Lecture Notes in Computer Science 12369

More information about this subseries at http://www.springer.com/series/7412

Andrea Vedaldi · Horst Bischof ·
Thomas Brox · Jan-Michael Frahm (Eds.)

Computer Vision – ECCV 2020

16th European Conference
Glasgow, UK, August 23–28, 2020
Proceedings, Part XXIV

Springer

Editors
Andrea Vedaldi
University of Oxford
Oxford, UK

Horst Bischof
Graz University of Technology
Graz, Austria

Thomas Brox
University of Freiburg
Freiburg im Breisgau, Germany

Jan-Michael Frahm
University of North Carolina at Chapel Hill
Chapel Hill, NC, USA

ISSN 0302-9743 ISSN 1611-3349 (electronic)
Lecture Notes in Computer Science
ISBN 978-3-030-58585-3 ISBN 978-3-030-58586-0 (eBook)
https://doi.org/10.1007/978-3-030-58586-0

LNCS Sublibrary: SL6 – Image Processing, Computer Vision, Pattern Recognition, and Graphics

This Springer imprint is published by the registered company Springer Nature Switzerland AG
The registered company address is: Gewerbestrasse 11, 6330 Cham, Switzerland

Foreword

Hosting the European Conference on Computer Vision (ECCV 2020) was certainly an exciting journey. From the 2016 plan to hold it at the Edinburgh International Conference Centre (hosting 1,800 delegates) to the 2018 plan to hold it at Glasgow's Scottish Exhibition Centre (up to 6,000 delegates), we finally ended with moving online because of the COVID-19 outbreak. While possibly having fewer delegates than expected because of the online format, ECCV 2020 still had over 3,100 registered participants.

Although online, the conference delivered most of the activities expected at a face-to-face conference: peer-reviewed papers, industrial exhibitors, demonstrations, and messaging between delegates. In addition to the main technical sessions, the conference included a strong program of satellite events with 16 tutorials and 44 workshops.

Furthermore, the online conference format enabled new conference features. Every paper had an associated teaser video and a longer full presentation video. Along with the papers and slides from the videos, all these materials were available the week before the conference. This allowed delegates to become familiar with the paper content and be ready for the live interaction with the authors during the conference week. The live event consisted of brief presentations by the oral and spotlight authors and industrial sponsors. Question and answer sessions for all papers were timed to occur twice so delegates from around the world had convenient access to the authors.

As with ECCV 2018, authors' draft versions of the papers appeared online with open access, now on both the Computer Vision Foundation (CVF) and the European Computer Vision Association (ECVA) websites. An archival publication arrangement was put in place with the cooperation of Springer. SpringerLink hosts the final version of the papers with further improvements, such as activating reference links and supplementary materials. These two approaches benefit all potential readers: a version available freely for all researchers, and an authoritative and citable version with additional benefits for SpringerLink subscribers. We thank Alfred Hofmann and Aliaksandr Birukou from Springer for helping to negotiate this agreement, which we expect will continue for future versions of ECCV.

August 2020

Vittorio Ferrari
Bob Fisher
Cordelia Schmid
Emanuele Trucco

Preface

Welcome to the proceedings of the European Conference on Computer Vision (ECCV 2020). This is a unique edition of ECCV in many ways. Due to the COVID-19 pandemic, this is the first time the conference was held online, in a virtual format. This was also the first time the conference relied exclusively on the Open Review platform to manage the review process. Despite these challenges ECCV is thriving. The conference received 5,150 valid paper submissions, of which 1,360 were accepted for publication (27%) and, of those, 160 were presented as spotlights (3%) and 104 as orals (2%). This amounts to more than twice the number of submissions to ECCV 2018 (2,439). Furthermore, CVPR, the largest conference on computer vision, received 5,850 submissions this year, meaning that ECCV is now 87% the size of CVPR in terms of submissions. By comparison, in 2018 the size of ECCV was only 73% of CVPR.

The review model was similar to previous editions of ECCV; in particular, it was double blind in the sense that the authors did not know the name of the reviewers and vice versa. Furthermore, each conference submission was held confidentially, and was only publicly revealed if and once accepted for publication. Each paper received at least three reviews, totalling more than 15,000 reviews. Handling the review process at this scale was a significant challenge. In order to ensure that each submission received as fair and high-quality reviews as possible, we recruited 2,830 reviewers (a 130% increase with reference to 2018) and 207 area chairs (a 60% increase). The area chairs were selected based on their technical expertise and reputation, largely among people that served as area chair in previous top computer vision and machine learning conferences (ECCV, ICCV, CVPR, NeurIPS, etc.). Reviewers were similarly invited from previous conferences. We also encouraged experienced area chairs to suggest additional chairs and reviewers in the initial phase of recruiting.

Despite doubling the number of submissions, the reviewer load was slightly reduced from 2018, from a maximum of 8 papers down to 7 (with some reviewers offering to handle 6 papers plus an emergency review). The area chair load increased slightly, from 18 papers on average to 22 papers on average.

Conflicts of interest between authors, area chairs, and reviewers were handled largely automatically by the Open Review platform via their curated list of user profiles. Many authors submitting to ECCV already had a profile in Open Review. We set a paper registration deadline one week before the paper submission deadline in order to encourage all missing authors to register and create their Open Review profiles well on time (in practice, we allowed authors to create/change papers arbitrarily until the submission deadline). Except for minor issues with users creating duplicate profiles, this allowed us to easily and quickly identify institutional conflicts, and avoid them, while matching papers to area chairs and reviewers.

Papers were matched to area chairs based on: an affinity score computed by the Open Review platform, which is based on paper titles and abstracts, and an affinity score computed by the Toronto Paper Matching System (TPMS), which is based on the paper's full text, the area chair bids for individual papers, load balancing, and conflict avoidance. Open Review provides the program chairs a convenient web interface to experiment with different configurations of the matching algorithm. The chosen configuration resulted in about 50% of the assigned papers to be highly ranked by the area chair bids, and 50% to be ranked in the middle, with very few low bids assigned.

Assignments to reviewers were similar, with two differences. First, there was a maximum of 7 papers assigned to each reviewer. Second, area chairs recommended up to seven reviewers per paper, providing another highly-weighed term to the affinity scores used for matching.

The assignment of papers to area chairs was smooth. However, it was more difficult to find suitable reviewers for all papers. Having a ratio of 5.6 papers per reviewer with a maximum load of 7 (due to emergency reviewer commitment), which did not allow for much wiggle room in order to also satisfy conflict and expertise constraints. We received some complaints from reviewers who did not feel qualified to review specific papers and we reassigned them wherever possible. However, the large scale of the conference, the many constraints, and the fact that a large fraction of such complaints arrived very late in the review process made this process very difficult and not all complaints could be addressed.

Reviewers had six weeks to complete their assignments. Possibly due to COVID-19 or the fact that the NeurIPS deadline was moved closer to the review deadline, a record 30% of the reviews were still missing after the deadline. By comparison, ECCV 2018 experienced only 10% missing reviews at this stage of the process. In the subsequent week, area chairs chased the missing reviews intensely, found replacement reviewers in their own team, and managed to reach 10% missing reviews. Eventually, we could provide almost all reviews (more than 99.9%) with a delay of only a couple of days on the initial schedule by a significant use of emergency reviews. If this trend is confirmed, it might be a major challenge to run a smooth review process in future editions of ECCV. The community must reconsider prioritization of the time spent on paper writing (the number of submissions increased a lot despite COVID-19) and time spent on paper reviewing (the number of reviews delivered in time decreased a lot presumably due to COVID-19 or NeurIPS deadline). With this imbalance the peer-review system that ensures the quality of our top conferences may break soon.

Reviewers submitted their reviews independently. In the reviews, they had the opportunity to ask questions to the authors to be addressed in the rebuttal. However, reviewers were told not to request any significant new experiment. Using the Open Review interface, authors could provide an answer to each individual review, but were also allowed to cross-reference reviews and responses in their answers. Rather than PDF files, we allowed the use of formatted text for the rebuttal. The rebuttal and initial reviews were then made visible to all reviewers and the primary area chair for a given paper. The area chair encouraged and moderated the reviewer discussion. During the discussions, reviewers were invited to reach a consensus and possibly adjust their ratings as a result of the discussion and of the evidence in the rebuttal.

After the discussion period ended, most reviewers entered a final rating and recommendation, although in many cases this did not differ from their initial recommendation. Based on the updated reviews and discussion, the primary area chair then made a preliminary decision to accept or reject the paper and wrote a justification for it (meta-review). Except for cases where the outcome of this process was absolutely clear (as indicated by the three reviewers and primary area chairs all recommending clear rejection), the decision was then examined and potentially challenged by a secondary area chair. This led to further discussion and overturning a small number of preliminary decisions. Needless to say, there was no in-person area chair meeting, which would have been impossible due to COVID-19.

Area chairs were invited to observe the consensus of the reviewers whenever possible and use extreme caution in overturning a clear consensus to accept or reject a paper. If an area chair still decided to do so, she/he was asked to clearly justify it in the meta-review and to explicitly obtain the agreement of the secondary area chair. In practice, very few papers were rejected after being confidently accepted by the reviewers.

This was the first time Open Review was used as the main platform to run ECCV. In 2018, the program chairs used CMT3 for the user-facing interface and Open Review internally, for matching and conflict resolution. Since it is clearly preferable to only use a single platform, this year we switched to using Open Review in full. The experience was largely positive. The platform is highly-configurable, scalable, and open source. Being written in Python, it is easy to write scripts to extract data programmatically. The paper matching and conflict resolution algorithms and interfaces are top-notch, also due to the excellent author profiles in the platform. Naturally, there were a few kinks along the way due to the fact that the ECCV Open Review configuration was created from scratch for this event and it differs in substantial ways from many other Open Review conferences. However, the Open Review development and support team did a fantastic job in helping us to get the configuration right and to address issues in a timely manner as they unavoidably occurred. We cannot thank them enough for the tremendous effort they put into this project.

Finally, we would like to thank everyone involved in making ECCV 2020 possible in these very strange and difficult times. This starts with our authors, followed by the area chairs and reviewers, who ran the review process at an unprecedented scale. The whole Open Review team (and in particular Melisa Bok, Mohit Unyal, Carlos Mondragon Chapa, and Celeste Martinez Gomez) worked incredibly hard for the entire duration of the process. We would also like to thank René Vidal for contributing to the adoption of Open Review. Our thanks also go to Laurent Charling for TPMS and to the program chairs of ICML, ICLR, and NeurIPS for cross checking double submissions. We thank the website chair, Giovanni Farinella, and the CPI team (in particular Ashley Cook, Miriam Verdon, Nicola McGrane, and Sharon Kerr) for promptly adding material to the website as needed in the various phases of the process. Finally, we thank the publication chairs, Albert Ali Salah, Hamdi Dibeklioglu, Metehan Doyran, Henry Howard-Jenkins, Victor Prisacariu, Siyu Tang, and Gul Varol, who managed to compile these substantial proceedings in an exceedingly compressed schedule. We express our thanks to the ECVA team, in particular Kristina Scherbaum for allowing open access of the proceedings. We thank Alfred Hofmann from Springer who again serve as the publisher. Finally, we thank the other chairs of ECCV 2020, including in particular the general chairs for very useful feedback with the handling of the program.

August 2020

Andrea Vedaldi
Horst Bischof
Thomas Brox
Jan-Michael Frahm

Organization

General Chairs

Vittorio Ferrari	Google Research, Switzerland
Bob Fisher	University of Edinburgh, UK
Cordelia Schmid	Google and Inria, France
Emanuele Trucco	University of Dundee, UK

Program Chairs

Andrea Vedaldi	University of Oxford, UK
Horst Bischof	Graz University of Technology, Austria
Thomas Brox	University of Freiburg, Germany
Jan-Michael Frahm	University of North Carolina, USA

Industrial Liaison Chairs

Jim Ashe	University of Edinburgh, UK
Helmut Grabner	Zurich University of Applied Sciences, Switzerland
Diane Larlus	NAVER LABS Europe, France
Cristian Novotny	University of Edinburgh, UK

Local Arrangement Chairs

Yvan Petillot	Heriot-Watt University, UK
Paul Siebert	University of Glasgow, UK

Academic Demonstration Chair

Thomas Mensink	Google Research and University of Amsterdam, The Netherlands

Poster Chair

Stephen Mckenna	University of Dundee, UK

Technology Chair

Gerardo Aragon Camarasa	University of Glasgow, UK

Tutorial Chairs

Carlo Colombo	University of Florence, Italy
Sotirios Tsaftaris	University of Edinburgh, UK

Publication Chairs

Albert Ali Salah	Utrecht University, The Netherlands
Hamdi Dibeklioglu	Bilkent University, Turkey
Metehan Doyran	Utrecht University, The Netherlands
Henry Howard-Jenkins	University of Oxford, UK
Victor Adrian Prisacariu	University of Oxford, UK
Siyu Tang	ETH Zurich, Switzerland
Gul Varol	University of Oxford, UK

Website Chair

Giovanni Maria Farinella	University of Catania, Italy

Workshops Chairs

Adrien Bartoli	University of Clermont Auvergne, France
Andrea Fusiello	University of Udine, Italy

Area Chairs

Lourdes Agapito	University College London, UK
Zeynep Akata	University of Tübingen, Germany
Karteek Alahari	Inria, France
Antonis Argyros	University of Crete, Greece
Hossein Azizpour	KTH Royal Institute of Technology, Sweden
Joao P. Barreto	Universidade de Coimbra, Portugal
Alexander C. Berg	University of North Carolina at Chapel Hill, USA
Matthew B. Blaschko	KU Leuven, Belgium
Lubomir D. Bourdev	WaveOne, Inc., USA
Edmond Boyer	Inria, France
Yuri Boykov	University of Waterloo, Canada
Gabriel Brostow	University College London, UK
Michael S. Brown	National University of Singapore, Singapore
Jianfei Cai	Monash University, Australia
Barbara Caputo	Politecnico di Torino, Italy
Ayan Chakrabarti	Washington University, St. Louis, USA
Tat-Jen Cham	Nanyang Technological University, Singapore
Manmohan Chandraker	University of California, San Diego, USA
Rama Chellappa	Johns Hopkins University, USA
Liang-Chieh Chen	Google, USA

Yung-Yu Chuang	National Taiwan University, Taiwan
Ondrej Chum	Czech Technical University in Prague, Czech Republic
Brian Clipp	Kitware, USA
John Collomosse	University of Surrey and Adobe Research, UK
Jason J. Corso	University of Michigan, USA
David J. Crandall	Indiana University, USA
Daniel Cremers	University of California, Los Angeles, USA
Fabio Cuzzolin	Oxford Brookes University, UK
Jifeng Dai	SenseTime, SAR China
Kostas Daniilidis	University of Pennsylvania, USA
Andrew Davison	Imperial College London, UK
Alessio Del Bue	Fondazione Istituto Italiano di Tecnologia, Italy
Jia Deng	Princeton University, USA
Alexey Dosovitskiy	Google, Germany
Matthijs Douze	Facebook, France
Enrique Dunn	Stevens Institute of Technology, USA
Irfan Essa	Georgia Institute of Technology and Google, USA
Giovanni Maria Farinella	University of Catania, Italy
Ryan Farrell	Brigham Young University, USA
Paolo Favaro	University of Bern, Switzerland
Rogerio Feris	International Business Machines, USA
Cornelia Fermuller	University of Maryland, College Park, USA
David J. Fleet	Vector Institute, Canada
Friedrich Fraundorfer	DLR, Austria
Mario Fritz	CISPA Helmholtz Center for Information Security, Germany
Pascal Fua	EPFL (Swiss Federal Institute of Technology Lausanne), Switzerland
Yasutaka Furukawa	Simon Fraser University, Canada
Li Fuxin	Oregon State University, USA
Efstratios Gavves	University of Amsterdam, The Netherlands
Peter Vincent Gehler	Amazon, USA
Theo Gevers	University of Amsterdam, The Netherlands
Ross Girshick	Facebook AI Research, USA
Boqing Gong	Google, USA
Stephen Gould	Australian National University, Australia
Jinwei Gu	SenseTime Research, USA
Abhinav Gupta	Facebook, USA
Bohyung Han	Seoul National University, South Korea
Bharath Hariharan	Cornell University, USA
Tal Hassner	Facebook AI Research, USA
Xuming He	Australian National University, Australia
Joao F. Henriques	University of Oxford, UK
Adrian Hilton	University of Surrey, UK
Minh Hoai	Stony Brooks, State University of New York, USA
Derek Hoiem	University of Illinois Urbana-Champaign, USA

Timothy Hospedales	University of Edinburgh and Samsung, UK
Gang Hua	Wormpex AI Research, USA
Slobodan Ilic	Siemens AG, Germany
Hiroshi Ishikawa	Waseda University, Japan
Jiaya Jia	The Chinese University of Hong Kong, SAR China
Hailin Jin	Adobe Research, USA
Justin Johnson	University of Michigan, USA
Frederic Jurie	University of Caen Normandie, France
Fredrik Kahl	Chalmers University, Sweden
Sing Bing Kang	Zillow, USA
Gunhee Kim	Seoul National University, South Korea
Junmo Kim	Korea Advanced Institute of Science and Technology, South Korea
Tae-Kyun Kim	Imperial College London, UK
Ron Kimmel	Technion-Israel Institute of Technology, Israel
Alexander Kirillov	Facebook AI Research, USA
Kris Kitani	Carnegie Mellon University, USA
Iasonas Kokkinos	Ariel AI, UK
Vladlen Koltun	Intel Labs, USA
Nikos Komodakis	Ecole des Ponts ParisTech, France
Piotr Koniusz	Australian National University, Australia
M. Pawan Kumar	University of Oxford, UK
Kyros Kutulakos	University of Toronto, Canada
Christoph Lampert	IST Austria, Austria
Ivan Laptev	Inria, France
Diane Larlus	NAVER LABS Europe, France
Laura Leal-Taixe	Technical University Munich, Germany
Honglak Lee	Google and University of Michigan, USA
Joon-Young Lee	Adobe Research, USA
Kyoung Mu Lee	Seoul National University, South Korea
Seungyong Lee	POSTECH, South Korea
Yong Jae Lee	University of California, Davis, USA
Bastian Leibe	RWTH Aachen University, Germany
Victor Lempitsky	Samsung, Russia
Ales Leonardis	University of Birmingham, UK
Marius Leordeanu	Institute of Mathematics of the Romanian Academy, Romania
Vincent Lepetit	ENPC ParisTech, France
Hongdong Li	The Australian National University, Australia
Xi Li	Zhejiang University, China
Yin Li	University of Wisconsin-Madison, USA
Zicheng Liao	Zhejiang University, China
Jongwoo Lim	Hanyang University, South Korea
Stephen Lin	Microsoft Research Asia, China
Yen-Yu Lin	National Chiao Tung University, Taiwan, China
Zhe Lin	Adobe Research, USA

Haibin Ling	Stony Brooks, State University of New York, USA
Jiaying Liu	Peking University, China
Ming-Yu Liu	NVIDIA, USA
Si Liu	Beihang University, China
Xiaoming Liu	Michigan State University, USA
Huchuan Lu	Dalian University of Technology, China
Simon Lucey	Carnegie Mellon University, USA
Jiebo Luo	University of Rochester, USA
Julien Mairal	Inria, France
Michael Maire	University of Chicago, USA
Subhransu Maji	University of Massachusetts, Amherst, USA
Yasushi Makihara	Osaka University, Japan
Jiri Matas	Czech Technical University in Prague, Czech Republic
Yasuyuki Matsushita	Osaka University, Japan
Philippos Mordohai	Stevens Institute of Technology, USA
Vittorio Murino	University of Verona, Italy
Naila Murray	NAVER LABS Europe, France
Hajime Nagahara	Osaka University, Japan
P. J. Narayanan	International Institute of Information Technology (IIIT), Hyderabad, India
Nassir Navab	Technical University of Munich, Germany
Natalia Neverova	Facebook AI Research, France
Matthias Niessner	Technical University of Munich, Germany
Jean-Marc Odobez	Idiap Research Institute and Swiss Federal Institute of Technology Lausanne, Switzerland
Francesca Odone	Università di Genova, Italy
Takeshi Oishi	The University of Tokyo, Tokyo Institute of Technology, Japan
Vicente Ordonez	University of Virginia, USA
Manohar Paluri	Facebook AI Research, USA
Maja Pantic	Imperial College London, UK
In Kyu Park	Inha University, South Korea
Ioannis Patras	Queen Mary University of London, UK
Patrick Perez	Valeo, France
Bryan A. Plummer	Boston University, USA
Thomas Pock	Graz University of Technology, Austria
Marc Pollefeys	ETH Zurich and Microsoft MR & AI Zurich Lab, Switzerland
Jean Ponce	Inria, France
Gerard Pons-Moll	MPII, Saarland Informatics Campus, Germany
Jordi Pont-Tuset	Google, Switzerland
James Matthew Rehg	Georgia Institute of Technology, USA
Ian Reid	University of Adelaide, Australia
Olaf Ronneberger	DeepMind London, UK
Stefan Roth	TU Darmstadt, Germany
Bryan Russell	Adobe Research, USA

Mathieu Salzmann	EPFL, Switzerland
Dimitris Samaras	Stony Brook University, USA
Imari Sato	National Institute of Informatics (NII), Japan
Yoichi Sato	The University of Tokyo, Japan
Torsten Sattler	Czech Technical University in Prague, Czech Republic
Daniel Scharstein	Middlebury College, USA
Bernt Schiele	MPII, Saarland Informatics Campus, Germany
Julia A. Schnabel	King's College London, UK
Nicu Sebe	University of Trento, Italy
Greg Shakhnarovich	Toyota Technological Institute at Chicago, USA
Humphrey Shi	University of Oregon, USA
Jianbo Shi	University of Pennsylvania, USA
Jianping Shi	SenseTime, China
Leonid Sigal	University of British Columbia, Canada
Cees Snoek	University of Amsterdam, The Netherlands
Richard Souvenir	Temple University, USA
Hao Su	University of California, San Diego, USA
Akihiro Sugimoto	National Institute of Informatics (NII), Japan
Jian Sun	Megvii Technology, China
Jian Sun	Xi'an Jiaotong University, China
Chris Sweeney	Facebook Reality Labs, USA
Yu-wing Tai	Kuaishou Technology, China
Chi-Keung Tang	The Hong Kong University of Science and Technology, SAR China
Radu Timofte	ETH Zurich, Switzerland
Sinisa Todorovic	Oregon State University, USA
Giorgos Tolias	Czech Technical University in Prague, Czech Republic
Carlo Tomasi	Duke University, USA
Tatiana Tommasi	Politecnico di Torino, Italy
Lorenzo Torresani	Facebook AI Research and Dartmouth College, USA
Alexander Toshev	Google, USA
Zhuowen Tu	University of California, San Diego, USA
Tinne Tuytelaars	KU Leuven, Belgium
Jasper Uijlings	Google, Switzerland
Nuno Vasconcelos	University of California, San Diego, USA
Olga Veksler	University of Waterloo, Canada
Rene Vidal	Johns Hopkins University, USA
Gang Wang	Alibaba Group, China
Jingdong Wang	Microsoft Research Asia, China
Yizhou Wang	Peking University, China
Lior Wolf	Facebook AI Research and Tel Aviv University, Israel
Jianxin Wu	Nanjing University, China
Tao Xiang	University of Surrey, UK
Saining Xie	Facebook AI Research, USA
Ming-Hsuan Yang	University of California at Merced and Google, USA
Ruigang Yang	University of Kentucky, USA

Kwang Moo Yi	University of Victoria, Canada
Zhaozheng Yin	Stony Brook, State University of New York, USA
Chang D. Yoo	Korea Advanced Institute of Science and Technology, South Korea
Shaodi You	University of Amsterdam, The Netherlands
Jingyi Yu	ShanghaiTech University, China
Stella Yu	University of California, Berkeley, and ICSI, USA
Stefanos Zafeiriou	Imperial College London, UK
Hongbin Zha	Peking University, China
Tianzhu Zhang	University of Science and Technology of China, China
Liang Zheng	Australian National University, Australia
Todd E. Zickler	Harvard University, USA
Andrew Zisserman	University of Oxford, UK

Technical Program Committee

Sathyanarayanan N. Aakur
Wael Abd Almgaeed
Abdelrahman Abdelhamed
Abdullah Abuolaim
Supreeth Achar
Hanno Ackermann
Ehsan Adeli
Triantafyllos Afouras
Sameer Agarwal
Aishwarya Agrawal
Harsh Agrawal
Pulkit Agrawal
Antonio Agudo
Eirikur Agustsson
Karim Ahmed
Byeongjoo Ahn
Unaiza Ahsan
Thalaiyasingam Ajanthan
Kenan E. Ak
Emre Akbas
Naveed Akhtar
Derya Akkaynak
Yagiz Aksoy
Ziad Al-Halah
Xavier Alameda-Pineda
Jean-Baptiste Alayrac
Samuel Albanie
Shadi Albarqouni
Cenek Albl
Hassan Abu Alhaija
Daniel Aliaga
Mohammad S. Aliakbarian
Rahaf Aljundi
Thiemo Alldieck
Jon Almazan
Jose M. Alvarez
Senjian An
Saket Anand
Codruta Ancuti
Cosmin Ancuti
Peter Anderson
Juan Andrade-Cetto
Alexander Andreopoulos
Misha Andriluka
Dragomir Anguelov
Rushil Anirudh
Michel Antunes
Oisin Mac Aodha
Srikar Appalaraju
Relja Arandjelovic
Nikita Araslanov
Andre Araujo
Helder Araujo
Pablo Arbelaez
Shervin Ardeshir
Sercan O. Arik
Anil Armagan
Anurag Arnab
Chetan Arora
Federica Arrigoni
Mathieu Aubry
Shai Avidan
Angelica I. Aviles-Rivero
Yannis Avrithis
Ismail Ben Ayed
Shekoofeh Azizi
Ioan Andrei Bârsan
Artem Babenko
Deepak Babu Sam
Seung-Hwan Baek
Seungryul Baek
Andrew D. Bagdanov
Shai Bagon
Yuval Bahat
Junjie Bai
Song Bai
Xiang Bai
Yalong Bai
Yancheng Bai
Peter Bajcsy
Slawomir Bak

Mahsa Baktashmotlagh
Kavita Bala
Yogesh Balaji
Guha Balakrishnan
V. N. Balasubramanian
Federico Baldassarre
Vassileios Balntas
Shurjo Banerjee
Aayush Bansal
Ankan Bansal
Jianmin Bao
Linchao Bao
Wenbo Bao
Yingze Bao
Akash Bapat
Md Jawadul Hasan Bappy
Fabien Baradel
Lorenzo Baraldi
Daniel Barath
Adrian Barbu
Kobus Barnard
Nick Barnes
Francisco Barranco
Jonathan T. Barron
Arslan Basharat
Chaim Baskin
Anil S. Baslamisli
Jorge Batista
Kayhan Batmanghelich
Konstantinos Batsos
David Bau
Luis Baumela
Christoph Baur
Eduardo Bayro-Corrochano
Paul Beardsley
Jan Bednavr'ik
Oscar Beijbom
Philippe Bekaert
Esube Bekele
Vasileios Belagiannis
Ohad Ben-Shahar
Abhijit Bendale
Róger Bermúdez-Chacón
Maxim Berman
Jesus Bermudez-cameo
Florian Bernard
Stefano Berretti
Marcelo Bertalmio
Gedas Bertasius
Cigdem Beyan
Lucas Beyer
Vijayakumar Bhagavatula
Arjun Nitin Bhagoji
Apratim Bhattacharyya
Binod Bhattarai
Sai Bi
Jia-Wang Bian
Simone Bianco
Adel Bibi
Tolga Birdal
Tom Bishop
Soma Biswas
Mårten Björkman
Volker Blanz
Vishnu Boddeti
Navaneeth Bodla
Simion-Vlad Bogolin
Xavier Boix
Piotr Bojanowski
Timo Bolkart
Guido Borghi
Larbi Boubchir
Guillaume Bourmaud
Adrien Bousseau
Thierry Bouwmans
Richard Bowden
Hakan Boyraz
Mathieu Brédif
Samarth Brahmbhatt
Steve Branson
Nikolas Brasch
Biagio Brattoli
Ernesto Brau
Toby P. Breckon
Francois Bremond
Jesus Briales
Sofia Broomé
Marcus A. Brubaker
Luc Brun
Silvia Bucci
Shyamal Buch
Pradeep Buddharaju
Uta Buechler
Mai Bui
Tu Bui
Adrian Bulat
Giedrius T. Burachas
Elena Burceanu
Xavier P. Burgos-Artizzu
Kaylee Burns
Andrei Bursuc
Benjamin Busam
Wonmin Byeon
Zoya Bylinskii
Sergi Caelles
Jianrui Cai
Minjie Cai
Yujun Cai
Zhaowei Cai
Zhipeng Cai
Juan C. Caicedo
Simone Calderara
Necati Cihan Camgoz
Dylan Campbell
Octavia Camps
Jiale Cao
Kaidi Cao
Liangliang Cao
Xiangyong Cao
Xiaochun Cao
Yang Cao
Yu Cao
Yue Cao
Zhangjie Cao
Luca Carlone
Mathilde Caron
Dan Casas
Thomas J. Cashman
Umberto Castellani
Lluis Castrejon
Jacopo Cavazza
Fabio Cermelli
Hakan Cevikalp
Menglei Chai
Ishani Chakraborty
Rudrasis Chakraborty
Antoni B. Chan

Kwok-Ping Chan
Siddhartha Chandra
Sharat Chandran
Arjun Chandrasekaran
Angel X. Chang
Che-Han Chang
Hong Chang
Hyun Sung Chang
Hyung Jin Chang
Jianlong Chang
Ju Yong Chang
Ming-Ching Chang
Simyung Chang
Xiaojun Chang
Yu-Wei Chao
Devendra S. Chaplot
Arslan Chaudhry
Rizwan A. Chaudhry
Can Chen
Chang Chen
Chao Chen
Chen Chen
Chu-Song Chen
Dapeng Chen
Dong Chen
Dongdong Chen
Guanying Chen
Hongge Chen
Hsin-yi Chen
Huaijin Chen
Hwann-Tzong Chen
Jianbo Chen
Jianhui Chen
Jiansheng Chen
Jiaxin Chen
Jie Chen
Jun-Cheng Chen
Kan Chen
Kevin Chen
Lin Chen
Long Chen
Min-Hung Chen
Qifeng Chen
Shi Chen
Shixing Chen
Tianshui Chen
Weifeng Chen
Weikai Chen
Xi Chen
Xiaohan Chen
Xiaozhi Chen
Xilin Chen
Xingyu Chen
Xinlei Chen
Xinyun Chen
Yi-Ting Chen
Yilun Chen
Ying-Cong Chen
Yinpeng Chen
Yiran Chen
Yu Chen
Yu-Sheng Chen
Yuhua Chen
Yun-Chun Chen
Yunpeng Chen
Yuntao Chen
Zhuoyuan Chen
Zitian Chen
Anchieh Cheng
Bowen Cheng
Erkang Cheng
Gong Cheng
Guangliang Cheng
Jingchun Cheng
Jun Cheng
Li Cheng
Ming-Ming Cheng
Yu Cheng
Ziang Cheng
Anoop Cherian
Dmitry Chetverikov
Ngai-man Cheung
William Cheung
Ajad Chhatkuli
Naoki Chiba
Benjamin Chidester
Han-pang Chiu
Mang Tik Chiu
Wei-Chen Chiu
Donghyeon Cho
Hojin Cho
Minsu Cho
Nam Ik Cho
Tim Cho
Tae Eun Choe
Chiho Choi
Edward Choi
Inchang Choi
Jinsoo Choi
Jonghyun Choi
Jongwon Choi
Yukyung Choi
Hisham Cholakkal
Eunji Chong
Jaegul Choo
Christopher Choy
Hang Chu
Peng Chu
Wen-Sheng Chu
Albert Chung
Joon Son Chung
Hai Ci
Safa Cicek
Ramazan G. Cinbis
Arridhana Ciptadi
Javier Civera
James J. Clark
Ronald Clark
Felipe Codevilla
Michael Cogswell
Andrea Cohen
Maxwell D. Collins
Carlo Colombo
Yang Cong
Adria R. Continente
Marcella Cornia
John Richard Corring
Darren Cosker
Dragos Costea
Garrison W. Cottrell
Florent Couzinie-Devy
Marco Cristani
Ioana Croitoru
James L. Crowley
Jiequan Cui
Zhaopeng Cui
Ross Cutler
Antonio D'Innocente

Rozenn Dahyot
Bo Dai
Dengxin Dai
Hang Dai
Longquan Dai
Shuyang Dai
Xiyang Dai
Yuchao Dai
Adrian V. Dalca
Dima Damen
Bharath B. Damodaran
Kristin Dana
Martin Danelljan
Zheng Dang
Zachary Alan Daniels
Donald G. Dansereau
Abhishek Das
Samyak Datta
Achal Dave
Titas De
Rodrigo de Bem
Teo de Campos
Raoul de Charette
Shalini De Mello
Joseph DeGol
Herve Delingette
Haowen Deng
Jiankang Deng
Weijian Deng
Zhiwei Deng
Joachim Denzler
Konstantinos G. Derpanis
Aditya Deshpande
Frederic Devernay
Somdip Dey
Arturo Deza
Abhinav Dhall
Helisa Dhamo
Vikas Dhiman
Fillipe Dias Moreira de Souza
Ali Diba
Ferran Diego
Guiguang Ding
Henghui Ding
Jian Ding
Mingyu Ding
Xinghao Ding
Zhengming Ding
Robert DiPietro
Cosimo Distante
Ajay Divakaran
Mandar Dixit
Abdelaziz Djelouah
Thanh-Toan Do
Jose Dolz
Bo Dong
Chao Dong
Jiangxin Dong
Weiming Dong
Weisheng Dong
Xingping Dong
Xuanyi Dong
Yinpeng Dong
Gianfranco Doretto
Hazel Doughty
Hassen Drira
Bertram Drost
Dawei Du
Ye Duan
Yueqi Duan
Abhimanyu Dubey
Anastasia Dubrovina
Stefan Duffner
Chi Nhan Duong
Thibaut Durand
Zoran Duric
Iulia Duta
Debidatta Dwibedi
Benjamin Eckart
Marc Eder
Marzieh Edraki
Alexei A. Efros
Kiana Ehsani
Hazm Kemal Ekenel
James H. Elder
Mohamed Elgharib
Shireen Elhabian
Ehsan Elhamifar
Mohamed Elhoseiny
Ian Endres
N. Benjamin Erichson
Jan Ernst
Sergio Escalera
Francisco Escolano
Victor Escorcia
Carlos Esteves
Francisco J. Estrada
Bin Fan
Chenyou Fan
Deng-Ping Fan
Haoqi Fan
Hehe Fan
Heng Fan
Kai Fan
Lijie Fan
Linxi Fan
Quanfu Fan
Shaojing Fan
Xiaochuan Fan
Xin Fan
Yuchen Fan
Sean Fanello
Hao-Shu Fang
Haoyang Fang
Kuan Fang
Yi Fang
Yuming Fang
Azade Farshad
Alireza Fathi
Raanan Fattal
Joao Fayad
Xiaohan Fei
Christoph Feichtenhofer
Michael Felsberg
Chen Feng
Jiashi Feng
Junyi Feng
Mengyang Feng
Qianli Feng
Zhenhua Feng
Michele Fenzi
Andras Ferencz
Martin Fergie
Basura Fernando
Ethan Fetaya
Michael Firman
John W. Fisher

Matthew Fisher
Boris Flach
Corneliu Florea
Wolfgang Foerstner
David Fofi
Gian Luca Foresti
Per-Erik Forssen
David Fouhey
Katerina Fragkiadaki
Victor Fragoso
Jean-Sébastien Franco
Ohad Fried
Iuri Frosio
Cheng-Yang Fu
Huazhu Fu
Jianlong Fu
Jingjing Fu
Xueyang Fu
Yanwei Fu
Ying Fu
Yun Fu
Olac Fuentes
Kent Fujiwara
Takuya Funatomi
Christopher Funk
Thomas Funkhouser
Antonino Furnari
Ryo Furukawa
Erik Gärtner
Raghudeep Gadde
Matheus Gadelha
Vandit Gajjar
Trevor Gale
Juergen Gall
Mathias Gallardo
Guillermo Gallego
Orazio Gallo
Chuang Gan
Zhe Gan
Madan Ravi Ganesh
Aditya Ganeshan
Siddha Ganju
Bin-Bin Gao
Changxin Gao
Feng Gao
Hongchang Gao
Jin Gao
Jiyang Gao
Junbin Gao
Katelyn Gao
Lin Gao
Mingfei Gao
Ruiqi Gao
Ruohan Gao
Shenghua Gao
Yuan Gao
Yue Gao
Noa Garcia
Alberto Garcia-Garcia
Guillermo Garcia-Hernando
Jacob R. Gardner
Animesh Garg
Kshitiz Garg
Rahul Garg
Ravi Garg
Philip N. Garner
Kirill Gavrilyuk
Paul Gay
Shiming Ge
Weifeng Ge
Baris Gecer
Xin Geng
Kyle Genova
Stamatios Georgoulis
Bernard Ghanem
Michael Gharbi
Kamran Ghasedi
Golnaz Ghiasi
Arnab Ghosh
Partha Ghosh
Silvio Giancola
Andrew Gilbert
Rohit Girdhar
Xavier Giro-i-Nieto
Thomas Gittings
Ioannis Gkioulekas
Clement Godard
Vaibhava Goel
Bastian Goldluecke
Lluis Gomez
Nuno Gonçalves
Dong Gong
Ke Gong
Mingming Gong
Abel Gonzalez-Garcia
Ariel Gordon
Daniel Gordon
Paulo Gotardo
Venu Madhav Govindu
Ankit Goyal
Priya Goyal
Raghav Goyal
Benjamin Graham
Douglas Gray
Brent A. Griffin
Etienne Grossmann
David Gu
Jiayuan Gu
Jiuxiang Gu
Lin Gu
Qiao Gu
Shuhang Gu
Jose J. Guerrero
Paul Guerrero
Jie Gui
Jean-Yves Guillemaut
Riza Alp Guler
Erhan Gundogdu
Fatma Guney
Guodong Guo
Kaiwen Guo
Qi Guo
Sheng Guo
Shi Guo
Tiantong Guo
Xiaojie Guo
Yijie Guo
Yiluan Guo
Yuanfang Guo
Yulan Guo
Agrim Gupta
Ankush Gupta
Mohit Gupta
Saurabh Gupta
Tanmay Gupta
Danna Gurari
Abner Guzman-Rivera

JunYoung Gwak
Michael Gygli
Jung-Woo Ha
Simon Hadfield
Isma Hadji
Bjoern Haefner
Taeyoung Hahn
Levente Hajder
Peter Hall
Emanuela Haller
Stefan Haller
Bumsub Ham
Abdullah Hamdi
Dongyoon Han
Hu Han
Jungong Han
Junwei Han
Kai Han
Tian Han
Xiaoguang Han
Xintong Han
Yahong Han
Ankur Handa
Zekun Hao
Albert Haque
Tatsuya Harada
Mehrtash Harandi
Adam W. Harley
Mahmudul Hasan
Atsushi Hashimoto
Ali Hatamizadeh
Munawar Hayat
Dongliang He
Jingrui He
Junfeng He
Kaiming He
Kun He
Lei He
Pan He
Ran He
Shengfeng He
Tong He
Weipeng He
Xuming He
Yang He
Yihui He
Zhihai He
Chinmay Hegde
Janne Heikkila
Mattias P. Heinrich
Stéphane Herbin
Alexander Hermans
Luis Herranz
John R. Hershey
Aaron Hertzmann
Roei Herzig
Anders Heyden
Steven Hickson
Otmar Hilliges
Tomas Hodan
Judy Hoffman
Michael Hofmann
Yannick Hold-Geoffroy
Namdar Homayounfar
Sina Honari
Richang Hong
Seunghoon Hong
Xiaopeng Hong
Yi Hong
Hidekata Hontani
Anthony Hoogs
Yedid Hoshen
Mir Rayat Imtiaz Hossain
Junhui Hou
Le Hou
Lu Hou
Tingbo Hou
Wei-Lin Hsiao
Cheng-Chun Hsu
Gee-Sern Jison Hsu
Kuang-jui Hsu
Changbo Hu
Di Hu
Guosheng Hu
Han Hu
Hao Hu
Hexiang Hu
Hou-Ning Hu
Jie Hu
Junlin Hu
Nan Hu
Ping Hu
Ronghang Hu
Xiaowei Hu
Yinlin Hu
Yuan-Ting Hu
Zhe Hu
Binh-Son Hua
Yang Hua
Bingyao Huang
Di Huang
Dong Huang
Fay Huang
Haibin Huang
Haozhi Huang
Heng Huang
Huaibo Huang
Jia-Bin Huang
Jing Huang
Jingwei Huang
Kaizhu Huang
Lei Huang
Qiangui Huang
Qiaoying Huang
Qingqiu Huang
Qixing Huang
Shaoli Huang
Sheng Huang
Siyuan Huang
Weilin Huang
Wenbing Huang
Xiangru Huang
Xun Huang
Yan Huang
Yifei Huang
Yue Huang
Zhiwu Huang
Zilong Huang
Minyoung Huh
Zhuo Hui
Matthias B. Hullin
Martin Humenberger
Wei-Chih Hung
Zhouyuan Huo
Junhwa Hur
Noureldien Hussein
Jyh-Jing Hwang
Seong Jae Hwang

Sung Ju Hwang
Ichiro Ide
Ivo Ihrke
Daiki Ikami
Satoshi Ikehata
Nazli Ikizler-Cinbis
Sunghoon Im
Yani Ioannou
Radu Tudor Ionescu
Umar Iqbal
Go Irie
Ahmet Iscen
Md Amirul Islam
Vamsi Ithapu
Nathan Jacobs
Arpit Jain
Himalaya Jain
Suyog Jain
Stuart James
Won-Dong Jang
Yunseok Jang
Ronnachai Jaroensri
Dinesh Jayaraman
Sadeep Jayasumana
Suren Jayasuriya
Herve Jegou
Simon Jenni
Hae-Gon Jeon
Yunho Jeon
Koteswar R. Jerripothula
Hueihan Jhuang
I-hong Jhuo
Dinghuang Ji
Hui Ji
Jingwei Ji
Pan Ji
Yanli Ji
Baoxiong Jia
Kui Jia
Xu Jia
Chiyu Max Jiang
Haiyong Jiang
Hao Jiang
Huaizu Jiang
Huajie Jiang
Ke Jiang
Lai Jiang
Li Jiang
Lu Jiang
Ming Jiang
Peng Jiang
Shuqiang Jiang
Wei Jiang
Xudong Jiang
Zhuolin Jiang
Jianbo Jiao
Zequn Jie
Dakai Jin
Kyong Hwan Jin
Lianwen Jin
SouYoung Jin
Xiaojie Jin
Xin Jin
Nebojsa Jojic
Alexis Joly
Michael Jeffrey Jones
Hanbyul Joo
Jungseock Joo
Kyungdon Joo
Ajjen Joshi
Shantanu H. Joshi
Da-Cheng Juan
Marco Körner
Kevin Köser
Asim Kadav
Christine Kaeser-Chen
Kushal Kafle
Dagmar Kainmueller
Ioannis A. Kakadiaris
Zdenek Kalal
Nima Kalantari
Yannis Kalantidis
Mahdi M. Kalayeh
Anmol Kalia
Sinan Kalkan
Vicky Kalogeiton
Ashwin Kalyan
Joni-kristian Kamarainen
Gerda Kamberova
Chandra Kambhamettu
Martin Kampel
Meina Kan
Christopher Kanan
Kenichi Kanatani
Angjoo Kanazawa
Atsushi Kanehira
Takuhiro Kaneko
Asako Kanezaki
Bingyi Kang
Di Kang
Sunghun Kang
Zhao Kang
Vadim Kantorov
Abhishek Kar
Amlan Kar
Theofanis Karaletsos
Leonid Karlinsky
Kevin Karsch
Angelos Katharopoulos
Isinsu Katircioglu
Hiroharu Kato
Zoltan Kato
Dotan Kaufman
Jan Kautz
Rei Kawakami
Qiuhong Ke
Wadim Kehl
Petr Kellnhofer
Aniruddha Kembhavi
Cem Keskin
Margret Keuper
Daniel Keysers
Ashkan Khakzar
Fahad Khan
Naeemullah Khan
Salman Khan
Siddhesh Khandelwal
Rawal Khirodkar
Anna Khoreva
Tejas Khot
Parmeshwar Khurd
Hadi Kiapour
Joe Kileel
Chanho Kim
Dahun Kim
Edward Kim
Eunwoo Kim
Han-ul Kim

Hansung Kim
Heewon Kim
Hyo Jin Kim
Hyunwoo J. Kim
Jinkyu Kim
Jiwon Kim
Jongmin Kim
Junsik Kim
Junyeong Kim
Min H. Kim
Namil Kim
Pyojin Kim
Seon Joo Kim
Seong Tae Kim
Seungryong Kim
Sungwoong Kim
Tae Hyun Kim
Vladimir Kim
Won Hwa Kim
Yonghyun Kim
Benjamin Kimia
Akisato Kimura
Pieter-Jan Kindermans
Zsolt Kira
Itaru Kitahara
Hedvig Kjellstrom
Jan Knopp
Takumi Kobayashi
Erich Kobler
Parker Koch
Reinhard Koch
Elyor Kodirov
Amir Kolaman
Nicholas Kolkin
Dimitrios Kollias
Stefanos Kollias
Soheil Kolouri
Adams Wai-Kin Kong
Naejin Kong
Shu Kong
Tao Kong
Yu Kong
Yoshinori Konishi
Daniil Kononenko
Theodora Kontogianni
Simon Korman
Adam Kortylewski
Jana Kosecka
Jean Kossaifi
Satwik Kottur
Rigas Kouskouridas
Adriana Kovashka
Rama Kovvuri
Adarsh Kowdle
Jedrzej Kozerawski
Mateusz Kozinski
Philipp Kraehenbuehl
Gregory Kramida
Josip Krapac
Dmitry Kravchenko
Ranjay Krishna
Pavel Krsek
Alexander Krull
Jakob Kruse
Hiroyuki Kubo
Hilde Kuehne
Jason Kuen
Andreas Kuhn
Arjan Kuijper
Zuzana Kukelova
Ajay Kumar
Amit Kumar
Avinash Kumar
Suryansh Kumar
Vijay Kumar
Kaustav Kundu
Weicheng Kuo
Nojun Kwak
Suha Kwak
Junseok Kwon
Nikolaos Kyriazis
Zorah Lähner
Ankit Laddha
Florent Lafarge
Jean Lahoud
Kevin Lai
Shang-Hong Lai
Wei-Sheng Lai
Yu-Kun Lai
Iro Laina
Antony Lam
John Wheatley Lambert
Xiangyuan lan
Xu Lan
Charis Lanaras
Georg Langs
Oswald Lanz
Dong Lao
Yizhen Lao
Agata Lapedriza
Gustav Larsson
Viktor Larsson
Katrin Lasinger
Christoph Lassner
Longin Jan Latecki
Stéphane Lathuilière
Rynson Lau
Hei Law
Justin Lazarow
Svetlana Lazebnik
Hieu Le
Huu Le
Ngan Hoang Le
Trung-Nghia Le
Vuong Le
Colin Lea
Erik Learned-Miller
Chen-Yu Lee
Gim Hee Lee
Hsin-Ying Lee
Hyungtae Lee
Jae-Han Lee
Jimmy Addison Lee
Joonseok Lee
Kibok Lee
Kuang-Huei Lee
Kwonjoon Lee
Minsik Lee
Sang-chul Lee
Seungkyu Lee
Soochan Lee
Stefan Lee
Taehee Lee
Andreas Lehrmann
Jie Lei
Peng Lei
Matthew Joseph Leotta
Wee Kheng Leow

Gil Levi
Evgeny Levinkov
Aviad Levis
Jose Lezama
Ang Li
Bin Li
Bing Li
Boyi Li
Changsheng Li
Chao Li
Chen Li
Cheng Li
Chenglong Li
Chi Li
Chun-Guang Li
Chun-Liang Li
Chunyuan Li
Dong Li
Guanbin Li
Hao Li
Haoxiang Li
Hongsheng Li
Hongyang Li
Houqiang Li
Huibin Li
Jia Li
Jianan Li
Jianguo Li
Junnan Li
Junxuan Li
Kai Li
Ke Li
Kejie Li
Kunpeng Li
Lerenhan Li
Li Erran Li
Mengtian Li
Mu Li
Peihua Li
Peiyi Li
Ping Li
Qi Li
Qing Li
Ruiyu Li
Ruoteng Li
Shaozi Li
Sheng Li
Shiwei Li
Shuang Li
Siyang Li
Stan Z. Li
Tianye Li
Wei Li
Weixin Li
Wen Li
Wenbo Li
Xiaomeng Li
Xin Li
Xiu Li
Xuelong Li
Xueting Li
Yan Li
Yandong Li
Yanghao Li
Yehao Li
Yi Li
Yijun Li
Yikang LI
Yining Li
Yongjie Li
Yu Li
Yu-Jhe Li
Yunpeng Li
Yunsheng Li
Yunzhu Li
Zhe Li
Zhen Li
Zhengqi Li
Zhenyang Li
Zhuwen Li
Dongze Lian
Xiaochen Lian
Zhouhui Lian
Chen Liang
Jie Liang
Ming Liang
Paul Pu Liang
Pengpeng Liang
Shu Liang
Wei Liang
Jing Liao
Minghui Liao
Renjie Liao
Shengcai Liao
Shuai Liao
Yiyi Liao
Ser-Nam Lim
Chen-Hsuan Lin
Chung-Ching Lin
Dahua Lin
Ji Lin
Kevin Lin
Tianwei Lin
Tsung-Yi Lin
Tsung-Yu Lin
Wei-An Lin
Weiyao Lin
Yen-Chen Lin
Yuewei Lin
David B. Lindell
Drew Linsley
Krzysztof Lis
Roee Litman
Jim Little
An-An Liu
Bo Liu
Buyu Liu
Chao Liu
Chen Liu
Cheng-lin Liu
Chenxi Liu
Dong Liu
Feng Liu
Guilin Liu
Haomiao Liu
Heshan Liu
Hong Liu
Ji Liu
Jingen Liu
Jun Liu
Lanlan Liu
Li Liu
Liu Liu
Mengyuan Liu
Miaomiao Liu
Nian Liu
Ping Liu
Risheng Liu

Sheng Liu
Shu Liu
Shuaicheng Liu
Sifei Liu
Siqi Liu
Siying Liu
Songtao Liu
Ting Liu
Tongliang Liu
Tyng-Luh Liu
Wanquan Liu
Wei Liu
Weiyang Liu
Weizhe Liu
Wenyu Liu
Wu Liu
Xialei Liu
Xianglong Liu
Xiaodong Liu
Xiaofeng Liu
Xihui Liu
Xingyu Liu
Xinwang Liu
Xuanqing Liu
Xuebo Liu
Yang Liu
Yaojie Liu
Yebin Liu
Yen-Cheng Liu
Yiming Liu
Yu Liu
Yu-Shen Liu
Yufan Liu
Yun Liu
Zheng Liu
Zhijian Liu
Zhuang Liu
Zichuan Liu
Ziwei Liu
Zongyi Liu
Stephan Liwicki
Liliana Lo Presti
Chengjiang Long
Fuchen Long
Mingsheng Long
Xiang Long
Yang Long
Charles T. Loop
Antonio Lopez
Roberto J. Lopez-Sastre
Javier Lorenzo-Navarro
Manolis Lourakis
Boyu Lu
Canyi Lu
Feng Lu
Guoyu Lu
Hongtao Lu
Jiajun Lu
Jiasen Lu
Jiwen Lu
Kaiyue Lu
Le Lu
Shao-Ping Lu
Shijian Lu
Xiankai Lu
Xin Lu
Yao Lu
Yiping Lu
Yongxi Lu
Yongyi Lu
Zhiwu Lu
Fujun Luan
Benjamin E. Lundell
Hao Luo
Jian-Hao Luo
Ruotian Luo
Weixin Luo
Wenhan Luo
Wenjie Luo
Yan Luo
Zelun Luo
Zixin Luo
Khoa Luu
Zhaoyang Lv
Pengyuan Lyu
Thomas Möllenhoff
Matthias Müller
Bingpeng Ma
Chih-Yao Ma
Chongyang Ma
Huimin Ma
Jiayi Ma
K. T. Ma
Ke Ma
Lin Ma
Liqian Ma
Shugao Ma
Wei-Chiu Ma
Xiaojian Ma
Xingjun Ma
Zhanyu Ma
Zheng Ma
Radek Jakob Mackowiak
Ludovic Magerand
Shweta Mahajan
Siddharth Mahendran
Long Mai
Ameesh Makadia
Oscar Mendez Maldonado
Mateusz Malinowski
Yury Malkov
Arun Mallya
Dipu Manandhar
Massimiliano Mancini
Fabian Manhardt
Kevis-kokitsi Maninis
Varun Manjunatha
Junhua Mao
Xudong Mao
Alina Marcu
Edgar Margffoy-Tuay
Dmitrii Marin
Manuel J. Marin-Jimenez
Kenneth Marino
Niki Martinel
Julieta Martinez
Jonathan Masci
Tomohiro Mashita
Iacopo Masi
David Masip
Daniela Massiceti
Stefan Mathe
Yusuke Matsui
Tetsu Matsukawa
Iain A. Matthews
Kevin James Matzen
Bruce Allen Maxwell
Stephen Maybank

Helmut Mayer
Amir Mazaheri
David McAllester
Steven McDonagh
Stephen J. Mckenna
Roey Mechrez
Prakhar Mehrotra
Christopher Mei
Xue Mei
Paulo R. S. Mendonca
Lili Meng
Zibo Meng
Thomas Mensink
Bjoern Menze
Michele Merler
Kourosh Meshgi
Pascal Mettes
Christopher Metzler
Liang Mi
Qiguang Miao
Xin Miao
Tomer Michaeli
Frank Michel
Antoine Miech
Krystian Mikolajczyk
Peyman Milanfar
Ben Mildenhall
Gregor Miller
Fausto Milletari
Dongbo Min
Kyle Min
Pedro Miraldo
Dmytro Mishkin
Anand Mishra
Ashish Mishra
Ishan Misra
Niluthpol C. Mithun
Kaushik Mitra
Niloy Mitra
Anton Mitrokhin
Ikuhisa Mitsugami
Anurag Mittal
Kaichun Mo
Zhipeng Mo
Davide Modolo
Michael Moeller
Pritish Mohapatra
Pavlo Molchanov
Davide Moltisanti
Pascal Monasse
Mathew Monfort
Aron Monszpart
Sean Moran
Vlad I. Morariu
Francesc Moreno-Noguer
Pietro Morerio
Stylianos Moschoglou
Yael Moses
Roozbeh Mottaghi
Pierre Moulon
Arsalan Mousavian
Yadong Mu
Yasuhiro Mukaigawa
Lopamudra Mukherjee
Yusuke Mukuta
Ravi Teja Mullapudi
Mario Enrique Munich
Zachary Murez
Ana C. Murillo
J. Krishna Murthy
Damien Muselet
Armin Mustafa
Siva Karthik Mustikovela
Carlo Dal Mutto
Moin Nabi
Varun K. Nagaraja
Tushar Nagarajan
Arsha Nagrani
Seungjun Nah
Nikhil Naik
Yoshikatsu Nakajima
Yuta Nakashima
Atsushi Nakazawa
Seonghyeon Nam
Vinay P. Namboodiri
Medhini Narasimhan
Srinivasa Narasimhan
Sanath Narayan
Erickson Rangel Nascimento
Jacinto Nascimento
Tayyab Naseer
Lakshmanan Nataraj
Neda Nategh
Nelson Isao Nauata
Fernando Navarro
Shah Nawaz
Lukas Neumann
Ram Nevatia
Alejandro Newell
Shawn Newsam
Joe Yue-Hei Ng
Trung Thanh Ngo
Duc Thanh Nguyen
Lam M. Nguyen
Phuc Xuan Nguyen
Thuong Nguyen Canh
Mihalis Nicolaou
Andrei Liviu Nicolicioiu
Xuecheng Nie
Michael Niemeyer
Simon Niklaus
Christophoros Nikou
David Nilsson
Jifeng Ning
Yuval Nirkin
Li Niu
Yuzhen Niu
Zhenxing Niu
Shohei Nobuhara
Nicoletta Noceti
Hyeonwoo Noh
Junhyug Noh
Mehdi Noroozi
Sotiris Nousias
Valsamis Ntouskos
Matthew O'Toole
Peter Ochs
Ferda Ofli
Seong Joon Oh
Seoung Wug Oh
Iason Oikonomidis
Utkarsh Ojha
Takahiro Okabe
Takayuki Okatani
Fumio Okura
Aude Oliva
Kyle Olszewski

Björn Ommer
Mohamed Omran
Elisabeta Oneata
Michael Opitz
Jose Oramas
Tribhuvanesh Orekondy
Shaul Oron
Sergio Orts-Escolano
Ivan Oseledets
Aljosa Osep
Magnus Oskarsson
Anton Osokin
Martin R. Oswald
Wanli Ouyang
Andrew Owens
Mete Ozay
Mustafa Ozuysal
Eduardo Pérez-Pellitero
Gautam Pai
Dipan Kumar Pal
P. H. Pamplona Savarese
Jinshan Pan
Junting Pan
Xingang Pan
Yingwei Pan
Yannis Panagakis
Rameswar Panda
Guan Pang
Jiahao Pang
Jiangmiao Pang
Tianyu Pang
Sharath Pankanti
Nicolas Papadakis
Dim Papadopoulos
George Papandreou
Toufiq Parag
Shaifali Parashar
Sarah Parisot
Eunhyeok Park
Hyun Soo Park
Jaesik Park
Min-Gyu Park
Taesung Park
Alvaro Parra
C. Alejandro Parraga
Despoina Paschalidou
Nikolaos Passalis
Vishal Patel
Viorica Patraucean
Badri Narayana Patro
Danda Pani Paudel
Sujoy Paul
Georgios Pavlakos
Ioannis Pavlidis
Vladimir Pavlovic
Nick Pears
Kim Steenstrup Pedersen
Selen Pehlivan
Shmuel Peleg
Chao Peng
Houwen Peng
Wen-Hsiao Peng
Xi Peng
Xiaojiang Peng
Xingchao Peng
Yuxin Peng
Federico Perazzi
Juan Camilo Perez
Vishwanath Peri
Federico Pernici
Luca Del Pero
Florent Perronnin
Stavros Petridis
Henning Petzka
Patrick Peursum
Michael Pfeiffer
Hanspeter Pfister
Roman Pflugfelder
Minh Tri Pham
Yongri Piao
David Picard
Tomasz Pieciak
A. J. Piergiovanni
Andrea Pilzer
Pedro O. Pinheiro
Silvia Laura Pintea
Lerrel Pinto
Axel Pinz
Robinson Piramuthu
Fiora Pirri
Leonid Pishchulin
Francesco Pittaluga
Daniel Pizarro
Tobias Plötz
Mirco Planamente
Matteo Poggi
Moacir A. Ponti
Parita Pooj
Fatih Porikli
Horst Possegger
Omid Poursaeed
Ameya Prabhu
Viraj Uday Prabhu
Dilip Prasad
Brian L. Price
True Price
Maria Priisalu
Veronique Prinet
Victor Adrian Prisacariu
Jan Prokaj
Sergey Prokudin
Nicolas Pugeault
Xavier Puig
Albert Pumarola
Pulak Purkait
Senthil Purushwalkam
Charles R. Qi
Hang Qi
Haozhi Qi
Lu Qi
Mengshi Qi
Siyuan Qi
Xiaojuan Qi
Yuankai Qi
Shengju Qian
Xuelin Qian
Siyuan Qiao
Yu Qiao
Jie Qin
Qiang Qiu
Weichao Qiu
Zhaofan Qiu
Kha Gia Quach
Yuhui Quan
Yvain Queau
Julian Quiroga
Faisal Qureshi
Mahdi Rad

Filip Radenovic
Petia Radeva
Venkatesh B. Radhakrishnan
Ilija Radosavovic
Noha Radwan
Rahul Raguram
Tanzila Rahman
Amit Raj
Ajit Rajwade
Kandan Ramakrishnan
Santhosh K. Ramakrishnan
Srikumar Ramalingam
Ravi Ramamoorthi
Vasili Ramanishka
Ramprasaath R. Selvaraju
Francois Rameau
Visvanathan Ramesh
Santu Rana
Rene Ranftl
Anand Rangarajan
Anurag Ranjan
Viresh Ranjan
Yongming Rao
Carolina Raposo
Vivek Rathod
Sathya N. Ravi
Avinash Ravichandran
Tammy Riklin Raviv
Daniel Rebain
Sylvestre-Alvise Rebuffi
N. Dinesh Reddy
Timo Rehfeld
Paolo Remagnino
Konstantinos Rematas
Edoardo Remelli
Dongwei Ren
Haibing Ren
Jian Ren
Jimmy Ren
Mengye Ren
Weihong Ren
Wenqi Ren
Zhile Ren
Zhongzheng Ren
Zhou Ren
Vijay Rengarajan
Md A. Reza
Farzaneh Rezaeianaran
Hamed R. Tavakoli
Nicholas Rhinehart
Helge Rhodin
Elisa Ricci
Alexander Richard
Eitan Richardson
Elad Richardson
Christian Richardt
Stephan Richter
Gernot Riegler
Daniel Ritchie
Tobias Ritschel
Samuel Rivera
Yong Man Ro
Richard Roberts
Joseph Robinson
Ignacio Rocco
Mrigank Rochan
Emanuele Rodolà
Mikel D. Rodriguez
Giorgio Roffo
Grégory Rogez
Gemma Roig
Javier Romero
Xuejian Rong
Yu Rong
Amir Rosenfeld
Bodo Rosenhahn
Guy Rosman
Arun Ross
Paolo Rota
Peter M. Roth
Anastasios Roussos
Anirban Roy
Sebastien Roy
Aruni RoyChowdhury
Artem Rozantsev
Ognjen Rudovic
Daniel Rueckert
Adria Ruiz
Javier Ruiz-del-solar
Christian Rupprecht
Chris Russell
Dan Ruta
Jongbin Ryu
Ömer Sümer
Alexandre Sablayrolles
Faraz Saeedan
Ryusuke Sagawa
Christos Sagonas
Tonmoy Saikia
Hideo Saito
Kuniaki Saito
Shunsuke Saito
Shunta Saito
Ken Sakurada
Joaquin Salas
Fatemeh Sadat Saleh
Mahdi Saleh
Pouya Samangouei
Leo Sampaio Ferraz Ribeiro
Artsiom Olegovich Sanakoyeu
Enrique Sanchez
Patsorn Sangkloy
Anush Sankaran
Aswin Sankaranarayanan
Swami Sankaranarayanan
Rodrigo Santa Cruz
Amartya Sanyal
Archana Sapkota
Nikolaos Sarafianos
Jun Sato
Shin'ichi Satoh
Hosnieh Sattar
Arman Savran
Manolis Savva
Alexander Sax
Hanno Scharr
Simone Schaub-Meyer
Konrad Schindler
Dmitrij Schlesinger
Uwe Schmidt
Dirk Schnieders
Björn Schuller
Samuel Schulter
Idan Schwartz

William Robson Schwartz
Alex Schwing
Sinisa Segvic
Lorenzo Seidenari
Pradeep Sen
Ozan Sener
Soumyadip Sengupta
Arda Senocak
Mojtaba Seyedhosseini
Shishir Shah
Shital Shah
Sohil Atul Shah
Tamar Rott Shaham
Huasong Shan
Qi Shan
Shiguang Shan
Jing Shao
Roman Shapovalov
Gaurav Sharma
Vivek Sharma
Viktoriia Sharmanska
Dongyu She
Sumit Shekhar
Evan Shelhamer
Chengyao Shen
Chunhua Shen
Falong Shen
Jie Shen
Li Shen
Liyue Shen
Shuhan Shen
Tianwei Shen
Wei Shen
William B. Shen
Yantao Shen
Ying Shen
Yiru Shen
Yujun Shen
Yuming Shen
Zhiqiang Shen
Ziyi Shen
Lu Sheng
Yu Sheng
Rakshith Shetty
Baoguang Shi
Guangming Shi
Hailin Shi
Miaojing Shi
Yemin Shi
Zhenmei Shi
Zhiyuan Shi
Kevin Jonathan Shih
Shiliang Shiliang
Hyunjung Shim
Atsushi Shimada
Nobutaka Shimada
Daeyun Shin
Young Min Shin
Koichi Shinoda
Konstantin Shmelkov
Michael Zheng Shou
Abhinav Shrivastava
Tianmin Shu
Zhixin Shu
Hong-Han Shuai
Pushkar Shukla
Christian Siagian
Mennatullah M. Siam
Kaleem Siddiqi
Karan Sikka
Jae-Young Sim
Christian Simon
Martin Simonovsky
Dheeraj Singaraju
Bharat Singh
Gurkirt Singh
Krishna Kumar Singh
Maneesh Kumar Singh
Richa Singh
Saurabh Singh
Suriya Singh
Vikas Singh
Sudipta N. Sinha
Vincent Sitzmann
Josef Sivic
Gregory Slabaugh
Miroslava Slavcheva
Ron Slossberg
Brandon Smith
Kevin Smith
Vladimir Smutny
Noah Snavely
Roger D. Soberanis-Mukul
Kihyuk Sohn
Francesco Solera
Eric Sommerlade
Sanghyun Son
Byung Cheol Song
Chunfeng Song
Dongjin Song
Jiaming Song
Jie Song
Jifei Song
Jingkuan Song
Mingli Song
Shiyu Song
Shuran Song
Xiao Song
Yafei Song
Yale Song
Yang Song
Yi-Zhe Song
Yibing Song
Humberto Sossa
Cesar de Souza
Adrian Spurr
Srinath Sridhar
Suraj Srinivas
Pratul P. Srinivasan
Anuj Srivastava
Tania Stathaki
Christopher Stauffer
Simon Stent
Rainer Stiefelhagen
Pierre Stock
Julian Straub
Jonathan C. Stroud
Joerg Stueckler
Jan Stuehmer
David Stutz
Chi Su
Hang Su
Jong-Chyi Su
Shuochen Su
Yu-Chuan Su
Ramanathan Subramanian
Yusuke Sugano

Masanori Suganuma
Yumin Suh
Mohammed Suhail
Yao Sui
Heung-Il Suk
Josephine Sullivan
Baochen Sun
Chen Sun
Chong Sun
Deqing Sun
Jin Sun
Liang Sun
Lin Sun
Qianru Sun
Shao-Hua Sun
Shuyang Sun
Weiwei Sun
Wenxiu Sun
Xiaoshuai Sun
Xiaoxiao Sun
Xingyuan Sun
Yifan Sun
Zhun Sun
Sabine Susstrunk
David Suter
Supasorn Suwajanakorn
Tomas Svoboda
Eran Swears
Paul Swoboda
Attila Szabo
Richard Szeliski
Duy-Nguyen Ta
Andrea Tagliasacchi
Yuichi Taguchi
Ying Tai
Keita Takahashi
Kouske Takahashi
Jun Takamatsu
Hugues Talbot
Toru Tamaki
Chaowei Tan
Fuwen Tan
Mingkui Tan
Mingxing Tan
Qingyang Tan
Robby T. Tan
Xiaoyang Tan
Kenichiro Tanaka
Masayuki Tanaka
Chang Tang
Chengzhou Tang
Danhang Tang
Ming Tang
Peng Tang
Qingming Tang
Wei Tang
Xu Tang
Yansong Tang
Youbao Tang
Yuxing Tang
Zhiqiang Tang
Tatsunori Taniai
Junli Tao
Xin Tao
Makarand Tapaswi
Jean-Philippe Tarel
Lyne Tchapmi
Zachary Teed
Bugra Tekin
Damien Teney
Ayush Tewari
Christian Theobalt
Christopher Thomas
Diego Thomas
Jim Thomas
Rajat Mani Thomas
Xinmei Tian
Yapeng Tian
Yingli Tian
Yonglong Tian
Zhi Tian
Zhuotao Tian
Kinh Tieu
Joseph Tighe
Massimo Tistarelli
Matthew Toews
Carl Toft
Pavel Tokmakov
Federico Tombari
Chetan Tonde
Yan Tong
Alessio Tonioni
Andrea Torsello
Fabio Tosi
Du Tran
Luan Tran
Ngoc-Trung Tran
Quan Hung Tran
Truyen Tran
Rudolph Triebel
Martin Trimmel
Shashank Tripathi
Subarna Tripathi
Leonardo Trujillo
Eduard Trulls
Tomasz Trzcinski
Sam Tsai
Yi-Hsuan Tsai
Hung-Yu Tseng
Stavros Tsogkas
Aggeliki Tsoli
Devis Tuia
Shubham Tulsiani
Sergey Tulyakov
Frederick Tung
Tony Tung
Daniyar Turmukhambetov
Ambrish Tyagi
Radim Tylecek
Christos Tzelepis
Georgios Tzimiropoulos
Dimitrios Tzionas
Seiichi Uchida
Norimichi Ukita
Dmitry Ulyanov
Martin Urschler
Yoshitaka Ushiku
Ben Usman
Alexander Vakhitov
Julien P. C. Valentin
Jack Valmadre
Ernest Valveny
Joost van de Weijer
Jan van Gemert
Koen Van Leemput
Gul Varol
Sebastiano Vascon
M. Alex O. Vasilescu

Subeesh Vasu
Mayank Vatsa
David Vazquez
Javier Vazquez-Corral
Ashok Veeraraghavan
Erik Velasco-Salido
Raviteja Vemulapalli
Jonathan Ventura
Manisha Verma
Roberto Vezzani
Ruben Villegas
Minh Vo
MinhDuc Vo
Nam Vo
Michele Volpi
Riccardo Volpi
Carl Vondrick
Konstantinos Vougioukas
Tuan-Hung Vu
Sven Wachsmuth
Neal Wadhwa
Catherine Wah
Jacob C. Walker
Thomas S. A. Wallis
Chengde Wan
Jun Wan
Liang Wan
Renjie Wan
Baoyuan Wang
Boyu Wang
Cheng Wang
Chu Wang
Chuan Wang
Chunyu Wang
Dequan Wang
Di Wang
Dilin Wang
Dong Wang
Fang Wang
Guanzhi Wang
Guoyin Wang
Hanzi Wang
Hao Wang
He Wang
Heng Wang
Hongcheng Wang
Hongxing Wang
Hua Wang
Jian Wang
Jingbo Wang
Jinglu Wang
Jingya Wang
Jinjun Wang
Jinqiao Wang
Jue Wang
Ke Wang
Keze Wang
Le Wang
Lei Wang
Lezi Wang
Li Wang
Liang Wang
Lijun Wang
Limin Wang
Linwei Wang
Lizhi Wang
Mengjiao Wang
Mingzhe Wang
Minsi Wang
Naiyan Wang
Nannan Wang
Ning Wang
Oliver Wang
Pei Wang
Peng Wang
Pichao Wang
Qi Wang
Qian Wang
Qiaosong Wang
Qifei Wang
Qilong Wang
Qing Wang
Qingzhong Wang
Quan Wang
Rui Wang
Ruiping Wang
Ruixing Wang
Shangfei Wang
Shenlong Wang
Shiyao Wang
Shuhui Wang
Song Wang
Tao Wang
Tianlu Wang
Tiantian Wang
Ting-chun Wang
Tingwu Wang
Wei Wang
Weiyue Wang
Wenguan Wang
Wenlin Wang
Wenqi Wang
Xiang Wang
Xiaobo Wang
Xiaofang Wang
Xiaoling Wang
Xiaolong Wang
Xiaosong Wang
Xiaoyu Wang
Xin Eric Wang
Xinchao Wang
Xinggang Wang
Xintao Wang
Yali Wang
Yan Wang
Yang Wang
Yangang Wang
Yaxing Wang
Yi Wang
Yida Wang
Yilin Wang
Yiming Wang
Yisen Wang
Yongtao Wang
Yu-Xiong Wang
Yue Wang
Yujiang Wang
Yunbo Wang
Yunhe Wang
Zengmao Wang
Zhangyang Wang
Zhaowen Wang
Zhe Wang
Zhecan Wang
Zheng Wang
Zhixiang Wang
Zilei Wang
Jianqiao Wangni

Anne S. Wannenwetsch
Jan Dirk Wegner
Scott Wehrwein
Donglai Wei
Kaixuan Wei
Longhui Wei
Pengxu Wei
Ping Wei
Qi Wei
Shih-En Wei
Xing Wei
Yunchao Wei
Zijun Wei
Jerod Weinman
Michael Weinmann
Philippe Weinzaepfel
Yair Weiss
Bihan Wen
Longyin Wen
Wei Wen
Junwu Weng
Tsui-Wei Weng
Xinshuo Weng
Eric Wengrowski
Tomas Werner
Gordon Wetzstein
Tobias Weyand
Patrick Wieschollek
Maggie Wigness
Erik Wijmans
Richard Wildes
Olivia Wiles
Chris Williams
Williem Williem
Kyle Wilson
Calden Wloka
Nicolai Wojke
Christian Wolf
Yongkang Wong
Sanghyun Woo
Scott Workman
Baoyuan Wu
Bichen Wu
Chao-Yuan Wu
Huikai Wu
Jiajun Wu
Jialin Wu
Jiaxiang Wu
Jiqing Wu
Jonathan Wu
Lifang Wu
Qi Wu
Qiang Wu
Ruizheng Wu
Shangzhe Wu
Shun-Cheng Wu
Tianfu Wu
Wayne Wu
Wenxuan Wu
Xiao Wu
Xiaohe Wu
Xinxiao Wu
Yang Wu
Yi Wu
Yiming Wu
Ying Nian Wu
Yue Wu
Zheng Wu
Zhenyu Wu
Zhirong Wu
Zuxuan Wu
Stefanie Wuhrer
Jonas Wulff
Changqun Xia
Fangting Xia
Fei Xia
Gui-Song Xia
Lu Xia
Xide Xia
Yin Xia
Yingce Xia
Yongqin Xian
Lei Xiang
Shiming Xiang
Bin Xiao
Fanyi Xiao
Guobao Xiao
Huaxin Xiao
Taihong Xiao
Tete Xiao
Tong Xiao
Wang Xiao
Yang Xiao
Cihang Xie
Guosen Xie
Jianwen Xie
Lingxi Xie
Sirui Xie
Weidi Xie
Wenxuan Xie
Xiaohua Xie
Fuyong Xing
Jun Xing
Junliang Xing
Bo Xiong
Peixi Xiong
Yu Xiong
Yuanjun Xiong
Zhiwei Xiong
Chang Xu
Chenliang Xu
Dan Xu
Danfei Xu
Hang Xu
Hongteng Xu
Huijuan Xu
Jingwei Xu
Jun Xu
Kai Xu
Mengmeng Xu
Mingze Xu
Qianqian Xu
Ran Xu
Weijian Xu
Xiangyu Xu
Xiaogang Xu
Xing Xu
Xun Xu
Yanyu Xu
Yichao Xu
Yong Xu
Yongchao Xu
Yuanlu Xu
Zenglin Xu
Zheng Xu
Chuhui Xue
Jia Xue
Nan Xue

Tianfan Xue
Xiangyang Xue
Abhay Yadav
Yasushi Yagi
I. Zeki Yalniz
Kota Yamaguchi
Toshihiko Yamasaki
Takayoshi Yamashita
Junchi Yan
Ke Yan
Qingan Yan
Sijie Yan
Xinchen Yan
Yan Yan
Yichao Yan
Zhicheng Yan
Keiji Yanai
Bin Yang
Ceyuan Yang
Dawei Yang
Dong Yang
Fan Yang
Guandao Yang
Guorun Yang
Haichuan Yang
Hao Yang
Jianwei Yang
Jiaolong Yang
Jie Yang
Jing Yang
Kaiyu Yang
Linjie Yang
Meng Yang
Michael Ying Yang
Nan Yang
Shuai Yang
Shuo Yang
Tianyu Yang
Tien-Ju Yang
Tsun-Yi Yang
Wei Yang
Wenhan Yang
Xiao Yang
Xiaodong Yang
Xin Yang
Yan Yang
Yanchao Yang
Yee Hong Yang
Yezhou Yang
Zhenheng Yang
Anbang Yao
Angela Yao
Cong Yao
Jian Yao
Li Yao
Ting Yao
Yao Yao
Zhewei Yao
Chengxi Ye
Jianbo Ye
Keren Ye
Linwei Ye
Mang Ye
Mao Ye
Qi Ye
Qixiang Ye
Mei-Chen Yeh
Raymond Yeh
Yu-Ying Yeh
Sai-Kit Yeung
Serena Yeung
Kwang Moo Yi
Li Yi
Renjiao Yi
Alper Yilmaz
Junho Yim
Lijun Yin
Weidong Yin
Xi Yin
Zhichao Yin
Tatsuya Yokota
Ryo Yonetani
Donggeun Yoo
Jae Shin Yoon
Ju Hong Yoon
Sung-eui Yoon
Laurent Younes
Changqian Yu
Fisher Yu
Gang Yu
Jiahui Yu
Kaicheng Yu
Ke Yu
Lequan Yu
Ning Yu
Qian Yu
Ronald Yu
Ruichi Yu
Shoou-I Yu
Tao Yu
Tianshu Yu
Xiang Yu
Xin Yu
Xiyu Yu
Youngjae Yu
Yu Yu
Zhiding Yu
Chunfeng Yuan
Ganzhao Yuan
Jinwei Yuan
Lu Yuan
Quan Yuan
Shanxin Yuan
Tongtong Yuan
Wenjia Yuan
Ye Yuan
Yuan Yuan
Yuhui Yuan
Huanjing Yue
Xiangyu Yue
Ersin Yumer
Sergey Zagoruyko
Egor Zakharov
Amir Zamir
Andrei Zanfir
Mihai Zanfir
Pablo Zegers
Bernhard Zeisl
John S. Zelek
Niclas Zeller
Huayi Zeng
Jiabei Zeng
Wenjun Zeng
Yu Zeng
Xiaohua Zhai
Fangneng Zhan
Huangying Zhan
Kun Zhan

Xiaohang Zhan
Baochang Zhang
Bowen Zhang
Cecilia Zhang
Changqing Zhang
Chao Zhang
Chengquan Zhang
Chi Zhang
Chongyang Zhang
Dingwen Zhang
Dong Zhang
Feihu Zhang
Hang Zhang
Hanwang Zhang
Hao Zhang
He Zhang
Hongguang Zhang
Hua Zhang
Ji Zhang
Jianguo Zhang
Jianming Zhang
Jiawei Zhang
Jie Zhang
Jing Zhang
Juyong Zhang
Kai Zhang
Kaipeng Zhang
Ke Zhang
Le Zhang
Lei Zhang
Li Zhang
Lihe Zhang
Linguang Zhang
Lu Zhang
Mi Zhang
Mingda Zhang
Peng Zhang
Pingping Zhang
Qian Zhang
Qilin Zhang
Quanshi Zhang
Richard Zhang
Rui Zhang
Runze Zhang
Shengping Zhang
Shifeng Zhang
Shuai Zhang
Songyang Zhang
Tao Zhang
Ting Zhang
Tong Zhang
Wayne Zhang
Wei Zhang
Weizhong Zhang
Wenwei Zhang
Xiangyu Zhang
Xiaolin Zhang
Xiaopeng Zhang
Xiaoqin Zhang
Xiuming Zhang
Ya Zhang
Yang Zhang
Yimin Zhang
Yinda Zhang
Ying Zhang
Yongfei Zhang
Yu Zhang
Yulun Zhang
Yunhua Zhang
Yuting Zhang
Zhanpeng Zhang
Zhao Zhang
Zhaoxiang Zhang
Zhen Zhang
Zheng Zhang
Zhifei Zhang
Zhijin Zhang
Zhishuai Zhang
Ziming Zhang
Bo Zhao
Chen Zhao
Fang Zhao
Haiyu Zhao
Han Zhao
Hang Zhao
Hengshuang Zhao
Jian Zhao
Kai Zhao
Liang Zhao
Long Zhao
Qian Zhao
Qibin Zhao
Qijun Zhao
Rui Zhao
Shenglin Zhao
Sicheng Zhao
Tianyi Zhao
Wenda Zhao
Xiangyun Zhao
Xin Zhao
Yang Zhao
Yue Zhao
Zhichen Zhao
Zijing Zhao
Xiantong Zhen
Chuanxia Zheng
Feng Zheng
Haiyong Zheng
Jia Zheng
Kang Zheng
Shuai Kyle Zheng
Wei-Shi Zheng
Yinqiang Zheng
Zerong Zheng
Zhedong Zheng
Zilong Zheng
Bineng Zhong
Fangwei Zhong
Guangyu Zhong
Yiran Zhong
Yujie Zhong
Zhun Zhong
Chunluan Zhou
Huiyu Zhou
Jiahuan Zhou
Jun Zhou
Lei Zhou
Luowei Zhou
Luping Zhou
Mo Zhou
Ning Zhou
Pan Zhou
Peng Zhou
Qianyi Zhou
S. Kevin Zhou
Sanping Zhou
Wengang Zhou
Xingyi Zhou

Yanzhao Zhou
Yi Zhou
Yin Zhou
Yipin Zhou
Yuyin Zhou
Zihan Zhou
Alex Zihao Zhu
Chenchen Zhu
Feng Zhu
Guangming Zhu
Ji Zhu
Jun-Yan Zhu
Lei Zhu
Linchao Zhu
Rui Zhu
Shizhan Zhu
Tyler Lixuan Zhu
Wei Zhu
Xiangyu Zhu
Xinge Zhu
Xizhou Zhu
Yanjun Zhu
Yi Zhu
Yixin Zhu
Yizhe Zhu
Yousong Zhu
Zhe Zhu
Zhen Zhu
Zheng Zhu
Zhenyao Zhu
Zhihui Zhu
Zhuotun Zhu
Bingbing Zhuang
Wei Zhuo
Christian Zimmermann
Karel Zimmermann
Larry Zitnick
Mohammadreza Zolfaghari
Maria Zontak
Daniel Zoran
Changqing Zou
Chuhang Zou
Danping Zou
Qi Zou
Yang Zou
Yuliang Zou
Georgios Zoumpourlis
Wangmeng Zuo
Xinxin Zuo

Additional Reviewers

Victoria Fernandez Abrevaya
Maya Aghaei
Allam Allam
Christine Allen-Blanchette
Nicolas Aziere
Assia Benbihi
Neha Bhargava
Bharat Lal Bhatnagar
Joanna Bitton
Judy Borowski
Amine Bourki
Romain Brégier
Tali Brayer
Sebastian Bujwid
Andrea Burns
Yun-Hao Cao
Yuning Chai
Xiaojun Chang
Bo Chen
Shuo Chen
Zhixiang Chen
Junsuk Choe
Hung-Kuo Chu
Jonathan P. Crall
Kenan Dai
Lucas Deecke
Karan Desai
Prithviraj Dhar
Jing Dong
Wei Dong
Turan Kaan Elgin
Francis Engelmann
Erik Englesson
Fartash Faghri
Zicong Fan
Yang Fu
Risheek Garrepalli
Yifan Ge
Marco Godi
Helmut Grabner
Shuxuan Guo
Jianfeng He
Zhezhi He
Samitha Herath
Chih-Hui Ho
Yicong Hong
Vincent Tao Hu
Julio Hurtado
Jaedong Hwang
Andrey Ignatov
Muhammad Abdullah Jamal
Saumya Jetley
Meiguang Jin
Jeff Johnson
Minsoo Kang
Saeed Khorram
Mohammad Rami Koujan
Nilesh Kulkarni
Sudhakar Kumawat
Abdelhak Lemkhenter
Alexander Levine
Jiachen Li
Jing Li
Jun Li
Yi Li
Liang Liao
Ruochen Liao
Tzu-Heng Lin
Phillip Lippe
Bao-di Liu
Bo Liu
Fangchen Liu

Hanxiao Liu
Hongyu Liu
Huidong Liu
Miao Liu
Xinxin Liu
Yongfei Liu
Yu-Lun Liu
Amir Livne
Tiange Luo
Wei Ma
Xiaoxuan Ma
Ioannis Marras
Georg Martius
Effrosyni Mavroudi
Tim Meinhardt
Givi Meishvili
Meng Meng
Zihang Meng
Zhongqi Miao
Gyeongsik Moon
Khoi Nguyen
Yung-Kyun Noh
Antonio Norelli
Jaeyoo Park
Alexander Pashevich
Mandela Patrick
Mary Phuong
Bingqiao Qian
Yu Qiao
Zhen Qiao
Sai Saketh Rambhatla
Aniket Roy
Amelie Royer
Parikshit Vishwas Sakurikar
Mark Sandler
Mert Bülent Sarıyıldız
Tanner Schmidt
Anshul B. Shah
Ketul Shah
Rajvi Shah
Hengcan Shi
Xiangxi Shi
Yujiao Shi
William A. P. Smith
Guoxian Song
Robin Strudel
Abby Stylianou
Xinwei Sun
Reuben Tan
Qingyi Tao
Kedar S. Tatwawadi
Anh Tuan Tran
Son Dinh Tran
Eleni Triantafillou
Aristeidis Tsitiridis
Md Zasim Uddin
Andrea Vedaldi
Evangelos Ververas
Vidit Vidit
Paul Voigtlaender
Bo Wan
Huanyu Wang
Huiyu Wang
Junqiu Wang
Pengxiao Wang
Tai Wang
Xinyao Wang
Tomoki Watanabe
Mark Weber
Xi Wei
Botong Wu
James Wu
Jiamin Wu
Rujie Wu
Yu Wu
Rongchang Xie
Wei Xiong
Yunyang Xiong
An Xu
Chi Xu
Yinghao Xu
Fei Xue
Tingyun Yan
Zike Yan
Chao Yang
Heran Yang
Ren Yang
Wenfei Yang
Xu Yang
Rajeev Yasarla
Shaokai Ye
Yufei Ye
Kun Yi
Haichao Yu
Hanchao Yu
Ruixuan Yu
Liangzhe Yuan
Chen-Lin Zhang
Fandong Zhang
Tianyi Zhang
Yang Zhang
Yiyi Zhang
Yongshun Zhang
Yu Zhang
Zhiwei Zhang
Jiaojiao Zhao
Yipu Zhao
Xingjian Zhen
Haizhong Zheng
Tiancheng Zhi
Chengju Zhou
Hao Zhou
Hao Zhu
Alexander Zimin

Contents – Part XXIV

Deep Novel View Synthesis from Colored 3D Point Clouds

Zhenbo Song[1(✉)], Wayne Chen[2], Dylan Campbell[2,3], and Hongdong Li[2,3]

[1] Nanjing University of Science and Technology, Nanjing, China
songzb@njust.edu.cn
[2] ANU–Australian National University, Canberra, Australia
[3] Australian Centre for Robotic Vision, Brisbane, Australia

Abstract. We propose a new deep neural network which takes a colored 3D point cloud of a scene as input, and synthesizes a photo-realistic image from a novel viewpoint. Key contributions of this work include a deep point feature extraction module, an image synthesis module, and a refinement module. Our PointEncoder network extracts discriminative features from the point cloud that contain both local and global contextual information about the scene. Next, the multi-level point features are aggregated to form multi-layer feature maps, which are subsequently fed into an ImageDecoder network to generate a synthetic RGB image. Finally, the output of the ImageDecoder network is refined using a RefineNet module, providing finer details and suppressing unwanted visual artifacts. W rotate and translate the 3D point cloud in order to synthesize new images from a novel perspective. We conduct numerous experiments on public datasets to validate the method in terms of quality of the synthesized views.

Keywords: Image synthesis · 3D point clouds · Virtual views

1 Introduction

This paper addresses the problem of how to render a dense photo-realistic RGB image of a static 3D scene from a novel viewpoint, only based on a set of sparse colored point clouds depicting the scene. The rendering pipeline is illustrated in Fig. 1. Traditional methods are often based on fitting point clouds to a piece-wise smooth mesh surface; they however suffer from the need of a strong scene prior and large amount of computation, despite which they can fail when the point clouds are too sparse or contain gross outliers, as is typical for real-world 3D range scans.

Practical uses of novel view synthesis include generating photo-realistic views from a real physical scene for immersive Augmented Reality applications. Structure from Motion (SfM) techniques have been applied to reconstruct 3D models

Electronic supplementary material The online version of this chapter (https://doi.org/10.1007/978-3-030-58586-0_1) contains supplementary material, which is available to authorized users.

A. Vedaldi et al. (Eds.): ECCV 2020, LNCS 12369, pp. 1–17, 2020.
https://doi.org/10.1007/978-3-030-58586-0_1

depicting the real scene. This way, the 3D models are represented as sparse set of 3D point clouds which are both computation and memory efficient, but falls short in visual appearance due to the very low density in discrete point samplings. This has motivated this paper to develop an efficient new method of dense novel view synthesis *directly* from sparse set of colored 3D point clouds.

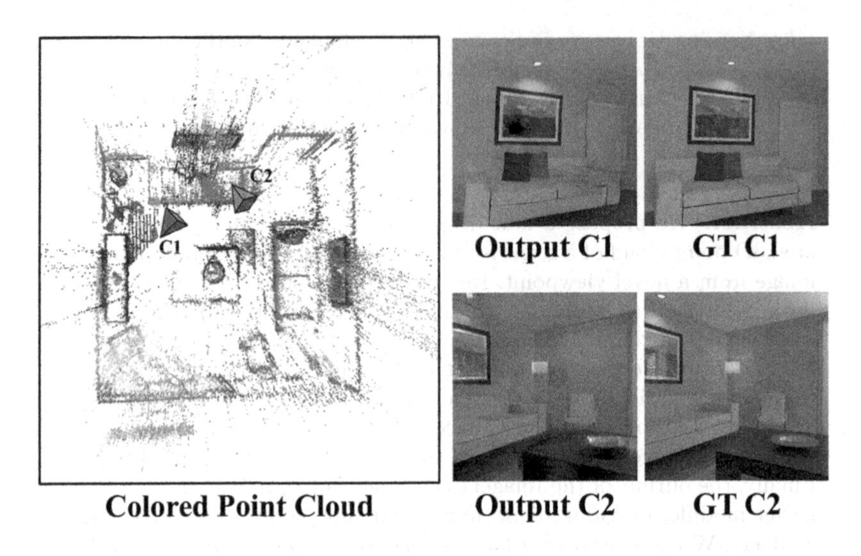

Fig. 1. Synthesizing novel views from colored 3D point clouds. The colored point cloud is generated from key frames of a video sequence using DSO [7]. Given two specific viewpoints C1 and C2 in the point cloud, our method synthesizes RGB images 'Output C1' and 'Output C2'. The corresponding ground-truth RGB images are labeled 'GT C1' and 'GT C2'.(Color figure online)

The most related recent work to this paper is *invsfm* [28], where the authors proposed a cascade of three U-Nets [17] to reveal scenes from SfM models. The input to their network is a sparse depth image with optional color and SIFT descriptors, that is, a projection of the SfM point cloud from a specific viewpoint. Their synthesized images are fairly convincing, but their pipeline does not take full advantage of the available 3D information. Projected point clouds lose adjacency information, with convolutions only sharing information between points that project nearby on the image. In contrast, the original point clouds retain this structural information, and convolutions share information between points that are nearby in 3D space. Moreover, a network trained on point cloud data is able to reason more intelligently about occlusion than one that takes a lossy z-buffering approach. Recently, point cloud processing has advanced considerably, with the development of PointNet [29] and PointNet++ [30] stimulating the field and leading to solid improvements in point cloud classification and segmentation. Additionally, generative adversarial networks (GAN) [10] have demonstrated the power of generating realistic images. Synthesizing both developments, *pc2px* [2] trained an image generator conditioned on the feature code of a object-level point cloud to render novel view images of the object. So far, directly generating images from scene-level point clouds remains an under-explored research area.

Inspired by these works, we develop a new type of U-Net, which encodes point clouds directly and decodes to 2D image space. We refer to the encoder as *PointEncoder* and the decoder as *ImageDecoder*. The main motivation for the design of this network is that we intend to make full use of all structural information in the point clouds, especially in the local regions of each point. Ideally, the 3D point features should help to recover better shapes and sharper edges in images. Meanwhile, we also use the associated RGB values for each point to enrich the 3D features with textural information. Consequently, our network is trained to generate RGB images from sparse colored point clouds. We further propose a network to refine the generated images and remove artifacts, called *RefineNet*. In summary, our contributions are:

1. a new image synthesis pipeline that generates images from novel viewpoints, given a sparse colored point cloud as input;
2. an encoder–decoder architecture that encodes 3D point clouds and decodes 2D images; and
3. a refinement network that improves the visual quality of the synthesized images.

Our approach generalizes effectively to a range of different real-world datasets, with good color consistency and shape recovery. We outperform the state-of-the-art method *invsfm* in two ways. Firstly, our network achieves better quantitative results, even with fewer points as input, for the scene revealing and novel view synthesis tasks. Secondly, our network has better qualitative visual results, with sharper edges and more complete shapes.

2 Related Work

There are two types of approaches for generating images from sparse point clouds: rendering after building a dense 3D model by point cloud upsampling or surface reconstruction; or directly recovering images using deep learning. In this section, we first review existing works on dense 3D model reconstruction and learning based image recovery. Then broader related topics are discussed such as novel view synthesis and image-to-image translation.

Dense 3D Model Reconstruction. Existing methods for building a dense 3D model from a point cloud can be grouped into two categories: point cloud upsampling, and surface reconstruction. PU-Net [38] and PU-GAN [20] are two deep learning based point cloud upsampling techniques. In these works, multi-level features for each point are learnt via deep neural networks. These features are further expanded in feature space and then split into a multitude of features to reconstruct a richer point cloud. Nevertheless, the upsampled point cloud is still not dense enough to enable image rendering. For mesh reconstruction, traditional algorithms often need strong priors including volumetric smoothing, structural repetition, part composition, and polygonal surface fitting [3]. Recently, some deep learning methods have been developed to address this problem. A 3D convolutional network called PointGrid, proposed by Le *et al.* [19], learns local approximation functions that help to reconstruct local shapes with better detail. However, reconstruction and storage of dense 3D models are not computationally efficient for practical applications.

Learning-Based Image Recovery. Instead of reconstructing the entire dense 3D model, some works synthesize images directly from sparse point clouds. A conditional GAN developed by Atienza [2] generates images from a deep point cloud code along with angles of camera viewpoints. Although the result does not outperform the state of the art, it shows more robustness towards downsampling and noise. Similarly, Milz *et al.* [11] adopt a GAN that conditions on an image projection. However, these two methods only work on object-level point clouds. In contrast, Pittaluga *et al.* [28] proposed a three stage neural network which recovers the source images of a SfM point cloud scene. The input to their network is a sparse depth image, that is, the projection of the point cloud onto the image plane with depth, color and a SIFT descriptor associated with each sparse 2D point. In contrast to these approaches, we focus on extracting the structural features of point clouds in 3D space and use them to generate better images.

Warping based Novel View Synthesis. Novel view synthesis from single or multiple images often requires a warping process to obtain a candidate image. Depth prediction is a typical strategy for warping. Liu *et al.* [21] regress pixel-wise depth and surface normal, then obtain the candidate image by warping with multiple surface homographies. Niklaus *et al.* [27] introduce a framework that inpaints the RGB image and depth map from a warped image so as to maintain space consistency. To achieve better depth estimation, multi-view images are applied in many methods, such as the use of multi-plane images (MPI) by Zhou *et al.* [39], and estimated depth volumes by Choi *et al.* [5] that leverage estimates of depth uncertainty from multiple views. The warped images using predicted depth maps often have only a few holes and missing pixels, which can be estimated using image completion networks. In comparison, our problem has much sparser inputs with significantly more missing data.

Image-to-Image Translation. Various methods [13,22,40] have succeeded in generating images from structural edges, changing the appearance style of existing images and synthesizing images from sketches. In our work, similar elements to these methods are used, such as an encoder–decoder architecture and adversarial training, for the task of pointset-to-image translation.

3 Method

Given a colored point cloud $\mathbf{P} \in \mathbb{R}^{N\times 6}$, where N is the number of points and each point has x, y, z coordinates and r, g, b color intensities, our goal is to generate an RGB image $\mathbf{I} \in \mathbb{R}^{H\times W\times 3}$ captured by a virtual camera from a specific viewpoint.

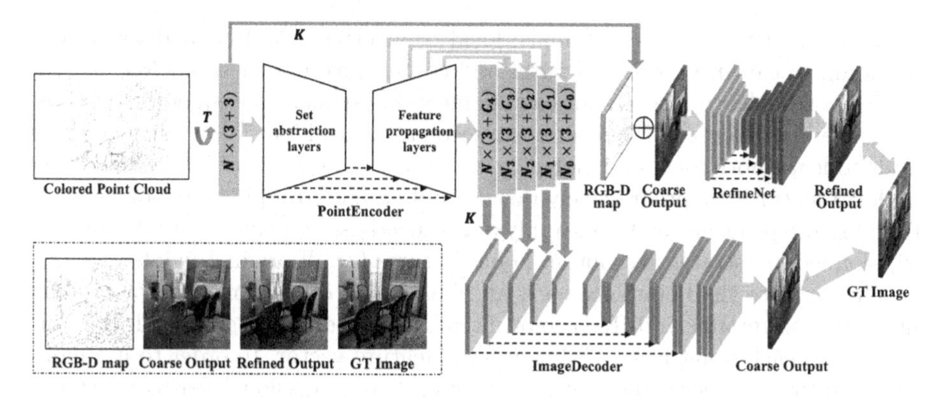

Fig. 2. Network architecture. The network has three modules with learnable parameters: a PointEncoder, an ImageDecoder, and a RefineNet. The PointEncoder has a PointNet++ structure [30] with set abstraction layers and feature propagation layers with skip connections. The ImageDecoder has a U-Net structure [17] but directly uses the projection maps from the PointEncoder. The RefineNet is a standard U-Net with an encoder–decoder structure which takes the coarse output from the ImageDecoder as input, alongside an additional RGB-D map. Visualizations of the different intermediate outputs are included at the bottom left.

The viewpoint is defined by the camera extrinsic parameters $\mathbf{T} \in SE(3)$ and the intrinsic parameters of typical pinhole camera model $\mathbf{K} \in \mathbb{R}^{3\times3}$. As shown in Fig. 2, our proposed view synthesis network has three main components: a PointEncoder, an ImageDecoder and a RefineNet. The first two networks together are the coarse image generator G_c and the RefineNet is the refined image generator G_r. We train a cascade of these two generators for pointset-to-image reconstruction and refinement, with an adversarial training strategy [10] using the discriminators D_c and D_r respectively. These discriminators use the PatchGAN [14] architecture and instance normalization [35] across all layers.

For the forward pass, the point cloud $\mathbf{P}$ is first rigidly transformed to $\mathbf{P}'$ by applying $\mathbf{T}$. The PointEncoder takes $\mathbf{P}'$ as an input to extract a set of point features in 3D space. These features are then associated onto feature map planes by projecting corresponding 3D points with the camera intrinsics $\mathbf{K}$. The ImageDecoder translates these feature maps into the image domain and produces a coarse RGB image of the final output size. Finally, the RefineNet produces a refined image using an encoder-decoder scheme, given the coarse image and an additional sparse RGB-D map.

3.1 Architecture

PointEncoder. Since point clouds are often sparse and the geometry and topology of the complete scene is unknown, it is difficult to generate photo-realistic images by rendering such point clouds. Thus to synthesize high-quality images, as much implicit structural information should be extracted from the point cloud as possible, such as surface normals, local connectivity, and color distribution. Intending to capture these structures and context, we use the PointNet++ [30] architecture to learn features for each point in 3D space. Consequently, our PointEncoder is composed of four set abstraction levels and four feature propagation levels to learn both local and global point features. Set abstraction layers generate local features by progressively downsampling and grouping the point cloud. Then feature propagation layers apply a distance-based feature interpolation and skip connection strategy to obtain point features for all original points.

The input to the PointEncoder is an $N \times (3 + 3)$ dimensional tensor consisting of 3D coordinates and RGB color intensities. After passing through the PointEncoder, each point has a C-dimensional feature vector. In order to use more point features, we save the features after each propagation level to construct a set of sub-pointsets with associated multi-scale point features. Specifically, after the i-th propagation level, we extract the point features $\mathbf{F}_i \in \mathbb{R}^{N_i \times (3+C_i)}$ of N_i subsampled points with 3D coordinates and C_i-dimensional feature channels. The final point feature set we adopt is denoted as $\mathbf{F} = \{\mathbf{F}_0, ..., \mathbf{F}_k\}$, where $N_k = N$ and $C_k = C$, (N_i, N_{i+1}) and (C_i, C_{i+1}) follow the rules of $N_i <= N_{i+1}$ and $C_i >= C_{i+1}$ respectively. Afterwards, $\mathbf{F}$ is projected and associated onto feature maps for the next step.

ImageDecoder. To decode the point features into an image, a bridge must be built between features in 3D space and features in image space. Considering the extraction process of point feature set $\mathbf{F}$, we observe that each 3D point in a sub-pointset represents a larger region in the original point cloud as the number of points in the subset gets smaller. As a result, feature vectors from smaller subsets contain richer contextual information than features from larger subsets. This is similar to how feature maps with less resolution but more channels in a convolutional neural network (CNN) encode information from a larger number of pixels. For the purpose of maintaining the scale consistency between the 3D space and image space, we project the point features to feature map planes with different resolutions according to their sub-pointset size. The ImageDecoder employs these feature maps and performs an upsampling and skip connection scheme like U-Net [31] until getting an image of the final output size.

More concretely, we project the point feature set $\mathbf{F}$ onto feature map planes $\mathbf{M} = \{\mathbf{M}_0, ..., \mathbf{M}_k\}$, where $\mathbf{M}_i \in \mathbb{R}^{H_i \times W_i \times C_i}$ corresponds to $\mathbf{F}_i$ and $\mathbf{M}_k$ has size $H \times W \times C$. Here, the generated feature maps $\mathbf{M}$ are regarded as a feature pyramid with the spatial dimension of its feature maps increasing by a scale of 2, as $H_{i+1} = 2H_i$ and $W_{i+1} = 2W_i$. In order to get the feature map $\mathbf{M}_i$, pixel coordinates in a map with size $H \times W$ are first calculated for all 3D points in $\mathbf{F}_i$ by perspective projection with camera intrinsics $\mathbf{K}$. After that, these pixel coordinates are rescaled in line with the size of $\mathbf{M}_i$ to associate the point features with it. If multiple points project to the same pixel, we retain the point closest to the camera. The ImageDecoder takes the feature pyramid $\mathbf{M}$ and decodes it into an RGB image.

RefineNet. By this stage in the network, we have generated an image from point features. However this is a coarse result and some problems remain unsolved.

One issue is that many 3D points are occluded by foreground surfaces in reality but still project onto the image plane even with z-buffering due to the sparsity of the point cloud. This brings deleterious features from non-visible regions of the point cloud onto the image plane. In addition, the PointEncoder predominately learns to reason about local shapes and structures, and so the color information is weakened. Accordingly, we propose the RefineNet module to estimate visibility implicitly and re-introduce the sparse color information.

As a standard U-Net architecture, the RefineNet receives a feature map of size $H \times W \times 7$, a concatenation of the coarse decoded image, the sparse RGB map, and the sparse depth map. The latter is used to analyse visibility in many geometric methods [1,4]. The sparse RGB-D image is obtained by associating the RGB values and z-value of the original point cloud onto a map of size $H \times W \times 4$ using the same projecting rules in the ImageDecoder. The output of the RefineNet is an RGB image with higher quality than the coarse image.

3.2 Training Loss

We employ Xavier initialization and then separately train the network in two independent adversarial steps. Firstly, the coarse generator G_c (the PointEncoder and ImageDecoder) is trained to generate coarse RGB images using ground-truth image supervision. After that, the parameters of G_c are fixed and the refined generator G_r (the RefineNet) is trained to refine the coarse images. Since the same loss function and ground truth supervision are utilized for both steps, we represent G_c and G_r together as G and discriminators D_c and D_r as D to simplify notation in next paragraphs. We notate the input for each step as x, which is a colored point cloud of size $N \times 6$ for G_c and a feature map of size $H \times W \times 7$ for G_r. The generator and discriminator G and D represent the functions that for G map from $\mathbb{R}^{N\times 6} \rightarrow \mathbb{R}^{H\times W\times 3}$ (or $\mathbb{R}^{H\times W\times 7} \rightarrow \mathbb{R}^{H\times W\times 3}$), and for D map from $\mathbb{R}^{H\times W\times 3} \rightarrow \mathbb{R}$.

For each step, the network is trained over a joint objective comprised of an ℓ_1 loss, an adversarial loss and a perceptual loss. Given the ground-truth image $\mathbf{I}_{\text{gt}} \in \mathbb{R}^{H\times W\times 3}$, the ℓ_1 loss and the adversarial loss are defined as

$$\mathcal{L}_{\ell_1} = ||\mathbf{I}_{\text{gt}} - G(x)||_1 \tag{1}$$

$$\mathcal{L}_{\text{adv}} = \log[D(\mathbf{I}_{\text{gt}})] + \log[1 - D(G(x))]. \tag{2}$$

A perceptual loss [8,15] is also used, which measures high-level perceptual and semantic distances between images. In our experiments, we use a feature reconstruction loss $\mathcal{L}_{\text{feat}}$ and a style loss $\mathcal{L}_{\text{style}}$ computed over different activation maps of the VGG-19 network [34] pre-trained on the ImageNet dataset [6]. The VGG-19 model is denoted as ϕ and the perceptual loss is computed using

$$\mathcal{L}_{\text{feat}} = \sum_{i=1}^{5} ||\phi_i(\mathbf{I}_{\text{gt}}) - \phi_i(G(x))||_1 \tag{3}$$

$$\mathcal{L}_{\text{style}} = \sum_{j=1}^{4} ||G_j^{\phi}(\mathbf{I}_{\text{gt}}) - G_j^{\phi}(G(x))||_1 \tag{4}$$

where ϕ_i generates the feature map after layers `relu1_1`, `relu2_1`, `relu3_1`, `relu4_1`, `relu5_1`; G_j^{ϕ} is a Gram matrix constructed from the feature map that is generated by ϕ_j, and ϕ_j corresponds to layers `relu2_2`, `relu3_4`, `relu4_4`, `relu5_2`. The Gram matrix treats each grid location of a feature map independently and captures information about relations between features themselves. While $\mathcal{L}_{\text{feat}}$ helps to preserve image content and overall spatial structure, $\mathcal{L}_{\text{style}}$ preserves stylistic features from the target image.

We manually set four hyperparameters as the coefficients of each loss term, and thus our overall loss function is given as follows

$$\mathcal{L}_G = \lambda_{\ell_1}\mathcal{L}_{\ell_1} + \lambda_{\text{adv}}\mathcal{L}_{\text{adv}} + \lambda_{\text{feat}}\mathcal{L}_{\text{feat}} + \lambda_{\text{style}}\mathcal{L}_{\text{style}}. \tag{5}$$

During training, the generator and discriminator are optimized together by applying alternating gradient updates.

4 Experiments

We evaluate our approach on several different datasets, including indoor and outdoor scenes, and on several different sources of 3D data. Specifically, we train our model on the SUN3D [37] dataset and then test it on two other indoor datasets, NYU-V2 [25] and ICL-NUIM [12], as well as the outdoor KITTI odometry dataset [9]. We also explore point clouds generated from different sources: depth measurements, COLMAP [32] and DSO [7]. We first compare the cascaded outputs of our proposed network, from the coarse and fine generators. Then we compare our approach with the state-of-the-art inverse SfM method [28], denoted *invsfm*, in terms of the synthesized image quality. We refer to the task of recovering views that were used to generate the input point clouds as *scene revealing*, and the task of recovering new views as *novel view synthesis*. Furthermore, to demonstrate the generalizability of our method, results on the KITTI dataset are reported, using point clouds generated by LiDAR sensors.

Training Data Preprocessing. SUN3D is a dataset of reconstructed spaces and provides RGB-D images and the ground-truth pose of each frame. By sampling from an RGB-D image, we can obtain a colored point cloud, which can be transformed to a novel view. Accordingly, we prepare the training data as a pair of RGB-D images and their relative pose, and then train our network with both current view inputs and novel view inputs. We use re-organized pairs of SUN3D data [36] to form a current-pointset–current-image pair and a current-pointset–novel-image pair for augmentation. In order to sample a sparse point cloud, we first sample 4096 pixels on each RGB image including feature points (ORB [24] or SIFT [23]), image edges and randomly sampled points. Then these pixels are inversely projected as a 3D point cloud, using the depth map and camera intrinsics, resulting in a colored point cloud.

Testing Data Preprocessing. We prepare two different types of point clouds for the evaluation of the scene revealing and novel view synthesis tasks. These two tasks have a significant difference: the scene revealing task intends to recover source images that participated in the generation of input pointsets, while the novel view synthesis task requires input pointsets generated from new views. As *invsfm* is the closest work to ours, for the scene revealing task we test our trained model on the SfM dataset they provide, which is processed from the NYU-V2 dataset [33] using COLMAP. As is typical for visual odometry or SLAM systems, 3D points are only triangulated from key frames. Therefore, we can evaluate the quality of novel view synthesis by using the remaining frames and the SfM pointset. In our experiment, we utilized DSO on the ICL-NUIM dataset to obtain pointsets and estimate the results for novel view synthesis. For unifying the size of input pointsets to $n \times 6$, we apply a sampling technique: randomly sampling when more than n points are in the field of view; and using nearest neighbor upsampling when there are fewer than n points.

Implementation Details. Our network is implemented using PyTorch and is trained with point clouds of size 4096×6 and images of size 256×256 using the Adam optimizer [16]. Since RefineNet is designed to perform image inpainting given a coarse input, we use the same empirical hyperparameter settings as EdgeConnect [26]: $\lambda_{\ell_1} = 1$, $\lambda_{\text{adv}} = \lambda_{\text{feat}} = 0.1$, and $\lambda_{\text{style}} = 250$. The learning rate of each generator starts at 10^{-4} and decreases to 10^{-6} during training until the objective converges. Discriminators are trained with a learning rate one tenth of the generators' rate.

Metrics. We measure the quality of the synthesized images using the following metrics: mean absolute error (MAE); structural similarity index (SSIM) with a window size of 11; and peak signal-to-noise ratio (PSNR). Here, a lower MAE and a higher SSIM or PSNR value indicates better results.

Runtime. The runtime for training on a single GTX 1080Ti GPU is 3d 21h 50 min for 30 K training examples and 50 epochs. For inference on a single TITAN XP GPU, the inference time is 0.038 s to synthesize a 256×256 image from an $N = 4096$ point cloud. In comparison, *invsfm* takes 0.068 s, almost double our inference time. For the PointEncoder/ImageDecoder/RefineNet, the inference time is divided up as 0.015/0.018/0.005 s.

4.1 Cascaded Outputs Comparison

In Fig. 3 we qualitatively compare the coarse and refined outputs of our two step generators where the size of input point clouds are all sampled to 4096. While the coarse results have good shape and patch reconstruction fidelity, the refined results recover colors and edges better. In addition, numerical comparison (Ours-coarse and Ours-refined) in Table 1 indicates that the RefineNet improves the results significantly. However, the performance of our coarse and refined outputs do not improve as the number of sampled points increases. The main reason is that there may not be that number of points in the field of view for many scenes

Fig. 3. Comparison of coarse and refined outputs. (Left to right) Input pointset, coarse output, refined output and ground-truth image. The input point clouds are sampled to a size of 4096. The coarse outputs reconstruct region shapes and patches while the refined outputs improve the color consistency and repair regions with artifacts.

and our upsampling strategy just replicates the points. Another reason is that we trained our model using 4096 points, thus the best performance is achieved when sampling the same number of points during testing. This reflects the capacity of our model for generating realistic images from very sparse pointsets. In our case, a 256×256 image is synthesized from only 4096 points, which is less than 6.25% of the pixels.

4.2 Scene Revealing Evaluation

To evaluate scene revealing performance, we utilize pointsets obtained from SFM on the NYU-V2 dataset. We make qualitative comparison of our approach with *invsfm* in Fig. 4 (first four columns), and additional results (last four columns) are reported using pointsets generated from RGB-D images. The results demonstrate that our work recovers sharper image edges and maintains better color consistency. With the 3D point features learnt by the PointEncoder, the network is able to generate more complete shapes, including small objects. In Table 1, quantitative results are given for comparison where our refined outputs achieve

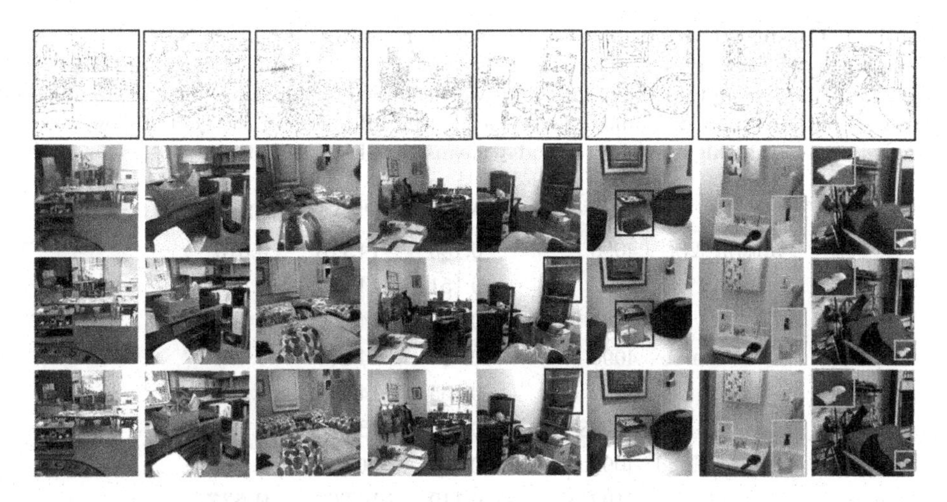

Fig. 4. Qualitative results for the scene revealing task on NYU-V2. (Top to bottom) Input pointset, *invsfm* results, our results, ground-truth images. Here our method uses 4096 sampled points while *invsfm* uses all points. The scenes are diverse and the point cloud sources differ: the first four are captured using SfM while the last four are sampled from RGB-D images. The first three columns show that our method generates sharper edges and better colors. Moreover, our results give better shape completion (red boxes) and finer small object reconstruction (green boxes).(Color figure online)

a notable improvement over *invsfm*. Even when using fewer input points, our approach has higher SSIM and PSNR scores as well as a lower MAE. It is also remarkable that our coarse results correspond closely to the results of *invsfm*, which reflects the effectiveness of our combination of the PointEncoder and ImageDecoder. Finally, the performance of our refined outputs remains stable with respect to the size of input pointsets, which indicates that our approach is robust to pointset density.

4.3 Novel View Synthesis Evaluation

Since the input of the network is a 3D pointset, synthesizing novel views of scenes can easily be achieved. As mentioned, we tested our proposed method on the non-keyframes of the DSO output. Note that non-keyframes are all aligned to specific poses in the pointsets, which thus can be seen as novel viewpoints with respect to the keyframes. We report the results of our model along with *invsfm*. Neither model is trained or fine-tuned on this dataset to ensure fair comparison. Qualitative results are displayed in Fig. 5 which shows that our model has advantages over *invsfm*. While the color effects of *invsfm* may partially fail in some cases, our approach recovers images with color consistency. The main characteristic of our model's ability to maintain shapes is also prominent here. Moreover, from the quantitative results in Table 2, we observe that our model

Table 1. Quantitative results for the scene revealing task on NYU-V2. The second column 'Max Points' refers to the size of the input point clouds, where 4096, 8192 and 12288 mean that the point clouds were sampled to this size using the sampling strategy outlined in Sect. 4, and >20000 means that all points in the field of view were used. ↑ means that higher is better and ↓ means that lower is better.

	Max Points	MAE ↓	PSNR ↑	SSIM ↑
invsfm [28]	4096	0.156	14.178	0.513
	8192	0.151	14.459	0.538
	>20000	0.150	14.507	0.544
Ours-coarse	4096	0.154	14.275	0.414
	8192	0.155	14.211	0.435
	12288	0.164	13.670	0.408
Ours-refined	4096	**0.117**	**16.439**	0.566
	8192	0.119	16.266	**0.577**
	12288	0.125	16.011	0.567

Table 2. Quantitative results for the novel view synthesis task on ICL-NUIM. Our method samples 4096 or 8192 3D points as input while *invsfm* takes all points in the view field. Our model achieves better results despite having many fewer input points.

	Max Points	MAE ↓	PSNR ↑	SSIM ↑
invsfm [28]	>20000	0.146	14.737	0.458
Ours-Coarse	4096	0.134	15.6	0.381
	8192	0.138	15.4	0.374
Ours-Refined	4096	**0.097**	**18.07**	0.579
	8192	0.101	17.75	**0.587**

outperforms *invsfm*, despite having fewer 3D points in the input pointset, by a greater margin than for the scene revealing task (Table 3).

4.4 Results on the KITTI Dataset

The LiDAR sensor and camera on the KITTI car are synchronized and calibrated with each other. While LiDAR provides accurate measurements of the 3D space, the camera captures the color and texture of a scene. By projecting the 3D pointset onto image planes, we can obtain the RGB values of each 3D point. Since the KITTI dataset also gives relative poses between frames in a sequence, novel view synthesis evaluation may be done on such a dataset. Figure 6 illustrates qualitative results for the scene revealing and view synthesis tasks. Although our model was not trained or fine-tuned on this dataset (or any outdoor dataset), it presents plausible results in that image colors, edges and basic shapes of objects are reconstructed effectively.

Fig. 5. Qualitative results for the novel view synthesis task on ICL-NUIM. (Top to bottom) Input pointset, *invsfm* results, our results, ground truth images. Here 4096 points are sampled for our method while *invsfm* takes all points. Our method constructs images with better color consistency (first two columns), sharper edges (red box), and finer-detail for small objects (green box).(Color figure online)

Table 3. Quantitative results on KITTI. We compare the output for scene revealing and view synthesis tasks on KITTI. Note that we did not train on any outdoor datasets, but our model is still able to generalize reasonably well to this data.

Type	**MAE** $\downarrow$	**PSNR** $\uparrow$	**SSIM** $\uparrow$
Scene revealing	0.154	13.8	0.514
Novel view synthesis	0.165	12.8	0.340

Fig. 6. Qualitative results on KITTI. (Top to bottom) Input pointset, scene revealing task results, novel view synthesis task results, ground truth images. The input pointsets are sampled to size 4096. Our model was not trained on any outdoor dataset, but still generates plausible images and recovers the shape of objects.

5 Conclusion

From the reported results above, it is clear that our pipeline has improved the performance over *invsfm*. This suggested it is possible to bypass a depthmap inpainting stage as used in *invsfm*. One possible explanation is that convolutions performed on the projected depthmap only share information between points that project nearby on the image, whereas processing directly on the point clouds removes this bias, sharing information between points that are nearby in 3D space. This difference in what is considered "nearby" is critical when reasoning about the geometric structure of a scene. It also means that the network is able to reason more intelligently about occlusion, beyond just z-buffering points that share a pixel. Indeed, the projection approach destroys information when multiple points project to the same pixel.

In this paper, we have demonstrated a deep learning solution to the view synthesis problem given a sparse colored 3D pointset as input. Our network is shown to perform satisfactorily in completing object shapes and reconstructing small objects, as well as maintaining color consistency. One limitation of the work is its sensitivity to outliers in the input pointset. Since outliers are common in many datasets, methods for filtering them from the point cloud could be investigated in future work, to improve the quality of the generated images. Our method assumes a static scene. A possible future extension is to synthesize novel views in a non-rigid dynamic scene [18].

Acknowledgements. This research was funded in part by the Australian Centre of Excellence for Robotic Vision (CE140100016), ARC-Discovery (DP 190102261) and ARC-LIEF (190100080) grants. The authors gratefully acknowledge GPUs donated by NVIDIA. We thank all anonymous reviewers and ACs for their comments." This work was completed when ZS was a visiting PhD student at ANU, and his visit was sponsored by the graduate school of Nanjing University of Science and Technology.

References

1. Alsadik, B., Gerke, M., Vosselman, G.: Visibility analysis of point cloud in close range photogrammetry. ISPRS Ann. Photogram. Remote Sens. Spat. Inf. Sci. **2**(5), 9 (2014)
2. Atienza, R.: A conditional generative adversarial network for rendering point clouds. In: Proceedings of the IEEE Conference on Computer Vision and Pattern Recognition Workshops, pp. 10–17 (2019)
3. Berger, M., et al.: State of the art in surface reconstruction from point clouds, April 2014
4. Biasutti, P., Bugeau, A., Aujol, J.F., Brédif, M.: Visibility estimation in point clouds with variable density. In: International Conference on Computer Vision Theory and Applications (VISAPP). Proceedings of the 14th International Conference on Computer Vision Theory and Applications, Prague, Czech Republic, February 2019
5. Choi, I., Gallo, O., Troccoli, A., Kim, M.H., Kautz, J.: Extreme view synthesis. In: Proceedings of the IEEE International Conference on Computer Vision, pp. 7781–7790 (2019)
6. Deng, J., Dong, W., Socher, R., Li, L.J., Li, K., Fei-Fei, L.: ImageNet: a large-scale hierarchical image database. In: 2009 IEEE Conference on Computer Vision and Pattern Recognition, pp. 248–255. IEEE (2009)
7. Engel, J., Koltun, V., Cremers, D.: Direct sparse odometry. IEEE Trans. Pattern Anal. Mach. Intell. **40**(3), 611–625 (2017)
8. Gatys, L.A., Ecker, A.S., Bethge, M.: Image style transfer using convolutional neural networks. In: Proceedings of the IEEE Conference on Computer Vision and Pattern Recognition, pp. 2414–2423 (2016)
9. Geiger, A., Lenz, P., Stiller, C., Urtasun, R.: Vision meets robotics: the kitti dataset. Int. J. Robot. Res. (IJRR) **32**, 1231–1237 (2013)
10. Goodfellow, I., et al.: Generative adversarial nets. In: Advances in Neural Information Processing Systems, pp. 2672–2680 (2014)
11. Milz, S., Simon, M., Fischer, K., Pöpperl, M., Gross, H.-M.: Points2Pix: 3D point-cloud to image translation using conditional GANs. In: Fink, G.A., Frintrop, S., Jiang, X. (eds.) DAGM GCPR 2019. LNCS, vol. 11824, pp. 387–400. Springer, Cham (2019). https://doi.org/10.1007/978-3-030-33676-9_27
12. Handa, A., Whelan, T., McDonald, J., Davison, A.J.: A benchmark for RGB-D visual odometry, 3D reconstruction and SLAM. In: 2014 IEEE International Conference on Robotics and Automation (ICRA), pp. 1524–1531. IEEE (2014)
13. Isola, P., Zhu, J.Y., Zhou, T., Efros, A.A.: Image-to-image translation with conditional adversarial networks. CoRR abs/1611.07004 (2016). http://arxiv.org/abs/1611.07004

14. Isola, P., Zhu, J.Y., Zhou, T., Efros, A.A.: Image-to-image translation with conditional adversarial networks. In: Proceedings of the IEEE Conference on Computer Vision and Pattern Recognition, pp. 1125–1134 (2017)
15. Johnson, J., Alahi, A., Fei-Fei, L.: Perceptual losses for real-time style transfer and super-resolution. In: Leibe, B., Matas, J., Sebe, N., Welling, M. (eds.) ECCV 2016. LNCS, vol. 9906, pp. 694–711. Springer, Cham (2016). https://doi.org/10.1007/978-3-319-46475-6_43
16. Kingma, D.P., Ba, J.: Adam: a method for stochastic optimization. In: Proceedings of the 3rd International Conference on Learning Representations (ICLR), May 2015
17. Kohl, S., et al.: A probabilistic U-Net for segmentation of ambiguous images. In: Advances in Neural Information Processing Systems, pp. 6965–6975 (2018)
18. Kumar, S., Dai, Y., Li, H.: Monocular dense 3D reconstruction of a complex dynamic scene from two perspective frames. In: International Conference on Computer Vision (2017)
19. Le, T., Duan, Y.: PointGrid: a deep network for 3D shape understanding. In: Proceedings of the IEEE Conference on Computer Vision and Pattern Recognition, pp. 9204–9214 (2018)
20. Li, R., Li, X., Fu, C.W., Cohen-Or, D., Heng, P.A.: PU-GAN: a point cloud upsampling adversarial network. In: Proceedings of the IEEE International Conference on Computer Vision, pp. 7203–7212 (2019)
21. Liu, M., He, X., Salzmann, M.: Geometry-aware deep network for single-image novel view synthesis. In: Proceedings of the IEEE Conference on Computer Vision and Pattern Recognition, pp. 4616–4624 (2018)
22. Liu, M.Y., Tuzel, O.: Coupled generative adversarial networks. In: Advances in Neural Information Processing Systems, pp. 469–477 (2016)
23. Lowe, D.G., et al.: Object recognition from local scale-invariant features. In: International Conference on Computer Vision, vol. 99, pp. 1150–1157 (1999)
24. Mur-Artal, R., Montiel, J.M.M., Tardos, J.D.: ORB-SLAM: a versatile and accurate monocular SLAM system. IEEE Trans. Robot. **31**(5), 1147–1163 (2015)
25. Silberman, N., Hoiem, D., Kohli, P., Fergus, R.: Indoor segmentation and support inference from RGBD images. In: Fitzgibbon, A., Lazebnik, S., Perona, P., Sato, Y., Schmid, C. (eds.) ECCV 2012. LNCS, vol. 7576, pp. 746–760. Springer, Heidelberg (2012). https://doi.org/10.1007/978-3-642-33715-4_54
26. Nazeri, K., Ng, E., Joseph, T., Qureshi, F., Ebrahimi, M.: EdgeConnect: generative image inpainting with adversarial edge learning. CoRR abs/1901.00212 (2019)
27. Niklaus, S., Mai, L., Yang, J., Liu, F.: 3D Ken Burns effect from a single image. ACM Trans. Graph. (TOG) **38**(6), 184 (2019)
28. Pittaluga, F., Koppal, S.J., Kang, S.B., Sinha, S.N.: Revealing scenes by inverting structure from motion reconstructions. In: Proceedings of the IEEE Conference on Computer Vision and Pattern Recognition, pp. 145–154 (2019)
29. Qi, C.R., Su, H., Mo, K., Guibas, L.J.: PointNet: deep learning on point sets for 3D classification and segmentation. In: Proceedings of the IEEE Conference on Computer Vision and Pattern Recognition, pp. 652–660 (2017)
30. Qi, C.R., Yi, L., Su, H., Guibas, L.J.: PointNet++: deep hierarchical feature learning on point sets in a metric space. In: Advances in Neural Information Processing Systems, pp. 5099–5108 (2017)
31. Ronneberger, O., Fischer, P., Brox, T.: U-Net: convolutional networks for biomedical image segmentation. In: Navab, N., Hornegger, J., Wells, W.M., Frangi, A.F. (eds.) MICCAI 2015. LNCS, vol. 9351, pp. 234–241. Springer, Cham (2015). https://doi.org/10.1007/978-3-319-24574-4_28

32. Schönberger, J.L., Frahm, J.M.: Structure-from-motion revisited. In: Conference on Computer Vision and Pattern Recognition (CVPR) (2016)
33. Silberman, N., Hoiem, D., Kohli, P., Fergus, R.: Indoor segmentation and support inference from RGBD images. In: Fitzgibbon, A., Lazebnik, S., Perona, P., Sato, Y., Schmid, C. (eds.) ECCV 2012. LNCS, vol. 7576, pp. 746–760. Springer, Heidelberg (2012). https://doi.org/10.1007/978-3-642-33715-4_54
34. Simonyan, K., Zisserman, A.: Very deep convolutional networks for large-scale image recognition. In: International Conference on Learning Representations (2015)
35. Ulyanov, D., Vedaldi, A., Lempitsky, V.: Improved texture networks: maximizing quality and diversity in feed-forward stylization and texture synthesis. In: Proceedings of the IEEE Conference on Computer Vision and Pattern Recognition, pp. 6924–6932 (2017)
36. Ummenhofer, B., et al.: Demon: depth and motion network for learning monocular stereo. In: IEEE Conference on Computer Vision and Pattern Recognition (CVPR) (2017). http://lmb.informatik.uni-freiburg.de//Publications/2017/UZUMIDB17
37. Xiao, J., Owens, A., Torralba, A.: Sun3D: a database of big spaces reconstructed using SFM and object labels. In: Proceedings of the IEEE International Conference on Computer Vision, pp. 1625–1632 (2013)
38. Yu, L., Li, X., Fu, C.W., Cohen-Or, D., Heng, P.A.: PU-Net: Point cloud upsampling network. In: Proceedings of the IEEE Conference on Computer Vision and Pattern Recognition, pp. 2790–2799 (2018)
39. Zhou, T., Tucker, R., Flynn, J., Fyffe, G., Snavely, N.: Stereo magnification: learning view synthesis using multiplane images. In: SIGGRAPH (2018)
40. Zhu, J.Y., Park, T., Isola, P., Efros, A.A.: Unpaired image-to-image translation using cycle-consistent adversarial networks. In: Proceedings of the IEEE International Conference on Computer Vision, pp. 2223–2232 (2017)

Consensus-Aware Visual-Semantic Embedding for Image-Text Matching

Haoran Wang[1], Ying Zhang[2], Zhong Ji[1(✉)], Yanwei Pang[1], and Lin Ma[2]

[1] School of Electrical and Information Engineering, Tianjin University, Tianjin, China
{haoranwang,Jizhong,pyw}@tju.edu.cn
[2] Tencent AI Lab, Shenzhen, China
yinggzhang@tencent.com, forest.linma@gmail.com

Abstract. Image-text matching plays a central role in bridging vision and language. Most existing approaches only rely on the image-text instance pair to learn their representations, thereby exploiting their matching relationships and making the corresponding alignments. Such approaches only exploit the superficial associations contained in the instance pairwise data, with no consideration of any external commonsense knowledge, which may hinder their capabilities to reason the higher-level relationships between image and text. In this paper, we propose a Consensus-aware Visual-Semantic Embedding (CVSE) model to incorporate the consensus information, namely the commonsense knowledge shared between both modalities, into image-text matching. Specifically, the consensus information is exploited by computing the statistical co-occurrence correlations between the semantic concepts from the image captioning corpus and deploying the constructed concept correlation graph to yield the consensus-aware concept (CAC) representations. Afterwards, CVSE learns the associations and alignments between image and text based on the exploited consensus as well as the instance-level representations for both modalities. Extensive experiments conducted on two public datasets verify that the exploited consensus makes significant contributions to constructing more meaningful visual-semantic embeddings, with the superior performances over the state-of-the-art approaches on the bidirectional image and text retrieval task. Our code of this paper is available at: https://github.com/BruceW91/CVSE.

Keywords: Image-text matching · Visual-semantic embedding · Consensus

1 Introduction

Vision and language understanding plays a fundamental role for human to perceive the real world, which has recently made tremendous progresses thanks to

H. Wang—Work done while Haoran Wang was a Research Intern with Tencent AI Lab.
H. Wang and Y. Zhang—Equal contribution.

A. Vedaldi et al. (Eds.): ECCV 2020, LNCS 12369, pp. 18–34, 2020.
https://doi.org/10.1007/978-3-030-58586-0_2

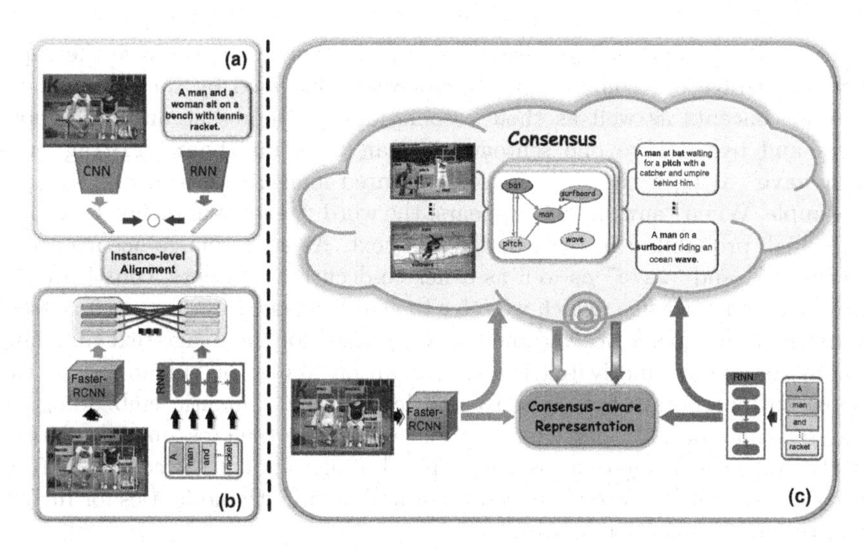

Fig. 1. The conceptual comparison between our proposed consensus-aware visual-semantic embedding (CVSE) approach and existing instance-level alignment based approaches. (a) instance-level alignment based on image and text global representation; (b) instance-level alignment exploiting the complicated fragment-level image-text matching; (c) our proposed CVSE approach.

the rapid development of deep learning. To delve into multi-modal data comprehending, this paper focuses on addressing the problem of image-text matching [27], which benefits a series of downstream applications, such as visual question answering [2,28], visual grounding [4,35,47], visual captioning [41,42,48], and scene graph generation [5]. Specifically, it aims to retrieve the texts (images) that describe the most relevant contents for a given image (text) query. Although thrilling progresses have been made, this task is still challenging due to the semantic discrepancy between image and text, which separately resides in heterogeneous representation spaces.

To tackle this problem, the current mainstream solution is to project the image and text into a unified joint embedding space. As shown in Fig. 1 (a), a surge of methods [10,21,30,43] employ the deep neural networks to extract the global representations of both images and texts, based on which their similarities are measured. However, these approaches failed to explore the relationships between image objects and sentence segments, leading to limited matching accuracy. Another thread of work [18,23] performs the fragment-level matching and aggregates their similarities to measure their relevance, as shown in Fig. 1 (b). Although complicated cross-modal correlations can be characterized, yielding satisfactory bidirectional image-text retrieval results, these existing approaches only rely on employing the image-text instance pair to perform cross-modal retrieval, which we name as instance-level alignment in this paper.

For human beings, besides the image-text instance pair, we have the capability to leverage our commonsense knowledge, expressed by the fundamental semantic concepts as well as their associations, to represent and align both images and texts. Take one sentence "`A man on a surfboard riding on a ocean wave`" along with its semantically-related image, shown in Fig. 1 (c), as an example. When "`surfboard`" appears, the word "`wave`" will incline to appear with a high probability in both image and text. As such, the co-occurrence of "`surfboard`" and "`wave`" as well as other co-occurred concepts, constitute the commonsense knowledge, which we refer to as *consensus*. However, such consensus information has not been studied and exploited for the image-text matching task. In this paper, motivated by this cognition ability of human beings, we propose to incorporate the consensus to learn visual-semantic embedding for image-text matching. In particular, we not only mine the cross-modal relationships between the image-text instance pairs, but also exploit the consensus from large-scale external knowledge to represent and align both modalities for further visual-textual similarity reasoning.

In this paper, we propose one Consensus-aware Visual-Semantic Embedding (CVSE) architecture for image-text matching, as depicted in Fig. 1 (c). Specifically, we first make the consensus exploitation by computing statistical co-occurrence correlations between the semantic concepts from the image captioning corpus and constructing the concept correlation graph to learn the consensus-aware concept (CAC) representations. Afterwards, based on the learned CAC representations, both images and texts can be represented at the consensus level. Finally, the consensus-aware representation learning integrates the instance-level and consensus-level representations together, which thereby serves to make the cross-modal alignment. Experiment results on public datasets demonstrate that the proposed CVSE model is capable of learning discriminative representations for image-text matching, and thereby boost the bidirectional image and sentence retrieval performances. Our contributions lie in three-fold.

- We make the first attempt to exploit the consensus information for image-text matching. As a departure from existing instance-level alignment based methods, our model leverages one external corpus to learn consensus-aware concept representations expressing the commonsense knowledge for further strengthening the semantic relationships between image and text.
- We propose a novel Consensus-aware Visual-Semantic Embedding (CVSE) model that unifies the representations of both modalities at the consensus level. And the consensus-aware concept representations are learned with one graph convolutional network, which captures the relationship between semantic concepts for more discriminative embedding learning.
- The extensive experimental results on two benchmark datasets demonstrate that our approach not only outperforms state-of-the-art methods for traditional image-text retrieval, but also exhibits superior generalization ability for cross-domain transferring.

2 Related Work

2.1 Knowledge Based Deep Learning

There has been growing interest in incorporating external knowledge to improve the data-driven neural network. For example, knowledge representation has been employed for image classification [31] and object recognition [7]. In the community of vision-language understanding, it has been explored in several contexts, including VQA [44] and scene graph generation [12]. In contrast, our CVSE leverages consensus knowledge to generate homogeneous high-level cross-modal representations and achieves visual-semantic alignment.

2.2 Image-Text Matching

Recently, there have been a rich line of studies proposed for addressing the problem of image-text matching. They mostly deploy the two-branch deep architecture to obtain the global [10,21,26,27,30,43] or local [17,18,23] representations and align both modalities in the joint semantic space. Mao *et al.* [30] adopted CNN and Recurrent Neural Network (RNN) to represent images and texts, followed by employing bidirectional triplet ranking loss to learn a joint visual-semantic embedding space. For fragment-level alignment, Karpathy *et al.* [18] measured global cross-modal similarity by accumulating local ones among all region-words pairs. Moreover, several attention-based methods [16,23,32,39] have been introduced to capture more fine-grained cross-modal interactions. To sum up, they mostly adhere to model the superficial statistical associations at instance level, whilst the lack of structured commonsense knowledge impairs their reasoning and inference capabilities for multi-modal data.

In contrast to previous studies, our CVSE incorporates the commonsense knowledge into the consensus-aware representations, thereby extracting the high-level semantics shared between image and text. The most relevant existing work to ours is [38], which enhances image representation by employing image scene graph as external knowledge to expand the visual concepts. Unlike [38], our CVSE is capable of exploiting the learned consensus-aware concept representations to uniformly represent and align both modalities at the consensus level. Doing so allows us to measure cross-modal similarity via disentangling higher-level semantics for both image and text, which further improves its interpretability.

3 Consensus-Aware Visual-Semantic Embedding

In this section, we elaborate on our Consensus-aware Visual-Semantic Embedding (CVSE) architecture for image-text matching (see Fig. 2). Different from instance-level representation based approaches, we first introduce a novel Consensus Exploitation module that leverages commonsense knowledge to capture the semantic associations among concepts. Then, we illustrate how to employ the Consensus Exploitation module to generate the consensus-level representation and combine it with the instance-level representation to represent both modalities. Lastly, the alignment objectives and inference method are represented.

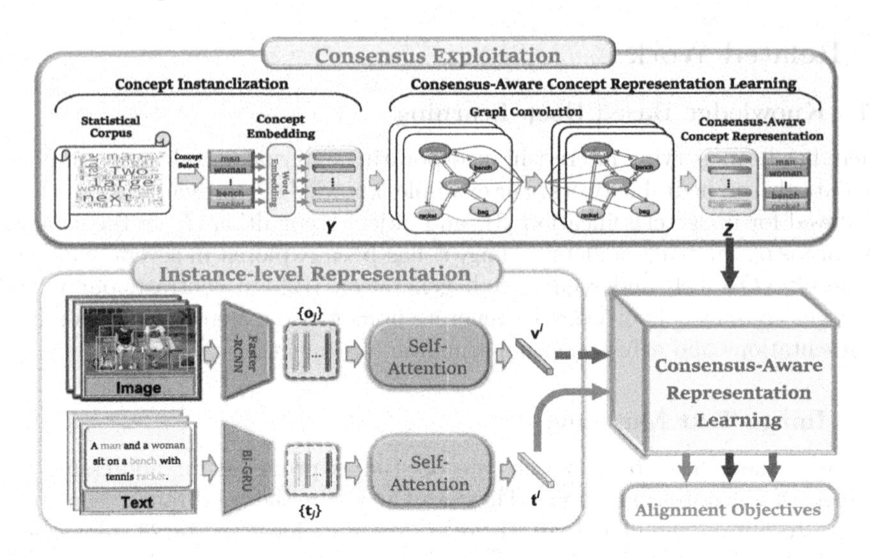

Fig. 2. The proposed CVSE model for image-text matching. Taking the fragment-level features of both modalities as input, it not only adopts the dual self-attention to generate the instance-level representations $\mathbf{v}^I$ and $\mathbf{t}^I$, but also leverages the consensus exploitation module to learn the consensus-level representations.

3.1 Exploit Consensus Knowledge to Enhance Concept Representations

As aforementioned, capturing the intrinsic associations among the concepts, which serves as the commonsense knowledge in human reasoning, can analogously supply high-level semantics for more accurate image-text matching. To achieve this, we construct a Consensus Exploitation (CE) module (see Fig. 2), which adopts graph convolution to propagate the semantic correlations among various concepts based on a correlation graph preserving their inter dependencies, which contributes to injecting more commonsense knowledge into the concept representation learning. It involves three key steps: (1) Concept instantiation, (2) Concept correlation graph building and, (3) Consensus-aware concept representation learning. The concrete details will be presented in the following.

Concept Instantiation. We rely on the image captioning corpus of the natural sentences to exploit the commonsense knowledge, which is represented as the semantic concepts and their correlations. Specifically, all the words in the corpus can serves as the candidate of semantic concepts. Due to the large scale of word vocabulary and the existence of some meaningless words, we follow [9,15] to remove the rarely appeared words from the word vocabulary. In particular, we select the words with top-q appearing frequencies in the concept vocabulary, which are roughly categorized into three types, *i.e.*, *Object*, *Motion*, and *Property*. For more detailed division principle, we refer readers to [13]. Moreover, according to the statistical frequency of the concepts with same type over the whole dataset, we restrict the ratio of the concepts with type of (*Object*, *Motion*, *Property*) to be (7:2:1). After that, we employ the glove [33] technique to instantialize these selected concepts, which is denoted as $\mathbf{Y}$.

Concept Correlation Graph Building. With the instantiated concepts, their co-occurrence relationship are examined to build one correlation graph and thereby exploit the commonsense knowledge. To be more specific, we construct a conditional probability matrix $\mathbf{P}$ to model the correlation between different concepts, with each element $\mathbf{P}_{ij}$ denoting the appearance probability of concept C_i when concept C_j appears:

$$\mathbf{P}_{ij} = \frac{\mathbf{E}_{ij}}{N_i} \tag{1}$$

where $\mathbf{E} \in \mathbb{R}^{q \times q}$ is the concept co-occurrence matrix, $\mathbf{E}_{ij}$ represents the co-occurrence times of C_i and C_j, and N_i is the occurrence times of C_i in the corpus. It is worth noting that $\mathbf{P}$ is an asymmetrical matrix, which allows us to capture the reasonable inter dependencies among various concepts rather than simple co-occurrence frequency.

Although the matrix $\mathbf{P}$ is able to capture the intrinsic correlation among the concepts, it suffers from several shortages. Firstly, it is produced by adopting the statistics of co-occurrence relationship of semantic concepts from the image captioning corpus, which may deviate from the data distribution of real scenario and further jeopardize its generalization ability. Secondly, the statistical patterns derived from co-occurrence frequency between concepts can be easily affected by the long-tail distribution, leading to biased correlation graph. To alleviate the above issues, we design a novel scale function, dubbed Confidence Scaling (CS) function, to rescale the matrix $\mathbf{E}$:

$$\mathbf{B}_{ij} = f_{CS}(\mathbf{E}_{ij}) = s^{\mathbf{E}_{ij} - u} - s^{-u}, \tag{2}$$

where s and u are two pre-defined parameters to determine the amplifying/shrinking rate for rescaling the elements of $\mathbf{E}$. Afterwards, to further prevent the correlation matrix from being over-fitted to the training data and improve its generalization ability, we also follow [6] to apply binary operation to the rescaled matrix $\mathbf{B}$:

$$\mathbf{G}_{ij} = \begin{cases} 0, & if \ \mathbf{B}_{ij} < \epsilon, \\ 1, & if \ \mathbf{B}_{ij} \geq \epsilon, \end{cases} \tag{3}$$

where $\mathbf{G}$ is the binarized matrix $\mathbf{B}$. ϵ denotes a threshold parameter filters noisy edges. Such scaling strategy not only assists us to focus on the more reliable co-occurrence relationship among the concepts, but also contributes to depressing the noise contained in the long-tailed data.

Consensus-Aware Concept Representation. Graph Convolutional Network (GCN) [3,20] is a multilayer neural network that operates on a graph and update the embedding representations of the nodes via propagating information based on their neighborhoods. Distinct from the conventional convolution operation that are implemented on images with Euclidean structures, GCN can learn the mapping function on graph-structured data. In this section, we employ the multiple stacked GCN layers to learn the concept representations (dubbed CGCN module), which introduces higher order neighborhoods information among the concepts to model their inter dependencies. More formally, given the instantiated concept representations $\mathbf{Y}$ and the concept correlation graph $\mathbf{G}$, the embedding feature of the l-th layer is calculated as

$$\mathbf{H}^{(l+1)} = \rho(\tilde{\mathbf{A}}\mathbf{H}^{(l)}\mathbf{W}^{(l)}) \tag{4}$$

where $\mathbf{H}^{(0)} = \mathbf{Y}$, $\tilde{\mathbf{A}} = \mathbf{D}^{-\frac{1}{2}}\mathbf{G}\mathbf{D}^{-\frac{1}{2}}$ denotes the normalized symmetric matrix and $\mathbf{W}^{l}$ represents the learnable weight matrix. ρ is a non-linear activation function, *e.g.*, ReLU function [22].

We take output of the last layer from GCN to acquire the final concept representations $\mathbf{Z} \in \mathbb{R}^{q \times d}$ with $\mathbf{z}_i$ denoting the generated embedding representation of concept C_i, and d indicating the dimensionality of the joint embedding space. Specifically, the i-th row vector of matrix $\mathbf{Z} = \{\mathbf{z}_1, ..., \mathbf{z}_q\}$, *i.e.* $\mathbf{z}_i$, represents the embedding representation for the i-th element of the concept vocabulary. For clarity, we name $\mathbf{Z}$ as consensus-aware concept (CAC) representations, which is capable of exploiting the commonsense knowledge to capture underlying interactions among various semantic concepts.

3.2 Consensus-Aware Representation Learning

In this section, we would like to incorporate the exploited consensus to generate the consensus-aware representation of image and text.

Instance-Level Image and Text Representations. As aforementioned, conventional image-text matching only rely on the individual image/text instance to yield the corresponding representations for matching, as illustrated in Fig. 2. Specifically, given an input image, we utilize a pre-trained Faster-RCNN [1,36] followed by a fully-connected (FC) layer to represent it by M region-level visual features $\mathbf{O} = \{\mathbf{o}_1, ..., \mathbf{o}_M\}$, whose elements are all F-dimensional vector. Given a sentence with L words, the word embedding is sequentially fed into a bi-directional GRU [37]. After that, we can obtain the word-level textual features $\{\mathbf{t}_1, ..., \mathbf{t}_L\}$ by performing mean pooling to aggregate the forward and backward hidden state vectors at each time step.

Afterwards, the self-attention mechanism [40] is used to concentrate on the informative portion of the fragment-level features to enhance latent embeddings for both modalities. Note that here we only describe the attention generation procedure of the visual branch, as it goes the same for the textual one. The region-level visual features $\{\mathbf{o}_1, ..., \mathbf{o}_M\}$ is used as the key and value items, while the global visual feature vector $\bar{\mathbf{O}} = \frac{1}{M}\sum_{m=1}^{M}\mathbf{o}_m$ is adopted as the query item for the attention strategy. As such, the self-attention mechanism refines

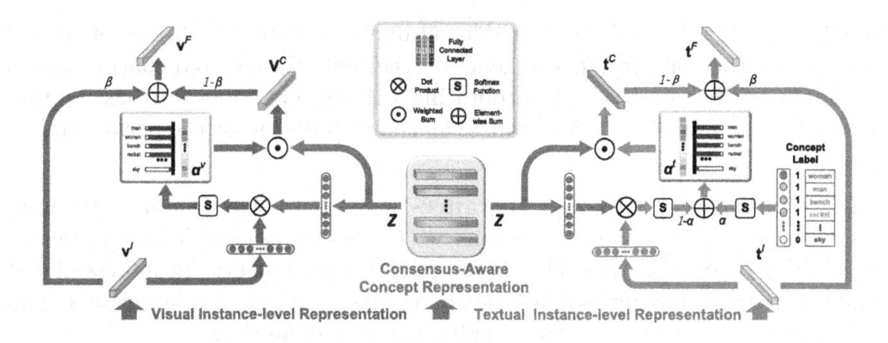

Fig. 3. Illustration of the consensus-level representation learning and the fusion between it and instance-level representation.

the instance-level visual representation as $\mathbf{v}^I$. With the same process on the word-level textual features $\{\mathbf{t}_1, ..., \mathbf{t}_L\}$, the instance-level textual representation is refined as $\mathbf{t}^I$.

Consensus-Level Image and Text Representations. In order to incorporate the exploited consensus, as shown in Fig. 3, we take the instance-level visual and textual representations ($\mathbf{v}^I$ and $\mathbf{t}^I$) as input to query from the CAC representations. The generated significance scores for different semantic concepts allow us to uniformly utilize the linear combination of the CAC representations to represent both modalities. Mathematically, the visual consensus-level representation $\mathbf{v}^C$ can be calculated as follows:

$$\begin{aligned} \mathbf{a}_i^v &= \frac{exp(\lambda \mathbf{v}^I \mathbf{W}^v \mathbf{z}_i^\top)}{\sum_{i=1}^q exp(\lambda \mathbf{v}^I \mathbf{W}^v \mathbf{z}_i^\top)}, \\ \mathbf{v}^C &= \sum_{i=1}^q \mathbf{a}_i^v \cdot \mathbf{z}_i, \end{aligned} \tag{5}$$

where $\mathbf{W}^v \in \mathbb{R}^{d \times d}$ is the learnable parameter matrix, $\mathbf{a}_i^v$ denotes the significance score corresponding to the semantic concept $\mathbf{z}_i$, and λ controls the smoothness of the softmax function.

For the text, due to the semantic concepts are instantiated from the textual statistics, we can annotate any given image-text pair via employing a set of concepts that appears in its corresponding descriptions. Formally, we refer to this multi-label tagging as concept label $\mathbf{L}^t \in \mathbb{R}^{q \times 1}$. Considering the consensus knowledge is explored from the textual statistics, we argue that it's reasonable to leverage the concept label as prior information to guide the consensus-level representation learning and alignment. Specifically, we compute the predicted concept scores $\mathbf{a}_i^t$ and consensus-level representation $\mathbf{t}^C$ as follows:

$$\begin{aligned} \mathbf{a}_j^t &= \alpha \, \frac{exp(\lambda \mathbf{L}_j^t)}{\sum_{j=1}^q exp(\lambda \mathbf{L}_j^t)} + (1-\alpha) \, \frac{exp(\lambda \mathbf{t}^I \mathbf{W}^t \mathbf{z}_j^\top)}{\sum_{j=1}^q exp(\lambda \mathbf{t}^I \mathbf{W}^t \mathbf{z}_j^\top)}, \\ \mathbf{t}^C &= \sum_{j=1}^q \mathbf{a}_j^t \cdot \mathbf{z}_j, \end{aligned} \tag{6}$$

where $\mathbf{W}^t \in \mathbb{R}^{d \times d}$ denotes the learnable parameter matrix. $\alpha \in [0,1]$ controls the proportion of the concept label to generate the textual predicted concept scores $\mathbf{a}_j^t$. We empirically find that incorporating the concept label into the textual consensus-level representation learning can significantly boost the performances.

Fusing Consensus-Level and Instance-Level Representations. We integrate the instance-level representations $\mathbf{v}^I(\mathbf{t}^I)$ and consensus-level representation $\mathbf{v}^C(\mathbf{t}^C)$ to comprehensively characterizing the semantic meanings of the visual and textual modalities. Empirically, we find that the simple weighted sum operation can achieve satisfactory results, which is defined as:

$$\begin{aligned} \mathbf{v}^F &= \beta \mathbf{v}^I + (1-\beta)\mathbf{v}^C, \\ \mathbf{t}^F &= \beta \mathbf{t}^I + (1-\beta)\mathbf{t}^C, \end{aligned} \tag{7}$$

where β is a tuning parameter controlling the ratio of two types of representations. And $\mathbf{v}^F$ and $\mathbf{t}^F$ respectively denote the combined visual and textual representations, dubbed consensus-aware representations.

3.3 Training and Inference

Training. During the training, we deploy the widely adopted bidirectional triplet ranking loss [10,11,21] to align the image and text:

$$\begin{aligned} \mathcal{L}_{rank}(\mathbf{v},\mathbf{t}) = \sum_{(\mathbf{v},\mathbf{t})} \{&\max[0, \gamma - s(\mathbf{v},\mathbf{t}) + s(\mathbf{v},\mathbf{t}^-)] \\ &+ \max[0, \gamma - s(\mathbf{t},\mathbf{v}) + s(\mathbf{t},\mathbf{v}^-)]\}, \end{aligned} \tag{8}$$

where γ is a predefined margin parameter, $s(\cdot,\cdot)$ denotes cosine distance function. Given the representations for a matched image-text pair $(\mathbf{v},\mathbf{t})$, its corresponding negative pairs are denoted as $(\mathbf{t},\mathbf{v}^-)$ and $(\mathbf{v},\mathbf{t}^-)$, respectively. The bidirectional ranking objectives are imposed on all three types of representations, including instance-level, consensus-level, and consensus-aware representations.

Considering that a matched image-text pair usually contains similar semantic concepts, we impose the Kullback Leibler (KL) divergence on the visual and textual predicted concept scores to further regularize the alignment:

$$\mathcal{D}_{KL}(\mathbf{a}^t \parallel \mathbf{a}^v) = \sum_{i=1}^{q} \mathbf{a}_i^t \, log(\frac{\mathbf{a}_i^t}{\mathbf{a}_i^v}), \tag{9}$$

In summary, the final training objectives of our CVSE model is defined as:

$$\mathcal{L} = \lambda_1 \mathcal{L}_{rank}(\mathbf{v^F},\mathbf{t^F}) + \lambda_2 \mathcal{L}_{rank}(\mathbf{v^I},\mathbf{t^I}) + \lambda_3 \mathcal{L}_{rank}(\mathbf{v^C},\mathbf{t^C}) + \lambda_4 \mathcal{D}_{KL}, \tag{10}$$

where $\lambda_1, \lambda_2, \lambda_3, \lambda_4$ aim to balance the weight of different loss functions.

Inference. During inference, we only deploy the consensus-aware representations $\mathbf{v}^F$ $(\mathbf{t}^F)$ and utilize cosine distance to measure their cross-modal similarity.

4 Experiments

4.1 Dataset and Settings

Datasets. Flickr30k [34] is an image-caption dataset containing 31,783 images, with each image annotated with five sentences. Following the protocol of [30], we split the dataset into 29,783 training, 1000 validation, and 1000 testing images. We report the performance evaluation of image-text retrieval on 1000 testing set. MSCOCO [24] is another image-caption dataset, totally including 123,287 images with each image roughly annotated with five sentence-level descriptions. We follow the public dataset split of [18], including 113,287 training images, 1000 validation images, and 5000 testing images. We report the experimental results on both 1 K testing set and 5 K testing set, respectively.

Evaluation Metrics. We employ the widely-used R@K as evaluation metric [10,21], which measures the fraction of queries for which the matched item is found among the top k retrieved results. We also report the "mR" criterion that averages all six recall rates of R@K, which provides a more comprehensive evaluation to testify the overall performance.

4.2 Implementation Details

All our experiments are implemented in PyTorch toolkit with a single NVIDIA Tesla P40 GPU. For representing visual modality, the amount of detected regions in each image is $M = 36$, and the dimensionality of region representation vectors is $F = 2048$. The dimensionality of word embedding space is set to 300. The dimensionality of joint space d is set to 1024. For the consensus exploitation, we adopt 300-dim GloVe [33] trained on the Wikipedia dataset to initialize the the semantic concepts. The size of the semantic concept vocabulary is $q = 300$. And two graph convolution layers are used, with the embedding dimensionality are set to 512 and 1024, respectively. Regarding the correlation matrix $\mathbf{G}$, we set $s = 5$ and $u = 0.02$ in Eq. (2), and $\epsilon = 0.3$ in Eq. (3). For image and text representation learning, we set $\lambda = 10$ in Eq. (5) and $\alpha = 0.35$ in Eq. (6), respectively. For the training objective, we empirically set $\beta = 0.75$ in Eq. (7) , $\gamma = 0.2$ in Eq. (8) and $\lambda_1, \lambda_2, \lambda_3, \lambda_4 = 3, 5, 1, 2$ in Eq. (10). Our CVSE model is trained by Adam optimizer [19] with mini-batch size of 128. The learning rate is set to be 0.0002 for the first 15 epochs and 0.00002 for the next 15 epochs. The dropout is also employed with a dropout rate of 0.4. Our code is available[1].

4.3 Comparison to State-of-the-art

The experimental results on the MSCOCO dataset are shown in Table 1. From Table 1, we can observe that our CVSE is obviously superior to the competitors in most evaluation metrics, which yield a result of 78.6% and 66.3% on

[1] https://github.com/BruceW91/CVSE.

Table 1. Comparisons of experimental results on MSCOCO 1K test set and Flickr30k test set.

Approach	MSCOCO dataset							Flickr30k dataset						
	Text retrieval			Image retrieval			mR	Text retrieval			Image retrieval			mR
	R@1	R@5	R@10	R@1	R@5	R@10		R@1	R@5	R@10	R@1	R@5	R@10	
DVSA [18]	38.4	69.9	80.5	27.4	60.2	74.8	39.2	22.2	48.2	61.4	15.2	37.7	50.5	58.5
m-RNN [30]	41.0	73.0	83.5	29.0	42.2	77.0	57.6	35.4	63.8	73.7	22.8	50.7	63.1	51.6
DSPE [43]	50.1	79.7	89.2	39.6	75.2	86.9	70.1	40.3	68.9	79.9	29.7	60.1	72.1	58.5
CMPM [49]	56.1	86.3	92.9	44.6	78.8	89	74.6	49.6	76.8	86.1	37.3	65.7	75.5	65.2
VSE++ [10]	64.7	–	95.9	52.0	–	92.0	–	52.9	–	87.2	39.6	–	79.5	–
PVSE [39]	69.2	91.6	96.6	55.2	86.5	93.7	–	–	–	–	–	–	–	
SCAN [23]	72.7	94.8	98.4	58.8	88.4	94.8	83.6	67.4	90.3	95.8	48.6	77.7	85.2	77.5
CAMP [45]	72.3	94.8	98.3	58.5	87.9	95.0	84.5	68.1	89.7	95.2	51.5	77.1	85.3	77.8
LIWE [46]	73.2	95.5	98.2	57.9	88.3	94.5	84.6	69.6	90.3	95.6	51.2	80.4	87.2	79.1
RDAN [14]	74.6	**96.2**	**98.7**	61.6	89.2	94.7	85.8	68.1	**91.0**	**95.9**	54.1	80.9	87.2	79.5
CVSE	**78.6**	95.0	97.5	**66.3**	**91.8**	**96.3**	**87.6**	**73.6**	90.4	94.4	**56.1**	**83.2**	**90.0**	**81.3**

R@1 for text retrieval and image retrieval, respectively. In particular, compared with the second best RDAN method, we achieve absolute boost (5.3%, 2.6%, 1.6%) on (R@1, R@5, R@10) for image retrieval. In contrast, we also find the performance of CVSE is slightly inferior to best method on R@5 and R@10 for text retrieval. However, as the most persuasive criteria, the mR metric of our CVSE still markedly exceeds other algorithms. Besides, some methods partially surpassed ours, such as SCAN [23] and RDAN [14] both exhaustively aggregating the local similarities over the visual and textual fragments, which leads to slow inference speed. By contrast, our CVSE just employs combined global representations so that substantially speeds up the inference stage. Therefore, considering the balance between effectiveness and efficiency, our CVSE still has distinct advantages over them.

The results on the Flickr30K dataset are presented in Table 1. It can be seen that our CVSE arrives at 81.3% on the criteria of "mR", which also outperforms all the state-of-the-art methods. Especially for image retrieval, the CVSE model surpasses the previous best method by (2.0%, 2.3%, 2.8%) on (R@1, R@5, R@10), respectively. The above results substantially demonstrate the effectiveness and necessity of exploiting the consensus between both modalities to align the visual and textual representations.

4.4 Ablation Studies

In this section, we perform several ablation studies to systematically explore the impacts of different components in our CVSE model. Unless otherwise specified, we validate the performance on the 1K test set of MSCOCO dataset.

Different Configuration of Consensus Exploitation. To start with, we explore how the different configurations of consensus exploitation module affects

Table 2. Effect of different configurations of CGCN module on MSCOCO Dataset.

Approaches	CGCN			Text retrieval			Image retrieval		
	Graph embedding	CS function	Concept label	R@1	R@5	R@10	R@1	R@5	R@10
CVSE$_{full}$	✓	✓	✓	**78.6**	**95.0**	97.5	**66.3**	91.8	**96.3**
CVSE$_{wo/GE}$		✓	✓	74.2	92.6	96.2	62.5	90.3	94.6
CVSE$_{wo/CS}$	✓		✓	77.4	93.8	97.0	65.2	**91.9**	95.8
CVSE$_{wo/CL}$	✓	✓		72.5	93.5	**97.7**	57.2	87.4	94.1

Table 3. Effect of different configurations of objective and inference scheme on MSCOCO Dataset.

Approaches	Objective		Inference scheme			Text retrieval			Image retrieval		
	Separate constraint	KL	Instance	Consensus	Fused	R@1	R@5	R@10	R@1	R@5	R@10
CVSE$_{wo/SC}$		✓			✓	74.9	91.1	95.7	63.6	90.8	94.2
CVSE$_{wo/KL}$	✓				✓	**77.3**	93.4	96.9	**65.6**	**92.8**	**95.9**
CVSE ($\beta = 1$)	✓	✓	✓			71.2	**93.8**	**97.4**	54.8	87.0	92.2
CVSE ($\beta = 0$)	✓	✓		✓		57.0	67.2	80.2	57.4	86.5	92.9

the performance of CVSE model. As shown in Table 2, it is observed that although we only adopt the glove word embedding as the CAC representations, the model (CVSE$_{wo/GC}$) can still achieve the comparable performance in comparison to the current leading methods. It indicates the semantic information contained in word embedding technique is still capable of providing weak consensus information to benefit image-text matching. Compared to the model (CVSE$_{wo/CS}$) where CS function in Eq. (2) is excluded, the CVSE model can obtain 1.2% and 1.1% performance gain on R@1 for text retrieval and image retrieval, respectively. Besides, we also find that if the concept label $\mathbf{L}^t$ is excluded, the performance of model (CVSE$_{wo/CL}$) evidently drops. We conjecture that this result is attributed to the consensus knowledge is collected from textual statistics, thus the textual prior information contained in concept label substantially contributes to enhancing the textual consensus-level representation so as to achieve more precise cross-modal alignment.

Different Configurations of Training Objective and Inference Strategy. We further explore how the different alignment objectives affect our performance. First, as shown in Table 3, when the separate ranking loss, *i.e.* $\mathcal{L}_{rank-I}$ and $\mathcal{L}_{rank-C}$ are both removed, the CVSE$_{wo/SC}$ model performs worse than our CVSE model, which validates the effectiveness of the two terms. Secondly, we find that the CVSE$_{wo/KL}$ produces inferior retrieval results, indicating the importance of $\mathcal{D}_{KL}$ for regularizing the distribution discrepancy of predicted concept scores between image and text, which again provides more interpretability that pariwise heterogeneous data should correspond to the approximate semantic concepts. Finally, we explore the relationship between instance-level features and consensus-level features for representing both modalities. Specifically, the CVSE ($\beta = 1$) denotes the CVSE model with $\beta = 1$ in Eq. (7), which employs

Table 4. Comparison results on cross-dataset generalization from MSCOCO to Flickr30k.

Approaches	Text retrieval			Image retrieval		
	R@1	R@5	R@10	R@1	R@5	R@10
RRF-Net [25]	28.8	53.8	66.4	21.3	42.7	53.7
VSE++ [10]	40.5	67.3	77.7	28.4	55.4	66.6
LVSE [8]	46.5	72.0	82.2	34.9	62.4	73.5
SCAN [23]	49.8	77.8	86.0	38.4	65.0	74.4
CVSE$_{wo/consensus}$	49.1	75.5	84.3	36.4	63.7	73.3
CVSE	**57.8**	**81.4**	**87.2**	**44.8**	**71.8**	**81.1**

instance-level representations alone. Similarly, the CVSE ($\beta = 0$) model refers to the CVSE that only adopts the consensus-level representations. Interestingly, we observe that deploying the representations from any single semantic level alone will yields inferior results compared to their combination. It substantially verifies the semantical complementarity between the instance-level and consensus-level representations is critical for achieving significant performance improvements.

4.5 Further Analysis

Consensus Exploitation for Domain Adaptation. To further verify the capacity of consensus knowledge, we test its generalization ability by conducting cross-dataset experiments, which was seldom investigated in previous studies whilst meaningful for evaluating the cross-modal retrieval performance in real scenario. Specifically, we conduct the experiment by directly transferring our model trained on MS-COCO to Flickr30k dataset. For comparison, except for two existing work [8,25] that provide the corresponding results, we additionally re-implement two previous studies [10,23] based on their open-released code. From Table 4, it's obvious that our CVSE outperforms all the competitors by a large margin. Moreover, compared to the baseline that only employs instance-level alignment (CVSE$_{wo/consensus}$), CVSE achieves compelling improvements. These results implicate the learned consensus knowledge can be shared between cross-domain heterogeneous data, which leads to significant performance boost.

The Visualization of Confidence Score for Concepts. In Fig. 4, we visualize the confidence score of concepts predicted by our CVSE. It can be seen that the prediction results are considerably reliable. In particular, some informative concepts that are not involved in the image-text pair can even be captured. For example, from Fig. 4(a), the associated concepts of "`traffic`" and "`buildings`" are also pinpointed for enhancing semantic representations.

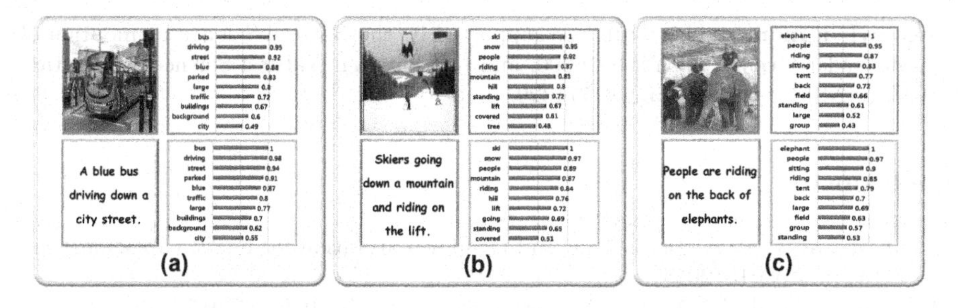

Fig. 4. The visualization results of predicted scores for top-10 concepts.

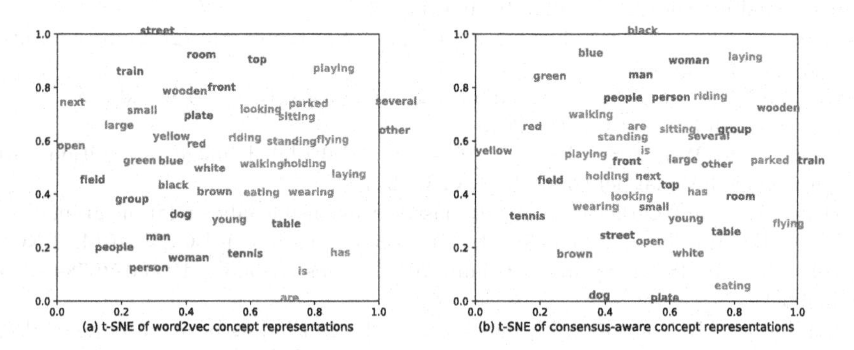

Fig. 5. The t-SNE results of word2vec based concept representations and our consensus-aware concepts representations. We randomly select 15 concepts per POS for performing visualization and annotate each POS with the same color.

The Visualization of Consensus-Aware Concept Representations. In Fig. 5, we adopt the t-SNE [29] to visualize the CAC representations. In contrast to the word2vec [33] based embedding features, the distribution of our CAC representations is more consistent with our common sense. For instance, the concepts with POS of *Motion*, such as "`riding`", is closely related to the concept of "`person`". Similarly, the concept of "`plate`" is closely associated with "`eating`". These results further verify the effectiveness of our consensus exploitation module in capturing the semantic associations among the concepts.

5 Conclusions

The ambiguous understanding of multi-modal data severely impairs the ability of machine to precisely associate images with texts. In this work, we proposed a Consensus-Aware Visual-Semantic Embedding (CVSE) model that integrates commonsense knowledge into the multi-modal representation learning for visual-semantic embedding. Our main contribution is exploiting the consensus knowledge to simultaneously pinpoint the high-level concepts and generate the unified consensus-aware concept representations for both image and text. We demonstrated the superiority of our CVSE for image-text retrieval by outperforming state-of-the-art models on widely used MSCOCO and Flickr30k datasets.

Acknowledgments. This work was supported by the Natural Science Foundation of Tianjin under Grant 19JCYBJC16000, and the National Natural Science Foundation of China (NSFC) under Grant 61771329.

References

1. Anderson, P., et al.: Bottom-up and top-down attention for image captioning and VQA. In: CVPR (2018)
2. Antol, S., et al.: VQA: Visual question answering. In: ICCV (2015)
3. Bruna, J., Zaremba, W., Szlam, A., LeCun, Y.: Spectral networks and locally connected networks on graphs. In: ICLR (2013)
4. Chen, J., Chen, X., Ma, L., Jie, Z., Chua, T.S.: Temporally grounding natural sentence in video (2018)
5. Chen, K., Gao, J., Nevatia, R.: Knowledge aided consistency for weakly supervised phrase grounding. In: CVPR (2018)
6. Chen, Z.M., Wei, X.S., Wang, P., Guo, Y.: Multi-label image recognition with graph convolutional networks. In: CVPR (2019)
7. Deng, J., et al.: Large-scale object classification using label relation graphs. In: Fleet, D., Pajdla, T., Schiele, B., Tuytelaars, T. (eds.) ECCV 2014. LNCS, vol. 8689, pp. 48–64. Springer, Cham (2014). https://doi.org/10.1007/978-3-319-10590-1_4
8. Engilberge, M., Chevallier, L., Pérez, P., Cord, M.: Finding beans in burgers: deep semantic-visual embedding with localization. In: CVPR (2018)
9. Fang, H., et al.: From captions to visual concepts and back. In: CVPR (2015)
10. Fartash, F., Fleet, D., Kiros, J., Fidler, S.: VSE++: improved visual-semantic embeddings. In: BMVC (2018)
11. Frome, A., et al.: Devise: a deep visual-semantic embedding model. In: NIPS (2013)
12. Gu, J., Zhao, H., Lin, Z., Li, S., Cai, J., Ling, M.: Scene graph generation with external knowledge and image reconstruction. In: CVPR (2019)
13. Hou, J., Wu, X., Zhao, W., Luo, J., Jia, Y.: Joint syntax representation learning and visual cue translation for video captioning. In: ICCV (2019)
14. Hu, Z., Luo, Y., Lin, J., Yan, Y., Chen, J.: Multi-level visual-semantic alignments with relation-wise dual attention network for image and text matching. In: IJCAI (2019)
15. Huang, Y., Wu, Q., Song, C., Wang, L.: Learning semantic concepts and order for image and sentence matching. In: CVPR (2018)
16. Ji, Z., Wang, H., Han, J., Pang, Y.: Saliency-guided attention network for image-sentence matching. In: ICCV (2019)
17. Karpathy, A., Joulin, A., Li, F.F.: Deep fragment embeddings for bidirectional image sentence mapping. In: NIPS (2014)
18. Karpathy, A., Li, F.F.: Deep visual-semantic alignments for generating image descriptions. In: CVPR (2015)
19. Kingma, D., Ba, J.: Adam: a method for stochastic optimization. In: ICLR (2014)
20. Kipf, T.N., Welling, M.: Semi-supervised classification with graph convolutional networks. In: ICLR (2016)
21. Kiros, R., Salakhutdinov, R., Zemel, R.: Unifying visual-semantic embeddings with multimodal neural language models. In: NIPS Workshop (2014)
22. Krizhevsky, A., Sutskever, I., Hinton, G.E.: ImageNet classification with deep convolutional neural networks. In: NIPS (2012)

23. Lee, K.H., Chen, X., Hua, G., Hu, H., He, X.: Stacked cross attention for image-text matching. In: ECCV (2018)
24. Lin, T.-Y., et al.: Microsoft COCO: common objects in context. In: Fleet, D., Pajdla, T., Schiele, B., Tuytelaars, T. (eds.) ECCV 2014. LNCS, vol. 8693, pp. 740–755. Springer, Cham (2014). https://doi.org/10.1007/978-3-319-10602-1_48
25. Liu, Y., Guo, Y., Bakker, E.M., Lew, M.S.: Learning a recurrent residual fusion network for multimodal matching. In: ICCV (2017)
26. Ma, L., Lu, Z., Shang, L., Li, H.: Multimodal convolutional neural networks for matching image and sentence. In: ICCV (2015)
27. Ma, L., Jiang, W., Jie, Z., Jiang, Y., Liu, W.: Matching image and sentence with multi-faceted representations. IEEE Trans. Circ. Syst. Video Technol. **30**(7), 2250–2261 (2020)
28. Ma, L., Lu, Z., Li, H.: Learning to answer questions from image using convolutional neural network (2016)
29. Maaten, L.V.D., Hinton, G.: Visualizing data using t-SNE. J. Mach. Learn. Res. **9**, 2579–2605 (2008)
30. Mao, J., Xu, W., Yang, Y., Wang, J., Huang, Z., Yuille, A.: Deep captioning with multimodal recurrent neural networks (M-RNN). In: ICLR (2015)
31. Marino, K., Salakhutdinov, R., Gupta, A.: The more you know: using knowledge graphs for image classification. In: CVPR (2017)
32. Nam, H., Ha, J., Kim, J.: Dual attention networks for multimodal reasoning and matching. In: CVPR (2017)
33. Pennington, J., Socher, R., Manning, C.D.: Glove: global vectors for word representation. In: EMNLP (2014)
34. Plummer, B., Wang, L., Cervantes, C., Caicedo, J., Hockenmaier, J., Lazebnik, S.: Flickr30k entities: collecting region-to-phrase correspondences for richer image-to-sentence models. In: ICCV (2015)
35. Plummer, B., Mallya, A., Cervantes, C., Hockenmaier, J., Lazebnik, S.: Phrase localization and visual relationship detection with comprehensive image-language cues. In: ICCV (2017)
36. Ren, S., He, K., Girshick, R., Sun, J.: Faster R-CNN: towards real-time object detection with region proposal networks. In: NIPS (2015)
37. Schuster, M., Paliwal, K.K.: Bidirectional recurrent neural networks. IEEE Trans. Sig. Process. **45**(11), 2673–2681 (1997)
38. Shi, B., Ji, L., Lu, P., Niu, Z., Duan, N.: Knowledge aware semantic concept expansion for image-text matching. In: IJCAI (2019)
39. Song, Y., Soleymani, M.: Polysemous visual-semantic embedding for cross-modal retrieval. In: CVPR (2019)
40. Vaswani, A., et al.: Attention is all you need. In: NIPS (2017)
41. Wang, B., Ma, L., Zhang, W., Liu, W.: Reconstruction network for video captioning (2018)
42. Wang, J., Jiang, W., Ma, L., Liu, W., Xu, Y.: Bidirectional attentive fusion with context gating for dense video captioning (2018)
43. Wang, L., Li, Y., Lazebnik, S.: Learning deep structure-preserving image-text embeddings. In: CVPR (2016)
44. Wang, P., Wu, Q., Shen, C., Dick, A., van den Hengel, A.: FVQA: Fact-based visual question answering. IEEE Trans. Pattern Anal. Mach. Intell. **40**(10), 2413–2427 (2018)
45. Wang, Z., et al.: Camp: Cross-modal adaptive message passing for text-image retrieval. In: ICCV (2019)

46. Wehrmann, J., Souza, D.M., Lopes, M.A., Barros, R.C.: Language-agnostic visual-semantic embeddings. In: ICCV (2019)
47. Yuan, Y., Ma, L., Wang, J., Liu, W., Zhu, W.: Semantic conditioned dynamtic modulation for temporal sentence grounding in videos (2019)
48. Zhang, W., Wang, B., Ma, L., Liu, W.: Reconstruct and represent video contents for captioning via reinforcement learning (2019). https://doi.org/10.1109/TPAMI.2019.2920899
49. Zhang, Y., Lu, H.: Deep cross-modal projection learning for image-text matching. In: ECCV (2018)

Spatial Hierarchy Aware Residual Pyramid Network for Time-of-Flight Depth Denoising

Guanting Dong, Yueyi Zhang(✉), and Zhiwei Xiong

University of Science and Technology of China, Hefei, China
gtdong@mail.ustc.edu.cn, {zhyuey,zwxiong}@ustc.edu.cn

Abstract. Time-of-Flight (ToF) sensors have been increasingly used on mobile devices for depth sensing. However, the existence of noise, such as Multi-Path Interference (MPI) and shot noise, degrades the ToF imaging quality. Previous CNN-based methods remove ToF depth noise without considering the spatial hierarchical structure of the scene, which leads to failures in obtaining high quality depth images from a complex scene. In this paper, we propose a Spatial Hierarchy Aware Residual Pyramid Network, called SHARP-Net, to remove the depth noise by fully exploiting the geometry information of the scene in different scales. SHARP-Net first introduces a Residual Regression Module, which utilizes the depth images and amplitude images as the input, to calculate the depth residual progressively. Then, a Residual Fusion Module, summing over depth residuals from all scales, is imported to refine the depth residual by fusing multi-scale geometry information. Finally, shot noise is further eliminated by a Kernel Prediction Network. Experimental results demonstrate that our method significantly outperforms state-of-the-art ToF depth denoising methods on both synthetic and realistic datasets. The source code is available at https://github.com/ashesknight/tof-mpi-remove.

Keywords: Time-of-Flight · Multi-Path Interference · Spatial hierarchy · Residual pyramid · Depth denoising

1 Introduction

Depth plays an important role in current research, especially in the field of computer vision. In the past decades, researchers have proposed various methods to obtain depth [22,29,30], among which Time-of-Flight (ToF) technology is becoming increasingly popular for depth sensing. Many successful consumer products, such as Kinect One [21], are equipped with ToF sensors, providing high quality depth image. These devices further promote many applications in

Electronic supplementary material The online version of this chapter (https://doi.org/10.1007/978-3-030-58586-0_3) contains supplementary material, which is available to authorized users.

A. Vedaldi et al. (Eds.): ECCV 2020, LNCS 12369, pp. 35–50, 2020.
https://doi.org/10.1007/978-3-030-58586-0_3

computer vision areas, for example scene understanding, action recognition and human-computer interaction. However, ToF depth images suffer from various noises, such as Multi-Path Interference (MPI) and shot noise, which limit the applicability of ToF imaging technologies.

ToF depth images are vulnerable to MPI noise, which originates in the fact that numerous multi-bounce lights are collected by one pixel during the exposure time. The existence of MPI breaks the key assumption that the receiving light is only reflected once in the scene and results in serious ToF depth error. Shot noise, a common and inevitable noise caused by sensor electronics, is another source of ToF depth noise. Figure 1 shows the depth error maps caused by shot noise and MPI noise respectively. It can be seen that both shot noise and MPI noise are widespread in ToF depth images but MPI noise is significantly intense in several regions such as corner and edge areas.

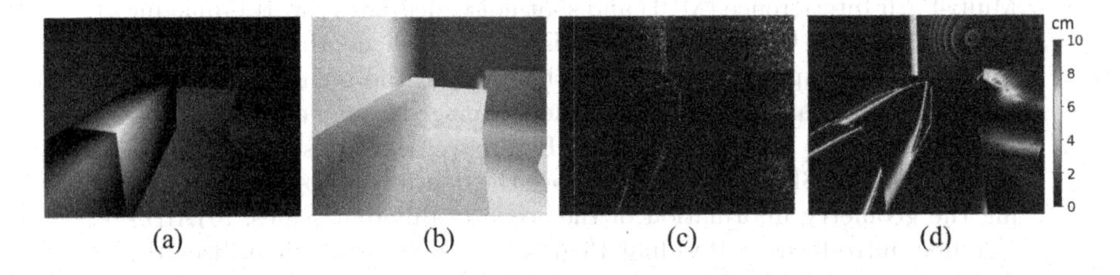

Fig. 1. (a) ToF amplitude image. (b) ToF ground truth depth. (c) Depth error map caused by shot noise. (d) Depth error map caused by MPI. The example comes from a synthetic dataset.

Recently, many Convolutional Neural Networks (CNN) based methods have been proposed for MPI removal in ToF sensors [17,23,26]. The fundamental theory of these CNN based methods is that the MPI noise of a pixel can be estimated as a linear combination of information from surrounding pixels. In the image space, CNN is a proper way to model this linear combination process with spatial convolution and achieves encouraging results. To fit the unknown parameters of convolution kernel, supervised learning is often utilized and the ground-truth labels without MPI of scenes are required. Since it is difficult to get the ground truth depth of realistic scenes, many synthetic ToF datasets are introduced for the training and testing of neural networks. Usually, these datasets consist of ToF depth images as well as corresponding amplitude images. Some datasets even contain the raw measurements of ToF sensors and color images, both of which are usually captured by the calibrated RGBD camera.

The large-scale datasets make it possible to learn the linear combination process of light transport through CNN based methods. However, the existing CNN based methods still have some limitations. Especially, the elimination of MPI noise for a complex scene is not satisfying. Specifically, in a complex scene, many objects with different shapes and sizes are located close to each other. In this case, each pixel of the ToF sensor may collect many light signals which are from various indirect light paths, which easily leads to intense MPI noise. Eliminating MPI noise in a complex scene still remains a challenging problem and needs more investigation.

A key observation is that in a scene, the objects usually have spatial hierarchical structures. For example, a showcase, a dog toy and the head of the dog toy can formulate a hierarchical relationship. In this case, the depth value of a point located at the surface of any object is usually affected by these three interrelated objects. In a complex scene with large-size shapes and detailed structures, there should be more diverse hierarchical relationships. And previous works have demonstrated that utilizing the hierarchical representations of the scene can lead to improvement in computer vision filed such as scene understanding [24,27], image embedding [4], image denoising [20], object detection [19], depth and 3D shape estimation [6,18]. Aforementioned works inspire us to explicitly utilize the spatial hierarchical relationships to improve the result of the MPI removal for ToF depth.

In this paper, we propose a Spatial Hierarchy Aware Residual Pyramid Network (SHARP-Net) to fully exploit scene structures in multiple scales for ToF depth denoising. The spatial hierarchical structure of the scene, in the forms of a feature pyramid with multiple scales, can provide a proper receptive field and more ample geometric relationships between the objects of the scene for the network, which improves the performance of noise removal.

Within SHARP-Net, a Residual Regression Module is first introduced, which consists of a feature extractor to build a feature pyramid and residual regression blocks to establish a depth residual pyramid in a coarse-to-fine manner. At upper levels of the residual pyramid, the depth residual maps represent MPI noise regressed by utilizing global geometry information. At lower levels, the depth residual maps describe subtle MPI effects by considering local scene structures. The Residual Regression Module pushes every level to utilize the available hierarchical relationships of the current level and deeply extracts the geometric information lying in the corresponding hierarchy of the scene. The geometric information obtained in different scales give excellent hints for estimating the MPI noise. Our proposed Residual Regression Module generates a depth residual map for each level, which is much different from the widely used U-Net structure. After going through Residual Regression Module, a depth residual pyramid is obtained to represent MPI estimation corresponding to the hierarchical structure of the scene. In order to further optimize the performance of SHARP-Net on both large-size shapes and detailed structures, we propose a Residual Fusion Module to explicitly choose predominant components by summing over the depth residuals from all scales. Finally, we employ a Depth Refinement Module, which is based on a Kernel Prediction Network, to remove shot noise and refine depth images.

Combining the Residual Regression Module, Residual Fusion Module and Depth Refinement Module, our SHARP-Net accurately removes noise for ToF depth images, especially MPI noise and shot noise. In short, we make the following contributions:

- We propose a Residual Regression Module to explicitly exploit the spatial hierarchical structure of the scene to accurately remove MPI noise and shot noise in large-size shapes and detailed structures simultaneously.
- We propose a Residual Fusion Module to selectively integrate the geometric information in different scales to further correct MPI noise, and introduce a Depth Refinement Module to effectively eliminate the shot noise.
- The proposed SHARP-Net significantly outperforms the state-of-the-art methods in the quantitative and qualitative comparison for ToF depth denoising on both the synthetic and realistic datasets.

2 Related Work

ToF imaging is affected by noise from different sources, such as shot noise and MPI noise [15,28]. Shot noise is caused by sensor electronics, which appears in all sensors. Shot noise removal for ToF sensors is well investigated. Traditional filtering algorithms, such as bilateral filtering, are able to eliminate shot noise effectively [2,16]. In contrast, MPI removal is a more difficult problem in ToF depth denoising. Many physics-based and learning-based MPI removal methods have been proposed.

For physics-based methods, Fuchs *et al.* conduct a series of studies to estimate MPI noise in the scene, from using single modulation frequency [9] to considering multiple albedos and reflections [10,14]. Feigin *et al.* propose a multi-frequency method to correct MPI through comparing the pixel-level changes of the raw measurements at different frequencies [5]. Gupta *et al.* study the impact of modulation frequencies on MPI and propose a phasor imaging method by emitting two signals with frequencies of great differences [12]. Freedman *et al.* propose a model based on a compressible backscattering representation to tackle the multi-path with more than two paths and achieve real-time processing speed [8].

For learning-based methods, Marco *et al.* exploit the transient imaging technology [13] to simulate the generation of MPI noise in ToF imaging process and produce a large dataset for ToF depth denoising. They also propose a two-stage deep neural network to refine ToF depth images [17]. Su *et al.* propose a deep end-to-end network for ToF depth denoising with raw correlation measurements as the input [26]. Guo *et al.* produce a large-scale ToF dataset FLAT, and introduce a kernel prediction network to remove MPI and shot noise [11]. To overcome the domain shift between the unlabelled realistic scene and the synthetic training dataset, Agresti *et al.* exploit an adversarial learning strategy, based on the generative adversarial network, to perform an unsupervised domain adaptation from the synthetic dataset to realistic scenes [2]. Qiu *et al.* take into account the corresponding RGB images provided by the RGB-D camera and propose a deep end-to-end network for camera alignment and ToF depth refinement [23].

Recently, residual pyramid methods have been adopted for a variety of computer vision tasks. For stereo matching, Song *et al.* build a residual pyramid to solve the degradation of depth images in tough areas, such as non-texture areas, boundary areas and tiny details [25]. For the monocular depth estimation,

Chen *et al.* propose a structure-aware residual pyramid to recover the depth image with high visual quality in a coarse-to-fine manner [6]. For image segmentation, Chen *et al.* propose a residual pyramid network to learn the main and residual segmentation in different scales [7]. For image super-resolution, Zheng *et al.* employ a joint residual pyramid network to effectively enlarge the receptive fields [31]. Our SHARP-Net refers to residual pyramid methods as well and achieve success in ToF depth denoising, which will be explained in detail of the following sections. To the best of our knowledge, SHARP-Net is the first work to apply residual pyramid to ToF depth denoising, which greatly surpasses existing methods by integrating spatial hierarchy.

3 ToF Imaging Model

In this section, we briefly introduce the mathematical models of ToF imaging and MPI.

With a single modulation frequency f_ω and four-step phase-shifted measurements r_i $(i = 1, 2, 3, 4)$, the depth d at each pixel is computed as

$$d = \frac{c}{4\pi f_\omega} \arctan\left(\frac{r_4 - r_2}{r_1 - r_3}\right), \tag{1}$$

where c is the speed of light in the vacuum. Under the ideal condition, it is assumed that a single light pulse is reflected only once in the scene and captured by a pixel (x, y) on the sensor. So the raw correlation measurement r_i can be modeled as

$$r_i(x, y) = \int_0^T s(t) b \cos(\omega t - \psi_i) dt, \tag{2}$$

where $s(t)$ is the received signal, $b \cos(\omega t - \psi_i)$ is the referenced periodic signal, ψ_i is the phase offset and T is the exposure temporal range.

In real world, MPI noise always exists. In this case, the received signal is changed to $\hat{s}(t)$, which can be described as

$$\hat{s}(t) = s(t) + \sum_{p \in P} s_p(t), \tag{3}$$

where P is the set of all the light paths p followed by indirectly received signals. Here indirectly received signals $s_p(t)$ represent the captured signals which are reflected multiple bounces after being emitted to the scene. The difference between $s(t)$ and $\hat{s}(t)$ further leads to a deviation to the depth d. In our proposed network, we call this deviation the depth residual. To better regress the depth residual, we bring in a residual pyramid to estimate MPI noise in multiple scales. At different levels of the pyramid, the deviation induced by the set P is regressed and further optimized by our network.

4 Spatial Hierarchy Aware Residual Pyramid Network

Our proposed Spatial Hierarchy Aware Residual Pyramid Network (SHARP-Net) consists of three parts: a Residual Regression Module as the backbone for multi-scale feature extraction, a Residual Fusion Module and a Depth Refinement Module to optimize the performance. The flowchart of SHARP-Net is shown in Fig. 2. The following subsections explain these three parts respectively.

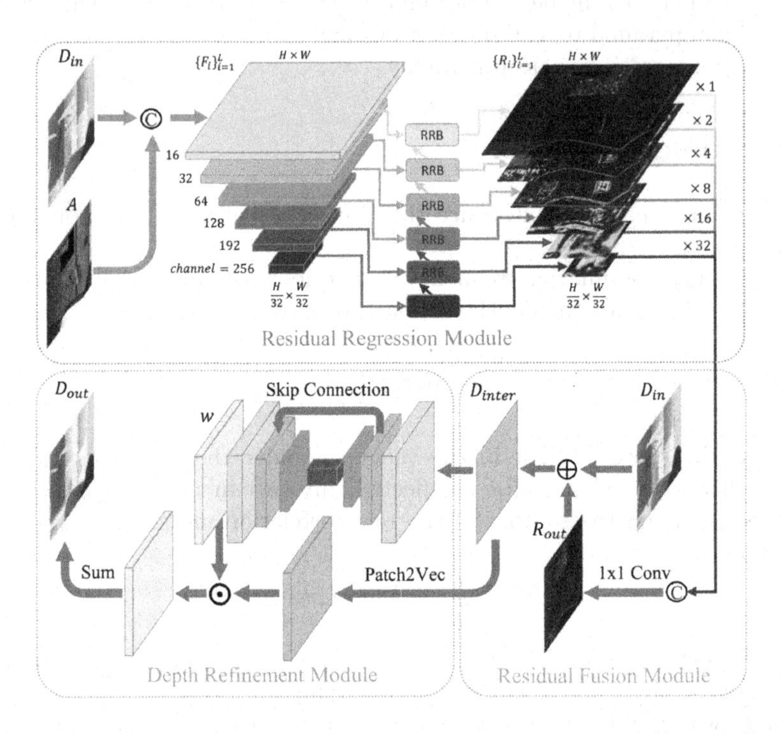

Fig. 2. Flowchart of Spatial Hierarchy Aware Residual Pyramid Network (SHARP-Net). Here ⊙ means the dot product operation, © is the concatenate operation, and ⊕ represents the addition operation. The 'Patch2Vec' represents the operation to reshape the neighbourhoods of each pixel to a vector.

4.1 Residual Regression Module

As the backbone of SHARP-Net, Residual Regression Module first introduces a feature encoder to extract a multi-scale feature pyramid $\{F_i\}_{i=1}^L$ from the combination of depth image D_{in} and amplitude image A, where F_i indicates the feature map extracted at the ith level, and L is the number of layers in the pyramid. When the size of the input image is $W \times H$, the size of feature maps at the ith level is $\frac{W}{2^{i-1}} \times \frac{H}{2^{i-1}} \times C_i$, where C_i is the number of output channels. In our network, we set $L = 6$ to keep the amount of parameters similar to that of state-of-the-art methods. The corresponding C_i are $16, 32, 64, 128, 192, 256$

respectively. From bottom to top, the feature pyramid gradually encodes the geometric information of the more detailed structure in the scene.

At each level, a Residual Regression Block, as shown in Fig. 3, is proposed to predict the depth residual map. The depth residual map from the lower level R_{i+1} is upsampled by the factor of 2 via bi-cubic interpolation, and then concatenated with the feature map at the current level. The new concatenated volume is the input of five sequential convolutional layers, which output the residual map R_i for the current level. Specifically, for the bottom level, the input of Residual Regression Block is only the feature map with size $\frac{W}{32} \times \frac{H}{32} \times 256$ because there is no depth residual map from the lower level. Different from the previous method [23] that directly regresses a residual map by sequentially up-sampling feature maps, our Residual Regression Module progressively regresses multi-scale residual maps in a coarse-to-fine manner by considering the hierarchical structures of the scene. The residual maps in lower resolutions depict depth noise existing in large-size shapes, while the residual map in higher resolutions focuses on depth noise existing in detailed structures. Finally, we get a residual pyramid $\{R_i\}_{i=1}^{L}$ consisting of the depth residual map at each level.

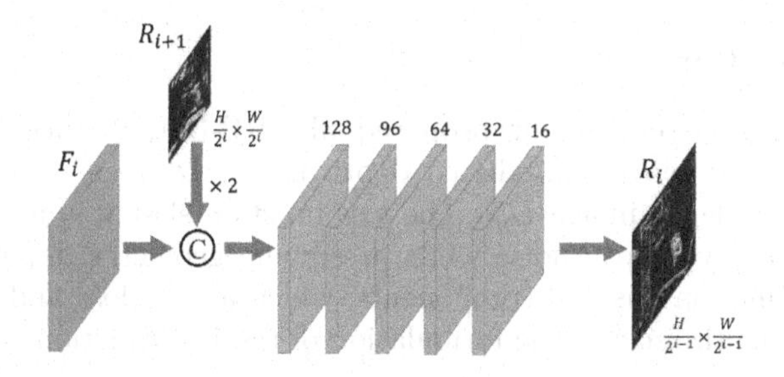

Fig. 3. Flowchart of residual regression block at the ith level

4.2 Residual Fusion Module

The uppermost level of the residual pyramid provides a depth residual map with the original resolution, which can be treated as an estimation of the depth error. However, depth residual map from a single level cannot fully utilize the geometry information of the scene. Although the uppermost level of the residual pyramid contains the information from all the levels below, after the convolutional operation, information from lower resolution levels may get lost. Thus, we propose a Residual Fusion Module to explicitly combine the depth residual maps in all scales. The depth residual map at each level is first upsampled to the original resolution via bi-cubic interpolation. Then all the upsampled depth residual maps are concatenated together. The new residual volume is the input of a 1×1 convolutional layer. After the convolutional operation, we get the final depth residual map R_{out}. The depth residual map is added to the original input depth image, by which the depth image is recovered as D_{inter}. The details of Residual Fusion Module are shown in Fig. 2.

4.3 Depth Refinement Module

After previous two modules, MPI noise is removed to a great extent. In the meantime, shot noise also gets alleviated, but not as much as MPI removal. The existence of shot noise still hinders the application of ToF depth sensing. To address this problem, we propose a Depth Refinement Module, which utilizes Kernel Prediction Network [3] to further remove shot noise.

Depth Refinement Module takes the intermediate depth image D_{inter} as the input, and employs a U-Net model with skip connection to generate a weight matrix. The weight matrix consists of a vectorized filter kernel for each pixel in the depth image. In our experiment, we set the kernel size k as 3 and the size of the weight matrix is $W \times H \times 9$. Next, we generate a patch matrix by vectoring a neighbourhood for each pixel in the depth image. We call the above operation 'Patch2Vec'. When the neighbourhood is a 3×3 area, it is easy to calculate that the size of the patch matrix is also $W \times H \times 9$. Then the weight matrix is multiplied element-wisely with the patch matrix, generating a 3D volume with the same size. By summing over the 3D volume, we finally get the refined depth image D_{out}. Figure 2 shows details of Depth Refinement Module as well.

4.4 Loss Function

To train the parameters in our proposed SHARP-Net, we need to compute the differences between the predicted depth image D_{out} and the corresponding ground truth depth image D_{gt}. The loss function should guide our network to accurately remove depth noise while preserving geometry details. Following [23], our loss function has two components, which are L_1 loss and its gradients on the refined depth image. The formulation of the loss function is depicted as

$$L = \frac{1}{N} \sum \left\| D_{out} - D_{gt} \right\|_1 + \lambda \left\| \nabla D_{out} - \nabla D_{gt} \right\|_1, \tag{4}$$

where $\|\cdot\|_1$ represents the L_1 norm, and N is the number of pixels. Here discrete Sobel operator is utilized to compute the gradients. In our experiments, we set $\lambda = 10$.

5 Experiments

5.1 Datasets

Our SHARP-Net is a supervised neural network to remove the noise for ToF depth images. To train all the parameters, we need ToF datasets with ground truth depth. To produce a suitable dataset, the mainstream method is applying the transient rendering technology to simulate the ToF imaging process while introducing MPI and shot noise [13]. Previous CNN based methods on ToF denoising have provided several synthetic datasets with thousands of scenes. In our experiments, we select two large-scale synthetic datasets ToF-FlyingThings3D (TFT3D) [23] and FLAT [11] for training and evaluation.

The TFT3D dataset contains 6250 different scenes such as living room and bathroom. We only utilize the ToF amplitude images and ToF depth images with resolution 640×480 as input for our proposed method. The FLAT dataset provides a total of 1929 scenes, which include the raw measurements and the corresponding ground truth depth. By using the pipeline released by the FLAT dataset, we convert the raw measurements to ToF depth images and ToF amplitude images with resolution 424×512. Furthermore, to evaluate the performance of SHARP-Net on realistic scenes, we also adopt the True Box dataset which is constructed by Agresti *et al.* in [2]. The ground truth depth of the True Box dataset is acquired by an active stereo system jointly calibrated with a ToF sensor. In total, there are 48 different scenes with resolution 239×320 on the dataset.

5.2 Data Pre-processing

We normalize the input depth images according to the range of depth value provided by the dataset, and filter out the pixels whose depth value are not within the range (0, 1]. For the convenience of experiments, we crop the images on the TFT3D dataset and FLAT dataset to size 384×512. For the True Box dataset, we crop the images to size 224×320. In addition, for the FLAT dataset, we exclude scenes without background following the experiment setting in [23]. For all the three datasets, we randomly select 20% scenes as the test set while the rest for training.

5.3 Training Settings

For the TFT3D dataset, the learning rate is set to be 4×10^{-4}, which is reduced 30% after every 2 epochs. We trained SHARP-Net for 40 epochs with a batch size of 2. For the FLAT dataset, we set the learning rate as 1×10^{-4} with conducting the rate decay. We train the SHARP-Net for 100 epochs with a batch size of 8. For the True Box dataset, the training settings are consistent with that of the TFT3D dataset. The network is implemented using TensorFlow framework [1] and trained using Adam optimizer. With four NVIDIA TITAN Xp graphics cards, the training process takes about 20 h for both TFT3D and FLAT datasets, less than half an hour for the True Box dataset.

5.4 Ablation Studies

SHARP-Net is a CNN based method with a 6-level Residual Regression Module as the backbone and two extra fusion and refinement modules. In order to validate the effectiveness of our proposed modules, we design experiments to compare SHARP-Net against its variants.

- WOFusRef: A variant of SHARP-Net without the Depth Refinement Module and the Residual Fusion Module.
- WORefine: A variant of SHARP-Net without the Depth Refinement Module.
- WOFusion: A variant of SHARP-Net without the Residual Fusion Module.

- FourLevel: A variant of SHARP-Net whose backbone has 4 levels.
- FiveLevel: A variant of SHARP-Net whose backbone has 5 levels.

For a fair comparison with FourLevel and FiveLevel, we need to ensure that the amount of parameters of these two variants are nearly the same with SHARP-Net. Therefore, we adjust the number of convolution kernel channels of the variants.

Table 1. Quantitative comparison with the variants of SHARP-Net on the TFT3D dataset.

Model	TFT3D dataset: MAE (cm)/Relative error				
	1st Quan.	2nd Quan.	3rd Quan.	4th Quan.	Overall
WOFusRef	0.12/7.7%	0.49/8.3%	1.08/9.4%	4.90/16.3%	1.69/13.8%
WORefine	0.12/7.7%	0.44/7.5%	0.97/8.4%	4.69/15.6%	1.55/12.7%
WOFusion	0.11/7.1%	0.42/7.2%	0.94/8.2%	5.01/16.7%	1.62/13.2%
FourLevel	0.15/9.6%	0.57/9.7%	1.24/10.7%	5.15/17.2%	1.78/14.5%
FiveLevel	0.12/7.7%	0.46/7.8%	1.00/8.8%	4.55/15.2%	1.53/12.5%
SHARP-Net	**0.09/5.8%**	**0.30/5.1%**	**0.67/5.8%**	**3.40/11.3%**	**1.19/9.7%**

For the quantitative comparison, we use two metrics, Mean Absolute Error (MAE) and relative error, to evaluate the performance. The MAE between the original noisy depth image and the ground truth depth image is depicted as the original MAE. Then, we define the relative error as the ratio of the MAE of each method to the MAE of the corresponding input. The overall and partial MAE/Relative Error at each error level are also calculated. Different denoising methods may have varying performances at different error levels. In our experiment, we adopt an evaluation method that is similar to the method in [23] to comprehensively evaluate our proposed SHARP-Net at different error levels. First, we calculate the per-pixel absolute error value between the input depth image and the ground truth. Then we sort all the per-pixel absolute errors in an ascending order. Next all the pixels in the test set are split into four quantiles (four error level sets). The difference between our evaluation method and the method in [23] is that we sort all the pixels in the test set instead of in a single image. This change makes our evaluation more reasonable because sorting in the whole test set eliminates the depth distinction over images. The pixels in the range of 0%–25% are classified into the 1st error level. In the same way, the pixels in the range of 25%–50% and 50%–75% and 75%–100% are classified into the 2nd, 3rd, and 4th error level. Pixels with depth value beyond the maximum depth for each dataset are considered as outlier here and excluded from any error level sets. Finally, we calculate the partial MAE and overall MAE for different error levels respectively.

For ablation studies, we just utilize the TFT3D dataset to compare our SHARP-Net against its variants. The overall MAE and partial MAE at each error level are reported in Table 1. From Table 1, it can be observed that SHARP-Net achieves the lowest MAE and relative error at all error levels. In addition, '4th Quan.' contributes the greatest share on the value of overall MAE compared with the remanent three quantiles. Comparing SHARP-Net with FourLevel and FiveLevel variants, we can see that at all error levels, MAE decreases as the total number of pyramid levels increases. This is because the network explicitly divides the scene into a more detailed hierarchical structure if the pyramid has more levels, which results in a more accurate estimation of MPI noise.

Comparing SHARP-Net with WORefine and WOFusion, it can be observed that the employment of Residual Fusion Module and Depth Refinement Module reduce the overall MAE by 26% and 23% respectively, which indicates the necessity of those two modules. Linking the comparison between WORefine and WOFusion at all error levels, it can be seen that either of the two modules facilitates the decline of MAE but the extent is limited. However, considering the difference between WOFusRef and SHARP-Net on the MAE and relative error, we conclude that utilizing these two modules together can greatly improve the performance of the noise removal at all error levels.

Table 2. Quantitative comparison with competitive ToF depth denoising methods on TFT3D, FLAT and True Box datasets.

Model	1st Quan.	2nd Quan.	3rd Quan.	4th Quan.	Overall
	TFT3D dataset: MAE (cm)/Relative error				
DeepToF	0.47/30.1%	1.56/26.6%	3.11/27.0%	9.01/30.0%	3.54/28.9%
ToF-KPN	0.19/12.2%	0.82/13.9%	1.87/16.2%	6.64/21.3%	2.38/19.4%
SHARP-Net	**0.09/5.8%**	**0.30/5.1%**	**0.67/5.8%**	**3.40/11.3%**	**1.19/9.7%**
	FLAT dataset: MAE (cm)/Relative error				
DeepToF	0.09/27.3%	0.44/33.6%	1.13/43.5%	2.74/37.8%	1.10/43.3%
ToF-KPN	0.08/24.2%	0.30/22.9%	0.66/25.4%	2.12/29.3%	0.79/31.1%
SHARP-Net	**0.04/12.1%**	**0.14/10.7%**	**0.32/12.3%**	**1.33/18.4%**	**0.46/18.1%**
	True Box dataset: MAE (cm)/Relative error				
DeepToF	0.31/42.5%	1.06/49.5%	2.15/52.9%	5.75/53.9%	2.32/52.7%
ToF-KPN	0.28/38.4%	0.87/40.6%	1.64/40.4%	4.51/42.3%	1.82/41.4%
SHARP-Net	**0.15/20.5%**	**0.47/21.9%**	**0.91/22.4%**	**3.02/28.3%**	**1.14/25.9%**

5.5 Results on Synthetic Datasets

To evaluate the performance of our proposed SHARP-Net, we compare it with two state-of-the-art ToF depth denoising methods DeepToF [17] and ToF-KPN [23]. The inputs of all selected methods are the concatenation of depth images and corresponding amplitude images. It should be noted that the original DeepToF is smaller than SHARP-Net in term of model size. For a fair comparison, we take the same strategy as [23] to replace the original DeepToF model with the U-Net backbone of ToF-KPN. The quantitative experimental results

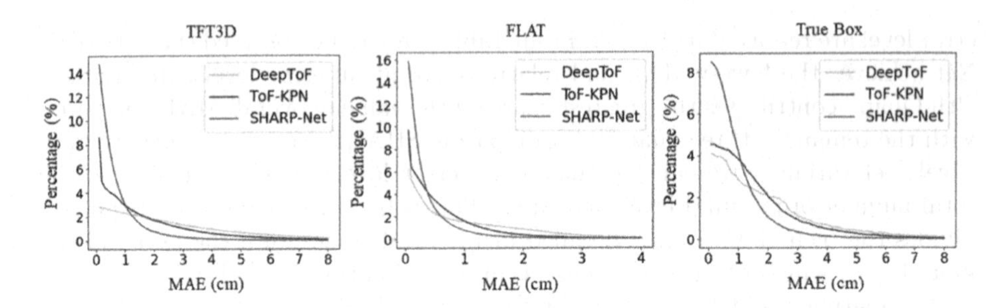

Fig. 4. The per-pixel error distribution curves of different methods on the TFT3D, FLAT and True Box datasets. The distribution curves of these three methods show that our proposed SHARP-Net obtains the optimal error distribution on all the datasets.

on the TFT3D and FLAT datasets are reported in Table 2. It can be seen that SHARP-Net achieves the lowest MAE and relative error at all error levels of the two synthetic datasets. The MAE between the input depth and ground truth depth is 12.24 cm and 2.54 cm for both TFT3D and FLAT datasets. After training on these datasets, SHARP-Net reduces the MAE to 1.19 cm and 0.46 cm in the test sets respectively.

The relative error is also a good indicator to measure the performance for different methods. From Table 2, it can be seen that the DeepToF method gives similar relative errors for all the four error levels, especially on the TFT3D dataset. Compared with two other methods, DeepToF's performance in terms of the relative error indicator is low. For ToF-KPN, the relative error increase as the error level increasing, which means ToF-KPN has better denoising performance for higher error level sets. For SHARP-Net, it can be seen that relative error is much smaller than the other two methods on the TFT3D dataset on the FLAT dataset, SHARP-Net is much better than DeepToF in term of relative error. Compared with ToF-KPN, SHARP-Net performs the same as ToF-KPN at the preceding three error levels, and outperform ToF-KPN at the highest error level.

For an intuitive comparison, in Fig. 4, we illustrate the per-pixel error distribution curves for all the methods on the TFT3D and FLAT datasets. It can be seen that after denoising by our SHARP-Net, the depth errors are mainly concentrated in the lower error region. In Fig. 5, we give several qualitative comparison results for SHARP-Net, ToF-KPN and DeepToF. It can be seen that the depth image corrected by our proposed method is more accurate, preserving more geometry structures in the scene. We observe that ToF-KPN performs better than DeepToF in removing the noise existing in detailed structures. However, the noise removal of ToF-KPN on large-size shapes is not adequate. In contrast, SHARP-Net demonstrates better results for large-size shapes and detailed structures simultaneously. In fact, our SHARP-Net also has some failure cases in depth denoising, for example low reflection areas and extremely complex geometry structures, which are the limitations of our method.

In Fig. 6, we compare the performance of all the methods along a scan line on a depth image selected from the TFT3D dataset. From the ToF amplitude image, we can observe that the scene is located in a living room. Many different objects

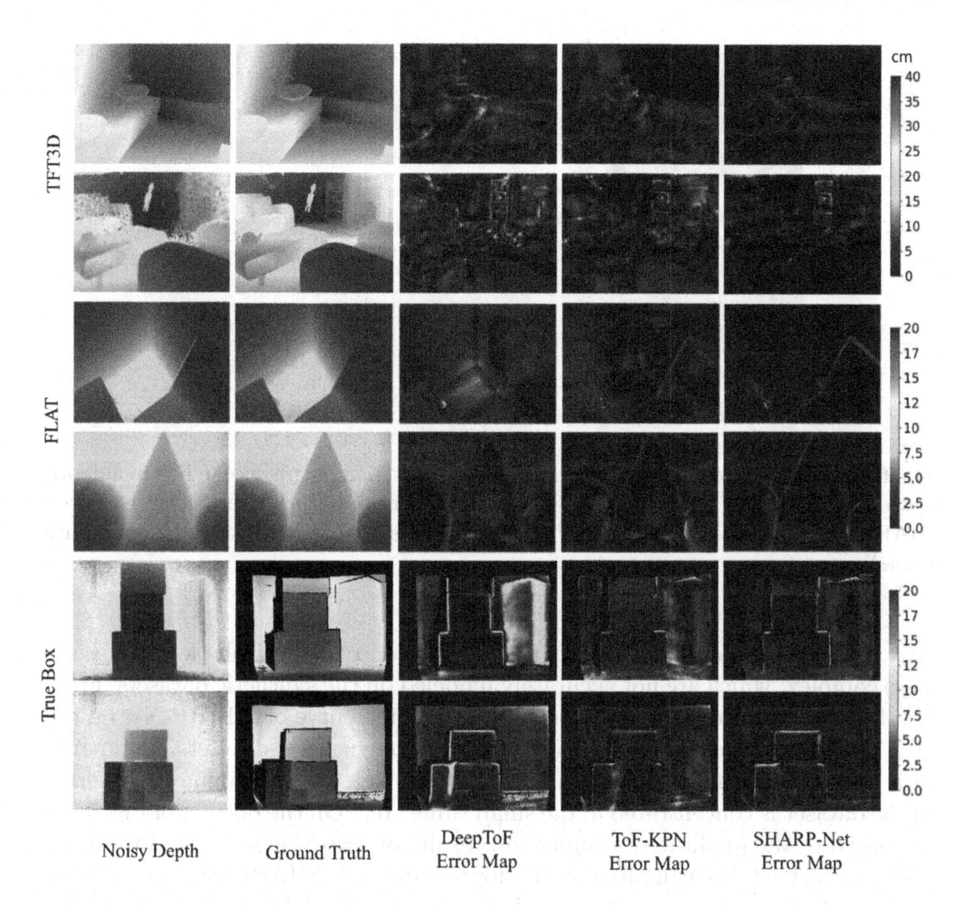

Fig. 5. Qualitative comparison on the TFT3D dataset, the FLAT dataset and the True Box dataset for ToF depth denoising. For each dataset, two scenes are selected for comparison. The colorbars in the right show the color scale for error maps with the unit in cm. (Color figure online)

appear in the living room and demonstrate complex hierarchical structures. The distinct depth variation along this scan line makes it suitable for this comparison. It can be seen that after depth denoising, the depth data corrected by SHARP-Net draw the closest line to the ground truth.

5.6 Results on the Realistic Dataset

Furthermore, we test our proposed SHARP-Net along with previous methods on a realistic dataset. All the tested models are retrained on the True Box training set. The experimental results for all the methods on the True Box dataset are also shown in Table 2. It can be seen that SHARP-Net surpasses other methods at all error levels. We also find that the relative errors at all error levels on the True Box dataset are significantly larger than those on the synthetic datasets. One reason

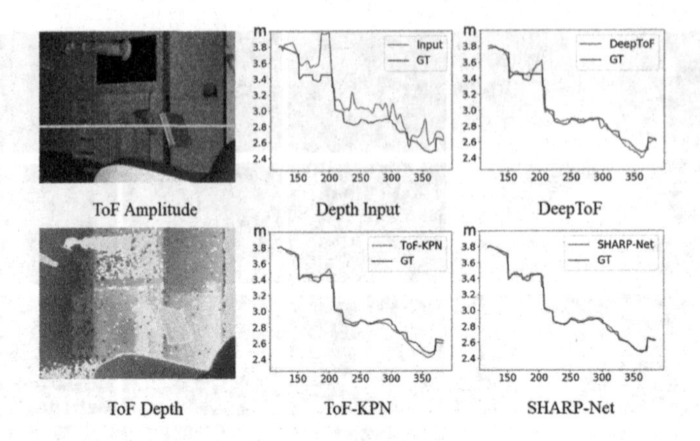

Fig. 6. Quantitative comparison with previous works along a green scan line in a depth image from the TFT3D dataset. 'GT' means the ground truth depth. Our proposed SHARP-Net demonstrates the best performance on depth denoising. (Color figure online)

may be that the noise generation mechanism for realistic ToF depth noise is more complex, which are not accurately modeled on the synthetic datasets. From Fig. 4, it can be observed that the error distribution curve of SHARP-Net on the True Box dataset is similar to those on the two synthetic datasets. Compared with other methods, after denoising by SHARP-Net, the remaining depth error on the dataset is concentrated in the small value area. On the bottom of Fig. 5, we demonstrate the qualitative comparison results on two scenes selected from the True Box test set. We can observe that for this dataset, SHARP-Net presents the best visual effects. Compared with other methods, SHARP-Net performs better in large-size shapes, especially in the background areas.

6 Conclusion

The Multi-Path Interference (MPI) seriously degrades the depth image captured by ToF sensors. In this work, we propose SHARP-Net, a Spatial Hierarchy Aware Residual Pyramid Network for ToF depth denoising. Our SHARP-Net progressively utilizes the spatial hierarchical structure of the scene to regress depth residual maps in different scales, obtaining a residual pyramid. A Residual Fusion Module is introduced to selectively fuse the residual pyramid by summing over the depth residual maps at all levels in the pyramid, and a Kernel Prediction Network based Depth Refinement Module is employed to further eliminate shot noise. Ablation studies validate the effectiveness of these modules. Experimental results demonstrate that our SHARP-Net greatly surpasses the state-of-the-art methods in both quantitative and qualitative comparison on synthetic and realistic datasets.

Acknowledgments. We acknowledge funding from National Key R&D Program of China under Grant 2017YFA0700800, and National Natural Science Foundation of China under Grants 61671419 and 61901435.

References

1. Abadi, M., et al.: Tensorflow: a system for large-scale machine learning. In: 12th Symposium on Operating Systems Design and Implementation, pp. 265–283 (2016)
2. Agresti, G., Schaefer, H., Sartor, P., Zanuttigh, P.: Unsupervised domain adaptation for ToF data denoising with adversarial learning. In: Proceedings of the IEEE Conference on Computer Vision and Pattern Recognition, pp. 5584–5593 (2019)
3. Bako, S., et al.: Kernel-predicting convolutional networks for denoising Monte Carlo renderings. ACM Trans. Graph. (TOG) **36**(4), 97 (2017)
4. Barz, B., Denzler, J.: Hierarchy-based image embeddings for semantic image retrieval. In: 2019 IEEE Winter Conference on Applications of Computer Vision (WACV), pp. 638–647. IEEE (2019)
5. Bhandari, A., Feigin, M., Izadi, S., Rhemann, C., Schmidt, M., Raskar, R.: Resolving multipath interference in Kinect: an inverse problem approach. In: 2014 IEEE SENSORS, pp. 614–617. IEEE (2014)
6. Chen, X., Chen, X., Zha, Z.: Structure-aware residual pyramid network for monocular depth estimation. In: Proceedings of the Twenty-Eighth International Joint Conference on Artificial Intelligence, IJCAI 2019, Macao, China, 10–16 August 2019, pp. 694–700 (2019)
7. Chen, X., Lou, X., Bai, L., Han, J.: Residual pyramid learning for single-shot semantic segmentation. IEEE Trans. Intell. Transp. Syst. **21**, 2990–3000 (2019)
8. Freedman, D., Smolin, Y., Krupka, E., Leichter, I., Schmidt, M.: SRA: fast removal of general multipath for ToF sensors. In: Fleet, D., Pajdla, T., Schiele, B., Tuytelaars, T. (eds.) ECCV 2014. LNCS, vol. 8689, pp. 234–249. Springer, Cham (2014). https://doi.org/10.1007/978-3-319-10590-1_16
9. Fuchs, S.: Multipath interference compensation in time-of-flight camera images. In: 2010 20th International Conference on Pattern Recognition, pp. 3583–3586. IEEE (2010)
10. Fuchs, S., Suppa, M., Hellwich, O.: Compensation for multipath in ToF camera measurements supported by photometric calibration and environment integration. In: Chen, M., Leibe, B., Neumann, B. (eds.) ICVS 2013. LNCS, vol. 7963, pp. 31–41. Springer, Heidelberg (2013). https://doi.org/10.1007/978-3-642-39402-7_4
11. Guo, Q., Frosio, I., Gallo, O., Zickler, T., Kautz, J.: Tackling 3D ToF artifacts through learning and the FLAT dataset. In: Ferrari, V., Hebert, M., Sminchisescu, C., Weiss, Y. (eds.) ECCV 2018. LNCS, vol. 11205, pp. 381–396. Springer, Cham (2018). https://doi.org/10.1007/978-3-030-01246-5_23
12. Gupta, M., Nayar, S.K., Hullin, M.B., Martin, J.: Phasor imaging: a generalization of correlation-based time-of-flight imaging. ACM Trans. Graph. (ToG) **34**(5), 156 (2015)
13. Jarabo, A., Marco, J., Muñoz, A., Buisan, R., Jarosz, W., Gutierrez, D.: A framework for transient rendering. ACM Trans. Graph. (ToG) **33**(6), 177 (2014)
14. Jiménez, D., Pizarro, D., Mazo, M., Palazuelos, S.: Modeling and correction of multipath interference in time of flight cameras. Image Vis. Comput. **32**(1), 1–13 (2014)

15. Jung, J., Lee, J.Y., Jeong, Y., Kweon, I.S.: Time-of-flight sensor calibration for a color and depth camera pair. IEEE Trans. Pattern Anal. Mach. Intell. **37**(7), 1501–1513 (2014)
16. Lenzen, F., Schäfer, H., Garbe, C.: Denoising time-of-flight data with adaptive total variation. In: Bebis, G., et al. (eds.) ISVC 2011. LNCS, vol. 6938, pp. 337–346. Springer, Heidelberg (2011). https://doi.org/10.1007/978-3-642-24028-7_31
17. Marco, J., et al.: DeepToF: off-the-shelf real-time correction of multipath interference in time-of-flight imaging. ACM Transactions on Graphics (ToG) **36**(6), 219 (2017)
18. Mo, K., et al.: StructureNet: hierarchical graph networks for 3D shape generation. arXiv preprint arXiv:1908.00575 (2019)
19. Nan, Y., Xiao, R., Gao, S., Yan, R.: An event-based hierarchy model for object recognition. In: 2019 IEEE Symposium Series on Computational Intelligence (SSCI), pp. 2342–2347. IEEE (2019)
20. Park, B., Yu, S., Jeong, J.: Densely connected hierarchical network for image denoising. In: Proceedings of the IEEE Conference on Computer Vision and Pattern Recognition Workshops (2019)
21. Payne, A., et al.: 7.6 a 512 × 424 CMOS 3D time-of-flight image sensor with multi-frequency photo-demodulation up to 130 MHz and 2 gs/s ADC. In: 2014 IEEE International Solid-State Circuits Conference Digest of Technical Papers (ISSCC), pp. 134–135. IEEE (2014)
22. Peng, J., Xiong, Z., Wang, Y., Zhang, Y., Liu, D.: Zero-shot depth estimation from light field using a convolutional neural network. IEEE Trans. Comput. Imaging **6**, 682–696 (2020)
23. Qiu, D., Pang, J., Sun, W., Yang, C.: Deep end-to-end alignment and refinement for time-of-flight RGB-D module. In: Proceedings of the IEEE International Conference on Computer Vision, pp. 9994–10003 (2019)
24. Shi, Y., Chang, A.X., Wu, Z., Savva, M., Xu, K.: Hierarchy denoising recursive autoencoders for 3D scene layout prediction. In: Proceedings of the IEEE Conference on Computer Vision and Pattern Recognition, pp. 1771–1780 (2019)
25. Song, X., Zhao, X., Hu, H., Fang, L.: EdgeStereo: a context integrated residual pyramid network for stereo matching. In: Jawahar, C.V., Li, H., Mori, G., Schindler, K. (eds.) ACCV 2018. LNCS, vol. 11365, pp. 20–35. Springer, Cham (2019). https://doi.org/10.1007/978-3-030-20873-8_2
26. Su, S., Heide, F., Wetzstein, G., Heidrich, W.: Deep end-to-end time-of-flight imaging. In: Proceedings of the IEEE Conference on Computer Vision and Pattern Recognition, pp. 6383–6392 (2018)
27. Yao, T., Pan, Y., Li, Y., Mei, T.: Hierarchy parsing for image captioning. In: Proceedings of the IEEE International Conference on Computer Vision, pp. 2621–2629 (2019)
28. Zanuttigh, P., Marin, G., Dal Mutto, C., Dominio, F., Minto, L., Cortelazzo, G.M.: Time-of-Flight and Structured Light Depth Cameras: Technology and Applications, pp. 978–983. Springer, Switzerland (2016). ISSBN
29. Zhang, S.: High-speed 3D shape measurement with structured light methods: a review. Opt. Lasers Eng. **106**, 119–131 (2018)
30. Zhang, Y., Xiong, Z., Wu, F.: Fusion of time-of-flight and phase shifting for high-resolution and low-latency depth sensing. In: 2015 IEEE International Conference on Multimedia and Expo (ICME), pp. 1–6. IEEE (2015)
31. Zheng, Y., Cao, X., Xiao, Y., Zhu, X., Yuan, J.: Joint residual pyramid for joint image super-resolution. J. Vis. Commun. Image Represent. **58**, 53–62 (2019)

Sat2Graph: Road Graph Extraction Through Graph-Tensor Encoding

Songtao He[1](✉), Favyen Bastani[1], Satvat Jagwani[1], Mohammad Alizadeh[1], Hari Balakrishnan[1], Sanjay Chawla[2], Mohamed M. Elshrif[2], Samuel Madden[1], and Mohammad Amin Sadeghi[3]

[1] Massachusetts Institute of Technology, Cambridge, USA
{songtao,favyen,satvat,alizadeh,hari,madden}@csail.mit.edu
[2] Qatar Computing Research Institute, Doha, Qatar
schawla@hbku.edu.qa, melshrif77@gmail.com
[3] University of Tehran, Tehran, Iran
m.a.sadeghi@gmail.com

Abstract. Inferring road graphs from satellite imagery is a challenging computer vision task. Prior solutions fall into two categories: (1) pixel-wise segmentation-based approaches, which predict whether each pixel is on a road, and (2) graph-based approaches, which predict the road graph iteratively. We find that these two approaches have complementary strengths while suffering from their own inherent limitations.

In this paper, we propose a new method, Sat2Graph, which combines the advantages of the two prior categories into a unified framework. The key idea in Sat2Graph is a novel encoding scheme, *graph-tensor encoding* (GTE), which encodes the road graph into a tensor representation. GTE makes it possible to train a simple, non-recurrent, supervised model to predict a rich set of features that capture the graph structure directly from an image. We evaluate Sat2Graph using two large datasets. We find that Sat2Graph surpasses prior methods on two widely used metrics, TOPO and APLS. Furthermore, whereas prior work only infers planar road graphs, our approach is capable of inferring stacked roads (e.g., overpasses), and does so robustly.

1 Introduction

Accurate and up-to-date road maps are critical in many applications, from navigation to self-driving vehicles. However, creating and maintaining digital maps is expensive and involves tedious manual labor. In response, automated solutions have been proposed to automatically infer road maps from different sources of data, including GPS tracks, aerial imagery, and satellite imagery. In this paper, we focus on extracting road network graphs from satellite imagery.

Although many techniques have been proposed [2,3,6,9,10,20–22,25,32,35, 36], extracting road networks from satellite imagery is still a challenging computer vision task due to the complexity and diversity of the road networks. Prior solutions fall into two categories: pixel-wise segmentation-based approaches and

A. Vedaldi et al. (Eds.): ECCV 2020, LNCS 12369, pp. 51–67, 2020.
https://doi.org/10.1007/978-3-030-58586-0_4

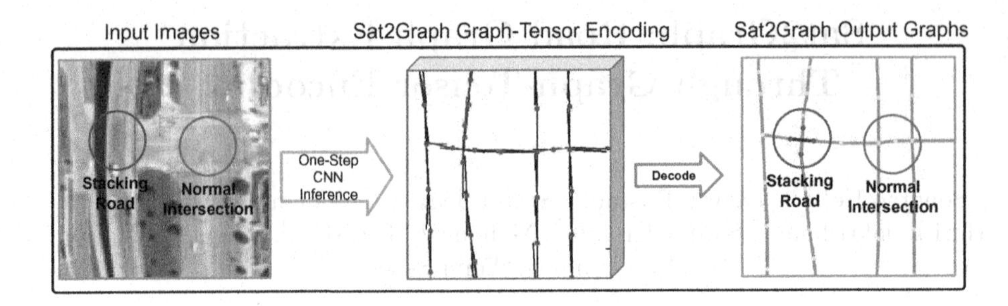

Fig. 1. Highlight of Sat2Graph.

graph-based approaches. Segmentation-based approaches assign a *roadness* score to each pixel in the satellite imagery. Then, they extract the road network graph using heuristic approaches. Here, the road segmentation acts as the intermediate representation of the road network graph. In contrast, graph-based approaches construct a road network graph directly from satellite imagery. Recently, Bastani *et al.* [2], as well as several follow-up works [10,21], utilize graph-based solutions that iteratively add vertices and edges to the partially constructed graph (Fig. 1).

We observe that the approaches in these two categories often tradeoff with each other. Segmentation-based approaches typically have a wider receptive field but rely on an intermediate non-graph representation and a post-processing heuristic (e.g., morphological thinning and line following) to extract road network graphs from this intermediate representation. The usage of the intermediate non-graph representation limits the segmentation-based approaches, and they often produce noisy and lower precision road networks compared with the graph-based methods as a result. To encourage the neural network model to focus more on the graph structure of road networks, recent work [3] proposes to train the road segmentation model jointly with road directions, and the approach achieves better road connectivity through this joint training strategy. However, a postprocessing heuristic is still needed.

In contrast, graph-based approaches [2,10,21] learn the graph structure directly. As a result, graph-based approaches yield road network graphs with better road connectivity compared with the original segmentation-based approach [2]. However, the graph generation process is often iterative, resulting in a neural network model that focuses more on local information rather than global information. To take more global information into account, recent work [10,21] proposes to improve the graph-based approaches with a sequential generative model, resulting in better performance compared with other state-of-art approaches.

Recent advancements [3,10,21] in segmentation-based approaches and graph-based approaches respectively are primarily focused on overcoming the inherent limitations of their baseline approaches, which are exactly from the same aspects that the methods in the competing baseline approach (i.e., from the other category) claim as advantages. Based on this observation, a natural question to ask is if it is possible to combine the segmentation-based approach and the graph-based approach into one unified approach that can benefit from the advantages of both?

Our answer to this question is a new road network extraction approach, Sat2Graph, which combines the inherent advantages of segmentation-based approaches and graph-based approaches into one simple, unified framework. To do this, we design a novel encoding scheme, *graph-tensor encoding* (GTE), to encode the road network graph into a tensor representation, making it possible to train a simple, non-recurrent, supervised model that predicts graph structures holistically from the input image.

In addition to the tensor-based network encoding, this paper makes two contributions:

1. Sat2Graph surpasses state-of-the-art approaches in a widely used topology-similarity metric at all precision-recall trade-off positions in an evaluation over a large city-scale dataset covering 720 km^2 area in 20 U.S. cities and the popular SpaceNet roads dataset [30].
2. Sat2Graph can naturally infer stacked roads, which prior approaches don't handle.

2 Related Work

Traditional Approaches. Extracting road networks from satellite imagery has long history [14,31]. Traditional approaches generally use heuristics and probabilistic models to infer road networks from imagery. For examples, Hinz *et al.* [20] propose an approach to create road networks through a complicated road model that is built using detailed knowledge about roads and the environmental context, such as the nearby buildings, vehicles and so on. Wegner *et al.* [32] propose to model the road network with higher-order conditional random fields (CRFs). They first segment the aerial images into super-pixels, then they connect these super-pixels based on the CRF model.

Segmentation-Based Approaches. With the increasing popularity of deep learning, researchers have used convolutional neural networks (CNN) to extract road network from satellite imagery [3,6,9,22,35,36]. For example, Cheng *et al.* [9] use an end-to-end cascaded CNN to extract road segmentation from satellite imagery. They apply a binary threshold to the road segmentation and use morphological thinning to extract the road center-lines. Then, a road network graph is produced through tracing the single-pixel-width road center-lines. Many other segmentation-based approaches proposed different improvements upon this basic graph extraction pipeline, including improved CNN backbones [6,36], improved post-processing strategy [22], improved loss functions [22,25], incorporating GAN [11,28,34], and joint training [3].

In contrast with existing segmentation-based approaches, Sat2Graph does not rely on the road segmentation as intermediate representation and learns the graph structure directly.

Graph-Based Approaches. Graph-based approaches construct a road network graph directly from satellite imagery. Recently, Bastani *et al.* [2] proposed RoadTracer, a graph-based approach to generate road network in an iterative way. The algorithm starts from a known location on the road map. Then, at each iteration, the algorithm uses a deep neural network to predict the next location to visit along the road through looking at the surrounding satellite imagery of the current location. Recent works [10,21] advanced the graph-based approach through applying sequential generative models (RNN) to generate road network iteratively. The usage of sequential models allows the graph generation model to take more context information into account compared with RoadTracer [2].

In contrast with existing graph-based approaches, Sat2Graph generates the road graphs in one shot (holistic). This allows Sat2Graph to easily capture the global information and make better coordination of vertex placement. The non-recurrent property of Sat2Graph also makes it easy to train and easy to extend (e.g., combine Sat2Graph with GAN). We think this simplicity of Sat2Graph is another advantage over other solutions.

Using Other Data Sources and Other Digital Map Inference Tasks. Extracting road networks from other data sources has also been extensively studied, e.g., using GPS trajectories collected from moving vehicles [1,5,8,12,13, 18,29]. Besides road topology inference, satellite imagery also enables inference of different map attributes, including high-definition road details [19,23,24], road safety [26] and road quality [7].

3 Sat2Graph

In this section, we present the details of our proposed approach - Sat2Graph. Sat2Graph relies on a novel encoding scheme that can encode the road network graph into a three-dimensional tensor. We call this encoding scheme Graph-Tensor Encoding (GTE). This graph-tensor encoding scheme allows us to train a simple, non-recurrent, neural network model to directly map the input satellite imagery into the road network graph (i.e., edges and vertices). As noted in the introduction, this graph construction strategy combines the advantages of segmentation-based and graph-based approaches.

3.1 Graph-Tensor Encoding (GTE)

We show our graph-tensor encoding (GTE) scheme in Fig. 2(a). For a road network graph $G = \{V, E\}$ that covers a W meters by H meters region, GTE uses a $\frac{W}{\lambda} \times \frac{H}{\lambda} \times (1 + 3 \cdot D_{max})$ 3D-tensor (denoted as T) to store the encoding of the graph. Here, the λ is the spatial resolution, i.e., one meter, which restricts the encoded graph in a way that no two vertices can be co-located within a $\lambda \times \lambda$ grid, and D_{max} is the maximum edges that can be encoded at each $\lambda \times \lambda$ grid.

The first two dimensions of T correspond to the two spatial axes in the 2D plane. We use the vector at each spatial location $u_{x,y} = [T_{x,y,1}, T_{x,y,2}, ...,$

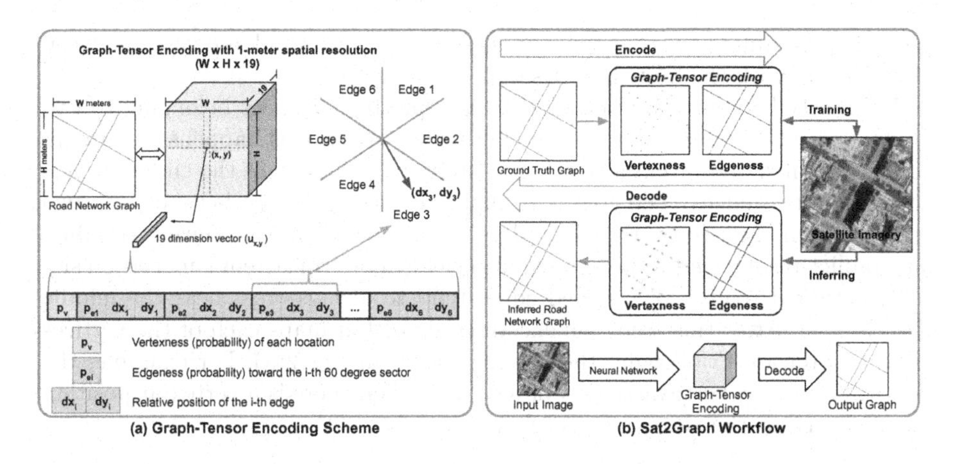

Fig. 2. Graph-Tensor Encoding and Sat2Graph workflow.

$T_{x,y,(1+3 \cdot D_{max})}]^T$ to encode the graph information. As shown in Fig. 2(a), the vector $u_{x,y}$ has $(1 + 3 \cdot D_{max})$ elements. Its first element $p_v \in [0, 1]$ (vertexness) encodes the probability of having a vertex at position (x, y). Following the first element are D_{max} 3-element groups, each of which encodes the information of a potential outgoing edge from position (x, y). For the i-th 3-element group, its first element $p_{e_i} \in [0, 1]$ (edgeness) encodes the probability of having an outgoing edge toward (dx_i, dy_i), i.e., an edge pointing from (x, y) to $(x + dx_i, y + dy_i)$. Here, we set D_{max} to six as we find that vertices with degree greater than six are very rare in road network graphs.

To reduce the number of possible different isomorphic encodings of the same input graph, GTE only uses the i-th 3-element group to encode edges pointing toward a $\frac{360}{D_{max}}$-degree sector from $(i - 1) \cdot \frac{360}{D_{max}}$ degrees to $i \cdot \frac{360}{D_{max}}$ degrees. We show this restriction and an example edge (in red color) in Fig. 2(a). This strategy imposes a new restriction on the encoded graphs – for each vertex in the encoded graph, there can only be at most one outgoing edge toward each $\frac{360}{D_{max}}$-degree sector. However, we find this restriction does not impact the representation ability of GTE for most road graphs. This is because the graphs encoded by GTE are *undirected.* We defer the discussion on this in Sect. 5.

Encode. Encoding a road network graph into GTE is straightforward. For road network extraction application, the encoding algorithm first interpolates the segment of straight road in the road network graph. It selects the minimum number of evenly spaced intermediate points so that the distance between consecutive points is under d meters. This interpolation strategy regulates the length of the edge vector in GTE, making the training process stable. Here, a small d value, e.g., $d < 5$, converts GTE back to the road segmentation, making GTE unable to represent stacking roads. A very large d value, e.g. $d = 50$, makes the GTE hard to approximate curvy roads. For these reasons, we think a d value between 15 to 25 can work the best. In our setup, we set d to 20.

For stacked roads, the interpolation may produce vertices belonging to two overlapped road segments at the same position. When this happens, we use an iterative conflict-resolution algorithm to shift the positions of the endpoint vertices of the two edges. The goal is to make sure the distance between any two vertices (from the two overlapping edges) is greater than 5 m. During training, this conflict-resolution pre-processing also yields more consistent supervision signal for stacked roads - overlapped edges tend to always cross near the middle of each edge. After this step, the encoding algorithm maps each of the vertices to the 3D-tensor T following the scheme shown in Fig. 2(a). For example, the algorithm sets the vertexness (p_v) of $u_{x,y}$ to 1 when there is a vertex at position (x, y), otherwise the vertexness is set to 0.

Decode. GTE's Decoding algorithm converts the predicted GTE (often noisy) of a graph back to the regular graph format (G = {V,E}). The decoding algorithm consists of two steps, (1) vertex extraction and, (2) edge connection. As both the vertexness predictions and edgeness predictions are real numbers between 0 and 1, we only consider vertices and edges with probability greater than a threshold (denoted as p_{thr}).

In the vertex extraction step, the decoding algorithm extracts the potential vertices through localizing the local maximas of the vertexness map (we show an example of this in Fig. 2(b)). The algorithm only considers the local maximas with vertexness greater than p_{thr}.

In the edge connection step, for each candidate vertex $v \in V$, the decoding algorithm connects its outgoing edges to other vertices. For the i-th edge of vertex $v \in V$, the algorithm computes its distance to all nearby vertices u through the following distance function,

$$\begin{aligned} d(v, i, u) = & \|(v_x + dx_i, v_y + dy_i) - (u_x, u_y)\| \\ & + w \cdot cos_{dist}((dx_i, dy_i), (u_x - v_x, u_y - v_y)), \end{aligned} \tag{1}$$

where $cos_{dist}(v_1, v_2)$ is the cosine distance of the two vectors, and w is the weight of the cosine distance in the distance function. Here, we set w to a large number, i.e., 100, to avoid incorrect connections. After computing this distance, the decoding algorithm picks up a vertex u' that minimizes the distance function $d(v, i, u)$, and adds an edge between v and u'. We set a maximum distance threshold, i.e., 15 m, to avoid incorrect edges being added to the graph when there are no good candidate vertices nearby.

3.2 Training Sat2Graph

We use cross-entropy loss (denoted as $\mathcal{L}_{CE}$) and L_2-loss to train Sat2Graph. The cross-entropy loss is applied to vertexness channel (p_v) and edgeness channels (p_{e_i} $i \in \{1, 2, ..., D_{max}\}$), and the L_2-loss is applied to the edge vector channels

$((dx_i, dy_i)\ i \in \{1, 2, ..., D_{max}\})$. GTE is inconsistent along long road segments. In this case, the same road structure can be mapped to different ground truth labels in GTE representation. Because of this inconsistency, we only compute the losses for edgeness and edge vectors at position (x, y) when there is a vertex at position (x, y) in the ground truth. We show the overall loss function below ($\hat{T}, \hat{p_v}, \hat{p_{e_i}}, \hat{d_{x_i}}, \hat{d_{y_i}}$ are from ground truth),

$$\begin{aligned} \mathcal{L}(T, \hat{T}) = & \sum_{(x,y)\in[1..W]\times[1..H]} \Bigg(\mathcal{L}_{CE}(p_v, \hat{p_v}) \\ & + \hat{T}_{x,y,1} \cdot \Big(\sum_{i=1}^{D_{max}} \big(\mathcal{L}_{CE}(p_{e_i}, \hat{p_{e_i}}) + \mathcal{L}_2((dx_i, dy_i), (\hat{dx_i}, \hat{dy_i}))\big)\Big)\Bigg) \end{aligned} \quad (2)$$

In Fig. 2(b), we show the training and inferring workflows of Sat2Graph. Sat2Graph is agnostic to the CNN backbones. In this paper, we choose to use the Deep Layer Aggregation (DLA) [33] segmentation architecture as our CNN backbone. We use residual blocks [17] for the aggregation function in DLA. The feasibility of training Sat2Graph with supervised learning is counter-intuitive because of the GTE's inconsistency. We defer the discussion of this to Sect. 5.

4 Evaluation

We now present experimental results comparing Sat2Graph to several state-of-the-art road-network generation systems.

4.1 Datasets

We conduct our evaluation on two datasets, one is a large city-scale dataset and the other is the popular SpaceNet roads dataset [30].

City-Scale Dataset. Our city-scale dataset covers 720 km^2 area in 20 U.S. cities. We collect road network data from OpenStreetMap [16] as ground truth and the corresponding satellite imagery through Google static map API [15]. The spatial resolution of the satellite imagery is set to one meter per pixel. This dataset enables us to evaluate the performance of different approaches at city scale, e.g., evaluating the quality of the shortest path crossing the entire downtown of a city on the inferred road graphs.

The dataset is organized as 180 tiles; each tile is a 2 km by 2 km square region. We randomly choose 15% (27 tiles) of them as a testing dataset and 5% (9 tiles) of them as a validation dataset. The remaining 80% (144 tiles) are used as training dataset.

SpaceNet Roads Dataset. Another dataset we used is the SpaceNet roads Dataset [30]. Because the ground truth of the testing data in the SpaceNet dataset is not public, we randomly split the 2549 tiles (non-empty) of the original training dataset into training (80%), testing (15%) and validating (5%) datasets. Each tile is a 0.4 km by 0.4 km square. Similar to the city-scale dataset, we resize the spatial resolution of the satellite imagery to one meter per pixel.

4.2 Baselines

We compare Sat2Graph with four different segmentation-based approaches and one graph-based approach.

Segmentation-Based Approaches. We use four different segmentation-based approaches as baselines.

1. *Seg-UNet:* Seg-UNet uses a simple U-Net [27] backbone to produce road segmentation from satellite imagery. The model is trained with cross-entropy loss. This scheme acts as the naive baseline as it is the most straightforward solution for road extraction.
2. *Seg-DRM* [22] *(ICCV-17):* Seg-DRM uses a stronger CNN backbone which contains 55 ResNet [17] layers to improve the road extraction performance. Meanwhile, Seg-DRM proposes to train the road segmentation model with soft-IoU loss to achieve better performance. However, we find training the Seg-DRM model with cross-entropy loss yields much better performance in terms of topology correctness. Thus, in our evaluation, we train the Seg-DRM model with cross-entropy loss.
3. *Seg-Orientation* [3] *(ICCV-19):* Seg-Orientation is a recent state-or-the-art approach which proposes to improve the road connectivity by joint learning of road orientation and road segmentation. Similar to Seg-DRM, we show the results of Seg-Orientation trained with cross-entropy loss as we find it performs better compared with soft-IoU loss.
4. *Seg-DLA:* Seg-DLA is our enhanced segmentation-based approach which uses the same CNN backbone as our Sat2Graph model. Seg-DLA, together with Seg-UNet, act as the baselines of an ablation study of Sat2Graph.

Graph-Based Approaches. For graph-based approaches, we compare our Sat2Graph solution with RoadTracer [2] (CVPR-18) by applying their code on our dataset. During inference, we use peaks in the segmentation output as starting locations for RoadTracer's iterative search.

4.3 Implementation Details

Data Augmentation: For all models in our evaluation, we augment the training dataset with random image brightness, hue and color temperature, random rotation of the tiles, and random masks on the satellite imagery.

Training: We implemented both Sat2Graph and baseline segmentation approaches using Tensorflow. We train the model on a V100 GPU for 300k iterations (about 120 epochs) with a learning rate starting from 0.001 and decreasing by 2x every 50k iterations. We train all models with the same receptive field, i.e., 352 by 352. We evaluate the performance on the validation dataset for each model every 5k iterations during training, and pick up the best model on the validation dataset as the converged model for each approach to avoid overfitting.

4.4 Evaluation Metrics

In the evaluation, we focus on the topology correctness of the inferred road graph rather than edge-wise correctness. This is because the topology correctness is often crucial in many real-world applications. For example, in navigation applications, a small missing road edge in the road graph could make two regions disconnected. This small missing road segment is a small error in terms of edge-wise correctness but a huge error in terms of topology correctness.

We evaluate the topology correctness of the inferred road graphs through two metrics, TOPO [4] and APLS [30]. Here, we describe the high level idea of these two metrics. Please refer to [4,30] for more details about these two metrics.

TOPO Metric: TOPO metric measures the similarity of sub-graphs sampled on the ground truth graph and the inferred graph from a seed location. The seed location is matched to the closest seed node on each graph. Here, given a seed node on a graph, the sub-graph contains all the nodes such that their distances (on the graph) to the seed node are less than a threshold, e.g., 300 m. For each seed location, the similarity between two sampled sub-graphs is quantified as precision, recall and F_1-score. The metric reports the average precision, recall and F_1-score over randomly sampled seed locations over the entire region.

The TOPO metric has different implementations. We implement the TOPO metric in a very strict way following the description in [18]. This strict implementation allows the metric to penalize detailed topology errors.

APLS Metric: APLS measures the quality of the shortest paths between two locations on the graph. For example, suppose the shortest path between two locations on the ground truth map is 200 m, but the shortest path between the same two locations on the inferred map is 20 m (a wrong shortcut), or 500 m, or doesn't exist. In these cases, the APLS metric yields a very low score, even though there might be only one incorrect edge on the inferred graph.

4.5 Quantitative Evaluation

Overall Quality. Each of the approaches we evaluated has one major hyper-parameter, which is often a probability threshold, that allows us to make different precision-recall trade-offs. We change this parameter for each approach to plot an precision-recall curve. We show the precision-recall curves for different approaches in Fig. 3. This precision-recall curve allows us to see the full picture of the capability of each approach. We also show the best achievable TOPO F_1-score and APLS score of each approach in Table 1 for reference.

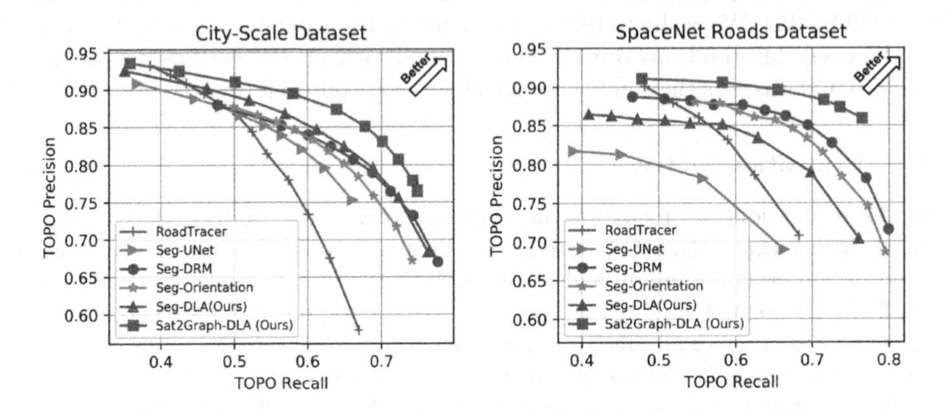

Fig. 3. TOPO metric precision-recall trade-off curves

From Fig. 3, we find an approach may not always better than another approach at different precision-recall position (TOPO metric). For examples, the graph-based approach RoadTracer performs better than others when the precision is high, whereas the segmentation-based approach DeepRoadMapper performs better when the recall is high.

Meanwhile, we find an approach may not always better than another approach on both TOPO and APLS. For example, in Table 1, RoadTracer has the best APLS score but the worst TOPO F_1-score in the five baselines on the city-scale dataset. This is because RoadTracer is good at coarse-grained road connectivity and the precision of inferred road graphs rather than recall. For example, in Fig. 4(a), RoadTracer is better compared with Seg-DRM and Seg-Orientation in terms of road connectivity when the satellite imagery is full of shadow.

In contrast, Sat2Graph surpasses all other approaches on APLS metric and at all TOPO precision-recall positions – for a given precision, Sat2Graph always has the best recall; and for a given recall, Sat2Graph always has the best precision. We think this is because Sat2Graph's graph-tensor encoding takes advantages from both the segmentation-based approaches and graph-based approaches, and allows Sat2Graph to infer stacking roads that none of the other approaches can handle. As an ablation study, we compare Sat2Graph-DLA with Seg-DLA (Seg-DLA uses the same CNN backbone as Sat2Graph-DLA). We find the superiority of Sat2Graph comes from the graph-tensor encoding rather than the stronger CNN backbone.

Benefit from GTE. In addition to the results shown in Table 1, we show the results of using GTE with other backbones. On our city-wide dataset, we find

Table 1. Comparison of the *best achievable* TOPO F_1-score and APLS score. We show the best TOPO F_1-score's corresponding precision and recall just for reference not for comparison. (All the values in this table are percentages)

Method	City-scale dataset				SpaceNet roads dataset			
	Prec.	Rec.	F_1	APLS	Prec.	Rec.	F_1	APLS
RoadTracer [2] (CVPR-18)	78.00	57.44	66.16	57.29	78.61	62.45	69.60	56.03
Seg-UNet	75.34	65.99	70.36	52.50	68.96	66.32	67.61	53.77
Seg-DRM [22] (ICCV-17)	76.54	71.25	73.80	54.32	82.79	72.56	77.34	62.26
Seg-orientation [3] (ICCV-19)	75.83	68.90	72.20	55.34	81.56	71.38	76.13	58.82
Seg-DLA (ours)	75.59	72.26	73.89	57.22	78.99	69.80	74.11	56.36
Sat2Graph-DLA (ours)	80.70	72.28	**76.26**	**63.14**	85.93	76.55	**80.97**	**64.43**

GTE can improve the TOPO F_1-score from 70.36% to 76.40% with the U-Net backbone and from 73.80% to 74.66% with the Seg-DRM backbone. Here, the improvement on Seg-DRM backbone is minor because Seg-DRM backbone has a very shallow decoder.

Sensitivity on w. In our decoding algorithm, we have a hyper-parameter w which is the weight of the cosine distance term in Eq. 1. In Table 2, we show how this parameter impacts the TOPO F_1-score on our city-wide dataset. We find the performance is robust to w - the F_1-scores are all greater than 76.2% with w in the range from 5 to 100.

Table 2. TOPO F_1 scores on our city-wide dataset with different w values.

Value of w	1	5	10	25	75	100	150
F_1-score	75.87%	76.28%	76.62%	76.72%	76.55%	76.26%	75.68%

Vertex Threshold and Edge Threshold. In our basic setup, we set the vertex threshold and the edge threshold of Sat2Graph to the same value. However, we can also use independent probability thresholds for vertices and edges. We evaluate this by choosing a fixed point and vary one probability threshold at a time. We find the vertex threshold dominates the performance and using a higher edge probability threshold (compared with the vertex probability) is helpful to achieve better performance.

Stacking Road. We evaluate the quality of the stacking road by matching the overpass/underpass crossing points between the ground truth graphs and the proposed graphs. In this evaluation, we find our approach has a precision of 83.11% (number of correct crossing points over the number of all proposed crossing points) and a recall of 49.81% (number of correct crossing points over the number of all ground-truth crossing points) on stacked roads. In fact only

0.37% of intersections are incorrectly predicted as overpasses/underpasses (false-positive rate). We find some small roads under wide highway roads are missing entirely. We think this is the reason for the low recall.

4.6 Qualitative Evaluation

Regular Urban Areas. In the regular urban areas (Fig. 4), we find the existing segmentation-based approach with a strong CNN backbone (Seg-DLA) and better data augmentation techniques has already been able to achieve decent results in terms of both precision and recall, even if the satellite imagery is full of shadows and occlusions. Compared with Sat2Graph, the most apparent remaining issue of the segmentation-based approach appears at parallel roads. We think the root cause of this issue is from the fundamental limitation of segmentation-based approaches—the road-segmentation intermediate representation. Sat2Graph eliminates this limitation through graph-tensor encoding, thereby, Sat2Graph is able to produce detailed road structures precisely even along closeby parallel roads.

Stacked Roads. We show the orthogonal superiority of Sat2Graph on stacked roads in Fig. 5. None of the existing approaches can handle stacked roads, whereas Sat2Graph can naturally infer stacked roads thanks to the graph-tensor

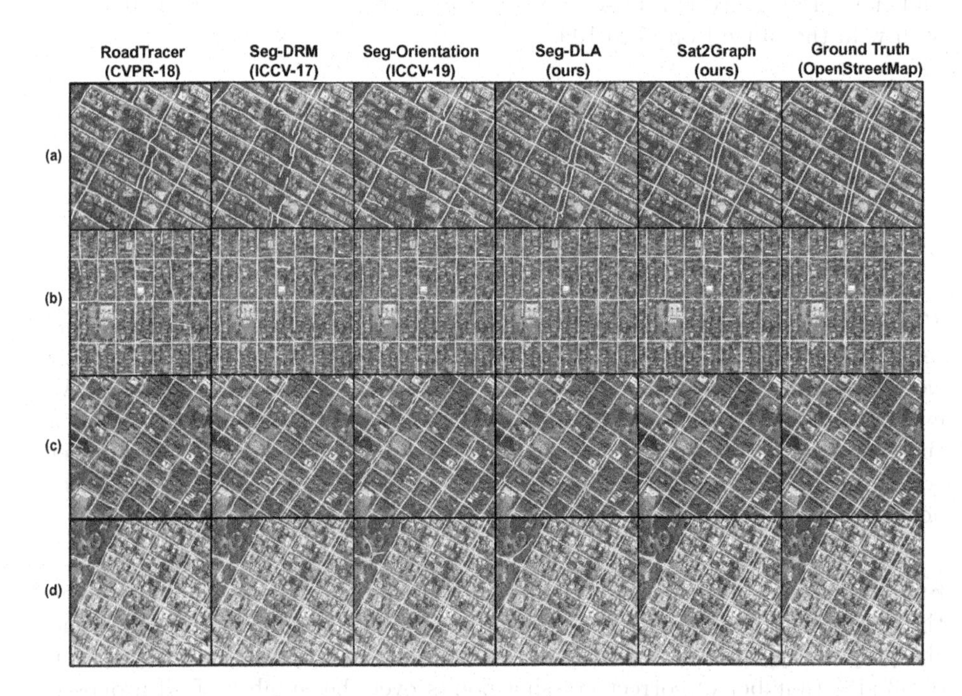

Fig. 4. Qualitative comparison in regular urban areas. We use the models that yield the best TOPO F_1 scores to create this visualization. (Color figure online)

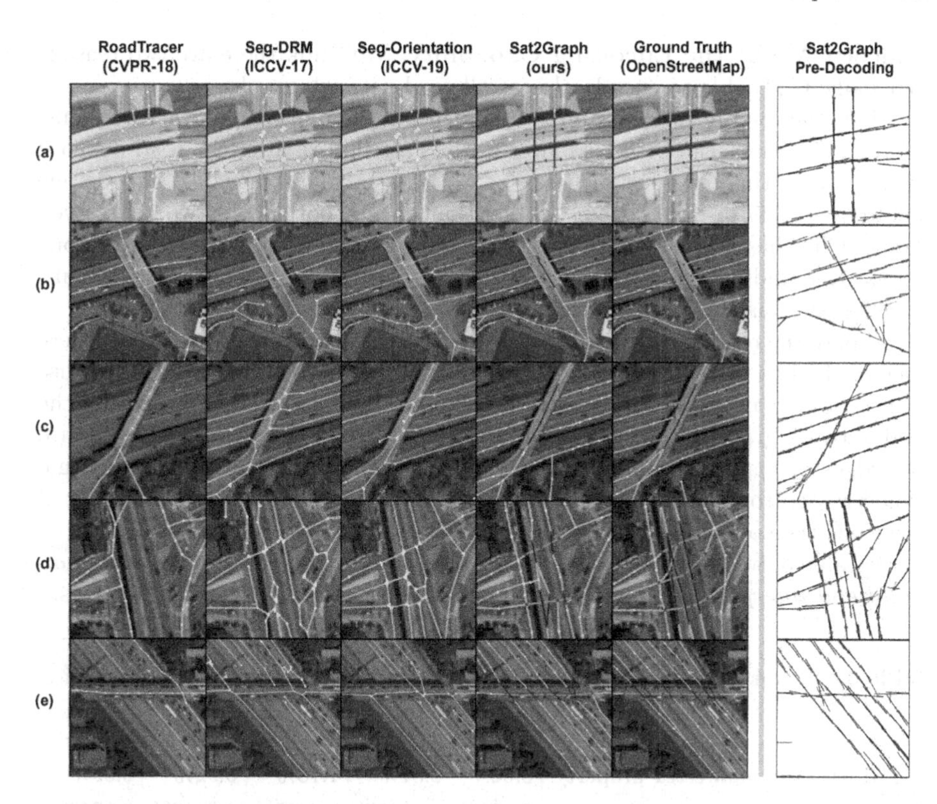

Fig. 5. Qualitative comparison for stacked roads. Sat2Graph robustly infers stacked roads in examples (a–c), but makes some errors in (d) and (e). Prior work infers only planar graphs and incorrectly captures road topology around stacked roads in all cases. We highlight edges that cross without connecting in green and blue. (Color figure online)

encoding. We find Sat2Graph may still fail to infer stacking roads in some complicated scenarios such as in Fig. 5(d–e). We think this can be further improved in a future work, such as adding discriminative loss to regulate the inferred road structure.

5 Discussion

There are two concerns regarding Sat2Graph: (1) it seems that the heavily restricted and non-lossless graph-tensor encoding may not be able to correctly represent all different road network graphs, and (2) training a model to output GTE representation with supervised learning seems impossible because the GTE representation is not consistent.

Concern About the Encoding Capability. We think there are two reasons that make GTE able to encode almost all road network graphs.

First, the road network graph is *undirected*. Although roads have directions, the road directions can be added later as road attributes, after the road network extraction. In this case, for each edge $e = (v_a, v_b)$, we only need to encode one link from v_a to v_b or from v_b to v_a, rather than encode both of the two links. Even though GTE has the $\frac{360}{D_{max}}$-degree sector restriction on outgoing edges from one vertex, this undirected-graph property makes it possible to encode very sharp branches such as the branch vertices between a highway and an exit ramp.

Second, the road network graph is *interpolatable*. There could be a case where none of the two links of an edge $e = (v_a, v_b)$ can be encoded into GTE because both v_a and v_b need to encode their other outgoing links. However, because the road network graph is interpolatable, we can always interpolate the edge e into two edges $e_1 = (v_a, v')$ and $e_2 = (v', v_b)$. After the interpolation, the original geometry and topology remain the same but we can use the additional vertex v' to encode the connectivity between v_a and v_b.

In Table 3, we show the ratios of edges that need to be fixed using the *undirected* and *interpolatable* properties in our dataset with different $D_{\max}$ values.

Table 3. The ratios of edges fixed using the undirected and interpolatable properties.

$D_{\max}$	3	4	5	6	8
Fixed with the *undirected* property	8.62%	2.81%	1.18%	0.92%	0.59%
Fixed with the *interpolatable* property	0.013%	0.0025%	0.0015%	0.0013%	0.0013%

Concern About Supervised Learning. Another concern with GTE is that for one input graph, there exist many different isomorphic encodings for it (e.g., there are many possible vertex interpolations on a long road segment.). These isomorphic encodings produce inconsistent ground truth labels. During training, this inconsistency of the ground truth can make it very hard to learn the right mapping through supervised learning.

However, counter-intuitively, we find Sat2Graph is able to learn through supervised learning and learn well. We find the key reason of this is because of the inconsistency of GTE representation doesn't equally impact the vertices and edges in a graph. For example, the locations of intersection vertices are always consistent in different isomorphic GTEs.

We find GTE has high label consistency for supervised learning at important places such as intersections and overpass/underpass roads. Often, these places are the locations where the challenges really come from. Although GTE has low consistency for long road segments, the topology of the long road segment is very simple and can still be corrected through GTE's decoding algorithm.

6 Conclusion

In this work, we have proposed a simple, unified road network extraction solution that combines the advantages from both segmentation-based approaches and graph-based approaches. Our key insight is a novel graph-tensor encoding scheme. Powered by this graph-tensor approach, Sat2Graph is able to surpass existing solutions in terms of topology-similarity metric at all precision-recall points in an evaluation over two large datasets. Additionally, Sat2Graph naturally infers stacked roads like highway overpasses that none of the existing approaches can handle.

References

1. Ahmed, M., Karagiorgou, S., Pfoser, D., Wenk, C.: A comparison and evaluation of map construction algorithms using vehicle tracking data. GeoInformatica **19**(3), 601–632 (2014)
2. Bastani, F., et al.: RoadTracer: automatic extraction of road networks from aerial images. In: Proceedings of the IEEE Conference on Computer Vision and Pattern Recognition, pp. 4720–4728 (2018)
3. Batra, A., Singh, S., Pang, G., Basu, S., Jawahar, C., Paluri, M.: Improved road connectivity by joint learning of orientation and segmentation. In: Proceedings of the IEEE Conference on Computer Vision and Pattern Recognition, pp. 10385–10393 (2019)
4. Biagioni, J., Eriksson, J.: Inferring road maps from global positioning system traces: survey and comparative evaluation. Transp. Res. Rec. **2291**(1), 61–71 (2012)
5. Biagioni, J., Eriksson, J.: Map inference in the face of noise and disparity. In: ACM SIGSPATIAL 2012 (2012)
6. Buslaev, A., Seferbekov, S.S., Iglovikov, V., Shvets, A.: Fully convolutional network for automatic road extraction from satellite imagery. In: CVPR Workshops, pp. 207–210 (2018)
7. Cadamuro, G., Muhebwa, A., Taneja, J.: Assigning a grade: accurate measurement of road quality using satellite imagery. arXiv preprint arXiv:1812.01699 (2018)
8. Cao, L., Krumm, J.: From GPS traces to a routable road map. In: ACM SIGSPATIAL, pp. 3–12 (2009)
9. Cheng, G., Wang, Y., Xu, S., Wang, H., Xiang, S., Pan, C.: Automatic road detection and centerline extraction via cascaded end-to-end convolutional neural network. IEEE Trans. Geosci. Remote Sens. **55**(6), 3322–3337 (2017)
10. Chu, H., et al.: Neural turtle graphics for modeling city road layouts. In: Proceedings of the IEEE International Conference on Computer Vision, pp. 4522–4530 (2019)
11. Costea, D., Marcu, A., Slusanschi, E., Leordeanu, M.: Creating roadmaps in aerial images with generative adversarial networks and smoothing-based optimization. In: Proceedings of the IEEE International Conference on Computer Vision Workshops, pp. 2100–2109 (2017)
12. Davies, J.J., Beresford, A.R., Hopper, A.: Scalable, distributed, real-time map generation. IEEE Pervasive Comput. **5**(4), 47–54 (2006)

13. Edelkamp, S., Schrödl, S.: Route planning and map inference with global positioning traces. In: Klein, R., Six, H.-W., Wegner, L. (eds.) Computer Science in Perspective. LNCS, vol. 2598, pp. 128–151. Springer, Heidelberg (2003). https://doi.org/10.1007/3-540-36477-3_10
14. Fortier, A., Ziou, D., Armenakis, C., Wang, S.: Survey of work on road extraction in aerial and satellite images. Center for Topographic Information Geomatics, Ontario, Canada. Technical report 241(3) (1999)
15. Google: Google Static Maps API. https://developers.google.com/maps/documentation/maps-static/intro. Accessed 21 Mar 2019
16. Haklay, M., Weber, P.: OpenStreetMap: user-generated street maps. IEEE Pervasive Comput. **7**(4), 12–18 (2008)
17. He, K., Zhang, X., Ren, S., Sun, J.: Deep residual learning for image recognition. In: Proceedings of the IEEE Conference on Computer Vision and Pattern Recognition, pp. 770–778 (2016)
18. He, S., et al.: RoadRunner: improving the precision of road network inference from GPS trajectories. In: ACM SIGSPATIAL (2018)
19. He, S., et al.: RoadTagger: robust road attribute inference with graph neural networks. arXiv preprint arXiv:1912.12408 (2019)
20. Hinz, S., Baumgartner, A.: Automatic extraction of urban road networks from multi-view aerial imagery. ISPRS J. Photogramm. Remote Sens. **58**(1–2), 83–98 (2003)
21. Li, Z., Wegner, J.D., Lucchi, A.: PolyMapper: extracting city maps using polygons. arXiv preprint arXiv:1812.01497 (2018)
22. Máttyus, G., Luo, W., Urtasun, R.: DeepRoadMapper: extracting road topology from aerial images. In: Proceedings of the IEEE International Conference on Computer Vision, pp. 3438–3446 (2017)
23. Mattyus, G., Wang, S., Fidler, S., Urtasun, R.: Enhancing road maps by parsing aerial images around the world. In: Proceedings of the IEEE International Conference on Computer Vision, pp. 1689–1697 (2015)
24. Máttyus, G., Wang, S., Fidler, S., Urtasun, R.: HD maps: fine-grained road segmentation by parsing ground and aerial images. In: Proceedings of the IEEE Conference on Computer Vision and Pattern Recognition, pp. 3611–3619 (2016)
25. Mosinska, A., Márquez-Neila, P., Koziński, M., Fua, P.: Beyond the pixel-wise loss for topology-aware delineation. In: The IEEE Conference on Computer Vision and Pattern Recognition (CVPR) (2018)
26. Najjar, A., Kaneko, S., Miyanaga, Y.: Combining satellite imagery and open data to map road safety. In: Thirty-First AAAI Conference on Artificial Intelligence (2017)
27. Ronneberger, O., Fischer, P., Brox, T.: U-Net: convolutional networks for biomedical image segmentation. In: Navab, N., Hornegger, J., Wells, W.M., Frangi, A.F. (eds.) MICCAI 2015. LNCS, vol. 9351, pp. 234–241. Springer, Cham (2015). https://doi.org/10.1007/978-3-319-24574-4_28
28. Shi, Q., Liu, X., Li, X.: Road detection from remote sensing images by generative adversarial networks. IEEE Access **6**, 25486–25494 (2017)
29. Stanojevic, R., Abbar, S., Thirumuruganathan, S., Chawla, S., Filali, F., Aleimat, A.: Robust road map inference through network alignment of trajectories. In: Proceedings of the 2018 SIAM International Conference on Data Mining. SIAM (2018)
30. Van Etten, A., Lindenbaum, D., Bacastow, T.M.: SpaceNet: a remote sensing dataset and challenge series. arXiv preprint arXiv:1807.01232 (2018)

31. Wang, W., Yang, N., Zhang, Y., Wang, F., Cao, T., Eklund, P.: A review of road extraction from remote sensing images. J. Traffic Transp. Eng. (Engl. Ed.) **3**(3), 271–282 (2016)
32. Wegner, J.D., Montoya-Zegarra, J.A., Schindler, K.: Road networks as collections of minimum cost paths. ISPRS J. Photogramm. Remote Sens. **108**, 128–137 (2015)
33. Yu, F., Wang, D., Shelhamer, E., Darrell, T.: Deep layer aggregation. In: Proceedings of the IEEE Conference on Computer Vision and Pattern Recognition, pp. 2403–2412 (2018)
34. Zhang, X., Han, X., Li, C., Tang, X., Zhou, H., Jiao, L.: Aerial image road extraction based on an improved generative adversarial network. Remote Sens. **11**(8), 930 (2019)
35. Zhang, Z., Liu, Q., Wang, Y.: Road extraction by deep residual U-Net. IEEE Geosci. Remote Sens. Lett. **15**(5), 749–753 (2018)
36. Zhou, L., Zhang, C., Wu, M.: D-LinkNet: LinkNet with pretrained encoder and dilated convolution for high resolution satellite imagery road extraction. In: CVPR Workshops, pp. 182–186 (2018)

Cross-Task Transfer for Geotagged Audiovisual Aerial Scene Recognition

Di Hu[1], Xuhong Li[1], Lichao Mou[2,3], Pu Jin[2], Dong Chen[4], Liping Jing[4], Xiaoxiang Zhu[2,3], and Dejing Dou[1](✉)

[1] Big Data Laboratory, Baidu Research, Beijing, China
{hudi04,lixuhong,doudejing}@baidu.com
[2] Technical University of Munich, Munich, Germany
{lichao.mou,pu.jin}@tum.de, {lichao.mou,xiaoxiang.zhu}@dlr.de
[3] German Aerospace Center, Cologne, Germany
[4] Beijing Key Lab of Traffic Data Analysis and Mining, Beijing Jiaotong University, Beijing, China
{chendong,lpjing}@bjtu.edu.cn

Abstract. Aerial scene recognition is a fundamental task in remote sensing and has recently received increased interest. While the visual information from overhead images with powerful models and efficient algorithms yields considerable performance on scene recognition, it still suffers from the variation of ground objects, lighting conditions etc. Inspired by the multi-channel perception theory in cognition science, in this paper, for improving the performance on the aerial scene recognition, we explore a novel audiovisual aerial scene recognition task using both images and sounds as input. Based on an observation that some specific sound events are more likely to be heard at a given geographic location, we propose to exploit the knowledge from the sound events to improve the performance on the aerial scene recognition. For this purpose, we have constructed a new dataset named *AuDio Visual Aerial sceNe reCognition datasEt* (ADVANCE). With the help of this dataset, we evaluate three proposed approaches for transferring the sound event knowledge to the aerial scene recognition task in a multimodal learning framework, and show the benefit of exploiting the audio information for the aerial scene recognition. The source code is publicly available for reproducibility purposes. (https://github.com/DTaoo/Multimodal-Aerial-Scene-Recognition)

Keywords: Cross-task transfer · Aerial scene classification · Geotagged sound · Multimodal learning · Remote sensing

1 Introduction

Scene recognition is a longstanding, hallmark problem in the field of computer vision, and it refers to assigning a scene-level label to an image based on its overall contents. Most scene recognition approaches in the community make use

A. Vedaldi et al. (Eds.): ECCV 2020, LNCS 12369, pp. 68–84, 2020.
https://doi.org/10.1007/978-3-030-58586-0_5

of ground images and have achieved remarkable performance. By contrast, overhead images usually cover larger geographical areas and are capable of offering more comprehensive information from a bird's eye view than ground images. Hence aerial scene recognition has received increased interest. The success of current state-of-the-art aerial scene understanding models can be attributed to the development of novel convolutional neural networks (CNNs) that aim at learning good visual representations from images.

Albeit successful, these models may not work well in some cases, particularly when they are directly used in worldwide applications, suffering the pervasive factors, such as different remote imaging sensors, lighting conditions, orientations, and seasonal variations. A study in neurobiology reveals that human perception usually benefits from the integration of both visual and auditory knowledge. Inspired by this investigation, we argue that aerial scenes' soundscapes are partially free of the aforementioned factors and can be a helpful cue for identifying scene categories (Fig. 1). This is based on an observation that the visual appearance of an aerial scene and its soundscape are closely connected. For instance, sound events like broadcasting, people talking, and perhaps whistling are likely to be heard in all train stations in the world, and cheering and shouting are expected to hear in most sports lands. However, incorporating the sound knowledge into a visual aerial scene recognition model and assessing its contributions to this task still remain underexplored. In addition, it is worth mentioning that with the now widespread availability of smartphones, wearable devices, and audio sharing platforms, geotagged audio data have been easily accessible, which enables us to explore the topic in this paper.

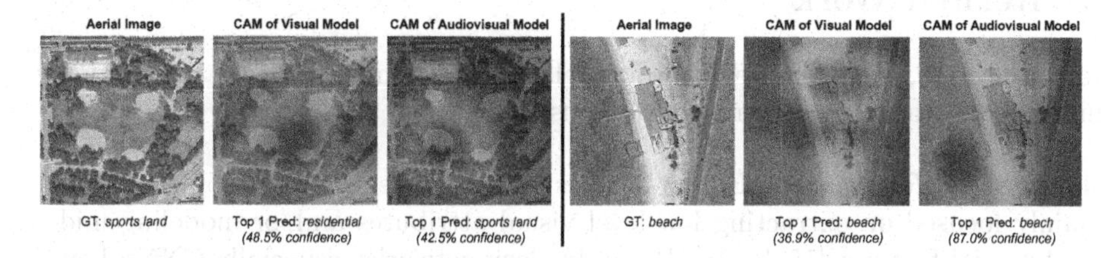

Fig. 1. Two examples showing aerial scenes' soundscapes could be a helpful cue for identifying their scene categories. More details of the audiovisual model please refer to Sect. 4.3. Here, we use the class activation mapping (CAM) technique to visualize what models are looking.

In this work, we are interested in the audiovisual aerial scene recognition task that simultaneously uses both visual and audio messages to identify the scene of a geographical region. To this end, we construct a new dataset, named *AuDio Visual Aerial sceNe reCognition datasEt* (ADVANCE), providing 5075 paired images and sound clips categorized to 13 scenes, which will be introduced in Sect. 3, for exploring the aerial scene recognition task. According to our preliminary experiments, simply concatenating representations from the two modalities is not helpful, slightly degrading the recognition performance compared to using a vision-based model. Knowing that sound events are related to scenes, this preliminary result indicates that the model cannot directly learn the underlying relation between the sound events and the scenes. So directly transferring the sound event knowledge to scene recognition may be the key to making progress. Following this direction, with the multimodal representations, we propose three approaches that can effectively exploit the audio knowledge to solve the aerial scene recognition task, which will be detailed in Sect. 4. We compare our proposed approaches with baselines in Sect. 5, showing the benefit of exploiting the sound event knowledge for the aerial scene recognition task.

Thereby, this work's contributions are threefold.

- The audiovisual perception of human beings gives us an incentive to investigate a novel audiovisual aerial scene recognition task. We are not aware of any previous work exploring this topic.
- We create an annotated dataset consisting of 5075 geotagged aerial image-sound pairs involving 13 scene classes. This dataset covers a large variety of scenes from across the world.
- We propose three approaches to exploit the audio knowledge, *i.e.*, preserving the capacity of recognizing sound events, constructing a mutual representation in order to learn the underlying relation between sound events and scenes, and directly learning this relation through the posterior probabilities of sound events given a scene. In addition, we validate the effectiveness of these approaches through extensive ablation studies and experiments.

2 Related Work

In this section, we briefly review some related works in aerial scene recognition, multimodal learning, and cross-task transfer.

Aerial Scene Recognition. Earlier studies on aerial scene recognition [23,24,32] mainly focused on extracting low-level visual attributes and/or modeling mid-level spatial features [15,16,28]. Recently, deep networks, especially CNNs, have achieved a large development in aerial scene recognition [4,5,20]. Moreover, some methods were proposed to solve the problem of the limited collection of aerial images by employing more efficient networks [19,33,37]. Although these methods have achieved great empirical success, they usually learn scene knowledge from the same modality, *i.e.*, image. Different from previous works, this paper mainly focuses on exploiting multiple modalities (*i.e.* image and sound) to achieve robust aerial scene recognition performance.

Multimodal Learning. Information in the real world usually comes as different modalities, with each modality being characterized by very distinct statistical properties, *e.g.*, sound and image [3]. An expected way to improve relevant task performance is by integrating the information from different modalities. In past decades, amounts of works have developed promising methods on the related topics, such as reducing the audio noise by introducing visual lip information for speech recognition [1,11], improving the performance of facial sentiment recognition by resorting to the voice signal [35]. Recently, more attention is paid to the task of learning to analyze real-world multimodal scenarios [12,13,21,34] and events [26,31]. These works have confirmed the advantages of multimodal learning. In this paper, we proposed to recognize the aerial scene by leveraging the bridge between scene and sound to help better understand aerial scenes.

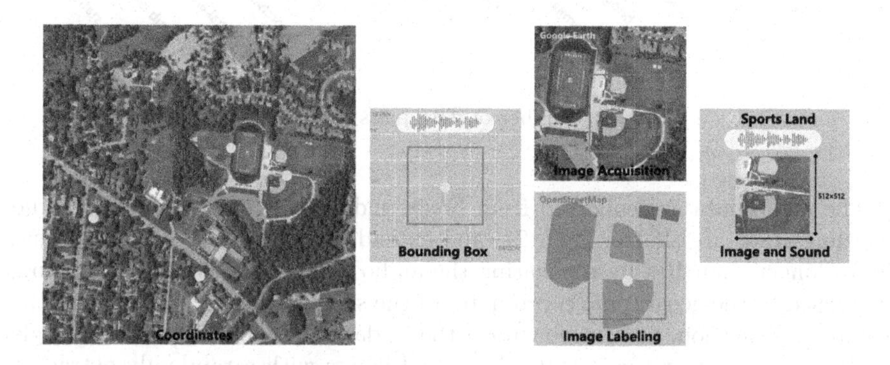

Fig. 2. The aerial images acquisition and labeling steps.

Cross-task Transfer. Transferring the learned knowledge from one task to another related task has been approved as an effective way for better data modeling and messages correlating [2,6,14]. Aytar et al. [2] proposed a teacher-student framework that transfers the discriminative knowledge of visual recognition to the representation learning task of sound modality via minimizing the differences in the distribution of categories. Imoto et al. [14] proposed a method for sound event detection by transferring the knowledge of scenes with soft labels. Gan et al. [8] transferred the visual object location knowledge for auditory localization learning. Salem et al. [25] proposed to transfer the sound clustering knowledge to the image recognition task by predicting the distribution of sound clusters from an overhead image, similarly work can be found in [22]. By contrast, this paper strives to exploit effective sound event knowledge to facilitate the aerial scene understanding task.

3 Dataset

To our knowledge, the audiovisual aerial scene recognition task has not been explored before. Salem *et al.* [25] established a dataset to explore the correlation between geotagged sound clips and overhead images. For further facilitating the research in this field, we construct a new dataset, with high-quality images and scene labels, named as ADVANCE[1], which in summary contains 5075 pairs of aerial images and sounds, classified into 13 classes.

[1] The dataset webpage: https://akchen.github.io/ADVANCE-DATASET/.

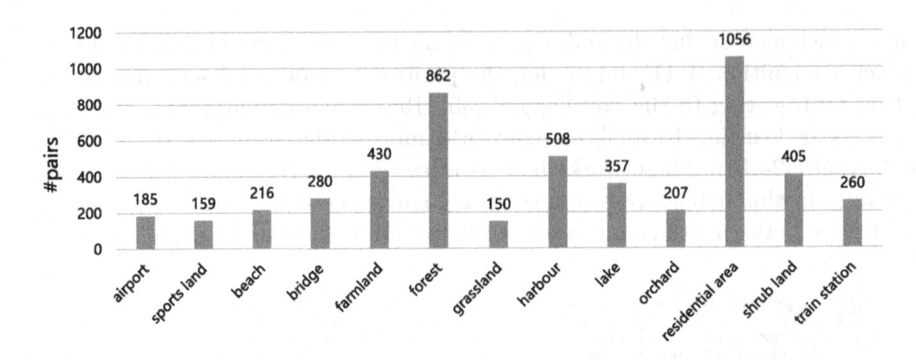

Fig. 3. Number of data pairs per class.

The audio data are collected from Freesound[2], where we remove the audio recordings that are shorter than 2 s, and extend those that are between 2 and 10 s to longer than 10 s by replicating the audio content. Each audio recording is attached to the geographic coordinates of the sound being recorded. From the location information, we can download the updated aerial images from Google Earth[3]. Then we pair the downloaded aerial image with a randomly extracted 10-s sound clip from the entire audio recording content. Finally, the paired data are labeled according to the annotations from OpenStreetMap[4], also using the attached geographic coordinates from the audio recording. Those annotations have manually been corrected and verified by participants in case that some of them are not up to date. The overview of the establishment is shown in Fig. 2.

Due to the inherent uneven distribution of scene classes, the collected data are strongly unbalanced, which makes difficult the training process. So, two extra steps are designed to alleviate the unbalanced-distribution problem. Firstly we filter out the scenes whose numbers of paired samples are less than 10, such as desert and the site of wind turbines. Then for scenes that have less than 100 samples, we apply a small offset to the original geographic coordinates in four directions. So, correspondingly, four new aerial images are generated from Google Earth and paired with the same audio recording, while for each image, a new 10-s sound clip is randomly extracted from the recording. Figure 3 reveals the final number of paired samples per class. Moreover, as shown in Fig. 4, the samples are distributed over the whole world, increasing the diversity of the aerial images and sounds.

4 Methodology

In this paper, we focus on the audiovisual aerial scene recognition task, based on two modalities, *i.e.*, image and audio. We propose to exploit the audio knowledge

[2] https://freesound.org/browse/geotags/.
[3] https://earthengine.google.com/.
[4] https://www.openstreetmap.org/.

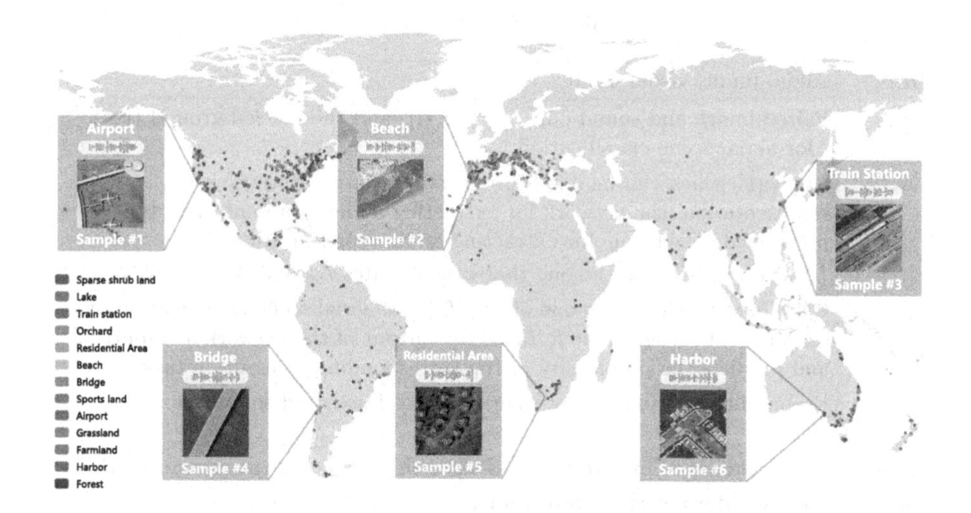

Fig. 4. Coordinates distribution and sample pairs of images and sound. Different scenes are represented by different color. Six sample pairs are displayed, which are composed of aerial images, sound and semantic labels. (Color figure online)

to better solve the aerial scene recognition task. In this section, we detail our proposed approaches for creating the bridge of knowledge transfer from sound event knowledge to the scene recognition in a multi-modality framework.

We take the notations from Table 1, note that the data $\boldsymbol{x}$ follows the empirical distribution μ of our built dataset ADVANCE. For the multimodal learning task with deep networks, we adopt the model architecture that concatenates representations from two deep convolutional networks on images and sound clips. So our main task, which is a supervised learning problem for aerial scene recognition, can be written as[5]

$$L_s = -\log\left[f_s(\boldsymbol{x}, N_{v+a})\right]_t \quad , \tag{1}$$

which is a cross-entropy loss with t-th class being the ground truth.

Furthermore, pre-training on related datasets helps accelerate the training process and improving the performance on the new dataset, especially on a relatively small dataset. For our task, the paired data samples are limited, and our preliminary experiments show that the two networks N_v and N_a benefit a lot from pre-training on the AID dataset [30] for classifying scenes from aerial images, and AudioSet [9] for recognizing 527 audio events from sound clips [29].

In the rest of this section, we formulate our proposed model architecture for addressing the multimodal scene recognition task, and present our idea of exploiting the audio knowledge following three directions: (1) avoid forgetting the audio knowledge during training by preserving the capacity of recognizing sound events; (2) construct a mutual representation that solves the main task

[5] For all loss functions, we omit the softmax activation function in f_s, the sigmoid activation function in f_e, and the expectation of $(\boldsymbol{x}, t)$ over μ for clarity.

Table 1. Main notations.

$\boldsymbol{a}$,$\boldsymbol{v}$	Audio input, visual input
$\boldsymbol{x}$,t	Paired image and sound clip, $\boldsymbol{x} = \{\boldsymbol{v}, \boldsymbol{a}\}$, and the labeled ground truth t for aerial scene classification
N_*	Network, which can be one of the network for extracting visual representation, the network for extracting audio representation, the pretrained (fixed) one for extracting audio representation, *i.e.*, $\{N_v, N_a, N_a^{(0)}\}$; also the one that concatenates N_v and N_a, *i.e.* N_{v+a}
f_*	Classifier, which can be one of $\{f_s, f_e\}$, for aerial scene classification or sound event recognition; f_* takes the output of the network as input, and predicts the probability of the corresponding recognition task
$\boldsymbol{s}$,$\boldsymbol{e}$	Probability distribution over aerial scene classes and sound event classes
s_k,s_t	k-th scene class' probability, and the t-th class being the ground truth
e_k	k-th sound event class' probability
$C(p, q)$	Binary KL divergence: $\log(\frac{p}{q}) + (1-p)\log(\frac{1-p}{1-q})$

and the sound event recognition task simultaneously, allowing the model to learn the underlying relation between sound events and scenes; (3) directly learn the relation between sound events and scenes. Our total objective function L is

$$L = L_s + \alpha L_\Omega \ , \tag{2}$$

where α controls the force of L_Ω, and L_Ω is one of the three approaches that are respectively presented in Sect. 4.1, 4.2 and 4.3, as illustrated in Fig. 5.

4.1 Preservation of Audio Source Knowledge

For our task of aerial scene recognition, audio knowledge is expected to be helpful since the scene information is related to sound events. While initializing the network by the pre-trained weights is an implicit way of transferring the knowledge to the main task, the audio source knowledge may easily be forgotten during fine-tuning. Without audio source knowledge, the model can hardly recognize the sound events, leading to a random confusion between sound events and scenes.

For preserving the knowledge, we propose to record the soft responses of target samples from the pre-trained model and retain them during fine-tuning. This simple but efficient approach [10] is named as knowledge distillation, for distilling the knowledge from an ensemble of models to a single model, and has also been used in domain adaptation [27] and lifelong learning [17]. All of them encourage to preserve the source knowledge by minimizing the KL divergence between the responses from the pre-trained model and the training model. For avoiding the saturated regions of the softmax, the pre-activations are divided by a large scalar, called temperature [10], to provide smooth responses, with which the knowledge can be easily transferred.

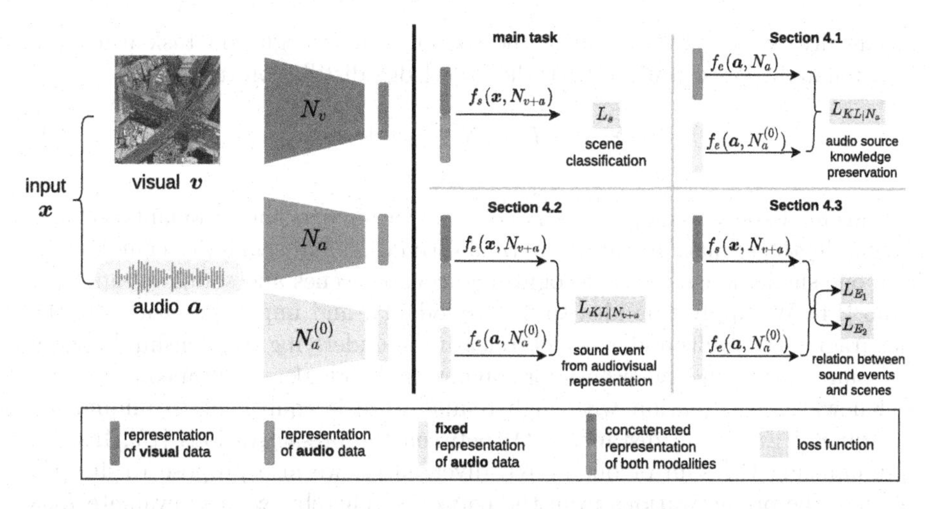

Fig. 5. Illustration of the main task and three cross-task transfer approaches (best viewed in color). We recall the notations: N_v, with trainable parameters, extracts visual representations, pretrained on the AID dataset; N_a, also with trainable parameters, extracts audio representations, pretrained on the AudioSet dataset; $N_a^{(0)}$, is the same as N_a except parameters being fixed; N_{v+a} simply applies both N_v and N_a. The classifier at the last layer of the network is presented by $f_{task}(input\ data, network)$, where the choice of *task* is $\{s$: scene classification, e : sound event recognition$\}$, *input data* is one of $\{\boldsymbol{v}, \boldsymbol{a}, \boldsymbol{x}\}$, and the set for *network* is $\{N_v, N_a, N_a^{(0)}, N_{v+a}\}$. On the left of this figure, our model takes a paired data sample $\boldsymbol{x}$ of an image $\boldsymbol{v}$ and a sound clip $\boldsymbol{a}$ as input, and extracts representations from different combinations of modalities and models (shown in different colors); On the right, the top-left block introduces our main task of aerial scene recognition, and the rest three blocks present the three cross-transfer approaches. (Color figure online)

However, for the reason that the audio event task is a multi-label recognition, $f_e(\boldsymbol{x}, N_*)$ is activated by the sigmoid function. The knowledge distillation technique is thus implemented by a sum of binary Kullback-Leibler divergences:

$$L_{KL|N_a} = \sum_i C(\ [f_e(\boldsymbol{a}, N_a^{(0)})]_i\ ||\ [f_e(\boldsymbol{a}, N_a)]_i\ )\ , \tag{3}$$

where $[f_e(\boldsymbol{a}, N_*)]_i$ indicates the probability of i-th sound event happening in sound clip $\boldsymbol{a}$, predicted by $N_a^{(0)}$ or N_a. This approach helps to preserve the audio knowledge from the source pretrained network from the AID dataset.

4.2 Audiovisual Representation for Multi-task

Different from the idea of preserving the knowledge within the audio modality, we encourage our multimodal model, along with the visual modality, to learn a mutual representation that recognizes scenes and sound events simultaneously.

Specifically, we optimize to solve the sound event recognition task using the concatenated representation, with the knowledge distillation technique:

$$L_{KL|N_{v+a}} = \sum_i C(\ [f_e(\boldsymbol{a}, N_a^{(0)})]_i \ ||\ [f_e(\boldsymbol{x}, N_{v+a})]_i\ ) \ . \tag{4}$$

This multi-task technique is very common within one single modality, such as solving depth estimation, surface normal estimation and semantic segmentation from one single image [7], or recognizing acoustic scenes and sound events from audio [14]. We apply this idea to multi-modality, and implement with Eq. (4), encouraging the multimodal model to learn the underlying relationship between the sound events and the scenes for solving the two tasks simultaneously.

Knowledge distillation with high temperature is equivalent to minimizing the squared Euclidean distance (SQ) between the pre-activations [10]. Instead of minimizing the sum of binary KL divergences, we also propose to directly compare the pre-activations from the networks. Thereby, we also evaluate L_{SQ} variant for Eq. 3 and 4 respectively:

$$\begin{aligned} L_{SQ|N_a} &= \left\| \hat{f}_e(\boldsymbol{a}, N_a^{(0)}) - \hat{f}_e(\boldsymbol{a}, N_a) \right\|_2^2 \ , \\ \check{L}_{SQ|N_{v+a}} &= \left\| \hat{f}_e(\boldsymbol{a}, N_a^{(0)}) - \hat{f}_e(\boldsymbol{x}, N_{v+a}) \right\|_2^2 \ , \end{aligned} \tag{5}$$

where $\hat{f}_e$ is the pre-activations, recalling that f_e is activated by sigmoid.

4.3 Sound Events in Different Scenes

The two previously proposed approaches are based on the multi-task learning framework, either using different or the same representations, in order to preserve the audio source knowledge or implicitly learn an underlying relation between aerial scenes and sound events. Here, we propose an explicit way for directly modeling the relation between scenes and sound events, and creating the bridge of transferring the knowledge between two modalities.

We employ the paired image-audio data samples from our built dataset as introduced in Sect. 3, analyze the happening sound events in each scene, and obtain the posteriors given one scene. Then instead of predicting the probability of sound events by the network, we estimate this probability distribution $p(\boldsymbol{e})$ with the help of posteriors $p(\boldsymbol{e}|s_k)$ and the predicted probability of scenes $p(\boldsymbol{s})$:

$$p(\boldsymbol{e}) = \sum_k p(s_k)\ p(\boldsymbol{e}|s_k) = \sum_k\ [f_s(\boldsymbol{x}, N_{v+a})]_k\ p(\boldsymbol{e}|s_k) \ , \tag{6}$$

where $p(s_k) = [f_s(\boldsymbol{x}, N_{v+a})]_k$ is the predicted probability of the k-th scene, and the posteriors $p(\boldsymbol{e}|s_k)$ is obtained by averaging $f_e(\boldsymbol{a}, N_a^{(0)})$ over all samples that belong to the scene s_k. This estimation $p(\boldsymbol{e})$ is in fact the compound distribution that marginalizes out the probability of scenes, while we search for the optimal scene probability distribution $p(\boldsymbol{s})$ (ideally one-hot) through aligning $p(\boldsymbol{e})$ with soft responses:

$$L_{E_1} = \sum_i C(\ [f_e(\boldsymbol{a}, N_a^{(0)})]_i \ || \ p(e_i)\ ) \ . \tag{7}$$

Besides estimating the probability of each sound event happening in a specific scene, we also investigate possible concomitant sound events. Some sound events may largely overlap under a given scene, and this coincidence can be used as a characteristic for recognizing scenes. We propose to extract this characteristic from $f_e(\boldsymbol{a}, N_a^{(0)})$ of all audio samples that belong to this specific scene.

We note $P(\boldsymbol{e}|s_k) \in \mathbb{R}^{n_k \times c}$ as the sound event probabilities of n_k samples in the scene s_k, where each row is each sample's probability of sound events in the scene s_k. Then with the Gram matrix $P(\boldsymbol{e}|s_k)^T P(\boldsymbol{e}|s_k)$, we extract the largest eigenvalue and the corresponding eigenvector $\boldsymbol{d}_k$ as the characteristic of $P(\boldsymbol{e}|s_k)$. This eigenvector $\boldsymbol{d}_k$ indicates the correlated sound events and quantifies their relevance in the scene s_k by the direction of this vector. We thus propose to align the direction of $\boldsymbol{d}_t$, the event relevance of the ground truth scene s_t, with the estimated $p(\boldsymbol{e})$ from Eq. 6:

$$L_{E_2} = \text{cosine}(\boldsymbol{d}_t, p(\boldsymbol{e})) \ . \tag{8}$$

Equation (7) and (8) have provided a way of explicitly building the connection between scenes and sound events. In the experiments, we use them together:

$$L_E = L_{E_1} + \beta L_{E_2} \ , \tag{9}$$

where β is a hyper-parameter controlling the importance of L_{E_2}.

5 Experiments

5.1 Implementation Details

Our built ADVANCE dataset is employed for evaluation, where 70% image-sound pairs are for training, 10% for validation, and 20% for testing. Note that, these three sub-sets do not share audiovisual pairs that are collected from the same coordinate. Before feeding the recognition model, we sub-sample the sound clips at 16 kHz. Then, following [29], the short-term Fourier transform is computed using a window size of 1024 and a hop length of 400. The generated spectrogram is then projected into the log-mel scale to obtain an audio matrix in $\mathbb{R}^{T \times F}$, where the time $T = 400$ and the frequency $F = 64$. Finally, we normalize each feature dimension to have zero mean and unit variance. The image data are all resized into 256 $\times$ 256, and horizontal flipping, color, and brightness jittering are used as data augmentation means.

In the network setting part, the visual pathway employs the AID pre-trained ResNet-101 for modeling the scene content [30] and the audio pathway adopts the AudioSet pre-trained ResNet-50 for modeling the sound content [29]. The whole network is optimized via an Adam optimizer with a weight decay rate 1e$-$4 and a relatively small learning rate 1e−5, as both backbones have been pre-trained from external knowledge. By using grid search strategy, the hyper-parameters of α and β are set as 0.1 and 0.001, respectively. We adopt the weighted-averaging precision, recall and F-score metrics for evaluation, which are more convincing when faced with uneven distribution of scene classes.

5.2 Aerial Scene Recognition

Figure 6 shows the recognition results of different learning approaches under the unimodal and multimodal scenario, from which we have four points should pay attention to. Firstly, according to the unimodal results, the sound data can provide a certain reference for different scene categories, although it is significantly worse than image-based results. Such phenomenon reminds us that we can take advantage of the audio information to improve recognition results further. Secondly, we recognize that simply using the information from both modalities does not bring benefits but slightly lowers the results (72.85 vs. 72.71 in F-score). This could be because the pre-trained knowledge for audio modality may be forgotten or the audio messages are not fully exploited just with the rough scene labels. Thirdly, when the sound event knowledge is transferred for the scene modeling, we have considerable improvements for all of the proposed approaches. The results of $L_{SQ|N_a}$ and $L_{KL|N_a}$ show that preserving audio event knowledge is an effective means for better exploiting audio messages for scene recognition, and the performance of $L_{SQ|N_{v+a}}$ and $L_{KL|N_{v+a}}$ demonstrates that transferring the unimodal knowledge of sound events to the multimodal network can help to learn better mutual representation of scene content across modalities. Fourthly, among all the compared approaches, our proposed L_E approach shows the best results, as it better imposes the sound event knowledge by imposing the underlying relation between scenes and sound events.

We use the CAM technique [36] to highlight the parts of the input image that make significant contributions to identifying the specific scene category. Figure 7 shows the comparison of the visualization results and the predicted probabilities of the ground-truth label among different approaches. By resorting to the sound event knowledge, as well as its association with scene information, our proposed model can better localize the salient area of the correct aerial scene and provide a higher predicted probability for the ground-truth category, e.g., the *harbour* and *bridge* class.

Apart from the multimodal setting, we have also conducted more experiments under the unimodal settings, shown in Table 2, for presenting the contributions from pre-trained models, and verifying the benefits from the sound event knowledge on the aerial scene recognition. For these unimodal experiments, we keep one modal data input and set the other to zeros. When only the audio data are considered, the sound event knowledge is transferred within the audio modality and thus $L_{SQ|N_a}$ is equivalent to $L_{SQ|N_{v+a}}$, similarly for the visual modality case. Comparing the results of randomly initializing the weights *i.e.* $L_s^{\dagger}$ and initializing with the pre-trained weights *i.e.* L_s, we find that initializing the network from the pre-trained model can significantly prompt the performance, which confirms

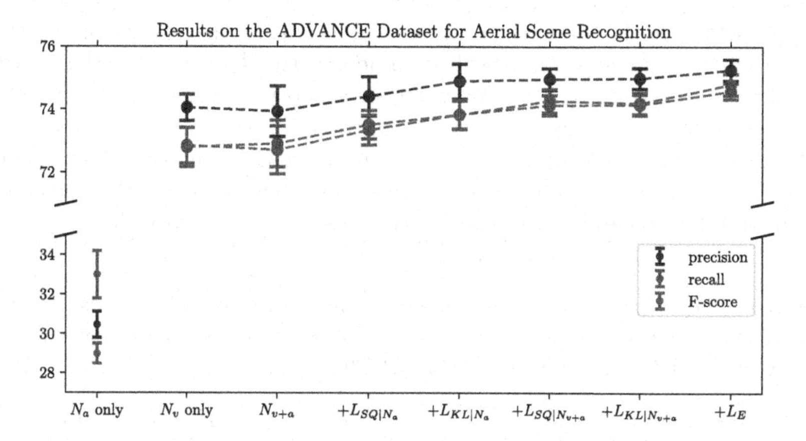

Fig. 6. Aerial scene recognition results on the ADVANCE dataset from 5 different runs, where the first approaches perform only the main loss function L_s, and the approaches with the symbol + mean they are respectively combined with L_s.

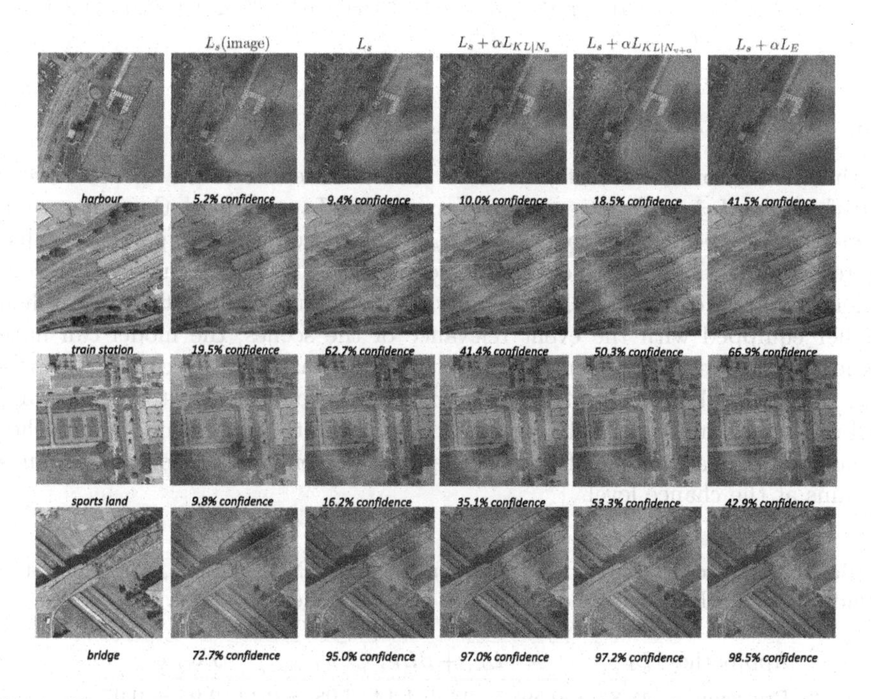

Fig. 7. The class activation map generated by different approaches for different categories, as well as the corresponding predict probabilities of ground-truth category. L_s(image) means the learning objective of L_s is just performed with image data.

that pre-training from a large-scale dataset benefits the learning task on the small datasets. Another remark from this table is that the results of the three proposed approaches show that both unimodal networks can take advantage of the sound event knowledge to achieve better scene recognition performance. It further validates the generalization of the proposed approaches, either in the multimodal or the unimodal input case. Compared with the multi-task framework of $L_{SQ|N_{v+a}}$ and $L_{KL|N_{v+a}}$, the L_E approach can better utilize the correlation between sound event and scene category via the statistical posteriors.

Table 2. Unimodal aerial scene recognition results on the ADVANCE dataset from 5 different runs, where † means random initialization and the approaches with the symbol + mean they are weightedly combined with L_s.

Modality	Approaches	L_s†	L_s	$+L_{SQ\|N_{v+a}}$	$+L_{KL\|N_{v+a}}$	$+L_E$
Sound	Precision	21.15 ± 0.68	30.46 ± 0.66	31.64 ± 0.65	30.00 ± 0.86	31.14 ± 0.30
	Recall	24.54± 0.67	32.99 ± 1.20	34.68 ± 0.49	34.29 ± 0.35	33.80 ± 1.03
	F-score	21.32 ± 0.42	28.99 ± 0.51	29.31 ± 0.71	28.51 ± 0.99	29.66 ± 0.13
Image	Precision	64.45 ± 0.97	74.05 ± 0.42	74.86 ± 0.94	74.36 ± 0.85	73.97 ± 0.39
	Recall	64.59 ± 1.12	72.79 ± 0.62	74.11 ± 0.89	73.40 ± 0.84	73.47 ± 0.52
	F-score	64.04 ± 1.07	72.85 ± 0.57	73.98 ± 0.92	73.52 ± 0.85	73.44 ± 0.45

5.3 Ablation Study

In this subsection, we directly validate the effectiveness of the scene-to-event transfer term L_{E_1} and the event relevance term L_{E_2}, without the supervision from the scene recognition objective of L_s. Table 3 shows the comparison results. By resorting to the scene-to-event transfer term, performing sound event recognition can reward the model the ability to distinguish different scenes. When further equipped with the event relevance of the scenes, the model can have higher performance. This demonstrates that cross-task transfer can indeed provide reasonable knowledge if the inherent correlation between these tasks are well exploited and utilized. By contrast, as the multi-task learning approaches do not well take advantage of this knowledge, the scene recognition performance remains at the chance level.

Table 3. Aerial scene recognition results on the ADVANCE dataset, where only the sound event knowledge is considered in the training stage.

Approaches	L_{E_1}	$L_{E_1} + \beta L_{E_2}$	$L_{KL\|N_{v+a}}$	$L_{SQ\|N_{v+a}}$
Precision	43.37 ± 0.59	54.23 ± 1.14	3.08 ± 0.14	2.95 ± 0.07
Recall	49.26 ± 0.36	52.57 ± 0.72	9.69 ± 0.43	9.28 ± 0.17
F-score	42.50 ± 0.42	48.65 ± 0.85	4.46 ± 0.20	4.24 ± 0.07

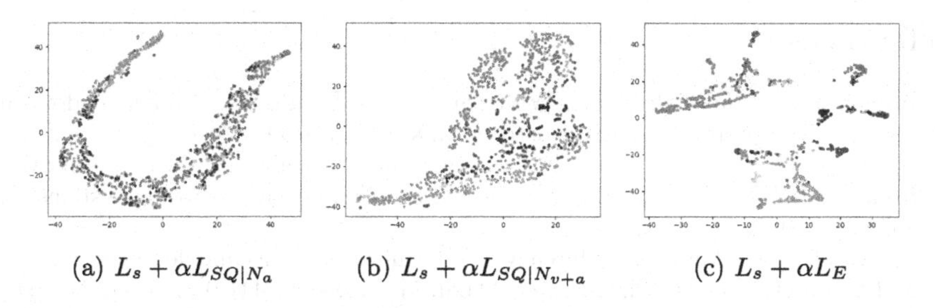

(a) $L_s + \alpha L_{SQ|N_a}$ (b) $L_s + \alpha L_{SQ|N_{v+a}}$ (c) $L_s + \alpha L_E$

Fig. 8. The aerial scene data embeddings indicated by the corresponding sound event distribution, where the points in different color mean in different scene categories. (Color figure online)

To better illustrate the correlation between aerial scenes and sound events, we further visualize the embedding results. Specifically, we use the well-trained cross-task transfer model to predict the sound event distribution on the testing set. Ideally, the sound event distribution can separate the scenes from each other, since each scene takes a different sound event distribution. Hence, we use t-SNE [18] to visualize the high-dimensional sound event distributions of different scenes. Figure 8 shows the visualization results, where the points in different color mean different scene categories. As $L_{SQ|N_a}$ is performed within the audio modality, the sound event knowledge cannot well transfer to the entire model, leading to the mixed scene distribution. By contrast, as $L_{SQ|N_{v+a}}$ transfers the sound event knowledge into the multimodal network, the predicted sound event distribution can separate different scenes to some extent. By introducing the correlation between scenes and events, *i.e.*, L_E, different scenes can be further disentangled, which confirms the feasibility and merits of cross task transfer.

6 Conclusions

In this paper, we explore a novel multimodal aerial scene recognition task that considers both visual and audio data. We have constructed a dataset consists of labeled paired audiovisual worldwide samples for facilitating the research on this topic. We propose to transfer the sound event knowledge to the scene recognition task for the reasons that the sound events are related to the scenes and that this underlying relation is not well exploited. Amounts of experimental results show the effectiveness of three proposed transfer approaches, confirming the benefit of exploiting the audio knowledge for the aerial scene recognition.

Acknowledgement. This work was supported in part by the National Natural Science Foundation of China under Grant 61822601 and 61773050; the Beijing Natural Science Foundation under Grant Z180006.

References

1. Assael, Y.M., Shillingford, B., Whiteson, S., De Freitas, N.: Lipnet: end-to-end sentence-level lipreading. arXiv preprint arXiv:1611.01599 (2016)
2. Aytar, Y., Vondrick, C., Torralba, A.: Soundnet: learning sound representations from unlabeled video. In: Advances in neural information processing systems. pp. 892–900 (2016)
3. Baltrušaitis, T., Ahuja, C., Morency, L.P.: Multimodal machine learning: a survey and taxonomy. IEEE Trans. Pattern Anal. Mach. Intell. **41**(2), 423–443 (2018)
4. Castelluccio, M., Poggi, G., Sansone, C., Verdoliva, L.: Land use classification in remote sensing images by convolutional neural networks. arXiv preprint arXiv:1508.00092 (2015)
5. Cheng, G., Yang, C., Yao, X., Guo, L., Han, J.: When deep learning meets metric learning: remote sensing image scene classification via learning discriminative CNNS. IEEE Trans. Geosci. Remote Sens. **56**(5), 2811–2821 (2018)
6. Ehrlich, M., Shields, T.J., Almaev, T., Amer, M.R.: Facial attributes classification using multi-task representation learning. In: Proceedings of the IEEE Conference on Computer Vision and Pattern Recognition Workshops, pp. 47–55 (2016)
7. Eigen, D., Fergus, R.: Predicting depth, surface normals and semantic labels with a common multi-scale convolutional architecture. In: Proceedings of the IEEE International Conference on Computer Vision, pp. 2650–2658 (2015)
8. Gan, C., Zhao, H., Chen, P., Cox, D., Torralba, A.: Self-supervised moving vehicle tracking with stereo sound. In: Proceedings of the IEEE International Conference on Computer Vision, pp. 7053–7062 (2019)
9. Gemmeke, J.F., et al.: Audio set: an ontology and human-labeled dataset for audio events. In: 2017 IEEE International Conference on Acoustics, Speech and Signal Processing (ICASSP), pp. 776–780. IEEE (2017)
10. Hinton, G., Vinyals, O., Dean, J.: Distilling the knowledge in a neural network. In: NIPS Deep Learning and Representation Learning Workshop (2015). http://arxiv.org/abs/1503.02531
11. Hu, D., Li, X., et al.: Temporal multimodal learning in audiovisual speech recognition. In: Proceedings of the IEEE Conference on Computer Vision and Pattern Recognition, pp. 3574–3582 (2016)
12. Hu, D., Nie, F., Li, X.: Deep multimodal clustering for unsupervised audiovisual learning. In: Proceedings of the IEEE Conference on Computer Vision and Pattern Recognition, pp. 9248–9257 (2019)
13. Hu, D., Wang, Z., Xiong, H., Wang, D., Nie, F., Dou, D.: Curriculum Audiovisual Learning, arXiv preprint arXiv:2001.09414 (2020)
14. Imoto, K., Tonami, N., Koizumi, Y., Yasuda, M., Yamanishi, R., Yamashita, Y.: Sound Event Detection by Multitask Learning of Sound Events and Scenes with Soft Scene Labels, arXiv preprint arXiv:2002.05848 (2020)
15. Kato, H., Harada, T.: Image reconstruction from bag-of-visual-words. In: Proceedings of the IEEE Conference on Computer Vision and Pattern Recognition, pp. 955–962 (2014)
16. Lazebnik, S., Schmid, C., Ponce, J.: Beyond bags of features: spatial pyramid matching for recognizing natural scene categories. In: 2006 IEEE Computer Society Conference on Computer Vision and Pattern Recognition (CVPR 2006), vol. 2, pp. 2169–2178. IEEE (2006)
17. Li, D., Chen, X., Zhang, Z., Huang, K.: Learning deep context-aware features over body and latent parts for person re-identification. In: IEEE Conference on Computer Vision and Pattern Recognition (CVPR), pp. 384–393 (2017)

18. Maaten, Lvd, Hinton, G.: Visualizing data using t-SNE. J. Mach. Learn. Res. **9**, 2579–2605 (2008)
19. Mou, L., Hua, Y., Zhu, X.X.: A relation-augmented fully convolutional network for semantic segmentation in aerial scenes. In: Proceedings of the IEEE Conference on Computer Vision and Pattern Recognition, pp. 12416–12425 (2019)
20. Nogueira, K., Penatti, O.A., Dos Santos, J.A.: Towards better exploiting convolutional neural networks for remote sensing scene classification. Pattern Recogn. **61**, 539–556 (2017)
21. Owens, A., Efros, A.A.: Audio-visual scene analysis with self-supervised multisensory features. In: Proceedings of the European Conference on Computer Vision (ECCV), pp. 631–648 (2018)
22. Owens, A., Wu, J., McDermott, J.H., Freeman, W.T., Torralba, A.: Learning sight from sound: ambient sound provides supervision for visual learning. Int. J. Comput. Vis. **126**, 1120–1137 (2018). https://doi.org/10.1007/s11263-018-1083-5
23. Risojević, V., Babić, Z.: Aerial image classification using structural texture similarity. In: 2011 IEEE International Symposium on Signal Processing and Information Technology (ISSPIT), pp. 190–195. IEEE (2011)
24. Risojević, V., Babić, Z.: Orientation difference descriptor for aerial image classification. In: 2012 19th International Conference on Systems, Signals and Image Processing (IWSSIP), pp. 150–153. IEEE (2012)
25. Salem, T., Zhai, M., Workman, S., Jacobs, N.: A multimodal approach to mapping soundscapes. In: IEEE International Geoscience and Remote Sensing Symposium (IGARSS) (2018)
26. Tian, Y., Shi, J., Li, B., Duan, Z., Xu, C.: Audio-visual event localization in unconstrained videos. In: Proceedings of the European Conference on Computer Vision (ECCV), pp. 247–263 (2018)
27. Tzeng, E., Hoffman, J., Darrell, T., Saenko, K.: Simultaneous deep transfer across domains and tasks. In: Proceedings of the IEEE International Conference on Computer Vision (ICCV), pp. 4068–4076 (2015)
28. Wang, J., Yang, J., Yu, K., Lv, F., Huang, T., Gong, Y.: Locality-constrained linear coding for image classification. In: 2010 IEEE Computer Society Conference on Computer Vision and Pattern Recognition, pp. 3360–3367. IEEE (2010)
29. Wang, Y.: Polyphonic sound event detection with weak labeling. PhD Thesis (2018)
30. Xia, G.S., et al.: Aid: a benchmark data set for performance evaluation of aerial scene classification. IEEE Trans. Geosci. Remote Sens. **55**(7), 3965–3981 (2017)
31. Xiao, F., Lee, Y.J., Grauman, K., Malik, J., Feichtenhofer, C.: Audiovisual slowfast networks for video recognition. arXiv preprint arXiv:2001.08740 (2020)
32. Yang, Y., Newsam, S.: Comparing SIFT descriptors and gabor texture features for classification of remote sensed imagery. In: 2008 15th IEEE International Conference on Image Processing, pp. 1852–1855. IEEE (2008)
33. Zhang, F., Du, B., Zhang, L.: Scene classification via a gradient boosting random convolutional network framework. IEEE Trans. Geosci. Remote Sens. **54**(3), 1793–1802 (2015)
34. Zhao, H., Gan, C., Rouditchenko, A., Vondrick, C., McDermott, J., Torralba, A.: The sound of pixels. In: Proceedings of the European Conference on Computer Vision (ECCV), pp. 570–586 (2018)
35. Zheng, W.L., Liu, W., Lu, Y., Lu, B.L., Cichocki, A.: Emotionmeter: a multimodal framework for recognizing human emotions. IEEE Trans. Cyber. **49**(3), 1110–1122 (2018)

36. Zhou, B., Khosla, A., Lapedriza, A., Oliva, A., Torralba, A.: Learning deep features for discriminative localization. In: Proceedings of the IEEE Conference on Computer Vision and Pattern Recognition, pp. 2921–2929 (2016)
37. Zou, Q., Ni, L., Zhang, T., Wang, Q.: Deep learning based feature selection for remote sensing scene classification. IEEE Geosci. Remote Sens. Lett. **12**(11), 2321–2325 (2015)

Polarimetric Multi-view Inverse Rendering

Jinyu Zhao(✉), Yusuke Monno, and Masatoshi Okutomi

Tokyo Institute of Technology, Tokyo, Japan
jzhao@ok.sc.e.titech.ac.jp

Abstract. A polarization camera has great potential for 3D reconstruction since the angle of polarization (AoP) of reflected light is related to an object's surface normal. In this paper, we propose a novel 3D reconstruction method called Polarimetric Multi-View Inverse Rendering (Polarimetric MVIR) that effectively exploits geometric, photometric, and polarimetric cues extracted from input multi-view color polarization images. We first estimate camera poses and an initial 3D model by geometric reconstruction with a standard structure-from-motion and multi-view stereo pipeline. We then refine the initial model by optimizing photometric rendering errors and polarimetric errors using multi-view RGB and AoP images, where we propose a novel polarimetric cost function that enables us to effectively constrain each estimated surface vertex's normal while considering four possible ambiguous azimuth angles revealed from the AoP measurement. Experimental results using both synthetic and real data demonstrate that our Polarimetric MVIR can reconstruct a detailed 3D shape without assuming a specific polarized reflection depending on the material.

Keywords: Multi-view reconstruction · Inverse rendering · Polarization

1 Introduction

Image-based 3D reconstruction has been studied for years and can be applied to various applications, e.g. model creation [15], localization [16], segmentation [20], and shape recognition [54]. There are two common approaches for 3D reconstruction: geometric reconstruction and photometric reconstruction. The geometric reconstruction is based on feature matching and triangulation using multi-view images. It has been well established as structure from motion (SfM) [8,50,57] for sparse point cloud reconstruction, often followed by dense reconstruction with multi-view stereo (MVS) [21–23]. On the other hand, the photometric reconstruction exploits shading information for each image pixel to derive dense surface normals. It has been well studied as shape from shading [13,59,61] and photometric stereo [25,27,58].

Electronic supplementary material The online version of this chapter (https://doi.org/10.1007/978-3-030-58586-0_6) contains supplementary material, which is available to authorized users.

A. Vedaldi et al. (Eds.): ECCV 2020, LNCS 12369, pp. 85–102, 2020.
https://doi.org/10.1007/978-3-030-58586-0_6

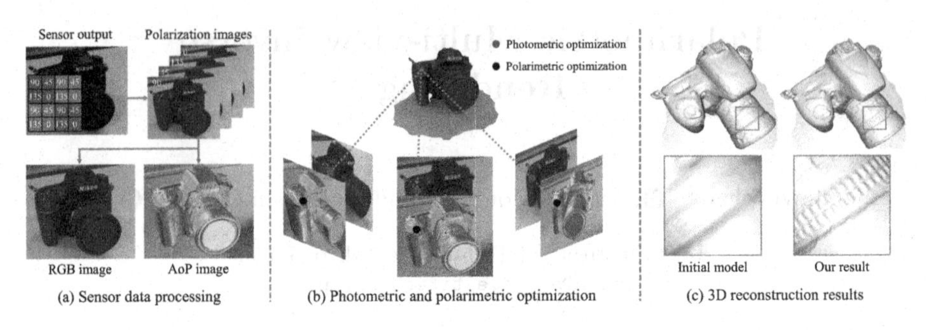

Fig. 1. Overview of our Polarimetric Multi-View Inverse Rendering (Polarimetric MVIR): (a) Color polarization sensor data processing to obtain the set of RGB and angle-of-polarization (AoP) images; (b) Using estimated camera poses and an initial model from SfM and MVS, Polarimetric MVIR optimizes photometric rendering errors and polarimetric errors by using multi-view RGB and AoP images; (c) Initial and our refined 3D model results.

There also exist other advanced methods combining the advantages of both approaches, e.g. multi-view photometric stereo [35,47] and multi-view inverse rendering (MVIR) [32]. These methods typically start with SfM and MVS for camera pose estimation and initial model reconstruction, and then refine the initial model, especially for texture-less surfaces, by utilizing shading cues.

Multi-view reconstruction using polarization images [18,60] has also received increasing attention with the development of one-shot polarization cameras using Sony IMX250 monochrome or color polarization sensor [38], e.g. JAI GO-5100MP-PGE [1] and Lucid PHX050S-Q [2] cameras. The use of polarimetric information has great potential for 3D reconstruction since the angle of polarization (AoP) of reflected light is related to the azimuth angle of the object's surface normal. One state-of-the-art method is Polarimetric MVS [18], which propagates initial sparse depth from SfM by using AoP images obtained by a polarization camera for creating a dense depth map for each view. Since there are four possible azimuth angles corresponding to one AoP measurement as detailed in Sect. 3, their depth propagation relies on the disambiguation of polarimetric ambiguities using the initial depth estimate by SfM.

In this paper, inspired by the success of MVIR [32] and Polarimetric MVS [18], we propose Polarimetric Multi-View Inverse Rendering (Polarimetric MVIR), which is a fully passive 3D reconstruction method exploiting all geometric, photometric, and polarimetric cues. We first estimate camera poses and an initial surface model based on SfM and MVS. We then refine the initial model by simultaneously using multi-view RGB and AoP images obtained from color polarization images (see Fig. 1) while estimating surface albedos and illuminations for each image. The key of our method is a novel global cost optimization framework for shape refinement. In addition to a standard photometric rendering term that evaluates RGB intensity errors (as in [32]), we introduce a novel polarimetric term that evaluates the difference between the azimuth angle of each estimated surface vertex's normal and four possible azimuth angles obtained from the corresponding AoP measurement. Our method takes all four possible ambiguous azimuth angles into account in the global optimization, instead of explicitly trying to solve the ambiguity as in Polarimetric MVS [18], which makes our method more robust to noise and mis-disambiguation. Experimental results using synthetic and real data demonstrate that, compared with existing MVS methods, MVIR, and Polarimetric MVS, Polarimetric MVIR can reconstruct a more detailed 3D model from unconstrained input images without any prerequisites for surface materials. Two main contributions of this work are summarized as below.

- We propose Polarimetric MVIR, which is the first 3D reconstruction method based on multi-view photometric and polarimetric optimization with an inverse rendering framework.
- We propose a novel polarimetric cost function that enables us to effectively constrain the surface normal of each vertex of the estimated surface mesh while considering the azimuth angle ambiguities as an optimization problem.

2 Related Work

In the past literature, a number of methods have been proposed for the geometric 3D reconstruction (e.g. SfM [8,50,57] and MVS [21–23]) and the photometric 3D reconstruction (e.g. shape from shading [13,59,61] and photometric stereo [25,27,58]). In this section, we briefly introduce the combined methods of geometric and photometric 3D reconstruction, and also polarimetric 3D reconstruction methods, which are closely related to our work.

Multi-view Geometric-Photometric 3D Reconstruction: The geometric approach is relatively robust to estimate camera poses and a sparse or dense point cloud, owing to the development of robust feature detection and matching algorithms [14,36]. However, it is weak in texture-less surfaces because sufficient feature correspondences cannot be obtained. In contrast, the photometric approach can recover fine details for texture-less surfaces by exploiting pixel-by-pixel shading information. However, it generally assumes a known or calibrated camera and lighting setup. Some advanced methods [39,56,57], including multi-view photometric stereo [35,47] and MVIR [32], combine the two approaches to take both advantages. These methods typically estimate camera poses and an initial model based on SfM and MVS, and then refine the initial model, especially in texture-less regions, by using shading cues from multiple viewpoints. Our Polarimetric MVIR is built on MVIR [32], which is an uncalibrated method and jointly estimates a refined shape, surface albedos, and each image's illumination.

Single-View Shape from Polarization (SfP): There are many SfP methods which estimate object's surface normals [10,26,29,43,44,52,55] based on the physical properties that AoP and degree of polarization (DoP) of reflected light are related to the azimuth and the zenith angles of the object's surface normal, respectively. However, existing SfP methods usually assume a specific surface material because of the material-dependent ambiguous relationship between AoP and the azimuth angle, and also the ambiguous relationship between DoP and the zenith angle. For instance, a diffuse polarization model is adopted in [10,26, 43,55], a specular polarization model is applied in [44], and dielectric material is considered in [29,52]. Some methods combine SfP with shape from shading or photometric stereo [9,37,43,46,52,62], where estimated surface normals from shading information are used as cues for resolving the polarimetric ambiguity. However, these methods require a calibrated lighting setup.

Multi-view Geometric-Polarimetric 3D Reconstruction: Some studies have shown that multi-view polarimetric information is valuable for surface normal estimation [11,24,41,42,48] and also camera pose estimation [17,19]. However, existing multi-view methods typically assume a specific material, e.g. diffuse objects [11,19], specular objects [41,42,48] and faces [24], to omit the polarimetric ambiguities. Recent two state-of-the-art methods, Polarimetric MVS [18] and Polarimetric SLAM [60], consider a mixed diffuse and specular reflection model to remove the necessity of known surface materials. These methods first disambiguate the ambiguity for AoP by using initial sparse depth cues from MVS or SLAM. Each viewpoint's depth map is then densified by propagating the sparse depth, where the disambiguated AoP values are used to find iso-depth contours along which the depth can be propagated. Although dense multi-view depth maps can be generated by the depth propagation, this approach relies on correct disambiguation which is not easy in general.

Advantages of Polarimetric MVIR: Compared to prior studies, our method has several advantages. First, it advances MVIR [32] by using polarimetric information while inheriting the benefits of MVIR. Second, similar to [18,60], our method is fully passive and does not require calibrated lighting and known surface materials. Third, polarimetric ambiguities are resolved as an optimization problem in shape refinement, instead of explicitly disambiguating them beforehand as in [18,60], which can avoid relying on the assumption that the disambiguation is correct. Finally, a fine shape can be obtained by simultaneously exploiting photometric and polarimetric cues, where multi-view AoP measurements are used for constraining each estimated surface vertex's normal, which is a more direct and natural way to exploit azimuth-angle-related AoP measurements for shape estimation.

3 Polarimetric Ambiguities in Surface Normal Prediction

3.1 Polarimetric Calculation

Unpolarized light becomes partially polarized after reflection by a certain object's surface. Consequently, under common unpolarized illumination, the intensity of reflected light observed by a camera equipped with a polarizer satisfies the following equation:

$$I(\phi_{pol}) = \frac{I_{max} + I_{min}}{2} + \frac{I_{max} - I_{min}}{2} \cos 2(\phi_{pol} - \phi), \quad (1)$$

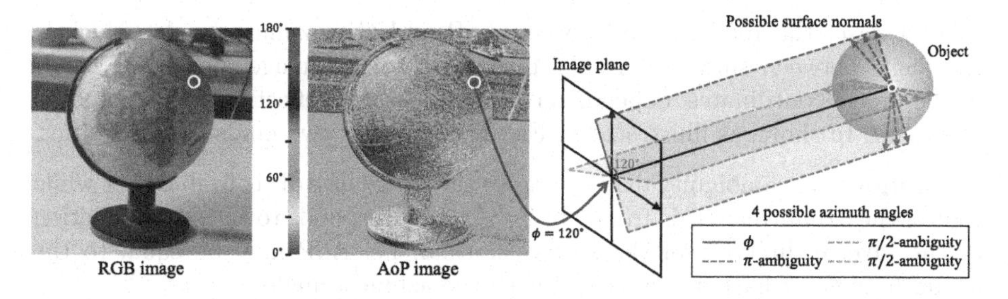

Fig. 2. Four possible azimuth angles ($\alpha = 30°,\ 120°,\ 210°$ and $300°$) corresponding to an observed AoP value ($\phi = 120°$). The transparent color plane shows the possible planes on which surface normal lies. Example possible surface normals are illustrated by the color dashed arrows on the object.

where I_{max} and I_{min} are the maximum and minimum intensities, respectively, ϕ_{pol} is the polarizer angle, and ϕ is the reflected light's AoP, which indicates reflection's direction of polarization. A polarization camera commonly observes the intensities of four polarization directions, i.e. I_0, I_{45}, I_{90}, and I_{135}. From those measurements, AoP can be calculated using the Stokes vector [53] as

$$\phi = \frac{1}{2}\tan^{-1}\frac{s_2}{s_1}, \tag{2}$$

where ϕ is the AoP, and s_1 and s_2 are the components of the Stokes vector

$$\mathbf{s} = \begin{bmatrix} s_0 \\ s_1 \\ s_2 \\ s_3 \end{bmatrix} = \begin{bmatrix} I_{max} + I_{min} \\ (I_{max} - I_{min})\cos(2\phi) \\ (I_{max} - I_{min})\sin(2\phi) \\ 0 \end{bmatrix} = \begin{bmatrix} I_0 + I_{90} \\ I_0 - I_{90} \\ I_{45} - I_{135} \\ 0 \end{bmatrix}, \tag{3}$$

where $s_3 = 0$ because circularly polarized light is not considered in this work.

3.2 Ambiguities

AoP of reflected light reveals information about the surface normal according to Fresnel equations, as depicted by Atkinson and Hancock [10]. There are two linear polarization components of the incident wave: s-polarized light and p-polarized light whose directions of polarization are perpendicular and parallel to the plane of incidence consisting of incident light and surface normal, respectively. For a dielectric, the reflection coefficient of s-polarized light is always greater than that of p-polarized light while the transmission coefficient of p-polarized light is always greater than that of s-polarized light. For a metal, the relationship are opposite. Consequently, the polarization direction of reflected light should be perpendicular or parallel to the plane of incidence according to the relationship between s-polarized and p-polarized light.

In this work, we consider a mixed polarization reflection model [12,18] which includes unpolarized diffuse reflection, polarized specular reflection (s-polarized light is stronger) and polarized diffuse reflection (p-polarized light is stronger).

In that case, the relationship between AoP and the azimuth angle, which is the angle between surface normal's projection to the image plane and x-axis in the image coordinates, depends on which polarized reflection's component is dominant. In short, as illustrated in Fig. 2, there exist two kinds of ambiguities.

π-ambiguity: π-ambiguity exists because the range of AoP is from 0 to π while that of the azimuth angle is from 0 to 2π. AoP corresponds to the same direction or the inverse direction of the surface normal, i.e. AoP may be equal to the azimuth angle or have π's difference with the azimuth angle.

$\pi/2$-ambiguity: It is difficult to decide whether polarized specular reflection or polarized diffuse reflection dominates without any prerequisites for surface materials. AoP has $\pi/2$'s difference with the azimuth angle when polarized specular reflection dominates, while it equals the azimuth angle or has π's difference with the azimuth angle when polarized diffuse reflection dominates. Therefore, there exists $\pi/2$-ambiguity in addition to π-ambiguity when determining the relationship between AoP and the azimuth angle.

As shown in Fig. 2, for the AoP value ($\phi = 120°$) for the pixel marked in red, there are four possible azimuth angles (i.e. $\alpha = 30°$, $120°$, $210°$ and $300°$) as depicted by the four lines on the image plane. The planes where the surface normal has to lie, which are represented by the four transparent color planes, are determined according to the four possible azimuth angles. The dashed arrows on the object show the examples of possible surface normals, which are constrained on the planes. In our method, the explained relationship between the AoP measurement and the possible azimuth angles is exploited to constrain the estimated surface vertex's normal.

4 Polarimetric Multi-view Inverse Rendering

4.1 Color Polarization Sensor Data Processing

To obtain input RGB and AoP images, we use a one-shot color polarization camera consisting of the 4×4 regular pixel pattern [38] as shown in Fig. 1(a), although our method is not limited to this kind of polarization camera. For every pixel, twelve values, i.e. 3 $(R, G, B) \times 4$ $(I_0, I_{45}, I_{90}, I_{135})$, are obtained by interpolating the raw mosaic data. As proposed in [45], pixel values for each direction in every 2×2 blocks are extracted to obtain Bayer-patterned data for that direction. Then, Bayer color interpolation [31] and polarization interpolation [40] are sequentially performed to obtain full-color-polarization data. As for the RGB images used for the subsequent processing, we employ unpolarized RGB component $\mathbf{I}_{min}$ obtained as $\mathbf{I}_{min} = (\mathbf{I}_0 + \mathbf{I}_{90})(1-\rho)/2$, where ρ is DoP and calculated by using the Stokes vector of Eq. (3) as $\rho = \sqrt{s_1^2 + s_2^2}/s_0$. Since using $\mathbf{I}_{min}$ can suppress the influence of specular reflection [10], it is beneficial for SfM and our photometric optimization. On the other hand, AoP values are calculated using Eqs. (2) and (3), where the intensities of four polarization directions (I_0, I_{45}, I_{90}, I_{135}) are obtained by averaging R, G, and B values for each direction.

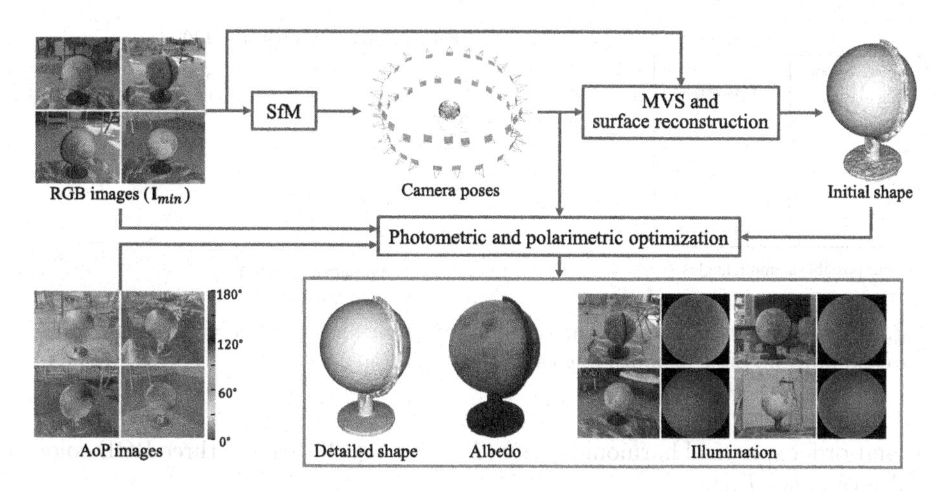

Fig. 3. The flowchart of our Polarimetric MVIR using multi-view RGB and AoP images.

4.2 Initial Geometric Reconstruction

Figure 3 shows the overall flow of our Poralimetric MVIR using multi-view RGB and AoP images. It starts with initial geometric 3D reconstruction as follows. SfM is firstly performed using the RGB images to estimate camera poses. Then, MVS and surface reconstruction are applied to obtain an initial surface model which is represented by a triangular mesh. The visibility of each vertex to each camera is then checked using the algorithm in [32]. Finally, to increase the number of vertices, the initial surface is subdivided by $\sqrt{3}$-subdivision [33] until the maximum pixel number in each triangular patch projected to visible cameras becomes smaller than a threshold.

4.3 Photometric and Polarimetric Optimization

The photometric and polarimetric optimization is then performed to refine the initial model while estimating each vertex's albedo and each image's illumination. The cost function is expressed as

$$\mathop{\arg\min}_{\mathbf{X},\mathbf{K},\mathbf{L}} E_{pho}(\mathbf{X},\mathbf{K},\mathbf{L}) + \tau_1 E_{pol}(\mathbf{X}) + \tau_2 E_{gsm}(\mathbf{X}) + \tau_3 E_{psm}(\mathbf{X},\mathbf{K}), \quad (4)$$

where E_{pho}, E_{pol}, E_{gsm}, and E_{psm} represent a photometric rendering term, a polarimetric term, a geometric smoothness term, and a photometric smoothness term, respectively. τ_1, τ_2, and τ_3 are weights to balance each term. Similar to MVIR [32], the optimization parameters are defined as below:
- $\mathbf{X} \in \mathbb{R}^{3\times n}$ is the vertex 3D coordinate, where n is the total number of vertices.
- $\mathbf{K} \in \mathbb{R}^{3\times n}$ is the vertex albedo, which is expressed in the RGB color space.
- $\mathbf{L} \in \mathbb{R}^{12\times p}$ is the scene illumination matrix, where p is the total number of images. Each image's illumination is represented by nine coefficients for the

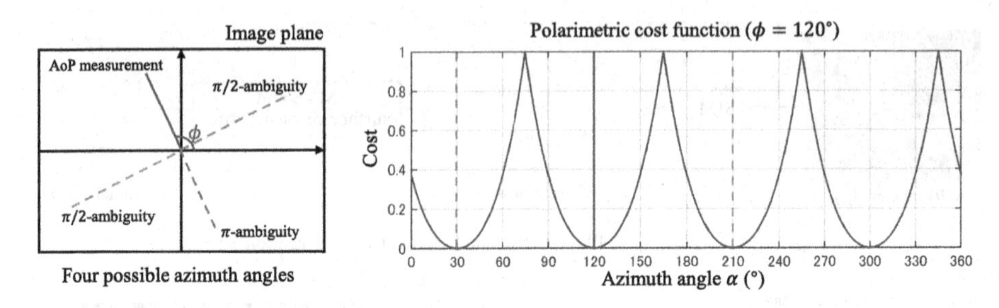

Fig. 4. An example of the polarimetric cost function ($\phi = 120°, k = 0.5$). Four lines correspond to possible azimuth angles as shown in Fig. 2.

second-order spherical harmonics basis $(L_0, \cdots, L_8)$ [49,57] and three RGB color scales (L_R, L_G, L_B).

Photometric Rendering Term: We adopt the same photometric rendering term as MVIR, which is expressed as

$$E_{pho}(\mathbf{X}, \mathbf{K}, \mathbf{L}) = \sum_i \sum_{c \in \mathcal{V}(i)} \frac{||\mathbf{I}_{i,c}(\mathbf{X}) - \hat{\mathbf{I}}_{i,c}(\mathbf{X}, \mathbf{K}, \mathbf{L})||^2}{|\mathcal{V}(i)|}, \tag{5}$$

which measures the pixel-wise intensity error between observed and rendered values. $\mathbf{I}_{i,c} \in \mathbb{R}^3$ is the observed RGB values of the pixel in c-th image corresponding to i-th vertex's projection and $\hat{\mathbf{I}}_{i,c} \in \mathbb{R}^3$ is the corresponding rendered RGB values. $\mathcal{V}(i)$ represents the visible camera set for i-th vertex. The perspective projection model is used to project each vertex to each camera. Suppose (K_R, K_G, K_B) and $(L_0, \cdots, L_8, L_R, L_G, L_B)$ represent the albedo for i-th vertex and the illumination for c-th image, where the indexes i and c are omitted for notation simplicity. The rendered RGB values are then calculated as

$$\hat{\mathbf{I}}_{i,c}(\mathbf{X}, \mathbf{K}, \mathbf{L}) = [K_R S(\mathbf{N}(\mathbf{X}), \mathbf{L}) L_R, K_G S(\mathbf{N}(\mathbf{X}), \mathbf{L}) L_G, K_B S(\mathbf{N}(\mathbf{X}), \mathbf{L}) L_B]^T, \tag{6}$$

where S is the shading calculated by using the second-order spherical harmonics illumination model [49,57] as

$$\begin{aligned} S(\mathbf{N}(\mathbf{X}), \mathbf{L}) = L_0 &+ L_1 N_y + L_2 N_z + L_3 N_x + L_4 N_x N_y + L_5 N_y N_z \\ &+ L_6 (N_z^2 - \frac{1}{3}) + L_7 N_x N_z + L_8 (N_x^2 - N_y^2), \end{aligned} \tag{7}$$

where $\mathbf{N}(\mathbf{X}) = [N_x, N_y, N_z]^T$ represents the vertex's normal vector, which is calculated as the average of adjacent triangular patch's normals. Varying illuminations for each image and spatially varying albedos are considered as in [32].

Polarimetric Term: To effectively constrain each estimated surface vertex's normal, we here propose a novel polarimetric term. Figure 4 shows an example of our polarimetric cost function for the case that the AoP measurement of the pixel corresponding to the vertex's projection equals 120°, i.e. $\phi = 120°$.

This example corresponds to the situation as shown in Fig. 2. In both figures, four possible azimuth angles derived from the AoP measurement are shown by blue solid, purple dashed, green dashed, and brown dashed lines on the image plane, respectively. These four possibilities are caused by both the π-ambiguity and the $\pi/2$-ambiguity introduced in Sect. 3.2. In the ideal case without noise, one of the four possible azimuth angles should be the same as the azimuth angle of (unknown) true surface normal.

Based on this principle, as shown in Fig. 4, our polarimetric term evaluates the difference between the azimuth angle of the estimated surface vertex's normal α and its closest possible azimuth angle from the AoP measurement (i.e. $\phi - \pi/2$, ϕ, $\phi + \pi/2$, or $\phi + \pi$). The cost function is mathematically defined as

$$E_{pol}(\mathbf{X}) = \sum_{i} \sum_{c \in \mathcal{V}(i)} \left(\frac{e^{-k\theta_{i,c}(\mathbf{X})} - e^{-k}}{1 - e^{-k}} \right)^2 / |\mathcal{V}(i)|, \tag{8}$$

where k is a parameter that determines the narrowness of the concave to assign the cost (see Fig. 4). $\theta_{i,c}$ is defined as

$$\theta_{i,c}(\mathbf{X}) = 1 - 4\eta_{i,c}(\mathbf{X})/\pi, \tag{9}$$

where $\eta_{i,c}$ is expressed as

$$\begin{aligned} \eta_{i,c}(\mathbf{X}) = \min(&|\alpha_{i,c}(\mathbf{N}(\mathbf{X})) - \phi_{i,c}(\mathbf{X}) - \pi/2|, |\alpha_{i,c}(\mathbf{N}(\mathbf{X})) - \phi_{i,c}(\mathbf{X})|, \\ &|\alpha_{i,c}(\mathbf{N}(\mathbf{X})) - \phi_{i,c}(\mathbf{X}) + \pi/2|, |\alpha_{i,c}(\mathbf{N}(\mathbf{X})) - \phi_{i,c}(\mathbf{X}) + \pi|). \end{aligned} \tag{10}$$

Here, $\alpha_{i,c}$ is the azimuth angle calculated by the projection of i-th vertex's normal to c-th image plane and $\phi_{i,c}$ is the corresponding AoP measurement.

Our polarimetric term mainly has two benefits. First, it enables us to constrain the estimated surface vertex's normal while simultaneously resolving the ambiguities based on the optimization using all vertices and all multi-view AoP measurements. Second, the concave shape of the cost function makes the normal constraint more robust to noise, which is an important property since AoP is susceptible to noise. The balance between the strength of the normal constraint and the robustness to noise can be adjusted by the parameter k.

Geometric Smoothness Term: The geometric smoothness term is applied to regularize the cost and to derive a smooth surface. This term is described as

$$E_{gsm}(\mathbf{X}) = \sum_{m} \left(\frac{\arccos\left(\mathbf{N}'_m(\mathbf{X}) \cdot \mathbf{N}'_{m_{avg}}(\mathbf{X})\right)}{\pi} \right)^q, \tag{11}$$

where $\mathbf{N}'_m$ represents the normal of m-th triangular patch, $\mathbf{N}'_{m_{avg}}$ represents the averaged normal of its adjacent patches, and q is a parameter to assign the cost. This term becomes small if the curvature of the surface is close to constant.

Photometric Smoothness Term: Changes of pixel values in each image may result from different albedos or shading since spatially varying albedos

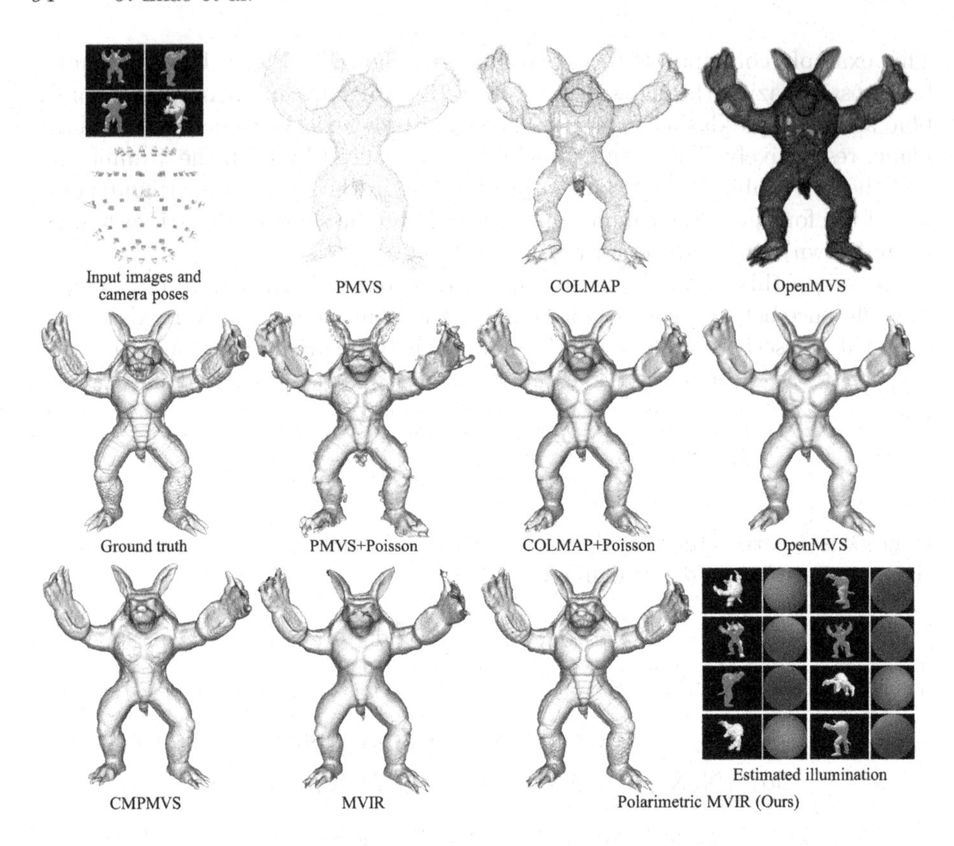

Fig. 5. Visual comparison for the Armadillo model

are allowed in our model. To regularize this uncertainty, the same photometric smoothness term as [32] is applied as

$$E_{psm}(\mathbf{X}, \mathbf{K}) = \sum_{i} \sum_{j \in \mathcal{A}(i)} w_{i,j}(\mathbf{X}) \left|\left|(\mathbf{K}_i - \mathbf{K}_j)\right|\right|^2, \tag{12}$$

where $\mathcal{A}(i)$ is the set of adjacent vertices of i-th vertex and $w_{i,j}$ is the weight for the pair of i-th and j-th vertices. A small weight is assigned, i.e. change of albedo is allowed, if a large chromaticity or intensity difference is observed between the corresponding pixels in the RGB image (see [32] for details). By this term, a smooth variation in photometric information is considered as the result of shading while a sharp variation is considered as the result of varying albedos.

5 Experimental Results

5.1 Implementation Details

We apply COLMAP [50] for SfM and OpenMVS [3] for MVS. The initial surface is reconstructed by the built-in surface reconstruction function of OpenMVS. The cost optimization of Eq. (4) is iterated three times by changing the weights as $(\tau_1, \tau_2, \tau_3) = (0.05, 1.0, 1.0)$, $(0.1, 1.0, 1.0)$, and $(0.3, 1.0, 1.0)$. For each iteration, the parameter q in Eq. (11) is changed as $q = 2.2$, 2.8, and 3.4, while the parameter k in Eq. (8) is set to 0.5 in all three iterations. By the three iterations, the surface normal constraint from AoP is gradually strengthened by allowing small normal variations to derive a fine shape while avoiding a local minimum. The non-linear optimization problem is solved by using Ceres solver [7].

5.2 Comparison Using Synthetic Data

Numerical evaluation was performed using four CG models (Armadillo, Stanford bunny, Dragon, and Buddha) available from Stanford 3D Scanning Repository [4]. Original 3D models were subdivided to provide enough number of vertices as ground truth. Since it is very difficult to simulate realistic polarization images and there are no public tools and datasets for polarimetric 3D reconstruction, we synthesized the RGB and the AoP inputs using Blender [5] as follows. Using spherically placed cameras, the RGB images were rendered under a point light source located at infinity and an environmental light uniformly contributing to the surface (see Fig. 5). For synthesizing AoP images, AoP for each pixel was obtained from the corresponding azimuth angle, meaning that there is no $\pi/2$-ambiguity, for the first experiment. The experiment was also conducted by randomly adding ambiguities and Gaussian noise to the azimuth angles.

We compared our Polarimetric MVIR with four representative MVS methods (PMVS [22], CMPMVS [28], MVS in COLMAP [51], OpenMVS [3]) and MVIR [32] using the same initial model as ours. Ground-truth camera poses are used to avoid the alignment problem among the models reconstructed from different methods. Commonly used metrics [6,34], i.e. accuracy which is the distance from each estimated 3D point to its nearest ground-truth 3D point and completeness which is the distance from each ground-truth 3D point to its nearest estimated 3D point, were used for evaluation. As estimated 3D points, the output point cloud was used for PMVS, COLMAP, and OpenMVS, while the output surface's vertices were used for CMPMVS, MVIR and our method.

Table 1 shows the comparison of the average accuracy and the average completeness for each model. The results show that our method achieves the best accuracy and completeness for all four models with significant improvements. Visual comparison for Armadillo is shown in Fig. 5, where the surfaces for PMVS and COLMAP were created using Poisson surface reconstruction [30] with our best parameter choice, while the surface for OpenMVS was obtained using its built-in function. We can clearly see that our method can recover more details than the other methods by exploiting AoP information. The visual comparison for the other models can be seen in our supplementary material.

Table 2 shows the numerical evaluation for our method when 50% random ambiguities and Gaussian noise with different noise levels were added to the azimuth-angle images. Note that top three rows are the results without any disturbance and same as those in Table 1, while the bottom five rows are the results of our method with ambiguity and noise added on AoP images. These results

Table 1. Comparisons of the average accuracy (Acc.) and completeness (Comp.) errors

		PMVS	CMPMVS	COLMAP	OpenMVS	MVIR	Ours
Armadillo	# of Vertices	60250	330733	268343	2045829	305555	305548
	Acc.($\times 10^{-2}$)	1.2634	0.6287	0.7436	0.8503	0.7667	**0.4467**
	Comp.($\times 10^{-2}$)	1.5261	0.8676	1.0295	0.6893	0.9311	**0.6365**
Bunny	# of Vertices	92701	513426	334666	2394638	399864	399863
	Acc.($\times 10^{-2}$)	1.0136	0.7766	0.7734	1.0222	0.7629	**0.5706**
	Comp.($\times 10^{-2}$)	1.3873	0.9581	1.6987	0.8466	0.8118	**0.6447**
Dragon	# of Vertices	88519	474219	399624	2820589	460888	460667
	Acc.($\times 10^{-2}$)	1.4321	0.8826	0.9001	1.0421	0.8563	**0.6258**
	Comp.($\times 10^{-2}$)	2.0740	1.4036	1.6606	1.3179	1.2237	**1.0222**
Buddha	# of Vertices	61259	338654	320539	2204122	348967	348691
	Acc.($\times 10^{-2}$)	1.7658	1.0565	0.9658	1.0878	1.0588	**0.7926**
	Comp.($\times 10^{-2}$)	2.4254	1.5666	2.0094	1.4859	1.3968	**1.1487**
Average	Acc.($\times 10^{-2}$)	1.3687	0.8361	0.8457	1.0006	0.8612	**0.6089**
	Comp.($\times 10^{-2}$)	1.8532	1.1990	1.5996	1.0849	1.0909	**0.8630**

Table 2. Numerical evaluation with 50% ambiguity and Gaussian noise

	Acc.($\times 10^{-2}$)	Comp.($\times 10^{-2}$)
CMPMVS (Best accuracy in the existing methods)	0.8361	1.1990
OpenMVS (Best completeness in the existing methods)	1.0006	1.0849
Ours (No ambiguity & $\sigma = 0°$)	**0.6089**	**0.8630**
Ours (50% ambiguity & $\sigma = 0°$)	0.6187	0.8727
Ours (50% ambiguity & $\sigma = 6°$)	0.6267	0.8820
Ours (50% ambiguity & $\sigma = 12°$)	0.6367	0.8892
Ours (50% ambiguity & $\sigma = 18°$)	0.6590	0.9115
Ours (50% ambiguity & $\sigma = 24°$)	0.7175	0.9643

demonstrate that our method is quite robust against the ambiguity and the noise and outperforms the best-performed existing methods even with the 50% ambiguities and a large noise level ($\sigma = 24°$).

5.3 Comparison Using Real Data

Figure 6 shows the visual comparison of the reconstructed 3D models using real images of a toy car (56 views) and a camera (31 views) captured under a normal lighting condition in the office using fluorescent light on the ceiling, and a statue (43 views) captured under outdoor daylight with cloudy weather. We captured the polarization images using Lucid PHX050S-Q camera [2]. We compared our method with CMPMVS, OpenMVS, and MVIR, which respectively provide the best accuracy, the best completeness, and the best balanced result among the existing methods shown in Table 1. The results of all compared methods and our albedo and illumination results can be seen in the supplementary material.

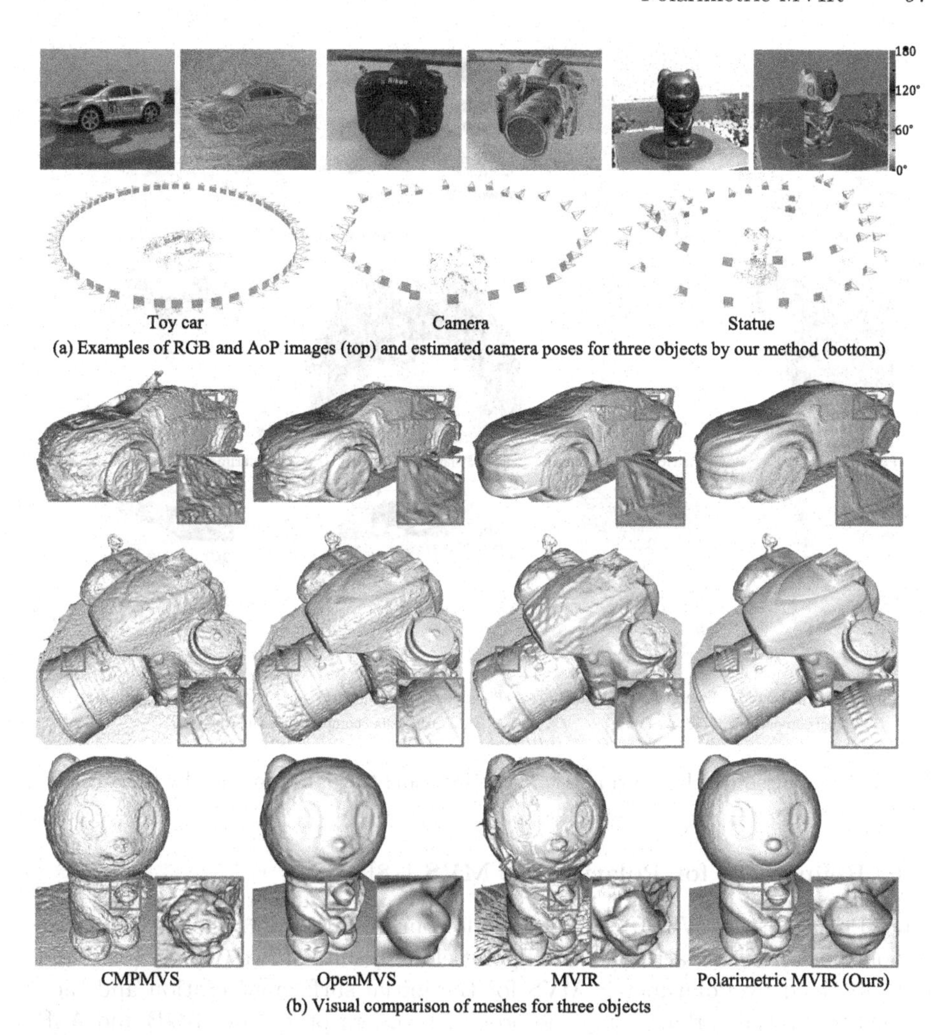

(a) Examples of RGB and AoP images (top) and estimated camera poses for three objects by our method (bottom)

(b) Visual comparison of meshes for three objects

Fig. 6. Visual comparison using real data for the three objects

The results of Fig. 6 show that CMPMVS can reconstruct fine details in relatively well-textured regions (e.g. the details of the camera lens), while it fails in texture-less regions (e.g. the front window of the car). OpenMVS can better reconstruct the overall shapes owing to the denser points, although some fine details are lost. MVIR performs well except for dark regions, where the shading information is limited (e.g. the top of the camera and the surface of the statue). On the contrary, our method can recover finer details and clearly improve the reconstructed 3D model quality by exploiting both photometric and polarimetric information, especially in regions such as the front body and the window of the toy car, and the overall surfaces of the camera and the statue.

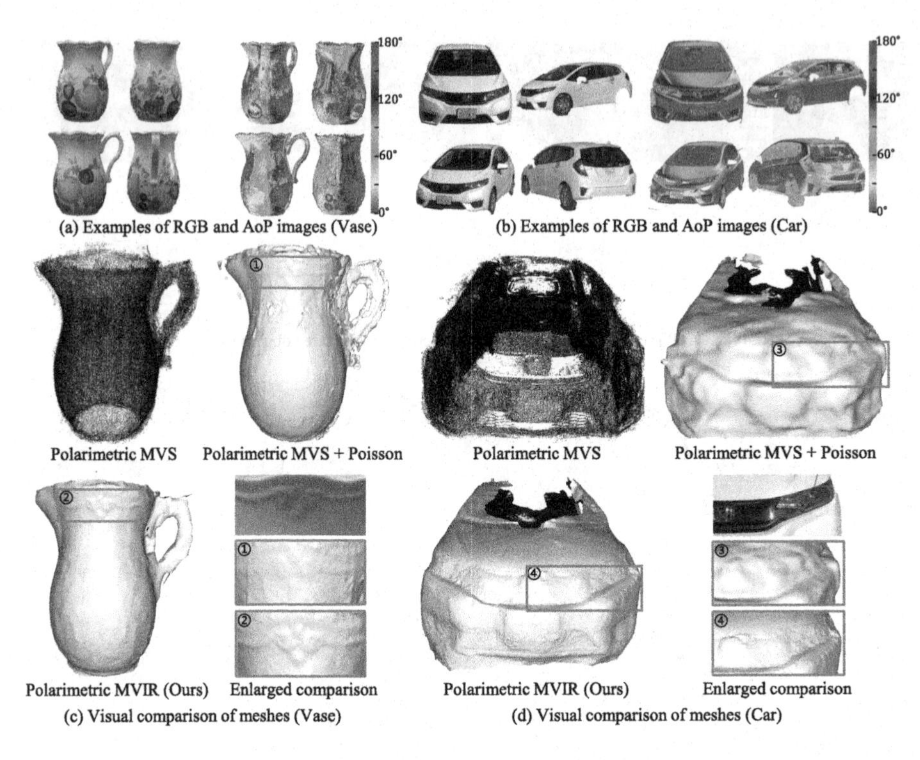

Fig. 7. Comparison with Polarimetric MVS [18] using the data provided by the authors.

5.4 Refinement for Polarimetric MVS [18]

Since Polarimetric MVS [18] can be used for our initial model to make better use of polarimetric information, we used point cloud results of two objects (vase and car) obtained by Polarimetric MVS for the initial surface generation and then refined the initial surface using the provided camera poses, and RGB and AoP images from 36 viewpoints as shown in Fig. 7 (a) and (b). As shown in Fig. 7 (c) and (d), Polarimetric MVS can provide dense point clouds, even for texture-less regions, by exploiting polarimetric information. However, there are still some outliers, which could be derived from AoP noise and incorrect disambiguation, and resultant surfaces are rippling. These artifacts are alleviated in our method (Polarimetric MVIR) by solving the ambiguity problem in our global optimization. Moreover, we can see that finer details are reconstructed using photometric shading information in our cost function.

6 Conclusions

In this paper, we have proposed Polarimetric MVIR, which can reconstruct a high-quality 3D model by optimizing multi-view photometric rendering errors and polarimetric errors. Polarimetric MVIR resolves the π- and $\pi/2$-ambiguities as an optimization problem, which makes the method fully passive and applicable to various materials. Experimental results have demonstrated that Polarimetric MVIR is robust to ambiguities and noise, and generates more detailed 3D models compared with existing state-of-the-art multi-view reconstruction methods.

Our Polarimetric MVIR has a limitation that it requires a reasonably good initial shape for its global optimization, which would encourage us to develop more robust initial shape estimation.

Acknowledgment. This work was partly supported by JSPS KAKENHI Grant Number 17H00744. The authors would like to thank Dr. Zhaopeng Cui for sharing the data of Polarimetric MVS.

References

1. https://www.jai.com/products/go-5100mp-pge
2. https://thinklucid.com/product/phoenix-5-0-mp-polarized-model/
3. https://github.com/cdcseacave/openMVS
4. http://graphics.stanford.edu/data/3Dscanrep/
5. https://www.blender.org/
6. Aanæs, H., Jensen, R.R., Vogiatzis, G., Tola, E., Dahl, A.B.: Large-scale data for multiple-view stereopsis. Int. J. Comput. Vis. **120**(2), 153–168 (2016). https://doi.org/10.1007/s11263-016-0902-9
7. Agarwal, S., Mierle, K., Others: Ceres solver. http://ceres-solver.org
8. Agarwal, S., Snavely, N., Simon, I., Seitz, S.M., Szeliski, R.: Building Rome in a day. In: Proceedings of IEEE International Conference on Computer Vision (ICCV), pp. 72–79 (2009)
9. Atkinson, G.A.: Polarisation photometric stereo. Comput. Vis. Image Underst. **160**, 158–167 (2017)
10. Atkinson, G.A., Hancock, E.R.: Recovery of surface orientation from diffuse polarization. IEEE Trans. Image Process. **15**(6), 1653–1664 (2006)
11. Atkinson, G.A., Hancock, E.R.: Shape estimation using polarization and shading from two views. IEEE Trans. Pattern Anal. Mach. Intell. **29**(11), 2001–2017 (2007)
12. Baek, S.H., Jeon, D.S., Tong, X., Kim, M.H.: Simultaneous acquisition of polarimetric SVBRDF and normals. ACM Trans. Graph. **37**(6), 268 (2018)
13. Barron, J.T., Malik, J.: Shape, illumination, and reflectance from shading. IEEE Trans. Pattern Anal. Mach. Intell. **37**(8), 1670–1687 (2014)
14. Bay, H., Ess, A., Tuytelaars, T., Van Gool, L.: Speeded-up robust features (SURF). Comput. Vis. Image Underst. **110**(3), 346–359 (2008)
15. Biehler, J., Fane, B.: 3D Printing with Autodesk: Create and Print 3D Objects with 123D. AutoCAD and Inventor. Pearson Education, London (2014)
16. Cao, S., Snavely, N.: Graph-based discriminative learning for location recognition. In: Proceedings of IEEE Conference on Computer Vision and Pattern Recognition (CVPR), pp. 700–707 (2013)
17. Chen, L., Zheng, Y., Subpa-Asa, A., Sato, I.: Polarimetric three-view geometry. In: Proceedings of European Conference on Computer Vision (ECCV), pp. 20–36 (2018)
18. Cui, Z., Gu, J., Shi, B., Tan, P., Kautz, J.: Polarimetric multi-view stereo. In: Proceedings of IEEE Conference on Computer Vision and Pattern Recognition (CVPR), pp. 1558–1567 (2017)

19. Cui, Z., Larsson, V., Pollefeys, M.: Polarimetric relative pose estimation. In: Proceedings of IEEE International Conference on Computer Vision (ICCV), pp. 2671–2680 (2019)
20. Dai, A., Nießner, M.: 3DMV: Joint 3D-multi-view prediction for 3D semantic scene segmentation. In: Proceedings of European Conference on Computer Vision (ECCV), pp. 452–468 (2018)
21. Furukawa, Y., Curless, B., Seitz, S.M., Szeliski, R.: Towards internet-scale multi-view stereo. In: Proceedings of IEEE Conference on Computer Vision and Pattern Recognition (CVPR), pp. 1434–1441 (2010)
22. Furukawa, Y., Ponce, J.: Accurate, dense, and robust multiview stereopsis. IEEE Trans. Pattern Anal. Mach. Intell. **32**(8), 1362–1376 (2009)
23. Galliani, S., Lasinger, K., Schindler, K.: Massively parallel multiview stereopsis by surface normal diffusion. In: Proceedings of IEEE International Conference on Computer Vision (ICCV), pp. 873–881 (2015)
24. Ghosh, A., Fyffe, G., Tunwattanapong, B., Busch, J., Yu, X., Debevec, P.: Multiview face capture using polarized spherical gradient illumination. ACM Trans. Graph. **30**(6), 129 (2011)
25. Haefner, B., Ye, Z., Gao, M., Wu, T., Quéau, Y., Cremers, D.: Variational uncalibrated photometric stereo under general lighting. In: Proceedings of IEEE International Conference on Computer Vision (ICCV), pp. 8539–8548 (2019)
26. Huynh, C.P., Robles-Kelly, A., Hancock, E.R.: Shape and refractive index from single-view spectro-polarimetric images. Int. J. Comput. Vis. **101**(1), 64–94 (2013). https://doi.org/10.1007/s11263-012-0546-3
27. Ikehata, S., Wipf, D., Matsushita, Y., Aizawa, K.: Photometric stereo using sparse Bayesian regression for general diffuse surfaces. IEEE Trans. Pattern Anal. Mach. Intell. **36**(9), 1816–1831 (2014)
28. Jancosek, M., Pajdla, T.: Multi-view reconstruction preserving weakly-supported surfaces. In: Proceedings of IEEE Conference on Computer Vision and Pattern Recognition (CVPR), pp. 3121–3128 (2011)
29. Kadambi, A., Taamazyan, V., Shi, B., Raskar, R.: Polarized 3D: high-quality depth sensing with polarization cues. In: Proceedings of IEEE International Conference on Computer Vision (ICCV), pp. 3370–3378 (2015)
30. Kazhdan, M., Hoppe, H.: Screened Poisson surface reconstruction. ACM Trans. Graph. **32**(3), 1–13 (2013)
31. Kiku, D., Monno, Y., Tanaka, M., Okutomi, M.: Beyond color difference: residual interpolation for color image demosaicking. IEEE Trans. Image Process. **25**(3), 1288–1300 (2016)
32. Kim, K., Torii, A., Okutomi, M.: Multi-view inverse rendering under arbitrary illumination and albedo. In: Leibe, B., Matas, J., Sebe, N., Welling, M. (eds.) ECCV 2016. LNCS, vol. 9907, pp. 750–767. Springer, Cham (2016). https://doi.org/10.1007/978-3-319-46487-9_46
33. Kobbelt, L.: $\sqrt{3}$-subdivision. In: Proceedings of Annual Conference on Computer Graphics and Interactive Techniques (SIGGRAPH), pp. 103–112 (2000)
34. Ley, A., Hänsch, R., Hellwich, O.: SyB3R: a realistic synthetic benchmark for 3D reconstruction from images. In: Leibe, B., Matas, J., Sebe, N., Welling, M. (eds.) ECCV 2016. LNCS, vol. 9911, pp. 236–251. Springer, Cham (2016). https://doi.org/10.1007/978-3-319-46478-7_15
35. Li, M., Zhou, Z., Wu, Z., Shi, B., Diao, C., Tan, P.: Multi-view photometric stereo: a robust solution and benchmark dataset for spatially varying isotropic materials. IEEE Trans. Image Process. **29**, 4159–4173 (2020)

36. Lowe, D.G.: Distinctive image features from scale-invariant keypoints. Int. J. Comput. Vis. **60**(2), 91–110 (2004). https://doi.org/10.1023/B:VISI.0000029664.99615.94
37. Mahmoud, A.H., El-Melegy, M.T., Farag, A.A.: Direct method for shape recovery from polarization and shading. In: Proceedings of IEEE International Conference on Image Processing (ICIP), pp. 1769–1772 (2012)
38. Maruyama, Y., et al.: 3.2-MP back-illuminated polarization image sensor with four-directional air-gap wire grid and 2.5-μm pixels. IEEE Trans. Electron Devices **65**(6), 2544–2551 (2018)
39. Maurer, D., Ju, Y.C., Breuß, M., Bruhn, A.: Combining shape from shading and stereo: a variational approach for the joint estimation of depth, illumination and albedo. In: Proceedings of British Machine Vision Conference (BMVC), p. 76 (2016)
40. Mihoubi, S., Lapray, P.J., Bigué, L.: Survey of demosaicking methods for polarization filter array images. Sensors **18**(11), 3688 (2018)
41. Miyazaki, D., Furuhashi, R., Hiura, S.: Shape estimation of concave specular object from multiview polarization. J. Electron. Imaging **29**(4), 041006 (2020)
42. Miyazaki, D., Shigetomi, T., Baba, M., Furukawa, R., Hiura, S., Asada, N.: Surface normal estimation of black specular objects from multiview polarization images. Opt. Eng. **56**(4), 041303 (2016)
43. Miyazaki, D., Tan, R.T., Hara, K., Ikeuchi, K.: Polarization-based inverse rendering from a single view. In: Proceedings of IEEE International Conference on Computer Vision (ICCV), vol. 2, pp. 982–987 (2003)
44. Morel, O., Meriaudeau, F., Stolz, C., Gorria, P.: Polarization imaging applied to 3D reconstruction of specular metallic surfaces. In: Proceedings of SPIE-IS&T Electronic Imaging (EI), vol. 5679, pp. 178–186 (2005)
45. Morimatsu, M., Monno, Y., Tanaka, M., Okutomi, M.: Monochrome and color polarization demosaicking using edge-aware residual interpolation. In: Proceedings of IEEE International Conference on Image Processing (ICIP). To appear (2020)
46. Ngo Thanh, T., Nagahara, H., Taniguchi, R.I.: Shape and light directions from shading and polarization. In: Proceedings of IEEE Conference on Computer Vision and Pattern Recognition (CVPR), pp. 2310–2318 (2015)
47. Park, J., Sinha, S.N., Matsushita, Y., Tai, Y.W., Kweon, I.S.: Robust multiview photometric stereo using planar mesh parameterization. IEEE Trans. Pattern Anal. Mach. Intell. **39**(8), 1591–1604 (2017)
48. Rahmann, S., Canterakis, N.: Reconstruction of specular surfaces using polarization imaging. In: Proceedings of IEEE Conference on Computer Vision and Pattern Recognition (CVPR), vol. 1, pp. 149–155 (2001)
49. Ramamoorthi, R., Hanrahan, P.: An efficient representation for irradiance environment maps. In: Proceedings of Annual Conference on Computer Graphics and Interactive Techniques (SIGGRAPH), pp. 497–500 (2001)
50. Schonberger, J.L., Frahm, J.M.: Structure-from-motion revisited. In: Proceedings of IEEE Conference on Computer Vision and Pattern Recognition (CVPR), pp. 4104–4113 (2016)
51. Schönberger, J.L., Zheng, E., Frahm, J.-M., Pollefeys, M.: Pixelwise view selection for unstructured multi-view stereo. In: Leibe, B., Matas, J., Sebe, N., Welling, M. (eds.) ECCV 2016. LNCS, vol. 9907, pp. 501–518. Springer, Cham (2016). https://doi.org/10.1007/978-3-319-46487-9_31
52. Smith, W., Ramamoorthi, R., Tozza, S.: Height-from-polarisation with unknown lighting or albedo. IEEE Trans. Pattern Anal. Mach. Intell. **41**(12), 2875–2888 (2019)

53. Stokes, G.G.: On the composition and resolution of streams of polarized light from different sources. Trans. Cambridge Philos. Soc. **9**, 399 (1851)
54. Su, H., Maji, S., Kalogerakis, E., Learned-Miller, E.: Multi-view convolutional neural networks for 3D shape recognition. In: Proceedings of IEEE International Conference on Computer Vision (ICCV), pp. 945–953 (2015)
55. Tozza, S., Smith, W.A., Zhu, D., Ramamoorthi, R., Hancock, E.R.: Linear differential constraints for photo-polarimetric height estimation. In: Proceedings of IEEE International Conference on Computer Vision (ICCV), pp. 2279–2287 (2017)
56. Wu, C., Liu, Y., Dai, Q., Wilburn, B.: Fusing multiview and photometric stereo for 3D reconstruction under uncalibrated illumination. IEEE Trans. Vis. Comput. Graph. **17**(8), 1082–1095 (2010)
57. Wu, C., Wilburn, B., Matsushita, Y., Theobalt, C.: High-quality shape from multi-view stereo and shading under general illumination. In: Proceedings of IEEE Conference on Computer Vision and Pattern Recognition (CVPR), pp. 969–976 (2011)
58. Wu, L., Ganesh, A., Shi, B., Matsushita, Y., Wang, Y., Ma, Y.: Robust photometric stereo via low-rank matrix completion and recovery. In: Proceedings of Asian Conference on Computer Vision (ACCV), pp. 703–717 (2010)
59. Xiong, Y., Chakrabarti, A., Basri, R., Gortler, S.J., Jacobs, D.W., Zickler, T.: From shading to local shape. IEEE Trans. Pattern Anal. Mach. Intell. **37**(1), 67–79 (2014)
60. Yang, L., Tan, F., Li, A., Cui, Z., Furukawa, Y., Tan, P.: Polarimetric dense monocular SLAM. In: Proceedings of IEEE Conference on Computer Vision and Pattern Recognition (CVPR), pp. 3857–3866 (2018)
61. Zhang, R., Tsai, P.S., Cryer, J.E., Shah, M.: Shape-from-shading: a survey. IEEE Trans. Pattern Anal. Mach. Intell. **21**(8), 690–706 (1999)
62. Zhu, D., Smith, W.A.: Depth from a Polarisation + RGB stereo pair. In: Proceedings of IEEE Conference on Computer Vision and Pattern Recognition (CVPR), pp. 7586–7595 (2019)

SideInfNet: A Deep Neural Network for Semi-Automatic Semantic Segmentation with Side Information

Jing Yu Koh[1(✉)], Duc Thanh Nguyen[2], Quang-Trung Truong[1], Sai-Kit Yeung[3], and Alexander Binder[1]

[1] Singapore University of Technology and Design, Singapore, Singapore
jingyu_koh@alumni.sutd.edu.sg
[2] Deakin University, Geelong, Australia
duc.nguyen@deakin.edu.au
[3] Hong Kong University of Science and Technology, Sai Kung District, Hong Kong

Abstract. Fully-automatic execution is the ultimate goal for many Computer Vision applications. However, this objective is not always realistic in tasks associated with high failure costs, such as medical applications. For these tasks, semi-automatic methods allowing minimal effort from users to guide computer algorithms are often preferred due to desirable accuracy and performance. Inspired by the practicality and applicability of the semi-automatic approach, this paper proposes a novel deep neural network architecture, namely SideInfNet that effectively integrates features learnt from images with side information extracted from user annotations. To evaluate our method, we applied the proposed network to three semantic segmentation tasks and conducted extensive experiments on benchmark datasets. Experimental results and comparison with prior work have verified the superiority of our model, suggesting the generality and effectiveness of the model in semi-automatic semantic segmentation.

Keywords: Semi-automatic semantic segmentation · Side information

1 Introduction

Most studies in Computer Vision tackle fully-automatic inference tasks which, ideally, perform automatically without human intervention. To achieve this, machine learning models are often well trained on rich datasets. However, these models may still fail in reality when dealing with unseen samples. A possible solution for this challenge is using assistive information provided by users, e.g.., user-provided brush strokes and bounding boxes [16]. Human input is also critical for tasks with high costs of failure. Examples include medical applications where predictions generated by computer algorithms have to be verified by human experts

Jing Yu Koh: Currently an AI Resident at Google.

Electronic supplementary material The online version of this chapter (https://doi.org/10.1007/978-3-030-58586-0_7) contains supplementary material, which is available to authorized users.

A. Vedaldi et al. (Eds.): ECCV 2020, LNCS 12369, pp. 103–118, 2020.
https://doi.org/10.1007/978-3-030-58586-0_7

before they can be used in treatment plans. In such cases, a semi-automatic approach that allows incorporation of easy-and-fast side information provided from human annotations may prove more reliable and preferable.

Semantic segmentation is an important Computer Vision problem aiming to associate each pixel in an image with a semantic class label. Recent semantic segmentation methods have been built upon deep neural networks [4,5,8,11]. However, these methods are not flexible to be extended with additional information from various sources, such as human annotations or multi-modal data. In addition, human interactions are not allowed seamlessly and conveniently.

In this paper, we propose SideInfNet, a general model that is capable of integrating domain knowledge learnt from domain data (e.g., images) with side information from user annotations or other modalities in an end-to-end fashion. SideInfNet is built upon a combination of advanced deep learning techniques. In particular, the backbone of SideInfNet is constructed from state-of-the-art convolutional neural network (CNN) based semantic segmentation models. To effectively calibrate the dense domain-dependent information against the spatially sparse side information, fractionally strided convolutions are added to the model. To speed up the inference process and reduce the computational cost while maintaining the quality of segmentation, adaptive inference gates are proposed to make the network's topology flexible and optimal. To the best of our knowledge, this combination presents a novel architecture for semi-automatic segmentation.

A key challenge in designing such a model is in making it generalize to different sparsity and modalities of side information. Existing work focuses on sparse pixel-wise side information, such as user-defined keypoints [13,19], and geotagged photos [7,22]. However, these methods may not perform optimally when the side information is non-uniformly distributed and/or poorly provided, e.g., brush strokes which can be drawn dense and intertwined. In [22], street-level panorama information is used as a source of side information. However, such knowledge is not available in tasks other than remote sensing, e.g., in tasks where the side information is provided as brush strokes. Furthermore, expensive nearest neighbor search is used for the kernel regression in [22], which we replace by efficient trainable fractionally strided convolutions. The Higher-Order Markov Random Field model proposed in [7] can be adapted to various side information types but is not end-to-end trainable. Compared with these works, SideInfNet provides superior performance in various tasks and on different datasets. Importantly, our model provides a principled compromise between fully-automatic and manual segmentation. The benefit gained by the model is well shown in tasks where there exists a mismatch between training and test distribution. A few brush strokes can drastically improve the performance on these tasks. We show the versatility of our proposed model on three tasks:

- **Zone segmentation** of low-resolution satellite imagery [7]. Geotagged street-level photographs from social media are used as side information.
- **BreAst Cancer Histology (BACH) segmentation** [2]. Whole-slide images are augmented with expert-created brush strokes to segment the slides into *normal*, *benign*, *in situ carcinoma* and *invasive carcinoma* regions.

- **Urban segmentation** of very high-resolution (VHR) overhead crops taken of the city of Zurich [21]. Brush annotations indicate geographic features and are augmented with imagery features to identify eight different urban and peri-urban classes from the Zurich Summer dataset [21].

2 Related Work

2.1 Interactive Segmentation

GrabCut [16] is a seminal work of interactive segmentation that operates in an unsupervised manner. The method allows users to provide interactions in the form of brush strokes and bounding boxes demarcating objects. Several methods have extended the GrabCut framework for both semantic segmentation and instance segmentation, e.g., [9,23]. However, these methods only support bounding box annotations and thus cannot be used in datasets containing irregular object shapes, e.g., non-rectangular zones in the Zurich Summer dataset [21].

Users can also provide prior and reliable cues to guide the segmentation process on-the-fly [3,10,14,17]. For instance, Perazzi et al. [14] proposed a CNN-based guidance method for segmenting user-defined objects from video data. In this work, users provide object bounding boxes or regions. It is also shown that increasing the number of user annotations led to improved segmentation quality. In a similar manner, Nagaraja et al. [17] tackled the task of object segmentation from video by combining motion cues and user annotations. In their work, users make scribbles to delineate the objects of interests. Experimental results verified the cooperation of sparse user annotations and motion cues, filling the gap between fully automatic and manual object segmentation. However, in the above methods, user annotations play a role as auxiliary cues but are not effectively incorporated (as features) into the segmentation process.

2.2 Semantic Segmentation with Side Information

One form of side information used in several segmentation problems is keypoint annotations. The effectiveness of oracle keypoints in human segmentation is illustrated in [19]. Similarly, in [13], a method for automatically learning keypoints was proposed. The keypoints are grouped into pose instances and used for instance segmentation of human subjects. The spatial layout of keypoints is important to represent meaningful human structures, but such constraint are not always held for other object types, such as cell masses in histopathology.

Literature has also demonstrated the advantages of using ground-level imagery as side information in remote sensing. For instance, in [12], multi-view imagery data, including aerial and ground images, were fused into a Markov Random Field (MRF) model to enhance the quality of fine-grained road segmentation. In [7], domain-dependent features from satellite images were learnt using CNNs while street-level photos were classified and considered as higher-orders in a Higher-Order MRF model. These methods are flexible to various CNN architectures but are not trainable in an end-to-end fashion.

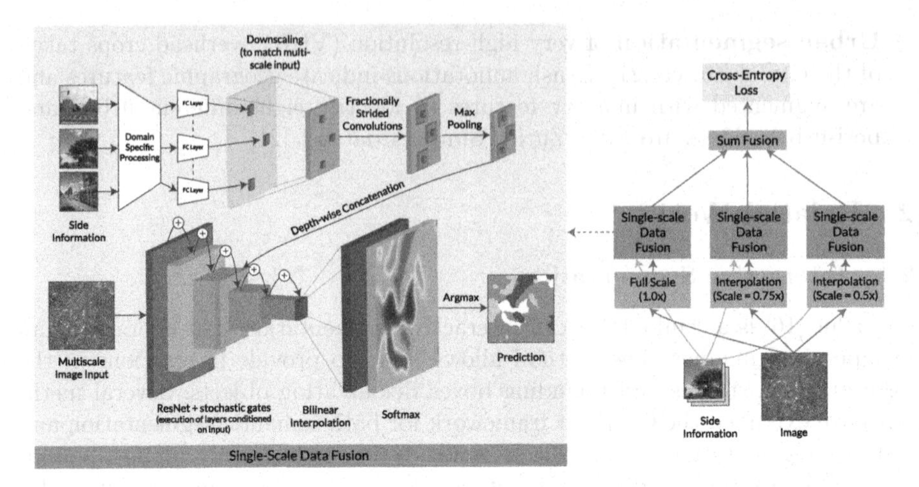

Fig. 1. Our proposed network architecture. A feature map of annotations is constructed based on the task. Our architecture for semantic segmentation is built on top of Deeplab-ResNet [5].

Workman et al. [22] proposed a model for fusing multi-view imagery data into a deep neural network for estimating geospatial functions land cover and land use. While this model is end-to-end, it has heavy computational requirements for its operation, e.g., for calculating and storing k nearest annotations, and thus may not be tractable for tasks with high density annotations. In addition, the model requires panorama knowledge to infer street-view photography.

3 SideInfNet

We propose SideInfNet, a novel neural network that fuses domain knowledge and user-provided side information in an end-to-end trainable architecture. SideInfNet allows the incorporation of multi-modal data, and is flexible with different annotation types and adaptive to various segmentation models. SideInfNet is built upon state-of-the-art semantic segmentation [5,11,15] and recent advances in adaptive neural networks [18,20]. This combination makes our model optimal while maintaining high quality segmentation results. For the sake of ease in presentation, we describe our method in the view of zone segmentation, a case study. However, our method is general and can be applied in different scenarios.

Zone segmentation aims to provide a zoning map for an aerial image, i.e., to identify the zone type for every pixel on the aerial image. Side information in this case includes street-level photos. These photos are captured by users and associated with geocodes that refer to their locations on the aerial image. Domain-dependent features are extracted from the input aerial image using some CNN-based semantic segmentation model (see Sect. 3.1). Side information features are then constructed from user-provided street-level photos (see Sect. 3.2 and Sect. 3.3). Associated geocodes in the street-level photos help to identify their locations in the receptive fields in the SideInfNet architecture where both domain-dependent and side information features are fused. To reduce the computational cost of the model while not sacrificing the quality of segmentation, adaptive inference gates are proposed to skip layers conditioned on input (see Sect. 3.4). Figure 1 illustrates the workflow of SideInfNet whose components are described in detail in the following subsections.

3.1 CNN-Based Semantic Segmentation

To extract domain-dependent features, we adopt the Deeplab-ResNet [5], a state-of-the-art CNN-based semantic segmentation. Deeplab-ResNet makes use of a series of dilated convolutional layers, with increasing rates to aggregate multi-scale features. To adapt Deeplab-ResNet into our framework, we retain the same architecture but extend the *conv2_3* layer with side information (see Sect. 3.2).

Specifically, the side information feature map is concatenated to the output of the *conv2_3* layer (see Fig. 1). As the original *conv2_3* layer outputs a feature map with 256 channels, concatenating the side information feature map results in a $\frac{H}{4} \times \frac{W}{4} \times (256 + d)$ dimensional feature map where H and W are the height and width of the input image, and d is the number of channels of the side information feature map. This extended feature map is the input to the next convolutional layer, *conv3_1*. We provide an ablation study on varying the dimension d in our supplementary material.

3.2 Side Information Feature Map Construction

Depending on applications, domain specific preprocessing may need to be applied to the side information. For instance, in the zone segmentation problem, we use the Places365-CNN in [24] to create vector representations for street-level photos (see details in Sect. 4.1). These vectors are then passed through a fully-connected layer returning d-dimensional vectors. Suppose that the input aerial image is of size $H \times W$. A side information feature map $\boldsymbol{x}^l$ of size $H \times W \times d$ can be created by initializing the d-dimensional vector at every location in $H \times W$ with the feature vector of the corresponding street-level photo, if one exists there. The feature vectors at locations that are not associated to any street-level photos are padded with zeros. Mapping image locations to street-level photos can be done using the associated geocodes of the street-level photos. Nearest neighbor interpolation is applied on the side information feature map to create multi-scale features. Features that fall in the same image locations (on the aerial image) due to downscaling are averaged. To make feature vectors consistent across scales and data samples, all feature vectors are normalized to the unit length.

There may exist misalignment in associating the side information features with their corresponding locations on the side information feature map. For instance, a brush stroke provided by a user may not well align with a true region. In the application of zoning, a street-level photo may not record the scene at the exact location where the photo is captured. Therefore, a direct reference of a street-level photo to a location on the feature map via the photo's geocode may not be a perfect association. However, one could expect that the side information could be propagated from nearby locations. To address this issue, we apply a series of fractionally-strided convolutions to the normalized feature map $\boldsymbol{x}^l$ to distribute the side information spatially. In our implementation, we use 3×3 kernels of ones, with stride length of 1 and padding of 1. After a single fractionally-strided convolution, side information features are distributed onto neighbouring 3×3 regions. We repeat this operation (denoted as f_c) n times and sum up all the feature maps to create the features for the next layer as follows,

$$\boldsymbol{x}^{l+1} = F(\boldsymbol{x}^l) = \sum_{i=1}^{n} w_i f_c^i(\boldsymbol{x}^l) \tag{1}$$

where w_i are learnable parameters and f_c^i is the i-th functional power of f_c, i.e.,

$$f_c^i(\boldsymbol{x}^l) = \begin{cases} f_c(\boldsymbol{x}^l), & i = 1 \\ f_c(f_c^{i-1}(\boldsymbol{x}^l)), & \text{otherwise} \end{cases} \tag{2}$$

The parameters w_i in (1) allow our model to learn the importance of spatial extent. We observe a decreasing pattern in w_i (i.e., $w_1 > w_2 > \cdots$) after training. This matches our intuition that information is likely to become less relevant with increased distances. The resulting feature map $\boldsymbol{x}^{l+1}$ represents a weighted sum of nearby feature vectors. We also normalize the feature vector at each location in the feature map by the number of the fractionally-strided convolutions used at that location. This has the effect of averaging overlapping features.

Lastly, we perform maxpooling to further downsample the side information feature map to fit with the counterpart domain-dependent feature map for feature fusion. We choose to perform feature fusion before the second convolutional block of Deeplab-ResNet, with the output of the *conv2_3* layer. We empirically found that this provided a good balance between computational complexity and segmentation quality. The output of the maxpooling layer is concatenated in the channels dimension to the output of the original layer (see Fig. 1). It is important to note that our proposed side information feature map construction method is general and can be applied alongside any CNN-based semantic segmentation architectures.

3.3 Fusion Weight Learning

As defined in (1), the output for each pixel (p, q) in the feature map $f_c^{i+1}(x^l)$ (after applying 3×3 fractionally-strided convolution of 1 s) can be described as:

$$f_c^{i+1}(x^l)_{p,q} = \sum_{j=1}^{3} \sum_{k=1}^{3} w_i x^l_{p-2+j,q-2+k}. \tag{3}$$

Gradient of the fusion weight w_i for each layer can be computed as,

$$\frac{\partial L}{\partial w_i} = \frac{\partial L}{\partial f_c^{i+1}(x^l)} \frac{\partial f_c^{i+1}(x^l)}{\partial w_i} = \sum_p \sum_q \frac{\partial L}{\partial f_c^{i+1}(x^l)_{p,q}} \sum_{j=1}^{3} \sum_{k=1}^{3} x^l_{p-2+n,q-2+n} \tag{4}$$

where $\frac{\partial L}{\partial f_c^{i+1}(x^l)}$ is back-propagated from the *conv2_3* layer.

For the fully-connected layers used for domain-specific processing (see Fig. 1), the layers are shared for each side-information instance. The shared weights w_{fc} can be learnt through standard back-propagation of a fully-connected layer:

$$\frac{\partial L}{\partial w_{\text{fc}}} = \frac{\partial L}{\partial f_c^1} \frac{\partial f_c^1}{\partial w_{\text{fc}}} \tag{5}$$

where $\frac{\partial L}{\partial f_c^1}$ is back-propagated from the first fusion layer (see (4)).

3.4 Adaptive Architecture

Inspired by advances in adaptive neural networks [18,20], we adopt adaptive inference graphs in SideInfNet. Adaptive inference graphs decide skip-connections in the network architecture using adaptive gates $\boldsymbol{z}^l$. Specifically, we define,

$$\boldsymbol{x}^{l+1} = \boldsymbol{x}^l + \boldsymbol{z}^l(h(\boldsymbol{x}^l)) \cdot F(\boldsymbol{x}^l) \tag{6}$$

where $\boldsymbol{z}^l(h(\boldsymbol{x}^l)) \in \{0, 1\}$ and h is some function that maps $\boldsymbol{x}^l \in H \times W \times d$ into a lower-dimensional space of $1 \times 1 \times d$. The gate $\boldsymbol{z}^l$ is conditioned on $\boldsymbol{x}^l$ and takes a binary decision (1 for "on" and 0 for "off").

Like [20], we set the early layers and the final classification layer of our model to always be executed, as these layers are critical for maintaining the accuracy. The gates are included in every other layer. We define the function h as,

$$h(\boldsymbol{x}^l) = \frac{1}{H \times W} \sum_{i=1}^{H} \sum_{j=1}^{W} \boldsymbol{x}_{i,j}^l \tag{7}$$

The feature map $h(\boldsymbol{x}^l)$ is passed into a multi-layer perceptron (MLP), which computes a relevance score to determine whether the layer l is executed. We also use a gate target rate t, that determines what fraction of layers should be activated. This is implemented as a mean squared error (MSE) loss and jointly optimized with the cross entropy loss. Each separate MLP determines whether its corresponding layer should be executed (contributing 1 to the total count), or not (contributing 0). Thus, the MSE loss encourages the overall learnt execution rate to be close to t. This is dynamic, i.e., more important layers would be executed more frequently and vice versa. For instance, a target rate $t = 0.8$ imposes a penalty on the loss function when the proportion of layers executed is greater or less than 80%. Our experimental results on this adaptive model are presented in Sect. 4, where we find that allowing a proportion of layers to be skipped helps improve segmentation quality.

4 Experiments and Results

In this section, we extensively evaluate our proposed SideInfNet in three different case studies. In each case study, we compare our method with its baseline and other existing works. We also evaluate our method under various levels of side information usage and with another CNN backbone.

4.1 Zone Segmentation

Experimental Setup: Like [7], we conducted experiments on three US cities: Boston (BOS), New York City (NYC), and San Francisco (SFO). Freely available satellite images hosted on Microsoft Bing Maps [6] were used. Ground-truth maps were retrieved at a service level of 12, which corresponds to a resolution of 38.2185 m per pixel. An example of the satellite imagery is shown in Fig. 2. We retrieved street-level photos from Mapillary [1], a service for sharing crowd-sourced geotagged photos. There were four zone types: *Residential*, *Commercial*, *Industrial* and *Others*. Table 1 summarizes the dataset used in this case study.

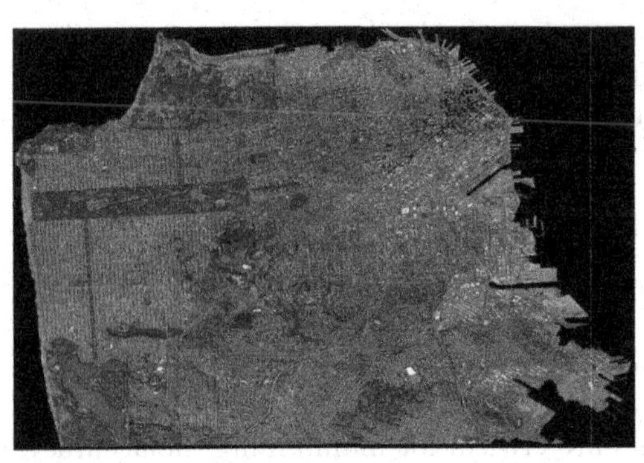

Fig. 2. Satellite image of San Francisco.

Table 1. Proportion of street-level photos (#photos).

Zone Type	City		
	BOS	NYC	SFO
Residential	25,607	16,395	50,116
Commercial	13,412	5,556	19,641
Industrial	2,876	9,327	15,219
Others	25,402	15,281	50,214

To extract side information features, we utilized the pre-trained model of Places365-CNN [24], which was designed for scene recognition. We fine-tuned the model on our data. During training the model, we froze the weights of the Places365-CNN and used this fine-tuned model to generate side information feature maps. We also applied a series of $n = 5$ fractionally-strided convolutions on feature maps generated from Places365-CNN. This acts as to distribute the side information from each geotagged photo 5 pixels in each cardinal direction.

Table 2. Segmentation performance on zoning. Best performances are highlighted.

Approach	Accuracy				mIOU			
	BOS	NYC	SFO	Mean	BOS	NYC	SFO	Mean
Deeplab-ResNet [5]	60.79%	59.58%	72.21%	64.19%	28.85%	23.77%	38.40%	30.34%
HO-MRF* [7]	59.52%	72.25%	73.93%	68.57%	31.92%	34.99%	46.53%	37.81%
Unified* [22]	67.91%	70.92%	75.92%	71.58%	40.51%	39.27%	55.36%	45.05%
SideInfNet	71.33%	71.08%	79.59%	74.00%	41.96%	39.59%	60.31%	47.29%

* Our implementation.

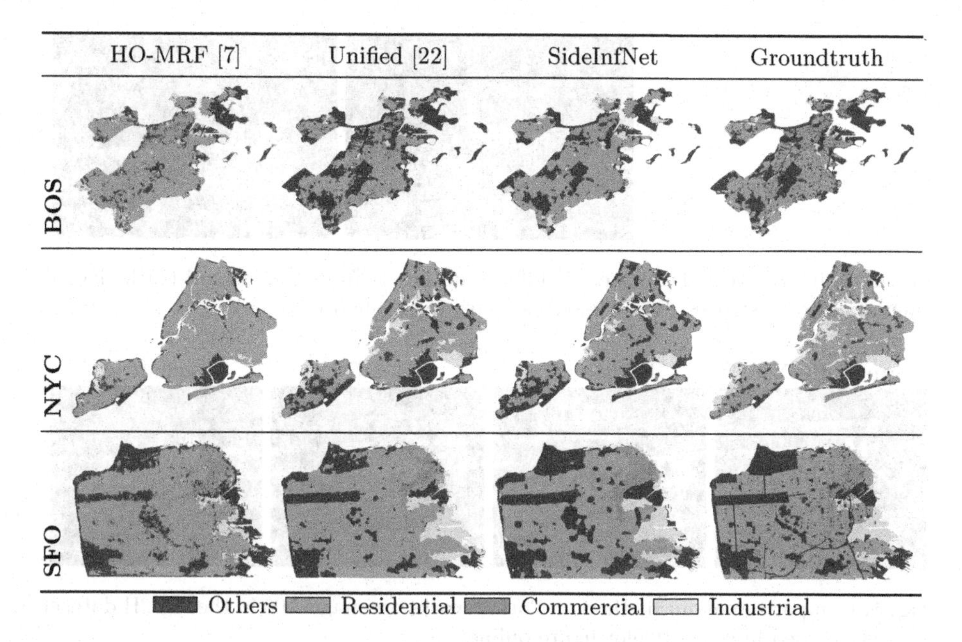

Fig. 3. Comparison of our method and previous works. Best viewed in color. (Color figure online)

Results: We evaluate our method and compare it with two recent works: Higher-Order Markov Random Field (HO-MRF) [7] and Unified model [22] using 3-fold cross validation, i.e., two cities are used for training and the other one is used for testing. To have a fair comparison, the same Places365-CNN model is used to extract side information in all methods. We also compare our method against the baseline Deeplab-ResNet, which directly performs semantic segmentation of satellite imagery without the use of geotagged photos.

Our results on both pixel accuracy and mean intersection over union (mIOU) are reported in Table 2. As shown in the table, our method significantly improves over its baseline, Deeplab-ResNet, proving the importance of side information. SideInfNet also outperforms prior work, with a relative improvement in pixel accuracy from the Unified model by 3.38% and from the HO-MRF by 7.92%. Improvement on mIOU scores is also significant, e.g., by 4.97% relative to the Unified model, and 25.07% relative to the HO-MRF model.

In addition to improved accuracy, our method offers several advantages over the previous works. First, compared with the HO-MRF [7], our method is trained end-to-end, allowing it to jointly learn optimal parameters for both semantic segmentation and side information feature extraction. Second, our method is efficient in computation. It simply performs a single forward pass through the network to produce segmentation results, opposed to iterative inference in the HO-MRF. Third, by using fractionally-strided convolutions, the complexity of our method is invariant to the side information density. This allows optimal

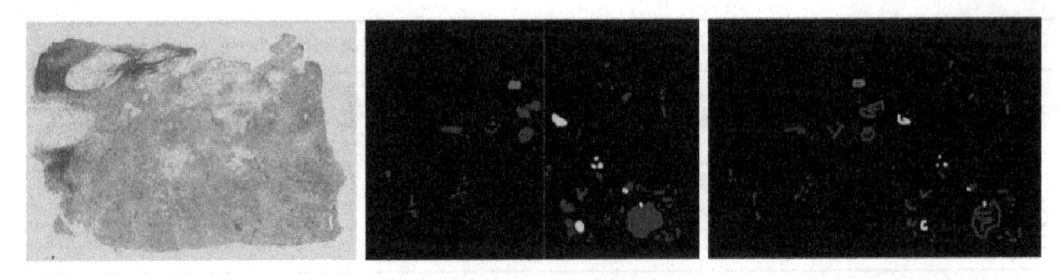

Fig. 4. Left: Whole-slide image. Middle: True labels from the ground-truth. Right: Simulated brush strokes. Best viewed in color. (Color figure online)

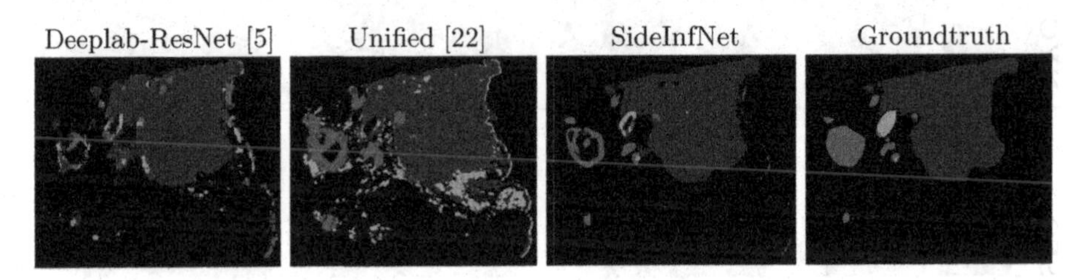

Fig. 5. Comparison of our method and other works on image A05 in the BACH dataset [2]. Best viewed in color. (Color figure online)

performance on regions with high density of side information. In contrast, the Unified model [22] requires exhaustive searches to determine nearest street-level photos for every pixel on satellite image and thus depend on the density of the street-level photos and the size of the satellite image.

We qualitatively show the segmentation results of our method and other works in Fig. 3. A clear drawback of the HO-MRF is that the results tend to be grainy, likely due to the sparsity of street-level imagery. In contrast, our method generally provides smoother results that form contiguous regions. Moreover, our method better captures fine grained details from street-level imagery.

4.2 BreAst Cancer Histology Segmentation

Experimental Setup: BACH (BreAst Cancer Histology) [2] is a dataset for breast cancer histology microscopy segmentation[1]. This dataset consists of high resolution whole-slide images that contain an entire sampled tissue. The whole-slide images were annotated by two medical experts, and images with disagreements were discarded. There are four classes: *normal*, *benign*, *in situ carcinoma* and *invasive carcinoma*. An example of a whole-slide image and its labels is shown in Fig. 4. As the *normal* class is considered background, it is not evaluated. Side information for BACH consists of expert brush stroke annotations,

[1] Data can be found at https://iciar2018-challenge.grand-challenge.org/. Due to the unavailability of the actual test set, we used slides A05 and A10 for testing, slide A02 for validation, and all other slides for training. This provides a fair class distribution, as not all slides contained all semantic classes.

indicating the potential presence of each class. In this case study, we use four different brush stroke colors to annotate the four classes.

BACH dataset does not include actual expert-annotated brush strokes. Therefore, to evaluate our method, we simulated expert annotations by using ground-truth labels in the dataset. Since the ground-truth was created by two experts, our brush strokes can be viewed as simulated rough expert input. To simulate situations where users have limited annotation time, we skipped annotating small regions that are likely to be omitted under time constraints. Figure 4 shows an example of our simulated brush strokes. In our experiments, we used slides A05 and A10 for testing, slide A02 for validation, and all other slides for training.

Table 3. Segmentation performance (mIOU) on BACH dataset. Best performances are highlighted.

Approach	A05	A10	Mean
Deeplab-ResNet [5]	34.08%	21.64%	27.86%
GrabCut [16]	30.20%	25.21%	27.70%
Unified* [22]	41.50%	17.23%	29.37%
SideInfNet	59.03%	35.45%	47.24%

* Our implementation.

Table 4. Segmentation performance on Zurich Summer dataset. Best performances are highlighted.

Approach	Accuracy	mIOU
Deeplab-ResNet [5]	73.20%	42.95%
GrabCut [16]	60.53%	26.89%
Unified* [22]	68.20%	42.09%
SideInfNet	78.97%	58.31%

* Our implementation.

Results: We evaluate three different methods: our proposed SideInfNet, Unified model [22], and GrabCut [16]. We were unable to run the HO-MRF model [7] on the BACH dataset due to the large size of the whole-slide images (note that the HO-MRF makes use of fully-connected MRF and thus is not computationally feasible under this context). In addition, since GrabCut is a binary segmentation method, to adapt this work to our case study, we ran the GrabCut model independently for each class. We report the performance of all the methods in Table 3. We also provide some qualitative results in Fig. 5.

Experimental results show that our method greatly outperforms previous works on BACH dataset. Furthermore, the Unified model [22] even performs worse than the baseline Deeplab-ResNet that used only whole-slide imagery. This suggests the limitation of the Unified model [22] in learning from dense annotations. Table 3 also confirms the role played by the side information (i.e., the Deeplab-ResNet vs SideInfNet). This aligns with our intuition, as we would expect that brush strokes provide stronger cues to guide the segmentation.

4.3 Urban Segmentation

Experimental Setup: The Zurich Summer v1.0 dataset [21] includes 20 very high resolution (VHR) overview crops taken from the city of Zurich, pansharpened to a PAN resolution of about 0.62 cm ground sampling distance (GSD). This is a much higher resolution compared to the low-resolution satellite imagery used

Fig. 6. Example satellite image, brush annotations, and ground-truth map from the Zurich Summer dataset [21]. Best viewed in color. (Color figure online)

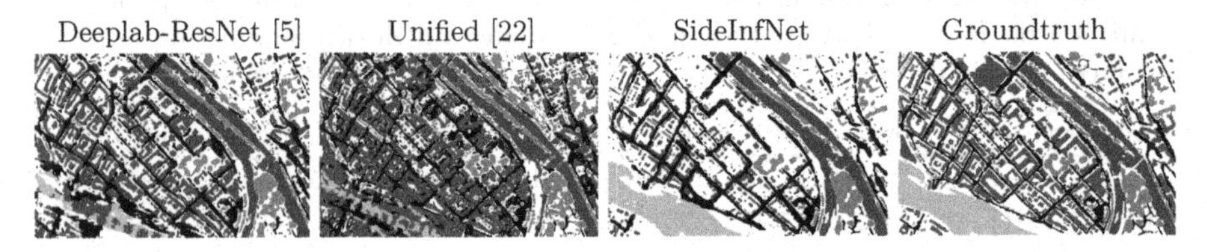

Fig. 7. Qualitative comparison of our method and other works on the Zurich Summer dataset [21]. Best viewed in color. (Color figure online)

in the zoning dataset. The Zurich Summer dataset contains eight different urban and periurban classes: Roads, Buildings, Trees, Grass, Bare Soil, Water, Railways and Swimming pools. Examples of satellite imagery, ground-truth labels, and brush annotations are shown in Fig. 6. Preprocessing steps and feature map construction are performed similarly to that of BACH. We also used rough brush strokes demarcating potential urban classes as side information.

Results: Our experimental results on the Zurich Summer dataset are summarized in Table 4. In general, similar trends with the BACH dataset are found, and our proposed method outperforms all prior works. Specifically, by using brush strokes, we are able to gain a relative improvement of 7.88% on accuracy and 35.76% on mIOU over the baseline Deeplab-ResNet. The Zurich dataset contains high-resolution satellite imagery, which suggests the usefulness of including brush annotations even with high fidelity image data. SideInfNet also outperforms the Unified model [22] with a relative improvement of 15.79% on accuracy and 38.53% on mIOU. This result proves the robustness of our method in dealing with dense annotations, which challenge the Unified model. GrabCut also under-performs due to its limitations as an unsupervised binary segmentation method. A qualitative comparison of our method with other works is also shown in Fig. 7.

Table 5. Performance of SideInfNet with varying side information.

Side information used	mIOU			Mean accuracy		
	Zoning [7]	BACH [2]	Zurich [21]	Zoning [7]	BACH [2]	Zurich [21]
100%	47.29%	47.24%	58.31%	74.00%	71.99%	78.97%
80%	40.27%	40.53%	52.32%	72.46%	68.60%	77.58%
60%	39.56%	34.16%	52.14%	72.39%	68.56%	76.33%
40%	37.70%	29.56%	49.49%	71.01%	64.87%	75.83%
20%	34.04%	26.15%	47.72%	68.11%	56.86%	74.29%
0%	28.11%	23.86%	45.98%	58.63%	60.48%	73.36%

4.4 Varying Levels of Side Information

In this experiment, we investigate the performance of our method when varying the availability of side information. To simulate various densities of brush strokes for an input image, we sample the original brush strokes (e.g., from 0% to 100% of the total number) and evaluate the segmentation performance of our method accordingly. The brush strokes could be randomly sampled. However, this approach may bias the spatial distribution of the brush strokes. To maintain the spatial distribution of the brush strokes for every sampling case, we perform k-means clustering on the original set of the brush strokes. For instance, if we wish to utilize a percentage p of the total brush strokes, and n brush strokes are present in total, we apply k-means algorithm with $k = \text{ceil}(np)$ on the centers of the brush strokes to spatially cluster the brush strokes into k groups. For each group, we select the brush stroke whose center is closest to the group's centroid. This step results in k brush strokes. We note that a similar procedure can be applied to sample street-level photos for zone segmentation.

We report the quantitative results of our method w.r.t varying side information in Table 5. In general, we observe a decreasing trend over the accuracy and mIOU as the proportion of side information decreases. This supports our hypothesis that side information is a key signal for improving segmentation accuracy. We also observe a trade off between human effort and segmentation accuracy. For instance, on the zone segmentation dataset [7], improvement over the baseline Deeplab-ResNet is achieved with as little as 20% of the original number of geotagged photos. This suggests that our proposed method can provide significant performance gains even with minimal human effort.

4.5 SideInfNet with Another CNN Backbone

To show the adaptability of SideInfNet, we experimented SideInfNet built with another CNN backbone. In particular, we adopted the VGG-19 as the backbone in our architecture. Note that VGG was also used in the Unified model [22]. To provide a fair comparison, we re-implemented both SideInfNet and Unified model with the same VGG architecture and evaluated both models using the

Table 6. Performance (mIOU) of SideInfNet with VGG.

Model	Zoning [7]	BACH [2]	Zurich [21]
SideInfNet-VGG	46.12%	49.53%	49.73%
Unified [22]	45.05%	29.37%	42.09%

same training/test split. We also utilized the original hyperparameters proposed in [22] in our implementation. We report the results of this experiment in Table 6.

Experimental results show that SideInfNet outperforms the Unified model [22] on all segmentation tasks when the same VGG backbone is used. These results confirm again the advantages of our method in feature construction and fusion.

5 Conclusion

This paper proposes SideInfNet, a novel end-to-end neural network for semi-automatic semantic segmentation with additional side information. Through extensive experiments on various datasets and modalities, we have shown the advantages of our method across a wide range of applications, including but not limited to remote sensing and medical image segmentation. In addition to being general, our method boasts improved accuracy and computational advantages over prior models. Lastly, our architecture is easily adapted to various semantic segmentation models and side information feature extractors.

The method proposed in this paper acts as a compromise between fully-automatic and manual segmentation. This is essential for many applications with high cost of failure, in which fully-automatic methods may not be widely accepted as of yet. Our model works well with dense brush stroke information, providing a quick and intuitive way for human experts to refine the model's outputs. In addition, our model also outperforms prior work on sparse pixel-wise annotations. By including side information to shape predictions, we are able to achieve an effective ensemble of human expertise and machine efficiency, producing both fast and accurate segmentation results.

Acknowledgement. – Duc Thanh Nguyen was partially supported by an internal SEBE 2019 RGS grant from Deakin University.

– Sai-Kit Yeung was partially supported by an internal grant from HKUST (R9429) and HKUST-WeBank Joint Lab.

– Alexander Binder was supported by the MoE Tier2 Grant MOE2016-T2-2-154, Tier1 grant TDMD 2016-2, SUTD grant SGPAIRS1811, TL grant RTDST1907012.

References

1. Mapillary AB. https://www.mapillary.com (2019). Accessed 01 Nov 2019
2. Aresta, G. et al.: Bach: Grand challenge on breast cancer histology images. Med. Image Anal. **56**, 122–139 (2019)
3. Caelles, S., Maninis, K.K., Pont-Tuset, J., Leal-Taixé, L., Cremers, D., Van Gool, L.: One-shot video object segmentation. In: Proceedings of the IEEE Conference on Computer Vision and Pattern Recognition, pp. 221–230 (2017)
4. Chen, L.C., Papandreou, G., Kokkinos, I., Murphy, K., Yuille, A.L.: Semantic image segmentation with deep convolutional nets and fully connected CRFs. arXiv preprint arXiv:1412.7062 (2014)
5. Chen, L.C., Papandreou, G., Kokkinos, I., Murphy, K., Yuille, A.L.: Deeplab: Semantic image segmentation with deep convolutional nets, atrous convolution, and fully connected CRFs. IEEE Trans. Pattern Anal. Mach. Intell. **40**(4), 834–848 (2018)
6. Corporation, M.: Bing maps tile system. https://msdn.microsoft.com/en-us/library/bb259689.aspx (2019). Accessed 01 Nov 2019
7. Feng, T., Truong, Q.T., Thanh Nguyen, D., Yu Koh, J., Yu, L.F., Binder, A., Yeung, S.K.: Urban zoning using higher-order Markov random fields on multi-view imagery data. In: Proceedings of the European Conference on Computer Vision (ECCV), pp. 614–630 (2018)
8. Girshick, R., Donahue, J., Darrell, T., Malik, J.: Rich feature hierarchies for accurate object detection and semantic segmentation. In: Proceedings of the IEEE Conference on Computer Vision and Pattern Recognition, pp. 580–587 (2014)
9. Göring, C., Fröhlich, B., Denzler, J.: Semantic segmentation using GrabCut. In: VISAPP, pp. 597–602 (2012)
10. Li, S., Seybold, B., Vorobyov, A., Fathi, A., Huang, Q., Jay Kuo, C.C.: Instance embedding transfer to unsupervised video object segmentation. In: Proceedings of the IEEE Conference on Computer Vision and Pattern Recognition, pp. 6526–6535 (2018)
11. Long, J., Shelhamer, E., Darrell, T.: Fully convolutional networks for semantic segmentation. In: Proceedings of the IEEE Conference on Computer Vision and Pattern Recognition, pp. 3431–3440 (2015)
12. Máttyus, G., Wang, S., Fidler, S., Urtasun, R.: HD maps: Fine-grained road segmentation by parsing ground and aerial images. In: Proceedings of the IEEE Conference on Computer Vision and Pattern Recognition, pp. 3611–3619 (2016)
13. Papandreou, G., Zhu, T., Chen, L.C., Gidaris, S., Tompson, J., Murphy, K.: Personlab: Person pose estimation and instance segmentation with a bottom-up, part-based, geometric embedding model. In: Proceedings of the European Conference on Computer Vision (ECCV), pp. 269–286 (2018)
14. Perazzi, F., Khoreva, A., Benenson, R., Schiele, B., Sorkine-Hornung, A.: Learning video object segmentation from static images. In: Proceedings of the IEEE Conference on Computer Vision and Pattern Recognition, pp. 2663–2672 (2017)
15. Ren, S., He, K., Girshick, R., Sun, J.: Faster R-CNN: Towards real-time object detection with region proposal networks. In: Advances in Neural Information Processing Systems, pp. 91–99 (2015)
16. Rother, C., Kolmogorov, V., Blake, A.: Grabcut: Interactive foreground extraction using iterated graph cuts. ACM Trans. Graphics (TOG) **23**, 309–314 (2004). ACM
17. Shankar Nagaraja, N., Schmidt, F.R., Brox, T.: Video segmentation with just a few strokes. In: Proceedings of the IEEE International Conference on Computer Vision, pp. 3235–3243 (2015)

18. Shazeer, N., Mirhoseini, A., Maziarz, K., Davis, A., Le, Q., Hinton, G., Dean, J.: Outrageously large neural networks: The sparsely-gated mixture-of-experts layer. arXiv preprint arXiv:1701.06538 (2017)
19. Tripathi, S., Collins, M., Brown, M., Belongie, S.: Pose2instance: Harnessing keypoints for person instance segmentation. arXiv preprint arXiv:1704.01152 (2017)
20. Veit, A., Belongie, S.: Convolutional networks with adaptive inference graphs. In: Proceedings of the European Conference on Computer Vision (ECCV), pp. 3–18 (2018)
21. Volpi, M., Ferrari, V.: Semantic segmentation of urban scenes by learning local class interactions. In: Proceedings of the IEEE Conference on Computer Vision and Pattern Recognition Workshops, pp. 1–9 (2015)
22. Workman, S., Zhai, M., Crandall, D.J., Jacobs, N.: A unified model for near and remote sensing. In: Proceedings of the IEEE International Conference on Computer Vision, pp. 2688–2697 (2017)
23. Xu, N., Price, B., Cohen, S., Yang, J., Huang, T.: Deep grabcut for object selection. arXiv preprint arXiv:1707.00243 (2017)
24. Zhou, B., Lapedriza, A., Xiao, J., Torralba, A., Oliva, A.: Learning deep features for scene recognition using places database. In: Advances in Neural Information Processing Systems, pp. 487–495 (2014)

Improving Face Recognition by Clustering Unlabeled Faces in the Wild

Aruni RoyChowdhury[1(✉)], Xiang Yu[2], Kihyuk Sohn[2], Erik Learned-Miller[1], and Manmohan Chandraker[2,3]

[1] University of Massachusetts, Amherst, USA
arunirc@cs.umass.edu
[2] NEC Labs America, San Jose, USA
xiangyu@nec-labs.com
[3] University of California, San Diego, USA

Abstract. While deep face recognition has benefited significantly from large-scale labeled data, current research is focused on leveraging unlabeled data to further boost performance, reducing the cost of human annotation. Prior work has mostly been in controlled settings, where the labeled and unlabeled data sets have no overlapping identities by construction. This is not realistic in large-scale face recognition, where one must contend with such overlaps, the frequency of which increases with the volume of data. Ignoring identity overlap leads to significant labeling noise, as data from the same identity is split into multiple clusters. To address this, we propose a novel identity separation method based on extreme value theory. It is formulated as an out-of-distribution detection algorithm, and greatly reduces the problems caused by overlapping-identity label noise. Considering cluster assignments as pseudo-labels, we must also overcome the labeling noise from clustering errors. We propose a modulation of the cosine loss, where the modulation weights correspond to an estimate of clustering uncertainty. Extensive experiments on both controlled and real settings demonstrate our method's consistent improvements over supervised baselines, e.g., 11.6% improvement on IJB-A verification.

1 Introduction

Deep face recognition has achieved impressive performance, benefiting from large-scale labeled data. Examples include DeepFace [38], which uses 4M labeled faces for training and FaceNet [32], which is trained on 200M labeled faces. Further improvements in recognition performance using traditional supervised learning

A. RoyChowdhury—now at Amazon, work done prior to joining, while interning at NEC Labs America.
K. Sohn—now at Google, work done prior to joining.

Electronic supplementary material The online version of this chapter (https://doi.org/10.1007/978-3-030-58586-0_8) contains supplementary material, which is available to authorized users.

A. Vedaldi et al. (Eds.): ECCV 2020, LNCS 12369, pp. 119–136, 2020.
https://doi.org/10.1007/978-3-030-58586-0_8

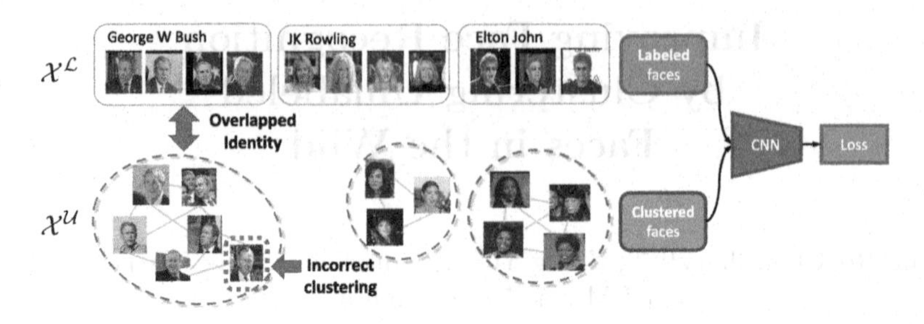

Fig. 1. Given a face recognition model trained on labeled faces ($\mathcal{X}^{\mathcal{L}}$), we wish to cluster unlabeled data ($\mathcal{X}^{\mathcal{U}}$) for additional training samples to further improve recognition performance. Key challenges include ***overlapping identities*** between labeled and unlabeled data (*George W Bush* images present in both $\mathcal{X}^{\mathcal{L}}$ and $\mathcal{X}^{\mathcal{U}}$) as well as noisy training labels arising from ***incorrect cluster assignments*** (a picture of *George Bush Sr.* is erroneously assigned to a cluster of *George W Bush* images).

may require tremendous annotation efforts to increase the labeled dataset volume, which is impractical, labor intensive and does not scale well. Therefore, exploiting unlabeled data to augment the labeled data, i.e., *semi-supervised learning*, is an attractive alternative. Preliminary work on generating pseudo-labels by clustering unlabeled faces has been shown to be effective in improving performance under controlled settings [37,42,44].

However, although learning from unlabeled data is a mature area and theoretically attractive, face recognition as a field has yet to adopt such methods in practical and realistic settings. There are several obstacles to directly applying such techniques that are peculiar to the setting of large-scale face recognition. First, there is a common assumption of semi-supervised face recognition methods is that there is *no class or identity overlap* between the unlabeled and the labeled data. This seemingly mild assumption, however, violates the basic premise of semi-supervised learning – that nothing is known about the labels of the unlabeled set. Thus, either practitioners must manually verify this property, meaning that the data is no longer *"truly" unlabeled*, or proceed under the assumption that identities are disjoint between labeled and unlabeled training datasets, which inevitably introduces labeling noise. When such overlapping identities are in fact present (Fig. 1), a significant price is paid in terms of performance, as demonstrated empirically in this work. A further practical concern in face recognition is the availability of massive labeled face datasets. Most current work using unlabeled faces focus on improving the performance of models trained with *limited labeled data* [42,44], and it is unclear if there are any benefits from using unlabeled datasets when baseline face recognition models are trained on *large-scale labeled data.*

In this paper, we present recipes for exploiting unlabeled data to further improve the performance of fully supervised state-of-the-art face recognition models, which are mostly trained on large-scale labeled datasets. We demonstrate that learning from unlabeled faces is indeed a practical avenue for improving deep face recognition, also addressing important practical challenges in the process – accounting for overlapping identities between labeled and unlabeled data, and attenuating the effect of noisy labels when training on pseudo-labeled data.

We begin with Face-GCN [42], a graph convolutional neural network (GCN) based face clustering method, to obtain pseudo-labels on unlabeled faces. To deal with the overlapping identity problem, we observe that the distribution of classification confidence on overlapping and disjoint identities is different – since our initial face feature is provided by a recognition engine trained on known identities, the confidence score of the overlapping identity images should be higher than those of non-overlapping identity images, as visualized in Fig. 3. Based on this observation, we approach the problem as *out-of-distribution detection* [9,18,20], and propose to parameterize the distribution of confidence scores as a mixture of Weibulls, motivated by extreme value theory. This results in an unsupervised procedure to separate overlapping identity samples from unlabeled data on-the-fly.

After resolving the overlapping identity caused label noise, the *systematic label noise* from the clustering algorithm remains, which is another prime cause for deteriorating performance in face recognition [40]. Instead of an additional complicated pruning step to discard noisy samples, *e.g.* as done in [42], we deal with the label noise during the re-training loop using the joint data of both labeled and clustered faces, by introducing a simple clustering uncertainty based attenuation on the training loss to reduce the effect of erroneous gradients caused by the noisy labeled data. This effectively smoothes the re-training procedure and has shown clear performance gains in our experiments. Our contributions are summarized as the following:

- To our knowledge, we are the first to tackle the practical issue of overlapping identities between labeled and unlabeled face data during clustering, formulated as an out-of-distribution detection.
- We successfully demonstrate that jointly leveraging large scale unlabeled data along with labeled data in a semi-supervised fashion can indeed significantly improve over supervised face recognition performance, i.e., substantial gains over a supervised CosFace [41] model across multiple public benchmarks.
- We introduce a simple and scalable uncertainty-modulated training loss into the semi-supervised learning setup, which is designed to compensate for the label noise introduced by the clustering procedure on unlabeled data.
- We provide extensive and ablative insights on both controlled and real-world settings, serving as a recipe for the semi-supervised face recognition or other large scale recognition problems.

2 Related Work

Face Clustering: Jain [12] provides a survey on classic clustering techniques. Most recent approaches [14,21,22,26,36] work on face features extracted from supervisedly-trained recognition engines. "Consensus-driven propagation" (CDP) [44] assigns pseudo-labels to unlabeled faces by forming a graph over the unlabeled samples. An ensemble of various network architectures provides multiple views of the unlabeled data, and an aggregation module decides on positive and negative pairs. Face-GCN [42] formulates the face clustering problem into a regression for cluster proposal purity, which can be fully supervised.

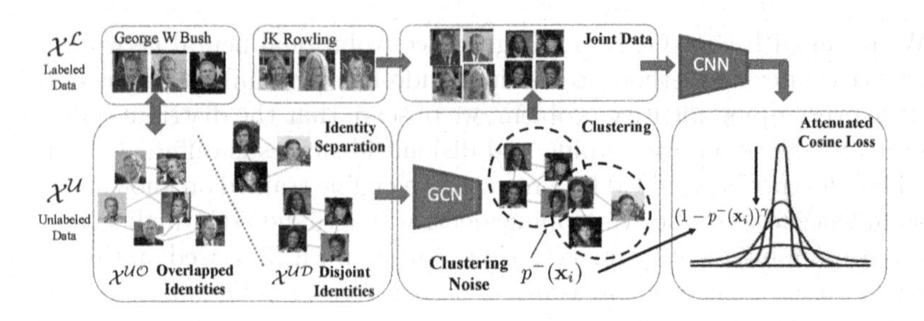

Fig. 2. Our approach trains a deep neural network [41] jointly on labeled faces $\mathcal{X}^{\mathcal{L}}$ and unlabeled faces $\mathcal{X}^{\mathcal{U}}$. Unlabeled samples with ***overlapping*** and ***disjoint*** identities w.r.t. $\mathcal{X}^{\mathcal{L}}$ are separated into $\mathcal{X}^{\mathcal{UO}}$ and $\mathcal{X}^{\mathcal{UD}}$, respectively (Sect. 3.1). The unlabeled faces in $\mathcal{X}^{\mathcal{UD}}$ are clustered using a ***graph conv-net*** or ***GCN*** (Sect. 3.2). Estimates of ***cluster uncertainty*** $p^-(\mathbf{x}_i)$ are used to modulate the cosine loss during re-training (Sect. 3.3).

Re-training the recognition engine with the clustered "pseudo-identities" and the original data improves performance, however, CDP [44] and Face-GCN re-training assumes the "pseudo-identities" and the original identities have no overlap, which does not always hold true. Meanwhile, their investigation stays in a controlled within-distribution setting using the MS-Celeb-1M dataset [7], which is far from realistic. In contrast, we demonstrate that these considerations are crucial to achieve gains for practical face recognition with truly large-scale labeled and unlabeled datasets.

Out-of-Distribution Detection: Extreme value distributions have been used in calibrating classification scores [29], classifier meta-analysis [31], open set recognition robust to adversarial images [2] and as normalization for score fusion from multiple biometric models [30], which is quite different from our problem. Recent approaches to out-of-distribution detection utilize the confidence of the predicted posteriors [9,20], while Lee *et al.* [18] use Mahalanobis distance-based classification along with gradient-based input perturbations. [18] outperforms the others, but does not scale to our setting – estimating per-subject covariance matrices is not feasible for the typical long-tailed class distribution in face recognition datasets.

Learning with Label Noise: Label-noise [24] has a significant effect on the performance of the face embeddings obtained from face recognition models trained on large datasets, as extensively studied in Wang *et al.* [40]. Indeed, even large scale human-annotated face datasets such as the well-known MS 1 Million (MS-1M) are shown to have some incorrect labeling, and gains in recognition performance can be attained by cleaning up the labeling [40]. Applying label-noise modeling to our problem of large-scale face recognition has its challenges – the labeled and unlabeled datasets are class-disjoint, a situation not considered by earlier methods [10,19,27]; having ~100k identities, typically long-tailed, make learning a label-transition matrix challenging [10,27]; label-noise from clustering pseudo-labels is typically structured and quickly memorized by a deep network, unlike the uniform-noise experiments in [1,39,45]. Our unsupervised label-noise estimation does not require a clean labeled dataset to learn a training curriculum unlike [13,28], and can thus be applied out-of-the-box.

3 Learning from Unlabeled Faces

Formally, let us consider samples $\mathcal{X} = \{\mathbf{x}_i\}_{i\in[n]}$, divided into two parts: $\mathcal{X}^{\mathcal{L}}$ and $\mathcal{X}^{\mathcal{U}}$ of sizes l and u respectively. Now $\mathcal{X}^{\mathcal{L}} := \{\mathbf{x}_1, \ldots, \mathbf{x}_l\}$ consist of faces that are provided with identity labels $\mathcal{Y}^{\mathcal{L}} := \{y_1, \ldots, y_l\}$, while we do not know the identities of the unlabeled faces $\mathcal{X}^{\mathcal{U}} := \{\mathbf{x}_{l+1}, \ldots, \mathbf{x}_{l+u}\}$. Our approach aims to improve the performance of a supervised face recognition model, trained on $(\mathcal{X}^{\mathcal{L}}, \mathcal{Y}^{\mathcal{L}})$, by first clustering the unlabeled faces $\mathcal{X}^{\mathcal{U}}$, then re-training on both labeled and unlabeled faces, using the cluster assignments on $\mathcal{X}^{\mathcal{U}}$ as pseudo-labels. Figure 2 visually summarizes the steps – **(1)** train a supervised face recognition model on $(\mathcal{X}^{\mathcal{L}}, \mathcal{Y}^{\mathcal{L}})$; **(2)** separate the samples in $\mathcal{X}^{\mathcal{U}}$ having overlapping identities with the labeled training set; **(3)** cluster the disjoint-identity unlabeled faces; **(4)** learn an unsupervised model for the likelihood of incorrect cluster assignments on the pseudo-labeled data; **(5)** re-train the face recognition model on labeled and pseudo-labeled faces, attenuating the training loss for pseudo-labeled samples using the estimated clustering uncertainty. In this section, we first describe the separation of overlapping identity samples from unlabeled data (Sect. 3.1), followed by an overview of the face clustering procedure (Sect. 3.2) and finally re-training the recognition model with an estimate of clustering uncertainty (Sect. 3.3).

3.1 Separating Overlapping Identities

Overlapping Identities. We typically have no control over the gathering of the unlabeled data $\mathcal{X}^{\mathcal{U}}$, so the same subject S may exist in labeled data (thus, be a class on which the baseline face recognition engine is trained) and also within our unlabeled dataset, *i.e.* $\mathcal{X}^{\mathcal{U}} = \mathcal{X}^{\mathcal{UO}} \cup \mathcal{X}^{\mathcal{UD}}$, where $\mathcal{X}^{\mathcal{UO}}$ and $\mathcal{X}^{\mathcal{UD}}$ denote the identity overlapped and identity disjoint subsets of $\mathcal{X}^{\mathcal{U}}$.

By default, the clustering will assign images of subject S in the unlabeled data as a new category. In this case, upon re-training with the additional pseudo-labeled data, the network will incorrectly learn to classify images of subject S into *two* categories. This is an important issue, since overlapping subjects can occur naturally in datasets collected from the Internet or recorded through passively mounted cameras, which to our knowledge has not been directly addressed by most recent pseudo-labeling methods [37,42,44].

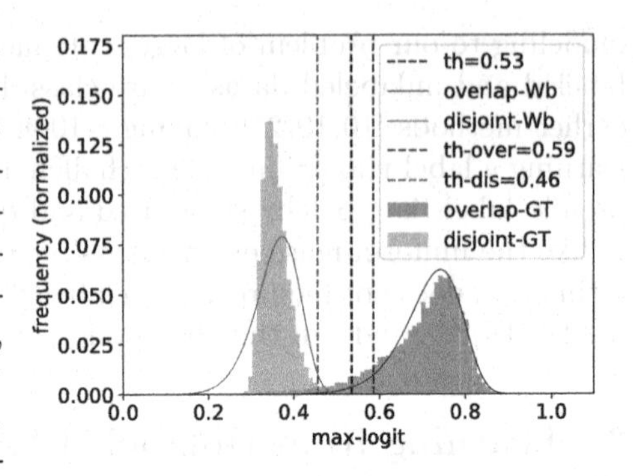

Fig. 3. Extreme value theory (EVT) provides a principled way of setting thresholds on the max-logits z_i for $\mathbf{x}_i \in \mathcal{X}^{\mathcal{U}}$ to separate *disjoint* and *overlapping* identities (red and blue vertical lines). An initial threshold is determined by Otsu's method (black vertical line). This plot uses splits from the MS-Celeb-1M dataset [7]. (Color figure online)

Out-of-Distribution Detection. The problem of separating unlabeled data into samples of disjoint and overlapping classes (w.r.t. the classes in the labeled data) can be regarded as an "out-of-distribution" detection problem. The intuition is that unlabeled samples with overlapping identities will have higher confidence scores from a face recognition engine, as the same labeled data is used to train the recognition model [9]. Therefore, we search for thresholds on the recognition confidence scores that can separate disjoint and overlapping identity samples. Note, since the softmax operation over several thousand categories can result in small values due to normalization, we use the *maximum logit* z_i for each sample $\mathbf{x}_i \in \mathcal{X}^{\mathcal{U}}$ as its confidence score. Since the z_i are the maxima over a large number of classes, we can draw upon results from extreme value theory (EVT) which state that the limiting distribution of the maxima of i.i.d random variables belongs to either the Gumbel, Fréchet or Weibull family [4]. Specifically, we model the z_i using the Weibull distribution,

$$f(z_i; \lambda, k) = \begin{cases} \frac{k}{\lambda}\left(\frac{z_i}{\lambda}\right)^{k-1} e^{-(z_i/\lambda)^k} & z_i \geq 0, \\ 0 & z_i < 0, \end{cases} \tag{1}$$

where $k > 0$ and $\lambda > 0$ denote the shape and scale parameters, respectively. We use Otsu's method [25] to obtain an initial threshold on the z_i values, then fit a two-component mixture of Weibulls, modeling the *identity-overlapping* and *identity-disjoint* sets $\mathcal{X}^{\mathcal{UO}}$ and $\mathcal{X}^{\mathcal{UD}}$, respectively. Selecting values corresponding to 95% confidence under each Weibull model provides thresholds for deciding if $\mathbf{x}_i \in \mathcal{X}^{\mathcal{UO}}$ or $\mathbf{x}_i \in \mathcal{X}^{\mathcal{UD}}$ with high confidence; we reject samples that fall outside of this interval. This approach does not require setting any hyper-parameters a priori, and can be applied to any new unlabeled dataset.

3.2 Clustering Faces with GCN

We use Face-GCN [42] to assign pseudo-labels for unlabeled faces in $\mathcal{X}^{\mathcal{UD}}$, which leverages a graph convolutional network (GCN) [15] for large-scale face clustering. We provide a brief overview of the approach for completeness. Based on features extracted from a pre-trained face recognition engine, a nearest-neighbor graph is constructed over all samples. By setting various thresholds on the edge weights of this graph, a set of connected components or cluster proposals are generated. During training, the aim is to regress the precision and recall of the cluster proposals arising from a single ground truth identity, motivated by object detection frameworks [8]. Since the proposals are generated based on labeled data, the Face-GCN is trained in a fully supervised way, unlike regular GCN training, which are typically trained with a classification loss, either for each node or an input graph as a whole. During testing, a "de-overlap" procedure uses predicted GCN scores for the proposals to partition an unlabeled dataset into a set of clusters. Please see [42] for further details.

3.3 Joint Data Re-training with Clustering Uncertainty

We seek to incorporate the uncertainty of whether a pseudo-labeled (*i.e.* clustered) sample was correctly labeled into the face recognition model re-training. Let a face drawn from the unlabeled dataset be $\mathbf{x}_i \in \mathcal{X}^{\mathcal{UD}}$. The feature representation for that face using the baseline supervised model is denoted as $\Phi(\mathbf{x}_i)$. Let cluster assignments obtained on $\mathcal{X}^{\mathcal{UD}}$ be $\{\mathcal{C}_1, \mathcal{C}_2, \dots, \mathcal{C}_K\}$, for K clusters. We train a logistic regression classifier to estimate $P(\mathcal{C}_k \mid \Phi(\mathbf{x}_i))$, for $k = 1, 2, \dots K$,

$$P(\mathcal{C}_k \mid \Phi(\mathbf{x}_i)) = \frac{\exp(\omega_k^\top \Phi(\mathbf{x}_i))}{\sum_j \exp(\omega_j^\top \Phi(\mathbf{x}_i))} \tag{2}$$

where ω_k are the classifier weights for the k-th cluster. Intuitively, we wish to determine how well a simple linear classifier on top of discriminative face descriptors can fit the cluster assignments. We compare the following uncertainty metrics: **(1)** *Entropy* of the posteriors across the K clusters, *i.e.* $\sum_k P(\mathcal{C}_k \mid \Phi(\mathbf{x}_i)) \log P(\mathcal{C}_k \mid \Phi(\mathbf{x}_i))$; **(2)** *Max-logit:* the largest logit value over the K clusters, **(3)** *Classification margin:* difference between the max and the second-max logit, indicating how easily a sample can flip between two clusters.

We consider two kinds of incorrect clustering corresponding to notions of precision and recall: **(1) Outliers**, samples whose identity does not belong to the identity of the cluster; **(2) Split-ID**, where samples from the same identity are spread over several clusters (Fig. 4(a)). In a controlled setting with known ground-truth identities, we validate our hypothesis that the uncertainty measures can distinguish between correct and incorrect cluster assignments (Fig. 4(b)). Note that Split-ID makes up the bulk of incorrectly-clustered samples, while outliers are about 10%. Figure 4(c) shows the distribution of class-margin on pseudo-labeled data on one split of the MS-1M dataset. Intuitively, samples that do not have a large classification margin are likely to be incorrect pseudo-labels, resulting in a bi-modal distribution – noisily labeled samples in one mode, and correctly labeled samples in the other. Notice that similar to overlapping v.s. disjoint identity, this is another distribution separation problem. A Weibull is fit to the lower portion of the distribution (orange curve), with an initial mode-separating threshold obtained from Otsu's method (black vertical line). The probability of sample $\mathbf{x}_i$ being incorrectly clustered is estimated by:

$$p^-(\mathbf{x}_i) = P(g(\mathbf{x}_i) \mid \theta^-_{Wb}), \tag{3}$$

where θ^-_{Wb} are the parameters of the learned Weibull model, $g(.)$ denotes the measure of uncertainty, *e.g.* class-margin. Note, ground-truth labels are not required for this estimation. We propose to associate the above uncertainty with the pseudo-labeled samples and set up a probabilistic face recognition loss.

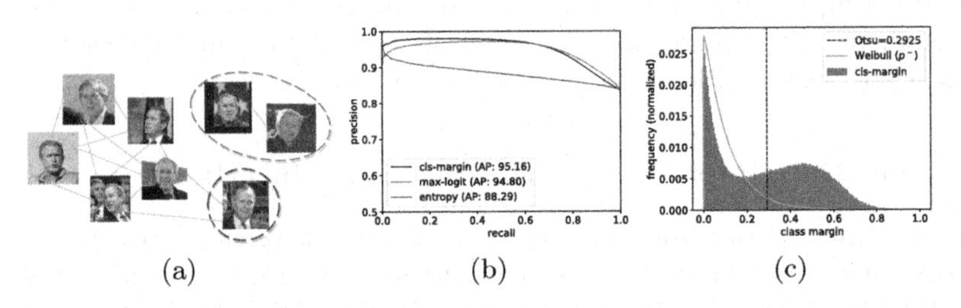

Fig. 4. Clustering uncertainty. (a) Illustration of incorrect pseudo-labels – an image of *George Bush Sr.* is included in a cluster of *George W Bush* images (outlier circled in blue); some *George W Bush* images are spread across multiple clusters ("split ID" circled in red). **(b)** Precision-recall curves showing Average Precision (AP) of predicting if a cluster assignment is correct using class-margin, max-logit and entropy. **(c)** Estimating clustering error $p^-(\mathbf{x}_i)$ from the distribution of class-margin (orange curve). (Color figure online)

The large margin cosine loss [41] is used for training:

$$\mathcal{L}(\mathbf{x}_i) = -\log \frac{\exp(\alpha(\mathbf{w}_j^\top \mathbf{f}_i - m))}{\exp(\alpha(\mathbf{w}_j^\top \mathbf{f}_i - m)) + \sum_{k \neq j} \exp(\alpha \mathbf{w}_k^\top \mathbf{f}_i)} \tag{4}$$

where $\mathbf{f}_i$ is the deep feature representation of the i-th training sample $\mathbf{x}_i$, $\mathbf{w}_j$ is the learned classifier weight for the j-th class, $m \in [0, 1]$ is an additive margin and α is a scaling factor; $\|\mathbf{f}_i\|$ and $\|\mathbf{w}_j\|$ are set to 1. For $\mathbf{x}_i \in \mathcal{X}^\mathcal{U}$, we modulate the training loss with the clustering uncertainty $p^-(\mathbf{x}_i)$, where γ controls the weighting curve shape:

$$\mathcal{L}^p(\mathbf{x}_i) = (1 - p^-(\mathbf{x}_i))^\gamma \mathcal{L}(\mathbf{x}_i), \tag{5}$$

4 Experiments

We augment supervised models trained on labeled data with additional pseudo-labeled data under various scenarios. We summarize the main findings first—

(i) the baseline supervised model benefits from additional pseudo-labeled training data; ***(ii)*** re-training on clustering without handling overlapping IDs can hurt performance, and our approach of separating overlaps is shown to be effective empirically; ***(iii)*** increasing diversity of training data by using unlabeled data from outside the distribution of the labeled set helps more than comparable amounts of within-domain unlabeled data; ***(iv)*** scaling up to using the entire MS-Celeb-1M [7] dataset (or MS1M for short) as labeled training set, as typically done by most deep face models, we see significant gains in performance *only* when the volume of unlabeled samples is comparable to the size of MS1M itself.

Experimental Setup. Table 1 summarizes the training data sources. The cleaned version of MS1M dataset contains 84,247 identities and 4,758,734 samples in total. Partitioning on the identities, the full MS1M dataset is split into 10 parts with approximately 8.4k identities and 470k samples per split. We create the following settings:

- ***Controlled disjoint*** *(Sect.* 4.1*):* Both labeled and unlabeled data are drawn from splits of MS1M (Table 1 *MS1M* splits 1 and 2, respectively). Thus, they have the same distribution and have no overlapping identities by construction, similar to the setting in [42]. We compare baseline clustering methods and the effect of clustering uncertainty on re-training the face recognition model.
- ***Controlled overlap*** *(Sect.* 4.2*):* we introduce simulated identity overlap between the two datasets (Table 1 *MS-Celeb-1M* splits *1-O* and *2-O*), showing the detrimental effect of naïvely clustering and re-training in this case, and the efficacy of our proposed approach.
- ***Semi-controlled*** *(Sect.* 4.3*):* we have limited labeled data (split-1 of MS1M) with unlabeled data from another dataset, i.e., VGGFace2 [3], containing 8.6k identities and 3.1 million images. This is closer to the realistic scenario, with potential identity overlaps and distribution shift between data sources.
- ***Uncontrolled*** *(Sect.* 4.5*):* close to the real-world setting, we use *all* the labeled data at our disposal (entire MS-Celeb-1M) and try to improve performance further by including unlabeled data from other datasets – VGGFace2 [3], IMDB-SenseTime [40], CASIA [43] & GlintAsian [5], by completely ignoring their ground truth labels. Note, this setting is not addressed in prior art on pseudo-labeling faces [42,44].

Evaluation. We report results on the following: verification accuracy on *Labeled Faces in the Wild* (LFW) [11,17] and *Celebrity Frontal to Profile* (CFP) [35]; identification at rank-1 and rank-5, and True Accept Rate (TAR) at False Accept Rates (FAR) of 1e−3 and 1e−4 on the challenging *IARPA Janus* benchmark (IJB-A) [16]. For clustering metrics we adopt the protocol used in [42].

Training Details. We use the high-performing CosFace model [41] for face recognition, with a 118-layer ResNet backbone, trained for 30 epochs on labeled data. Re-training is done from scratch with identical settings. Face-GCN uses the publicly available code of GCN-D [42]. For further details please refer to the supplementary materials.

Table 1. Statistics for training datasets.

Dataset	#IDs	Images
MS-Celeb-full	84k	4.7M
MS-Celeb-split-1	8.4k	505k
MS-Celeb-split-2	8.4k	467k
MS-Celeb-split-1-O	16.8k	729k
MS-Celeb-split-2-O	16.8k	705k
VGGFace2 [3]	8.6k	3.1M
CASIA-WebFace [43]	10.5k	455k
IMDB-SenseTime [40]	51k	1M
GlintAsian [5]	94k	2.8M

Table 2. Controlled: Face clustering baselines. Comparing the GCN-based method with standard clustering algorithms. The GCN is trained on MS-Celeb-1M split 1, tested on split 2.

Method	Prec	Rec	F1	#Clstr
K-means	55.77	87.56	68.14	5k
FastHAC	99.32	64.66	78.32	117k
DBSCAN	99.62	46.83	63.71	352k
GCN	95.87	79.43	86.88	45k
GCN-iter2	97.94	87.28	92.30	32k

4.1 Controlled Disjoint: MS-Celeb-1M Splits

In controlled setting, Split-1 of MS-Celeb-1M is used as the labeled dataset to train the face recognition model in a fully supervised fashion. The face clustering module is also trained in a supervised way on the labeled Split-1 data. The unlabeled data is from Split-2 of MS-Celeb-1M: ground truth labels are ignored, features are extracted on all the samples and the trained GCN model provides the cluster assignments.

Clustering. The performance of various clustering methods are summarized in Table 2, i.e., ***K-means*** [34], ***FastHAC*** [23] and ***DBSCAN*** [6,33], with optimal hyper-parameter settings[1]. The GCN is clearly better than the baseline clustering approaches. GCN typically provides an over-clustering of the actual number of identities – the precision is comparably higher than the recall (95.87% versus 79.43%), indicating high purity per cluster, but samples from the same identity end up being spread out across multiple clusters ("split ID").

Re-training. The results are summarized in Table 3. Re-training CosFace on labeled Split-1 and pseudo-labeled Split-2 data (*+GCN*) improves over training on just the labeled Split-1 (*Baseline GT-1*) across the benchmarks. The performance is upper-bounded when perfect labels are available on Split-2 (*+GT-2*). Note that re-training on cluster assignments from simpler methods like K-Means and HAC also improve over the baseline.

Re-train w/iterative Clustering. We perform a second iteration of clustering, using the re-trained CosFace model as feature extractor. The re-trained CosFace model has more discriminative features, resulting in better clustering (Table 2 *GCN-iter2* versus *GCN*). However, another round of re-training CosFace on these cluster-assignments yields smaller gains (Table 3 *+GCN-iter2* v.s. *+GCN*).

Insights. With limited labeled data, training on clustered faces significantly improves recognition performance. Simpler clustering methods like K-means are

[1] K-means: K = 5k, FastHAC: dist = 0.85, DBSCAN: minsize = 2, eps = 0.8.

also shown to improve recognition performance – if training Face-GCN is not practical, off-the-shelf clustering algorithms can also provide pseudo-labels. A second iteration gives small gains, indicating diminishing returns.

Table 3. Controlled disjoint: Re-training CosFace on the union of labeled and pseudo-labeled data (*+GCN*), pseudo-label on second iteration (*+GCN-iter-2*), with an upper bound from ground truth (*GT-2*). ↑ indicates improvement from baseline.

Model	LFW	↑	CFP-fp	↑	IJBA-idt.	↑	IJBA-vrf.	↑
					Rank-1, 5		FAR@1e−3,−4	
Baseline GT-1	99.20	–	92.37	–	92.66, 96.42	–	80.23, 69.64	–
+K-means	99.47	0.27	94.11	1.74	93.80, 96.79	1.14, 0.37	87.03, 78.00	6.80, 8.36
+FastHAC	99.42	0.22	93.56	0.90	93.84, **96.81**	1.18, **0.39**	84.78, 75.21	4.55, 5.57
+GCN	99.48	0.28	**95.51**	**3.14**	94.11, 96.55	1.45, 0.13	87.60, 77.67	7.37, 7.93
+GCN-*iter-2*	**99.57**	**0.37**	94.14	1.77	**94.46**, 96.40	**1.80**, −0.02	**88.00, 78.78**	**7.77, 9.14**
+GT-2 (bound)	99.58	0.38	95.56	3.19	95.24, 97.24	2.58, 0.82	89.45, 81.02	9.22, 11.38

Table 4. Controlled overlaps: Re-training with overlapping identity unlabeled data.

Model	LFW	↑	CFP-fp	↑	IJBA-idt.	↑	IJBA-vrf.	↑
					Rank-1, 5		FAR@1e−3,−4	
Baseline	99.45	–	95.17	–	94.52, 96.60	–	87.36, 75.06	–
+GCN (naive)	99.37	−0.08	93.17	−2.0	93.72, 96.65	−0.80, 0.05	87.02, 79.39	−0.34, 4.33
+GCN (disjoint)	99.57	0.12	95.01	−0.16	**94.83**, 96.98	**0.31**, 0.38	89.29, 82.64	1.93, 7.58
+GCN (overlap)	**99.58**	**0.13**	94.30	−0.87	94.47, 96.64	−0.05, 0.04	86.93, 78.42	−0.43, 3.36
+GCN (both)	**99.58**	**0.13**	**95.36**	**0.19**	94.81, **97.05**	0.29, **0.45**	**89.43, 82.86**	**2.07, 7.80**

4.2 Controlled Overlap: Overlapping Identities

We simulate the real-world overlapping-identity scenario mentioned in Sect. 3.1 to empirically observe its impact on the "pseudo-labeling by clustering" pipeline. We create two subsets of MS1M with around 16k identities each, having about 8.5k overlapping identities (suffix "*O*" for *overlaps* in Table 1). The labeled subset $\mathcal{X}^L$ contains around 720k samples (Split-1-O). The unlabeled subset, Split-2-O, contains approximately 467k *disjoint-identity* ($\mathcal{X}^{UD}$) and 224k *overlapping-identity* ($\mathcal{X}^{UO}$) samples.

Disjoint/Overlap. Modeling the disjoint/overlapping identity separation as an out-of-distribution problem is an effective approach, especially on choosing the max-logit score as the feature for OOD. A simple Otsu's threshold provides acceptably low error rates, i.e., **6.2%** false positive rate and **0.69%** false negative rate, while using 95% confidence intervals from Weibulls, we achieve much lower error rates of **2.3%** FPR and **0.50%** FNR.

Table 5. Disjoint/overlap clustering. Results of clustering the entire unlabeled $\mathcal{X}^{\mathcal{U}}$ ("Split-2-O") and clustering the estimated ID-disjoint portion $\mathcal{X}^{\mathcal{UD}}$.

Data	Prec.	Rec.	F1	#IDs	#Clstr	#Img
$\mathcal{X}^{\mathcal{U}}$	98.7	84.8	91.2	16.8k	60k	693k
$\mathcal{X}^{\mathcal{UD}}$	98.8	85.2	91.5	11.7k	39k	464k

Table 6. Semi-controlled: clustering. Comparing performance on "within-domain" splits of MS-Celeb-1M vs. VGGFace2 data.

Train	Test	Prec.	Rec.	F1	#clstr
Split-1	Split-2	95.87	79.43	86.88	45k
Split-1	VGG2	97.65	59.62	74.04	614k
Full	VGG2	98.88	72.76	83.83	224k

Clustering. Table 5 shows the results from clustering all the unlabeled data (*Naive*) versus separating out the identity disjoint portion of the unlabeled data and then clustering (*Disjoint*). On both sets of unlabeled samples, the GCN clustering achieves high precision and fairly high recall, indicating that the clusters we use in re-training the face recognition engine are of good quality.

Re-training. The results are shown in Table 4. Naively re-training on the additional pseudo-labels clearly hurts performance (*Baseline* v.s. *GCN(naive)*). Adding pseudo-labels from the *disjoint* data improves over the baseline across the benchmarks. Merging the *overlapping* samples with their estimated identities in the labeled data is done based on the softmax outputs of the baseline model, causing improvements in some cases (*e.g.* LFW and IJBA verif.) but degrading performance in others (*e.g.* IJBA ident. and YTF). Merging overlapping identities as well as clustering disjoint identities also shows improvements over the baseline across several benchmarks.

Insights. Overlapping identities with the labeled training set clearly has a detrimental effect when retraining and must be accounted for when merging unlabeled data sources – the choice of modeling max-logit scores for this separation is shown to be simple and effective. Overall, discarding overlapping samples from re-training, and clustering *only* the disjoint samples, appears to be a better strategy. Adding pseudo-labeled data for classes that exist in the labeled set seems to have limited benefits, versus adding more identities.

Table 7. Semi-controlled: MS-Celeb-1M split 1 and VGGFace2. Note that similar volume of pseudo-labeled data from MS-Celeb-1M split 2 (*+MS1M-GCN-2*) gives lower benefits compared to data from VGGFace2 (*+VGG-GCN*) in challenging settings like IJB-A verification at FAR = 1e−4, IJB-A identification Rank-1.

Model	LFW	↑	CFP-fp	↑	IJBA-idt.	↑	IJBA-vrf.	↑
					Rank-1, 5		FAR@1e−3,−4	
MS1M-GT-1	99.20	–	92.37	–	92.66, 96.42	–	80.23, 69.64	–
+MS1M-GCN-2	99.48	0.28	**95.51**	**3.14**	94.11, 96.55	1.45, 0.13	87.60, 77.67	7.37, 12.03
+VGG-GCN (ours)	**99.55**	**0.35**	94.60	2.23	**94.72, 96.97**	**2.06, 0.55**	**88.12, 82.48**	**7.89, 12.84**
+VGG-GT (bound)	99.70	0.50	97.81	5.44	96.93, 98.25	4.27, 1.83	93.20, 84.67	12.97, 15.03

4.3 Semi-controlled: Limited Labeled, Large-Scale Unlabeled Data

MS-Celeb-1M Split 1 forms the labeled data, while the unlabeled data is from VGGFace2 (Table 1). We simply discard VGGFace2 samples estimated to have overlapping identities with MS-Celeb-1M Split-1. Out of the total 3.1M samples, about 2.9M were estimated to be identity-disjoint with MS-Celeb-1M Split-1.

Clustering. The same GCN model trained on Split-1 of MS-Celeb-1M in Sect. 4.1 is used to obtain cluster assignments on VGGFace2. Table 6 compares the clustering on MS-Celeb-1M Split2 (controlled) v.s. the current setting. The F-score on VGGFace2 is reasonable – 74.04%, but lower than the F-score on Split-2 MS-Celeb-1M (86.88%) – we are no longer dealing with within-dataset unlabeled data.

Re-training. To keep similar volumes of labeled and pseudo-labeled data we randomly select 50 images per cluster from the largest 8.5k clusters of VGGFace2. Re-training results are in Table 7. We generally see benefits from VGGFace2 data over both *baseline* and *MS1M-split-2*: YTF: 93.82% → 94.64% → **95.14%**, IJBA idnt. rank-1: 92.66% → 94.11% → **94.72%**, IJBA verif. at FAR 1e−4: 69.635% → 77.665% → **82.484%**. When the full VGGFace2 labeled dataset is used to augment MS1M-split-1, *VGG-GT(full)*, we get the upper bound performance.

Insights. Ensuring the *diversity* of unlabeled data is important, in addition to other concerns like clustering accuracy and data volume: pseudo-labels from VGGFace2 benefit more than using more data from within MS1M.

4.4 Soft Labels for Clustering Uncertainty

Table 8 shows results of re-training the face recognition model with our proposed cluster-uncertainty weighted loss (Sect. 3.3) on the pseudo-labeled samples (*GCN-soft*). We set $\gamma = 1$ (ablation in supplemental). We empirically find that incorporating this cluster uncertainty into the training loss improves results in both controlled and large-scale settings (3 out of 4 evaluation protocols). In the controlled setting, the soft pseudo-labels, MS1M-GCN-*soft*, improves over MS1M-GCN (hard cluster assignments) on challenging IJB-A protocols (77.67% → **78.78%** @FAR 1e−4) and is slightly better on LFW. In the large-scale setting, comparing VGG-GCN and VGG-GCN-*soft*, we again see significant improvements on IJB-A (81.85% → **90.16%** @FAR 1e−4) and gains on the LFW benchmark as well. Qualitative analyses of clustering errors and uncertainty estimates $p^-(\mathbf{x}_i)$ are included in the supplemental.

4.5 Uncontrolled: Large-Scale Labeled and Unlabeled Data

The earlier cases either had limited labeled data, unlabeled data from an identical distribution as the labeled data by construction, or both aspects together. Now, the *entire* MS-Celeb-1M is used as labeled training data for training the baseline CosFace model as well as the GCN. We gradually add several well-known face

Table 8. Effect of Cluster Uncertainty: Re-training CosFace with the proposed clustering uncertainty (*GCN-soft*) shows improvements in both controlled (*MS1M-GT-split1*) and large-scale settings, *MS1M-GT-full* (CosFace [41]).

Model	LFW	↑	CFP-fp	↑	IJBA-idt.	↑	IJBA-vrf.	↑
					Rank-1, 5		FAR@1e−3,−4	
MS1M-GT-*split1*	99.20	–	92.37	–	92.66, 96.42	–	80.23, 69.64	–
+MS1M-GCN (ours)	99.48	0.28	**95.51**	**3.14**	94.11, 96.55	1.45, 0.13	87.60, 77.67	7.37, 12.03
+MS1M-GCN-*soft* (ours)	**99.50**	**0.30**	94.71	2.34	**94.76, 97.10**	**2.10, 0.68**	**87.97, 79.43**	**7.74, 9.79**
MS1M-GT-*full* (CosFace)	99.70	-	**98.10**	-	95.47, 97.04	-	92.82, 80.68	-
+VGG-GCN (ours)	99.73	0.03	97.63	−0.47	95.87, 97.45	0.40, 0.41	93.88, 81.85	1.06, 1.17
+VGG-GCN-*soft* (ours)	**99.75**	**0.05**	97.57	−0.53	**96.37, 97.70**	**0.90, 0.66**	**93.94, 90.16**	**1.12, 9.48**

Table 9. Uncontrolled: pseudo-labels. Showing the clusters and samples in the uncontrolled setting with full-MS1M and unlabeled data of increasingly larger volume – (1) VGG2 [3]; (2) merging CASIA [43] & IMDB-SenseTime [40] with VGG2; (3) merging GlintAsian [5] with all the above.

Dataset:	VGG2	+(CASIA, IMDB)	+Glint
True classes	8631	57,271	149,824
Clusters	224,466	452,598	719,722
Samples	1,257,667	2,133,286	3,673,517
Prec	98.88	91.35	88.16
Rec	72.76	77.53	66.93
F-score	83.83	83.88	76.09

Table 10. Uncontrolled: re-training. Merging unlabeled training samples with the entire MS-Celeb-1M labeled data consistently surpasses the fully-supervised MS1M-GT-*full* (CosFace [41]) trained on the entire labeled MS-Celeb-1M dataset.

Model	LFW	↑	CFP-fp	↑	IJBA-idt.	↑	IJBA-vrf.	↑
					Rank-1, 5		FAR@1e−3,−4	
MS1M-GT-*full* (CosFace)	99.70	–	98.10	–	95.47, 97.04	–	92.82, 80.68	–
+VGG-GCN (ours)	**99.73**	**0.03**	97.63	−0.47	95.87, 97.45	0.40, 0.41	93.88, 81.85	1.06, 1.17
+CASIA-IMDB (ours)	**99.73**	**0.03**	97.81	−0.29	96.66, 97.89	1.19, 0.85	93.79, 89.58	0.97, 8.90
+GlintAsian (ours final)	**99.73**	**0.03**	**98.24**	**0.14**	**96.94, 98.21**	**1.47, 1.17**	**94.89, 92.29**	**2.07, 11.61**

recognition datasets (ignoring their labels) to MS-Celeb-1M labeled samples during re-training (Table 9)[2]. Along with more data, these datasets bring in more *varied* or *diverse* samples (analysis in supplemental).

Re-training. The re-training results are shown in Table 10. As expected, we get limited benefits from adding moderate amounts of unlabeled data when the baseline model is trained on a large labeled dataset like MS-Celeb-1M. When incorporating data from only VGGFace2, there are improvements on LFW (99.7% → 99.73%), and on IJBA, ident. (95.47% → 95.87%) and verif. (80.68% →

[2] In particular, we estimated a 40% overlap in identities between MS-Celeb and VGG2.

81.85%). There are however some instances of decreased performance on the smaller scale dataset CFP-fp. When the volume of unlabeled data is of comparable magnitude (4.7M labeled versus 3.6M unlabeled) by merging all the other datasets (VGGFace2, CASIA, IMDB-SenseTime and GlintAsian), we get a clear advantage on the challenging IJBA benchmarks (rank-1 identification: 95.47% → **96.94%**, verification TAR at FAR 1e−4: 80.68% → **92.29%**).

Insights. The crucial factors in improving face recognition when we have access to all available labeled data from MS1M appear to be *both* diversity and volume – it is only when we merged unlabeled data from all the other data sources, reaching comparable number of samples to MS1M, that we could improve over the performance attained from training on just the ground-truth labels of MS1M, suggesting that current high-performing face recognition models can benefit from even larger training datasets. While acquiring datasets of such scale purely through manual annotation is prohibitively expensive and labor-intensive, using pseudo-labels is shown to be a feasible alternative.

5 Conclusion

The pseudo-labeling approach described in this paper provides a recipe for improving fully supervised face recognition, i.e., CosFace, leveraging large unlabeled sources of data to augment an existing labeled dataset. The experimental results show consistent performance gains across various scenarios and provide insights into the practice of large-scale face recognition with unlabeled data – **(1)** we require comparable volumes of labeled and unlabeled data to see significant performance gains, especially when several million labeled samples are available; **(2)** overlapped identities between labeled and unlabeled data is a major concern and needs to be handled in real-world scenarios; **(3)** along with large amounts of unlabeled data, greater gains are observed if the new data shows certain domain gap w.r.t. the labeled set; **(4)** incorporating scalable measures of clustering uncertainty on the pseudo-labels is helpful in dealing with label noise. Overall, learning from unlabeled faces is shown to be an effective approach to further improve face recognition performance.

Acknowledgments. This research was partly sponsored by the AFRL and DARPA under agreement FA8750-18-2-0126. The U.S. Government is authorized to reproduce and distribute reprints for Governmental purposes notwithstanding any copyright notation thereon. The views and conclusions contained herein are those of the authors and should not be interpreted as necessarily representing the official policies or endorsements, either expressed or implied, of the AFRL and DARPA or the U.S. Government.

References

1. Arazo, E., Ortego, D., Albert, P., O'Connor, N.E., McGuinness, K.: Unsupervised label noise modeling and loss correction. In: International Conference on Machine Learning (ICML) (June 2019)

2. Bendale, A., Boult, T.E.: Towards open set deep networks. In: Proceedings of the IEEE Conference on Computer Vision and Pattern Recognition, pp. 1563–1572 (2016)
3. Cao, Q., Shen, L., Xie, W., Parkhi, O.M., Zisserman, A.: VGGFace2: a dataset for recognising faces across pose and age. In: IEEE FG (2018)
4. De Haan, L., Ferreira, A.: Extreme Value Theory: An Introduction. Springer, New York (2007). https://doi.org/10.1007/0-387-34471-3
5. DeepGlint: Glint asian. http://trillionpairs.deepglint.com/overview. Accessed 11 Nov 2019
6. Ester, M., Kriegel, H.P., Sander, J., Xu, X., et al.: A density-based algorithm for discovering clusters in large spatial databases with noise. In: KDD, vol. 96, pp. 226–231 (1996)
7. Guo, Y., Zhang, L., Hu, Y., He, X., Gao, J.: MS-Celeb-1M: a dataset and benchmark for large-scale face recognition. In: Leibe, B., Matas, J., Sebe, N., Welling, M. (eds.) ECCV 2016. LNCS, vol. 9907, pp. 87–102. Springer, Cham (2016). https://doi.org/10.1007/978-3-319-46487-9_6
8. He, K., Gkioxari, G., Dollar, P., Girshick, R.: Mask R-CNN. In: ICCV (2017)
9. Hendrycks, D., Gimpel, K.: A baseline for detecting misclassified and out-of-distribution examples in neural networks. In: ICLR (2017)
10. Hendrycks, D., Mazeika, M., Wilson, D., Gimpel, K.: Using trusted data to train deep networks on labels corrupted by severe noise. In: Advances in Neural Information Processing Systems, pp. 10456–10465 (2018)
11. Huang, G.B., Mattar, M., Berg, T., Learned-Miller, E.: Labeled faces in thewild: A database for studying face recognition in unconstrained environments. Technical report, University of Massachusetts, Amherst (2007)
12. Jain, A.K.: Data clustering: 50 years beyond K-means. Pattern Recogn. Lett. **31**(8), 651–666 (2010)
13. Jiang, L., Zhou, Z., Leung, T., Li, L.J., Fei-Fei, L.: MentorNet: Learning data-driven curriculum for very deep neural networks on corrupted labels. arXiv preprint arXiv:1712.05055 (2017)
14. Jin, S., Su, H., Stauffer, C., Learned-Miller, E.: End-to-end face detection and cast grouping in movies using Erdos-Renyi clustering. In: ICCV (2017)
15. Kipf, T.N., Welling, M.: Semi-supervised classification with graph convolutional networks. In: ICLR (2017)
16. Klare, B.F., et al.: Pushing the frontiers of unconstrained face detection and recognition: IARPA Janus benchmark A. In: Proceedings of the IEEE Conference on Computer Vision and Pattern Recognition, pp. 1931–1939 (2015)
17. Learned-Miller, E., Huang, G.B., RoyChowdhury, A., Li, H., Hua, G.: Labeled faces in the wild: a survey. In: Kawulok, M., Celebi, M.E., Smolka, B. (eds.) Advances in Face Detection and Facial Image Analysis, pp. 189–248. Springer, Cham (2016). https://doi.org/10.1007/978-3-319-25958-1_8
18. Lee, K., Lee, K., Lee, H., Shin, J.: A simple unified framework for detecting out-of-distribution samples and adversarial attacks. In: Advances in Neural Information Processing Systems, pp. 7167–7177 (2018)
19. Li, Y., Yang, J., Song, Y., Cao, L., Luo, J., Li, L.J.: Learning from noisy labels with distillation. In: ICCV, pp. 1928–1936 (2017)
20. Liang, S., Li, Y., Srikant, R.: Enhancing the reliability of out-of-distribution image detection in neural networks. In: ICLR (2017)
21. Lin, W.A., Chen, J.C., Castillo, C.D., Chellappa, R.: Deep density clustering of unconstrained faces. In: Proceedings of the IEEE Conference on Computer Vision and Pattern Recognition, pp. 8128–8137 (2018)

22. Lin, W.A., Chen, J.C., Chellappa, R.: A proximity-aware hierarchical clustering of faces. In: 2017 12th IEEE International Conference on Automatic Face & Gesture Recognition, FG 2017, pp. 294–301. IEEE (2017)
23. Müllner, D., et al.: fastcluster: fast hierarchical, agglomerative clustering routines for R and Python. J. Stat. Softw. **53**(9), 1–18 (2013)
24. Natarajan, N., Dhillon, I.S., Ravikumar, P.K., Tewari, A.: Learning with noisy labels. In: Advances in Neural Information Processing Systems, pp. 1196–1204 (2013)
25. Otsu, N.: A threshold selection method from gray-level histograms. IEEE Trans. Syst. Man Cybern. **9**(1), 62–66 (1979)
26. Otto, C., Wang, D., Jain, A.K.: Clustering millions of faces by identity. IEEE Trans. Pattern Anal. Mach. Intell. **40**(2), 289–303 (2017)
27. Patrini, G., Rozza, A., Menon, A.K., Nock, R., Qu, L.: Making deep neural networks robust to label noise: a loss correction approach. In: Proceedings of the IEEE Conference on Computer Vision and Pattern Recognition (CVPR), pp. 2233–2241 (2017)
28. Ren, M., Zeng, W., Yang, B., Urtasun, R.: Learning to reweight examples for robust deep learning. arXiv preprint arXiv:1803.09050 (2018)
29. Rudd, E.M., Jain, L.P., Scheirer, W.J., Boult, T.E.: The extreme value machine. IEEE Trans. Pattern Anal. Mach. Intell. **40**(3), 762–768 (2017)
30. Scheirer, W., Rocha, A., Micheals, R., Boult, T.: Robust fusion: extreme value theory for recognition score normalization. In: Daniilidis, K., Maragos, P., Paragios, N. (eds.) ECCV 2010. LNCS, vol. 6313. Springer, Heidelberg (2010). https://doi.org/10.1007/978-3-642-15558-1_35
31. Scheirer, W.J., Rocha, A., Micheals, R.J., Boult, T.E.: Meta-recognition: the theory and practice of recognition score analysis. IEEE Trans. Pattern Anal. Mach. Intell. **33**(8), 1689–1695 (2011)
32. Schroff, F., Kalenichenko, D., Philbin, J.: FaceNet: a unified embedding for face recognition and clustering. In: Proceedings of the IEEE Conference on Computer Vision and Pattern Recognition, pp. 815–823 (2015)
33. Schubert, E., Sander, J., Ester, M., Kriegel, H.P., Xu, X.: DBSCAN revisited, revisited: why and how you should (still) use DBSCAN. ACM Trans. Database Syst. (TODS) **42**(3), 19 (2017)
34. Sculley, D.: Web-scale K-means clustering. In: Proceedings of the 19th International Conference on World Wide Web, pp. 1177–1178. ACM (2010)
35. Sengupta, S., Chen, J.C., Castillo, C., Patel, V.M., Chellappa, R., Jacobs, D.W.: Frontal to profile face verification in the wild. In: 2016 IEEE Winter Conference on Applications of Computer Vision (WACV), pp. 1–9. IEEE (2016)
36. Shi, Y., Otto, C., Jain, A.K.: Face clustering: representation and pairwise constraints. IEEE Trans. Inf. Forensics Secur. **13**(7), 1626–1640 (2018)
37. Sohn, K., Shang, W., Yu, X., Chandraker, M.: Unsupervised domain adaptation for distance metric learning. In: ICLR (2019)
38. Taigman, Y., Yang, M., Ranzato, M., Wolf, L.: DeepFace: closing the gap to human-level performance in face verification. In: CVPR (2014)
39. Toneva, M., Sordoni, A., Combes, R.T., Trischler, A., Bengio, Y., Gordon, G.J.: An empirical study of example forgetting during deep neural network learning. In: ICLR (2019)
40. Wang, F., et al.: The devil of face recognition is in the noise. In: Ferrari, V., Hebert, M., Sminchisescu, C., Weiss, Y. (eds.) ECCV 2018. LNCS, vol. 11213, pp. 780–795. Springer, Cham (2018). https://doi.org/10.1007/978-3-030-01240-3_47

41. Wang, H., et al.: CosFace: large margin cosine loss for deep face recognition. In: Proceedings of the IEEE Conference on Computer Vision and Pattern Recognition, pp. 5265–5274 (2018)
42. Yang, L., Zhan, X., Chen, D., Yan, J., Loy, C.C., Lin, D.: Learning to cluster faces on an affinity graph. In: Proceedings of the IEEE Conference on Computer Vision and Pattern Recognition (CVPR) (2019)
43. Yi, D., Lei, Z., Liao, S., Li, S.Z.: Learning face representation from scratch. arXiv preprint arXiv:1411.7923 (2014)
44. Zhan, X., Liu, Z., Yan, J., Lin, D., Loy, C.C.: Consensus-driven propagation in massive unlabeled data for face recognition. In: Ferrari, V., Hebert, M., Sminchisescu, C., Weiss, Y. (eds.) ECCV 2018. LNCS, vol. 11213, pp. 576–592. Springer, Cham (2018). https://doi.org/10.1007/978-3-030-01240-3_35
45. Zhang, H., Cisse, M., Dauphin, Y.N., Lopez-Paz, D.: mixup: Beyond empirical risk minimization. arXiv preprint arXiv:1710.09412 (2017)

NeuRoRA: Neural Robust Rotation Averaging

Pulak Purkait(✉), Tat-Jun Chin, and Ian Reid

The University of Adelaide, Adelaide, SA 5005, Australia
pulak.isi@gmail.com
https://github.com/pulak09/NeuRoRA

Abstract. Multiple rotation averaging is an essential task for structure from motion, mapping, and robot navigation. The conventional methods for this task seek parameters of the absolute orientations that agree best with the observed noisy measurements according to a robust cost function. These robust cost functions are highly nonlinear and are designed based on certain assumptions about the noise and outlier distributions. In this work, we aim to build a neural network that learns the noise patterns from the data and predict/regress the model parameters from the noisy relative orientations. The proposed network is a combination of two networks: (1) a view-graph cleaning network, which detects outlier edges in the view-graph and rectifies noisy measurements; and (2) a fine-tuning network, which fine-tunes an initialization of absolute orientations bootstrapped from the cleaned graph, in a single step. The proposed combined network is very fast, moreover, being trained on a large number of synthetic graphs, it is more accurate than the conventional iterative optimization methods.

Keywords: Robust rotation averaging · Message passing neural networks

1 Introduction

Recently, we have witnessed a surge of interest in applying neural networks in various computer vision and robotics problems, such as, single-view depth estimation [16], absolute pose regression [28] and 3D point-cloud classification [35]. However, we still rely on robust optimizations at different steps of geometric problems, for example, robot navigation and mapping. The reason is that neural networks have not yet proven to be effective in solving constrained optimization problems. Some classic examples of the test-time geometric optimization include rotation averaging [8,14,15,23,26,45] (a.k.a. rotation synchronization [5,38,44]), pose-graph optimization [30,42], local bundle adjustment [33]

Electronic supplementary material The online version of this chapter (https://doi.org/10.1007/978-3-030-58586-0_9) contains supplementary material, which is available to authorized users.

A. Vedaldi et al. (Eds.): ECCV 2020, LNCS 12369, pp. 137–154, 2020.
https://doi.org/10.1007/978-3-030-58586-0_9

and global structure from motion [46]. These optimization methods estimate the model parameters that agree best with the observed noisy measurements by minimizing a robust (typically non-convex) cost function. Often, these loss functions are designed based on certain assumptions about the sensor noise and outlier distributions. However, the observed noise distribution in a real-world application could be far from those assumptions. A few such examples of noise patterns in real datasets are displayed in Fig. 2. Furthermore the nature and the structure of the objective loss function is the same for different problem instances of a specific task. Nonetheless, existing methods optimize the loss function for each instance. Moreover, an optimization during test-time could be slow for a target task involving a large number of parameters, and often forestalls a real-time solution to the problem.

In this work, with the advancement of machine learning, we address the following question: *"can we learn the noise patterns in data, given thousands of different problem instances of a specific task, and regress the target parameters instead of optimizing them during test-time?"* The answer is affirmative for some specific applications, and we propose a learning framework that exceeds baseline optimization methods for a geometric problem. We choose *multiple rotation averaging* (MRA) as a target application to validate our claim.

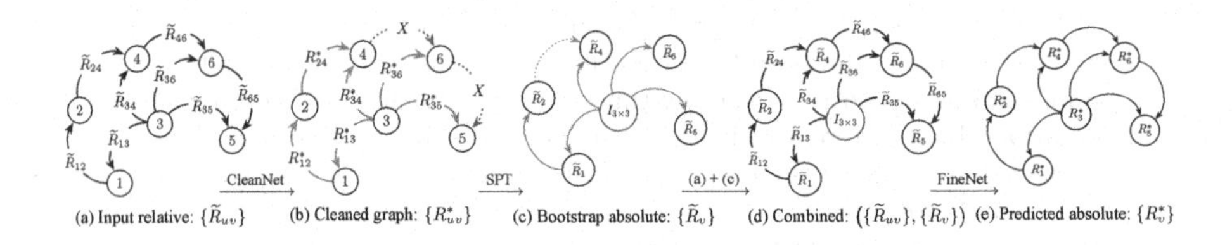

(a) Input relative: $\{\widetilde{R}_{uv}\}$ (b) Cleaned graph: $\{R^{*}_{uv}\}$ (c) Bootstrap absolute: $\{\widetilde{R}_{v}\}$ (d) Combined: $(\{\widetilde{R}_{uv}\}, \{\widetilde{R}_{v}\})$ (e) Predicted absolute: $\{R^{*}_{v}\}$

Fig. 1. The proposed method NeuRoRA is a two-step approach: in the first step a graph-based network (CleanNet) is utilized to clean the view-graph by removing outliers and rectifying noisy measurements. An initialization from the cleaned view-graph, instantiated from a shortest path tree (SPT), is then further fine-tuned using a separate graph-based network (FineNet). The notations are outlined in Table 1.

In MRA, the task is to estimate the absolute orientations of cameras given some of their pairwise noisy relative orientations defined on a view-graph. There are a different number of cameras for each problem instance of MRA, and usually sparsely connected to each other. Further, the observed relative orientations are often corrupted by outliers. The conventional methods for this task [5,8, 14,23,44] optimize the parameters of the absolute orientations of the cameras that are most compatible (up to a robust cost) with the observed noisy relative orientations.

We propose a neural network for robust MRA. Our network is a combination of two simple four-layered message-passing neural networks defined on the view-graphs, summarized in Fig. 1. We name our method *Neural Robust Rotation Averaging*, which is abbreviated as NeuRoRA in the rest of the manuscript.

Contribution and Findings

- A graph-based neural network NeuRoRA is proposed as an alternative to conventional optimizations for MRA.
- NeuRoRA requires *no* explicit optimization at test-time and hence it is **much faster** (10–50× on CPUs and 500–2000× on GPUs) than the baselines.
- The proposed NeuRoRA is **more accurate** than the conventional optimization methods of MRA. The mean/median orientation error of the predicted absolute orientations by the proposed method is 1.45°/0.74°, compared to 2.17°/1.25° by an optimization method [8].
- Being **a small size** network, NeuRoRA is fast and can easily be deployed to real-world applications (network size <0.5 Mb).

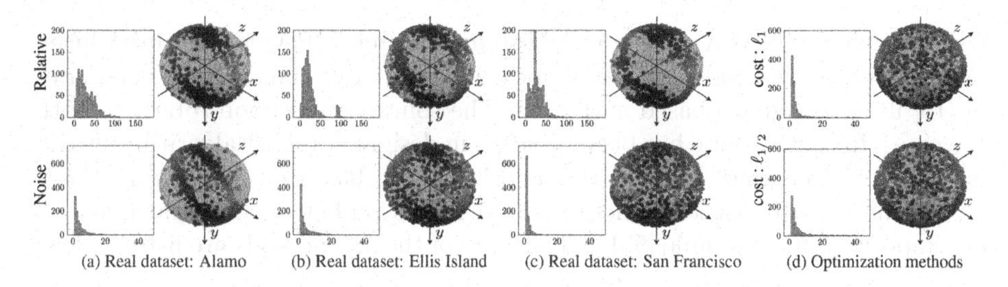

(a) Real dataset: Alamo (b) Real dataset: Ellis Island (c) Real dataset: San Francisco (d) Optimization methods

Fig. 2. Here we qualitatively illustrate that the noise distribution in real data diverges considerably from the noise assumptions baked into most optimization methods. We plot the angle and axes of observed **relative** orientations (first row) and the same of **noise** (second row) in real datasets (for clarity only 10^3 random samples) are displayed. The noise orientation is calculated from the ground-truth absolute and the observed relative orientations. The view-graphs of (a)–(b) are shared by [46] and (c) is shared by [12]. We plotted histograms of the magnitudes of the angles (in degrees) and the axes of the orientations. Notice that the axes of the sampled relative and noise orientations for the real data in (a)–(c) are not uniformly distributed on the unit ball. The sampled noise orientations (somewhat vertical axes) are far from the typical distribution assumptions regarded by optimization algorithms. Samples from such noise distributions (ℓ_1 [44] and $\ell_{1/2}$ [8]) are shown in (d).

2 Related Works

We separate the related methods into two separate sections—**(i)** learning based methods as an alternative to optimizations, **(ii)** relevant optimizations specific to MRA, and **(iii)** other related optimization methods.

(i) Learning to optimize. A neural network is proposed as an alternative to non-linear least square optimizations in [11] for camera tracking and mapping. It exploits the least square structure of the problem and uses a recurrent network to compute updated steps of the optimization variables. In a similar direction, [31] relaxes the assumptions made by inverse compositional algorithms for dense image alignment by incorporating data-driven priors. [40] proposes a bundle adjustment layer that learns to predict the dampening parameter of the Levenberg-Marquardt algorithm to optimize depth and camera parameters. In contrast to the direct optimization-based methods that explicitly use regularizers to solve an ill-posed problem, [1] implicitly learn the prior through a data-driven method. Aoki *et al.* [3] proposed an iterative algorithm based on PointNet [35] for point-cloud registration as an alternative to direct optimization. Learning to predict an approximate solution to combinatorial optimization problem over graphs, *e.g.* minimum vertex cover, traveling salesman problem, etc., is proposed in [29]. Learning methods to optimize general black-box functions [10] have also received a lot of attention recently. These conventional learning-based methods are tailored to some specific problems, where in this work, we are interested in an alternative learning-based solution for geometric problems, *e.g.* SFM.

(ii) Robust optimization for rotation averaging. MRA was first introduced in [18] where a linear solution was proposed using quaternion averaging and later in [19] using Lie group based averaging. The solutions were non-robust in both the cases. Recently, there has been progress in designing robust algorithms [9,22] for rotation averaging. Most of the algorithms are based on iterative methods for optimizing a robust loss function. MRA is also exploited using sparse matrix decomposition, for example [5,44]. The state of the art methods are listed below:

- Chatterjee and Govindu [8] fine-tune an initialization by first performing an iterative ℓ_1 minimization, followed by another iterative reweighted least squares with a more robust loss function $\ell_{\frac{1}{2}}$.
- Hartley *et al.* [22] propose a straight-forward method. It fine-tunes an initialization by the Weiszfeld algorithm of ℓ_1 averaging [7]. At every iteration, the absolute orientations of each camera are updated by the median of those computed from its neighbors.
- Arrigoni *et al.* [5] formulate the problem as a low-rank and sparse matrix decomposition and utilizes existing decomposition algorithms that caters for missing data, outliers and noise in the pairwise observations.
- Wang *et al.* [44] employ the alternating direction method to minimize a robust cost function involving the sum of unsquared deviations.

Rotation averaging is surveyed recently in a vast amount of literature [4,6,34,43].

(iii) Other related optimization methods. DISCO [12] employs a two-step approach. In the first step, a loopy belief propagation is used for an initial estimation of the absolute orientations which are fine-tuned by Levenberg-Marquardt method in the second step. The problem of detecting outliers in the view-graph has been extensively studied in the literature [13,20,27,32,36,47]. Optimizing/cleaning the view-graph for sfm also proposed in [21,37,39].

Huang *et al.* [25] proposed a neural network solving the pairwise matching problem (*c.f.* page 2, 2nd col) accurately. The key component of [25] is a network that takes two 3D scans and a relative transformation between them as input and outputs a score indicating the goodness of the scan alignment which is iteratively employed to fine-tune the absolute pose. Therefore, [25] is only valid (and tailored) for alignments of multiple scans.

3 Multiple Rotation Averaging

Consider N cameras with M pairwise relative orientation measurements forming a directed view-graph $\mathcal{G} = (\mathcal{V}, \mathcal{E})$. A vertex $\mathcal{V}_v \in \mathcal{V}$ corresponds to the absolute orientation $\widehat{R}_v$ (to be estimated) of the vth camera and an edge $\mathcal{E}_{uv} \in \mathcal{E}$ corresponds to the observed relative orientation $\widetilde{R}_{uv}$ from uth camera to vth camera. Conventionally, the task is to estimate the absolute orientations $\{\widehat{R}_v\}$, with respect to a global reference of orientations, such that the estimated orientations are most consistent with the observed noisy relative orientation measurements, *i.e.* $\widetilde{R}_{uv} \approx \widehat{R}_v \widehat{R}_u^{-1}, \forall \mathcal{E}_{uv} \in \mathcal{E}$. Further, the observed measurements are corrupted by outliers, *i.e.* some of the orientations $\widetilde{R}_{uv}$ are far from $\widehat{R}_v \widehat{R}_u^{-1}$. Conventionally, the solution is obtained by minimizing a robust cost function that penalizes the discrepancy between observed noisy relative orientations $\{\widetilde{R}_{uv}\}$ and the estimated relative orientations $\{R^*_{uv}\} := \{R^*_v {R^*_u}^{-1}\}$. The corresponding optimization problem can then be expressed as

$$\underset{\{R^*_v\}}{\arg\min} \sum_{\mathcal{E}_{uv} \in \mathcal{E}} \rho\Big(d\big(R^*_{uv}, \widetilde{R}_{uv}\big)\Big) \tag{1}$$

where $\rho(.)$ is a robust cost and $d(.,.)$ is a distance measure between the orientations. The nature of the above optimization is a typical complex multi-variable nonlinear optimization problem with thousands of variables (for thousands of cameras) and there seems to be no direct method (closed-form solution) minimizing the above cost even without outliers [23].

The Choice of Distance Measure $d(\widetilde{R}, R)$. There are three commonly used distance measurements in the rotation group SO(3): (i) the geodesic or angle metric $d_\theta = \angle(\widetilde{R}, R)$, (ii) the chordal metric $d_C = \|\widetilde{R} - R\|_F$ and (iii) the quaternion metric $d_Q = \min\{\|q_{\widetilde{R}} - q_R\|, \|q_{\widetilde{R}} + q_R\|\}$ where q_R and $q_{\widetilde{R}}$ are quaternion representations of R and $\widetilde{R}$ respectively, and $\|.\|_F$ is the Frobenius norm. The metrics d_C and d_Q are proven to be $2\sqrt{2}\sin(d_\theta/2)$ and $2\sin(d_\theta/4)$ respectively [23], thus, all the metrics are the same to the first order. In our implementation, we employ the quaternion representations (with non-negative scalars).

The Choice of Robust Cost $\rho(.)$. In practical applications, *e.g.* robot navigation, the agent usually ends up with some corrupt measurements (outliers), due to symmetric and repetitive man-made structures, in addition to the sensor noise. To estimate the absolute orientations of the cameras that are immune to those outliers, the conventional methods optimize a robust cost $\rho(.)$ as discussed above. An exhaustive list of such robust functions can be found in [8].

The noise and outliers in the observed relative orientations is assumed to follow some distributions subject to the cost function with mean identity orientation [23,44]. However, in real data, we observe very different noise distributions and a few such examples are shown in Fig. 2. Further, optical axis of most of the cameras are horizontal and hence the axes of the relative orientations are vertical. By training a neural network to perform the task, our aim is for the neural network to capture these patterns while predicting the absolute orientations.

4 Learning to Predict Absolute Orientations

Let $\mathcal{D} := \{\mathcal{G}\}$ be a dataset of ground-truth view-graphs. Each view-graph $\mathcal{G} := (\mathcal{V}, \mathcal{E})$ contains a noisy relative orientation measurement $\widetilde{R}_{uv}$ for each edge $\mathcal{E}_{uv} \in \mathcal{E}$ and a ground-truth absolute orientation $\widehat{R}_v$ for each camera $\mathcal{V}_v \in \mathcal{V}$. The desired neural network learns a mapping Φ that takes noisy relative measurements $\{\widetilde{R}_{uv}\}$ as input and predicts the absolute orientations $\{R_v^\Phi\} := \Phi\Big(\{\widetilde{R}_{uv}\};\ \Theta\Big)$ as output, where Θ is the set of network parameters. To train the parameters of such network, one could minimize the discrepancy between the ground-truth $\widehat{R}_{uv} := \widehat{R}_v \widehat{R}_u^{-1}$ and the estimated $R_{uv}^\Phi := R_v^\Phi {R_u^\Phi}^{-1}$ relative orientations (*cf.* Eq. (1)), *i.e.*

$$\arg\min_{\Theta} \sum_{\mathcal{G}\in\mathcal{D}} \sum_{\mathcal{E}_{uv}\in\mathcal{E}} d(R_{uv}^\Phi, \widehat{R}_{uv}) \tag{2}$$

In contrast to (1), where conventional methods optimize the orientation parameters for each instance of the view-graph $\mathcal{G} \in \mathcal{D}$, here in (2), the network parameters are optimized during training that learn the mapping Φ effectively from observed relative orientations $\{\widetilde{R}_{uv}\}$ to the target absolute orientations $\{\widehat{R}_v\}$, *i.e.* $\{\widehat{R}_v\} \approx \Phi\Big(\{\widetilde{R}_{uv}\};\ \Theta\Big)$ over the entire dataset of view-graphs $\mathcal{D}$.

Direct Training of Φ and Gauge Freedom. For an arbitrary orientation R,

$$R_{uv}^* := R_v^* {R_u^*}^{-1} = (R_v^* R)(R_u^* R)^{-1}, \quad \forall \mathcal{E}_{uv} \in \mathcal{E} \tag{3}$$

Therefore, $\{R_v^*\}$ and $\{R_v^* R\}$ essentially represent the same solution to the MRA problem (1) and there is a gauge freedom of degree 3. The mapping Φ is thus one-to-many as $\{R_v^\Phi\}$ and $\{R_v^\Phi R\}$ correspond to the same cost (2). This gauge freedom makes it difficult to train such a network. Further, one could choose a direct cost (no associated gauge freedom) to learn an one-to-one mapping Φ, *e.g.*

$$\arg\min_{\Theta} \sum_{\mathcal{G}\in\mathcal{D}} \sum_{\mathcal{V}_v\in\mathcal{V}} d(R_v^\Phi, \widehat{R}_v) \tag{4}$$

where the reference orientation is fixed according to the ground-truth. Again, $\{\widehat{R}_v\}$ and $\{\widehat{R}_v R\}$ represent the same ground-truth where the reference orientations are fixed at different directions. One could fix the issue by fixing the reference orientation to the orientation of the first camera in all the view-graphs in $\mathcal{D}$. However, in a graph (set representation), the nodes are permutation invariant. Thus the choice of the first camera, and hence the reference orientation, is arbitrary. Therefore, one needs to pass the reference orientation or the index of the first camera (possibly via a binary encoding) to the network as an additional input to be able to train such a network. However, we employ an alternative strategy adopted from the conventional optimization methods [8,23], *i.e.* initialize a solution of the absolute orientations under a fixed reference and pass the initialization to the network to fine-tune the solution. The network gets the reference orientation as an additional input via initialization (see Fig. 1(d)) and regress the parameters, *i.e.* $\{\widehat{R}_v\} \approx \Phi\Big(\{\widetilde{R}_{uv}\}, \{\widetilde{R}_v\};\ \Theta\Big)$. Further, we train the network by minimizing a combined cost where the 1st term (2) enforces the consistency over the entire graph and the 2nd term (4) enforces a unique solution, *i.e.*

$$\arg\min_{\Theta} \sum_{\mathcal{G}\in\mathcal{D}} \Big(\sum_{\mathcal{E}_{uv}\in\mathcal{E}} d(R^{\Phi}_{uv}, \widehat{R}_{uv}) + \beta \sum_{\mathcal{V}_v\in\mathcal{V}} d(R^{\Phi}_{v}, \widehat{R}_{v}) \Big) \tag{5}$$

where β is a weight parameter. Note that the reference orientation are now fixed at the orientation of a certain camera c in the initialization $\{\widetilde{R}_v\}$ as well as in the ground-truth absolute orientations $\{\widehat{R}_v\}$. Although, the choice of c is not critical, in practice, the camera c with most neighboring cameras is chosen as the reference, *i.e.* $\widetilde{R}_c = \widehat{R}_c = I_{3\times 3}$.

The above mapping Φ is now one-to-one. However, it requires an initialization $\{\widetilde{R}_v\}$ as an additional input. Conventional methods initialize the absolute orientations using a spanning tree of the view graph. However even a single outlier in that spanning tree can lead to a very poor initialization, so it is very important to identify these outliers beforehand. Further, noise in the relative orientation along each edge of the spanning tree will also propagate at the subsequent nodes while computing the initial absolute orientations. Thus, we first clean the view-graph by removing the outliers and rectifying the noisy measurements, and then bootstrap an initialization from the cleaned view-graph.

Cleaning the View-Graph. Given the local structure in the view-graph, *i.e.* measurements of all the edges that the pair of adjacent nodes $\{\mathcal{V}_u, \mathcal{V}_v\}$ are connected to (and possibly subsequent edges), an outlier edge $\mathcal{E}_{uv}$ can be detected. To be specific, chaining the relative orientations along a cycle in the local structure of the view-graph forms an orientation close to the identity orientation and an indication of an outlier in the cycle otherwise. The presence of an outliers in multiple such cycles through the current edge indicates that the edge to be an outlier. Instead of designing such explicit algorithms, we use another neural network to clean the graph. The proposed method can be summarized as follows:

- A graph-based network is employed to clean the view-graph by removing outlier measurements and rectifying noisy measurements (see Sect. 4.2).
- The cleaned view-graph is then utilized to initialize the absolute orientations (see Sect. 4.3).
- The initialization is fine-tuned using a separate network (see Sect. 4.4).

For clarity of the rest of the paper, the notations are outlined in Table 1.

4.1 The Network Design Choice

Generalizing convolution operators to irregular domains, such as graphs, is typically expressed as neighborhood aggregation or a message-passing scheme. The proposed network is built using such Message-Passing Neural Networks (MPNN) [17], directly operating on view-graphs $\mathcal{G}$. A MPNN is defined in terms of message functions $m_v^{(t)}$ and update functions $\gamma^{(t)}$ that run for T time-steps (layers). At each time-step, the hidden state $h_v^{(t)}$ at each node (feature) in the graph is updated according to

$$h_v^{(t)} = \gamma^{(t)}\left(h_v^{(t-1)}, m_v^{(t)}\right) \tag{6}$$

where $m_v^{(t)}$ is the condensed message at node v, coming from the neighboring nodes $u \in \mathcal{N}_v$, and can be expressed as follows:

$$m_v^{(t)} = \square_{\mathcal{V}_u \in \mathcal{N}_v} \phi^{(t)}\left(h_v^{(t-1)}, h_u^{(t-1)}, e_{uv}\right) \tag{7}$$

where $\square$ denotes a differentiable, permutation invariant symmetric function, *e.g. mean*, *soft-max*, etc.; $\gamma^{(t)}$ and $\phi^{(t)}$ are concatenation operations followed by 1-D convolutions and ReLUs; e_{uv} is the edge feature of the edge $\mathcal{E}_{uv}$, $h_{u\to v}^{(t)} := \phi^{(t)}\left(h_v^{(t-1)}, h_u^{(t-1)}, e_{uv}\right)$ is the accumulated message for the edge $\mathcal{E}_{uv}$ at time-step (t); and $\mathcal{N}_v$ is the set of all neighboring cameras connected to $\mathcal{V}_v$. A diagram of the elements involved in computing the next-level features is shown in Fig. 3.

Table 1. The notations and symbols used in the manuscript

Orientation parameters in the view-graph			
$\widetilde{R}_{uv} \vert R^*_{uv}$	: Observed|Noise-rectified relative	$\widetilde{R}_v \vert R^{\Phi}_v \vert R^*_v$	: Initial|Refined|predicted absolute
$\widehat{R}_{uv} \vert \widehat{R}_v$	: Ground-truth relative|absolute	$\{R_{uv}\} \vert \{R_v\}$	: Set of all relative|absolute
The network parameters and symbols			
$\alpha^*_{uv} \vert \widehat{\alpha}_{uv}$	: Predicted|Ground-truth outlier-score	$h_v^{(t)} \vert m_v^{(t)}$	: Features|Message at node v
$\phi^{(t)} \vert \gamma^{(t)}$	: Feature update|Accumulated message	lp_1, lp_2, lp_3	: Single layers of linear perceptrons

4.2 View-Graph Cleaning Network

The view-graph cleaning network (CleanNet) is built on a MPNN. The input to CleanNet is a noisy view-graph and the output is a clean one, *i.e.* the network takes noisy relative orientations $\widetilde{R}_{uv}$ as the edge features e_{uv} and predicts the noise-rectified relative orientations R^*_{uv} from the accumulated message $h_{u\to v}^{(T)} := \phi^{(T)}\left(h_v^{(T-1)}, h_u^{(T-1)}, e_{uv}\right)$ at the last layer. It also predicts a score α^*_{uv} depicting the probability of the edge $\mathcal{E}_{uv}$ to be an outlier. *i.e.*

$$R^*_{uv} = lp_1\left(h_{u\to v}^{(T)}\right) \star \widetilde{R}_{uv} \quad \text{and} \quad \alpha^*_{uv} = lp_2\left(h_{u\to v}^{(T)}\right) \tag{8}$$

where $lp_1(.)$ and $lp_2(.)$ are single-layered linear perceptrons that map the accumulated messages to the edge noise orientation and outlier score respectively.

'$\star$' is the matrix multiplication. The hidden states are initialized by null vectors, *i.e.* $h_v^{(0)} = \emptyset$. Note that instead of directly estimating the rectified orientations, we predict the noise in the relative orientation measurements, which are then multiplied to obtain the rectified orientations. The loss is chosen as the weighted combination of mean orientation error $\mathcal{L}_{mre}$ of the rectified R^*_{uv} and ground-truth $\widehat{R}_{uv} := \widehat{R}_v \widehat{R}_u^{-1}$ relative orientations, and mean binary cross-entropy error $\mathcal{L}_{bce}$ of the predicted α^*_{uv} and the ground-truth outlier score $\widehat{\alpha}_{uv}$, *i.e.*

$$\mathcal{L} = \sum_{\mathcal{G} \in \mathcal{D}} \sum_{\mathcal{E}_{uv} \in \mathcal{E}} \Big(\mathcal{L}_{mre}\big(R^*_{uv}, \widehat{R}_{uv}\big) + \lambda \mathcal{L}_{bce}\big(\alpha^*_{uv}, \widehat{\alpha}_{uv}\big) \Big) \tag{9}$$

where λ is a weight parameter (fixed as $\lambda = 10$). We formulate the orientations using unit quaternions and the predictions are normalized accordingly. The error in the prediction is also normalized by the degree of the node, *i.e.*

$$\mathcal{L}_{mre}\big(R^*_{uv}, \widehat{R}_{uv}\big) = \frac{1}{|\mathcal{N}_v||\mathcal{N}_u|} d_Q\Big(\frac{R^*_{uv}}{\|R^*_{uv}\|_2}, \widehat{R}_{uv}\Big) \tag{10}$$

Experimentally, we observed the above loss produces superior performance than the standard discrepancy loss (2). Note that the ground-truth outlier score $\widehat{\alpha}_{uv}$ is generated based on the amount of noise in the relative orientations. Specifically, if the amount of noise in the relative orientation $\widetilde{R}_v^{-1} \widehat{R}_{uv} \widetilde{R}_u > 20°$, the ground-truth edge label is assigned as an outlier, *i.e.* $\widehat{\alpha}_{uv} = 1$ and $\widehat{\alpha}_{uv} = 0$ otherwise.

An edge $\mathcal{E}_{uv}$ is marked as an outlier edge if the predicted outlier score α^*_{uv} is greater than a predefined threshold ϵ. In all of our experiments, we choose the threshold $\epsilon = 0.75$[1]. A cleaned view-graph $\mathcal{G}^*$ is then generated by removing outlier edges from $\mathcal{G}$ and replacing noisy relative orientations $\widetilde{R}_{uv}$ by the rectified orientations R^*_{uv}. Note that the cleaned graph $\mathcal{G}^*$ is only employed to bootstrap an initialization of the absolute orientations.

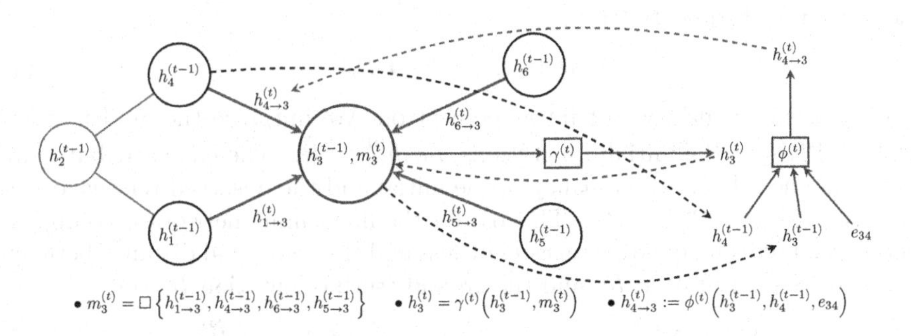

Fig. 3. An illustration of computing next level features of a message-passing network.

[1] The choice is not critical in the range $\epsilon \in [0.35, 0.8]$.

4.3 Bootstrapping Absolute Orientations

Hartley *et al.* [22] proposed generating a spanning tree by setting the camera with the maximum number of neighbors as the root and recursively adding adjacent cameras without forming a cycle. The reference orientation is fixed at the camera at the root of the spanning tree. The orientations of the rest of the cameras in the tree are computed by propagating away the rectified orientations R^*_{uv} from the root node along the edges, *i.e.* $\widetilde{R}_v = R^*_{uv}\widetilde{R}_u$.

As discussed before, the noise in the relative orientation along each edge $R^*_{uv}\widehat{R}^{-1}_{uv}$ propagates at the subsequent nodes while computing the initial absolute orientations of the cameras. Therefore, the spanning tree that minimizes the sum of depths of all the nodes (a.k.a. shortest path tree [41]) is the best spanning tree for the initialization. Starting with a root node, a shortest path tree could be computed by greedily connecting nodes to each neighboring node in the breadth-first order. The best shortest path tree can be found by applying the same procedure with each one of the nodes as a root node (time complexity $\mathcal{O}(n^2)$) [24]. However, we employed the procedure just once (time complexity $\mathcal{O}(n)$) with the root at the node with the maximum number of adjacent nodes (similar to Hartley *et al.* [22]) and observed similar results as with the best spanning tree. The reference orientation of the initialization and the ground-truth is fixed at the root of the tree. This procedure is very fast and it takes only a fraction of a second for a large view-graph with thousands of cameras. We abbreviate this procedure as SPT and it is the default initializer in all of our experiments.

4.4 Fine-Tuning Network

The fine-tuning network (FineNet) is again built on a MPNN. It takes the initial absolute orientations $\{\widetilde{R}_v\}$ and the relative orientation measurements $\{\widetilde{R}_{uv}\}$ as inputs, and predicts the refined absolute orientations $\{R^*_v\}$ as the output. The refined orientations are estimated from the hidden states $h_v^{(T)}$ of the nodes at the last layer of the network, *i.e.*

$$R^*_v = lp_3\big(h_v^{(T)}\big) \star \widetilde{R}_v \tag{11}$$

where lp_3 is a single layer of linear perceptron. We initialize the hidden states of the MPNN by the initial orientations, *i.e.* $h_v^{(0)} = \widetilde{R}_v$. The edge attributes are chosen as the relative discrepancy of the initial and the observed relative orientations, *i.e.* $e_{uv} = \widetilde{R}_v^{-1}\widetilde{R}_{uv}\widetilde{R}_u$. The loss for the fine-tuning network is computed as the weighted sum of edge consistency loss and the rotational distance between the predicted orientation $\widetilde{R}_v$ and the ground-truth orientation $\widetilde{R}_v$, *i.e.*

$$\mathcal{L} = \sum_{\mathcal{G}\in\mathcal{D}}\Big(\sum_{\mathcal{E}_{uv}\in\mathcal{E}} \mathcal{L}_{mre}\big(R^*_{uv}, \widehat{R}_{uv}\big) + \frac{\beta}{|\mathcal{N}_v|}\sum_{\mathcal{V}_v\in\mathcal{V}} d_Q\big(\frac{R^*_v}{\|R^*_v\|}, \widehat{R}_v\big)\Big) \tag{12}$$

where $\mathcal{L}_{mre}$ is chosen as the quaternion distance (10). This is a combination of two loss functions chosen according to (5). We value consistency of the entire graph (enforced via relative orientations in the first term) over individual accuracy (second term), and so choose $\beta = 0.1$.

4.5 Training

The view-graph cleaning network and the fine-tuning network are trained separately. For each edge $\mathcal{E}_{uv}$ in the view-graph with observed orientation $\widetilde{R}_{uv}$, an additional edge $\mathcal{E}_{vu}$ is included in the view-graph in the opposite direction with orientation $\widetilde{R}_{vu} := \widetilde{R}_{uv}^{-1}$. This will ensure the messages flow in both directions of an edge. In both of the above networks, the parameters are chosen as: the number of time-steps $T = 4$, the permutation invariant function $\square$ as the *mean*, and the length of the message $m_v^{(t)}$ and hidden state $h_v^{(t)}$ are 32.

Network Parameters Θ: The parameters are involved with: (i) 1-D convolutions of $\gamma^{(t)}$ in (6) and $\phi^{(t)}$ in (7), and (ii) linear perceptrons lp_1, lp_2 in (8) and lp_3 in (11). With the above hyper-parameters (*i.e.* time-steps, length of the messages, etc.), the total number of parameters of NeuRoRA becomes ≈49.8K. Note that increasing the hyper-parameters could lead to a bigger network (more parameters) and that results a slower performance. A small size network is much faster but is not capable of predicting accurate outputs for larger view-graphs. We have tried different network sizes and found the current network size is a good balance between speed and accuracy.

Architecture Setup: The networks are implemented in `PyTorch Toolbox`[2] and trained on a GTX 1080 Ti GPU with a learning rate of 0.5×10^{-4} and weight decay 10^{-4}. Each of CleanNet and FineNet are trained for 250 epochs (takes ∼4–6 h) to learn the network parameters. To prevent the networks from over-fitting on the training dataset, we randomly drop 25% of the edges of each view-graph along with observed noisy relative orientations in each epoch. During testing all the edges were kept active. The parameters that yielded the minimum validation loss were kept for evaluation. All the baselines including the proposed networks were evaluated on an Intel Core i7 CPU.

5 Results

Experiments were carried out on synthetic as well as real datasets to demonstrate the true potential of our approach.

Baseline Methods. We evaluated NeuRoRA against the following baseline methods (also described in Sect. 2):

- Chatterjee and Govindu [8]: the latest implementation with the default parameters and cost function in the shared scripts[3] were employed. We also employed their evaluation strategy to compare the predicted orientations.
- Weiszfeld algorithm [22]: the algorithm is straightforward but computationally expensive and we only ran it for 50 iterations of ℓ_1 averaging.
- Arrigoni *et al.* [5]: the authors shared the code with the optimal parameters[4].

[2] https://pytorch-geometric.readthedocs.io.
[3] http://www.ee.iisc.ac.in/labs/cvl/research/rotaveraging/.
[4] http://www.diegm.uniud.it/fusiello/demo/gmf/.

– Wang and Singer [44]: this is employed with a publicly available scripts[5].

We also ran the graph cleaning network (CleanNet) followed by bootstrapping initial orientation (using SPT) as a baseline CleanNet-SPT, and ran SPT on the noisy graph followed by fine-tuning network (FineNet) as another baseline SPT-FineNet. Note that the proposed network NeuRoRA takes CleanNet-SPT as an initialization and then fine-tunes the initialization in a single step by FineNet. NeuRoRA-$v2$ is a variation of the proposed method where an initialization from CleanNet-SPT is fine-tuned in two steps of FineNet, *i.e.* the output of FineNet in the first step is fed as an initialization of FineNet in the second step.

Synthetic Dataset. We carefully designed a synthetic dataset that closely resembles the real-world datasets. Since the amount of noise in observed relative measurements changes with the sensor type (*e.g.* camera device), the structure of the connections in the view-graphs and the outlier ratios are varied with the scene (Fig. 2). A single view-graph was generated as follows: (1) the number of cameras were sampled in the range 250–1000 and their orientations were generated randomly on a horizontal plane (yaw only), (2) pairwise edges and corresponding relative orientations were randomly introduced between the cameras that amounted to (10–30)% of all possible pairs, (3) the relative orientations were then corrupted by a noise with a std σ where σ is chosen uniformly in the range ($5°$–$30°$) once for the entire view-graph, and the directions are chosen randomly on the vertical plane (to emulate realistic distributions 2), and (4) the relative orientations were further corrupted by (0–30)% of outliers with random orientations. Our synthetic dataset consisted of 1200 sparse view-graphs. The dataset was divided into training (80%), validation (10%), and testing (10%).

The results are furnished in Table 2. The average angular error on all the view-graphs in the dataset is displayed. The proposed method NeuRoRA performs remarkably well compared to the baselines in terms of accuracy and speed. NeuRoRA-$v2$ further improves the results. Overall, Chatterjee [8] performs well but the performance does not improve with a better initialization. Unlike Wang [44], Weiszfeld [22] improves the performance with a better initialization given by CleanNet-SPT, but, it can not improve the solution further given an even better initialization by NeuRoRA. Notice that the proposed NeuRoRA is three orders of magnitude faster with a GPU than the baseline methods.

Real Dataset. We summarize the real datasets and display in Table 3. There are a total of 19 publicly available view-graphs with observed noisy relative orientations and the ground-truth absolute orientations. The ground-truth orientations were obtained by applying incremental bundle adjustment [2] on the view-graphs. The TNotreDame dataset is shared by Chatterjee *et al.* [8][6]. The Artsquad and SanFrancisco datasets are provided by DISCO [12][7]. The rest of the view-graphs are publicly shared by 1DSFM [46][8]. The ground-truth orienta-

[5] https://github.com/huangqx/map_synchronization.
[6] http://www.ee.iisc.ac.in/labs/cvl/research/rotaveraging/.
[7] http://vision.soic.indiana.edu/projects/disco/.
[8] http://www.cs.cornell.edu/projects/1dsfm/.

Table 2. Results of Rotation averaging on a test synthetic dataset. The average angular error on all the view-graphs in our dataset is displayed. The proposed method NeuRoRA is remarkably faster than the baselines while producing better results. There is no GPU implementations of [5,8,22,44] available, thus the runtime comparisons on cuda are excluded. Note that NeuRoRA takes only **0.0016 s** on average on a GPU.

Baseline methods	mn	md	cpu			mn	md	cpu	
Chatterjee [8]	2.17°	1.25°	5.38 s	(1×)	Arrigoni [5]	2.92°	1.42°	8.20 s	(0.65×)
Weiszfeld [22]	3.35°	1.02°	50.92 s	(0.11×)	Wang [44]	2.77°	1.40°	9.75 s	(0.55×)
Proposed methods									
CleanNet-SPT + [8]	2.11°	1.26°	5.41 s	(0.99×)	NeuRoRA	**1.45°**	**0.74°**	**0.21 s**	(**24**×)
CleanNet-SPT + [22]	1.74°	1.01°	50.36 s	(0.11×)	NeuRoRA-$v2$	**1.30°**	**0.68°**	**0.30 s**	(**18**×)
Other methods									
CleanNet-SPT	2.93°	1.47°	**0.11 s**	(**47**×)	SPT-FineNet	3.00°	1.57°	**0.11 s**	(**47**×)
CleanNet-SPT + [44]	2.77°	1.40°	9.86 s	(0.53×)	SPT-FineNet + [44]	2.78°	1.40°	9.86 s	(0.53×)
SPT-FineNet + [8]	2.12°	1.26°	5.41 s	(0.99×)	SPT-FineNet + [22]	1.78°	1.01°	50.36 s	(0.11×)
NeuRoRA + [8]	2.11°	1.26°	5.51 s	(0.97×)	NeuRoRA + [22]	1.73°	1.01°	50.46 s	(0.10×)

mn: mean of the angular error; md: median of the angular error; cpu: the average runtime of the method; MethodA + MethodB: MethodB is initialized by the solution of MethodA.

tions are available for some of those cameras (indicated in parenthesis) and the training, validation and testing are performed only on those cameras.

Due to limited availability of real datasets for training, we employed network parameters pre-trained on the above synthetic dataset and further fine-tuned on the real datasets in round-robin fashion (leave one out). Such evaluation protocol is employed because we did not want to divide the sequences into training and testing sequences that might favor one particular method. The finetuning is done for each round of the round-robin using the real-data apart from the held-out test sequence. Overall, the proposed NeuRoRA outperformed the baselines for this task in terms of accuracy and efficiency (Table 3). The Artsquad and SanFrancisco datasets have different orientation patterns as shared from a different source [12]. In particular SanFrancisco dataset is captured along a road which is significantly different from others. Thus, the performance of NeuRoRA falls short to Chatterjee [8] and Wang [44] only on those two sequences, but, is still better than Weiszfeld [22] and Arrigoni [5]. Nonetheless, the proposed NeuRoRA is much faster than others.

Robustness Check. In this experiment we study the generalization capability of NeuRoRA. To check the individual effects of different sensor settings, we generate a number of synthetic dataset varying **(i)** #cameras **(ii)** #edges, **(iii)** amount of noise and outliers, and **(iv)** planar/random camera motion. NeuRoRA is then trained on one of such datasets and evaluated on the others. Each dataset consists of 1000 view-graphs (large ones contain only 100). Results are furnished in Table 4. Notice that NeuRoRa generalizes well across dataset changes except when the network is trained on planar cameras and tested on random. We therefore advice to use two separate networks for planar and nonplanner scenes. Notice that Chatterjee [8] demands a large memory for large

Table 3. Results of MRA on real datasets. The proposed method NeuRoRA is much faster than the baselines while producing overall similar or better results. The number of cameras, for which ground-truths are available, is shown within parenthesis.

Datasets			Chatterjee [8]			Weiszfeld [22]			Arrigoni [5]			Wang [44]			NeuRoRA		
Name	#cameras	#edges	mn	md	cpu	mn	md	cpu	mn	md	cpu	mn	md	cpu	mn	md	cpu
Alamo	627(577)	49.5%	**4.2**	**1.1**	20.5 s	4.9	1.4	84.0 s	6.2	1.6	2.7 s	5.3	1.4	20.6 s	4.9	1.2	**2.2 s**
EllisIsland	247(227)	66.8%	2.8	**0.5**	2.5 s	4.4	1.0	8.9 s	3.9	1.2	**0.2 s**	3.6	1.1	2.6 s	**2.6**	0.6	0.4 s
GendrmMarkt	742(677)	17.5%	37.6	7.7	11.1 s	29.4	9.6	53.7 s	41.6	13.3	8.9 s	32.6	6.1	12.5 s	**4.5**	**2.9**	**0.5 s**
MadridMetrop	394(341)	30.7%	6.9	1.2	3.2 s	7.5	2.7	14.5 s	6.0	1.7	0.9 s	5.0	1.4	3.6 s	**2.5**	**1.1**	**0.2 s**
MontrealNotre	474(450)	46.8%	1.5	**0.5**	9.1 s	2.1	0.7	41.5 s	4.8	0.9	2.9 s	2.0	0.8	10.1 s	**1.2**	0.6	**1.0 s**
NYCLibrary	376(332)	29.3%	3.0	1.3	4.8 s	3.8	2.1	14.4 s	3.9	1.5	1.4 s	2.9	1.4	3.2 s	**1.9**	**1.1**	**0.2 s**
NotreDame	553(553)	68.1%	3.5	**0.6**	23.3 s	4.7	0.8	80.8 s	3.9	1.0	4.2 s	3.5	0.9	19.5 s	**1.6**	**0.6**	**2.0 s**
PiazzaDelP	354(338)	39.5%	4.0	0.8	3.3 s	4.8	1.3	16.7 s	10.8	1.2	0.6 s	6.2	1.1	3.6 s	**3.0**	**0.7**	**0.4 s**
Piccadilly	2508(2152)	10.2%	6.9	2.9	449.0 s	26.4	7.5	∼20 m	22.0	9.7	43.7 s	10.1	3.9	118.1 s	**4.7**	**1.9**	**5.9 s**
RomanForum	1134(1084)	10.9%	3.1	1.5	20.2 s	4.8	1.8	115.0 s	13.2	8.2	16.8 s	4.6	3.5	19.6 s	**2.3**	**1.3**	**1.3 s**
TowerLondon	508(472)	18.5%	3.9	2.4	1.9 s	4.7	2.9	17.1 s	4.6	1.8	3.9 s	2.9	1.5	3.6 s	**2.6**	**1.4**	**0.3 s**
Trafalgar	5433(5058)	4.6%	**3.5**	**2.0**	858.4 s	15.6	11.3	∼92 m	48.6	13.2	167.4 s	17.2	16.0	319.2 s	5.3	2.2	**15.5 s**
UnionSquare	930(789)	5.9%	9.3	3.9	6.8 s	40.9	10.3	42.8 s	9.2	4.4	12.1 s	6.8	3.2	4.1 s	**5.9**	**2.0**	**0.6 s**
ViennaCath	918(836)	24.6%	8.2	1.2	48.1 s	11.7	1.9	158.3 s	19.3	2.39	6.0 s	10.1	1.8	25.7 s	**3.9**	**1.5**	**2.1 s**
Yorkminster	458(437)	26.5%	3.5	1.6	4.0 s	5.7	2.0	32.0 s	4.5	1.6	2.5 s	3.5	1.3	4.9 s	**2.5**	**0.9**	**0.4 s**
Acropolis	463(463)	10.7%	1.1	0.7	1.5 s	**0.6**	**0.3**	15.0 s	2.7	1.7	3.2 s	2.4	1.6	1.7 s	0.8	0.5	**0.2 s**
ArtsQuad	5530(4978)	1.4%	**4.8**	3.5	116.1 s	34.4	23.1	∼32 m	35.2	15.8	189.4 s	6.0	**3.2**	73.9 s	27.5	7.3	**5.0 s**
SanFran	7866(7866)	0.3%	**3.6**	**3.4**	15.2 s	18.8	16.4	∼22 m	66.8	43.9	354.7 s	89.2	75.5	27.2 s	17.6	12.6	**2.6 s**
TNotreDame	715(715)	25.3%	**1.0**	**0.4**	10.6 s	1.4	0.6	72.5 s	2.4	0.9	5.7 s	1.7	0.8	14.8 s	1.7	0.7	**1.4 s**

mn: mean of the angular error (in deg); md: median of the angular error (in deg); cpu: the runtime of the method on a cpu (s: in sec, m: in minute); entries with >20° or >120 s are marked in red.

Table 4. Robustness check: results of Rotation averaging on multiple datasets where NeuRoRA is trained on one and evaluated on the other datasets.

Robustness	Training datasets				Evaluation datasets				NeuRoRA			Chatterjee [8]		
	$\lvert\mathcal{V}\rvert$	$\lvert\mathcal{E}\rvert$	E&O	P	$\lvert\mathcal{V}\rvert$	$\lvert\mathcal{E}\rvert$	E&O	P	mn	md	cpu	mn	md	cpu
#cameras ($\lvert\mathcal{V}\rvert$)	1000	25.0%	30°&10%	✓	**250**	25.0%	30°&10%	✓	**1.1°**	**0.9°**	**0.1 s**	1.8°	1.7°	0.3 s
	250	25.0%	30°&10%	✓	**5000**	**2.5%**	30°&10%	✓	**1.1°**	**1.0°**	**4.9 s**	1.4°	1.3°	~12 m
	250	25.0%	30°&10%	✓	**10000**	**2.5%**	30°&10%	✓	**0.7°**	**0.6°**	**18.7 s**	Out of memory		
	250	25.0%	30°&10%	✓	**25000**	**2.5%**	30°&10%	✓	**0.6°**	**0.5°**	**142.6 s**	Out of memory		
#edges ($\lvert\mathcal{E}\rvert$)	1000	25.0%	30°&10%	✓	1000	**2.5%**	30°&10%	✓	**2.4°**	**2.1°**	**0.1 s**	3.0°	2.8°	3.3 s
	1000	**2.5%**	30°&10%	✓	1000	25.0%	30°&10%	✓	**0.5°**	**0.4°**	**2.5 s**	0.9°	0.8°	43.2 s
Noise & outliers (E&O)	1000	25.0%	30°&10%	✓	1000	25.0%	**10°&5%**	✓	0.4°	**0.3°**	**2.5 s**	**0.3°**	**0.3°**	31.6 s
	1000	25.0%	**10°& 5%**	✓	1000	25.0%	30°&10%	✓	**0.6°**	**0.5°**	**2.5 s**	0.9°	0.8°	43.2 s
Planar (P)	1000	25.0%	30°&10%	✓	1000	25.0%	30°&10%	✗	2.2°	1.6°	**2.5 s**	**0.9°**	**0.8°**	26.3 s
	1000	25.0%	30°&10%	✗	1000	25.0%	30°&10%	✓	**0.9°**	**0.7°**	**2.5 s**	**0.9°**	0.8°	26.3 s

E: noise with a std chosen uniformly in the range (0°–E°); O: percentage of outliers; P: flag for planar cameras.

view-graphs and failed to execute on a system of 64 Gb of RAM. We tested our method on view-graphs upto 25K vertices and 24M edges on the same system.

6 Discussion

We have proposed a graph-based neural network for absolute orientation regression of a number of cameras from their observed relative orientations. The proposed network is exceptionally faster than the strong optimization-based baselines while producing better results on most datasets. The outstanding performance of the current work and the relevant neural networks for test-time optimization leads to the following question: *"can we then replace all the optimizations in robotics/computer vision by a suitable neural network-based regression?"* The answer is obviously *No*. For instance, if an optimization at test-time requires solving a simpler convex cost with a few parameters to optimize, a naive gradient descent will find the globally optimal parameters, while a network-based regression would only estimate sub-optimal parameters. To date, neural nets have been proven to be consistently better at solving pattern recognition problems than solving a constraint optimization problems. A few neural network-based solutions are proposed recently that can exploit the patterns in the data while solving a test-time optimization. Therefore the current work also opens up many questions related to the right tool for a specific application.

Acknowledgement. We gratefully acknowledge the support of the Australian Research Council through the Centre of Excellence for Robotic Vision, CE140100016 and the Australian Research Council Discovery Project DP200101675.

References

1. Adler, J., Öktem, O.: Solving ill-posed inverse problems using iterative deep neural networks. Inverse Prob. **33**(12), 124007 (2017)
2. Agarwal, S., Snavely, N., Seitz, S.M., Szeliski, R.: Bundle adjustment in the large. In: Daniilidis, K., Maragos, P., Paragios, N. (eds.) ECCV 2010. LNCS, vol. 6312, pp. 29–42. Springer, Heidelberg (2010). https://doi.org/10.1007/978-3-642-15552-9_3
3. Aoki, Y., Goforth, H., Srivatsan, R.A., Lucey, S.: PointNetLK: robust & efficient point cloud registration using PointNet. In: Proceedings of CVPR, pp. 7163–7172 (2019)
4. Arrigoni, F., Fusiello, A.: Synchronization problems in computer vision with closed-form solutions. IJCV **128**(1), 26–52 (2020)
5. Arrigoni, F., Rossi, B., Fragneto, P., Fusiello, A.: Robust synchronization in SO(3) and SE(3) via low-rank and sparse matrix decomposition. CVIU **174**, 95–113 (2018)
6. Carlone, L., Tron, R., Daniilidis, K., Dellaert, F.: Initialization techniques for 3D SLAM: a survey on rotation estimation and its use in pose graph optimization. In: Proceedings of ICRA, pp. 4597–4604 (2015)
7. Chandrasekaran, R., Tamir, A.: Open questions concerning Weiszfeld's algorithm for the Fermat-Weber location problem. Math. Program. **44**(1–3), 293–295 (1989)

8. Chatterjee, A., Govindu, V.M.: Robust relative rotation averaging. TPAMI **40**(4), 958–972 (2017)
9. Chatterjee, A., Madhav Govindu, V.: Efficient and robust large-scale rotation averaging. In: Proceedings of ICCV, pp. 521–528 (2013)
10. Chen, Y., et al.: Learning to learn without gradient descent by gradient descent. In: Proceedings of ICML, pp. 748–756. JMLR.org (2017)
11. Clark, R., Bloesch, M., Czarnowski, J., Leutenegger, S., Davison, A.J.: Learning to solve nonlinear least squares for monocular stereo. In: Ferrari, V., Hebert, M., Sminchisescu, C., Weiss, Y. (eds.) ECCV 2018. LNCS, vol. 11212, pp. 291–306. Springer, Cham (2018). https://doi.org/10.1007/978-3-030-01237-3_18
12. Crandall, D., Owens, A., Snavely, N., Huttenlocher, D.: Discrete-continuous optimization for large-scale structure from motion. In: Proceedings of CVPR, pp. 3001–3008 (2011)
13. Enqvist, O., Kahl, F., Olsson, C.: Non-sequential structure from motion. In: Proceedings of ICCV Workshops, pp. 264–271 (2011)
14. Eriksson, A., Olsson, C., Kahl, F., Chin, T.J.: Rotation averaging and strong duality. In: Proceedings of CVPR, pp. 127–135 (2018)
15. Fredriksson, J., Olsson, C.: Simultaneous multiple rotation averaging using Lagrangian duality. In: Lee, K.M., Matsushita, Y., Rehg, J.M., Hu, Z. (eds.) ACCV 2012. LNCS, vol. 7726, pp. 245–258. Springer, Heidelberg (2013). https://doi.org/10.1007/978-3-642-37431-9_19
16. Garg, R., Vijay Kumar, B.G., Carneiro, G., Reid, I.: Unsupervised CNN for single view depth estimation: geometry to the rescue. In: Leibe, B., Matas, J., Sebe, N., Welling, M. (eds.) ECCV 2016. LNCS, vol. 9912, pp. 740–756. Springer, Cham (2016). https://doi.org/10.1007/978-3-319-46484-8_45
17. Gilmer, J., Schoenholz, S.S., Riley, P.F., Vinyals, O., Dahl, G.E.: Neural message passing for quantum chemistry. In: Proceedings of ICML, pp. 1263–1272. JMLR.org (2017)
18. Govindu, V.M.: Combining two-view constraints for motion estimation. In: Proceedings of CVPR, vol. 2, p. II (2001)
19. Govindu, V.M.: Lie-algebraic averaging for globally consistent motion estimation. In: Proceedings of CVPR, vol. 1, p. I (2004)
20. Govindu, V.M.: Robustness in motion averaging. In: Narayanan, P.J., Nayar, S.K., Shum, H.-Y. (eds.) ACCV 2006. LNCS, vol. 3852, pp. 457–466. Springer, Heidelberg (2006). https://doi.org/10.1007/11612704_46
21. Guibas, L.J., Huang, Q., Liang, Z.: A condition number for joint optimization of cycle-consistent networks. In: Proceedings of NeurIPS, pp. 1005–1015 (2019)
22. Hartley, R., Aftab, K., Trumpf, J.: L1 rotation averaging using the Weiszfeld algorithm. In: Proceedings of CVPR, pp. 3041–3048 (2011)
23. Hartley, R., Trumpf, J., Dai, Y., Li, H.: Rotation averaging. IJCV **103**(3), 267–305 (2013)
24. Hassin, R., Tamir, A.: On the minimum diameter spanning tree problem. Inf. Process. Lett. **53**(2), 109–111 (1995)
25. Huang, X., Liang, Z., Zhou, X., Xie, Y., Guibas, L.J., Huang, Q.: Learning transformation synchronization. In: Proceedings of CVPR, pp. 8082–8091 (2019)
26. Huynh, D.Q.: Metrics for 3D rotations: comparison and analysis. J. Math. Imaging Vis. **35**(2), 155–164 (2009)
27. Jiang, N., Cui, Z., Tan, P.: A global linear method for camera pose registration. In: Proceedings of ICCV, pp. 481–488 (2013)
28. Kendall, A., Grimes, M., Cipolla, R.: PoseNet: a convolutional network for real-time 6-DOF camera relocalization. In: Proceedings of ICCV, pp. 2938–2946 (2015)

29. Khalil, E., Dai, H., Zhang, Y., Dilkina, B., Song, L.: Learning combinatorial optimization algorithms over graphs. In: Proceedings of NIPS, pp. 6348–6358 (2017)
30. Kümmerle, R., Grisetti, G., Strasdat, H., Konolige, K., Burgard, W.: g2o: a general framework for graph optimization. In: Proceedings of ICRA, pp. 3607–3613 (2011)
31. Lv, Z., Dellaert, F., Rehg, J.M., Geiger, A.: Taking a deeper look at the inverse compositional algorithm. In: Proceedings of CVPR, pp. 4581–4590 (2019)
32. Moulon, P., Monasse, P., Marlet, R.: Global fusion of relative motions for robust, accurate and scalable structure from motion. In: Proceedings of ICCV, pp. 3248–3255 (2013)
33. Mouragnon, E., Lhuillier, M., Dhome, M., Dekeyser, F., Sayd, P.: Generic and real-time structure from motion using local bundle adjustment. Image Vis. Comput. **27**(8), 1178–1193 (2009)
34. Özyeşil, O., Voroninski, V., Basri, R., Singer, A.: A survey of structure from motion*. Acta Numerica **26**, 305–364 (2017)
35. Qi, C.R., Su, H., Mo, K., Guibas, L.J.: PointNet: deep learning on point sets for 3D classification and segmentation. In: Proceedings of CVPR, pp. 652–660 (2017)
36. Shah, R., Chari, V., Narayanan, P.J.: View-graph selection framework for SfM. In: Ferrari, V., Hebert, M., Sminchisescu, C., Weiss, Y. (eds.) ECCV 2018. LNCS, vol. 11209, pp. 553–568. Springer, Cham (2018). https://doi.org/10.1007/978-3-030-01228-1_33
37. Shen, T., Zhu, S., Fang, T., Zhang, R., Quan, L.: Graph-based consistent matching for structure-from-motion. In: Leibe, B., Matas, J., Sebe, N., Welling, M. (eds.) ECCV 2016. LNCS, vol. 9907, pp. 139–155. Springer, Cham (2016). https://doi.org/10.1007/978-3-319-46487-9_9
38. Singer, A.: Angular synchronization by eigenvectors and semidefinite programming. Appl. Comput. Harmonic Anal. **30**(1), 20–36 (2011)
39. Sweeney, C., Sattler, T., Hollerer, T., Turk, M., Pollefeys, M.: Optimizing the viewing graph for structure-from-motion. In: Proceedings of ICCV, pp. 801–809 (2015)
40. Tang, C., Tan, P.: BA-Net: dense bundle adjustment network. In: ICLR 2019 (2019)
41. Tarjan, R.E.: Sensitivity analysis of minimum spanning trees and shortest path trees. Inf. Process. Lett. **14**(1), 30–33 (1982)
42. Tron, R., Vidal, R.: Distributed image-based 3-D localization of camera sensor networks. In: Proceedings of CDC, pp. 901–908 (2009)
43. Tron, R., Zhou, X., Daniilidis, K.: A survey on rotation optimization in structure from motion. In: Proceedings of CVPR Workshops, pp. 77–85 (2016)
44. Wang, L., Singer, A.: Exact and stable recovery of rotations for robust synchronization. Inf. Infer. J. IMA **2**(2), 145–193 (2013)
45. Wilson, K., Bindel, D., Snavely, N.: When is rotations averaging hard? In: Leibe, B., Matas, J., Sebe, N., Welling, M. (eds.) ECCV 2016. LNCS, vol. 9911, pp. 255–270. Springer, Cham (2016). https://doi.org/10.1007/978-3-319-46478-7_16
46. Wilson, K., Snavely, N.: Robust global translations with 1DSfM. In: Fleet, D., Pajdla, T., Schiele, B., Tuytelaars, T. (eds.) ECCV 2014. LNCS, vol. 8691, pp. 61–75. Springer, Cham (2014). https://doi.org/10.1007/978-3-319-10578-9_5
47. Zach, C., Klopschitz, M., Pollefeys, M.: Disambiguating visual relations using loop constraints. In: Proceedings of CVPR, pp. 1426–1433 (2010)

SG-VAE: Scene Grammar Variational Autoencoder to Generate New Indoor Scenes

Pulak Purkait[1](✉), Christopher Zach[2], and Ian Reid[1]

[1] Australian Institute of Machine Learning and School of Computer Science, The University of Adelaide, Adelaide, SA 5005, Australia
pulak.isi@gmail.com

[2] Chalmers University of Technology, 41296 Goteborg, Sweden

Abstract. Deep generative models have been used in recent years to learn coherent latent representations in order to synthesize high-quality images. In this work, we propose a neural network to learn a generative model for sampling consistent indoor scene layouts. Our method learns the co-occurrences, and appearance parameters such as shape and pose, for different objects categories through a grammar-based auto-encoder, resulting in a compact and accurate representation for scene layouts. In contrast to existing grammar-based methods with a user-specified grammar, we construct the grammar automatically by extracting a set of production rules on reasoning about object co-occurrences in training data. The extracted grammar is able to represent a scene by an augmented parse tree. The proposed auto-encoder encodes these parse trees to a latent code, and decodes the latent code to a parse tree, thereby ensuring the generated scene is always valid. We experimentally demonstrate that the proposed auto-encoder learns not only to generate valid scenes (i.e. the arrangements and appearances of objects), but it also learns coherent latent representations where nearby latent samples decode to similar scene outputs. The obtained generative model is applicable to several computer vision tasks such as 3D pose and layout estimation from RGB-D data.

Keywords: Scene grammar · Indoor scene synthesis · VAE

1 Introduction

Recently proposed approaches for deep generative models have seen great success in producing high quality RGB images [7,16,17,24] and continuous latent representations from images [12]. Our work aims to learn coherent latent representations for generating natural indoor scenes comprising different object categories

Electronic supplementary material The online version of this chapter (https://doi.org/10.1007/978-3-030-58586-0_10) contains supplementary material, which is available to authorized users.

A. Vedaldi et al. (Eds.): ECCV 2020, LNCS 12369, pp. 155–171, 2020.
https://doi.org/10.1007/978-3-030-58586-0_10

and their respective appearances (*i.e.* pose and shape). Such a learned representation has direct use for various computer vision and scene understanding tasks, including **(i)** 3D scene-layout estimation [27], **(ii)** 3D visual grounding [3,31], **(iii)** Visual Question Answering [1,18], and **(iv)** robot navigation [19].

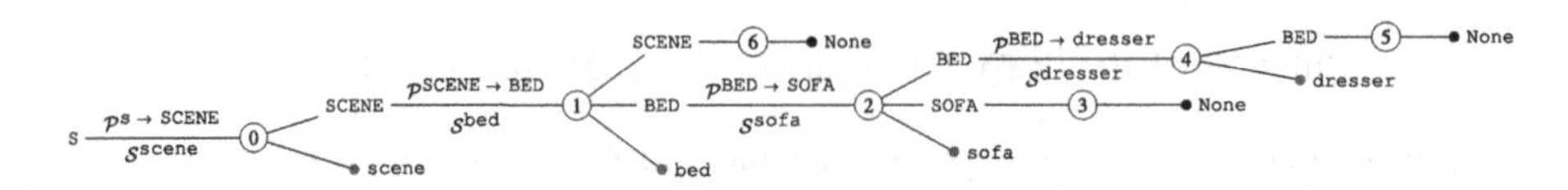

Fig. 1. An example of parse tree obtained by applying the CFG to a scene comprising `bed, sofa, dresser`. The sequence of production rules ⓪–⑥ are marked in order. The attributes of production rules are displayed above and below of the rules.

Developing generative models for such discrete domains has been explored in a limited number of works [9,23,34]. These works utilize prior knowledge of indoor scenes by manually defining attributed grammars. However, the number of rules in such grammars can be prohibitively large for real indoor environments, and consequently, these methods are evaluated only on synthetic data with a small number of objects. Further, the Monte Carlo based inference method can be intractably slow: up to 40 min [23] or one hour [9] to estimate a single layout. Deep generative models for discrete domains have been proposed in [6] (employing sequential representations) and in [14] (based on formal grammars). In our work, we extend [14] by integrating object attributes, such as pose and shape of objects in a scene. Further, the underlying grammar is often defined manually [14,23], but we propose to extract suitable grammar rules from training data automatically. The main components of our approach are thus:

- a scene grammar variational autoencoder (SG-VAE) that captures the appearances (*i.e.* pose and shape) of objects in the same 3D spatial configurations in a compact latent code (Sect. 2);
- a context free grammar that explains causal relationships among objects which frequently co-occur, automatically extracted from training data (Sect. 3);
- the practicality of the learned latent space is also demonstrated for a computer vision task.

Our SG-VAE is fast and has the ability to represent the scene in a coherent latent space as shown in Sect. 4.

2 Deep Generative Model for Scene Generation

The proposed method is influenced by the Grammar Variational Autoencoder [14], so we begin with a brief description of that prior art.

The Grammar VAE takes a valid string (in their case a chemical formula) and begins by parsing it into a set of production rules. These rules are represented as

1-hot binary vectors and encoded compactly to a latent code by the VAE. Latent codes can then be sampled and decoded to production rules and corresponding valid strings. More specifically, each production rule is represented by a 1-hot vector of size N, where N is the total number of rules, *i.e.* $N = |\mathcal{R}|$ (where $\mathcal{R}$ is the set of rules). The maximum size T of the sequence is fixed in advance. Thus the scene is represented by a sequence $\mathcal{X} \in \{0,1\}^{N \times T}$ of 1-hot vectors (note that when fewer than T rules are needed, a dummy/null rule is used to pad the sequence up to length T ensuring that the input to the autoencoder is always the same size). $\mathcal{X}$ is then encoded to a continuous (low)-dimensional latent posterior distribution $\mathcal{N}(\boldsymbol{\mu}(\mathcal{X}), \boldsymbol{\Sigma}(\mathcal{X}))$. The decoding network, which is a recurrent network, maps latent vectors to a set of unnormalized log probability vectors (logits) corresponding to the production rules. To convert from the output logits to a valid sequence of production rules, each logit vector is considered in turn. The max output in the logit vector gives a 1-hot encoding of a production rule, but only some sequences of rules are valid. To avoid generating a rule that is inconsistent with the rules that have preceded it, invalid rules are masked out of the logit and the max is taken over only unmasked elements. This ensures that the Grammar VAE only ever generates valid outputs. Further details of the Grammar VAE can be found in [14].

Adapting this idea to the case of generating scenes requires that we incorporate not only valid co-occurrences of objects, but also valid attributes such as absolute pose (3D location and orientation) and shape (3D bounding boxes) of the objects in the scene. More specifically, our proposed SG-VAE is adapted from the Grammar VAE in the following ways:

- The object attributes, *i.e.* absolute pose and shape of the objects are estimated while inferring the production rules.
- The SG-VAE is moreover designed to generate valid 3D scenes which adhere not only to the rules of grammar, but also generate valid poses.

2.1 Scene-Grammar Variational Autoencoder

We represent the objects in indoor scenes explicitly by a set of production rules, so that the entire arrangement—*i.e.* the occurrences and appearances (*i.e.* pose and shape) of the objects in a scene—is guaranteed to be consistent during inference. Nevertheless we also aim to capture the advantages of deep generative models in admitting a compact representation that can be rapidly decoded. While a standard VAE would implicitly *encourage* decoded outputs to be scene-like, our proposed solution extends the Grammar VAE [14] to explicitly enforce an underlying grammar, while still possessing the aforementioned advantages of deep generative models. For example, given an appearance of an object *bed*, the model finds strong evidence for co-occurrence of another indoor object, *e.g. dresser*. Furthermore, given the attributes (3D pose and bounding boxes) of one object (*bed*), the attributes of the latter (*dresser*) can be inferred.

The model comprises two parts: **(i)** a context free grammar (CFG) that represents valid configurations of objects; **(ii)** a Variational Autoencoder (VAE) that maps a sequence of production rules (*i.e.* a valid scene) to a low dimensional latent space, and decodes a latent vector to a sequence of production rules which in turn define a valid scene.

2.2 CFG of Indoor Scenes

A context-free grammar can be defined by a 4-tuple of sets $G = (S, \Sigma, \mathcal{V}, \mathcal{R})$ where S is a distinct non-terminal symbol known as start symbol; Σ is the finite set of non-terminal symbols; $\mathcal{V}$ is the set of terminal symbols; and $\mathcal{R}$ is the set of production rules. Note that in a CFG, the left hand side is always a non-terminal symbol. A set of all valid configurations $\mathcal{C}$ derived from the production rules defined by the CFG G is called a language. In contrast to [23] where the grammar is pre-specified, we propose a data-driven algorithm to generate a set of production rules that constitutes a CFG.

We select a few objects and associate a number of non-terminals. Only those objects that lead to co-occurrence of other objects also exist as non-terminals (described in detail in Sect. 3.2). A valid production rule is thus *"an object category, corresponding to a non-terminal, generates another object category"*. For clarity, non-terminals are denoted in upper-case with the object name. For example `BED` and `bed` are the non-terminal and the terminal symbols corresponding to the object category *bed*. Thus occurrence of a non-terminal `BED` leads to occurrence of the immediate terminal symbol `bed` and possibly further occurrences of other terminal symbols that *bed* co-occurs with, *e.g.* `dresser`. Thus, a set of rules {`S` $\longrightarrow$ `scene SCENE; SCENE` $\longrightarrow$ `bed BED SCENE; BED` $\longrightarrow$ `bed BED; BED` $\longrightarrow$ `dresser BED; BED` $\longrightarrow$ `None; SCENE` $\longrightarrow$ `None`} can be defined accordingly. Note that an additional object category *scene* is incorporated to represent the shape and size of the room. The learned scene grammar is composed of following rules:

(R1) *involving start symbol* `S`: generates the terminal `scene` and non-terminal `SCENE` that represents the indoor scene layout with attributes as the room size and room orientation, *e.g.* `S` $\rightarrow$ `scene SCENE;`. This rule ensures generating a room first.
(R2) *involving non-terminal* `SCENE`: generates a terminal and a non-terminal corresponding to an object category, *e.g.* `SCENE` $\rightarrow$ `bed BED SCENE;`.
(R3) *generating a terminal object category*: a non-terminal generates a terminal corresponding to another object category, *e.g.* `BED` $\rightarrow$ `dresser BED;`.
(R4) *involving* `None`: non-terminal symbols assigned to `None`, *e.g.* `BED` $\rightarrow$ `None;`.

`None` is an empty object and corresponding rule is a dummy rule indicating that the generation of the non-terminal is complete and the parser is now ready to handle the next non-terminal in the stack. The proposed method to deduce a CFG from data is described in detail in Sect. 3.

Note that the above CFG creates a necessary but not sufficient description. For example, a million dresser and a bed in a bedroom is a valid configuration by the grammar. Likewise, the relative orientation and shape are not included in the grammar, therefore a scene consisting of couple of small beds on a huge pillow is also a valid scene under the grammar. However, these issues are handled further by the co-occurrence distributions learned by the autoencoder.

2.3 The VAE Network

Let $\mathcal{D}$ be a set of scenes comprising multiple objects. Let $\mathcal{S}_i^j$ be the (bounding box) shape parameters and $\mathcal{P}_i^j = (T_i^j;\ \gamma_i^j)$ be the (absolute) pose parameters of jth object in the ith scene where T_i^j is the center and γ_i^j is the (yaw) angle corresponding to the direction of the object in the horizontal plane, respectively. Note that an object bounding box is aligned with gravity, thus there is only one degree of freedom in its orientation. The world co-ordinates are aligned with the camera co-ordinate frame.

The pose and shape attributes $\Theta^{j\to k} = (\mathcal{P}_i^{j\to k}, \mathcal{S}_i^k)$ are associated with a production rule in which a non-terminal X_j yields a terminal X_k. The pose parameters $\mathcal{P}_i^{j\to k}$ of the terminal object X_k are computed w.r.t. the non-terminal object X_j on the left of the production rule. *i.e.* $\mathcal{P}_i^{j\to k} = \left(\mathcal{P}_i^j\right)^{-1}\mathcal{P}_i^k$. The absolute poses of the objects are determined by chaining the relative poses on the path from the root node to the terminal node in the parse tree (see Fig. 1). Note that pose and shape attributes of the production rules corresponding to `None` object are fixed to zero.

The VAE must encode and decode both production rules (1-hot vectors) and the corresponding pose and shape parameters. We achieve this by having separate initial branches of the encoder into which the attributes $\Theta^{j\to k}$, and the 1-hot vectors are passed. Features from the 1-hot encoding branch and the pose-shape branch are then concatenated after a number of 1D convolutional layers. These concatenated features undergo further 1D convolutional layers before being flattened and mapped to the latent space (thereby predicting $\boldsymbol{\mu}$ and $\boldsymbol{\Sigma}$ of $\mathcal{N}(\boldsymbol{\mu}, \boldsymbol{\Sigma})$). The decoding network is a recurrent network consisting of a stack of GRUs, that takes samples $\boldsymbol{z} \sim \mathcal{N}(\boldsymbol{\mu}, \boldsymbol{\Sigma})$ (employing reparameterization trick [12]) and outputs logits (corresponding to the production rules) and corresponding attributes $\Theta^{j\to k}$. Logits corresponding to invalid production rules are masked out.

The reconstruction loss of our SG-VAE consists of two parts: **(i)** a cross entropy loss corresponding to the 1-hot encoding of the production rules—note that *soft-max* is computed only on the components after mask-out—and **(ii)** a mean squared error loss corresponding to the production rule attributes (but omitting the terms of `None` objects). Thus, the loss is given as follows:

$$\mathcal{L}_{total}(\phi, \theta; \mathcal{X}, \Theta) = \mathcal{L}_{vae}(\phi, \theta; \mathcal{X}) + \lambda_1\Big(\mathcal{L}_{pose}(\phi, \theta; \mathcal{P}) + \lambda_2\mathcal{L}_{shape}(\phi, \theta; \mathcal{S})\Big) \quad (1)$$

where $\mathcal{L}_{vae}$ is the autoencoder loss [14], and $\mathcal{L}_{pose}$ and $\mathcal{L}_{shape}$ are mean squared error loss corresponding to pose and shape parameters, respectively; ϕ, and θ are the encoder and decoder parameters of the autoencoder that we optimize; $(\mathcal{X}, \Theta)$ are the set of training examples comprising 1-hot encoders and rule attributes. Instead of directly regressing the orientation parameter, the respective *sines* and *cosines* are regressed. Our choice is $\lambda_1 = 10$ and $\lambda_2 = 1$ in all experiments.

3 Discovery of the Scene Grammar

In much previous work a grammar is manually specified. However in this work we aim to discover a suitable grammar for scene layouts in a data-driven manner. It comprises two parts. First we generate a causal graph of all pairwise relationships discovered in the training data, as described in more detail in Sect. 3.1. Second we prune this causal graph by removing all but the dominant discovered relationships, as described in Sect. 3.2.

3.1 Data-Driven Relationship Discovery

We aim to discover causal relationships of different objects that reflects the influence of the appearance (*i.e.* pose and shape) of one object to another. We learn the relationship using hypothesis testing, with each successful hypothesis added to a causal graph (directed) $\mathcal{G} : (\mathcal{V},\ \mathcal{E})$ where the vertex set $\mathcal{V} = \{X_1, \ldots, X_n\}$ is the set of different object categories, and edge set $\mathcal{E}$ is the set of causal relationships. An edge $(X_j \circ\!\!\rightarrow X_{j'}) \in \mathcal{E}$ corresponds to a direct causal influence on occurrence of the object X_j to the object $X_{j'}$. We conduct separate **(i)** appearance based and **(ii)** co-occurrence based testing for causal relationships between a pair of object categories as set out below.

Algorithm 1: χ^2-test for conditional independence check

Input: Co-occurrences O of the objects $X_j, X_{j'}, X_k$

Output: **True** if $X_j \perp\!\!\!\perp X_{j'} \mid X_k$ and **False** Otherwise

1 $$\chi^2 = \sum_{j,j',k \in (\{0,1\})^3} \frac{\left(O_{j,j',k} - \frac{O_{j,k} O_{j',k}}{O_k}\right)^2}{\frac{O_{j,k} O_{j',k}}{O_k}},$$

$O_{j,j',k}$: frequency of occurrences of (j, j', k),
$O_{j,k}$: frequency of occurrences of (j, k),
O_k : frequency of occurrences of k,
N : number of scenes

2 Compute p-value from cumul. χ^2 distrib. with above χ^2 value and d.o.f. ; /* D.o.f is 2 */

3 **return** p-value $< \tau$ (we choose $\tau = 0.05$)

(i) Testing for Dependency Based on Co-occurrences. We seek to capture loose associations (*e.g.* sofa and TV) and determine if these associations have a potentially causal nature. To do so, for each pair of object categories we then consider whether these categories are dependent, given a third category. This is performed using the Chi-squared (χ^2) test described below. If the dependence persists across all possible choices of the third category, we conclude that the dependence is not induced by another object, therefore a potentially causal link should exist between them. This exhaustive series of tests is $O(Nm^3)$, where N is the number of scenes and m is the number of categories (in our case, 84). However it is performed offline and only once. This procedure creates an undirected graph with links between pairs where a causal relationship is hypothesized to exist. To establish the direction of causation—*i.e.* turn the undirected graph into a directed one, we use Pearl's Inductive Causation algorithm [20] and the procedure is summarized in Algorithm 2 of the supplementary.

In more detail the χ^2-test checks for conditional independence of a pair of object categories $\{X_j, X_{j'}\}$ given an additional object category $X_k \in \mathcal{V} \setminus \{X_j, X_{j'}\}$. The probabilities required for the test are obtained from the relative frequencies of the objects and their co-occurrences in the dataset. Algorithm 1 describes this in detail. By way of example, *pillow* and *blanket* might co-occur in a substantial number of scenes, however, their co-occurrences are influenced by a third object category *bed*. In this case, the pairwise relationship between *pillow* and *blanket* is determined to be independent, given the presence of *bed*, so no link between *pillow* and *blanket* is created.

(ii) Testing for Dependency Based on Shape and Pose. In addition to the conditional co-occurrence captured above, we also seek to capture covering/enclosing and supporting relationships—which are defined by the shape and pose of the objects as well as the categories—in the causal graph. More precisely we hypothesize a causal relationship between object categories A and B if:

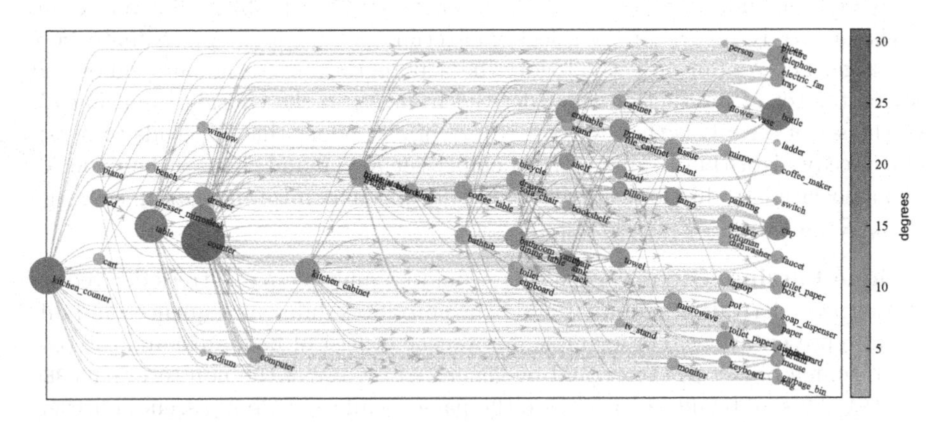

Fig. 2. The above graph is generated by the modified IC algorithm on SUNRGBD-3D Dataset. An arrowhead of an edge indicates the direction of causation and the color of a node indicates its degree.

- object category A is found to support object category B (*i.e.* their relative poses and shapes are such that, within a threshold, one is above and touching the other), or,
- object category B is enclosed/covered by another larger object category A (again this is determined using a threshold on the objects' relative shapes and poses).

We accept a hypothesis and establish the causal relationship (by entering a suitable edge into the causal graph) if at least 30% of the co-occurrences of these object categories in the dataset agree with the hypothesis. The final directed causal relational graph $\mathcal{G}$ is the union of the causal graphs generated by the above tests. Note that we do not consider any dependencies that would lead to a cycle in the graph [4]. The result of the procedure is also displayed in Fig. 2.

3.2 Creating a CFG from the Causal Graph

We now need to create a Context Free Grammar from the causal graph. The CFG is characterised by non-terminal symbols that generate other symbols. Suppose we choose a particular node in the causal graph (*i.e.* an object category) and assume it is non-terminal. By tracing the full set of directed edges in the causal graph emanating from this node we create a set of production rules. This non-terminal and associated rules are then tested against the dataset to determine how many scenes are explained (formally "covered") by the rules. A good choice of non-terminals will lead to good coverage. Our task then is to determine an optimal set of non-terminals and associated rules to give the best coverage of the full dataset of scenes.

Note that finding such a set is a combinatorial hard problem. Therefore, we devise a greedy algorithm to select non-terminals and find approximate best coverage. Let X_j, an object category, be a potential non-terminal symbol and $\mathcal{R}_j$ be the set of production rules derived from X_j in the causal graph. Let C_j be the set of terminals that $\mathcal{R}_j$ covers (essentially nodes that X_j leads to in $\mathcal{G}$). Our greedy algorithm begins with an empty set $\mathcal{R} = \emptyset$ and chooses the node X_j and associated production rule set $\mathcal{R}_j$ to add that maximize the *gain* in coverage $\mathcal{G}_{gain}(\mathcal{R}_j, \mathcal{R}) = \frac{1}{|\mathcal{R}_j|} \sum_{I_i \in \mathcal{I} \setminus \mathcal{C}} |Y_i|/|I_i|$.

Unique Parsing. Given a scene there could be multiple parse trees derived by the leftmost derivation grammar and hence produces different sequences and different representations. For example, for a scene consist of `bed`, `sofa` and `pillow`, the terminal object `pillow` could be generated by any of the non-terminals `BED` or `SOFA`. This ambiguity can confuse the parser while encoding a scene. Further, different orderings of multiple occurrences of an object lead to different parse trees. We consider following parsing rules to remove the ambiguity:

- Fix the order of the object categories in the order of precedence defined by the grammar.
- Multiple occurrences of an object are sorted in the appearance w.r.t. the preceded object in the anti-clockwise direction starting from the object making minimum angle to the orientation of the preceded object. One such example is shown in Fig. 3. Note that this ordering is only required during training.

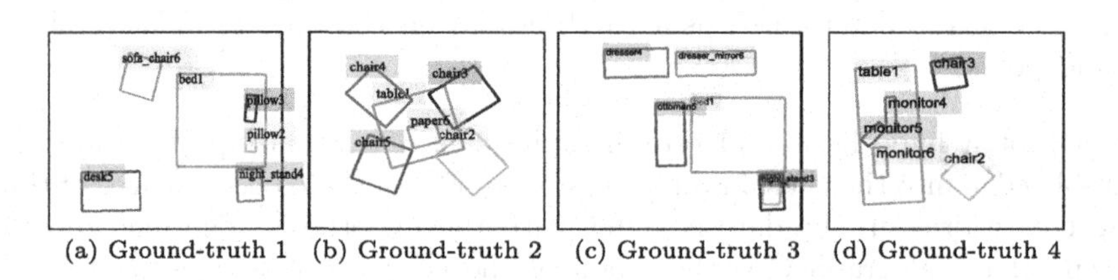

(a) Ground-truth 1 (b) Ground-truth 2 (c) Ground-truth 3 (d) Ground-truth 4

Fig. 3. The parsing order is displayed by a numeral concatenated with the object name. (a) the objects are sorted in the order of precedence defined by the grammar, and (b) multiple chairs are sorted in the appearance w.r.t. the table in anti-clockwise direction starting from the bottom right corner. (c)–(d) More examples of the object order in the ground-truth samples from SUN RGB-D dataset [26].

4 Experiments

Dataset. We evaluate the proposed method on SUN RGB-D Dataset [26] consisting of 10, 335 real scenes with 64, 595 3D bounding boxes and about 800 object categories. The dataset is a collection of multiple datasets [10,25,32], and is highly unbalanced: *e.g.* a single object category *chair* corresponds to about 31% of all the bounding boxes and 38% of total object categories occur just once in the entire dataset. We consider object categories appearing at least 10 times in the dataset for evaluation. Further, very similar object categories are merged, yielding 84 object categories and 62, 485 bounding boxes.

Table 1. Results of 3D bounding box reconstruction of some of the frequent objects under a valid reconstruction with IoU >0.25 [21]. The pose estimation results are furnished within braces (angular errors in degrees, displacement errors in meters).

Objects	chair	bed	table	ktchn_cntr	piano
SG-VAE	**92.3(5.88°,0.13m)**	100.0(**2.80°**,**0.08 m**)	**96.2**(**2.84°**,**0.08 m**)	100.0(1.69°,0.08 m)	**67.1**(**6.93°**,**0.11 m**)
BL1	75.4 (8.30°, **0.14 m**)	98.5 (5.70°, **0.09 m**)	93.8 (3.70°, **0.08 m**)	**100.0**(**1.62°**,**0.06 m**)	48.7 (10.3°, **0.12 m**)
BL2 [6]	33.7 (28.1°, 0.32 m)	75.1 (7.12°, 0.45 m)	90.2 (7.69°, 0.33 m)	83.2 (5.96°, 0.09 m)	0.80 (45.1°, 0.43 m)
BL3[14]+[34]	29.2 (41.8°, 0.26 m)	88.7 (35.7°, 0.58 m)	72.9 (56.7°, 0.34 m)	100.0 (52.5°, 0.78 m)	26.3 (17.4°, 0.33 m)

We separated 10% of the data at random for validation. On average there are 4.29 objects per image and maximum number of objects in an image is considered to be 15. This also provides the upper bound of the length of the sequence generated by the grammar. Note that the dataset is the intersection of the given dataset and the scene language, *i.e.* the possible set of scenes generated by the CFG.

Table 2. IoU of room layout estimation

Methods	SG-VAE	BL1	BL2 [6]	BL3 [14]+[34]
Grammar	✓	✓		✓
Pose & Shape	✓	✓	✓	
IoU	**0.6240**	0.5673	0.2964	0.5119

Baseline Methods for Evaluation (Ablation Studies). To evaluate the individual effects of (i) output of the decoder structure, (ii) usage of grammar, and (iii) the pose and shape attributes, the following baselines are chosen:

(BL1) *Variant of SG-VAE*: In contrast to the proposed SG-VAE where attributes of each rule are directly concatenated with 1-hot encoding of the rule, in this variant separate attributes for each rule type are predicted by the decoder and rest are filled with zeros.

(BL2) *No Grammar VAE* [6]: No grammar is considered in this baseline. The 1-hot encodings correspond to the object type is concatenated with the absolute pose of the objects (in contrast to rule-type and relative pose in SG-VAE) respectively.

(BL3) *Grammar VAE* [14] + *Make home* [34]: The Grammar VAE is incorporated with our extracted grammar to sample a set of coherent objects and [34] is used to arrange them. Sampled 10 times and solution corresponding to best IoU w.r.t. groundtruth is employed. The details of the above baselines are provided in the supplementary.

All the baselines including SG-VAE are implemented in `python 2.7 (Tensorflow)` and trained on a GTX 1080 Ti GPU.

Evaluation Metrics. The relative poses of individual objects in a scene are accumulated to compute their absolute poses which are then combined with the shape parameters to compute the scene layout. The reconstructed scene layouts are then compared against the groundtruth layouts.

- **3D bounding box reconstruction.** We employed IoU to measure the shape similarity of the bounding boxes. Reconstructed bounding boxes with IoU >0.25 are considered as true positives. The results are reported in Table 1.
- **3D Pose estimation.** The average pose error is considered over two separate metrices: **(i)** angular error in degrees, **(ii)** displacement error in meters (see Table 1).
- **Room Layout Estimation.** The evaluation is conducted as the IoU of the *occupied* space between the groundtruth and the predicted layouts (see Table 2). The intersection is computed only over the true positives.

Interpolation in Latent Space. Two distinct scenes are encoded into the latent space, *e.g.*, $\mathcal{N}(\boldsymbol{\mu}_1, \boldsymbol{\Sigma}_1)$ and $\mathcal{N}(\boldsymbol{\mu}_2, \boldsymbol{\Sigma}_2)$ and new scenes are then synthesized from interpolated vectors of the means, *i.e.* from $\alpha\boldsymbol{\mu}_1 + (1-\alpha)\boldsymbol{\mu}_2$. We performed the experiment on a set of random pairs chosen from the test dataset. The results are shown in Fig. 4. Notice that the decoder behaves gracefully w.r.t. perturbations of the latent code and always yields a valid and realistic scene.

4.1 Comparison with Baselines on Other Datasets

The conventional indoor scene synthesis methods (for example, Grains [16], Human-centric [23] (HC), fast-synth [24] (FS) etc.) are tailored to and trained on SUNCG dataset [28]. The dataset consists of synthetic scenes generated by graphic designers. Moreover, the dataset is no longer publicly available (along with the meta-files).[1] Therefore the following evaluation protocols are employed to assess the performance of different methods.

Comparison on Synthetic Scene Quality. To conduct a qualitative evaluation, we employ a classifier (based on Pointnet [22]) to predict a scene layout to

[1] Due to the legal dispute around SUNCG [28] we include our results for SUNCG (conducted on our internal copy) in Table 3 only for illustrative purposes.

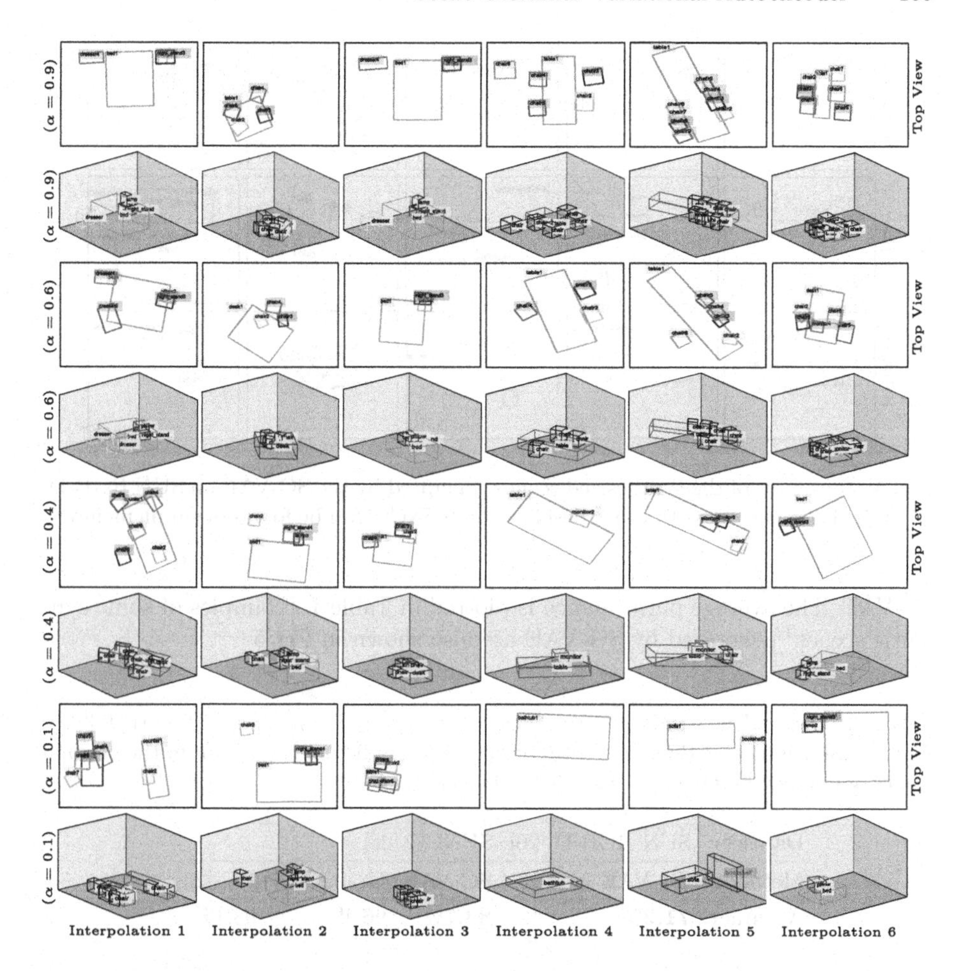

Fig. 4. Synthetic scenes decoded from linear interpolations $\alpha\boldsymbol{\mu}_1 + (1-\alpha)\boldsymbol{\mu}_2$ of the means $\boldsymbol{\mu}_1$ and $\boldsymbol{\mu}_2$ of the latent distributions of two separate scenes. The generated scenes are valid in terms of the co-occurrences of the object categories and their shapes and poses (more examples can be found in the supplementary). The room-size and the camera view-point are fixed for better visualization. Best viewed electronically.

be an original or generated by a scene synthesis method. If the generated scenes are very similar to the original scenes, the classifier performs poorly (lower accuracy) and indicates the efficacy of the synthesis method. The classifier takes a scene layout of multiple objects, individually represented by the concatenation of 1-hot code and the attributes, as input and predicts a binary label according to the scene-type. The classifier is trained and tested on a dataset of $2K$ original and synthetic scenes (50% training and 50% testing). Note that SG-VAE is trained on the synthetic data generated by the interpolations of latent vectors (some examples are shown in Fig. 4) and real data of SUN RGB-D [26]. Lower accuracy of the classifier validates the superior performance of the proposed

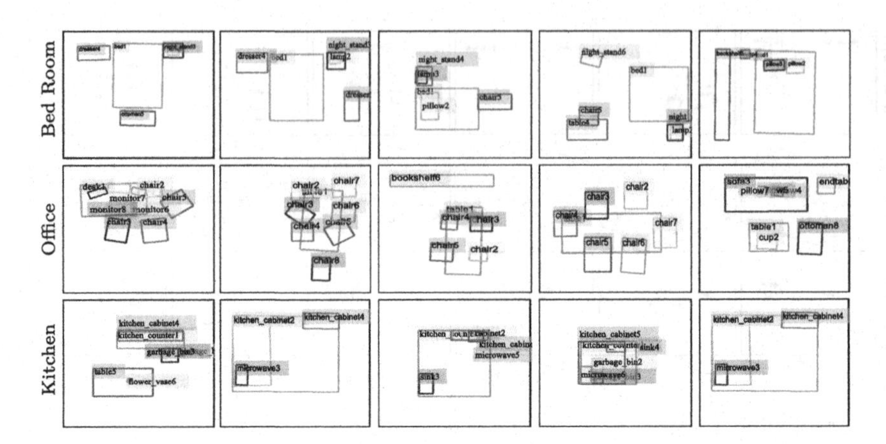

Fig. 5. Top-views of the synthesized scenes generated by the SG-VAE on SUN RGB-D. A detailed comparison with other baselines on SUNCG can be found in supplementary.

SG-VAE. The average performance is plotted in Table 3. Examples of some synthetic scenes[2] generated by SG-VAE are also shown in Fig. 5.

Table 3. Original vs. synthetic classification accuracy (trained with Pointnet [22]): The accuracy indicates that the synthetic scenes are indistinguishable from the original scenes and hence lower (closer to 50%) is better.

Datasets	SUN RGB-D [26]	SUNCG [28]		
Methods	SG-VAE	SG-VAE	Grains [16]	HC [23]
Accuracy	**71.3%**	**83.7%**	96.4%	98.1%

Runtime Comparison. All the methods are evaluated on a single CPU and the runtime is displayed in Table 4. Note that the decoder of the proposed SG-VAE takes only ~1 ms to generate the parse tree and the rest of the time is consumed by the renderer (generating bounding boxes). The proposed method is almost two orders of magnitude faster than the other scene synthesis methods.

Table 4. Average time required to generate a single scene.

Methods	SG-VAE	Grains [16]	FS [24]	HC [23]
Avg. runtime	**8.5 ms**	1.2×10^2 ms	1.8×10^3 ms	2.4×10^5 ms

[2] We thank the authors of Grains [16] and HC [23] for sharing the code. More results and the proposed SG-VAE for SUNCG are in the supplementary material.

4.2 Scene Layout Estimation from the RGB-D Image

The task is to predict the 3D scene layout given an RGB-D image. Typically, the state of the art methods are based on sophisticated region proposals and subsequent processing [21,27]. With this experiment, we aim to demonstrate the potential use of the latent representation learned by the proposed auto-encoder for a computer vision task, and therefore we employ a simple approach at this point. We (linearly) map deep features (extracted from images by a DNN [38]) to the latent space of the scene-grammar autoencoder. The decoder subsequently generates a 3D scene configuration with associated bounding boxes and object labels from the projected latent vector. Since during the deep feature extraction and the linear projection, the spatial information of the bounding boxes are lost, the predicted scene layout is then combined with a bounding box detection to produce the final output.

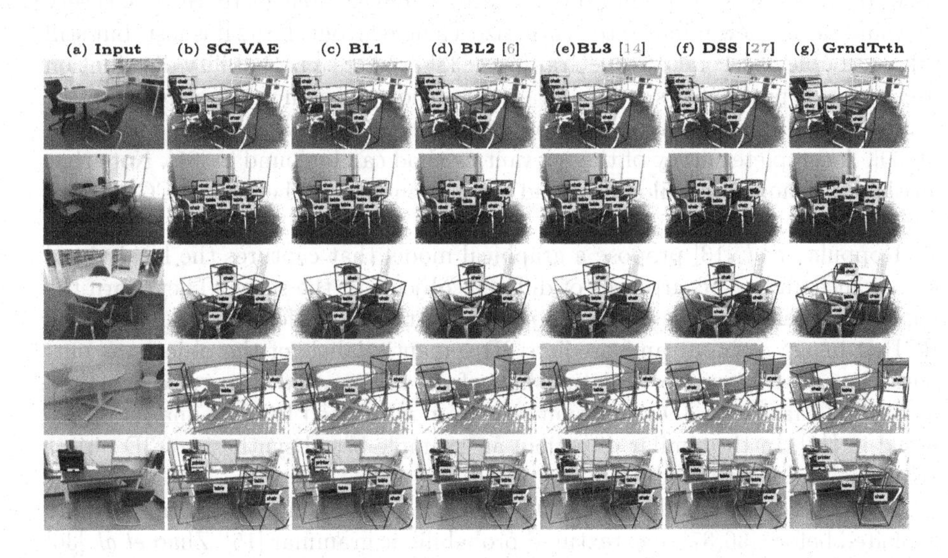

Fig. 6. A few results on SUNRGB Dataset inferred from the RGB-D images.

The bounding box detector of DSS [27] is employed and the scores of the detection are updated based on our reconstruction as follows: the score (confidence of the prediction) of a detected bounding box is doubled if a similar bounding box (in terms of shape and pose) of the same category is reconstructed by our method. A 3D non-maximum suppression is applied to the modified scores to get the final scene layout. The details can be found in the supplementary.

We selected the average IoU for room layout estimation as the evaluation metric, and the results are presented in Table 5. The proposed method and other grammar-based baselines improve the scene layout estimation from the same by sophisticated methods such as deep sliding shapes [27]. Furthermore, the proposed method tackles the problem in a much simpler and faster way. Thus, it can be

Table 5. IoU for RGBD to room layout estimation

Methods	SG-VAE	BL1	BL2 [6]	BL3 [14]+[34]	DSS [27]
IoU	**0.4387**	0.4315	0.4056	0.4259	0.4070

employed to any 3D scene layout estimation method with very little overhead (*e.g.* a few ms in addition to 5.6 s of [27]). Results on some test images where SG-VAE produces better IoUs are displayed in Fig. 6.

5 Related Works

The most relevant method to ours is Grains [16]. It requires training separate networks for each of the room-types—bedroom, office, kitchen etc. HC [23] is very slow and takes a few minutes to synthesize a single layout. FS [24] is fast, but still takes a couple of seconds. SceneGraphNet [33] predicts a probability distribution over object types that fits well in a target location given an incomplete layout. A similar graph-based method is proposed in [29] and CNN-based method proposed in [30]. A complete survey of the relevant method can be found in [35]. Note that all these methods are tailored to and trained on the synthetic SUNCG dataset which is currently unavailable.

Koppula *et al.* [13] propose a graphical model that captures the local visual appearance and co-occurrences of different objects in the scene. They learn the appearance relationships among objects from the visual features that takes an RGB-D image as input and predicts 3D semantic labels of the objects as output. The pair-wise support relationships of the indoor objects are also exploited in [8,25]. Learning to 3D scene synthesis from annotated RGB-D images is proposed in [11]. In the similar direction, an example-based synthesis of 3D object arrangements is proposed in [5].

Grammar-based models for 3D scene reconstruction have been partially exploited before [36,37], *e.g.* textured probabilistic grammar [15]. Zhao *et al.* [36] proposed handcoded grammar to its terminal symbols (line segments) and later extended to different functional groups in [37]. Choi *et al.* [2] proposed a 3D geometric phrase model that estimates a scene layout with multiple object interactions. Note that all the above methods are based on hand-coded production rules, in contrast, the proposed method exploits a self-supervision to yield the production rules of the grammar.

6 Conclusion

We proposed a grammar-based autoencoder SG-VAE for generating natural indoor scene layouts containing multiple objects. By construction the output of SG-VAE always yields a valid configuration (w.r.t. the grammar) of objects, which was also experimentally confirmed. We demonstrated that the obtained latent representation of an SG-VAE has desirable properties such as the ability to interpolate between latent states in a meaningful way. The latent space of SG-VAE can also be easily adapted to computer vision problems (*e.g.* 3D scene layout estimation from RGB-D images). Nevertheless, we believe that there is potential in leveraging the latent space of SG-VAEs to the other tasks, *e.g.* fine-tuning the latent space for a consistent layout over multiple cameras which is part of the future work.

Acknowledgement. We gratefully acknowledge the support of the Australian Research Council through the Centre of Excellence for Robotic Vision, CE140100016, and the Wallenberg AI, Autonomous Systems and Software Program (WASP) funded by the Knut and Alice Wallenberg Foundation.

References

1. Anderson, P., et al.: Bottom-up and top-down attention for image captioning and visual question answering. In: Proceedings of CVPR, pp. 6077–6086 (2018)
2. Choi, W., Chao, Y.W., Pantofaru, C., Savarese, S.: Understanding indoor scenes using 3D geometric phrases. In: Proceedings of CVPR, pp. 33–40 (2013)
3. Deng, C., Wu, Q., Wu, Q., Hu, F., Lyu, F., Tan, M.: Visual grounding via accumulated attention. In: Proceedings of CVPR, pp. 7746–7755 (2018)
4. Dor, D., Tarsi, M.: A simple algorithm to construct a consistent extension of a partially oriented graph (1992)
5. Fisher, M., Ritchie, D., Savva, M., Funkhouser, T., Hanrahan, P.: Example-based synthesis of 3D object arrangements. ACM Trans. Graph. (TOG) **31**(6), 1–11 (2012)
6. Gómez-Bombarelli, R., et al.: Automatic chemical design using a data-driven continuous representation of molecules. ACS Cent. Sci. **4**(2), 268–276 (2018)
7. Goodfellow, I., et al.: Generative adversarial nets. In: Proceedings of NIPS, pp. 2672–2680 (2014)
8. Guo, R., Hoiem, D.: Support surface prediction in indoor scenes. In: Proceedings of ICCV, pp. 2144–2151 (2013)
9. Huang, S., Qi, S., Zhu, Y., Xiao, Y., Xu, Y., Zhu, S.-C.: Holistic 3D scene parsing and reconstruction from a single RGB image. In: Ferrari, V., Hebert, M., Sminchisescu, C., Weiss, Y. (eds.) ECCV 2018. LNCS, vol. 11211, pp. 194–211. Springer, Cham (2018). https://doi.org/10.1007/978-3-030-01234-2_12
10. Janoch, A., et al.: A category-level 3D object dataset: putting the kinect to work. In: Fossati, A., Gall, J., Grabner, H., Ren, X., Konolige, K. (eds.) Consumer Depth Cameras for Computer Vision. ACVPR, pp. 141–165. Springer, London (2013). https://doi.org/10.1007/978-1-4471-4640-7_8
11. Kermani, Z.S., Liao, Z., Tan, P., Zhang, H.: Learning 3D scene synthesis from annotated RGB-D images. Comput. Graph. Forum **35**, 197–206 (2016). Wiley Online Library
12. Kingma, D.P., Welling, M.: Auto-encoding variational Bayes. In: Proceedings of ICLR, pp. 469–477 (2014)
13. Koppula, H.S., Anand, A., Joachims, T., Saxena, A.: Semantic labeling of 3D point clouds for indoor scenes. In: Proceedings of NIPS, pp. 244–252 (2011)
14. Kusner, M.J., Paige, B., Hernández-Lobato, J.M.: Grammar variational autoencoder. In: Proceedings of ICML, pp. 1945–1954. JMLR.org (2017)

15. Li, D., Hu, D., Sun, Y., Hu, Y.: 3D scene reconstruction using a texture probabilistic grammar. Multimedia Tools Appl. **77**(21), 28417–28440 (2018)
16. Li, M., et al.: GRAINS: generative recursive autoencoders for indoor scenes. ACM Trans. Graph. (TOG) **38**(2), 12 (2019)
17. Liu, M.Y., Tuzel, O.: Coupled generative adversarial networks. In: Proceedings of NIPS, pp. 469–477 (2016)
18. Lu, J., Yang, J., Batra, D., Parikh, D.: Hierarchical question-image co-attention for visual question answering. In: Proceedings of NIPS, pp. 289–297 (2016)
19. Meyer, J.A., Filliat, D.: Map-based navigation in mobile robots: II. A review of map-learning and path-planning strategies. Cogn. Syst. Res. **4**(4), 283–317 (2003)
20. Pearl, J., Verma, T.S.: A theory of inferred causation. In: Studies in Logic and the Foundations of Mathematics, vol. 134, pp. 789–811. Elsevier (1995)
21. Qi, C.R., Chen, X., Litany, O., Guibas, L.J.: ImVoteNet: boosting 3D object detection in point clouds with image votes. In: Proceedings of CVPR, pp. 4404–4413 (2020)
22. Qi, C.R., Su, H., Mo, K., Guibas, L.J.: PointNet: deep learning on point sets for 3D classification and segmentation. In: Proceedings of CVPR, pp. 652–660 (2017)
23. Qi, S., Zhu, Y., Huang, S., Jiang, C., Zhu, S.C.: Human-centric indoor scene synthesis using stochastic grammar. In: Proceedings of CVPR, pp. 5899–5908 (2018)
24. Ritchie, D., Wang, K., Lin, Y.: Fast and flexible indoor scene synthesis via deep convolutional generative models. In: Proceedings of CVPR, pp. 6182–6190 (2019)
25. Silberman, N., Hoiem, D., Kohli, P., Fergus, R.: Indoor segmentation and support inference from RGBD images. In: Fitzgibbon, A., Lazebnik, S., Perona, P., Sato, Y., Schmid, C. (eds.) ECCV 2012. LNCS, vol. 7576. Springer, Heidelberg (2012). https://doi.org/10.1007/978-3-642-33715-4_54
26. Song, S., Lichtenberg, S.P., Xiao, J.: Sun RGB-D: a RGB-D scene understanding benchmark suite. In: Proceedings of CVPR, pp. 567–576 (2015)
27. Song, S., Xiao, J.: Deep sliding shapes for amodal 3D object detection in RGB-D images. In: Proceedings of CVPR, pp. 808–816 (2016)
28. Song, S., Yu, F., Zeng, A., Chang, A.X., Savva, M., Funkhouser, T.: Semantic scene completion from a single depth image. In: Proceedings of CVPR, pp. 1746–1754 (2017)
29. Wang, K., Lin, Y.A., Weissmann, B., Savva, M., Chang, A.X., Ritchie, D.: PlanIT: planning and instantiating indoor scenes with relation graph and spatial prior networks. ACM Trans. Graph. (TOG) **38**(4), 1–15 (2019)
30. Wang, K., Savva, M., Chang, A.X., Ritchie, D.: Deep convolutional priors for indoor scene synthesis. ACM Trans. Graph. (TOG) **37**(4), 1–14 (2018)
31. Xiao, F., Sigal, L., Jae Lee, Y.: Weakly-supervised visual grounding of phrases with linguistic structures. In: Proceedings of CVPR, pp. 5945–5954 (2017)
32. Xiao, J., Owens, A., Torralba, A.: SUN3D: a database of big spaces reconstructed using SfM and object labels. In: Proceedings of ICCV, pp. 1625–1632 (2013)
33. Zhou, Y., While, Z., Kalogerakis, E.: SceneGraphNet: neural message passing for 3D indoor scene augmentation. In: Proceedings of ICCV (2019)
34. Yu, L.F., Yeung, S.K., Tang, C.K., Terzopoulos, D., Chan, T.F., Osher, S.J.: Make it home: automatic optimization of furniture arrangement. ACM Trans. Graph. (TOG) **30**, 86 (2011)
35. Zhang, S.H., Zhang, S.K., Liang, Y., Hall, P.: A survey of 3D indoor scene synthesis. J. Comput. Sci. Technol. **34**(3), 594–608 (2019)
36. Zhao, Y., Zhu, S.C.: Image parsing with stochastic scene grammar. In: Proceedings of NIPS, pp. 73–81 (2011)

37. Zhao, Y., Zhu, S.C.: Scene parsing by integrating function, geometry and appearance models. In: Proceedings of CVPR, pp. 3119–3126 (2013)
38. Zhou, B., Lapedriza, A., Xiao, J., Torralba, A., Oliva, A.: Learning deep features for scene recognition using places database. In: Proceedings of NIPS, pp. 487–495 (2014)

Unsupervised Learning of Optical Flow with Deep Feature Similarity

Woobin Im[1], Tae-Kyun Kim[1,2], and Sung-Eui Yoon[1](✉)

[1] School of Computing, KAIST, Daejeon, South Korea
iwbn@kaist.ac.kr, sungeui@kaist.edu

[2] Department of Electrical and Electronic Engineering, Imperial College London, London, UK
tk.kim@imperial.ac.uk

Abstract. Deep unsupervised learning for optical flow has been proposed, where the loss measures image similarity with the warping function parameterized by estimated flow. The census transform, instead of image pixel values, is often used for the image similarity. In this work, rather than the handcrafted features i.e. census or pixel values, we propose to use deep self-supervised features with a novel similarity measure, which fuses multi-layer similarities. With the fused similarity, our network better learns flow by minimizing our proposed feature separation loss. The proposed method is a polarizing scheme, resulting in a more discriminative similarity map. In the process, the features are also updated to get high similarity for matching pairs and low for uncertain pairs, given estimated flow. We evaluate our method on FlyingChairs, MPI Sintel, and KITTI benchmarks. In quantitative and qualitative comparisons, our method effectively improves the state-of-the-art techniques.

Keywords: Unsupervised · Self-supervised · Optical flow · Deep feature · Similarity

1 Introduction

In computer vision, optical flow estimation is a fundamental step towards motion understanding. It describes the velocity of each point in the 3D world as the projection of points to the 2D motion field. Thanks to its effective motion description, it has been largely used for many applications, e.g., video recognition [7,31], frame interpolation [4,36], and inpainting [38], to name a few.

Recently, end-to-end deep networks for optical flow estimation [9,16,28,32] have made architectural progresses, resulting in significant improvements in terms of flow accuracy, efficiency, and generalization capability. For supervised optical flow training, large-scale synthetic datasets, e.g., FlyingChairs [9], have

Electronic supplementary material The online version of this chapter (https://doi.org/10.1007/978-3-030-58586-0_11) contains supplementary material, which is available to authorized users.

A. Vedaldi et al. (Eds.): ECCV 2020, LNCS 12369, pp. 172–188, 2020.
https://doi.org/10.1007/978-3-030-58586-0_11

been primarily used. While there are real-world datasets including Middlebury [4] and KITTI [10,26], their sizes are limited to few hundred images and each of them is limited to a specific scenario. This limitation is mainly due to the extremely prohibitive cost to obtain or manually label accurate matching points in thousands of video frames in the wild.

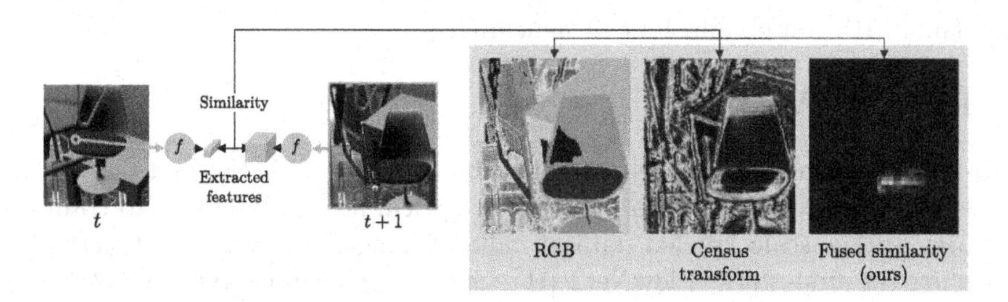

Fig. 1. This figure shows different similarity maps of the reference point at time step t to all pixels in the target image at $t+1$; red means higher similarity. We compare the similarity computed by our deep feature against ones computed by census transform and RGB. The fused similarity shows improved discriminative response, while the census transform tends to be sensitive to local edge appearance and RGB shows high similarity with similar colors. For simple visualization, we compute similarity in the spatial domain (Color figure online)

To train deep networks without ground-truth flows, unsupervised approaches have been proposed. In principle, unsupervised methods exploit an assumption that two matching points have similar features and learn to generate flows maximizing the similarity. In this line of research, choosing appropriate features is critical for accurate optical flow estimation. The early work [19,29] applies RGB pixel values and image gradients as the feature, and recently it has been shown that the census transform [39] is highly effective for optical flow learning [22,25].

Another interesting aspect of the unsupervised optical flow networks is that a network learns more than just the loss it is trained with. A recent work [22] found that configuring the loss function only with the data term does work without the smoothness term. This observation implies that the network feature learns meaningful patterns in moving objects only with the photometric constancy assumption. A similar observation is also found in the literature on self-supervision [11,27], where deep features learn semantic patterns by conducting simple unsupervised tasks like a jigsaw puzzle or guessing rotation.

In this work, we learn self-supervised network features and use them to improve the unsupervised optical flow. To learn from self-features, we propose to use the similarity based on the product fusion of multi-layer features (Sect. 3.2). We visualize similarity maps computed by different features in Fig. 1. Our fused similarity demonstrates discriminative matching points highlighting the matching pair, while lessening unmatched areas. On the other hand, the similarity map (e.g., computed by the cosine) with RGB or the census transform shows many high-response points across the whole area. This is mainly because those features only encode patterns in a local area and do not represent semantic meanings. We propose three loss functions utilizing the feature similarity for optical flow training (Sect. 3.3). Across various quantitative and qualitative validations, we demonstrate the benefits of the proposed feature similarity and the loss function (Sect. 4). When compared to other deep unsupervised methods, our method achieves state-of-the-art results under various measures across FlyingChairs, MPI Sintel, and KITTI benchmarks.

2 Related Work

End-to-end Supervised Deep Methods. FlowNet [9] is the first end-to-end framework that exploits a deep network for optical flow estimation. To train the network, a large-scale labeled dataset called FlyingChairs was constructed [9]. Following the first work, FlowNet2 [16], SpyNet [28], and PWC-Net [32] have made architectural progresses, resulting in significant improvements in terms of flow accuracy, efficiency, and generalization capability.

Datasets. Large-scale synthetic datasets are available for supervised optical flow training including FlyingChairs [9]. Following FlyingChairs, FlyingThings3D [24] and FlyingChairs-Occ [15] datasets are incorporated into the collection with improved reality and additional information. Real-world datasets annotated with a rich optical flow are lacking. Middlebury [4] and KITTI [10,26] are the most commonly used ones. However, not only the scenes they have are limited to few hundreds of images, but also they are constrained to specific scenarios: indoor static objects for Middlebury and driving for KITTI. This limitation is mainly due to the prohibitive cost to obtain or manually label accurate matching points in thousands of video frames in the wild.

End-to-end Unsupervised Deep Methods. To make use of deep networks for optical flow without expensive ground-truth flows, deep unsupervised approaches have been proposed. Earlier methods [19,29] brought ideas from the classical variational methods, which adopt the energy functional containing the data and smoothness terms into loss functions of deep learning. The loss function in unsupervised methods can be calculated using the warping technique [17].

In unsupervised optical flow learning, how to filter out unreliable signals from its loss is critical for achieving better results. In terms of the data term, the census transform [39] has been proven to be effective in deep unsupervised optical flow [22,25]. Similarly, occlusion handling [25,34] can eliminate possibly occluded points from the loss calculation, where no target pixels exist due to occlusion. Rather than just giving up supervision on occluded ones, hallucinated occlusion [22,23] can be helpful to give meaningful loss for those occluded points. Additionally, training with multiple frames improves the noisy loss signal. Janai et al. [18] use the constant velocity assumption and Liu et al. [23] build multiple cost-volumes to give more information to a model.

Deep Features for Matching. In the fields of matching and tracking keypoints or objects, deep network features have been used for robust matching [5,33,42]. At a high level, matching techniques are related to optical flow estimation [2,3,13,30,35,37]. However, addressing all the pixels and their matching in a dense manner (cf. sparse keypoints) is challenging. To the best of our knowledge, deep features have not been exploited in objective functions for optical flow estimation. On the other hand, there has been successful exploitation of deep features in pixel-wise tasks, e.g., depth estimation [41] and generation by warping [14]. In our evaluations, the Zhan's method [41], which was proposed for stereo matching, or the losses more direct to optimize the deep features, e.g., the triplet losses, have shown poor performance in unsupervised optical flow estimation. Given the high level of noise due to occlusion, deformation, motion blur, and the unsupervised settings of optical flow, the level of optimization, i.e., the loss function for features, needs to be carefully designed.

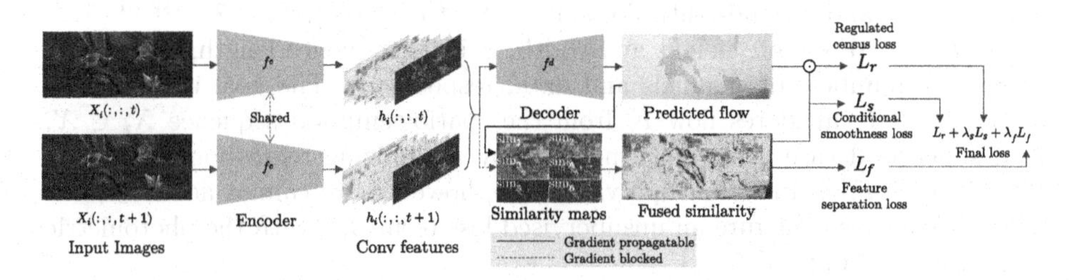

Fig. 2. This shows the overview of our method, which is end-to-end trainable for both optical flow and self-supervised deep features

3 Our Approach

In this work, we successfully learn and exploit deep features for improving unsupervised optical flow estimation. The proposed framework simultaneously improves the features and optical flow while adding a small additional cost for training and no extra cost at runtime. For effective training, we build new loss functions (Sect. 3.3) utilizing the fused similarity (Sect. 3.2). The feature separation loss separates certain and uncertain matchings by our fused similarity like a contrastive loss. Noting that even the best features can fail in cases like occlusions, the separation mechanism effectively improves resulting flows by discouraging to match uncertain pixels, as well as encouraging to improve certain ones. Additionally, the regulated census loss and conditional smoothness loss make use of the census features and the smoothness constraint adaptively considering the deep similarity to compensate for the low precision of deep features.

Figure 2 illustrates our method. Our method is an end-to-end trainable network, which takes a sequence of images as an input for optical flow estimation; we use PWC-Net [32] structure as a base model of the encoder and the decoder, and train it using self-features, i.e., spatial conv features. In the training phase, we add a similarity branch in which similarities of predicted flows are calculated and fused; the resulting fused similarity is actively used in our training process. The aggregated similarity over layers resolves disagreement among multiple-layer features. We apply three loss functions to be minimized upon the fused similarity: feature separation loss, regulated census loss, and conditional smoothness loss. In the learning process, both the encoder and decoder are initialized using the conventional unsupervised optical flow loss. Then, the proposed feature-based losses are minimized. The method converges under different initialization and parameter settings.

3.1 Background on Unsupervised Optical Flow Learning

The learning-based optical flow method commonly works on a dataset to train a model that has a set of spatio-temporal images $\mathcal{X} = \{X_1, X_2, \ldots, X_N\}$, $X_i \in \mathbb{R}^{H \times W \times T \times C}$, and ground-truth flows $\mathcal{Y} = \{F_1, F_2, \ldots F_N\}$, $F_i \in \mathbb{R}^{H \times W \times T-1 \times 2}$, where H and W denote height and width, T is its sequence length, and C and N are the numbers of channels and data, respectively. The goal is to train a model f_θ that calculates flow $\hat{F}_i$ from the spatio-temporal sequence $X_i \in \mathcal{X}$. In a supervised case, we train a network by minimizing regression loss: $L_s = \frac{1}{N}\sum_i^N ||F_i - \hat{F}_i||_2^2$. In an unsupervised case, however, we cannot access $\mathcal{Y}$, but only $\mathcal{X}$. We thus configure an unsupervised loss term, L_p, with the photometric consistency assumption:

$$L_p = \frac{1}{N}\sum_i^N \sum_{(x,y,t)\in\Omega} \Psi(X_i(x,y,t) - X_i(x+u, y+v, t+1)), \tag{1}$$

where Ω contains all the spatio-temporal coordinates, $(u,v) = \hat{F}_i(x,y,t)$ is an estimated flow at (x,y,t), and Ψ is the robust penalty function [22]; $\Psi(x) = (|x|+\epsilon)^q$. Note that $X(x+u, y+v, \cdot)$ includes the warping operation using bilinear sampling [17], which supports back-propagation for end-to-end optimization. In this paper, we use the census transform in L_p with the same configuration used in [22] for all experiments unless otherwise stated.

Occlusion handling is performed by checking consistency [25] between forward and backward flows. The estimated occlusion mask is denoted with $\hat{C}_i^o(x,y,t)$, which is 1 if $\hat{F}_i(x,y,t)$ is not occluded, otherwise 0. This geometrically means the backward flow vector should be the inverse of the forward one if it is not occluded. Our loss terms use the occlusion map as previous work [25].

In this work, we use the data distillation loss [22] for occluded pixels:

$$L_d = \frac{1}{N}\sum_i^N \sum_{(x,y,t)\in\Omega} \Psi(\hat{F}_i^s(x,y,t) - \hat{F}_i^t(x,y,t))M_f(x,y,t), \tag{2}$$

where $\hat{F}_i^t$ is a flow from a teacher model, $\hat{F}_i^s$ is a flow from a student model, and M_f is a valid mask. In short, the teacher model processes inputs without artificial synthetic occlusion, and the student model gets inputs with the generated occlusion. The student flow then learns from the teacher flow.

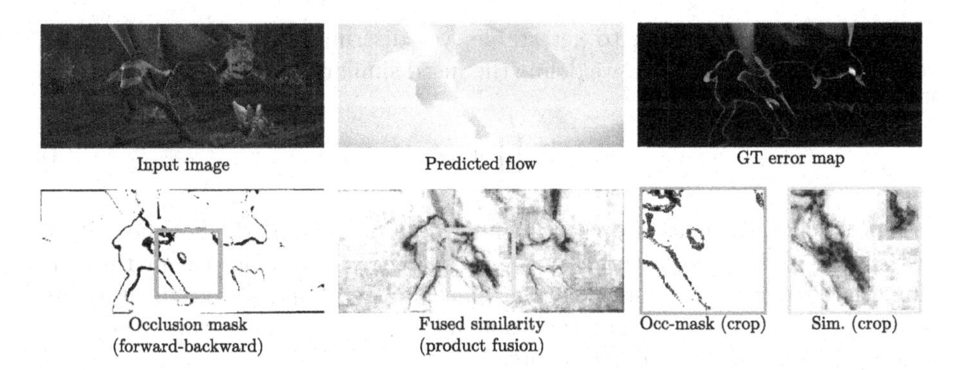

Fig. 3. We visualize the occlusion mask ($\hat{C}^o$) and the similarity map (sim_f). By comparing the ground-truth error map and the fused similarity, we can observe that the similarity is low when the error is high. The bottom row shows occlusion mask and the similarity. Since the occlusion mask does not consider matching confidence, it does not represent how confident the matching is. On the other hand, our fused similarity marks whether the predicted flow is confident. Furthermore, we can use back-propagation since the similarity is differentiable.

3.2 Feature Similarity from Multiple Layers

We use a model f_θ that estimates optical flow, where the model can be decomposed into an encoder $f^e_{\theta_e}$ and a decoder $f^d_{\theta_d}$, such that $f_\theta(X_i) = f^d_{\theta_d}(f^e_{\theta_e}(X_i))$; $\theta = \theta_e \cup \theta_d$. The encoder f^e is a function that outputs L-layered features $h_i = (h^1_i, h^2_i, \dots, h^L_i)$; a lower numbered layer indicates a shallower layer in the deep network (Fig. 3).

Given a matching $\hat{F}(x, y, t)$, we define the similarity between $X_i(x, y, t)$ and $X_i(x+u, y+v, t+1)$ in a layer l to be:

$$\text{sim}_l(x, y, t; h^l_i, \hat{F}_i) = \frac{h^l_i(x, y, t) \cdot h^l_i(x+u, y+v, t+1) + 1}{2}, \tag{3}$$

where $(u, v) = \hat{F}_i(x, y, t)$. Since we use l2-normalization for h^l_i, the function $\text{sim}(\cdot)$ is equivalent to the normalized cosine similarity between the reference feature and the target feature. Note that, for flows going out of the frame, no gradient is propagated during training. Additionally, we update encoder weights by indirect gradient, i.e., back-propagation through the decoder, while ignoring direct gradient from sim_l to h^l_i. That strategy is chosen because the direct gradient easily bypasses the decoder, and the encoder easily suffers from overfitting, in the end, the output flow is downgraded.

Fused Similarity. To utilize features from all layers, we propose to fuse multi-layer feature similarities $(\text{sim}_1, \text{sim}_2, \dots, \text{sim}_L)$. In CNNs, lower layer features tend to show high response for low-level details, e.g. edge, color, etc., while higher layers focus on objects [40]. As a result, similarity response of a lower layer usually bursts around the whole image, while similarity of a deeper layer has few modes. Therefore, to get stable, yet discriminative feature similarity response as shown in Fig. 1, we define the fused similarity as product of multiple features:

$$\mathrm{sim}_f(\cdot) = \prod_{l=1}^{L} \mathrm{sim}_l(\cdot; h_i^l, \hat{F}_i). \tag{4}$$

For efficient calculation during training, we downsample the flow field $\hat{F}_i$ to the size of each layer feature using area interpolation before calculating the similarity; we use area interpolation, since it can propagate gradient to all source points, while other interpolation methods, e.g., bilinear interpolation, only update few nearest source points.

3.3 Learning Optical Flow with Feature Similarity

In this section, we propose three loss functions to effectively use the similarity map for optical flow estimation.

Feature Separation Loss. Given deep similarity, a model can learn flow by simply maximizing $\mathrm{sim}(\cdots; h_i, \hat{F}_i)$, since larger similarity possibly means better matching solution. However, this simple approach in practice leads to worse results, because matchings between pixels under occlusion do not get better, even as we increase the similarity. In other words, maximizing the similarity for these points makes the flows incorrectly matched to random pixels with higher similarity.

To address this matching issue with uncertainty, we suppress flows with lower similarity by minimizing their similarity further, while refining flows with higher similarity by maximizing their similarity. First, we define a similarity threshold k, which we separate the values from:

$$k = \frac{1}{2}(k_{noc} + k_{occ}), \tag{5}$$

where k_{noc} and k_{occ} are average similarities of non-occluded pixels and occluded pixels: $k_{noc} = \frac{\sum_{\Omega}(\mathrm{sim}_f \cdot \hat{C}_i)}{\sum_{\Omega}(\hat{C}_i)}$, $k_{occ} = \frac{\sum_{\Omega}(\mathrm{sim}_f \cdot (1-\hat{C}_i))}{\sum_{\Omega}(1-\hat{C}_i)}$. Since occlusion is an effective criterion to set the boundary value, so that other kinds of difficulties can also be covered as shown in the experiments (Fig. 6).

We then formulate the feature separating loss term as:

$$L_f = \frac{1}{N} \sum_{i}^{N} \sum_{(x,y,t) \in \Omega} -(\mathrm{sim}_f(x,y,t) - k)^2. \tag{6}$$

L_f is a quadratic loss function encouraging the similarity to be far from k, which serves as a boundary value that decides the direction of update. In other words, it suppresses uncertain flows, i.e. $\mathrm{sim}_f(x,y) < k$, down towards 0, and certain flows, i.e. $\mathrm{sim}_f(x,y) > k$, up towards 1. This can be also interpreted as minimizing entropy in semi-supervised learning [12]; we make a network output more informative by regularizing the similarity to be at each polar. A similar approach separating feature similarity from a different domain, image retrieval, has been shown to be effective [21].

One may concern that minimizing the similarity of uncertain flows, i.e., $\text{sim}_f < k$, can lead to an arbitrary matching solution. However, the product operation in the fused similarity (Eq. 4) makes sim_l with a higher similarity relatively retained, while changing smaller similarity much faster. Given any a, b s.t. $1 \le a, b \le L$, whose similarity is not 0, one can derive the following equation:

$$\frac{\partial L_f}{\partial \text{sim}_a(x,y,t)} = \frac{\text{sim}_b(x,y,t)}{\text{sim}_a(x,y,t)} \left(\frac{\partial L_f}{\partial \text{sim}_b(x,y,t)} \right). \tag{7}$$

That is, the scale difference between the two gradients is proportional to the fractional ratio of them, which can grow much faster when the denominator becomes smaller in the scale of the multiplicative inverse. As a result, L_f is minimized by the smaller similarity approaching zero; higher layer similarities can be preserved to prevent arbitrary matching.

Regulated Census Loss. Since L_f (Eq. 6) is fully self-regulated, using only L_f for training can mislead the network itself. Thus, by modifying the well-known unsupervised loss (Eq. 1), we additionally use a regulated census loss, L_r, controlled by similarity:

$$L_r = \frac{1}{N} \sum_i^N \sum_{(x,y,t) \in \Omega} \Psi(\cdot) \hat{C}_i^o(x,y,t) \text{sim}_f(x,y,t). \tag{8}$$

The warping operation used in Eq. 1 is the bilinear sampling. This can only take into account four nearest pixels, making it difficult to address the pixel position far from the estimation, as also pointed by Wang et al. [34]. Therefore, the unsupervised loss (Eq. 1) does not give a correct direction when the current estimation is far from the desired target flow; whatever the loss is, it would be a noise in that case. In contrast, deep features have larger receptive fields with global context, so does the fused similarity. Thus, we can suppress the noise signal by multiplying the similarity; the similarity is designed to indicate whether the current estimation is near the desired target point.

Conditional Smoothness Loss. We use the smoothness prior for spatial locations with low similarity. In general, using the smoothness prior for all pixels degrades the accuracy because the flow-field is blurred. Meanwhile, if clear matching is not found, the smoothness constraint can help by being harmonized with surrounding flows. We thus define our smoothness prior loss considering similarity as:

$$L_s = \frac{1}{N} \sum_i^N \sum_{(x,y,t) \in \Omega} (|\nabla u|^2 + |\nabla v|^2) M_l(x,y,t), \tag{9}$$

$$M_l(x,y,t) = \begin{cases} 1, & \text{if } \text{sim}_f(\cdot) < k, \\ 0, & \text{otherwise.} \end{cases} \tag{10}$$

Loss for Training. We jointly use the aforementioned losses to train our model with stochastic gradient descent. Our final loss function is sum of these loss functions:

$$L = L_r + \lambda_f L_f + \lambda_s L_s + \lambda_d L_d, \tag{11}$$

where λs are weight parameters, and L_d is the data distillation loss defined in Eq. 2.

4 Experimental Results

Our network structure is based on PWC-Net [32], which is a deep network that contains warping, cost-volume, and context network to cover large displacements. We train the network from scratch using Adam optimizer [20] for stochastic gradient descent. In all experiment, we set the mini-batch size to 4.

Training Procedure. Overall, we follow the training process of DDFlow [22] to initialize the network. To train the model, we first pretrain our network with FlyingChairs [9] and finetune it with each target dataset. For pretraining, we use the conventional photometric loss with RGB (Eq. 1) for 200k steps and additional 300k steps with occlusion handling. The resulting weights become the base network parameters for the following experiments. Next, using each target dataset, we finetune the base network. From this stage, we use the census transform for photometric loss. We train the model for 200k steps with occlusion handling. We then apply the final loss L (Eq. 11). We run first 1k steps without L_d, i.e. $\lambda_d = 0$. Then, the teacher network (Sect. 3.1) is fixed to use L_d and continues training using all the losses to 50k steps. We set hyper-parameters to $\lambda_s = 10^{-4}$ and $\lambda_f = 4$. We follow $\lambda_d = 1$ from the previous work [22].

Data Augmentation. For better generalization, we augment the training data using random cropping, random flipping, random channel swapping and color jittering; color jittering includes random brightness and saturation. We normalize the input RGB value into $[-0.5, 0.5]$.

FlyingChairs. The FlyingChairs [9] is a synthetic dataset created by combining chair and background images. The dataset consisting of 20 k pairs of images has a given train/test split, thus we use its training set for training our network and evaluate our model on its test set.

MPI Sintel. MPI Sintel [6] is a rendered dataset, originally from an open-source movie, Sintel. For training, we use 1k images from the training set and upload our results of the test set to its benchmark server for evaluation. We use both versions of rendering, i.e., clean and final, for training.

KITTI. KITTI [10] has a driving scene from the real world. This dataset has only 200 pairs of images with ground-truth flows. We thus train our model using unlabeled images from a multi-view extension set of KITTI without duplicated images in the benchmark training or testing sets, following the previous work [34].

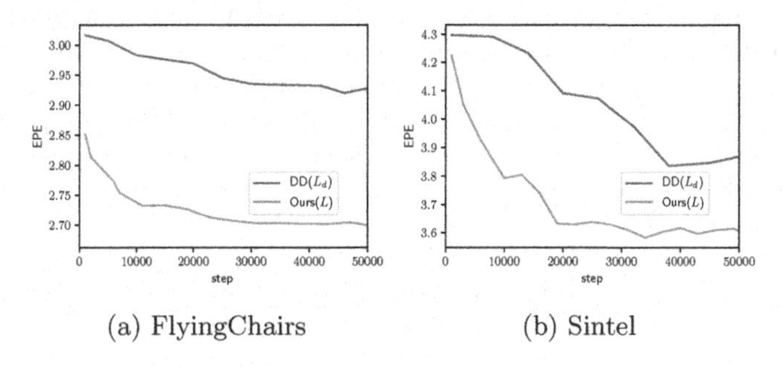

(a) FlyingChairs (b) Sintel

Fig. 4. Training graphs of end-point-error (EPE) on two datasets. For DD(L_d), we use census, occlusion handling and L_d, which is the same setting to [22]. For ours(L), we use the full loss function L (Eq. 11). Ours performs consistently better during training

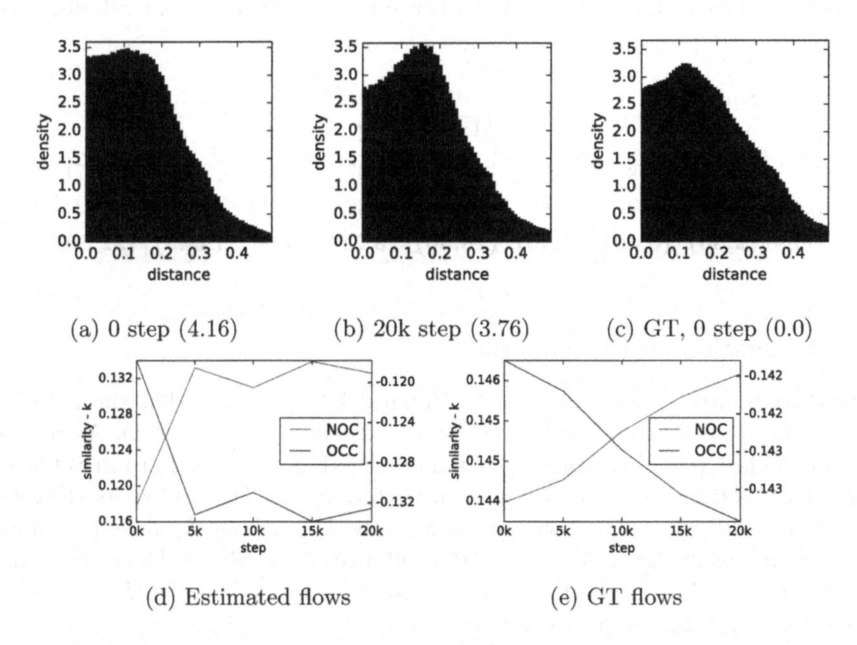

(a) 0 step (4.16) (b) 20k step (3.76) (c) GT, 0 step (0.0)

(d) Estimated flows (e) GT flows

Fig. 5. (a–c) We measure the distance from the boundary k (Eq. 5) to fused similarity on Sintel Final dataset. (a–c) show distance distributions of different steps; we show EPE inside parenthesis. During training, our feature separation loss pushes fused similarity away from the boundary k; note that the greater the distance is, the more separation we have. (d–e) show average $\text{sim}_f - k$ for occluded/non-occluded pixels. Note that occluded pixels show negative value. We measure the distance using ground-truth (GT) flows to observe the effect mainly on the encoder feature excluding the effect from the decoder side.

Table 1. Ablation study on various settings. Average end-point error (EPE) is used as a metric. For Sintel, the results are evaluated over all (ALL), non-occluded (NOC), and occluded (OCC) pixels. Results in parenthesis are achieved by testing the model using the data that the model is trained on. In left columns we show types of losses we use: occlusion handling ($\hat{C}$), data-distillation [22] (L_d), and ours: feature separation (L_f) and regulated census (L_r). The first two rows show the performance of our pretrained network trained with FlyingChairs

Base feature	$\hat{C}$	L_d	L_f	L_r	FlyingChairs	Sintel Clean			Sintel Final		
					ALL	ALL	NOC	OCC	ALL	NOC	OCC
RGB					4.01	5.61	3.01	38.64	6.44	3.76	40.45
RGB	✓				3.64	4.40	2.12	33.33	5.42	3.02	36.01
Census	✓				3.10	(3.22)	**(1.26)**	(28.14)	(4.37)	(2.25)	(31.25)
Census	✓	✓			2.93	(3.15)	(1.49)	(24.35)	(3.86)	(2.11)	(26.16)
Census	✓	✓	✓		2.87	(3.25)	(1.46)	(25.95)	(4.15)	(2.27)	(28.01)
Census	✓	✓		✓	2.81	(2.91)	(1.29)	(23.40)	(3.62)	(1.95)	(24.98)
Census	✓	✓	✓	✓	**2.69**	**(2.86)**	(1.28)	**(22.85)**	**(3.57)**	**(1.94)**	**(24.38)**

Table 2. Average EPE on different displacements. ALL: all pixels. *a*-*b*: pixels displaced within $[a, b)$

	Sintel clean				Sintel final			
	ALL	0–10	10–40	40+	ALL	0–10	10–40	40+
L_p	(3.22)	(0.59)	(3.86)	(20.45)	(4.37)	(0.83)	(5.41)	(27.17)
L_d	(3.15)	(0.59)	(4.02)	(19.51)	(3.86)	(0.71)	(5.06)	(23.63)
Ours	**(2.86)**	**(0.49)**	**(3.45)**	**(18.36)**	**(3.57)**	**(0.64)**	**(4.49)**	**(22.36)**

4.1 Evaluation on Benchmarks

Ablation Study. The results on benchmark datasets show that the network better learns to estimate flows with the feature similarity (Table 1). As can be seen in the last row of the table, our final method ($L_d + L_f + L_r$) works better in most cases than the other settings on both datasets. Due to low localization precision of deep features, however, using only L_f without L_r does not much improve the result. Interestingly, L_r that adaptively regulates the conventional census loss is highly effective. When L_r is combined together with L_f, the combined loss ($L_f + L_r$) performs the best.

Table 3. Average EPE depending on loss weighting parameters

(a) Smoothness weight

λ_s	10^{-2}	10^{-3}	10^{-4}	10^{-5}
FlyingChairs	3.27	2.84	2.85	**2.83**
Sintel Final	(4.41)	(4.34)	**(4.16)**	(4.41)

(b) Separation weight

λ_f	2.0	3.0	4.0	5.0
FlyingChairs	2.95	2.90	**2.85**	3.21
Sintel Final	(4.33)	(4.31)	**(4.16)**	(4.36)

Table 4. Comparison to state-of-the-art deep unsupervised optical flow methods. Results in parentheses indicates it is evaluated using data it is trained on. We report average end-point-error for most categories and percentage of erroneous pixels for KITTI testset calculated from benchmark server. Best results in red and second best results in blue.

Method	Chairs	Sintel Clean		Sintel Final		KITTI 2015	
	Test	Train	Test	Train	Test	Train	Test(Fl)
BackToBasic [19]	5.3	–	–	–	–	–	–
DSTFlow-ft [29]	5.52	(6.16)	10.41	(6.81)	11.27	16.79	39%
OccAwareFlow-best [34]	3.30	(4.03)	7.95	(5.95)	9.15	8.88	31.2%
UnFlow-CSS-ft [25]	–	–	–	(7.91)	10.22	8.10	23.30%
MultiFrameOccFlow-ft [18]	–	(3.89)	7.23	(5.52)	8.81	6.59	22.94%
DDFlow-ft [22]	2.97	(2.92)	6.18	(3.98)	7.40	5.72	14.29%
SelFlow-ft-Sintel [23]	–	(2.88)†	6.56†	(3.87)†	6.57†	4.84	14.19%
Ours-Chairs	2.69	3.66	–	4.67	–	16.99	–
Ours-ft-Sintel	3.01	(2.86)	5.92	(3.57)	6.92	12.75	–
Ours-ft-KITTI	4.32	5.49	–	7.24	–	5.19	13.38%

†: Pretrained on the original Sintel movie

In Sintel where we report EPE on NOC and OCC, L_f improves better on OCC than on NOC. It implies that the suppression part in L_f takes effect, which discourages matching with higher similarity for uncertain flows. Table 2 shows that our method effectively covers various ranges of displacements.

We have also tested direct feature learning by the triplet loss [8], it did not improve the baseline accuracy DDFlow [22] i.e. $L_d + L_p$ alone in the experiments. The proposed losses and learning strategy are crucial to unsupervised learning of feature and optical flow.

Loss Parameter. We show results with different parameter settings in Table 3. Our smoothness constraint helps the network refine flows with lower similarity, when set to lower weight values; with a high weight value, the flow field becomes over-smoothed. With respect to λ_f, a weight greater than 4 can deteriorate the learning. As can be seen in Fig. 4, our final loss (Eq. 11) converges well and it achieves higher performance than the baseline of using L_d.

Analysis on Similarity. During training, our method gradually improves flow and encoder feature. Figure 5 illustrates the fused similarity with respect to each training step. Figure 5a shows the distribution before we apply our feature separation loss. After 20k steps of training with our method, the distribution (Fig. 5b) becomes similar to ground-truth distribution (Fig. 5c). Figure 5d–5e plots average difference, i.e., $(\text{sim}_f - k)$, of each types of pixels, where we observe the feature similarity becomes more discriminative gradually; average similarities of non-occluded pixels and occluded pixels become higher and lower respectively. The last result using GT flows shows solely on the feature factor (encoder) without flow estimation factor (decoder), which confirms the updated, more discriminative features.

Qualitative Results. Training with our feature similarity loss makes the flow and similarity much discriminative. During training, the similarity map changes in a way that higher similarity becomes higher and lower similarity goes lower. As a result, the network can improve its prediction by reinforcing clear matching and suppressing uncertain matching. In Fig. 6, we visualize how the similarity loss can be beneficial to flow learning. In the examples, our loss effectively improves the flow estimation by pushing similarity to each polar; that is why our method performs well in uncertain regions. In Fig. 7, we compare ours with SelFlow [23]. In the first two examples, the uncertain snow regions are the most challenging part, so that the state-of-the-art SelFlow fails in such regions. On the other hand, our method is able to handle such regions effectively, since the similarity loss suppresses the flows in that regions.

Quantitative Comparison to State-of-the-art. We compare our method with existing deep unsupervised methods in Table 4. Our method effectively improves the baseline framework [22] and shows competitive results against other unsupervised methods. For Sintel, SelFlow gets a better result in the Sintel Final testset; it is trained on 10 k additional Sintel movie frames, while our model uses 1 k frames from the MPI Sintel dataset. Since our method can be used jointly with hallucinated occlusion and multiple-frame schemes from SelFlow [23], we expect a much stronger unsupervised model if the two are combined. In the real dataset KITTI, our method effectively improves over our baseline model

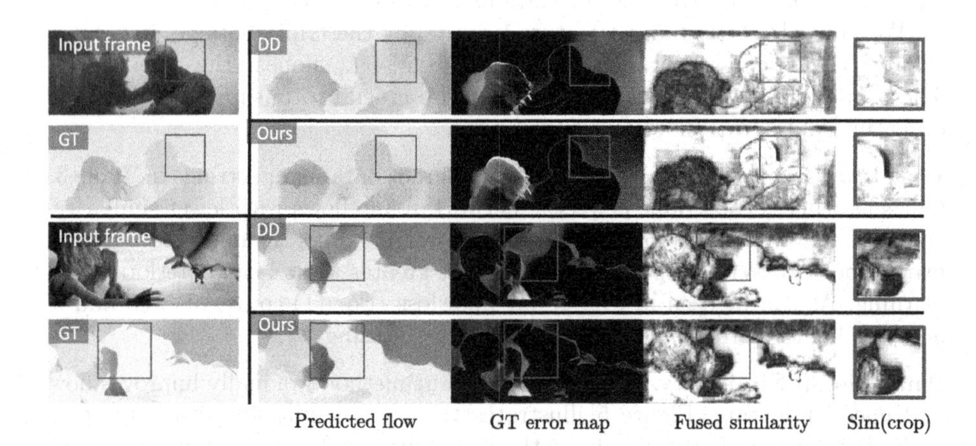

Fig. 6. Fused similarity of models trained w/o and w/the feature similarity loss, denoted by **DD** and **ours**, respectively. The similarity loss suppresses similarity of the background snow texture behind the fighting man, resulting in the similarity map in the second row. In the second example, the sky region is expanded since the simlilarity increases by the similarity loss. As a result, the flow field effectively separates the brown wing of the dragon and the sky in brown (Color figure online)

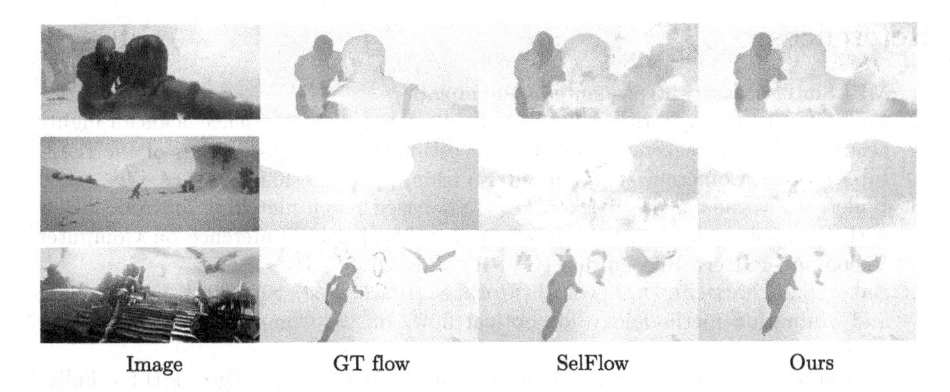

Fig. 7. Comparison to SelFlow [23] on the Sintel testset. We retrieve the resulting images and the visualizations of ground-truth from its benchmark website [1]; it provides only twelve test samples for each method

(DDFlow), and reduces the percentage of erroneous pixels to 13.38% in the test benchmark. Overall, our approach achieves top-1 or top-2 consistently across different benchmarks. This demonstrates the robustness of our approach and benefits of utilizing deep self-supervised features and fused similarity.

Failure Cases. Most unsupervised methods tend to estimate motion in a larger area than it really is, and it occurs more frequently for smaller and faster objects. In contrast, since our method estimates the similarity of a matching and refines it with the similarity, it sometimes reduces flows for small and fast-moving objects. In the last example in Fig. 7, ours fails to catch the movement of few birds flying fast over the stairs.

5 Conclusion

In this paper, we have shown that learning flow from self-supervised features is feasible and effective with our fused similarity and feature separation loss. This allows the network to better learn flows and features from a sequence of images in the unsupervised way. We observe that, using the feature separation loss, the flows are updated to make the fused similarity more discriminative, while suppressing uncertain flows and reinforcing clear flows. The experiments show that the proposed method achieves competitive results in both qualitative and quantitative evaluation. The promising results confirm that, without labels, the self-supervised features can be used to improve itself. This kind of self-regulation techniques would be proven more effective under semi-supervised settings, where part of labels are available.

Acknowledgment. This research was supported by Next-Generation Information Computing Development Program through the National Research Foundation of Korea(NRF) funded by the Ministry of Science, ICT (NRF-2017M3C4A7066317).

References

1. MPI Sintel dataset. http://sintel.is.tue.mpg.de/
2. Bailer, C., Taetz, B., Stricker, D.: Flow fields: dense correspondence fields for highly accurate large displacement optical flow estimation. In: Proceedings of the IEEE International Conference on Computer Vision, pp. 4015–4023 (2015)
3. Bailer, C., Varanasi, K., Stricker, D.: CNN-based patch matching for optical flow with thresholded hinge embedding loss. In: The IEEE Conference on Computer Vision and Pattern Recognition (CVPR), July 2017
4. Baker, S., Scharstein, D., Lewis, J., Roth, S., Black, M.J., Szeliski, R.: A database and evaluation methodology for optical flow. Int. J. Comput. Vis. **92**(1), 1–31 (2011)
5. Bertinetto, L., Valmadre, J., Henriques, J.F., Vedaldi, A., Torr, P.H.S.: Fully-convolutional siamese networks for object tracking. In: Hua, G., Jégou, H. (eds.) ECCV 2016. LNCS, vol. 9914, pp. 850–865. Springer, Cham (2016). https://doi.org/10.1007/978-3-319-48881-3_56
6. Butler, D.J., Wulff, J., Stanley, G.B., Black, M.J.: A naturalistic open source movie for optical flow evaluation. In: Fitzgibbon, A., Lazebnik, S., Perona, P., Sato, Y., Schmid, C. (eds.) ECCV 2012. LNCS, vol. 7577, pp. 611–625. Springer, Heidelberg (2012). https://doi.org/10.1007/978-3-642-33783-3_44
7. Carreira, J., Zisserman, A.: Quo vadis, action recognition? A new model and the kinetics dataset. In: The IEEE Conference on Computer Vision and Pattern Recognition (CVPR), July 2017
8. Dong, X., Shen, J.: Triplet loss in siamese network for object tracking. In: Ferrari, V., Hebert, M., Sminchisescu, C., Weiss, Y. (eds.) ECCV 2018. LNCS, vol. 11217, pp. 472–488. Springer, Cham (2018). https://doi.org/10.1007/978-3-030-01261-8_28
9. Dosovitskiy, A., et al.: FlowNet: learning optical flow with convolutional networks. In: Proceedings of the IEEE International Conference on Computer Vision, pp. 2758–2766 (2015)
10. Geiger, A., Lenz, P., Urtasun, R.: Are we ready for autonomous driving? The kitti vision benchmark suite. In: 2012 IEEE Conference on Computer Vision and Pattern Recognition, pp. 3354–3361. IEEE (2012)
11. Gidaris, S., Singh, P., Komodakis, N.: Unsupervised representation learning by predicting image rotations. In: International Conference on Learning Representations (2018). https://openreview.net/forum?id=S1v4N2l0-
12. Grandvalet, Y., Bengio, Y.: Semi-supervised learning by entropy minimization. In: Advances in Neural Information Processing Systems, pp. 529–536 (2005)
13. Güney, F., Geiger, A.: Deep discrete flow. In: Lai, S.-H., Lepetit, V., Nishino, K., Sato, Y. (eds.) ACCV 2016. LNCS, vol. 10114, pp. 207–224. Springer, Cham (2017). https://doi.org/10.1007/978-3-319-54190-7_13
14. Han, X., Hu, X., Huang, W., Scott, M.R.: ClothFlow: a flow-based model for clothed person generation. In: Proceedings of the IEEE International Conference on Computer Vision, pp. 10471–10480 (2019)
15. Hur, J., Roth, S.: Iterative residual refinement for joint optical flow and occlusion estimation. In: Proceedings of the IEEE Conference on Computer Vision and Pattern Recognition, pp. 5754–5763 (2019)
16. Ilg, E., Mayer, N., Saikia, T., Keuper, M., Dosovitskiy, A., Brox, T.: FlowNet 2.0: evolution of optical flow estimation with deep networks. In: Proceedings of the IEEE conference on Computer Vision and Pattern Recognition, pp. 2462–2470 (2017)

17. Jaderberg, M., et al.: Spatial transformer networks. In: Advances in Neural Information Processing Systems, pp. 2017–2025 (2015)
18. Janai, J., Güney, F., Ranjan, A., Black, M., Geiger, A.: Unsupervised learning of multi-frame optical flow with occlusions. In: Ferrari, V., Hebert, M., Sminchisescu, C., Weiss, Y. (eds.) ECCV 2018. LNCS, vol. 11220, pp. 713–731. Springer, Cham (2018). https://doi.org/10.1007/978-3-030-01270-0_42
19. Yu, J.J., Harley, A.W., Derpanis, K.G.: Back to basics: unsupervised learning of optical flow via brightness constancy and motion smoothness. In: Hua, G., Jégou, H. (eds.) ECCV 2016. LNCS, vol. 9915, pp. 3–10. Springer, Cham (2016). https://doi.org/10.1007/978-3-319-49409-8_1
20. Kingma, D.P., Ba, J.: Adam: a method for stochastic optimization. arXiv preprint arXiv:1412.6980 (2014)
21. Liu, C.,et al.: Guided similarity separation for image retrieval. In: Advances in Neural Information Processing Systems, pp. 1554–1564 (2019)
22. Liu, P., King, I., Lyu, M.R., Xu, J.: DDFlow: learning optical flow with unlabeled data distillation. In: Proceedings of the AAAI Conference on Artificial Intelligence, vol. 33(01), pp. 8770–8777, July 2019. https://doi.org/10.1609/aaai.v33i01.33018770, https://aaai.org/ojs/index.php/AAAI/article/view/4902
23. Liu, P., Lyu, M., King, I., Xu, J.: SelFlow: self-supervised learning of optical flow. In: Proceedings of the IEEE Conference on Computer Vision and Pattern Recognition, pp. 4571–4580 (2019)
24. Mayer, N., et al.: A large dataset to train convolutional networks for disparity, optical flow, and scene flow estimation. In: Proceedings of the IEEE Conference on Computer Vision and Pattern Recognition, pp. 4040–4048 (2016)
25. Meister, S., Hur, J., Roth, S.: Unflow: unsupervised learning of optical flow with a bidirectional census loss. In: Thirty-Second AAAI Conference on Artificial Intelligence (2018)
26. Menze, M., Geiger, A.: Object scene flow for autonomous vehicles. In: Proceedings of the IEEE Conference on Computer Vision and Pattern Recognition, pp. 3061–3070 (2015)
27. Noroozi, M., Favaro, P.: Unsupervised learning of visual representations by solving jigsaw puzzles. In: Leibe, B., Matas, J., Sebe, N., Welling, M. (eds.) ECCV 2016. LNCS, vol. 9910, pp. 69–84. Springer, Cham (2016). https://doi.org/10.1007/978-3-319-46466-4_5
28. Ranjan, A., Black, M.J.: Optical flow estimation using a spatial pyramid network. In: Proceedings of the IEEE Conference on Computer Vision and Pattern Recognition, pp. 4161–4170 (2017)
29. Ren, Z., Yan, J., Ni, B., Liu, B., Yang, X., Zha, H.: Unsupervised deep learning for optical flow estimation. In: Thirty-First AAAI Conference on Artificial Intelligence (2017)
30. Revaud, J., Weinzaepfel, P., Harchaoui, Z., Schmid, C.: EpicFlow: edge-preserving interpolation of correspondences for optical flow. In: Proceedings of the IEEE Conference on Computer Vision and Pattern Recognition, pp. 1164–1172 (2015)
31. Simonyan, K., Zisserman, A.: Two-stream convolutional networks for action recognition in videos. In: Advances in Neural Information Processing Systems, pp. 568–576 (2014)
32. Sun, D., Yang, X., Liu, M.Y., Kautz, J.: PWC-Net: CNNs for optical flow using pyramid, warping, and cost volume. In: Proceedings of the IEEE Conference on Computer Vision and Pattern Recognition, pp. 8934–8943 (2018)
33. Ufer, N., Ommer, B.: Deep semantic feature matching. In: Proceedings of the IEEE Conference on Computer Vision and Pattern Recognition, pp. 6914–6923 (2017)

34. Wang, Y., Yang, Y., Yang, Z., Zhao, L., Wang, P., Xu, W.: Occlusion aware unsupervised learning of optical flow. In: Proceedings of the IEEE Conference on Computer Vision and Pattern Recognition, pp. 4884–4893 (2018)
35. Weinzaepfel, P., Revaud, J., Harchaoui, Z., Schmid, C.: DeepFlow: large displacement optical flow with deep matching. In: Proceedings of the IEEE International Conference on Computer Vision, pp. 1385–1392 (2013)
36. Werlberger, M., Pock, T., Unger, M., Bischof, H.: Optical flow guided TV-L^1 video interpolation and restoration. In: Boykov, Y., Kahl, F., Lempitsky, V., Schmidt, F.R. (eds.) EMMCVPR 2011. Optical flow guided tv-l 1 video interpolation and restoration., vol. 6819, pp. 273–286. Springer, Heidelberg (2011). https://doi.org/10.1007/978-3-642-23094-3_20
37. Xu, J., Ranftl, R., Koltun, V.: Accurate optical flow via direct cost volume processing. In: The IEEE Conference on Computer Vision and Pattern Recognition (CVPR), July 2017
38. Xu, R., Li, X., Zhou, B., Loy, C.C.: Deep flow-guided video inpainting. In: The IEEE Conference on Computer Vision and Pattern Recognition (CVPR), June 2019
39. Zabih, R., Woodfill, J.: Non-parametric local transforms for computing visual correspondence. In: Eklundh, J.-O. (ed.) ECCV 1994. LNCS, vol. 801, pp. 151–158. Springer, Heidelberg (1994). https://doi.org/10.1007/BFb0028345
40. Zeiler, M.D., Fergus, R.: Visualizing and understanding convolutional networks. In: Fleet, D., Pajdla, T., Schiele, B., Tuytelaars, T. (eds.) ECCV 2014. LNCS, vol. 8689, pp. 818–833. Springer, Cham (2014). https://doi.org/10.1007/978-3-319-10590-1_53
41. Zhan, H., Garg, R., Saroj Weerasekera, C., Li, K., Agarwal, H., Reid, I.: Unsupervised learning of monocular depth estimation and visual odometry with deep feature reconstruction. In: Proceedings of the IEEE Conference on Computer Vision and Pattern Recognition, pp. 340–349 (2018)
42. Zhang, Z., Peng, H.: Deeper and wider siamese networks for real-time visual tracking. In: Proceedings of the IEEE Conference on Computer Vision and Pattern Recognition, pp. 4591–4600 (2019)

Blended Grammar Network for Human Parsing

Xiaomei Zhang[1,2](✉), Yingying Chen[1,2,3], Bingke Zhu[1,2], Jinqiao Wang[1,2,4], and Ming Tang[1]

[1] National Laboratory of Pattern Recognition, Institute of Automation, CAS, Beijing 100190, China
{xiaomei.zhang,yingying.chen,bingke.zhu,jqwang,tangm}@nlpr.ia.ac.cn
[2] School of Artificial Intelligence, University of Chinese Academy of Sciences, Beijing 100049, China
[3] ObjectEye Inc., Beijing, China
[4] NEXWISE Co., Ltd., Guangzhou, China

Abstract. Although human parsing has made great progress, it still faces a challenge, i.e., how to extract the whole foreground from similar or cluttered scenes effectively. In this paper, we propose a Blended Grammar Network (BGNet), to deal with the challenge. BGNet exploits the inherent hierarchical structure of a human body and the relationship of different human parts by means of grammar rules in both cascaded and paralleled manner. In this way, conspicuous parts, which are easily distinguished from the background, can amend the segmentation of inconspicuous ones, improving the foreground extraction. We also design a Part-aware Convolutional Recurrent Neural Network (PCRNN) to pass messages which are generated by grammar rules. To train PCRNNs effectively, we present a blended grammar loss to supervise the training of PCRNNs. We conduct extensive experiments to evaluate BGNet on PASCAL-Person-Part, LIP, and PPSS datasets. BGNet obtains state-of-the-art performance on these human parsing datasets.

1 Introduction

Human parsing aims to segment human images into multiple human parts of fine-grained semantics and benefits a detailed understanding of images. It has many applications, such as person re-identification [8], human behavior analysis [37], clothing style recognition and retrieval [40], clothing category classification [35], to name a few. With the rapid development of electronic commerce and online shopping, human parsing has been attracting much attention [13,14,17, 19,20,23,28,34,38,39,41,45]. However, it is still a challenging task and faces the difficulty in accurate extraction of the whole foreground from similar, cluttered scenes, and blurred images. Figure 1 provides two examples where there exist similar appearances to the foreground in the background or cluttered scenes.

In order to deal with the problem, HAZA [38] and Joint [39] introduced human detection into their human parsing algorithms to abate the interference of

A. Vedaldi et al. (Eds.): ECCV 2020, LNCS 12369, pp. 189–205, 2020.
https://doi.org/10.1007/978-3-030-58586-0_12

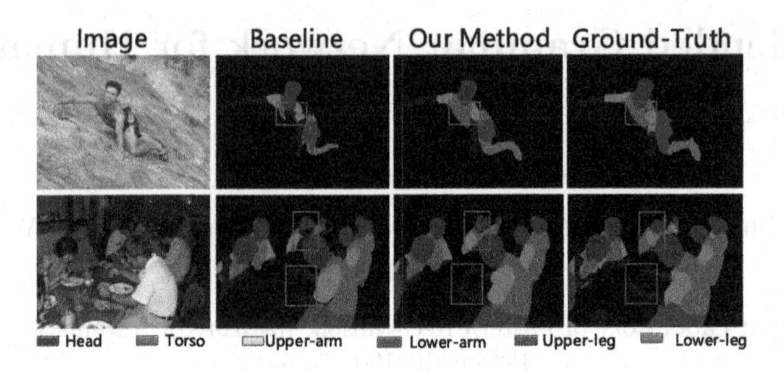

Fig. 1. Examples of the challenge in human parsing. The original images and ground-truth come from PASCAL-Person-Part dataset [5]. There are similar appearances to the foreground in the background. Second column: the baseline fails to extract the whole foreground. Third column: our method has better performance in extracting foreground from similar or cluttered scenes.

the similar or cluttered background. Nevertheless, their methods strongly depend on the detectors, and still need to extract the whole body of a human from the cluttered background around the bounding boxes of detection, even if the detection is correct. If the detectors fail to locate a human body, that body will be treated as the cluttered background, or vice versa. This may in turn increase the background interference for human parsing. Additionally, MuLA [29] and LIP [18] combined human pose estimation and human parsing to improve the foreground extraction. Nevertheless, human pose estimation does not focus on assigning a unique label to a pixel. The assistance against the cluttered background by means of the human pose estimation is limited.

In this paper, we propose a Blended Grammar Network (BGNet) to deal with the above problem by exploiting the inherent hierarchical structure of a human body and the relationship of different human parts. We design a Part-aware Convolutional Recurrent Neural Network (PCRNN) to model grammar rules in our BGNet, which can adaptively extract features to improve accuracy. A blended grammar loss is designed to supervise the training of PCRNNs.

Our BGNet, which is developed to exploit the inherent hierarchical structure of a human body and capture the relationship of different human parts, is based on two insights. One comes from the visual psychology which claimed that humans often pay attention to conspicuous parts first, such as head, and then tend to other parts [11,33] when observing a person. And the other is reported by [44], which verified that conspicuous parts, such as head, are relatively more separable and can more easily be distinguished from the background. According to these insights, the grammar rules of BGNet leverage conspicuous parts (e.g., torso, head) distinguished from the background easily to amend the segmentation of inconspicuous ones (e.g., low-leg, low-arm), improving the foreground extraction. And we use grammar rules to progressively explore the inherent hierarchical structure of a human body in both cascaded and paralleled manner, as shown in Fig. 2. Specifically, a latter grammar rule inherits the outputs of its former rule as one of its inputs in a cascaded manner, and two grammar rules take the outputs of the same grammar rule as their one input in a paralleled manner. For example, $Rule_1^c$, $Rule_2^h$ and $Rule_3^h$ connect in a cascaded manner, meanwhile, $Rule_2^h$ and $Rule_4^v$ connect in parallel. In this way, BGNet combines the advantages of both cascaded and paralleled architectures, thus further improving the accuracy of foreground extraction.

We propose PCRNN to represent grammar rules and model messages passing, which can adaptively extract features to improve accuracy. PCRNN consists of two stages to effectively generate reliable features of parts, as shown in Fig. 4. The first stage of grammar rules (e.g., $Rule_2^h$) uses two inputs, the results of the former rule (e.g., the outputs of the $Rule_1^c$) and features of its corresponding part (e.g., the features of upper-arm), to model the relationship among human parts. The second stage extracts features by adaptively selecting the results (S_r) of the first stage and generates features of its corresponding parts (e.g., head, torso and upper-arm in $Rule_2^h$). To train PCRNNs effectively, we design a blended grammar loss which uses a less-to-more manner with increasing parts in PCRNNs. The supervision of each PCRNN is the ground truth of its corresponding human parts.

Extensive experiments show that our network achieves new state-of-the-art results consistently on three public benchmarks, PASCAL-Person-Part [5], LIP [14] and PPSS [27]. Our method outperforms the best competitors by 3.08%, 2.42%, and 4.59% on PASCAL-Person-Part, LIP, and PPSS in terms of mIoU, respectively. In summary, our contributions are in three folds:

1. We propose a novel Blended Grammar Network (BGNet) to improve the extraction accuracy of the foreground out of the cluttered background in both cascaded and paralleled manner. And grammar rules of BGNet use the conspicuous parts to amend the inconspicuous ones, improving the foreground extraction.
2. We design a Part-aware Convolutional Recurrent Neural Network (PCRNN) to pass messages across BGNet and a novel deep blended grammar loss to supervise the training of PCRNNs. With the grammar loss, the PCRNN effectively represents the relationship of human parts.
3. The proposed BGNet achieves new state-of-the-art results consistently on three public human parsing benchmarks, including PASCAL-Person-Part, LIP, and PPSS.

2 Related Work

Human Parsing. Many research efforts have been devoted to human parsing [14,17,19,20,23,28,38,39,45]. Chen *et l.* [4] proposed an attention mechanism that learns to weight the multi-scale features at each pixel location softly. Xia *et al.* [38] proposed HAZN for object part parsing, which adapted to the local scales of objects and parts by detection methods. Human pose estimation and semantic part segmentation were two complementary tasks [18], in which the former provided an object-level shape prior to regularize part segments while the latter constrained the variation of pose location. Ke *et al.* [12] proposed Graphonomy, which incorporated hierarchical graph transfer learning upon the conventional parsing network to predict all labels. Wang *et al.* [36] combined neural networks with the compositional hierarchy of human bodies for efficient and complete human parsing. Different from the above methods, our BGNet exploits the inherent hierarchical structure of a human body and the relationship of different human parts by means of grammar rules.

Grammar Model. Grammar models are powerful tools for modeling high-level human knowledge in some tasks of human pose estimation [7], Clothing Category Classification [35] and so on. Qi *et al.* [31] presented the stochastic grammar to predict human activities. Grammar models allow an expert injects domain-specific knowledge into the algorithms, avoiding local ambiguities [1,30]. Wang *et al.* [35] designed the fashion grammar to capture the relations among a few joints and focused on the local context through short grammar rule, in which grammar layers connected in parallel. Different from the above methods, first, our BGNet uses a blended architecture of both parallelled and cascaded connections among grammar rules, and latter rules inherit and refine results of former rules. Second, our grammar rules exploit the relations of different human parts to capture the local and global context because human parsing needs not only the local context but also the global context. Third, our PCRNN passes messages unidirectionally. The purpose of our PCRNNs is to carry out the message passing from conspicuous parts (e.g., torso, head) to the segmentation of inconspicuous parts (e.g., low-leg).

Supervision Mechanism. Various types of supervision approaches have been popular in the tasks of human parsing [18,45], semantic segmentation [9,10,21, 42,43] and human pose estimation [16]. Liang *et al.* [18] introduced a structure-sensitive loss to evaluate the quality of the predicted parsing results from a joint structure perspective. Zhu *et al.* [45] proposed a component-aware region convolution structure to segment human parts with hierarchical supervision. Zhao *et al.* [43] designed an auxiliary loss in the backbone. Zhao *et al.* [42] developed an image cascade network (ICNet) with cascade label guidance. Ke *et al.* [16] proposed multi-scale supervision to strengthen contextual feature learning in matching body keypoints by combining feature heatmaps across scales. Different from them, our deep blended grammar loss uses the less-to-more manner with an increasing number of parts in rules.

3 Blended Grammar Network

The overall architecture of BGNet is shown in Fig. 2. We the feature extractor to generate original features. Then, a convolutional layer is applied to features to generate the corresponding coarse predictions of human parts, which are learned under the supervision from the ground-truth segmentation. The coarse prediction is sent into grammar rules to exploit the inherent hierarchical structure of a human body and the relationship of different human parts. We design PCRNN

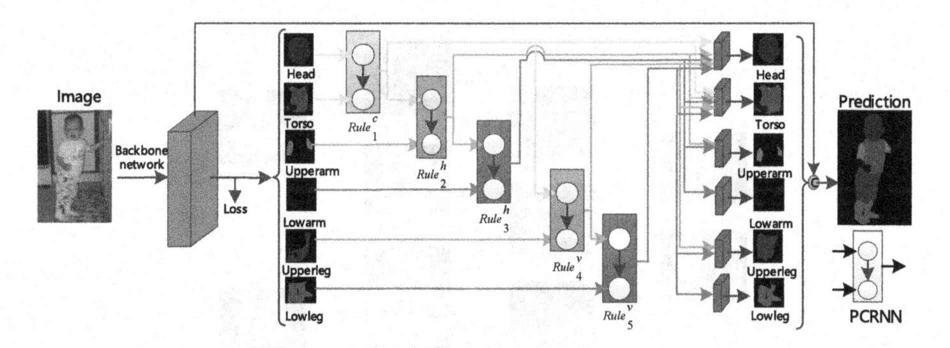

Fig. 2. Overview of the proposed Blended Grammar Network (BGNet). An input image goes through the backbone network to generate its original features and the corresponding coarse predictions of human parts. Then grammar rules are applied to exploit the inherent hierarchical structure of a human body and the relationship of different human parts. Outputs of grammar rules and original features are concatenated to get the final fine prediction. Every grammar rule is represented by one PCRNN. © indicates the concatenation operation.

to model grammar rules, and every grammar rule is represented by one PCRNN. Finally, the outputs of the feature extractor and PCRNN are concatenated to obtain the final fine prediction.

3.1 Grammar

Dependency grammars [30] have been widely used in natural language processing for syntactic parsing. It has a root node S and set of n other nodes $\{A_1, ..., A_n\}$ with rules like

$$\begin{aligned} &S \rightarrow A_1|A_2 \cdots |A_n, \\ &A_i \rightarrow a_i|a_iA_j|A_ja_i, \forall i = 1, 2..., n, j{\neq}i, \end{aligned} \tag{1}$$

where "|" denotes "or" function, "→" denotes the flow of information, root node S can transit to any other nodes once, and then each node A_i can terminate as a_i or transit to another node A_j to the left or right side. It is seen that S can also indirectly transmit to any other node through an intermediate node.

Our Grammar. Inspired by dependency grammars, we define a novel progressive grammar model of grammar rules, to segment human parts, as shown in Fig. 2. Grammar rules of BGNet set conspicuous parts as root node to amend the segmentation of inconspicuous parts by means of their relationship, improving the extraction accuracy of the foreground out of the cluttered background. The more conspicuous parts are used in more rules. For example, head is the most conspicuous part [44], thus head is integrated into all grammar rules.

We consider a total of six human parts to constitute grammar rules which are tried during preparing the paper, as shown in Tabel 1. According to the performance and human anatomical and anthropomorphic constraints, we use five grammar rules, $Rule^c_1$, $Rule^h_2$, $Rule^h_3$, $Rule^v_4$ and $Rule^v_5$, to progressively exploit

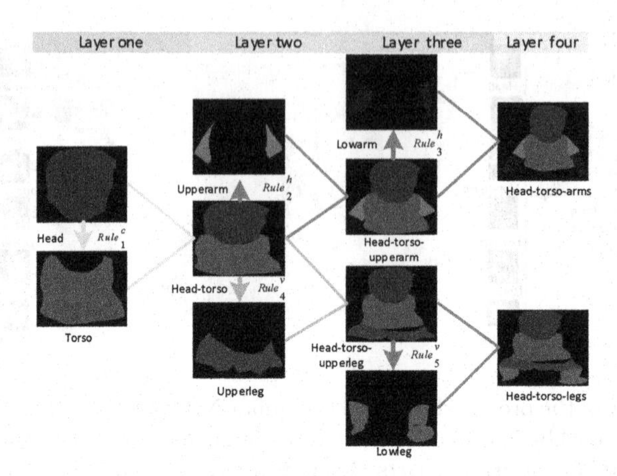

Fig. 3. Message across our BGNet. Lines and arrows with different colors and their connective human parts represent different grammar rules.

the inherent hierarchical structure of a human body. These five grammar rules describe central, horizontal, and vertical relation, respectively. Due to different datasets with different kinds of labels of human parts, the grammar rules may be adjusted slightly. The five rules are:

$$\begin{aligned} &Rule_1^c : head \rightarrow torso, \\ &Rule_2^h : head \rightarrow torso \rightarrow upperarm, \\ &Rule_3^h : head \rightarrow torso \rightarrow upperarm \rightarrow lowarm, \\ &Rule_4^v : head \rightarrow torso \rightarrow upperleg, \\ &Rule_5^v : head \rightarrow torso \rightarrow upperleg \rightarrow lowleg. \end{aligned} \tag{2}$$

There is a progressive relation among the five grammar rules. For example, $Rule_2^h$ is the growth of $Rule_1^c$, and $Rule_3^h$ is the growth of $Rule_2^h$. Thus, Eq. 2 can be represented by the following expression,

$$\begin{aligned} &Rule_1^c : head \rightarrow torso, \\ &Rule_2^h : Rule_1^c \rightarrow upperarm, \\ &Rule_3^h : Rule_2^h \rightarrow lowarm, \\ &Rule_4^v : Rule_1^c \rightarrow upperleg, \\ &Rule_5^v : Rule_4^v \rightarrow lowleg. \end{aligned} \tag{3}$$

3.2 Blended Grammar Network Structure

As shown in Fig. 2, BGNet is a blended architecture among grammar rules. It combines the advantages of both cascaded and paralleled architectures. i.e., the latter rule inherits the results of its former rule in a cascaded way, meanwhile, $Rule_2^h$ and $Rule_4^v$ all take the results of $Rule_1^c$ as inputs in parallel because they have the tight relationship with $Rule_1^c$. In the cascaded way, the former results provide valuable context to the latter layers, meanwhile, the latter layers take features of a corresponding part as input to further refine the results of the former layer. In the paralleled way, BGNet can make full use of the relationship of human parts.

Figure 3 provides a more vivid representation of the message passing in the blended architecture. $Rule_1^c$ is on the first layer, which exploits the relation between the head and torso and generates the features of head-torso in the middle of the second layer. In the next layer, both $Rule_2^h$ and $Rule_4^v$ leverage the features of head-torso to further explore the relation of upper-arm and upper-leg, respectively. Similarly, $Rule_3^h$ and $Rule_5^v$ repeat the growth by inheriting results from $Rule_2^h$ and $Rule_4^v$, respectively.

Because every grammar rule generates features of its corresponding human parts, thus, different parts have the different number of features. In order to obtain the same number of features of each part, we concatenate the features generated by all relevant grammar rules of a part. Then we apply a convolutional layer on the concatenating features to generate its final prediction. These are

$$\begin{aligned} M^p &= concat(\{M_r^p\}_{r=1}^5), \\ P^p &= \mathcal{W}^p * M^p + b^p, \end{aligned} \tag{4}$$

where M_r^p denotes the feature of part p of the rule r, $concat(\cdot)$ denotes the concatenating function, M^p denote all features of part p and P^p denotes the feature of part p, $\mathcal{W}^p$ refers to weights, b^p refers to bias.

Note that the predictions of all parts and the outputs of the baseline F are concatenated, that is,

$$F^b = concat(\{P^p\}_{p=1}^n, F), \tag{5}$$

where n denotes the number of parts.

Different Architectures of Grammar Network. To prove the capability of our blended architecture, we introduce another architecture named Paralleled Grammar Network (PGNet), in which grammar rules connect in parallel. Two architectures of networks take the outputs of the baseline as their inputs, and use PCRNNs to pass messages. In PGNet, there are also five grammar rules and 5 PCRNNs to pass messages. However, the paralleled connection among grammar rules are designed in PGNet, and the results of the former rules can not be inherited by the latter rules. The results show that our blended architecture has better performance on human parsing than the paralleled architecture.

3.3 Part-Aware CRNN

We design Part-aware CRNN (PCRNN) to pass messages which are generated by grammar rules in BGNet, as shown in Fig. 4. Each grammar rule of BGNet is represented by one PCRNN. PCRNN has the capacity to model the relationship among human parts. What is more, PCRNN can preserve the spatial semantic information via the convolutional operations.

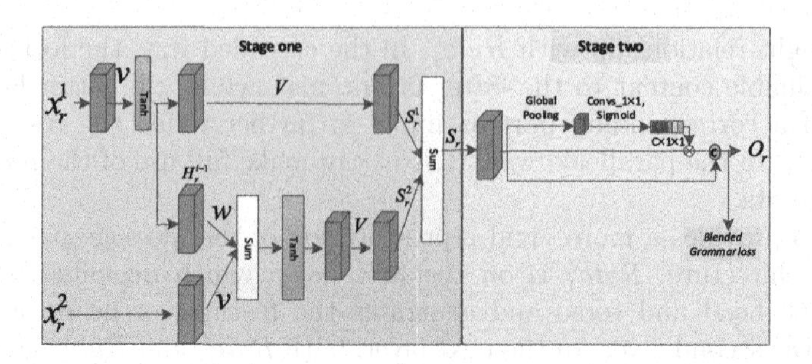

Fig. 4. Architecture of the proposed PCRNN. © indicates the concatenation operation.

PCRNN has two inputs, including the inherited results generated by former rule and features of its corresponding part. It consists of two stages. The first stage models the relationship among human parts. The second stage extracts features by adaptively selecting the results of the first stage. And at the end of it, a convolutional layer is applied to generate the corresponding coarse predictions of human parts under the supervision of the blended grammar loss. Then we concatenate all features of one part. The functions are

$$\begin{aligned} H_r^i &= tanh(\mathcal{V} * \chi_r^i + \mathcal{W} * H_r^{i-1} + b), \\ S_r^i &= V * \mathcal{H}_r^i + c, \\ S_r &= \mathcal{W}_r * (S_r^1 + S_r^2) + b_r, \\ O_r &= concat(sigmoid(gp(S_r)) \circ S_r, S_r), \\ M_r^p &= \mathcal{S}(O_r), \end{aligned} \tag{6}$$

where χ_r^i denotes the input i of the rule r, i refers to 1,2 and r is from 1 to 5, H_r^{i-1} denote the information of the input $i-1$ of the rule r, S_r^i denotes the middle result of the input i, S_r denotes the summation of S_r^i, O_r denotes the output of the rule r, S denotes the sliced operation, M_r^p denotes the feature of corresponding part p of the rule r, '$\circ$' denotes the channel-wise multiplication, $\mathcal{V}$, $\mathcal{W}$, $\mathcal{W}_r$ and V are weights, b, b_r and c are bias. When $i = 1$, H_r^{i-1} does not exist.

The first input of PCRNN is updated by the results of its former one,

$$\chi_{r+1}^1 \leftarrow O_r, \tag{7}$$

where χ_{r+1}^1 denotes the first input of the rule $r+1$.

Our PCRNN passes messages unidirectionally. The purpose of our PCRNNs is to carry out the message passing from conspicuous parts (e.g., torso, head) to the segmentation of inconspicuous parts (e.g., low-leg, low-arm). Exploiting inconspicuous parts to amend conspicuous ones is generally unreliable because distinguishing inconspicuous parts from their background is much harder than distinguishing conspicuous ones. We design B-PCRNN which passes messages back and forth to improve performance directionally. Experimental results on B-PCRNN and PCRNN in Table 1 and show that PCRNN has better performance.

Blended Grammar Loss. Every PCRNN has its corresponding blended grammar loss which locates at the end of it. The supervision is the ground truth of its corresponding human parts. For example, the supervision of $Rule_1^c$ is head and torso, the next layer, the supervision of $Rule_2^h$ is head, torso and upper-arm, etc. It is seen that our grammar loss uses a less-to-more manner with the increasing number of parts in rules.

$$L_r = -\frac{1}{MN}\sum_{i=1}^{MN}\sum_{k=1}^{K}(y_i = k)log(p_{i,k}), \tag{8}$$

where M and N is the height and width of the input image, and K is the number of the categories in rule r, y is a binary indicator (0 or 1) if categories label k is the correct classification for observation i, and p predicts probability observation i of categories k.

3.4 Loss Function

In our BGNet, we design a novel deep blended grammar loss to supervise the training of PCRNNs, termed L_r. For the deep blended grammar loss, we utilize softmax cross-entropy loss.

Following PSPNet [43], BGNet employs two deep auxiliary losses, one locates at the end of the baseline and the other is applied after the twenty-second block of the fourth stage of ResNet101, i.e., the res4b22 residue block, which is named as L_{aux1} and L_{aux2}, respectively. The loss at the end of our method is named as $L_{softmax}$. The total loss can be formulated as:

$$L = \lambda L_{softmax} + \lambda_1 L_{aux1} + \lambda_2 L_{aux2} + \sum_{r=1}^{n} L_r, \tag{9}$$

where we fix the hyper-parameters $\lambda = 0.6$, $\lambda_1 = 0.1$, and $\lambda_2 = 0.3$ in our experiments, r denotes rules in BGNet, n denotes the number of rules and it is 5 in our parsing network. We experiment with setting the auxiliary loss weight λ_1 and λ_2 between 0 and 1, respectively. Then, we set the loss at the end of our method λ between 0 and 1. $\lambda = 0.6$, $\lambda_1 = 0.1$ and $\lambda_2 = 0.3$ yield the best results.

4 Experiments

4.1 Datasets

PASCAL-Person-Part. PASCAL-Person-Part [5] has multiple person appearances in an unconstrained environment. Each image has 7 labels: background, head, torso, upper-arm, lower-arm, upper-leg and lower-leg. We use the images containing human for training (1716 images) and validation (1817 images).

LIP. LIP dataset [14] contains 50,462 images in total, including 30,362 for training, 10,000 for testing and 10,000 for validation. LIP defines 19 human parts (clothes) labels, including hat, hair, sunglasses, upper-clothes, dress, coat, socks, pants, gloves, scarf, skirt, jumpsuits, face, right-arm, left-arm, right-leg, left-leg, right-shoe and left-shoe, and a background class. We use its training set to train our network and its validated set to test our network.

PPSS. PPSS dataset [27] includes 3,673 annotated samples, which are divided into a training set of 1,781 images and a testing set of 1,892 images. It defines seven human parts, including hair, face, upper-clothes, low-clothes, arms, legs and shoes. Collected from 171 surveillance videos, the dataset can reflect the occlusion and illumination variation in the real scene.

Evaluation Metrics. We evaluate the mean pixel Intersection-over-Union (mIoU) of our network in experiments.

4.2 Implementation Details

As for the baseline, we use the FCN-like ResNet-101 [15] (pre-trained on ImageNet [32]). In addition, the PPM module [43] is applied for extracting more effective features with multi-scale context. Following PSPNet [43], the classification layer and last two pooling layers are removed and the dilation rate of the convolution layers after the removed pooling layers are set to 2 and 4 respectively. Thus, the output feature is 8× smaller than the input image.

We train all the models using stochastic gradient descent (SGD) solver, momentum is 0.9 and weight decay is 0.0005. As for these three datasets (PASCAL-Person-Part, LIP and PPSS), we resize images to 512 × 512, 473 × 473, and 512 × 512 as the input size, respectively, the batch sizes are 8, 12, and 8, respectively, the epochs of three datasets are 100, 120, 120, respectively. We do not use OHEM. For data augmentation, we apply the random scaling (from 0.5 to 1.5) and left-right flipping during training. In the inference process, we test images on the multi-scale to acquire a multi-scale context.

4.3 Ablation Study

In this section, we conduct several experiments to analyze the effect of each component in our BGNet on PASCAL-Person-Part [5]. The grammar rules for PASCAL-Person-Part are represented in Eq. 3.

Grammar Rule and PCRNN. To show some empirical details of designing the five grammar rules and evaluate the performance of the PCRNN, we conduct experiments by four settings without blended grammar loss in BGNet, as illustrated in Tabel 1. First, the common convolutional layers have an approximate quantity of parameters compared with 5 PCRNNs, named Baseline + CCL. Second, we reverse the order of all the five grammar rules, such as changing the $Rule_1^c$ as $torso \rightarrow head$, named Baseline + R-PCRNN. Third, we also adopt B-PCRNN [35] to express the grammar rules, named Baseline + B-PCRNN. Fourth, PCRNN is proposed in our paper, named Baseline + PCRNN. We have tried more grammar rules to parse a human body while preparing the paper. For

Table 1. Ablation study for our network. The results are obtained on the validation set of PASCAL-Person-Part [5]. CCL denotes the common convolutional layers with an approximate quantity of parameters compared with 5 PCRNNs. R-PCRNN denotes the reversible order of grammar rules. B-PCRNN is proposed by [35]. PCRNN is proposed in our paper. BGL denotes our novel deep blended grammar loss. DA denotes data augmentation. MS denotes multi-scale testing.

Method	CCL	R-PCRNN	B-PCRNN	PCRNN	BGL	DA	MS	Ave.
Baseline								66.61
Baseline + CCL	✓							67.04
Baseline + R-PCRNN		✓						67.50
Baseline + B-PCRNN			✓					68.1
Baseline + PCRNN				✓				70.35
Baseline + PCRNN + BGL				✓	✓			72.46
Baseline + PCRNN + BGL + DA				✓	✓	✓		73.13
Baseline + PCRNN + BGL + DA + MS				✓	✓	✓	✓	74.42

Table 2. These five grammar rules influence each part category. From the numbers marked with different colors, we show that each grammar rule can improve the accuracy of human parts. The results are obtained on the validation set of PASCAL-Person-Part.

Method	Rule-1	Rule-2	Rule-3	Rule-4	Rule-5	Head	Torso	U-arms	L-arms	U-legs	L-legs	Background	Ave.
Baseline						86.95	70.72	58.22	55.17	51.22	47.09	96.11	66.61
(a)	✓					87.98	72.93	60.03	57.57	54.3	50.71	96.12	68.52
(b)	✓	✓				88.02	73.65	64.12	62.72	56.15	53.73	96.1	70.64
(c)	✓	✓	✓			88.11	73.58	64.51	63.22	57.39	55.6	96.13	71.22
(d)	✓	✓	✓	✓		88.67	73.8	66.73	66.08	58.15	55	96.18	71.93
(e)	✓	✓	✓	✓	✓	88.74	75.68	67.09	64.99	58.14	56.18	96.37	72.46

example, $torso \rightarrow upper - arm$,$upper - arm \rightarrow low - arm$, and so on. But the improvement over these five grammar rules can be negligible. Therefore, these five grammar rules are the best choice.

Because the conspicuous parts, which are easily distinguished from the background, can amend the inconspicuous ones, improving the foreground extraction. Exploiting inconspicuous parts to amend conspicuous ones is generally unreliable because distinguishing inconspicuous parts from their background is much harder than distinguishing conspicuous ones. Thus, PCRNN is better than B-PCRNN [35] and has the best performance compared with other settings.

In Table 2, (a)–(e) presents the performance of BGNet with adding grammar rules. It is seen that each grammar rule can improve the accuracy of the grammar model. The five grammar rules improve the accuracy of every part, as shown in Table 2. From the numbers marked with different colors, it can be seen that $Rule_1$ improves head by 1.03% and torso by 2.21%, compared with the baseline, $Rule_2$

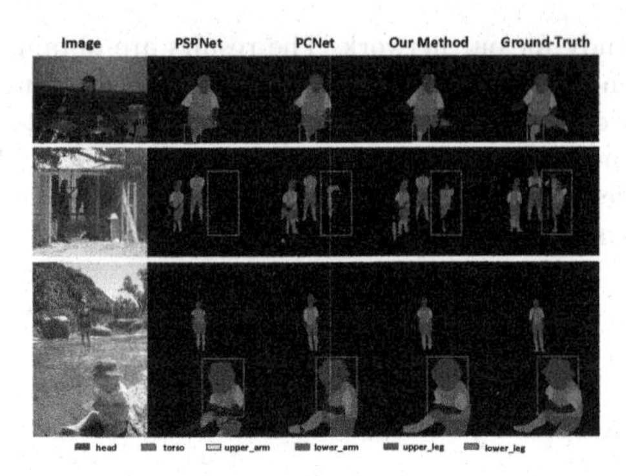

Fig. 5. Qualitative comparison among our method and state-of-the-art approaches on PASCAL-Person-Part [5] dataset. In the first two rows, our method extracts more complete foregrounds from cluttered scenes. And in the last two rows, our method segments different human parts more accurately, such as head and upper-arm.

improves upper-arm by 4.09%, compared with (a), $Rule_3$ improves low-arm by 0.5%, compared with (b), $Rule_4$ improves upper-legs by 0.76%, compared with (c), and $Rule_5$ improves low-legs by 1.18%, compared with (d).

Computation Comparison. We experiment by adding common convolutional layers on the top of the baseline, named Baseline + CCL. The computation of Baseline + CCL is similar to our framework. Our framework yields a result of 72.46% on PASCAL-Person-Part, exceeding Baseline + CCL 67.04% by 4.42%. The result shows that our method improves performance due to our algorithm rather than extra computational overhead.

Different Architectures. Blended Grammar Network (BGNet) and Paralleled Grammar Network (PGNet) are two architectures of the grammar model. BGNet yields a result of 74.42% and exceeds paralleled PGNet 71.02% by 3.4%, showing our blended architecture has better performance than the paralleled architecture.

4.4 Comparison with State-of-the-Art

Results on PASCAL-Person-Part Dataset. To further demonstrate the effectiveness of BGNet, we compare it with state-of-the-art methods on the validation set of PASCAL-Person-Part. As shown in Table 3, our BGNet outperforms other methods on all categories. Furthermore, BGNet achieves the state-of-the-art performance, i.e., 74.42%, and outperforms the previous best one by 3.08%.

Qualitative Comparison. The qualitative comparison of results on PASCAL-Person-Part [5] is visualized in Fig. 5. From the first row, we find that our method has better performance in extracting the foreground from cluttered scenes compared with the PSPNet [43] and PCNet [45]. In the second row, PSPNet misses some parts of human bodies, and PCNet only can segment a few parts. However, most of the parts can be segmented by our network. For the head, upper-arm

Table 3. Performance comparison in terms of mean pixel Intersection-over-Union (mIoU) (%) with the state-of-the-art methods on PASCAL-Person-Part [5].

Method	Head	Torso	U-arms	L-arms	U-legs	L-legs	Background	Ave.
HAZA [38]	80.76	60.50	45.65	43.11	41.21	37.74	93.78	57.54
LIP [14]	83.26	62.40	47.80	45.58	42.32	39.48	94.68	59.36
MMAN [28]	82.58	62.83	48.49	47.37	42.80	40.40	94.92	59.91
MuLA [29]	–	–	–	–	–	–	–	65.1
PCNet [45]	86.81	69.06	55.35	55.27	50.21	48.54	96.07	65.90
Holistic [17]	–	–	–	–	–	–	–	66.3
WSHP [6]	87.15	72.28	57.07	56.21	52.43	50.36	97.72	67.60
PGN [13]	90.89	75.12	55.83	64.61	55.42	41.47	95.33	68.40
RefineNet [21]	–	–	–	–	–	–	–	68.6
Learning [36]	88.02	72.91	64.31	63.52	55.61	54.96	96.02	70.76
Graphonomy [12]	–	–	–	–	–	–	–	71.14
DPC [2]	88.81	74.54	63.85	63.73	57.24	54.55	96.66	71.34
CDCL [22]	86.39	74.70	68.32	65.98	59.86	58.70	95.79	72.82
BGNet (ours)	**90.18**	**77.44**	**68.93**	**67.15**	**60.79**	**59.27**	**97.12**	**74.42**

Table 4. Performance comparison in terms of mean pixel Intersection-over-Union (mIoU) (%) with state-of-the-art methods on LIP [14].

Method	hat	hair	glov	sung	clot	dress	coat	sock	pant	suit	scarf	skirt	face	l-arm	r-arm	l-leg	r-leg	l-sh	r-sh	bkg	Ave
FCN-8s [25]	39.79	58.96	5.32	3.08	49.08	12.36	26.82	15.66	49.41	6.48	0.00	2.16	62.65	29.78	36.63	28.12	26.05	17.76	17.70	78.02	28.29
DeepLabV2 [3]	56.48	65.33	29.98	19.67	62.44	30.33	51.03	40.51	69.00	22.38	11.29	20.56	70.11	49.25	52.88	42.37	35.78	33.81	32.89	84.53	41.64
Attention [4]	58.87	66.78	23.32	19.48	63.20	29.63	49.70	35.23	66.04	24.73	12.84	20.41	70.58	50.17	54.03	38.35	37.70	26.20	27.09	84.00	42.92
DeepLab-ASPP [3]	56.48	65.33	29.98	19.67	62.44	30.33	51.03	40.51	69.00	22.38	11.29	20.56	70.11	49.25	52.88	42.37	35.78	33.81	32.89	84.53	44.03
LIP [14]	59.75	67.25	28.95	21.57	65.30	29.49	51.92	38.52	68.02	24.48	14.92	24.32	71.01	52.64	55.79	40.23	38.80	28.08	29.03	84.56	44.73
ASN [26]	56.92	64.34	28.07	17.78	64.90	30.85	51.90	39.75	71.78	25.57	7.97	17.63	70.77	53.53	56.70	49.58	48.21	34.57	33.31	84.01	45.41
MMAN [28]	57.66	66.63	30.70	20.02	64.15	28.39	51.98	41.46	71.03	23.61	9.65	23.20	68.54	55.30	58.13	51.90	52.17	38.58	39.05	84.75	46.81
JPPNet [18]	63.55	70.20	36.16	23.48	68.15	31.42	55.65	44.56	72.19	28.39	18.76	25.14	73.36	61.97	63.88	58.21	57.99	44.02	44.09	86.26	51.37
CE2P [23]	65.29	72.54	39.09	32.73	69.46	32.52	56.28	49.67	74.11	27.23	14.19	22.51	75.50	65.14	66.59	60.10	58.59	46.63	46.12	87.67	53.10
BraidNet [24]	66.8	72.0	42.5	32.1	69.8	33.7	57.4	49.0	74.9	32.4	19.3	27.2	74.9	65.5	67.9	**60.2**	59.6	47.4	47.9	88.0	54.4
BGNet (ours)	**69.18**	**73.14**	**44.27**	**34.16**	**72.32**	**36.13**	**60.69**	**50.93**	**76.56**	**38.78**	**31.38**	**33.86**	**76.21**	**66.53**	**68.04**	59.83	**60.06**	**47.96**	**48.01**	**88.30**	**56.82**

in the last row, ours performs well on these small parts and large parts in the image compared with the other methods.

Results on LIP Dataset. According to PASCAL-Person-Part, we define 6 human parts to constitute grammar rules, which are head, upper-clothes, arm, upper-leg, low-leg and shoes. The region of the head is generated by merging parsing labels of hat, hair, sunglasses and face. Similarly, upper-clothes, coat, dress, jumpsuits and scarf are merged to be upper-clothes, right-arm, left-arm and gloves for arm, pants and skirt for upper-leg. The rest regions can also be obtained by correcting labels. Due to different datasets with different kinds of

Table 5. In the first two rows, performance comparison of model trained on LIP to test the PPSS [27]. In the bottom half, performance comparison in terms of mean pixel Intersection-over-Union (mIoU) (%) with the state-of-the-art methods on PPSS.

Method	Hair	Face	U-cloth	arms	L-cloth	Legs	Background	Ave
MMAN [28]	53.1	50.2	69.0	29.4	55.9	21.4	85.7	52.1
BGNet (ours)	59.36	57.15	63.94	42.68	59.96	27.68	86.09	56.69
DDN [27]	35.5	44.1	68.4	17.0	61.7	23.8	80.0	47.2
ASN [26]	51.7	51.0	65.9	29.5	52.8	20.3	83.8	50.7
BGNet (ours)	**70.67**	**62.31**	**82.59**	**48.12**	**72.61**	**29.82**	**92.97**	**65.44**

labels of human parts, the grammar rules may be adjusted slightly. The five grammar rules for LIP dataset are:

$$\begin{aligned}
&Rule_1^c : head \rightarrow upperclothes,\\
&Rule_2^h : Rule_1^c \rightarrow arm,\\
&Rule_3^v : Rule_1^c \rightarrow upperleg,\\
&Rule_4^v : Rule_3^v \rightarrow lowleg,\\
&Rule_5^v : Rule_4^v \rightarrow shoes.
\end{aligned}$$

We compare our method with previous networks on the validation set, which are FCN-8s [25], Attention [4], LIP [14], BraidNet [24] and so on. As shown in Table 4, our method outperforms all priors. Our proposed framework yields 56.82% in terms of mIoU on the LIP. Compared with the best methods, ours exceeds it 2.42%.

Results on PPSS Dataset. Similar to LIP [14], we merge hair and face into the head and the grammar rules may be adjusted slightly. The five grammar rules for PPSS dataset are:

$$\begin{aligned}
&Rule_1^c : head \rightarrow upperclothes,\\
&Rule_2^h : Rule_1^c \rightarrow arms,\\
&Rule_3^v : Rule_1^c \rightarrow lowcloth,\\
&Rule_4^v : Rule_3^v \rightarrow legs,\\
&Rule_5^v : Rule_4^v \rightarrow shoes.
\end{aligned}$$

We compare our method with some methods on the testing set, DDN [4], ASN [14] and MMAN [28]. In the first two rows of Table 5, we deploy the model trained on LIP [14] to the testing set of the PPSS [27] without any fine-tuning, to evaluate the generalization ability of we proposed model, which is similar to MMAN. We merge the fine-grained labels of LIP into coarse-grained human parts defined in PPSS. From Table 5, our method outperforms MMAN by 4.59%. We also train our method on the training set of PPSS dataset, whose results in segmentation on the testing set achieve further improvement. Our proposed framework achieves 65.44% in terms of Mean IoU on PPSS dataset. Compared with ASN, our method exceeds 14.74%.

5 Conclusion

In this work, we propose a Blended Grammar Network (BGNet) to improve the extraction accuracy of the foreground from the cluttered background. BGNet exploits the inherent hierarchical structure of a human body and the relationship of human parts by means of grammar rules in both cascaded and paralleled manner. Then, we design the Part-aware Convolutional Recurrent Neural Network (PCRNN) to pass messages across BGNet which can adaptively extract features to improve accuracy. To train PCRNN effectively, we develop a blended grammar loss to supervise the training of PCRNNs, which uses a less-to-more manner with increasing parts in grammar rules. Finally, extensive experiments show that BGNet improves the performance of the baseline models on three datasets significantly. These results on three datasets prove that our framework works well on different kinds of datasets.

Acknowledgement. This work was supported by Research and Development Projects in the Key Areas of Guangdong Province (No.2019B010153001), and National Natural Science Foundation of China (No.61772527, 61976210, 61806200, 61702510 and 61876086). Thanks Prof. Si Liu and Wenkai Dong for their help on paper writing.

References

1. Amit, Y., Trouvé, A.: POP: Patchwork of parts models for object recognition. IJCV **75**(2), 267–282 (2007). https://doi.org/10.1007/s11263-006-0033-9
2. Chen, L.C., et al.: Searching for efficient multi-scale architectures for dense image prediction. In: NeurIPS, pp. 8699–8710 (2018)
3. Chen, L.C., Papandreou, G., Kokkinos, I., Murphy, K., Yuille, A.L.: DeepLab: semantic image segmentation with deep convolutional nets, atrous convolution, and fully connected CRFs. IEEE TPAMI **40**(4), 834–848 (2018)
4. Chen, L.C., Yang, Y., Wang, J., Xu, W., Yuille, A.L.: Attention to scale: scale-aware semantic image segmentation. In: CVPR, June 2016
5. Chen, X., Mottaghi, R., Liu, X., Fidler, S., Urtasun, R., Yuille, A.: Detect what you can: detecting and representing objects using holistic models and body parts. In: CVPR, pp. 1979–1986 (2014)
6. Fang, H.S., Lu, G., Fang, X., Xie, J., Tai, Y.W., Lu, C.: Weakly and semi supervised human body part parsing via pose-guided knowledge transfer. arXiv preprint arXiv:1805.04310 (2018)
7. Fang, H., Xu, Y., Wang, W., Liu, X., Zhu, S.C.: Learning pose grammar to encode human body configuration for 3D pose estimation. In: AAAI (2018)
8. Farenzena, M., Bazzani, L., Perina, A., Murino, V., Cristani, M.: Person re-identification by symmetry-driven accumulation of local features. In: CVPR, pp. 2360–2367 (2010)
9. Fu, J., et al.: Dual attention network for scene segmentation. In: Proceedings of the IEEE Conference on Computer Vision and Pattern Recognition, pp. 3146–3154 (2019)

10. Fu, J., Liu, J., Wang, Y., Lu, H.: Densely connected deconvolutional network for semantic segmentation. In: 2017 IEEE International Conference on Image Processing (ICIP), pp. 3085–3089. IEEE (2017)
11. Garland-Thomson, R.: Staring: How We Look. Oxford University Press, Oxford (2009)
12. Gong, K., Gao, Y., Liang, X., Shen, X., Wang, M., Lin, L.: Graphonomy: universal human parsing via graph transfer learning. In: CVPR (2019)
13. Gong, K., Liang, X., Li, Y., Chen, Y., Yang, M., Lin, L.: Instance-level human parsing via part grouping network. In: Proceedings of the European Conference on Computer Vision (ECCV), pp. 770–785 (2018)
14. Gong, K., Liang, X., Zhang, D., Shen, X., Lin, L.: Look into person: self-supervised structure-sensitive learning and a new benchmark for human parsing. In: CVPR, vol. 2, p. 6 (2017)
15. He, K., Zhang, X., Ren, S., Sun, J.: Deep residual learning for image recognition. In: CVPR, pp. 770–778 (2016)
16. Ke, L., Chang, M.C., Qi, H., Lyu, S.: Multi-scale structure-aware network for human pose estimation. In: ECCV, pp. 713–728 (2018)
17. Li, Q., Arnab, A., Torr, P.H.: Holistic, instance-level human parsing. arXiv preprint arXiv:1709.03612 (2017)
18. Liang, X., Gong, K., Shen, X., Lin, L.: Look into person: joint body parsing & pose estimation network and a new benchmark. IEEE TPAMI **41**, 871–885 (2018)
19. Liang, X., Lin, L., Shen, X., Feng, J., Yan, S., Xing, E.P.: Interpretable structure-evolving LSTM. In: CVPR, pp. 2175–2184 (2017)
20. Liang, X., Shen, X., Xiang, D., Feng, J., Lin, L., Yan, S.: Semantic object parsing with local-global long short-term memory. In: CVPR, pp. 3185–3193 (2016)
21. Lin, G., Milan, A., Shen, C., Reid, I.: RefineNet: multi-path refinement networks for high-resolution semantic segmentation. In: CVPR, pp. 1925–1934 (2017)
22. Lin, K., Wang, L., Luo, K., Chen, Y., Liu, Z., Sun, M.T.: Cross-domain complementary learning with synthetic data for multi-person part segmentation. arXiv preprint arXiv:1907.05193 (2019)
23. Liu, T., et al.: Devil in the details: towards accurate single and multiple human parsing. In: AAAI (2019)
24. Liu, X., Zhang, M., Liu, W., Song, J., Mei, T.: BraidNet: braiding semantics and details for accurate human parsing. In: Proceedings of the 27th ACM International Conference on Multimedia, pp. 338–346. ACM (2019)
25. Long, J., Shelhamer, E., Darrell, T.: Fully convolutional networks for semantic segmentation. In: CVPR, pp. 3431–3440 (2015)
26. Luc, P., Couprie, C., Chintala, S., Verbeek, J.: Semantic segmentation using adversarial networks. arXiv preprint arXiv:1611.08408 (2016)
27. Luo, P., Wang, X., Tang, X.: Pedestrian parsing via deep decompositional network. In: CVPR, pp. 2648–2655 (2014)
28. Luo, Y., Zheng, Z., Zheng, L., Guan, T., Yu, J., Yang, Y.: Macro-micro adversarial network for human parsing. In: ECCV (2018)
29. Nie, X., Feng, J., Yan, S.: Mutual learning to adapt for joint human parsing and pose estimation. In: ECCV, pp. 502–517 (2018)
30. Park, S., Nie, B.X., Zhu, S.C.: Attribute and-or grammar for joint parsing of human pose, parts and attributes. IEEE TPAMI **40**(7), 1555–1569 (2018)
31. Qi, S., Huang, S., Wei, P., Zhu, S.C.: Predicting human activities using stochastic grammar. In: CVPR (2017)
32. Russakovsky, O., et al.: ImageNet large scale visual recognition challenge. Int. J. Comput. Vis. **115**(3), 211–252 (2015). https://doi.org/10.1007/s11263-015-0816-y

33. Thorpe, S., Fize, D., Marlot, C.: Speed of processing in the human visual system. Nature **381**(6582), 520 (1996)
34. Wang, P., Shen, X., Lin, Z., Cohen, S., Price, B., Yuille, A.: Joint object and part segmentation using deep learned potentials. In: ICCV, pp. 1573–1581 (2015)
35. Wang, W., Xu, Y., Shen, J., Zhu, S.: Attentive fashion grammar network for fashion landmark detection and clothing category classification. In: CVPR (2018)
36. Wang, W., Zhang, Z., Qi, S., Shen, J., Pang, Y., Shao, L.: Learning compositional neural information fusion for human parsing. In: Proceedings of the IEEE International Conference on Computer Vision (ICCV) (2019)
37. Wang, Y., Duan, T., Liao, Z., Forsyth, D.: Discriminative hierarchical part-based models for human parsing and action recognition. J. Mach. Learn. Res. **13**(1), 3075–3102 (2012)
38. Xia, F., Wang, P., Chen, L.C., Yuille, A.L.: Zoom better to see clearer: human and object parsing with hierarchical auto-zoom net. In: ICCV, pp. 648–663 (2015)
39. Xia, F., Wang, P., Chen, X., Yuille, A.: Joint multi-person pose estimation and semantic part segmentation. In: CVPRW, pp. 6080–6089 (2017)
40. Yamaguchi, K., Kiapour, M.H., Berg, T.L.: Paper doll parsing: retrieving similar styles to parse clothing items. In: ICCV, pp. 3519–3526 (2013)
41. Zhang, X., Chen, Y., Zhu, B., Wang, J., Tang, M.: Part-aware context network for human parsing. In: Proceedings of the IEEE/CVF Conference on Computer Vision and Pattern Recognition, pp. 8971–8980 (2020)
42. Zhao, H., Qi, X., Shen, X., Shi, J., Jia, J.: ICNet for real-time semantic segmentation on high-resolution images. In: ECCV, pp. 405–420 (2018)
43. Zhao, H., Shi, J., Qi, X., Wang, X., Jia, J.: Pyramid scene parsing network. In: CVPR (2017)
44. Zhou, B., Khosla, A., Lapedriza, A., Oliva, A., Torralba, A.: Learning deep features for discriminative localization. In: CVPR, pp. 2921–2929 (2016)
45. Zhu, B., Chen, Y., Tang, M., Wang, J.: Progressive cognitive human parsing. In: AAAI (2018)

P²Net: Patch-Match and Plane-Regularization for Unsupervised Indoor Depth Estimation

Zehao Yu[1,2], Lei Jin[1,2], and Shenghua Gao[1,3](✉)

[1] ShanghaiTech Universtiy, Shanghai, China
{yuzh,jinlei,gaoshh}@shanghaitech.edu.cn
[2] DGene Inc, Shanghai, China
[3] Shanghai Engineering Research Center of Intelligent Vision and Imaging, Shanghai, China

https://github.com/svip-lab/Indoor-SfMLearner

Abstract. This paper tackles the unsupervised depth estimation task in indoor environments. The task is extremely challenging because of the vast areas of non-texture regions in these scenes. These areas could overwhelm the optimization process in the commonly used unsupervised depth estimation framework proposed for outdoor environments. However, even when those regions are masked out, the performance is still unsatisfactory. In this paper, we argue that the poor performance suffers from the non-discriminative point-based matching. To this end, we propose P²Net. We first extract points with large local gradients and adopt patches centered at each point as its representation. Multiview consistency loss is then defined over patches. This operation significantly improves the robustness of the network training. Furthermore, because those textureless regions in indoor scenes (*e.g.*, wall, floor, roof, *etc.*) usually correspond to planar regions, we propose to leverage superpixels as a plane prior. We enforce the predicted depth to be well fitted by a plane within each superpixel. Extensive experiments on NYUv2 and ScanNet show that our P²Net outperforms existing approaches by a large margin.

Keywords: Unsupervised depth estimation · Patch-based representation · Multiview photometric consistency · Piece-wise planar loss

1 Introduction

Depth estimation, as a fundamental problem in computer vision, bridges the gap between 2D images and 3D world. Lots of supervised depth estimation methods [7, 10,30] have been proposed with the recent trend in convolution neural networks

Z. Yu and L. Jin—Equal Contribution.

Electronic supplementary material The online version of this chapter (https://doi.org/10.1007/978-3-030-58586-0_13) contains supplementary material, which is available to authorized users.

A. Vedaldi et al. (Eds.): ECCV 2020, LNCS 12369, pp. 206–222, 2020.
https://doi.org/10.1007/978-3-030-58586-0_13

(CNNs). However, capturing a large number of images in different scenes with accurate ground truth depth requires expensive hardware and time [4,15,38,41,43]. To overcome the above challenges, another line of work [14,16,46,55] focuses on unsupervised depth estimation that only uses either stereo videos or monocular videos as training data. The key supervisory signal in these work is the appearance consistency between the real view and the view synthesized based on the estimated scene geometry and ego-motion of the camera. Bilinear interpolation [20] based warping operation allows the training process to be fully differentiable.

While recent works of unsupervised depth estimation [50,54,56] have demonstrated impressive results on outdoor datasets, the same training process may easily collapse [53] on indoor datasets such as NYUv2 [41] or ScanNet [4]. The primary reason is that indoor environments contain large non-texture regions where the photometric consistency (the main supervisory signal in unsupervised learning) is unreliable. In such regions, the predicted depth might decay to infinite, while the synthesized view still has a low photometric error. Similar problems [16,17,32,50] are also observed on outdoor datasets, especially in road regions. While the propotion of such regions is small on outdoor datasets, which would only lead to degradation in performance, the large non-texture regions on indoor scenarios can easily overwhelm the whole training process.

An intuitive try would be to mask out all the non-texture regions during the loss calculation. However, as the experimental results will demonstrate, merely ignoring the gradients from these non-texture regions still leads to inferior results. The reason is that we are minimizing per pixel (point) based multi-view photometric consistency error in the training process, where each point should be matched correctly across different views. Such point-based representation is not discriminative enough for matching in indoor scenes, since many other pixels in images could have the same intensity values. This operation could easily result in false matching. Taking inspiration from traditional multi-view stereo approaches [12,39] that represent a point with a local patch, we propose to replace point-based representation with a patch-based representation to increase the discriminative ability in the matching process. Specifically, points with large local gradients are selected as our keypoints. We assume the same depth for pixels within a local window around every keypoint. We then project these local patches to different views with the predicted depth map and camera motion, and minimize multi-view photometric consistency error over the patches. Compared to point-based representation, our patch-based solution leads to a more distinctive characterization that produces more representative gradients with a wider basin of convergence.

Finally, to handle the rest large non-texture regions in indoor scenes, we draw inspiration from the recent success of work [11,29,51] that leverages the plane prior for indoor scene reconstruction. We make the assumption that homogeneous-colored regions, for example, walls, can be approximated with a plane. Here we adopt a similar strategy with the previous work [2,3] that approximates the planar regions with superpixels. Specifically, we first extract planar regions by superpixels [9], then use a planar consistency loss to enforce the predicted depth in these regions can be well fitted by a plane, *i.e.*, low plane-fitting error within each superpixel. This allows our network to produce a more robust result.

Compared with MovingIndoor [53], a pioneer work on unsupervised indoor depth estimation that requires to first establish sparse correspondences between consecutive frames, and then propagates the sparse flows to the entire image, our P^2Net is direct, and no pre-matching process is required. Therefore, there is no concern for falsely matched pairs that might misguide the training of the network. Further, the supervisory signal of MovingIndoor [53] comes from the consistency between the synthesized optical flow and the predicted flow of the network. Such indirect supervision might also lead to a sub-optimal result. Our P^2Net instead supervises the network from two aspects: local patches for textured regions and planar consistency for the non-texture regions.

Our contributions can be summarized as follows: i) we propose to extract discriminative keypoints with large local gradients and use patches centered at each point as its representation. ii) patch-match: A patch-based warping process that assumes the same depth for pixels within a local patch is proposed for a more robust matching. iii) plane-regularization: we propose to use superpixels to represent those homogeneous-texture or non-texture piece-wise planar regions and regularize the depth consistency within each superpixel. On the one hand, our P^2Net leverages the discriminative patch-based representation that improves the matching robustness. On the other hand, our P^2Net encodes the piece-wise planar prior into the network. Consequently, our approach is more suitable for indoor scene depth estimation. Extensive experiments on widely-used indoor datasets NYUv2 [41] and ScanNet [4] demonstrate that P^2Net outperforms state-of-the-art by a large margin.

2 Related Work

2.1 Supervised Depth Estimation

A vast amount of research has been done in the field of supervised depth estimation. With the recent trend in convolution neural networks (CNNs), many different deep learning based approaches have been proposed. Most of them frame the problem as a per-pixel regression problem. Particularly, Eigen et al. [5] propose a multi-scale coarse-to-fine approach. Laina et al. [25] improve the performance of depth estimation by introducing a fully convolutional architecture with several up-convolution blocks. Kim et al. [22] use conditional random fields to refine the depth prediction. Recently, Fu et al. [10] treat the problem from an ordinal regression perspective. With a carefully designed discretization strategy and an ordinal loss, their method is able to achieve new state-of-the-art results in supervised depth estimation. Other work focus on combining depth estimation with semantic segmentation [21,52] and surface norm estimation [6,34]. Yin et al. [49] show that high-order 3D geometric constraints, the so-called virtual normal, can further improve depth prediction accuracy. However, all of these methods rely on vast amounts of labeled data, which is still a large cost in both hardware and time.

2.2 Unsupervised Depth Estimation

Unsupervised learning of depth estimation has been proposed to ease the demand for large-scale labeled training data. One line of work exploits stereo images or videos [14,16,46] as training data and trains a network to minimize the photometric error between synthesized view and real view. Godard et al. [16] introduce a left-right disparity consistency as regularization. Another line of work learns depth from monocular video sequences. Zhou et al. [55] introduce a separate network to predict camera motion between input images. Their method learns to estimate depth and ego-motion simultaneously. Later work also focus on joint-learning by minimizing optical flow errors [37,50], or combining SLAM pipelines into deep networks [40,44]. However, none of the above approaches produce satisfactory results on indoor datasets. MovingIndoor [53] is the first to study unsupervised depth estimation in indoor scenes. The authors propose an optical flow estimation network, SFNet, initialized with sparse flows from matching results of SURF [1]. The dense optical flows are used as the supervisory signal for the learning of the depth and pose. By contrast, we propose to supervise the training with a more discriminative patch-based multi-view photometric consistency error and regularize the depth within homogeneous-color regions with a planar consistency loss. Our method is direct, and no pre-matching process is required. Therefore, there is no concern for falsely matched pairs that might misguide the training of the network.

2.3 Piece-Wise Planar Scene Reconstruction

Piece-wise planar reconstruction is an active research topic in multi-view 3D reconstruction [11,13], SLAM [2,3] and has drawn increasing attention recently [28,29,48,51]. Traditional methods [12,13] generate plane hypotheses by fitting planes to triangulated 3D points, then assign hypotheses to each pixel via a global optimization. Concha and Civera [2,3] used superpixels [9] to describe non-texture region in a monocular dense SLAM system. Their method has shown impressive reconstruction results. Raposo et al. [36] proposed πMatch, a vSLAM pipeline with plane features to for a piecewise planar reconstruction. In their more recent work [35], they recovered structure and motion from planar regions and combined these estimations into stereo algorithms. Together with Deep CNNs, Liu et al. [29] learn to infer plane parameters and associate each pixel to a plane in a supervised manner. Yang and Zhou [48] learn a similar network with only depth supervision. Following work [28,51] further formulate the planar reconstruction problem as an instance segmentation problem and have shown significant improvements. Inspired by these work, we incorporate the planar prior for homogeneous-color regions into our unsupervised framework and propose a planar consistency loss to regularize the depth map in such regions in the training phrase.

3 Method

3.1 Overview

Our goal is to learn a depth estimator for indoor environments with only monocular videos. Following recent success on unsupervised depth estimation [55], our

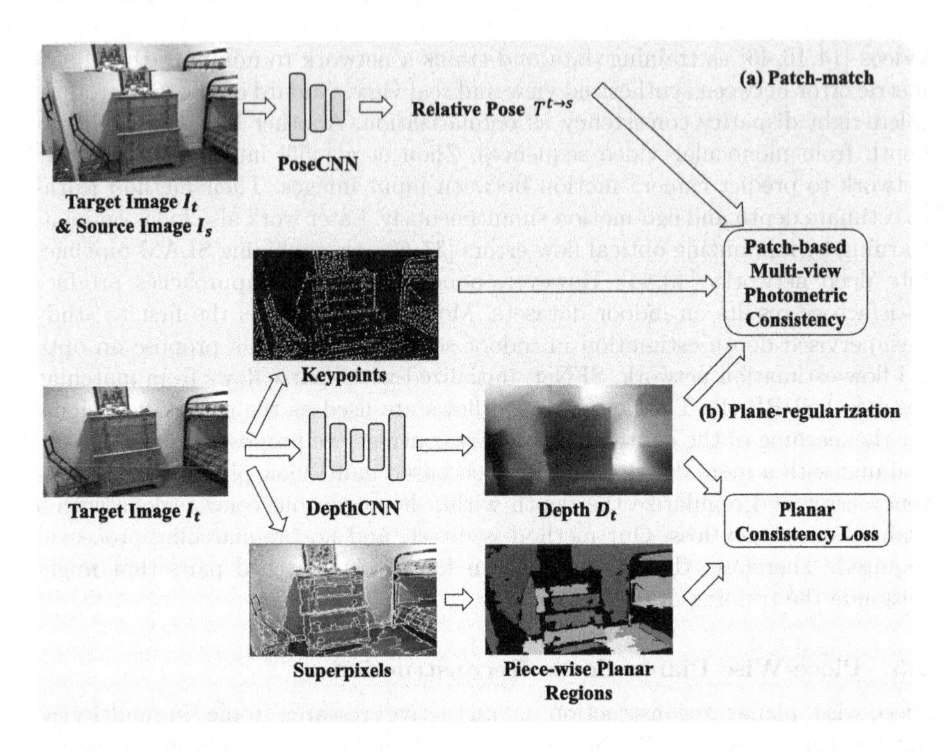

Fig. 1. Overall network architecture. Given input images, DepthCNN predicts the corresponding depth for the target image I_t, PoseCNN outputs the relative pose from the source to the target view. Our P^{2}Net consists of two parts: a) **Patch-match Module**: We warp the selected pixels along with their local neighbors with a patch-based warping module. b) **Plane-regularization Module**: We enforce depth consistency in large superpixel regions.

P^{2}Net contains two learnable modules: DepthCNN and PoseCNN. DepthCNN takes a target view image I_t as input and outputs its corresponding depth D_t. PoseCNN takes a source view image I_s and a target view image I_t as input and predicts the relative pose $T_{t \to s}$ between two consecutive frames. A commonly used strategy is to first synthesize a novel view I_t' with the predicted depth map D_t and camera motion $T_{t \to s}$, and minimize the photometric consistency error between the synthesized view I_t' and its corresponding real view I_t. However, the training process soon collapses when directly applying this strategy to indoor scenarios.

Our observation is that the large non-texture regions in indoor scenes might easily overwhelm the whole training process. Therefore, we propose to select representative keypoints that have large local variances. However, representing a point with a single intensity value, as done in previous unsupervised learning frameworks [16,17], is non-discriminative and may result in false matching. To address this problem, we propose a **Patch-match Module**, a patch-based representation that combines a point with the local window centered at that point to increase their discriminative abilities and minimize patch-based multi-view photometric consistency error. To handle the large non-texture regions, we propose a **Plane-regularization Module** to extract homogeneous-color regions using large superpixels and enforce that the predicted depth map within a superpixel may be approximated by a plane. The overview of our P^{2}Net is depicted in Fig. 1.

3.2 Keypoints Extraction

Different from outdoor scenes, the large proportion of the non-texture regions in indoor scenes can easily overwhelm the training process, leading to trivial solutions where DepthCNN always predicts an infinity depth, and PoseCNN always gives an identity rotation. Thus, only points within textured regions should be kept in the training process to avoid the network being stuck in such trivial results. Here, we adopt the points selection strategy from Direct Sparse Odometry (DSO) [8] for its effectiveness and efficiency. Points from DSO are sampled from pixels that have large intensity gradients. Examples of extracted DSO keypoints are shown in Fig. 3.

A critical benefit of our direct method over matching based approaches [53] is that we do not need to pre-compute the matching across images, which itself is a challenging problem. As a result, our points need to be extracted from the target image once only. No hand-crafted descriptor for matching is needed.

3.3 Patch-Based Multi-view Photometric Consistency Error

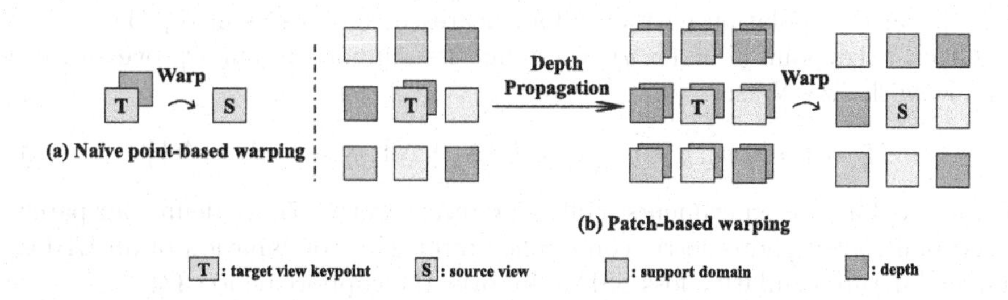

Fig. 2. Two types of warping operations. a) Naive point-based warping. b) Our proposed patch-based warping. Note that we are defining pixels over its support domain and warp the entire window. Combining support domains into the pixel leads to more robust representations. Best viewed in color. (Color figure online)

With the extracted keypoints from the previous step, we can simply define a photometric consistency error by comparing the corresponding pixels' values. However, such point-based representation is not representative enough and may easily cause false matching because there are many pixels with the same intensity values in an image. In traditional sparse SLAM pipelines [8], to overcome the above challenge, a support domain Ω_{p_i} is defined over each point p_i's local window. Photometric loss is then accumulated over each support domain Ω_{p_i} instead of a single isolated point. This operation would lead to more robust results as the extracted keypoints combined with their support domains are becoming much more unique.

Inspired from the above operation, here we propose a patch-based warping process as in Fig. 2. Specifically, we extract DSO keypoints p_i^t from the target view t, the original point-based warping process first back-projects the keypoints to the source view I_s with:

$$p_i^{t \longrightarrow s} = KT^{t \longrightarrow s} D(p_i) K^{-1} p_i^t \tag{1}$$

where K denotes the camera intrinsic parameters, $T^{t \longrightarrow s}$ the relative pose between the source view I_s and the target view I_t, and $D(p_i)$ the depth of point p_i. Then we sample the intensity values with bilinear interpolation [20] at $p_i^{t \longrightarrow s}$ in the source view.

On the contrast, our approach assumes a same depth within each pixel's local window $\Omega_{p_i}^t$. Then, for every extracted keypoint, we warp the point together with its local support region $\Omega_{p_i}^t$ with the exact same depth. Our warping process can thus be described as:

$$\Omega_{p_i}^{t \longrightarrow s} = KT^{t \longrightarrow s} D(p_i) K^{-1} \Omega_{p_i}^t \tag{2}$$

where $\Omega_{p_i}^t$ and $\Omega_{p_i}^{t \longrightarrow s}$ denotes the support domains of the point p_i in the target view and the source view, respectively. From a SLAM perspective, we characterize each point over its support region, such patch-based approaches makes the representation of each point more distinctive and robust. From a deep learning perspective, our operation allows a larger region of valid gradients compared to the bilinear interpolation with only four nearest neighbors as in Eq. (1).

Given a keypoint $p = (x, y)$, we define its support region Ω_p over a local window with size N as:

$$\Omega_p = \{(x + x_p, y + y_p), x_p \in \{-N, 0, N\}, y_p \in \{-N, 0, N\}\} \tag{3}$$

N is set to 3 in our experiments. Following recent work [17], we define our patch-based multi-view photometric consistency error as a combination of an L1 loss and a structure similarity loss SSIM [45] over the support region Ω_{p_i}:

$$L_{SSIM} = SSIM(I_t \left[\Omega_{p_i}^t\right], I_s \left[\Omega_{p_i}^{t \longrightarrow s}\right]) \tag{4}$$

$$L_{L1} = ||I_t \left[\Omega_{p_i}^t\right] - I_s \left[\Omega_{p_i}^{t \longrightarrow s}\right] ||_1 \tag{5}$$

$$L_{ph} = \alpha L_{SSIM} + (1 - \alpha) L_{L1} \tag{6}$$

where $I_t [p]$ denotes pixel values at p in image I_t via a bilinear interpolation, and $\alpha = 0.85$ a weighting factor. Note that when more than one source images are used in the photometric loss, we follow [17] to select the one with the minimum L_{ph} for robustness purpose.

Fig. 3. Examples of input images, their corresponding keypoints, superpixels and piece-wise planar regions obtained from large superpixels.

3.4 Planar Consistency Loss

Finally, to further constrain the large non-texture regions in indoor scenes, we propose to enforce piecewise planar constraints into our network. Our assumption is that, most of the homogeneous-color regions are planar regions, and we can assume a continuous depth that satisfies the planar assumptions within these regions. Following representative work on reconstruction of indoor scenes [2,3], we adopt the Felzenszwalb superpixel segmentation [9] in our approach. The segmentation algorithm follows a greedy approach and segments areas with low gradients, and hence produces more planar regions. Examples with images, superpixels segmentation and piece-wise planar regions determined by superpixels, are demonstrated in Fig. 3. We can see that our assumption is reasonable, since indoor scenes generally consists of many man-made objects, like floor, walls, roof, *etc.* Further, previous work also shows the good performance of indoor scene reconstruction with a piece-wise planar assumption in [28,29,51].

Specifically, given an input image I, we first extract superpixels from the image and only keep regions larger than 1000 pixels. An intuition is that the planar regions, like walls, floor, the surface of a table, are more likely to be within a larger area. Given an extracted superpixel SPP_m and its corresponding depth $D(p_n)$ from an image, where p_n enumerates all the pixels within SPP_m, we first backproject all the points p_n back to 3D space,

$$p_n^{3D} = D(p_n)K^{-1}p_n, p_n \subseteq SPP_m \tag{7}$$

where p_n^{3D} denotes the corresponding point of p_n in 3D world. We define the plane in 3D following [29,51] as

$$A_m^\top p_n^{3D} = \mathbf{1} \tag{8}$$

where A_m is plane parameter of SPP_m.

We use a least square method to fit the plane parameters A_m. Mathematically, we form two data matrices Y_m and P_n, where $Y_m = \mathbf{1} = [1\,1\,...\,1]^\top$, $P_n = [p_1^{3D}\; p_2^{3D}\; ...\; p_n^{3D}]^\top$:

$$P_n A_m = Y_m \tag{9}$$

Then A_m can be computed with a closed-form solution:

$$A_m = \left(P_n^\top P_n + \epsilon E\right)^{-1} P_n^\top Y_m. \tag{10}$$

where E is an identity matrix, and ϵ a small scalar for numerical stability. After obtaining the plane parameters, We can then retrieve our fitted planar depth for each pixel within the superpixel SPP_m as $D^{'}(p_n) = (A_m^\top K^{-1} p_n)^{-1}$. We then add another constraint to enforce a low plane-fitting error within each superpixel:

$$L_{spp} = \sum_{m=1}^{M} \sum_{n=1}^{N} |D(p_n) - D^{'}(p_n)| \tag{11}$$

Here M denotes the number of superpixels, and N number of pixels in each superpixel.

3.5 Loss Function

We also adopt an edge-aware smoothness term L_{sm} over the entire depth map as that in [16,17]:

$$L_{sm} = |\partial_x d_t^*|\, e^{-|\partial_x I_t|} + |\partial_y d_t^*|\, e^{-|\partial_y I_t|}, \tag{12}$$

where ∂_x denotes the gradients along the x direction, ∂_y along the y direction and $d_t^* = d_t/\overline{d_t}$ is the normalized depth.

Our overall loss function is defined as:

$$L = L_{ph} + \lambda_1 L_{sm} + \lambda_2 L_{spp} \tag{13}$$

where λ_1 is set to 0.001, λ_2 is set to 0.05 in our experiments.

4 Experiments

4.1 Implementation Details

We implement our solution under the PyTorch [33] framework. Following the pioneer work on unsupervised depth estimation in outdoor scenes, we use the same encoder-decoder architecture as that in [17] with separate ResNet18s [18] pretrained on ImageNet as our backbones, the same PoseCNN as that in [17]. Adam [23] is adopted as our optimizer. The network is trained for a total of 41 epochs with a batch size of 12. Initial learning rate is set to $1e-4$ for the first 25 epochs. Then we decay it once by 0.1 for the next 10 epochs. We adopt random flipping and color augmentation during training. All images are resized to

Table 1. Performance comparison on the NYUv2 dataset. We report results of depth supervised approaches in the first block, plane supervised results in the second block, unsupervised results in the third and fourth block, and the supervised upper bound of our approach denoted as ResNet18 in the final block. PP denotes the final result with left-right fliping augmentation in evaluation. Our approach achieves state-of-the-art performance among the unsupervised ones. ↓ indicates the lower the better, ↑ indicates the higher the better.

Methods	Supervised	rms ↓	rel ↓	log10 ↓	$\delta < 1.25$ ↑	$\delta < 1.25^2$ ↑	$\delta < 1.25^3$ ↑
Make3D [38]	✓	1.214	0.349	-	0.447	0.745	0.897
Liu et al. [31]	✓	1.200	0.350	0.131	-	-	-
Ladicky et al. [24]	✓	1.060	0.335	0.127	-	-	-
Li et al. [26]	✓	0.821	0.232	0.094	0.621	0.886	0.968
Liu et al. [30]	✓	0.759	0.213	0.087	0.650	0.906	0.976
Li et al. [27]	✓	0.635	0.143	0.063	0.788	0.958	0.991
Xu et al. [47]	✓	0.586	0.121	0.052	0.811	0.954	0.987
DORN [10]	✓	0.509	0.115	0.051	0.828	0.965	0.992
Hu et al. [19]	✓	0.530	0.115	0.050	0.866	0.975	0.993
PlaneNet [29]	✓	0.514	0.142	0.060	0.827	0.963	0.990
PlaneReg [51]	✓	0.503	0.134	0.057	0.827	0.963	0.990
MovingIndoor [53]	×	0.712	0.208	0.086	0.674	0.900	0.968
Monov2 [17]	×	0.617	0.170	0.072	0.748	0.942	0.986
P^{2}Net (3 frames)	×	**0.599**	**0.159**	**0.068**	**0.772**	**0.942**	**0.984**
P^{2}Net (5 frames)	×	0.561	0.150	0.064	0.796	0.948	0.986
P^{2}Net (5 frames PP)	×	**0.553**	**0.147**	**0.062**	**0.801**	**0.951**	**0.987**
ResNet18	✓	0.591	0.138	0.058	0.823	0.964	0.989

288×384 pixels during training. Predicted depth are up-sampled back to the original resolution during testing. Since unsupervised monocular depth estimation exists scale ambiguity, we adopt the same median scaling strategy as that in [17,55] for evaluation. A larger baseline is also beneficial for training, and we use a 3-frame (one target frame, 2 source frames) input in our ablation experiments and report the final results with a 5-frame (one target frame, 4 source frames) input. Besides the standard DSO keypoints, we also draw points randomly to have a fixed number of 3K points from one image.

4.2 Datasets

We evaluate our P^{2}Net on two publicly available datasets of indoor scenes, including NYU Depth V2 [41] and ScanNet [4].

NYU Depth V2. NYU Depth V2 consists of a total 582 indoor scenes. We adopt the same train split of 283 scenes following previous work on indoor depth estimation [53] and provide our results on the official test set with the standard depth evaluation criteria. We sample the training set at 10 frames interval as our target views and use ± 10, ± 20 frames as our source views. This leaves us around 20K unique images, a number much less than the 180K images used in the previous work of unsupervised indoor depth estimation [53]. We undistort the input image as in [42] and crop 16 black pixels from the border region.

We compare with MovingIndoor [53], the pioneer work on unsupervised indoor depth estimation and Monov2 [17], a state-of-the-art unsupervised depth estimation method on outdoor datasets. Quantitative results are provided in Table 1. Our method achieves the best result. We further provide some visualization of our predicted depth in Fig. 4. GeoNet collapsed during training as we inspected. Compared to MovingIndoor [53], our method preserves much more details owing to the patch-based multi-view consistency module. A supervised upper bound, denoted as ResNet18, is also provided here by replacing the backbone network in [19] with ours.

Results for surface normal estimation are provided in Table 2. We compare with other methods that fits norm from the point clouds. Not only is our result the best among the unsupervised ones, it is also close to supervised results like DORN [10]. We visualize some results of our method for surface normal estimation in Fig. 5.

ScanNet. ScanNet [4] contains around 2.5M images captured in 1513 scenes. While there is no current official train/test split on ScanNet for depth estimation, we randomly pick 533 testing images from diverse scenes. We directly evaluate our models pretrained on NYUv2 under a transfer learning setting to test the generalizability of our approach. We showcase some of the prediction results in Fig. 4. We achiever better result as reported in Table 3.

Table 2. Surface normal evaluation on NYUv2. PP denotes the final result with left-right fliping augmentation in evaluation.

Methods	Supervised	Mean ↓	11.2° ↑	22.5° ↑	30° ↑
GeoNet [34]	✓	36.8	15.0	34.5	46.7
DORN [10]	✓	36.6	15.7	36.5	49.4
MovingIndoor [53]	×	43.5	10.2	26.8	37.9
Monov2 [17]	×	43.8	10.4	26.8	37.3
P^2Net (3 frames)	×	38.8	11.5	31.8	44.8
P^2Net (5 frames)	×	36.6	15.0	36.7	49.0
P^2Net (5 frames pp)	×	**36.1**	**15.6**	**37.7**	**50.0**

4.3 Ablation Experiments

Patch-Match and Plane-Regularization. For our baseline, we first calculate the variance within a local region for each pixel. This servers as our texture/non-texture region map. Photometric loss is directly multiplied by the map. This represents the most straightforward case when only point-based supervision is provided. We report the numbers in the first row of Table 4. Then we add our proposed Patch-match module and report the results in the second line, the

Fig. 4. Depth visualization on NYUv2 (first 6 rows) and ScanNet (last 2 rows). We trained our model on NYUv2 and directly transfer the weights to ScanNet without fine-tunning. From left right: input image, results of MovingIndoor [53], our results and ground truth depth. GeoNet would collapse on indoor datasets due to the large non-texture regions. Compared to MovingIndoor [53], our methods preserve more details.

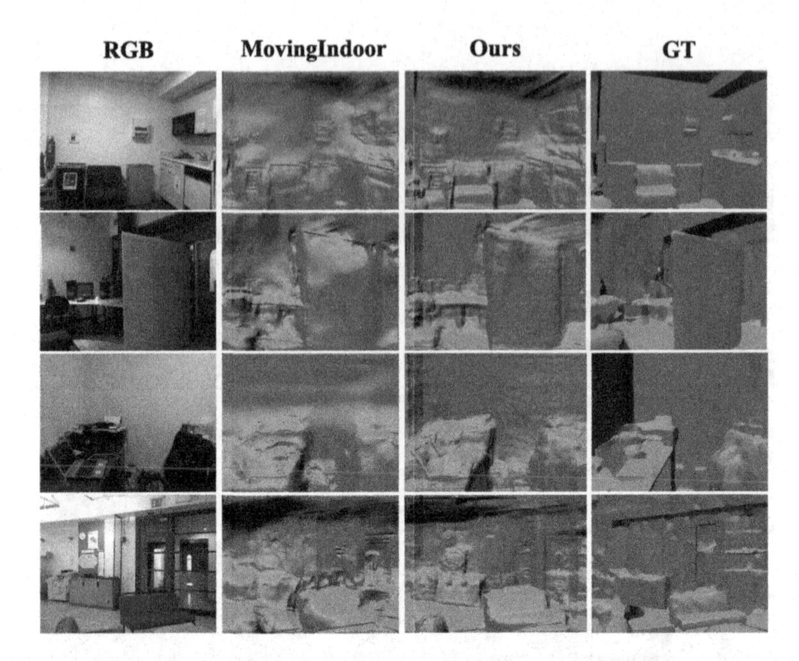

Fig. 5. Visualization of fitted surface norm from 3D point clouds on the NYUv2 dataset. From left to right: input image, results of MovingIndoor [53], ours and ground truth normal. Our method produces more smooth results in planar regions.

Table 3. Performance comparison on transfer learning. Results are evaluated directly with NYUv2 pretrained models on ScanNet. Our model still achieves the best result.

Methods	rms ↓	rel ↓	log10 ↓	$\delta < 1.25$ ↑	$\delta < 1.25^2$ ↑	$\delta < 1.25^3$ ↑
MovingIndoor [53]	0.483	0.212	0.088	0.650	0.905	0.976
Monov2 [17]	0.458	0.200	0.083	0.672	0.922	0.981
P^{2}Net	**0.420**	**0.175**	**0.074**	**0.740**	**0.932**	**0.982**

Plane-regularization module in the fourth line. Experiments demonstrate the effectiveness of our proposed modules.

Different Keypoint Types. Here, we demonstrate that our method is not limited to some specific type of keypoint detectors. We replace DSO with a blob region detector SURF [1]. We achieve similar results as reported in line two and three in Table 4.

Camera Pose. Following previous work [42] on predicting depth from videos, we provide our camera pose estimation results on the ScanNet dataset, consisting a total of 2000 pairs of images from diverse scenes. Note that since our method is monocular, there exists scale ambiguity in our predictions. Hence, we follow [42] and rescale our translation during evaluation. Results are reported in Table 5. Our method performs better than MovingIndoor [53].

Table 4. Ablation study of our proposed module on the NYUv2 dataset.

Keypoint	Patch match	Plane regularization	rms ↓	rel ↓	$\delta < 1.25$ ↑	$\delta < 1.25^2$ ↑	$\delta < 1.25^3$ ↑
-			0.786	0.240	0.628	0.884	0.962
DSO	✓		0.612	0.166	0.758	0.945	0.985
SURF	✓		0.622	0.169	0.750	0.941	0.986
DSO	✓	✓	**0.599**	**0.159**	**0.772**	**0.942**	**0.984**

Table 5. Results on camera pose.

Method	rot(deg)	tr(deg)	tr(cm)
Moving [53]	1.96	39.17	1.40
Monov2 [17]	2.03	41.12	**0.83**
P^2Net	**1.86**	**35.11**	0.89

Table 6. Results on KITTI.

Method	rel↓	rms ↓	$\delta < 1.25$ ↑
Moving [53]	0.130	5.294	-
P^2Net	**0.126**	**5.140**	**0.862**
Monov2 [17]	0.115	4.863	0.877

Results on Outdoor Scenes. Here we also provide our results on the KITTI benchmark in Table 6. We trained and evaluated our results on the same subset as in [17]. Our method outperforms another unsupervised indoor depth estimation approach MovingIndoor. Different from indoor scenes, the main challenge in outdoor scenes are moving objects (like cars) and occlusions, which seldom occur in indoor scenes. Our method does not take such priors into consideration. On the contrast, Monov2 is specially designed to handle these cases.

5 Conclusion

This paper propose P^2Net that leverages patches and superpixels for unsupervised depth estimation task in indoor scenes. Extensive experiments validate the effectiveness of our P^2Net. Here for simplicity we adopt the fronto-parallel assumption. One possible solution could be to first pretrain the network and calculate normal from depth. Then we can combine normal into the training process.

Acknowledgements. The work was supported by National Key R&D Program of China (2018AAA0100704), NSFC #61932020, and ShanghaiTech-Megavii Joint Lab. We would also like to thank Junsheng Zhou from Tsinghua University for detailed commons of reproducing his work and some helpful discussions.

References

1. Bay, H., Tuytelaars, T., Van Gool, L.: SURF: speeded up robust features. In: Leonardis, A., Bischof, H., Pinz, A. (eds.) ECCV 2006, Part I. LNCS, vol. 3951, pp. 404–417. Springer, Heidelberg (2006). https://doi.org/10.1007/11744023_32
2. Concha, A., Civera, J.: Using superpixels in monocular SLAM. In: ICRA (2014)

3. Concha, A., Civera, J.: DPPTAM: dense piecewise planar tracking and mapping from a monocular sequence. In: 2015 IEEE/RSJ International Conference on Intelligent Robots and Systems (IROS), pp. 5686–5693. IEEE (2015)
4. Dai, A., Chang, A.X., Savva, M., Halber, M., Funkhouser, T., Nießner, M.: ScanNet: richly-annotated 3D reconstructions of indoor scenes. In: CVPR (2017)
5. Eigen, D., Puhrsch, C., Fergus, R.: Prediction from a single image using a multiscale deep network. In: NIPS (2014)
6. Eigen, D., Fergus, R.: Predicting depth, surface normals and semantic labels with a common multi-scale convolutional architecture. In: ICCV (2015)
7. Eigen, D., Puhrsch, C., Fergus, R.: Depth map prediction from a single image using a multi-scale deep network. In: NIPS, pp. 2366–2374 (2014)
8. Engel, J., Koltun, V., Cremers, D.: Direct sparse odometry. IEEE Trans. Pattern Anal. Mach. Intell. **40**(3), 611–625 (2017)
9. Felzenszwalb, P.F., Huttenlocher, D.P.: Efficient graph-based image segmentation. Int. J. Comput. Vision **59**(2), 167–181 (2004)
10. Fu, H., Gong, M., Wang, C., Batmanghelich, K., Tao, D.: Deep ordinal regression network for monocular depth estimation. In: CVPR (2018)
11. Furukawa, Y., Curless, B., Seitz, S.M., Szeliski, R.: Manhattan-world stereo. In: CVPR (2009)
12. Furukawa, Y., Ponce, J.: Accurate, dense, and robust multiview stereopsis. IEEE Trans. Pattern Anal. Mach. Intell. **32**(8), 1362–1376 (2009)
13. Gallup, D., Frahm, J.M., Pollefeys, M.: Piecewise planar and non-planar stereo for urban scene reconstruction. In: 2010 IEEE Computer Society Conference on Computer Vision and Pattern Recognition (2010)
14. Garg, R., B.G., V.K., Carneiro, G., Reid, I.: Unsupervised CNN for single view depth estimation: geometry to the rescue. In: Leibe, B., Matas, J., Sebe, N., Welling, M. (eds.) ECCV 2016, Part VIII. LNCS, vol. 9912, pp. 740–756. Springer, Cham (2016). https://doi.org/10.1007/978-3-319-46484-8_45
15. Geiger, A., Lenz, P., Urtasun, R.: Are we ready for autonomous driving? the KITTI vision benchmark suite. In: CVPR (2012)
16. Godard, C., Mac Aodha, O., Brostow, G.J.: Unsupervised monocular depth estimation with left-right consistency. In: CVPR (2017)
17. Godard, C., Mac Aodha, O., Firman, M., Brostow, G.J.: Digging into self-supervised monocular depth estimation. In: ICCV (2019)
18. He, K., Zhang, X., Ren, S., Sun, J.: Deep residual learning for image recognition. In: CVPR (2016)
19. Hu, J., Ozay, M., Zhang, Y., Okatani, T.: Revisiting single image depth estimation: toward higher resolution maps with accurate object boundaries. In: WACV (2019)
20. Jaderberg, M., Simonyan, K., Zisserman, A., et al.: Spatial transformer networks. In: NIPS (2015)
21. Jiao, J., Cao, Y., Song, Y., Lau, R.: Look deeper into depth: monocular depth estimation with semantic booster and attention-driven loss. In: Ferrari, V., Hebert, M., Sminchisescu, C., Weiss, Y. (eds.) ECCV 2018, Part XV. LNCS, vol. 11219, pp. 55–71. Springer, Cham (2018). https://doi.org/10.1007/978-3-030-01267-0_4
22. Kim, S., Park, K., Sohn, K., Lin, S.: Unified depth prediction and intrinsic image decomposition from a single image via joint convolutional neural fields. In: Leibe, B., Matas, J., Sebe, N., Welling, M. (eds.) ECCV 2016, Part VIII. LNCS, vol. 9912, pp. 143–159. Springer, Cham (2016). https://doi.org/10.1007/978-3-319-46484-8_9
23. Kingma, D.P., Ba, J.: Adam: A method for stochastic optimization. arXiv preprint arXiv:1412.6980 (2014)

24. Ladicky, L., Shi, J., Pollefeys, M.: Pulling things out of perspective. In: CVPR (2014)
25. Laina, I., Rupprecht, C., Belagiannis, V., Tombari, F., Navab, N.: Deeper depth prediction with fully convolutional residual networks. In: 3DV (2016)
26. Li, B., Shen, C., Dai, Y., Van Den Hengel, A., He, M.: Depth and surface normal estimation from monocular images using regression on deep features and hierarchical CRFs. In: CVPR (2015)
27. Li, J., Klein, R., Yao, A.: A two-streamed network for estimating fine-scaled depth maps from single RGB images. In: ICCV (2017)
28. Liu, C., Kim, K., Gu, J., Furukawa, Y., Kautz, J.: PlaneRCNN: 3D plane detection and reconstruction from a single image. In: CVPR (2019)
29. Liu, C., Yang, J., Ceylan, D., Yumer, E., Furukawa, Y.: PlaneNet: piece-wise planar reconstruction from a single RGB image. In: CVPR (2018)
30. Liu, F., Shen, C., Lin, G., Reid, I.: Learning depth from single monocular images using deep convolutional neural fields. IEEE Trans. Pattern Anal. Mach. Intell. **38**(10), 2024–2039 (2015)
31. Liu, M., Salzmann, M., He, X.: Discrete-continuous depth estimation from a single image. In: CVPR (2014)
32. Luo, C., et al.: Every pixel counts++: Joint learning of geometry and motion with 3D holistic understanding. arXiv preprint arXiv:1810.06125 (2018)
33. Paszke, A., et al.: PyTorch: an imperative style, high-performance deep learning library. In: NIPS (2019)
34. Qi, X., Liao, R., Liu, Z., Urtasun, R., Jia, J.: GeoNet: geometric neural network for joint depth and surface normal estimation. In: CVPR (2018)
35. Raposo, C., Antunes, M., Barreto, J.P.: Piecewise-planar stereoscan: sequential structure and motion using plane primitives. IEEE Trans. Pattern Anal. Mach. Intell. **40**(8), 1918–1931 (2018)
36. Raposo, C., Barreto, J.P.: πMatch: monocular vSLAM and piecewise planar reconstruction using fast plane correspondences. In: Leibe, B., Matas, J., Sebe, N., Welling, M. (eds.) ECCV 2016, Part VIII. LNCS, vol. 9912, pp. 380–395. Springer, Cham (2016). https://doi.org/10.1007/978-3-319-46484-8_23
37. Ren, Z., Yan, J., Ni, B., Liu, B., Yang, X., Zha, H.: Unsupervised deep learning for optical flow estimation. In: AAAI (2017)
38. Saxena, A., Sun, M., Ng, A.Y.: Make3D: learning 3D scene structure from a single still image. IEEE Trans. Pattern Anal. Mach. Intell. **31**(5), 824–840 (2008)
39. Schönberger, J.L., Zheng, E., Frahm, J.-M., Pollefeys, M.: Pixelwise view selection for unstructured multi-view stereo. In: Leibe, B., Matas, J., Sebe, N., Welling, M. (eds.) ECCV 2016, Part III. LNCS, vol. 9907, pp. 501–518. Springer, Cham (2016). https://doi.org/10.1007/978-3-319-46487-9_31
40. Shi, Y., Zhu, J., Fang, Y., Lien, K., Gu, J.: Self-supervised learning of depth and ego-motion with differentiable bundle adjustment. arXiv preprint arXiv:1909.13163 (2019)
41. Silberman, N., Hoiem, D., Kohli, P., Fergus, R.: Indoor segmentation and support inference from RGBD images. In: Fitzgibbon, A., Lazebnik, S., Perona, P., Sato, Y., Schmid, C. (eds.) ECCV 2012, Part V. LNCS, vol. 7576, pp. 746–760. Springer, Heidelberg (2012). https://doi.org/10.1007/978-3-642-33715-4_54
42. Teed, Z., Deng, J.: DeepV2D: Video to depth with differentiable structure from motion. arXiv preprint arXiv:1812.04605 (2018)
43. Vasiljevic, I., et al.: DIODE: A Dense Indoor and Outdoor DEpth Dataset. CoRR abs/1908.00463 (2019)

44. Wang, C., Miguel Buenaposada, J., Zhu, R., Lucey, S.: Learning depth from monocular videos using direct methods. In: CVPR (2018)
45. Wang, Z., Bovik, A.C., Sheikh, H.R., Simoncelli, E.P.: Image quality assessment: from error visibility to structural similarity. IEEE Trans. Image Process. **13**(4), 600–612 (2004)
46. Xie, J., Girshick, R., Farhadi, A.: Deep3D: fully automatic 2D-to-3D video conversion with deep convolutional neural networks. In: Leibe, B., Matas, J., Sebe, N., Welling, M. (eds.) ECCV 2016, Part IV. LNCS, vol. 9908, pp. 842–857. Springer, Cham (2016). https://doi.org/10.1007/978-3-319-46493-0_51
47. Xu, D., Ricci, E., Ouyang, W., Wang, X., Sebe, N.: Multi-scale continuous CRFs as sequential deep networks for monocular depth estimation. In: CVPR (2017)
48. Yang, F., Zhou, Z.: Recovering 3D planes from a single image via convolutional neural networks. In: Ferrari, V., Hebert, M., Sminchisescu, C., Weiss, Y. (eds.) ECCV 2018, Part X. LNCS, vol. 11214, pp. 87–103. Springer, Cham (2018). https://doi.org/10.1007/978-3-030-01249-6_6
49. Yin, W., Liu, Y., Shen, C., Yan, Y.: Enforcing geometric constraints of virtual normal for depth prediction. In: ICCV (2019)
50. Yin, Z., Shi, J.: GeoNet: unsupervised learning of dense depth, optical flow and camera pose. In: CVPR (2018)
51. Yu, Z., Zheng, J., Lian, D., Zhou, Z., Gao, S.: Single-image piece-wise planar 3D reconstruction via associative embedding. In: CVPR (2019)
52. Zhang, Z., Cui, Z., Xu, C., Jie, Z., Li, X., Yang, J.: Joint task-recursive learning for semantic segmentation and depth estimation. In: Ferrari, V., Hebert, M., Sminchisescu, C., Weiss, Y. (eds.) ECCV 2018, Part X. LNCS, vol. 11214, pp. 238–255. Springer, Cham (2018). https://doi.org/10.1007/978-3-030-01249-6_15
53. Zhou, J., Wang, Y., Qin, K., Zeng, W.: Moving indoor: unsupervised video depth learning in challenging environments. In: ICCV (2019)
54. Zhou, J., Wang, Y., Qin, K., Zeng, W.: Unsupervised high-resolution depth learning from videos with dual networks. In: ICCV (2019)
55. Zhou, T., Brown, M., Snavely, N., Lowe, D.G.: Unsupervised learning of depth and ego-motion from video. In: CVPR (2017)
56. Zou, Y., Luo, Z., Huang, J.-B.: DF-Net: unsupervised joint learning of depth and flow using cross-task consistency. In: Ferrari, V., Hebert, M., Sminchisescu, C., Weiss, Y. (eds.) ECCV 2018, Part V. LNCS, vol. 11209, pp. 38–55. Springer, Cham (2018). https://doi.org/10.1007/978-3-030-01228-1_3

Efficient Attention Mechanism for Visual Dialog that Can Handle All the Interactions Between Multiple Inputs

Van-Quang Nguyen[1(✉)], Masanori Suganuma[1,2], and Takayuki Okatani[1,2]

[1] Grad School of Information Sciences, Tohoku University, Sendai, Japan
{quang,suganuma,okatani}@vision.is.tohoku.ac.jp
[2] RIKEN Center for AIP, Tokyo, Japan

Abstract. It has been a primary concern in recent studies of vision and language tasks to design an effective attention mechanism dealing with interactions between the two modalities. The Transformer has recently been extended and applied to several bi-modal tasks, yielding promising results. For visual dialog, it becomes necessary to consider interactions between three or more inputs, i.e., an image, a question, and a dialog history, or even its individual dialog components. In this paper, we present a neural architecture named *Light-weight Transformer for Many Inputs* (LTMI) that can efficiently deal with all the interactions between multiple such inputs in visual dialog. It has a block structure similar to the Transformer and employs the same design of attention computation, whereas it has only a small number of parameters, yet has sufficient representational power for the purpose. Assuming a standard setting of visual dialog, a layer built upon the proposed attention block has less than one-tenth of parameters as compared with its counterpart, a natural Transformer extension. The experimental results on the VisDial datasets validate the effectiveness of the proposed approach, showing improvements of the best NDCG score on the VisDial v1.0 dataset from 57.59 to 60.92 with a single model, from 64.47 to 66.53 with ensemble models, and even to 74.88 with additional finetuning.

Keywords: Visual dialog · Attention · Multimodality

1 Introduction

Recently, an increasing amount of attention has been paid to problems lying at the intersection of the vision and language domains. Many pilot tasks in this intersecting region have been designed and introduced to the research community, together with datasets. Visual dialog has been developed aiming at a higher level of vision-language interactions [7], as compared with VQA (visual question answering)

Electronic supplementary material The online version of this chapter (https://doi.org/10.1007/978-3-030-58586-0_14) contains supplementary material, which is available to authorized users.

A. Vedaldi et al. (Eds.): ECCV 2020, LNCS 12369, pp. 223–240, 2020.
https://doi.org/10.1007/978-3-030-58586-0_14

[2] and VCR (visual commonsense reasoning). It extends VQA to multiple rounds; given an image and a history of question-answer pairs about the image, an agent is required to answer a new question. For example, to answer the question *'What color are they?'*, the agent needs to understand the context from a dialog history to know what *'they'* refers to and look at the relevant image region to find out a color.

In recent studies of vision-language tasks, a primary concern has been to design an attention mechanism that can effectively deal with interactions between the two modalities. In the case of visual dialog, it becomes further necessary to consider interactions between an image, a question, and a dialog history or additionally multiple question-answer pairs in the history. Thus, the key to success will be how to deal with such interactions between three and more entities. Following a recent study [36], we will use the term *utility* to represent each of these input entities for clarity, since the term *modality* is inconvenient to distinguish between the question and the dialog history.

Existing studies have considered attention from one utility to another based on different hypotheses, such as "question → history → image" path in [18,28], and "question → image → history → question" path in [12,43], etc. These methods cannot take all the interactions between utilities into account, although the missing interactions could be crucial. Motivated by this, a recent study tries to capture all the possible interactions by using a factor graph [36]. However, building the factor graph is computationally inefficient, which seemingly hinders the method from unleashing the full potential of modeling all the interactions, especially when the dialog history grows long.

The Transformer [41] has become a standard neural architecture for various tasks in the field of natural language processing, especially since the huge success of its pretrained model, BERT [11]. Its basic mechanism has recently been extended to the bi-modal problems of vision and language, yielding promising results [6,13,26,27,47]. Then, it appears to be natural to extend it further to deal with many-to-many utility interactions. However, it is not easy due to several reasons. As its basic structure is designed to be deal with self-attention, even in the simplest case of bi-modality, letting X and Y be the two utilities, there are four patterns of attention, $X \rightarrow Y$, $Y \rightarrow X$, $X \rightarrow X$, and $Y \rightarrow Y$; we need an independent Transformer block for each of these four. When extending this to deal with many-to-many utility interactions, the number of the blocks and thus of their total parameters increases proportionally with the square of the number of utilities, making it computationally expensive. Moreover, it is not apparent how to aggregate the results from all the interactions.

To cope with this, we propose a neural architecture named *Light-weight Transformer for Many Inputs* (LTMI) that can deal with all the interactions between many utilities. While it has a block structure similar to the Transformer and shares the core design of attention computation, it differs in the following two aspects. One is the difference in the implementation of multi-head attention. Multi-head attention in the Transformer projects the input feature space linearly to multiple lower-dimensional spaces, enabling to handle multiple attention maps, where the linear mappings are represented with learnable parameters. In the proposed model, we instead split the input feature space to subspaces mechanically according to its indexes, removing all the learnable parameters from the attention computation.

The other difference from the Transformer is that LTMI is designed to receive multiple utilities and compute all the interactions to one utility from all the others including itself. This yields the same number of attended features as the input utilities, which are then concatenated in the direction of the feature space dimensions and then linearly projected back to the original feature space. We treat the parameters of the last linear projection as only learnable parameters in LTMI. This design makes it possible to retain sufficient representational power with a much fewer number of parameters, as compared with a natural extension of the Transformer block to many utilities. By using the same number of blocks in parallel as the number of utilities, we can deal with all the interactions between the utilities; see Fig. 2 for example. Assuming three utilities and the feature space dimensionality of 512, a layer consisting of LTMI has 2.38M parameters, whereas its counterpart based on naive Transformer extension has 28.4M parameters.

2 Related Work

2.1 Attention Mechanisms for Vision-Language Tasks

Attention mechanisms are currently indispensable to build neural architectures for vision-language tasks, such as VQA [4,16,20,29,31,45,48,49] and visual grounding [10,46,52], etc. Inspired by the recent success of the Transformer for language tasks [11,41], several studies have proposed its extensions to bi-modal vision-language tasks [6,13,26,27,40,47]. Specifically, for VQA, it is proposed to use intra-modal and inter-modal attention blocks and stack them alternately to fuse question and image features [13]; it is also proposed to use a cascade of modular co-attention layers that compute the self-attention and guided-attention of question and image features [47]. The method of pretraining a Transformer model used in BERT [11] is employed along with Transformer extension to bi-modal tasks for several vision-language tasks [6,26,27]. They first pretrain the models on external datasets, such as COCO Captions [5] or Conceptual Captions dataset [38], and then fine-tune them on several target tasks.

2.2 Visual Dialog

The task of visual dialog has recently been proposed by two groups of researchers concurrently [7,9]. De Vries et al. introduced the GuessWhat?! dataset, which is built upon goal-oriented dialogs held by two agents to identify unknown objects in an image through a set of yes/no questions [9]. Das et al. released the VisDial dataset, which is built upon dialogs consisting of pairs of a question and an answer about an image that are provided in the form of natural language texts [7]. Kottur et al. recently introduced CLEVR-Dialog as the diagnostic dataset for visual dialog [23].

Most of the existing approaches employ an encoder-decoder architecture [39]. They can be categorized into the following three groups by the design of the encoder: i) fusion-based methods, e.g., LF [7] and HRE [7], which fuses the inputs by their concatenation followed by the application of a feed-forward or recurrent network, and Synergistic [14], which fuses the inputs at multiple stages; ii) attention-based methods that compute attended features of the input image, question, and history utilities, e.g., MN [7], CoAtt [43], HCIAE [28], Synergistic [14], ReDAN [12], FGA [36], and CDF [19]; ReDAN compute the attention over several reasoning steps, FGA models all the interactions over many utilities via a factor graph; iii) methods that attempt to resolve visual co-reference, e.g., RvA [32] and CorefNMN [22], which use neural modules to form an attention mechanism, DAN [18], which employs a network having two attention modules, and AMEM [37], which utilizes a memory mechanism for attention. As for the decoder, there are two designs: i) discriminative decoders that rank the candidate answers using the cross-entropy loss [7] or the n-pair loss [28]; and ii) generative decoders that yield an answer by using a MLE loss [7], weighted likelihood estimation [50], or a combination with adversarial learning [28,43], which trains a discriminator on both positive and negative answers, then transferring it to the generator with auxiliary adversarial learning.

Other approaches include GNN [51], which models relations in a dialog by an unknown graph structure; the employment of reinforcement learning [3,8]; and HACAN [44] which adopts policy gradient to learn the impact of history by intentionally imposing the wrong answer into dialog history. In [30,42], pre-trained vision-language models are adopted, which consist of many Transformer blocks with hundreds of millions parameters, leading to some performance gain. Qi et al. [34] present model-agnostic principles for visual dialog to maximize performance.

3 Efficient Attention Mechanism for Many Utilities

3.1 Attention Mechanism of Transformer

As mentioned earlier, the Transformer has been applied to several bi-modal vision-language tasks, yielding promising results. The Transformer computes and uses attention from three types of inputs, Q (query), K (key), and V (value). Its computation is given by

$$\mathcal{A}(Q, K, V) = \text{softmax}\left(\frac{QK^\top}{\sqrt{d}}\right)V, \tag{1}$$

where Q, K, and V are all collection of features, each of which is represented by a d-dimensional vector. To be specific, $Q = [q_1, \ldots, q_M]^\top \in \mathbb{R}^{M \times d}$ is a collection of M features; similarly, K and V are each a collection of N features, i.e., $K, V \in \mathbb{R}^{N \times d}$. In Eq. (1), V is attended with the weights computed from the similarity between Q and K.

The above computation is usually multi-plexed in the way called multi-head attention. It enables to use a number of attention distributions in parallel, aiming

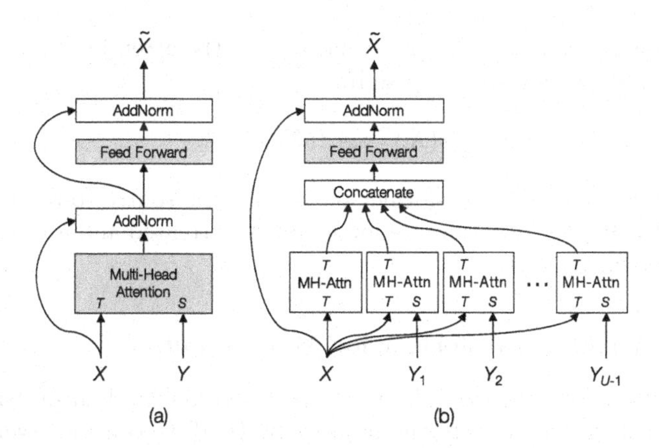

Fig. 1. (a) Source-to-target attention for bi-modal problems implemented by the standard Transformer block; the source Y is attended by weights computed from the similarity between the target X and Y. (b) The proposed block that can deal with many utilities; the source features $\{Y_1, \ldots, Y_{U-1}\}$ are attended by weights computed between them and the target X. Shaded boxes have learnable weights

at an increase in representational power. The outputs of H 'heads' are concatenated, followed by linear transformation with learnable weights $W^O \in \mathbb{R}^{d \times d}$ as

$$\mathcal{A}^{\mathrm{M}}(Q, K, V) = \left[\mathrm{head}_1, \cdots, \mathrm{head}_H\right] W^O. \tag{2}$$

Each head is computed as follows:

$$\mathrm{head}_h = \mathcal{A}(QW_h^Q, KW_h^K, VW_h^V), \;\; h = 1, \ldots, H, \tag{3}$$

where W_h^Q, W_h^K, $W_h^V \in \mathbb{R}^{d \times d_H}$ each are learnable weights inducing a linear projection from the feature space of d-dimensions to a lower space of $d_H (= d/H)$-dimensions. Thus, one attentional block $\mathcal{A}^{\mathrm{M}}(Q, K, V)$ has the following learnable weights:

$$(W_1^Q, W_1^K, W_1^V), \cdots, (W_H^Q, W_H^K, W_H^V) \text{ and } W^O. \tag{4}$$

3.2 Application to Bi-modal Tasks

While Q, K, and V in NLP tasks are of the same modality (i.e., language), the above mechanism has been extended to bi-modality and applied to vision-language tasks in recent studies [6,13,26,27,40,47]. They follow the original idea of the Transformer, considering attention from source features Y to target features X as

$$\mathcal{A}_Y(X) = \mathcal{A}^{\mathrm{M}}(X, Y, Y). \tag{5}$$

In MCAN [47], language feature is treated as the source and visual feature is as the target. In [26] and others [6,13,27,40], co-attention, i.e., attention in the both directions, is considered. Self-attention, i.e., the attention from features to themselves, is given as a special case by

$$\mathcal{A}_X(X) = \mathcal{A}^{\mathrm{M}}(X, X, X). \tag{6}$$

In the above studies, the Transformer block with the source-to-target attention and that with the self-attention are independently treated and are stacked, e.g., alternately or sequentially.

3.3 Light-Weight Transformer for Many Inputs

Now suppose we wish to extend the above attention mechanism to a greater number of utilities[1]; we denote the number by U. If we consider every possible source-target pairs, there are $U(U-1)$ cases in total, as there are U targets, for each of which $U-1$ sources exist. Then we need to consider attention computation $\mathcal{A}_Y(X)$ over $U-1$ sources Y's for each target X. Thus, the straightforward extension of the above attention mechanism to U utilities needs $U(U-1)$ times the number of parameters listed in Eq. (4). If we stack the blocks, the total number of parameters further increases proportionally.

To cope with this, we remove all the weights from Eq. (5). To be specific, for each head $h(=1,\ldots,H)$, we choose and freeze (W_h^Q, W_h^K, W_h^V) as

$$W_h^Q = W_h^K = W_h^V = [\underbrace{O_{d_H},\cdots,O_{d_H}}_{(h-1)d_H}, I_{d_H}, \underbrace{O_{d_H},\cdots,O_{d_H}}_{(H-h)d_H}]^\top, \tag{7}$$

where O_{d_H} is a $d_H \times d_H$ zero matrix and I_{d_H} is a $d_H \times d_H$ identity matrix. In short, the subspace for each head is determined to be one of H subspaces obtained by splitting the d-dimensional feature space with its axis indexes. Besides, we set $W^O = I$, which is the linear mapping applied to the concatenation of the heads' outputs. Let $\bar{\mathcal{A}}_Y(X)$ denote this simplified attention mechanism.

Now let the utilities be denoted by $\{X, Y_1, \ldots, Y_{U-1}\}$, where $X \in \mathbb{R}^{M\times d}$ is the chosen target and others $Y_i \in \mathbb{R}^{N_i \times d}$ are the sources. Then, we compute all the source-to-target attention as $\bar{\mathcal{A}}_{Y_1}(X), \cdots, \bar{\mathcal{A}}_{Y_{U-1}}(X)$. In the standard Transformer block (or rigorously its natural extensions to bi-modal problems), the attended features are simply added to the target as $X + \mathcal{A}_Y(X)$, followed by normalization and subsequent computations. To recover some of the loss in representational power due to the simplification yielding $\bar{\mathcal{A}}_Y(X)$, we propose a different approach to aggregate $\bar{\mathcal{A}}_{Y_1}(X), \cdots, \bar{\mathcal{A}}_{Y_{U-1}}(X)$ and X. Specifically, we concatenate all the source-to-target attention plus the self-attention $\bar{\mathcal{A}}_X(X)$ from X to X as

$$X_{\mathrm{concat}} = [\bar{\mathcal{A}}_X(X), \bar{\mathcal{A}}_{Y_1}(X), \cdots, \bar{\mathcal{A}}_{Y_{U-1}}(X)], \tag{8}$$

[1] As we stated in Introduction, we use the term *utility* here to mean a collection of features.

where $X_{\text{concat}} \in \mathbb{R}^{M \times Ud}$. We then apply linear transformation to it given by $W \in \mathbb{R}^{Ud \times d}$ and $b \in \mathbb{R}^d$ with a single fully-connected layer, followed by the addition of the original X and layer normalization as

$$\tilde{X} = \text{LayerNorm}(\text{ReLU}(X_{\text{concat}}W + \mathbf{1}_M \cdot b^\top) + X), \tag{9}$$

where $\mathbf{1}_M$ is M-vector with all ones. With this method, we aim at recovery of representational power as well as the effective aggregation of information from all the utilities.

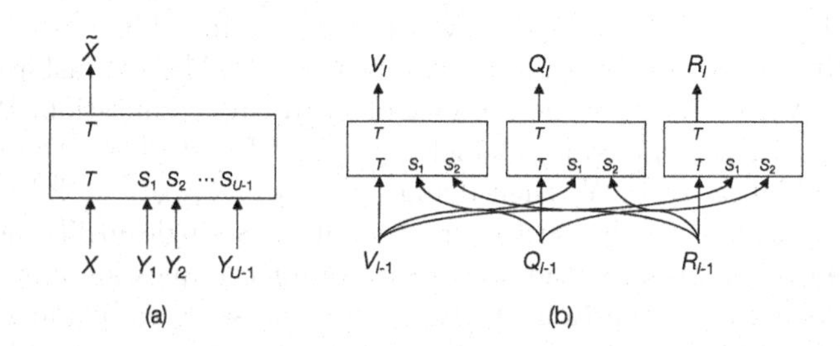

Fig. 2. (a) Simplified symbol of the proposed block shown in Fig. 1(b). (b) Its application to Visual Dialog

3.4 Interactions Between All Utilities

We have designed a basic block (Fig. 1(b)) that deals with attention from many sources to a single target. We wish to consider all possible interactions between all the utilities, not a single utility being the only target. To do this, we use U basic blocks to consider all the source-to-target attention. Using the basic block as a building block, we show how an architecture is designed for visual dialog having three utilities, visual features V, question features Q, and dialog history features R, in Fig. 2(b).

4 Implementation Details for Visual Dialog

4.1 Problem Definition

The problem of Visual Dialog is stated as follows. An agent is given the image of a scene and a dialog history containing T entities, which consists of a caption and question-answer pairs at $T-1$ rounds. Then, the agent is further given a new question at round T along with 100 candidate answers for it and requested to answer the question by choosing one or scoring each of the candidate answers.

4.2 Representation of Utilities

We first extract features from an input image, a dialog history, and a new question at round T to obtain their representations. For this, we follow the standard method employed in many recent studies. For the image utility, we use the bottom-up mechanism [1], which extracts region-level image features using the Faster-RCNN [35] pre-trained on the Visual Genome dataset [24]. For each region (i.e., a bounding box = an object), we combine its CNN feature and geometry to get a d-dimensional vector v_i $(i = 1, \ldots, K)$, where K is the predefined number of regions. We then define $V = [v_1, v_2, \cdots, v_K]^\top \in \mathbb{R}^{K \times d}$. For the question utility, after embedding each word using an embedding layer initialized by pretrained GloVe vectors, we use two-layer Bi-LSTM to transform them to q_i $(i = 1, \ldots, N)$, where N is the number of words in the question. We optionally use the positional embedding widely used in NLP studies. We examine its effects in an ablation test. We then define $Q = [q_1, \ldots, q_N]^\top \in \mathbb{R}^{N \times d}$. For the dialog history utility, we choose to represent it as a single utility here. Thus, each of its entities represents the initial caption or the question-answer pair at one round. As with the question utility, we use the same embedding layer and a two-layer Bi-LSTM together with the positional embeddings for the order of dialog rounds to encode them with a slight difference in formation of an entity vector r_i $(i = 1, \ldots, T)$, where T is the number of Q&A plus the caption. We then define $R = [r_1, \ldots, r_T]^\top \in \mathbb{R}^{T \times d}$. More details are provided in the supplementary material.

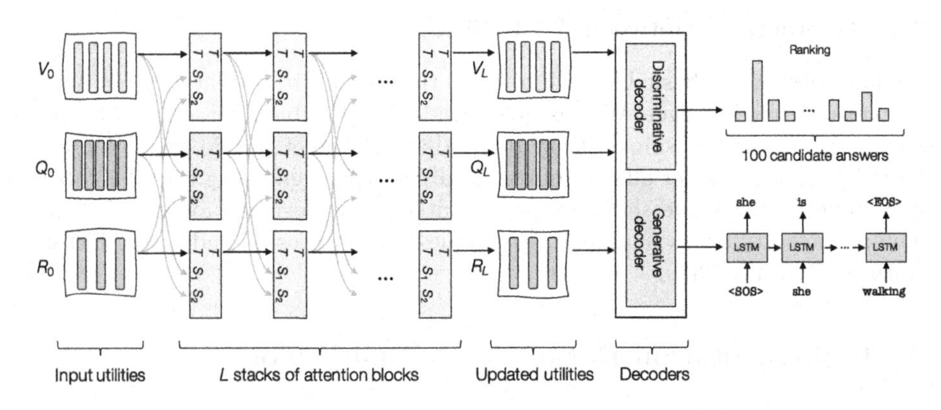

Fig. 3. The entire network built upon the proposed LTMI for Visual Dialog

4.3 Overall Network Design

Figure 3 shows the entire network. It consists of an encoder and a decoder. The encoder consists of L stacks of the proposed attention blocks; a single stack has U blocks in parallel, as shown in Fig. 2(b). We set $V_0 = V$, $Q_0 = Q$, and

$R_0 = R$ as the inputs of the first stack. After the l-th stack, the representations of the image, question, and dialog history utilities are updated as V_l, Q_l, and R_l, respectively. In the experiments, we apply dropout with the rate of 0.1 to the linear layer inside every block. There is a decoder(s) on top of the encoder. We consider a discriminative decoder and a generative decoder, as in previous studies. Their design is explained below.

4.4 Design of Decoders

Decoders receive the updated utility representations, V_L, Q_L, and R_L at their inputs. We convert them independently into d-dimensional vectors c_V, c_Q, and c_R, respectively. This conversion is performed by a simple self-attention computation. We take c_V as an example here. First, attention weights over the entities of V_L are computed by a two-layer network as

$$a_V = \mathrm{softmax}(\mathrm{ReLU}(V_L W_1 + \mathbf{1}_K b_1^\top) W_2 + \mathbf{1}_K b_2), \tag{10}$$

where $W_1 \in \mathbb{R}^{d\times d}$, $W_2 \in \mathbb{R}^{d\times 1}$, $b_1 \in \mathbb{R}^d$, $b_2 \in \mathbb{R}^1$, and $\mathbf{1}_K$ is K-vector with all ones. Then, c_V is given by

$$c_V = \sum_{i=1}^{K} v_{L,i}^\top a_{V,i}, \tag{11}$$

where $v_{L,i}$ is the i-th row vector of V_L and $a_{V,i}$ is the i-th attention weight (a scalar). The others, i.e., c_Q and c_R, can be obtained similarly.

These vectors are integrated and used by the decoders. In our implementation for visual dialog, we found that c_R does not contribute to better results; thus we use only c_V and c_Q. Note that this does not mean the dialog utility R is not necessary; it is interacted with other utilities inside the attention computation, contributing to the final prediction. The two d-vectors c_V and c_Q are concatenated as $[c_V^\top, c_Q^\top]^\top$, and this is projected to d-dimensional space, yielding a context vector $c \in \mathbb{R}^d$.

We design the discriminative and generative decoders following the previous studies. Receiving c and the candidate answers, the two decoders compute the score of each candidate answer in different ways. See details in the supplementary material.

4.5 Multi-task Learning

We observe in our experiments that accuracy is improved by training the entire network using the two decoders simultaneously. This is simply done by minimizing the sum of the losses, $\mathcal{L}_D$ for the discriminative one and $\mathcal{L}_G$ for the generative one (we do not use weights on the losses):

$$\mathcal{L} = \mathcal{L}_D + \mathcal{L}_G. \tag{12}$$

The increase in performance may be attributable to the synergy of learning two tasks while sharing the same encoder. Details will be given in Sect. 5.3.

5 Experimental Results

5.1 Experimental Setup

Dataset. We use the VisDial v1.0 dataset in our experiments which consists of the train 1.0 split (123,287 images), the val 1.0 split (2,064 images), and test v1.0 split (8,000 images). Each image has a dialog composed of 10 question-answer pairs along with a caption. For each question-answer pair, 100 candidate answers are given. The val v1.0 split and 2,000 images of the train v1.0 split are provided with dense annotations (i.e., relevance scores) for all candidate answers. Although the test v1.0 split was also densely annotated, the information about the ground truth answers and the dense annotations are not publicly available. Additionally, we evaulate the method on the Audio Visual Scene-aware Dialog Dataset [15]; the results are shown in the supplementary.

Evaluation Metrics. From the visual dialog challenge 2018, normalized discounted cumulative gain (NDCG) has been used as the principal metric to evaluate methods on the VisDial v1.0 dataset. Unlike other classical retrieval metrics such as R@1, R@5, R@10, mean reciprocal rank (MRR), and mean rank, which are only based on a single ground truth answer, NDCG is computed based on the relevance scores of all candidate answers for each question, which can properly handle the case where each question has more than one correct answer, such as *'yes it is'* and *'yes'*; such cases do occur frequently.

Other Configurations. We employ the standard method used by many recent studies for the determination of hyperparameters etc. For the visual features, we detect $K = 100$ objects from each image. For the question and history features, we first build the vocabulary composed of 11,322 words that appear at least five times in the training split. The captions, questions, and answers are truncated or padded to 40, 20, and 20 words, respectively. Thus, $N = 20$ for the question utility Q. T for the history utilities varies depending on the number of dialogs. We use pre-trained 300-dimensional GloVe vectors [33] to initialize the embedding layer, which is shared for all the captions, questions, and answers.

For the attention blocks, we set the dimension of the feature space to $d = 512$ and the number of heads H in each attention block to 4. We mainly use models having two stacks of the proposed attention block. We train our models on the VisDial v0.9 and VisDial v1.0 dataset using the Adam optimizer [21] with 5 epochs and 15 epochs respectively. The learning rate is warmed up from 1×10^{-5} to 1×10^{-3} in the first epoch, then halved every 2 epochs. The batch size is set to 32 for the both datasets.

5.2 Comparison with State-of-the-Art Methods

Compared Methods. We compare our method with previously published methods on the VisDial v0.9 and VisDial v1.0 datasets, including LF, HRE, MN [7], LF-Att, MN-Att (with attention) [7], SAN [45], AMEM [37], SF [17],

HCIAE [28] and Sequential CoAttention model (CoAtt) [43], Synergistic [14], FGA [36], GNN [51], RvA [32], CorefNMN [22], DAN [18], and ReDAN [12], all of which were trained without using external datasets or data imposition. Unless noted otherwise, the results of our models are obtained from the output of discriminative decoders.

Table 1. Comparison of the performances of different methods on the validation set of VisDial v1.0 with discriminative and generative decoders.

Model	Discriminative						Generative					
	NDCG↑	MRR↑	R@1↑	R@5↑	R@10↑	Mean↓	NDCG↑	MRR↑	R@1↑	R@5↑	R@10↑	Mean↓
MN [7]	55.13	60.42	46.09	78.14	88.05	4.63	56.99	47.83	38.01	57.49	64.08	18.76
CoAtt [43]	57.72	62.91	48.86	80.41	89.83	4.21	59.24	49.64	40.09	59.37	65.92	17.86
HCIAE [28]	57.75	62.96	48.94	80.5	89.66	4.24	59.70	49.07	39.72	58.23	64.73	18.43
ReDAN [12]	59.32	64.21	50.6	81.39	90.26	4.05	60.47	50.02	40.27	59.93	66.78	17.4
LTMI	**62.72**	62.32	48.94	78.65	87.88	4.86	**63.58**	50.74	40.44	61.61	69.71	14.93

Results on the val v1.0 Split. We first compare single-model performance on the val v1.0 split. We select here MN, CoAtt, HCIAE, and ReDAN for comparison, as their performances from the both decoders in all metrics are available in the literature. To be specific, we use the accuracy values reported in [12] for a fair comparison, in which these methods are reimplemented using the bottom-up-attention features. Similar to ours, all these methods employ the standard design of discriminative and generative decoders as in [7]. Table 1 shows the results. It is seen that our method outperforms all the compared methods on the NDCG metric with large margins regardless of the decoder type. Specifically, as compared with ReDAN, the current state-of-the-art on the VisDial v1.0 dataset, our model has improved NDCG from 59.32 to 62.72 and from 60.47 to 63.58 with discriminative and generative decoders, respectively.

Results on the Test-Standard v1.0 Split. We next consider performance on the test-standard v1.0 split. In our experiments, we encountered a phenomenon that accuracy values measured by NDCG and other metrics show a trade-off relation (see the supplementary material for details), depending much on the choice of metrics (i.e., NDCG or others) for judging convergence at the training time. This is observed in the results reported in [12] and is attributable to the inconsistency between the two types of metrics. Thus, we show two results here, the one obtained using NDCG for judging convergence and the one using MRR for it; the latter is equivalent to performing early stopping.

Table 2(a) shows single-model performances on the blind test-standard v1.0 split. With the outputs from the discriminative decoder, our model gains improvement of 3.33pp in NDCG from the best model. When employing the aforementioned early stopping, our model achieves at least comparable or better performance in other metrics as well.

Many previous studies report the performance of an ensemble of multiple models. To make a comparison, we create an ensemble of 16 models with some

Table 2. Comparison in terms of (a) single- and (b) ensemble-model performance on the blind test-standard v1.0 split of the VisDial v1.0 dataset and in terms of (c) the number of parameters of the attention mechanism. The result obtained by early stopping on MRR metric is denoted by ⋆ and those with fine-tuning on dense annotations are denoted by †.

a) Performance of single models

Model	**NDCG** ↑	MRR ↑	R@1 ↑	R@5 ↑	R@10 ↑	Mean ↓
LF [7]	45.31	55.42	40.95	72.45	82.83	5.95
HRE [7]	45.46	54.16	39.93	70.45	81.50	6.41
MN [7]	47.50	55.49	40.98	72.30	83.30	5.92
MN-Att [7]	49.58	56.90	42.42	74.00	84.35	5.59
LF-Att [7]	49.76	57.07	42.08	74.82	85.05	5.41
FGA [36]	52.10	63.70	49.58	**80.97**	88.55	4.51
GNN [51]	52.82	61.37	47.33	77.98	87.83	4.57
CorefNMN [22]	54.70	61.50	47.55	78.10	88.80	4.40
RvA [32]	55.59	63.03	49.03	80.40	89.83	4.18
Synergistic [14]	57.32	62.20	47.90	80.43	89.95	4.17
DAN [18]	57.59	63.20	49.63	79.75	89.35	4.30
LTMI⋆	59.03	**64.08**	**50.20**	80.68	**90.35**	**4.05**
LTMI	**60.92**	60.65	47.00	77.03	87.75	4.90

b) Performance of ensemble models

Model	**NDCG** ↑	MRR ↑	R@1 ↑	R@5 ↑	R@10 ↑	Mean ↓
FGA [36]	52.10	67.30	53.40	85.28	92.70	3.54
Synergistic [14]	57.88	63.42	49.30	80.77	90.68	3.97
DAN [18]	59.36	64.92	51.28	81.60	90.88	3.92
ReDAN [12]	64.47	53.73	42.45	64.68	75.68	6.63
LTMI	66.53	63.19	49.18	80.45	89.75	4.14
P1_P2[34]†	74.91	49.13	36.68	62.98	78.55	7.03
VD-BERT[42]†	**75.13**	50.00	38.28	60.93	77.28	6.90
LTMI†	74.88	**52.14**	**38.93**	**66.60**	**80.65**	**6.53**

c) Num. of attention parameters and the metrics scores

Model	# params	MRR↑	**NDCG**↑
DAN [18]	12.6M	63.20	57.59
RvA [32]	11.9M	63.03	55.59
Naive Transformer	56.8M	62.09	55.10
LTMI⋆ (MRR-based)	4.8M	**64.08**	59.92
LTMI (Q, V)	4.8M	60.65	60.92
LTMI (Q, V, R)	4.8M	60.76	**61.12**

differences, from initialization with different random seeds to whether to use sharing weights across attention blocks or not, the number of attention blocks (i.e. $L = 2, 3$), and the number of objects in the image (i.e. $K = 50, 100$). Aiming at achieving the best performance, we also enrich the image features by incorporating the class label and attributes of each object in an image, which are also obtained from the pretrained Faster-RCNN model. Details are given in the supplementary material. We take the average of the outputs (probability distributions) from the discriminative decoders of these models to rank the candidate answers. Furthermore, we also test fine-tuning each model with its discriminative decoder on the available dense annotations from the train v1.0 and val v1.0, where the cross-entropy loss with soft labels (i.e. relevance scores) is minimized for two epochs. Table 2(b) shows the results. It is observed that our ensemble model (w/o the fine-tuning) achieves the best NDCG = 66.53 in all the ensemble models.

With optional fine-tuning, our ensemble model further gains a large improvement in NDCG, resulting in the third place in the leaderboard. The gap in NDCG to the first place (VD-BERT) is only 0.25pp, while our model yields performance that is better in all the other metrics, i.e, by 2.14pp, 5.67pp, and 3.37pp in MRR, R@5, and R@10, respectively, and 5.36% reduction in Mean.

Table 2(c) shows the number of parameters of the multi-modal attention mechanism employed in the recent methods along with their NDCG scores on the VisDial v1.0 test-standard set. We exclude the parameters of the networks computing the input utilities and the decoders, as they are basically shared among these methods. 'Naive Transformer' consists of two stacks of transformer blocks with simple extension to three utilities as mentioned in Sect. 1. The efficiency of our models can be observed. Note also that the gap between (Q, V) and (Q, V, R) is small, contrary to the argument in [34].

Table 3. Ablation study on the components of our method on the val v1.0 split of VisDial dataset. ↑ indicates the higher the better.

(a)

Component	Details	A-NDCG ↑	D-NDCG ↑	G-NDCG ↑
Number of attention blocks	1	65.37	62.06	62.95
	2	**65.75**	**62.72**	**63.58**
	3	65.42	62.48	63.22
Self-Attention	No	65.38	61.76	63.31
	Yes	**65.75**	**62.72**	**63.58**
Attended features aggregation	Add	64.12	60.28	61.49
	Concat	**65.75**	**62.72**	**63.58**
Shared Attention weights	No	**65.75**	**62.72**	**63.58**
	Yes	65.57	62.50	63.24

(b)

Component	Details	A-NDCG ↑	D-NDCG ↑	G-NDCG ↑
Context feature aggregation	[Q]	65.12	61.50	63.19
	[Q, V]	**65.75**	**62.72**	**63.58**
	[Q, V, R]	65.53	62.37	63.38
Decoder Type	Gen	-	-	62.35
	Disc	-	61.80	-
	Both	**65.75**	**62.72**	**63.58**
The number of objects in an image	36	65.25	62.40	63.08
	50	65.24	62.29	63.12
	100	**65.75**	**62.72**	**63.58**
Positional and spatial embeddings	No	65.18	61.84	62.96
	Yes	**65.75**	**62.72**	**63.58**

5.3 Ablation Study

To evaluate the effect of each of the components of our method, we perform the ablation study on the val v1.0 split of VisDial dataset. We evaluate here the accuracy of the discriminative decoder and the generative decoder separately. We denote the former by D-NDCG and the latter by G-NDCG, and the accuracy of their averaged model by A-NDCG (i.e., averaging the probability distributions over the candidate answers obtained by the discriminative and generative decoders). The results are shown in Table 3(a–b).

The first block of Table 3(a) shows the effect of the number of stacks of the proposed attention blocks. We observe that the use of two to three stacks achieves good performance on all three measures. More stacks did not bring further improvement, and thus are omitted in the table.

The second block of Table 3(a) shows the effect of self-attention, which computes the interaction within a utility, i.e., $\bar{\mathcal{A}}_X(X)$. We examine this because it can be removed from the attention computation. It is seen that self-attention does contribute to good performance. The third block shows the effects of how to aggregate the attended features. It is seen that their concatenation yields better performance than their simple addition. The fourth block shows the impact of sharing the weights across the stacks of the attention blocks. If the weights can be shared as in [25], it contributes a further decrease in the number of parameters. We observe that the performance does drop if weight sharing is employed, but the drop is not very large.

The first block of Table 3(b) shows the effect of how to aggregate the context features c_V, c_Q, and c_R in the decoder(s), which are obtained from the outputs of our encoder. As mentioned above, the context vector c_R of the dialog history does not contribute to the performance. However, the context vector c_v of the image is important for achieving the best performance. The second block of

Table 3(b) shows the effects of simultaneously training the both decoders (with the entire model). It is seen that this contributes greatly to the performance; this indicates the synergy of learning two tasks while sharing the encoder, resulting better generalization as compared with those trained with a single decoder.

We have also confirmed that the use of fewer objects leads to worse results. Besides, the positional embedding for representing the question and history utilities as well as the spatial embedding (i.e., the bounding box geometry of objects) for image utility representation have a certain amount of contribution.

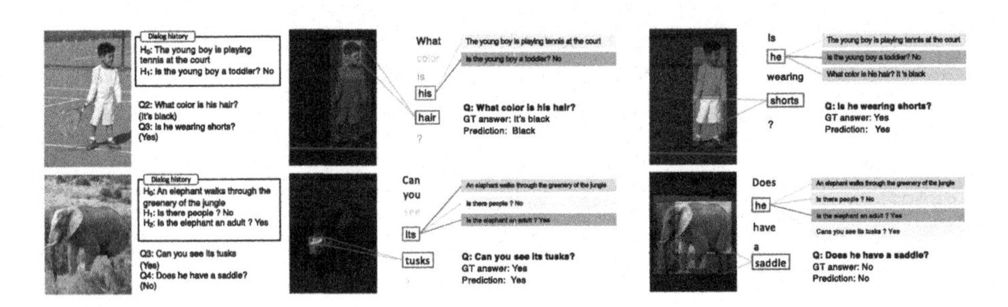

Fig. 4. Examples of visualization for the attention weights generated in our model at two Q&A rounds on two images. See Sect. 5.4 for details.

5.4 Visualization of Generated Attention

Figure 4 shows attention weights generated in our model on two rounds of Q&A on two images. We show here two types of attention. One is the self-attention weights used to compute the context vectors c_V and c_Q. For c_V, the attention weights a_V are generated over image regions (i.e., bounding boxes), as in Eq. (10). Similarly, for c_Q, the attention weights are generated over question words. These two sets of attention weights are displayed by brightness of the image bounding-boxes and darkness of question words, respectively, in the center and the rightmost columns. It can be observed from these that the relevant regions and words are properly highlighted at each Q&A round.

The other attention we visualize is the source-to-target attention computed inside the proposed block. We choose here the image-to-question attention $\bar{A}_V(Q)$ and the history-to-question attention $\bar{A}_R(Q)$. For each, we compute the average of the attention weights over all the heads computed inside the block belonging to the upper stack. In Fig. 4, the former is displayed by the red boxes connected between an image region and a question word; only the region with the largest weight is shown for the target word; the word with the largest self-attention weight is chosen for the target. The history-to-question attention is displayed by the Q&As highlighted in blue color connected to a selected question word that is semantically ambiguous, e.g., *'its'*, *'he'*, and *'his'*. It is seen that the model performs proper visual grounding for the important words, *'hair'*, *'shorts'*, and *'tusks'*. It is also observed that the model properly resolves the co-reference for the words, *'he'* and *'its'*.

6 Summary and Conclusion

In this paper, we have proposed LTMI (Light-weight Transformer for Many Inputs) that can deal with all the interactions between multiple input utilities in an efficient way. As compared with other methods, the proposed architecture is much simpler in terms of the number of parameters as well as the way of handling inputs (i.e., their equal treatment), and nevertheless surpasses the previous methods in accuracy; it achieves the new state-of-the-art results on the VisDial datasets, e.g., high NDCG scores on the VisDial v1.0 dataset. Thus, we believe our method can be used as a simple yet strong baseline.

Acknowledgments. This work was partly supported by JSPS KAKENHI Grant Number JP15H05919 and JP19H01110.

References

1. Anderson, P., et al.: Bottom-up and top-down attention for image captioning and visual question answering. In: Proceedings of the IEEE Conference on Computer Vision and Pattern Recognition, pp. 6077–6086 (2018)
2. Antol, S., et al.: VQA: visual question answering. In: Proceedings of the IEEE International Conference on Computer Vision, pp. 2425–2433 (2015)
3. Chattopadhyay, P., et al.: Evaluating visual conversational agents via cooperative human-AI games. In: Proceedings of AAAI Conference on Human Computation and Crowdsourcing (2017)
4. Chen, K., Wang, J., Chen, L.C., Gao, H., Xu, W., Nevatia, R.: ABC-CNN: an attention based convolutional neural network for visual question answering. arXiv preprint arXiv:1511.05960 (2015)
5. Chen, X., et al.: Microsoft COCO captions: data collection and evaluation server. arXiv preprint arXiv:1504.00325 (2015)
6. Chen, Y.C., et al.: UNITER: learning universal image-text representations. arXiv preprint arXiv:1909.11740 (2019)
7. Das, A., et al.: Visual dialog. In: Proceedings of the IEEE Conference on Computer Vision and Pattern Recognition, pp. 326–335 (2017)
8. Das, A., Kottur, S., Moura, J.M., Lee, S., Batra, D.: Learning cooperative visual dialog agents with deep reinforcement learning. In: Proceedings of the IEEE International Conference on Computer Vision, pp. 2951–2960 (2017)
9. De Vries, H., Strub, F., Chandar, S., Pietquin, O., Larochelle, H., Courville, A.: Guesswhat?! visual object discovery through multi-modal dialogue. In: Proceedings of the IEEE Conference on Computer Vision and Pattern Recognition, pp. 5503–5512 (2017)
10. Deng, C., Wu, Q., Wu, Q., Hu, F., Lyu, F., Tan, M.: Visual grounding via accumulated attention. In: Proceedings of the IEEE Conference on Computer Vision and Pattern Recognition, pp. 7746–7755 (2018)

11. Devlin, J., Chang, M.W., Lee, K., Toutanova, K.: BERT: pre-training of deep bidirectional transformers for language understanding. arXiv preprint arXiv:1810.04805 (2018)
12. Gan, Z., Cheng, Y., Kholy, A.E., Li, L., Liu, J., Gao, J.: Multi-step reasoning via recurrent dual attention for visual dialog. In: Proceedings of the Conference of the Association for Computational Linguistics, pp. 6463–6474 (2019)
13. Gao, P., et al.: Dynamic fusion with intra-and inter-modality attention flow for visual question answering. In: Proceedings of the IEEE Conference on Computer Vision and Pattern Recognition, pp. 6639–6648 (2019)
14. Guo, D., Xu, C., Tao, D.: Image-question-answer synergistic network for visual dialog. In: Proceedings of the IEEE Conference on Computer Vision and Pattern Recognition, pp. 10434–10443 (2019)
15. Hori, C., et al.: End-to-end audio visual scene-aware dialog using multimodal attention-based video features. In: ICASSP 2019–2019 IEEE International Conference on Acoustics, Speech and Signal Processing (ICASSP), pp. 2352–2356 (2019)
16. Ilievski, I., Yan, S., Feng, J.: A focused dynamic attention model for visual question answering. arXiv preprint arXiv:1604.01485 (2016)
17. Jain, U., Lazebnik, S., Schwing, A.G.: Two can play this game: visual dialog with discriminative question generation and answering. In: Proceedings of the IEEE Conference on Computer Vision and Pattern Recognition, pp. 5754–5763 (2018)
18. Kang, G.C., Lim, J., Zhang, B.T.: Dual attention networks for visual reference resolution in visual dialog. In: Proceedings of the Conference on Empirical Methods in Natural Language Processing, pp. 2024–2033 (2019)
19. Kim, H., Tan, H., Bansal, M.: Modality-balanced models for visual dialogue. arXiv preprint arXiv:2001.06354 (2020)
20. Kim, J.H., Jun, J., Zhang, B.T.: Bilinear attention networks. In: Advances in Neural Information Processing Systems, pp. 1564–1574 (2018)
21. Kingma, D.P., Ba, J.: Adam: a method for stochastic optimization. arXiv preprint arXiv:1412.6980 (2014)
22. Kottur, S., Moura, J.M.F., Parikh, D., Batra, D., Rohrbach, M.: Visual coreference resolution in visual dialog using neural module networks. In: Ferrari, V., Hebert, M., Sminchisescu, C., Weiss, Y. (eds.) ECCV 2018. LNCS, vol. 11219, pp. 160–178. Springer, Cham (2018). https://doi.org/10.1007/978-3-030-01267-0_10
23. Kottur, S., Moura, J.M., Parikh, D., Batra, D., Rohrbach, M.: CLEVR-dialog: a diagnostic dataset for multi-round reasoning in visual dialog. arXiv preprint arXiv:1903.03166 (2019)
24. Krishna, R., et al.: Visual genome: connecting language and vision using crowdsourced dense image annotations. Int. J. Comput. Vis. **123**(1), 32–73 (2017)
25. Lan, Z., Chen, M., Goodman, S., Gimpel, K., Sharma, P., Soricut, R.: ALBERT: a lite BERT for self-supervised learning of language representations. arXiv preprint arXiv:1909.11942 (2019)
26. Li, L.H., Yatskar, M., Yin, D., Hsieh, C.J., Chang, K.W.: VisualBERT: a simple and performant baseline for vision and language. arXiv preprint arXiv:1908.03557 (2019)
27. Lu, J., Batra, D., Parikh, D., Lee, S.: ViLBERT: pretraining task-agnostic visiolinguistic representations for vision-and-language tasks. arXiv preprint arXiv:1908.02265 (2019)
28. Lu, J., Kannan, A., Yang, J., Parikh, D., Batra, D.: Best of both worlds: transferring knowledge from discriminative learning to a generative visual dialog model. In: Advances in Neural Information Processing Systems, pp. 314–324 (2017)

29. Lu, J., Yang, J., Batra, D., Parikh, D.: Hierarchical question-image co-attention for visual question answering. In: Advances in Neural Information Processing Systems, pp. 289–297 (2016)
30. Murahari, V., Batra, D., Parikh, D., Das, A.: Large-scale pretraining for visual dialog: a simple state-of-the-art baseline. arXiv preprint arXiv:1912.02379 (2019)
31. Nguyen, D.K., Okatani, T.: Improved fusion of visual and language representations by dense symmetric co-attention for visual question answering. In: Proceedings of the IEEE Conference on Computer Vision and Pattern Recognition, pp. 6087–6096 (2018)
32. Niu, Y., Zhang, H., Zhang, M., Zhang, J., Lu, Z., Wen, J.R.: Recursive visual attention in visual dialog. In: Proceedings of the IEEE Conference on Computer Vision and Pattern Recognition, pp. 6679–6688 (2019)
33. Pennington, J., Socher, R., Manning, C.: Glove: global vectors for word representation. In: Proceedings of the Conference on Empirical Methods in Natural Language Processing, pp. 1532–1543 (2014)
34. Qi, J., Niu, Y., Huang, J., Zhang, H.: Two causal principles for improving visual dialog. In: Proceedings of the IEEE/CVF Conference on Computer Vision and Pattern Recognition, pp. 10860–10869 (2020)
35. Ren, S., He, K., Girshick, R., Sun, J.: Faster R-CNN: towards real-time object detection with region proposal networks. In: Advances in Neural Information Processing Systems, pp. 91–99 (2015)
36. Schwartz, I., Yu, S., Hazan, T., Schwing, A.G.: Factor graph attention. In: Proceedings of the IEEE Conference on Computer Vision and Pattern Recognition, pp. 2039–2048 (2019)
37. Seo, P.H., Lehrmann, A., Han, B., Sigal, L.: Visual reference resolution using attention memory for visual dialog. In: Advances in Neural Information Processing Systems, pp. 3719–3729 (2017)
38. Sharma, P., Ding, N., Goodman, S., Soricut, R.: Conceptual captions: a cleaned, hypernymed, image alt-text dataset for automatic image captioning. In: Proceedings of the Annual Meeting of the Association for Computational Linguistics, pp. 2556–2565 (2018)
39. Sutskever, I., Vinyals, O., Le, Q.V.: Sequence to sequence learning with neural networks. In: Advances in Neural Information Processing Systems, pp. 3104–3112 (2014)
40. Tan, H., Bansal, M.: LXMERT: learning cross-modality encoder representations from transformers. In: Proceedings of the Conference on Empirical Methods in Natural Language Processing (2019)
41. Vaswani, A., et al.: Attention is all you need. In: Advances in Neural Information Processing Systems, pp. 5998–6008 (2017)
42. Wang, Y., Joty, S., Lyu, M.R., King, I., Xiong, C., Hoi, S.C.: VD-BERT: a unified vision and dialog transformer with BERT. arXiv preprint arXiv:2004.13278 (2020)
43. Wu, Q., Wang, P., Shen, C., Reid, I., van den Hengel, A.: Are you talking to me? Reasoned visual dialog generation through adversarial learning. In: Proceedings of the IEEE Conference on Computer Vision and Pattern Recognition, pp. 6106–6115 (2018)
44. Yang, T., Zha, Z.J., Zhang, H.: Making history matter: history-advantage sequence training for visual dialog. In: Proceedings of the IEEE International Conference on Computer Vision, pp. 2561–2569 (2019)
45. Yang, Z., He, X., Gao, J., Deng, L., Smola, A.: Stacked attention networks for image question answering. In: Proceedings of the IEEE Conference on Computer Vision and Pattern Recognition, pp. 21–29 (2016)

46. Yu, L., et al.: MAttNet: modular attention network for referring expression comprehension. In: Proceedings of the IEEE Conference on Computer Vision and Pattern Recognition, pp. 1307–1315 (2018)
47. Yu, Z., Yu, J., Cui, Y., Tao, D., Tian, Q.: Deep modular co-attention networks for visual question answering. In: Proceedings of the IEEE Conference on Computer Vision and Pattern Recognition, pp. 6281–6290 (2019)
48. Yu, Z., Yu, J., Fan, J., Tao, D.: Multi-modal factorized bilinear pooling with co-attention learning for visual question answering. In: Proceedings of the IEEE International Conference on Computer Vision, pp. 1821–1830 (2017)
49. Yu, Z., Yu, J., Xiang, C., Fan, J., Tao, D.: Beyond bilinear: generalized multimodal factorized high-order pooling for visual question answering. IEEE Trans. Neural Netw. Learn. Syst. **29**(12), 5947–5959 (2018)
50. Zhang, H., et al.: Generative visual dialogue system via weighted likelihood estimation. In: Proceedings of the International Joint Conference on Artificial Intelligence, pp. 1025–1031 (2019)
51. Zheng, Z., Wang, W., Qi, S., Zhu, S.C.: Reasoning visual dialogs with structural and partial observations. In: Proceedings of the IEEE Conference on Computer Vision and Pattern Recognition, pp. 6669–6678 (2019)
52. Zhuang, B., Wu, Q., Shen, C., Reid, I., van den Hengel, A.: Parallel attention: a unified framework for visual object discovery through dialogs and queries. In: Proceedings of the IEEE Conference on Computer Vision and Pattern Recognition, pp. 4252–4261 (2018)

Adaptive Mixture Regression Network with Local Counting Map for Crowd Counting

Xiyang Liu[1], Jie Yang[2], Wenrui Ding[3(✉)], Tieqiang Wang[4], Zhijin Wang[2], and Junjun Xiong[2]

[1] School of Electronic and Information Engineering, Beihang University, Beijing, China
xiyangliu@buaa.edu.cn
[2] Shunfeng Technology (Beijing) Co., Ltd, Beijing, China
jieyang2@sfmail.sf-express.com
[3] Institute of Unmanned Systems, Beihang University, Beijing, China
ding@buaa.edu.cn
[4] Institute of Automation, Chinese Academy of Sciences, Beijing, China

Abstract. The crowd counting task aims at estimating the number of people located in an image or a frame from videos. Existing methods widely adopt density maps as the training targets to optimize the point-to-point loss. While in testing phase, we only focus on the differences between the crowd numbers and the global summation of density maps, which indicate the inconsistency between the training targets and the evaluation criteria. To solve this problem, we introduce a new target, named local counting map (LCM), to obtain more accurate results than density map based approaches. Moreover, we also propose an adaptive mixture regression framework with three modules in a coarse-to-fine manner to further improve the precision of the crowd estimation: scale-aware module (SAM), mixture regression module (MRM) and adaptive soft interval module (ASIM). Specifically, SAM fully utilizes the context and multi-scale information from different convolutional features; MRM and ASIM perform more precise counting regression on local patches of images. Compared with current methods, the proposed method reports better performances on the typical datasets. The source code is available at https://github.com/xiyang1012/Local-Crowd-Counting.

Keywords: Crowd counting · Local counting map · Adaptive mixture regression network

This work is done when Xiyang Liu is an intern at Shunfeng Technology.

Electronic supplementary material The online version of this chapter (https://doi.org/10.1007/978-3-030-58586-0_15) contains supplementary material, which is available to authorized users.

A. Vedaldi et al. (Eds.): ECCV 2020, LNCS 12369, pp. 241–257, 2020.
https://doi.org/10.1007/978-3-030-58586-0_15

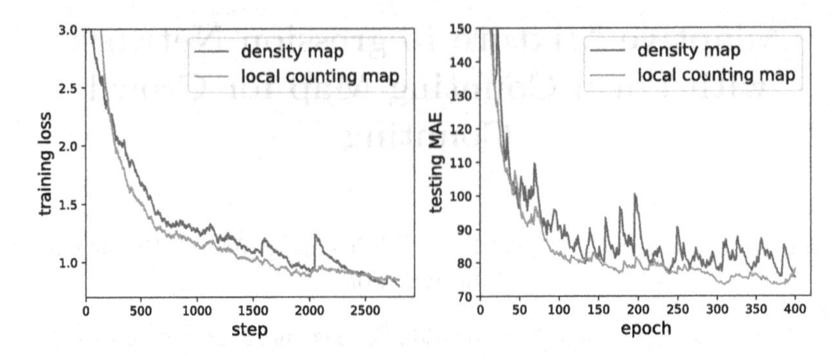

Fig. 1. Training loss curves (*left*) and testing loss curves (*right*) between the two networks sharing VGG16 as the backbone with different regression targets, density map and local counting map on ShanghaiTech Part A dataset. The network trained with the local counting map has the lower error and more stable performance on the testing dataset than the one with the density map

1 Introduction

The main purpose of visual crowd counting is to estimate the numbers of people from static images or frames. Different from pedestrian detection [12,15,18], crowd counting datasets only provide the center points of heads, instead of the precise bounding boxes of bodies. So most of the existing methods draw the density map [11] to calculate crowd number. For example, CSRNet [13] learned a powerful convolutional neural network (CNN) to get the density map with the same size as the input image. Generally, for an input image, the ground truth of its density map is constructed via a Gaussian convolution with a fixed or adaptive kernel on the center points of heads. Finally, the counting result can be represented via the summation of the density map.

In recent years, benefit from the powerful representation learning ability of deep learning, crowd counting researches mainly focus on CNN based methods [1,3,20,25,36] to generate high-quality density maps. The mean absolute error (MAE) and mean squared error (MSE) are adopted as the evaluation metrics of crowd counting task. However, we observed an inconsistency problem for the density map based methods: the training process minimizes the L_1/L_2 error of the density map, which actually represents a point-to-point loss [6], while the evaluation metrics in the testing stage only focus on the differences between the ground-truth crowd numbers and the overall summation of the density maps. Therefore, the model with minimum training error of the density map does not ensure the optimal counting result when testing.

To draw this issue, we introduce a new learning target, named local counting map (LCM), in which each value represents the crowd number of a local patch rather than the probability value indicating whether has a person or not in the density map. In Sect. 3.1, we prove that LCM is closer to the evaluation metric than the density map through a mathematical inequality deduction. As shown in Fig. 1, LCM markedly alleviates the inconsistency problem brought by

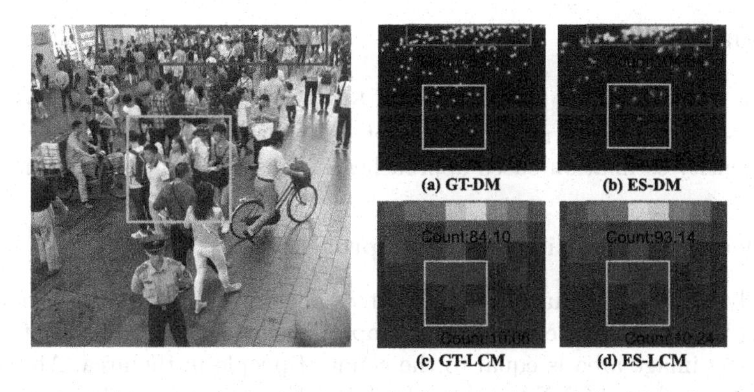

Fig. 2. An intuitive comparison between the local counting map (LCM) and the density map (DM) on local areas. LCM has more accurate estimation counts on both dense (the red box) and sparse (the yellow box) populated areas. (*GT-DM*: ground-truth of DM; *ES-DM*: estimation of DM; *GT-LCM*: ground-truth of LCM; *ES-LCM*: estimation of LCM) (Color figure online)

the density map. We also give an intuitive example to illustrate the prediction differences of LCM and density map. As shown in Fig. 2, the red box represents the dense region and the yellow one represents the sparse region. The prediction of density map is not reliable in dense areas, while LCM has more accurate counting results in these regions.

To further improve the counting performance, we propose an adaptive mixture regression framework to give an accurate estimation of crowd numbers in a coarse-to-finer manner. Specifically, our approach mainly includes three modules: 1) scale-aware module (SAM) to fully utilize the context and multi-scale information contained in feature maps from different layers for estimation; 2) mixture regression module (MRM) and 3) adaptive soft interval module (ASIM) to perform precise counting regression on local patches of images.

In summary, the main contributions in this work are in the followings:

- We introduce a new learning target LCM, which alleviates the inconsistency problem between training targets and evaluation criteria, and reports better counting performance compared with the density map.
- We propose an adaptive mixture regression framework in a coarse-to-finer manner, which fully utilizes the context and multi-scale information from different convolutional features and performs more accurate counting regression on local patches.

The rest of the paper is described as follows: Sect. 2 reviews the previous work of crowd counting; Sect. 3 details our method; Sect. 4 presents the experimental results on typical datasets; Sect. 5 concludes the paper.

2 Related Work

Recently, CNN based approaches have become the focus of crowd counting researches. According to regression targets, they can be classified into two categories: density estimation based approaches and direct counting regression ones.

2.1 Density Estimation Based Approaches

The early work [11] defined the concept of density map and transformed the counting task to estimate the density map of an image. The integral of density map in any image area is equal to the count of people in the area. Afterwards, Zhang *et al.* [35] used CNN to regress both the density map and the global count. It laid the foundation for subsequent works based on CNN methods. To improve performance, some methods aimed at improving network structures. MCNN [36] and Switch-CNN [2] adopted multi-column CNN structures for mapping an image to its density map. CSRNet [13] removed multi-column CNN and used dilated convolution to expand the receptive field. SANet [3] introduced a novel encoder-decoder network to generate high-resolution density maps. HACNN [26] employed attention mechanisms at various CNN layers to selectively enhance the features. PaDNet [29] proposed a novel end-to-end architecture for pan-density crowd counting. Other methods aimed at optimizing the loss function. ADMG [30] produced a learnable density map representation. SPANet [6] put forward MEP loss to find the pixel-level subregion with high discrepancy to the ground truth. Bayesian Loss [17] presented a Bayesian loss to adopt a more reliable supervision on the count expectation at each annotated point.

2.2 Direct Counting Regression Approaches

Counting regression approaches directly estimate the global or local counting number of an input image. This idea was first adopted in [5], which proposed a multi-output regressor to estimate the counts of people in spatially local regions for crowd counting. Afterwards, Shang *et al.* [21] made a global estimation for a whole image, and adopted the counting number constraint as a regularization item. Lu *et al.* [16] regressed a local count of the sub-image densely sampled from the input, and merged the normalized local estimations to a global prediction. Paul *et al.* [19] proposed the redundant counting, and it was generated with the square kernel instead of the Gaussian kernel adopted by the density map. Chattopadhyay *et al.* [4] employed a divide and conquer strategy while incorporating context across the scene to adapt the subitizing idea to counting. Stahl *et al.* [28] adopted a local image divisions method to predict global image-level counts without using any form of local annotations. S-DCNet [32] exploited a spatial divide and conquer network that learned from closed set and generalize to open set scenarios.

Though many approaches have been proposed to generate high-resolution density maps or predict global and local counts, the robust crowd counting of diverse scenes remains hard. Different with previous methods, we firstly introduce a novel regression target, and then adopt an adaptive mixture regression network in a coarse-to-fine manner for better crowd counting.

3 Proposed Method

In this section, we first introduce LCM in details and prove its superiority compared with the density map in Sect. 3.1. After that, we describe SAM, MRM and ASIM of the adaptive mixture regression framework in Sects. 3.2, 3.3 and 3.4, respectively. The overview of our framework is shown in Fig. 3.

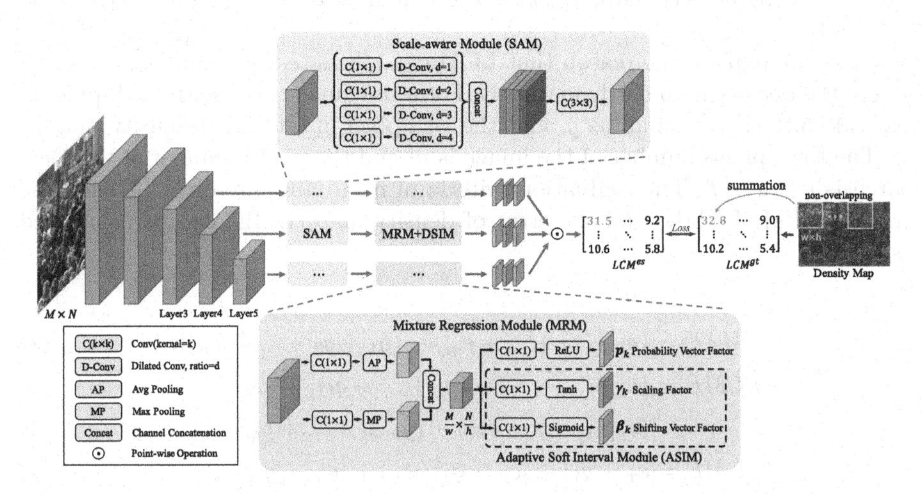

Fig. 3. The overview of our framework mainly including three modules: 1) scale-aware module (SAM), used to enhance multi-scale information of feature maps via multi-column dilated convolution; 2) mixture regression module (MRM) and 3) adaptive soft interval module (ASIM), used to regress feature maps to the probability vector factor $\boldsymbol{p}_k$, the scaling factor γ_k and the shifting vector factors $\boldsymbol{\beta}_k$ of the k-th mixture, respectively. We adopt the feature maps of layers 3, 4 and 5 as the inputs of SAM. The local counting map (LCM) is calculated according to parameters $\{\boldsymbol{p}_k, \gamma_k, \boldsymbol{\beta}_k\}$ and point-wise operation in Eq. (8). For an input $M \times N$ image and the $w \times h$ patch size, the output of the entire framework is a $\frac{M}{w} \times \frac{N}{h}$ LCM

3.1 Local Counting Map

For a given image containing n heads, the ground-truth annotation can be described as $GT(p) = \sum_{i=1}^{n} \delta(p - p_i)$, where p_i is the pixel position of the i-th head's center point. Generally, the generation of the density map is based on a fixed or adaptive Gaussian kernel G_σ, which is described as $D(p) = \sum_{i=1}^{n} \delta(p - p_i) * G_\sigma$. In this work, we fix the spread parameter σ of the Gaussian kernel as 15.

Each value in LCM represents the crowd number of a local patch, rather than a probability value indicating whether has a person or not in the density map. Because heads may be at the boundary of two patches in the process of regionalizing an image, it's unreasonable to divide people directly. Therefore, we generate LCM by summing the density map patch-by-patch. Then, the crowd number of local patch in the ground-truth LCM is not discrete value, but continuous value calculated based on the density map. The LCM can be described as the result of the non-overlapping sliding convolution operation as follows:

$$LCM = D * \mathbf{1}_{(w,h)}, \tag{1}$$

where D is the density map, $\mathbf{1}_{(w,h)}$ is the matrix of ones and (w, h) is the local patch size.

Next, we explain the reason that LCM can alleviate the inconsistency problem of the density map mathematically. For a test image, we set the i-th pixel in ground-truth density map as g_i and the i-th pixel in estimated density map as e_i. The total pixels number of the image is m and the pixels number of the local patch is $t = w \times h$. The evaluation criteria of mean absolute error (MAE), the error of LCM (LCME) and the error of density map (DME) can be calculated as follows:

$$\begin{aligned} MAE =& \left|(e_1 + e_2 + ... + e_m) - (g_1 + g_2 + ... + g_m)\right|, & (2)\\ LCME =& \left|(e_1 + ... + e_t) - (g_1 + ... + g_t)\right| + ...+ \\ & \left|(e_{m-t} + ... + e_m) - (g_{m-t} + ... + g_m)\right|, & (3)\\ DME =& \left|e_1 - g_1\right| + \left|e_2 - g_2\right| + ... + \left|e_m - g_m\right|. & (4) \end{aligned}$$

According to absolute inequality theorem, we can get the relationship among them:

$$MAE \leq LCME \leq DME. \tag{5}$$

When $t = 1$, we have LCME = DME. When $t = m$, we get LCME = MAE. LCME provides a general form of loss function adopted for crowding counting. No matter what value t takes, LCME proves to be a closer bound of MAE than DME theoretically.

On the other side, we clarify the advantages of LCME for training, compared with DME and MAE. 1) DME mainly trains the model to generate probability responses pixel-by-pixel. However, pixel-level position labels generated by a Gaussian kernel may be low-quality and inaccurate for training, due to severe occlusions, large variations of head size, shape and density, etc. There is also a gap between the training loss DME and the evaluation criteria MAE. So the model with minimum training DME does not ensure the optimal counting result when testing with MAE. 2) MAE means direct global counting from an entire image. But global counting is an open-set problem and the crowd number ranges from 0 to ∞, the MAE optimization makes the regression range greatly uncertain. Meanwhile, global counting would ignore all spatial annotated information, which couldn't provide visual density presentations of the prediction results. 3) LCM provides a more reliable training label than the density map, which discards the inaccurate pixel-level position information of density maps and focuses on the count values of local patches. LCME also lessens the gap between DME and MAE. Therefore, we adopt LCME as the training loss rather than MAE or DME.

3.2 Scale-Aware Module

Due to the irregular placement of cameras, the scales of heads in an image are usually very polytropic, which brings great challenge to crowd counting task. To deal with this problem, we propose scale-aware module (SAM) to enhance the multi-scale feature extraction capability of the network. The previous works, such as L2SM [33] and S-DCNet [32], mainly focused on the fusion of feature maps from different CNN layers and acquire multi-scale information through feature pyramid network structure. Different from them, the proposed SAM achieves multi-scale information enhancement only on a single layer feature map and performs this operation at different convolutional layers to bring rich information to subsequent regression modules.

For fair comparisons, we treat VGG16 as the backbone network for CNN-based feature extraction. As shown in Fig. 3, we enhance the feature maps of layers 3, 4 and 5 of the backbone through SAM, respectively. SAM first compresses the channel of feature map via 1×1 convolution. Afterwards, the compressed feature map is processed through dilated convolution with different expansion ratios of 1, 2, 3 and 4 to perceive multi-scale features of heads. The extracted multi-scale feature maps are fused via channel-wise concatenation operation and 3×3 convolution. The size of final feature map is consistent with the input one.

3.3 Mixture Regression Module

Given an testing image, the crowd numbers of different local patches vary a lot, which means great uncertainty on the estimation of local counting. Instead of taking the problem as a hard regression in TasselNet [16], we model the estimation as the probability combination of several intervals. We propose the MRM module to make the local regression more accurate via a coarse-to-fine manner.

First, we discuss the case of coarse regression. For a certain local patch, we assume that the patch contains the upper limit of the crowd as C_m. Thus, the number of people in this patch is considered to be $[0, C_m]$. We equally divide $[0, C_m]$ into s intervals and the length of each interval is $\frac{C_m}{s}$. The vector $\boldsymbol{p} = [p_1, p_2, ..., p_s]^T$ represents the probability of s intervals, and the vector $\boldsymbol{v} = [v_1, v_2, ..., v_s]^T = [\frac{1 \cdot C_m}{s}, \frac{2 \cdot C_m}{s}, ..., C_m]^T$ represents the value of s intervals.

Then the counting number C_p of a local patch in coarse regression can be obtained as followed:

$$C_p = \boldsymbol{p}^T \boldsymbol{v} = \sum_{i=1}^{s} p_i \cdot v_i = \sum_{i=1}^{s} p_i \cdot \frac{i \cdot C_m}{s} = C_m \sum_{i=1}^{s} \frac{p_i \cdot i}{s}. \tag{6}$$

Next, we discuss the situation of fine mixture regression. We assume that the fine regression is consisted of K mixtures. Then, the interval number of the k-th mixture is s_k. The vector $\boldsymbol{p}$ of the k-th mixture is $\boldsymbol{p_k} = [p_{k,1}, p_{k,2}, ..., p_{k,s}]^T$ and the vector $\boldsymbol{v}$ is $\boldsymbol{v_k} = [v_{k,1}, v_{k,2}, ..., v_{k,s}]^T = [\frac{1 \cdot C_m}{\prod_{j=1}^{k} s_j}, \frac{2 \cdot C_m}{\prod_{j=1}^{k} s_j}, ..., \frac{s_k \cdot C_m}{\prod_{j=1}^{k} s_j}]^T$. The counting number C_p of a local patch in mixture regression can be calculated as followed:

$$C_p = \sum_{k=1}^{K} \boldsymbol{p_k}^T \boldsymbol{v_k} = \sum_{k=1}^{K} (\sum_{i=1}^{s_k} p_{k,i} \cdot \frac{i_k \cdot C_m}{\prod_{j=1}^{k} s_j}) = C_m \sum_{k=1}^{K} \sum_{i=1}^{s_k} \frac{p_{k,i} \cdot i_k}{\prod_{j=1}^{k} s_j}. \tag{7}$$

To illustrate the operation of MRM clearly, we take the regression with three mixtures ($K = 3$) for example. For the first mixture, the length of each interval is C_m/s_1. The interval is roughly divided, and the network learns a preliminary estimation of the degree of density, such as sparse, medium, or dense. As the deeper feature in the network contains richer semantic information, we adopt the feature map of layer 5 for the first mixture. For the second and third mixtures, the length of each interval is $C_m/(s_1 \times s_2)$ and $C_m/(s_1 \times s_2 \times s_3)$, respectively. Based on the fine estimation of the second and third mixtures, the network performs more accurate and detailed regression. Since the shallower features in the network contain detailed texture information, we exploit the feature maps of layer 4 and layer 3 for the second and third mixtures of counting regression, respectively.

3.4 Adaptive Soft Interval Module

In Sect. 3.3, it is very inflexible to directly divide the regression interval into several non-overlapping intervals. The regression of value at hard-divided interval boundary will cause a significant error. Therefore, we propose ASIM, which can shift and scale interval adaptively to make the regression process smooth.

For shifting process, we add an extra interval shifting vector factor $\boldsymbol{\beta_k} = [\beta_{k,1}, \beta_{k,2}, ..., \beta_{k,s}]^T$ to represent interval shifting of the i-th interval of the k-th mixture, and the index of the k-th mixture $\bar{i}_k$ can be updated to $\bar{i}_k = i_k + \beta_{k,i}$.

For scaling process, similar to the shifting process, we add an additional interval scaling factor γ to represent interval scaling of each mixture, and the interval number of the k-th mixture $\bar{s}_k$ can be updated to $\bar{s}_k = s_k(1 + \gamma_k)$.

The network can get the output parameters $\{\boldsymbol{p_k}, \gamma_k, \boldsymbol{\beta_k}\}$ for an input image. Based on Eq. (7) and the given parameters C_m and s_k, we can update the mixture regression result C_p to:

$$C_p = C_m \sum_{k=1}^{K} \sum_{i=1}^{s_k} \frac{p_{k,i} \cdot \bar{i}_k}{\prod_{j=1}^{k} \bar{s}_j} = C_m \sum_{k=1}^{K} \sum_{i=1}^{s_k} \frac{p_{k,i} \cdot (i_k + \beta_{k,i})}{\prod_{j=1}^{k} [s_j(1+\gamma_k)]}. \tag{8}$$

Now, we detail the specific implementation of MRM and ASIM. As shown in Fig. 3, for the feature maps from SAM, we downsample them to size $\frac{M}{w} \times \frac{N}{h}$ by following a two-stream model (1×1 convolution and avg pooling, 1×1 convolution and max pooling) and channel-wise concatenation operation. In this way, we can get the fused feature map from the two-stream model to avoid excessive information loss caused via down-sampling. With linear mapping via 1×1 convolution and different activation functions (ReLU, Tanh and Sigmoid), we get regression factors $\{\boldsymbol{p}_k, \gamma_k, \boldsymbol{\beta}_k\}$, respectively. We should note that, $\{\boldsymbol{p}_k, \gamma_k, \boldsymbol{\beta}_k\}$ are the output of MRM and ASIM modules, only related to the input image. LCM is calculated according to parameters $\{\boldsymbol{p}_k, \gamma_k, \boldsymbol{\beta}_k\}$ and point-wise operation in Eq. (8). Crowd number can be calculated via global summation over the LCM. The entire network can be trained end-to-end. The target of network optimization is L_1 distance between the estimated LCM (LCM^{es}) and the ground-truth LCM (LCM^{gt}), which is defined as $Loss = \|LCM^{es} - LCM^{gt}\|_1$.

4 Experiments

In this section, we first introduce four public challenging datasets and the essential implementation details in our experiments. After that, we compare our method with state-of-the-art methods. Finally, we conduct extensive ablation studies to prove the effectiveness of each component of our method.

4.1 Datasets

We evaluate our method on four publicly available crowd counting benchmark datasets: ShanghaiTech [36] Part A and Part B, UCF-QNRF [8] and UCF-CC-50 [7]. These datasets are introduced as follows.

ShanghaiTech. The ShanghaiTech dataset [36] is consisted of two parts: Part A and Part B, with a total of 330,165 annotated heads. Part A is collected from the Internet and represents highly congested scenes, where 300 images are used for training and 182 images for testing. Part B is collected from shopping street surveillance camera and represents relatively sparse scenes, where 400 images are used for training and 316 images for testing.

UCF-QNRF. The UCF-QNRF dataset [8] is a large crowd counting dataset with 1535 high resolution images and 1.25 million annotated heads, where 1201 images are used for training and 334 images for testing. It contains extremely dense scenes where the maximum crowd count of an image can reach 12865. We resize the long side of each image within 1920 pixels to reduce cache occupancy, due to the large resolution of images in the dataset.

UCF-CC-50. The UCF-CC-50 dataset [7] is an extremely challenging dataset, containing 50 annotated images of complicated scenes collected from the Internet. In addition to different resolutions, aspect ratios and perspective distortions, this dataset also has great variants of crowd numbers, varying from 94 to 4543.

4.2 Implementation Details

Evaluation Metrics. We adopt mean absolute error (MAE) and mean squared error (MSE) as metrics to evaluate the accuracy of crowd counting estimation, which are defined as:

$$MAE = \frac{1}{N}\sum_{i=1}^{N}|C_i^{es} - C_i^{gt}|, \quad MSE = \sqrt{\frac{1}{N}\sum_{i=1}^{N}(C_i^{es} - C_i^{gt})^2}, \tag{9}$$

where N is the total number of testing images, C_i^{es} (*resp.* C_i^{gt}) is the estimated (*resp.* ground-truth) count of the i-th image, which can be calculated by summing the estimated (*resp.* ground-truth) LCM of the i-th image.

Data Augmentation. In order to ensure our network can be sufficiently trained and keep good generalization, we randomly crop an area of $m \times m$ pixels from the original image for training. For the ShanghaiTech Part B and UCF-QNRF datasets, m is set to 512. For the ShanghaiTech Part A and UCF-CC-50 datasets, m is set to 384. Random mirroring is also performed during training. In testing, we use the original image to infer without crop and resize operations. For the fair comparison with the previous typical work CSRNet [13] and SANet [3], we does not add the scale augmentation during the training and test stages.

Training Details. Our method is implemented with PyTorch. All experiments are carried out on a server with an Intel Xeon 16-core CPU (3.5 GHz), 64 GB RAM and a single Titan Xp GPU. The backbone of network is directly adopted from convolutional layers of VGG16 [24] pretrained on ImageNet, and the other convolutional layers employ random Gaussian initialization with a standard deviation of 0.01. The learning rate is initially set to $1e^{-5}$. The training epoch is set to 400 and the batch size is set to 1. We train our networks with Adam optimization [10] by minimizing the loss function.

4.3 Comparisons with State of the Art

The proposed method exhibits outstanding performance on all the benchmarks. The quantitative comparisons with state-of-the-art methods on four datasets are presented in Table 1 and Table 2. In addition, we also tell the visual comparisons in Fig. 6.

ShanghaiTech. We compare the proposed method with multiple classic methods on ShanghaiTech Part A & Part B dataset and it has significant performance improvement. On Part A, our method improves 9.69% in MAE and 14.47% in

Table 1. Comparisons with state-of-the-art methods on ShanghaiTech Part A and Part B datasets

Dataset	Part A		Part B	
Method	MAE	MSE	MAE	MSE
MCNN [36]	110.2	173.2	26.4	41.3
Switch-CNN [2]	90.4	135.0	21.6	33.4
CP-CNN [25]	73.6	106.4	20.1	30.1
CSRNet [13]	68.2	115.0	10.6	16.0
SANet [3]	67.0	104.5	8.4	13.6
PACNN [22]	66.3	106.4	8.9	13.5
SFCN [31]	64.8	107.5	7.6	13.0
Encoder-Decoder [9]	64.2	109.1	8.2	12.8
CFF [23]	65.2	109.4	7.2	12.2
Bayesian Loss [17]	62.8	101.8	7.7	12.7
SPANet+CSRNet [6]	62.4	99.5	8.4	13.2
RANet [34]	59.4	102.0	7.9	12.9
PaDNet [29]	**59.2**	**98.1**	8.1	12.2
Ours	61.59	98.36	**7.02**	**11.00**

MSE compared with CSRNet, improves 8.07% in MAE and 5.42% in MSE compared with SANet. On Part B, our method improves 33.77% in MAE and 31.25% in MSE compared with CSRNet, improves 16.43% in MAE and 19.12% in MSE compared with SANet.

UCF-QNRF. We then compare the proposed method with other related methods on the UCF-QNRF dataset. To the best of our knowledge, UCF-QNRF is currently the largest and most widely distributed crowd counting dataset. Bayesian Loss [17] achieves 88.7 in MAE and 154.8 in MSE, which currently maintains the highest accuracy on this dataset, while our method improves 2.37% in MAE and 1.68% in MSE, respectively.

UCF-CC-50. We also conduct experiments on the UCF-CC-50 dataset. The crowd numbers in images vary from 96 to 4633, bringing a great challenging for crowd counting. We follow the 5-fold cross validation as [7] to evaluate our method. With a small amount of training images, our network can still converge well in this dataset. Compared with the latest method Bayesian Loss [17], our method improves 19.76% in MAE and 18.18% in MSE and achieves the state-of-the-art performance.

4.4 Ablation Studies

In this section, we perform ablation studies on ShanghaiTech dataset and demonstrate the roles of several modules in our approach.

Table 2. Comparisons with state-of-the-art methods on UCF-QNRF and UCF-CC-50 datasets

Dataset	UCF-QNRF		UCF-CC-50	
Method	MAE	MSE	MAE	MSE
MCNN [36]	277	426	377.6	509.1
Switch-CNN [2]	228	445	318.1	439.2
Composition Loss [8]	132	191	–	–
Encoder-Decoder [9]	113	188	249.4	354.5
RANet [34]	111	190	239.8	319.4
S-DCNet [32]	104.4	176.1	204.2	301.3
SFCN [31]	102.0	171.4	214.2	318.2
DSSINet [14]	99.1	159.2	216.9	302.4
MBTTBF [27]	97.5	165.2	233.1	300.9
PaDNet [29]	96.5	170.2	185.8	278.3
Bayesian Loss [17]	88.7	154.8	229.3	308.2
Ours	**86.6**	**152.2**	**184.0**	**265.8**

Table 3. An quantitative comparison with two different targets on testing datasets between LCM and density map

Target	Part A		Part B	
	MAE	MSE	MAE	MSE
Density map	72.98	114.89	9.79	14.40
Local counting map	**69.52**	**110.23**	**8.96**	**13.51**

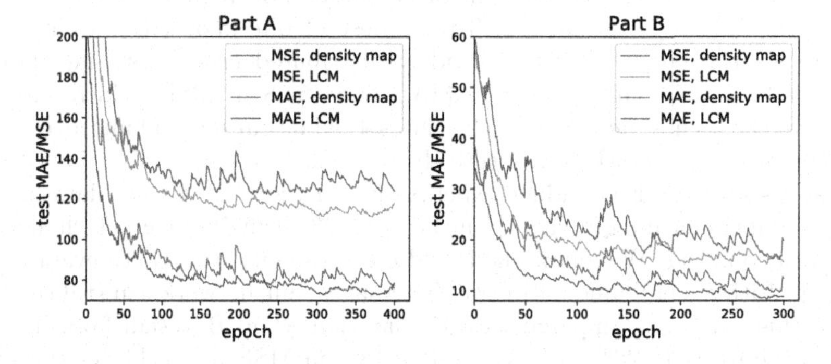

Fig. 4. The curves of testing loss for different regression targets LCM and density map. LCM has lower error and smoother convergence curves on both MAE and MSE than density map

Effect of Regression Target. We analyze the effects of different regression targets firstly. As shown in Table 3, the LCM we introduced has better

Table 4. Ablation study on different combinations of modules including MRM, ASIM and SAM in the regression framework

Module	Part A		Part B	
	MAE	MSE	MAE	MSE
LCM	69.52	110.23	8.96	13.51
MRM	65.24	104.81	7.79	12.55
MRM+ASIM	63.85	102.48	7.56	11.98
MRM+ASIM+SAM	**61.59**	**98.36**	**7.02**	**11.00**

performance than the density map, with 4.74% boost in MAE and 4.06% boost in MSE on Part A, 8.47% boost in MAE and 6.18% boost in MSE on Part B. As shown in Fig. 4, LCM has more stable and lower MAE & MSE testing curves. It indicates that LCM alleviates the inconsistency problem between the training target and the evaluation criteria to bring performance improvement. Both of them adopt VGG16 as the backbone networks without other modules.

Effect of Each Module. To validate the effectiveness of several modules, we train our model with four different combinations: 1) VGG16+LCM (Baseline); 2) MRM; 3) MRM+ASIM; 4) MRM+ASIM+SAM. As shown in Table 4, MRM improves the MAE from 69.52 to 65.24 on Part A and from 8.96 to 7.79 on Part B, compared with our baseline direct LCM regression. With ASIM, it improves the MAE from 65.24 to 63.85 on Part A and from 7.79 to 7.56 on Part B. With SAM, it improves the MAE from 63.85 to 61.59 on Part A and from 7.56 to 7.02 on Part B, respectively. The combination of MRM+ASIM+SAM achieves the best performance, 61.59 in MAE and 98.36 in MSE on Part A, 7.02 in MAE and 11.00 in MSE on Part B.

Table 5. The effects of different local patch sizes with MRM module

Size	Part A		Part B	
	MAE	MSE	MAE	MSE
16×16	70.45	114.12	9.41	13.93
32×32	69.28	109.24	8.68	13.44
64×64	**65.24**	**104.81**	**7.79**	**12.55**
128×128	67.73	105.15	7.93	12.78

Effect of Local Patch Size. We analyze the effects of different local patch sizes on regression results with MRM. As shown in Table 5, the performance gradually improves with local patch size increasing and it slightly drops until 128×128 patch size. Our method gets the best performance with 64×64 patch size on Part A and Part B. When the local patch size is too small, the heads information that local patch can represent is too limited, and it is difficult to

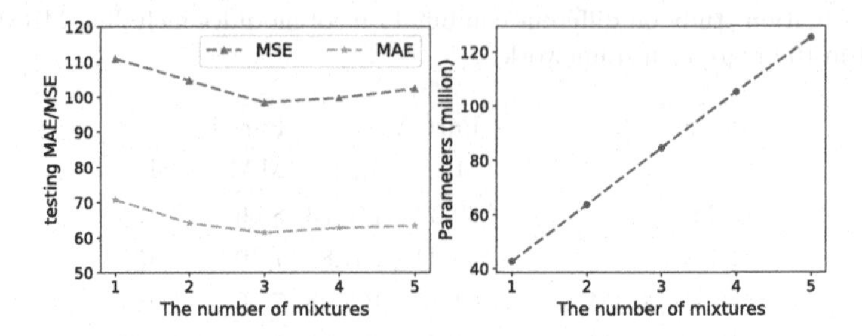

Fig. 5. The left figure shows the MAE & MSE performance of different mixtures number on ShanghaiTech Part A. The right one shows the relationship between the number of mixtures and the parameters of mixtures

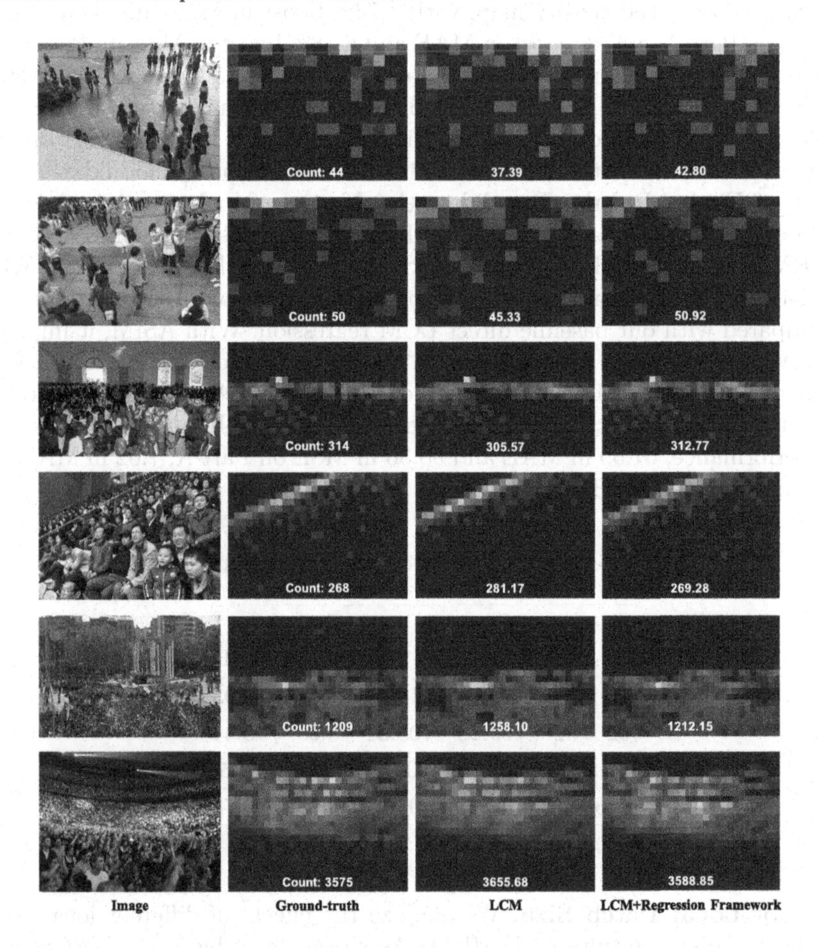

Fig. 6. From top to bottom, we exhibit several example images with different densities from sparse, medium to dense. The second, third and fourth columns display ground-truth LCM, LCM generated with the baseline model (VGG16+LCM) and LCM generated with our proposed regression framework, respectively.

map weak features to the counting value. When the local patch size is 1×1, the regression target changes from LCM to the density map. When the local patch size is too large, the counting regression range will also expand, making it difficult to perform fine and accurate estimation.

Effect of Mixtures Number K. We measure the performance of adaptive mixture regression network with different mixture numbers K. As shown in Fig. 5, the testing error firstly drops and then slightly arises with the increasing number of K. On the one hand, smaller K (e.g., $K = 1$) means single division and it will involve a coarse regression on the local patch. On the other hand, larger K (e.g., $K = 5$) means multiple divisions. It's obviously unreasonable when we divide each interval via undersize steps, such as 0.1 and 0.01. The relationship between the number of mixtures and model parameters is shown in Fig. 5. To achieve a proper balance between the accuracy and computational complexity, we take $K = 3$ as the mixtures number in experiments.

5 Conclusion

In this paper, we introduce a new learning target named local counting map, and show its feasibility and advantages in local counting regression. Meanwhile, we propose an adaptive mixture regression framework in a coarse-to-fine manner. It reports marked improvements in counting accuracy and the stability of the training phase, and achieves the start-of-the-art performances on several authoritative datasets. In the future, we will explore better ways of extracting context and multi-scale information from different convolutional layers. Additionally, we will explore other forms of local area supervised learning approaches to further improve crowd counting performance.

References

1. Babu Sam, D., Sajjan, N.N., Venkatesh Babu, R., Srinivasan, M.: Divide and Grow: capturing huge diversity in crowd images with incrementally growing CNN. In: Proceedings of the IEEE Conference on Computer Vision and Pattern Recognition (2018)
2. Babu Sam, D., Surya, S., Venkatesh Babu, R.: Switching convolutional neural network for crowd counting. In: Proceedings of the IEEE Conference on Computer Vision and Pattern Recognition (2017)
3. Cao, X., Wang, Z., Zhao, Y., Su, F.: Scale aggregation network for accurate and efficient crowd counting. In: Ferrari, V., Hebert, M., Sminchisescu, C., Weiss, Y. (eds.) ECCV 2018. LNCS, vol. 11209, pp. 757–773. Springer, Cham (2018). https://doi.org/10.1007/978-3-030-01228-1_45
4. Chattopadhyay, P., Vedantam, R., Selvaraju, R.R., Batra, D., Parikh, D.: Counting everyday objects in everyday scenes. In: Proceedings of the IEEE Conference on Computer Vision and Pattern Recognition (2017)
5. Chen, K., Loy, C.C., Gong, S., Xiang, T.: Feature mining for localised crowd counting. In: Proceedings of the British Machine Vision Conference (2012)

6. Cheng, Z.Q., Li, J.X., Dai, Q., Wu, X., Hauptmann, A.G.: Learning spatial awareness to improve crowd counting. In: Proceedings of the International Conference on Computer Vision (2019)
7. Idrees, H., Saleemi, I., Seibert, C., Shah, M.: Multi-source multi-scale counting in extremely dense crowd images. In: Proceedings of the IEEE Conference on Computer Vision and Pattern Recognition (2013)
8. Idrees, H., et al.: Composition loss for counting, density map estimation and localization in dense crowds. In: Ferrari, V., Hebert, M., Sminchisescu, C., Weiss, Y. (eds.) ECCV 2018. LNCS, vol. 11206, pp. 544–559. Springer, Cham (2018). https://doi.org/10.1007/978-3-030-01216-8_33
9. Jiang, X., et al.: Crowd counting and density estimation by trellis encoder-decoder networks. In: Proceedings of the IEEE Conference on Computer Vision and Pattern Recognition (2019)
10. Kingma, D.P., Ba, J.: Adam: a method for stochastic optimization. arXiv preprint arXiv:1412.6980 (2014)
11. Lempitsky, V., Zisserman, A.: Learning to count objects in images. In: Proceedings of the Conference and Workshop on Neural Information Processing Systems (2010)
12. Li, J., Liang, X., Shen, S., Xu, T., Feng, J., Yan, S.: Scale-aware fast R-CNN for pedestrian detection. IEEE Tran. Multimedia **20**(4), 985–996 (2017)
13. Li, Y., Zhang, X., Chen, D.: CSRNet: dilated convolutional neural networks for understanding the highly congested scenes. In: Proceedings of the IEEE Conference on Computer Vision and Pattern Recognition (2018)
14. Liu, L., Qiu, Z., Li, G., Liu, S., Ouyang, W., Lin, L.: Crowd counting with deep structured scale integration network. In: Proceedings of the IEEE International Conference on Computer Vision (2019)
15. Liu, W., Liao, S., Ren, W., Hu, W., Yu, Y.: High-level semantic feature detection: a new perspective for pedestrian detection. In: Proceedings of the IEEE Conference on Computer Vision and Pattern Recognition (2019)
16. Lu, H., Cao, Z., Xiao, Y., Zhuang, B., Shen, C.: TasselNet: counting maize tassels in the wild via local counts regression network. Plant Methods **13**(1), 79 (2017)
17. Ma, Z., Wei, X., Hong, X., Gong, Y.: Bayesian loss for crowd count estimation with point supervision. In: Proceedings of the International Conference on Computer Vision (2019)
18. Mao, J., Xiao, T., Jiang, Y., Cao, Z.: What can help pedestrian detection? In: Proceedings of the IEEE Conference on Computer Vision and Pattern Recognition (2017)
19. Paul Cohen, J., Boucher, G., Glastonbury, C.A., Lo, H.Z., Bengio, Y.: Count-ception: counting by fully convolutional redundant counting. In: Proceedings of the International Conference on Computer Vision (2017)
20. Sam, D.B., Babu, R.V.: Top-down feedback for crowd counting convolutional neural network. In: Thirty-second AAAI Conference on Artificial Intelligence (2018)
21. Shang, C., Ai, H., Bai, B.: End-to-end crowd counting via joint learning local and global count. In: Proceedings of the International Conference on Image Processing (2016)
22. Shi, M., Yang, Z., Xu, C., Chen, Q.: Revisiting perspective information for efficient crowd counting. In: Proceedings of the IEEE Conference on Computer Vision and Pattern Recognition (2019)
23. Shi, Z., Mettes, P., Snoek, C.G.M.: Counting with focus for free. In: Proceedings of the International Conference on Computer Vision (2019)
24. Simonyan, K., Zisserman, A.: Very deep convolutional networks for large-scale image recognition. arXiv preprint arXiv:1409.1556 (2014)

25. Sindagi, V.A., Patel, V.M.: Generating high-quality crowd density maps using contextual pyramid CNNs. In: Proceedings of the International Conference on Computer Vision (2017)
26. Sindagi, V.A., Patel, V.M.: HA-CNN: hierarchical attention-based crowd counting network. IEEE Trans. Image Process. **29**, 323–335 (2019)
27. Sindagi, V.A., Patel, V.M.: Multi-level bottom-top and top-bottom feature fusion for crowd counting. In: Proceedings of the IEEE International Conference on Computer Vision (2019)
28. Stahl, T., Pintea, S.L., van Gemert, J.C.: Divide and count: generic object counting by image divisions. IEEE Trans. Image Process. **28**(2), 1035–1044 (2018)
29. Tian, Y., Lei, Y., Zhang, J., Wang, J.Z.: PaDNet: pan-density crowd counting. IEEE Trans. Image Process. **29**, 2714–2727 (2020)
30. Wan, J., Chan, A.: Adaptive density map generation for crowd counting. In: Proceedings of the International Conference on Computer Vision (2019)
31. Wang, Q., Gao, J., Lin, W., Yuan, Y.: Learning from synthetic data for crowd counting in the wild. In: Proceedings of the IEEE Conference on Computer Vision and Pattern Recognition (2019)
32. Xiong, H., Lu, H., Liu, C., Liu, L., Cao, Z., Shen, C.: From open set to closed set: counting objects by spatial divide-and-conquer. In: Proceedings of the International Conference on Computer Vision (2019)
33. Xu, C., Qiu, K., Fu, J., Bai, S., Xu, Y., Bai, X.: Learn to scale: generating multipolar normalized density maps for crowd counting. In: Proceedings of the International Conference on Computer Vision (2019)
34. Zhang, A., et al.: Relational attention network for crowd counting. In: Proceedings of the IEEE International Conference on Computer Vision (2019)
35. Zhang, C., Li, H., Wang, X., Yang, X.: Cross-scene crowd counting via deep convolutional neural networks. In: Proceedings of the IEEE Conference on Computer Vision and Pattern Recognition (2015)
36. Zhang, Y., Zhou, D., Chen, S., Gao, S., Ma, Y.: Single-image crowd counting via multi-column convolutional neural network. In: Proceedings of the IEEE Conference on Computer Vision and Pattern Recognition (2016)

BIRNAT: Bidirectional Recurrent Neural Networks with Adversarial Training for Video Snapshot Compressive Imaging

Ziheng Cheng[1], Ruiying Lu[1], Zhengjue Wang[1], Hao Zhang[1], Bo Chen[1(✉)], Ziyi Meng[2,3], and Xin Yuan[4(✉)]

[1] National Laboratory of Radar Signal Processing, Xidian University, Xi'an, China
zhcheng@stu.xidian.edu.cn, ruiyinglu_xidian@163.com, zhengjuewang@163.com, zhanghao_xidian@163.com, bchen@mail.xidian.edu.cn

[2] Beijing University of Posts and Telecommunications, Beijing, China
mengziyi@bupt.edu.cn

[3] New Jersey Institute of Technology, Newark, NJ, USA

[4] Nokia Bell Labs, Murray Hill, NJ, USA
xyuan@bell-labs.com

Abstract. We consider the problem of video snapshot compressive imaging (SCI), where multiple high-speed frames are coded by different masks and then summed to a single measurement. This measurement and the modulation masks are fed into our Recurrent Neural Network (RNN) to reconstruct the desired high-speed frames. Our end-to-end sampling and reconstruction system is dubbed **BI**directional **R**ecurrent **N**eural networks with **A**dversarial **T**raining (BIRNAT). To our best knowledge, this is the first time that recurrent networks are employed to SCI problem. Our proposed BIRNAT outperforms other deep learning based algorithms and the state-of-the-art optimization based algorithm, DeSCI, through exploiting the underlying correlation of sequential video frames. BIRNAT employs a deep convolutional neural network with Resblock and feature map self-attention to reconstruct the first frame, based on which bidirectional RNN is utilized to reconstruct the following frames in a sequential manner. To improve the quality of the reconstructed video, BIRNAT is further equipped with the adversarial training besides the mean square error loss. Extensive results on both simulation and real data (from two SCI cameras) demonstrate the superior performance of our BIRNAT system. The codes are available at https://github.com/BoChenGroup/BIRNAT.

Keywords: Snapshot compressive imaging · Compressive sensing · Deep learning · Convolutional neural networks · Recurrent Neural Network

---

Electronic supplementary material The online version of this chapter (https://doi.org/10.1007/978-3-030-58586-0_16) contains supplementary material, which is available to authorized users.

A. Vedaldi et al. (Eds.): ECCV 2020, LNCS 12369, pp. 258–275, 2020.
https://doi.org/10.1007/978-3-030-58586-0_16

1 Introduction

Videos are essentially sequential images (frames). Due to the high redundancy in these frames, a video codec [25] can achieve a high (>100) compression rate for a high-definition video. Two potential problems exist in this conventional sampling plus compression framework: i) the high-dimensional data has to be captured and saved, which requires a significant amount of memory and power; ii) the codec, though efficient, introduces latency for the following transmission. To address the first challenge, one novel idea is to build an *optical encoder*, *i.e.*, compressing the video during capture. Inspired by the compressive sensing (CS) [5,6], video snapshot compressive imaging (SCI) [13,23,36] was proposed aiming to provide a promising solution of this *optical encoder*. The underlying principle is to modulate the video frames with a higher speed than the capture rate of the camera. With knowledge of modulation, high-speed video frames can be reconstructed from each single measurement by using advanced algorithms [53]. It has been shown that 148 frames can be recovered from a single measurement in the coded aperture compressive temporal imaging (CACTI) system [23]. With this optical encoder in hand, another challenge, namely an *efficient decoder* is also required to make the video SCI system being practical. Previous algorithms are usually based on iterative optimization, which needs a long time (even hours [22]) to provide a good result. Inspired by deep learning, there are some research attempting to employ convolutional neural networks (CNNs) to reconstruct the high-speed scene from the SCI measurements [15,26,34,35,51,54]. Though the testing speed is promising (tens of milliseconds), none of them can outperform the state-of-the-art optimization algorithm, namely DeSCI [22] on both simulation and real data. Please refer to Fig. 1 for a brief comparison.

Bearing these concerns in mind, in order to achieve high-quality reconstructed videos in a short time, this paper aims to develop an end-to-end deep network to reconstruct high quality images for video SCI, specifically, by investigating the *spatial correlation* via an attention based CNN with Resblock (AttRes-CNN) and *temporal correlation* via a Bidirectional Recurrent Neural Network.

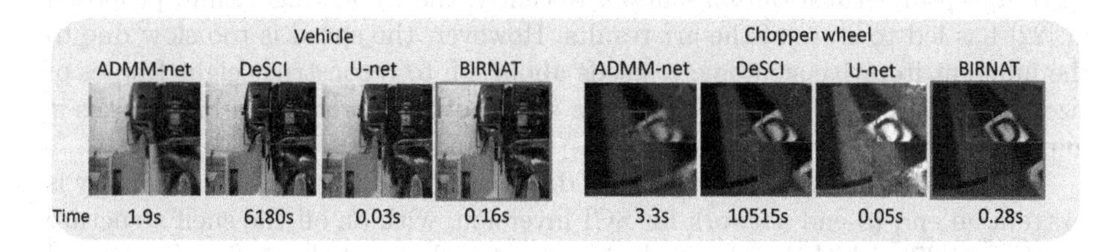

Fig. 1. Selected reconstructed frames using state-of-the-art methods, where DeSCI is an optimization algorithm, ADMM-net and U-net are based on CNNs and BIRNAT (ours) is based on RNNs. Left: simulation data `Vehicle`, the evaluation metric PSNR (in dB) is 23.62 (ADMM-net), 27.04 (DeSCI), 26.43 (U-net) and **27.84 (BIRNAT)**. Right: Real data from CACTI [23]. The testing time is reported at the bottom row.

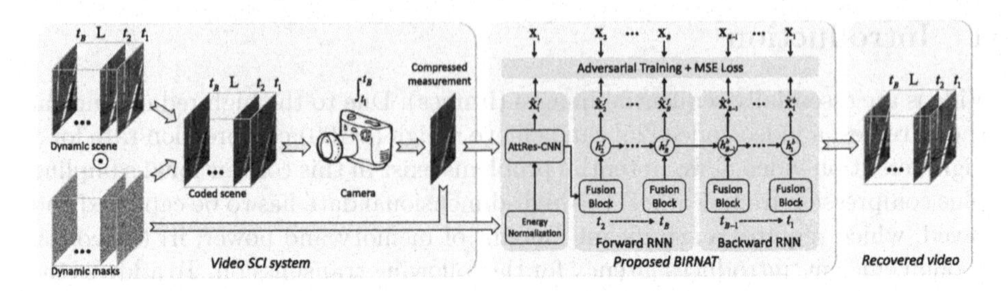

Fig. 2. Principle of video SCI (left) and the proposed BIRNAT for reconstruction (middle). A dynamic scene, shown as a sequence of images at different timestamps ($[t_1, t_2, \ldots, t_B]$, top-left), passes through a dynamic aperture (bottom-left), which imposes individual coding patterns. The coded frames after the aperture are then integrated over time on a camera, forming a single-frame compressed measurement (middle). This measurement along with the dynamic masks are fed into our BIRNAT to reconstruct the time series (right) of the dynamic scene.

1.1 Video Snapshot Compressive Imaging

As shown in Fig. 2, in video SCI, a dynamic scene, modeled as a time-series of two-dimensional (2D) images, passes through a dynamic aperture which applies timestamp-specified spatial coding. In specific, the value of each timestamp-specified spatial coding is superposed by a random pattern and thus the spatial coding of each two timestamps are different (a shifting binary pattern was used in [23]) from each other. The coded frames after the aperture are then integrated over time on a camera, forming a *compressed coded measurement.* Given the coding pattern for each frame, the time series of the scene can be reconstructed from the compressed measurement through iterative optimization based algorithms, which have been developed extensively before. Based on this idea, different video SCI systems have been built in the literature. The modulation approach can be categorized into spatial light modulator (SLM) (including digital micromirror device (DMD)) [13,35,36,40] and physical mask [23,55]. However, one common bottleneck to preclude the wide applications of SCI is the *slow reconstruction speed and poor reconstruction quality.* Recently, the DeSCI algorithm, proposed in [22] has led to state-of-the-art results. However, the speed is too slow due to the inherent iterative strategy; it needs about 2 h to reconstruct eight frames of size 256×256 pixels (Fig. 1 left) from a snapshot measurement, which makes it impractical for real applications.

Motivated by the recent advances of deep learning, one straightforward way is to train an end-to-end network for SCI inversion, with an off-the-shelf structure like U-net [37], which has been used as the backbone of the design for several inverse problems [1,27,28,30,35]. This was also our first choice but it turned out that a single U-net cannot lead to good results as shown in Fig. 1 since it fails to consider the inherent temporal correlation within in video frames for video SCI. Aiming to fill this research gap, in this paper, we propose a Recurrent Neural Network (RNN) based network dubbed **BI**directional **R**ecurrent **N**eural networks with **A**dversarial **T**raining (BIRNAT) for video SCI reconstruction.

1.2 Related Work

For SCI problems, the well established algorithms include TwIST [2], GAP-TV [52] and GMM [16,49], where different priors are used. As mentioned before, the DeSCI algorithm [22] has led to state-of-the-art results for video SCI. DeSCI applies the weighted nuclear norm minimization [10] of nonlocal similar patches in the video frames into the alternating direction method of multipliers [3] regime. Inspired by the recent advances of deep learning on image restoration [47,59], researchers have started using deep learning in computational imaging [15,21,26,30,35,48,57]. Deep fully-connected neural network was used for video CS in [15] and most recently, a deep tensor ADMM-net was proposed in [26] for video SCI problem. A joint optimization and reconstruction network was trained in [51] for video CS. The coding patterns used in [15] is a repeated pattern of a small block; this is not practical in real imaging systems and only simulation results were shown therein. The real data quality shown in [51] is low. The deep tensor ADMM-net [26] employs deep-unfolding technique [38,50] and limited results were shown.

To fill the gap of speed and quality for video SCI reconstruction, this paper develop an RNN based network. Intuitively, the desired high-speed video frames are strongly correlated and a network to fully exploit this correlation should improve the reconstructed video quality. RNNs, originally developed to capture temporal correlations for text and speech, e.g., [9,14], are becoming increasingly popular for video tasks, such as deblurring [32], super-resolution [11,31] and object segmentation [45]. Although these works achieve high performance in their tasks, how to use RNN to build a unified structure for SCI problems still remains challenging.

1.3 Contributions and Organization of This Paper

In a nutshell, we build a new reconstruction framework (BIRNAT) for video SCI and specific contributions are summarized as follows:

1) We build an end-to-end deep learning based reconstruction regime for video SCI reconstruction and use *RNN* to exploit the temporal correlation.
2) A CNN with Resblock [12] is proposed to reconstruct the first frame as a reference for the reconstruction of following frames by RNN. Considering the limitation of convolution in CNN only extracting the local dependencies, we equip it with a *self-attention* module to capture the global (non-local) spatial dependencies, resulting in AttRes-CNN.
3) Given the reconstruction of the first frame, a *Bidirectional RNN* is developed to sequentially infer the following frames, where the *backward RNN* refines the results of the *forward RNN* to improve the reconstructed video further.
4) This *dual-stage* framework is *jointly trained* via combining mean square error (MSE) loss and *adversarial training* [7] to achieve good results.
5) We apply our model on the six benchmark simulation datasets and it produces 0.59 dB higher PSNR than DeSCI on average. We further verify our BIRNAT

on real datasets captured by the CACTI camera and another camera [35]. It shows competitive, sometimes higher performance than DeSCI but with >30,000 times shorter inference time.

The rest of this paper is organized as follows. Section 2 present the mathematical model of video SCI. The proposed BIRNAT is developed in Sect. 3. Simulation and real data results are reported in Sect. 4 and Sect. 5 concludes the entire paper.

2 Mathematical Model of Video SCI

Recalling Fig. 2, we assume that B high-speed frames $\{\mathbf{X}_k\}_{k=1}^{B} \in \mathbb{R}^{n_x \times n_y}$ are modulated by the coding patterns $\{\mathbf{C}_k\}_{k=1}^{B} \in \mathbb{R}^{n_x \times n_y}$, correspondingly. The measurement $\mathbf{Y} \in \mathbb{R}^{n_x \times n_y}$ is given by

$$\mathbf{Y} = \textstyle\sum_{k=1}^{B} \mathbf{X}_k \odot \mathbf{C}_k + \mathbf{G}\,, \tag{1}$$

where $\odot$ denotes the Hadamard (element-wise) product and $\mathbf{G}$ represents the noise. For all B pixels (in the B frames) at position (i,j), $i = 1,\ldots,n_x$; $j = 1,\ldots,n_y$, they are collapsed to form one pixel in the snapshot measurement as

$$y_{i,j} = \textstyle\sum_{k=1}^{B} c_{i,j,k} x_{i,j,k} + g_{i,j}\,. \tag{2}$$

Define $\boldsymbol{x} = \left[\boldsymbol{x}_1^\top,\ldots,\boldsymbol{x}_B^\top\right]$, where $\boldsymbol{x}_k = \mathrm{vec}(\mathbf{X}_k)$, and let $\mathbf{D}_k = \mathrm{diag}(\mathrm{vec}(\mathbf{C}_k))$, for $k = 1,\ldots,B$, where vec( ) vectorizes the matrix inside ( ) by stacking the columns and diag( ) places the ensured vector into the diagonal of a diagonal matrix. We thus have the vector formulation of the sensing process of video SCI:

$$\boldsymbol{y} = \boldsymbol{\Phi}\boldsymbol{x} + \boldsymbol{g}\,, \tag{3}$$

where $\boldsymbol{\Phi} \in \mathbb{R}^{n \times nB}$ is the sensing matrix with $n = n_x n_y$, $\boldsymbol{x} \in \mathbb{R}^{nB}$ is the desired signal, and $\boldsymbol{g} \in \mathbb{R}^{n}$ again denotes the vectorized noise. Unlike traditional CS [5], the sensing matrix considered here is not a dense matrix. In SCI, the matrix $\boldsymbol{\Phi}$ in (3) has a very special structure and can be written as

$$\boldsymbol{\Phi} = [\mathbf{D}_1,\ldots,\mathbf{D}_B]\,, \tag{4}$$

where $\{\mathbf{D}_k\}_{k=1}^{B}$ are diagonal matrices. Therefore, the compressive sampling rate in SCI is equal to $1/B$. It has recently been proved that high quality reconstruction is achievable when $B > 1$ [18,19].

3 Proposed Network for Reconstruction

Having obtained the measurement $\mathbf{Y}$ and coding patterns $\{\mathbf{C}_k\}_{k=1}^{B}$, BIRNAT is developed to *predict* the high-speed frames $\{\widehat{\mathbf{X}}_k\}_{k=1}^{B}$, which are also regarded as the *reconstructions of real high-speed frames* $\{\mathbf{X}_k\}_{k=1}^{B}$. In this section, we will introduce each module of the proposed BIRNAT, including a novel *measurement preprocessing* method in Sect. 3.1, an attentional resblock based CNN to reconstruct the first (reference) frame in Sect. 3.2, and a bidirectional RNN to sequentially reconstruct the following frames in Sect. 3.3. Combining adversarial training and MSE loss, BIRNAT is trained end-to-end as described in Sect. 3.4.

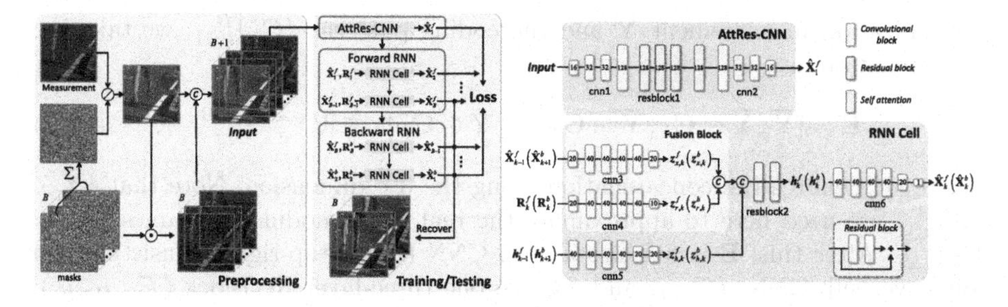

Fig. 3. Left: the proposed preprocessing approach to normalize the measurement. We fed the concatenation of normalization measurement $\overline{\mathbf{Y}}$ and $\{\overline{\mathbf{Y}} \odot \mathbf{C}_k\}_{k=1}^B$ into the proposed BIRNAT. Middle: the specific structure of BIRNAT including i) the attention based CNN (AttRes-CNN) to reconstruct the first frame $\widehat{\mathbf{X}}_1^f$; ii) forward RNN to recurrently reconstruct the following frames $\{\widehat{\mathbf{X}}_k^f\}_{k=2}^B$; iii) backward RNN to perform reverse-order reconstruction $\{\widehat{\mathbf{X}}_k^b\}_{k=B-1}^1$. Right: details of AttRes-CNN and RNN cell. C denotes concatenation along the channel dimension. The numbers in the AttRes-CNN and RNN cell denote the numbers of channels in each feature map.

3.1 Measurement Energy Normalization

Recapping the definition of measurement $\mathbf{Y}$ in (1), it is a weighted ($\{\mathbf{C}_k\}_{k=1}^B$) summation of the high-speed frames $\{\mathbf{X}_k\}_{k=1}^B$. As a result, $\mathbf{Y}$ is usually a *non-energy-normalized* image. For example, some pixels in $\mathbf{Y}$ may gather only one- or two-pixel energy from $\{\mathbf{X}_k\}_{k=1}^B$, while some ones may gather $B-1$ or B. Thus, it is not suitable to directly feed $\mathbf{Y}$ into a network, which motivates us to develop a measurement energy normalization method depicted in Fig. 3 (left).

To be concrete, we first sum all coding patterns $\{\mathbf{C}_k\}_{k=1}^B$ to achieve the energy normalization matrix $\mathbf{C}'$ as

$$\mathbf{C}' = \textstyle\sum_{k=1}^B \mathbf{C}_k\,, \tag{5}$$

where each element in $\mathbf{C}'$ describes how many corresponding pixels of $\{\mathbf{X}_k\}_{k=1}^B$ are integrated into the measurement $\mathbf{Y}$. Then we normalize the measurement $\mathbf{Y}$ by $\mathbf{C}'$ to obtain the energy-normalization measurement $\overline{\mathbf{Y}}$ as

$$\overline{\mathbf{Y}} = \mathbf{Y} \oslash \mathbf{C}', \tag{6}$$

where $\oslash$ denotes the matrix dot (element-wise) division. From Fig. 3 and the definition of $\overline{\mathbf{Y}}$, it can be observed obviously that $\overline{\mathbf{Y}}$ owns more visual information than $\mathbf{Y}$. Meanwhile, $\overline{\mathbf{Y}}$ can be regarded as an approximate average of the high-speed frames $\{\mathbf{X}_k\}_{k=1}^B$, preserving the motionless information such as background and motion trail information.

3.2 AttRes-CNN

In order to initiate RNN, a reference frame is required. Towards this end, we propose a ResBlock [12] based deep CNN for the first frame ($\widehat{\mathbf{X}}_1$) reconstruction. Aiming to fuse all the visual information in hand including our proposed normalization measurement $\overline{\mathbf{Y}}$ and the coding patterns $\{\mathbf{C}_k\}_{k=1}^B$, we take the concatenation as:

$$\mathbf{E} = [\overline{\mathbf{Y}}, \overline{\mathbf{Y}} \odot \mathbf{C}_1, \overline{\mathbf{Y}} \odot \mathbf{C}_2 \dots, \overline{\mathbf{Y}} \odot \mathbf{C}_B]_3 \in \mathbb{R}^{n_x \times n_y \times (B+1)}, \tag{7}$$

where $[\,]_3$ denotes the concatenation along the 3rd dimension. Note that $\{\overline{\mathbf{Y}} \odot \mathbf{C}_k\}_{k=1}^B$ are used here to approximate the real mask-modulated frames $\{\mathbf{X}_k \odot \mathbf{C}_k\}_{k=1}^B$. After this, $\mathbf{E}$ is fed into a deep CNN (Fig. 3 top-right) consisting two four-layer sub-CNNs ($\mathcal{F}_{cnn1}$ and $\mathcal{F}_{cnn2}$), one three-layer ResBlock ($\mathcal{F}_{resblock1}$), and one self-attention module [43] ($\mathcal{F}_{atten}$) as

$$\widehat{\mathbf{X}}_1 = \mathcal{F}_{cnn2}(\mathbf{L}_3), \;\; \mathbf{L}_3 = \mathcal{F}_{atten}(\mathbf{L}_2), \;\; \mathbf{L}_2 = \mathcal{F}_{resblock1}(\mathbf{L}_1), \;\; \mathbf{L}_1 = \mathcal{F}_{cnn1}(\mathbf{E}), \tag{8}$$

where, $\mathcal{F}_{cnn1}$ is used to fuse different visual information in $\mathbf{E}$ to achieve feature $\mathbf{L}_1$; $\mathcal{F}_{resblock1}$ is employed to further capture the spatial correlation when going deeper, and also to alleviate the gradient vanishing; $\mathcal{F}_{cnn2}$, whose structure is mirror symmetry with $\mathcal{F}_{cnn1}$, is used to reconstruct the first frame $\widehat{\mathbf{X}}_1$ of the desired video, and $\mathcal{F}_{atten}$ is developed to capture long-range dependencies (*e.g.*, non-local similarity), discussed as follows.

Self-attention Module. Note that the traditional CNN is only able to capture local dependencies since the convolution operator in CNN has a local receptive field, while in images/videos, non-local similarity [4] is generally used to improve the restoration performance. To explore the non-local information in networks, we employ a self-attention module [44] to capture the long range dependencies [30] among regions to assist our first frame reconstruction.

We perform the self-attention over the pixels of feature map output from $\mathcal{F}_{resblock1}$, denoted by $\mathbf{L}_2 \in \mathbb{R}^{h_x \times h_y \times b}$, where h_x, h_y and b represents the length, width and number of channel in the feature map $\mathbf{L}_2$, respectively. By imposing 1×1 convolution on $\mathbf{L}_2$, we obtain the query $\mathbf{Q}$, key $\mathbf{K}$ and value $\mathbf{V}$ matrix as

$$\mathbf{Q} = \boldsymbol{w}_1 * \mathbf{L}_2, \quad \mathbf{K} = \boldsymbol{w}_2 * \mathbf{L}_2, \quad \mathbf{V} = \boldsymbol{w}_3 * \mathbf{L}_2, \tag{9}$$

where $\{\boldsymbol{w}_1, \boldsymbol{w}_2\} \in \mathbb{R}^{1 \times 1 \times b \times b'}$ and $\boldsymbol{w}_3 \in \mathbb{R}^{1 \times 1 \times b \times b}$ with the fourth dimension representing the number of filters (b' for $\{\boldsymbol{w}_1, \boldsymbol{w}_2\}$ and b for $\boldsymbol{w}_3$), $\{\mathbf{Q}, \mathbf{K}\} \in \mathbb{R}^{h_x \times h_y \times b'}$, $\mathbf{V} \in \mathbb{R}^{h_x \times h_y \times b}$, $*$ represents convolutional operator. $\mathbf{Q}$, $\mathbf{K}$ and $\mathbf{V}$ are then reshaped to $\mathbf{Q}' \in \mathbb{R}^{h_{xy} \times b'}$, $\mathbf{K}' \in \mathbb{R}^{h_{xy} \times b'}$ and $\mathbf{V}' \in \mathbb{R}^{h_{xy} \times b}$, where $h_{xy} = h_x \times h_y$, which means we treat each pixel in the feature map $\mathbf{L}_2$ as a "token", whose feature is $1 \times b'$. After that we construct the attention map $\mathbf{A} \in \mathbb{R}^{h_{xy} \times h_{xy}}$ with element $a_{j,i}$ defined by $a_{j,i} = \frac{exp(s_{i,j})}{\sum_{j=1}^{h_{xy}} exp(s_{i,j})}$, where $s_{i,j}$ is the element in the matrix $\mathbf{S} = \mathbf{Q}'\mathbf{K}'^T \in \mathbb{R}^{h_{xy} \times h_{xy}}$. Here $a_{j,i}$ represents that the extent of the model depends on the ith location when generating the jth region. Having obtained the attention map $\mathbf{A}$, we can impose it on the value matrix $\mathbf{V}'$ to achieve the self-attention feature map $\mathbf{L}_3'$ as

$$\mathbf{L}_3' = reshape(\mathbf{A}\mathbf{V}') \in \mathbb{R}^{h_x \times h_y \times b}, \tag{10}$$

where *reshape*() reshapes the 2D matrix $\mathbf{AV}' \in \mathbb{R}^{h_{xy} \times b'}$ to the 3D matrix. Lastly, we multiply the self-attention feature map $\mathbf{L}_3'$ by a *scale learnable* parameter λ and add it back to the input feature map $\mathbf{L}_2$ [30], leading to the final result

$$\mathbf{L}_3 = \mathbf{L}_2 + \lambda \mathbf{L}_3'. \tag{11}$$

Recapping the reconstruction process of the first frame in (8), it can be regarded as a nonlinear combination of $\overline{\mathbf{Y}}$ and $\{\overline{\mathbf{Y}} \odot \mathbf{C}_k\}_{k=1}^B$. After obtaining the first frame $\widehat{\mathbf{X}}_1$, we use it as a base to reconstruct the following frames by our next proposed sequential model. Therefore, it is important to build the ResBlocks based deep CNN to obtain a good reference frame.

3.3 Bidirectional Recurrent Reconstruction Network

After getting the first frame $\widehat{\mathbf{X}}_1$ via the AttRes-CNN, we now propose a *bidirectional RNN* to perform the reconstruction of the following frames $\{\widehat{\mathbf{X}}_k\}_{k=2}^B$ in a sequel manner. The overall structure of BIRNAT is described in Fig. 3, and we give detailed discussion below.

The Forward RNN: The forward RNN takes $\widehat{\mathbf{X}}_1$ as the initial input, fusing different visual information at corresponding frames to sequentially output the forward reconstruction of other frames $\{\widehat{\mathbf{X}}_k^f\}_{k=2}^B$ (the superscript f denotes 'forward'). For simplicity, in the following description, we take the frame k as an example to describe the RNN cell, which is naturally extended to each frame.

Specifically, at frame k where $k = 2, \cdots, B$, a fusion block, including two parallel six-layer CNNs $\mathcal{F}_{cnn3}$ and $\mathcal{F}_{cnn4}$, is used to fuse the visual information of the reconstruction at the $(k-1)^{th}$ frame $\widehat{\mathbf{X}}_{k-1}^f$, and a reference image at the kth frame $\mathbf{R}_k$ as

$$\boldsymbol{z}_{i,k}^f = \left[\boldsymbol{z}_{x,k}^f, \boldsymbol{z}_{r,k}^f\right]_3, \quad \boldsymbol{z}_{x,k}^f = \mathcal{F}_{cnn3}(\widehat{\mathbf{X}}_{k-1}^f), \quad \boldsymbol{z}_{r,k}^f = \mathcal{F}_{cnn4}(\mathbf{R}_k^f), \tag{12}$$

where $\widehat{\mathbf{X}}_{k-1}^f$ and $\mathbf{R}_k^f$ are fed into each CNN-based feature extractor to achieve $\boldsymbol{z}_{x,k}^f$ and $\boldsymbol{z}_{r,k}^f$ respectively, which are then concatenated as the fused image feature $\boldsymbol{z}_{i,k}^f$. The reference image at the kth frame, $\mathbf{R}_k^f$, is acquired by

$$\mathbf{R}_k^f = \left[\overline{\mathbf{Y}}, \mathbf{Y} - \textstyle\sum_{t=1}^{k-1} \mathbf{C}_t \odot \widehat{\mathbf{X}}_t^f - \sum_{t=k+1}^{B} \mathbf{C}_t \odot \overline{\mathbf{Y}}\right]_3. \tag{13}$$

Recalling the definition of measurement $\mathbf{Y}$ in (1), the second item in (13) can be seen as an approximation of $\mathbf{C}_k \odot \mathbf{X}_k$. This is due to the reason that the predicted frames $\widehat{\mathbf{X}}_t^f$ before k and our proposed normalization measurement $\overline{\mathbf{Y}}$ after k are used to approximate the corresponding real frames $\mathbf{X}_k$. Basically, considering the approximation of $\widehat{\mathbf{X}}_k^f$ should be more accurate than $\overline{\mathbf{Y}}$, the second item in (13) is going closer to the real $\mathbf{C}_k \odot \mathbf{X}_k$. This is one of the motivation that we build the backward RNN in the following subsection. Furthermore, comparing the second item on the current frame, the first item $\overline{\mathbf{Y}}$ in (13) contains more consistent visual information over the consecutive frames, such as background. Thus, we put $\overline{\mathbf{Y}}$ in the $\mathbf{R}_k^f$ to help the model reconstruct smoother video frames.

Having obtained $\boldsymbol{z}_{i,k}^f$, it is concatenated with the features $\boldsymbol{z}_{h,k}^f$ extracted from the hidden units $\boldsymbol{h}_{k-1}^f$ (we initialize $\boldsymbol{h}_1$ with zero), to get the fused features $\boldsymbol{g}_k^f$

$$\boldsymbol{g}_k^f = [\boldsymbol{z}_{i,k}^f, \boldsymbol{z}_{h,k}^f]_3, \quad \boldsymbol{z}_{h,k}^f = \mathcal{F}_{cnn5}(\boldsymbol{h}_{k-1}^f), \tag{14}$$

where $\mathcal{F}_{cnn5}$ is another cnn-based feature extractor. After that, $\boldsymbol{g}_k^f$ is fed into a two-layer ResBlock to achieve the hidden units $\boldsymbol{h}_k^f$ at frame k as

$$\boldsymbol{h}_k^f = \mathcal{F}_{resblock2}(\boldsymbol{g}_k^f), \tag{15}$$

which is then used to generate the forward reconstruction $\widehat{\mathbf{X}}_k^f$ by a CNN as

$$\widehat{\mathbf{X}}_k^f = \mathcal{F}_{cnn6}(\boldsymbol{h}_k^f), \tag{16}$$

where $\mathcal{F}_{cnn6}$ is a six-layer CNN. As a result, the current reconstructed frame $\widehat{\mathbf{X}}_k^f$ and hidden units $\boldsymbol{h}_k^f$ are transported to the same cell to sequentially generate the next frame, until we get the last reconstructed frame $\widehat{\mathbf{X}}_B^f$. Finally, we can get the reconstruction of forward RNN $\{\widehat{\mathbf{X}}_k^f\}_{k=1}^B$ (we regard the construction of first frame $\widehat{\mathbf{X}}_1$ from CNN in (8) as $\widehat{\mathbf{X}}_1^f$).

Although the forward RNN is able to achieve appealing results (refer to Table 1), it ignores the sequential information in a reverse order, which has been widely used in natural language processing [17]. Besides, we observe that the performance of forward RNN improves as k goes from 1 to B. We attribute it to the following two reasons: i) the latter frame uses more information from reconstructed frames; ii) the approximation of the second item in (13) are more accurate. Based on these observations, we add the backward RNN to improve the performance of reconstruction further, especially for the front frames.

The Backward RNN: The backward RNN takes $\widehat{\mathbf{X}}_B^f$ and $\boldsymbol{h}_B^f$ as input to sequentially output the backward reconstruction of each frame $\{\widehat{\mathbf{X}}_k^b\}_{k=B-1}^1$ (the superscript b denotes the backward). At frame k, the structure of backward RNN cell is similar to the forward one, with a little difference on the inputs of each cell. Referring to Fig. 3 and the description of the forward RNN above, in the following, we only discuss the difference between backward and forward RNN.

The first difference is the second item in (12). Due to the opposite order to the forward RNN, at frame k, the backward RNN will use the reconstruction of frame $k+1$. The corresponding networks of (12) for backward RNN are thus changed to

$$\boldsymbol{z}_{i,k}^b = [\boldsymbol{z}_{x,k}^b, \boldsymbol{z}_{r,k}^b]_3, \quad \boldsymbol{z}_{x,k}^b = \mathcal{F}_{cnn3}(\widehat{\mathbf{X}}_{k+1}^b), \quad \boldsymbol{z}_{r,k}^b = \mathcal{F}_{cnn4}(\mathbf{R}_k^b). \tag{17}$$

The second difference is the definition of backward reference image $\mathbf{R}_k^b$. According to the definition of $\mathbf{R}_k^f$ in (13) at frame k, since the reconstruction of frames after k are not obtained, we have to use the normalization measurement $\overline{\mathbf{Y}}$ to approximate them. In backward RNN, it is natural to use each reconstruction from forward RNN $\{\widehat{\mathbf{X}}_k^f\}_{k=1}^B$ directly as

$$\mathbf{R}_k^b = \left[\overline{\mathbf{Y}}, \mathbf{Y} - \textstyle\sum_{t=B,t\neq k}^{1} \mathbf{C}_t \odot \widehat{\mathbf{X}}_t^f\right]_3, \tag{18}$$

where the first item $\overline{\mathbf{Y}}$ is retained to memory its visual information and help the backward RNN to improve the performance.

The networks used in forward and backward RNN do not share the parameters but have the same structure. Another important difference is that the hidden units $\boldsymbol{h}_1^f$ are set to zeros in the forward RNN, while the hidden units $\boldsymbol{h}_B^b$ are set to $\boldsymbol{h}_B^f$ in the backward RNN. This change builds a closer connection between forward and backward RNN and provides more information for backward RNN.

3.4 Optimization

BIRNAT contains four modules: i) the measurement energy normalization, ii) AttRes-CNN, iii) the forward RNN and iv) the backward RNN. Except for i), other modules have their corresponding parameters. Specifically, all learnable parameters in BIRNAT are denoted by $\boldsymbol{\Theta} = \{\mathbf{W}^c, \mathbf{W}^f, \mathbf{W}^b\}$, where $\mathbf{W}^c = \{\mathbf{W}_{cnn1}^c, \mathbf{W}_{cnn2}^c, \mathbf{W}_{resblock1}^c, \mathbf{W}_{attn}^c\}$ are the parameters of the AttRes-CNN; $\mathbf{W}^f = \{\mathbf{W}_{cnn3}^f, \mathbf{W}_{cnn4}^f, \mathbf{W}_{cnn5}^f, \mathbf{W}_{cnn6}^f, \mathbf{W}_{resblock2}^f\}$ are the parameters of forward RNN; $\mathbf{W}^b = \{\mathbf{W}_{cnn3}^b, \mathbf{W}_{cnn4}^b, \mathbf{W}_{cnn5}^b, \mathbf{W}_{cnn6}^b, \mathbf{W}_{resblock2}^b\}$ are the parameters of backward RNN. In the following, we will introduce how to jointly learn them at the training stage and use the well-learned parameters at the testing stage.

Learning Parameters at the Training Stage. At the training stage, besides measurement and the coding patterns $\{\mathbf{Y}_n, \{\mathbf{C}_{n,k}\}_{k=1}^B\}_{n=1}^N$ for N training videos, the real frames $\{\{\mathbf{X}_{n,k}\}_{k=1}^B\}_{n=1}^N$ are also provided as the supervised signal. In order to minimize the reconstruction error of all the frames, the mean square error is used as the loss function

$$\mathcal{L} = \textstyle\sum_{n=1}^N \alpha\mathcal{L}_n^f + \mathcal{L}_n^b, \tag{19}$$

$$\mathcal{L}_n^f = \textstyle\sum_{k=1}^B ||\widehat{\mathbf{X}}_{n,k}^f - \mathbf{X}_{n,k}||_2^2, \qquad \mathcal{L}_n^b = \textstyle\sum_{k=B-1}^1 ||\widehat{\mathbf{X}}_{n,k}^b - \mathbf{X}_{n,k}||_2^2, \tag{20}$$

where $\mathcal{L}_n^f$ and $\mathcal{L}_n^b$ represent the MSE loss of forward and backward RNN, respectively, and α is a trade-off parameter, which is set to 1 in our experiments.

To further improve the quality of each reconstructed frames and make the generated video smoother, we introduce the adversarial training [8] in addition to the MSE loss in (19). To be more specific, the input video frames $\{\mathbf{X}_{n,k}\}_{n=1,k=1}^{N,B}$ are treated as "real" samples, while the reconstructed frames $[\{\widehat{\mathbf{X}}_{n,k}^b\}_{n=1,k=1}^{N,B-1}, \{\widehat{\mathbf{X}}_{n,B}^f\}_{n=1}^N]_3$, generated from previous networks, are assumed as the "fake" samples. The adversarial training loss can be formulated as

$$\mathcal{L}_g = \mathbb{E}_{\mathbf{X}}[\log D(\mathbf{X})] + \mathbb{E}_{\mathbf{Y}}[\log(1 - D(G(\mathbf{Y}, \{\mathbf{C}_k\}_{k=1}^B)))], \tag{21}$$

where G is the generator which outputs reconstructed video frames, and D is the discriminator that has same structure with [29]. As a result, the final loss function of our model is

$$\mathcal{L} = \sum_{n=1}^{N}(\alpha\mathcal{L}_n^f + \mathcal{L}_n^b) + \beta\mathcal{L}_g, \tag{22}$$

where β is a trade-off parameter. In the experiments, β is set to 0.001.

Performing SCI Reconstruction at the Testing Stage. During testing, with the well-learned network parameters $\boldsymbol{\Theta}$, we can achieve the frames $\{\widehat{\mathbf{X}}_k^f\}_{k=1}^B$ and $\{\widehat{\mathbf{X}}_k^b\}_{k=B-1}^1$. Considering the advantages of backward RNN that uses a good visual features generated by the forward RNN, we use the reconstructed frame 1 to $B-1$ from backward RNN, and frame B from forward RNN to construct the final reconstruction of our system, that is $[\{\widehat{\mathbf{X}}_k^b\}_{k=1}^{B-1}, \widehat{\mathbf{X}}_B^f]_3$. Note that our proposed BIRNAT can also be used in other SCI systems [24,39,41,42,56,58].

4 Experiments

In this section, we compare BIRNAT with several state-of-the-art methods on both simulation and real datasets.

4.1 Training, Testing Datasets and Experimental Settings

Datasets: Considering the following two reasons: i) the video SCI reconstruction task does not have a specific training set; ii) the SCI imaging technology is suitable for any scene, we choose the dataset `DAVIS2017` [33], originally used in video object segmentation task as the training set. We first evaluate BIRNAT on six simulation datasets including `Kobe`, `Runner`, `Drop`, `Traffic` [22], `Aerial` and `Vehicle` [26]. After that, we also evaluate BIRNAT on several real datasets captured by real video SCI cameras [23,35].

Table 1. The average results of PSNR in dB (left entry) and SSIM (right entry) and running time per measurement/shot in seconds by different algorithms on 6 datasets. Best results are in **red and bold**, second best results are blue underlined.

Dataset	Kobe		Traffic		Runner		Drop		Aerial		Vehicle		Average		Running time
GAP-TV [52]	26.45	0.8448	20.89	0.7148	28.81	0.9092	34.74	0.9704	25.05	0.8281	24.82	0.8383	26.79	0.8576	4.2
DeSCI [22]	**33.25**	**0.9518**	28.72	0.9250	**38.76**	0.9693	**43.22**	**0.9925**	25.33	0.8603	27.04	0.9094	32.72	0.9347	6180
U-net [35]	27.79	0.8071	24.62	0.8403	34.12	0.9471	36.56	0.9494	27.18	0.8690	26.43	0.8817	29.45	0.8824	0.0312
PnP-FFDNet [54]	30.50	0.9256	24.18	0.8279	32.15	0.9332	40.70	0.9892	25.27	0.8291	25.42	0.8493	29.70	0.8924	3.0
BIRNAT w/o SA& AT& BR	31.06	0.9158	27.17	0.9198	36.62	0.9674	40.67	0.9802	28.40	0.9103	27.24	0.9125	31.86	0.9343	0.0856
BIRNAT w/o SA& AT	32.18	0.9168	28.93	0.9298	38.06	0.9716	42.10	0.9889	28.95	0.9092	27.68	0.9173	32.98	0.9389	0.1489
BIRNAT w/o AT	32.66	0.9490	29.30	0.9418	38.25	0.9748	42.08	0.9914	28.98	0.9163	27.79	0.9234	33.18	0.9494	0.1512
BIRNAT w/o SA	32.27	0.9341	28.99	0.9391	38.44	0.9753	42.22	0.9916	**29.00**	**0.9170**	27.74	0.9233	33.11	0.9467	0.1489
BIRNAT	32.71	0.9504	**29.33**	**0.9422**	38.70	**0.9760**	42.28	0.9918	28.99	0.9166	**27.84**	**0.9274**	**33.31**	**0.9507**	0.1647

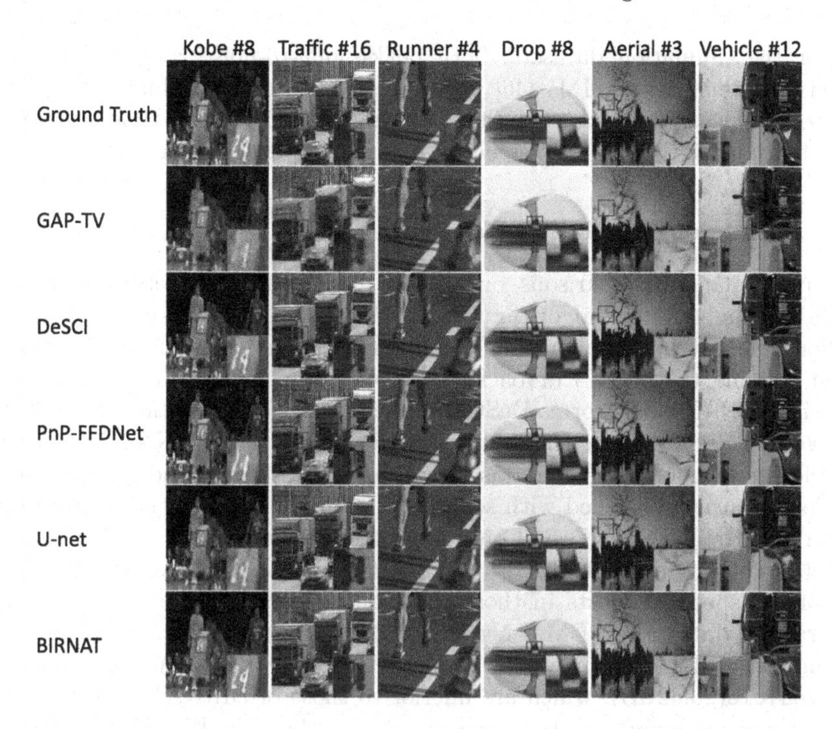

Fig. 4. Reconstructed frames of GAP-TV, DeSCI, U-net and BIRNAT on six simulated video SCI datasets. Please watch the full video in the SM for details.

Implementation Details of BIRNAT: Following the setting in [22], eight ($B = 8$) sequential frames are modulated by the shifting binary masks $\{\mathbf{C}_k\}_{k=1}^{B}$ and then collapsed into a single measurement $\mathbf{Y}$. We randomly crop patch cubes $256 \times 256 \times 8$ from original scenes in `DAVIS2017` and obtain 26,000 training data pairs with data augmentation. Our model is trained for 100 epochs in total. Starting with the initial learning rate of 3×10^{-4}, we reduce the learning rate by 10% every 10 epochs, and it costs about 3 day for training the entire network. The Adam optimizer [20] is employed for the optimization. All experiments are run on the NVIDIA RTX 8000 GPU based on PyTorch. The detailed architecture for BIRNAT is given in the supplementary material (SM).

Counterparts and Performance Metrics: As introduced above, various methods have been proposed for SCI reconstruction. Hereby we compare our model with three competitive counterparts. The first one GAP-TV [52] is a widely used efficient baseline with decent performance. The second one DeSCI [22] currently produces state-of-the-art results. For the results of other algorithms, please refer to [22]. In order to compare with the deep learning based methods, we repurposed U-net to SCI tasks as in [35], where the CNN is employed to capture local correlations in an end-to-end manner. We further compare with the most recent plug-and-play (PnP) algorithm proposed in [54].

For the simulation datasets, both peak-signal-to-noise ratio (PSNR) and structural similarity (SSIM) [46] are used as metrics to evaluate the performance. Besides, to see whether they can be applied to a real-time system, we give the running time of reconstructing the video at the testing stage.

4.2 Results on Simulation Datasets

The performance comparisons on the six benchmark datasets are given in Table 1, using different algorithms, *i.e.* GAP-TV, DeSCI, U-net and various versions of BIRNAT without self-attention (denoted as 'w/o SA') or adversarial training ('w/o AT') or backward RNN ('w/o BR'). It can be observed that: i) BIRNAT outperforms DeSCI on the `Traffic`(0.61 dB), `Aerial`(3.66 dB) and `Vehicle`(0.80 dB) by the metric PSNR. Obviously, BIRNAT can provide superior performance on the datasets with complex background, owning to the non-local features obtained with self-attention and the sequential dependencies constructed by RNN; ii) DeSCI only improved a little bit over BIRNAT on the Kobe(0.54 dB), Runner(0.06 dB) and Drop(0.94 dB), since there are high-speed motions of specific objects in those three datasets, which are rarely found in the training data. The ADMM-net [26] used different training sets for different testing sets and it only shows the results on `Kobe`(30.15 dB), `Aerial`(26.85 dB) and `Vehicle`(23.62 dB), which are inferior to those of BIRNAT. BIRNAT gets leading average performance on these six datasets both for PSNR and SSIM; iii) BIRNAT achieves 30000 times speedups over DeSCI at the testing stage; iv) the attention mechanism and adversarial training are beneficial to performance.

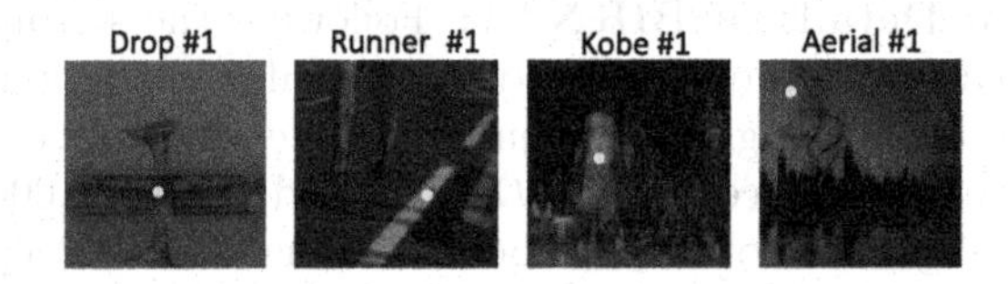

Fig. 5. Selected attention maps of the first frame. Yellow points denote the pixels randomly selected from each image, and red areas denote the active places. (Color figure online)

Figure 4 shows selected reconstructed frames of BIRNAT on these six datasets compared with GAP-TV, DeSCI and our repurposed U-net. We can observe that while DeSCI smooths out the details in the reconstructed video, BIRNAT provides sharper borders and finer details, owning to the better interpolation with both spatial and temporal information extracted by CNN and Bidirectional RNN. To further explore the influence of attention mechanism, we illustrate the attention map in Fig. 5, where we plot the attended active areas (highlighted red color) of a randomly selected pixel. It can be seen that those non-local regions in red color are corresponding to the highly semantically related areas. These attention-aware features can provide long range spatial dependencies among pixels, which is helpful for the first frame reconstruction and gives a better basement for generating the following frames.

4.3 Results on Real SCI Data

We now apply BIRNAT to real data captured by the SCI cameras [23,35] to verify the its robustness. The `Wheel` snapshot measurement of size 256×256 pixels encodes 14 ($B = 14$) high-speed videos. The mask is the shifting random mask with the pixel shifts determined by the pre-set translation of the printed film. The `Domino` and `Water Balloon` snapshot measurement of size 512×512 pixels encodes 10 frames, in which the mask is controlled by a DMD [35]. The real captured data have noise inside and thus the SCI for real data is more challenging. As shown in Fig. 6, the reconstructed video by BIRNAT shows finer and complete details compared with other methods, with a significant saving on the reconstruction time during testing compared to DeSCI. This indicates the applicability and efficiency of our algorithm in real applications.

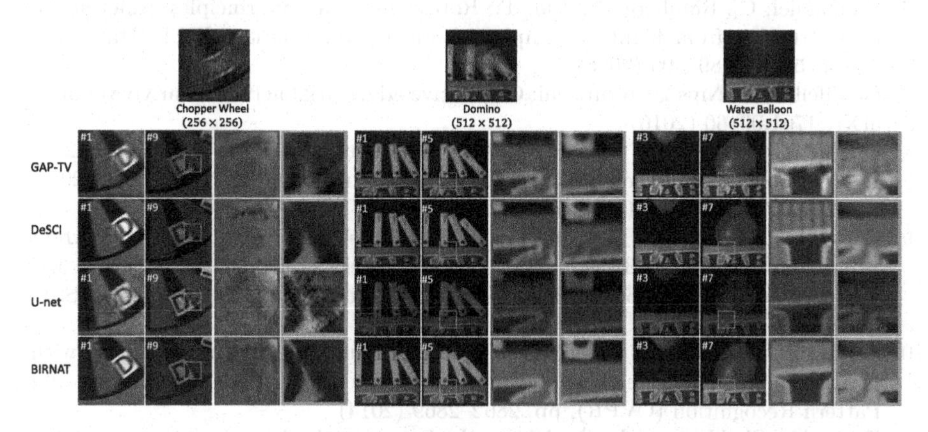

Fig. 6. Real data `Wheel` (left), `Domino` (middle) and `Water Balloon` (right): results of GAP-TV, DeSCI, U-net and BIRNAT. Please refer to more real data results in the SM.

5 Conclusions

In this paper, we have proposed a bidirectional RNN with adversarial training for snapshot compressive imaging system, called BIRNAT. We employ a dual-stage framework, where the first frame is reconstructed through an attention ResBlock based deep CNN, and then the following frames are sequentially inferred by RNN. The experimental results on both simulation and real-world SCI camera data have demonstrated that the proposed method achieves superior performance and outperforms current state-of-the-art algorithms.

Acknowledgement. B. Chen acknowledges the support of the Program for Oversea Talent by Chinese Central Government, the 111 Project (No. B18039), and NSFC (61771361) and Shaanxi Innovation Team Project.

References

1. Barbastathis, G., Ozcan, A., Situ, G.: On the use of deep learning for computational imaging. Optica **6**(8), 921–943 (2019)
2. Bioucas-Dias, J., Figueiredo, M.: A new TwIST: Two-step iterative shrinkage/thresholding algorithms for image restoration. IEEE Trans. Image Process. **16**(12), 2992–3004 (2007)
3. Boyd, S., Parikh, N., Chu, E., Peleato, B., Eckstein, J.: Distributed optimization and statistical learning via the alternating direction method of multipliers. Found. Trends Mach. Learn. **3**(1), 1–122 (2011)
4. Buades, A., Coll, B., Morel, J.M.: A non-local algorithm for image denoising. In: 2005 IEEE Computer Society Conference on Computer Vision and Pattern Recognition, CVPR 2005, vol. 2, pp. 60–65. IEEE (2005)
5. Donoho, D.L.: Compressed sensing. IEEE Trans. Inf. Theor. **52**(4), 1289–1306 (2006)
6. Emmanuel, C., Romberg, J., Tao, T.: Robust uncertainty principles: exact signal reconstruction from highly incomplete frequency information. IEEE Trans. Inf. Theor. **52**(2), 489–509 (2006)
7. Goodfellow, I.: Nips 2016 tutorial: Generative adversarial networks. arXiv preprint arXiv:1701.00160 (2016)
8. Goodfellow, I.J., et al.: Generative adversarial nets. In: Proceedings of the 27th International Conference on Neural Information Processing Systems, NIPS 2014, vol. 2, pp. 2672–2680 (2014)
9. Graves, A., Mohamed, A., Hinton, G.: Speech recognition with deep recurrent neural networks. In: 2013 IEEE International Conference on Acoustics, Speech and Signal Processing, pp. 6645–6649 (May 2013). https://doi.org/10.1109/ICASSP.2013.6638947
10. Gu, S., Zhang, L., Zuo, W., Feng, X.: Weighted nuclear norm minimization with application to image denoising. In: IEEE Conference on Computer Vision and Pattern Recognition (CVPR), pp. 2862–2869 (2014)
11. Haris, M., Shakhnarovich, G., Ukita, N.: Recurrent back-projection network for video super-resolution. In: The IEEE Conference on Computer Vision and Pattern Recognition (CVPR) (June 2019)
12. He, K., Zhang, X., Ren, S., J, S.: Deep residual learning for image recognition. In: CVPR (2016)
13. Hitomi, Y., Gu, J., Gupta, M., Mitsunaga, T., Nayar, S.K.: Video from a single coded exposure photograph using a learned over-complete dictionary. In: 2011 International Conference on Computer Vision, pp. 287–294. IEEE (2011)
14. Huang, Y., Wang, W., Wang, L.: Video super-resolution via bidirectional recurrent convolutional networks. IEEE Trans. Pattern Anal. Mach. Intell. **40**(4), 1015–1028 (2018). https://doi.org/10.1109/TPAMI.2017.2701380
15. Iliadis, M., Spinoulas, L., Katsaggelos, A.K.: Deep fully-connected networks for video compressive sensing. Digit. Sig. Proc. **72**, 9–18 (2018). https://doi.org/10.1016/j.dsp.2017.09.010
16. Yang, J., et al.: Video compressive sensing using Gaussian mixture models. IEEE Trans. Image Process. **23**(11), 4863–4878 (2014)

17. Jaeger, H.: A tutorial on training recurrent neural networks, covering BPPT, RTRL, EKF and the "echo state network" approach (2005)
18. Jalali, S., Yuan, X.: Snapshot compressed sensing: performance bounds and algorithms. IEEE Trans. Inf. Theor. **65**(12), 8005–8024 (2019). https://doi.org/10.1109/TIT.2019.2940666
19. Jalali, S., Yuan, X.: Compressive imaging via one-shot measurements. In: IEEE International Symposium on Information Theory (ISIT) (2018)
20. Kingma, D., Ba, J.: Adam: a method for stochastic optimization. In: ICLR (2015)
21. Kulkarni, K., Lohit, S., Turaga, P., Kerviche, R., Ashok, A.: ReconNet: non-iterative reconstruction of images from compressively sensed random measurements. In: CVPR (2016)
22. Liu, Y., Yuan, X., Suo, J., Brady, D., Dai, Q.: Rank minimization for snapshot compressive imaging. IEEE Trans. Pattern Anal. Mach. Intell. **41**(12), 2990–3006 (2019)
23. Llull, P., et al.: Coded aperture compressive temporal imaging. Opt. Exp. **21**(9), 10526–10545 (2013). https://doi.org/10.1364/OE.21.010526
24. Llull, P., Yuan, X., Carin, L., Brady, D.J.: Image translation for single-shot focal tomography. Optica **2**(9), 822–825 (2015)
25. Lu, G., Ouyang, W., Xu, D., Zhang, X., Cai, C., Gao, Z.: DVC: an end-to-end deep video compression framework. In: CVPR (2019)
26. Ma, J., Liu, X., Shou, Z., Yuan, X.: Deep tensor ADMM-Net for snapshot compressive imaging. In: IEEE/CVF Conference on Computer Vision (ICCV) (2019)
27. Meng, Z., Ma, J., Yuan, X.: End-to-end low cost compressive spectral imaging with spatial-spectral self-attention. In: Vedaldi, A., Bischof, H., Brox, T., Frahm, J.M. (eds.) ECCV 2020. LNCS, vol. 12368, pp. 187–204. Springer, Cham (2020). https://doi.org/10.1007/978-3-030-58592-1_12
28. Meng, Z., Qiao, M., Ma, J., Yu, Z., Xu, K., Yuan, X.: Snapshot multispectral endomicroscopy. Opt. Lett. **45**(14), 3897–3900 (2020)
29. Mescheder, L., Nowozin, S., Geiger, A.: Which training methods for GANs do actually converge? In: International Conference on Machine Learning (ICML) (2018)
30. Miao, X., Yuan, X., Pu, Y., Athitsos, V.: λ-Net: reconstruct hyperspectral images from a snapshot measurement. In: IEEE/CVF Conference on Computer Vision (ICCV) (2019)
31. Mikolov, T., Karafiát, M., Burget, L., Cernockỳ, J., Khudanpur, S.: Recurrent neural network based language model. In: INTERSPEECH, vol. 2, p. 3 (2010)
32. Nah, S., Son, S., Lee, K.M.: Recurrent neural networks with intra-frame iterations for video deblurring. In: The IEEE Conference on Computer Vision and Pattern Recognition (CVPR) (June 2019)
33. Pont-Tuset, J., Perazzi, F., Caelles, S., Arbelaez, P., Sorkine-Hornung, A., Gool, L.V.: The 2017 DAVIS challenge on video object segmentation. CoRR abs/1704.00675 (2017). http://arxiv.org/abs/1704.00675
34. Qiao, M., Liu, X., Yuan, X.: Snapshot spatial-temporal compressive imaging. Opt. Lett. **45**(7), 1659–1662 (2020)
35. Qiao, M., Meng, Z., Ma, J., Yuan, X.: Deep learning for video compressive sensing. APL Photonics **5**(3), 030801 (2020). https://doi.org/10.1063/1.5140721
36. Reddy, D., Veeraraghavan, A., Chellappa, R.: P2c2: programmable pixel compressive camera for high speed imaging. In: CVPR 2011, pp. 329–336. IEEE (2011)
37. Ronneberger, O., Fischer, P., Brox, T.: U-Net: convolutional networks for biomedical image segmentation. In: Navab, N., Hornegger, J., Wells, W.M., Frangi, A.F. (eds.) MICCAI 2015. LNCS, vol. 9351, pp. 234–241. Springer, Cham (2015). https://doi.org/10.1007/978-3-319-24574-4_28

38. Roux, J.R.L., Weninger, J.: Deep unfolding: Model-based inspiration of novel deep architectures (2014)
39. Sun, Y., Yuan, X., Pang, S.: High-speed compressive range imaging based on active illumination. Opt. Exp. **24**(20), 22836–22846 (2016)
40. Sun, Y., Yuan, X., Pang, S.: Compressive high-speed stereo imaging. Opt. Exp. **25**(15), 18182–18190 (2017). https://doi.org/10.1364/OE.25.018182
41. Tsai, T.H., Llull, P., Yuan, X., Carin, L., Brady, D.J.: Spectral-temporal compressive imaging. Opt. Lett. **40**(17), 4054–4057 (2015)
42. Tsai, T.H., Yuan, X., Brady, D.J.: Spatial light modulator based color polarization imaging. Opt. Exp. **23**(9), 11912–11926 (2015)
43. Vaswani, A., et al.: Attention is all you need. In: Guyon, I., et al. (eds.) Advances in Neural Information Processing Systems, vol. 30, pp. 5998–6008. Curran Associates, Inc. (2017). http://papers.nips.cc/paper/7181-attention-is-all-you-need.pdf
44. Vaswani, A., et al.: Attention is all you need. In: Advances in Neural Information Processing Systems, pp. 5998–6008 (2017)
45. Ventura, C., Bellver, M., Girbau, A., Salvador, A., Marques, F., Giro-i Nieto, X.: RVOS: end-to-end recurrent network for video object segmentation. In: The IEEE Conference on Computer Vision and Pattern Recognition (CVPR) (June 2019)
46. Wang, Z., Bovik, A.C., Sheikh, H.R., Simoncelli, E.P., et al.: Image quality assessment: from error visibility to structural similarity. IEEE Trans. Image Process. **13**(4), 600–612 (2004)
47. Xie, J., Xu, L., Chen, E.: Image denoising and inpainting with deep neural networks. In: Pereira, F., Burges, C.J.C., Bottou, L., Weinberger, K.Q. (eds.) Advances in Neural Information Processing Systems, vol. 25, pp. 341–349. Curran Associates, Inc. (2012). http://papers.nips.cc/paper/4686-image-denoising-and-inpainting-with-deep-neural-networks.pdf
48. Xu, K., Ren, F.: CSVideoNet: A real-time end-to-end learning framework for high-frame-rate video compressive sensing. arXiv: 1612.05203 (December 2016)
49. Yang, J., Liao, X., Yuan, X., Llull, P., Brady, D.J., Sapiro, G., Carin, L.: Compressive sensing by learning a Gaussian mixture model from measurements. IEEE Trans. Image Process. **24**(1), 106–119 (2015)
50. Yang, Y., Sun, J., Li, H., Xu, Z.: Deep ADMM-Net for compressive sensing MRI. In: Lee, D.D., Sugiyama, M., Luxburg, U.V., Guyon, I., Garnett, R. (eds.) Advances in Neural Information Processing Systems, vol. 29, pp. 10–18. Curran Associates, Inc. (2016)
51. Yoshida, M., et al.: Joint optimization for compressive video sensing and reconstruction under hardware constraints. In: Ferrari, V., Hebert, M., Sminchisescu, C., Weiss, Y. (eds.) ECCV 2018. LNCS, vol. 11214, pp. 649–663. Springer, Cham (2018). https://doi.org/10.1007/978-3-030-01249-6_39
52. Yuan, X.: Generalized alternating projection based total variation minimization for compressive sensing. In: 2016 IEEE International Conference on Image Processing (ICIP), pp. 2539–2543 (September 2016)
53. Yuan, X., Brady, D., Katsaggelos, A.K.: Snapshot compressive imaging: Theory, algorithms and applications. IEEE Sig. Process. Mag. (2020)
54. Yuan, X., Liu, Y., Suo, J., Dai, Q.: Plug-and-play algorithms for large-scale snapshot compressive imaging. In: The IEEE/CVF Conference on Computer Vision and Pattern Recognition (CVPR) (June 2020)
55. Yuan, X., et al.: Low-cost compressive sensing for color video and depth. In: IEEE Conference on Computer Vision and Pattern Recognition (CVPR), pp. 3318–3325 (2014). https://doi.org/10.1109/CVPR.2014.424

56. Yuan, X., Pang, S.: Structured illumination temporal compressive microscopy. Biomed. Opt. Exp. **7**, 746–758 (2016)
57. Yuan, X., Pu, Y.: Parallel lensless compressive imaging via deep convolutional neural networks. Opt. Exp. **26**(2), 1962–1977 (2018)
58. Yuan, X., Tsai, T.H., Zhu, R., Llull, P., Brady, D., Carin, L.: Compressive hyperspectral imaging with side information. IEEE J. Sel. Top. Sig. Process. **9**(6), 964–976 (2015)
59. Zhang, K., Zuo, W., Chen, Y., Meng, D., Zhang, L.: Beyond a Gaussian denoiser: residual learning of deep CNN for image denoising. IEEE Trans. Image Process. **26**(7), 3142–3155 (2017). https://doi.org/10.1109/TIP.2017.2662206

Ultra Fast Structure-Aware Deep Lane Detection

Zequn Qin, Huanyu Wang, and Xi Li(✉)

College of Computer Science and Technology, Zhejiang University, Hangzhou, China
zequnqin@gmail.com, {huanyuhello,xilizju}@zju.edu.cn

Abstract. Modern methods mainly regard lane detection as a problem of pixel-wise segmentation, which is struggling to address the problem of challenging scenarios and speed. Inspired by human perception, the recognition of lanes under severe occlusion and extreme lighting conditions is mainly based on contextual and global information. Motivated by this observation, we propose a novel, simple, yet effective formulation aiming at extremely fast speed and challenging scenarios. Specifically, we treat the process of lane detection as a row-based selecting problem using global features. With the help of row-based selecting, our formulation could significantly reduce the computational cost. Using a large receptive field on global features, we could also handle the challenging scenarios. Moreover, based on the formulation, we also propose a structural loss to explicitly model the structure of lanes. Extensive experiments on two lane detection benchmark datasets show that our method could achieve the state-of-the-art performance in terms of both speed and accuracy. A light weight version could even achieve 300+ frames per second with the same resolution, which is at least 4x faster than previous state-of-the-art methods. Our code is available at https://github.com/cfzd/Ultra-Fast-Lane-Detection.

Keywords: Lane detection · Fast formulation · Structural loss · Anchor

1 Introduction

With a long research history in computer vision, lane detection is a fundamental problem and has a wide range of applications [8] (e.g., ADAS and autonomous driving). For lane detection, there are two kinds of mainstream methods, which

This work is supported by key scientific technological innovation research project by Ministry of Education, Zhejiang Provincial Natural Science Foundation of China under Grant LR19F020004, Baidu AI Frontier Technology Joint Research Program, and Zhejiang University K.P. Chao's High Technology Development Foundation.

Electronic supplementary material The online version of this chapter (https://doi.org/10.1007/978-3-030-58586-0_17) contains supplementary material, which is available to authorized users.

A. Vedaldi et al. (Eds.): ECCV 2020, LNCS 12369, pp. 276–291, 2020.
https://doi.org/10.1007/978-3-030-58586-0_17

are traditional image processing methods [1,2,28] and deep segmentation methods [11,21,22]. Recently, deep segmentation methods have made great success in this field because of great representation and learning ability. There are still some important and challenging problems to be addressed.

As a fundamental component of autonomous driving, the lane detection algorithm is heavily executed. This requires an extremely low computational cost of lane detection. Besides, present autonomous driving solutions are commonly equipped with multiple camera inputs, which typically demand lower computational cost for every camera input. In this way, a faster pipeline is essential to lane detection. For this purpose, SAD [9] is proposed to solve this problem by self-distilling. Due to the dense prediction property of SAD, which is based on segmentation, the method is computationally expensive.

Another problem of lane detection is called *no-visual-clue*, as shown in Fig. 1. Challenging scenarios with severe occlusion and extreme lighting conditions correspond to another key problem of lane detection. In this case, the lane detection urgently needs higher-level semantic analysis of lanes. Deep segmentation methods naturally have stronger semantic representation ability than conventional image processing methods, and become mainstream. Furthermore, SCNN [22] addresses this problem by proposing a message passing mechanism between adjacent pixels, which significantly improves the performance of deep segmentation methods. Due to the dense pixel-wise communication, this kind of message passing requires a even more computational cost.

Also, there exists a phenomenon that the lanes are represented as segmented binary features rather than lines or curves. Although deep segmentation methods dominate the lane detection fields, this kind of representation makes it difficult to explicitly utilize the prior information like rigidity and smoothness of lanes.

Fig. 1. Illustration of difficulties in lane detection. Most of challenging scenarios are severely occluded or distorted with various lighting conditions, resulting in little or no visual clues of lanes can be used for lane detection.

With the above motivations, we propose a novel lane detection formulation aiming at extremely fast speed and solving the *no-visual-clue* problem. Meanwhile, based on the proposed formulation, we present a structural loss to explicitly utilize prior information of lanes. Specifically, our formulation is proposed to **select locations of lanes at predefined rows of the image using global features** instead of segmenting every pixel of lanes based on a local receptive field, which significantly reduces the computational cost. The illustration of location selecting is shown in Fig. 2.

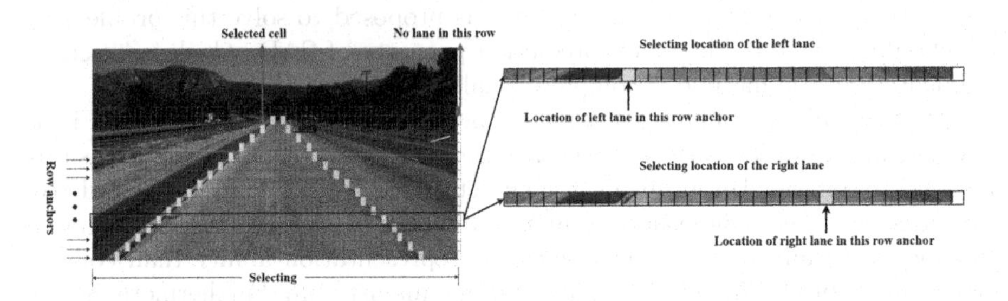

Fig. 2. Illustration of selecting on the left and right lane. In the right part, the selecting of a row is shown in detail. Row anchors are the predefined row locations, and our formulation is defined as horizontally selecting on each of row anchor. On the right of the image, a background gridding cell is introduced to indicate no lane in this row.

For the *no-visual-clue* problem, our method could also achieve good performance because our formulation is conducting the procedure of selecting in rows based on global features. With the aid of global features, our method has a receptive field of the whole image. Compared with segmentation based on a limited receptive field, visual clues and messages from different locations can be learned and utilized. In this way, our new formulation could solve the speed and the *no-visual-clue* problems simultaneously. Moreover, based on our formulation, lanes are represented as selected locations on different rows instead of the segmentation map. Hence, we can directly utilize the properties of lanes like rigidity and smoothness by optimizing the relations of selected locations, i.e., the structural loss. The contribution of this work can be summarized in three parts:

- We propose a novel, simple, yet effective formulation of lane detection aiming at extremely fast speed and solving the *no-visual-clue* problem. Compared with deep segmentation methods, our method is selecting locations of lanes instead of segmenting every pixel and works on the different dimensions, which is ultra fast. Besides, our method uses global features to predict, which has a larger receptive field than the segmentation formulation. In this way, the *no-visual-clue* problem can also be addressed.
- Based on the proposed formulation, we present a structural loss which explicitly utilizes prior information of lanes. To the best of our knowledge, this is the first attempt at optimizing such information explicitly in deep lane detection methods.

- The proposed method achieves the state-of-the-art performance in terms of both accuracy and speed on the challenging CULane dataset. A light weight version of our method could even achieve 300+ FPS with a comparable performance with the same resolution, which is at least 4 times faster than previous state-of-the-art methods.

2 Related Work

Traditional Methods. Traditional approaches usually solve the lane detection problem based on visual information. The main idea of these methods is to take advantage of visual clues through image processing like the HSI color model [25] and edge extraction algorithms [27,29]. When the visual information is not strong enough, tracking is another popular post-processing solution [13,28]. Besides tracking, Markov and conditional random fields [16] are also used as post-processing methods. With the development of machine learning, some methods [6,15,20] that adopt algorithms like template matching and support vector machines are proposed.

Deep Learning Models. With the development of deep learning, some methods [11,12] based on deep neural networks show the superiority in lane detection. These methods usually use the same formulation by treating the problem as a semantic segmentation task. For instance, VPGNet [17] proposes a multi-task network guided by vanishing points for lane and road marking detection. To use visual information more efficiently, SCNN [22] utilizes a special convolution operation in the segmentation module. It aggregates information from different dimensions via processing sliced features and adding them together one by one, which is similar to the recurrent neural networks. Some works try to explore light weight methods for real-time applications. Self-attention distillation (SAD) [9] is one of them. It applies an attention distillation mechanism, in which high and low layers' attentions are treated as teachers and students, respectively.

Besides the mainstream segmentation formulation, other formulations like Sequential prediction and clustering are also proposed. In [18], a long short-term memory (LSTM) network is adopted to deal with the long line structure of lanes. With the same principle, Fast-Draw [24] predicts the direction of lanes at each lane point, and draws them out sequentially. In [10], the problem of lane detection is regarded as clustering binary segments. The method proposed in [30] also uses a clustering approach to detect lanes. Different from the 2D view of previous works, a lane detection method in 3D formulation [4] is proposed to solve the problem of non-flatten ground.

3 Method

In this section, we describe the details of our method, including the new formulation and lane structural losses. Besides, a feature aggregation method for high-level semantics and low-level visual information is also depicted.

3.1 New Formulation for Lane Detection

As described in the introduction section, fast speed and the *no-visual-clue* problems are important for lane detection. Hence, how to effectively handle these problems is key to good performance. In this section, we show the derivation of our formulation by tackling the speed and the *no-visual-clue* problem. For a better illustration, Table 1 shows some notations used hereinafter.

Table 1. Notation.

Variable	Type	Definition
H	Scalar	Height of image
W	Scalar	Width of image
h	Scalar	Number of row anchors
w	Scalar	Number of gridding cells
C	Scalar	Number of lanes
X	Tensor	The global features of image
f	Function	The classifier for selecting lane locations
$P \in R^{C \times h \times (w+1)}$	Tensor	Group prediction
$T \in R^{C \times h \times (w+1)}$	Tensor	Group target
$Prob \in R^{C \times h \times w}$	Tensor	Probability of each location
$Loc \in R^{C \times h}$	Matrix	Locations of lanes

Definition of Our Formulation. In order to cope with the problems above, we propose to formulate lane detection to a **row-based selecting method based on global image features**. In other words, our method is selecting the correct locations of lanes on each predefined row using the global features. In our formulation, lanes are represented as a series of horizontal locations at predefined rows, i.e., row anchors. In order to represent locations, the first step is gridding. On each row anchor, the location is divided into many cells. In this way, the detection of lanes can be described as selecting certain cells over predefined row anchors, as shown in Fig. 3(a).

Suppose the maximum number of lanes is C, the number of row anchors is h and the number of gridding cells is w. Suppose X is the global image feature and f^{ij} is the classifier used for selecting the lane location on the i-th lane, j-th row anchor. Then the prediction of lanes can be written as:

$$P_{i,j,:} = f^{ij}(X), \text{ s.t. } i \in [1, C], j \in [1, h], \tag{1}$$

in which $P_{i,j,:}$ is the $(w+1)$-dimensional vector represents the probability of selecting $(w+1)$ gridding cells for the i-th lane, j-th row anchor. Suppose $T_{i,j,:}$ is the one-hot label of correct locations. Then, the optimization of our formulation corresponds to:

$$L_{cls} = \sum_{i=1}^{C} \sum_{j=1}^{h} L_{CE}(P_{i,j,:}, T_{i,j,:}), \tag{2}$$

in which L_{CE} is the cross entropy loss. We use an extra dimension to indicate the absence of lane, so our formulation is composed of $(w + 1)$-dimensional instead of w-dimensional classifications.

From Eq. 1 we can see that our method predicts the probability distribution of all locations on each row anchor based on global features. As a result, the correct location can be selected based on the probability distribution.

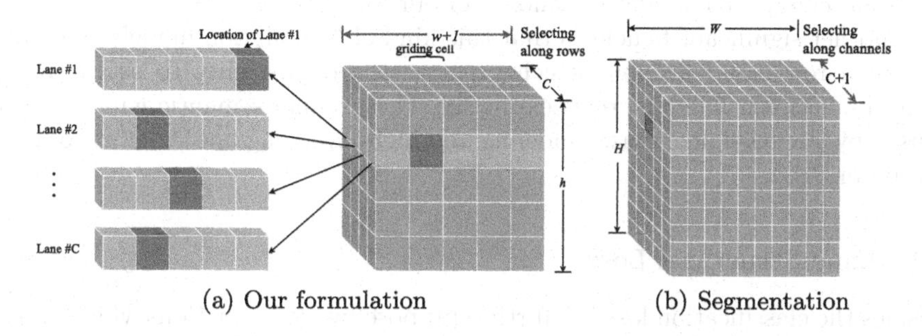

Fig. 3. Illustration of our formulation and segmentation. Our formulation is selecting locations (grids) on rows, while segmentation is classifying every pixel. The dimensions used for classifying are also different, which is marked in red. Besides, our formulation uses global features as input, which has larger receptive field than segmentation. (Color figure online)

How the Formulation Achieves Fast Speed. The differences between our formulation and segmentation are shown in Fig. 3. It can be seen that our formulation is much simpler than the commonly used segmentation. Suppose the image size is $H \times W$. In general, the number of predefined row anchors and gridding size are far less than the size of an image, that is to say, $h \ll H$ and $w \ll W$. In this way, the original segmentation formulation needs to conduct $H \times W$ classifications that are $(C + 1)$-dimensional, while our formulation only needs to solve $C \times h$ classification problems that are $(w+1)$-dimensional. In this way, the scale of computation can be reduced considerably because the computational cost of our formulation is $C \times h \times (w+1)$ while the one for segmentation is $H \times W \times (C + 1)$. For example, using the common settings of the CULane dataset [22], the ideal computational cost of our method is 1.7×10^4 calculations and the one for segmentation is 1.15×10^6 calculations. The computational cost is significantly reduced and our formulation could achieve extremely fast speed.

How the Formulation Handles the No-Visual-Clue Problem. In order to handle the *no-visual-clue* problem, utilizing information from other locations is important because *no-visual-clue* means no information at the target location. For example, a lane is occluded by a car, but we could still locate the lane by information from other lanes, road shape, and even car direction. In this way, utilizing information from other locations is key to solve the *no-visual-clue* problem, as shown in Fig. 1.

From the perspective of the receptive field, our formulation has a receptive field of the whole image, which is much bigger than segmentation methods. The context information and messages from other locations of the image can be utilized to address the *no-visual-clue* problem. From the perspective of learning, prior information like shape and direction of lanes can also be learned using structural loss based on our formulation, as shown in Sect. 3.2. In this way, the *no-visual-clue* problem can be handled in our formulation.

Another significant benefit is that this kind of formulation models lane location in a row-based fashion, which gives us the opportunity to establish the relations between different rows explicitly. The original semantic gap, which is caused by low-level pixel-wise modeling and high-level long line structure of lane, can be bridged.

3.2 Lane Structural Loss

Besides the classification loss, we further propose two loss functions which aim at modeling location relations of lane points. In this way, the learning of structural information can be encouraged.

The first one is derived from the fact that lanes are continuous, that is to say, the lane points in adjacent row anchors should be close to each other. In our formulation, the location of the lane is represented by a classification vector. So the continuous property is realized by constraining the distribution of classification vectors over adjacent row anchors. In this way, the similarity loss function can be:

$$L_{sim} = \sum_{i=1}^{C} \sum_{j=1}^{h-1} \|P_{i,j,:} - P_{i,j+1,:}\|_1 , \tag{3}$$

in which $P_{i,j,:}$ is the prediction on the j-th row anchor and $\|\cdot\|_1$ represents L_1 norm.

Another structural loss function focuses on the shape of lanes. Generally speaking, most of the lanes are straight. Even for the curve lane, the majority of it is still straight due to the perspective effect. In this work, we use the second-order difference equation to constrain the shape of the lane, which is zero for the straight case.

To consider the shape, the location of the lane on each row anchor needs to be calculated. The intuitive idea is to obtain locations from the classification prediction by finding the maximum response peak. For any lane index i and row anchor index j, the location $Loc_{i,j}$ can be represented as:

$$Loc_{i,j} = \underset{k}{\operatorname{argmax}}\ P_{i,j,k} , \ \text{s.t.} \ \ k \in [1, w] \tag{4}$$

in which k is an integer representing the location index. It should be noted that we do not count in the background gridding cell and the location index k only ranges from 1 to w, instead of $w+1$.

However, the *argmax* function is not differentiable and can not be used with further constraints. Besides, in the classification formulation, classes have no apparent order and are hard to set up relations between different row anchors. To solve this problem, we propose to use the expectation of predictions as an approximation of location. We use the softmax function to get the probability of different locations:

$$Prob_{i,j,:} = softmax(P_{i,j,1:w}), \tag{5}$$

in which $P_{i,j,1:w}$ is a w-dimensional vector and $Prob_{i,j,:}$ represents the probability at each location. For the same reason as Eq. 4, background gridding cell is not included and the calculation only ranges from 1 to w. Then, the expectation of locations can be written as:

$$Loc_{i,j} = \sum_{k=1}^{w} k \cdot Prob_{i,j,k} \tag{6}$$

in which $Prob_{i,j,k}$ is the probability of the i-th lane, the j-th row anchor, and the k-th location. The benefits of this localization method are twofold. The first one is that the expectation function is differentiable. The other is that this operation recovers the continuous location with the discrete random variable.

According to Eq. 6, the second-order difference constraint can be written as:

$$\begin{aligned} L_{shp} = \sum_{i=1}^{C}\sum_{j=1}^{h-2} \|&(Loc_{i,j} - Loc_{i,j+1}) \\ &- (Loc_{i,j+1} - Loc_{i,j+2})\|_1, \end{aligned} \tag{7}$$

in which $Loc_{i,j}$ is the location on the i-th lane, the j-th row anchor. The reason why we use the second-order difference instead of the first-order difference is that the first-order difference is not zero in most cases. So the network needs extra parameters to learn the distribution of the first-order difference of lane location. Moreover, the constraint of the second-order difference is relatively weaker than that of the first-order difference, thus resulting in less influence when the lane is not straight. Finally, the overall structural loss can be:

$$L_{str} = L_{sim} + \lambda L_{shp}, \tag{8}$$

in which λ is the loss coefficient.

3.3 Feature Aggregation

In Sect. 3.2, the loss design mainly focuses on the intra-relations of lanes. In this section, we propose an auxiliary feature aggregation method that performs

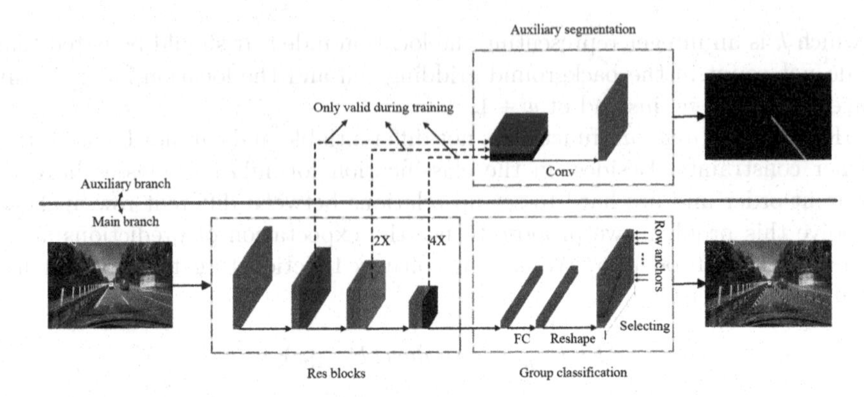

Fig. 4. Overall architecture. The auxiliary branch is shown in the upper part, which is only valid when training. The feature extractor is shown in the blue box. The classification-based prediction and auxiliary segmentation task are illustrated in the green and orange boxes, respectively. (Color figure online)

on the global context and local features. An auxiliary segmentation task utilizing multi-scale features is proposed to model local features. It should be noted that our method only uses the auxiliary segmentation task in the training phase, and it would be removed in the testing phase. We use cross entropy as our auxiliary segmentation loss. In this way, the overall loss of our method can be written as:

$$L_{total} = L_{cls} + \alpha L_{str} + \beta L_{seg}, \tag{9}$$

in which L_{seg} is the segmentation loss, α and β are loss coefficients. The overall architecture can be seen in Fig. 4.

4 Experiments

In this section, we demonstrate the effectiveness of our method with extensive experiments. The following sections mainly focus on three aspects: 1) Experimental settings. 2) Ablation studies. 3) Results on two lane detection datasets.

4.1 Experimental Setting

Datasets. To evaluate our method, we conduct experiments on two widely used benchmark datasets: TuSimple [26] and CULane [22] datasets. TuSimple dataset is collected with stable lighting conditions in highways. CULane dataset consists of nine different scenarios. The detailed information can be seen in Table 2.

Evaluation metrics. For TuSimple dataset, the main evaluation metric is accuracy. The accuracy is calculated by:

$$accuracy = \frac{\sum_{clip} C_{clip}}{\sum_{clip} S_{clip}}, \tag{10}$$

in which C_{clip} and S_{clip} are the number of correct predictions and ground truth.

For CULane, each lane is treated as a 30-pixel-width line. The intersection-over-union (IoU) is computed between ground truth and predictions. F1-measure is taken as the evaluation metric and formulated as follows:

$$F1 - measure = \frac{2 \times Precision \times Recall}{Precision + Recall}, \tag{11}$$

where $Precision = \frac{TP}{TP+FP}$, $Recall = \frac{TP}{TP+FN}$, TP is the true positive, FP is the false positive, and FN is the false negative.

Table 2. Datasets description

Dataset	#Frame	Train	Validation	Test	Resolution	#Lane	Environment
TuSimple	6,408	3,268	358	2,782	1280×720	≤ 5	Highway
CULane	133,235	88,880	9,675	34,680	1640×590	≤ 4	Urban and highway

Implementation Details. For both datasets, we use the row anchors that are defined by the dataset. Specifically, the row anchors of Tusimple range from 160 to 710 with a step of 10. The counterpart of CULane ranges from 260 to 530. The number of gridding cells is set to 100 on the Tusimple dataset and 150 on the CULane dataset. The corresponding ablation study on the Tusimple dataset can be seen in Sect. 4.2.

In the optimizing process, images are resized to 288×800 following [22]. We use Adam [14] to train our model with cosine decay learning rate strategy [19] initialized with 4e−4. Loss coefficients λ, α and β in Eq. 8 and 9 are all set to 1. The batch size is set to 32, and the total number of training epochs is set 100 for TuSimple dataset and 50 for CULane dataset. All models are trained and tested with pytorch [23] and nvidia GTX 1080Ti GPU.

Data Augmentation. Due to the inherent structure of lanes, a classification-based network could easily over-fit the training set and show poor performance on the validation set. To prevent this phenomenon and gain generalization ability, we use an augmentation method composed of rotation, vertical and horizontal shift. Besides, to preserve the lane structure, the lane is extended or cropped till the boundary of the image. The results of augmentation can be seen in Fig. 5.

4.2 Ablation Study

In this section, we verify our method with several ablation studies. The experiments are all conducted with the same settings as Sect. 4.1.

Effects of Number of Gridding Cells. As described in Sect. 3.1, we use gridding and selecting to establish the relations between structural information in lanes and classification-based formulation. We use different numbers of gridding

(a) Original anaotation (b) Augmentated result

Fig. 5. Demonstration of augmentation. The lane on the right image is extended to maintain the lane structure, which is marked with red ellipse. (Color figure online)

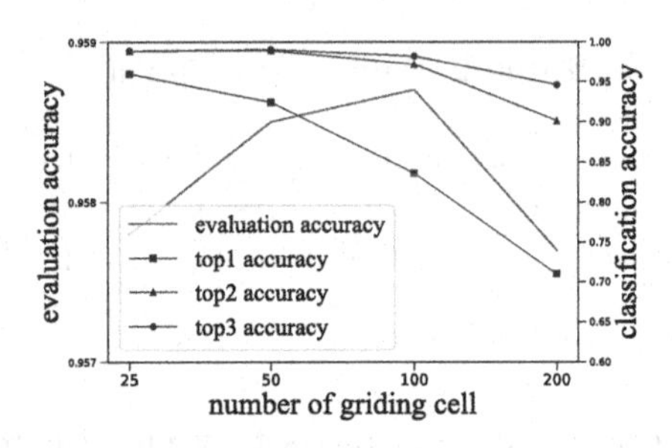

Fig. 6. Performance under different numbers of gridding cells. Evaluation accuracy is the metric of Tusimple, and classification accuracy is the standard accuracy.

cells to demonstrate the effects on our method. We divide the image using 25, 50, 100 and 200 cells in columns. The results can be seen in Fig. 6.

With the increase of the number of gridding cells, we can see that both top1, top2 and top3 classification accuracy drops gradually. It is because more gridding cells require finer-grained and harder classification. However, the evaluation accuracy is not strictly monotonic. Although a smaller number of gridding cells means higher classification accuracy, the localization error would be larger, since the gridding cell is too large to represent precise location. In this work, we choose 100 as our number of gridding cells on the Tusimple Dataset.

Effectiveness of Localization Methods. Since our method formulates the lane detection as a group classification problem, one natural question is what are the differences between classification and regression. To test in an regression manner, we replace the classification head with a similar regression head. We use four experimental settings, which are respectively REG, REG Norm, CLS and CLS Exp. CLS means the classification method, while REG means the regression method. CLS Exp is the classification method with Eq. 6. The REG Norm setting is a variant of REG, which normalizes the learning target within $[0, 1]$.

The results are shown in Table 3. We can see that classification with the expectation could gain better performance than the standard method. This result also proves the analysis in Eq. 6 that the expectation based localization is more

Table 3. Comparison between classification and regression on the Tusimple dataset. REG and REG Norm are regression methods. The ground truth of REG Norm is normalized. CLS is classification method and CLS Exp is the one with Eq. 6.

Type	REG	REG Norm	CLS	CLS Exp
Accuracy	71.59	67.24	95.77	95.87

precise than *argmax* operation. Meanwhile, classification-based methods could consistently outperform the regression-based methods.

Effectiveness of the Proposed Modules. To verify the effectiveness of the proposed modules, we conduct both qualitative and quantitative experiments.

First, we show the quantitative results of our modules. As shown in Table 4, the experiments of different module combinations are carried out.

Table 4. Experiments of the proposed modules on Tusimple benchmark with Resnet-34 backbone. Baseline stands for conventional segmentation formulation.

Baseline	New formulation	Structural loss	Feature aggregation	Accuracy
✓				92.84
	✓			95.64 (+2.80)
	✓	✓		95.96 (+3.12)
	✓		✓	95.98 (+3.14)
	✓	✓	✓	96.06 (+3.22)

From Table 4, we can see that the new formulation gains significant performance improvement compared with segmentation formulation. Besides, both lane structural loss and feature aggregation could enhance the performance.

Second, we illustrate the effectiveness of lane similarity loss in Eq. 3. The results are shown in Fig. 7. We can see that similarity loss makes the classification prediction smoother and thus gains better performance.

4.3 Results

In this section, we show the results on the Tusimple and the CULane datasets. In these experiments, Resnet-18 and Resnet-34 [7] are used as backbone models.

For the Tusimple dataset, seven methods are used for comparison, including Res18-Seg [3], Res34-Seg [3], LaneNet [21], EL-GAN [5], SCNN [22] and SAD [9]. Both Tusimple evaluation accuracy and runtime are compared in this experiment. The runtime of our method is recorded with the average time for 100 runs. The results are shown in Table 5.

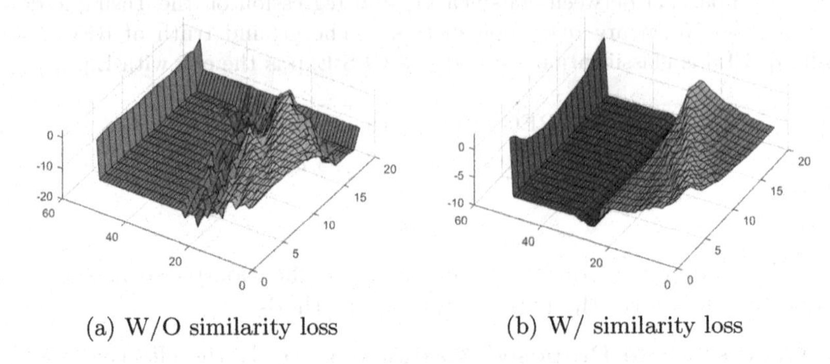

(a) W/O similarity loss (b) W/ similarity loss

Fig. 7. Qualitative comparison of similarity loss. The predicted distributions of group classification of the same lane are shown. Figure (a) shows the visualization of distribution without similarity loss, while Fig. (b) shows the counterpart with similarity loss.

From Table 5, we can see that our method achieves comparable performance with state-of-the-art methods while our method could run extremely fast. Compared with SCNN, our method could infer 41.7 times faster. Even compared with the second-fastest network SAD, our method is still more than 2 times faster.

Table 5. Comparison with other methods on TuSimple test set.

Method	Accuracy	Runtime (ms)	Multiple
Res18-Seg [3]	92.69	25.3	5.3x
Res34-Seg [3]	92.84	50.5	2.6x
LaneNet [21]	96.38	19.0	7.0x
EL-GAN [5]	96.39	>100	<1.3x
SCNN [22]	96.53	133.5	1.0x
SAD [9]	**96.64**	13.4	10.0x
Res34-Ours	96.06	5.9	22.6x
Res18-Ours	95.87	**3.2**	**41.7x**

For the CULane dataset, four methods, including Seg [3], SCNN [22], FastDraw [24] and SAD [9], are used for comparison. F1-measure and runtime are compared. The results can be seen in Table 6.

It is observed in Table 6 that our method achieves the best performance in terms of both accuracy and speed. It proves the effectiveness of the proposed formulation and structural loss on these challenging scenarios because our method could utilize global and structural information to address the *no-visual-clue* and speed problem. The fastest model of our formulation achieves 322.5 FPS with a resolution of 288×800, which is the same as other compared methods.

The visualizations of our method on the Tusimple and CULane datasets are shown in Fig. 8. We can see our method performs well under various conditions.

Table 6. Comparison of F1-measure and runtime on CULane testing set with IoU threshold = 0.5. For crossroad, only false positives are shown. The less, the better.

Category	R50-Seg [3]	SCNN [22]	FD-50 [24]	R34-SAD	SAD [9]	Res18-Ours	Res34-Ours
Normal	87.4	90.6	85.9	89.9	90.1	87.7	**90.7**
Crowded	64.1	69.7	63.6	68.5	68.8	66.0	**70.2**
Night	60.6	66.1	57.8	64.6	66.0	62.1	**66.7**
No-line	38.1	43.4	40.6	42.2	41.6	40.2	**44.4**
Shadow	60.7	66.9	59.9	67.7	65.9	62.8	**69.3**
Arrow	79.0	84.1	79.4	83.8	84.0	81.0	**85.7**
Dazzlelight	54.1	58.5	57.0	59.9	**60.2**	58.4	59.5
Curve	59.8	64.4	65.2	66.0	65.7	57.9	**69.5**
Crossroad	2505	1990	7013	1960	1998	**1743**	2037
Total	66.7	71.6	-	70.7	70.8	68.4	**72.3**
Runtime (ms)	-	133.5	-	50.5	13.4	3.1	5.7
Multiple	-	1.0x	-	2.6x	10.0x	43.0x	23.4x
FPS	-	7.5	-	19.8	74.6	322.5	175.4

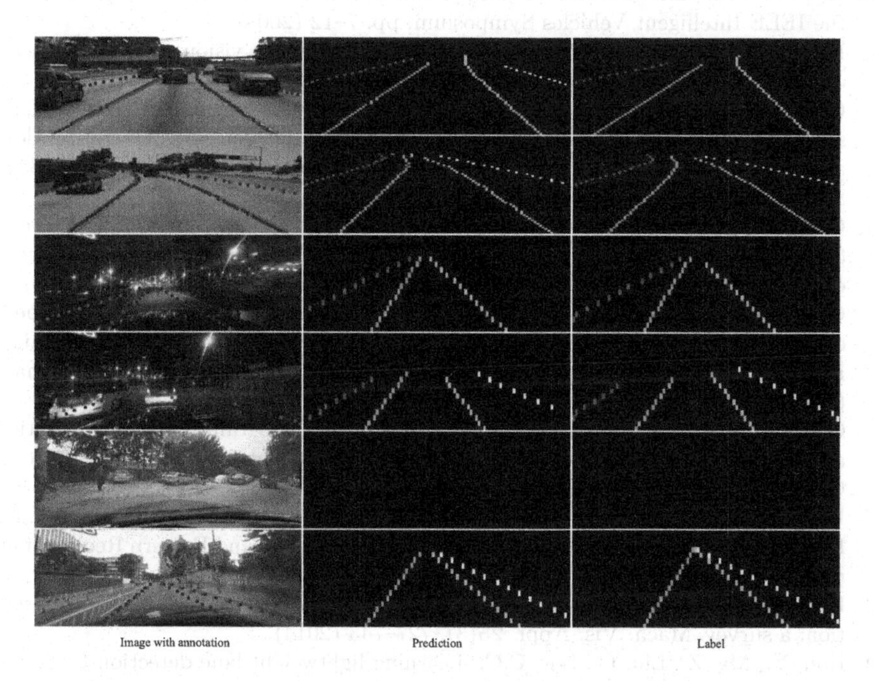

Fig. 8. Visualization results. The first two rows are results on the Tusimple dataset and the rest rows are on the CULane dataset. From left to right, the results are image, prediction and label. In the image, predictions are marked in blue and ground truth are marked in red. Because our method only predicts on the predefined row anchors, the scales of images and labels in the vertical direction are not identical. (Color figure online)

5 Conclusion

In this paper, we have proposed a novel formulation with structural loss and achieves remarkable speed and accuracy. The proposed formulation regards lane detection as a problem of row-based selecting using global features. In this way, the problem of speed and *no-visual-clue* can be addressed. Besides, structural loss used for explicitly modeling of lane prior information is also proposed. The effectiveness of our formulation and structural loss are well justified with both qualitative and quantitative experiments. Especially, our model using Resnet-34 backbone could achieve state-of-the-art accuracy and speed. A light weight Resnet-18 version of our method could even achieve 322.5 FPS with a comparable performance at the same resolution.

References

1. Aly, M.: Real time detection of lane markers in urban streets. In: Proceedings of the IEEE Intelligent Vehicles Symposium, pp. 7–12 (2008)
2. Bertozzi, M., Broggi, A.: GOLD: a parallel real-time stereo vision system for generic obstacle and lane detection. IEEE Trans. Image Process. **7**(1), 62–81 (1998)
3. Chen, L.C., Papandreou, G., Kokkinos, I., Murphy, K., Yuille, A.L.: DeepLab: semantic image segmentation with deep convolutional nets, atrous convolution, and fully connected CRFs. IEEE Trans. Pattern Anal. Mach. Intell. **40**(4), 834–848 (2017)
4. Garnett, N., Cohen, R., Pe'er, T., Lahav, R., Levi, D.: 3D-LaneNet: end-to-end 3D multiple lane detection. In: Proceedings of the IEEE International Conference on Computer Vision, pp. 2921–2930 (2019)
5. Ghafoorian, M., Nugteren, C., Baka, N., Booij, O., Hofmann, M.: EL-GAN: embedding loss driven generative adversarial networks for lane detection. In: Leal-Taixé, L., Roth, S. (eds.) ECCV 2018. LNCS, vol. 11129, pp. 256–272. Springer, Cham (2019). https://doi.org/10.1007/978-3-030-11009-3_15
6. Gonzalez, J.P., Ozguner, U.: Lane detection using histogram-based segmentation and decision trees. In: Proceedings of the IEEE Intelligent Transportation Systems Conference, pp. 346–351 (2000)
7. He, K., Zhang, X., Ren, S., Sun, J.: Deep residual learning for image recognition. In: Proceedings of the IEEE Conference on Computer Vision and Pattern Recognition, pp. 770–778 (2016)
8. Hillel, A.B., Lerner, R., Levi, D., Raz, G.: Recent progress in road and lane detection: a survey. Mach. Vis. Appl. **25**(3), 727–745 (2014)
9. Hou, Y., Ma, Z., Liu, C., Loy, C.C.: Learning lightweight lane detection CNNs by self attention distillation. In: Proceedings of the IEEE International Conference on Computer Vision, pp. 1013–1021 (2019)
10. Hsu, Y.C., Xu, Z., Kira, Z., Huang, J.: Learning to cluster for proposal-free instance segmentation. In: Proceedings of the International Joint Conference on Neural Networks, pp. 1–8 (2018)
11. Huval, B., et al.: An empirical evaluation of deep learning on highway driving. arXiv preprint arXiv:1504.01716 (2015)

12. Kim, J., Lee, M.: Robust lane detection based on convolutional neural network and random sample consensus. In: Loo, C.K., Yap, K.S., Wong, K.W., Teoh, A., Huang, K. (eds.) ICONIP 2014. LNCS, vol. 8834, pp. 454–461. Springer, Cham (2014). https://doi.org/10.1007/978-3-319-12637-1_57
13. Kim, Z.: Robust lane detection and tracking in challenging scenarios. IEEE Trans. Intell. Transp. Syst. **9**(1), 16–26 (2008)
14. Kingma, D.P., Ba, J.: Adam: A method for stochastic optimization. arXiv preprint arXiv:1412.6980 (2014)
15. Kluge, K., Lakshmanan, S.: A deformable-template approach to lane detection. In: Proceedings of the Intelligent Vehicles Symposium, pp. 54–59 (1995)
16. Krähenbühl, P., Koltun, V.: Efficient inference in fully connected CRFs with Gaussian edge potentials. In: Advances in Neural Information Processing Systems, pp. 109–117 (2011)
17. Lee, S., et al.: VPGNet: vanishing point guided network for lane and road marking detection and recognition. In: Proceedings of the IEEE International Conference on Computer Vision (2017)
18. Li, J., Mei, X., Prokhorov, D., Tao, D.: Deep neural network for structural prediction and lane detection in traffic scene. IEEE Trans. Neural Netw. Learn. Syst. **28**(3), 690–703 (2016)
19. Loshchilov, I., Hutter, F.: Sgdr: Stochastic gradient descent with warm restarts. arXiv preprint arXiv:1608.03983 (2016)
20. Mandalia, H.M., Salvucci, M.D.D.: Using support vector machines for lane-change detection. Proc. Hum. Factors Ergon. Soc. Annu. Meet. **49**, 1965–1969 (2005)
21. Neven, D., De Brabandere, B., Georgoulis, S., Proesmans, M., Van Gool, L.: Towards end-to-end lane detection: an instance segmentation approach. In: Proceedings of the IEEE Intelligent Vehicles Symposium, pp. 286–291 (2018)
22. Pan, X., Shi, J., Luo, P., Wang, X., Tang, X.: Spatial as deep: spatial CNN for traffic scene understanding. In: Proceedings of the AAAI Conference on Artificial Intelligence, pp. 7276–7283 (2018)
23. Paszke, A., et al.: Automatic differentiation in PyTorch (2017)
24. Philion, J.: FastDraw: addressing the long tail of lane detection by adapting a sequential prediction network. In: Proceedings of the IEEE Conference on Computer Vision and Pattern Recognition, pp. 11582–11591 (2019)
25. Sun, T.Y., Tsai, S.J., Chan, V.: HSI color model based lane-marking detection. In: Proceedings of the IEEE Intelligent Transportation Systems Conference, pp. 1168–1172 (2006)
26. TuSimple: TuSimple benchmark. https://github.com/TuSimple/tusimple-benchmark. Accessed Nov 2019
27. Wang, Y., Shen, D., Teoh, E.K.: Lane detection using spline model. Pattern Recogn. Lett. **21**(8), 677–689 (2000)
28. Wang, Y., Teoh, E.K., Shen, D.: Lane detection and tracking using b-snake. Image Vis. Comput. **22**(4), 269–280 (2004)
29. Yu, B., Jain, A.K.: Lane boundary detection using a multi-resolution Hough transform. Proc. Int. Conf. Image Process. **2**, 748–751 (1997)
30. Yuenan, H.: Agnostic lane detection. arXiv preprint arXiv:1905.03704 (2019)

Cross-Identity Motion Transfer for Arbitrary Objects Through Pose-Attentive Video Reassembling

Subin Jeon[1(✉)], Seonghyeon Nam[1], Seoung Wug Oh[1], and Seon Joo Kim[1,2]

[1] Yonsei University, Sinchon-dong, South Korea
{subinjeon,shnnam,sw.oh,seonjookim}@yonsei.ac.kr
[2] Facebook, Menlo Park, USA

Abstract. We propose an attention-based networks for transferring motions between arbitrary objects. Given a source image(s) and a driving video, our networks animate the subject in the source images according to the motion in the driving video. In our attention mechanism, dense similarities between the learned keypoints in the source and the driving images are computed in order to retrieve the appearance information from the source images. Taking a different approach from the well-studied warping based models, our attention-based model has several advantages. By reassembling non-locally searched pieces from the source contents, our approach can produce more realistic outputs. Furthermore, our system can make use of multiple observations of the source appearance (e.g. front and sides of faces) to make the results more accurate. To reduce the training-testing discrepancy of the self-supervised learning, a novel cross-identity training scheme is additionally introduced. With the training scheme, our networks is trained to transfer motions between different subjects, as in the real testing scenario. Experimental results validate that our method produces visually pleasing results in various object domains, showing better performances compared to previous works.

Keywords: Motion transfer · Generative adversarial network · Video to video translation

1 Introduction

Motion transfer is a task of transferring motion between different subjects. In other words, it generates a video conditioned on the appearance extracted from source image(s) and the motion patterns from a driving video. Here, the appearance means an identity of a subject observed in source image(s) and the motion

Electronic supplementary material The online version of this chapter (https://doi.org/10.1007/978-3-030-58586-0_18) contains supplementary material, which is available to authorized users.

A. Vedaldi et al. (Eds.): ECCV 2020, LNCS 12369, pp. 292–308, 2020.
https://doi.org/10.1007/978-3-030-58586-0_18

patterns refer to a sequence of poses that change continuously. Motion transfer has long been studied for its practical applications such as video editing and virtual/augmented reality. Recent progress in deep generative model has shown the capability of generating highly realistic images, and it has led to vigorous attempts to transfer motions in learning basis [3,6,25,32,34,35].

The majority of previous learning-based approaches relies on pre-computed (or given) landmark annotations to represent poses [6,19,28,34]. With an accurate pose representation, those methods have achieved impressive results especially in the human face and body domains where the keypoint extraction is well-studied. However, the dependency on pre-computed keypoints limits their application. For example, [19,28] are only applicable to specific domains (e.g. face and body), and [6,34] work only for specific subjects.

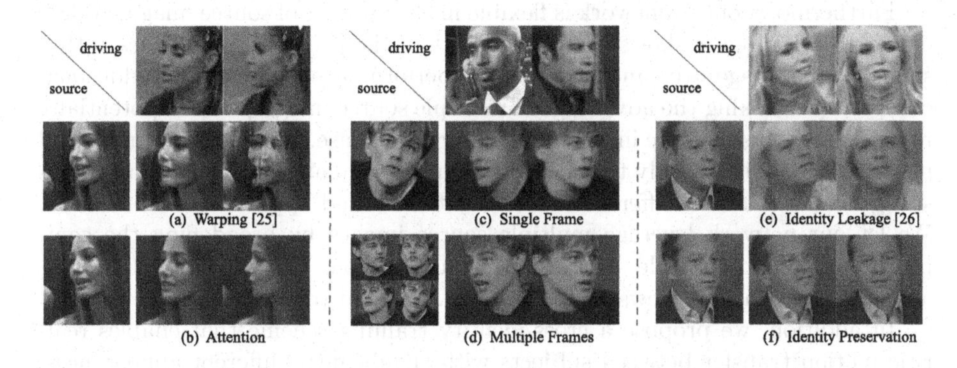

Fig. 1. (a) Previous motion transfer methods is based on spatial warping of single image. (b) We propose a non-local search based approach that handles a wide range of motion. (c) A single source image contains insufficient appearance data. (d) Our method can utilize multiple sources with various views. (e) Self-supervision causes identity-leakage in motion transfer. (f) We propose a cross-identity training which is effective in preserving identities of a source image.

To overcome the above issue, several attempts have been made to transfer object motions without using pre-defined pose representations [25,26,35] recently. They loosened the constraint on requiring the pre-defined keypoints by employing the self-supervised learning scheme for the pose representation [9], enabling applications beyond specific object categories. To learn to extract representations and transfer motion simultaneously, previous methods commonly take warping-based approach to synthesize images in various poses from a source image. However, we argue that the warping operation is not the best option for this task due to several limitations. In this paper, we highlight those limitations and propose a novel attention-based approach for arbitrary motion transfer.

While the warping operation works well for handling small motions, it has some challenges in achieving realistic synthesis. First, it is difficult to model large and complex motions. For example, it is more difficult to create a side view than to raise head slightly from the source image with a front view, as illustrated in Fig. 1(a). Second, it is hard to make use of more than one source image. Most of previous works cannot use more than one source image by its design [19,25,28], and some methods rely on heuristics to combine results obtained with several source images [35]. Observations from different poses and views are essential in completely recovering occluded parts in various views. As shown in Fig. 1(d), using multiple sources is beneficial in synthesizing various views.

We propose a novel attention-based method to transfer a variety of motion from arbitrary object categories. The main idea of our method is to reassemble the source image by finding proper visual appearance in the source image for each part of the driving pose. Compared with the warping-based approaches, our method produces more realistic output as it non-locally searches the best source content to synthesize every parts of an image.

Furthermore, our framework is flexible in the number of source images. While our approach already shows better results than the previous approaches with a single source image, we can maximize the performance by providing additional observations. Taking the advantage of multiple source images, we can potentially resolve the occlusion issue in the single observation case. Previous warping-based methods are designed only to use single source image and they focus on synthesizing unseen parts or different views with the limited appearance information. We let our network leverage multiple source images by broadening the comparison region to multiple source images and it shows the capability to utilize appearance in various views and poses (Fig. 1(c) and (d)).

In addition, we propose a cross-identity training scheme that enables realistic motion transfer between subjects with significantly different appearances. Previously, motion transfer networks are usually trained by transferring poses in the same subject even though the actual task is to transfer poses between different subjects. While this training trick is required to learn the networks in an unsupervised way, it is not appropriate for the real testing scenario. For example, models trained in this way often fails to maintain the personal shape that is unique for each subject independent from its poses (Fig. 1(e)). With the proposed training scheme, the network can learn to perform the motion transfer between different subjects, as in the real testing, without additional supervision. Furthermore, our cross-identity reconstruction loss encourages the network to preserve the identity-specific property (e.g. the shape of a human face) after the motion transfer.

To sum up, our contribution can be summarized as follows:

- We propose a pose-attentive video reassembling network for motion transfer of arbitrary objects. Our network implicitly learns the pose representation specialized for transferring motion without annotation. Through the non-local search based reassembling of visual pieces, we overcome the limitation of warping based motion transfers.
- Our framework can naturally process and take advantage of multiple images or video as the source, which helps to synthesize more realistic outputs.

- We propose a cross-identity training method to preserve identity-relevant properties such as object shape while changing the pose. This training strategy matches the testing scenarios, therefore, is more realistic compared to previous works that train on same subjects.

2 Related Work

Remarkable achievements in the generative modelling [12] have given rise to interests in video generation, leading to extensive development of video generation methods [8,24,29–31,37]. As a type of video generation, motion transfer refers to the task of synthesizing motion extracted from a driving video on different subjects. In this section, we address major approaches of motion transfer.

Video-to-Video Translation. Motion transfer has been addressed from the perspective of video-to-video translation. This class of works [3,6,34] proposed methods to learn a mapping between two videos in order to convert a domain of videos into another. In those frameworks, the generator memorizes diverse aspects of the source appearance using a sequence of frames in the source video and learns to transfer the appearance on the driving video. These methods have the advantage of utilizing the rich appearance information of diverse pose and view in the source video, making realistic results. However, they require time-consuming training per each subject using a large amount of source images related to the subject to train the subject-specific generator.

Pose Guided Image Generation. This class refers to the image generation conditioned on a source image and a driving pose. In order to represent the pose to condition on, they use keypoint annotations or off-the-shelf keypoint detector. Using the keypoint annotations and a single source image, many works [2,10,17,19,20,27,28] attempted to generate the source subject in a novel pose by warping a single source image. This approach has strength in transferring motion on new subjects at the test time, but experiences difficulties caused by insufficient appearance information in the single source image. Recently, multi-source based methods [13,18,32,36] have emerged to overcome the drawback of a single source image. These works are similar to our work in that they can transfer motion on new subjects with multiple source images. However, we tackle more challenging case because we need to extract pose from images directly and generalize to various object categories, whereas this stream of works is restricted to a specific object category (human body or face).

Motion Transfer for Arbitrary Objects. This class of works learns to extract pose representation implicitly from driving frames to transfer motion, extending object scopes to arbitrary objects. X2Face [35] and Monkey-net [25] transfer motion by learning to warp a source image to pose of a driving image in unsupervised way. These warping-based method showed superiority in transferring local motion, but experienced difficulties in synthesizing motion when the movement becomes larger and more complex. To solve the large motion problem, Siarohin [26] et al. enhanced Monkey-net by taking the local movement into consideration using local affine transformation of keypoints. Our method also falls in to this category of transferring motion on arbitrary objects. Different from previous works, we propose a non-local motion matching mechanism to capture drastic motion changes in order to overcome the limitation of previous warping-based approaches.

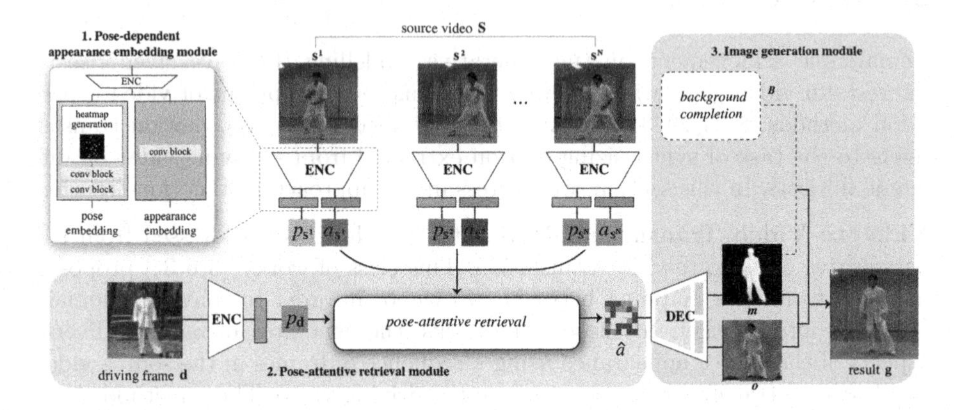

Fig. 2. Overview of our work. It consists of (1) pose-dependent appearance embedding module, (2) pose-attentive retrieval module, and (3) image generation module. The pose-dependent appearance embedding module takes frames in the source video to embed (pose, appearance) pairs. Pose attention block retrieves appearance correspondent to the driving pose. The driving pose is extracted using the shared pose embedding network. Finally, image generation module generates the results.

3 Method

3.1 Overview

Given a source video $\mathbf{S} = \{\mathbf{s}^n\}_{n=1,2,\ldots,N}$ and a driving video $\mathbf{D} = \{\mathbf{d}^t\}_{t=1,2,\ldots,T}$, our task is to synthesize a video where the motion of the foreground object is similar to that of $\mathbf{D}$ while the appearance is same as that of $\mathbf{S}$. Note that we deal with both cases of the single source frame $N = 1$ and the multiple source frames $N > 1$.

The overall architecture is illustrated in Fig. 2. Our model consists of following modules: (1) pose-dependent appearance embedding module, (2) pose-attentive retrieval module, and (3) image generation module. The pose-dependent appearance embedding module takes S as input, and extracts the pose and the appearance representations for each frame in S independently, using a shared embedding network. The pose attentive retrieval module extracts the pose from the driving frame, and searches for proper appearance features in the source frames using our spatio-temporal attention mechanism. Finally, our decoder synthesizes outputs using the appearance features and one of the source images for the background.

Our network is trained using a large number of videos containing subjects in the same object category. At each iteration, the network is trained using two videos with different identities by minimizing the self-reconstruction loss and the cross-identity reconstruction loss. After training, it generates output video frame conditioned on S and each frame d^t in the driving video. In the following, we describe our method in detail.

3.2 Network Architecture

Pose-Dependent Appearance Embedding. Given a source video S, we extract poses P_S and their corresponding appearance feature A_S. Each s^n, n-th frame of S, is processed independently through the shared embedding network, as illustrated in the upper side of Fig. 2. It starts by putting a single image s^n into the shared encoder. The features extracted from the encoder, then, goes through two parallel streams, for the pose and the appearance representation, respectively. In the pose stream, it first reduces the number of channels to K with a convolutional layer. Here, K implicitly determines the number of keypoints to be extracted by the network. Each channel is condensed to represent a key point in a spatial domain by using the method proposed in [16]. In order to capture a local relationship between keypoints, the extracted keypoints are processed by a couple of convolutional layers. In case of the appearance stream, the appearance representation is embedded through a convolutional layer from the encoder feature. Each s^n gets a (p_{s^n}, a_{s^n}) pair, resulting in $P_S = \{p_{s^n}\}_{n=1,2,\ldots,N}$ and $A_S = \{a_{s^n}\}_{n=1,2,\ldots,N}$, which are concatenated in time-dimension respectively.

Pose-Attentive Retrieval Block. Given P^S, A^S, and a driving frame d, the goal is to synthesize a figure with the same identity as in the source video but in the pose of the subject in the driving frame. To this end, a computational block that matches poses between frames to retrieve appearance information for the synthesis is designed. We take insights from a attention mechanism in the previous memory networks [22,23]. However, different from [22,23] that matches visual features, we explicitly design our module to match pose keypoints with a different motivation.

To match between poses, the driving pose p_d is first extracted through the same pose embedding network used for the source video. Then, similarities between the source poses P_S and the driving pose p_d are computed to determine where to retrieve relevant appearance information from the source appearance A_S. Specifically, soft weights are computed in a non-local manner by comparing every pixel locations in the source pose embedding P_S with every pixels of the driving pose embedding p_d. Finally, the appearance $\hat{a}$ correspondent to p_d is reassembled by taking a weighted summation of the source appearance features A_S with regard to the soft weights. Our appearance retrieval operation can be summarized as:

$$\hat{a}^i = \frac{1}{Z} \sum_{\forall j} f(p_d^i, P_S^j) A_S^j, \tag{1}$$

where $Z = \sum_{\forall j} f(p_d^i, P_S^j)$ is a normalization factor, and i and j indicate the pixel location of the driving and the source pose features, respectively. At first, it computes the attention map using the similarity function f, The similarity function f is defined as follow:

$$f(p_d^i, P_S^j) = \exp(p_d^i \cdot P_S^j), \tag{2}$$

where $\cdot$ is dot-product.

Image Generation. Contrary to the object parts that change according to the driving pose, the background should remain still regardless of the variation of poses. Thus, the image generation module should focus on generating the object parts while the background are directly obtained from the source images. To this end, the decoder produces two outputs by dividing the output branches; one to create the object parts o and the other to create a mask for the object m. m is a single channel probability map obtained by applying the sigmoid function. We compute the background image B by feeding one of source images to a simple encoder-decoder network to inpaint the occluded part. Formally, we get the result by

$$g = m \odot o + (1 - m) \odot B, \tag{3}$$

where $\odot$ represents element-wise multiplication.

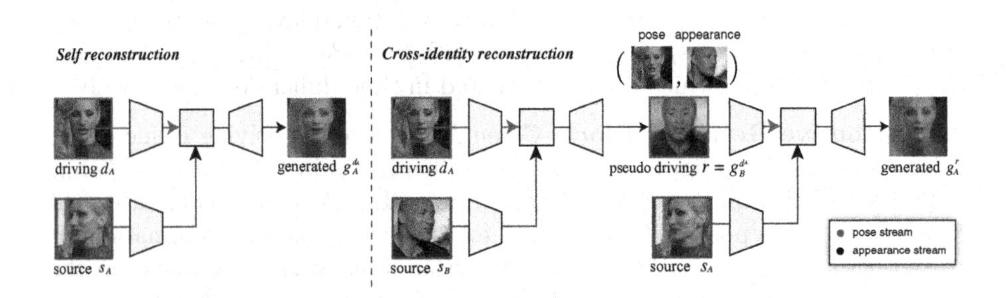

Fig. 3. Self-reconstruction and cross-identity reconstruction training scheme.

3.3 Training and Losses

Our generator is trained using two main losses: self-reconstruction loss and cross-identity reconstruction loss. The distinction between them depends on whether they use the same identity of objects on the source and the driving images, as illustrated in Fig. 3.

Self-reconstruction Loss. If source images s_A and the driving image d_A have the same identity A, the generated image $g_A^{d_A}$ has the identity of A and the pose from d_A, which is equal to d_A. Therefore the network can be trained by using d_A as the ground-truth image of $g_A^{d_A}$ in a self-supervised way. Note that we split a single video of an identity into source images $s_{identity}$ and driving image $d_{identity}$, and there is no overlap between them.

We impose the reconstruction loss on the pair of $(g_A^{d_A}, d_A)$ to train the overall network. Additionally, we add adversarial loss for the quality of generated image. Thus the self-reconstruction loss is formulated as follows:

$$\begin{aligned} L_{self} = & L_{GAN}(G, D) + \lambda_{FM} L_{FM}(G, D) \\ & + \lambda_{VGG} L_{VGG}(g_A^{d_A}, d_A). \end{aligned} \tag{4}$$

Here, we give adversarial loss $L_{GAN}(G, D)$ using the least-square GAN formulation [21]. $L_{FM}(G, D)$ represents the discriminator feature-matching loss [33] and $L_{VGG}(g_A^{d_A}, d_A)$ is the perceptual loss using the pretrained VGG19 network.

This objective is commonly used to train previous motion transfer networks in a self-supervised way [25,35]. However, using this self-reconstruction alone can possibly cause severe issues in the real testing, where subjects in the source and the driving image are different. For example, only working with samples from the same video, the model can simply learn to assume the source and the driving subjects to have the same identity. Thus the model may not be prepared for dealing with subjects that show a significant appearance difference. We address this issue by additionally imposing a novel cross-identity reconstruction loss that simulates the real testing cases, where the identity of subjects in the source and the driving are different.

Cross-Identity Reconstruction Loss. Consider transferring motion between different identities, which is our real task. To train a network in this scenario, it requires a lot of pairs of different subjects in the same posture. However, obtaining such pairs of images is not only time-consuming, but also results in annotations that involve subjective judgments. We overcome the challenge by two consecutive generation procedures between different subjects. The key idea behind our method is that we synthesize an image that has a different identity but the same pose as the one in the source video, then we use it as a pseudo-driving frame to transform an arbitrary frame in the source. In this case, the output image should be the same as the original source frame with the same pose since the poses of both the driving and the source image are identical. By minimizing the pixel distance, the network is enforced to preserve the identity-relevant content in the output. Although our method shares a similar concept with existing self-supervision methods [3,38], our method presents a novel approach for tackling the cross-identity motion transfer in that it provides a way of supervision using different identities through the pseudo-driving image.

Figure 3 illustrates the procedure of the cross-identity reconstruction scheme. First, we synthesize a pseudo-driving frame using two different videos. Specifically, we use d_A and s_B as the driving and the source image, respectively. The generator synthesizes the output $r = g_B^{d_A}$, where the object B mimics the pose in d_A. Then, we perform another generation using the output r as the pseudo-driving image and s_A as the source images. In this case, the generator outputs g_A^r that has the identity of A and the pose of r. Because the pose of r is d_A, g_A^r and d_A can be expected to be identical. We minimize a reconstruction loss between this pair of images with an adversarial loss similar to the self-reconstruction:

$$
\begin{aligned} L_{cross} = &L_{GAN}(G, D) + \lambda_{FM} L_{FM}(G, D) \\ &+\lambda_{VGG} L_{VGG}(g_A^r, d_A). \end{aligned} \tag{5}
$$

Full Objective. Our final training objective is described as

$$
L_{total} = L_{self} + L_{cross}, \tag{6}
$$

which is optimized by solving a minimax optimization problem on an adversarial learning framework.

3.4 Inference

At the inference time, our network takes a driving video and source images as input. Note that there is no limitation to the number of source images. At first, it performs the pose-dependent appearance embedding on all of the source images. Each source image is processed individually using the shared network. Regardless of the number of source frames used at the training time, our network can fully leverage diverse observations from all the source images. The resulting image is generated based on a frame in the driving video using the pose-attentive retrieval module. To transfer motions throughout the driving video, every frame in the driving video is forwarded one-by-one.

4 Experiments

4.1 Experimental Setup

Datasets. We conduct experiments on diverse object categories using following datasets. **VoxCeleb2** [7] dataset is a large dataset of human speaking videos downloaded from YouTube. VoxCeleb2 contains 150,480 videos of 6,112 celebrities with 145,569/4,911 train/test split. **Thai-Chi-HD** [26] contains 3,334 videos of Thai-chi performers. Among them, we use 3,049 videos for training and 285 videos for testing, following the official split. **BAIR robot pushing dataset** [11] is a collection of videos about robot arms moving and pushing various objects on the table. It provides 42,880 training and 128 test videos. Note that the motion contained in the 3 datasets is different from each other, which is face, body, and machine, respectively. By using them, we show the effectiveness of our method on a variety of objects.

Baselines. We compare our network with unsupervised motion transfer methods, which are able to expand on arbitrary objects. Particularly for VoxCeleb2, we additionally compare our methods with multi-source motion transfer methods that use facial landmarks detected from a external landmark detector. To the best of our knowledge, we are not aware of any multi-source motion transfer methods designed for arbitrary objects. All of our baselines are listed below:

- X2Face [35]: It transfers motion by warping objects (especially face) in the direction of x and y axis. Also, they utilize multiple source images by averaging all of the appearance features.
- Monkey-net [25]: It uses keypoints learned in self-supervised way to extract motion from general object categories. It transfers motion by estimating optical flows between keypoints.
- First-order [26]: To express more complex and large motion, it uses keypoints and affine transformation for each keypoint.
- NeuralHead [36]: It transfers the appearance of source frames into a set of face landmarks. It fine-tunes subject-specific generator parameters using few shots of source images at test time.

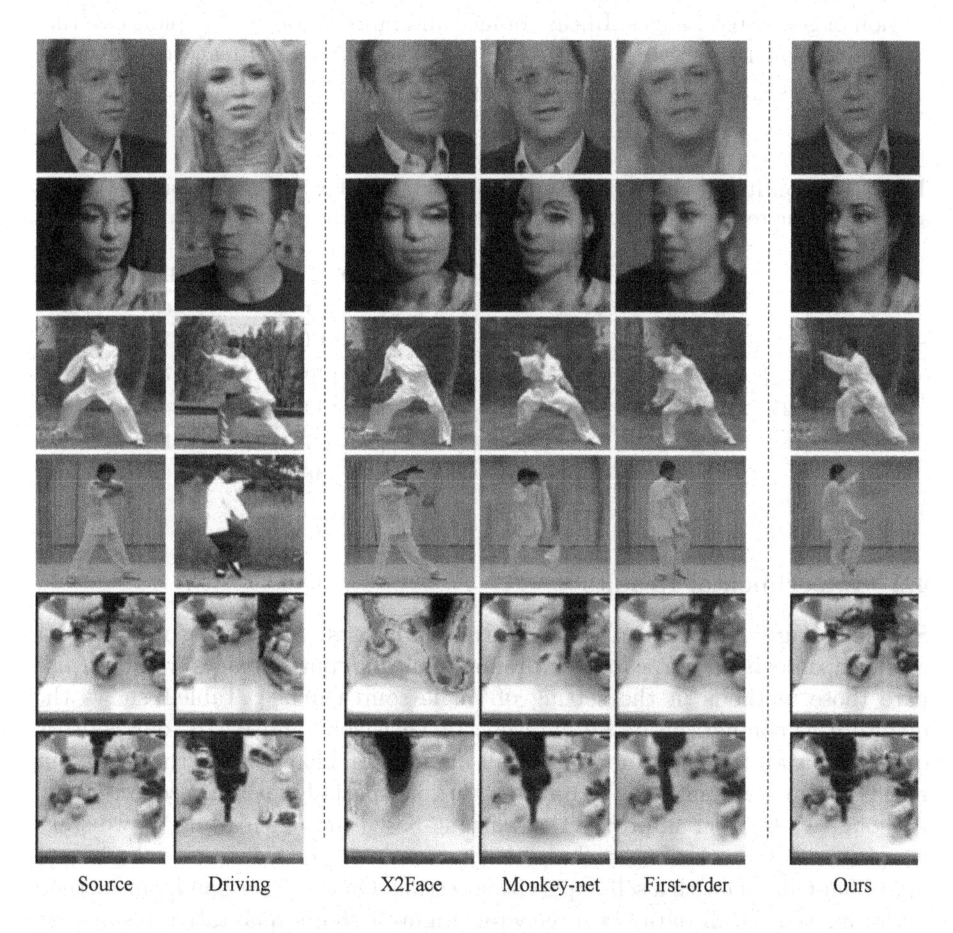

Fig. 4. Qualitative results of single source animation. We conduct experiments on VoxCeleb2 [7], Thai-Chi-HD [26], and BAIR [11] datasets. Source images and driving images represent appearance and pose to condition on, respectively.

Metrics. We evaluate motion transfer methods with the following criteria. We follow metrics proposed in Monkey-net [25] and extend the evaluation scope by applying them on self reconstruction results and the motion transfer on different identity.

- Fréchet Inception Distance (FID) [15]: This score indicates the quality of generated images. It is computed by comparing the feature statistics of generated and real images to estimate how realistic the generated images are.
- Average Keypoint Distance (AKD): It represents the average distance between keypoints of generated and driving videos, and evaluates whether the motion of driving video is transferred or not. We extract face and body keypoints using external keypoint detection networks [4,5].
- Average Euclidean Distance (AED): It shows the degree of identity preservation of generated images. In case of face and human body, we exploit external face recognition network [1,14] to embed identity-related features and compute average euclidean distance between them.

Table 1. Quantitative comparison on various datasets. For all metrics, smaller values are better. The red and blue color indicate the first and the second ranks, respectively.

	VoxCeleb2				Thai-Chi-HD				BAIR	
	Self		Cross		Self		Cross		Self	Cross
	FID	AKD	FID	AED	FID	AKD	FID	AED	FID	FID
X2Face [35]	48.63	22.23	55.46	0.69	94.99	4.88	101.74	1.42	375.75	379.60
Monkey-net [25]	**27.98**	18.26	**52.49**	**0.66**	**30.43**	2.85	101.64	2.67	**24.28**	75.57
First-order [26]	35.39	**2.52**	78.21	0.78	18.55	1.47	37.38	2.15	31.80	**52.39**
Ours	17.24	1.90	36.39	0.62	35.60	**2.74**	**48.33**	**2.02**	20.75	34.81

4.2 Experimental Results

Single Image Animation of Arbitrary Objects. As existing methods for arbitrary objects are not capable of taking multiple source images, we first compare those methods in the setting of single source image. Table 1 shows the quantitative comparison on all datasets. In VoxCeleb2 and BAIR, our network outperforms all baselines methods in all metrics. In Thai-Chi-HD, however, our results are worse than those of First-order [26] especially for self-reconstruction. Although First-order improves the warping-based approach using a two-stage warping with hallucinating occluded background, its generalization is limited to only Thai-Chi-HD with a self-supervision setting. On the other hand, our method performs well on all datasets in general. Figure 4 shows qualitative results. As can be seen, our method shows the effectiveness in transferring various types of motion. Warping-based methods work for the small movement change, e.g. leg regions of the third row in Fig. 4. However, they have difficulties in synthesizing complex scenes with challenging viewpoint changes.

Comparison with Multi-source Methods. We compare results of using multiple frames. For fair comparison with NeuralHead [36], we use the test set and the results provided by the authors of NeuralHead [36]. The test set is composed of 3 subsets with 1, 8, and 32 source frames in a self-reconstruction setting.

Table 2 represents the quantitative comparison. As can be seen, our method significantly outperforms X2Face and is competitive to NeuralHead. This is because NeuralHead uses keypoints extracted from the pre-trained landmark detector and produces sharper images, while our method learns to estimate the pose without supervision. However, NeuralHead is specifically designed for facial videos with external landmarks, therefore the applicability to arbitrary objects is limited. Note that our method outperforms both baselines on AED. It indicates that our method better preserves the identity of the source due to our cross-identity training scheme, which is also observed in Fig. 5.

Table 2. Quantitative results of few-shot motion transfer methods on VoxCeleb2.

	X2Face [35]			NeuralHead [36]			Ours		
	1	8	32	1	8	32	1	8	32
FID	97.37	110.29	127.32	50.10	45.88	**44.85**	56.19	60.17	60.80
AKD	113.79	96.18	142.09	4.48	3.71	**3.45**	7.47	4.26	4.09
AED	0.67	0.80	0.79	0.54	0.45	0.43	0.46	0.33	**0.29**

Table 3. Ablation study on the cross-identity training scheme.

	Self			Cross	
	FID	AKD	AED	FID	AED
w/o cross	18.49	5.79	0.39	55.11	0.76
w/ cross	17.24	1.90	0.38	36.39	0.62

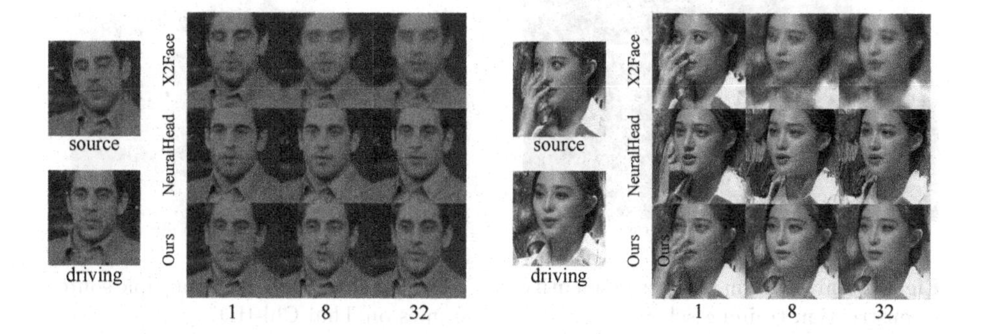

Fig. 5. Comparison with face few-shot motion transfer methods. We compare results with varying number of source images.

4.3 Human Evaluation

We conducted a user study using Amazon Mechanical Turk (AMT) to compare our results with previous warping based motion transfer methods [25,26,35] subjectively. We randomly selected 50 results of VoxCeleb2 [7], Thai-chi-HD [26] and

BAIR robot pushing [11] datasets. We used a single frame for source, and uniformly sampled 32 frames from the driving video. Source images and corresponding driving videos are matched randomly to make pairs of distinct identities. In the user study, users ranked the result videos according to the following criteria: (1) whether the identity of the source image is preserved or not, and (2) whether the motion of the driving video is transferred or not. Each question was rated by 10 people and we averaged the total ranks. The results are shown in Table 4. As can be seen, our results are preferred over previous warping-based methods on VoxCeleb2 and BAIR with a large margin. Even though our method achieves lower quantitative scores than First-order [26] on Thai-Chi-HD in Table 1, the user study shows that our results are preferred by users as much as those of First-order.

Table 4. Human evaluation on AMT. The scores indicate average ranking values among 4 methods.

	X2Face [35]	Monkey-net [25]	First-order [26]	Ours
VoxCeleb2	3.116	2.884	2.414	**1.586**
Thai-Chi-HD	3.438	3.048	1.792	**1.722**
BAIR	3.58	2.104	2.48	**1.836**

Table 5. Ablation study on pose-attentive retrieval module.

Method	Self		Cross	
	FID	AKD	FID	AED
ConcatAppearance	17.57	4.21	63.52	0.92
ConcatNearest	23.69	4.27	52.13	0.87
ConcatAverage	51.35	103.78	52.67	1.35
Ours	**17.24**	**1.90**	**36.39**	**0.62**

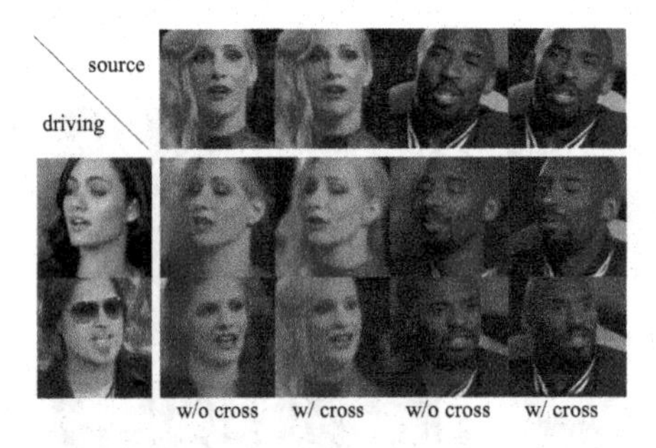

Fig. 6. Ablation on the cross-identity reconstruction training scheme.

Fig. 7. Ablation on using multiple source frames on Thai-Chi-HD.

4.4 Analysis

Efficacy of Cross-Identity Reconstruction Loss. We conduct ablation study on the cross-identity training scheme using VoxCeleb2 dataset by transferring motion between people of different genders and ages. The quantitative and qualitative results are shown in Table 3 and Fig. 6. As represented in the first row of each example in Fig. 6, the self-reconstruction training can disentangle shape and appearance. However, it is not able to further disentangle identity-specific shape such as face shape or skeleton. For example, results without the

Fig. 8. Visualization of keypoints and attention. We visualize learned keypoints on Thai-Chi-HD and BAIR datasets in the first row. The second row represents activated attention maps according to driving images.

cross-identity reconstruction training show the tendency that the identity of source images is not preserved for all the driving videos. On the other hand, the identity is well preserved across diverse driving videos with the cross-identity training scheme.

Moreover, the results using only the self-reconstruction training scheme show implausible output in regions surrounding the face such as hair, neck, and shoulder. We attribute it to the fact that the self-reconstruction training scheme focuses on the facial region, which is the most different part between frames in the same video. With cross-identity training, our network is enforced to consider all parts of a person, thereby further improves the quality. The efficacy of our cross-identity reconstruction training scheme is also demonstrated in Table 3. When adding our cross-identity reconstruction, both image quality and identity preservation are improved.

Efficacy of Pose-Attentive Retrieval Module. For a deeper analysis on pose-attentive retrieval module, we replace the module with following baselines. **ConcatAppearance** concatenates pose and appearance features. It can take only a single source image as input. **ConcatNearest** takes multiple source images as input and then selects a single image that has the most similar pose with the driving pose using Euclidean distance of the keypoints. **ConcatAverage** takes the average of feature maps of multiple images. As can be seen in Table 5, ConcatAppearance lowers the quality, especially in AKD. ConcatNearest could not improve the quality, as well. Moreover, ConcatAverage deteriorates performance.

Using Multiple Source Images at Inference Time. Figure 7 demonstrates the effectiveness of utilizing multiple source images. The driving images are shown on the first row, and the second and the third row show the output images when 1 or 4 source images are given, respectively. With a single source image, reconstructing occluded parts is difficult as shown in the head region in the second row. On the other hand, exploiting multiple source images alleviates the issue by copying and pasting the occluded region from the source images. The performance gain is also shown quantitatively in Table 2.

Visualization. To understand the learned representation of our method, we visualize the learned keypoints and attention in Fig. 8. The first row illustrates 10 keypoints extracted from the given frames. In the second row, we visualize the attention map in the source images corresponding to the query points in the driving pose.

5 Conclusion

In this paper, we presented a pose-attentive video reassembling method to tackle the problem of cross-identity motion transfer method for arbitrary objects. Unlike existing warping-based approaches, our pose-attentive reassembling method better handles a wide range of motion and is also able to exploit multiple sources images to robustly reconstruct occluded parts. Furthermore, our cross-identity training scheme disentangles identity-relevant content from object pose to preserve identity in the setting of cross-identity motion transfer.

Acknowledgement. This work was conducted by Center for Applied Research in Artificial Intelligence (CARAI) grant funded by DAPA and ADD (UD190031RD).

References

1. Amos, B., Ludwiczuk, B., Satyanarayanan, M., et al.: OpenFace: A general-purpose face recognition library with mobile applications. CMU School of Computer Science (June 2016)
2. Balakrishnan, G., Zhao, A., Dalca, A.V., Durand, F., Guttag, J.: Synthesizing images of humans in unseen poses. In: Proceedings of the IEEE Conference on Computer Vision and Pattern Recognition, pp. 8340–8348 (2018)
3. Bansal, A., Ma, S., Ramanan, D., Sheikh, Y.: Recycle-GAN: unsupervised video retargeting. In: Ferrari, V., Hebert, M., Sminchisescu, C., Weiss, Y. (eds.) ECCV 2018. LNCS, vol. 11209, pp. 122–138. Springer, Cham (2018). https://doi.org/10.1007/978-3-030-01228-1_8
4. Bulat, A., Tzimiropoulos, G.: How far are we from solving the 2D & 3D face alignment problem? (and a dataset of 230,000 3D facial landmarks). In: Proceedings of the IEEE International Conference on Computer Vision, pp. 1021–1030 (2017)
5. Cao, Z., Simon, T., Wei, S.E., Sheikh, Y.: Realtime multi-person 2D pose estimation using part affinity fields. In: Proceedings of the IEEE Conference on Computer Vision and Pattern Recognition, pp. 7291–7299 (2017)
6. Chan, C., Ginosar, S., Zhou, T., Efros, A.A.: Everybody dance now. In: Proceedings of the IEEE International Conference on Computer Vision, pp. 5933–5942 (2019)
7. Chung, J.S., Nagrani, A., Zisserman, A.: VoxCeleb2: deep speaker recognition. Proc. Interspeech **2018**, 1086–1090 (2018)
8. Denton, E., Fergus, R.: Stochastic video generation with a learned prior. arXiv preprint arXiv:1802.07687 (2018)
9. Denton, E.L., et al.: Unsupervised learning of disentangled representations from video. In: Advances in Neural Information Processing Systems, pp. 4414–4423 (2017)

10. Ding, H., Sricharan, K., Chellappa, R.: ExprGAN: facial expression editing with controllable expression intensity. In: 32nd AAAI Conference on Artificial Intelligence (2018)
11. Ebert, F., Finn, C., Lee, A.X., Levine, S.: Self-supervised visual planning with temporal skip connections. arXiv preprint arXiv:1710.05268 (2017)
12. Goodfellow, I., et al.: Generative adversarial nets. In: Advances in Neural Information Processing Systems, pp. 2672–2680 (2014)
13. Ha, S., Kersner, M., Kim, B., Seo, S., Kim, D.: MarioNETte: few-shot face reenactment preserving identity of unseen targets. In: Proceedings of the AAAI Conference on Artificial Intelligence (2020)
14. Hermans, A., Beyer, L., Leibe, B.: In defense of the triplet loss for person re-identification. arXiv preprint arXiv:1703.07737 (2017)
15. Heusel, M., Ramsauer, H., Unterthiner, T., Nessler, B., Hochreiter, S.: GANs trained by a two time-scale update rule converge to a local Nash equilibrium. In: Advances in Neural Information Processing Systems, pp. 6626–6637 (2017)
16. Jakab, T., Gupta, A., Bilen, H., Vedaldi, A.: Unsupervised learning of object landmarks through conditional image generation. In: Bengio, S., Wallach, H., Larochelle, H., Grauman, K., Cesa-Bianchi, N., Garnett, R. (eds.) Advances in Neural Information Processing Systems, vol. 31, pp. 4016–4027. Curran Associates, Inc. (2018). http://papers.nips.cc/paper/7657-unsupervised-learning-of-object-landmarks-through-conditional-image-generation.pdf
17. Kulkarni, T.D., Whitney, W.F., Kohli, P., Tenenbaum, J.: Deep convolutional inverse graphics network. In: Advances in Neural Information Processing Systems, pp. 2539–2547 (2015)
18. Lathuilière, S., Sangineto, E., Siarohin, A., Sebe, N.: Attention-based fusion for multi-source human image generation. In: The IEEE Winter Conference on Applications of Computer Vision, pp. 439–448 (2020)
19. Ma, L., Jia, X., Sun, Q., Schiele, B., Tuytelaars, T., Van Gool, L.: Pose guided person image generation. In: Advances in Neural Information Processing Systems, pp. 406–416 (2017)
20. Ma, L., Sun, Q., Georgoulis, S., Van Gool, L., Schiele, B., Fritz, M.: Disentangled person image generation. In: Proceedings of the IEEE Conference on Computer Vision and Pattern Recognition, pp. 99–108 (2018)
21. Mao, X., Li, Q., Xie, H., Lau, R.Y., Wang, Z., Paul Smolley, S.: Least squares generative adversarial networks. In: Proceedings of the IEEE International Conference on Computer Vision, pp. 2794–2802 (2017)
22. Oh, S.W., Lee, J.Y., Xu, N., Kim, S.J.: Video object segmentation using space-time memory networks. In: Proceedings of the IEEE International Conference on Computer Vision, pp. 9226–9235 (2019)
23. Oh, S.W., Lee, S., Lee, J.Y., Kim, S.J.: Onion-peel networks for deep video completion. In: Proceedings of the IEEE International Conference on Computer Vision, pp. 4403–4412 (2019)
24. Saito, M., Matsumoto, E., Saito, S.: Temporal generative adversarial nets with singular value clipping. In: Proceedings of the IEEE International Conference on Computer Vision, pp. 2830–2839 (2017)
25. Siarohin, A., Lathuilière, S., Tulyakov, S., Ricci, E., Sebe, N.: Animating arbitrary objects via deep motion transfer. In: Proceedings of the IEEE Conference on Computer Vision and Pattern Recognition, pp. 2377–2386 (2019)

26. Siarohin, A., Lathuilière, S., Tulyakov, S., Ricci, E., Sebe, N.: First order motion model for image animation. In: Advances in Neural Information Processing Systems, vol. 32, pp. 7137–7147. Curran Associates, Inc. (2019). http://papers.nips.cc/paper/8935-first-order-motion-model-for-image-animation.pdf
27. Siarohin, A., Sangineto, E., Lathuilière, S., Sebe, N.: Deformable gans for pose-based human image generation. In: Proceedings of the IEEE Conference on Computer Vision and Pattern Recognition, pp. 3408–3416 (2018)
28. Tran, L., Yin, X., Liu, X.: Disentangled representation learning GAN for pose-invariant face recognition. In: Proceedings of the IEEE Conference on Computer Vision and Pattern Recognition, pp. 1415–1424 (2017)
29. Tulyakov, S., Liu, M.Y., Yang, X., Kautz, J.: MoCoGAN: decomposing motion and content for video generation. In: Proceedings of the IEEE Conference on Computer Vision and Pattern Recognition, pp. 1526–1535 (2018)
30. Villegas, R., Yang, J., Zou, Y., Sohn, S., Lin, X., Lee, H.: Learning to generate long-term future via hierarchical prediction. In: Proceedings of the 34th International Conference on Machine Learning, vol. 70, pp. 3560–3569. JMLR.org (2017)
31. Vondrick, C., Pirsiavash, H., Torralba, A.: Generating videos with scene dynamics. In: Advances In Neural Information Processing Systems, pp. 613–621 (2016)
32. Wang, T.C., Liu, M.Y., Tao, A., Liu, G., Catanzaro, B., Kautz, J.: Few-shot video-to-video synthesis. In: Advances in Neural Information Processing Systems, pp. 5014–5025 (2019)
33. Wang, T.C., Liu, M.Y., Zhu, J.Y., Tao, A., Kautz, J., Catanzaro, B.: High-resolution image synthesis and semantic manipulation with conditional GANs. In: Proceedings of the IEEE Conference on Computer Vision and Pattern Recognition, pp. 8798–8807 (2018)
34. Wang, T.C., et al.: Video-to-video synthesis. In: Advances in Neural Information Processing Systems, pp. 1152–1164 (2018)
35. Wiles, O., Koepke, A.S., Zisserman, A.: X2Face: a network for controlling face generation using images, audio, and pose codes. In: Ferrari, V., Hebert, M., Sminchisescu, C., Weiss, Y. (eds.) ECCV 2018. LNCS, vol. 11217, pp. 690–706. Springer, Cham (2018). https://doi.org/10.1007/978-3-030-01261-8_41
36. Zakharov, E., Shysheya, A., Burkov, E., Lempitsky, V.: Few-shot adversarial learning of realistic neural talking head models. In: Proceedings of the IEEE International Conference on Computer Vision, pp. 9459–9468 (2019)
37. Zhao, L., Peng, X., Tian, Yu., Kapadia, M., Metaxas, D.: Learning to forecast and refine residual motion for image-to-video generation. In: Ferrari, V., Hebert, M., Sminchisescu, C., Weiss, Y. (eds.) ECCV 2018. LNCS, vol. 11219, pp. 403–419. Springer, Cham (2018). https://doi.org/10.1007/978-3-030-01267-0_24
38. Zhu, J.Y., Park, T., Isola, P., Efros, A.A.: Unpaired image-to-image translation using cycle-consistent adversarial networks. In: Proceedings of the IEEE International Conference on Computer Vision, pp. 2223–2232 (2017)

Domain Adaptive Object Detection via Asymmetric Tri-Way Faster-RCNN

Zhenwei He and Lei Zhang[(✉)]

Learning Intelligence and Vision Essential (LiVE) Group,
School of Microelectronics and Communication Engineering,
Chongqing University, Chongqing, China
{hzw,leizhang}@cqu.edu.cn
http://www.leizhang.tk/

Abstract. Conventional object detection models inevitably encounter a performance drop as the domain disparity exists. Unsupervised domain adaptive object detection is proposed recently to reduce the disparity between domains, where the source domain is label-rich while the target domain is label-agnostic. The existing models follow a parameter shared siamese structure for adversarial domain alignment, which, however, easily leads to the collapse and out-of-control risk of the source domain and brings negative impact to feature adaption. The main reason is that the labeling unfairness (asymmetry) between source and target makes the parameter sharing mechanism unable to adapt. Therefore, in order to avoid the source domain collapse risk caused by parameter sharing, we propose an asymmetric tri-way Faster-RCNN (ATF) for domain adaptive object detection. Our ATF model has two distinct merits: 1) A ancillary net supervised by source label is deployed to learn ancillary target features and simultaneously preserve the discrimination of source domain, which enhances the structural discrimination (object classification vs. bounding box regression) of domain alignment. 2) The asymmetric structure consisting of a chief net and an independent ancillary net essentially overcomes the parameter sharing aroused source risk collapse. The adaption safety of the proposed ATF detector is guaranteed. Extensive experiments on a number of datasets, including Cityscapes, Foggy-cityscapes, KITTI, Sim10k, Pascal VOC, Clipart and Watercolor, demonstrate the SOTA performance of our method.

Keywords: Object detection · Transfer learning · Deep learning

1 Introduction

Object detection is one of the significant tasks in computer vision, which has extensive applications in video surveillance, self-driving, face analysis, medical imaging, etc. Motivated by the development of CNNs, researchers have made the object detection models fast, reliable, and precise [8,13,15,24,26,30]. However, in real-world scenarios, object detection models face enormous challenges due

A. Vedaldi et al. (Eds.): ECCV 2020, LNCS 12369, pp. 309–324, 2020.
https://doi.org/10.1007/978-3-030-58586-0_19

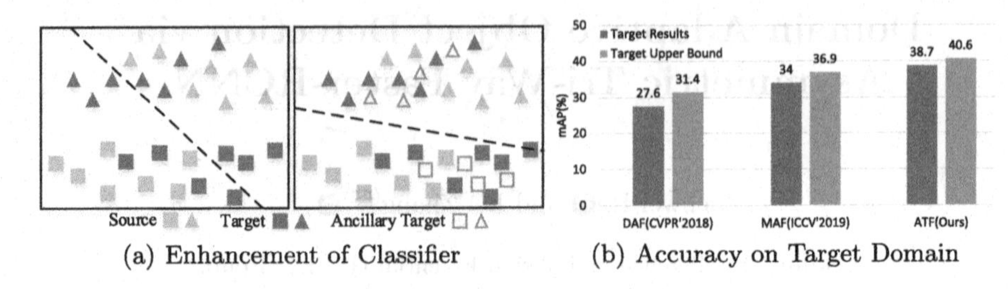

(a) Enhancement of Classifier (b) Accuracy on Target Domain

Fig. 1. The motivation of our approach. (a) shows the learning effect of the classification decision boundary for both source and target data without/with the ancillary target features. (b) presents the target domain detection performance (mAP) based on different transfer models and the upper bound of the target performance based on their deep features and groundtruth target labels.

to the diversity of application environments such as different weathers, backgrounds, scenes, illuminations, and object appearances. The unavoidable environment change causes the domain shift circumstance for object detection models. Nevertheless, conventional object detection models are domain constrained, which can not take into account the domain shift happened in open conditions and noticeable performance degradation is resulted due to the environmental changes. One way to avoid the influence of domain shift is to train the detector by domain/scene specific data, but labeling the training data of each domain is time-consuming and impractical. For reducing the labeling cast and getting a domain adaptive detector, in this paper, the unsupervised domain adaptive object detection task is addressed by transferring the label-rich source domain to the label-agnostic target domain. Since the training label of the target domain is unnecessary, there is no extra annotation cast for the target domain. More importantly, detectors are more generalizable to the environmental change benefitting from the co-training between domains.

Very recently, unsupervised domain adaptive object detection is proposed to mitigate the domain shift problem by transferring the knowledge from the sematic related source domain to target domain [4,17,20,22]. Most of cross-domain object detection models learn the domain invariant features with transfer learning ideas. Inspired by the pioneer of this field [4,17], the detector can be domain-invariant if only the source and target domains are sufficiently confused. No suspicion that a domain-invariant feature representation can enable the object detector to be domain adaptive. However, the domain-invariant detector does not guarantee good object classification and bounding box (bbox) regression. This is mainly due to the lack of domain specific data. Relying solely on labeled source data and unlabeled target data to entirely eliminate the domain disparity is actually not an easy task. This is motivated in Fig. 1(a) (left), where we use a toy classifier as an example. We can see that as the features from different domains are not fully aligned, the decision boundary learned with labeled source data and unlabeled target data at hand can not correctly classify the samples from the target domain. Thus, we have an instinct thought to implicitly learn ancillary target data for domain-invariant class discriminative and bbox regressive features. Specifically, to better characterize the domain-invariant detector, as is shown in Fig. 1(a) (right), we propose to learn the ancillary target features with a specialized module, which we call "ancillary net". We see that the target process features contribute to the new decision boundary of the classifier such that the target features are not only domain invariant but class separable.

Most of the CNN based domain adaption algorithms aim to learn a transferable feature representation [5,10,33,40]. They utilize the feature learning ability of CNN to extract domain-invariant representation for the source and target domains. A popular strategy to learn transferable features with CNN is adversarial training just like the generative adversarial net (GAN) [14]. The adversarial training strategy is a two-player gaming, in which a discriminator is trained to distinguish different domains, while a generator is trained to extract domain-invariant features to fool the discriminator. However, adversarial learning also has some risks. As indicated by [25], forcing the features to be domain-invariant may inevitably distort the original distribution of domain data, and the structural discrimination (intra-class compactness vs. inter-class separability) between two domains may be destroyed. This is mainly because the target data is completely unlabeled.

Similarly, distribution distortion also occurs in cross-domain object detection. Since the target data is unlabeled and the model is trained only with source labels, the learned source features can be discriminative and reliable, while the discrimination of the target features is vulnerable and untrustworthy. However, most existing models such as DAF [4] and MAF [17] default that the source and target domains share the same network with parameter sharing. A forthcoming problem of parameter sharing network is that aligning the reliable source features toward the unreliable target features may enhance the risk of source domain collapse and eventually deteriorate the structural discrimination of the model. It will inevitably bring a negative impact to object classification and bbox regression of the detector. According to the domain adaption theory in [1], the expected target risk $\epsilon_T(h)$ is upper bounded by the empirical source risk $\epsilon_S(h)$, domain discrepancy $d_{\mathcal{A}}$ and shared error $\lambda = \epsilon_T(h^*) + \epsilon_S(h^*)$ of the ideal hypothesis h^* for both domains. Therefore, effectively controlling the source risk and avoiding the collapse of the source domain is particularly important for the success of a domain adaptive detector. In this paper, we propose an **A**symmetric **T**ri-way structure to enhance the transferability of **F**aster-RCNN, which is called ATF and consists of a chief net and an ancillary net, as is shown in Fig. 2. The asymmetry originates from that the ancillary net is independent of the parameter shared chief net. Because the independent ancillary net is only trained by the labeled source data, the asymmetry can largely avoid source collapse and feature distortion during transfer.

Our model inclines to preserve the discrimination of source features and simultaneously guide the structural transfer of target features. One evidence is shown in Fig. 1(b), in which we implement the domain adaptive detectors, such as DAF [4], MAF [17] and ATF from Cityscapes [6] dataset to the Foggy

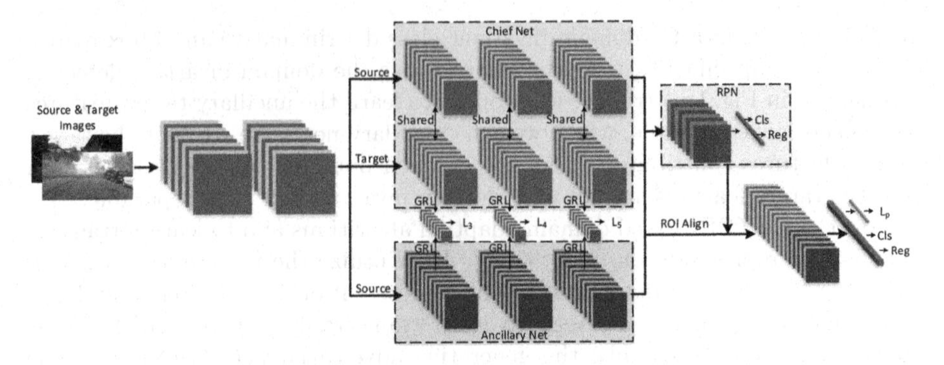

Fig. 2. The net structure of our ATF. Our ATF has three streams upon the backbone. The first two streams with shared parameters are the Chief net (orange color). Another stream is the ancillary net (blue color) which is independent of the chief net. All streams are fed into the same RPN. A ROI-Align layer is deployed to get the pooled features of all streams. The pooled features are used for final detection. The chief net is trained by ancillary data guided domain adversarial loss and source label guided detection loss. The ancillary net is trained with source label guided detection loss. (Color figure online)

Cityscapes [34], respectively. Additionally, in order to observe the upper target performance, we also use the features from the networks and train the detector with the ground truth target label. We can see that our ATF achieves both higher performance (11.1% and 4.7% resp.) than DAF (CVPR'18) and MAF (ICCV'19). In summary, this paper has two distinct merits. 1) We propose a source only guided ancillary net in order to learn ancillary target data for reducing the domain bias and model bias. 2) We propose an asymmetric tri-way structure, which overcomes the adversarial model collapse of the parameter shared network, i.e. the out-of-control risk in the source domain.

2 Related Work

Object Detection. Object detection is an essential task of computer vision, which has been studied for many years. Boosted by the development of deep convolutional neural networks, object detection has recently achieved significant advances [3,23,29,38]. Object detection models can be roughly categorized into two types: one-stage and two-stage detection. One-stage object detection is good at computational efficiency, which attains real-time object detection. SSD [26] is the first one-stage object detector. Although SSD has a similar way to detect objects as RPN [30], it uses multiple layers for various scales. YOLO [28] outputs sparse detection results with high computation speed. Recently, RetinaDet [24] addresses the unbalance of foreground and background with the proposed focal loss. Two-stage detectors generate region proposals for the detection, for example, Faster-RCNN [30] introduced the RPN for the proposal generation. FPN [23] utilized multi-layers for the detection of different scales. NAS-FPN [12] learn the net structure of FPN to achieve better performance. In this paper, we select the Faster-RCNN as the base network for domain adaptive detection.

Domain Adaption. Domain adaption is proposed for bridging the gap between different domains. The domain adaption problem has been investigated for several different computer vision tasks, such as image classification and object segmentation [5,32,33,39]. Inspired by the accomplishment of deep learning, early domain adaptive models minimize the disparity estimate between different domains such as maximum mean discrepancy (MMD) [27,37]. Recently, several domain adaptive models were proposed based on adversarial learning. The two-player game between the feature extractor and domain discriminator promotes the confusion between different domains. Ganin *et al.* [9] proposed the gradient reverse layer (GRL), which reverses the gradient backpropagation for adversarial learning. For feature regularization, Cicek *et al.* introduced a regularization method base on the VAT. Besides the adversarial of extractor and discriminator, Saito *et al.* [32] employed two distinct classifiers to reduce the domain disparity. In our ATF, an unsupervised domain adaption mechanism is exploited.

Domain Adaptive Object Detection. Domain adaptive or cross-domain object detection has been raised very recently for the unconstrained scene. This task was firstly proposed by Chen *et al.* [4], which addresses the domain shift on both image-level and instance-level. This work indicated that if the features are domain-invariant, the detector can achieve better performance on the target domain. After that, several models are also proposed for domain adaptive detection. In [31], low-level features and high-level features are treated with strong and weak alignment, respectively. Kim *et al.* [21] introduces a new loss function and exploits the self-training strategy to improve the detection of the target domain. He and Zhang proposed a MAF [17] model, in which a hierarchical structure is designed to reduce the domain disparity at different scales. In addition to the domain adaption guided detectors, the mean teacher is introduced to get the pseudo labels for the target domain [2]. Besides that, Khodabandeh *et al.* [20] train an extra classifier and use KL-divergence to learn a model for correction.

3 The Proposed ATF Approach

In this section, we introduce the details of our Asymmetric Tri-way Faster-RCNN (ATF) model. For convenience, the fully labeled source domain is marked as $D_s = \{(x_i^s, b_i^s, y_i^s)\}_i^{n_s}$, where x_i^s stands the image, b_i^s is the coordinate of bounding boxes, y_i^s is the category label and n_s is the number of samples. The unlabeled target domain is marked as $D_t = \{(x_i^t)\}_i^{n_t}$, where n_t denotes the number of samples. Our task is to transfer the semantic knowledge from D_s to D_t and achieve successful detection in the target domain.

3.1 Network Architecture of ATF

Our proposed ATF model is based on the Faster-RCNN [30] detection framework. In order to overcome the out-of-control risk of source domain in the conventional symmetric Siamese network structure with shared parameters, we introduce an asymmetric tri-way network as the backbone. Specifically, the images from the source or target domain are fed into the first two convolution blocks. On top of that, we divide the structure into three streams. As shown in Fig. 2, the first two streams with the shared parameters in orange color are the Chief net. Features from the source and target data are fed into the two streams, respectively. The third stream with blue color is the proposed Ancillary net, which is parametrically independent of the chief net. That is, the ancillary net has different parameters from the chief net. The source only data is fed into the ancillary net during the training phase. Three streams of the network share the same region proposal network (RPN) as Faster-RCNN does. We pool the features of all streams based on the proposals with the ROI-Align layer. Finally, we get the detection results on top of the network with the pooled features.

An overview of our network structure is illustrated in Fig. 2. For training the ATF model, we design the adversarial domain confusion strategy, which is established between the chief net and ancillary net. The training loss of our ATF consists of two kinds of losses: domain adversarial confusion (**Dac**) loss in the chief net for bounding the domain discrepancy $d_{\mathcal{A}}$ and the source labeled guided detection loss (**Det**) in the ancillary net for bounding the empirical source risk $\epsilon_S(h)$. So, the proposed model is easily trained end-to-end.

3.2 Principle of the Chief Net

Domain discrepancy is the primary factor that leads to performance degradation in cross-domain object detection. In order to reduce the domain discrepancy $d_{\mathcal{A}}$, we introduce the domain adversarial confusion (Dac) mechanism which bridges the gap between the chief net (target knowledge) and the ancillary net (source knowledge). The features from the ancillary net should have a similar distribution to the target stream features from the chief net. Considering that object detection refers to two stages, i.e., image-level feature learning (global) and proposal-level feature learning (local), we propose two alignment modules based on the Dac mechanism, i.e.g.lobal domain alignment with Dac and local domain alignment with Dac.

Global Domain Alignment with Dac. Obviously, the Dac guided global domain alignment focuses on the low-level convolutional blocks between the chief net (target stream) and the ancillary net in ATF. Let x_i^s and x_i^t be two images from the source and target domains, respectively. The feature maps of the $k^{th}(k=3,4,5)$ block of the chief net and ancillary net as shown in Fig. 2 are defined as $F_c(x_i^t, \theta_c^k)$ and $F_a(x_i^s, \theta_a^k)$, respectively. d is the binary domain label, and $d=1$ for source domain and 0 for target domain. The Dac based global domain alignment loss for k^{th} block ($\mathcal{L}_{G-Dac}^k$) can be written as:

$$\mathcal{L}_{G-Dac}^k = -\sum_{u,v}((1-d)\log(D_k(F_c(x_i^t, \theta_c^k)^{(u,v)}, \theta_d^k)+d\log(D_k(F_a(x_i^s, \theta_a^k)^{(u,v)}, \theta_d^k)) \tag{1}$$

where the (u, v) stands for the pixel coordinate of the feature map. D_k is the discriminator of kth block. θ_c^k, θ_a^k, and θ_d^k are the parameters of chief net, ancillary net and discriminator in the kth block, respectively. In principle, the discriminator D tries to minimize $\mathcal{L}_{G-Dac}^k(D)$ to distinguish the features from different domains. With the gradient reversal layer (GRL) [9], the chief net F_c and ancillary net F_a try to maximize $\mathcal{L}_{G-Dac}^k(F_c, F_a)$. Then, the global features between the chief net (target stream) and the ancillary net are confused (aligned).

Local Domain Alignment with Dac. The domain alignment on the global image level is still not enough for the local object based detector. Therefore, we propose to further align the local object-level features across domains. As shown in Fig. 2, the features pooled by the ROI-Align layer stand for the local part of an image, including foreground and background. Similar to the global domain alignment module, we confuse (align) the local object-level features pooled from the target stream of the chief net and the ancillary net. Suppose the pooled features from the ancillary net to be f_a and features from the chief net to be f_c, the Dac based local domain alignment loss ($\mathcal{L}_{L-Dac}$) on the local object-level features is formulated as:

$$\mathcal{L}_{L-Dac} = -\frac{1}{N}\sum_{n}((1-d)\log(D_l(F_l(f_c^n, \theta_f), \theta_d) + d\log(D_l(F_l(f_a^n, \theta_f), \theta_d)) \quad (2)$$

where the D_l is the local domain discriminator, F_l is the local backbone network, θ_l and θ_d are the parameters of the backbone and the discriminator, respectively. We implement the adversarial learning with GRL [9]. The discriminator tries to minimize $\mathcal{L}_{L-Dac}(D_l)$ while the local backbone network tries to maximize $\mathcal{L}_{L-Dac}(F_l)$ for local domain confusion.

3.3 Principle of the Ancillary Net

As discussed above, the chief net aims to bound the domain discrepancy d_A. In this section, we propose to bound the empirical source risk ϵ_S by using the ancillary net. The reason why we construct a specialized ancillary net for bounding the source risk rather than using the source stream of the chief net has been elaborated. The main reason is that the chief net is parameter shared, so the empirical risk of source stream in the chief net is easily out-of-control due to the unlabeled problem of the target domain. Because the source domain is sufficiently labeled with object categories and bounding boxes in each image, the source risk ϵ_S of the ancillary net is easy to be bounded by the classification loss and regression loss of detector. In our implementation, the detection loss for the chief net is reused for the supervision of the ancillary net.

From the principles of the chief net and the ancillary net, we know that the ancillary net is trained to generate features that have a similar distribution to the target stream in the chief net as shown in Eqs. (1) and (2). That is, the ancillary net adjusts the features learned by the target stream of the chief net to adapt to the source data trained detector. Meanwhile, the ancillary net is restricted by the classifier and regressor of the source detector, such that the structural discrimination is preserved in the source domain. Therefore, with the domain alignment and source risk minimization, the expected task risk can be effectively bounded for domain adaptive object detection.

3.4 Training Loss of Our ATF

The proposed asymmetric tri-way Faster-RCNN (ATF) contains the two loss functions, the detection based source risk loss and the domain alignment loss. The detection loss function for both chief and ancillary nets is shown as:

$$\mathcal{L}_{Det} = \mathcal{L}_{cls}(x_i^s, b_i^s, y_i^s) + \mathcal{L}_{reg}(x_i^s, b_i^s, y_i^s) \tag{3}$$

where $\mathcal{L}_{cls}$ is the softmax based cross-entropy loss and $\mathcal{L}_{reg}$ is the smooth-L_1 loss, which are standard detection losses for bounding the empirical source risk. In summary, by revisiting the Eqs. (1), (2) and (3), the total loss function for training our model can be written as:

$$\mathcal{L}_{ATF} = \mathcal{L}_{Det} + \alpha(\mathcal{L}_{L-Dac} + \sum_{k=3}^{5} \mathcal{L}_{G-Dac}^{k}) \tag{4}$$

where the α is a hyper-parameter to adjust the weight of domain alignment loss. The model is easily trained end-to-end with Stochastic Gradient Descent (SGD). Overview of our ATF can be observed in Fig. 2.

4 Experiments

In this section, we evaluate our approach on several different datasets, including Cityscapes [6], Foggy Cityscapes [34], KITTI [11], SIM10k [19], Pascal VOC [7], Clipart [18] and Watercolor [18]. We compare our results with state-of-the-art methods to show the effectiveness of our model.

4.1 Implementation Details

The base network of our ATF model is VGG-16 [36] or Res-101 [16] in experiments. We follow the same experimental setting as [4]. The ImageNet pre-trained model is used for initialization. For each iteration, one labeled source sample and one unlabeled target sample are fed into ATF. In test phase, the chief net is used to get the detection results. For all datasets, we report the average precisions (AP, %) and mean average precisions (mAP, %) with a threshold of 0.5.

4.2 Datasets

Cityscapes: Cityscapes [6] captures high-quality video for in different cities for automotive vision. The dataset includes 5000 manually selected images from 27 cities, which are collected with a similar weather condition. These images are annotated with dense pixel-level image annotation. Although the Cityscapes dataset is labeled for the semantic segmentation task, we generate the bounding box based on the pixel-level as [4] did.

Foggy Cityscapes: Foggy Cityscapes [34] dataset simulates the foggy weather based on the Cityscapes. The pixel-level labels of Cityscapes can be inherited by the Foggy Cityscapes such that we can generate the bounding box. In our experiments, the validation set of the Foggy Cityscapes is used for testing.

KITTI: KITTI [11] dataset is collected by the autonomous driving platform. Images of the dataset are manually selected in several different scenes. The dataset includes 14999 images and 80256 bounding boxes for the detection task. Only the training set of KITTI is used for our experiment.

SIM10K: Images of SIM10K [19] are generated by the engine of Grand Theft Auto V (GTA V). The dataset simulates different scenes, such as different time or weathers. SIM10K contains 10000 images with 58701 bounding boxes of car. All images of the dataset are used for training.

Pascal VOC: Pascal VOC [7] is a famous object detection dataset. This dataset contains 20 categories with bounding boxes. The image scale of the dataset is diverse. In our experiment, the training and validation split of VOC07 and 12 are used as the training set, which results in about 15k images.

Clipart and Watercolor: The Clipart and Watercolor [18] are constructed by the Amazon Mechanical Turk, which is introduced for the domain adaption detection task. Similar to the Pascal VOC, the Clipart contains 1000 images and 20 categories. Watercolor has 2000 images of 6 categories. Half of the datasets are introduced for training while the remaining is used for the test.

4.3 Cross-Domain Detection in Different Visibility and Cameras

Domain Adaption Across Different Visibility. Visibility change caused by weather can shift the data distribution. In this part, we evaluate our ATF with the cityscapes [6] and the foggy cityscapes [34] datasets. We treat Cityscapes as source domain and Foggy Cityscapes as target domain. Our model uses VGG16 as the base net in the experiment. We introduce the source only trained Faster-RCNN (without adaptation), DAF [4], MAF [17], Strong-Weak [31], Diversify&match(D&match) [22], Noisy Labeling(NL) [20], and SCL [35] for the comparison. Our ATF is trained for 18 epochs in the experiment, where the learning rate is set as 0.001 and changes to 0.0001 in the 12th epoch.

The results are presented in Table 1. We can see that our ATF achieves 38.7% mAP, which outperforms all the compared models. Due to the lack of domain specific data, the models which only concentrate on the feature alignment can not work well, such as MAF [17], Strong-Weak [31] and SCL [35]. With the ancillary target features from the ancillary net to reduce the domain shift and bias, our model gets preferable performance. Additionally, our ATF model also outperforms the pseudo label based model [20], in which it has to generate and update the pseudo labels with features extracted by a source only trained extractor, where the target features are untrustworthy. The unreliable feature based pseudo labels can not lead to a precise target model. In order to prove

Table 1. The cross-domain detection results from Cityscapes to Foggy Cityscapes.

Methods	Person	Rider	Car	Truck	Bus	Train	Mcycle	Bcycle	mAP
Faster-RCNN	24.1	33.1	34.3	4.1	22.3	3.0	15.3	26.5	20.3
DAF(CVPR'18) [4]	25.0	31.0	40.5	22.1	35.3	20.2	20.0	27.1	27.6
MAF(ICCV'19) [17]	28.2	39.5	43.9	23.8	39.9	33.3	29.2	33.9	34.0
Strong-Weak [31]	29.9	42.3	43.5	24.5	36.2	32.6	30.0	35.3	34.3
D& Match [22]	30.8	40.5	44.3	27.2	38.4	34.5	28.4	32.2	34.6
NL /w res101 [20]	**35.1**	42.2	49.2	30.1	45.3	27.0	26.9	36.0	36.5
SCL [35]	31.6	44.0	44.8	**30.4**	41.8	**40.7**	**33.6**	36.2	37.9
ATF (1-block)	33.3	43.6	44.6	24.3	39.6	10.5	27.2	35.6	32.3
ATF (2-blocks)	34.0	46.0	49.1	26.4	**46.5**	14.7	30.7	37.5	35.6
ATF (ours)	34.6	**47.0**	**50.0**	23.7	43.3	38.7	33.4	**38.8**	**38.7**

Table 2. The results of domain adaptive object detection on Cityscapes and KITTI.

Tasks	Faster-RCNN	DAF [4]	MAF [17]	S-W [31]	SCL [35]	ATF (ours)
K to C	30.2	38.5	41.0	37.9	41.9	**42.1**
C to K	53.5	64.1	72.1	71.0	72.7	**73.5**

the effectiveness of ancillary net, we conduct ablation studies that reduce the convolutional blocks of ancillary net. The results of 1-block (the 3rd and 4th blocks are removed) and 2-blocks (the 3rd blocks is removed) from the ancillary net are shown in Table 1. As we reduce the convolutional block of the ancillary net, the performance drops. The merit of the proposed ancillary net is validated.

Domain Adaption Across Different Cameras. The camera change is another important factor leading to the domain shift in real-world application scenarios. In this experiment, we employ the Cityscapes (C) [34] and KITTI (K) [11] as the source and target domains, respectively. The source only trained Faster-RCNN (without adaption), DAF [4], MAF [17], and Strong-Weak [31] are implemented for comparisons. The AP of car on the target domain is computed for the test. The experimental results are presented in Table 2.

In Table 2, K to C represents that the KITTI is used as the source domain, while the Cityscapes is used as the target domain and vice versa. We can observe that our ATF achieves the best performance on both K to C and C to K tasks among all the compared models, which testify the effectiveness of our model in alleviating the domain shift problem caused by the change of cameras.

4.4 Cross-Domain Detection on Large Domain Shift

In this section, we concentrate on the domains with large domain disparity, especially, from the real image to the comical or artistic images. We employ the Pascal VOC [7] dataset as the source domain, which contains the real image. The Clipart or Watercolor [18] is exploited as the target domain. The backbone for the experiments is the ImageNet pretrained ResNet-101. We train our model for

Table 3. The cross-domain detection results from Pascal VOC to Clipart.

Methods	Aero	Bike	Bird	Boat	Bottle	Bus	Car	Cat	Chair	Cow	
Faster-RCNN	35.6	52.5	24.3	23.0	20.0	43.9	32.8	10.7	30.6	11.7	
DAF [4]	15.0	34.6	12.4	11.9	19.8	21.1	23.2	3.1	22.1	26.3	
BDC-Faster	20.2	46.4	20.4	19.3	18.7	41.3	26.5	6.4	33.2	11.7	
WST-BSR [21]	28.0	64.5	23.9	19.0	21.9	**64.3**	**43.5**	16.4	**42.2**	25.9	
Strong-Weak [31]	26.2	48.5	32.6	33.7	38.5	54.3	37.1	18.6	34.8	58.3	
MAF [17]	38.1	61.1	25.8	**43.9**	40.3	41.6	40.3	9.2	37.1	48.4	
SCL [35]	**44.7**	50.0	**33.6**	27.4	**42.2**	55.6	38.3	**19.2**	37.9	69.0	
ATF (ours)	41.9	**67.0**	27.4	36.4	41.0	48.5	42.0	13.1	39.2	**75.1**	
Methods	Table	Dog	Horse	Mbike	Person	Plant	Sheep	Sofa	Train	TV	mAP
Faster-RCNN	13.8	6.0	36.8	45.9	48.7	41.9	16.5	7.3	22.9	32.0	27.8
DAF [4]	10.6	10.0	19.6	39.4	34.6	29.3	1.0	17.1	19.7	24.8	19.8
BDC-Faster	26.0	1.7	36.6	41.5	37.7	44.5	10.6	20.4	33.3	15.5	25.6
WST-BSR [21]	30.5	7.9	25.5	**67.6**	54.5	36.4	10.3	**31.2**	**57.4**	43.5	35.7
Strong-Weak [31]	17.0	12.5	33.8	65.5	**61.6**	52.0	9.3	24.9	54.1	**49.1**	38.1
MAF [17]	24.2	13.4	36.4	52.7	57.0	**52.5**	18.2	24.3	32.9	39.3	36.8
SCL [35]	30.1	**26.3**	34.4	67.3	61.0	47.9	21.4	26.3	50.1	47.3	41.5
ATF (ours)	**33.4**	7.9	**41.2**	56.2	61.4	50.6	**42.0**	25.0	53.1	39.1	**42.1**

8 epochs with the learning rate of 0.001 and change the learning rate to 0.0001 in the 6th epoch to ensure convergence.

Transfer from Pascal VOC to Clipart. The Clipart [18] contains the comical images which has the same 20 categories as Pascal VOC [7]. In this experiment, we introduce the source only Faster RCNN, DAF [4], WST-BSR [21], MAF [17], Strong-Weak [31] and SCL [35] for the comparison. The results are shown in Table 3. Our ATF achieves 42.1% mAP and outperforms all models.

Table 4. The cross-domain detection results from Pascal VOC to Watercolor.

Methods	Bike	Bird	Car	Cat	Dog	Person	mAP
Faster-RCNN	68.8	46.8	37.2	32.7	21.3	60.7	44.6
DAF [4]	75.2	40.6	48.0	31.5	20.6	60.0	46.0
BDC-Faster	68.6	48.3	47.2	26.5	21.7	60.5	45.5
WST-BSR [21]	75.6	45.8	**49.3**	34.1	30.3	64.1	49.9
MAF [17]	73.4	55.7	46.4	36.8	28.9	60.8	50.3
Strong-Weak [31]	**82.3**	55.9	46.5	32.7	**35.5**	66.7	53.3
ATF (ours)	78.8	**59.9**	47.9	**41.0**	34.8	**66.9**	**54.9**

Transfer from Pascal VOC to Watercolor. The Watercolor [18] dataset contains 6 categories which are the same as the Pascal VOC [7]. In this experiment, the source only trained Faster-RCNN, DAF [4], WST-BSR [21], MAF [17]

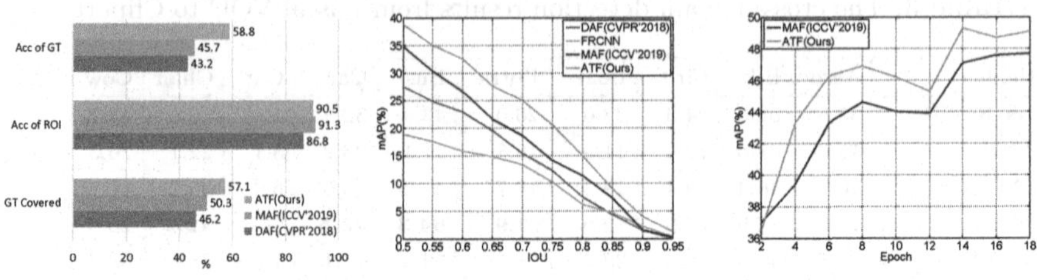

(a) Accuracy of Classifiers (b) Performance v.s IOUs (c) Performance v.s Epoch

Fig. 3. Analysis of our model. (a) We test the accuracy of the classifier in the detector. **Acc of GT**: We crop the ground truth of the image and use the classifier on R-CNN to classify them. **Acc of ROI**: The accuracy of RCNN's classifier with the ROIs generated by RPN. **GT Covered**: The proportion of ground truth seen by RPN. (b) The performance change with different IOUs. Better viewed in color version. (c) The performance of source domain with different epoches. (Color figure online)

and Strong-Weak [31] are introduced for comparison. The results are shown in Table 4, from which we can observe that our proposed ATF achieves the best performance among all compared models and the advantage is further proved.

4.5 Cross-Domain Detection from Synthetic to Real

Domain adaption from the synthetic scene to the real scene is an important application scenario for domain adaptive object detection. In this section, we introduce the SIM10k [19] as the synthetic scene and the Cityscapes [34] as the real scene. The source only trained Faster-RCNN, DAF [4], MAF [17], Strong-Weak [31], and SCL [35] are introduced for comparisons. The AP of the car is computed for the evaluation of the experiment which is presented in Table 5. We observe that our model outperforms all the compared models. The superiority of the proposed ATF is further demonstrated for the cross-domain object detection.

Table 5. The results of domain adaptive object detection on SIM10k and Cityscapes.

Methods	F-RCNN	DAF [4]	MAF [17]	S-W [31]	SCL [35]	ATF (ours)
AP (%)	34.6	38.9	41.1	40.1	42.6	**42.8**

4.6 Analysis and Discussion

In this section, we will implement some experiments to analyze our ATF model with four distinct aspects, including parameter sensitivity, accuracy of classifiers, IOU v.s. detection performance and visualization.

Table 6. Parameter Sensitivity on α.

α	0.3	0.5	0.7	0.9	1.1	1.3
mAP (%)	36.7	37.5	38.7	38.7	38.4	38.0

Parameter Sensitivity on α . In this part, we show the sensitivity of parameter α in Eq. (3). α controls the power of domain adaption. We conduct the cross-domain experiments from Cityscapes to Foggy Cityscapes. The sensitivity of α is shown in Table 6. When $\alpha = 0.7$, our model achieves the best performance.
The Accuracy of Classifiers. Our model enhances the training of the detector with the ancillary target feature. In this part, we analyze the classifier in the detector with different models. First, we static the number of ground truth boxes covered by the ROIs. If the IOU between the ROI and ground truth is higher than 0.5, we think the corresponding ROI is predicted by the RPN (region proposal network). Second, we compute the accuracy of the RCNN classifier with ROIs from the RPN. Last, we crop the ground truth of the testing set and use the RCNN classifier to predict their label and get the accuracy. The DAF [4], MAF [17] and our model are tested with Cityscapes and Foggy Cityscapes datasets. The results are shown in Fig. 3(a). We observe that the RPN of our model finds 57.1% of the ground truth, which is the best result of all compared models. The RCNN classifiers from all three tested models achieve above 90% accuracy when classifying generated ROIs. However, when the cropped ground truth samples are fed into the model, the accuracy of the RCNN classifier sharply drops. The ground truth samples missed by RPN are also misjudged by the RCNN in the experiments. Therefore, we experimentally find that it is very important to improve the recall of RPN in cross-domain object detection task. In ATF, benefited by the asymmetric structure which enhances the structural discrimination for the detector and new decision boundary contributed by ancillary target features, our model achieves a better recall in RPN.
Detection Performance w.r.t. Different IOUs. The IOU threshold is an important parameter in the test phase. In the previous experiments, the IOU threshold is set as 0.5 by default. In this part, we test our model with different IOUs. The source only Faster-RCNN, DAF [4], MAF [17], and our ATF are implemented for comparison. We conduct the experiments on the Cityscapes [6] and Foggy Cityscapes [34] datasets. The results are shown in Fig. 3(b), where the IOU threshold is increased from 0.5 to 0.95. The performance drops as the IOU threshold increases in the experiment. Our model achieves the best performance on all tested IOU thresholds.
Source Performance w.r.t. Epoches in Monitoring Source Risk. In this part, we monitor the training process of adaption from Cityscapes [6] to Foggy Cityscapes [34]. The mAP (%) on the test set of Cityscapes is shown in Fig. 3(c). Our ATF achieves higher mAP compared to the parameter shared MAF [17] during the training phase. Benefited by the asymmetric structure, our ATF can well prevent the collapse of the source domain and preserve the structural

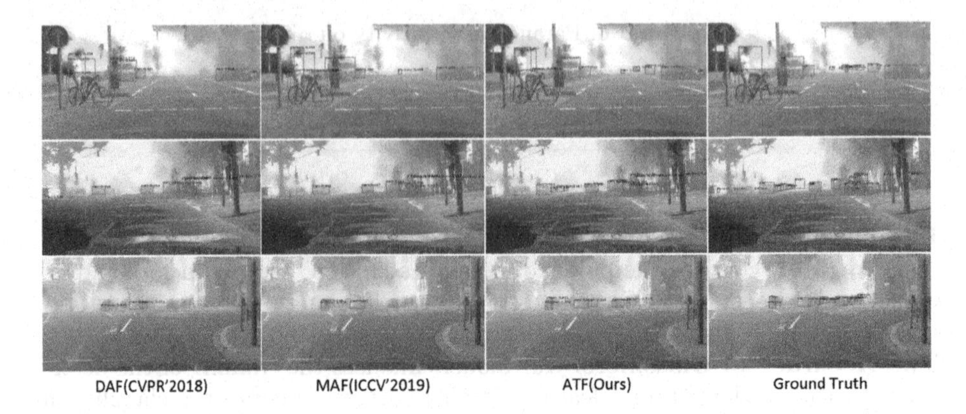

Fig. 4. The visualization results on the target domain (Foggy Cityscapes [34]).

discrimination of source features. This experiment fully proves that parameter sharing of network will deteriorate the empirical source risk $\epsilon_S(h)$, which then leads to high target risk $\epsilon_T(h)$. Thus, a safer adaption of our ATF than the parameter shared MAF is verified.

Visualization of Domain Adaptive Detection. Figure 4 shows some qualitative object detection results of several models on the Foggy Cityscapes dataset [34], i.e. target domain. The state-of-the-art models, DAF [4] and MAF [17], are also presented. We can clearly observe that our ATF shows the best domain adaptive detection results and better matches the ground-truth.

5 Conclusions

In this paper, we propose an asymmetric tri-way network (ATF) to address the out-of-control problem of parameter shared Siamese transfer network for unsupervised domain adaptive object detection. In ATF, an independent network, i.e. the ancillary net, supervised by source labels, is proposed without parameter sharing. Our model has two contributions: 1) Since the domain disparity is hard to be eliminated in parameter shared siamese network, we propose the asymmetric structure to enhance the training of the detector. The asymmetry can well alleviate the labeling unfairness between source and target. 2) The proposed ancillary net enables the structural discrimination preservation of source feature distribution, which to a large extent promotes the feature reliability of the target domain. Our model is easy to be implemented for training the chief net and ancillary net. We conduct extensive experiments on a number of benchmark datasets and state-of-the-art results are obtained by our ATF.

Acknowledgement. This work was supported by the National Science Fund of China under Grants (61771079) and Chongqing Youth Talent Program.

References

1. Ben-David, S., Blitzer, J., Crammer, K., Pereira, F.: Analysis of representations for domain adaptation. In: NeurIPS (2006)
2. Cai, Q., Pan, Y., Ngo, C.W., Tian, X., Duan, L., Yao, T.: Exploring object relation in mean teacher for cross-domain detection. In: CVPR, pp. 11457–11466 (2019)
3. Cai, Z., Vasconcelos, N.: Cascade R-CNN: delving into high quality object detection. In: CVPR, pp. 6154–6162 (2018)
4. Chen, Y., Li, W., Sakaridis, C., Dai, D., Van Gool, L.: Domain adaptive faster R-CNN for object detection in the wild. In: CVPR, pp. 3339–3348 (2018)
5. Chen, Y.C., Lin, Y.Y., Yang, M.H., Huang, J.B.: CrDoCo: pixel-level domain transfer with cross-domain consistency. In: CVPR, pp. 1791–1800 (2019)
6. Cordts, M., et al.: The cityscapes dataset for semantic urban scene understanding. In: CVPR, pp. 3213–3223 (2016)
7. Everingham, M., Van Gool, L., Williams, C.K., Winn, J., Zisserman, A.: The Pascal visual object classes (voc) challenge. IJCV **88**(2), 303–338 (2010)
8. Fu, C.Y., Liu, W., Ranga, A., Tyagi, A., Berg, A.C.: DSSD: Deconvolutional single shot detector. arXiv preprint arXiv:1701.06659 (2017)
9. Ganin, Y., Lempitsky, V.: Unsupervised domain adaptation by backpropagation. arXiv preprint arXiv:1409.7495 (2014)
10. Ganin, Y., et al.: Domain-adversarial training of neural networks. JMLR **17**(1), 2096–2030 (2016)
11. Geiger, A., Lenz, P., Urtasun, R.: Are we ready for autonomous driving? The KITTI vision benchmark suite. In: CVPR, pp. 3354–3361. IEEE (2012)
12. Ghiasi, G., Lin, T.Y., Le, Q.V.: NAS-FPN: learning scalable feature pyramid architecture for object detection. In: CVPR, pp. 7036–7045 (2019)
13. Girshick, R.: Fast R-CNN. In: ICCV, pp. 1440–1448 (2015)
14. Goodfellow, I., et al.: Generative adversarial nets. In: NeurIPS, pp. 2672–2680 (2014)
15. He, K., Gkioxari, G., Dollar, P., Girshick, R.: Mask R-CNN. TPAMI **42**, 386–397 (2018)
16. He, K., Zhang, X., Ren, S., Sun, J.: Deep residual learning for image recognition. In: CVPR, pp. 770–778 (2016)
17. He, Z., Zhang, L.: Multi-adversarial faster-RCNN for unrestricted object detection. In: ICCV, pp. 6668–6677 (2019)
18. Inoue, N., Furuta, R., Yamasaki, T., Aizawa, K.: Cross-domain weakly-supervised object detection through progressive domain adaptation. In: CVPR, pp. 5001–5009 (2018)
19. Johnson-Roberson, M., Barto, C., Mehta, R., Sridhar, S.N., Rosaen, K., Vasudevan, R.: Driving in the matrix: can virtual worlds replace human-generated annotations for real world tasks? arXiv preprint arXiv:1610.01983 (2016)
20. Khodabandeh, M., Vahdat, A., Ranjbar, M., Macready, W.G.: A robust learning approach to domain adaptive object detection. In: ICCV, pp. 480–490 (2019)
21. Kim, S., Choi, J., Kim, T., Kim, C.: Self-training and adversarial background regularization for unsupervised domain adaptive one-stage object detection. In: ICCV, pp. 6092–6101 (2019)
22. Kim, T., Jeong, M., Kim, S., Choi, S., Kim, C.: Diversify and match: a domain adaptive representation learning paradigm for object detection. In: CVPR, pp. 12456–12465 (2019)

23. Lin, T.Y., Dollár, P., Girshick, R., He, K., Hariharan, B., Belongie, S.: Feature pyramid networks for object detection. In: CVPR, pp. 2117–2125 (2017)
24. Lin, T.Y., Goyal, P., Girshick, R., He, K., Dollár, P.: Focal loss for dense object detection. In: ICCV, pp. 2980–2988 (2017)
25. Liu, H., Long, M., Wang, J., Jordan, M.: Transferable adversarial training: A general approach to adapting deep classifiers. In: ICML, pp. 4013–4022 (2019)
26. Liu, W., et al.: SSD: single shot multibox detector. In: Leibe, B., Matas, J., Sebe, N., Welling, M. (eds.) ECCV 2016. LNCS, vol. 9905, pp. 21–37. Springer, Cham (2016). https://doi.org/10.1007/978-3-319-46448-0_2
27. Long, M., Zhu, H., Wang, J., Jordan, M.I.: Unsupervised domain adaptation with residual transfer networks. In: NeurIPS, pp. 136–144 (2016)
28. Redmon, J., Divvala, S., Girshick, R., Farhadi, A.: You only look once: unified, real-time object detection. In: CVPR, pp. 779–788 (2016)
29. Redmon, J., Farhadi, A.: YOLOv3: An incremental improvement. arXiv preprint arXiv:1804.02767 (2018)
30. Ren, S., He, K., Girshick, R., Sun, J.: Faster R-CNN: Towards real-time object detection with region proposal networks. TPAMI **6**, 1137–1149 (2017)
31. Saito, K., Ushiku, Y., Harada, T., Saenko, K.: Strong-weak distribution alignment for adaptive object detection. In: CVPR, pp. 6956–6965 (2019)
32. Saito, K., Watanabe, K., Ushiku, Y., Harada, T.: Maximum classifier discrepancy for unsupervised domain adaptation. In: CVPR, pp. 3723–3732 (2018)
33. Saito, K., Yamamoto, S., Ushiku, Y., Harada, T.: Open set domain adaptation by backpropagation. In: Ferrari, V., Hebert, M., Sminchisescu, C., Weiss, Y. (eds.) ECCV 2018. LNCS, vol. 11209, pp. 156–171. Springer, Cham (2018). https://doi.org/10.1007/978-3-030-01228-1_10
34. Sakaridis, C., Dai, D., Van Gool, L.: Semantic foggy scene understanding with synthetic data. IJCV **126**(9), 973–992 (2018)
35. Shen, Z., Maheshwari, H., Yao, W., Savvides, M.: SCL: Towards accurate domain adaptive object detection via gradient detach based stacked complementary losses. arXiv preprint arXiv:1911.02559 (2019)
36. Simonyan, K., Zisserman, A.: Very deep convolutional networks for large-scale image recognition. arXiv preprint arXiv:1409.1556 (2014)
37. Sun, B., Saenko, K.: Deep CORAL: correlation alignment for deep domain adaptation. In: Hua, G., Jégou, H. (eds.) ECCV 2016. LNCS, vol. 9915, pp. 443–450. Springer, Cham (2016). https://doi.org/10.1007/978-3-319-49409-8_35
38. Vu, T., Jang, H., Pham, T.X., Yoo, C.: Cascade RPN: delving into high-quality region proposal network with adaptive convolution. In: NeurIPS, pp. 1430–1440 (2019)
39. Wang, Q., Breckon, T.P.: Unsupervised domain adaptation via structured prediction based selective pseudo-labeling. arXiv preprint arXiv:1911.07982 (2019)
40. Xu, R., Li, G., Yang, J., Lin, L.: Larger norm more transferable: an adaptive feature norm approach for unsupervised domain adaptation. In: ICCV, pp. 1426–1435 (2019)

Exclusivity-Consistency Regularized Knowledge Distillation for Face Recognition

Xiaobo Wang[1], Tianyu Fu[1], Shengcai Liao[2], Shuo Wang[1], Zhen Lei[3,4](✉), and Tao Mei[1]

[1] JD AI Research, Beijing, China
wangxiaobo2015cbsr@gmail.com, tmei@jd.com
[2] Inception Institute of Artificial Intelligence (IIAI), Abu Dhabi, UAE
[3] CBSR & NLPR, Institute of Automation, Chinese Academy of Science, Beijing, China
zlei@nlpr.ia.ac.cn
[4] School of Artificial Intelligence, University of Chinese Academy of Science, Beijing, China

Abstract. Knowledge distillation is an effective tool to compress large pre-trained Convolutional Neural Networks (CNNs) or their ensembles into models applicable to mobile and embedded devices. The success of which mainly comes from two aspects: the designed student network and the exploited knowledge. However, current methods usually suffer from the low-capability of mobile-level student network and the unsatisfactory knowledge for distillation. In this paper, we propose a novel position-aware exclusivity to encourage large diversity among different filters of the same layer to alleviate the low-capability of student network. Moreover, we investigate the effect of several prevailing knowledge for face recognition distillation and conclude that the knowledge of feature consistency is more flexible and preserves much more information than others. Experiments on a variety of face recognition benchmarks have revealed the superiority of our method over the state-of-the-arts.

Keywords: Face recognition · Knowledge distillation · Weight exclusivity · Feature consistency

1 Introduction

Convolutional neural networks (CNNs) have gained impressive success in the recent advanced face recognition systems [12,13,24,44,45,47,48,53]. However, the performance advantages are driven at the cost of training and deploying resource-intensive networks with millions of parameters. As face recognition shifts toward mobile and embedded devices, the computational cost of large

X. Wang and T. Fu—Equal contribution

A. Vedaldi et al. (Eds.): ECCV 2020, LNCS 12369, pp. 325–342, 2020.
https://doi.org/10.1007/978-3-030-58586-0_20

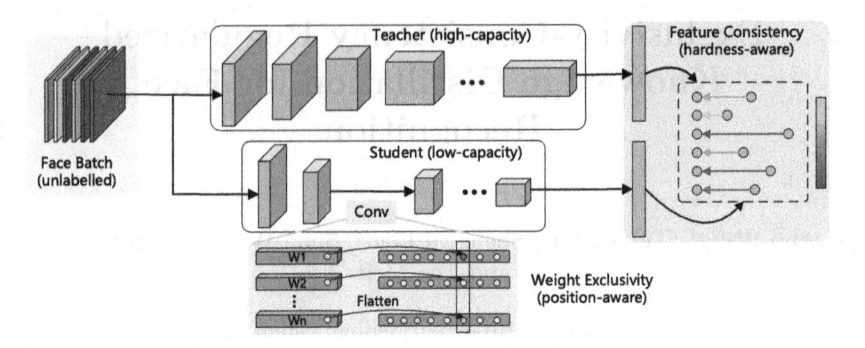

Fig. 1. Overview of our exclusivity-consistency regularized knowledge distillation. The target student network is trained with unlabelled face data by the position-aware weight exclusivity and the hardness-aware feature consistency.

CNNs prevents them from being deployed to these devices. It motivates research of developing compact yet still discriminative models. Several directions such as model pruning, model quantization and knowledge distillation have been suggested to make the model smaller and cost-efficient. Among them, knowledge distillation is being actively investigated. The distillation process aims to learn a compact network (student) by utilizing the knowledge of a larger network or its ensemble (teacher) as supervision. Unlike other compression methods, it can downsize a network regardless of the structural difference between teacher and student.

For face recognition model compression, there have been several attempts [9,10,16,18,19,26,35,40,51,52] in literatures to distil large CNNs, so as to make their deployments easier. Hinton *et al.* [16] propose the first knowledge distillation based on the soften probability consistency, where a temperature parameter is introduced in the softmax function to disclose the similarity structure of data. Wang *et al.* [40] use both the soften and one-hot probability consistency knowledge for face recognition and alignment. Luo *et al.* [31] propose a neuron selection method by leveraging the essential characteristics (domain knowledge) of the learned face representation. Karlekar *et al.* [19] simultaneously exploit the one-hot probability consistency and the feature consistency for knowledge transfer between different face resolutions. Yan *et al.* [52] employ the one-hot probability consistency to guide the network training and design a recursive knowledge distillation strategy to relieve the discrepancy between the teacher and student models. Peng *et al.* [35] use the knowledge of probability consistency to transfer not only the instance-level information, but also the correlation between instances. Although current knowledge distillation methods can achieve more promising results than directly training the mobile-level student network, most of them are limited because of the low-capability of pre-defined student network and the inflexible probability consistency knowledge.

In practice, the common dilemma is that we only have a teacher model at hand and do not know how it was trained (including training sets, loss functions and training strategies of teacher *etc.*). The task of knowledge distillation is to distil a mobile-level student model from the pre-given teacher. However, as the student network is much more smaller than the teacher, it usually suffer from low-capability for achieving a good performance. Moreover, what kind of knowledge should be used under such dilemma is an open issue. To address these problems, this paper proposes a novel exclusivity-consistency regularized knowledge distillation namely EC-KD, to simultaneously exploit the weight exclusivity and the feature consistency into one framework for face recognition model compression. Figure 1 gives an illustration of our proposed EC-KD. To sum up, the contributions of this paper can be summarized as follows:

- We propose a novel position-aware exclusivity regularization to encourage large diversity among different filters of the same convolutional layer to alleviate the low-capability of student network.
- We investigate several knowledge for face recognition model distillation and demonstrate that the knowledge of feature consistency is more flexible and powerful than others in face recognition. Moreover, a hardness-aware feature consistency term is developed for fitting the teacher knowledge well.
- We conduct extensive experiments on a variety of face recognition benchmarks, including LFW [17], CALFW [59], CPLFW [60], SLLFW [8], AgeDB [32], CFP [38], RFW [43], MegaFace [20], Trillion-Pairs [6] and IQIYI-VID [30] have verified the superiority of our approach over the state-of-the-arts. Our code is available at http://www.cbsr.ia.ac.cn/users/xiaobowang/.

2 Related Work

Knowledge Distillation. Many studies have been conducted since Hinton *et al.* [16] proposed the first knowledge distillation based on the soften class probabilities. Romero *et al.* [36] used the hidden layer response of a teacher network as a hint for a student network to improve knowledge distillation. Zagoruyko and Komodakis *et al.* [56] found the area of activated neurons in a teacher network and transferred the activated area to a student network. Luo *et al.* [31] resorted to the top hidden layer as the knowledge and used the attributes to select the important neurons. Karlekar *et al.* [19] simultaneously exploited one-hot labels and feature vectors for the knowledge transfer between different face resolutions. Heo *et al.* utilized an adversarial attack to discover supporting samples [14] and focused on the transfer of activation boundaries formed by hidden neurons [15] for knowledge distillation. Some studies [1,2,4,23,40,54] extended knowledge distillation to other applications.

Deep Face Recognition. Face recognition is an essential open-set metric learning problem, which is different from the closed-set image classification. Specifically, rather than the traditional softmax loss, face recognition is usually supervised by margin-based softmax losses [7,24,28,42,46,47], metric learning losses [37] or both [39]. Moreover, the training set used in face recognition is often with larger identities than image classification. To achieve better performance, large

CNNs like ResNet [7] or AttentionNet [46] are usually employed, which makes them hard to deploy on mobile and embedded devices. Some works [3,50] start to design small networks, but the balance between inference time and performance is unsatisfactory, which motivates us to use the knowledge distillation tool for further model compression.

3 Proposed Formulation

3.1 Weight Exclusivity

It is well-known that larger CNNs exhibit higher capability than smaller ones. For face recognition model compression, this phenomenon is more obvious. To this end, we need take some steps to improve the capability of the target student network. In this paper, we try to exploit the diverse information among different filters. To achieve this, several methods [5,27,29] have been proposed. However, they are all value-aware criteria and require to normalize the filters (fixed magnitude), which is in contradiction to the weight decay (dynamic magnitude) thus may not address the diversity well. Alternatively, we define a novel position-aware exclusivity. Specifically, assume that all filters in a convolutional layer are a tensor $\mathcal{W} \in \mathbb{R}^{N \times M \times K_1 \times K_2}$, where N and M are the numbers of filters and input channels, K_1 and K_2 are the spatial height and width of the filters, respectively. Usually, $K_1 = K_2 = K$. Suppose the tensor $\mathcal{W} \in \mathbb{R}^{N \times M \times K_1 \times K_2}$ is reshaped as vectors $\boldsymbol{W} = [\boldsymbol{w}_1, \ldots, \boldsymbol{w}_N]^T \in \mathbb{R}^{N \times D}$, where $D = MK_1K_2$. We define a new measure of diversity, *i.e.*, Exclusivity.

Definition 1 ***(Weight Exclusivity).*** *Exclusivity between two filter vectors* $\boldsymbol{w}_i \in \mathbb{R}^{1 \times D}$ *and* $\boldsymbol{w}_j \in \mathbb{R}^{1 \times D}$ *is defined as* $\mathcal{H}(\boldsymbol{w}_i, \boldsymbol{w}_j) := \|\boldsymbol{w}_i \odot \boldsymbol{w}_j\|_0 = \sum_{k=1}^{D} (\boldsymbol{w}_i(k) \cdot \boldsymbol{w}_j(k) \neq 0)$, *where the operator* $\odot$ *designates the Hadamard product (i.e., element-wise product), and* $\|\cdot\|_0$ *is the* ℓ_0*-norm.*

From the definition, we can observe that the exclusivity encourages two filter vectors to be as diverse as possible. Ideally, if the position k of $\boldsymbol{w}_i$ (*i.e.*, $\boldsymbol{w}_i(k)$) is not equal to zero, then the exclusivity term encourages the same position k of $\boldsymbol{w}_j$ (*i.e.*, $\boldsymbol{w}_j(k)$) to be zero. In other words, the same position from different filters are competing to survive and the winner positions are set to large values while the loser ones are set to zeros. Consequently, we can say that the defined exclusivity term is *position-aware*. Compared with *value-aware* regularizations, *e.g.*, the orthonormal regularization [29] to minimize $\|\boldsymbol{W}\boldsymbol{W}^T - \boldsymbol{I}\|_F^2$ and the hyperspherical diversity [27], our *position-aware* exclusivity has the following two advantages. One is that *value-aware* criteria are often based on the normalized weights (*i.e.*, fixed magnitude by setting $\|\boldsymbol{w}_i\|_2^2 = 1$), which is in contradiction to the weight decay (*i.e.*, dynamic magnitude by regularizing the norm of weights) thus may not address the diversity well in practice. Our *position-aware* exclusivity has no such restriction. The other one is that our weight exclusivity can be seamlessly incorporated into the traditional weight decay (please see the Sect. 3.3). Nevertheless, the non-convexity and discontinuity of ℓ_0-norm make our exclusivity hard to optimize. Fortunately, it is known that ℓ_1-norm is the tightest convex relaxation of ℓ_0-norm [49], thus we have the following relaxed exclusivity.

Definition 2 ***(Relaxed Weight Exclusivity)***. *Relaxed exclusivity between two filters* $\boldsymbol{w}_i \in \mathbb{R}^{1\times D}$ *and* $\boldsymbol{w}_j \in \mathbb{R}^{1\times D}$ *is defined as* $\mathcal{H}(\boldsymbol{w}_i, \boldsymbol{w}_j) := \|\boldsymbol{w}_i \odot \boldsymbol{w}_j\|_1 = \sum_{k=1}^{D} |\boldsymbol{w}_i(k)| \cdot |\boldsymbol{w}_j(k)|$*, where* $|\cdot|$ *is the absolute value.*

Consequently, our final **weight exclusivity** is formulated as:

$$\mathcal{L}_{\mathrm{WE}}(\mathcal{W}) := \sum_{1\leq j\neq i\leq N} \|\boldsymbol{w}_i \odot \boldsymbol{w}_j\|_1 = \sum_{1\leq j\neq i\leq N} \sum_{k=1}^{D} |\boldsymbol{w}_i(k)| \cdot |\boldsymbol{w}_j(k)|. \tag{1}$$

3.2 Feature Consistency

In face recognition knowledge distillation, the common dilemma is that we only have a teacher model at hand and do not know how it was trained (including training sets, loss functions and training strategies *etc.*). But the task is to obtain a student network with satisfactory performance as well as can be applicable to mobile and embedded devices. As a result, we have the following cases:

One-Hot Labels. If the training set of student network is well-labelled, we can directly train the target student network with one-hot labels. Obviously, this manner does not utilize the knowledge of teacher.

Probability Consistency (PC). Let's denote the final softmax output as $\boldsymbol{z}$, the soft label for teacher model T can be defined as $P_{\mathrm{T}}^{\mathcal{T}} = \mathrm{softmax}(\boldsymbol{z}_{\mathrm{T}}/\tau)$, where τ is the temperature parameter. Similarly, the soft label for student network S is $P_{\mathrm{S}}^{\mathcal{T}} = \mathrm{softmax}(\boldsymbol{z}_{\mathrm{S}}/\tau)$. Prevailing approaches usually exploit the popular **probability consistency** as follows:

$$\mathcal{L}_{\mathrm{PC}} := \mathcal{L}(P_{\mathrm{T}}^{\mathcal{T}}, P_{\mathrm{S}}^{\mathcal{T}}) = \mathcal{L}(\mathrm{softmax}(\boldsymbol{z}_{\mathrm{T}}/\tau), \mathrm{softmax}(\boldsymbol{z}_{\mathrm{S}}/\tau)), \tag{2}$$

where $\mathcal{L}$ is the cross entropy loss between $P_{\mathrm{T}}^{\mathcal{T}}$ and $P_{\mathrm{S}}^{\mathcal{T}}$. However, the formulation of PC is inflexible due to the potential discrepancies between teacher and student networks. For example, 1) If the training classes of teacher are different from student's or the teacher model was pre-trained by metric learning losses (*e.g.*, contrastive or triplet losses), $P_{\mathrm{T}}^{\mathcal{T}} = \mathrm{softmax}(\boldsymbol{z}_{\mathrm{T}}/\tau)$ can not be computed. 2) If the training set of student network contains noisy labels, the performance is not guaranteed because of the unreliable $P_{\mathrm{S}}^{\mathcal{T}} = \mathrm{softmax}(\boldsymbol{z}_{\mathrm{S}}/\tau)$. To sum up, all these point to: the probability consistency knowledge is not flexible and powerful for face recognition.

Feature Consistency (FC). In face recognition, we can also use the feature layer as hint to train the student network. The **feature consistency** can be formulated as follows:

$$\mathcal{L}_{\mathrm{FC}} := \mathcal{H}(F_{\mathrm{S}}, F_{\mathrm{T}}) = ||F_{\mathrm{S}} - F_{\mathrm{T}}||, \tag{3}$$

Algorithm 1: Exclusivity-Consistency Regularized Knowledge Distillation (EC-KD)

Input: Unlabelled training data; Pre-trained teacher model
;
Initialization: $e = 1$; Randomly initialize $\boldsymbol{\Theta}^{\mathrm{S}}$;
while $e \leq E$ **do**
  Shuffle the unlabelled training set $\mathcal{S}$ and fetch mini-batch $\mathcal{S}_m$;
  Forward: Compute the re-weighted matrix $\boldsymbol{G}$ and the final loss (Eq. (5));
  Backward: Compute the gradient $\frac{\partial \mathcal{L}_{\text{EC-KD}}}{\partial \boldsymbol{W}} = \frac{\partial \mathcal{L}_{\text{HFC}}}{\partial \boldsymbol{W}} + \lambda_1(\boldsymbol{G} \odot \boldsymbol{W})$ and
  Update the student model $\boldsymbol{\Theta}^{\mathrm{S}}$ by SGD.
end
Output: Target student model $\boldsymbol{\Theta}^{\mathrm{S}}$.

where $\mathcal{H}$ is the L2 loss, F_{S} and F_{T} are the features from student and teacher. From the formulation, it can be concluded that FC is flexible for training because it is not restricted by the discrepancies between the unknown teacher and the target student. Moreover, to make full use of feature consistency knowledge, we further develop a hardness-aware one. Intuitively, for face samples that are far away from their teachers, they should be emphasized. As a consequence, we define a re-weighted softmax function, $s_i = \frac{e^{\mathcal{H}_i}}{\sum_{j=1}^{m} e^{\mathcal{H}_j}}$, where m is the batch size and our **hardness-aware feature consistency** is simply formulated as:

$$\mathcal{L}_{\text{HFC}} := (1 + s_i)\mathcal{H}(F_{\mathrm{S}}, F_{\mathrm{T}}). \tag{4}$$

3.3 Exclusivity-Consistency Regularized Knowledge Distillation

Based on the above analysis, we prefer to simultaneously harness the weight exclusivity and the feature consistency, *i.e.*, Eqs. (1) and (4), together with the weight decay result in our final Exclusivity-Consistency Regularized Knowledge Distillation (EC-KD):

$$\mathcal{L}_{\text{EC-KD}} = \mathcal{L}_{\text{HFC}} + \lambda_1 \underbrace{\|\boldsymbol{W}\|_F^2}_{\text{weight decay}} + \underbrace{\lambda_2 \sum_{1 \leq j \neq i \leq N} \sum_{k=1}^{D} |\boldsymbol{w}_i(k)| \cdot |\boldsymbol{w}_j(k)|}_{\text{weight exclusivity}} \tag{5}$$

where λ_1 and λ_2 are the trade-off parameters. Typically, weight decay is a unary cost to regularize the norm of filters, while weight exclusivity is a pairwise cost to promote the direction of the filters. Therefore, they are complementary to each other. For simplicity, we empirically set $\lambda_2 = 2\lambda_1$. In consequence, the weight decay and the weight exclusivity can be seamlessly formulated as $\Phi(\boldsymbol{W}) :=$

$$\|\boldsymbol{W}\|_F^2 + 2 \sum_{1 \leq j \neq i \leq N} \sum_{k=1}^{D} |\boldsymbol{w}_i(k)| \cdot |\boldsymbol{w}_j(k)| = \sum_{k=1}^{D} \left(\sum_{i=1}^{N} |\boldsymbol{w}_i(k)| \right)^2 = \|\boldsymbol{W}\|_{1,2}^2 \tag{6}$$

As observed from Eq. (6), it can be split into a set of smaller problems. For each column $\boldsymbol{w}_{\cdot k}$ of $\boldsymbol{W}$, according to [21], the gradient of $\Phi(\boldsymbol{W})$ with respect to $\boldsymbol{w}_{\cdot k}$ is computed as follows:

$$\frac{\partial \Phi(\boldsymbol{W})}{\partial \boldsymbol{w}_{\cdot k}} = \mathbf{g}_{\cdot k} \odot \boldsymbol{w}_{\cdot k} = \left[\frac{\|\boldsymbol{w}_{\cdot k}\|_1}{|\boldsymbol{w}_{\cdot k}(1)| + \epsilon}, \ldots, \frac{\|\boldsymbol{w}_{\cdot k}\|_1}{|\boldsymbol{w}_{\cdot k}(N)| + \epsilon}\right]^T \odot \boldsymbol{w}_{\cdot k}, \tag{7}$$

where $\epsilon \to 0^+$ (a small constant) is introduced to avoid zero denominators. Since both $\boldsymbol{g}_{\cdot k}$ and $\boldsymbol{w}_{\cdot k}$ depend on $\boldsymbol{w}_{\cdot k}$, we employ an efficient re-weighted algorithm to iteratively update $\boldsymbol{g}_{\cdot k}$ and $\boldsymbol{w}_{\cdot k}$. In each iteration, for the forward, we compute the re-weighted matrix $\boldsymbol{G} = [\boldsymbol{g}_{\cdot 1}, \ldots, \boldsymbol{g}_{\cdot D}] \in \mathbb{R}^{N \times D}$, while for the backward, we update the weight by using the gradient $\frac{\partial \Phi(\boldsymbol{W})}{\partial \boldsymbol{W}} = \boldsymbol{G} \odot \boldsymbol{W}$. For clarity, the whole scheme of our framework is summarized in Algorithm 1.

Table 1. Face datasets for training and test. (P) and (G) refer to the probe and gallery set, respectively. (V) refers to video clips.

	Datasets	#Identities	Images
Training	CASIA-WebFace-R [55]	9,809	0.39M
Test	LFW [17]	5,749	13,233
	CALFW [59]	5,749	12,174
	CPLFW [60]	5,749	11,652
	SLLFW [8]	5,749	13,233
	AgeDB [32]	568	16,488
	CFP [38]	500	7,000
	RFW [43]	11,430	40,607
	MegaFace [20]	530(P)	1M(G)
	Trillion-Pairs [6]	5,749(P)	1.58M(G)
	IQIYI-VID [30]	4,934	565,372(V)

4 Experiments

4.1 Datasets

Training Set. We use CASIA-WebFace [55] as the training set of student models unless otherwise specified. Specifically, we use the publicly available one[1].

Test Set. We use ten face recognition benchmarks, including LFW [17], CALFW [59], CPLFW [60], SLLFW[8], AgeDB [32], CFP [38], RFW[43], MegaFace [20, 33], Trillion-Pairs [6] and IQIYI-VID [30] as the test sets. LFW contains 13,233 web-collected images from 5,749 different identities. CALFW [59] was collected

[1] https://github.com/ZhaoJ9014/face.evoLVe.PyTorch.

by crowdsourcing efforts to seek the pictures of people in LFW with age gap as large as possible on the Internet. CPLFW [60] is similar to CALFW, but from the perspective of pose difference. SLLFW [8] selects 3,000 similar-looking negative face pairs from the original LFW image collection. AgeDB [32] contains images annotated with accurate to the year, noise-free labels. CFP [38] consists of collected images of celebrities in frontal and profile views. RFW [43] is a benchmark for measuring racial bias, which consists of four test subsets, namely Caucasian, Indian, Asian and African. MegaFace [33] aims at evaluating the face recognition performance at the million scale of distractors. Trillion-Pairs [6] is a benchmark for testing the face recognition performance with trillion scale of distractors. IQIYI-VID [30] contains multiple video clips from IQIYI variety shows, films and television dramas.

Dataset Overlap Removal. In face recognition, it is very important to perform open-set evaluation [7,28,42], *i.e.*, there should be no overlapping identities between training set and test set. To this end, we need to carefully remove the overlapped identities between the employed training set and the test sets. For the overlap identities removal tool, we use the publicly available script provided by [42] to check whether if two names are of the same person. As a result, we remove 766 identities from the training set of the CASIA-WebFace. For clarity, we denote the refined training set as CASIA-WebFace-R. Important statistics of the datasets are summarized in Table 1. To be rigorous, all the experiments in this paper are based on the refined training dataset.

4.2 Experimental Settings

Data Processing. We detect the faces by adopting the FaceBoxes detector [57,58] and localize five landmarks (two eyes, nose tip and two mouth corners) through a simple 6-layer CNN [11]. The detected faces are cropped and resized to 144×144, and each pixel (ranged between [0,255]) in RGB images is normalized by subtracting 127.5 and divided by 128. For all the training faces, they are horizontally flipped with probability 0.5 for data augmentation.

Pre-trained Teacher. There are many kinds of network architectures [3,28,41] and several loss functions [7,46] for face recognition. Without loss of generality, we use SEResNet50-IR [7] as the teacher model, which was trained by SV-AM-Softmax loss [46]. For all the experiments in this paper, the teacher is pre-given and frozen. Here we provide the details of teacher to the competitors KD [16], FitNet [36], AB[15], BBS [14] and ONE [22].

Student. We use MobileFaceNet [3] and its variants as the student model. The feature dimension of student is 512.

Training. All the student models are trained from scratch, with the batch size of 256 on 4 P40 GPUs parallelly. All experiments in this paper are implemented by PyTorch [34]. The weight decay λ_1 is empirically set to 0.0005 and the momentum is 0.9. The learning rate is initially 0.1 and divided by 10 at the 9, 18, 26 epochs, and we finish the training process at 30 epochs.

Test. We use the learned student network to extract face features. For the evaluation metric, cosine similarity is utilized. We follow the unrestricted with labeled outside data protocol [17] to report the performance on LFW, CALFW, CPLFW, SLLFW, AgeDB, CFP and RFW. Moreover, we also report the BLUFR (TPR@FAR=1e-4) protocol [25] on LFW. On Megaface and Trillion-Pairs, both face identification and verification are conducted by ranking and thresholding the scores. Specifically, for face identification (Id.), the Cumulative Matching Characteristics (CMC) curves are adopted to evaluate the Rank-1 accuracy. For face verification (Veri.), the Receiver Operating Characteristic (ROC) curves at different false alarm rates are adopted. On IQIYI-VID, the MAP@100 is adopted as the evaluation indicator. MAP (Mean Average Precision) refers to the average accuracy rate of the videos of person ID retrieved in the test set for each person ID (as the query) in the training set.

Table 2. Performance (%) comparison of different knowledge.

	LFW	BLUFR	MF-Id	MF-Veri.
SEResNet50-IR(T)	99.21	96.41	86.14	88.32
MobileFaceNet(S)	99.08	93.65	80.27	85.20
PC [Eq. (2)]	98.48	84.17	69.33	68.07
FC [Eq. (3)]	99.11	95.57	83.96	86.56
PC + Loss [46]	99.01	93.71	81.74	85.48
FC + Loss [46]	99.15	94.29	81.90	84.72
HFC [Eq. (4)]	**99.20**	**95.59**	**84.19**	**87.86**

4.3 Ablation Study and Exploratory Experiments

Feature Consistency vs. Other Knowledge. In this part, we use MobileFaceNet [3] as the student network. For the adopted knowledge, we compare the soften probability consistency (PC) (*i.e.*, Eq. (2)), feature consistency (FC) (*i.e.*, Eq. (3)) and their combinations with the softmax-based loss [46]. The results in Table 2 show that simply using the knowledge of soften probability consistency is not enough. It should combine with the softmax-based loss (one-hot labels) to achieve satisfactory performance. While the simple knowledge of feature consistency can achieve higher performance, which reveals that it preserves much more information than probability consistency. We also observe that the improvement by combining the knowledge of feature consistency with the softmax-based loss is limited. Moreover, from the results of our hardness-aware feature consistency (HFC), we can see that the feature consistency knowledge should be emphasized.

Effect of Filter Numbers. We further evaluate the feature consistency knowledge with different number of filters. Specifically, we change the filter numbers of all convolutional layers from the student network (*i.e.*, MobileFaceNet) to 2, 1/2, 1/4 and 1/8 times size. The performance is reported in Table 3, from which

Table 3. Comparison of different filter numbers.

#Filters	Model size	Flops	Infer time	LFW	MF-Id	MF-Veri
(2×)128	16 MB	1.44G	158 ms	99.34	87.19	90.82
(Orig.)64	4.8 MB	0.38G	84 ms	99.11	83.96	87.57
(1/2)32	1.7 MB	0.11G	49 ms	98.55	74.32	78.71
(1/4)16	648 KB	0.03G	34 ms	97.60	52.60	58.69
(1/8)8	304 KB	0.01G	28 ms	94.29	25.32	27.04

we can conclude that smaller networks usually exhibit lower capability for face recognition. To achieve a good balance between different factors, we employ the (1/2)MobileFaceNet as the student network in the following experiments unless otherwise specified.

Table 4. Performance (%) comparison of different diversities.

	LFW	BLUFR	MF-Id	MF-Veri
SEResNet50-IR(T)	99.21	96.41	86.14	88.32
(1/2)MobileFaceNet(S)	98.64	89.32	69.54	74.93
$\text{FC}_{\text{wd}=0}$ [Eq. (3)]	97.43	64.26	51.65	24.50
$\text{FC}_{\text{wd}=5\text{e}-4}$ [Eq. (3)]	98.55	90.29	74.32	78.71
FC+Orth. [29]	98.30	86.19	67.80	73.86
FC+MHE [27]	**98.86**	90.44	74.58	79.39
FC+Exclusivity	98.63	**91.07**	**75.29**	**79.56**

Position-Aware Exclusivity vs. Value-Aware Regularizations. Promoting orthogonality (Orth.) [29] or minimum hyperspherical energy (MHE) [27] among filters has been a popular choice for encouraging diversity. However, their assumption of normalized weights is limited and usually lead to very slow convergence. From the values in Table 4, we can see that the weight decay (*e.g.*, wd = 5e−4 vs. wd = 0) to control the norm of filters is not negligible. Moreover, the diversity to determine the direction of filters is also important. The experiments show that our position-aware exclusivity can achieve better diversity and result in higher performance than previous value-aware regularizations.

Convergence. For our EC-KD method, we alternatively update the re-weighted matrix $\boldsymbol{G}$ and the filters $\boldsymbol{W}$. Specifically, we compute the re-weighted matrix $\boldsymbol{G}$ in the forward and update the filters $\boldsymbol{W}$ in the backward, which is different from the standard stochastic gradient descent (SGD) algorithm. Although the convergence of our method is not easy to be theoretically analyzed, it would be intuitive to see its empirical behavior. Here, we give the loss changes as the

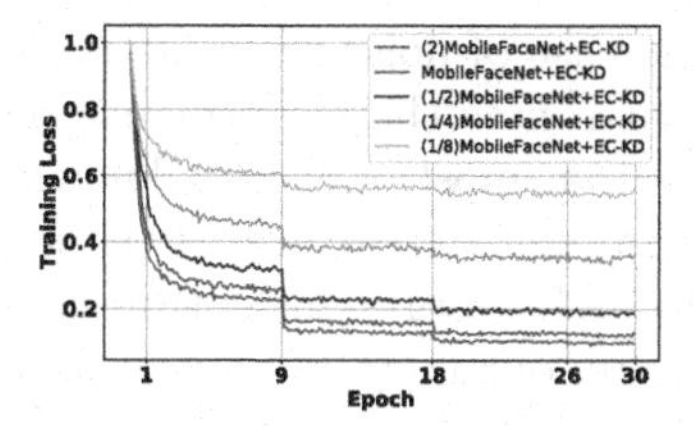

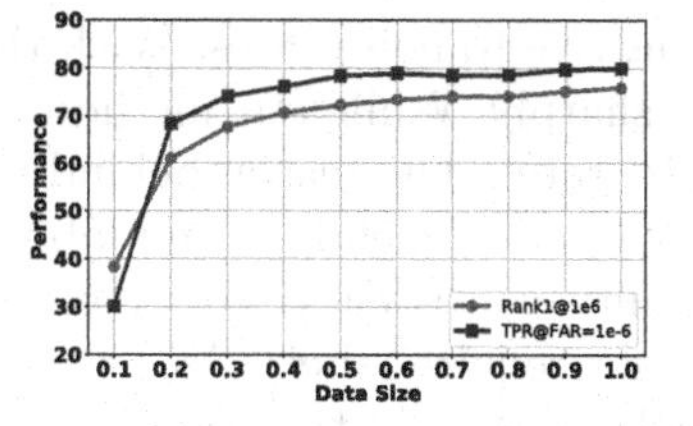

Fig. 2. Left: Convergence of EC-KD. **Right**: Effect of different data size.

number of epochs increases. From the curves in the left of Fig. 2, it can be observed that our method has a good behavior of convergence under various models. Moreover, we can also observe that smaller networks suffer from low-capability and their loss values are generally higher than the larger ones.

Effect of Data Size. Since the training set of student network can be unlabelled, we evaluate the performance gain from using different percentages of unlabelled training data in our EC-KD algorithm. Various percentages (from 10% to 100%, with step 10%) of unlabelled data are randomly emerged for training the target student network. As can be seen from the curves in the right of Fig. 2, at the beginning, the performance on MegaFace goes up as the amount of unlabelled data increases, but the improvement is minor when the percentage of data is large enough. So we conclude that our method can benefit from the small scale training sets and thus can reduce the training time.

Table 5. Performance (%) comparison of different methods with different noise rates.

	LFW	BLUFR	MF-Id.	MF-Veri.
SEResNet50-IR(T)	99.21	96.41	86.14	88.32
(1/2)MobileFaceNet(S)	98.64	89.32	69.54	74.93
KD [16] (symmetric = 0.1)	98.30	77.77	64.09	41.42
EC-KD (symmetric = 0.1)	**98.93**	**91.11**	**75.94**	**79.87**
KD [16] (symmetric = 0.2)	94.98	40.77	30.82	5.83
EC-KD (symmetric = 0.2)	**98.91**	**91.87**	**75.78**	**80.45**
KD [16] (symmetric = 0.3)	93.56	27.18	25.53	2.15
EC-KD (symmetric = 0.3)	**98.89**	**91.03**	**75.98**	**79.64**

Effect of Noisy Labels. To validate the robustness of our method under the case that the training set of student network contains noisy labels, in this experiments, we use the training set CASIA-WebFace-R with different synthetic noise rates to train the student network. The symmetric noise is generated by randomly selecting a label with equal probabilities among all the classes [45]. From the values in Table 5, we can see that the probability consistency method KD [16] is very sensitive to noise rates. With the increase of noise rates, its performance decreases sharply. While our method can guarantee the performance regardless of the noise rates. The reason behind this is that most of current knowledge distillation methods depend on well-labelled training set because of the knowledge of probability consistency. Our method EC-KD resorts to feature consistency and dose not require labels. As a consequence, our method is insensitive to the noisy labels existing in the training set.

Table 6. Performance (%) comparison of different methods by using new training sets.

Training set	Method	LFW	BLUFR	MF-Id.	MF-Veri.
CASIA-WebFace-R	SEResNet50-IR(T)	99.21	96.41	86.14	88.32
IQIYI-VID-Training	(1/2)MobileFaceNet(S)	86.08	66.68	29.62	26.90
	EC-KD (**Ours**)	**98.41**	**85.18**	**64.27**	**67.47**

Generalization Capability. In practice, it is hard to know how the teacher network was pre-trained. More frequently, we only have the teacher model at hand. In this case, we may face the situation that the training set of student is different from teacher's. For example, the teacher is pre-trained by CASIA-WebFace-R dataset but we can only get a new dataset (*e.g.*, IQIYI-VID-Training [30]) to train the target student network. As shown in Table 6, it can be seen that directly training the student network from scratch (*i.e*, (1/2)MobileFaceNet) is hard to boost the performance. Current knowledge distillation methods like KD [16] are unable to train the student network because the training classes of teacher and student are different. In contrast, our method EC-KD can not only be used for training the student network, but also be used to effectively transfer the useful knowledge and achieve higher performance.

4.4 Comparison to State-of-the-Art Methods

Results on LFW, CALFW, CPLFW, SLLFW, AgeDB and CFP. Table 7 shows the results of different approaches on LFW [17], CALFW [59], CPLFW [60], SLLFW [8], AgeDB [32] and CFP [38] sets. The bold values in each column represent the best result. From the numbers, we observe that most of the knowledge distillation methods are better than simply training the student network from scratch (*i.e.*, (1/2)MobileFaceNet). Among all the competitors, the Selection [31], AB [15] and BSS [14] seem to show better generalization ability than others. For our method, we boost about 2% average improvement over the baseline (1/2)MobileFaceNet. Although we cannot beat the competitors on each test set, we achieve the best average (Avg.) performance on these six test sets than the best competitor Selection [31] because of our position-aware weight exclusivity and hardness-aware feature consistency.

Table 7. Performance (%) of different knowledge distillation methods on LFW, CALFW, CPLFW, SLLFW, AgeDB and CFP.

Method	LFW	CALFW	CPLFW	SLLFW	AgeDB	CFP	**Avg.**
SEResNet50-IR	99.21	90.78	84.06	97.76	93.98	93.38	93.23
(1/2)MobileFaceNet	98.64	87.79	78.03	94.39	89.91	86.52	89.21
KD [16]	98.81	89.35	76.38	95.13	90.95	85.11	89.28
FitNet [36]	**99.06**	89.33	77.51	95.41	91.21	87.01	89.92
Selection [31]	98.66	89.06	79.83	95.15	91.50	89.30	90.58
AB [15]	97.54	85.93	74.30	92.78	92.06	89.95	88.76
BSS [14]	98.98	89.18	77.28	95.51	91.78	85.14	89.64
ONE [22]	98.41	88.36	78.06	94.21	89.73	86.08	89.14
EC-KD (**Ours**)	98.96	**89.39**	**80.98**	**95.58**	**92.33**	**90.20**	**91.22**

Results on RFW. Table 8 displays the performance comparison of all the methods on the RFW test set. The results exhibit the same trends that emerged on previous test sets. Concretely, most of the knowledge distillation methods are consistently better than directly training the student network from scratch. For instance, the classical Knowledge Distillation (KD) [16] boost about 2.3% average performance than the baseline (1/2)MobileFaceNet. While for our proposed EC-KD, it can further boost the performance. Specifically, it achieves about 1.5% average improvement on the four subsets of RFW than the KD [16]. From the experiments, we can conclude the effectiveness of our weight exclusivity and hardness-aware feature consistency knowledge.

Results on MegaFace and Trillion-Pairs. Table 9 gives the identification and verification results on MegaFace [33] and Trillion-Pairs challenge. In particular, compared with the baseline, *i.e.*, (1/2)MobileFaceNet, most of the competi-

Table 8. Performance (%) of different methods on RFW.

Method	Caucasian	Indian	Asian	African	**Avg.**
SEResNet50-IR(T)	92.66	88.50	84.00	83.50	87.16
(1/2)MobileFaceNet(S)	84.16	78.33	80.83	77.66	80.24
KD [16]	88.16	80.66	81.16	**80.33**	82.57
FitNet [36]	87.50	82.16	81.83	78.00	82.37
Selection [31]	88.83	79.83	78.83	77.50	81.24
AB [15]	83.33	75.83	74.16	71.16	76.12
BSS [14]	89.16	79.66	81.83	77.16	81.95
ONE [22]	86.83	77.66	80.33	78.33	80.78
EC-KD (**Ours**)	**91.33**	**82.83**	**82.83**	79.49	**84.12**

Table 9. Performance (%) of different methods on MegaFace and Trillion-Pairs.

Method	MF-Id	MF-Veri	TP-Id	TP-Veri	**Avg.**
SEResNet50-IR(T)	86.14	88.32	33.08	32.09	59.90
(1/2)MobileFaceNet(S)	69.54	74.93	16.57	4.77	41.45
KD [16]	69.86	71.67	3.60	1.08	36.55
FitNet [36]	71.74	74.12	5.96	1.24	38.26
Selection [31]	74.50	79.08	16.31	15.65	46.38
AB [15]	**77.42**	75.24	17.13	16.41	46.55
BSS [14]	71.56	73.72	5.25	1.08	37.90
ONE [22]	68.63	74.07	16.60	11.34	42.66
EC-KD (**Ours**)	75.89	**79.87**	**17.67**	**16.96**	**47.59**

tors have shown their strong abilities to fit the teacher knowledge and usually achieve better performance on MegaFace. While on Trillion-Pairs, these methods usually fail. The reason may be that the knowledge of probability consistency is hard to preserve at the small rank and very low false alarm rate, due to its high-dimensionality. For feature consistency knowledge, it has been proved that it can preserve much more information than probability consistency knowledge. Moreover, with the weight exclusivity to exploit the diverse information among different filters, our EC-KD can keep the performance, even at the small rank or very low false alarm rate (*e.g.* 0.54% for Id. rate and 0.55 % for Veri. rate over the best competitor AB [15] on Trillion-Pairs).

Table 10. Performance (%) of different methods on IQIYI-VID test set.

Training set	Method	MAP(%)
CASIA-WebFace-R	SEResNet50-IR(T)	45.83
	(1/2)MobileFaceNet(S)	26.43
CASIA-WebFace-R	KD [16]	21.26
	FitNet [36]	21.74
	Selection [31]	32.76
	AB [15]	23.51
	BSS [14]	21.37
	ONE [22]	25.45
	EC-KD (**Ours**)	33.28
IQIYI-VID-Training [30]	EC-KD (**Ours**)	**38.03**

Results on IQIYI-VID. Table 10 shows the mean of the average accuracy rate (MAP) of different methods on IQIYI-VID test set. The performance is not high because of the domain gap between the training set CASIA-WebFace-R and the test set IQIYI-VID. Specifically, directly training the student network from scratch can only reach MAP = 26.43%. Moreover, we empirically find that most of the competitors degrade the performance in this case. While for our method, despite the large domain gap between training and test sets, we still keep the useful knowledge and can achieve higher performance (about 7% improvement over the baseline (1/2)MobileFaceNet). These experiments have also shown the generalization capability of our method. Moreover, if we can collect the training set that are similar to the test set (without needing the labeled training data), we can further improve the performance (38.03% vs. 33.28%).

5 Conclusion

In this paper, we have exploited a new measurement of diversity, *i.e.*, exclusivity, to improve the low-capability of student network. Different from the weight decay, which is a unary cost to regularize the norm of filters, our weight exclusivity is a pairwise cost to promote the direction of the filters. Therefore, these two branches are complementary to each other. Moreover, we have demonstrated that the feature consistency knowledge is more flexible and preserves much more information than others for face recognition distillation. Incorporating the weight exclusivity and the hardness-aware feature consistency together gives birth to a new knowledge distillation, namely EC-KD. Extensive experiments on a variety of face recognition benchmarks have validated the effectiveness and generalization capability of our method.

Acknowledgement. This work was supported in part by the National Key Research & Development Program (No. 2020YFC2003901), Chinese National Natural Science Foundation Projects #61872367 and partially supported by Beijing Academy of Artificial Intelligence (BAAI).

References

1. Aguinaldo, A., Chiang, P.Y., Gain, A., Patil, A., Pearson, K., Feizi, S.: Compressing gans using knowledge distillation. arXiv preprint arXiv:1902.00159 (2019)
2. Chen, G., Choi, W., Yu, X., Han, T., Chandraker, M.: Learning efficient object detection models with knowledge distillation. In: Advances in Neural Information Processing Systems, pp. 742–751 (2017)
3. Chen, S., Liu, Y., Gao, X., Han, Z.: MobileFaceNets: efficient CNNs for accurate real-time face verification on mobile devices. In: Zhou, J., et al. (eds.) CCBR 2018. LNCS, vol. 10996, pp. 428–438. Springer, Cham (2018). https://doi.org/10.1007/978-3-319-97909-0_46
4. Chen, Y., Wang, N., Zhang, Z.: Darkrank: accelerating deep metric learning via cross sample similarities transfer. In: Thirty-Second AAAI Conference on Artificial Intelligence (2018)

5. Cogswell, M., Ahmed, F., Girshick, R., Zitnick, L., Batra, D.: Reducing overfitting in deep networks by decorrelating representations. arXiv preprint arXiv:1511.06068 (2015)
6. DeepGlint (2018). http://trillionpairs.deepglint.com/overview
7. Deng, J., Guo, J., Xue, N., Zafeiriou, S.: Arcface: additive angular margin loss for deep face recognition. In: Proceedings of the IEEE Conference on Computer Vision and Pattern Recognition, pp. 4690–4699 (2019)
8. Deng, W., Hu, J., Zhang, N., Chen, B., Guo, J.: Fine-grained face verification: FGLFW database, baselines, and human-DCMN partnership. Pattern Recogn. **66**, 63–73 (2017)
9. Duong, C.N., Luu, K., Quach, K.G., Le, N.: Shrinkteanet: million-scale lightweight face recognition via shrinking teacher-student networks. arXiv preprint arXiv:1905.10620 (2019)
10. Feng, Y., Wang, H., Hu, R., Yi, D.T.: Triplet distillation for deep face recognition. arXiv preprint arXiv:1905.04457 (2019)
11. Feng, Z.H., Kittler, J., Awais, M., Huber, P., Wu, X.J.: Wing loss for robust facial landmark localisation with convolutional neural networks. In: Proceedings of the IEEE Conference on Computer Vision and Pattern Recognition, pp. 2235–2245 (2018)
12. Guo, J., Zhu, X., Lei, Z., Li, S.Z.: Face synthesis for eyeglass-robust face recognition. In: Zhou, J., et al. (eds.) CCBR 2018. LNCS, vol. 10996, pp. 275–284. Springer, Cham (2018). https://doi.org/10.1007/978-3-319-97909-0_30
13. Guo, J., Zhu, X., Zhao, C., Cao, D., Lei, Z., Li, S.Z.: Learning meta face recognition in unseen domains. In: Proceedings of the IEEE/CVF Conference on Computer Vision and Pattern Recognition, pp. 6163–6172 (2020)
14. Heo, B., Lee, M., Yun, S., Choi, J.Y.: Knowledge distillation with adversarial samples supporting decision boundary. arXiv preprint arXiv:1805.05532, vol. 3 (2018)
15. Heo, B., Lee, M., Yun, S., Choi, J.Y.: Knowledge transfer via distillation of activation boundaries formed by hidden neurons. arXiv preprint arXiv:1811.03233 (2018)
16. Hinton, G., Vinyals, O., Dean., J.: Distilling the knowledge in a neural network. In: arXiv preprint arXiv:1503.02531 (2015)
17. Huang, G., Ramesh, M., Miller., E.: Labeled faces in the wild: a database for studying face recognition in unconstrained enviroments. Technical report (2007)
18. Jin, X., et al.: Knowledge distillation via route constrained optimization. arXiv preprint arXiv:1904.09149 (2019)
19. Karlekar, J., Feng, J., Wong, Z.S., Pranata, S.: Deep face recognition model compression via knowledge transfer and distillation. arXiv preprint arXiv:1906.00619 (2019)
20. Kemelmacher-Shlizerman, I., Seitz, S.M., Miller, D., Brossard, E.: The megaface benchmark: 1 million faces for recognition at scale. In: Proceedings of the IEEE Conference on Computer Vision and Pattern Recognition, pp. 4873–4882 (2016)
21. Kong, D., Fujimaki, R., Liu, J., Nie, F., Ding, C.: Exclusive feature learning on arbitrary structures via ℓ_{1, 2}-norm. In: Advances in Neural Information Processing Systems, pp. 1655–1663 (2014)
22. Lan, X., Zhu, X., Gong, S., Lan, X.: Knowledge distillation by on-the-fly native ensemble. arXiv preprint arXiv:1806.04606 (2018)
23. Li, Q., Jin, S., Yan, J.: Mimicking very efficient network for object detection. In: Proceedings of the IEEE Conference on Computer Vision and Pattern Recognition, pp. 6356–6364 (2017)

24. Liang, X., Wang, X., Lei, Z., Liao, S., Li, S.Z.: Soft-margin softmax for deep classification. In: Liu, D., Xie, S., Li, Y., Zhao, D., El-Alfy, E.S. (eds.) ICONIP 2017. LNCS, vol. 10635, pp. 413–421. Springer, Cham (2017). https://doi.org/10.1007/978-3-319-70096-0_43
25. Liao, S., Lei, Z., Yi, D., Li, S.Z.: A benchmark study of large-scale unconstrained face recognition. In: International Conference on Biometrics (2014)
26. Lin, R., et al.: Regularizing neural networks via minimizing hyperspherical energy. In: Proceedings of the IEEE/CVF Conference on Computer Vision and Pattern Recognition, pp. 6917–6927 (2020)
27. Liu, W., et al.: Learning towards minimum hyperspherical energy. In: Advances in Neural Information Processing Systems, pp. 6222–6233 (2018)
28. Liu, W., Wen, Y., Yu, Z., Li, M., Raj, B., Song, L.: Sphereface: deep hypersphere embedding for face recognition. In: Proceedings of the IEEE Conference on Computer Vision and Pattern Recognition, pp. 212–220 (2017)
29. Liu, W., et al.: Deep hyperspherical learning. In: Advances in Neural Information Processing Systems, pp. 3950–3960 (2017)
30. Liu, Y., et al.: iQIYI-VID: a large dataset for multi-modal person identification. arXiv preprint arXiv:1811.07548 (2018)
31. Luo, P., Zhu, Z., Liu, Z., Wang, X., Tang, X.: Face model compression by distilling knowledge from neurons. In: Thirtieth AAAI Conference on Artificial Intelligence (2016)
32. Moschoglou, S., Papaioannou, A., Sagonas, C., Deng, J., Kotsia, I., Zafeiriou, S.: Agedb: the first manually collected, in-the-wild age database. In: Proceedings of the IEEE Conference on Computer Vision and Pattern Recognition Workshops, pp. 51–59 (2017)
33. Nech, A., Kemelmacher-Shlizerman, I.: Level playing field for million scale face recognition. In: Proceedings of the IEEE Conference on Computer Vision and Pattern Recognition, pp. 7044–7053 (2017)
34. Paszke, A., et al.: Automatic differentiation in pytorch (2017)
35. Peng, B., et al.: Correlation congruence for knowledge distillation. In: Proceedings of the IEEE International Conference on Computer Vision, pp. 5007–5016 (2019)
36. Romero, A., Ballas, N., Kahou., S.: Fitnets: hints for thin deep nets. arXiv preprint arXiv:1412.6550 (2014)
37. Schroff, F., Kalenichenko, D., Philbin, J.: Facenet: a unified embedding for face recognition and clustering. In: Proceedings of the IEEE Conference on Computer Vision and Pattern Recognition, pp. 815–823 (2015)
38. Sengupta, S., Chen, J.C., Castillo, C., Patel, V.M., Chellappa, R., Jacobs, D.W.: Frontal to profile face verification in the wild. In: 2016 IEEE Winter Conference on Applications of Computer Vision (WACV), pp. 1–9. IEEE (2016)
39. Sun, Y., Wang, X., Tang, X.: Deeply learned face representations are sparse, selective, and robust. In: Proceedings of the IEEE Conference on Computer Vision and Pattern Recognition, pp. 2892–2900 (2015)
40. Wang, C., Lan, X., Zhang, Y.: Model distillation with knowledge transfer from face classification to alignment and verification. arXiv preprint arXiv:1709.02929 (2017)
41. Wang, F., et al.: The devil of face recognition is in the noise. In: Ferrari, V., Hebert, M., Sminchisescu, C., Weiss, Y. (eds.) ECCV 2018. LNCS, vol. 11213, pp. 780–795. Springer, Cham (2018). https://doi.org/10.1007/978-3-030-01240-3_47
42. Wang, F., Cheng, J., Liu, W., Liu, H.: Additive margin softmax for face verification. IEEE Signal Process. Lett. **25**(7), 926–930 (2018)

43. Wang, M., Deng, W., Hu, J., Peng, J., Tao, X., Huang, Y.: Racial faces in-the-wild: reducing racial bias by deep unsupervised domain adaptation. arXiv:1812.00194 (2018)
44. Wang, X., Shuo, W., Cheng, C., Shifeng, Z., Tao, M.: Loss function search for face recognition. In: Proceedings of the 37-th International Conference on Machine Learning (2020)
45. Wang, X., Wang, S., Wang, J., Shi, H., Mei, T.: Co-mining: deep face recognition with noisy labels. In: Proceedings of the IEEE International Conference on Computer Vision, pp. 9358–9367 (2019)
46. Wang, X., Wang, S., Zhang, S., Fu, T., Shi, H., Mei, T.: Support vector guided softmax loss for face recognition. arXiv:1812.11317 (2018)
47. Wang, X., Zhang, S., Lei, Z., Liu, S., Guo, X., Li, S.Z.: Ensemble soft-margin softmax loss for image classification. arXiv preprint arXiv:1805.03922 (2018)
48. Wang, X., Zhang, S., Wang, S., Fu, T., Shi, H., Mei, T.: Mis-classified vector guided softmax loss for face recognition. arXiv preprint arXiv:1912.00833 (2019)
49. Wright, J., Yang, A.Y., Ganesh, A., Sastry, S.S., Ma, Y.: Robust face recognition via sparse representation. IEEE Trans. Pattern Anal. Mach. Intell. **31**(2), 210–227 (2008)
50. Wu, X., He, R., Sun, Z., Tan, T.: A light CNN for deep face representation with noisy labels. IEEE Trans. Inf. Forensics Secur. **13**(11), 2884–2896 (2018)
51. Xie, P., Singh, A., Xing, E.P.: Uncorrelation and evenness: a new diversity-promoting regularizer. In: International Conference on Machine Learning, pp. 3811–3820 (2017)
52. Yan, M., Zhao, M., Xu, Z., Zhang, Q., Wang, G., Su, Z.: Vargfacenet: an efficient variable group convolutional neural network for lightweight face recognition. In: Proceedings of the IEEE International Conference on Computer Vision Workshops (2019)
53. Yao, T., Pan, Y., Li, Y., Mei, T.: Exploring visual relationship for image captioning. In: Ferrari, V., Hebert, M., Sminchisescu, C., Weiss, Y. (eds.) Computer Vision – ECCV 2018. LNCS, vol. 11218, pp. 711–727. Springer, Cham (2018). https://doi.org/10.1007/978-3-030-01264-9_42
54. Yao, T., Pan, Y., Li, Y., Qiu, Z., Mei, T.: Boosting image captioning with attributes. In: Proceedings of the IEEE International Conference on Computer Vision, pp. 4894–4902 (2017)
55. Yi, D., Lei, Z., Liao, S., Li., S.Z.: Learning face representation from scratch. arXiv:1411.7923. (2014)
56. Zagoruyko, S., Komodakis, N.: Paying more attention to attention: improving the performance of convolutional neural networks via attention transfer. arXiv preprint arXiv:1612.03928 (2016)
57. Zhang, S., Wang, X., Lei, Z., Li, S.Z.: Faceboxes: a CPU real-time and accurate unconstrained face detector. Neurocomputing **364**, 297–309 (2019)
58. Zhang, S., Zhu, X., Lei, Z., Shi, H., Wang, X., Li, S.Z.: Faceboxes: a CPU real-time face detector with high accuracy. In: 2017 IEEE International Joint Conference on Biometrics (IJCB), pp. 1–9. IEEE (2017)
59. Zheng, T., Deng, W., Hu, J., Hu, J.: Cross-age lfw: a database for studying cross-age face recognition in unconstrained environments. arXiv:1708.08197 (2017)
60. Zheng, T., Deng, W., Zheng, T., Deng, W.: Cross-pose lfw: a database for studying crosspose face recognition in unconstrained environments. Technical report (2018)

Learning Camera-Aware Noise Models

Ke-Chi Chang[1,2], Ren Wang[1], Hung-Jin Lin[1], Yu-Lun Liu[1], Chia-Ping Chen[1], Yu-Lin Chang[1], and Hwann-Tzong Chen[2(✉)]

[1] MediaTek Inc., Hsinchu, Taiwan
[2] National Tsing Hua University, Hsinchu, Taiwan
htchen@cs.nthu.edu.tw

Abstract. Modeling imaging sensor noise is a fundamental problem for image processing and computer vision applications. While most previous works adopt statistical noise models, real-world noise is far more complicated and beyond what these models can describe. To tackle this issue, we propose a data-driven approach, where a generative noise model is learned from real-world noise. The proposed noise model is camera-aware, that is, different noise characteristics of different camera sensors can be learned simultaneously, and a single learned noise model can generate different noise for different camera sensors. Experimental results show that our method quantitatively and qualitatively outperforms existing statistical noise models and learning-based methods. The source code and more results are available at https://arcchang1236.github.io/CA-NoiseGAN/.

Keywords: Noise model · Denoising · GANs · Sensor

1 Introduction

Modeling imaging sensor noise is an important task for many image processing and computer vision applications. Besides low-level applications such as image denoising [4,9,25,26], many high-level applications, such as detection or recognition [10,17,20,21], can benefit from a better noise model.

Many existing works assume statistical noise models in their applications. The most common and simplest one is signal-independent additive white Gaussian noise (AWGN) [25]. A combination of Poisson and Gaussian noise, containing both signal-dependent and signal-independent noise, is shown to be a better fit for most camera sensors [7,9].

However, the behavior of real-world noise is very complicated. Different noise can be induced at different stages of an imaging pipeline. Real-world noise includes but is not limited to photon noise, read noise, fixed-pattern noise, dark current noise, row/column noise, and quantization noise. Thus simple statistical noise models can not well describe the behavior of real-world noise.

Recently, several learning-based noise models are proposed to better represent the complexity of real-world noise in a data-driven manner [1,5,12]. In this paper,

Electronic supplementary material The online version of this chapter (https://doi.org/10.1007/978-3-030-58586-0_21) contains supplementary material, which is available to authorized users.

A. Vedaldi et al. (Eds.): ECCV 2020, LNCS 12369, pp. 343–358, 2020.
https://doi.org/10.1007/978-3-030-58586-0_21

we propose a learning-based generative model for signal-dependent synthetic noise. The synthetic noise generated by our model is perceptually more realistic than existing statistical models and other learning-based methods. When used to train a denoising network, better denoising quality can also be achieved.

Moreover, the proposed method is camera-aware. Different noise characteristics of different camera sensors can be learned simultaneously by a single generative noise model. Then this learned noise model can generate different synthetic noise for different camera sensors respectively.

Our main contributions are summarized as follows:

- propose a learning-based generative model for camera sensor noise
- achieve camera awareness by leveraging camera-specific Poisson-Gaussian noise and a camera characteristics encoding network
- design a novel feature matching loss for signal-dependent patterns, which leads to significant improvement of visual quality
- outperform state-of-the-art noise modeling methods and improve image denoising performance.

2 Related Work

Image denoising is one of the most important applications and benchmarks in noise modeling. Similar to the recent success of deep learning in many vision tasks, deep neural networks also dominate recent advances of image denoising.

DnCNN [25] shows that a residual neural network can perform blind denoising well and obtains better results than previous methods on additive white Gaussian noise (AWGN). However, a recent denoising benchmark DND [19], consisting of real photographs, found that the classic BM3D method [6] outperforms DnCNN on real-world noise instead. The main reason is that real-world noise is more complicated than AWGN, and DnCNN failed to generalize to real-world noise because it was trained only with AWGN.

Instead of AWGN, CBDNet [9] and Brooks *et al.* [4] adopt Poisson-Gaussian noise and demonstrate significant improvement on the DND benchmark. Actually, they adopt an approximated version of Poisson-Gaussian noise by a heteroscedastic Gaussian distribution:

$$n \sim \mathcal{N}(0, \delta_{\text{shot}} I + \delta_{\text{read}}), \tag{1}$$

where n is the noise sampling, I is the intensity of a noise-free image, and δ_{shot} and δ_{read} denote the Poisson and Gaussian components, respectively. Moreover, δ_{shot} and δ_{read} for a specific camera sensor can be obtained via a calibration process [14]. The physical meaning of these two components corresponds to the signal-dependent and signal-independent noise of a specific camera sensor.

Recently, several learning-based noise modeling approaches have been proposed [1,5,12]. GCBD [5] is the first GAN-based noise modeling method. Its generative noise model, however, takes only a random vector as input but does not take the intensity of the clean image into account. That means the generated noise is not signal-dependent. Different characteristics between different camera sensors are not considered either. The synthetic noise is learned and imposed on sRGB images, rather than the raw images. These are the reasons why GCBD didn't deliver promising denoising performance on the DND benchmark [19].

GRDN [12] is another GAN-based noise modeling method. Their model was trained with paired data of clean images and real noisy images of smartphone cameras, provided by NTIRE 2019 Real Image Denoising Challenge [3], which is a subset of the SIDD benchmark [2]. In addition to a random seed, the input of the generative noise model also contained many conditioning signals: the noise-free image, an identifier indicating the camera sensor, ISO level, and shutter speed. Although GRDN can generate signal-dependent and camera-aware noise, the denoising network trained with this generative noise model only improved slightly. The potential reasons are two-fold: synthetic noise was learned and imposed on sRGB images, not raw images; a plain camera identifier is too simple to represent noise characteristics of different camera sensors.

Noise Flow [1] applied a flow-based generative model that maximizes the likelihood of real noise on raw images, and then exactly evaluated the noise modeling performance qualitatively and quantitatively. To do so, the authors proposed using Kullback-Leibler divergence and negative log-likelihood as the evaluation metrics. Both training and evaluation were conducted on SIDD [2]. To our knowledge, Noise Flow is the first deep learning-based method that demonstrates significant improvement in both noise modeling and image denoising capabilities. However, they also fed only a camera identifier into a gain layer to represent complex noise characteristics of different camera sensors.

3 Proposed Method

Different from most existing works, the proposed learning-based approach aims to model noise characteristics for each camera sensor. Figure 1 shows an overview of our framework, which comprises two parts: the Noise-Generating Network and the Camera-Encoding Network. The Noise-Generating Network, introduced in Sect. 3.1, learns to generate synthetic noise according to the content of a clean input image and the characteristics of a target camera. The target camera characteristics are extracted via the Camera-Encoding Network from noisy images captured by that target camera, which is illustrated in Sect. 3.2. Finally, Sect. 3.3 shows how to train these two networks in an end-to-end scheme.

3.1 Noise-Generating Network

As depicted in the upper part of Fig. 1, a clean image $\mathbf{I}_{C_i^s}$ from the sth camera and the initial synthetic noise $\tilde{\mathbf{n}}_{\text{init}}$ are fed into a noise generator G and then transformed into various feature representations through convolutional layers. At the last layer, the network produces a residual image $R(\tilde{\mathbf{n}}_{\text{init}}|\mathbf{I}_{C_i^s})$ that approximates the difference between real noise $\mathbf{n} \sim \mathbb{P}_r$ and $\tilde{\mathbf{n}}_{\text{init}}$, where $\mathbb{P}_r$ indicates the

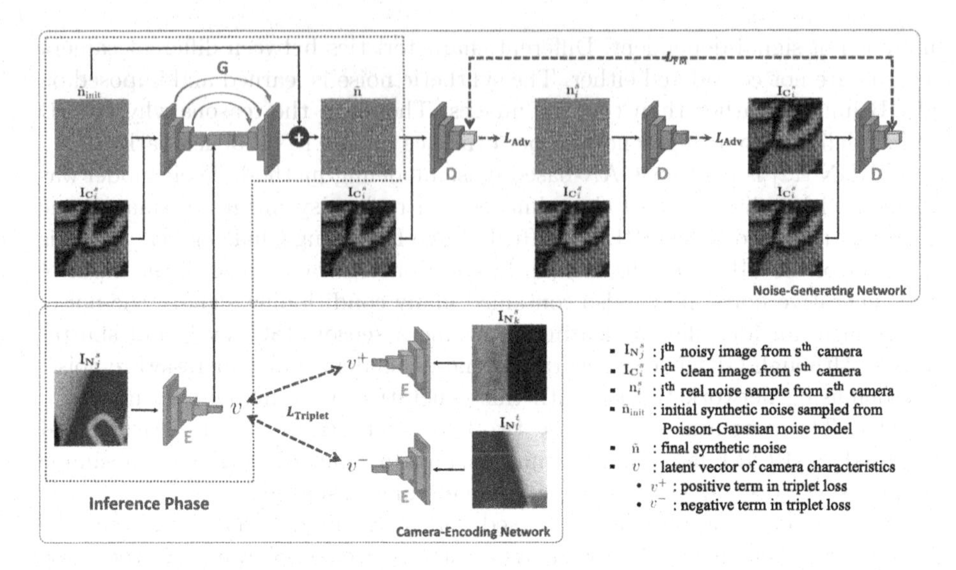

Fig. 1. An overview of our noise-modeling framework. The proposed architecture comprises two sub-networks: the Noise-Generating Network and the Camera-Encoding Network. First, a clean image $\mathbf{I}_{C_i^s}$ and the initial synthetic noise $\tilde{\mathbf{n}}_{\text{init}}$ sampled from Poisson-Gaussian noise model are fed into the generator G. In addition, a latent vector $\boldsymbol{v}$ provided by the camera encoder E, which represents the camera characteristics, is concatenated with the features of the middle layers of G. Eventually, the final synthetic noise $\tilde{\mathbf{n}}$ is generated by G. To jointly train G and E, a discriminator D is introduced for the adversarial loss L_{Adv} and the feature matching loss L_{FM}. Moreover, a triplet loss L_{Triplet} is proposed to let the latent space of $\boldsymbol{v}$ be more reliable

real noise distribution. Ideally, we can generate realistic synthetic noise $\tilde{\mathbf{n}} \approx \mathbf{n}$ from the estimated residual image as

$$\tilde{\mathbf{n}} = G(\tilde{\mathbf{n}}_{\text{init}} | \mathbf{I}_{C_i^s}) = \tilde{\mathbf{n}}_{\text{init}} + R(\tilde{\mathbf{n}}_{\text{init}} | \mathbf{I}_{C_i^s}). \tag{2}$$

To achieve this objective, we adopt adversarial learning for making the generated noise distribution $\mathbb{P}_g$ fit $\mathbb{P}_r$ as closely as possible. A discriminator D is used to measure the distance between distributions by distinguishing real samples from fake ones, such that G can minimize the distance through an adversarial loss L_{Adv}. Therefore, we need to collect pairs of clean images and real noise $(\mathbf{I}_{C_i^s}, \mathbf{n}_i^s)$ as the real samples.

A real noise sample $\mathbf{n}_i^s$ can be acquired by subtracting $\mathbf{I}_{C_i^s}$ from the corresponding noisy image $\mathbf{I}_{N_i^s}$, *i.e.*, $\mathbf{n}_i^s = \mathbf{I}_{N_i^s} - \mathbf{I}_{C_i^s}$. Note that a clean image could have many corresponding noisy images because noisy images can be captured at different ISOs to cover a wide range of noise levels. For simplicity, we let i denote not only the scene but also the shooting settings of a noisy image.

In addition to measuring the distance in adversarial learning, the discriminator D also plays another role in our framework. It is observed that some signal-dependent patterns like spots or stripes are common in real noise; hence we propose a feature matching loss L_{FM} and treat D as a feature extractor. The feature matching loss forces the generated noise $\tilde{\mathbf{n}}$ and the clean image $\mathbf{I}_{C_i^s}$ to share similar high-level features because we assume these signal-dependent patterns should be the most salient traits in clean images.

It is worthwhile to mention that a noise model should be capable of generating a variety of reasonable noise samples for the same input image and noise level. GANs usually take a random vector sampled from Gaussian distribution as the input of the generator to ensure this stochastic property. In most cases, this random vector is not directly relevant to the main task. However, our goal is exactly to generate random noise, which implies that this random vector could be treated as the initial synthetic noise. Moreover, Gaussian distribution can be replaced with a more representative statistical noise model. For this reason, we apply Poisson-Gaussian noise model to the initial synthetic noise $\tilde{\mathbf{n}}_{\text{init}}$ as in (1):

$$\tilde{\mathbf{n}}_{\text{init}} \sim \mathcal{N}(0, {\delta_{\text{shot}}}_i^s \mathbf{I}_{C_i^s} + {\delta_{\text{read}}}_i^s)\,, \tag{3}$$

where ${\delta_{\text{shot}}}_i^s$ and ${\delta_{\text{read}}}_i^s$ are the Poisson and the Gaussian component for $\mathbf{I}_{N_i^s}$, respectively. Note that these two parameters not only describe the preliminary noise model for the sth camera but also control the noise level of $\tilde{\mathbf{n}}_{\text{init}}$ and $\tilde{\mathbf{n}}$.

3.2 Camera-Encoding Network

FUNIT [16] has shown that encoding the class information is helpful to specify the class domain for an input image. Inspired by their work, we would like to encode the camera characteristics in an effective representation. Since ${\delta_{\text{shot}}}_i^s$ and ${\delta_{\text{read}}}_i^s$ are related to the sth camera in (3), the generator G is actually aware of the camera characteristics from $\tilde{\mathbf{n}}_{\text{init}}$. However, this awareness is limited to the assumption of the Poisson-Gaussian noise model. We, therefore, propose a novel Camera-Encoding Network to overcome this problem.

As depicted in the lower part of Fig. 1, a noisy image $\mathbf{I}_{N_j^s}$ is fed into a camera encoder E and then transformed into a latent vector $\boldsymbol{v} = E(\mathbf{I}_{N_j^s})$. After that, the latent vector $\boldsymbol{v}$ is concatenated with the middle layers of G. Thus, the final synthetic noise is rewritten as

$$\tilde{\mathbf{n}} = G(\tilde{\mathbf{n}}_{\text{init}}|\mathbf{I}_{C_i^s}, \boldsymbol{v}) = \tilde{\mathbf{n}}_{\text{init}} + R(\tilde{\mathbf{n}}_{\text{init}}|\mathbf{I}_{C_i^s}, \boldsymbol{v}). \tag{4}$$

We consider $\boldsymbol{v}$ as a representation for the characteristics of the sth camera and expect G can generate more realistic noise with this latent vector.

Aiming at this goal, the camera encoder E must have the ability to extract the core information for each camera, regardless of the content of input images. Therefore, a subtle but important detail here is that we feed the jth noisy image rather than the ith noisy image into E, whereas G takes the ith clean image as its input. Specifically, the jth noisy image is randomly selected from the data of the sth camera. Consequently, E has to provide latent vectors beneficial to the generated noise but ignoring the content of input images.

Additionally, some regularization should be imposed on $\boldsymbol{v}$ to make the latent space more reliable. FUNIT calculates the mean over a set of class images to provide a representative class code. Nevertheless, this approach assumes that the latent space consists of hypersphere manifolds. Apart from FUNIT, we use a triplet loss L_{Triplet} as the regularization. The triplet loss is used to minimize the intra-camera distances while maximizing the inter-camera distances, which allows the latent space to be more robust to image content. The detailed formulation will be shown in the next section.

One more thing worth clarifying is why the latent vector $\boldsymbol{v}$ is extracted from the noisy image $\mathbf{I}_{N_j^s}$ rather than the real noise sample $\mathbf{n}_j^s$. The reason is out of consideration for making data preparation easier in the inference phase, which is shown as the violet block in Fig. 1. Collecting paired data $(\mathbf{I}_{C_j^s}, \mathbf{I}_{N_j^s})$ to acquire $\mathbf{n}_j^s$ is cumbersome and time-consuming in real world. With directly using noisy images to extract latent vectors, there is no need to prepare a large number of paired data during the inference phase.

3.3 Learning

To jointly train the aforementioned networks, we have briefly introduced three loss functions: 1) the adversarial loss L_{Adv}, 2) the feature matching loss L_{FM}, and 3) the triplet loss L_{Triplet}. In this section, we describe the formulations for these loss functions in detail.

Adversarial Loss. GANs are well-known for reducing the divergence between the generated data distribution and real data distribution in the high-dimensional image space. However, there are several GAN frameworks for achieving this goal. Among these frameworks, we choose WGAN-GP [8] to calculate the adversarial loss L_{Adv}, which minimizes Wasserstein distance for stabilizing the training. The L_{Adv} is thus defined as

$$L_{\text{Adv}} = -\mathop{\mathbb{E}}_{\tilde{\mathbf{n}} \sim \mathbb{P}_g} [D(\tilde{\mathbf{n}}|\mathbf{I}_C)], \tag{5}$$

where D scores the realness of the generated noise. In more depth, scores are given at the scale of patches rather than whole images because we apply a PatchGAN [11] architecture to D. The advantage of using this architecture is that it prefers to capture high-frequency information, which is associated with the characteristics of noise.

On the other hand, the discriminator D is trained by

$$L_D = \mathop{\mathbb{E}}_{\tilde{\mathbf{n}} \sim \mathbb{P}_g} [D(\tilde{\mathbf{n}}|\mathbf{I}_C)] - \mathop{\mathbb{E}}_{\mathbf{n} \sim \mathbb{P}_r} [D(\mathbf{n}|\mathbf{I}_C)] + \lambda_{\text{gp}} \mathop{\mathbb{E}}_{\hat{\mathbf{n}} \sim \mathbb{P}_{\hat{\mathbf{n}}}} [(\|\nabla_{\hat{\mathbf{n}}} D(\hat{\mathbf{n}}|\mathbf{I}_C)\|_2 - 1)^2], \tag{6}$$

where λ_{gp} is the weight of gradient penalty, and $\mathbb{P}_{\hat{\mathbf{n}}}$ is the distribution sampling uniformly along straight lines between paired points sampled from $\mathbb{P}_g$ and $\mathbb{P}_r$.

Feature Matching Loss. In order to regularize the training for GANs, some works [16,23] apply the feature matching loss and extract features through the discriminator networks. Following these works, we propose a feature matching loss L_{FM} to encourage G to generate signal-dependent patterns in synthetic noise. The L_{FM} is then calculated as

$$L_{\mathrm{FM}} = \mathop{\mathbb{E}}_{\tilde{\mathbf{n}} \sim \mathbb{P}_g} \left[\| D_f(\tilde{\mathbf{n}}|\mathbf{I}_C) - D_f(\mathbf{I}_C|\mathbf{I}_C) \|_1 \right], \tag{7}$$

where D_f denotes the feature extractor constructed by removing the last layer from D. Note that D_f is not optimized by L_{FM}.

Triplet Loss. The triplet loss was first proposed to illustrate the triplet relation in embedding space by [22]. We use the triplet loss to let the latent vector $\boldsymbol{v} = E(\mathbf{I}_{N_j^s})$ be more robust to the content of noisy image. Here we define the positive term $\boldsymbol{v}^+$ as the latent vector also extracted from the sth camera, and the negative term $\boldsymbol{v}^-$ is from a different camera on the contrary. In particular, $\boldsymbol{v}^+$ and $\boldsymbol{v}^-$ are obtained by encoding the randomly selected noisy images $\mathbf{I}_{N_k^s}$ and $\mathbf{I}_{N_l^t}$, respectively. Note that $\mathbf{I}_{N_k^s}$ is not restricted to any shooting setting, which means the images captured with different shooting settings of the same camera are treated as positive samples. The objective is to minimize the intra-camera distances while maximizing the inter-camera distances. The triplet loss L_{Triplet} is thus given by

$$L_{\mathrm{Triplet}} = \mathop{\mathbb{E}}_{\boldsymbol{v}, \boldsymbol{v}^+, \boldsymbol{v}^- \sim \mathbb{P}_e} \left[\max(0, \left\| \boldsymbol{v} - \boldsymbol{v}^+ \right\|_2 - \left\| \boldsymbol{v} - \boldsymbol{v}^- \right\|_2 + \alpha) \right], \tag{8}$$

where $\mathbb{P}_e$ is the latent space distribution and α is the margin between positive and negative pairs.

Full Loss. The full objective of the generator G is combined as

$$L_{\mathrm{G}} = L_{\mathrm{Adv}} + \lambda_{\mathrm{FM}} L_{\mathrm{FM}} + \lambda_{\mathrm{Triplet}} L_{\mathrm{Triplet}}, \tag{9}$$

where λ_{FM} and $\lambda_{\mathrm{Triplet}}$ control the relative importance for each loss term.

4 Experimental Results

In this section, we first describe our experiment settings and the implementation details. Then, Sect. 4.1 shows the quantitative and qualitative results. Sect. 4.2 presents extensive ablation studies to justify our design choices. The effectiveness and robustness of the Camera-Encoding Network are evaluated in Sect. 4.3.

Dataset. We train and evaluate our method on Smartphone Image Denoising Dataset (SIDD) [2], which consists of approximately 24,000 pairs of real noisy-clean images. The images are captured by five different smartphone cameras: Google Pixel, iPhone 7, Samsung Galaxy S6 Edge, Motorola Nexus 6, and LG G4. These images are taken in ten different scenes and under a variety of lighting conditions and ISOs. SIDD is currently the most abundant dataset available for real noisy and clean image pairs.

Table 1. Quantitative evaluation of different noise models. Our proposed noise model yields the best Kullback-Leibler divergence (D_{KL}). Relative improvements of our method over other baselines are shown in parentheses

	Gaussian	Poisson-Gaussian	Noise flow	Ours
D_{KL}	0.54707 (99.5%)	0.01006 (74.7%)	0.00912 (72.0%)	**0.00159**

Implementation Details. We apply Bayer preserving augmentation [15] to all SIDD images, including random cropping and horizontal flipping. At both training and testing phases, the images are cropped into 64×64 patches. Totally 650,000 pairs of noisy-clean patches are generated. Then we randomly select 500,000 pairs as the training set and 150,000 pairs as the test set. The scenes in the training set and test set are mutually exclusive to prevent overfitting. Specifically, the scene indices of the test set are 001, 002 and 008, and the remaining indices are used for the training set.

To synthesize the initial synthetic noise $\tilde{\mathbf{n}}_{\mathrm{init}}$, we set the Poisson component $\delta_{\mathrm{shot}}{}_i^s$ and Gaussian component $\delta_{\mathrm{read}}{}_i^s$ in (3) to the values provided by SIDD, which are estimated using the method proposed by [14]. The weight of gradient penalty of L_D in (6) is set to $\lambda_{\mathrm{gp}} = 10$, and the margin of L_{Triplet} in (8) is set to $\alpha = 0.2$. The loss weights of L_G in (9) are set to $\lambda_{\mathrm{FM}} = 1$ and $\lambda_{\mathrm{Triplet}} = 0.5$.

We use the Adam optimizer [13] in all of our experiments, with an initial learning rate of 0.0002, $\beta_1 = 0.5$, and $\beta_2 = 0.999$. Each training batch contains 64 pairs of noisy-clean patches. The generator G, discriminator D, and camera encoder E are jointly trained to convergence with 300 epochs. It takes about 3 days on a single GeForce GTX 1080 Ti GPU.

All of our experiments are conducted on linear raw images. Previous works have shown that many image processing methods perform better in Bayer RAW domain than in sRGB domain [19]. For noise modeling or image denoising, avoiding non-linear transforms (such as gamma correction) or spatial operations (such as demosaicking) is beneficial because we can prevent noise characteristics from being dramatically changed by these operations.

Methods in Comparison. We compare our method with two mostly-used statistical models: Gaussian noise model and Poisson-Gaussian noise model, and one state-of-the-art learning-based method: Noise Flow [1].

4.1 Quantitative and Qualitative Results

To perform the quantitative comparison, we adopt the Kullback-Leibler divergence (D_{KL}) as suggested in [1]. Table 1 shows the average D_{KL} between real noise and synthetic noise generated by different noise models. Our method achieves the smallest average Kullback-Leibler divergence, which means that our method can synthesize more realistic noise than existing methods.

Table 2. Ablation study of our model. L_{adv}: the adversarial loss, L_{FM}: the feature matching loss, E: the Camera-Encoding Network, L_{Triplet}: the triplet loss. The Kullback-Leibler divergence D_{KL} is measured in different settings

L_{Adv}	√	√	√	√
L_{FM}		√	√	√
E			√	√
L_{Triplet}				√
D_{KL}	0.01445	0.01374	0.01412	**0.00159**

Figure 2 shows the synthetic noise generated in linear RAW domain by all noise models and then processed by the camera pipeline toolbox provided by SIDD [2]. Each two consecutive rows represent an image sample for different ISOs (indicated by a number) and different lighting conditions (L and N denote low and normal lighting conditions respectively). Our method can indeed generate synthetic noise that is more realistic and perceptually closer to the real noise.

4.2 Ablation Studies

In this section, we perform ablation studies to investigate how each component contributes to our method, including the feature matching loss L_{FM}, the Camera-Encoding Network E, the triplet loss L_{Triplet}, and the initial synthetic noise $\tilde{\mathbf{n}}_{\text{init}}$. The results are shown in Table 2 and 3.

Feature Matching Loss L_{FM}. Figure 3 shows that the feature matching loss is effective in synthesizing signal-dependent noise patterns and achieving better visual quality. With the feature matching loss L_{FM}, the network is more capable of capturing low-frequency signal-dependent patterns. As shown in Table 2, the Kullback-Leibler divergence can also be improved from 0.01445 to 0.01374.

Camera-Encoding Network and Triplet Loss. The Camera-Encoding Network is designed to represent the noise characteristics of different camera sensors. However, simply adding a Camera-Encoding Network alone provides no advantage (0.01374 → 0.01412), as shown in Table 2. The triplet loss is essential to learn effective camera-specific latent vectors, and the KL divergence can be significantly reduced from 0.01412 to 0.00159.

The camera-specific latent vectors can also be visualized in the t-SNE space [18]. As shown in Fig. 4, the Camera-Encoding Network can effectively extract camera-specific latent vectors from a single noisy image with the triplet loss.

Initial Synthetic Noise $\tilde{\mathbf{n}}_{\text{init}}$. Table 3 shows the average KL divergence when using Gaussian or Poisson-Gaussian noise as the initial noise $\tilde{\mathbf{n}}_{\text{init}}$.

Table 3. Ablation study of the initial synthetic noise $\tilde{\mathbf{n}}_{\text{init}}$. Using Poisson-Gaussian as initial synthetic noise model performs better than using Gaussian

$\tilde{\mathbf{n}}_{\text{init}}$	Gaussian	Poisson-Gaussian
D_{KL}	0.06265	**0.00159**

Table 4. Analysis of noisy images from different cameras. The comparison of Kullback-Leibler divergence for different cameras of the noisy image, where $\tilde{\mathbf{n}}_A = G(\tilde{\mathbf{n}}_{\text{init}}|\mathbf{I}_{C_i^s}, E(\mathbf{I}_{N_j^s}))$ and $\tilde{\mathbf{n}}_B = G(\tilde{\mathbf{n}}_{\text{init}}|\mathbf{I}_{C_i^s}, E(\mathbf{I}_{N_k^t}))$

	$(\tilde{\mathbf{n}}_A\|\mathbf{n}_i^s)$	$(\tilde{\mathbf{n}}_A\|\mathbf{n}_j^s)$	$(\tilde{\mathbf{n}}_B\|\mathbf{n}_i^s)$
D_{KL}	0.00159	0.17921	0.01324

The KL divergence severely degrades from 0.00159 to 0.06265 if we use Gaussian noise instead of Poisson-Gaussian noise. This result shows that using a better synthetic noise as initial and predicting a residual to refine it can yield better-synthesized noise.

4.3 Robustness Analysis of the Camera-Encoding Network

To further verify the behavior and justify the robustness of the Camera-Encoding Network, we design several experiments with different input conditions.

Comparing Noise for Different Imaging Conditions or Different Cameras. Given a clean image $\mathbf{I}_{C_i^s}$ and a noisy image $\mathbf{I}_{N_j^s}$ also from the sth camera, our noise model should generate noise $\tilde{\mathbf{n}}_A = G(\tilde{\mathbf{n}}_{\text{init}}|\mathbf{I}_{C_i^s}, E(\mathbf{I}_{N_j^s}))$. The Kullback-Leibler divergence $D_{\text{KL}}(\tilde{\mathbf{n}}_A\|\mathbf{n}_i^s)$ between the generated noise and the corresponding real noise should be very small (0.00159 in Table 4). On the other hand, $D_{\text{KL}}(\tilde{\mathbf{n}}_A\|\mathbf{n}_j^s)$ between the generated noise and a non-corresponding real noise should be quite large (0.17921 in Table 4), owing to the different imaging conditions, even though the real noise $\mathbf{n}_j^s$ is from the same s^{th} camera.

If the latent vector is extracted by a noisy image of the tth camera instead of the sth camera, the generated noise becomes $\tilde{\mathbf{n}}_B = G(\tilde{\mathbf{n}}_{\text{init}}|\mathbf{I}_{C_i^s}, E(\mathbf{I}_{N_k^t}))$. Because the latent vector is from a different camera, we expect that $D_{\text{KL}}(\tilde{\mathbf{n}}_A\|\mathbf{n}_i^s) < D_{\text{KL}}(\tilde{\mathbf{n}}_B\|\mathbf{n}_i^s)$. Table 4 also verifies these results.

Analysis of Different Noisy Images from the Same Camera. Another important property of the Camera-Encoding Network is that it must capture camera-specific characteristics from a noisy image, and the extracted latent vector should be irrelevant to the image content of the input noisy image. To verify this, we randomly select five different noisy images from the same camera. These different noisy images are fed into the Camera-Encoding Network, while other inputs for the Noise Generating Network are kept fixed. Because these noisy

Table 5. Analysis of different noisy images from the same camera. The Kullback-Leibler divergence results from five randomly selected noisy images but fixed inputs for the generator

Noisy image sets	1st	2nd	3rd	4th	5th
D_{KL}	0.00159	0.00180	0.00183	0.00163	0.00176

images are from the same camera, the generated noise should be robust and consistent. Table 5 shows that the D_{KL} between the generated noise and real noise remains low for different noisy images.

5 Application to Real Image Denoising

5.1 Real-World Image Denoising

We conduct real-world denoising experiments to further compare different noise models. For all noise models, we follow Noise Flow [1] to use the same 9-layer DnCNN network [25] as the baseline denoiser. Learning-based noise models (Noise Flow and ours) are trained with SIDD dataset. We then train a denoiser network with synthetic training pairs generated by each noise model separately.

Table 6 shows the average PSNR and SSIM [24] on the test set. The denoisers trained with statistical noise models (Gaussian and Poisson-Gaussian) are worse than those trained with learning-based noise models (Noise Flow and Ours), which also outperform the denoiser trained with real data only (the last row of Table 6). This is because the amount of synthetic data generated by noise models is unlimited, while the amount of real data is fixed.

Our noise model outperforms Noise Flow in terms of both PSNR and SSIM while using more training data for training noise models leads to better denoising performance. Table 6 also shows that using both real data and our noise model results in further improved PSNR and SSIM.

5.2 Camera-Specific Denoising Networks

To verify the camera-aware ability of our method, we train denoiser networks with our generative noise models, which are trained with and without the Camera-Encoding Network (and with and without the triplet loss) respectively.

For our noise model without the Camera-Encoding Network, we train a single generic denoiser network for all cameras. For our noise models with the Camera-Encoding Network, we train camera-specific denoiser networks with and without the triplet loss for each camera. The denoising performance is shown in Table 7.

The results show that the Camera-Encoding Network with the triplet loss can successfully capture camera-specific noise characteristics and thus enhance the performance of camera-specific denoiser networks.

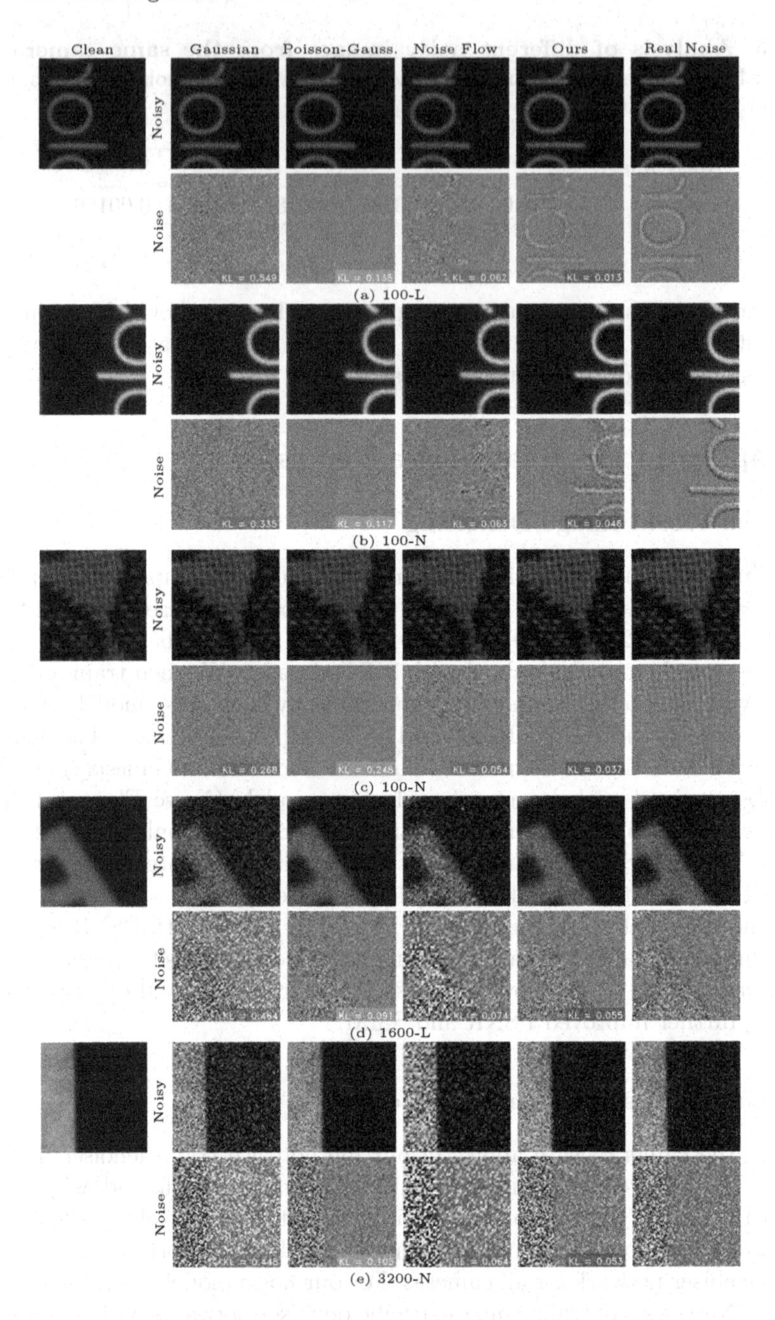

Fig. 2. Visualization of noise models. The synthetic noise samples of several noise modeling methods on different clean images with different ISO/lighting conditions are illustrated Quantitatively, the proposed method outperforms others in terms of D_{KL} measurement with real noise distribution. Furthermore, ours have many clear structures that fit the texture of clean images and real noise. Note that the noise value is scaled up for better visualization purpose

Table 6. Real-world image denoising. The denoising networks using our noise model outperform those using existing statistical noise models and learning-based models. Red indicates the best and blue indicates the second best performance (While training using both synthetic and real data, Ours + Real, synthetic and real data are sampled by a ratio of 5 : 1 in each mini-batch)

Noise model	# of training data for noise model	# of *real* training data for denoiser	PSNR	SSIM
Gaussian	-	-	43.63	0.968
Poisson-Gaussian	-	-	44.99	0.982
Noise Flow [1]	100k	-	47.49	0.991
	500k	-	48.52	0.992
Ours	100k	-	47.97	0.992
	500k	-	48.71	0.993
Ours + Real	100k	100k	47.93	0.994
	500k	500k	48.72	0.994
Real only	-	100k	47.08	0.989
	-	500k	48.30	0.994

Table 7. Real-world image denoising using camera-aware noise model. Grouping the proposed Camera-Encoding Network and triplet loss L_{Triplet} can extract camera-specific latent vectors and thus improve camera-specific denoiser networks

Model	PSNR on test cameras				
	IP	GP	S6	N6	G4
w/o $(E + L_{\text{Triplet}})$	57.4672	44.5180	40.0183	44.7954	51.8048
with E, w/o L_{Triplet}	49.8788	45.7755	40.4976	41.8447	51.8139
with $(E + L_{\text{Triplet}})$	**58.6073**	**45.9624**	**41.8881**	**46.4726**	**53.2610**

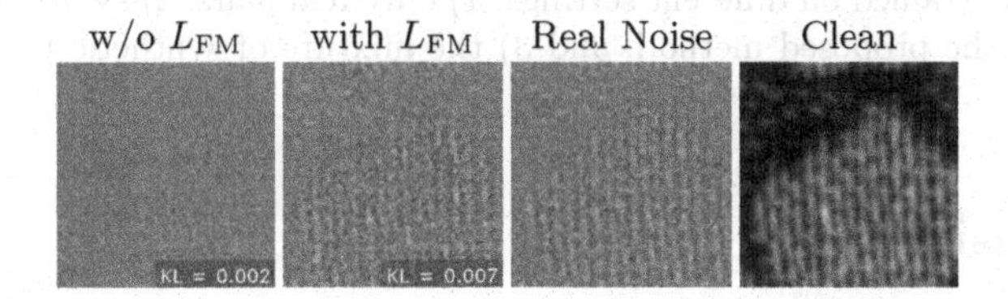

Fig. 3. Visualization of synthetic noise with and without feature matching loss. With the feature matching loss L_{FM}, the generated noise is highly correlated to the image content. Hence, they have more distinct structures resembling the texture of clean images. Besides, the D_{KL} measurements are slightly improved

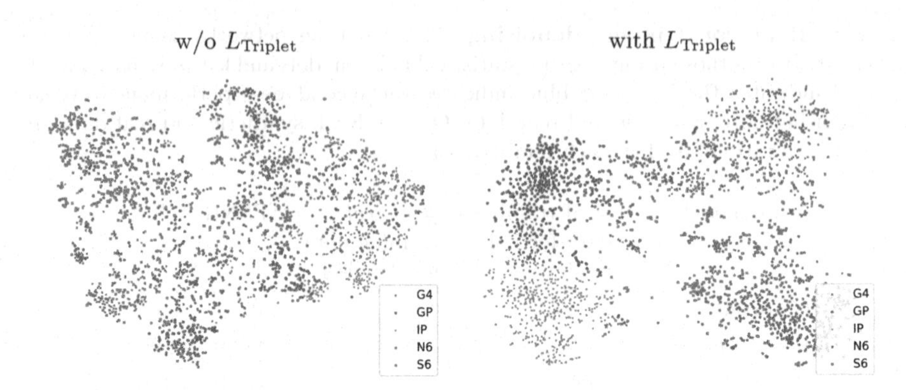

Fig. 4. Ablation study on the distributions of latent vectors from a camera encoder trained with or without L_{Triplet}. We project the encoded latent vectors $\boldsymbol{v}$ of noisy images from five different cameras with t-SNE. The Camera-Encoding Network trained with L_{Triplet} can effectively group the characteristics of different cameras

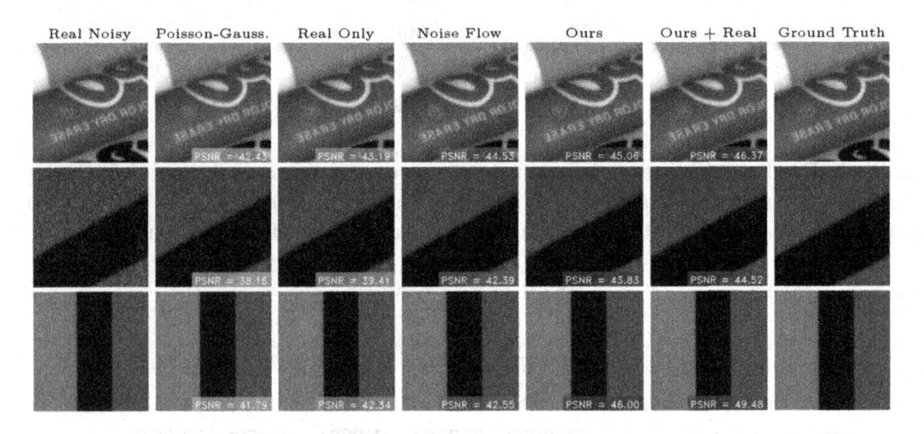

Fig. 5. Results of denoisers trained on different noise models. We compare the denoised results trained on different settings, 1) only real pairs, 2) synthetic pairs with Noise Flow or the proposed method, and 3) the mixture of synthetic pairs from ours and real pairs

6 Conclusions

We have presented a novel learning-based generative method for real-world noise. The proposed noise model outperforms existing statistical models and learning-based methods quantitatively and qualitatively. Moreover, the proposed method can capture different characteristics of different camera sensors in a single noise model. We have also demonstrated that the real-world image denoising task can benefit from our noise model. As for future work, modeling real-world noise with few-shot or one-shot learning could be a possible direction. This could reduce the burden of collecting real data for training a learning-based noise model.

References

1. Abdelhamed, A., Brubaker, M.A., Brown, M.S.: Noise flow: noise modeling with conditional normalizing flows. In: Proceedings of the IEEE International Conference on Computer Vision, pp. 3165–3173 (2019)
2. Abdelhamed, A., Lin, S., Brown, M.S.: A high-quality denoising dataset for smartphone cameras. In: Proceedings of the IEEE Conference on Computer Vision and Pattern Recognition, pp. 1692–1700 (2018)
3. Abdelhamed, A., Timofte, R., Brown, M.S., et al.: Ntire 2019 challenge on real image denoising: Methods and results. In: The IEEE/CVF Conference on Computer Vision and Pattern Recognition Workshops (CVPRW), June 2019
4. Brooks, T., Mildenhall, B., Xue, T., Chen, J., Sharlet, D., Barron, J.T.: Unprocessing images for learned raw denoising. In: Proceedings of the IEEE Conference on Computer Vision and Pattern Recognition, pp. 11036–11045 (2019)
5. Chen, J., Chen, J., Chao, H., Yang, M.: Image blind denoising with generative adversarial network based noise modeling. In: Proceedings of the IEEE Conference on Computer Vision and Pattern Recognition, pp. 3155–3164 (2018)
6. Dabov, K., Foi, A., Katkovnik, V., Egiazarian, K.: Image denoising with block-matching and 3D filtering. In: Image Processing: Algorithms and Systems, Neural Networks, and Machine Learning, vol. 6064, p. 606414. International Society for Optics and Photonics (2006)
7. Foi, A., Trimeche, M., Katkovnik, V., Egiazarian, K.: Practical Poissonian-Gaussian noise modeling and fitting for single-image raw-data. Trans. Image Proc. **17**(10), 1737–1754 (2008). https://doi.org/10.1109/TIP.2008.2001399
8. Gulrajani, I., Ahmed, F., Arjovsky, M., Dumoulin, V., Courville, A.C.: Improved training of Wasserstein GANs. In: Advances in Neural Information Processing Systems, pp. 5767–5777 (2017)
9. Guo, S., Yan, Z., Zhang, K., Zuo, W., Zhang, L.: Toward convolutional blind denoising of real photographs. In: Proceedings of the IEEE Conference on Computer Vision and Pattern Recognition, pp. 1712–1722 (2019)
10. He, K., Gkioxari, G., Dollár, P., Girshick, R.: Mask R-CNN. In: Proceedings of the IEEE International Conference on Computer Vision, pp. 2961–2969 (2017)
11. Isola, P., Zhu, J.Y., Zhou, T., Efros, A.A.: Image-to-image translation with conditional adversarial networks. In: Proceedings of the IEEE Conference on Computer Vision and Pattern Recognition, pp. 1125–1134 (2017)
12. Kim, D.W., Ryun Chung, J., Jung, S.W.: GRDN: grouped residual dense network for real image denoising and GAN-based real-world noise modeling. In: Proceedings of the IEEE Conference on Computer Vision and Pattern Recognition Workshops (2019)
13. Kingma, D.P., Ba, J.: Adam: a method for stochastic optimization. arXiv preprint arXiv:1412.6980 (2014)
14. Liu, C., Szeliski, R., Kang, S.B., Zitnick, C.L., Freeman, W.T.: Automatic estimation and removal of noise from a single image. IEEE Trans. Pattern Anal. Mach. Intell. **30**(2), 299–314 (2007)
15. Liu, J., et al.: Learning raw image denoising with bayer pattern unification and Bayer preserving augmentation. In: Proceedings of the IEEE Conference on Computer Vision and Pattern Recognition Workshops (2019)
16. Liu, M.Y., et al.: Few-shot unsupervised image-to-image translation. In: Proceedings of the IEEE International Conference on Computer Vision, pp. 10551–10560 (2019)

17. Liu, W., et al.: SSD: single shot MultiBox detector. In: Leibe, B., Matas, J., Sebe, N., Welling, M. (eds.) ECCV 2016. LNCS, vol. 9905, pp. 21–37. Springer, Cham (2016). https://doi.org/10.1007/978-3-319-46448-0_2
18. Maaten, L., Hinton, G.: Visualizing data using t-SNE. J. Mach. Learn. Res. **9**, 2579–2605 (2008)
19. Plotz, T., Roth, S.: Benchmarking denoising algorithms with real photographs. In: Proceedings of the IEEE Conference on Computer Vision and Pattern Recognition, pp. 1586–1595 (2017)
20. Redmon, J., Divvala, S., Girshick, R., Farhadi, A.: You only look once: unified, real-time object detection. In: Proceedings of the IEEE Conference on Computer Vision and Pattern Recognition, pp. 779–788 (2016)
21. Ren, S., He, K., Girshick, R., Sun, J.: Faster R-CNN: towards real-time object detection with region proposal networks. In: Advances in Neural Information Processing Systems, pp. 91–99 (2015)
22. Schroff, F., Kalenichenko, D., Philbin, J.: Facenet: a unified embedding for face recognition and clustering. In: Proceedings of the IEEE Conference on Computer Vision and Pattern Recognition, pp. 815–823 (2015)
23. Wang, T.C., Liu, M.Y., Zhu, J.Y., Tao, A., Kautz, J., Catanzaro, B.: High-resolution image synthesis and semantic manipulation with conditional GANs. In: Proceedings of the IEEE Conference on Computer Vision and Pattern Recognition, pp. 8798–8807 (2018)
24. Wang, Z., Bovik, A.C., Sheikh, H.R., Simoncelli, E.P.: Image quality assessment: from error visibility to structural similarity. IEEE Trans. Image Process. **13**(4), 600–612 (2004)
25. Zhang, K., Zuo, W., Chen, Y., Meng, D., Zhang, L.: Beyond a Gaussian denoiser: residual learning of deep CNN for image denoising. IEEE Trans. Image Process. **26**(7), 3142–3155 (2017)
26. Zhang, K., Zuo, W., Zhang, L.: FFDNet: toward a fast and flexible solution for CNN-based image denoising. IEEE Trans. Image Process. **27**(9), 4608–4622 (2018)

Towards Precise Completion of Deformable Shapes

Oshri Halimi[1(✉)], Ido Imanuel[1(✉)], Or Litany[2(✉)], Giovanni Trappolini[3(✉)], Emanuele Rodolà[3(✉)], Leonidas Guibas[2(✉)], and Ron Kimmel[1(✉)]

[1] Technion - Israel Institute of Technology, Haifa, Israel
oshri.halimi@gmail.com, ido.imanuel@gmail.com, ron@cs.technion.ac.il
[2] Stanford University, Stanford, USA
or.litany@gmail.com
[3] Sapienza University of Rome, Rome, Italy
giovanni.trappolini@uniroma1.it, rodola@di.uniroma1.it

Abstract. According to Aristotle, *"the whole is greater than the sum of its parts"*. This statement was adopted to explain human perception by the Gestalt psychology school of thought in the twentieth century. Here, we claim that when observing a part of an object which was previously acquired as a whole, one could deal with both partial correspondence and shape completion in a holistic manner. More specifically, given the geometry of a full, articulated object in a given pose, as well as a partial scan of the same object in a different pose, we address the *new* problem of matching the part to the whole while simultaneously reconstructing the new pose from its partial observation. Our approach is data-driven and takes the form of a Siamese autoencoder without the requirement of a consistent vertex labeling at inference time; as such, it can be used on unorganized point clouds as well as on triangle meshes. We demonstrate the practical effectiveness of our model in the applications of single-view deformable shape completion and dense shape correspondence, both on synthetic and real-world geometric data, where we outperform prior work by a large margin.

Keywords: Shape completion · 3D deep learning · Shape analysis

1 Introduction

One of Aristotle's renowned sayings declares *"the whole is greater than the sum of its parts"*. This fundamental observation was narrowed down to human perception of planar shapes by the Gestalt psychology school of thought in the

O. Halimi and I. Imanuel—Equal contribution.

Electronic supplementary material The online version of this chapter (https://doi.org/10.1007/978-3-030-58586-0_22) contains supplementary material, which is available to authorized users.

A. Vedaldi et al. (Eds.): ECCV 2020, LNCS 12369, pp. 359–377, 2020.
https://doi.org/10.1007/978-3-030-58586-0_22

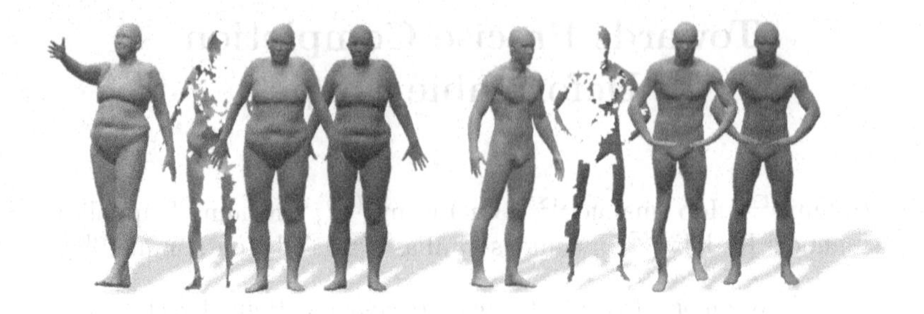

Fig. 1. Left to right: input reference shape, input part, output completion, and the ground truth full model.

twentieth century. A guiding idea of Gestalt theory is the principle of *reification*, arguing that human perception contains more spatial information than can be extracted from the sensory stimulus, and thus giving rise to the view that the mind generates the additional information based on verbatim acquired patterns. Here, we adopt this line of thought in the context of non-rigid shape completion. Specifically, we argue that given access to a complete shape in one pose, one can accurately complete partial views of that shape at any other pose (Fig. 1).

3D data acquisition using depth sensors is often done from a single view point, resulting in an incomplete point cloud. Many downstream applications require completing the partial observations and recovering the full shape. Based on this need, the task of shape completion has been extensively studied in the literature. The required fidelity of completion, however, is task dependent. In fact, in many cases even an approximate completion would be satisfying. For example, completing a car captured from one side by assuming the occluded side is symmetric would be perfectly acceptable for the purpose of obstacle avoidance in autonomous navigation, even if in reality the other side of that car has a large dent. In other cases, however, e.g. when capturing a person for telepresence or medical procedure purposes, it is crucial that the completion is exact, and no hallucination of shape details takes place. Clearly, this requirement is only viable given access to additional measurements or prior information. Here, we wish to focus on the latter case, which we coin as *precise shape completion*. In particular, provided a complete **non-rigid** shape in one pose, we require a solution for completing a partial view of the same shape in a different pose that is **accurate**, **fast**, and can handle **single-view partiality** resulting from self-occlusion. In this work we make a first attempt to address this specific setting, as opposed to the ubiquitous regime of precise *pose*-reconstruction, and as such, we focus solely on types of partialities that induce mild pose ambiguity.

Existing methods for rigid and non-rigid shape completion from partial scans fall largely into two categories: generative and alignment based. Generative methods have proven to be very powerful in completing shapes by learning to match the class distribution. However, they inherently aim at solving an ill-posed problem. Namely, they assume access only to the partial observation at inference time, and thus are incapable of performing *precise* shape completion of shapes unseen at train time. Non-rigid registration methods can take a full shape and align it to a partial observation and thus fit our prescribed setting. However, state of art methods are slow, and can usually handle only mild partiality. Here we propose a new method for precise completion of a partial non-rigid shape in an arbitrary (target) pose, given the full shape in a different (source) pose. Our method is fast, accurate and can handle severe partiality. Based on a deep neural network for point clouds, we learn a function that encodes the partial and full shapes, and outputs the complete shape at the target pose. By providing the full shape, our completion achieves much higher accuracy than existing methods. Since completion in done a single feed-forward pass our solution is orders of magnitude faster than competing methods. In addition, our generated training set of rendered partial views and their corresponding complete shapes covers a broad range of plausible human poses, appearances and partialities which allows our method to gracefully generalize to unseen instances. Finally, our solution effortlessly recovers dense correspondences between the partial and full shapes that considerably improves state of the art performance on the FAUST projection benchmark.

Our main contributions can be summarized as follows:

1. We introduce a deep Siamese architecture to tackle *precise* non-rigid shape completion;
2. Our solution is significantly faster, more accurate and can handle more severe partialities than previous methods.
3. The recovered correspondences achieves state-of-the-art performance in partial shape correspondence.

2 Related Work

Roughly speaking, there exist three approaches that address the challenge of reconstructing the geometry of an articulated shape from its partial scan, namely, partial non-rigid registration of surfaces, surface registration to a known skeleton, and shape completion of a given partial surface. While the first approach is the closest to our setting, none of these approaches has yet provided a good solution to the described application. Currently, state-of-the-art partial nonrigid shape registration/alignment [26,45,47] methods do not handle the significant partiality one often obtains when using commodity depth sensors, and their processing time even for a moderate-size point clouds of few thousand vertices vary between few minutes at best, to a few hours. In our experimental section, we compare with the most efficient methods belonging to this class.

The methods that can handle substantial partiality usually rely on some modification of the iterative closest point (ICP) algorithm and often have difficulties in handling large deformations between the full and the partial shape [50,66]. Another existing approach for the non-rigid alignment problem is to use an explicit deformation model such as skeleton rigging. These methods often completely ignore the detailed geometric and textural information in the actual scanned surface. Moreover, they rely on a rigged model for the full template, which is a limiting assumption when the full model is not restricted to a standard pose. The shape completion setting, as explained below, does not accommodate the full shape and therefore hallucinates details by construction, resulting in inferior completions, as shown by our results and ablation sections. We now turn to review some of the above approaches in more detail.

Shape Completion. Recovering a complete shape from partial or noisy measurements is a longstanding research problem that comes in many flavors. In an early attempt to use one pose in order to geometrically reconstruct another, Devir *et al.* [21] considered mapping a model shape in a given pose onto a noisy version of the shape in a different pose. Elad and Kimmel were the first to treat shapes as metric spaces [22,23]. They matched shapes by comparing second order moments of embedding the intrinsic metric into a Euclidean one via classical scaling. In the context of deformable shapes, early efforts focused on completion based on geometric priors [36] or reoccurring patterns [13,38,40,62]. These methods are not suited for severe partiality. For such cases model-based techniques are quite popular, for example, category-specific parametric morphable models that can be fitted to the partial data [1,5,24,44,65]. Model-based shape completion was demonstrated for key-points input [2], and was recently proven to be quite useful for recovering 3D body shapes from 2D images [28,70,71,78]. Parametric morphable models [5], coupled with axiomatic image formation models were used to train a network to reconstruct face geometry from images [57,58,64]. Still, much less attention has been given to the task of fitting a model to a partial 3D point cloud. Recently, Jiang *et al.* [34] tackled this problem using a skeleton-aware architecture. However, their approach works well when full coverage of the underlying shape is given. [63] proposed a real time solution based on a reinforcement learning agent controlled by a GAN network. [32] reconstructed a 3D completion by generating and back-projecting multi-view depth maps. [75] focused on the ambiguity in completion from a single view, and suggested to address it using adversarially learned shape priors. Finally, [67] suggested a weakly supervised approach and showed performance on realistic data.

Nonrigid Partial Shape Matching. Dense non-rigid shape correspondence [15,17,19,29,37,41,60] is a key challenge in 3D computer vision and graphics, and has been widely explored in the last few years. A particularly challenging setting arises whenever one of the two shapes has missing geometry. Bronstein et al. [9–14] dealt with partial matching of articulated objects in various scenarios, including pruning of the intrinsic structure while accounting for cuts. This setting has been tackled with moderate success in a few recent papers [42,56,59], however, it largely remains an open problem whenever the partial shape exhibits severe artifacts or large, irregular missing parts. In this paper we tackle precisely this setting, demonstrating unprecedented performance on a variety of real-world and synthetic datasets.

Deep Learning of Surfaces. Following the success of convolutional neural networks on images in the recent years, the geometry processing community has been rapidly adopting and designing computational modules suited for such

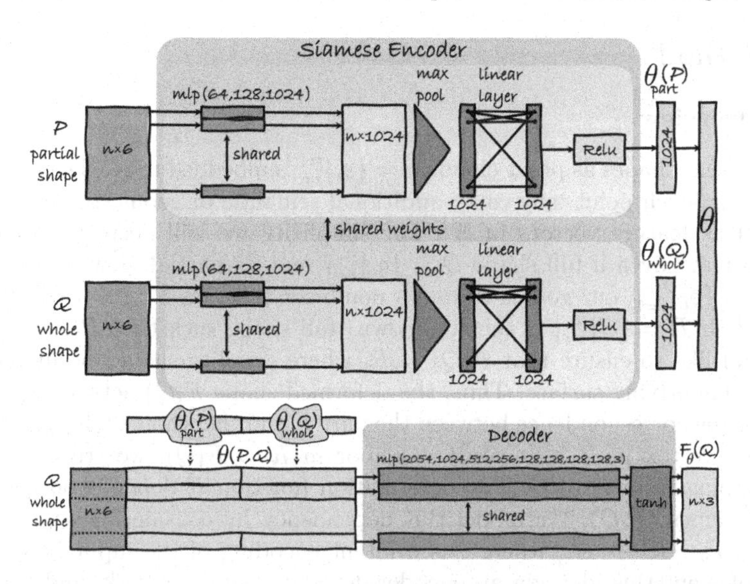

Fig. 2. Network Architecture. Siamese encoder architecture at the top, and the decoder (generator) architecture at the bottom. A shape is provided to the encoder as a list of 6D points, representing the spatial and unit normal coordinates. The latent codes of the input shapes $\theta_{part}(P)$ and $\theta_{whole}(Q)$ are concatenated to form a latent code θ representing the input pair. Based on this latent code, the decoder deforms the full shape by operating on each of its points with the same function. The result is the deformed full shape $F_\theta(\mathrm{Q})$.

data. The main challenge is that unlike images, geometric structures like surfaces come in many types of representations, and each requires a unique handling. Early efforts focused on a simple extension from a single image to multi-view representations [68,74]. Another natural extension are 3D CNNs on volumetric grids [76]. A host of techniques for mesh processing were developed as part of a research branch termed *geometric deep learning* [16]. These include graph-based methods [30,72,73], intrinsic patch extraction [8,48,49], and spectral techniques [29,41]. Point cloud networks [54,55] have recently gained much attention. Offering a light-weight computation restricted to sparse points with a sound geometric explanation [35], these networks have shown to provide a good compromise between complexity and accuracy, and are dominating the field of 3D object detection [53,77], semantic segmentation [3,25], and even temporal point cloud processing [18,43]. For generative methods, recent implicit and parametric methods have demonstrated promising results [27,51]. Following the success of encoding non-rigid shape deformations using a point cloud network [26], here, we also choose to use a point cloud representation. Importantly, while the approach presented in [26] predicts alignment of two shapes, it is not designed to handle severe partiality, and assumes a fixed template for the source shape. Instead, we show how to align arbitrary input shapes and focus on such a partiality.

3 Method

3.1 Overview

We represent shapes as point clouds $S = \{s_i\}_{i=1}^{n_s}$ embedded in $\mathbb{R}^3$. Depending on the setting, each point may carry additional semantic or geometric information encoded as feature vectors in $\mathbb{R}^d$. For simplicity we will keep $d = 3$ in our formulation. Given a full shape $Q = \{q_i\}_{i=1}^{n_q}$ and its partial view in a different pose $P = \{p_i\}_{i=1}^{n_p}$, our goal is to find a nonlinear function $F : \mathbb{R}^3 \rightarrow \mathbb{R}^3$ aligning Q to P[1]. If $R = \{r_i\}_{i=1}^{n_r}$ is the (unknown) full shape such that $P \subset R$, ideally we would like to ensure that $F(Q) = R$, where equality should be understood as same underlying surface. Thus, the deformed shape $F(Q)$ acts as a proxy to solve for the correspondence between the part P and the whole Q. By calculating for every vertex in P its nearest neighbor in $R \approx F(Q)$, we trivially obtain the mapping from P to Q. The deformation function F depends on the input pair of shapes (P, Q). We model this dependency by considering a parametric function $F_\theta : \mathbb{R}^3 \rightarrow \mathbb{R}^3$, where θ is a latent encoding of the input pair (P, Q). We implement this idea via an encoder-decoder neural network, and learn the space of parametric deformations from example pairs of partial and complete shapes, together with full uncropped versions of the partial shapes, serving as the ground truth completion. Our network is composed of an encoder E and a generator F_θ. The encoder takes as input the pair (P, Q) and embeds it into a latent code θ. To map points from Q to their new location, we feed them to the generator along with the latent code. Our network architecture shares a common factor with 3D-CODED architecture [26], namely the deformation of one shape based on the latent code of the another. However [26] uses a fixed template and is therefore only suited for no or mild partiality, as the template cannot make up for lost shape details in the part. Our pipeline on the other hand, is designed to merge two sources of information into the reconstructed model, resulting in an accurate reconstruction under extreme partiality. In the supplementary we perform an analysis where we train our network in a fixed-template setting, similar to 3D-CODED and demonstrate the advantage of our paradigm. In what follows we first describe each module, and then give details on the training procedure and the loss function. We refer to Fig. 2 for a schematic illustration of our learning model.

3.2 Encoder

We encode P and Q using a Siamese pair of single-shape encoders, each producing a global shape descriptor (respectively θ_{part} and θ_{whole}). The two codes are then concatenated so as to encode the information of the specific pair of shapes, $\theta = [\theta_{part}, \theta_{whole}]$. Considering the specific architecture of the single-shape encoder, we think about the encoder network as a channel transforming

[1] In our setting, we assume that the pose can be inferred from the partial shape (*e.g.*, an entirely missing limb would make the prediction ambiguous), hence the deformation function F is well defined.

geometric information to a vector representation. We would like to utilize architectures which have been empirically proven to encode the 3D surface with the least loss of information, thus enabling the decoder to convert the resulting latent code θ to an accurate spatial deformation F_θ. Encouraged by recent methods [26,27] that showed detailed reconstruction using PointNet [54], we also adopt it as our backbone encoder. We provide the encoder 6 input channels, representing the vertex location and the vertex normal field. We justify this design choice in Sect. 3.6. Specifically, our encoder passes all 6D points of the input shape through the same "mlp" units, of hidden dimensions $(64, 128, 1024)$. Here, the term "mlp", carries the same meaning as in PointNet, i.e. multi-layer perceptron, with ReLU activation, and batch normalization layers. After a max-pool operation over the input points, we receive a single 1024-dimensional vector. Finally, we apply a linear layer of size 1024 and a ReLu activation function. Hence, each shape in the input pair is represented by a latent code $\theta_{whole}, \theta_{part}$ of size 1024 respectively. We concatenate these to a joint representation θ of size 2048.

3.3 Generator

Given the code θ, representing the partial and full shapes, the generator has to predict the deformation function F_θ to be applied to the full shape Q. We realize F_θ as a Multi-Layer Perceptron (MLP) that maps an input point q_i on the full shape Q, to its corresponding output point r_i on the ground truth completed shape. The MLP operates pointwise on the tuple (q_i, θ), with θ kept fixed. The result is the destination location $F_\theta(q_i) \in \mathbb{R}^3$, for each input point of the full shape Q. This generator architecture allows, in principle, to calculate the output reconstruction in a flexible resolution, by providing the generator a full shape with some desired output resolution. In detail, the generator consists of 9 layers of hidden dimensions $(2054, 1024, 512, 256, 128, 128, 128, 128, 3)$, followed by a hyperbolic tangent activation function. The output of the decoder is the 3D coordinates. In addition, we can compute a normal field based on the vertex coordinates, making the overall output of the decoder a 6D point. The normal is calculated using the known connectivity of the full shape Q, from our training dataset. Thus, the reconstruction loss in the below section could be generalized and defined using the normal output channel, as well. In the implementation section, we ablate this design choice, and show it leads to a performance improvement.

3.4 Loss Function

The loss definition should reflect the visual plausibility of the reconstructed shape. Measuring such a quality analytically is a challenging problem worth studying on itself. Yet, in this paper we adopt a naive measurement of the Euclidean proximity between the ground-truth and the reconstruction. Formally, we define the loss as,

$$\mathcal{L}(P,Q,R) = \sum_{i=0}^{n_q} \left\| F_{\theta(P,Q)}(q_i) - r_i \right\|^2, \quad (1)$$

where $r_i = \pi^*(q_i) \in R$ is the matched point of $q_i \in Q$, given by the ground-truth mapping $\pi^* : Q \to R$.

3.5 Training Procedure

We train our model using samples from datasets of human shapes. These contain 3D models of different subjects in various poses. The datasets are described in detail in Sect. 4.1. Each training sample is a triplet (P, Q, R) of a partial shape P, a full shape in a different pose Q and a ground truth completion R. The shapes Q and R are sampled from the same subject in two different poses. To receive P we render a depth map of R, at a viewpoint of zero elevation and a random azimuth angle in the range 0° and 360°. These projections approximate the typical partiality pattern of depth sensors. Note that despite the large missing region, these projections largely retain the pose, making the reconstruction task well-defined. We also analysed different types of projections, such as projections from different elevation angles. This analysis is provided in the supplementary. The training examples $(P_n, Q_n, R_n)_{n=1}^{N}$ were provided in batches to the Siamese Network, where N is the size of the train set. Each input pair is fed to the encoder to receive the latent code $\theta(P_n, Q_n)$ and the reconstruction $F_{\theta(P_n,Q_n)}(Q_n)$ is determined by the generator. This reconstruction is subsequently compared against the ground-truth reconstruction R_n using the loss in Eq. (1).

3.6 Implementation Considerations

Our implementation is available at https://github.com/OshriHalimi/precise_shape_completion. The network was trained using the PyTorch [52] ADAM optimizer with a learning rate of 0.001 and a momentum of 0.9. Each training batch contained 10 triplet examples (P, Q, R). The network was trained for 50 epochs, each containing 1000 batches. The input shapes, Q and R, are were centered, such that their center of mass lies at the origin.

Surace Normals. In practice we found it helpful to include normal vectors as additional input features, making each input point 6D. The normal vector field is especially helpful for disambiguating contact points of the surface allowing prevention of contradicting requirements of the estimated deformation function. The input normals were computed using the connectivity for the mesh inputs, and approximated using Hoppe's method [31] for point clouds, as described in the experimental section. We note that at training, we always had access to the mesh connectivity and Hoppe's method was only applied on real scans, at inference time. Additionally, in the loss evaluation, we found that by considering also surface normals, in addition to point coordinates, fine details are better preserved. Therefore, in Eq. (1), we defined r_i as the concatenation of the coordinates and unit normal vector at each point: $(\boldsymbol{x}_{ri}, \alpha\boldsymbol{n}_{ri}) \in \mathbb{R}^6$. We used a scale factor of $\alpha = 0.1$, for the normal vector. To conclude, we used the surface normals in two places: (A) as additional channels for the input shapes, and (B) in the loss definition. To quantify the contribution of each design choice we ran all 4 configurations on FAUST dataset [6]. The relative improvement w.r.t not using normals at all is as follows: A+\B−: 4.6%; A− \B+: −3.3%; A+\B+ (as in the paper): 13%. Our experiments indicate that setting A+ is consistently helpful in disambiguating contact points, and that the chosen setting A+\B+, is the best performing.

ICP Refinement. Empirically, the network reconstruction is often slightly shifted from the source partial scan. To recover the partial correspondence via a nearest neighbor query, it is crucial that the alignment be as exact as possible, and therefore we apply a rigid Iterative Closest Point algorithm [4], as refinement, choosing the moving input as the partial shape, and the fixed input as the network reconstruction. Since the initial alignment is already adequate, this step is both stable and fast.

Activation Function. Studying the displacement field statistics between all pose pairs in our training datasets, we observed that the maximal coordinate displacement is bounded by 1.804 m, and relatively symmetric. Accordingly, in the generator module, we used the activation 2*tanh(x) - a symmetric function, bounded in the range $[-2, 2]$, akin to [26].

4 Experiments

The proposed method tackles two important tasks in nonrigid shape analysis: shape completion and partial shape matching. We emphasize the graceful handling of severe partiality resulting from range scans. In contrast, prior efforts either addressed one of these tasks or attempted to address both at mild partiality conditions. Here, we describe the different datasets used and then evaluate our method on both tasks. Finally, we show performance on real scanned data.

4.1 Datasets

We utilize two datasets of human shapes for training and evaluation, FAUST [6] and AMASS [46]. In addition we use raw scans from Dynamic FAUST [7] for testing purposes only. FAUST was generated by fitting SMPL parametric body model [44] to raw scans. It is a relatively small set of 10 subjects posing at 10 poses each. Following training and evaluation protocols from previous works (e.g. [41]), we kept the same train/test split, and for each of these sets, we generated 10 projected views per model, using pyRender [33]. AMASS, on the other hand, is currently the largest and most diverse dataset of human shapes designed specifically for deep learning applications. It was generated by unifying 15 archived datasets of marker-based optical motion capture (mocap) data. Each mocap sequence was converted to a sequence of rigged meshes using SMPL+H model [61]. Consequently, AMASS provides a richer resource for evaluating generalization. We generated a large set of single-view projections by sampling every 100th frame of all provided sequences. We then used pyRender [33] to render each shape from 10 equally spaced azimuth angles, keeping elevation at zero. Keeping the data splits prescribed by [46], our dataset comprises a total of $110K$, $10K$, and $1K$ full shapes for train, validation and test, respectively; and 10 times that in partial shapes. Note that at train time we randomly mix and match full shapes and their parts which drastically increases the effective set size.

4.2 Methods in Comparison

The problem of deformable shape completion was recently studied by Litany *et al.* [39]. In their work, completion is achieved via optimization in a learned shape space. Different from us, their task is completion from a partial view without explicit access to a full model. This is an important distinction as it means missing parts can only be hallucinated. In contrast, we assume the shape details are provided but are not in the correct pose. Moreover, their solution requires a preliminary step of solving partial matching to a template model, which by itself is a hard problem. Here, we solve for it jointly with the alignment. The optimization at inference time also makes their solution quite slow. Instead we output a result in a single feed forward fashion. 3D-CODED [26] performs template alignment to an input shape in two stages: fast inference and slow refinement. It is designed for inputs which are either full or has mild partiality. Here we evaluate the performance of their network predictions under significant partiality. In the refinement step we use directional Chamfer distance, as suggested by the authors in the partial case. FARM [47] is another alignment-based solution that has shown impressive results on shape completion and dense correspondences. It builds on the SMPL [44] human body model due to its compact parameterization, yet, we found it to be very slow to converge (up to 30 min for a single shape) and prone to getting trapped in local minima. We also tried to compare with a recent nonrigid registration method [45] that aligns a given full source point cloud to a partial target point cloud. However, this method didn't converge on our moderate size point clouds (< 7000 vertices) even within 48 h, therefore we do not report on this method. 3D-EPN [20] is a rigid shape completion method. Based on a 3D-CNN, it accepts a voxelized signed distance field as input, and outputs that of a completed shape. Results are then converted to a mesh via computation of an isosurface. Comparison with classic Poisson reconstruction [36] is also provided. It serves as a naïve baseline as it has access only to the partial input. Lacking a single good measure of completion quality, we provide 5 different ones (see Tables 1 and 2). Each measurement highlights a different aspect of the predicted completion. We report the root mean square error (RMSE) of the Euclidean distance between each point on the reconstructed shape and its ground truth mapping. We report this measure for predictions with well defined correspondence to the true reconstruction. We also report the RMSE of two directional Chamfer distances: ground-truth to prediction, and vice versa. The former measures coverage of the target shape by the prediction and the later

Table 1. FAUST Shape Completion. Comparison of different methods with respect to errors in vertex position and shape volume.

	Euclidean distance	Volumetric err.	Chamfer GT → Recon.	Chamfer Recon. → GT	Full Chamfer
Poisson [36]	–	24.8 ± 23.2	7.3	3.64	10.94
3D-EPN [20]	–	89.7 ± 33.8	4.52	4.87	9.39
3D-CODED [26]	35.50	21.8 ± 0.3	11.15	38.49	49.64
FARM [47]	35.77	43.08 ± 20.4	9.5	3.9	13.4
Litany *et al.* [39]	7.07	9.24 ± 8.62	2.84	2.9	5.74
Ours	**2.94**	$\mathbf{7.05 \pm 3.45}$	**2.42**	**1.95**	**4.37**

penalizes prediction outliers. We report the sum of both as full the Chamfer distance. Finally, we report volumetric error as the absolute volume difference divided by the ground truth volume. Please note that the results reported in [39] as "Euclidean distance error" are reported differently in our Table 1 and 2. We confirm that the column named "Euclidean Distance Error" in [39] is, in fact, a directional Chamfer distance from GT to reconstruction. We, therefore, reported that error in the appropriate column and added a computation of the Euclidean distance.

4.3 Single View Completion

We evaluate our method on the task of deformable shape completion on FAUST and AMASS.

FAUST Projections. We follow the evaluation protocol proposed in [39] and summarize the completion results of our method and prior art in Table 1. As can be seen, our network generates a much more accurate completion. Contrary to optimization-based methods [26,39,47] which are very slow at inference, our feed-forward network performs inference in less than a second. To better appreciate the quality of our reconstructions, in Fig. 4 we visualize completions predicted by various methods. Note how our method accurately preserves fine details that were lost in previous methods. In the supplementary, we analyse the reconstruction error as a function of proximity between the source and the target pose, as well as provide additional completion results.

Table 2. AMASS Shape Completion. Comparison of different methods with respect to errors in vertex position and shape volume.

	Euclidean distance	Volumetric err.	Chamfer GT → Recon.	Chamfer Recon. → GT	Full chamfer
3D-CODED [26]	36.14	–	13.65	35.35	49
FARM [47]	27.75	49.42 ± 29.12	11.17	5.14	16.31
Ours	**6.58**	$\mathbf{27.62 \pm 15.27}$	**4.86**	**3.06**	**7.92**

AMASS Projections. Using our test set of partial shapes from AMASS (generated as described in 4.1), we compare our method with two recent methods based on shape alignment: 3D-CODED [26], and FARM [47]. As described in 4.2, 3D-CODED is a learning-based method that uses a fixed template and is not designed to handle severe partiality. FARM, on the other hand, is an optimization method built for the same setting as ours. We summarize the results in Table 2. As can be seen, our method outperforms the two baselines by a large margin in all reported metrics. Note that on some of the examples (about 30%) FARM crashed during the optimization. We therefore only report the errors on its successful runs. Visualizations of several completions are shown in Fig. 3. Additional completions are visualized in the supplementary.

4.4 Non-rigid Partial Correspondences

Finding dense correspondences between a full shape and its deformed parts is still an active research topic. Here we propose a solution in the form of alignment between the full shape and the partial shape, allowing for the recovery of the correspondence by a simple nearest neighbor search. As before, we evaluate this task on both FAUST and AMASS.

FAUST Projections. On the FAUST projections dataset, we compare with two alignment-based methods, FARM and 3D-CODED. We also compare with 3 methods designed to only recover correspondences, that is, without performing shape completion: MoNet [49], and two 3-layered Euclidean CNN baselines, trained on either SHOT [69] descriptors or depth maps. Results are reported in Fig. 5. As in the single view completion experiment, the test set consists of 200 shapes: 2 subjects at 10 different poses with 10 projected views each. The direct matching baselines solve a labeling problem, assigning each input vertex a matching index in a fixed template shape. Differently, 3D-CODED deforms a fixed template and recovers correspondence by a nearest neighbor query for each input vertex using a one-sided Chamfer distance, as suggested in [26]. Our method and FARM both require a complete shape as input, which we chose as the null pose of each of the test examples. Due to slow convergence and unstable behavior of FARM we only kept 20 useful matching results on which we report the performance. As seen in Fig. 5, our method outperforms prior art by a significant margin. This result is particularly interesting since it demonstrates that even though we solve an alignment problem, which is strictly harder than correspondence, we receive better results than methods that specialize in the latter. At the same time, looking at the poor performance demonstrated by the other alignment-based methods, we conclude that simply solving an alignment problem is not enough and the details of our method and training scheme allow for a substantial difference. Qualitative correspondence results are visualized in the supplementary.

AMASS Projections. As FAUST is limited in variability, we further test our method on the recently published AMASS dataset. On the task of partial correspondence, we compare with FARM [47] and 3D-CODED [26] for which code

Fig. 3. AMASS Shape Completion. At the top from left to right: full shape Q, partial shape P, ground truth completion R. At the bottom from left to right: reconstructions of FARM [47], 3D-CODED [26] and ours.

Fig. 4. FAUST Shape Completion. At the top from left to right: full shape Q, partial shape P, ground truth completion R. At the bottom from left to right: reconstructions from FARM [47], 3D-EPN [20], Poisson [36], 3D-CODED [26], Litany *et al.* [39] and ours.

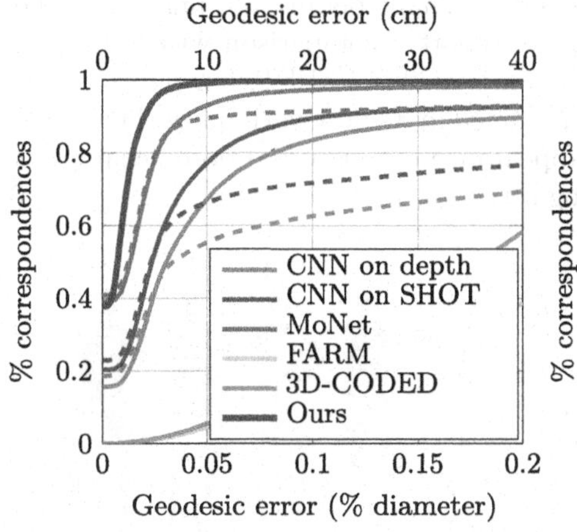

Fig. 5. Partial correspondence error, FAUST dataset. Same color dashed and solid lines indicate performance before and after refinement, respectively.

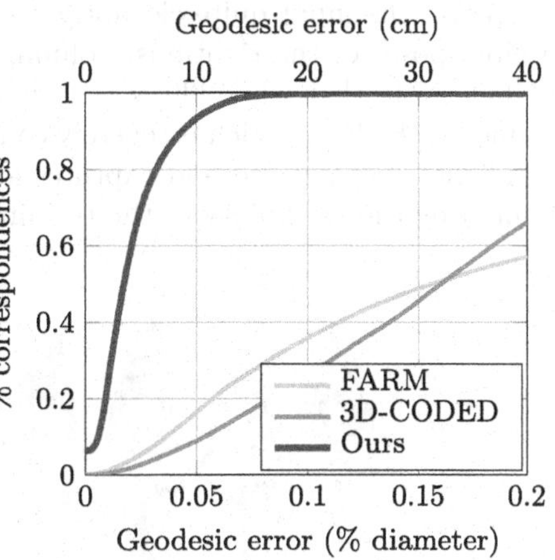

Fig. 6. Partial correspondence error, AMASS dataset.

was available online. We report the correspondence error graphs in Fig. 6. For evaluation we used 200 pairs of partial and full shapes chosen randomly (but consistently between different methods). Specifically, for each of the 4 subjects in AMASS test set we randomized 50 pairs of full poses: one was taken as the full shape Q and one was projected to obtain the partial shape P, using the full unprojected version as the ground truth completion R. As with FAUST, we report the error curve of FARM taking the average of only the successful runs. As can be observed, our method outperforms both methods by a large margin. Qualitative correspondence results are visualized in the supplementary.

4.5 Real Scans

To evaluate our method in real-world conditions, we test it on raw measurements taken during the preparation of the Dynamic FAUST [7] dataset. This use case nicely matches our setting: these are partial scans of a subject for which we have a complete reference shape at a different pose. As preprocessing we compute point normals for the input scan using the method presented in [31]. The point cloud and the reference shape are subsequently inserted into a network pretrained on FAUST. The template, raw scan, and our reconstruction are shown, from left to right, in Fig. 7. We show our result both as the recovered point cloud as well as the recovered mesh using the template triangulation. As apparent from the figure, this is a challenging test case as it introduces several properties not seen at test time: a point cloud without connectivity leads to noisier normals, scanner noise, different point density and extreme partiality (note the missing bottom half of the shapes). Despite all these, the proposed network was able to recover the input quite elegantly, preserving shape details and mimicking the desired pose. In the rightmost column, we report a comparison with Litany *et al.* [39]. Note that while [39] was trained on Dynamic FAUST, our network was trained on FAUST which is severely constrained in its pose variability. The result highlights that our method captures appearance details while pose accuracy is limited by the variability of the training set.

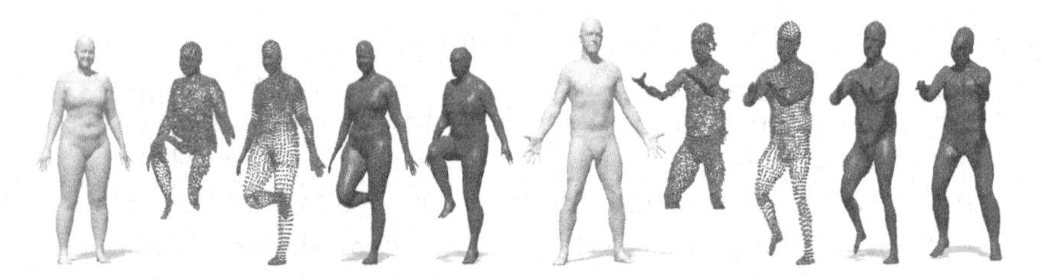

Fig. 7. Completion from real scans from the Dynamic Faust dataset [7]. From left to right: Input reference shape; input raw scan; our completed shape as a point cloud; and as mesh; completion from Litany *et al.* [39].

5 Conclusions

We proposed an alignment-based solution to the problem of shape completion from range scans. Different from most previous works, we focus on the setting where a complete shape is given, but is at a different pose than that of the scan. Our data-driven solution is based on learning the space of distortions, linking scans at various poses to whole shapes in other poses. As a result, at test time we can accurately align unseen pairs of parts and whole shapes at different poses.

Acknowledgements. We gratefully thank Rotem Cohen for contributing to the article visualizations. This work was supported by the Israel Ministry of Science and Technology grant number 3-14719, the Technion Hiroshi Fujiwara Cyber Security Research Center and the Israel Cyber Directorate, the Vannevar Bush Faculty Fellowship, the SAIL-Toyota Center for AI Research, and by Amazon Web Services. Giovanni Trappolini and Emanuele Rodolà are supported by the ERC Starting Grant No. 802554 (SPECGEO) and the MIUR under grant "Dipartimenti di eccellenza 2018-2022" of the Department of Computer Science of Sapienza University.

References

1. Allen, B., Curless, B., Popović, Z., Hertzmann, A.: Learning a correlated model of identity and pose-dependent body shape variation for real-time synthesis. In: Proceedings of the 2006 ACM SIGGRAPH/Eurographics Symposium on Computer Animation, pp. 147–156. Eurographics Association (2006)
2. Anguelov, D., Srinivasan, P., Koller, D., Thrun, S., Rodgers, J., Davis, J.: Scape: shape completion and animation of people. ACM Trans. Graph. **24**(3), 408–416 (2005)
3. Ben-Shabat, Y., Lindenbaum, M., Fischer, A.: 3D point cloud classification and segmentation using 3D modified fisher vector representation for convolutional neural networks. arXiv preprint arXiv:1711.08241 (2017)
4. Besl, P.J., McKay, N.D.: Method for registration of 3-D shapes. In: Sensor Fusion IV: Control Paradigms and Data Structures, vol. 1611, pp. 586–606. International Society for Optics and Photonics (1992)
5. Blanz, V., Vetter, T.: A morphable model for the synthesis of 3D faces. In: Proceedings of Computer Graphics and Interactive Techniques, pp. 187–194 (1999)
6. Bogo, F., Romero, J., Loper, M., Black, M.J.: FAUST: dataset and evaluation for 3D mesh registration. In: Proceedings of CVPR (2014)
7. Bogo, F., Romero, J., Pons-Moll, G., Black, M.J.: Dynamic FAUST: registering human bodies in motion. In: IEEE Conference on Computer Vision and Pattern Recognition (CVPR), July 2017
8. Boscaini, D., Masci, J., Rodolà, E., Bronstein, M.: Learning shape correspondence with anisotropic convolutional neural networks. In: Advances in Neural Information Processing Systems, pp. 3189–3197 (2016)
9. Bronstein, A.M., Bronstein, M.M., Bruckstein, A., Kimmel, R.: Matching two-dimensional articulated shapes using generalized multidimensional scaling. In: Proceedings of Articulated Motion and Deformable Objects (AMDO) (2006)
10. Bronstein, A.M., Bronstein, M.M., Kimmel, R.: Expression-invariant 3D face recognition. In: Kittler, J., Nixon, M.S. (eds.) AVBPA 2003. LNCS, vol. 2688, pp. 62–70. Springer, Heidelberg (2003). https://doi.org/10.1007/3-540-44887-X_8
11. Bronstein, A.M., Bronstein, M.M., Kimmel, R.: Three-dimensional face recognition. Int. J. Comput. Vis. **64**(1), 5–30 (2005)
12. Bronstein, A.M., Bronstein, M.M., Kimmel, R.: Face2face: an isometric model for facial animation. In: Conference on Articulated Motion and Deformable Objects (AMDO) (2006)
13. Bronstein, A.M., Bronstein, M.M., Kimmel, R.: Robust expression-invariant face recognition from partially missing data. In: Leonardis, A., Bischof, H., Pinz, A. (eds.) ECCV 2006. LNCS, vol. 3953, pp. 396–408. Springer, Heidelberg (2006). https://doi.org/10.1007/11744078_31

14. Bronstein, A.M., Bronstein, M.M., Kimmel, R.: Expression-invariant representations of faces. IEEE Trans. Image Process. **16**(1), 188–197 (2007)
15. Bronstein, A.M., Bronstein, M.M., Kimmel, R.: Generalized multidimensional scaling: a framework for isometry-invariant partial surface matching. PNAS **103**(5), 1168–1172 (2006)
16. Bronstein, M.M., Bruna, J., LeCun, Y., Szlam, A., Vandergheynst, P.: Geometric deep learning: going beyond Euclidean data. IEEE Signal Process. Mag. **34**(4), 18–42 (2017)
17. Chen, Q., Koltun, V.: Robust nonrigid registration by convex optimization. In: Proceedings of ICCV (2015)
18. Choy, C., Gwak, J., Savarese, S.: 4D spatio-temporal convnets: Minkowski convolutional neural networks. arXiv preprint arXiv:1904.08755 (2019)
19. Cosmo, L., Panine, M., Rampini, A., Ovsjanikov, M., Bronstein, M.M., Rodolà, E.: Isospectralization, or how to hear shape, style, and correspondence. In: Proceedings of the IEEE Conference on Computer Vision and Pattern Recognition, pp. 7529–7538 (2019)
20. Dai, A., Qi, C.R., Nießner, M.: Shape completion using 3D-encoder-predictor CNNs and shape synthesis. arXiv:1612.00101 (2016)
21. Devir, Y., Rosman, G., Bronstein, A. M. Bronstein, M.M., Kimmel, R.: On reconstruction of non-rigid shapes with intrinsic regularization. In: Proceedings of Workshop on Nonrigid Shape Analysis and Deformable Image Alignment (NORDIA) (2009)
22. Elad, A., Kimmel, R.: Bending invariant representations for surfaces. In: Proceedings of CVPR 2001, Hawaii, December 2001
23. Elad, A., Kimmel, R.: On bending invariant signatures for surfaces. IEEE Trans. Pattern Anal. Mach. Intell. (PAMI) **25**(10), 1285–1295 (2003)
24. Gerig, T., et al.: Morphable face models-an open framework. arXiv preprint arXiv:1709.08398 (2017)
25. Graham, B., Engelcke, M., van der Maaten, L.: 3D semantic segmentation with submanifold sparse convolutional networks. In: Proceedings of the IEEE Conference on Computer Vision and Pattern Recognition, pp. 9224–9232 (2018)
26. Groueix, T., Fisher, M., Kim, V.G., Russell, B.C., Aubry, M.: 3D-CODED: 3D correspondences by deep deformation. In: Ferrari, V., Hebert, M., Sminchisescu, C., Weiss, Y. (eds.) ECCV 2018. LNCS, vol. 11206, pp. 235–251. Springer, Cham (2018). https://doi.org/10.1007/978-3-030-01216-8_15
27. Groueix, T., Fisher, M., Kim, V.G., Russell, B.C., Aubry, M.: Atlasnet: A papier-Mâchè approach to learning 3D surface generation. arXiv preprint arXiv:1802.05384 (2018)
28. Guler, R.A., Kokkinos, I.: Holopose: Holistic 3D human reconstruction in-the-wild. In: Proceedings of the IEEE Conference on Computer Vision and Pattern Recognition, pp. 10884–10894 (2019)
29. Halimi, O., Litany, O., Rodolà, E., Bronstein, A.M., Kimmel, R.: Unsupervised learning of dense shape correspondence. In: Proceedings of the IEEE Conference on Computer Vision and Pattern Recognition, pp. 4370–4379 (2019)
30. Hanocka, R., Hertz, A., Fish, N., Giryes, R., Fleishman, S., Cohen-Or, D.: MeshCNN: a network with an edge. ACM Trans. Graph. (TOG) **38**(4), 90 (2019)
31. Hoppe, H., DeRose, T., Duchamp, T., McDonald, J., Stuetzle, W.: Surface reconstruction from unorganized points, vol. 26. ACM (1992)
32. Hu, T., Han, Z., Shrivastava, A., Zwicker, M.: Render4completion: synthesizing multi-view depth maps for 3D shape completion. In: Proceedings of the IEEE International Conference on Computer Vision Workshops (2019)

33. Huang, J., Zhou, Y., Funkhouser, T., Guibas, L.: Framenet: learning local canonical frames of 3D surfaces from a single RGB image. arXiv preprint arXiv:1903.12305 (2019)
34. Jiang, H., Cai, J., Zheng, J.: Skeleton-aware 3D human shape reconstruction from point clouds. In: Proceedings of the IEEE International Conference on Computer Vision, pp. 5431–5441 (2019)
35. Joseph-Rivlin, M., Zvirin, A., Kimmel, R.: Mom^{e}Net: flavor the moments in learning to classify shapes. In: Proceedings of IEEE International Conference on Computer Vision (CVPR) Workshops (2019)
36. Kazhdan, M., Hoppe, H.: Screened Poisson surface reconstruction. TOG **32**(3), 29 (2013)
37. Kim, V.G., Lipman, Y., Funkhouser, T.A.: Blended intrinsic maps. Trans. Graph. **30**, 4 (2011)
38. Korman, S., Ofek, E., Avidan, S.: Peeking template matching for depth extension. In: Proceedings of CVPR (2015)
39. Litany, O., Bronstein, A., Bronstein, M., Makadia, A.: Deformable shape completion with graph convolutional autoencoders. In: CVPR (2018)
40. Litany, O., Remez, T., Bronstein, A.: Cloud dictionary: sparse coding and modeling for point clouds. arXiv:1612.04956 (2016)
41. Litany, O., Remez, T., Rodolà, E., Bronstein, A.M., Bronstein, M.M.: Deep functional maps: Structured prediction for dense shape correspondence. In: Proceedings of ICCV, vol. 2, p. 8 (2017)
42. Litany, O., Rodolà, E., Bronstein, A.M., Bronstein, M.M.: Fully spectral partial shape matching. Comput. Graph. Forum **36**(2), 247–258 (2017)
43. Liu, X., Yan, M., Bohg, J.: Meteornet: deep learning on dynamic 3D point cloud sequences. In: Proceedings of the IEEE International Conference on Computer Vision, pp. 9246–9255 (2019)
44. Loper, M., Mahmood, N., Romero, J., Pons-Moll, G., Black, M.J.: SMPL: a skinned multi-person linear model. ACM Trans. Graph. (TOG) **34**(6), 248 (2015)
45. Ma, J., Wu, J., Zhao, J., Jiang, J., Zhou, H., Sheng, Q.Z.: Nonrigid point set registration with robust transformation learning under manifold regularization. IEEE Trans. Neural Netw. Learn. Syst. **30**(12), 3584–3597 (2018)
46. Mahmood, N., Ghorbani, N., Troje, N.F., Pons-Moll, G., Black, M.J.: Amass: archive of motion capture as surface shapes. In: The IEEE International Conference on Computer Vision (ICCV), October 2019. https://amass.is.tue.mpg.de
47. Marin, R., Melzi, S., Rodolà, E., Castellani, U.: Farm: functional automatic registration method for 3D human bodies. In: Computer Graphics Forum. Wiley Online Library (2018)
48. Masci, J., Boscaini, D., Bronstein, M., Vandergheynst, P.: Geodesic convolutional neural networks on Riemannian manifolds. In: Proceedings of the IEEE International Conference on Computer Vision Workshops, pp. 37–45 (2015)
49. Monti, F., Boscaini, D., Masci, J., Rodolà, E., Svoboda, J., Bronstein, M.M.: Geometric deep learning on graphs and manifolds using mixture model CNNs. In: IEEE Conference on Computer Vision and Pattern Recognition (CVPR), pp. 5425–5434. IEEE (2017)
50. Newcombe, R.A., Fox, D., Seitz, S.M.: Dynamicfusion: reconstruction and tracking of non-rigid scenes in real-time. In: Proceedings of the IEEE Conference on Computer Vision and Pattern Recognition, pp. 343–352 (2015)
51. Park, J.J., Florence, P., Straub, J., Newcombe, R., Lovegrove, S.: Deepsdf: learning continuous signed distance functions for shape representation. arXiv preprint arXiv:1901.05103 (2019)

52. Paszke, A., et al.: Automatic differentiation in pytorch (2017)
53. Qi, C.R., Litany, O., He, K., Guibas, L.J.: Deep hough voting for 3d object detection in point clouds. arXiv preprint arXiv:1904.09664 (2019)
54. Qi, C.R., Su, H., Mo, K., Guibas, L.J.: Pointnet: deep learning on point sets for 3D classification and segmentation. In: Proceedings of CVPR (2017)
55. Qi, C.R., Yi, L., Su, H., Guibas, L.J.: Pointnet++: deep hierarchical feature learning on point sets in a metric space. arXiv:1706.02413 (2017)
56. Rampini, A., Tallini, I., Ovsjanikov, M., Bronstein, A.M., Rodolà, E.: Correspondence-free region localization for partial shape similarity via hamiltonian spectrum alignment. arXiv preprint arXiv:1906.06226 (2019)
57. Richardson, E., Sela, M., Or-El, R., Ron, K.: Learning detailed face reconstruction from a single image. In: IEEE Conference on Computer Vision and Pattern Recognition (CVPR), Hawaii, Honolulu (2017)
58. Richardson, E., Sela, M., Ron, K.: 3D face reconstruction by learning from synthetic data. In: 4th International Conference on 3D Vision (3DV) Stanford University, CA, USA (2016)
59. Rodolà, E., Cosmo, L., Bronstein, M.M., Torsello, A., Cremers, D.: Partial functional correspondence. Comput. Graph. Forum **36**(1), 222–236 (2017)
60. Rodolà, E., Rota Bulo, S., Windheuser, T., Vestner, M., Cremers, D.: Dense nonrigid shape correspondence using random forests. In: Proceedings of the IEEE Conference on Computer Vision and Pattern Recognition, pp. 4177–4184 (2014)
61. Romero, J., Tzionas, D., Black, M.J.: Embodied hands: modeling and capturing hands and bodies together. ACM Trans. Graph. (Proc. SIGGRAPH Asia) **36**(6) (2017)
62. Sarkar, K., Varanasi, K., Stricker, D.: Learning quadrangulated patches for 3D shape parameterization and completion. arXiv:1709.06868 (2017)
63. Sarmad, M., Lee, H.J., Kim, Y.M.: RL-GAN-Net: a reinforcement learning agent controlled gan network for real-time point cloud shape completion. In: Proceedings of the IEEE Conference on Computer Vision and Pattern Recognition, pp. 5898–5907 (2019)
64. Sela, M., Richardson, E., Kimmel, R.: Unrestricted facial geometry reconstruction using image-to-image translation. In: International Conference on Computer Vision (ICCV), Venice, Italy (2017)
65. Shtern, A., Sela, M., Kimmel, R.: Fast blended transformations for partial shape registration. J. Math. Imaging Vis. **60**(6), 913–928 (2018)
66. Slavcheva, M., Baust, M., Cremers, D., Ilic, S.: Killingfusion: Non-rigid 3D reconstruction without correspondences. In: Proceedings of the IEEE Conference on Computer Vision and Pattern Recognition, pp. 1386–1395 (2017)
67. Stutz, D., Geiger, A.: Learning 3D shape completion from laser scan data with weak supervision. In: Proceedings of the IEEE Conference on Computer Vision and Pattern Recognition, pp. 1955–1964 (2018)
68. Su, H., Maji, S., Kalogerakis, E., Learned-Miller, E.: Multi-view convolutional neural networks for 3d shape recognition. In: Proceedings of CVPR (2015)
69. Tombari, F., Salti, S., Di Stefano, L.: Unique signatures of histograms for local surface description. In: International Conference on Computer Vision (ICCV), pp. 356–369 (2010)
70. Varol, G., et al.: BodyNet: volumetric inference of 3D human body shapes. In: Ferrari, V., Hebert, M., Sminchisescu, C., Weiss, Y. (eds.) ECCV 2018. LNCS, vol. 11211, pp. 20–38. Springer, Cham (2018). https://doi.org/10.1007/978-3-030-01234-2_2

71. Varol, G., et al.: Learning from synthetic humans. In: Proceedings of the IEEE Conference on Computer Vision and Pattern Recognition, pp. 109–117 (2017)
72. Verma, N., Boyer, E., Verbeek, J.: Dynamic filters in graph convolutional networks. arXiv:1706.05206 (2017), http://arxiv.org/abs/1706.05206
73. Wang, Y., Sun, Y., Liu, Z., Sarma, S.E., Bronstein, M.M., Solomon, J.M.: Dynamic graph CNN for learning on point clouds. ACM Trans. Graph. (TOG) **38**(5), 146 (2019)
74. Wei, L., Huang, Q., Ceylan, D., Vouga, E., Li, H.: Dense human body correspondences using convolutional networks. In: Proceedings of CVPR (2016)
75. Wu, J., Zhang, C., Zhang, X., Zhang, Z., Freeman, W.T., Tenenbaum, J.B.: Learning shape priors for single-view 3D completion and reconstruction. In: Ferrari, V., Hebert, M., Sminchisescu, C., Weiss, Y. (eds.) ECCV 2018. LNCS, vol. 11215, pp. 673–691. Springer, Cham (2018). https://doi.org/10.1007/978-3-030-01252-6_40
76. Wu, Z., et al.: 3D shapenets: a deep representation for volumetric shapes. In: Proceedings of CVPR (2015)
77. Xu, D., Anguelov, D., Jain, A.: Pointfusion: deep sensor fusion for 3d bounding box estimation. In: Proceedings of the IEEE Conference on Computer Vision and Pattern Recognition, pp. 244–253 (2018)
78. Zanfir, A., Marinoiu, E., Sminchisescu, C.: Monocular 3d pose and shape estimation of multiple people in natural scenes-the importance of multiple scene constraints. In: Proceedings of the IEEE Conference on Computer Vision and Pattern Recognition, pp. 2148–2157 (2018)

Iterative Distance-Aware Similarity Matrix Convolution with Mutual-Supervised Point Elimination for Efficient Point Cloud Registration

Jiahao Li[1](✉), Changhao Zhang[2], Ziyao Xu[3], Hangning Zhou[3], and Chi Zhang[3]

[1] Washington University in St. Louis, St. Louis, USA
jiahao.li@wustl.edu
[2] Xi'an Jiaotong University, Xi'an, China
cvchanghao@gmail.com
[3] Megvii Inc., Beijing, China
{xuziyao,zhouhangning,zhangchi}@megvii.com

Abstract. In this paper, we propose a novel learning-based pipeline for partially overlapping 3D point cloud registration. The proposed model includes an iterative distance-aware similarity matrix convolution module to incorporate information from both the feature and Euclidean space into the pairwise point matching process. These convolution layers learn to match points based on joint information of the entire geometric features and Euclidean offset for each point pair, overcoming the disadvantage of matching by simply taking the inner product of feature vectors. Furthermore, a two-stage learnable point elimination technique is presented to improve computational efficiency and reduce false positive correspondence pairs. A novel mutual-supervision loss is proposed to train the model without extra annotations of keypoints. The pipeline can be easily integrated with both traditional (e.g. FPFH) and learning-based features. Experiments on partially overlapping and noisy point cloud registration show that our method outperforms the current state-of-the-art, while being more computationally efficient.

Keyword: Point Cloud Registration

1 Introduction

Point cloud registration is an important task in computer vision, which aims to find a rigid body transformation to align one 3D point cloud (source) to another (target). It has a variety of applications in computer vision, augmented reality and virtual reality, such as pose estimation and 3D reconstruction. The

Electronic supplementary material The online version of this chapter (https://doi.org/10.1007/978-3-030-58586-0_23) contains supplementary material, which is available to authorized users.

A. Vedaldi et al. (Eds.): ECCV 2020, LNCS 12369, pp. 378–394, 2020.
https://doi.org/10.1007/978-3-030-58586-0_23

most widely used traditional registration method is Iterative Closest Point (ICP) [3], which is only suitable for estimating small rigid transformation. However, in many real world applications, this assumption does not hold. The task of registering two point clouds with large rotation and translation is called global registration. Some global registration methods [42,43] are proposed to overcome the limitation of ICP, but are usually very slow compared to ICP.

In recent years, deep learning models have dominated the field of computer vision [8,9,13,15,31]. Many computer vision tasks are proven to be solved better using data-driven methods based on neural networks. Recently, some learning-based neural network methods for point cloud registration are proposed [1,35,36]. They are capable of dealing with large rotation angles, and are typically much faster than traditional global registration methods. However, they still have major drawbacks. For example, DCP [35] assumes that all the points in the source point cloud have correspondences in the target point cloud. Although promising, learning-based point cloud registration methods are far from perfect.

In this paper, we propose the **Iterative Distance-Aware Similarity Matrix Convolution Network (IDAM)**, a novel learnable pipeline for accurate and efficient point cloud registration. The intuition for IDAM is that while many registration methods use local geometric features for point matching, ICP uses the distance as the only criterion for matching. We argue that incorporating both geometric and distance features into the iterative matching process can resolve ambiguity and have better performance than using either of them. Moreover, point matching involves computing a similarity score, which is usually computed using the inner product or $L2$ distance between feature vectors. This simple matching method does not take into consideration the interaction of features of different point pairs. We propose to use a learned module to compute the similarity score based on the entire concatenated features of the two points of interest. These two intuition can be realized using a single learnable **similarity matrix convolution** module that accepts pairwise inputs in both the feature and Euclidean space.

Another major problem for global registration methods is efficiency. To reduce computational complexity, we propose a novel **two-stage point elimination** technique to keep a balance between performance and efficiency. The first point elimination step, **hard point elimination**, independently filters out the majority of individual points that are not likely to be matched with confidence. The second step, **hybrid point elimination**, eliminates correspondence pairs instead of individual points. It assigns low weights to those pairs that are probable to be false positives while solving the absolute orientation problem. We design a novel **mutual-supervision loss** to train these learned point elimination modules. This loss allows the model to be trained end-to-end without extra annotations of keypoints. This two-stage elimination process makes our method significantly faster than the current state-of-art global registration methods.

Our learned registration pipeline is compatible with both learning-based and traditional point cloud feature extraction methods. We show by experiments that our method performs well with both FPFH [27] and Graph Neural Network (GNN) [17,37,40] features. We compare our model to other point cloud registration methods, showing that the power of learning is not only restricted to feature extraction, but is also critical for the registration process.

2 Related Work

Local Registration. The most widely used traditional local registration method is Iterative Closest Point (ICP) [3]. It finds for each point in the source the closest neighbor in the target as the correspondence. Trimmed ICP [5] extends ICP to handle partially overlapping point clouds. Other methods [4,26,28] are mostly variants to the vanilla ICP.

Global Registration. The most important non-learning global registration methods is RANSAC [6]. Usually FPFH [27] or SHOT [33] feature extraction methods are used with RANSAC. However, RANSAC is very slow compared to ICP. Fast Global Registration (FGR) [43] uses FPFH features and an alternating optimization technique to speed up global registration. Go-ICP [42] adopts a brute-force branch-and-bound strategy to find the rigid transformation. There are also other methods [11,14,20,41] that utilize a variety of optimization techniques.

Data-Driven Registration. PointNetLK [1] pioneers the recent learning-based registration methods. It adapts PointNet [22] and the Lucas & Kanade [18] algorithm into a single trainable recurrent deep neural network. Deep Closest Point (DCP) [35] proposed to use a transformer network based on DGCNN [37] to extract features, and train the network end-to-end by back-propagating through the SVD layer. PRNet [36] tries to extend DCP to an iterative pipeline and deals with partially overlapping point cloud registration.

Learning on Point Cloud. Recently a large volume of research papers apply deep learning techniques for learning on point clouds. Volumetric methods [21,45] apply discrete 3D convolution on the voxel representation. OctNet [24] and O-CNN [34] try to design efficient high-resolution 3D convolution using the sparsity property of point clouds. Other methods [16,19,32,38] try to directly define convolution in the continuous Euclidean space, or convert the point clouds to a new space for implementing easy convolution-like operations [25,30]. Contrary to the effort of adapting convolution to point clouds, PointNet [22] and PointNet++ [23], which use simple permutation invariant pooling operations to aggregate information from individual points, are widely used recently due to their simplicity. [17,37] view point clouds as graphs with neighbors connecting to each other, and apply graph neural networks (GNN) [40] to extract features.

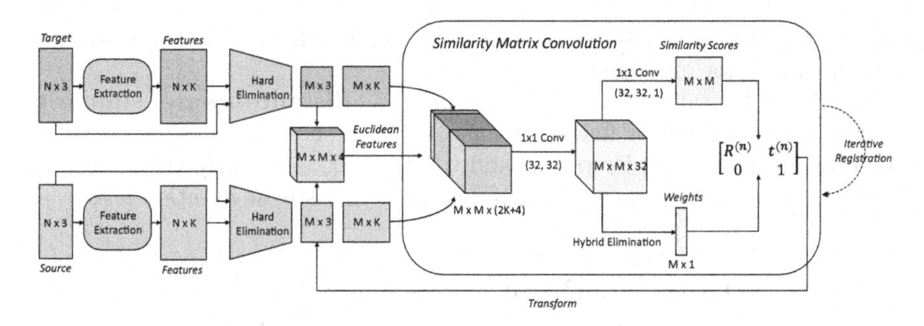

Fig. 1. The overall architecture of the IDAM registration pipeline. Details of hard point elimination and hybrid point elimination are demonstrated in Fig. 2.

3 Model

This section describes the proposed IDAM point cloud registration model. The diagram of the whole pipeline is shown in Fig. 1. The details of each component is explained in the following sections.

3.1 Notation

Here we introduce some notation that will be used throughout the paper. The problem of interest is that for a given source point cloud $\mathcal{S}$ of $N_\mathcal{S}$ points and a target point cloud $\mathcal{T}$ of $N_\mathcal{T}$ points, we need to find the ground truth rigid body transformation $(\mathbf{R}^*, \mathbf{t}^*)$ that aligns $\mathcal{S}$ to $\mathcal{T}$. Let $\mathbf{p}_i \in \mathcal{S}$ denote the ith point in the source, and $\mathbf{q}_j \in \mathcal{T}$ the jth point in the target.

3.2 Similarity Matrix Convolution

To find the rigid body transformation $\mathbf{R}^*$ and $\mathbf{t}^*$, we need to find a set of point correspondences between the source and target point clouds. Most of the existing methods achieve this by using the inner product (or $L2$ distance) of the point features as a measure of similarity, and directly pick the ones with the highest (or lowest for $L2$) response.

However, this has two shortcomings. First of all, one point in $\mathcal{S}$ may have multiple possible correspondences in $\mathcal{T}$, and one-shot matching is not ideal since the points chosen as correspondences may not be the correct ones due to randomness. Inspired by ICP, we argue that incorporating distance information between points into an iterative matching processing can alleviate this problem, since after an initial registration, correct point correspondences are more likely to be closer to each other.

The second drawback of direct feature similarity computation is that it has limited power of identifying the similarity between two points because the way of matching is the same for different pairs. Instead, we have a learned network that accepts the whole feature vectors and outputs the similarity scores. This way, the network takes into consideration the combinations of features from two points in a pair for matching.

Based on the above intuition, we propose **distance-aware similarity matrix convolution** for finding point correspondences. Suppose we have the geometric features $\mathbf{u}^{\mathcal{S}}(i)$ for $\mathbf{p}_i \in \mathcal{S}$ and $\mathbf{u}^{\mathcal{T}}(j)$ for $\mathbf{q}_j \in \mathcal{T}$, both with dimension K. We form the **distance-augmented feature tensor** at iteration n as

$$\mathbf{T}^{(n)}(i,j) = [\mathbf{u}^{\mathcal{S}}(i); \mathbf{u}^{\mathcal{T}}(j); \|\mathbf{p}_i - \mathbf{q}_j\|; \frac{\mathbf{p}_i - \mathbf{q}_j}{\|\mathbf{p}_i - \mathbf{q}_j\|}] \tag{1}$$

where $[\cdot;\cdot]$ denotes concatenation. The $(2K+4)$-dimensional vector at the (i,j) location of $\mathbf{T}^{(n)}$ is a combination of the geometric and Euclidean features for the point pair $(\mathbf{p}_i, \mathbf{q}_j)$. The 4-dimensional Euclidean features comprise the distance between $\mathbf{p}_i$ and $\mathbf{q}_j$, and the unit vector pointing from $\mathbf{q}_j$ to $\mathbf{p}_i$. Each augmented feature vector in $\mathbf{T}^{(n)}$ encodes the joint information of the local shapes of the two points and their current relative position, which are useful for computing similarity scores at each iteration.

The distance-augmented feature tensor $\mathbf{T}^{(n)}$ can be seen as a $(2K+4)$-channel 2D image. To extract a similarity score for each point pair, we apply a series of 1×1 2D convolution on $\mathbf{T}^{(n)}$ that outputs a single channel image of the same spatial size at the last layer. This is equivalent to applying a multi-layer perceptron on the augmented feature vector at each position. Then we apply a Softmax function on each row of the single channel image to get the **similarity matrix**, denoted as $\mathbf{S}^{(n)}$. $\mathbf{S}^{(n)}(i,j)$ represents the "similarity score" (the higher the more similar) for $\mathbf{p}_i$ and $\mathbf{q}_j$. Each row of $\mathbf{S}^{(n)}$ defines a normalized probability distribution over all the points in $\mathcal{T}$ for some $\mathbf{p} \in \mathcal{S}$. As a result, $\mathbf{S}^{(n)}(i,j)$ can also be interpreted as the probability of $\mathbf{q}_j$ being the correspondence of $\mathbf{p}_i$. The 1×1 convolutions learn their weights using the **point matching loss** described in Sect. 3.4. They learn to take into account the interaction between the shape and distance information to output a more accurate similarity score compared to simple inner product.

To find the correspondence pairs, we take the argmax of each row of $\mathbf{S}^{(n)}$. The results are a set of correspondence pairs $\{(\mathbf{p}_i, \mathbf{p}'_i) \mid \forall \mathbf{p}_i \in \mathcal{S}\}$, with which we solve the following optimization problem to find the estimated rigid transformation $(\mathbf{R}^{(n)}, \mathbf{t}^{(n)})$

$$\mathbf{R}^{(n)}, \mathbf{t}^{(n)} = \underset{\mathbf{R},\mathbf{t}}{\operatorname{argmin}} \sum_i \|\mathbf{R}\mathbf{p}_i + \mathbf{t} - \mathbf{p}'_i\|^2 \tag{2}$$

This is a classical absolute orientation problem [10], which can be efficiently solved with the orthogonal Procrustes algorithm [2,7] using Singular Value Decomposition (SVD). $\mathbf{R}^{(n)}$ and $\mathbf{t}^{(n)}$ are then used to transform the source point cloud to a new position before entering the next iteration. The final estimate for $(\mathbf{R}^*, \mathbf{t}^*)$ is the composition of the intermediate $(\mathbf{R}^{(n)}, \mathbf{t}^{(n)})$ for all the iterations.

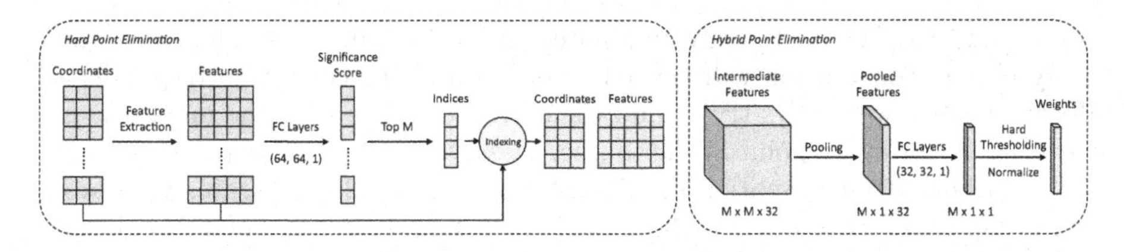

Fig. 2. Comparison of hard point elimination and hybrid point elimination. Hard point elimination filters points based on the features extracted independently for each point, while hybrid point elimination utilizes the joint information of the point pairs to compute weights for the orthogonal Procrustes algorithm.

3.3 Two-Stage Point Elimination

Although similarity matrix convolution is powerful in terms of matching, it is computationally expensive to apply convolution on the large $N_{\mathcal{S}} \times N_{\mathcal{T}} \times (2K+4)$ tensor, because $N_{\mathcal{S}}$ and $N_{\mathcal{T}}$ are typically more than a thousand. However, if we randomly down-sample the point clouds, the performance of the model would degrade drastically since many points no longer have correspondences. To tackle this dilemma, we propose a **two-stage point elimination** process. It consists of **hard point elimination** and **hybrid point elimination** (Fig. 2), which targets on improving efficiency and accuracy respectively. While manually labelling keypoints for point clouds is not practical, we propose a **mutual-supervision loss**, that uses the information in the similarity matrices $\mathbf{S}^{(n)}$ to supervise the point elimination modules. The details of the mutual-supervision loss is described in Sect. 3.4. In this section, we present the point elimination process for inference.

Hard Point Elimination. To reduce the computational burden of similarity matrix convolution, we first propose the **hard point elimination** (Fig. 2 Left). Given the extracted local shape features for each point, we apply a multi-layer perceptron on the feature vector, and output a **significance score**. A high score means a more prominent point, such as a corner point, that can be matched with high confidence later (see the Appendix for visualization). It filters out those points in the "flat" regions that are ambiguous during matching. This process is done on individual points, and does not take into account the point pair information as in similarity matrix convolution. As a result, it is efficient to compute the significance score. We preserve the M points for each point cloud with highest significance scores, and eliminate the remaining points. In our network, we choose $M = \lceil \frac{N}{6} \rceil$, where N can be $N_{\mathcal{S}}$ or $N_{\mathcal{T}}$. Denote the set of points in $\mathcal{S}$ preserved by hard point elimination as $\mathcal{B}_{\mathcal{S}}$, and that for the target as $\mathcal{B}_{\mathcal{T}}$.

Hybrid Point Elimination. While hard point elimination improves the efficiency significantly, it has negative effect on the performance of the model. The correct corresponding point in the target point cloud for a point in the source point cloud may be mistakenly eliminated in hard elimination. Therefore, similarity matrix convolution will never be able to find the correct correspondence. However, since we always try to find the correspondence with the maximal similarity score for every point in the source, these "negative" point pairs can make the rigid body transformation obtained by solving Eq. 2 inaccurate. This problem is especially severe when the model is dealing with two point clouds that only partially overlap with each other. In this case, even without any elimination, some points will not have any correspondence whatsoever.

To alleviate this problem, we propose a **hybrid point elimination** (Fig. 2 Right) process, applied after similarity matrix convolution. Hybrid point elimination is a mixture of both hard elimination and soft elimination, and operates on point pairs instead of individual points. It uses a permutation-invariant pooling operation to aggregate information across all possible correspondences for a given point in the source, and outputs the **validity score**, for which a higher score means higher probability of having a true correspondence. Formally, let $\mathbf{F}$ be the intermediate output (see Fig. 1) of the similarity matrix convolution of shape $M \times M \times K'$. Hybrid point elimination first computes the validity score

$$v(i) = \sigma(f(\bigoplus_j(\mathbf{F}(i,j)))) \tag{3}$$

where $\sigma(\cdot)$ is the sigmoid function, $\bigoplus$ is an element-wise permutation invariant pooling method, such as "mean" or "max", and f is a multi-layer perceptron that takes the pooled features as input and outputs the scores. This permutation invariant pooling technique is used in a variety point cloud processing [22,23] and graph neural network [17,37] models. Following [22,23] we use element-wise max for $\bigoplus$. This way, we have a validity score for each point in the source, and thus for each point pair. It can be seen as the probability that a correspondence pair is correct.

With this validity score, we then compute the **hybrid elimination weights**. The weight for the ith point pair is defined as

$$w_i = \frac{v(i) \cdot \mathbb{1}[\![v(i) \geq \text{median}_k(v(k))]\!]}{\sum_i v(i) \cdot \mathbb{1}[\![v(i) \geq \text{median}_k(v(k))]\!]} \tag{4}$$

where $\mathbb{1}[\![\cdot]\!]$ is the indicator function. What this weighting process does is that it gives 0 weight to those points with lowest validity scores (hard elimination), and weighs the rest proportionally to the validity scores (soft elimination). With this elimination weight vector, we can obtain the $(\mathbf{R}^{(n)}, \mathbf{t}^{(n)})$ with a slightly different objective function from Eq. 2

$$\mathbf{R}^{(n)}, \mathbf{t}^{(n)} = \underset{\mathbf{R},\mathbf{t}}{\text{argmin}} \sum_i w_i \|\mathbf{R}\mathbf{p}_i + \mathbf{t} - \mathbf{p}'_i\|^2 \tag{5}$$

This can still be solved using SVD with little overhead. Ideally, the hybrid point elimination can eliminate those point pairs that are not correct due to noise and incompletion, giving better performance on estimating $\mathbf{R}^{(n)}$ and $\mathbf{t}^{(n)}$ (see the Appendix for visualization).

3.4 Mutual-Supervision Loss

In this section, we describe in detail the **mutual-supervision loss** that is used to train the network. With this loss, we can train the similarity matrix convolution, along with the two-stage elimination module, without extra annotations of keypoints. The loss is the sum of three parts, which will be explained in the following.

Note that training on all the points during each forward-backward loop is inefficient and unnecessary. However, since hard point elimination does not function properly during training yet, we do not have direct access to $\mathcal{B}_\mathcal{S}$ and $\mathcal{B}_\mathcal{T}$ (see the definitions in Sect. 3.3). Therefore, we need some way to sample points from the source and the target for training. This sampling technique is described in Sect. 3.5. In this section we accept that as given, and abuse the notation $\mathcal{B}_\mathcal{S}$ for the **source sampled set** and $\mathcal{B}_\mathcal{T}$ for the **target sampled set**, which both contain the M sampled points for training. Let $\mathbf{p}_i$ denote the ith point in $\mathcal{B}_\mathcal{S}$ and $\mathbf{q}_j$ denote the jth point in $\mathcal{B}_\mathcal{T}$

Point Matching Loss. The point matching loss is used to supervise the similarity matrix convolution. It is a standard cross-entropy loss. The point matching loss for the nth iteration is defined as

$$\mathcal{L}_{\text{match}}^{(n)}(\mathcal{S},\mathcal{T},\mathbf{R}^*,\mathbf{t}^*) = \frac{1}{M}\sum_{i=1}^{M} -\log(\mathbf{S}^{(n)}(i,j^*)) \cdot \mathbb{1}[\![\|\mathbf{R}^*\mathbf{p}_i + \mathbf{t}^* - \mathbf{q}_{j^*}\|^2 \leq r^2]\!] \tag{6}$$

where

$$j^* = \underset{1\leq j\leq M}{\text{argmin}}\, \|\mathbf{R}^*\mathbf{p}_i + \mathbf{t}^* - \mathbf{q}_j\|^2 \tag{7}$$

is the index of the point in the target sampled set $\mathcal{B}_\mathcal{T}$ that is closest to the ith point in the source sampled set $\mathcal{B}_\mathcal{S}$ under the ground truth transformation. r is hyper-parameter that controls the minimal radius within which two points are considered close enough. If the distance of $\mathbf{p}_i$ and $\mathbf{q}_{j^*}$ is larger than r, they can not be seen as correspondences, and no supervision signal is applied on them. This happens frequently when the model is dealing with partially overlapping point clouds. The total point matching loss is the average of those for all the iterations.

Negative Entropy Loss. This loss is used for training hard point elimination. The problem for training hard point elimination is that we do not have direct access to annotations of keypoints. Therefore, we propose to use a **mutual supervision** technique, which uses the result of the point matching loss to supervise hard point elimination. This mutual supervision is based on the intuition that if a point $\mathbf{p}_i \in \mathcal{B}_\mathcal{S}$ is a prominent point (high significance score), the probability distribution defined by the ith row of $\mathbf{S}^{(n)}$ should have low entropy because it is confident in matching. On the other hand, the supervision on the similarity matrices has no direct relationship to hard point elimination. Therefore, the **negative entropy** of the probability distribution can be seen as a supervision signal for the significance scores. Mathematically, the **negative entropy loss** for the nth iteration can be defined as

$$\mathcal{L}_{\text{hard}}^{(n)}(\mathcal{S}, \mathcal{T}, \mathbf{R}^*, \mathbf{t}^*) = \frac{1}{M} \sum_{i=1}^{M} |s(i) - \sum_{j=1}^{M} \mathbf{S}^{(n)}(i,j) \log(\mathbf{S}^{(n)}(i,j))|^2 \tag{8}$$

where $s(i)$ is the significant score for the ith point in $\mathcal{B}_\mathcal{S}$. Although this loss can be defined for any iteration, we only use the one for first iteration, because in the early stages of registration the shape features are more important than the Euclidean features. We want the hard point elimination module learns to filter points only based on shape information. We cut the gradient flow from the negative entropy loss to $\mathbf{S}^{(n)}$ to prevent interference with the training of similarity matrix convolution.

Hybrid Elimination Loss. A similar mutual supervision idea can also be used for training the hybrid point elimination. The difference is that hybrid elimination takes into account the point pair information, while hard point elimination only looks at individual points. As a result, the mutual supervision signal is much more obvious for hybrid point elimination. We simply use the probability that there exists a point in $\mathcal{B}_\mathcal{T}$ which is the correspondence of point $\mathbf{p}_i \in \mathcal{B}_\mathcal{S}$ as the supervision signal for v_i (validity score). Instead of computing the probability explicitly, the **hybrid elimination loss** for the nth iteration is defined as

$$\mathcal{L}_{\text{hybrid}}^{(n)}(\mathcal{S}, \mathcal{T}, \mathbf{R}^*, \mathbf{t}^*) = \frac{1}{M} \sum_{i=1}^{M} -\mathbb{I}_i \cdot \log(v_i) - (1 - \mathbb{I}_i) \cdot \log(1 - v_i) \tag{9}$$

where

$$\mathbb{I}_i = \mathbb{1}[\![\|\mathbf{R}^*\mathbf{p}_i + \mathbf{t}^* - \mathbf{q}_{\text{argmax}_j \mathbf{S}^{(n)}(i,j)}\|^2 \leq r^2]\!] \tag{10}$$

In effect, this loss assigns a positive label 1 to those points in $\mathcal{B}_\mathcal{S}$ that correctly finds its correspondence, and a negative label 0 to those that do not. In the long run, those point pairs with high probability of correct matching will have higher validity scores.

3.5 Balanced Sampling for Training

In this section, we describe a balanced sampling technique to sample points for training our network. We first sample $\lceil \frac{M}{2} \rceil$ points from $\mathcal{S}$ with the following unnormalized probability distribution

$$p_{\text{pos}}(i) = \mathbb{1}[\![(\min_{\mathbf{q} \in \mathcal{T}} \|\mathbf{R}^*\mathbf{p}_i + \mathbf{t}^* - \mathbf{q}\|^2) \leq r^2]\!] + \epsilon \tag{11}$$

where $\epsilon = 10^{-6}$ is some small number. This sampling process aims to randomly sample "positive" points from $\mathcal{S}$, in the sense that they indeed have correspondences in the target. It introduces the ϵ to avoid errors when encountering the singularity cases where no points in the source have correspondences in the target.

Similarly, we sample $(M - \lceil \frac{M}{2} \rceil)$ "negative" points from $\mathcal{S}$ using the unnormalized distribution

$$p_{\text{neg}}(i) = \mathbb{1}[\![(\min_{\mathbf{q} \in \mathcal{T}} \|\mathbf{R}^* \mathbf{p}_i + \mathbf{t}^* - \mathbf{q}\|^2) > r^2]\!] + \epsilon \tag{12}$$

This way, we have a set $\mathcal{B}_\mathcal{S}$ of points of size M, with both positive and negative instances. To sample points from the target, we simply find the closest points of each point from $\mathcal{B}_\mathcal{S}$ in the target

$$\mathcal{B}_\mathcal{T} = \{\operatorname*{argmin}_{\mathbf{q}} \|\mathbf{R}^* \mathbf{p}_i + \mathbf{t}^* - \mathbf{q}\| \mid i \in \mathcal{B}_\mathcal{S}\} \tag{13}$$

This balanced sampling technique randomly samples points from $\mathcal{S}$ and $\mathcal{T}$, while keeping a balance between points that have correspondences and points that do not.

4 Experiments

This section shows the experimental results to demonstrate the performance and efficiency of our method. We also conduct ablation study to show the effectiveness of each component of our model.

4.1 Experimental Setup

We train our model with the Adam [12] optimizer for 40 epochs. The initial learning rate is 1×10^{-4}, and is multiplied by 0.1 after 30 epochs. We use a weight decay of 1×10^{-3} and no Dropout [29]. We use the FPFH implementation from the Open3D [44] library and a very simple graph neural network (GNN) for feature extraction. The details of the GNN architecture are described in the supplementary material. For both FPFH and GNN features, the number of iterations is set to 3.

Following [36], all the experiments are done on the ModelNet40 [39] dataset. ModelNet40 includes 9843 training shapes and 2468 testing shapes from 40 object categories. For a given shape, we randomly sample 1024 points to form a point cloud. For each point cloud, we randomly generate rotations within $[0°, 45°]$ and translation in $[-0.5, 0.5]$. The original point cloud is used as the source, and the transformed point cloud as the target. To generate partially overlap point clouds, we follow the same method as [36], which fixes a random point far away from the two point clouds, and preserve 768 points closest to the far point for each point cloud.

We compare our method to ICP, Go-ICP, FGR, FPFH+RANSAC, PointNetLK, DCP and PRNet. All the data-driven methods are trained on the same training set. We use the same metrics as [35,36] to evaluate all these methods. For the rotation matrix, the root mean square error (RMSE($\mathbf{R}$)) and mean absolute error (MAE($\mathbf{R}$)) in degrees are used. For the translation vector, the root mean square error (RMSE($\mathbf{t}$)) and mean absolute error (MAE($\mathbf{t}$)) are used.

4.2 Results

In this section, we show the results for three different experiments to demonstrate the effectiveness and robustness of our method. These experimental settings are the same as those in [36]. We also include in the supplementary material some visualization results for these experiments.

Unseen Shapes. First, we train our model on the training set of ModelNet40 and evaluate on the test set. Both the training set and test set of ModelNet40 contain point clouds from all the 40 categories. This experiment evaluates the ability to generalize to unseen point clouds. Table 1 shows the results.

Table 1. Results for testing on point clouds of unseen shapes in ModelNet40.

Model	RMSE($\mathbf{R}$)	MAE($\mathbf{R}$)	RMSE($\mathbf{t}$)	MAE($\mathbf{t}$)
ICP	33.68	25.05	0.29	0.25
FPFH+RANSAC	2.33	1.96	**0.015**	0.013
FGR	11.24	2.83	0.030	0.008
Go-ICP	14.0	3.17	0.033	0.012
PointNetLK	16.74	7.55	0.045	0.025
DCP	6.71	4.45	0.027	0.020
PRNet	3.20	1.45	0.016	0.010
FPFH+IDAM	**2.46**	**0.56**	0.016	**0.003**
GNN+IDAM	2.95	0.76	0.021	0.005

We can see that local registration method ICP performs poorly because the initial rotation angles are large. FPFH+RANSAC is the best performing traditional method, which is comparable to many learning-based methods. Note that both RANSAC and FGR use FPFH methods, and our method with FPFH features outperforms both of them. Neural network models have a good balance between performance and efficiency. Our method outperforms all the other methods with both hand-crafted (FPFH) and learned (GNN) features.

Surprisingly, FPFH+IDAM has better performance than GNN+IDAM. One possibility is that the GNN overfits to the point clouds in the training set, and does not generalize well to unseen shapes. However, as will be shown in later sections, GNN+IDAM is more robust to noise and also more efficient that FPFH+IDAM.

Unseen Categories. In the second experiment, we use the first 20 categories in the training set of ModelNet40 for training, and evaluate on the other 20 categories on the test set. This experiment tests the capability to generalize to point clouds of unseen categories. The results are summarized in Table 2. We can see that without training on the testing categories, all the learning-based methods perform worse consistently. Traditional methods are not affected that much as expected. Based on different evaluation metrics, FPFH+RANSAC and FPFH+IDAM are the best performing methods.

Table 2. Results for testing on point clouds of unseen categories in ModelNet40.

Model	RMSE(**R**)	MAE(**R**)	RMSE(**t**)	MAE(**t**)
ICP	34.89	25.46	0.29	0.25
FPFH+RANSAC	**2.11**	1.82	**0.015**	0.013
FGR	9.93	1.95	0.038	0.007
Go-ICP	12.53	2.94	0.031	0.010
PointNetLK	22.94	9.66	0.061	0.033
DCP	9.77	6.95	0.034	0.025
PRNet	4.99	2.33	0.021	0.015
FPFH+IDAM	3.04	**0.61**	0.019	**0.004**
GNN+IDAM	3.42	0.93	0.022	0.005

Gaussian Noise. In the last experiment, we add random Gaussian noise with standard deviation 0.01 to all the shapes, and repeat the first experiment (unseen shapes). The random noise is clipped to $[-0.05, 0.05]$. As shown in Table 3, both traditional methods and IDAM based on FPFH features perform much worse than the noise-free case. This demonstrates that FPFH is not very robust to noise. The performance of data-driven methods are comparable to the noise-free case, thanks to the powerful feature extraction networks. Our method based on GNN features has the best performance compared to others.

4.3 Efficiency

We test the speed of our method, and compare it to ICP, FGR, FPFH+RANSAC, PointNetLK, DCP and PRNet. We use the Open3D implementation of ICP, FGR and FPFH+RANSAC, and the official implementation of PointNetLK, DCP and PRNet released by the authors. The experiments are done on a machine with 2 Intel Xeon Gold 6130 CPUs and a single Nvidia GeForce RTX 2080 Ti GPU. We use a batch size of 1 for all the neural network based models. The speed is measured in seconds per frame.

We test the speed on point clouds with 1024, 2048 and 4096 points, and the results are summarized in Table 4. It can be seen that neural network based

Table 3. Results for testing on point clouds of unseen shapes in ModelNet40 with Gaussian noise.

Model	RMSE(**R**)	MAE(**R**)	RMSE(**t**)	MAE(**t**)
ICP	35.07	25.56	0.29	0.25
FPFH+RANSAC	5.06	4.19	0.021	0.018
FGR	27.67	13.79	0.070	0.039
Go-ICP	12.26	2.85	0.028	0.029
PointNetLK	19.94	9.08	0.057	0.032
DCP	6.88	4.53	0.028	0.021
PRNet	4.32	2.05	**0.017**	0.012
FPFH+IDAM	14.21	7.52	0.067	0.042
GNN+IDAM	**3.72**	**1.85**	0.023	**0.011**

methods are generally faster than traditional methods. When the number of points is small, IDAM with GNN features is only slower than DCP. But as the number of points increases, IDAM+GNN is much faster than all the other methods. Although FPFH+RANSAC has the best performance among non-learning methods, it is also the slowest. Note that our method with FPFH features is 2× to 5× faster than the other two methods (FGR and RANSAC) which also use FPFH.

Table 4. Comparison of speed of different models. IDAM(G) and IDAM(F) represent GNN+IDAM and FPFH+IDAM respectively. RANSAC also uses FPFH. Speed is measured in seconds-per-frame.

	IDAM(G)	IDAM(F)	ICP	FGR	RANSAC	PointNetLK	DCP	PRNet
1024 points	0.026	0.050	0.095	0.123	0.159	0.082	0.015	0.022
2048 points	0.038	0.078	0.185	0.214	0.325	0.085	0.030	0.048
4096 points	0.041	0.175	0.368	0.444	0.685	0.098	0.084	0.312

4.4 Ablation Study

In this section, we present the results of ablation study of IDAM to show the effectiveness of each component. We examine three key components of our model: distance-aware similarity matrix convolution (denoted as SM), hard point elimination (HA) and hybrid point elimination (HB). We use BS to denote the model that does not contain any of the three components mentioned above. Since hard point elimination is necessary for similarity matrix convolution due to memory constraints, we replace it with random point elimination in BS. We use inner-product of features in BS when similarity matrix convolution is disabled. As a result, BS is just a simple model that uses the inner-product of features as similarity scores to find correspondences, and directly solves the absolute orientation problem (Eq. 2). We add the components one by one and compare their performance for GNN features. We conducted the experiments under the settings of "unseen categories" as described in Sect. 4.2. The results are summarized in Table 5.

It can be seen that even with random sampling, similarity matrix convolution already outperforms the baseline (BS) by a large margin. The two-stage point elimination (HA and HB) further boosts the performance significantly.

Table 5. Comparison of the performance of different model choices for IDAM. These experiments examine the effectiveness of similarity matrix convolution (SM), hard point elimination (HA) and hybrid point elimination (HB).

Model	RMSE($\mathbf{R}$)	MAE($\mathbf{R}$)	RMSE($\mathbf{t}$)	MAE($\mathbf{t}$)
BS	7.77	5.33	0.055	0.047
BS+SM	5.08	3.58	0.056	0.042
BS+HA+SM	4.31	2.89	0.029	0.019
BS+HA+SM+HB	**3.42**	**0.93**	**0.022**	**0.005**

5 Conclusions

In this paper, we propose a novel data-driven pipeline named IDAM for partially overlapping 3D point cloud registration. We present a novel distance-aware similarity matrix convolution to augment the network's ability of finding correct correspondences in each iteration. Moreover, a novel two-stage point elimination method is proposed to improve performance while reducing computational complexity. We design a mutual-supervised loss for training IDAM end-to-end without extra annotations of keypoints. Experiments show that our method performs better than the current state-of-the-art point cloud registration methods and is robustness to noise.

Acknowledgements. This work was supported in part by the National Key Research and Development Program of China under Grant 2017YFA0700800.

References

1. Aoki, Y., Goforth, H., Srivatsan, R.A., Lucey, S.: Pointnetlk: robust & efficient point cloud registration using pointnet. In: Proceedings of the IEEE Conference on Computer Vision and Pattern Recognition, pp. 7163–7172 (2019)
2. Arun, K.S., Huang, T.S., Blostein, S.D.: Least-squares fitting of two 3-D point sets. IEEE Trans. Pattern Anal. Mach. Intell. **5**, 698–700 (1987)
3. Besl, P.J., McKay, N.D.: Method for registration of 3-D shapes. In: Sensor Fusion IV: Control Paradigms and Data Structures, vol. 1611, pp. 586–606. International Society for Optics and Photonics (1992)

4. Bouaziz, S., Tagliasacchi, A., Pauly, M.: Sparse iterative closest point. In: Computer Graphics Forum, vol. 32, pp. 113–123. Wiley Online Library (2013)
5. Chetverikov, D., Svirko, D., Stepanov, D., Krsek, P.: The trimmed iterative closest point algorithm. In: Object Recognition Supported by User Interaction for Service Robots, vol. 3, pp. 545–548. IEEE (2002)
6. Fischler, M.A., Bolles, R.C.: Random sample consensus: a paradigm for model fitting with applications to image analysis and automated cartography. Commun. ACM **24**(6), 381–395 (1981)
7. Golub, G.H., Van Loan, C.F.: Matrix Computations, vol. 3. JHU Press, Baltimore (2012)
8. He, K., Zhang, X., Ren, S., Sun, J.: Deep residual learning for image recognition. In: Proceedings of the IEEE Conference on Computer Vision and Pattern Recognition, pp. 770–778 (2016)
9. He, K., Zhang, X., Ren, S., Sun, J.: Identity mappings in deep residual networks. In: Leibe, B., Matas, J., Sebe, N., Welling, M. (eds.) ECCV 2016. LNCS, vol. 9908, pp. 630–645. Springer, Cham (2016). https://doi.org/10.1007/978-3-319-46493-0_38
10. Horn, B.K.: Closed-form solution of absolute orientation using unit quaternions. Josa A **4**(4), 629–642 (1987)
11. Izatt, G., Dai, H., Tedrake, R.: Globally optimal object pose estimation in point clouds with mixed-integer programming. In: Amato, N.M., Hager, G., Thomas, S., Torres-Torriti, M. (eds.) Robotics Research. SPAR, vol. 10, pp. 695–710. Springer, Cham (2020). https://doi.org/10.1007/978-3-030-28619-4_49
12. Kingma, D.P., Ba, J.: Adam: a method for stochastic optimization. arXiv preprint arXiv:1412.6980 (2014)
13. Krizhevsky, A., Sutskever, I., Hinton, G.E.: Imagenet classification with deep convolutional neural networks. In: Advances in Neural Information Processing Systems, pp. 1097–1105 (2012)
14. Le, H.M., Do, T.T., Hoang, T., Cheung, N.M.: SDRSAC: semidefinite-based randomized approach for robust point cloud registration without correspondences. In: Proceedings of the IEEE Conference on Computer Vision and Pattern Recognition, pp. 124–133 (2019)
15. LeCun, Y., Bengio, Y., Hinton, G.: Deep learning. Nature **521**(7553), 436–444 (2015)
16. Li, Y., Bu, R., Sun, M., Wu, W., Di, X., Chen, B.: PointCNN: convolution on x-transformed points. In: Advances in Neural Information Processing Systems, pp. 820–830 (2018)
17. Liu, Y., Fan, B., Xiang, S., Pan, C.: Relation-shape convolutional neural network for point cloud analysis. In: Proceedings of the IEEE Conference on Computer Vision and Pattern Recognition, pp. 8895–8904 (2019)
18. Lucas, B.D., Kanade, T., et al.: An iterative image registration technique with an application to stereo vision (1981)
19. Mao, J., Wang, X., Li, H.: Interpolated convolutional networks for 3D point cloud understanding. In: Proceedings of the IEEE International Conference on Computer Vision, pp. 1578–1587 (2019)
20. Maron, H., Dym, N., Kezurer, I., Kovalsky, S., Lipman, Y.: Point registration via efficient convex relaxation. ACM Trans. Graph. (TOG) **35**(4), 1–12 (2016)
21. Maturana, D., Scherer, S.: Voxnet: A 3D convolutional neural network for real-time object recognition. In: 2015 IEEE/RSJ International Conference on Intelligent Robots and Systems (IROS), pp. 922–928. IEEE (2015)

22. Qi, C.R., Su, H., Mo, K., Guibas, L.J.: Pointnet: deep learning on point sets for 3D classification and segmentation. In: Proceedings of the IEEE Conference on Computer Vision and Pattern Recognition, pp. 652–660 (2017)
23. Qi, C.R., Yi, L., Su, H., Guibas, L.J.: Pointnet++: deep hierarchical feature learning on point sets in a metric space. In: Advances in Neural Information Processing Systems, pp. 5099–5108 (2017)
24. Riegler, G., Osman Ulusoy, A., Geiger, A.: OctNet: learning deep 3D representations at high resolutions. In: Proceedings of the IEEE Conference on Computer Vision and Pattern Recognition, pp. 3577–3586 (2017)
25. Rippel, O., Snoek, J., Adams, R.P.: Spectral representations for convolutional neural networks. In: Advances in Neural Information Processing Systems, pp. 2449–2457 (2015)
26. Rusinkiewicz, S., Levoy, M.: Efficient variants of the ICP algorithm. In: Proceedings Third International Conference on 3-D Digital Imaging and Modeling, pp. 145–152. IEEE (2001)
27. Rusu, R.B., Blodow, N., Beetz, M.: Fast point feature histograms (FPFH) for 3D registration. In: 2009 IEEE International Conference on Robotics and Automation, pp. 3212–3217. IEEE (2009)
28. Segal, A., Haehnel, D., Thrun, S.: Generalized-ICP. In: Robotics: Science and Systems, vol. 2, p. 435. Seattle, WA (2009)
29. Srivastava, N., Hinton, G., Krizhevsky, A., Sutskever, I., Salakhutdinov, R.: Dropout: a simple way to prevent neural networks from overfitting. J. Mach. Learn. Res. **15**(1), 1929–1958 (2014)
30. Su, H., et al.: Splatnet: sparse lattice networks for point cloud processing. In: Proceedings of the IEEE Conference on Computer Vision and Pattern Recognition, pp. 2530–2539 (2018)
31. Szegedy, C., Vanhoucke, V., Ioffe, S., Shlens, J., Wojna, Z.: Rethinking the inception architecture for computer vision. In: Proceedings of the IEEE Conference on Computer Vision and Pattern Recognition, pp. 2818–2826 (2016)
32. Thomas, H., Qi, C.R., Deschaud, J.E., Marcotegui, B., Goulette, F., Guibas, L.J.: KPConv: flexible and deformable convolution for point clouds. In: Proceedings of the IEEE International Conference on Computer Vision, pp. 6411–6420 (2019)
33. Tombari, F., Salti, S., Di Stefano, L.: Unique signatures of histograms for local surface description. In: Daniilidis, K., Maragos, P., Paragios, N. (eds.) ECCV 2010. LNCS, vol. 6313, pp. 356–369. Springer, Heidelberg (2010). https://doi.org/10.1007/978-3-642-15558-1_26
34. Wang, P.S., Liu, Y., Guo, Y.X., Sun, C.Y., Tong, X.: O-CNN: octree-based convolutional neural networks for 3D shape analysis. ACM Trans. Graph. (TOG) **36**(4), 1–11 (2017)
35. Wang, Y., Solomon, J.M.: Deep closest point: Learning representations for point cloud registration. In: Proceedings of the IEEE International Conference on Computer Vision, pp. 3523–3532 (2019)
36. Wang, Y., Solomon, J.M.: PRNet: self-supervised learning for partial-to-partial registration. In: Advances in Neural Information Processing Systems, pp. 8812–8824 (2019)
37. Wang, Y., Sun, Y., Liu, Z., Sarma, S.E., Bronstein, M.M., Solomon, J.M.: Dynamic graph CNN for learning on point clouds. ACM Trans. Graph. (TOG) **38**(5), 1–12 (2019)
38. Wu, W., Qi, Z., Fuxin, L.: PointConv: deep convolutional networks on 3D point clouds. In: Proceedings of the IEEE Conference on Computer Vision and Pattern Recognition, pp. 9621–9630 (2019)

39. Wu, Z., et al.: 3D shapenets: a deep representation for volumetric shapes. In: Proceedings of the IEEE Conference on Computer Vision and Pattern Recognition, pp. 1912–1920 (2015)
40. Wu, Z., Pan, S., Chen, F., Long, G., Zhang, C., Yu, P.S.: A comprehensive survey on graph neural networks. arXiv preprint arXiv:1901.00596 (2019)
41. Yang, H., Carlone, L.: A polynomial-time solution for robust registration with extreme outlier rates. arXiv preprint arXiv:1903.08588 (2019)
42. Yang, J., Li, H., Jia, Y.: Go-ICP: Solving 3D registration efficiently and globally optimally. In: Proceedings of the IEEE International Conference on Computer Vision, pp. 1457–1464 (2013)
43. Zhou, Q.-Y., Park, J., Koltun, V.: Fast global registration. In: Leibe, B., Matas, J., Sebe, N., Welling, M. (eds.) ECCV 2016. LNCS, vol. 9906, pp. 766–782. Springer, Cham (2016). https://doi.org/10.1007/978-3-319-46475-6_47
44. Zhou, Q.Y., Park, J., Koltun, V.: Open3D: a modern library for 3D data processing. arXiv preprint arXiv:1801.09847 (2018)
45. Zhou, Y., Tuzel, O.: VoxelNet: end-to-end learning for point cloud based 3D object detection. In: Proceedings of the IEEE Conference on Computer Vision and Pattern Recognition, pp. 4490–4499 (2018)

Pairwise Similarity Knowledge Transfer for Weakly Supervised Object Localization

Amir Rahimi[1(✉)], Amirreza Shaban[2(✉)], Thalaiyasingam Ajanthan[1], Richard Hartley[1,3], and Byron Boots[4]

[1] ANU, ACRV, Canberra, Australia
amir.rahimi@anu.edu.au
[2] Georgia Tech, Atlanta, Georgia
ashaban@uw.edu
[3] Google Research, Mountain View, USA
[4] University of Washington, Seattle, USA

Abstract. Weakly Supervised Object Localization (WSOL) methods only require image level labels as opposed to expensive bounding box annotations required by fully supervised algorithms. We study the problem of learning localization model on target classes with weakly supervised image labels, helped by a fully annotated source dataset. Typically, a WSOL model is first trained to predict class generic objectness scores on an off-the-shelf fully supervised source dataset and then it is progressively adapted to learn the objects in the weakly supervised target dataset. In this work, we argue that learning only an objectness function is a weak form of knowledge transfer and propose to learn a classwise pairwise similarity function that directly compares two input proposals as well. The combined localization model and the estimated object annotations are jointly learned in an alternating optimization paradigm as is typically done in standard WSOL methods. In contrast to the existing work that learns pairwise similarities, our approach optimizes a unified objective with convergence guarantee and it is computationally efficient for large-scale applications. Experiments on the COCO and ILSVRC 2013 detection datasets show that the performance of the localization model improves significantly with the inclusion of pairwise similarity function. For instance, in the ILSVRC dataset, the Correct Localization (CorLoc) performance improves from 72.8% to 78.2% which is a new state-of-the-art for WSOL task in the context of knowledge transfer.

Keywords: Weakly supervised object localization · Transfer learning · Multiple instance learning · Object detection

A. Rahimi and A. Shaban—Authors contributed equally.

Electronic supplementary material The online version of this chapter (https://doi.org/10.1007/978-3-030-58586-0_24) contains supplementary material, which is available to authorized users.

A. Vedaldi et al. (Eds.): ECCV 2020, LNCS 12369, pp. 395–412, 2020.
https://doi.org/10.1007/978-3-030-58586-0_24

1 Introduction

Weakly Supervised Object Localization (WSOL) methods have gained a lot of attention in computer vision [1–9]. Despite their supervised counterparts [10–14] that require the object class and their bounding box annotations, WSOL methods only require the image level labels indicating presence or absence of object classes. In spite of major improvements [1,5] in this area of research, there is still a large performance gap between weakly supervised and fully supervised object localization algorithms. In a successful attempt, WSOL methods are adopted to use an already annotated object detection dataset, called source dataset, to improve the weakly supervised learning performance in new classes [4,15]. These approaches learn transferable knowledge from the source dataset and use it to speed up learning new categories in the weakly supervised setting.

Multiple Instance Learning (MIL) methods like MI-SVM [16] are the predominant methods in weakly supervised object localization [1,5,6]. Typically, images are decomposed into bags of object proposals and the problem is posed as selecting one proposal from each bag that contains an object class. MIL methods take advantage of alternating optimization to progressively learn a classwise objectness (unary) function and the optimal selection in re-training and re-localization steps, respectively. Typically, the source dataset is used to learn an initial generic objectness function which is used to steer the selection toward objects and away from background proposals [4,15,17–20]. However, solely learning an objectness measure is a sub-optimal form of knowledge transfer as it can only discriminate objects from background proposals, while it is unable to discriminate between different object classes. Deselaers *et al.* [7] propose to additionally learn a pairwise similarity function from the fully annotated dataset and frame WOSL as a graph labeling problem where nodes represent bags and each proposal corresponds to one label for the corresponding node. The edges which reflect the cost of wrong pairwise labeling are derived from the learned pairwise similarities. Additionally, they propose an ad-hoc algorithm to progressively adapt the scoring functions to learn the weakly supervised classes using alternating re-training and re-localization steps. Unlike the alternating optimization in MIL, re-training and re-localization steps in [7] does not optimize a unified objective and therefore the convergence of their method could not be guaranteed. Despite good performance on medium scale problems, this method is less popular especially in large scale problems where computing all the pairwise similarities is intractable.

In this work, we adapt the localization model in MIL to additionally learn a pairwise similarity function and use a two-step alternating optimization to jointly learn the augmented localization model and the optimal selection. In the re-training step, the pairwise and unary functions are learned given the current selected proposals for each class. In the re-localization step, the selected proposals are updated given the current pairwise and unary similarity functions. We show that with a properly chosen localization loss function, the objective in the re-localization step can be equivalently expressed as a graph labeling problem very similar to the model in [7]. We use the computationally effective iterated conditional modes (ICM) graph inference algorithm [21] in the re-localization step which updates the selection of one bag in each iteration. Unfortunately, the ICM algorithm is prone to local minimum and its performance is highly dependent on the quality of its initial conditions. Inspired by the recent work on few-shot object localization [22], we divide the dataset into smaller mini-problems and solve each mini-problem individually using TRWS [23]. We combine the solutions of these mini-problems to initialize the ICM algorithm. Surprisingly, we observe that initializing ICM with the optimal selection from mini-problems of small sizes considerably improves the convergence point of ICM.

Our work addresses the main disadvantages of graph labeling algorithm in [7]. First, we formulate learning pairwise and unary functions and updating the optimal proposal selections with graph labeling within a two-step alternating optimization framework where each step is optimizing a unified objective and the convergence is guaranteed. Second, we propose a computationally efficient graph inference algorithm which uses a novel initialization method combined with ICM updates in the re-localization step. Our experiments show our method significantly improves the performance of MIL methods in large-scale COCO [24] and ILSVRC 2013 detection [25] datasets. Particularly, our method sets a new state-of-the-art performance of 78.2% correct localization [7] for the WSOL task in the ILSVRC 2013 detection dataset.[1]

2 Related Work

We review the MIL based algorithms among other branches in WSOL [17,19]. These approaches exploit alternating optimization to learn a detector and the optimal selection jointly. The algorithm iteratively alternates between re-localizing the objects given the current detector and re-training the detector given the current selection. In the recent years, alternating optimization scheme combined with deep neural networks has been the state-of-the-art in WSOL [1,2,26]. However, due to the non-convexity of its objective function, this method is prone to local minimum which typically leads to sub-optimal results [27,28] e.g. selecting the salient parts instead of the whole object. Addressing this issue has been the main focus of research in WSOL in the recent years [1,5,29]. In multi-fold [5], weakly supervised dataset is split into separate training and testing folds to avoid overfitting. Kumar *et al.* [29] propose an iterative self-paced learning algorithm that gradually learns from easy to hard samples to avoid getting stuck in bad local optimum points. Wan *et al.* [1] propose a continuation MIL algorithm to smooth out the non-convex loss function in order to alleviate the local optimum problem in a systematic way.

Transfer learning is another way to improve WSOL performance. These approaches utilize the information in a fully annotated dataset to learn an improved object detector on a weakly supervised dataset [4,15,18,20]. They leverage the common visual information between object classes to improve the localization performance in the target weakly supervised dataset. In a standard

[1] Source code is available on https://github.com/AmirooR/Pairwise-Similarity-knowledge-Transfer-WSOL.

knowledge transfer framework, the fully annotated dataset is used to learn a class agnostic objectness measure. This measure is incorporated during the alternating optimization step to steer the detector toward objects and away from the background [4]. Although the objectness measure is a powerful metric in differentiating between background and foreground, it fails to discriminate between different object classes. Several works have utilized pairwise similarity measure for improving WSOL [7,19,22]. Shaban *et al.* [22] use a relation network to predict pairwise similarity between pairs of proposals in the context of few-shot object co-localization. Deselaers *et al.* [7] frame WSOL as a graph labeling problem with pairwise and unary potentials and progressively adapt the potential functions to learn weakly supervised classes. Tang *et al.* [19] utilize the pairwise similarity between proposals to capture the inter-class diversity for the co-localization task. Hayder *et al.* [30,31] use pairwise learning for object co-detection.

3 Problem Description and Background

We review the standard dataset definition and optimization method for the weakly supervised object localization problem [1,4,5,7].

Dataset and Notation. Suppose each image is decomposed into a collection of object proposals which form a bag $\mathcal{B} = \{\mathbf{e}_i\}_{i=1}^m$ where an object proposal $\mathbf{e}_i \in \mathbb{R}^d$ is represented by a d-dimensional feature vector. We denote $y(\mathbf{e}) \in \mathcal{C} \cup \{c_\varnothing\}$ the label for object proposal $\mathbf{e}$. In this definition $\mathcal{C}$ is a set of object classes and $c_\varnothing$ denotes the background class. Given a class $c \in \mathcal{C}$ we can also define the binary label

$$y_c(\mathbf{e}) = \begin{cases} 1 \text{ if } y(\mathbf{e}) = c \\ 0 \text{ otherwise.} \end{cases} \tag{1}$$

With this notation a dataset is a set of bags along with the labels. For a weakly supervised dataset, only bag-level labels that denote the presence/absence of objects in a given bag are available. More precisely, the label for bag $\mathcal{B}$ is written as $\mathcal{Y}(\mathcal{B}) = \{c \mid \exists \mathbf{e} \in \mathcal{B} \text{ s.t. } y(\mathbf{e}) = c \in \mathcal{C}\}$. Let $Y_c(\mathcal{B}) \in \{0,1\}$ denote the binary bag label which indicates the presence/absence of class c in bag $\mathcal{B}$.

Given a weakly supervised dataset $\mathcal{D}_\mathcal{T} = \{\mathcal{T}, \mathcal{Y}_\mathcal{T}\}$ called the target dataset, with $\mathcal{T} = \{\mathcal{B}_j\}_{j=1}^N$ and corresponding bag labels $\mathcal{Y}_\mathcal{T} = \{\mathcal{Y}(\mathcal{B})\}_{\mathcal{B}\in\mathcal{T}}$, the goal is to estimate the latent proposal unary labeling[2] $\mathbf{y}_c$ for all object classes $c \in \mathcal{C}_\mathcal{T}$ in the target set.

For ease of notation, we also introduce a pairwise labeling function between pairs of proposals. The pairwise labeling function $r : \mathbb{R}^d \times \mathbb{R}^d \to \{0,1\}$ is designated to output 1 when two object proposals belong to the same object class and 0 otherwise, *i.e.*,

$$r(\mathbf{e}, \mathbf{e}') = \begin{cases} 1 \text{ if } y(\mathbf{e}) = y(\mathbf{e}') \neq c_\varnothing \\ 0 \text{ otherwise.} \end{cases} \tag{2}$$

[2] Notice, the labeling is a function defined over a finite set of variables, which can be treated as a vector. Here, $\mathbf{y}_c$ denotes the vector of labels $y_c(\mathbf{e})$ for all proposals $\mathbf{e}$.

Likewise, given a class c, two proposals are related under the class conditional pairwise labeling function $r_c : \mathbb{R}^d \times \mathbb{R}^d \to \{0,1\}$ if they both belong to class c. Similar to the unary labeling, since the pairwise labeling function is also defined over a finite set of variables, it can be seen as a vector. Unless we use the word vector or function, the context will determine whether we use the unary or pairwise labeling as a vector or a function. We use the "hat" notation to refer to the estimated (pseudo) unary or pairwise labeling by the weakly supervised learning algorithm.

Multiple Instance Learning (MIL). In standard MIL [16], the problem is solved by jointly learning a unary score function $\psi_c^{\mathrm{U}} : \mathbb{R}^d \to \mathbb{R}$ (typically represented by a neural network) and a feasible (pseudo) labeling $\hat{\mathbf{y}}_c$ that minimize the empirical unary loss

$$\mathcal{L}_c^{\mathrm{U}}(\boldsymbol{\psi}_c^{\mathrm{U}}, \hat{\mathbf{y}}_c \mid \mathcal{T}) = \sum_{\mathcal{B}\in\mathcal{T}} \sum_{\mathbf{e}\in\mathcal{B}} \ell(\psi_c^{\mathrm{U}}(\mathbf{e}), \hat{y}_c(\mathbf{e})), \tag{3}$$

where the loss function $\ell : \mathbb{R} \times \{0,1\} \to \mathbb{R}$ measures the incompatibility between predicted scores $\psi_c^{\mathrm{U}}(\mathbf{e})$ and the pseudo labels $\hat{y}_c(\mathbf{e})$. Here, likewise to the labeling, we denote the class score for all the proposals as a vector $\boldsymbol{\psi}_c^{\mathrm{U}}$. Note that the unary labeling $\hat{\mathbf{y}}_c$ is feasible if exactly one proposal has label 1 in each positive bag, and every other proposal has label 0 [5]. To this end, the set of feasible labeling $\mathcal{F}$ can be defined as

$$\mathcal{F} = \left\{ \hat{\mathbf{y}}_c \mid \hat{y}_c(\mathbf{e}) \in \{0,1\}, \sum_{\mathbf{e}\in\mathcal{B}} \hat{y}_c(\mathbf{e}) = Y_c(\mathcal{B}), \forall \mathcal{B} \in \mathcal{T} \right\}. \tag{4}$$

Finally, the problem is framed as minimizing the loss over all possible vectors $\boldsymbol{\psi}_c^{\mathrm{U}}$ (*i.e.*, unary functions represented by the neural network) and the feasible labels $\hat{\mathbf{y}}_c$

$$\begin{aligned} &\min_{\boldsymbol{\psi}_c^{\mathrm{U}}, \hat{\mathbf{y}}_c} \mathcal{L}_c^{\mathrm{U}}(\boldsymbol{\psi}_c^{\mathrm{U}}, \hat{\mathbf{y}}_c \mid \mathcal{T}), \\ &\text{s.t. } \hat{\mathbf{y}}_c \in \mathcal{F}. \end{aligned} \tag{5}$$

Optimization. This objective is typically minimized in an iterative two-step alternating optimization paradigm [32]. The optimization process starts with some initial value of the parameters and labels, and iteratively alternates between *re-training* and *re-localization* steps until convergence. In the re-training step, the parameters of the unary score function ψ_c^{U} are optimized while the labels $\hat{\mathbf{y}}_c$ are fixed. In the re-localization step, proposal labels are updated given the current unary scores. The optimization in the re-localization step is equivalent to assigning positive label to the proposal with the highest unary score within each positive bag and label 0 to all other proposals [16]. Formally, label of the proposal $\mathbf{e} \in \mathcal{B}$ in bag $\mathcal{B}$ is updated as

$$\hat{y}_c(\mathbf{e}) = \begin{cases} 1 \text{ if } Y_c(\mathcal{B}) = 1 \text{ and } \mathbf{e} = \operatorname{argmax}_{\mathbf{e}'\in\mathcal{B}} \psi_c^{\mathrm{U}}(\mathbf{e}') \\ 0 \text{ otherwise.} \end{cases} \tag{6}$$

Knowledge Transfer. In this paper, we also assume having access to an auxiliary fully annotated dataset $\mathcal{D}_{\mathcal{S}}$ (source dataset) with object classes in $\mathcal{C}_{\mathcal{S}}$ which is a disjoint set from the target dataset classes, *i.e.*, $\mathcal{C}_{\mathcal{T}} \cap \mathcal{C}_{\mathcal{S}} = \varnothing$. In the standard practice [4,18,20], the source dataset is used to learn a class agnostic unary score $\psi^{\mathrm{U}} : \mathbb{R}^d \rightarrow \mathbb{R}$ which measures how likely the input proposal $\mathbf{e}$ tightly encloses a foreground object. Then, the unary score vector used in Eq. 6 is adapted to $\boldsymbol{\psi}_c^{\mathrm{U}} \leftarrow \lambda \boldsymbol{\psi}_c^{\mathrm{U}} + (1-\lambda)\boldsymbol{\psi}^{\mathrm{U}}$ for some $0 \leq \lambda \leq 1$. This steers the labeling toward choosing proposals that contain complete objects. Although the class agnostic unary score function ψ^{U} is learned on the source classes, since objects share common properties, it transfers to the unseen classes in the target set.

4 Proposed Method

In addition to learning the unary scores, we also learn a classwise pairwise similarity function $\psi_c^{\mathrm{P}} : \mathbb{R}^d \times \mathbb{R}^d \rightarrow \mathbb{R}$ that estimates the pairwise labeling between pairs of proposals. That is for the target class c, pairwise similarity score $\psi_c^{\mathrm{P}}(\mathbf{e}, \mathbf{e}')$ between two input proposals $\mathbf{e}, \mathbf{e}' \in \mathbb{R}^d$ has a high value if two proposals are related, *i.e.*, $\hat{r}_c(\mathbf{e}, \mathbf{e}') = 1$ and a low value otherwise. We define the empirical pairwise similarity loss to measure the incompatibility between pairwise similarity function predictions and the pairwise labeling $\hat{\mathbf{r}}_c$

$$\mathcal{L}_c^{\mathrm{P}}(\boldsymbol{\psi}_c^{\mathrm{P}}, \hat{\mathbf{r}}_c | \mathcal{T}) = \sum_{\substack{\mathcal{B},\mathcal{B}' \in \mathcal{T} \\ \mathcal{B} \neq \mathcal{B}'}} \sum_{\substack{\mathbf{e} \in \mathcal{B} \\ \mathbf{e}' \in \mathcal{B}'}} \ell(\psi_c^{\mathrm{P}}(\mathbf{e}, \mathbf{e}'), \hat{r}_c(\mathbf{e}, \mathbf{e}')), \tag{7}$$

where $\boldsymbol{\psi}_c^{\mathrm{P}}$ denotes the vector of the pairwise similarities of all pairs of proposals, and $\ell : \mathbb{R} \times \{0, 1\} \rightarrow \mathbb{R}$ is the loss function. We define the overall loss as the weighted sum of the empirical pairwise similarity and the unary loss

$$\mathcal{L}_c(\boldsymbol{\psi}_c, \hat{\mathbf{z}}_c | \mathcal{T}) = \alpha \mathcal{L}_c^{\mathrm{P}}(\boldsymbol{\psi}_c^{\mathrm{P}}, \hat{r}_c | \mathcal{T}) + \mathcal{L}_c^{\mathrm{U}}(\boldsymbol{\psi}_c^{\mathrm{U}}, \hat{y}_c | \mathcal{T}), \tag{8}$$

where $\boldsymbol{\psi}_c = \left[\boldsymbol{\psi}_c^{\mathrm{U}}, \boldsymbol{\psi}_c^{\mathrm{P}}\right]$ is the vector of unary and pairwise similarity scores combined, and $\hat{\mathbf{z}}_c = [\hat{\mathbf{y}}_c, \hat{\mathbf{r}}_c]$ denotes the concatenation of unary and pairwise labeling vectors, and $\alpha > 0$ controls the importance of the pairwise similarity loss.

We employ alternating optimization to jointly optimize the loss over the parameters of the scoring functions ψ_c^{U} and ψ_c^{P} (re-training) and labelings $\hat{\mathbf{z}}_c$ (re-localization). In re-training, the objective function is optimized to learn the pairwise similarity and the unary scoring functions from the pseudo labels. In re-localization, we use the current scores to update the labelings.

Training the model with fixed labels, *i.e.* re-training step, is straightforward and can be implemented within any common neural network framework. We use sigmoid cross entropy loss in both empirical unary and pairwise similarity losses

$$\ell(x, y) = -(1-y)\log(1-\sigma(x)) - y\log(\sigma(x)), \tag{9}$$

where $x \in \mathbb{R}$ is the predicted logit, $y \in \{0, 1\}$ is the label, and $\sigma : \mathbb{R} \rightarrow \mathbb{R}$ denotes the sigmoid function $\sigma(x) = 1/(1+\exp(-x))$. The choice of the loss function directly affects the objective function in the re-localization step. As we will show later, since sigmoid cross entropy loss is a linear function of label y it leads to a *linear objective function* in the re-localization step. To speed up the re-training step, we train pairwise similarity and unary scoring functions for all the classes together by optimizing the total loss

$$\mathcal{L}(\boldsymbol{\psi} \mid \hat{\mathbf{z}}, \mathcal{T}) = \sum_{c \in \mathcal{C}_\mathcal{T}} \mathcal{L}_c(\boldsymbol{\psi}_c, \hat{\mathbf{z}}_c \mid \mathcal{T}), \tag{10}$$

where $\boldsymbol{\psi} = [\boldsymbol{\psi}_c]_{c \in \mathcal{C}_\mathcal{T}}$ and $\hat{\mathbf{z}} = [\hat{\mathbf{z}}_c]_{c \in \mathcal{C}_\mathcal{T}}$ are the concatenation of respective vectors for all classes. Note that we learn the parameters of the scoring functions that minimize the loss, while $\hat{\mathbf{z}}$ remains fixed in this step. Since the dataset is large, we employ Stochastic Gradient Descent (SGD) with momentum for optimization. Additionally, we subsample proposals in each bag by sampling 3 proposals with foreground and 7 proposal with background label in each training iteration.

4.1 Re-localization

In this step, we minimize the empirical loss function in Eq. 8 over the feasible labeling $\hat{\mathbf{z}}_c$ for the given model parameters. We first define feasible labeling set $\mathcal{A}$ and simplify the objective function to an equivalent, simple linear form. Then, we discuss algorithms to optimize the objective function in the large scale settings.

For $\hat{\mathbf{z}}_c$ to be feasible, labeling should be feasible, *i.e.*, $\hat{\mathbf{y}}_c \in \mathcal{F}$ and pairwise labeling $\hat{\mathbf{r}}_c$ should also be consistent with the unary labeling. For dataset $\mathcal{D}_\mathcal{T}$ and target class c, this constraint set, known as the *local polytope* in the MRF literature [33], is expressed as

$$\mathcal{A} = \left\{ \hat{\mathbf{z}}_c \left| \begin{array}{ll} \sum_{\mathbf{e} \in \mathcal{B}} \hat{y}_c(\mathbf{e}) = Y_c(\mathcal{B}) & \mathcal{B} \in \mathcal{T} \\ \sum_{\mathbf{e} \in \mathcal{B}} \hat{r}_c(\mathbf{e}, \mathbf{e}') = \hat{y}_c(\mathbf{e}') & \mathcal{B}, \mathcal{B}' \in \mathcal{T}, \mathcal{B}' \neq \mathcal{B}, \mathbf{e}' \in \mathcal{B}' \\ \hat{r}_c(\mathbf{e}, \mathbf{e}'), \hat{y}_c(\mathbf{e}) \in \{0, 1\} & c \in \mathcal{C}, \text{for all } \mathbf{e} \text{ and } \mathbf{e}' \end{array} \right. \right\}. \tag{11}$$

Next, we simplify the loss function in the re-localization step. Let $\mathcal{T}_c = \{\mathcal{B} \mid \mathcal{B} \in \mathcal{T}, c \in \mathcal{Y}(\mathcal{B})\}$ and $\mathcal{T}_{\bar{c}} = \mathcal{T} \setminus \mathcal{T}_c$ denote the set of positive and negative bags with respect to class c. The loss function in Eq. 8 can be decomposed into three parts

$$\begin{aligned} \mathcal{L}_c(\boldsymbol{\psi}_c, \hat{\mathbf{z}}_c | \mathcal{T}) = {} & \mathcal{L}_c(\boldsymbol{\psi}_c, \hat{\mathbf{z}}_c | \mathcal{T}_c) + \mathcal{L}_c(\boldsymbol{\psi}_c, \hat{\mathbf{z}}_c | \mathcal{T}_{\bar{c}}) \\ & + \sum_{\substack{\mathbf{e} \in \mathcal{B} \in \mathcal{T}_c \\ \mathbf{e}' \in \mathcal{B}' \in \mathcal{T}_{\bar{c}}}} \ell(\psi_c^{\mathrm{P}}(\mathbf{e}, \mathbf{e}'), \hat{r}_c(\mathbf{e}, \mathbf{e}')) + \ell(\psi_c^{\mathrm{P}}(\mathbf{e}', \mathbf{e}), \hat{r}_c(\mathbf{e}', \mathbf{e})), \end{aligned}$$

were the first two terms are the loss function in Eq. 8 defined over the positive set $\mathcal{T}_c$ and negative set $\mathcal{T}_{\bar{c}}$, and last term is the loss defined by the pairwise similarities between these two sets. Since for any feasible labeling all the proposals in negative bags has label 0 and remain fixed, only the value of $\mathcal{L}_c(\boldsymbol{\psi}_c, \hat{\mathbf{z}}_c | \mathcal{T}_c)$ changes within $\mathcal{A}$ and other terms are constant. Furthermore, by observing that for sigmoid cross entropy loss in Eq. 9 we have $\ell(x, y) = \ell(x, 0) - yx$, for $y \in [0, 1]$,[3] we can further break down $\mathcal{L}_c(\psi_c, \hat{\mathbf{z}}_c | \mathcal{T}_c)$ as

$$\mathcal{L}_c(\psi_c, \hat{\mathbf{z}}_c \mid \mathcal{T}_c) = \mathcal{L}_c(\psi_c, \mathbf{0} \mid \mathcal{T}_c) \\ \underbrace{-\alpha \sum_{\substack{\mathcal{B}, \mathcal{B}' \in \mathcal{T}_c \\ \mathcal{B} \neq \mathcal{B}'}} \sum_{\substack{\mathbf{e} \in \mathcal{B} \\ \mathbf{e}' \in \mathcal{B}'}} \psi_c^{\mathrm{P}}(\mathbf{e}, \mathbf{e}') \hat{r}_c(\mathbf{e}, \mathbf{e}') - \sum_{\mathcal{B} \in \mathcal{T}} \sum_{\mathbf{e} \in \mathcal{B}} \psi_c^{\mathrm{U}}(\mathbf{e}) \hat{y}_c(\mathbf{e})}_{\mathcal{L}_{\mathrm{reloc}}(\hat{\mathbf{z}}_c | \psi_c, \mathcal{T}_c)}, \tag{12}$$

where $\mathbf{0}$ is zero vector of the same dimension as $\hat{\mathbf{z}}_c$. Since the first term is constant with respect to $\hat{\mathbf{z}}_c = [\hat{\mathbf{y}}_c, \hat{\mathbf{r}}_c]$, re-localization can be equivalently done by optimizing $\mathcal{L}_{\mathrm{reloc}}(\hat{\mathbf{z}}_c \mid \psi_c, \mathcal{T}_c)$ over the feasible set $\mathcal{A}$

$$\begin{aligned} \min_{\hat{\mathbf{z}}_c} & -\alpha \hat{\mathbf{r}}_c^\top \psi_c^{\mathrm{P}} - \hat{\mathbf{y}}_c^\top \psi_c^{\mathrm{U}}, \\ & \text{s.t. } \hat{\mathbf{z}}_c \in \mathcal{A}, \end{aligned} \tag{13}$$

where we use the equivalent vector form to represent the re-localization loss in Eq. 12. The re-localization optimization is an Integer Linear Program (ILP) and has been widely studied in literature [34]. The optimization can be equivalently expressed as a graph labeling problem with pairwise and unary potentials [35]. In the equivalent graph labeling problem, each bag is represented by a node in the graph where each proposal of the bag corresponds to a label of that node, and pairwise and unary potentials are equivalent to the negative pairwise similarity and negative unary scores in our problem. We discuss different graph inference methods and their limitations and present a practical method for large-scale settings.

Inference. Finding an optimal solution $\hat{\mathbf{z}}_c^*$ that minimizes the loss function defined in Eq. 13 is NP-hard and thus not feasible to compute exactly, except in small cases. Loopy belief propagation [36], TRWS [23], and AStar [37], are among the many inference algorithms used for approximate graph labeling problem. Unfortunately, finding an approximate labeling quickly becomes impractical as the size of $\mathcal{T}_c$ increases, since the dimension of $\hat{\mathbf{z}}_c$ increases quadratically with the numbers of bags in $\mathcal{T}_c$ due to dense pairwise connectivity. Due to this limitation, we employ an older well-known iterated conditional modes (ICM) algorithm for optimization [21]. In each iteration, ICM only updates one unary label in $\hat{\mathbf{y}}_c$ along with the pairwise labels that are related to this unary label while all the other elements of $\hat{\mathbf{z}}_c$ are fixed. The block that gets updated in each iteration is shown in Fig. 1. ICM generates monotonically non-increasing objective values and is computationally efficient. However, since ICM performs coordinate descent type updates and the problem in Eq. 13 is neither convex nor differentiable as the constraint set is discrete, ICM is prone to get stuck at a local minimum and its solution significantly depends on the quality of the initial labeling.

[3] See Appendix for the proof.

Algorithm 1: Re-localization

Input: Dataset $\mathcal{D}_{\mathcal{T}}$, batch size K, #epochs E
Output: Optimal unary labeling $\hat{\mathbf{y}}^*$
for $c \in \mathcal{C}_{\mathcal{T}}$ **do**
  $T \leftarrow round(\frac{|\mathcal{T}_c|}{K})$, $\hat{\mathbf{y}}_c \leftarrow \mathbf{0}$
  for $t \leftarrow 1$ ***to*** T **do**
    `// Sample next mini-problem`
    $\mathcal{X} \sim \mathcal{T}_c$
    `// Solve mini-problem with TRWS [23]`
    $[\bar{\mathbf{y}}_c^*, \bar{\mathbf{r}}_c^*] \leftarrow \text{argmin}_{\bar{\mathbf{z}}_c} -\alpha \bar{\mathbf{r}}_c^\top \bar{\boldsymbol{\psi}}_c^{\text{P}} - \bar{\mathbf{y}}_c^\top \bar{\boldsymbol{\psi}}_c^{\text{U}}$ s.t. $\bar{\mathbf{z}}_c \in \bar{\mathcal{A}}$
    Update corresponding block of $\hat{\mathbf{y}}_c$ with $\bar{\mathbf{y}}^*$
  `// Finetune for` E `epochs`
  $\hat{\mathbf{y}}_c^* \leftarrow \texttt{ICM}(\hat{\mathbf{y}}_c, E)$
return $\{\hat{\mathbf{y}}_c^*\}_{c \in \mathcal{C}_{\mathcal{T}}}$

Recent work [22] has shown that using accurate pairwise and unary functions learned on the source dataset, the re-localization method performs reasonably well by only looking at few bags. Motivated by this, we divide the full size problem into a set of disjoint mini-problems, solve each mini-problem efficiently using TRWS inference algorithm, and use these results to initialize the ICM algorithm.

The initialization algorithm samples a mini-problem $\mathcal{X} \in \mathcal{T}_c$ and optimizes the re-localization problem $\mathcal{L}_{\text{reloc}}(\bar{\mathbf{z}}_c \mid \bar{\boldsymbol{\psi}}_c, \mathcal{X})$ where vectors $\bar{\mathbf{z}}_c$ and $\bar{\boldsymbol{\psi}}_c$ are parts of vectors $\hat{\mathbf{z}}_c$ and $\boldsymbol{\psi}_c$ that are within the mini-problem defined by $\mathcal{X}$ (see Fig. 1). This process is repeated until all the bags in the dataset are covered. The complete re-localization step is illustrated in Algorithm 1.

Next, we analysis the time complexity of the re-localization step. We practically observed that computing the pairwise similarity scores is the computation bottleneck, thus we analyze the time complexities in terms of the number of pairwise similarity scores each algorithm computes. Let $M = \max_{c \in \mathcal{C}_{\mathcal{T}}} |\mathcal{T}_c|$ denotes the maximum number of positive bags, and $B = \max_{\mathcal{B} \in \mathcal{T}} |\mathcal{B}|$ be the maximum bag size. To solve the exact optimization in Eq. 13, we need to compute the vector $\boldsymbol{\psi}_c$ with $\mathcal{O}(B^2M^2)$ elements. On the other hand, each iteration of ICM only computes $\mathcal{O}(BM)$ pairwise similarity scores. We additionally compute a total of $\mathcal{O}(MKB^2)$ pairwise similarity scores for the initialization where K is the size of the mini-problem. Thus, ICM algorithm would be asymptotically more efficient than the exact optimization in terms of total number of pairwise similarity scores it computes, if it is run for $\Omega(MB)$ iterations or $E = \Omega(B)$ epochs. We practically observe that by initializing ICM with the result of the proposed initialization scheme it convergences in few epochs.

Even though Eq. 13 is similar to the DenseCRF formulation [38], the pairwise potentials are not amenable to the efficient filtering method [39] which is the backbone of DenseCRF methods [38,40]. Therefore, it is intractable to use any existing sophisticated optimization algorithm except for ICM and Mean-Field [41]. Nevertheless, our block-wise application of TRWS provides an effective initialization for ICM. We additionally experimented with the block version of ICM [35] but it performs similarly while being slower.

Fig. 1. ICM iteration (left) and initialization (right) graphical models. In both graphs, each node represents a bag (with B proposals) within a dataset with $|\mathcal{T}_c| = 9$ bags. **Left:** ICM updates the unary label of the selected node (shown in green). Edges show all the pairwise labels that gets updated in the process. Since the unary labeling of other nodes are fixed each blue edge represents B elements in vector $\hat{\mathbf{r}}_c$. **Right:** For initialization we divide the dataset into smaller mini-problems (with size $K = 3$ in this example) and solve each of them individually. Each edge represents B^2 pairwise scores that need to be computed. (Color figure online)

4.2 Knowledge Transfer

To transfer knowledge from the fully annotated source set $\mathcal{D}_\mathcal{S}$, we first learn *class generic* pairwise similarity $\psi^\mathrm{P} : \mathbb{R}^d \times \mathbb{R}^d \to \mathbb{R}$ and unary $\psi^\mathrm{U} : \mathbb{R}^d \to \mathbb{R}$ functions from the source set. Since the labels are available for all the proposals in the source set, learning the pairwise and unary functions is straightforward. We simply use stochastic gradient descent (SGD) to optimize the loss

$$\mathcal{L}^T(\psi^\mathrm{P}, \psi^\mathrm{U} | \mathcal{S}, \mathbf{r}, \mathbf{o}) = \alpha \sum_{\substack{\mathcal{B},\mathcal{B}' \in \mathcal{S} \\ \mathcal{B} \neq \mathcal{B}'}} \sum_{\substack{\mathbf{e} \in \mathcal{B} \\ \mathbf{e}' \in \mathcal{B}'}} \ell(\psi^\mathrm{P}(\mathbf{e}, \mathbf{e}'), r(\mathbf{e}, \mathbf{e}'))) + \sum_{\mathcal{B} \in \mathcal{S}} \sum_{\mathbf{e} \in \mathcal{B}} \ell(\psi^\mathrm{U}(\mathbf{e}), o(\mathbf{e})), \tag{14}$$

where $o(\mathbf{e}) \in \{0, 1\}$ is class generic objectness label, *i.e.*,

$$o(\mathbf{e}) = \begin{cases} 1 \text{ if } y(\mathbf{e}) \neq c_\varnothing \\ 0 \text{ otherwise,} \end{cases} \tag{15}$$

and relation function $r : \mathbb{R}^d \times \mathbb{R}^d \to \mathbb{R}$ is defined by Eq. 6. Here we do not use hat notation since groundtruth proposal labels are available for the source dataset $\mathcal{D}_\mathcal{S}$. We skip the details as the loss in Eq. 14 has a similar structure to the re-training loss. Note that in general the class generic functions ψ^U and ψ^P and class specific functions ψ_c^U and ψ_c^P use different feature sets extracted from different networks. Having learned these functions, we adapt both pairwise similarity and score vectors in the re-localization step in Algorithm 1 as

$$\begin{aligned} \psi_c^\mathrm{P} &\leftarrow (1 - \lambda_1)\psi_c^\mathrm{P} + \lambda_1 \psi^\mathrm{P} \\ \psi_c^\mathrm{U} &\leftarrow (1 - \lambda_2)\psi_c^\mathrm{U} + \lambda_2 \psi^\mathrm{U}, \end{aligned}$$

where $0 \leq \lambda_1, \lambda_2 \leq 1$ controls the weight of transferred and adaptive functions in pairwise similarity and unary functions respectively.

We start the alternating optimization with a *warm-up* re-localization step where only the learned class generic pairwise and unary functions above are used in the re-localization algorithm, *i.e.*, $\lambda_1, \lambda_2 = 1$. The warm-up re-localization step provides high quality pseudo labels to the first re-training step and speeds up the convergence of the alternating optimization algorithm.

4.3 Network Architectures

Proposal and Feature Extraction. Following the experiment protocol in [4], we use a Faster-RCNN [42] model trained on the source dataset $\mathcal{D}_\mathcal{S}$ to extract region proposals from each image. We keep the box features in the last layer of Faster-RCNN as transferred features to be used in the class generic score functions. Following [4,15,43], we extract AlexNet [44] feature vectors from each proposal as input to the class specific scoring functions ψ_c^{U} and ψ_c^{P}.

Scoring Functions. Let $\mathbf{e}$ and $\mathbf{e}'$ denote features in $\mathbb{R}^d$ extracted from two image proposals. Linear layers are employed to model the class generic unary function ψ^{U} and all the classwise unary functions ψ_c^{U} i.e. $\psi_c^{\mathrm{U}}(\mathbf{e}) = \mathbf{w}_c^\top \mathbf{e} + b_c$ where $\mathbf{w}_c \in \mathbb{R}^d$ is the weight and $b_c \in \mathbb{R}$ is the bias parameter. We borrow the relation network architecture from [22] to model the pairwise similarity functions ψ^{P} and ψ_c^{P}. The details of the relation network architecture are discussed in the Appendix.

5 Experiments

We evaluate the main applicability of our technique on different weakly supervised datasets and analyze how each part affects the final results in our method. We report the widely accepted Correct Localization (CorLoc) metric [7] for the object localization task as our evaluation metric.

5.1 COCO 2017 Dataset

We employ a split of COCO 2017 [24] dataset to evaluate the effect of different initialization strategies and our pairwise retraining and re-localization steps. The dataset has 80 classes in total. We take the same split of [22,45] with 63 source $\mathcal{C}_\mathcal{S}$ and 17 target $\mathcal{C}_\mathcal{T}$ classes and follow [22] to create the source and target splits to create source and target datasets with 111,085 and 8,245 images, respectively.

Similar to [22], we use Faster-RCNN [42] with ResNet 50 [46] backbone as our proposal generator and feature extractor for knowledge transfer. We keep the top $B = 100$ proposals generated by Faster-RCNN for experiments on the COCO 2017.

We first study different approaches for initializing the ICM method in the re-localization step. Then, we present the result of the full proposed method and compare it with other baselines.

Initialization Scheme. Since the ICM algorithm is sensitive to initialization, we devise the following experiment to evaluate different initialization methods. To limit total running time of the experiment, we only do this evaluation in the warm-up re-localization step. We start by training class generic unary and pairwise similarity scoring functions on the source dataset $\mathcal{D}_\mathcal{S}$. Next, we initialize the labeling of the images in $\mathcal{D}_\mathcal{T}$ using the following initialization strategies:

- Random: randomly select a proposal from each bag.
- Objectness: select the proposal with the highest unary score from each bag.
- Proposed initialization method: Proposed initialization method discussed in Sect. 4.1. We conduct the experiment with different mini-problem sizes $K \in \{2, 4, 8, 64\}$. We use TRWS [23] algorithm for inference in each mini-problem.

Finally, we perform ICM with each of the initialization methods. Figure 2 shows the CorLoc and Energy vs. time plots as well as the computation time for different initialization methods. As expected, $K = 64$ exhibits the best initialization performance. However, ICM converges to similar energy when $4 \leq K \leq 64$ is used in the initialization method. In the extreme case with mini-problem of size $K = 2$, ICM converges to a worse local minimum in terms of CorLoc and energy value. Surprisingly, random initialization converges to the same result as objectness and $K = 2$. We also tried initializing ICM with the proposal that covers the complete image as it is the initialization scheme that is commonly used in MIL alternating optimization algorithms [4,5]. But it performs significantly worse than the other initialization methods.

These results highlight the importance of initialization in ICM inference. Note that increasing K beyond 64 might provide a better initialization to ICM and increase the results further but it quickly becomes impractical as the time plot in Fig. 2 illustrates. As a rule of thumb, one should increase the mini-problem size as far as time and computational resources allow.

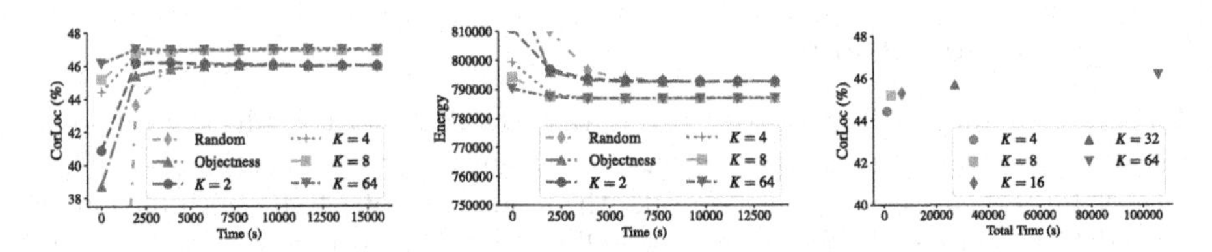

Fig. 2. Left: ICM CorLoc@0.5(%) vs. time for different initialization methods. See initialization schemes for definition of each initialization method. Markers indicate start of a new epoch. ICM inference convergences in 2 epochs and demonstrates its best performance when is initialized with the proposed initialization method. **Middle:** Energy vs. time for different initialization methods. The energies in the plot are computed by summing over energies of all classes. **Right:** Runtime vs. CorLoc(%) comparison of the proposed initialization scheme with various mini-problem sizes. We observe a quadratic time increase with respect to the mini-problem size.

Full Pipeline. Here, we conduct an experiment to determine the importance of learning pairwise similarities on the COCO dataset. We compare our full method with the unary method which only learns and uses unary scoring functions during, warm-up, re-training and re-localization steps. This method is analogous to [4]. The difference is that it uses cross entropy loss and SGD training instead of

Support Vector Machine used in [4]. Also, we do not employ hard-negative mining after each re-training step. For this experiment, we use mini-problems of size $K = 4$ for initializing ICM. We run both methods for 5 iterations of alternating optimization on the target dataset. Our method achieves **48.3%** compared to 39.4% CorLoc@0.5 of the unary method. This clearly shows the effectiveness of our pairwise similarity learning.

5.2 ILSVRC 2013 Detection Dataset

We closely follow the experimental protocol of [4,15,43] to create source and target datasets on ILSVRC 2013 [25] detection dataset. The dataset has 200 categories with full bounding box annotations. We use the first 100 alphabetically ordered classes as source categories $\mathcal{C}_\mathcal{S}$ and the remaining 100 classes as target categories $\mathcal{C}_\mathcal{T}$. The dataset is divided into source training set $\mathcal{D}_\mathcal{S}$ with 63k images, target set $\mathcal{D}_\mathcal{T}$ with 65k images, and a target test set with 10k images. We report CorLoc of different algorithms on $\mathcal{D}_\mathcal{T}$. Similar to previous works [4,15,43], we additionally train a detector from the output of our method on target set $\mathcal{D}_\mathcal{T}$, and evaluate it on the target test set.

For a fair comparison, we use a similar proposal generator and multi-fold strategy as [4]. We use Faster-RCNN [42] with Inception-Resnet [47] backbone trained on source dataset $\mathcal{D}_\mathcal{S}$ for object bounding box generation. The experiment on COCO suggests a small mini-problem size K would be sufficient to achieve good performance in the re-localization step. We use $K = 8$ to balance the time and accuracy in this experiment.

Baselines and Results. We compare our method with two knowledge transfer techniques [4,15] for WSOL. In addition, we demonstrate the results of the following baselines that only use unary scoring function:

- Warm-up (unary): To see the importance of learning pairwise similarities in knowledge transfer, we perform the warm-up re-localization with only the transferred unary scores ψ^{U}. This can be achieved by simply selecting the box with the highest unary score within each bag. We compare this results with the result of the warm-up step which uses both pairwise and unary scores in knowledge transfer.
- Unary: Standard MIL objective in Eq. 5 which only learns labeling and the unary scoring function.

We compare these results with our full pipeline which starts with a warm-up re-localization step followed by alternating re-training and re-localization steps. The results are illustrated in Table 1. Compared to Uijlings *et al.* [4], our method improves the CorLoc@0.5 performance on the target set by 4% and mAP@0.5 on the target test set by 4.8%. Warm-up re-localization improves CorLoc performance of warm-up (unary) by 4.9% with transferring a pairwise similarity measure from the source classes. Note that the result of warm-up step without any re-training performs on par with the Uijlings *et al.* [4] MIL method. The

Table 1. Performance of different methods on ILSVRC 2013. Proposal generators and their backbone models are shown in the second and third column. Total time is shown in "Training+Inference" format. CorLoc is reported on the target set. The last column shows the performance of an object detector trained on the target set and evaluated on the target test set. *The first 3 methods use RCNN detector with AlexNet backbone while other methods utilize Faster-RCNN detector with Inception-Resnet backbone.

Method	Proposal generator	Backbone	CorLoc@0.5	CorLoc@0.7	Time(hours)	mAP@0.5
LSDA [15]	Selective search [48]	AlexNet [44]	28.8	-	-	18.1*
Tang *et al.* [43]	Selective search [48]	AlexNet [44]	-	-	-	20.0*
Uijlings *et al.* [4]	SSD [12]	Inception-V3 [49]	70.3	58.8	-	23.3*
Uijlings *et al.* [4]	Faster-RCNN	Inception-Resnet	74.2	61.7	-	36.9
Warm-up (unary)	Faster-RCNN	Inception-Resnet	68.9	59.5	0	-
Warm-up	Faster-RCNN	Inception-Resnet	73.8	62.3	5+3	-
Unary	Faster-RCNN	Inception-Resnet	72.8	62.0	13+2	38.1
Full (ours)	Faster-RCNN	Inception-Resnet	**78.2**	**65.5**	65+13	**41.7**
Supervised [4]						46.2

CorLoc performance at the stricter IoU>0.7 also shows similar results. Some of the success cases are shown in Fig. 3.

Compared to [4], our implementation of the MIL method performs worse with IoU threshold 0.5 but better with stricter threshold 0.7. We believe the reason is having a different loss function and hard-negative mining in [4].

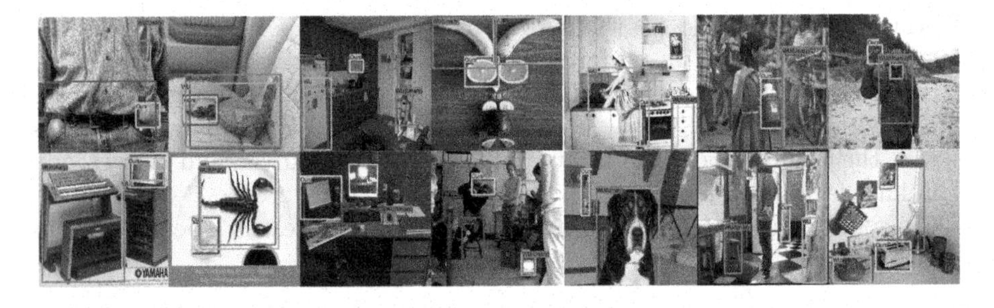

Fig. 3. Success cases on ILSVRC 2013 dataset. Unary method that relies on the objectness function tends to select objects from source classes that have been seen during training. Note that "banana", "dog", and "chair" are samples from source classes. Bounding boxes are tagged with method names. "GT" and "WU" stand for groundtruth and warm-up respectively. See Appendix for a larger set of success and failure cases.

6 Conclusion

We study the problem of learning localization models on target classes from weakly supervised training images, helped by a fully annotated source dataset. We adapt MIL localization model by adding a classwise pairwise similarity module that learns to directly compare two input proposals. Similar to the standard MIL approach, we learn the augmented localization model and annotations jointly by two-step alternating optimization. We represent the re-localization step as a graph labeling problem and propose a computationally efficient inference algorithm for optimization. Compared to the previous work [7] that uses pairwise similarities for this task, the proposed method is represented in alternating optimization framework with convergence guarantee and is computationally efficient in large-scale settings. The experiments show that learning pairwise similarity function improves the performance of WSOL over the standard MIL.

Acknowledgement. We would like to thank Jasper Uijlings for providing helpful instructions and their original models and datasets. We gratefully express our gratitude to Judy Hoffman for her advice on improving the experiments. We also thank the anonymous reviewers for their helpful comments to improve the paper. This research is supported in part by the Australia Research Council Centre of Excellence for Robotics Vision (CE140100016).

References

1. Wan, F., Liu, C., Ke, W., Ji, X., Jiao, J., Ye, Q.: C-MIL: continuation multiple instance learning for weakly supervised object detection. In: Proceedings of the IEEE Conference on Computer Vision and Pattern Recognition, pp. 2199–2208 (2019)
2. Gao, J., Wang, J., Dai, S., Li, L.J., Nevatia, R.: NOTE-RCNN: NOise tolerant ensemble RCNN for semi-supervised object detection. In: Proceedings of the IEEE International Conference on Computer Vision, pp. 9508–9517 (2019)
3. Arun, A., Jawahar, C., Kumar, M.P.: Dissimilarity coefficient based weakly supervised object detection. In: Proceedings of the IEEE Conference on Computer Vision and Pattern Recognition, pp. 9432–9441 (2019)
4. Uijlings, J., Popov, S., Ferrari, V.: Revisiting knowledge transfer for training object class detectors. In: Proceedings of the IEEE Conference on Computer Vision and Pattern Recognition, pp. 1101–1110 (2018)
5. Cinbis, R.G., Verbeek, J., Schmid, C.: Weakly supervised object localization with multi-fold multiple instance learning. IEEE Trans. Pattern Anal. Mach. Intell. **39**(1), 189–203 (2016)
6. Bilen, H., Pedersoli, M., Tuytelaars, T.: Weakly supervised object detection with posterior regularization. In: British Machine Vision Conference, vol. 3 (2014)
7. Deselaers, T., Alexe, B., Ferrari, V.: Localizing objects while learning their appearance. In: Daniilidis, K., Maragos, P., Paragios, N. (eds.) ECCV 2010. LNCS, vol. 6314, pp. 452–466. Springer, Heidelberg (2010). https://doi.org/10.1007/978-3-642-15561-1_33
8. Tang, P., Wang, X., Bai, X., Liu, W.: Multiple instance detection network with online instance classifier refinement. In: Proceedings of the IEEE Conference on Computer Vision and Pattern Recognition (2017)

9. Tang, P., et al.: PCL: proposal cluster learning for weakly supervised object detection. IEEE Trans. Pattern Anal. Mach. Intell. **42**(1), 176–191 (2018)
10. He, K., Gkioxari, G., Dollár, P., Girshick, R.: Mask R-CNN. In: Proceedings of the IEEE International Conference on Computer Vision, pp. 2961–2969 (2017)
11. Redmon, J., Divvala, S., Girshick, R., Farhadi, A.: You only look once: unified, real-time object detection. In: Proceedings of the IEEE Conference on Computer Vision and Pattern Recognition, pp. 779–788 (2016)
12. Liu, W., et al.: SSD: single shot multibox detector. In: Leibe, B., Matas, J., Sebe, N., Welling, M. (eds.) ECCV 2016. LNCS, vol. 9905, pp. 21–37. Springer, Cham (2016). https://doi.org/10.1007/978-3-319-46448-0_2
13. Singh, B., Najibi, M., Davis, L.S.: SNIPER: efficient multi-scale training. In: Advances in Neural Information Processing Systems, pp. 9310–9320 (2018)
14. Lin, T.Y., Goyal, P., Girshick, R., He, K., Dollár, P.: Focal loss for dense object detection. In: Proceedings of the IEEE International Conference on Computer Vision, pp. 2980–2988 (2017)
15. Hoffman, J., et al.: Large scale visual recognition through adaptation using joint representation and multiple instance learning. J. Mach. Learn. Res. **17**(1), 4954–4984 (2016)
16. Andrews, S., Tsochantaridis, I., Hofmann, T.: Support vector machines for multiple-instance learning. In: Advances in Neural Information Processing Systems, pp. 577–584 (2003)
17. Bilen, H., Vedaldi, A.: Weakly supervised deep detection networks. In: Proceedings of the IEEE Conference on Computer Vision and Pattern Recognition, pp. 2846–2854 (2016)
18. Rochan, M., Wang, Y.: Weakly supervised localization of novel objects using appearance transfer. In: Proceedings of the IEEE Conference on Computer Vision and Pattern Recognition, pp. 4315–4324 (2015)
19. Tang, K., Joulin, A., Li, L.J., Fei-Fei, L.: Co-localization in real-world images. In: Proceedings of the IEEE Conference on Computer Vision and Pattern Recognition, pp. 1464–1471 (2014)
20. Guillaumin, M., Ferrari, V.: Large-scale knowledge transfer for object localization in imagenet. In: Proceedings of the IEEE Conference on Computer Vision and Pattern Recognition, pp. 3202–3209. IEEE (2012)
21. Besag, J.: On the statistical analysis of dirty pictures. J. Roy. Stat. Soc.: Ser. B (Methodol.) **48**(3), 259–279 (1986)
22. Shaban, A., Rahimi, A., Bansal, S., Gould, S., Boots, B., Hartley, R.: Learning to find common objects across few image collections. In: Proceedings of the IEEE International Conference on Computer Vision, pp. 5117–5126 (2019)
23. Kolmogorov, V.: Convergent tree-reweighted message passing for energy minimization. IEEE Trans. Pattern Anal. Mach. Intell. **28**(10), 1568–1583 (2006)
24. Lin, T.-Y., et al.: Microsoft COCO: common objects in context. In: Fleet, D., Pajdla, T., Schiele, B., Tuytelaars, T. (eds.) ECCV 2014. LNCS, vol. 8693, pp. 740–755. Springer, Cham (2014). https://doi.org/10.1007/978-3-319-10602-1_48
25. Russakovsky, O., et al.: Imagenet large scale visual recognition challenge. Int. J. Comput. Vis. **115**(3), 211–252 (2015)
26. Zhu, Y., Zhou, Y., Ye, Q., Qiu, Q., Jiao, J.: Soft proposal networks for weakly supervised object localization. In: Proceedings of the IEEE International Conference on Computer Vision, pp. 1841–1850 (2017)
27. Bilen, H., Pedersoli, M., Tuytelaars, T.: Weakly supervised object detection with convex clustering. In: Proceedings of the IEEE Conference on Computer Vision and Pattern Recognition, pp. 1081–1089 (2015)

28. Wan, F., Wei, P., Jiao, J., Han, Z., Ye, Q.: Min-entropy latent model for weakly supervised object detection. In: Proceedings of the IEEE Conference on Computer Vision and Pattern Recognition, pp. 1297–1306 (2018)
29. Kumar, M.P., Packer, B., Koller, D.: Self-paced learning for latent variable models. In: Advances in Neural Information Processing Systems, pp. 1189–1197 (2010)
30. Hayder, Z., Salzmann, M., He, X.: Object co-detection via efficient inference in a fully-connected CRF. In: Fleet, D., Pajdla, T., Schiele, B., Tuytelaars, T. (eds.) ECCV 2014. LNCS, vol. 8691, pp. 330–345. Springer, Cham (2014). https://doi.org/10.1007/978-3-319-10578-9_22
31. Hayder, Z., He, X., Salzmann, M.: Structural kernel learning for large scale multiclass object co-detection. In: Proceedings of the IEEE International Conference on Computer Vision, pp. 2632–2640. IEEE (2015)
32. Ortega, J.M., Rheinboldt, W.C.: Iterative Solution of Nonlinear Equations in Several Variables, vol. 30. SIAM, Philadelphia (1970)
33. Wainwright, M.J., Jordan, M.I., et al.: Graphical models, exponential families, and variational inference. Found. Trends® Mach. Learn. **1**(1–2), 1–305 (2008)
34. Schrijver, A.: Theory of Linear and Integer Programming. John Wiley & Sons, Hoboken (1998)
35. Savchynskyy, B., et al.: Discrete graphical models–an optimization perspective. Found. Trends® Comput. Graph. Vis. **11**(3–4), 160–429 (2019)
36. Weiss, Y., Freeman, W.T.: On the optimality of solutions of the max-product belief-propagation algorithm in arbitrary graphs. IEEE Trans. Inf. Theory **47**(2), 736–744 (2001)
37. Bergtholdt, M., Kappes, J., Schmidt, S., Schnörr, C.: A study of parts-based object class detection using complete graphs. IJCV **87**(1–2), 93 (2010)
38. Krähenbühl, P., Koltun, V.: Efficient inference in fully connected CRFs with Gaussian edge potentials. In: Advances in Neural Information Processing Systems, pp. 109–117 (2011)
39. Adams, A., Baek, J., Davis, M.A.: Fast high-dimensional filtering using the permutohedral lattice. Comput. Graph. Forum **29**, 753–762 (2010). Wiley Online Library
40. Ajanthan, T., Desmaison, A., Bunel, R., Salzmann, M., Torr, P.H., Pawan Kumar, M.: Efficient linear programming for dense CRFs. In: Proceedings of the IEEE Conference on Computer Vision and Pattern Recognition, pp. 3298–3306 (2017)
41. Blake, A., Kohli, P., Rother, C.: Markov Random Fields for Vision and Image Processing. MIT Press, Cambridge (2011)
42. Ren, S., He, K., Girshick, R., Sun, J.: Faster R-CNN: towards real-time object detection with region proposal networks. In: Advances in Neural Information Processing Systems, pp. 91–99 (2015)
43. Tang, Y., Wang, J., Gao, B., Dellandréa, E., Gaizauskas, R., Chen, L.: Large scale semi-supervised object detection using visual and semantic knowledge transfer. In: Proceedings of the IEEE Conference on Computer Vision and Pattern Recognition, pp. 2119–2128 (2016)
44. Krizhevsky, A., Sutskever, I., Hinton, G.E.: Imagenet classification with deep convolutional neural networks. In: Advances in Neural Information Processing Systems, pp. 1097–1105 (2012)
45. Bansal, A., Sikka, K., Sharma, G., Chellappa, R., Divakaran, A.: Zero-shot object detection. In: Ferrari, V., Hebert, M., Sminchisescu, C., Weiss, Y. (eds.) ECCV 2018. LNCS, vol. 11205, pp. 397–414. Springer, Cham (2018). https://doi.org/10.1007/978-3-030-01246-5_24

46. He, K., Zhang, X., Ren, S., Sun, J.: Deep residual learning for image recognition. In: Proceedings of the IEEE Conference on Computer Vision and Pattern Recognition, pp. 770–778 (2016)
47. Szegedy, C., Ioffe, S., Vanhoucke, V., Alemi, A.A.: Inception-v4, inception-resnet and the impact of residual connections on learning. In: Thirty-First AAAI Conference on Artificial Intelligence (2017)
48. Uijlings, J.R., Van De Sande, K.E., Gevers, T., Smeulders, A.W.: Selective search for object recognition. Int. J. Comput. Vis. **104**(2), 154–171 (2013)
49. Szegedy, C., Vanhoucke, V., Ioffe, S., Shlens, J., Wojna, Z.: Rethinking the inception architecture for computer vision. In: Proceedings of the IEEE Conference on Computer Vision and Pattern Recognition, pp. 2818–2826 (2016)

Environment-Agnostic Multitask Learning for Natural Language Grounded Navigation

Xin Eric Wang[1,2](✉), Vihan Jain[3], Eugene Ie[3], William Yang Wang[2], Zornitsa Kozareva[3], and Sujith Ravi[3,4]

[1] University of California, Santa Cruz, USA
xwang366@ucsc.edu
[2] University of California, Santa Barbara, USA
[3] Google Research, Mountain View, USA
[4] Amazon, Seattle, USA

Abstract. Recent research efforts enable study for natural language grounded navigation in photo-realistic environments, e.g., following natural language instructions or dialog . However, existing methods tend to overfit training data in seen environments and fail to generalize well in previously unseen environments. To close the gap between seen and unseen environments, we aim at learning a generalized navigation model from two novel perspectives: (1) we introduce a multitask navigation model that can be seamlessly trained on both Vision-Language Navigation (VLN) and Navigation from Dialog History (NDH) tasks, which benefits from richer natural language guidance and effectively transfers knowledge across tasks; (2) we propose to learn environment-agnostic representations for the navigation policy that are invariant among the environments seen during training, thus generalizing better on unseen environments. Extensive experiments show that environment-agnostic multitask learning significantly reduces the performance gap between seen and unseen environments, and the navigation agent trained so outperforms baselines on unseen environments by 16% (relative measure on success rate) on VLN and 120% (goal progress) on NDH. Our submission to the CVDN leaderboard establishes a new state-of-the-art for the NDH task on the holdout test set. Code is available at https://github.com/google-research/valan.

Keywords: Vision-and-language navigation · Natural language grounding · Multitask learning · Agnostic learning

X. E. Wang and V. Jain—Equal contribution.
X. E. Wang and S. Ravi—Work done at Google.

Electronic supplementary material The online version of this chapter (https://doi.org/10.1007/978-3-030-58586-0_25) contains supplementary material, which is available to authorized users.

A. Vedaldi et al. (Eds.): ECCV 2020, LNCS 12369, pp. 413–430, 2020.
https://doi.org/10.1007/978-3-030-58586-0_25

1 Introduction

Navigation in visual environments by following natural language guidance [18] is a fundamental capability of intelligent robots that simulate human behaviors, because humans can easily reason about the language guidance and navigate efficiently by interacting with the visual environments. Recent efforts [3,9,36,42] empower large-scale learning of natural language grounded navigation that is situated in photo-realistic simulation environments.

Nevertheless, the generalization problem commonly exists for these tasks, especially indoor navigation: the agent usually performs poorly on unknown environments that have never been seen during training. One of the leading causes of such behavior is data scarcity, as it is expensive and time-consuming to extend either visual environments or natural language guidance. The number of scanned houses for indoor navigation is limited due to high expense and privacy concerns. Besides, unlike vision-only navigation tasks [23,25,30–32,50] where episodes can be exhaustively sampled in simulation, natural language grounded navigation is supported by human demonstrated interaction in natural language. It is impractical to fully collect all the samples for individual tasks.

Therefore, it is essential though challenging to efficiently learn a more generalized policy for natural language grounded navigation tasks from existing data [48,49]. In this paper, we study how to resolve the generalization and data scarcity issues from two different angles. First, previous methods are trained for one task at the time, so each new task requires training a new agent instance from scratch that can only solve the one task on which it was trained. In this work, we propose a generalized multitask model for natural language grounded navigation tasks such as Vision-Language Navigation (VLN) and Navigation from Dialog History (NDH), aiming to efficiently transfer knowledge across tasks and effectively solve all the tasks simultaneously with one agent.

Furthermore, even though there are thousands of trajectories paired with language guidance, the underlying house scans are restricted. For instance, the popular Matterport3D environment [6] contains only 61 unique house scans in the training set. The current models perform much better in seen environments by taking advantage of the knowledge of specific houses they have acquired over multiple task completions during training, but fail to generalize to houses not seen during training. To overcome this shortcoming, we propose an environment-agnostic learning method to learn a visual representation that is invariant to specific environments but can still support navigation. Endowed with the learned environment-agnostic representations, the agent is further prevented from the overfitting issue and generalizes better on unseen environments.

To the best of our knowledge, we are the first to introduce natural language grounded multitask and environment-agnostic training regimes and validate their effectiveness on VLN and NDH tasks. Extensive experiments demonstrate that our environment-agnostic multitask navigation model can not only efficiently execute different language guidance in indoor environments but also outperform the single-task baseline models by a large margin on both tasks. Besides, the performance gap between seen and unseen environments is significantly reduced. Furthermore, our leaderboard submission for the NDH task establishes a new state-of-the-art outperforming the existing best agent by more than 66% on the primary metric of goal progress on the holdout test set.

2 Background

Vision-and-Language Navigation. As depicted in Fig. 1, Vision-and-Language Navigation [3,7] task requires an embodied agent to navigate in photo-realistic environments to carry out natural language instructions. For a given path, the associated natural language instructions describe the step-by-step guidance from the starting position to the target position. The agent is spawned at an initial pose $p_0 = (v_0, \phi_0, \theta_0)$, which includes the spatial location, heading and elevation angles. Given a natural language instruction $X = \{x_1, x_2, ..., x_n\}$, the agent is expected to perform a sequence of actions $\{a_1, a_2, ..., a_T\}$ and arrive at the target position v_{tar} specified by the language instruction X. In this work, we consider the VLN task defined for Room-to-Room (R2R) [3] dataset, which contains instruction-trajectory pairs across 90 different indoor environments (houses). The instructions for a given trajectory in the dataset on an average contain 29 words. Previous VLN methods have studied various aspects to improve the navigation performance, such as planning [46], data augmentation [14,15,40], cross-modal alignment [20,45], progress estimation [28], error correction [22,29], interactive language assistance [33,34] etc. This work tackles VLN via multitask learning and environment-agnostic learning, which is orthogonal to all these prior arts.

Navigation from Dialog History. Different from Visual Dialog [10] that involves dialog grounded in a single image, the recently introduced Cooperative Vision-and-Dialog Navigation (CVDN) dataset [42] includes interactive language assistance for indoor navigation, which consists of over 2,000 embodied, human-human dialogs situated in photo-realistic home environments. The task of Navigation from Dialog History (NDH) demonstrated in Fig. 1, is defined as: given a target object t_0 and a dialog history between humans cooperating to perform the task, the embodied agent must infer navigation actions towards the goal room that contains the target object. The dialog history is denoted as $<t_0, Q_1, A_1, Q_2, A_2, ..., Q_i, A_i>$, including the target object t_0, the questions Q and answers A till the turn i ($0 \leq i \leq k$, where k is the total number of Q-A turns from the beginning to the goal room). The agent, located in p_0, is trying to move closer to the goal room by inferring from the dialog history that happened before. The dialog for a given trajectory lasts 6 utterances (3 question-answer exchanges) and is 82 words long on an average.

Multitask Learning. The basis of multitask learning is the notion that tasks can serve as mutual sources of inductive bias for each other [5]. When multiple tasks are trained jointly, multitask learning causes the learner to prefer the hypothesis that explains all the tasks simultaneously, leading to more generalized solutions. Multitask learning has been successful in natural language processing [8], speech recognition [11], computer vision [17], drug discovery [37], and Atari games [41]. The deep reinforcement learning methods that

Fig. 1. While the NDH task (left) requires an agent to navigate using dialog history between two human players - a navigator (N) who is trying to find the goal room with the help of an oracle (O), the VLN task (right) requires navigating using instructions written by human annotators.

have become very popular for training models on natural language grounded navigation tasks [19,20,40,45] are known to be data inefficient. In this work, we introduce multitask reinforcement learning for such tasks to improve data efficiency by positive transfer across related tasks.

Agnostic Learning. A few studies on agnostic learning have been proposed recently. For example, Model-Agnostic Meta-Learning (MAML) [13] aims to train a model on a variety of learning tasks and solve a new task using only a few training examples. Liu *et al.* [27] proposes a unified feature disentangler that learns domain-invariant representation across multiple domains for image translation. Other domain-agnostic techniques are also proposed for supervised [26] and unsupervised domain adaption [35,38]. In this work, we pair the environment classifier with a gradient reversal layer [16] to learn an environment-agnostic representation that can be better generalized on unseen environments in a zero-shot fashion where no adaptation is involved.

Distributed Actor-Learner Navigation Learning Framework. To train models for the various language grounded navigation tasks like VLN and NDH, we use the VALAN framework [24], a distributed actor-learner learning infrastructure. The framework is inspired by IMPALA [12] and uses its off-policy correction method called V-trace to scale reinforcement learning methods to thousands of machines efficiently. The framework additionally supports a variety of supervision strategies essential for navigation tasks such as teacher-forcing [3], student-forcing [3] and mixed supervision [42]. The framework is built using TensorFlow [1] and supports ML accelerators (GPU, TPU).

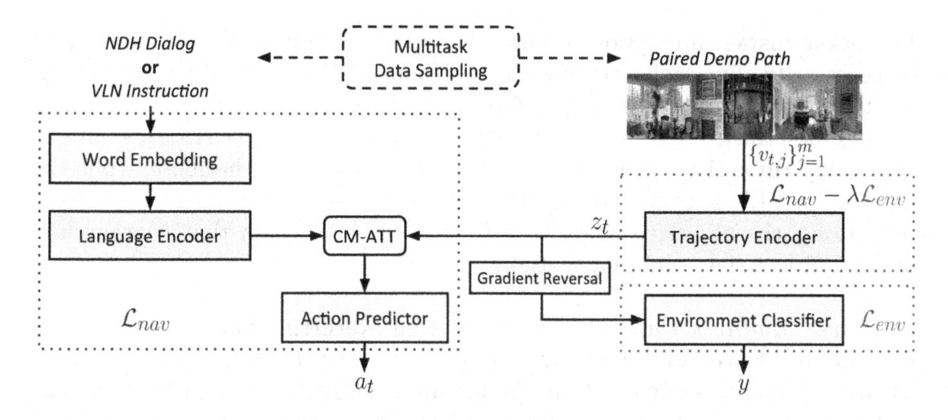

Fig. 2. Overview of environment-agnostic multitask learning.

3 Environment-Agnostic Multitask Learning

3.1 Overview

Our environment-agnostic multitask navigation model is illustrated in Fig. 2. First, we adapt the reinforced cross-modal matching (RCM) model [45] and make it seamlessly transfer across tasks by sharing all the learnable parameters for both NDH and VLN, including joint word embedding layer, language encoder, trajectory encoder, cross-modal attention module (CM-ATT), and action predictor. Furthermore, to learn the environment-agnostic representation $\boldsymbol{z}_t$, we equip the navigation model with an environment classifier whose objective is to predict which house the agent is. However, note that between trajectory encoder and environment classifier, a gradient reversal layer [16] is introduced to reverse the gradients back-propagated to the trajectory encoder, making it learn representations that are environment-agnostic and thus more generalizable in unseen environments. During training, the environment classifier is minimizing the environment classification loss $\mathcal{L}_{env}$, while the trajectory encoder is maximizing $\mathcal{L}_{env}$ and minimizing the navigation loss $\mathcal{L}_{nav}$. The other modules are optimized with the navigation loss $\mathcal{L}_{nav}$ simultaneously. Below we introduce multitask reinforcement learning and environment-agnostic representation learning. A more detailed model architecture is presented in Sect. 4.

3.2 Multitask Reinforcement Learning

In this section, we describe how we adapted the RCM agent model to learn the two tasks of VLN and NDH simultaneously. It is worth noting that even though both the VLN and NDH tasks use the same Matterport3D indoor environments [6], there are significant differences in the motivations and the overall objectives of the two tasks. While the natural language descriptions associated with the paths in the VLN task are step-by-step instructions to follow the ground-truth paths, the descriptions of the paths in the NDH task are series of question-answer interactions (dialog) between two human players which need not necessarily align sequentially with the ground-truth paths. This difference in the style of the two tasks also manifests in their respective datasets—the average path description length and average path length in the NDH task's dataset are roughly three times that of the VLN task's dataset. Furthermore, while the objective in VLN is to find the exact goal node in the environment (i.e., point navigation), the objective in NDH is to find the goal room that contains the specified object (i.e., room navigation).

Interleaved Multitask Data Sampling. To avoid overfitting individual tasks, we adopt an interleaved multitask data sampling strategy to train the model. Particularly, each data sample within a mini-batch can be from either task, so that the VLN instruction-trajectory pairs and NDH dialog-trajectory pairs are interleaved in a mini-batch though they may have different learning objectives.

Reward Shaping. Following prior art [45,46], we first implement a discounted cumulative reward function R for the VLN and NDH tasks:

$$R(s_t, a_t) = \sum_{t'=t}^{T} \gamma^{t'-t} r(s_{t'}, a_{t'}) \tag{1}$$

where γ is the discounted factor. For the VLN task, we choose the immediate reward function such that the agent is rewarded at each step for getting closer to (or penalized for getting further from) the target location. At the end of the episode, the agent receives a reward only if it terminated successfully. Formally,

$$r(s_{t'}, a_{t'}) = \begin{cases} d(s_{t'}, v_{tar}) - d(s_{t'+1}, v_{tar}) & \text{if } t' < T \\ \mathbb{1}[d(s_T, v_{tar}) \leq d_{th}] & \text{if } t' = T \end{cases} \tag{2}$$

where $d(s_t, v_{tar})$ is the distance between state s_t and the target location v_{tar}, $\mathbb{1}[.]$ is the indicator function and d_{th} is the maximum distance from v_{tar} that the agent is allowed to terminate for success.

Different from VLN, the NDH task is essentially room navigation instead of point navigation because the agent is expected to reach a room that contains the target object. Suppose the goal room is occupied by a set of nodes $\{v_i\}_1^N$, we replace the distance function $d(s_t, v_{tar})$ in Eq. 2 with the minimum distance to the goal room $d_{room}(s_t, \{v_i\}_1^N)$ for NDH:

$$d_{room}(s_t, \{v_i\}_1^N) = \min_{1 \leq i \leq N} d(s_t, v_i) \tag{3}$$

Navigation Loss. Since human demonstrations are available for both VLN and NDH tasks, we use behavior cloning to constrain the learning algorithm to model state-action spaces that are most relevant to each task. Following previous works [45], we also use reinforcement learning to aid the agent's ability to recover from erroneous actions in unseen environments. During navigation model training, we adopt a mixed training strategy of reinforcement learning and behavior cloning, so the navigation loss function is:

$$\mathcal{L}_{nav} = -\mathbb{E}_{a_t \sim \pi}[R(s_t, a_t) - b] - \mathbb{E}[\log \pi(a_t^* | s_t)] \tag{4}$$

where we use REINFORCE policy gradients [47] and supervised learning gradients to update the policy π. b is the estimated baseline to reduce the variance and a_t^* is the human demonstrated action.

3.3 Environment-Agnostic Representation Learning

To further improve the navigation policy's generalizability, we propose to learn a latent environment-agnostic representation that is invariant among seen environments. The objective is to not learn the intricate environment-specific features that are irrelevant to general navigation (e.g. unique house appearances), preventing the model from overfitting to specific seen environments. We can reformulate the navigation policy as

$$\pi(a_t | s_t) = p(a_t | \boldsymbol{z}_t, s_t) p(\boldsymbol{z}_t | s_t) \tag{5}$$

where $\boldsymbol{z}_t$ is a latent representation.

As shown in Fig. 2, $p(a_t | \boldsymbol{z}_t, s_t)$ is modeled by the policy module (including CM-ATT and action predictor) and $p(\boldsymbol{z}_t | s_t)$ is modeled by the trajectory encoder. In order to learn the environment-agnostic representation, we employ an environment classifier and a gradient reversal layer [16]. The environment classifier is parameterized to predict the house identity, so its loss function $\mathcal{L}_{env}$ is defined as

$$\mathcal{L}_{env} = -\mathbb{E}[\log p(y = y^* | \boldsymbol{z}_t)] \tag{6}$$

where y^* is the ground-truth house label. The gradient reversal layer has no parameters. It acts as an identity transform during forward-propagation, but multiplies the gradient by $-\lambda$ and passes it to the trajectory encoder during back-propagation. Therefore, in addition to minimizing the navigation loss $\mathcal{L}_{nav}$, the trajectory encoder is also maximizing the environment classification loss $\mathcal{L}_{env}$. While the environment classifier is minimizing the classification loss conditioned on the latent representation $\boldsymbol{z}_t$, the trajectory encoder is trying to increase the classifier's entropy, resulting in an adversarial learning objective.

4 Model Architecture

Language Encoder. The natural language guidance (instruction or dialog) is tokenized and embedded into n-dimensional space $\boldsymbol{X} = \{\boldsymbol{x}_1, \boldsymbol{x}_2, ..., \boldsymbol{x}_n\}$ where the word vectors $\boldsymbol{x}_i$ are initialized randomly. The vocabulary is restricted to tokens that occur at least five times in the training instructions (the vocabulary used when jointly training VLN and NDH tasks is the union of the two tasks' vocabularies.). All out-of-vocabulary tokens are mapped to a single out-of-vocabulary identifier. The token sequence is encoded using a bi-directional LSTM [39] to create $\boldsymbol{H}^{\boldsymbol{X}}$ following:

$$\boldsymbol{H}^{\boldsymbol{X}} = [\boldsymbol{h}_1^{\boldsymbol{X}}; \boldsymbol{h}_2^{\boldsymbol{X}}; ...; \boldsymbol{h}_n^{\boldsymbol{X}}], \quad \boldsymbol{h}_t^{\boldsymbol{X}} = \sigma(\overrightarrow{\boldsymbol{h}}_t^{\boldsymbol{X}}, \overleftarrow{\boldsymbol{h}}_t^{\boldsymbol{X}}) \tag{7}$$

$$\overrightarrow{\boldsymbol{h}}_t^{\boldsymbol{X}} = LSTM(\boldsymbol{x}_t, \overrightarrow{\boldsymbol{h}}_{t-1}^{\boldsymbol{X}}), \quad \overleftarrow{\boldsymbol{h}}_t^{\boldsymbol{X}} = LSTM(\boldsymbol{x}_t, \overleftarrow{\boldsymbol{h}}_{t+1}^{\boldsymbol{X}}) \tag{8}$$

where $\overrightarrow{\boldsymbol{h}}_t^{\boldsymbol{X}}$ and $\overleftarrow{\boldsymbol{h}}_t^{\boldsymbol{X}}$ are the hidden states of the forward and backward LSTM layers at time step t respectively, and the σ function is used to combine $\overrightarrow{\boldsymbol{h}}_t^{\boldsymbol{X}}$ and $\overleftarrow{\boldsymbol{h}}_t^{\boldsymbol{X}}$ into $\boldsymbol{h}_t^{\boldsymbol{X}}$.

Trajectory Encoder. Similar to benchmark models [14,20,45], at each time step t, the agent perceives a 360-degree panoramic view at its current location. The view is discretized into k view angles ($k = 36$ in our implementation, 3 elevations by 12 headings at 30-degree intervals). The image at view angle i, heading angle ϕ and elevation angle θ is represented by a concatenation of the pre-trained CNN image features with the 4-dimensional orientation feature [sin ϕ; cos ϕ; sin θ; cos θ] to form $\boldsymbol{v}_{t,i}$. The visual input sequence $\boldsymbol{V} = \{\boldsymbol{v}_1, \boldsymbol{v}_2, ..., \boldsymbol{v}_m\}$ is encoded using a LSTM to create $\boldsymbol{H}^{\boldsymbol{V}}$ following:

$$\boldsymbol{H}^{\boldsymbol{V}} = [\boldsymbol{h}_1^{\boldsymbol{V}}; \boldsymbol{h}_2^{\boldsymbol{V}}; ...; \boldsymbol{h}_m^{\boldsymbol{V}}], \quad \text{where } \boldsymbol{h}_t^{\boldsymbol{V}} = LSTM(\boldsymbol{v}_t, \boldsymbol{h}_{t-1}^{\boldsymbol{V}}) \tag{9}$$

$\boldsymbol{v}_t = \text{Attention}(\boldsymbol{h}_{t-1}^{\boldsymbol{V}}, \boldsymbol{v}_{t,1..k})$ is the attention-pooled representation of all view angles using previous agent state $\boldsymbol{h}_{t-1}$ as the query. We use the dot-product attention [43] hereafter.

Policy Module. The policy module comprises of cross-modal attention (CM-ATT) unit as well as an action predictor. The agent learns a policy π_θ over parameters θ that maps the natural language instruction $\boldsymbol{X}$ and the initial visual scene $\boldsymbol{v}_1$ to a sequence of actions $[a_1, a_2, ..., a_n]$. The action space which is common to VLN and NDH tasks consists of navigable directions from the current location. The available actions at time t are denoted as $\boldsymbol{u}_{t,1..l}$, where $\boldsymbol{u}_{t,j}$ is the representation of the navigable direction j from the current location obtained similarly to $\boldsymbol{v}_{t,i}$. The number of available actions, l, varies per location, since graph node connectivity varies. Following Wang *et al.* [45], the model predicts the probability p_d of each navigable direction d using a bilinear dot product:

$$p_d = \text{softmax}([\boldsymbol{h}_t^{\boldsymbol{V}}; \boldsymbol{c}_t^{\text{text}}; \boldsymbol{c}_t^{\text{visual}}]\boldsymbol{W}_c(\boldsymbol{u}_{t,d}\boldsymbol{W}_u)^T) \tag{10}$$

where $\boldsymbol{c}_t^{\text{text}} = \text{Attention}(\boldsymbol{h}_t^{\boldsymbol{V}}, \boldsymbol{h}_{1..n}^{\boldsymbol{X}})$ and $\boldsymbol{c}_t^{\text{visual}} = \text{Attention}(\boldsymbol{c}_t^{\text{text}}, \boldsymbol{v}_{t,1..k})$. $\boldsymbol{W}_c$ and $\boldsymbol{W}_u$ are learnable parameters.

Environment Classifier. The environment classifier is a two-layer perceptron with a SoftMax layer as the last layer. Given the latent representation $\boldsymbol{z}_t$ (which is $\boldsymbol{h}_t^{\boldsymbol{V}}$ in our setting), the classifier generates a probability distribution over the house labels.

5 Experiments

5.1 Experimental Setup

Implementation Details. We use a 2-layer bi-directional LSTM for the instruction encoder, where the size of LSTM cells is 256 in each direction. The inputs to the encoder are 300-dimensional embeddings initialized randomly. For the visual encoder, we use a 2-layer LSTM with a cell size of 512. The encoder inputs are image features derived as mentioned in Sect. 4. The cross-modal attention layer size is 128 units. The environment classifier has one hidden layer of size 128 units, followed by an output layer of size equal to the number of classes. The negative gradient multiplier λ in the gradient reversal layer is empirically tuned and fixed at a value of 1.3 for all experiments. During training, some episodes in the batch are identical to available human demonstrations in the training dataset, where the objective is to increase the agent's likelihood of choosing human actions (behavioral cloning [4]). The rest of the episodes are constructed by sampling from the agent's own policy. For the NDH task, we deploy mixed supervision similar to Thomason *et al.* [42], where the navigator's or oracle's path is selected as ground-truth depending on if the navigator was successful in reaching the correct end node following the question-answer exchange with the oracle or not. In the experiments, unless otherwise stated, we use the entire dialog history from the NDH task for model training. *All the reported results in subsequent studies are averages of at least three independent runs.*

Evaluation Metrics. The agents are evaluated on two datasets, namely *Validation Seen* that contains new paths from the training environments and *Validation Unseen* that contains paths from previously unseen environments. The evaluation metrics for VLN task are as follows: *Path Length (PL)* measures the total length of the predicted path; *Navigation Error (NE)* measures the distance between the last nodes in the predicted and the reference paths; *Success Rate (SR)* measures how often the last node in the predicted path is within some threshold distance of the last node in the reference path; *Success weighted by Path Length (SPL)* [2] measures Success Rate weighted by the normalized Path Length; and *Coverage weighted by Length Score (CLS)* [21] measures predicted path's conformity to the reference path weighted by length score. For the NDH task, the agent's progress is defined as a reduction (in meters) from the distance to the goal region at the agent's first position versus at its last position [42].

5.2 Environment-Agnostic Multitask Learning

Table 1 shows the results of training the navigation model using environment-agnostic learning (*EnvAg*) as well as multitask learning (*MT-RCM*). First, both learning methods independently help the agent learn more generalized navigation policy, as is evidenced by a significant reduction in agent's performance gap between seen and unseen environments (better visualized with Fig. 3). For instance, the performance gap in goal progress on the NDH task drops from 3.85 m to 0.92 m using multitask learning, and the performance gap in success

Table 1. The agent's performance under different training strategies. The single-task RCM (ST-RCM) model is independently trained and tested on VLN or NDH tasks. The standard deviation across 3 independent runs is reported.

Fold	Model	NDH	VLN				
		Progress ↑	PL	NE ↓	SR ↑	SPL ↑	CLS ↑
Val Seen	seq2seq [42]	5.92					
	RCM [45][a]		12.08	**3.25**	**67.60**	-	-
	Ours						
	ST-RCM	**6.49** ± 0.95	10.75 ± 0.26	5.09 ± 0.49	52.39 ± 3.58	48.86 ± 3.66	63.91 ± 2.41
	ST-RCM + EnvAg	6.07 ± 0.56	11.31 ± 0.26	4.93 ± 0.49	52.79 ± 3.72	48.85 ± 3.71	63.26 ± 2.31
	MT-RCM	5.28 ± 0.56	10.63 ± 0.10	5.09 ± 0.05	**56.42** ± 1.21	**49.67** ± 1.07	**68.28** ± 0.16
	MT-RCM + EnvAg	5.07 ± 0.45	11.60 ± 0.30	**4.83** ± 0.12	53.30 ± 0.71	49.39 ± 0.74	64.10 ± 0.16
Val Unseen	seq2seq [42]	2.10					
	RCM [45]		15.00	6.02	40.60	-	-
	Ours						
	ST-RCM	2.64 ± 0.06	10.60 ± 0.27	6.10 ± 0.06	42.93 ± 0.21	38.88 ± 0.20	54.86 ± 0.92
	ST-RCM + EnvAg	3.15 ± 0.29	11.36 ± 0.27	5.79 ± 0.06	44.40 ± 2.14	40.30 ± 2.12	55.77 ± 1.31
	MT-RCM	4.36 ± 0.17	10.23 ± 0.14	**5.31** ± 0.18	46.20 ± 0.55	**44.19** ± 0.64	54.99 ± 0.87
	MT-RCM + EnvAg	**4.65** ± 0.20	12.05 ± 0.23	5.41 ± 0.20	**47.22** ± 1.00	41.80 ± 1.11	**56.22** ± 0.87

[a] The equivalent RCM model without intrinsic reward is used as the benchmark.

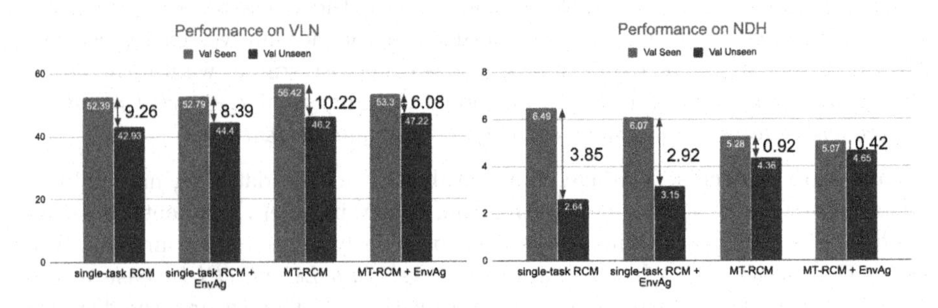

Fig. 3. Visualizing performance gap between seen and unseen environments for VLN (success rate) and NDH (progress) tasks.

rate on the VLN task drops from 9.26% to 8.39% using environment-agnostic learning. Second, the two techniques are complementary—the agent's performance when trained with both the techniques simultaneously improves on unseen environments compared to when trained separately. Finally, we note here that *MT-RCM + EnvAg* outperforms the baseline goal progress of 2.10 m [42] on NDH validation unseen dataset by more than 120%. At the same time, it outperforms the equivalent RCM baseline [45] of 40.6% success rate by more than 16% (relative measure) on VLN validation unseen dataset.

To further validate our results on NDH task, we evaluated the *MT-RCM + EnvAg* agent on the test set of NDH dataset which is held out as the CVDN challenge.[1] Table 2 shows that our submission to the leaderboard with *MT-RCM* + **EnvAg** establishes a new state-of-the-art on this task outperforming the existing best agent by more than 66%.

[1] https://evalai.cloudcv.org/web/challenges/challenge-page/463/leaderboard/1292.

Table 2. Comparison on CVDN Leaderboard Test Set. Note that the metric *Progress* is the same as *dist_to_end_reduction*.

	Agent	Progress ↑
Baselines	Random	0.83
	Shortest Path Agent (*upper bound*)	9.76
Leaderboard Submissions	Seq2Seq [42]	2.35
	MT-RCM + EnvAg	**3.91**

5.3 Multitask Learning

We then conduct studies to examine cross-task transfer using multitask learning alone. First, we experiment multitasking learning with access to different parts of the dialog—the target object t_o, the last oracle answer A_i, the prefacing navigator question Q_i, and the full dialog history. Table 3 shows the results of jointly training *MT-RCM* model on VLN and NDH tasks. (1) *Does VLN complement NDH?* Yes, consistently. On NDH Val Unseen, *MT-RCM* consistently benefits from following shorter paths with step-by-step instructions in VLN for all kinds of dialog inputs. It shows that VLN can serve as an essential task to boost learning of primitive action-and-instruction following and therefore support more complicated navigation tasks like NDH. (2) *Does NDH complement VLN?* Yes, under certain conditions. From the results on VLN Val Unseen, we can observe that MT-RCM with only target objects as the guidance performs equivalently or slightly worse than *VLN-RCM*, showing that extending visual paths alone (even with final targets) is not helpful in VLN. But we can see a consistent and gradual increase in the success rate of *MT-RCM* on the VLN task as it is trained on paths with richer dialog history from the NDH task. This shows that the agent benefits from more fine-grained information about the path implying the importance given by the agent to the language instructions in the task. (3) Multitask learning improves the generalizability of navigation models: the seen-unseen performance gap is narrowed. (4) As a side effect, results of different dialog inputs on NDH Val Seen *versus* Unseen verify the essence of language guidance in generalizing navigation to unseen environments.

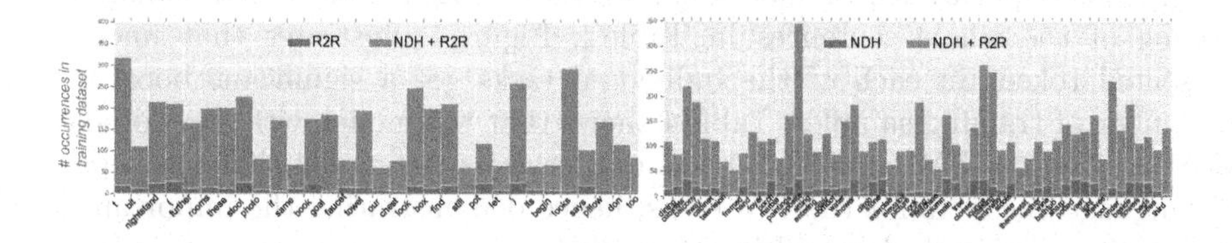

Fig. 4. Selected tokens from the vocabulary for VLN (left) and NDH (right) tasks which gained more than 40 additional occurrences in the training dataset due to joint-training.

Table 3. Comparison of agent performance when trained separately *vs.* jointly on VLN and NDH tasks.

Fold	Model	NDH Evaluation					VLN Evaluation				
		Inputs for NDH				Progress	PL	NE	SR	SPL	CLS
		t_o	A_i	Q_i	$A_{1:i-1}; Q_{1:i-1}$	↑		↓	↑	↑	↑
Val Seen	NDH-RCM	✓				**6.97**					
		✓	✓			**6.92**					
		✓	✓	✓		**6.47**					
		✓	✓	✓	✓	**6.49**					
	VLN-RCM						10.75	5.09	52.39	48.86	63.91
	MT-RCM	✓				3.00	11.73	4.87	54.56	52.00	65.64
		✓	✓			5.92	11.12	4.62	54.89	**52.62**	66.05
		✓	✓	✓		5.43	10.94	**4.59**	54.23	52.06	66.93
		✓	✓	✓	✓	5.28	10.63	5.09	**56.42**	49.67	**68.28**
Val Unseen	NDH-RCM	✓				1.25					
		✓	✓			2.69					
		✓	✓	✓		2.69					
		✓	✓	✓	✓	2.64					
	VLN-RCM						10.60	6.10	42.93	38.88	54.86
	MT-RCM	✓				**1.69**	13.12	5.84	42.75	38.71	53.09
		✓	✓			**4.01**	11.06	5.88	42.98	40.62	54.30
		✓	✓	✓		**3.75**	11.08	5.70	44.50	39.67	54.95
		✓	✓	✓	✓	**4.36**	10.23	**5.31**	**46.20**	**44.19**	**54.99**

Table 4. Comparison of agent performance when language instructions are encoded by separate *vs.* shared encoder for VLN and NDH tasks.

Language Encoder	Val Seen						Val Unseen					
	NDH	VLN					NDH	VLN				
	Progress ↑	PL	NE ↓	SR ↑	SPL ↑	CLS ↑	Progress ↑	PL	NE ↓	SR ↑	SPL ↑	CLS ↑
Shared	**5.28**	10.63	5.09	**56.42**	**49.67**	**68.28**	**4.36**	10.23	**5.31**	**46.20**	**44.19**	**54.99**
Separate	5.17	11.26	**5.02**	52.38	48.80	64.19	4.07	11.72	6.04	43.64	39.49	54.57

Besides, we show multitask learning results in better language grounding through more appearance of individual words in Fig. 4 and shared semantic encoding of the whole sentences in Table 4. Figure 4 illustrates that under-represented tokens in each of the individual tasks get a significant boost in the number of training samples. Table 4 shows that the model with shared language encoder for NDH and VLN tasks outperforms the model that has separate language encoders for the two tasks, hence demonstrating the importance of parameter sharing during multitask learning.

Furthermore, we observed that the agent's performance improves significantly when trained on a mixture of VLN and NDH paths even when the size of the training dataset is fixed, advancing the argument that multitask learning on NDH and VLN tasks complements the agent's learning. More details of the ablation studies can be found in the Appendix.

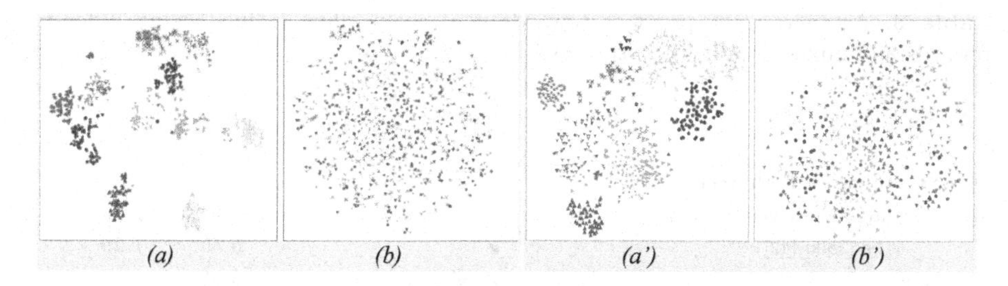

Fig. 5. t-SNE visualization of trajectory encoder's output for VLN task across 11 different color-coded seen (*a*,*b*) and unseen (*a'*,*b'*) environments. The depicted representations in *(a)* and *(a')* are learned with environment-aware objective while those in *(b)* and *(b')* are learned with environment-agnostic objective (Color figure online).

Table 5. Environment-agnostic *versus* environment-aware learning.

(a) Comparison on NDH.

Model	Val Seen Progress ↑	Val Unseen Progress ↑
RCM	6.49	2.64
EnvAware	**8.38**	1.81
EnvAg	6.07	**3.15**

(b) Comparison on VLN.

Model	Val Seen					Val Unseen				
	PL	NE ↓	SR ↑	SPL ↑	CLS ↑	PL	NE ↓	SR ↑	SPL ↑	CLS ↑
RCM	10.75	5.09	52.39	48.86	63.91	10.60	6.10	42.93	38.88	54.86
EnvAware	10.30	**4.36**	**57.59**	**54.05**	**68.49**	10.13	6.30	38.83	35.65	54.79
EnvAg	11.31	4.93	52.79	48.85	63.26	11.36	**5.79**	**44.40**	**40.30**	**55.77**

5.4 Environment-Agnostic Learning

From Table 1, it can be seen that both VLN and NDH tasks benefit from environment-agnostic learning independently. To further examine the generalization property of environment-agnostic learning, we train a model with the opposite objective—learn to correctly predict the navigation environments by removing the gradient reversal layer (*environment-aware learning*). The results in Table 5 demonstrate that environment-aware learning leads to overfitting on the training dataset as the performance on environments seen during training consistently increases for both the tasks. In contrast, environment-agnostic learning leads to a more generalized navigation policy that performs better on unseen environments. Figure 5 further shows that due to environment-aware learning, the model learns to represent visual inputs from the same environment closer to each other while the representations of different environments are farther from each other resulting in a clustering learning effect. On the other hand, environment-agnostic learning leads to more general representation across different environments, which results in better performance on unseen environments.

5.5 Reward Shaping for NDH Task

As discussed in Sect. 3.2, we conducted studies to shape the reward for the NDH task. Table 6 presents the results of training the agent with access to different parts of the dialog history. The results demonstrate that the agents rewarded

Table 6. Average agent progress towards goal room when trained using different rewards and mixed supervision strategy.

	Model	Inputs				Goal Progress (m)	
		t_0	A_i	Q_i	$A_{1:i-1}; Q_{1:i-1}$	Val Seen	Val Unseen
Baselines	Shortest-Path Agent					9.52	9.58
	Random Agent					0.42	1.09
	Seq2Seq [42]	✓				5.71	**1.29**
		✓	✓			6.04	2.05
		✓	✓	✓		6.16	1.83
		✓	✓	✓	✓	5.92	2.10
Ours	NDH-RCM (distance to goal location)	✓				4.18	0.42
		✓	✓			4.96	2.34
		✓	✓	✓		4.60	2.25
		✓	✓	✓	✓	5.02	2.58
	NDH-RCM (distance to goal room)	✓				**6.97**	1.25
		✓	✓			**6.92**	**2.69**
		✓	✓	✓		**6.47**	**2.69**
		✓	✓	✓	✓	**6.49**	**2.64**

for getting closer to the goal room consistently outperform the agents rewarded for getting closer to the exact goal location. This proves that using a reward function better aligned with the NDH task's objective yields better performance than other reward functions.

6 Conclusion

In this work, we presented an environment-agnostic multitask learning framework to learn generalized policies for agents tasked with natural language grounded navigation. We applied the framework to train agents that can simultaneously solve two popular and challenging tasks in the space: Vision-and-Language Navigation and Navigation from Dialog History. We showed that our approach effectively transfers knowledge across tasks and learns more generalized environment representations. As a result, the trained agents not only close down the performance gap between seen and unseen environments but also outperform the single-task baselines on both tasks by a significant margin. Furthermore, the studies show the two approaches of multitask learning and environment-agnostic learning independently benefit the agent learning and complement each other. There are possible future extensions to our work—*MT-RCM* can further be adapted to other language-grounded navigation datasets (e.g., Touchdown [7], TalkTheWalk [44], StreetLearn [31]); and complementary techniques like environmental dropout [40] can be combined with environment-agnostic learning to learn more general representations.

References

1. Abadi, M., et al.: Tensorflow: a system for large-scale machine learning. In: 12th USENIX Symposium on Operating Systems Design and Implementation (OSDI 2016), pp. 265–283 (2016). https://www.usenix.org/system/files/conference/osdi16/osdi16-abadi.pdf
2. Anderson, P., et al.: On evaluation of embodied navigation agents (2018). arXiv:1807.06757 [cs.AI]
3. Anderson, P., et al.: Vision-and-language navigation: interpreting visually-grounded navigation instructions in real environments. In: Proceedings of the IEEE Conference on Computer Vision and Pattern Recognition, pp. 3674–3683 (2018)
4. Bain, M., Sammut, C.: A framework for behavioural cloning. In: Machine Intelligence 15, Intelligent Agents, St. Catherine's College, Oxford, July 1995, pp. 103–129. Oxford University, Oxford (1999). http://dl.acm.org/citation.cfm?id=647636.733043
5. Caruana, R.: Multitask learning: a knowledge-based source of inductive bias. In: Proceedings of the Tenth International Conference on Machine Learning, pp. 41–48. Morgan Kaufmann (1993)
6. Chang, A., et al.: Matterport3D: learning from RGB-D data in indoor environments. In: International Conference on 3D Vision (3DV) (2017)
7. Chen, H., Suhr, A., Misra, D., Snavely, N., Artzi, Y.: Touchdown: natural language navigation and spatial reasoning in visual street environments. In: Proceedings of the IEEE Conference on Computer Vision and Pattern Recognition, pp. 12538–12547 (2019)
8. Collobert, R., Weston, J.: A unified architecture for natural language processing: deep neural networks with multitask learning. In: Proceedings of the 25th International Conference on Machine Learning, ICML 2008, pp. 160–167. ACM, New York (2008). https://doi.org/10.1145/1390156.1390177
9. Das, A., Datta, S., Gkioxari, G., Lee, S., Parikh, D., Batra, D.: Embodied question answering. In: Proceedings of the IEEE Conference on Computer Vision and Pattern Recognition Workshops. pp. 2054–2063 (2018)
10. Das, A., et al.: Visual dialog. In: Proceedings of the IEEE Conference on Computer Vision and Pattern Recognition (CVPR) (2017)
11. Deng, L., Hinton, G., Kingsbury, B.: New types of deep neural network learning for speech recognition and related applications: an overview. In: 2013 IEEE International Conference on Acoustics, Speech and Signal Processing, pp. 8599–8603 (2013). https://doi.org/10.1109/ICASSP.2013.6639344
12. Espeholt, L., et al.: IMPALA: scalable distributed deep-RL with importance weighted actor-learner architectures. In: Dy, J., Krause, A. (eds.) Proceedings of the 35th International Conference on Machine Learning. Proceedings of Machine Learning Research, PMLR, Stockholmsmässan, Stockholm Sweden, 10–15 July 2018, vol. 80, pp. 1407–1416. http://proceedings.mlr.press/v80/espeholt18a.html
13. Finn, C., Abbeel, P., Levine, S.: Model-agnostic meta-learning for fast adaptation of deep networks. In: Proceedings of the 34th International Conference on Machine Learning, vol. Volume 70, pp. 1126–1135. JMLR. org (2017)
14. Fried, D., et al.: Speaker-follower models for vision-and-language navigation. In: Neural Information Processing Systems (NeurIPS) (2018)
15. Fu, T.J., Wang, X., Peterson, M., Grafton, S., Eckstein, M., Wang, W.Y.: Counterfactual vision-and-language navigation via adversarial path sampling. arXiv preprint arXiv:1911.07308 (2019)

16. Ganin, Y., Lempitsky, V.: Unsupervised domain adaptation by backpropagation. In: Proceedings of the 32Nd International Conference on International Conference on Machine Learning, ICML 2015, vol. 37, pp. 1180–1189. JMLR.org (2015). http://dl.acm.org/citation.cfm?id=3045118.3045244
17. Girshick, R.: Fast R-CNN. In: 2015 IEEE International Conference on Computer Vision (ICCV), pp. 1440–1448 (2015). https://doi.org/10.1109/ICCV.2015.169
18. Hemachandra, S., Duvallet, F., Howard, T.M., Roy, N., Stentz, A., Walter, M.R.: Learning models for following natural language directions in unknown environments. In: 2015 IEEE International Conference on Robotics and Automation (ICRA), pp. 5608–5615. IEEE (2015)
19. Huang, H., Jain, V., Mehta, H., Baldridge, J., Ie, E.: Multi-modal discriminative model for vision-and-language navigation. In: Proceedings of the Combined Workshop on Spatial Language Understanding (SpLU) and Grounded Communication for Robotics (RoboNLP), pp. 40–49. Association for Computational Linguistics, Minneapolis (2019). https://doi.org/10.18653/v1/W19-1605, https://www.aclweb.org/anthology/W19-1605
20. Huang, H., et al.: Transferable representation learning in vision-and-language navigation. In: Proceedings of the IEEE/CVF International Conference on Computer Vision (ICCV), vol. 178 (2019)
21. Jain, V., Magalhães, G., Ku, A., Vaswani, A., Ie, E., Baldridge, J.: Stay on the path: instruction fidelity in vision-and-language navigation. In: ACL (2019)
22. Ke, L., et al.: Tactical rewind: self-correction via backtracking in vision-and-language navigation. In: Proceedings of the IEEE Conference on Computer Vision and Pattern Recognition, pp. 6741–6749 (2019)
23. Kolve, E., et al.: AI2-THOR: An Interactive 3D Environment for Visual AI. arXiv (2017)
24. Lansing, L., Jain, V., Mehta, H., Huang, H., Ie, E.: VALAN: vision and language agent navigation. ArXiv abs/1912.03241 (2019)
25. Li, J., et al.: Unsupervised reinforcement learning of transferable meta-skills for embodied navigation. In: Proceedings of the IEEE/CVF Conference on Computer Vision and Pattern Recognition, pp. 12123–12132 (2020)
26. Li, Y., Baldwin, T., Cohn, T.: What's in a Domain? Learning Domain-Robust Text Representations using Adversarial Training. In: NAACL-HLT (2018)
27. Liu, A.H., Liu, Y.C., Yeh, Y.Y., Wang, Y.C.F.: A unified feature disentangler for multi-domain image translation and manipulation. In: Advances in Neural Information Processing Systems, pp. 2590–2599 (2018)
28. Ma, C.Y., et al.: Self-monitoring navigation agent via auxiliary progress estimation. arXiv preprint arXiv:1901.03035 (2019)
29. Ma, C.Y., Wu, Z., AlRegib, G., Xiong, C., Kira, Z.: The regretful agent: heuristic-aided navigation through progress estimation. In: Proceedings of the IEEE Conference on Computer Vision and Pattern Recognition, pp. 6732–6740 (2019)
30. Savva, M., et al.: Habitat: a platform for embodied AI research. In: Proceedings of the IEEE/CVF International Conference on Computer Vision (ICCV) (2019)
31. Mirowski, P., et al.: Learning to navigate in cities without a map. In: Bengio, S., Wallach, H., Larochelle, H., Grauman, K., Cesa-Bianchi, N., Garnett, R. (eds.) Advances in Neural Information Processing Systems 31, pp. 2419–2430. Curran Associates, Inc. (2018). http://papers.nips.cc/paper/7509-learning-to-navigate-in-cities-without-a-map.pdf
32. Mirowski, P.W., et al.: Learning to navigate in complex environments. ArXiv abs/1611.03673 (2016)

33. Nguyen, K., Daumé III, H.: Help, anna! visual navigation with natural multimodal assistance via retrospective curiosity-encouraging imitation learning. arXiv preprint arXiv:1909.01871 (2019)
34. Nguyen, K., Dey, D., Brockett, C., Dolan, B.: Vision-based navigation with language-based assistance via imitation learning with indirect intervention. In: Proceedings of the IEEE Conference on Computer Vision and Pattern Recognition, pp. 12527–12537 (2019)
35. Peng, X., Huang, Z., Sun, X., Saenko, K.: Domain agnostic learning with disentangled representations. In: Proceedings of the 36th International Conference on Machine Learning, ICML 2019, Long Beach, California, USA, 9–15 June 2019, pp. 5102–5112 (2019)
36. Qi, Y., et al.: Reverie: remote embodied visual referring expression in real indoor environments. In: Proceedings of the IEEE/CVF Conference on Computer Vision and Pattern Recognition, pp. 9982–9991 (2020)
37. Ramsundar, B., Kearnes, S.M., Riley, P., Webster, D., Konerding, D.E., Pande, V.S.: Massively multitask networks for drug discovery. ArXiv abs/1502.02072 (2015)
38. Romijnders, R., Meletis, P., Dubbelman, G.: A domain agnostic normalization layer for unsupervised adversarial domain adaptation. In: 2019 IEEE Winter Conference on Applications of Computer Vision (WACV), pp. 1866–1875. IEEE (2019)
39. Schuster, M., Paliwal, K.K.: Bidirectional recurrent neural networks. IEEE Trans. Signal Process. **45**, 2673–2681 (1997)
40. Tan, H., Yu, L., Bansal, M.: Learning to navigate unseen environments: back translation with environmental dropout. In: Proceedings of the 2019 Conference of the North American Chapter of the Association for Computational Linguistics: Human Language Technologies, vol. 1 (Long and Short Papers), pp. 2610–2621. Association for Computational Linguistics, Minneapolis (2019). https://doi.org/10.18653/v1/N19-1268
41. Teh, Y., et al.: Distral: robust multitask reinforcement learning. In: Advances in Neural Information Processing Systems, pp. 4496–4506 (2017)
42. Thomason, J., Murray, M., Cakmak, M., Zettlemoyer, L.: Vision-and-dialog navigation. In: Conference on Robot Learning (CoRL) (2019)
43. Vaswani, A., et al.: Attention is all you need. In: Advances in Neural Information Processing Systems, pp. 5998–6008 (2017)
44. de Vries, H., Shuster, K., Batra, D., Parikh, D., Weston, J., Kiela, D.: Talk the walk: navigating New York City through grounded dialogue. arXiv preprint arXiv:1807.03367 (2018)
45. Wang, X., et al.: Reinforced cross-modal matching and self-supervised imitation learning for vision-language navigation. In: Proceedings of the IEEE Conference on Computer Vision and Pattern Recognition, pp. 6629–6638 (2019)
46. Wang, X., Xiong, W., Wang, H., Wang, W.Y.: Look before you leap: bridging model-free and model-based reinforcement learning for planned-ahead vision-and-language navigation. In: Ferrari, V., Hebert, M., Sminchisescu, C., Weiss, Y. (eds.) ECCV 2018. LNCS, vol. 11220, pp. 38–55. Springer, Cham (2018). https://doi.org/10.1007/978-3-030-01270-0_3
47. Williams, R.J.: Simple statistical gradient-following algorithms for connectionist reinforcement learning. Mach. Learn. **8**(3), 229–256 (1992). https://doi.org/10.1007/BF00992696
48. Wu, Y., Wu, Y., Gkioxari, G., Tian, Y.: Building generalizable agents with a realistic and rich 3D environment. arXiv preprint arXiv:1801.02209 (2018)

49. Wu, Y., Wu, Y., Tamar, A., Russell, S., Gkioxari, G., Tian, Y.: Learning and planning with a semantic model. arXiv preprint arXiv:1809.10842 (2018)
50. Xia, F., R. Zamir, A., He, Z.Y., Sax, A., Malik, J., Savarese, S.: Gibson env: real-world perception for embodied agents. In: 2018 IEEE Conference on Computer Vision and Pattern Recognition (CVPR). IEEE (2018)

TPFN: Applying Outer Product Along Time to Multimodal Sentiment Analysis Fusion on Incomplete Data

Binghua Li[1,2], Chao Li[1], Feng Duan[2(✉)], Ning Zheng[1], and Qibin Zhao[1]

[1] RIKEN Center for Advanced Intelligence Project (AIP), Tokyo, Japan
{chao.li,ning.zheng,qibin.zhao}@riken.jp

[2] College of Artificial Intelligence, Nankai University, Tianjin, China
nkvhua@outlook.com, duanf@nankai.edu.cn

Abstract. Multimodal sentiment analysis (MSA) has been widely investigated in both computer vision and natural language processing. However, studies on the imperfect data especially with missing values are still far from success and challenging, even though such an issue is ubiquitous in the real world. Although previous works show the promising performance by exploiting the low-rank structures of the fused features, only the first-order statistics of the temporal dynamics are concerned. To this end, we propose a novel network architecture termed Time Product Fusion Network (TPFN), which *takes the high-order statistics over both modalities and temporal dynamics into account*. We construct the fused features by the outer product along adjacent time-steps, such that richer modal and temporal interactions are utilized. In addition, we claim that the low-rank structures can be obtained by regularizing the Frobenius norm of latent factors instead of the fused features. Experiments on CMU-MOSI and CMU-MOSEI datasets show that TPFN can compete with state-of-the art approaches in multimodal sentiment analysis in cases of both random and structured missing values.

Keywords: Multimodal sentiment analysis · Multimodal learning · Matrix/tensor decomposition · Incomplete data

1 Introduction

Multimodal learning is recently one of the increasingly popular yet challenging tasks in both computer vision (CV) [1,12,14], and natural language processing (NLP) [5,52]. As its important application, multimodal sentiment analysis (MSA) is to predict the categories or intensity of the sentiment by jointly mining the data with various modalities, *e.g.*, visual, acoustic and language [44,50,51].

B. Li and C. Li—Equal Contribution.

Electronic supplementary material The online version of this chapter (https://doi.org/10.1007/978-3-030-58586-0_26) contains supplementary material, which is available to authorized users.

A. Vedaldi et al. (Eds.): ECCV 2020, LNCS 12369, pp. 431–447, 2020.
https://doi.org/10.1007/978-3-030-58586-0_26

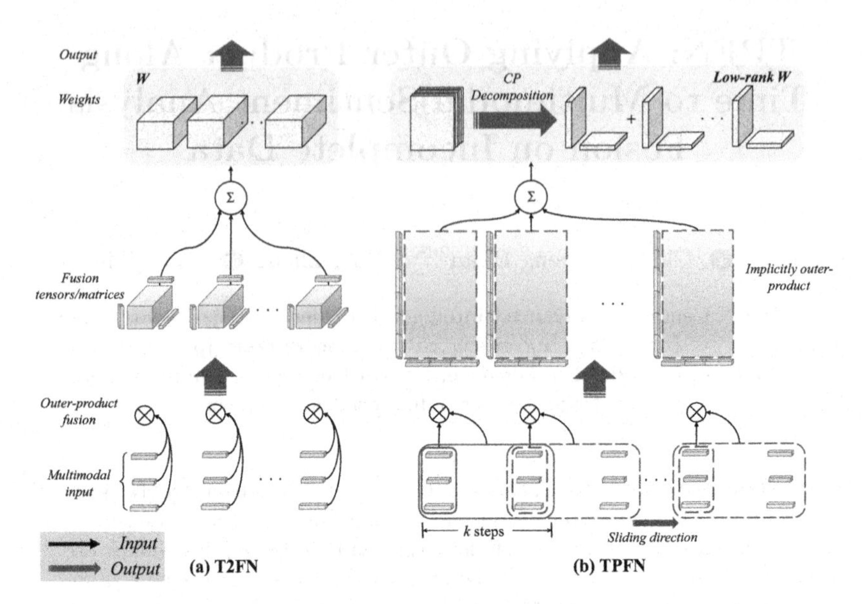

Fig. 1. Comparison between T2FN [23] and TPFN (our method). In the proposed model, we concatenate the features along adjacent k time-steps. In the fusion phase, the out product is implicitly applied to avoid the additional computational/storage consumption, and the corresponding weight tensor is also assumed with the low-rank CP format.

Although approaches on MSA have been well developed in an ideal situation, the imperfectness of the data is still a challenge we have to face in the real world, especially when there are missing values at unknown locations caused by mismatched modalities or sensor failures [35,40]. Recently, Liang *et al.* [23] demonstrates that the representation obtained from incomplete data has the low-rank structure in MSA, proposing a low-rank regularization based model termed temporal tensor fusion network (T2FN) against the bias caused by missing values. Despite the method achieves robust implementation against the imperfect data, there are still two aspects we can go further: 1) T2FN fully captures the interactions among all modalities of the data yet fails to exploit the data dynamics across the temporal domain and 2) the outer product is explicitly applied in the model, which would lead to heavy consumption of memory resources when increasing the time-steps or feature dimension.

To this end, we propose a novel model termed Time Product Fusion Network (TPFN) for the issue of incomplete MSA. Compared to previous works, TPFN 1) *captures additional interactions among different time-steps* to exploit higher-order statistics of temporal dynamics, and 2) *alleviates the issue of the unacceptable model size* by approximating the weight tensor using the well-known CANDECOMP/PARAFAC (CP) decomposition [21]. For the latter, we theoretically claim that the low-rank structure of the fused features can be controlled by the Frobenius norm of the latent factors. It allows us to avoid the explicit calculation of the outer product used in [23]. Figure 1 gives the schematic diagram to demonstrate the difference between T2FN and the proposed TPFN. As shown in Fig. 1, we construct the interaction among features by implicitly exploiting the outer-product along the adjacent time-steps. As for the learnable weights in the model, we apply the low-rank CP format to reducing the required model size.

1.1 Related Works

Multimodal Sentiment Analysis. The studies on sentiment analysis (SA) are started from 2000s [34], and widely discussed in the NLP community [7,24,36,39]. Its extension, multimodal sentiment analysis (MSA), recently attracts more attention as the visual and audio features are capable of significantly improving the prediction performance compared to the conventional SA [31]. As a multimodal learning task, the core issue on the MSA study is how to efficiently fuse the features from multiple modalities [15,31,43,45,52]. In the existing methods, the outer-product-based fusion strategy shows impressive performance with very neat model [2,19,23,26,28,49]. The basic idea behind those methods is to exploit the outer-product to obtain the high-order statistics of features. The fused features reflect rich information about the interaction among multiple modalities. Table 1 compares several important characteristics of the state-of-the-art methods based on outer-product. As shown in Table 1, only T2FN [23] and our method can deal with the imperfect data, while other methods generally assume that both the training and test datasets are ideal. Compared to T2FN, our method takes the *high-order statistics of temporal dynamics* into account and further applies matrix/tensor decomposition on weights to reducing the number of parameters. Although there have been several studies on developing robust models for multimodal learning [6,25,29,40], they cannot trivially applied to the MSA task due to the difference of specific tasks and data.

Table 1. Comparison of the outer-product-based methods. Each column corresponds one characteristic of the methods, where "*Decomposition*" indicates whether exploiting matrix/tensor decomposition for dimension reduction and "*Low-rankness*" means whether the low-rankness of features is considered in the model.

Methods	Decomposition	Temporal dynamics	Imperfect data	Low-rankness
TFN [49]	✗	✗	✗	✗
LMF [26]	✓	✗	✗	✗
HFFN [28]	✗	✗	✗	✗
HPFN [19]	✓	✓ (high-order)	✗	✗
T2FN [23]	✗	✓ (low-order)	✓	✓
TPFN(ours)	✓	✓ (high-order)	✓	✓

Low-Rankness in Robust Learning. Low-rank approximation is a collection of well-known methods to cope with the issue on the imperfect data like noise

[13,32] and incompleteness [18,22,42,48]. The existing low-rank approximation methods can be roughly split into two categories: (a) matrix/tensor decomposition [8,17,41] and (b) nuclear norm minimization [11,16,27]. On the other side, in the studies on artificial neural networks (ANN), the low-rank assumption is generally imposed on weights for model compression [33,46] or for alleviating the overfitting issue [3,4]. Except for T2FN [23], there is seldom work in ANN to apply the low-rank regularization *directly* to feature maps for robust learning. Inspired by the prior arts, we also apply the low-rank regularization to coping with the imperfect data issue. However, we highlight the difference from T2FN that in this work the low-rank structure is obtained by regularizing the Frobenius norm of the latent factors instead of on the fused features as T2FN does. The new trick allows us to avoid the additional requirement on computation and storage of the model.

2 Preliminaries

To be self-contained of the paper, we concisely review the necessary multilinear algebra and operations, which play important roles to understand our model.

Notation. We use italic letters like a, b, A, B to denote scalars, boldface lowercase letters like $\mathbf{a}, \mathbf{b}, ...$ to denote vectors, boldface capital letters $\mathbf{A}$, $\mathbf{B}$, ... to denote matrices and calligraphic letters like $\mathcal{A}, \mathcal{B}, ...$ for tensors of arbitrary order. Time series are denoted by the underlined italic capital letters, *e.g.*., $\underline{A} = \{\mathcal{A}_1, \dots, \mathcal{A}_T\}$. The operation "$\circ$" denotes the element-wise product, "$\bigotimes$" denotes outer product of vectors and "$\cdot$" denotes the matrix-tensor product. More details on multi-linear operations are given in [10] and the references therein.

CP Format. In our model, we apply the well-known (CP) decomposition [8,17] to representing the weights in the last layer of the proposed network. Specifically, CP decomposition is to represent a tensor as a finite sum of rank-1 factors constructed by the outer products of vectors. Given an pth-order tensor $\mathcal{W}$ and the factor matrices $\mathbf{W}^{(p)}, p \in [P]$, the CP decomposition of $\mathcal{W}$ is given by:

$$\mathcal{W} = \sum_{r=1}^{R} \bigotimes_{p=1}^{P} \mathbf{w}_r^{(p)}, \tag{1}$$

where R denotes the CP-rank of $\mathcal{W}$[1], and $\mathbf{w}_r^{(p)}$ denotes the r-th column of $\mathbf{W}^{(p)}$. Note that if $P = 2$ then Eq. (1) is degenerated into the trivial product of two factor matrices, *i.e.*,

$$\mathcal{W} = \sum_{r=1}^{R} \bigotimes_{p=1}^{2} \mathbf{w}_r^{(p)} = \mathbf{W}^{(1)}\mathbf{W}^{(2),\top}, \tag{2}$$

[1] Without ambiguity, we also use the notion of CP-rank to represent the number of rank-1 factors used in matrix/tensor approximation.

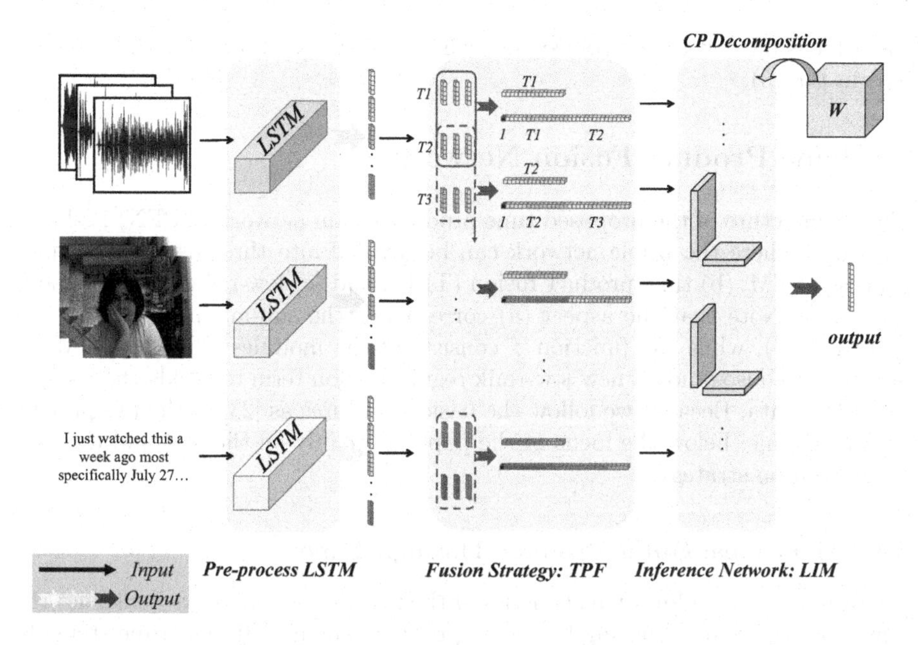

Fig. 2. Details of our TPFN with $k = 2$ and stride as 1. Three Modules are involved in our model: The Pre-process LSTM, the Time Product Fusion Module and the Low-rank Inference Module. Low-rank regularization is applied on the multimodal representation

where $\cdot^\top$ denotes the transpose operation of a matrix. In the rest of the paper, we also need to exploit the tensor nuclear norm $\|\mathcal{W}\|_*$, the dual norm of tensor spectral norm, to introduce a more efficient low-rank regularization into the model. As introduced in [37], the tensor nuclear norm is a good surrogate of tensor rank. It therefore implies that bounding the nuclear norm generally results in a low-rank regularization.

Problem Setting. In this paper, we consider the MSA task as a multimodal learning problem. Specifically, we assume a sample in the task to be a triple $(\underline{A}, \underline{V}, \underline{L})$, where $\underline{A} = \{\mathcal{A}_1, \ldots, \mathcal{A}_T\}$, $\underline{V} = \{\mathcal{V}_1, \ldots, \mathcal{V}_T\}$ and $\underline{L} = \{\mathcal{L}_1, \ldots, \mathcal{L}_T\}$ denote the time series of the length T *w.r.t.* the acoustic, visual and language data, respectively. The goal of our work is hence to learn a mapping from the multimodal data to an output associated with a specific task, such as classification or regression. Mathematically, we would like to learn a composite function

$$\widehat{\mathbf{y}} = f\left(\phi_a(\underline{A}), \phi_v(\underline{V}), \phi_l(\underline{L})\right), \tag{3}$$

where ϕ_a, ϕ_v and ϕ_l denotes the sub-mappings from the raw data to the features, and the function f includes the fusion phase and the mapping from the fused features to targets. As in the previous works of MSA, ϕ_a, ϕ_v and ϕ_l generally consist of deep neural networks like CNN [20] and RNN [30] to embed the raw data into the feature space. We also adopt the same architecture yet put the main focus on developing robust and efficient fusion method, *i.e.* the function $f(\cdot)$ in Eq. (3).

3 Time Product Fusion Network

The architecture of the proposed time product fusion network (TPFN) is shown in Fig. 2, where the whole network can be divided into three aspects: (a) pre-process LSTM, (b) time product fusion (TPF), and (c) low-rank inference module (LIM). Note that the aspect (a) corresponds the sub-mappings ϕ_a, ϕ_v and ϕ_l in Eq. (3), while the function f consists of the modules TPF and LIM. In addition, we also study a new low-rank regularization term to tackle the issue of imperfect data. Because we follow the basic structures as [23] in the pre-process LSTM module, below the focus of the paper is mainly on the rest and the new regularization strategy.

3.1 Main Idea: Outer Product Through Time

To investigate the additional statistics of the feature dynamics, we construct the fusion operation by imposing the outer products among adjacent time-steps. In the MSA task, assume that we obtain the multimodal features per T time-steps by the sub-nets ϕ_a, ϕ_v and ϕ_l in Eq. (3), where the features are denotes by the matrices $\mathbf{A} \in \mathbb{R}^{d_a \times T}$ for acoustic, $\mathbf{V} \in \mathbb{R}^{d_v \times T}$ for visual, and $\mathbf{V} \in \mathbb{R}^{d_v \times T}$ for the language modal, respectively. Let $\mathbf{a}_t$, $\mathbf{v}_t$, and $\mathbf{l}_t$ be the t-th column of $\mathbf{A}$, $\mathbf{V}$ and $\mathbf{L}$, respectively, then we first concatenate the features from all modalities for the given time-step $t \in [T]$:

$$\mathbf{z}_t^\top = [\mathbf{a}_t^\top, \mathbf{v}_t^\top, \mathbf{l}_t^\top] \in \mathbb{R}^{1 \times L}, \tag{4}$$

where $L = d_a + d_v + d_l$. As shown in Eq. (4), all features in the t-th step are involved in the concatenated vector $\mathbf{z}_t$. Next, to model the interaction across time-steps, we construct enhanced vectors *w.r.t.* $\mathbf{z}_t$ by further concatenating the adjacent time-steps, *i.e.*,

$$\mathbf{z}_t^{e,\top} = [1, \mathbf{z}_t^\top, ..., \mathbf{z}_{t+k-1}^\top] \in \mathbb{R}^{1 \times (kL+1)}, k > 0, t \in [T], \tag{5}$$

where the element "1" is also padded to retain the intra-modal correlation for each modality. Using Eq. (4) and (5), the tth-step ingredient of the fused features is therefore calculated as

$$\mathbf{M}_t = \mathbf{z}_t \otimes \mathbf{z}_t^e = \mathbf{z}_t \otimes [1, \mathbf{z}_t, \ldots, \mathbf{z}_{t+k-1}] \in \mathbb{R}^{L \times (kL+1)}. \tag{6}$$

Note that $\mathbf{M}_t$ can be divided into three chunks: $\mathbf{z}_t \bigotimes 1$, $\mathbf{z}_t \bigotimes \mathbf{z}_t$, and $\mathbf{z}_t \bigotimes \mathbf{z}_{t+i}$. The first chunk keeps the *unimodal information* by the product with the identity. The second chunk is to model the *inter-modality interaction* in the local step, and the third attempts to explore the *dynamics across time.*

More interestingly, the calculation of $\mathbf{M}_t$ for all $t \in [T - k + 1]$ is equivalent to sliding a window of the size k along the temporal domain. As shown in

Fig. 2, we first collect the features from all modalities together by concatenation, and then use a window of the size k and stride 1 to slide the concatenated features from top to bottom. For each step, we extract the features contained in the window and employ the outer-product to fuse the features. With the time window sliding, all intra-modality and inter-modality dynamics will be involved. As a special case, if k is set to 1, *i.e.* $\mathbf{z}_e^{\mathrm{e}\top} = [1, \mathbf{z}_t^\top]$, only the interaction within a local time-step would be captured.

At last, we conduct the pooling by summation on all $\mathbf{M}_t$'s to obtain the final fused features. If we define the factor matrices $\mathbf{Z} = [\mathbf{z}_1, \mathbf{z}_2 \ldots, \mathbf{z}_{T-k+1}]$ and $\mathbf{Z}^{\mathrm{e}} = [\mathbf{z}_1^{\mathrm{e}}, \mathbf{z}_2^{\mathrm{e}}, ..., \mathbf{z}_{T-k+1}^{\mathrm{e}}]$, then the fused feature matrix can be written as

$$\mathbf{M} = \sum_{t=1}^{N} \mathbf{M}_t = \mathbf{Z}\mathbf{Z}^{e,\top} \in \mathbb{R}^{L\times(kL+1)}, \tag{7}$$

where $N = T - k + 1$. In summary, the feature matrix $\mathbf{M}$ in our model reflects the interaction across not only multiple modalities but also time steps, which supplies more information to tackle the issue of imperfect data compared to T2FN [23].

However, it leads to troubles if we directly use $\mathbf{M}$ in Eq. (7) as the inputs of the sequential layers. It is because the size of $\mathbf{M}$ quadratically grows when increasing L, which equals the sum of dimension of features through all modalities. To alleviate such an issue, we will show in the following section that the acceptable feature size can be obtained by leveraging the inherent low-rank structure of $\mathbf{M}$.

3.2 Low-Rank Inference Module

Below, we introduce the low-rank inference module (LIM), which not only maps the "fused features" to the final output but also avoids using numerous parameters in the model.

Consider a collection of affine functionals $g_i(\cdot)$, $i \in [d_o]$, each of which maps a feature matrix to a scalar. Given i, the functional can be thus written as

$$g_i(\mathbf{M}; \mathbf{W}_i, b_i) = \langle \mathbf{W}_i, \mathbf{M} \rangle + b_i, \tag{8}$$

where $\langle \cdot, \cdot \rangle$ denotes the trivial inner product, and $\mathbf{W}_i \in \mathbb{R}^{L\times(kL+1)}$ and b_i denote the weight and bias in the context of neural networks, respectively. Note that a large size of the feature matrix $\mathbf{M}$ also leads to the weight $\mathbf{W}_i$ with the same size, which would be unaffordable in practice. To tackle the issue, we decompose $\mathbf{W}_i$, $\forall i$ as the aforementioned CP format of the rank equaling R. As given in Eq. (1) and (2), the weights can be decomposed as

$$\mathbf{W}_i = \mathbf{W}_i^{(1)} \mathbf{W}_i^{(2),\top} = \sum_{r=1}^{R} \mathbf{w}_{i,r}^{(1)} \otimes \mathbf{w}_{i,r}^{(2)}, \tag{9}$$

where $\mathbf{w}_{i,r}^{(j)}, j = 1,2$ denotes the r-th column of $\mathbf{W}_i^{(j)}$. Note from Eq. (7) and (9) that both $\mathbf{W}_i$ and $\mathbf{M}$ can be expressed by summation on outer products of two vectors shaped $\mathbb{R}^L$ and $\mathbb{R}^{kL+1}$. Hence we modify Eq. (8) as

$$\begin{aligned}
\widehat{y}_i = g_i\left(\mathbf{M}; \mathbf{W}_i, b_i\right) &= \left\langle \sum_{r=1}^{R} \mathbf{w}_{i,r}^{(1)} \otimes \mathbf{w}_{i,r}^{(2)}, \sum_{t=1}^{N} \mathbf{z}_t \otimes \mathbf{z}_t^e \right\rangle + b_i \\
&= \sum_{r=1}^{R}\sum_{t=1}^{N} tr\left((\mathbf{w}_{i,r}^{(1)} \otimes \mathbf{w}_{i,r}^{(2)})^\top (\mathbf{z}_t \otimes \mathbf{z}_t^e) \right) + b_i \\
&= \sum_{r=1}^{R}\sum_{t=1}^{N} (\mathbf{w}_{i,r}^{(1)\top} \cdot \mathbf{z}_t)(\mathbf{w}_{i,r}^{(2)\top} \cdot \mathbf{z}_t^e) + b_i \\
&= \left\langle \underbrace{\mathbf{W}_i^{(1),\top}\mathbf{Z}}_{R\times N}, \underbrace{\mathbf{W}_i^{(2),\top}\mathbf{Z}^e}_{R\times N} \right\rangle + b_i
\end{aligned}, \qquad (10)$$

where $tr(\cdot)$ denotes the trace function, and the equality in the third line is obtained due to the cyclic property. At last, $\widehat{\mathbf{y}}^\top = [\widehat{y}_1, \widehat{y}_2, \ldots, \widehat{y}_{d_o}]^\top$ gives the final output of the proposed network. As shown in Eq. (10), $\widehat{y}_i$ is obtained by calculating the inner product between $\mathbf{W}_i^{(1),\top}\mathbf{Z}$ and $\mathbf{W}_i^{(2),\top}\mathbf{Z}^e$, of which the size $R \times N$ would generally far smaller than the size $L \times (kL+1)$ *w.r.t.* $\mathbf{M}$, especially when L is large (it usually happens when dealing with multimodal data including visual and linguistic features). As for the computational complexity, Eq. (10) results in $\mathcal{O}(kRNL)$, while totally $\mathcal{O}(kNL^2)$ is needed if we directly calculate the feature matrix $\mathbf{M}$ and use Eq. (8) to obtain the output. In the experimental section, we will also empirically prove that the proposed TPFN incorporates more interactions across the temporal domain yet with fewer parameters due to our low-rank inference module.

3.3 Low-Rank Regularization

As mentioned in T2FN [23], the existence of missing values would increase the rank of the fused feature matrix $\mathbf{M}$. Inspired by the claim, we also impose the low-rank regularization into the model to handle the issue of imperfect data. However, unlike regularizing the Frobenius norm of $\mathbf{M}$ directly, we argue that the rank can be bounded by the norm of its latent factor matrix $\mathbf{Z}$.

To do so, we first introduce the key lemma, which appears as Lemma 1 in [38] and is popularly used in collaborative filtering [47]. Assuming a matrix $\mathbf{X} \in \mathbb{R}^{m\times n}$, we have

Lemma 1 (from [38]**).** *Assume matrix* $\mathbf{X} \in \mathbb{R}^{m\times n}$*, which can be represented by arbitrary decomposition* $\mathbf{X} = \mathbf{U}\mathbf{V}^\top$*, then we have*

$$\|\mathbf{X}\|_* = \min_{\mathbf{U},\mathbf{V}} \|\mathbf{U}\|_F \|\mathbf{V}\|_F, \quad s.t. \mathbf{X} = \mathbf{U}\mathbf{V}^\top. \qquad (11)$$

Lemma 1 implies that the nuclear norm of the matrix $\mathbf{X}$ is upper bounded by the product of Frobenius norm of its factor matrices. Using Lemma 1, we propose the claim that the nuclear norm of $\mathbf{M}$ defined in Eq. (7) is upper bounded by the Frobenius norm of $\mathbf{Z}$. Specifically,

Claim 1. *Define the matrices* $\mathbf{M}$ *and* $\mathbf{Z}$ *as above, then the following inequality holds:*

$$\|\mathbf{M}\|_* \leq \sqrt{N + k\|\mathbf{Z}\|_F}\|\mathbf{Z}\|_F. \tag{12}$$

The proof is trivial by exploiting the relation $\|\mathbf{Z}^{\mathrm{e}}\|_F^2 \leq N + k\|\mathbf{Z}\|_F^2$. Since the matrix nuclear norm is the convex envelope of matrix rank [37], Claim 1 implies that the rank of the feature matrix $\mathbf{M}$ would be "controlled" by the Frobenius norm of its factor matrix $\mathbf{Z}$. More importantly, Claim 1 allow us to avoid explicitly calculating the fused matrix $\mathbf{M}$ for the low-rank regularization yet the model still results in the robustness against the imperfect data. Therefore, in our model we multiply $\|\mathbf{Z}\|_F$ with a norm factor λ (tuning parameter) and add it to the loss function as regularization to train the network.

4 Experiments

Dataset. We evaluate our method on two datasets: CMU-MOSI [53] and CMU-MOSEI [54]. *CMU-MOSI* is a multimodal sentiment analysis dataset containing 93 videos, which are then split into 2,199 short video clips by sentence. *CMU-MOSEI* is a dataset that can be applied to both emotion recognition and sentiment analysis, and it is the largest dataset on multimodal sentiment analysis at present. It contains 23453 labeled movies collected from 1000 different speakers in YouTube, covering 250 hot topics. CMU-MOSI and CMU-MOSEI datasets we use are pre-trained by the methods in [9] and [54], respectively[2]. Dataset statistics are shown in Table 2.

Table 2. Dataset statistics of **CMU-MOSI** and **CMU-MOSEI**. Number of samples and size of features for each modality are listed.

	Number of Samples			Size of Features		
	Train	**Val**	**Test**	**Acoustic**	**Visual**	**Language**
MOSEI	15,290	2,291	4,832	74	35	300
MOSI	1,284	229	686	5	20	300

Incompleteness Modelling. Like [23], we exploit two strategies for dropping the data to simulate the incompleteness, *i.e.* random drop and structured drop.

[2] See http://immortal.multicomp.cs.cmu.edu/raw_datasets/processed_data/.

Table 3. Comparison of classification accuracy (ACC-2) and the total number of parameters used in the model for the *CMU-MOSI* task. **RD** and **SD** represent the random and structural drop task, respectively. The "**Low**", "**Medium**" and "**High**" columns correspond the missing percentage equaling 0.1, 0.5, 0.9, respectively. The "**Params**" column shows the number of parameters used in the methods.

Task	Method	Low	Medium	High	Params
RD	TFN	0.7361	0.7172	0.4475	759,424
	LMF	0.7346	0.7317	0.5218	**2,288**
	HPFN	0.7565	0.6982	0.5568	4,622,039
	T2FN	0.7769	0.7113	0.5962	19,737
	TPFN/reg(ours)	0.7638	0.7594	0.5845	8,652
	TPFN(ours)	**0.7915**	**0.7609**	**0.6559**	19,488
SD	TFN	0.7317	0.6880	0.5758	390,784
	LMF	0.7346	0.7128	0.5976	**792**
	HPFN	0.7463	0.7186	0.6151	1,168,247
	T2FN	0.7478	0.7142	0.6137	19,737
	TPFN/reg(ours)	**0.7682**	0.7288	0.6151	11,360
	TPFN(ours)	0.7594	**0.7434**	**0.6516**	7,344

Given the missing percentage $p \in \{0.0, 0.1, ..., 0.9\}$, we *i.i.d.* drop the entries of the data at random for the former, and randomly remove the whole time step for the latter.

General Setting. We select the window size k from $\{1, 2, 3, 4, 5\}$, and keep the stride equalling 1. The CP-rank r is tuned from $\{4, 8, 12, 16, 24, 32\}$ and regularization parameter λ (if has) is tuned from $\{0.0, 0.0001, 0.001, 0.003, 0.01\}$. The hidden size of the pre-process LSTM is selected from $\{8, 16\}$, $\{4, 8, 16\}$, $\{64, 128\}$ for acoustic, visual and language, respectively. We train our method for 200 epochs in all experiments and employ early stop when the model does achieve the minimum loss on valid set for over 20 times. Adam Optimizer is used in our paper and the learning rate is tuned from $\{0.0003, 0.001, 0.003\}$.

Goal. The aim of our experiments include two aspects: First, we demonstrate the effectiveness of our methods on the incomplete multimodal sentiment analysis task, comparing with the results by the current state-of-the-art (SOTA) approaches; Second, we discuss the impact of tuning parameters such as the window size k, regularization parameter λ and the rank r on performance.

4.1 Performance on MOSI and MOSEI

For comparison, we implement TFN [49], LMF [26], HPFN [19] and T2FN [23] as baselines to evaluate the performance of our TPFN method. We also show the performance of TPFN without regularization (TPFN/reg), demonstrating that the low-rank regularization does improve the performance on the incomplete data. Table 3 and 4 show the classification accuracy (ACC-2) of the methods on CMU-MOSI and CMU-MOSEI, respectively. We select the missing percentage $p = 0.1, 0.5, 0.9$ to represent the low, medium and high incompleteness strength, respectively. Also, we show in Fig. 3 the performance change of our method under a full range of missing percentage p on CMU-MOSI dataset.

Results on Accuray. Overall, our methods obtain the superior performance among all methods. Similar results goes in CMU-MOSEI. Figure 3 also shows that our method can maintain a relatively stable performance as missing percentage increases. Meanwhile, it is shown that the low-rank regularization is helpful for the task on incomplete multimodal data since TPFN without the reg term performs worse than the one equipped with the regularization. We can also see from the "**Params**" that TFPN uses less number of parameters than T2FN in the experiment. Although LMF used the least number of parameters, our methods significantly outperforms LMF specifically when the missing percentage is high.

Table 4. Comparison of classification accuracy (ACC-2) and the total number of parameters used in the model for the *CMU-MOSEI* task. **RD** and **SD** represent the random and structural drop task, respectively. The "**Low**", "**Medium**" and "**High**" columns correspond the missing percentage equaling 0.1, 0.5, 0.9, respectively. The "**Params**" column shows the number of parameters used in the methods.

Task	Method	Low	Medium	High	Params
RD	TFN	0.7195	0.7193	0.6705	1,353,856
	LMF	0.7307	0.7233	0.6684	**1,208**
	HPFN	0.7371	0.7189	0.7119	1,295,895
	T2FN	0.7394	**0.7382**	0.7104	18,785
	TPFN/reg(ours)	0.7375	0.7297	0.7156	14,240
	TPFN(ours)	**0.7411**	0.7367	**0.7334**	16,842
SD	TFN	0.7295	0.7121	0.6968	759,424
	LMF	0.7313	0.7067	0.7057	**1,304**
	HPFN	0.7311	0.7245	0.7003	1,296,423
	T2FN	0.7350	0.7295	0.7173	9,945
	TPFN/reg(ours)	**0.7437**	0.7301	0.7007	7,056
	TPFN(ours)	0.7386	**0.7382**	**0.7301**	5,796

4.2 Effect of Time Window Size k

To investigate how the performance changes as the time window size k varies, we keep all parameters in Sect. 4.1 except for k, and conduct the random and structured drop task on CMU-MOSI. The results are given in Fig. 4.

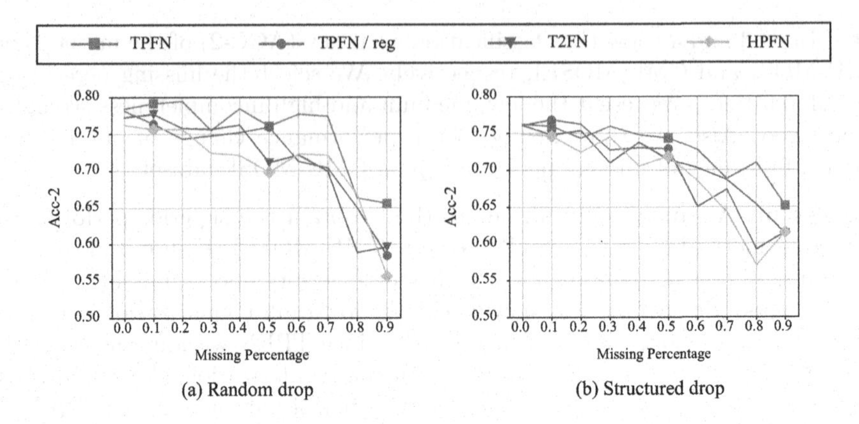

Fig. 3. Classification accuracy on CMU-MOSI in the full range of the missing percentage. The marker points corresponds to the results shown in Table 3.

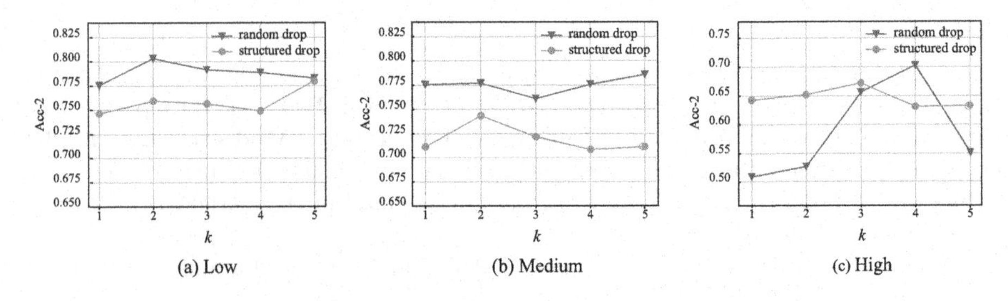

Fig. 4. Comparison on CMU-MOSI as k varies in the random drop task and the structured drop task.

It can be observed when the missing percentage p is small, that the accuracy floats within a limited range as k varies in $[1, 2, 3, 4, 5]$, which indicates that our method is relatively robust with respect to k values in those cases. While in the case with high missing percentage, a suitable value of k can remarkably improve the performance in both tasks. We infer that choosing suitable k can help the method to "see" more information when the modalities get sparse. Assume that $\mathbf{a}_t$ is missing yet $\mathbf{v}_t$ and $\mathbf{l}_t$ are not, setting $k = 1$ implies that the visual and language modality can hardly interact with the acoustic one in the t-th step, while a bigger k incorporates additional information of modalities from the adjacent time-steps (such as $\mathbf{a}_{t+1}$), allowing exploring the inter-modality dynamics. Generally, $k = 1$ means the dynamics across time is neglected, while more interactions would be taken into account for a larger k. Note that the performance tends to degradation when the given k is too large as shown in Fig. 4. We conjecture that the domination by dynamics across time weakens the local dynamics. As Eq. (6) shows, only $\mathbf{z}_t \bigotimes 1$ and $\mathbf{z}_t \bigotimes \mathbf{z}_t$ contain the interaction within time-steps. When k increasing, more correlations between series are involved, leading to the weakness on local dynamics.

4.3 Effect of Regularization Parameter λ

We keep all parameters used in Sect. 4.1 unchanged except for λ, and conduct the random and structured drop task on CMU-MOSI. Results are shown in Fig. 5.

As shown in Fig. 5, TPFN is stable with the change of λ, yet a suitable value of λ is indeed able to improve the performance compared to the one without the low-rank regularization ($\lambda = 0$). As our aforementioned discussion, the regularization on $\|\mathbf{Z}\|_F$ is able to bound the low-rank structure of the fused features, and therefore results in the "restored" features, which is closer to the clean features than their incomplete counterparts. However, when λ gets too large, the loss function would be dominated by the regularization terms, and therefore degrades the performance as the result. Note that the significant improvement by the regularization term appears when the missing percentage is high. It is because the low-rank prior could guide the method to revise the bias by the severe incompleteness of the features.

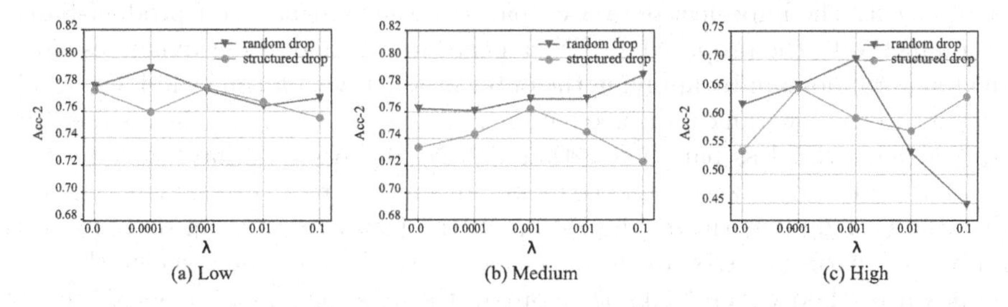

Fig. 5. Comparison on CMU-MOSI as λ varies in the random drop task and the structured drop task.

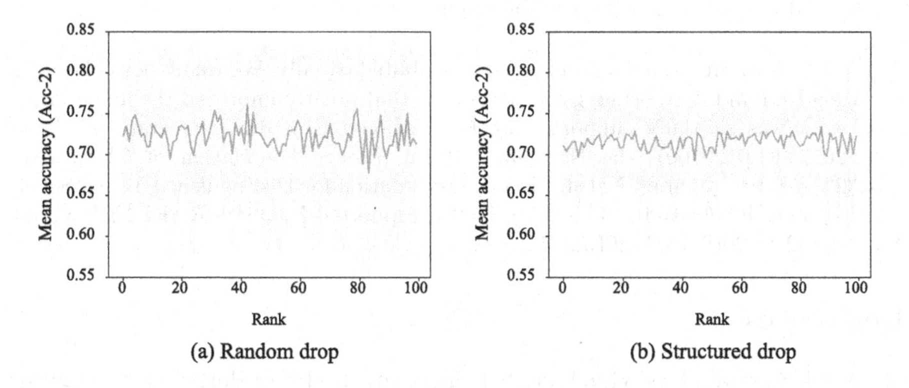

Fig. 6. Performance on CMU-MOSI as the rank r varies. The results are obtained by averaging the all missing percentage and varying the CP-rank r from 1 to 100.

4.4 Discussion on CP-rank

To discuss how the CP-rank effects the performance, we also keep all parameters in Sect. 4.1 except for r, and conduct the random and structured drop tasks on CMU-MOSI. We use the mean value of accuracy obtained from 10 different missing percentages to illustrate the overall performance, and the experimental results are shown in Fig. 6, where the CP-rank r varies from 1 to 100.

It is shown that TPFN is robust with respect to the CP-rank r for both the random and structural drop tasks. In other words, the performance seems not to be remarkably influenced as the rank increases. The similar phenomenon has been discussed in the previous study on LMF [26]. More supplemental materials and codes are available in the webpage https://qibinzhao.github.io.

5 Conclusions

The main focus of this paper is on a new fusion strategy for multimodal sentiment analysis with the imperfect data. Compared to the existing outer-product-based fusion methods, the proposed TPFN can capture the high-order dynamics along the temporal domain by applying the outer product within time windows. Additionally, in the low-rank inference module, our method achieves the competitive performance with less parameters than T2FN [23]. Also, we have introduced a new low-rank regularization for the model. In contrast to T2FN, we have claimed that the Frobenius norm regularization on the factor matrix can obtain a low-rank fused feature matrix. In the experiments, We have not only shown that the proposed method outperforms the state-of-the-arts, but also further discussed how the window size k, the regularization parameter λ and the CP-rank r affect the performance, showing that a moderate determination of tuning parameters are helpful for the task with the incomplete data.

Acknowledgment. Binghua and Chao contributed equally. We thank our colleagues Dr. Ming Hou and Zihao Huang for discussions that greatly improved the manuscript. This work was partially supported by the National Key R&D Program of China (No. 2017YFE0129700), the National Natural Science Foundation of China (No. 61673224) and the Tianjin Natural Science Foundation for Distinguished Young Scholars (No. 18JCJQJC46100). This work is also supported by JSPS KAKENHI (Grant No. 20H04249, 20H04208, 20K19875).

References

1. Antol, S., et al.: Vqa: visual question answering. In: Proceedings of ICCV (2015)
2. Barezi, E.J., Fung, P.: Modality-based factorization for multimodal fusion. arXiv preprint arXiv:1811.12624 (2018)
3. Ben-Younes, H., Cadene, R., Cord, M., Thome, N.: Mutan: multimodal tucker fusion for visual question answering. In: Proceedings of ICCV (2017)
4. Ben-Younes, H., Cadene, R., Thome, N., Cord, M.: Block: bilinear superdiagonal fusion for visual question answering and visual relationship detection. In: Proceedings of AAAI (2019)

5. Busso, C., Bulut, M., Lee, C.C., Kazemzadeh, A., Mower, E., Kim, S., Chang, J.N., Lee, S., Narayanan, S.S.: Iemocap: interactive emotional dyadic motion capture database. Lang. Resour. Eval. **42**(4), 335 (2008)
6. Cai, L., Wang, Z., Gao, H., Shen, D., Ji, S.: Deep adversarial learning for multi-modality missing data completion. In: Proceedings of SIGKDD (2018)
7. Cambria, E., Poria, S., Bajpai, R., Schuller, B.: Senticnet 4: a semantic resource for sentiment analysis based on conceptual primitives. In: Proceedings of COLING (2016)
8. Carroll, J.D., Chang, J.J.: Analysis of individual differences in multidimensional scaling via an n-way generalization of "eckart-young" decomposition. Psychometrika **35**(3), 283–319 (1970)
9. Chen, M., Wang, S., Liang, P.P., Baltrušaitis, T., Zadeh, A., Morency, L.P.: Multimodal sentiment analysis with word-level fusion and reinforcement learning. In: Proceedings of ICMI (2017)
10. Cichocki, A., et al.: Tensor networks for dimensionality reduction and large-scale optimization: Part 1 low-rank tensor decompositions. Found. Trends Mach. Learn. **9**(4—-5), 249–429 (2016)
11. Dong, J., Zheng, H., Lian, L.: Low-rank laplacian-uniform mixed model for robust face recognition. In: Proceedings of CVPR (2019)
12. Duong, C.T., Lebret, R., Aberer, K.: Multimodal classification for analysing social media. arXiv preprint arXiv:1708.02099 (2017)
13. Fan, H., Chen, Y., Guo, Y., Zhang, H., Kuang, G.: Hyperspectral image restoration using low-rank tensor recovery. EEE J. Sel. Top. Appl. Earth Obs. Remote Sens. **10**(10), 4589–4604 (2017)
14. Goyal, Y., Khot, T., Summers-Stay, D., Batra, D., Parikh, D.: Making the v in vqa matter: elevating the role of image understanding in visual question answering. In: Proceedings of CVPR (2017)
15. Gu, Y., Li, X., Chen, S., Zhang, J., Marsic, I.: Speech intention classification with multimodal deep learning. In: Canadian Conference on Artificial Intelligence. pp. 260–271 (2017)
16. Guo, J., Zhou, Z., Wang, L.: Single image highlight removal with a sparse and low-rank reflection model. In: Proceedings of ECCV (2018)
17. Harshman, R.A., et al.: Foundations of the parafac procedure: Models and conditions for an" explanatory" multimodal factor analysis. UCLA Working Phonetics Paper (1970)
18. He, W., Yao, Q., Li, C., Yokoya, N., Zhao, Q.: Non-local meets global: an integrated paradigm for hyperspectral denoising. In: Proceedings of CVPR (2019)
19. Hou, M., Tang, J., Zhang, J., Kong, W., Zhao, Q.: Deep multimodal multilinear fusion with high-order polynomial pooling. In: Proceedings of NeurIPS (2019)
20. Jia, Y., et al.: Caffe: Convolutional architecture for fast feature embedding. In: Proceedings of MM (2014)
21. Kolda, T.G., Bader, B.W.: Tensor decompositions and applications. SIAM review **51**(3), 455–500 (2009)
22. Li, C., He, W., Yuan, L., Sun, Z., Zhao, Q.: Guaranteed matrix completion under multiple linear transformations. In: Proceedings of CVPR (2019)
23. Liang, P.P., et al.: Learning representations from imperfect time series data via tensor rank regularization. arXiv preprint arXiv:1907.01011 (2019)
24. Liu, B., Zhang, L.: A Survey of Opinion Mining and Sentiment Analysis. In: Aggarwal, C., Zhai, C., (eds.) Mining Text Data. Springer, Boston, MA (2012) https://doi.org/10.1007/978-1-4614-3223-4_13

25. Liu, H., Lin, M., Zhang, S., Wu, Y., Huang, F., Ji, R.: Dense auto-encoder hashing for robust cross-modality retrieval. In: Proceedings of MM (2018)
26. Liu, Z., Shen, Y., Lakshminarasimhan, V.B., Liang, P.P., Zadeh, A., Morency, L.P.: Efficient low-rank multimodal fusion with modality-specific factors. arXiv preprint arXiv:1806.00064 (2018)
27. Lu, C., Peng, X., Wei, Y.: Low-rank tensor completion with a new tensor nuclear norm induced by invertible linear transforms. In: Proceedings of CVPR (2019)
28. Mai, S., Hu, H., Xing, S.: Divide, conquer and combine: hierarchical feature fusion network with local and global perspectives for multimodal affective computing. In: Proceedings of ACL (2019)
29. Miech, A., Laptev, I., Sivic, J.: Learning a text-video embedding from incomplete and heterogeneous data. arXiv preprint arXiv:1804.02516 (2018)
30. Mikolov, T., Chen, K., Corrado, G., Dean, J.: Efficient estimation of word representations in vector space. arXiv preprint arXiv:1301.3781 (2013)
31. Morency, L.P., Mihalcea, R., Doshi, P.: Towards multimodal sentiment analysis: Harvesting opinions from the web. In: Proceedings of ICMI (2011)
32. Nimishakavi, M., Jawanpuria, P.K., Mishra, B.: A dual framework for low-rank tensor completion. In: Proceedings of NeurIPS (2018)
33. Pan, Y., et al.: Compressing recurrent neural networks with tensor ring for action recognition. In: Proceedings of AAAI (2019)
34. Pang, B., Lee, L., et al.: Opinion mining and sentiment analysis. Found. Trends Inf. Ret. **2**(1—-21), 1–135 (2008)
35. Pham, H., Liang, P.P., Manzini, T., Morency, L.P., Póczos, B.: Found in translation: learning robust joint representations by cyclic translations between modalities. In: Proceedings of AAAI (2019)
36. Poria, S., Cambria, E., Winterstein, G., Huang, G.B.: Sentic patterns: dependency-based rules for concept-level sentiment analysis. Know.-Based Syst. **69**, 45–63 (2014)
37. Recht, B., Fazel, M., Parrilo, P.A.: Guaranteed minimum-rank solutions of linear matrix equations via nuclear norm minimization. SIAM review **52**(3), 471–501 (2010)
38. Srebro, N., Shraibman, A.: Rank, trace-norm and max-norm. In: Proceedings of COLT (2005)
39. Taboada, M., Brooke, J., Tofiloski, M., Voll, K., Stede, M.: Lexicon-based methods for sentiment analysis. CL **37**(2), 267–307 (2011)
40. Tran, L., Liu, X., Zhou, J., Jin, R.: Missing modalities imputation via cascaded residual autoencoder. In: Proceedings of CVPR (2017)
41. Tucker, L.R.: Some mathematical notes on three-mode factor analysis. Psychometrika **31**(3), 279–311 (1966)
42. Wang, A., Li, C., Jin, Z., Zhao, Q.: Robust tensor decomposition via orientation invariant tubal nuclear norms. In: Proceedings of AAAI (2020)
43. Wang, H., Meghawat, A., Morency, L.P., Xing, E.P.: Select-additive learning: improving generalization in multimodal sentiment analysis. In: Proceedings of ICME (2017)
44. Wang, Y., Shen, Y., Liu, Z., Liang, P.P., Zadeh, A., Morency, L.P.: Words can shift: dynamically adjusting word representations using nonverbal behaviors. In: Proceedings of AAAI (2019)
45. Wöllmer, M., Weninger, F., Knaup, T., Schuller, B., Sun, C., Sagae, K., Morency, L.P.: Youtube movie reviews: sentiment analysis in an audio-visual context. IEEE Intell. Syst. **28**(3), 46–53 (2013)

46. Yang, Y., Krompass, D., Tresp, V.: Tensor-train recurrent neural networks for video classification. In: Proceedings ICML (2017)
47. Yu, K., Zhu, S., Lafferty, J., Gong, Y.: Fast nonparametric matrix factorization for large-scale collaborative filtering. In: Proceedings of SIGIR (2009)
48. Yuan, L., Li, C., Mandic, D., Cao, J., Zhao, Q.: Tensor ring decomposition with rank minimization on latent space: an efficient approach for tensor completion. In: Proceedings of AAAI (2019)
49. Zadeh, A., Chen, M., Poria, S., Cambria, E., Morency, L.P.: Tensor fusion network for multimodal sentiment analysis. arXiv preprint arXiv:1707.07250 (2017)
50. Zadeh, A., Liang, P.P., Mazumder, N., Poria, S., Cambria, E., Morency, L.P.: Memory fusion network for multi-view sequential learning. In: Proceedings of AAAI (2018)
51. Zadeh, A., Liang, P.P., Poria, S., Vij, P., Cambria, E., Morency, L.P.: Multi-attention recurrent network for human communication comprehension. In: Proc. AAAI (2018)
52. Zadeh, A., Zellers, R., Pincus, E., Morency, L.P.: Mosi: multimodal corpus of sentiment intensity and subjectivity analysis in online opinion videos. arXiv preprint arXiv:1606.06259 (2016)
53. Zadeh, A., Zellers, R., Pincus, E., Morency, L.P.: Multimodal sentiment intensity analysis in videos: facial gestures and verbal messages. IEEE Intell. Syst. **31**(6), 82–88 (2016)
54. Zadeh, A.B., Liang, P.P., Poria, S., Cambria, E., Morency, L.P.: Multimodal language analysis in the wild: cmu-mosei dataset and interpretable dynamic fusion graph. In: Proceedings of ACL (2018)

ProxyNCA++: Revisiting and Revitalizing Proxy Neighborhood Component Analysis

Eu Wern Teh[1,2](✉), Terrance DeVries[1,2], and Graham W. Taylor[1,2]

[1] University of Guelph, Guelph, ON, Canada
{eteh,terrance,gwtaylor}@uoguelph.ca
[2] Vector Institute, Toronto, ON, Canada

Abstract. We consider the problem of distance metric learning (DML), where the task is to learn an effective similarity measure between images. We revisit ProxyNCA and incorporate several enhancements. We find that low temperature scaling is a performance-critical component and explain why it works. Besides, we also discover that Global Max Pooling works better in general when compared to Global Average Pooling. Additionally, our proposed fast moving proxies also addresses small gradient issue of proxies, and this component synergizes well with low temperature scaling and Global Max Pooling. Our enhanced model, called ProxyNCA++, achieves a 22.9% point average improvement of Recall@1 across four different zero-shot retrieval datasets compared to the original ProxyNCA algorithm. Furthermore, we achieve state-of-the-art results on the CUB200, Cars196, Sop, and InShop datasets, achieving Recall@1 scores of 72.2, 90.1, 81.4, and 90.9, respectively.

Keywords: Metric learning · Zero-shot learning · Image retrieval

1 Introduction

Distance Metric Learning (DML) is the task of learning effective similarity measures between examples. It is often applied to images, and has found numerous applications such as visual products retrieval [1,16,23], person re-identification [30,37], face recognition [22], few-shot learning [14,28], and clustering [11]. In this paper, we focus on DML's application on zero-shot image retrieval [13,17,23,33], where the task is to retrieve images from previously unseen classes.

Proxy-Neighborhood Component Analysis (ProxyNCA) [17] is a proxy-based DML solution that consists of updatable proxies, which are used to represent class distribution. It allows samples to be compared with these proxies instead of one another to reduce computation. After the introduction of ProxyNCA, there

Electronic supplementary material The online version of this chapter (https://doi.org/10.1007/978-3-030-58586-0_27) contains supplementary material, which is available to authorized users.

A. Vedaldi et al. (Eds.): ECCV 2020, LNCS 12369, pp. 448–464, 2020.
https://doi.org/10.1007/978-3-030-58586-0_27

are very few works that extend ProxyNCA [21,35], making it less competitive when compared with recent DML solutions [13,32,36].

Our contributions are the following: First, we point out the difference between NCA and ProxyNCA, and propose to use proxy assignment probability which aligns ProxyNCA with NCA [7]. Second, we explain why low temperature scaling works and show that it is a performance-critical component of ProxyNCA. Third, we explore different global pooling strategies and find out that Global Max Pooling (GMP) outperforms the commonly used Global Average Pooling (GAP), both for ProxyNCA and other methods. Fourth, we suggest using faster moving proxies that compliment well with both GMP and low temperature scaling, which also address the small gradient issue due to L^2-Normalization of proxies. Our enhanced ProxyNCA, which we called ProxyNCA++, has a 22.9 percentage points of improvement over ProxyNCA on average for Recall@1 across four different zero-shot retrieval benchmarks (performance gains are highlighted in Fig. 1). In addition, we also achieve state-of-the-art performance on all four benchmark datasets across all categories.

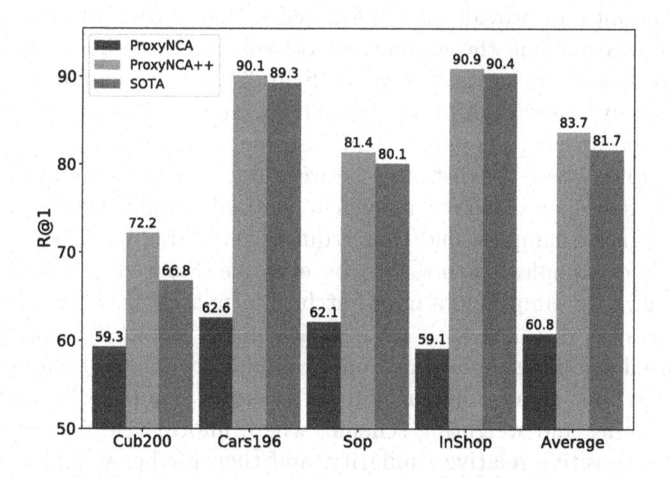

Fig. 1. A summary of the average performance on Recall@1 for all datasets. With our proposed enhancements, we improve upon the original ProxyNCA by 22.9pp, and outperform current state-of-the-art models by 2.0pp on average.

2 Related Work

The core idea of Distance Metric Learning (DML) is to learn an embedding space where similar examples are attracted, and dissimilar examples are repelled. To restrict the scope, we limit our review to methods that consider image data. There is a large body of work in DML, and it can be traced back to the 90s, where Bromley et al. [2] designed a Siamese neural network to verify signatures.

Later, DML was used in facial recognition, and dimensionality reduction in the form of a contrastive loss [4,9], where pairs of similar and dissimilar images are selected, and the distance between similar pairs of images is minimized while the distance between dissimilar images is maximized.

Like contrastive loss, which deals with the actual distance between two images, triplet loss optimizes the relative distance between a positive pair (an anchor image and an image similar to the anchor image) and a negative pair (an anchor image and an image dissimilar to the anchor image) [3]. In addition to contrastive and triplet loss, there is a long line of work which proposes new loss functions, such as angular loss [31], histogram loss [26], margin-based loss [33], and hierarchical triplet loss [6]. Wang et al. [32] categorize this group as paired-based DML.

One weakness of paired-based methods is the sampling process. First, the number of possible pairs grows polynomially with the number of data points, which increases the difficulty of finding an optimal solution. Second, if a pair or triplet of images is sampled randomly, the average distance between two samples is approximately $\sqrt{2}$-away [33]. In other words, a randomly sampled image is highly redundant and provides less information than a carefully chosen one.

In order to overcome the weaknesses of pair-based methods, several works have been proposed in the last few years. Schroff et al. [22] explore a curriculum learning strategy where examples are selected based on the distances of samples to the anchored images. They use a semi-hard negative mining strategy to select negative samples where the distances between negative pairs are at least greater than the positive pairs. However, such a method usually generates very few semi-hard negative samples, and thus requires very large batches (on the order of thousands of samples) in order to be effective. Song et al. [23] propose to utilize all pair-wise samples in a mini-batch to form triplets, where each positive pair compares its distance with all negative pairs. Wu et al. [33] proposed a distance-based sampling strategy, where examples are sampled based on inverse n-dimensional unit sphere distances from anchored samples. Wang et al. [32] propose a mining and weighting scheme, where informative pairs are sampled by measuring positive relative similarity, and then further weighted using self-similarity and negative relative similarity.

Apart from methods dedicated to addressing the weakness of pair-based DML methods, there is another line of work that tackles DML via class distribution estimation. The motivation for this camp of thought is to compare samples to proxies, and in doing so, reduce computation. One method that falls under this line of work is the Magnet Loss [19], in which samples are associated with a cluster centroid, and at each training batch, samples are attracted to cluster centroids of similar classes and repelled by cluster centroids of different classes. Another method in this camp is ProxyNCA [17], where proxies are stored in memory as learnable parameters. During training, each sample is pushed towards its proxy while repelling against all other proxies of different classes. ProxyNCA is discussed in greater detail in Sect. 3.2.

Similar to ProxyNCA, Zhai et al. [36] design a proxy-based solution that emphasizes on the Cosine distance rather than the Euclidean squared distance. They also use layer norm in their model to improve robustness against poor weight initialization of new parameters and introduces class balanced sampling during training, which improves their retrieval performance. In our work, we also use these enhancements in our architecture.

Recently, a few works in DML have explored ensemble techniques. Opitz et al. [18] train an ensemble DML by reweighting examples using online gradient boosting. The downside of this technique is that it is a sequential process. Xuan et al. [35] address this issue by proposing an ensemble technique where ensemble models are trained separately on randomly combined classes. Sanakoyeu et al. [21] propose a unique divide-and-conquer strategy where the data is divided periodically via clustering based on current combined embedding during training. Each cluster is assigned to a consecutive chunk of the embedding, called learners, and they are randomly updated during training. Apart from ensemble techniques, there is recent work that attempts to improve DML in general. Jacob et al. [13] discover that DML approaches that rely on Global Average Pooling (GAP) potentially suffer from the scattering problem, where features learned with GAP are sensitive to outlier. To tackle this problem, they propose HORDE, which is a high order regularizer for deep embeddings that computes higher-order moments of features.

3 Methods

In this section, we revisit NCA and ProxyNCA and discuss six enhancements that improve the retrieval performance of ProxyNCA. The enhanced version, which we call ProxyNCA++, is shown in Fig. 2.

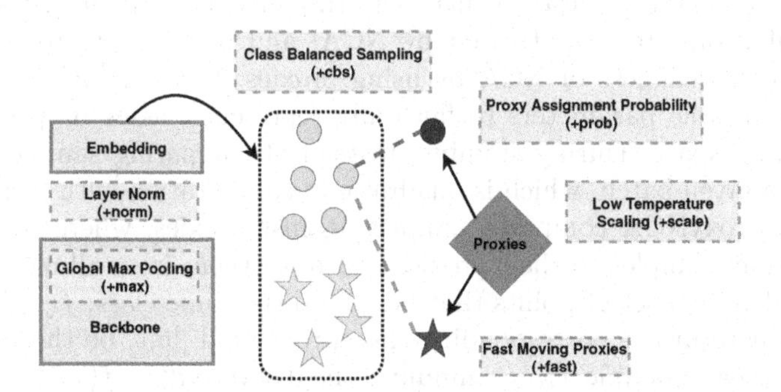

Fig. 2. We show an overview of our architecture, ProxyNCA++, which consists of the original building blocks of ProxyNCA and six enhancements, which are shown in the dashed boxes. ProxyNCA consists of a pre-trained backbone model, a randomly initialized embedding layer, and randomly initialized proxies. The six enhancements in ProxyNCA++ are proxy assignment probability (+prob), low temperature scaling (+scale), class balanced sampling (+cbs), layer norm (+norm), global max pooling (+max) and fast-moving proxies (+fast).

3.1 Neighborhood Component Analysis (NCA)

Neighborhood Component Analysis (NCA) is a DML algorithm that learns a Mahalanobis distance for k-nearest neighbors (KNN). Given two points, x_i and x_j, Goldberg et al. [7] define p_{ij} as the assignment probability of x_i to x_j:

$$p_{ij} = \frac{-d(x_i, x_j)}{\sum_{k \notin i} -d(x_i, x_k)} \tag{1}$$

where $d(x_i, x_k)$ is Euclidean squared distance computed on some learned embedding. In the original work, it was parameterized as a linear mapping, but nowadays, the method is often used with nonlinear mappings such as feedforward or convolutional neural networks. Informally, p_{ij} is the probability that points i and j are said to be "neighbors".

The goal of NCA is to maximize the probability that points assigned to the same class are neighbors, which, by normalization, minimizes the probability that points in different classes are neighbors:

$$L_{\text{NCA}} = -\log\left(\frac{\sum_{j \in C_i} \exp(-d(x_i, x_j))}{\sum_{k \notin C_i} \exp(-d(x_i, x_k))}\right). \tag{2}$$

Unfortunately, the computation of NCA loss grows polynomially with the number of samples in the dataset. To speed up computation, Goldberg et al. use random sampling and optimize the NCA loss with respect to the small batches of samples.

3.2 ProxyNCA

ProxyNCA is a DML method which performs metric learning in the space of class distributions. It is motivated by NCA, and it attempts to address the computational weakness of NCA by using proxies. In ProxyNCA, *proxies* are stored as learnable parameters to faithfully represent classes by prototypes in an embedding space. During training, instead of comparing samples with one another in a given batch, which is quadratic in computation with respect to the batch size, ProxyNCA compares samples against proxies, where the objective aims to attract samples to their proxies and repel them from all other proxies.

Let C_i denote a set of points that belong to the same class, $f(a)$ be a proxy function that returns a corresponding class proxy, and $||a||_2$ be the L^2-Norm of vector a. For each sample x_i, we minimize the distance $d(x_i, f(x_i))$ between the sample, x_i and its own proxy, $f(x_i)$ and maximize the distance $d(x_i, f(z))$ of that sample with respect to all other proxies Z, where $f(z) \in Z$ and $z \notin C_i$.

$$L_{\text{ProxyNCA}} = -\log\left(\frac{\exp\left(-d(\frac{x_i}{||x_i||_2}, \frac{f(x_i)}{||f(x_i)||_2})\right)}{\sum_{f(z) \in Z} \exp\left(-d(\frac{x_i}{||x_i||_2}, \frac{f(z)}{||f(z)||_2})\right)}\right). \tag{3}$$

3.3 Aligning with NCA by Optimizing Proxy Assignment Probability

Using the same motivation as NCA (Equation 1), we propose to optimize the proxy assignment probability, P_i. Let A denote the set of all proxies. For each x_i, we aim to maximize P_i.

$$P_i = \frac{\exp\left(-d(\frac{x_i}{||x_i||_2}, \frac{f(x_i)}{||f(x_i)||_2})\right)}{\sum_{f(a)\in A}\exp\left(-d(\frac{x_i}{||x_i||_2}, \frac{f(a)}{||f(a)||_2})\right)} \quad (4)$$

$$L_{\text{ProxyNCA++}} = -\log(P_i) \quad (5)$$

Since P_i is a probability score that must sum to one, maximizing P_i for a proxy also means there is less chance for x_i to be assigned to other proxies. In addition, maximizing P_i also preserves the original ProxyNCA properties where x_i is attracted toward its own proxy $f(x_i)$ while repelling proxies of other classes, Z. It is important to note that in ProxyNCA, we maximize the distant ratio between $-d(x_i, y_j)$ and $\sum_{f(z)\in Z} -d(x_i, f(z))$, while in ProxyNCA++, we maximize the proxy assignment probability, P_i, a subtle but important distinction. Table 8 shows the effect of proxy assignment probability to ProxyNCA and its enhancements.

3.4 About Temperature Scaling

Temperature scaling is introduced in [12], where Hinton et al. use a high temperature ($T > 1$) to create a softer probability distribution over classes for knowledge distillation purposes. Given a logit y_i and a temperature variable T, a temperature scaling is defined as $q_i = \frac{\exp(y_i/T)}{\sum_j \exp(y_j/T)}$. By incorporating temperature scaling to the loss function of ProxyNCA++ in Eq. 4, the new loss function has the following form:

$$L_{\text{ProxyNCA++}} = -\log\left(\frac{\exp\left(-d(\frac{x_i}{||x_i||_2}, \frac{f(x_i)}{||f(x_i)||_2}) * \frac{1}{T}\right)}{\sum_{f(a)\in A}\exp\left(-d(\frac{x_i}{||x_i||_2}, \frac{f(a)}{||f(a)||_2}) * \frac{1}{T}\right)}\right) \quad (6)$$

When $T = 1$, we have a regular Softmax function. As T gets larger, the output of the softmax function will approach a uniform distribution. On the other hand, as T gets smaller, it leads to a peakier probability distribution. Low temperature scaling ($T < 1$) is used in [34] and [36]. In this work, we attempt to explain why low-temperature scaling works by visualizing its effect on synthetic data. In Fig. 3, as T gets smaller, the decision boundary is getting more refined and can classify the samples better. In other words, as T becomes smaller, the model can overfit to the problem better and hence generating better decision boundaries.

In Fig. 4 (a), we show a plot of R@1 score with respect to temperature scale on the CUB200 dataset. The highest test average R@1 happens at $T = \frac{1}{9}$. Lowering T beyond this point will allow the model to overfit more to the training set and to make it less generalizable. Hence, we see a drop in test performance. Table 9 shows the effect of low temperature scaling to ProxyNCA and its enhancements.

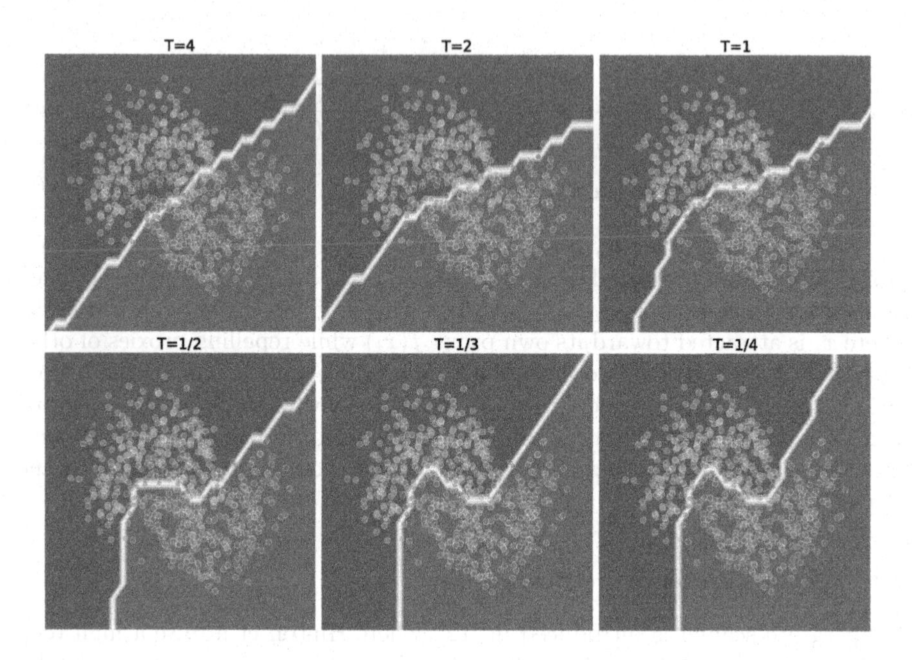

Fig. 3. The effect of temperature scaling on the decision boundary of a Softmax Classifier trained on the two moons synthetic dataset

3.5 About Global Pooling

In DML, the de facto global pooling operation used by the community is Global Average Pooling (GAP). In this paper, we investigate the effect of global pooling of spatial features on zero-shot image retrieval. We propose the use of Global K-Max Pooling [5] to interpolate between GAP and Global Max Pooling (GMP). Given a convolution feature map of $M \times M$ dimension with E channels, $g \in \mathbb{R}^{M \times M \times E}$ and a binary variable, $h_i \in \{0, 1\}$, Global K-Max Pooling is defined as:

$$\text{Global } k\text{-Max}(g_\epsilon) = \max_h \frac{1}{k} \sum_{i=1}^{M^2} h_i \cdot g_\epsilon \text{ , s.t.} \sum_{i=1}^{M^2} h_i = k, \forall \epsilon \in E \tag{7}$$

When $k = 1$, we have GMP, and when $k = M^2$, we have GAP. Figure 4 (b) is a plot of Recall@1 with different k value of Global K-Max Pooling on the CUB200 dataset. There is a negative correlation of 0.98 between k and

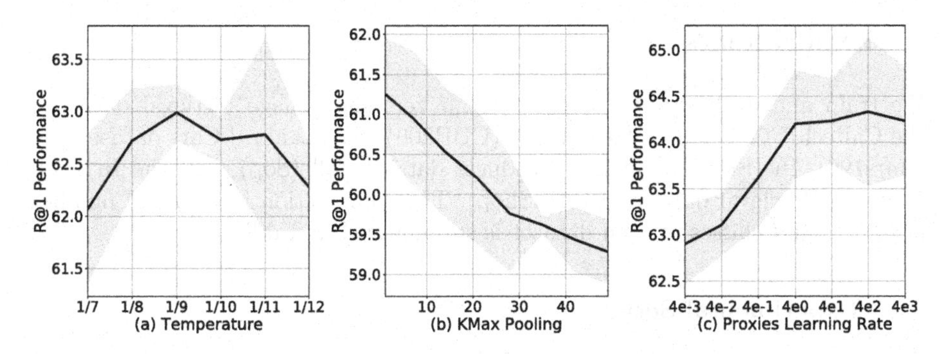

Fig. 4. We show three plots of R@1 with different (a) temperature scales, (b) k values for K-Max Pooling and (c) proxy learning rates on on CUB200 [29]. The shaded areas represent one standard deviation of uncertainty.

Recall@1 performance, which shows that a lower k value results in better retrieval performance.

3.6 About Fast Moving Proxies

In ProxyNCA, the proxies, the embedding layer, and the backbone model all share the same learning rate. We hypothesize that the proxies should be moving faster than the embedding space in order to represent the class distribution better. However, in our experiments, we discovered that the gradient of proxies is smaller than the gradient of the embedding layer and backbone model by three orders of magnitude, and this is caused by the L^2-Normalization of proxies. To mitigate this problem, we use a higher learning rate for the proxies.

From our ablation studies in Table 9, we observe that fast moving proxies synergize better with low temperature scaling and Global Max Pooling. We can see a 1.4pp boost in R@1 if we combine fast proxies and low temperature scaling. There is also a 2.1pp boost in the retrieval performance if we combine fast proxies, low temperature scaling, and Global Max Pooling. Figure 4 (c) is a plot of Recall@1 with different proxy learning rates on CUB200.

3.7 Layer Norm (Norm) and Class Balanced Sampling (CBS)

The use of layer normalization [27] without affine parameters is explored by Zhai et al. [36]. Based on our experiments, we also find that this enhancement helps to boost performance. Besides, we also use a class balanced sampling strategy in our experiments, where we have more than one instance per class in each training batch. To be specific, for every batch of size N_b, we only sample N_c classes from which we then randomly select $\lfloor N_b/N_c \rfloor$ examples. This sampling strategy commonly appears in pair-based DML approaches [23,32,33] as a baseline and Zhai et al. is the first paper that uses it in a proxy-based DML method.

4 Experiments

We train and evaluate our model on four zero-shot image retrieval datasets: the Caltech-UCSD Birds dataset [29] (CUB200), the Stanford Cars dataset [15] (Cars196), the Stanford Online Products dataset [23] (Sop), and the In Shop Clothing Retrieval dataset [16] (InShop). The composition in terms of number of images and classes of each dataset is summarized in Table 1.

4.1 Experimental Setup

For each dataset, we use the first half of the original training set as our training set and the second half of the original training set as our validation set. In all of our experiments, we use a two-stage training process. We first train our models on the training set and then use the validation set to perform hyper-parameter tuning (e.g., selecting the best epoch for early stopping, learning rate, etc.). Next, we train our models with the fine-tuned hyper-parameters on the combined training and validation sets (i.e., the complete original training set).

Table 1. We show the composition of all four zero-shot image retrieval datasets considered in this work. In addition, we also report the learning rates, the batch size, and cbs (class balanced sampling) instances for each dataset during training. The number of classes for the Sop and InShop datasets is large when compared to CUB200 and Cars196 dataset. However, the number of instances per class is very low for the Sop and InShop datasets. In general, ProxyNCA does not require a large batch size when compared to pairs-based DML methods. To illustrate this, we also show the batch sizes used in [32], which is current state-of-the-art among pairs-based methods. Their technique requires a batch size which is several times larger than ProxyNCA++.

	Images	Classes	Avg	Batch size (ours)	Batch size (MS [32])	Base lr	Proxy lr	cbs
CUB200	11,788	200	58	32	80	4e–3	4e2	4
Cars196	16,185	196	82	32	–	4e–3	4e2	4
Sop	120,053	22,634	5	192	1000	2.4e–2	2.4e2	3
InShop	52,712	11,967	4	192	-	2.4e–2	2.4e3	3

We use the same learning rate for both stages of training. We also set the number of proxies to be the same as the number of classes in the training set. For our experiments with fast proxies, we use a different learning rate for proxies (see Table 1 for details). We also use a temperature value of $\frac{1}{9}$ across all datasets.

In the first stage of training, we use the "reduce on loss plateau decay" annealing [8] to control the learning rate of our model based on the recall performance (R@1) on the validation set. We set the patience value to four epochs in our experiments. We record the epochs where the learning rate is reduced and also save the best epochs for early stopping on the second stage of training.

In all of our experiments, we leverage the commonly used ImageNet [20] pre-trained Resnet50 [10] model as our backbone (see Table 2 for commonly used backbone architectures). Features are extracted after the final convolutional block of the model and are reduced to a spatial dimension of 1×1 using a global pooling operation. This procedure results in a 2048 dimensional vector, which is fed into a final embedding layer. In addition, we also experiment with various embedding sizes. We observe a gain in performance as we increase the size of the embedding. It is important to note that not all DML techniques yield better performance as embedding size increases. For some techniques such as [23,32], a larger embedding size hurts performance.

Table 2. Commonly used backbone architectures for zero-shot image retrieval, with associated ImageNet Top-1 Error % for each architecture

Architecture	Abbreviation	Top-1 Error (%)
Resnet18 [10]	R18	30.24
GoogleNet [25]	I1	30.22
Resnet50 [10]	R50	23.85
InceptionV3 [24]	I3	22.55

During training, we scale the original images to a random aspect ratio (0.75 to 1.33) before applying a crop of random size (0.08 to 1.0 of the scaled image). After cropping, we resize the images to 256×256. We also perform random horizontal flipping for additional augmentation. During testing, we resize the images to 288×288 and perform a center crop of size 256×256.

4.2 Evaluation

We evaluate retrieval performance based on two evaluation metrics: (a) Recall@K (R@K) and (b) Normalized Mutual Information, $\text{NMI}(\Omega, \mathbb{C}) = \frac{2 * I(\Omega, \mathbb{C})}{H(\Omega) + H(\mathbb{C})}$, where Ω represents ground truth label, $\mathbb{C}$ represents the set of clusters computed by K-means, I stands for mutual information and H stands for entropy. The purpose of NMI is to measure the purity of the cluster on unseen data.

Using the same evaluation protocols detailed in [13,16,17,32], we evaluate our model using unseen classes on four datasets. The InShop dataset [16] is slightly different than all three other datasets. There are three groupings of data: training set, query set, and gallery set. The query and gallery set have the same classes, and these classes do not overlap with the training set. Evaluation is done based on retrieval performance on the gallery set.

Tables 3, 4, 5, and 6 show the results of our experiments[1]. For each dataset, we report the results of our method, averaged over five runs. We also report the

[1] For additional experiments on different crop sizes, please refer to the corresponding supplementary materials

standard deviation of our results to account for uncertainty. Additionally, we also show the results of ProxyNCA++ trained with smaller embedding sizes (512, 1024). Our ProxyNCA++ model outperforms ProxyNCA and all other state-of-the-art methods in all categories across all four datasets. Note, our model trained with a 512-dimensional embedding also outperform all other methods in the same embedding space except for The InShop dataset [16], where we tie in the R@1 category.

Table 3. Recall@k for k = 1,2,4,8 and NMI on CUB200-2011 [29]

R@k	1	2	4	8	NMI	Arch	Emb
ProxyNCA [17]	49.2	61.9	67.9	72.4	59.5	I1	128
Margin [33]	63.6	74.4	83.1	90.0	69.0	R50	128
MS [32]	65.7	77.0	86.3	91.2	-	I3	512
HORDE [13]	66.8	77.4	85.1	91.0	-	I3	512
NormSoftMax [36]	61.3	73.9	83.5	90.0	-	R50	512
NormSoftMax [36]	65.3	76.7	85.4	91.8	-	R50	2048
ProxyNCA	59.3 ± 0.4	71.2 ± 0.3	80.7 ± 0.2	88.1 ±0.3	63.3 ± 0.5	R50	2048
ProxyNCA++	69.0 ± 0.8	79.8 ± 0.7	87.3 ± 0.7	92.7 ± 0.4	73.9 ± 0.5	R50	512
ProxyNCA++	70.2 ± 1.6	80.7 ± 1.4	88.0 ± 0.9	93.0 ± 0.4	74.2 ± 1.0	R50	1024
ProxyNCA++ (–max, –fast)	69.1 ± 0.5	79.6 ± 0.4	87.3 ± 0.3	92.7 ± 0.2	73.3 ± 0.7	R50	2048
ProxyNCA++	**72.2 ± 0.8**	**82.0 ± 0.6**	**89.2 ± 0.6**	**93.5 ± 0.4**	**75.8± 0.8**	R50	2048

Table 4. Recall@k for k = 1,2,4,8 and NMI on CARS196 [15]

R@k	1	2	4	8	NMI	Arch	Emb
ProxyNCA [17]	73.2	82.4	86.4	88.7	64.9	I1	128
Margin [33]	79.6	86.5	91.9	95.1	69.1	R50	128
MS [32]	84.1	90.4	94.0	96.1	-	I3	512
HORDE [13]	86.2	91.9	95.1	97.2	–	I3	512
NormSoftMax [36]	84.2	90.4	94.4	96.9	–	R50	512
NormSoftMax [36]	89.3	94.1	96.4	98.0	–	R50	2048
ProxyNCA	62.6 ± 9.1	73.6 ± 8.6	82.2 ± 6.9	88.9 ±4.8	53.8 ± 7.0	R50	2048
ProxyNCA++	86.5 ± 0.4	92.5 ± 0.3	95.7 ± 0.2	97.7 ± 0.1	73.8 ± 1.0	R50	512
ProxyNCA++	87.6 ± 0.3	93.1 ± 0.1	96.1 ± 0.2	97.9 ± 0.1	75.7 ± 0.3	R50	1024
ProxyNCA++ (-max, -fast)	87.9 ± 0.2	93.2 ± 0.2	96.1 ± 0.2	97.9 ± 0.1	76.0 ± 0.5	R50	2048
ProxyNCA++	**90.1±0.2**	**94.5±0.2**	**97.0±0.2**	**98.4±0.1**	**76.6±0.7**	R50	2048

4.3 Ablation Study

In Table 7, we perform an ablation study on the performance of our proposed methods using the CUB200 dataset. The removal of the low temperature scaling component gives the most significant drop in R@1 performance (-10.8pt).

Table 5. Recall@k for k = 1,10,100,1000 and NMI on Stanford Online Products [23].

R@k	1	10	100	1000	Arch	Emb
ProxyNCA [17]	73.7	-	-	-	I1	128
Margin [33]	72.7	86.2	93.8	98.0	R50	128
MS [32]	78.2	90.5	96.0	98.7	I3	512
HORDE [13]	80.1	91.3	96.2	98.7	I3	512
NormSoftMax [36]	78.2	90.6	96.2	-	R50	512
NormSoftMax [36]	79.5	91.5	96.7	-	R50	2048
ProxyNCA	62.1 ± 0.4	76.2 ± 0.4	86.4 ± 0.2	93.6 ± 0.3	R50	2048
ProxyNCA++	80.7 ± 0.5	92.0 ± 0.3	96.7 ± 0.1	98.9 ± 0.0	R50	512
ProxyNCA++	80.7 ± 0.4	92.0 ± 0.2	96.7 ± 0.1	98.9 ± 0.0	R50	1024
ProxyNCA++(-max, -fast)	72.1 ± 0.2	85.4 ± 0.1	93.0 ± 0.1	96.7 ± 0.2	R50	2048
ProxyNCA++	**81.4 ± 0.1**	**92.4 ± 0.1**	**96.9 ± 0.0**	**99.0 ± 0.0**	R50	2048

Table 6. Recall@k for k = 1,10,20,30,40 on the In-Shop Clothing Retrieval dataset [23]

R@k	1	10	20	30	40	Arch	Emb
MS [32]	89.7	97.9	98.5	98.8	99.1	I3	512
HORDE [13]	90.4	97.8	98.4	98.7	98.9	I3	512
NormSoftMax [36]	88.6	97.5	98.4	98.8	–	R50	512
NormSoftMax [36]	89.4	97.8	98.7	99.0	–	R50	2048
ProxyNCA	59.1±0.7	80.6±0.6	84.7±0.3	86.7±0.4	88.1±0.5	R50	2048
ProxyNCA++	90.4 ± 0.2	98.1 ± 0.1	98.8 ± 0.0	99.0 ±0.1	99.2±0.0	R50	512
ProxyNCA++	90.4 ± 0.4	98.1 ± 0.1	98.8 ± 0.1	99.1 ±0.1	99.2±0.1	R50	1024
ProxyNCA++ (-max, -fast)	82.5 ± 0.3	93.5 ± 0.1	95.4 ± 0.2	96.3 ±0.0	96.8±0.0	R50	2048
ProxyNCA++	**90.9 ± 0.3**	**98.2 ± 0.0**	**98.9 ± 0.0**	**99.1 ± 0.0**	**99.4± 0.0**	R50	2048

This is followed by Global Max Pooling (–3.2pt), Layer Normalization (–2.6pt), Class Balanced Sampling (–2.6pt), Fast proxies (–1.9pt) and Proxy Assignment Probability (–1.1pt).

We compare the effect of the Global Max Pooling (GMP) and the Global Average Pooling (GAP) on other metric learning methodologies [13,22,32,33] in Table 11 on CUB200 dataset. The performance of all other models improves when GAP is replaced with GMP, with the exception of HORDE [13]. In HORDE, Jacob et al. [13] include both the pooling features as well as the higher-order moment features in the loss calculation. We speculate that since this method is designed to reduce the effect of outliers, summing max-pooled features canceled out the effect of higher-order moment features, which may have lead to sub-optimal performance.

Table 7. An ablation study of ProxyNCA++ and its enhancements on CUB200 [29].

R@k	1	2	4	8	NMI
ProxyNCA++ (Emb: 2048)	72.2 ± 0.8	82.0 ± 0.6	89.2 ± 0.6	93.5 ± 0.4	75.8 ± 0.8
-scale	61.4 ± 0.4	72.4 ± 0.5	81.5 ± 0.3	88.4 ± 0.5	64.8 ± 0.4
-max	69.0 ± 0.6	80.3 ± 0.5	88.1 ± 0.4	93.1 ± 0.1	74.3 ± 0.4
-norm	69.6 ± 0.3	80.5 ± 0.5	88.0 ± 0.2	93. 0± 0.2	75.2 ± 0.4
-cbs	69.6 ± 0.6	80.1 ± 0.3	87.7 ± 0.3	92.8 ± 0.2	73.4 ± 0.3
-fast	70.3 ± 0.9	80.6 ± 0.4	87.7 ± .5	92.5 ± 0.3	73.5 ± 0.9
-prob	71.1 ± 0.7	81.1 ± 0.3	87.9 ± 0.3	92.6 ± 0.3	73.4 ± 0.8

Table 8. An ablation study of the effect of Proxy Assignment Probability (+prob) to ProxyNCA and its enhancements on CUB200 [29].

R@1	Without prob	With prob
ProxyNCA (Emb: 2048)	59.3 ± 0.4	59.0 ± 0.4
+scale	62.9 ± 0.4	63.4 ± 0.6
+scale +norm	65.3 ± 0.7	65.7 ± 0.8
+scale +max	65.1 ± 0.3	66.2 ± 0.3
+scale +norm +cbs	67.2 ± 0.8	69.1 ± 0.5
+scale +norm +cbs +max	68.8 ± 0.7	70.3 ± 0.9
+scale +norm +cbs +max +fast	71.1 ± 0.7	72.2 ± 0.8

Table 9. An ablation study of the effect of low temperature scaling to ProxyNCA and its enhancements on CUB200 [29]. Without low temperature scaling, three out of six enhancements (in red) get detrimental results when they are applied to ProxyNCA.

R@1	Without scale	With scale
ProxyNCA (Emb: 2048)	59.3 ± 0.4	62.9 ± 0.4
+cbs	54.8 ± 6.2	64.0 ± 0.4
+prob	59.0 ± 0.4	63.4 ± 0.6
+norm	60.2 ± 0.6	65.3 ± 0.7
+max	61.3 ± 0.7	65.1 ± 0.3
+fast	56.3 ± 0.8	64.3 ± 0.8
+max +fast	60.3 ± 0.5	67.2 ± 0.5
+norm +prob +cbs	60.4 ± 0.7	69.1 ± 0.5
+norm +prob +cbs +max	61.2 ± 0.7	70.3 ± 0.9
+norm +prob +cbs +max +fast	61.4 ± 0.4	72.2 ± 0.8

Table 10. An ablation study of ProxyNCA the effect of Global Max Pooling to ProxyNCA and its enhancements on CUB200 [29]. We can see a 2.1pp improvement on average after replacing GAP with GMP.

R@1	Global Average Pooling	Global Max Pooling
ProxyNCA (Emb: 2048)	59.3 ± 0.4	61.3 ± 0.7
+cbs	54.8 ± 6.2	55.5 ± 6.2
+prob	59.0 ± 0.4	61.2 ± 0.7
+norm	60.2 ± 0.6	60.9 ± 0.9
+scale	62.9 ± 0.4	65.1 ± 0.3
+fast	56.3 ± 0.8	60.3 ± 0.5
+scale +fast	64.3 ± 0.8	67.2 ± 0.5
+norm +prob +cbs	60.4 ± 0.7	61.2 ± 0.7
+norm +prob +cbs +fast	56.2 ± 0.9	61.4 ± 0.4
+norm +prob +cbs +scale	69.1 ± 0.5	70.3 ± 0.9
+norm +prob +cbs +scale +fast	69.0 ± 0.6	72.2 ± 0.8

Table 11. Comparing the effect of Global Max Pooling and Global Average Pooling on the CUB200 dataset for a variety of methods.

Method	Pool	R@1	Arch	Emb
WithoutTraining	avg	45.0	R50	2048
	max	**53.1**	R50	2048
Margin [33]	avg	63.3	R50	128
	max	**64.3**	R50	128
Triplet-Semihard sampling [22]	avg	60.5	R50	128
	max	**61.6**	R50	128
MS [32]	avg	64.9	R50	512
	max	**68.5**	R50	512
MS [32]	avg	65.1	I3	512
	max	**66.1**	I3	512
Horde (Contrastive Loss) [13]	avg	**65.1**	I3	512
	max	63.1	I3	512

5 Conclusion

We revisit ProxyNCA and incorporate several enhancements. We find that low temperature scaling is a performance-critical component and explain why it works. Besides, we also discover that Global Max Pooling works better in general when compared to Global Average Pooling. Additionally, our proposed fast moving proxies also addresses small gradient issue of proxies, and this component synergizes well with low temperature scaling and Global Average pooling.

The new and improved ProxyNCA, which we call ProxyNCA++, outperforms the original ProxyNCA by 22.9% points on average across four zero-shot image retrieval datasets for Recall@1. In addition, we also achieve state-of-art results on all four benchmark datasets for all categories.

References

1. Bell, S., Bala, K.: Learning visual similarity for product design with convolutional neural networks. ACM Trans. Graph. **34**(4), 1–10 (2015)
2. Bromley, J., Guyon, I., LeCun, Y., Säckinger, E., Shah, R.: Signature verification using a "siamese" time delay neural network. In: Proceedings of the 6th International Conference on Neural Information Processing Systems, NIPS 1993, pp. 737–744, San Francisco, CA, USA (1993)
3. Chechik, G., Sharma, V., Shalit, U., Bengio, S.: Large scale online learning of image similarity through ranking. J. Mach. Learn. Res. **11**, 1109–1135 (2010)
4. Chopra, S., Hadsell, R., LeCun, Y.: Learning a similarity metric discriminatively, with application to face verification. In: 2005 IEEE Computer Society Conference on Computer Vision and Pattern Recognition CVPR 2005. vol. 1, pp. 539-546 IEEE (2005)
5. Thibaut, D., Nicolas, T., Matthieu, C.: Weldon: weakly supervised learning of deep convolutional neural networks. In: Proceedings of the IEEE Conference on Computer Vision and Pattern Recognition, pp. 4743–4752 (2016)
6. Weifeng, G.: Deep metric learning with hierarchical triplet loss. In: The European Conference on Computer Vision (ECCV) (2018)
7. Jacob, G., Geoffrey, E.H., Sam, T.R., Ruslan, R.S.: Neighbourhood components analysis. In: Advances in Neural Information Processing Systems, pp. 513–520 (2005)
8. Goodfellow, I., Yoshua, B., Aaron, C.: Deep Learning. MIT Press (2016) http://www.deeplearningbook.org
9. Raia, H., Sumit, C., Yann, L.: Dimensionality reduction by learning an invariant mapping. In: 2006 IEEE Computer Society Conference on Computer Vision and Pattern Recognition (CVPR 2006), vol. 2, pp. 1735–1742. IEEE (2006)
10. Kaiming, H., Xiangyu, Z., Shaoqing, R., Jian, S.: Deep residual learning for image recognition. In: The IEEE Conference on Computer Vision and Pattern Recognition (CVPR) (2016)
11. Hershey, J. R., Chen, Z., Le Roux, J., Watanabe, S.: Deep clustering: Discriminative embeddings for segmentation and separation. In: 2016 IEEE International Conference on Acoustics, Speech and Signal Processing (ICASSP), pp. 31–35, (2016)
12. Geoffrey, H., Oriol, V., Jeff, D.: Distilling the knowledge in a neural network. arXiv preprint arXiv:1503.02531, 2015
13. Pierre, J., David, P., Histace, A., Edouard, K.: Metric learning with horde: High-order regularizer for deep embeddings. arXiv preprint arXiv:1908.02735 (2019)
14. Gregory, K.: Siamese neural networks for one-shot image recognition. In: ICML Deep Learning Workshop (2015)
15. Jonathan, K., Michael, S., Jia, D., Li, F-F.: 3d object representations for fine-grained categorization. In: 4th International IEEE Workshop on 3D Representation and Recognition (3dRR-13), Sydney, Australia (2013)

16. Ziwei, L., Ping, L., Shi, Q., Xiaogang, W., Xiaoou, T.: Deepfashion: Powering robust clothes recognition and retrieval with rich annotations. In: Proceedings of the IEEE Conference on Computer Vision and Pattern Recognition, pp. 1096–1104, (2016)
17. Yair, M.-A., Alexander, T., Thomas, K. L., Sergey, I., Saurabh, S.: No fuss distance metric learning using proxies. In: Proceedings of the IEEE International Conference on Computer Vision, pp. 360–368 (2017)
18. Michael, O., Georg, W., Horst, P., Horst, B.: Bier - boosting independent embeddings robustly. In: The IEEE International Conference on Computer Vision (ICCV) (2017)
19. Oren, R., Manohar, P., Piotr, D., Lubomir, B.: Metric learning with adaptive density discrimination. arXiv preprint arXiv:1511.05939 (2015)
20. Russakovsky, O., et al.: ImageNet large scale visual recognition challenge. Int. J. Comput. Vis. **115**(3), 211–252 (2015). https://doi.org/10.1007/s11263-015-0816-y
21. Artsiom, S., Vadim, T., Uta, B., Bjorn, O.: Divide and conquer the embedding space for metric learning. In: Proceedings of the IEEE Conference on Computer Vision and Pattern Recognition, pp. 471–480, (2019)
22. Florian, S., Dmitry, K., James, P.: Facenet: a unified embedding for face recognition and clustering. In: The IEEE Conference on Computer Vision and Pattern Recognition (CVPR) (2015)
23. Hyun, O.S., Yu, X., Stefanie, J., Silvio, S.: Deep metric learning via lifted structured feature embedding. In: IEEE Conference on Computer Vision and Pattern Recognition (CVPR) (2016)
24. Chopra, S., Hadsell, R., LeCun, Y.: Learning a similarity metric discriminatively, with application to face verification. In: 2005 IEEE Computer Society Conference on Computer Vision and Pattern Recognition (CVPR 2005). vol. 1, pp. 539-546. IEEE (2005)
25. Szegedy, C., et al.: Going deeper with convolutions. In: 2015 IEEE Conference on Computer Vision and Pattern Recognition (CVPR), pp. 1–9 (2015)
26. Evgeniya, U., Victor, L.: Learning deep embeddings with histogram loss. In: Lee, D.D., Sugiyama, M., Luxburg, U.V., Guyon, I., Garnett, R., (eds.) Advances in Neural Information Processing Systems 29, pp. 4170–4178. Curran Associates Inc (2016)
27. Ashish, V.: Attention is all you need. In: Advances in Neural Information Processing Systems, pp. 5998–6008 (2017)
28. Vinyals, O., Blundell, C., Lillicrap, T., Wierstra, D.: Matching networks for one shot learning. In: Proceedings of the 30th International Conference on Neural Information Processing Systems, NIPS 2016, pp. 3637–3645, USA, Curran Associates Inc (2016)
29. Catherine, W., Steve, B., Peter, W., Pietro, P., Serge, B.: The caltech-ucsd birds-200-2011 dataset (2011)
30. Wang, G., Yuan, Y., Chen, X., Li, J., Zhou, X.: Learning discriminative features with multiple granularities for person re-identification. In: Proceedings of the 26th ACM International Conference on Multimedia, MM 2018, pp. 274–282, New York, USA, ACM (2018)
31. Jian, W., Feng, Z., Shilei, W., Xiao, L., Yuanqing, L.: Deep metric learning with angular loss. In: The IEEE International Conference on Computer Vision (ICCV) (2017)
32. Xun, W., Xintong, H., Weilin, H., Dengke, D., Matthew, R.S.: Multi-similarity loss with general pair weighting for deep metric learning. In: Proceedings of the IEEE Conference on Computer Vision and Pattern Recognition, pp. 5022–5030 (2019)

33. Chao-Yuan, W., Manmatha, R., Alexander, J.S., Philipp, K.: Sampling matters in deep embedding learning. In: Proceedings of the IEEE International Conference on Computer Vision, pp. 2840–2848 (2017)
34. Zhirong, W., Alexei, A.E., Stella, X.Y.: Improving generalization via scalable neighborhood component analysis. In Proceedings of the European Conference on Computer Vision (ECCV), pp. 685–701 (2018)
35. Hong, X., Richard, S., Robert, P.: Deep randomized ensembles for metric learning. In: Proceedings of the European Conference on Computer Vision (ECCV), pp. 723–734 (2018)
36. Zhai, A., Wu, H.Y.: Classification is a strong baseline for deep metric learning (2019)
37. Feng, Z., et al.: Pyramidal person re-identification via multi-loss dynamic training. In: The IEEE Conference on Computer Vision and Pattern Recognition (CVPR) (2019)

Learning with Privileged Information for Efficient Image Super-Resolution

Wonkyung Lee, Junghyup Lee, Dohyung Kim, and Bumsub Ham(✉)

Yonsei University, Seoul, South Korea
bumsub.ham@yonsei.ac.kr

Abstract. Convolutional neural networks (CNNs) have allowed remarkable advances in single image super-resolution (SISR) over the last decade. Most SR methods based on CNNs have focused on achieving performance gains in terms of quality metrics, such as PSNR and SSIM, over classical approaches. They typically require a large amount of memory and computational units. FSRCNN, consisting of few numbers of convolutional layers, has shown promising results, while using an extremely small number of network parameters. We introduce in this paper a novel distillation framework, consisting of teacher and student networks, that allows to boost the performance of FSRCNN drastically. To this end, we propose to use ground-truth high-resolution (HR) images as privileged information. The encoder in the teacher learns the degradation process, subsampling of HR images, using an imitation loss. The student and the decoder in the teacher, having the same network architecture as FSRCNN, try to reconstruct HR images. Intermediate features in the decoder, affordable for the student to learn, are transferred to the student through feature distillation. Experimental results on standard benchmarks demonstrate the effectiveness and the generalization ability of our framework, which significantly boosts the performance of FSRCNN as well as other SR methods. Our code and model are available online: https://cvlab.yonsei.ac.kr/projects/PISR.

Keywords: Privileged information · Super-resolution · Distillation

1 Introduction

Single image super-resolution (SISR) aims at reconstructing a high-resolution (HR) image from a low-resolution (LR) one, which has proven useful in various tasks including object detection [3], face recognition [17,63], medical imaging [16], and information forensics [35]. With the great success of deep learning,

W. Lee, J. Lee and D. Kim—Equal contribution.

Electronic supplementary material The online version of this chapter (https://doi.org/10.1007/978-3-030-58586-0_28) contains supplementary material, which is available to authorized users.

A. Vedaldi et al. (Eds.): ECCV 2020, LNCS 12369, pp. 465–482, 2020.
https://doi.org/10.1007/978-3-030-58586-0_28

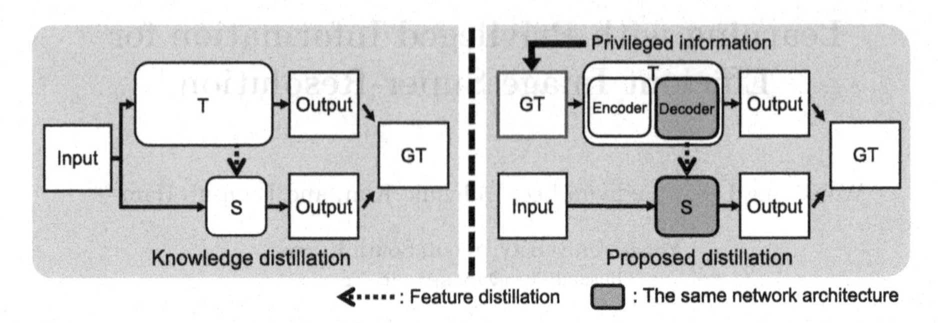

Fig. 1. Compressing networks using knowledge distillation (left) transfers the knowledge from a large teacher model (T) to a small student model (S), with the same input, *e.g.*, LR images in the case of SISR. Differently, the teacher in our framework (right) takes the ground truth (*i.e.*, HR image) as an input, exploiting it as privileged information, and transfers the knowledge via feature distillation. (Best viewed in color.) (Color figure online)

SRCNN [10] first introduces convolutional neural networks (CNNs) for SISR, outperforming classical approaches by large margins. After that, CNN-based SR methods focus on designing wider [34,49,62] or deeper [20,29,32,39,60,61] network architectures for the performance gains. They require a high computational cost and a large amount of memory, and thus implementing them directly on a single chip for, *e.g.*, televisions and mobile phones, is extremely hard without neural processing units and off-chip memory.

Many works introduce cost-effective network architectures [1,11,18,19,27,30,45] to reduce the computational burden and/or required memory, using recursive layers [30,45] or additional modules specific for SISR [1,27]. Although they offer a good compromise in terms of PSNR and speed/memory, specially-designed or recursive architectures may be difficult to implement on hardware devices. Network pruning [19] and parameter quantization [18], typically used for network compression, are alternative ways for efficient SR networks, where the pruning removes redundant connections of nodes and the quantization reduces bit-precision of weights or activations. The speedup achieved by the pruning is limited due to irregular memory accesses and poor data localizations [53], and the performance of the network quantization is inherently bound by that of a full-precision model. Knowledge distillation is another way of model compression, where a large model (*i.e.*, a teacher network) transfers a softened version of the output distribution (*i.e.*, logits) [23] or intermediate feature representations [2,14,22,43] to a small one (*i.e.*, a student network), which has shown the effectiveness in particular for the task of image classification. Generalized distillation [37] goes one step further, allowing a teacher to make use of extra (privileged) information at training time, and assisting the training process of a student network with the complementary knowledge [15,24].

We present in this paper a simple yet effective framework for an efficient SISR method. The basic idea is that ground-truth HR images can be thought of as privileged information (Fig. 1), which has not been explored in both SISR and privileged learning. It is true that the HR image includes the complementary information (*e.g.*, high-frequency components) of LR images, but current SISR methods have used it just to penalize an incorrect reconstruction at the end of CNNs. On the contrary, our approach to using HR images as privileged information allows to extract the complementary features and leverage them explicitly for the SISR task. To implement this idea, we introduce a novel distillation framework where teacher and student networks try to reconstruct HR image but using different inputs (*i.e.*, ground-truth HR and corresponding LR images for the teacher and the student, respectively), which is clearly different from the conventional knowledge distillation framework (Fig. 1). Specifically, the teacher network has an hourglass architecture consisting of an encoder and a decoder. The encoder extracts compact features from HR images while encouraging them to imitate LR counterparts using an imitation loss. The decoder, which has the same network architecture as the student, reconstructs the HR images again using the compact features. Intermediate features in the decoder are then transferred to the student via feature distillation, such that the student learns the knowledge (*e.g.*, high frequencies or fine details of HR inputs) of the teacher trained with the privileged data (*i.e.*, HR image). Note that our framework is useful in that the student can be initialized with the network parameters of the decoder, which allows to transfer the reconstruction capability of the teacher to the student. We mainly exploit FSRCNN [11] as the student network, since it has a hardware-friendly architecture (*i.e.*, a stack of convolutional layers) and the number of parameters is extremely small compared to other CNN-based SR methods. Experimental results on standard SR benchmarks demonstrate the effectiveness of our approach, which boosts the performance of FSRCNN without any additional modules. To the best of our knowledge, our framework is the first attempt to leverage the privileged information for SISR. The main contributions of our work can be summarized as follows:

- We present a novel distillation framework for SISR that leverages the ground truth (*i.e.*, HR images) as privileged information to transfer the important knowledge of the HR images to a student network.
- We propose to use an imitation loss to train a teacher network, making it possible to distill the knowledge a student is able to learn.
- We demonstrate that our approach boosts the performance of the current SISR methods, significantly, including FSRCNN [11], VDSR [29], IDN [27], and CARN [1]. We show an extensive experimental analysis with ablation studies.

2 Related Work

SISR. Early works on SISR design image priors to constrain the solution space [9,28,55], and leverage external datasets to learn the relationship between HR and LR images [6,13,44,47,56], since lots of HR images can be reconstructed from a single LR image. CNNs have allowed remarkable advances in

SISR. Dong *et al.* pioneer the idea of exploiting CNNs for SISR, and propose SRCNN [10] that learns a mapping function directly from input LR to output HR images. Recent methods using CNNs exploit a much larger number of convolutional layers. Sparse [32,34,39] or dense [20,49,62] skip connections between them prevent a gradient vanishing problem, achieving significant performance gains over classical approaches. More recently, efficient networks for SISR in terms of memory and/or runtime have been introduced. Memory-efficient SR methods [30,33,45,46] reduce the number of network parameters by reusing them recursively. They further improve the reconstruction performance using residual units [45], memory [46] or feedback [33] modules but at the cost of runtime. Runtime-efficient methods [1,11,26,27] on the other hand are computationally cheap. They use cascaded [1] or multi-branch [26,27] architectures, or exploit group convolutions [8,54]. The main drawback of such SR methods is that their hardware implementations are difficult due to the network architectures specially-designed for the SR task. FSRCNN [11] reduces both runtime and memory. It uses typical convolutional operators with a small number of filters and feature channels, except the deconvolution layer at the last part of the network. Although FSRCNN has a hardware-friendly network architecture, it is largely outperformed by current SR methods.

Feature Distillation. The purpose of knowledge distillation is to transfer the representation ability of a large model (teacher) to a small one (student) for enhancing the performance of the student model. It has been widely used to compress networks, typically for classification tasks. In this framework, the softmax outputs of a teacher are regarded as soft labels, providing informative clues beyond discrete labels [23]. Recent methods extend this idea to feature distillation, which transfers intermediate feature maps [2,43], their transformations [22,58], the differences of features before and after a stack of layers [57], or pairwise relations within feature maps [36]. In particular, the variational information distillation (VID) method [2] transfers the knowledge by maximizing the mutual information between feature maps of teacher and student networks. We exploit VID for feature distillation, but within a different framework. Instead of sharing the same inputs (*i.e.*, LR images) with the student, our teacher network inputs HR images, that contain the complementary information of LR images, to take advantage of privileged information.

Closely related to ours, SRKD [14] applies the feature distillation technique to SISR in order to compress the size of SR network, where a student is trained to have similar feature distributions to those of a teacher. Following the conventional knowledge distillation, the student and teacher networks in SRKD use the same inputs of LR images. This is clearly different from our method in that our teacher takes ground-truth HR images as inputs, allowing to extract more powerful feature representations for image reconstruction.

Generalized Distillation. Learning using privileged information [50,51] is a machine learning paradigm that uses extra information, which requires an

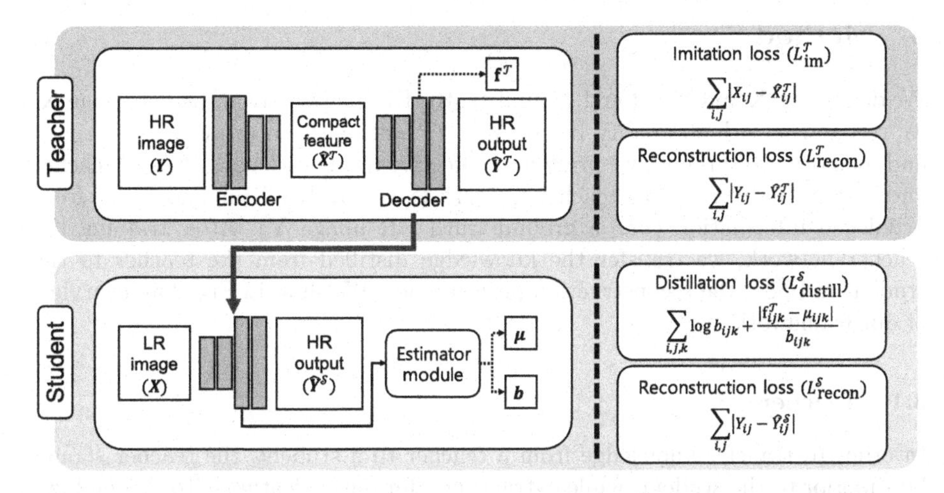

Fig. 2. Overview of our framework. A teacher network inputs a HR image $\mathbf{Y}$ and extracts a compact feature representation $\hat{\mathbf{X}}^{\mathcal{T}}$ using an encoder. The decoder in the network then reconstructs a HR output $\hat{\mathbf{Y}}^{\mathcal{T}}$. To train the teacher network, we use imitation $L^{\mathcal{T}}_{\text{im}}$ and reconstruction $L^{\mathcal{T}}_{\text{recon}}$ losses. After training the teacher, a student network is initialized with weights of the decoder in the teacher network (red line), and restores a HR output $\hat{\mathbf{Y}}^{\mathcal{S}}$ from a LR image $\mathbf{X}$. Note that the student network and the decoder share the same network architecture. The estimator module takes intermediate feature maps of the student network, and outputs location and scale maps, $\boldsymbol{\mu}$ and $\mathbf{b}$, respectively. To train the student network, we exploit a reconstruction loss $L^{\mathcal{S}}_{\text{recon}}$ together with a distillation loss $L^{\mathcal{S}}_{\text{distill}}$ using the intermediate representation $\mathbf{f}^{\mathcal{T}}$ of the teacher network and the parameter maps of $\boldsymbol{\mu}$ and $\mathbf{b}$. See text for details. (Best viewed in color.) (Color figure online)

additional cost, at training time, but with no accessibility to it at test time. In a broader context, generalized distillation [37] covers both feature distillation and learning using privileged information. The generalized distillation enables transferring the privileged knowledge of a teacher to a student. For example, the works of [15,24] adopt the generalized distillation approach for object detection and action recognition, where depth images are used as privileged information. In the framework, a teacher is trained to extract useful features from depth images. They are then transferred to a student which takes RGB images as inputs, allowing the student to learn complementary representations from privileged information. Our method belongs to generalized distillation, since we train a teacher network with ground-truth HR images, which can be viewed as privileged information, and transfer the knowledge to a student network. Different from previous methods, our method does not require an additional cost for privileged information, since the ground truth is readily available at training time.

3 Method

We denote by $\mathbf{X}$ and $\mathbf{Y}$ LR and ground-truth HR images. Given the LR image $\mathbf{X}$, we reconstruct a high-quality HR output $\hat{\mathbf{Y}}^{\mathcal{S}}$ efficiently in terms of both speed and memory. To this end, we present an effective framework consisting of teacher and student networks. The teacher network learns to distill the knowledge from privileged information (*i.e.*, a ground-truth HR image $\mathbf{Y}$). After training the teacher network, we transfer the knowledge distilled from the teacher to the student to boost the reconstruction performance. We show in Fig. 2 an overview of our framework.

3.1 Teacher

In order to transfer knowledge from a teacher to a student, the teacher should be superior to the student, while extracting informative features. To this end, we treat ground-truth HR images as privileged information, and exploit an *intelligent teacher* [50]. As will be seen in our experiments, the network architecture of the teacher influences the SR performance significantly. As the teacher network inputs ground-truth HR images, it may not be able to extract useful features, and just learn to copy the inputs for the reconstruction of HR images, regardless of its capacity. Moreover, a large difference for the number of network parameters or the performance gap between the teacher and the student discourages the distillation process [7,41]. To reduce the gap while promoting the teacher to capture useful features, we exploit an hourglass architecture for the teacher network. It projects the HR images into a low-dimensional feature space to generate compact features, and reconstructs the original HR images from them, such that the teacher learns to extract better feature representations for an image reconstruction task. Specifically, the teacher network consists of an encoder $G^{\mathcal{T}}$ and a decoder $F^{\mathcal{T}}$. Given a pair of LR and HR images, the encoder $G^{\mathcal{T}}$ transforms the input HR image $\mathbf{Y}$ into the feature representation $\hat{\mathbf{X}}^{\mathcal{T}}$ in a low-dimensional space:

$$\hat{\mathbf{X}}^{\mathcal{T}} = G^{\mathcal{T}}(\mathbf{Y}), \tag{1}$$

where the feature representation of $\hat{\mathbf{X}}^{\mathcal{T}}$ has the same size as the LR image. The decoder $F^{\mathcal{T}}$ reconstructs the HR image $\hat{\mathbf{Y}}^{\mathcal{T}}$ using the compact feature $\hat{\mathbf{X}}^{\mathcal{T}}$:

$$\hat{\mathbf{Y}}^{\mathcal{T}} = F^{\mathcal{T}}(\hat{\mathbf{X}}^{\mathcal{T}}). \tag{2}$$

For the decoder, we use the same architecture as the student network. It allows the teacher to have a similar representational capacity as the student, which has proven useful in [41].

Loss. To train the teacher network, we use reconstruction and imitation losses, denoted by $L^{\mathcal{T}}_{\text{recon}}$ and $L^{\mathcal{T}}_{\text{im}}$, respectively. The reconstruction term computes the mean absolute error (MAE) between the HR image $\mathbf{Y}$ and its reconstruction $\hat{\mathbf{Y}}^{\mathcal{T}}$ defined as:

$$L_{\text{recon}}^{\mathcal{T}} = \frac{1}{HW} \sum_{i=1}^{H} \sum_{j=1}^{W} |Y_{ij} - \hat{Y}_{ij}^{\mathcal{T}}|, \tag{3}$$

where H and W are height and width of the HR image, respectively, and we denote by Y_{ij} an intensity value of $\mathbf{Y}$ at position (i, j). It encourages the encoder output (*i.e.*, compact feature $\hat{\mathbf{X}}^{\mathcal{T}}$) to contain useful information for the image reconstruction and forces the decoder to reconstruct the HR image again using the compact feature. The imitation term restricts the representational power of the encoder, making the output of the encoder close to the LR image. Concretely, we define this term as the MAE between the LR image $\mathbf{X}$ and the encoder output $\hat{\mathbf{X}}^{\mathcal{T}}$:

$$L_{\text{im}}^{\mathcal{T}} = \frac{1}{H'W'} \sum_{i=1}^{H'} \sum_{j=1}^{W'} |X_{ij} - \hat{X}_{ij}^{\mathcal{T}}|, \tag{4}$$

where H' and W' are height and width of the LR image, respectively. This facilitates an initialization of the student network that takes the LR image $\mathbf{X}$ as an input. Note that our framework avoids the trivial solution that the compact feature becomes the LR image since the network parameters in the encoder are updated by both the imitation and reconstruction terms. The overall objective is a sum of reconstruction and imitation terms, balanced by the parameter $\lambda^{\mathcal{T}}$:

$$L_{\text{total}}^{\mathcal{T}} = L_{\text{recon}}^{\mathcal{T}} + \lambda^{\mathcal{T}} L_{\text{im}}^{\mathcal{T}}. \tag{5}$$

3.2 Student

A student network has the same architecture as the decoder $F^{\mathcal{T}}$ in the teacher, but uses a different input. It takes a LR image $\mathbf{X}$ as an input and generates a HR image $\hat{\mathbf{Y}}^{\mathcal{S}}$:

$$\hat{\mathbf{Y}}^{\mathcal{S}} = F^{\mathcal{S}}(\mathbf{X}). \tag{6}$$

We initialize the weights of the student network with those of the decoder in the teacher. This transfers the reconstruction capability of the teacher to the student and provides a good starting point for optimization. Note that several works [15,24] point out that how to initialize network weights is crucial for the performance of a student. We adopt FSRCNN [11], a hardware-friendly SR architecture, as the student network $F^{\mathcal{S}}$.

Loss. Although the network parameters of the student $F^{\mathcal{S}}$ and the decoder $F^{\mathcal{T}}$ in the teacher are initially set to the same, the features extracted from them are different due to the different inputs. Besides, these parameters are not optimized with input LR images. We further train the student network $F^{\mathcal{S}}$ with a reconstruction loss $L_{\text{recon}}^{\mathcal{S}}$ and a distillation loss $L_{\text{distill}}^{\mathcal{S}}$. The reconstruction term is similarly defined as Eq. (3) using the ground-truth HR image and its reconstruction from the student network, dedicating to the SISR task:

$$L_{\text{recon}}^{\mathcal{S}} = \frac{1}{HW} \sum_{i=1}^{H} \sum_{j=1}^{W} |Y_{ij} - \hat{Y}_{ij}^{\mathcal{S}}|. \tag{7}$$

The distillation term focuses on transferring the knowledge of the teacher to the student. Overall, we use the following loss to train the student network:

$$L_{\text{total}}^{\mathcal{S}} = L_{\text{recon}}^{\mathcal{S}} + \lambda^{\mathcal{S}} L_{\text{distill}}^{\mathcal{S}}, \tag{8}$$

where $\lambda^{\mathcal{S}}$ is a distillation parameter. In the following, we describe the distillation loss in detail.

We adopt the distillation loss proposed in the VID method [2], which maximizes mutual information between the teacher and the student. We denote by $\mathbf{f}^{\mathcal{T}}$ and $\mathbf{f}^{\mathcal{S}}$ the intermediate feature maps of the teacher and student networks, respectively, having the same size of $C \times H' \times W'$, where C is the number of channels. We define mutual information $I(\mathbf{f}^{\mathcal{T}}; \mathbf{f}^{\mathcal{S}})$ as follows:

$$I(\mathbf{f}^{\mathcal{T}}; \mathbf{f}^{\mathcal{S}}) = H(\mathbf{f}^{\mathcal{T}}) - H(\mathbf{f}^{\mathcal{T}} | \mathbf{f}^{\mathcal{S}}), \tag{9}$$

where we denote by $H(\mathbf{f}^{\mathcal{T}})$ and $H(\mathbf{f}^{\mathcal{T}} | \mathbf{f}^{\mathcal{S}})$ marginal and conditional entropies, respectively. To maximize the mutual information, we should minimize the conditional entropy $H(\mathbf{f}^{\mathcal{T}} | \mathbf{f}^{\mathcal{S}})$. However, an exact optimization w.r.t the weights of the student is intractable, as it involves an integration over a conditional probability $p(\mathbf{f}^{\mathcal{T}} | \mathbf{f}^{\mathcal{S}})$. The variational information maximization technique [4] instead approximates the conditional distribution $p(\mathbf{f}^{\mathcal{T}} | \mathbf{f}^{\mathcal{S}})$ using a parametric model $q(\mathbf{f}^{\mathcal{T}} | \mathbf{f}^{\mathcal{S}})$, such as the Gaussian or Laplace distributions, making it possible to find a lower bound of the mutual information $I(\mathbf{f}^{\mathcal{T}}; \mathbf{f}^{\mathcal{S}})$. Using this technique, we maximize the lower bound of mutual information $I(\mathbf{f}^{\mathcal{T}}; \mathbf{f}^{\mathcal{S}})$ for feature distillation. As the parametric model $q(\mathbf{f}^{\mathcal{T}} | \mathbf{f}^{\mathcal{S}})$, we use a multivariate Laplace distribution with parameters of location and scale, $\boldsymbol{\mu} \in \mathbb{R}^{C \times H' \times W'}$ and $\mathbf{b} \in \mathbb{R}^{C \times H' \times W'}$, respectively. We define the distillation loss $L_{\text{distill}}^{\mathcal{S}}$ as follows:

$$L_{\text{distill}}^{\mathcal{S}} = \frac{1}{CH'W'} \sum_{i=1}^{C} \sum_{j=1}^{H'} \sum_{k=1}^{W'} \log b_{ijk} + \frac{|f_{ijk}^{\mathcal{T}} - \mu_{ijk}|}{b_{ijk}}, \tag{10}$$

where we denote by μ_{ijk} the element of $\boldsymbol{\mu}$ at the position (i, j, k). This minimizes the distance between the features $\mathbf{f}^{\mathcal{T}}$ of the teacher and the location map $\boldsymbol{\mu}$. The scale map $\mathbf{b}$ controls the extent of distillation. For example, when the student does not benefit from the distillation, the scale parameter b_{ijk} increases in order to reduce the extent of distillation. This is useful for our framework where the teacher and student networks take different inputs, since it adaptively determines the features the student is affordable to learn from the teacher. The term $\log b_{ijk}$ prevents a trivial solution where the scale parameter goes to infinite. We estimate these maps of $\boldsymbol{\mu}$ and $\mathbf{b}$ from the features of the student $\mathbf{f}^{\mathcal{S}}$. Note that other losses designed for feature distillation can also be used in our framework (See the supplementary material).

Estimator Module. We use a small network to estimate the parameters of location $\boldsymbol{\mu}$ and scale $\mathbf{b}$ in Eq. (10). It consists of location and scale branches, where each takes the features of the student $\mathbf{f}^{\mathcal{S}}$ and estimates the location and scale maps, separately. Both branches share the same network architecture of two 1×1 convolutional layers and a PReLU [21] between them. For the scale branch, we add the softplus function ($\zeta(x) = \log(1 + e^{x})$) [12] at the last layer, forcing the scale parameter to be positive. Note that the estimation module is used only at training time.

4 Experiments

4.1 Experimental Details

Implementation Details. The encoder in the teacher network consists of 4 blocks of convolutional layers followed by a PReLU [21]. All the layers, except the second one, perform convolutions with stride 1. In the second block, we use the convolution with stride s (*i.e.*, a scale factor) to downsample the size of the HR image to that of the LR image. The kernel sizes of the first two and the last two blocks are 5×5 and 3×3, respectively. The decoder in the teacher and the student network have the same architecture as FSRCNN [11] consisting of five components: Feature extraction, shrinking, mapping, expanding, and deconvolution modules. We add the estimator module for location and scale maps on top of the expanding module in the student network. We use these maps together with the output features of the expanding module in the decoder to compute the distillation loss. We set the hyperparameters for losses using a grid search on the DIV2K dataset [48], and choose the ones ($\lambda^{\mathcal{T}} = 10^{-4}$, $\lambda^{\mathcal{S}} = 10^{-6}$) that give the best performance. We implement our framework using `PyTorch` [42].

Training. To train our network, we use the training split of DIV2K [48] corresponding 800 pairs of LR and HR images, where the LR images are synthesized by bicubic downsampling. We randomly crop HR patches of size 192×192 from the HR images. LR patches are cropped from the corresponding LR images according to the scale factor. For example, LR patches of size 96×96 are used for the scale factor of 2. We use data augmentation techniques, including random rotation and horizontal flipping. The teacher network is trained with random initialization. We train our model with a batch size of 16 about 1000k iterations over the training data. We use the Adam [31] with $\beta_1 = 0.9$ and $\beta_2 = 0.999$. As a learning rate, we use 10^{-3} and reduce it until 10^{-5} using a cosine annealing technique [38].

Evaluation. We evaluate our framework on standard benchmarks including Set5 [5], Set14 [59], B100 [40], and Urban100 [25]. Following the experimental protocol in [34], we use the peak signal to noise ratio (PSNR) and the structural similarity index (SSIM) [52] on the luminance channel as evaluation metrics.

Table 1. Average PSNR of student and teacher networks, trained with variants of our framework, on the Set5 [5] dataset. We use FSRCNN [11], reproduced by ourselves using the DIV2K [48] dataset without distillation, as the baseline in the first row. We denote by VID_G and VID_L VID losses [2] with the Gaussian and Laplace distributions, respectively. The performance gains of each component over the baseline are shown in the parentheses. The number in bold indicates the best performance and underscored one is the second best.

Hourglass architecture	Weight transfer	L_{im}^{T}	L_{distill}^{S}	Student PSNR	Teacher PSNR
–	–	–	–	37.15 (baseline)	–
✗	–	–	MAE	37.19 (+0.04)	57.60
✓	✗	✗	MAE	37.22 (+0.07)	37.70
✓	✓	✗	MAE	37.23 (+0.08)	37.70
✓	✓	✓	MAE	37.27 (+0.12)	37.65
✓	✓	✓	VID_G [2]	<u>37.31</u> (+0.16)	37.65
✓	✓	✓	VID_L [2]	**37.33** (+0.18)	37.65

4.2 Ablation Studies

We present an ablation analysis on each component of our framework. We report quantitative results in terms of the average PSNR on Set5 [5] with the scale factor of 2. The results on a large dataset (*i.e.*, B100 [40]) can be seen in the supplementary material. We show in Table 1 the average PSNR for student networks trained with variants of our framework. The results of the baseline in the first row are obtained using FSRCNN [11]. From the second row, we can clearly see that feature distillation boosts the PSNR performance. For the teacher network in the second row, we use the same network architecture as FSRCNN except for the deconvolution layers. In contrast to FSRCNN, the teacher inputs HR images, and thus we replace the deconvolution layer with a convolutional layer, preserving the size of the inputs. We can see from the third row that a teacher network with an hourglass architecture improves the student performance. The hourglass architecture limits the performance of the teacher and degrades the performance (*e.g.*, a 19.9 dB decrease compared to that of the teacher in the second row), reducing the performance gap between the teacher and the student. This allows the feature distillation to be more effective, thus the student of the third row performs better (37.22 dB) than that of the second row (37.19 dB), which can also be found in recent works [7,41]. The fourth row shows that the student network benefits from initializing the network weights with those of the decoder in the teacher, since this provides a good starting point for learning, and transfers the reconstruction capability of the teacher. From the fifth row, we observe that an imitation loss further improves the PSNR performance, making it easier for the student to learn features from the teacher. The next two rows show that the VID loss [2], especially with the Laplace distribution (VID_L), provides better results than the MAE, and combining all components gives the best performance. The distillation loss based on the MAE forces the feature maps

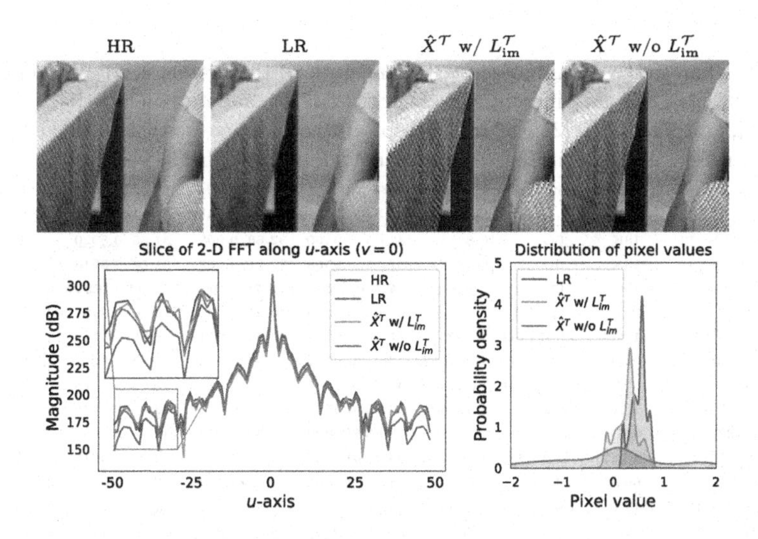

Fig. 3. Analysis on compact features in spatial (top) and frequency (bottom left) domains and the distribution of pixel values (bottom right). To visualize the compact features in the frequency domain, we apply the 2D Fast Fourier Transform (FFT) to the image, obtaining its magnitude spectrum. It is then sliced along the u-axis. (Best viewed in color.) (Color figure online)

of the student and teacher networks to be the same. This strong constraint on the feature maps is, however, problematic in our case, since we use different inputs for the student and teacher networks. The VID method allows the student to learn important features adaptively. We also compare the performance of our framework and a typical distillation approach with different losses in the supplementary material.

4.3 Analysis on Compact Features

In Fig. 3, we show an analysis on compact features in spatial and frequency domains. Compared to the LR image, the compact features $\hat{\mathbf{X}}^{\mathcal{T}}$ show high-frequency details regardless of whether the imitation loss $L_{\text{im}}^{\mathcal{T}}$ is used or not. This can also be observed in the frequency domain – The compact features contain more high-frequency components than the LR image, and the magnitude spectrums of them are more similar to that of the HR image especially for high-frequency components. By taking these features as inputs, the decoder in the teacher shows the better performance than the student (Table 1) despite the fact that they have the same architecture. This demonstrates that the compact features extracted from the ground truth contain useful information for reconstructing the HR image, encouraging the student to reconstruct more accurate results via feature distillation. In the bottom right of the Fig. 3, we can see that the pixel distributions of the LR image and the compact feature are largely different without the imitation loss, discouraging the weight transfer to the student.

Table 2. Quantitative comparison with the state of the art on SISR. We report the average PSNR/SSIM for different scale factors (2×, 3×, and 4×) on Set5 [5], Set14 [59], B100 [40], and Urban100 [25]. *: models reproduced by ourselves using the DIV2K [48] dataset without distillation; Ours: student networks of our framework.

Scale	Methods	Param.	MultiAdds	Runtime	Set5 [5] PSNR/SSIM	Set14 [59] PSNR/SSIM	B100 [40] PSNR/SSIM	Urban100 [25] PSNR/SSIM
2	FSRCNN [11]	13K	6.0G	0.83 ms	37.05/0.9560	32.66/0.9090	31.53/0.8920	29.88/0.9020
	FSRCNN*	13K	6.0G	0.83 ms	37.15/0.9568	32.71/0.9095	31.58/0.8913	30.05/0.9041
	FSRCNN (Ours)	13K	6.0G	0.83 ms	**37.33/0.9576**	**32.79/0.9105**	**31.65/0.8926**	**30.24/0.9071**
	Bicubic Int.	–	–	–	33.66/0.9299	30.24/0.8688	29.56/0.8431	26.88/0.8403
	DRCN [30]	1,774K	17,974.3G	239.93 ms	37.63/0.9588	33.04/0.9118	31.85/0.8942	30.75/0.9133
	DRRN [45]	297K	6,796.9G	105.76 ms	37.74/0.9591	33.23/0.9136	32.05/0.8973	31.23/0.9188
	MemNet [46]	677K	2,662.4G	21.06 ms	37.78/0.9597	33.28/0.9142	32.08/0.8978	31.31/0.9195
	CARN [1]	1,592K	222.8G	8.43 ms	37.76/0.9590	33.52/0.9166	32.09/0.8978	31.92/0.9256
	IDN [27]	591K	136.5G	7.01 ms	37.83/0.9600	33.30/0.9148	32.08/0.8985	31.27/0.9196
	SRFBN [33]	3,631K	1,126.7G	108.52 ms	38.11/0.9609	33.82/0.9196	32.29/0.9010	32.62/0.9328
	IMDN [26]	694K	159.6G	6.97 ms	38.00/0.9605	33.63/0.9177	32.19/0.8996	32.17/0.9283
3	FSRCNN [11]	13K	5.0G	0.72 ms	33.18/0.9140	29.37/0.8240	28.53/0.7910	26.43/0.8080
	FSRCNN*	13K	5.0G	0.72 ms	33.15/0.9157	29.45/0.8250	28.52/0.7895	26.49/0.8089
	FSRCNN (Ours)	13K	5.0G	0.72 ms	**33.31/0.9179**	**29.57/0.8276**	**28.61/0.7919**	**26.67/0.8153**
	Bicubic Int.	–	–	–	30.39/0.8682	27.55/0.7742	27.21/0.7385	24.46/0.7349
	DRCN [30]	1,774K	17,974.3G	239.19 ms	33.82/0.9226	29.76/0.8311	28.80/0.7963	27.15/0.8276
	DRRN [45]	297K	6,796.9G	98.58 ms	34.03/0.9244	29.96/0.8349	28.95/0.8004	27.53/0.8378
	MemNet [46]	677K	2,662.4G	11.33 ms	34.09/0.9248	30.00/0.8350	28.96/0.8001	27.56/0.8376
	CARN [1]	1,592K	118.8G	3.86 ms	34.29/0.9255	30.29/0.8407	29.06/0.8034	28.06/0.8493
	IDN [27]	591K	60.6G	3.62 ms	34.11/0.9253	29.99/0.8354	28.95/0.8013	27.42/0.8359
	SRFBN [33]	3,631K	500.8G	76.74 ms	34.70/0.9292	30.51/0.8461	29.24/0.8084	28.73/0.8641
	IMDN [26]	703K	71.7G	5.36 ms	34.36/0.9270	30.32/0.8417	29.09/0.8046	28.17/0.8519
4	FSRCNN [11]	13K	4.6G	0.67 ms	30.72/0.8660	27.61/0.7550	26.98/0.7150	24.62/0.7280
	FSRCNN*	13K	4.6G	0.67 ms	30.89/0.8748	27.72/0.7599	27.05/0.7176	24.76/0.7358
	FSRCNN (Ours)	13K	4.6G	0.67 ms	**30.95/0.8759**	**27.77/0.7615**	**27.08/0.7188**	**24.82/0.7393**
	Bicubic Int	–	–	–	28.42/0.8104	26.00/0.7027	25.96/0.6675	23.14/0.6577
	DRCN [30]	1,774K	17,974.3G	243.62 ms	31.53/0.8854	28.02/0.7670	27.23/0.7233	25.14/0.7510
	DRRN [45]	297K	6,796.9G	57.09 ms	31.68/0.8888	28.21/0.7721	27.38/0.7284	25.44/0.7638
	MemNet [46]	677K	2,662.4G	8.55 ms	31.74/0.8893	28.26/0.7723	27.40/0.7281	25.50/0.7630
	CARN [1]	1,592K	90.9G	3.16 ms	32.13/0.8937	28.60/0.7806	27.58/0.7349	26.07/0.7837
	IDN [27]	591K	34.1G	3.08 ms	31.82/0.8903	28.25/0.7730	27.41/0.7297	25.41/0.7632
	SRFBN [33]	3,631K	281.7G	48.39ms	32.47/0.8983	28.81/0.7868	27.72/0.7409	26.60/0.8015
	IMDN [26]	715K	41.1G	4.38 ms	32.21/0.8948	28.58/0.7811	27.56/0.7353	26.04/0.7838

The imitation loss $L_{\text{im}}^{\mathcal{T}}$ alleviates this problem by encouraging the distributions of the LR image and the compact feature to be similar.

4.4 Results

Quantitative Comparison. We compare in Table 2 the performance of our student model with the state of the art, particularly for efficient SISR methods [1,10,11,26,27,30,33,45,46]. For a quantitative comparison, we report the average PSNR and SSIM [52] for upsampling factors of 2, 3, and 4, on standard benchmarks [5,25,40,59]. We also report the number of model parameters and operations (MultiAdds), required to reconstruct a HR image of size 1280 × 720, and present the average runtime of each method measured on the Set5 [5] using

Table 3. Quantitative results of student networks using other SR methods. We report the average PSNR for different scale factors (2×, 3×, and 4×) on Set5 [5] and B100 [40]. *: models reproduced by ourselves using the DIV2K [48] dataset; Ours: student networks of our framework.

Methods	2× Set5/B100	3× Set5/B100	4× Set5/B100
FSRCNN-L*	37.59/31.90	33.76/28.81	31.47/27.29
FSRCNN-L (Ours)	**37.65/31.92**	**33.85/28.83**	**31.52/27.30**
VDSR [29]	37.53/31.90	33.67/28.82	31.35/27.29
VDSR*	37.64/31.96	33.80/28.83	31.37/27.25
VDSR (Ours)	**37.77/32.00**	**33.85/28.86**	**31.51/27.29**
IDN [27]	37.83/32.08	34.11/28.95	31.82/27.41
IDN*	37.88/32.12	34.22/29.02	**32.03**/27.49
IDN (Ours)	**37.93/32.14**	**34.31/29.03**	32.01/**27.51**
CARN [1]	37.76/32.09	34.29/29.06	32.13/27.58
CARN*	37.75/32.02	34.08/28.94	31.77/27.44
CARN (Ours)	**37.82/32.08**	**34.10/28.95**	**31.83/27.45**

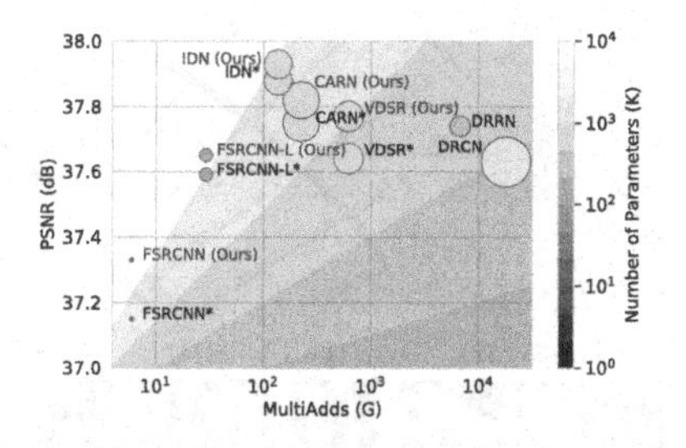

Fig. 4. Trade-off between the number of operations and the average PSNR on Set5 [5] (2×). The size of the circle and background color indicate the number of parameters and the efficiency of the model (white: high, black: low), respectively. (Best viewed in color.) (Color figure online)

the same machine with a NVIDIA Titan RTX GPU. From this table, we can observe two things: (1) Our student model trained with the proposed framework outperforms FSRCNN [11] by a large margin, consistently for all scale factors, even both have the same network architecture. It demonstrates the effectiveness of our approach to exploiting ground-truth HR images as privileged information; (2) The model trained with our framework offers a good compromise in terms of PSNR/SSIM and the number of parameters/operations/runtimes. For example, DRCN [30] requires 1,774K parameters, 17,974.3G operations and average runtime of 233.93 ms to achieve the average PSNR of 30.75 dB on Urban100 [25] for a factor of 2. On the contrary, our framework further boosts FSRCNN without modifying the network architecture, achieving the average PSNR of 30.24 dB with 13K parameters/6.0G operations only, while taking 0.83 ms for inference.

In Table 3, we show the performances of student networks, adopting the architectures of other SR methods, trained with our framework using the DIV2K dataset [48]. We reproduce their models (denoted by *) using the same training setting but without distillation. The FSRCNN-L has the same components as FSRCNN [11] but with much more parameters (126K vs. 13K), where the numbers of filters in feature extraction and shrinking components are both 56, and the mapping module consists of 4 blocks of convolutional layers. Note that the multi-scale learning strategy in the CARN [1] is not used for training the network, and thus the performance is slightly lower than the original one. We can see that all the SISR methods benefit from our framework except for IDN [27] for the scale factor of 4 on Set5. In particular, the performances of the variant of FSRCNN [10] and VDSR [29] are significantly boosted through our framework. Additionally, our framework further improves the performances of the cost-effective SR methods [1,27], which are specially-designed to reduce the number of parameters and operations while improving the reconstruction performance. Considering the performance gains of recent SR methods, the results are

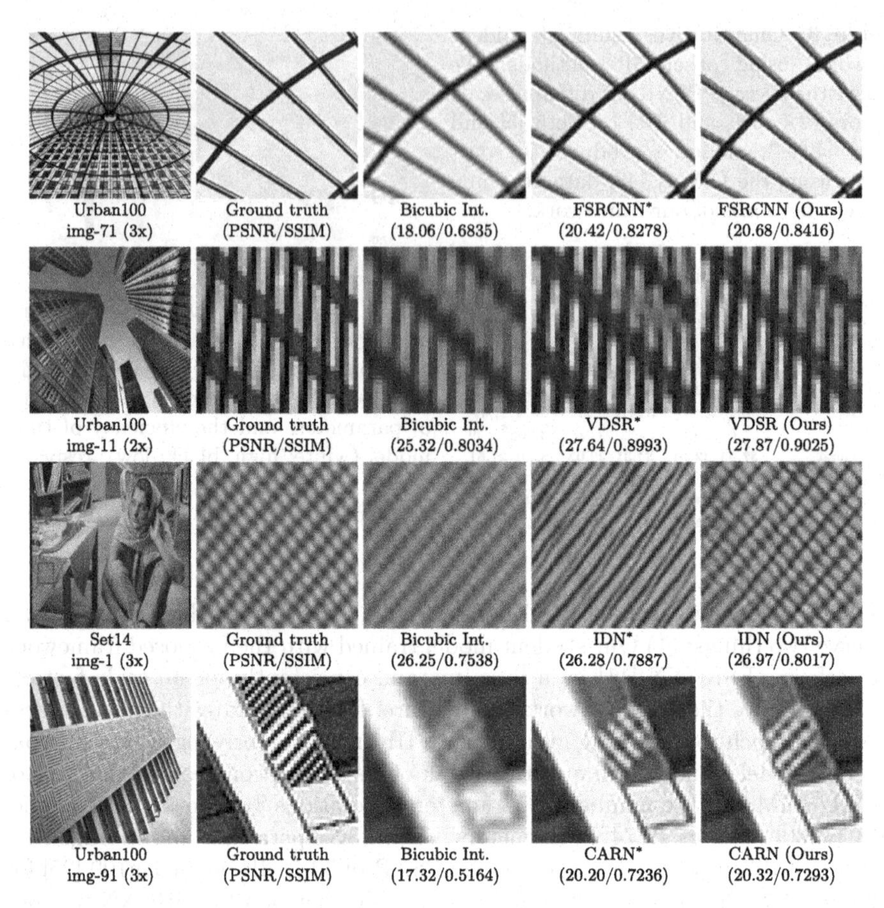

Fig. 5. Visual comparison of reconstructed HR images (2× and 3×) on Urban100 [25] and Set14 [59]. We report the average PSNR/SSIM in the parentheses. (Best viewed in color.) (Color figure online)

significant, demonstrating the effectiveness and generalization ability of our framework. For example IDN [27] and SRFBN [33] outperform the second-best methods by 0.05 dB and 0.02 dB, respectively, in terms of PSNR on Set5 [5] for a factor of 2. We visualize in Fig. 4 the performance comparison of student networks using various SR methods and the state of the art in terms of the number of operations and parameters. It confirms once more the efficiency of our framework.

Qualitative Results. We show in Fig. 5 reconstruction examples on the Urban100 [25] and Set14 [59] datasets using the student networks. We can clearly see that the student models provide better qualitative results than their baselines. In particular, our models remove artifacts (*e.g.*, the borders around the sculpture in the first row) and reconstruct small-scale structures (*e.g.*, windows in the second row and the iron railings in the last row) and textures (*e.g.*, the patterns of the tablecloth in the third row). More qualitative results can be seen in the supplementary material.

5 Conclusion

We have presented a novel distillation framework for SISR leveraging ground-truth HR images as privileged information. The detailed analysis on each component of our framework clearly demonstrates the effectiveness of our approach. We have shown that the proposed framework substantially improves the performance of FSRCNN as well as other methods. In future work, we will explore distillation losses specific to our model to further boost the performance.

Acknowledgement. This research was supported by the Samsung Research Funding & Incubation Center for Future Technology (SRFC-IT1802-06).

References

1. Ahn, N., Kang, B., Sohn, K.-A.: Fast, accurate, and lightweight super-resolution with cascading residual network. In: Ferrari, V., Hebert, M., Sminchisescu, C., Weiss, Y. (eds.) ECCV 2018. LNCS, vol. 11214, pp. 256–272. Springer, Cham (2018). https://doi.org/10.1007/978-3-030-01249-6_16
2. Ahn, S., Hu, S.X., Damianou, A., Lawrence, N.D., Dai, Z.: Variational information distillation for knowledge transfer. In: CVPR (2019)
3. Bai, Y., Zhang, Y., Ding, M., Ghanem, B.: SOD-MTGAN: small object detection via multi-task generative adversarial network. In: Ferrari, V., Hebert, M., Sminchisescu, C., Weiss, Y. (eds.) ECCV 2018. LNCS, vol. 11217, pp. 210–226. Springer, Cham (2018). https://doi.org/10.1007/978-3-030-01261-8_13
4. Barber, D., Agakov, F.V.: The IM algorithm: a variational approach to information maximization. In: NIPS (2003)
5. Bevilacqua, M., Roumy, A., Guillemot, C., Alberi-Morel, M.L.: Low-complexity single-image super-resolution based on nonnegative neighbor embedding. In: BMVC (2012)
6. Chang, H., Yeung, D.Y., Xiong, Y.: Super-resolution through neighbor embedding. In: CVPR (2004)
7. Cho, J.H., Hariharan, B.: On the efficacy of knowledge distillation. In: ICCV (2019)
8. Chollet, F.: Xception: deep learning with depthwise separable convolutions. In: CVPR (2017)
9. Dai, S., Han, M., Xu, W., Wu, Y., Gong, Y., Katsaggelos, A.K.: SoftCuts: a soft edge smoothness prior for color image super-resolution. IEEE TIP **18**(5), 969–981 (2009)
10. Dong, C., Loy, C.C., He, K., Tang, X.: Image super-resolution using deep convolutional networks. IEEE TPAMI **38**(2), (2015)
11. Dong, C., Loy, C.C., Tang, X.: Accelerating the super-resolution convolutional neural network. In: Leibe, B., Matas, J., Sebe, N., Welling, M. (eds.) ECCV 2016. LNCS, vol. 9906, pp. 391–407. Springer, Cham (2016). https://doi.org/10.1007/978-3-319-46475-6_25

12. Dugas, C., Bengio, Y., Bélisle, F., Nadeau, C., Garcia, R.: Incorporating second-order functional knowledge for better option pricing. In: NIPS (2001)
13. Freeman, W.T., Jones, T.R., Pasztor, E.C.: Example-based super-resolution. IEEE CG&A **22**(2), 56–65 (2002)
14. Gao, Q., Zhao, Y., Li, G., Tong, T.: Image super-resolution using knowledge distillation. In: ACCV (2018)
15. Garcia, N.C., Morerio, P., Murino, V.: Modality distillation with multiple stream networks for action recognition. In: Ferrari, V., Hebert, M., Sminchisescu, C., Weiss, Y. (eds.) ECCV 2018. LNCS, vol. 11212, pp. 106–121. Springer, Cham (2018). https://doi.org/10.1007/978-3-030-01237-3_7
16. Greenspan, H.: Super-resolution in medical imaging. Comput. J. **52**(1), 43–63 (2008)
17. Gunturk, B.K., Batur, A.U., Altunbasak, Y., Hayes, M.H., Mersereau, R.M.: Eigenface-domain super-resolution for face recognition. IEEE TIP **12**(5), 597–606 (2003)
18. Han, S., Mao, H., Dally, W.: Deep compression: compressing deep neural networks with pruning, trained quantization and Huffman coding. In: ICLR (2016)
19. Han, S., Pool, J., Tran, J., Dally, W.: Learning both weights and connections for efficient neural network. In: NIPS (2015)
20. Haris, M., Shakhnarovich, G., Ukita, N.: Deep back-projection networks for super-resolution. In: CVPR (2018)
21. He, K., Zhang, X., Ren, S., Sun, J.: Delving deep into rectifiers: surpassing human-level performance on imagenet classification. In: ICCV (2015)
22. Heo, B., Kim, J., Yun, S., Park, H., Kwak, N., Choi, J.Y.: A comprehensive overhaul of feature distillation. In: ICCV (2019)
23. Hinton, G., Vinyals, O., Dean, J.: Distilling the knowledge in a neural network. In: NIPS Workshop (2014)
24. Hoffman, J., Gupta, S., Darrell, T.: Learning with side information through modality hallucination. In: CVPR (2016)
25. Huang, J.B., Singh, A., Ahuja, N.: Single image super-resolution from transformed self-exemplars. In: CVPR (2015)
26. Hui, Z., Gao, X., Yang, Y., Wang, X.: Lightweight image super-resolution with information multi-distillation network. In: ACMMM (2019)
27. Hui, Z., Wang, X., Gao, X.: Fast and accurate single image super-resolution via information distillation network. In: CVPR (2018)
28. Sun, J., Xu, Z., Shum, H.-Y.: Image super-resolution using gradient profile prior. In: CVPR (2008)
29. Kim, J., Kwon Lee, J., Mu Lee, K.: Accurate image super-resolution using very deep convolutional networks. In: CVPR (2016)
30. Kim, J., Kwon Lee, J., Mu Lee, K.: Deeply-recursive convolutional network for image super-resolution. In: CVPR (2016)
31. Kingma, D.P., Ba, J.: Adam: a method for stochastic optimization. In: ICLR (2015)
32. Ledig, C., et al.: Photo-realistic single image super-resolution using a generative adversarial network. In: CVPR (2017)
33. Li, Z., Yang, J., Liu, Z., Yang, X., Jeon, G., Wu, W.: Feedback network for image super-resolution. In: CVPR (2019)
34. Lim, B., Son, S., Kim, H., Nah, S., Mu Lee, K.: Enhanced deep residual networks for single image super-resolution. In: CVPR Workshop (2017)
35. Lin, W.S., Tjoa, S.K., Zhao, H.V., Liu, K.R.: Digital image source coder forensics via intrinsic fingerprints. IEEE TIFS **4**(3), 460–475 (2009)

36. Liu, Y., Chen, K., Liu, C., Qin, Z., Luo, Z., Wang, J.: Structured knowledge distillation for semantic segmentation. In: CVPR (2019)
37. Lopez-Paz, D., Bottou, L., Schölkopf, B., Vapnik, V.: Unifying distillation and privileged information. In: ICLR (2016)
38. Loshchilov, I., Hutter, F.: SGDR: stochastic gradient descent with warm restarts. In: ICLR (2017)
39. Mao, X., Shen, C., Yang, Y.B.: Image restoration using very deep convolutional encoder-decoder networks with symmetric skip connections. In: NIPS (2016)
40. Martin, D., Fowlkes, C., Tal, D., Malik, J., et al.: A database of human segmented natural images and its application to evaluating segmentation algorithms and measuring ecological statistics. In: ICCV (2001)
41. Mirzadeh, S.I., Farajtabar, M., Li, A., Ghasemzadeh, H.: Improved knowledge distillation via teacher assistant: bridging the gap between student and teacher. In: AAAI (2020)
42. Paszke, A., et al.: Automatic differentiation in PyTorch (2017)
43. Romero, A., Ballas, N., Kahou, S.E., Chassang, A., Gatta, C., Bengio, Y.: FitNets: hints for thin deep nets. In: ICLR (2015)
44. Schulter, S., Leistner, C., Bischof, H.: Fast and accurate image upscaling with super-resolution forests. In: CVPR (2015)
45. Tai, Y., Yang, J., Liu, X.: Image super-resolution via deep recursive residual network. In: CVPR (2017)
46. Tai, Y., Yang, J., Liu, X., Xu, C.: MemNet: a persistent memory network for image restoration. In: ICCV (2017)
47. Timofte, R., De, V., Gool, L.V.: Anchored neighborhood regression for fast example-based super-resolution. In: ICCV (2013)
48. Timofte, R., Agustsson, E., Van Gool, L., Yang, M.H., Zhang, L.: NTIRE 2017 challenge on single image super-resolution: methods and results. In: CVPR Workshop (2017)
49. Tong, T., Li, G., Liu, X., Gao, Q.: Image super-resolution using dense skip connections. In: ICCV (2017)
50. Vapnik, V., Izmailov, R.: Learning using privileged information: similarity control and knowledge transfer. JMLR **16**, 2023–2049 (2015)
51. Vapnik, V., Vashist, A.: A new learning paradigm: learning using privileged information. Neural Netw. **22**(5–6), 544–557 (2009)
52. Wang, Z., et al.: Image quality assessment: from error visibility to structural similarity. IEEE TIP **13**(4), 600–612 (2004)
53. Wen, W., Wu, C., Wang, Y., Chen, Y., Li, H.: Learning structured sparsity in deep neural networks. In: NIPS (2016)
54. Xie, S., Girshick, R., Dollar, P., Tu, Z., He, K.: Aggregated residual transformations for deep neural networks. In: CVPR (2017)
55. Yan, Q., Xu, Y., Yang, X., Nguyen, T.Q.: Single image super-resolution based on gradient profile sharpness. IEEE TIP **24**(10), 3187–3202 (2015)
56. Yang, J., Wright, J., Huang, T., Ma, Y.: Image super-resolution as sparse representation of raw image patches. In: CVPR (2008)
57. Yim, J., Joo, D., Bae, J., Kim, J.: A gift from knowledge distillation: fast optimization, network minimization and transfer learning. In: CVPR (2017)
58. Zagoruyko, S., Komodakis, N.: Paying more attention to attention: improving the performance of convolutional neural networks via attention transfer. In: ICLR (2017)
59. Zeyde, R., Elad, M., Protter, M.: On single image scale-up using sparse-representations. In: Curves and Surfaces (2010)

60. Zhang, K., Zuo, W., Chen, Y., Meng, D., Zhang, L.: Beyond a Gaussian denoiser: residual learning of deep CNN for image denoising. IEEE TIP **26**, 3142–3155 (2017)
61. Zhang, K., Zuo, W., Gu, S., Zhang, L.: Learning deep CNN denoiser prior for image restoration. In: CVPR (2017)
62. Zhang, Y., Tian, Y., Kong, Y., Zhong, B., Fu, Y.: Residual dense network for image super-resolution. In: CVPR (2018)
63. Zou, W.W., Yuen, P.C.: Very low resolution face recognition problem. IEEE TIP **21**(1), (2011)

Joint Visual and Temporal Consistency for Unsupervised Domain Adaptive Person Re-identification

Jianing Li and Shiliang Zhang(✉)

Department of Computer Science, School of EE&CS, Peking University, Beijing 100871, China
{ljn-vmc,slzhang.jdl}@pku.edu.cn

Abstract. Unsupervised domain adaptive person Re-IDentification (ReID) is challenging because of the large domain gap between source and target domains, as well as the lackage of labeled data on the target domain. This paper tackles this challenge through jointly enforcing visual and temporal consistency in the combination of a local one-hot classification and a global multi-class classification. The local one-hot classification assigns images in a training batch with different person IDs, then adopts a Self-Adaptive Classification (SAC) model to classify them. The global multi-class classification is achieved by predicting labels on the entire unlabeled training set with the Memory-based Temporal-guided Cluster (MTC). MTC predicts multi-class labels by considering both visual similarity and temporal consistency to ensure the quality of label prediction. The two classification models are combined in a unified framework, which effectively leverages the unlabeled data for discriminative feature learning. Experimental results on three large-scale ReID datasets demonstrate the superiority of proposed method in both unsupervised and unsupervised domain adaptive ReID tasks. For example, under unsupervised setting, our method outperforms recent unsupervised domain adaptive methods, which leverage more labels for training.

Keywords: Domain adaption · Person re-identification · Convolution neural networks

1 Introduction

Person Re-Identification (ReID) aims to identify a probe person in a camera network by matching his/her images or video sequences and has many promising applications like smart surveillance and criminal investigation. Recent years have witnessed the significant progresses on supervised person ReID in discriminative feature learning from labeled person images [14,17,23,27,32,38] and videos [11–13]. However, supervised person ReID methods rely on a large amount of labeled data which is expensive to annotate. Deep models trained on the source domain

A. Vedaldi et al. (Eds.): ECCV 2020, LNCS 12369, pp. 483–499, 2020.
https://doi.org/10.1007/978-3-030-58586-0_29

suffer substantial performance drop when transferred to a different target domain. Those issues make it hard to deploy supervised ReID models in real applications.

To tackle this problem, researchers focus on unsupervised learning [5,29,39], which could take advantage of abundant unlabeled data for training. Compared with supervised learning, unsupervised learning relieves the requirement for expensive data annotation, hence shows better potential to push person ReID towards real applications. Recent works define unsupervised person ReID as a transfer learning task, which leverages labeled data on other domains. Related works can be summarized into two categories, *e.g.*, (1) using Generative Adversarial Network (GAN) to transfer the image style from labeled source domain to unlabeled target domain while preserving identity labels for training [31,39,41], or (2) pre-training a deep model on source domain, then clustering unlabeled data in target domain to estimate pseudo labels for training [5,34]. The second category has significantly boosted the performance of unsupervised person ReID. However, there is still a considerable performance gap between supervised and unsupervised person ReID. The reason may be because many persons share similar appearance and the same person could exhibit different appearances, leading to unreliable label estimation. Therefore, more effective ways to utilize the unlabeled data should still be investigated.

This work targets to learn discriminative features for unlabeled target domain through generating more reliable label predictions. Specifically, reliable labels can be predicted from two aspects. First, since each training batch samples a small number of images from the training set, it is likely that those images are sampled from different persons. We thus could label each image with a distinct person ID and separate them from each other with a classification model. Second, it is not reliable to estimate labels on the entire training set with only visual similarity. We thus consider both visual similarity and temporal consistency for multi-class label prediction, which is hence utilized to optimize the inter and intra class distances. Compared with previous methods, which only utilize visual similarity to cluster unlabeled images [5,34], our method has potential to exhibit better robustness to visual variance. Our temporal consistency is inferred based on the video frame number, which can be easily acquired without requiring extra annotations or manual alignments.

The above intuitions lead to two classification tasks for feature learning. The local classification in each training batch is conducted by a Self-Adaptive Classification (SAC) model. Specially, in each training batch, we generate a self-adaptive classifier from image features and apply one-hot label to separate images from each other. The feature optimization in the entire training set is formulated as a multi-label classification task for global optimization. We propose the Memory-based Temporal-Guided Cluster (MTC) to predict multi-class labels based on both visual similarity and temporal consistency. In other words, two images are assigned with the same label if they a) share large visual similarity and b) share enough temporal consistency.

Inspired by [30], we compute the temporal consistency based on the distribution of time interval between two cameras, *i.e.*, interval of frame numbers of two images. For example, when we observe a person appears in camera i at time t, according to the estimated distribution, he/she would have high possibility to be recorded by camera j at time $t + \Delta t$, and has low possibility will be recorded by another camera k. This cue would effectively filter hard negative samples with similar visual appearance, as well as could be applied in ReID to reduce the search space. To further ensure the accuracy of clustering result, MTC utilizes image features stored in the memory bank. Memory bank is updated with augmented features after each training iteration to improve feature robustness.

The two classification models are aggregated in a unified framework for discriminative feature learning. Experiments on three large-scale person ReID datasets show that, our method exhibits substantial superiority to existing unsupervised and domain adaptive ReID methods. For example, we achieve rank1 accuracy of 79.5% on Market-1501 with unsupervised training, and achieve 86.8% after unsupervised domain transfer, respectively.

Our promising performance is achieved with the following novel components. 1) The SAC model efficiently performs feature optimization in each local training batch by assigning images with different labels. 2) The MTC method performs feature optimization in the global training set by predicting labels with visual similarity and temporal consistency. 3) Our temporal consistency does not require any extra annotations or manual alignments, and could be utilized in both model training and ReID similarity computation. To the best of our knowledge, this is an early unsupervised person ReID work utilizing temporal consistency for label prediction and model training.

2 Related Work

This work is closely related to unsupervised domain adaptation and unsupervised domain adaptive person ReID. This section briefly summarizes those two categories of works.

Unsupervised Domain Adaptation (UDA) has been extensively studied in image classification. The aim of UDA is to align the domain distribution between source and target domains. A common solution of UDA is to define and minimize the domain discrepancy between source and target domain. Gretton *et al.* [9] project data samples into a reproducing kernel Hilbert space and compute the difference of sample means to reduce the Maximum Mean Discrepancy (MMD). Sun *et al.* [28] propose to learn a transformation to align the mean and covariance between two domains in the feature space. Pan *et al.* [24] propose to align each class in source and target domain through Prototypical Networks. Adversarial learning is also widely used to minimize domain shift. Ganin *et al.* [6] propose a Gradient Reversal Layer (GRL) to confuse the feature learning model and make it can't distinguish the features from source and target domains. DRCN [7] takes a similar approach but also performs multi-task learning to reconstruct target domain images. Different from domain adaption in person ReID, traditional UDA mostly assumes that the source domain and target domain share same classes. However, in person ReID, different domain commonly deals with different persons, thus have different classes.

Unsupervised Domain Adaptive Person ReID: Early methods design hand craft features for person ReID [8,20]. Those methods can be directly adapted to unlabeled dataset, but show unsatisfactory performance. Recent works propose to train deep models on labeled source domain and then transfer to unlabeled target domain. Yu *et al.* [34] use the labeled source dataset as a reference to learn soft labels. Fu *et al.* [5] cluster the global and local features to estimate pseudo labels, respectively. Generative Adversarial Network (GAN) is also applied to bridge the gap across cameras or domains. Wei *et al.* [31] transfer images from the source domain to target domain while reserving the identity labels for training. Zhong *et al.* [40] apply CycleGAN [42] to generate images under different camera styles for data augmentation. Zhong *et al.* [39] introduce the memory bank [33] to minimize the gap between source and target domains.

Most existing methods only consider visual similarity for feature learning on unlabeled data, thus are easily influenced by the large visual variation and domain bias. Different from those works, we consider visual similarity and temporal consistency for feature learning. Compared with existing unsupervised domain adaptive person ReID methods, our method exhibits stronger robustness and better performance. As shown in our experiments, our approach outperforms recent ReID methods under both unsupervised and unsupervised domain adaptive settings. To the best of our knowledge, this is an early attempt to jointly consider visual similarity and temporal consistency in unsupervised domain adaptive person ReID. Another person ReID work [30] also uses temporal cues. Different with our work, it focuses on supervised training and only uses temporal cues in the ReID stage for re-ranking.

3 Proposed Method

3.1 Formulation

For any query person image q, the person ReID model is expected to produce a feature vector to retrieve the image g containing the same person from a gallery set. In other words, the ReID model should guarantee q share more similar feature with g than with other images. Therefore, learning a discriminative feature extractor is critical for person ReID.

In unsupervised domain adaptive person ReID, we have an unlabeled target domain $T = \{t_i\}_{i=1}^{N_T}$ containing N_T person images. Additionally, a labeled source domain $S = \{s_i, y_i\}_{i=1}^{N_S}$ containing N_S labeled person images is provided as an auxiliary training set, where y_i is the identity label associated with the person image s_i. The goal of domain adaptive person ReID is to learn a discriminative feature extractor $\mathrm{f}(\cdot)$ for T, using both S and T.

The training of $\mathrm{f}(\cdot)$ can be conducted by minimizing the training loss on both source and target domains. With person ID labels, the training on S can be considered as a classification task by minimizing the cross-entropy loss, *i.e.*,

$$\mathcal{L}_{src} = -\frac{1}{N_S}\sum_{i=1}^{N_S} \log\ \mathrm{P}(y_i|s_i), \tag{1}$$

where $\mathrm{P}(y_i|s_i)$ is the predicted probability of sample s_i belonging to class y_i. This supervised learning ensures the performance of $\mathrm{f}(\cdot)$ on source domain.

To gain discriminative power of $\mathrm{f}(\cdot)$ to the target domain, we further compute training loss with predicted labels on T. First, because each training batch samples $n_T, n_T \ll N_T$ images from T, it is likely that n_T images are sampled from different persons. We thus simply label each image t_i in the mini-batch with a distinct person ID label, *i.e.*, an one-hot vector $\boldsymbol{l}_i$ with $\boldsymbol{l}_i[j] = 1$ only if $i = j$. A Self-Adaptive Classification (SAC) model is adopted to separate images of different persons in the training batch. The objective of SAC can be formulated as minimizing the classification loss, *i.e.*,

$$\mathcal{L}_{local} = \frac{1}{n_T}\sum_{i=1}^{n_T} \mathrm{L}(\mathcal{V} \times \mathrm{f}(t_i), \boldsymbol{l}_i), \tag{2}$$

where n_T denotes the number of images in a training batch. $\mathrm{f}(\cdot)$ produces a d-dim feature vector. $\mathcal{V}$ stores n_T d-dim vectors as the classifier. $\mathcal{V} \times \mathrm{f}(t_i)$ computes the classification score, and $\mathrm{L}(\cdot)$ computes the loss by comparing classification scores and one-hot labels. Details of classifier $\mathcal{V}$ will be given in Sect. 3.2.

Besides the local optimization in each training batch, we further predict labels on the entire T and perform a global optimization. Since each person may have multiple images in T, we propose the Memory-based Temporal-guide Cluster (MTC) to predict a multi-class label for each image. For an image t_i, MTC predicts its multi-class label $\boldsymbol{m}_i$, where $\boldsymbol{m}_i[j] = 1$ only if t_i and t_j are regarded as containing the same person.

Predicted label $\boldsymbol{m}_i$ allows for a multi-label classification on T. We introduce a memory bank $\mathcal{K} \in \mathbf{R}^{N_T \times d}$ to store N_T image features as a N_T-class classifier [39]. The multi-label classification loss is computed by classifying image feature $\mathrm{f}(t_i)$ with the memory bank $\mathcal{K}$, then comparing the classification scores with multi-class label $\boldsymbol{m}_i$. The multi-label classification loss on T can be represented as

$$\mathcal{L}_{global} = \frac{1}{N_T}\sum_{i=1}^{N_T} \mathrm{L}(\mathcal{K} \times \mathrm{f}(t_i), \boldsymbol{m}_i), \tag{3}$$

where $\mathcal{K} \times \mathrm{f}(t_i)$ produces the classification score. The memory bank $\mathcal{K}$ is updated after each training iteration as

$$\mathcal{K}[i]^t = (1-\alpha)\mathcal{K}[i]^{t-1} + \alpha\,\mathrm{f}(t_i), \tag{4}$$

where the superscript t denotes the training epoch, α is the updating rate. Detailed of MTC and $\boldsymbol{m}_i$ computation will be presented in Sect. 3.3.

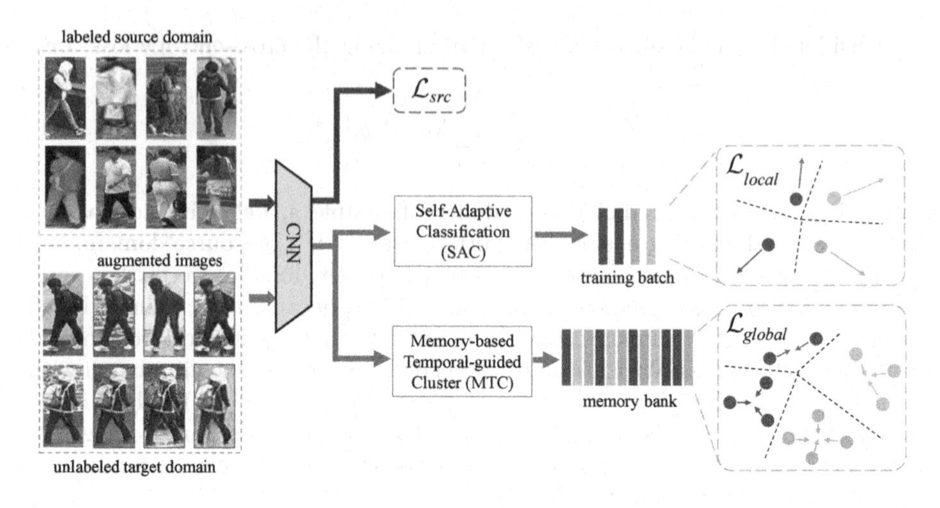

Fig. 1. Overview of the proposed framework for unsupervised domain adaptive ReID model training. $\mathcal{L}_{src}$ is computed on the source domain. SAC computes $\mathcal{L}_{local}$ in each training batch. MTC computes $\mathcal{L}_{global}$ on the entire target domain. SAC and MTC predict one-hot label and multi-class label for each image, respectively. Without $\mathcal{L}_{src}$, our framework works as unsupervised training.

By combining the above losses computed on S and T, the overall training loss of our method can be formulated as,

$$\mathcal{L} = \mathcal{L}_{src} + w_1\mathcal{L}_{local} + w_2\mathcal{L}_{global}, \tag{5}$$

where w_1 and w_2 are loss weights.

The accuracy of predicted labels, *i.e.*, $\boldsymbol{l}$ and $\boldsymbol{m}$ is critical for the training on T. The accuracy of $\boldsymbol{l}$ can be guaranteed by setting batch size $n_T \ll N_T$, and using careful sampling strategies. To ensure the accuracy of $\boldsymbol{m}$, MTC considers both visual similarity and temporal consistency for label prediction.

We illustrate our training framework in Fig. 1, where $\mathcal{L}_{local}$ can be efficiently computed within each training batch by classifying a few images. $\mathcal{L}_{global}$ is a more powerful supervision by considering the entire training set T. The combination of $\mathcal{L}_{local}$ and $\mathcal{L}_{global}$ utilizes both temporal and visual consistency among unlabeled data and guarantees strong robustness of the learned feature extractor $\mathrm{f}(\cdot)$. The following parts proceed to introduces the computation of $\mathcal{L}_{local}$ in SAC, and $\mathcal{L}_{local}$ in MTC, respectively.

3.2 Self-adaptive Classification

SAC classifies unlabeled data in each training batch. As shown in Eq. (2), the key of SAC is the classifier $\mathcal{V}$. For a batch consisting of n_T images, the classifier $\mathcal{V}$ is defined as a $n_T \times d$ sized tensor, where the i-th d-dim vector represents the classifiers for the i-th image. To enhance its robustness, $\mathcal{V}$ is calculated based on features of original images and their augmented duplicates.

Specifically, for an image t_i in training batch, we generate k images $t_i^{(j)}$ ($j = 1, 2, ..., k$) with image argumentation. This enlarges the training batch to $n_T \times (k+1)$ images belonging to n_T categories. The classifier $\mathcal{V}$ is computed as,

$$\mathcal{V} = [v_1, v_2, ...v_{n_T}] \in \mathbf{R}^{n_T \times d}, \; v_i = \frac{1}{k+1}(\mathrm{f}(t_i) + \sum_{j=1}^{k} \mathrm{f}(t_i^{(j)})), \tag{6}$$

where v_i is the averaged feature of t_i and its augmented images. It can be inferred that, the robustness of $\mathcal{V}$ enhances as $\mathrm{f}(\cdot)$ gains more discriminative power. We thus call $\mathcal{V}$ as a self-adapted classifier.

Data augmentation is critical to ensure the robustness of $\mathcal{V}$ to visual variations. We consider each camera as a style domain and adopt CycleGAN [42] to train camera style transfer models [40]. For each image under a specific camera, we totally generate $C-1$ images with different styles, where C is the camera number in the target domain. We set $k < C-1$. Therefore, each training batch randomly selects k augmented images for training.

Based on classifier $\mathcal{V}$ and the one-hot label $\boldsymbol{l}$, the $\mathcal{L}_{local}$ of SAC can be formulated as the cross-entropy loss, *i.e.*,

$$\mathcal{L}_{local} = -\frac{1}{n_T \times (k+1)} \sum_{i=1}^{n_T} (\log(\mathrm{P}(i|t_i) + \sum_{j=1}^{k} \log(\mathrm{P}(i|t_i^{(j)})), \tag{7}$$

where $\mathrm{P}(i|t_i)$ is the probability of image t_i being classified to label i, *i.e.*,

$$\mathrm{P}(i|t_i) = \frac{\exp(v_i^T \cdot \mathrm{f}(t_i)/\beta_1)}{\sum_{n=1}^{n_T} \exp(v_n^T \cdot \mathrm{f}(t_i)/\beta_1)} \tag{8}$$

where β_1 is a temperature factor to balance the feature distribution.

$\mathcal{L}_{local}$ can be efficiently computed on n_T images. Minimizing $\mathcal{L}_{local}$ enlarges the feature distance of images in the same training batch, meanwhile decreases the feature distance of augmented images in the same category. It thus boosts the discriminative power of $\mathrm{f}(\cdot)$ on T.

3.3 Memory-Based Temporal-Guided Cluster

MTC predicts the multi-class label $\boldsymbol{m}_i$ for image t_i through clustering images in T, *i.e.*, images inside the same cluster are assigned with the same label. The clustering is conducted based on the pair-wise similarity considering both visual similarity and temporal consistency of two images.

Visual similarity can be directly computed using the feature extractor $\mathrm{f}(\cdot)$ or the features stored in the memory bank $\mathcal{K}$. Using $\mathrm{f}(\cdot)$ requires to extract features for each image in T, which introduces extra time consumption. Meanwhile, the features in $\mathcal{K}$ can be enhanced by different image argumentation strategies, making them more robust. We hence use features in $\mathcal{K}$ to compute the visual similarity between two images t_i and t_j, *i.e.*,

$$\mathrm{vs}(t_i, t_j) = \mathrm{cosine}(\mathcal{K}[i], \mathcal{K}[j]), \tag{9}$$

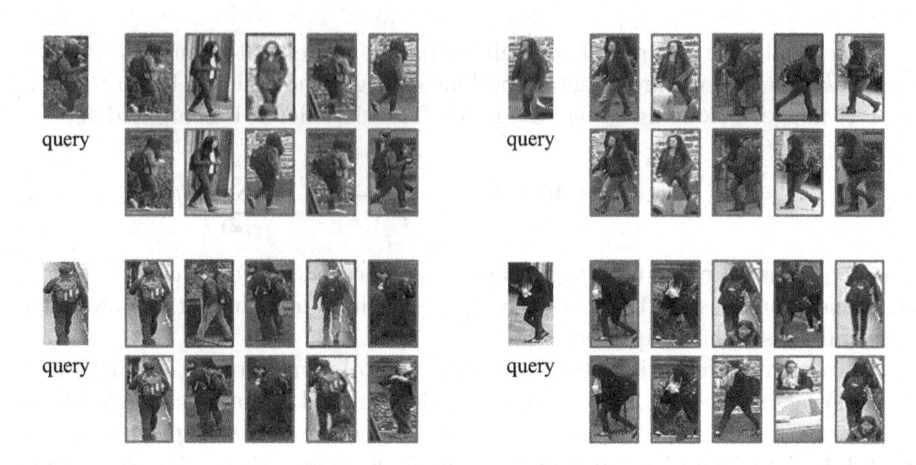

Fig. 2. Illustration of person ReID results on DukeMTMC-reID dataset. Each example shows the top-5 retrieved images by visual similarity (first tow) and joint similarity computed in Eq. (12) (second row). The true match is annotated by the green bounding box and false match is annotated by the red bounding box. (Color figure online)

where vs(·) computes the visual similarity with cosine distance.

Temporal consistency is independent to visual features and is related to the camera id and frame id of each person image. Suppose we have two images t_i from camera a and t_j from camera b with frame IDs fid_i and fid_j, respectively. The temporal consistency between t_i and t_j can be computed as,

$$\mathrm{ts}(t_i, t_j) = \mathrm{H}_{(a,b)}(fid_i - fid_j), \tag{10}$$

where $\mathrm{H}_{(a,b)}(\cdot)$ is a function for camera pair (a, b). It estimates the temporal consistency based on frame id interval of t_i and t_j, which reflects the time interval when they are recorded by cameras a and b.

$\mathrm{H}_{(a,b)}(\cdot)$ can be estimated based on a histogram $\bar{\mathrm{H}}_{(a,b)}(int)$, which shows the probability of appearing identical person at camera a and b for frame id interval int. $\bar{\mathrm{H}}_{(a,b)}(int)$ can be easily computed on datasets with person ID labels. To estimate it on unlabeled T, we first cluster images in T with visual similarity in Eq. (9) to acquire pseudo person ID labels. Suppose $n_{(a,b)}$ is the total number of image pairs containing identical person in camera a and b. The value of int-th bin in histogram, *i.e.*, $\bar{\mathrm{H}}_{(a,b)}(int)$ is computed as,

$$\bar{\mathrm{H}}_{(a,b)}(int) = n_{(a,b)}^{int} / n_{(a,b)}, \tag{11}$$

where $n_{(a,b)}^{int}$ is the number of image pairs containing identical person in camera a and b with frame id intervals int.

For a dataset with C cameras, $C(C-1)/2$ histograms will be computed. We finally use Gaussian function to smooth the histogram and take the smoothed histogram as $\mathrm{H}_{(a,b)}(\cdot)$ for temporal consistency computation.

Our final pair-wise similarity is computed based on vs(·) and ts(·). Because those two similarities have different value ranges, we first normalize them, then perform the fusion. This leads to the joint similarity function J(·), *i.e.*,

$$\mathrm{J}(t_i, t_j) = 1/(1 + \lambda_0 e^{-\gamma_0 \,\mathrm{vs}(t_i,t_j)}) \times 1/(1 + \lambda_1 e^{-\gamma_1 \,\mathrm{ts}(t_i,t_j)}), \tag{12}$$

where λ_0 and λ_1 are smoothing factors, γ_0 and γ_1 are shrinking factors.

Equation (12) computes more reliable similarities between images than either Eq. (9) or Eq. (10). J(·) can also be used in person ReID for query-gallery similarity computation. Figure 2 compares some ReID results achieved by visual similarity and joint similarity, respectively. It can be observed that, the joint similarity is more discriminative than the visual similarity.

We hence cluster images in target domain T based on J(·) and assign the multi-class label for each image. For an image t_i, its multi-class label $\boldsymbol{m}_i[j] = 1$ only if t_i and t_j are in the same cluster. Based on $\boldsymbol{m}$, the $\mathcal{L}_{global}$ on T can be computed as,

$$\mathcal{L}_{global} = -\frac{1}{N_T} \sum_{i=1}^{N_T} \sum_{j=1}^{N_T} \boldsymbol{m}_i[j] \times \log \bar{\mathrm{P}}(j|t_i)/|\boldsymbol{m}_i|_1, \tag{13}$$

where $|\cdot|_1$ computes the L-1 norm. $\bar{\mathrm{P}}(j|t_i)$ denotes the probability of image t_i being classified to the j-th class in multi-label classification, *i.e.*,

$$\bar{\mathrm{P}}(j|t_i) = \frac{\exp(\mathcal{K}[j]^T \cdot \mathrm{f}(t_i)/\beta_2)}{\sum_{n=1}^{N_T} \exp(\mathcal{K}[n]^T \cdot \mathrm{f}(t_i)/\beta_2)}, \tag{14}$$

where β_2 is the temperature factor. The following section proceeds to discuss the effects of parameters and conduct comparisons with recent works.

4 Experiment

4.1 Dataset

We evaluate our methods on three widely used person ReID datasets, *e.g.*, Market1501 [36], DukeMTMC-ReID [26,37], and MSMT17 [31], respectively.

Market1501 consists of 32,668 images of 1,501 identities under 6 cameras. The dataset is divided into training and test sets, which contains 12,936 images of 751 identities and 19,732 images of 750 identities, respectively.

DukeMTMC-ReID is composed of 1,812 identities and 36,411 images under 8 cameras. 16,522 images of 702 pedestrians are used for training. The other identities and images are included in the testing set.

MSMT17 is currently the largest image person ReID dataset. MSMT17 contains 126,441 images of 4,101 identities under 15 cameras. The training set of MSMT17 contains 32,621 bounding boxes of 1,041 identities, and the testing set contains 93,820 bounding boxes of 3,060 identities.

We follow the standard settings in previous works [5,39] for training in domain adaptive person ReID and unsupervised person ReID, respectively. Performance is evaluated by the Cumulative Matching Characteristic (CMC) and mean Average Precision (mAP). We use **JVTC** to denote our method.

Table 1. Evaluation of individual components of JVTC.

Dataset	DukeMTMC → Market1501					Market1501 → DukeMTMC				
Method	mAP	r1	r5	r10	r20	mAP	r1	r5	r10	r20
Supervised	69.7	86.3	94.3	96.5	97.6	61.0	80.2	89.1	91.9	94.2
Direct transfer	18.2	42.1	60.7	67.9	74.8	16.6	31.8	48.4	55.0	61.7
Baseline	46.6	77.4	89.5	93.0	95.1	43.6	66.1	77.7	81.7	84.8
SAC	41.8	64.5	76.0	79.6	92.3	37.5	59.4	74.1	78.3	81.4
MTC	56.4	79.8	91.0	93.9	95.9	51.1	71.3	81.1	84.3	86.3
JVTC	61.1	83.8	93.0	95.2	96.9	56.2	75.0	85.1	88.2	90.4
JVTC+	**67.2**	**86.8**	**95.2**	**97.1**	**98.1**	**66.5**	**80.4**	**89.9**	**92.2**	**93.7**

4.2 Implementation Details

We adopt ResNet50 [10] as the backbone and add a 512-dim embedding layer for feature extraction. We initialize the backbone with the model pre-trained on ImageNet [2]. All models are trained and finetuned with PyTorch. Stochastic Gradient Descent (SGD) is used to optimize our model. Input images are resized to 256 × 128. The mean value is subtracted from each (B, G, and R) channel. The batch size is set as 128 for both source and target domains. Each training batch in the target domain contains 32 original images and each image has 3 augmented duplicates, *i.e.*, we set $k=3$.

The temperature factor β_1 is set as 0.1 and β_2 is set as 0.05. The smoothing factors and shrinking factors λ_0, λ_1, γ_0 and γ_1 in Eq. (12) are set as $1, 2, 5$ and 5, respectively. The initial learning rate is set as 0.01, and is reduced by ten times after 40 epoches. The multi-class label $\boldsymbol{m}$ are updated every 5 epochs based on visual similarity initially, and the joint similarity is introduced at 30-th epoch. Only local loss $\mathcal{L}_{local}$ is applied at the initial epoch. The $\mathcal{L}_{global}$ is applied at the 10-th epoch. The training is finished after 100 epoches. The memory updating rate α starts from 0 and grows linearly to 1. The loss weights w_1 and w_2 are set as 1 and 0.2, respectively. DBSCAN [4] is applied for clustering.

4.3 Ablation Study

Evaluation of Individual Components: This section investigates the effectiveness of each component in our framework, *e.g.*, the SAC and MTC. We summarize the experimental results in Table 1. In the table, "Supervised" denotes training deep models with labeled data on the target domain, and testing on the testing set. "Direct transfer" denotes directly using the model trained on source domain for testing. "Baseline" uses memory bank for multi-label classification, but predicts multi-class label only based on visual similarity. "SAC" is implemented based on "Direct transfer" by applying SAC model for one-hot classification. "MTC" utilizes both visual similarity and temporal consistency

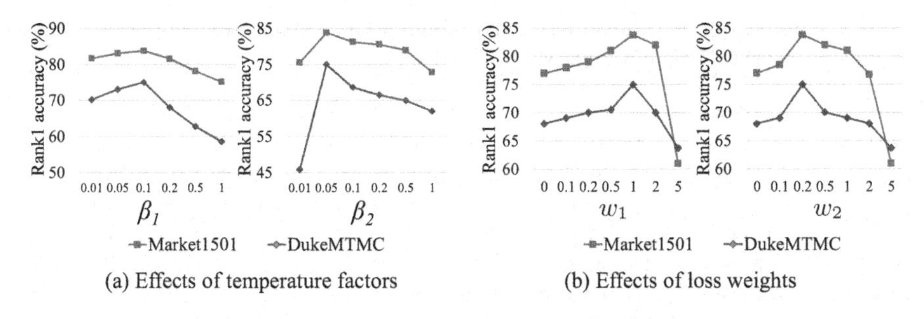

Fig. 3. Influences of temperature factors β_1 and β_2 in (a), and loss weights w_1, w_2 in (b). Experiments are conducted on Market1501 and DukeMTMC-reID.

for multi-class label prediction. "JVTC" combines SAC and MTC. "JVTC+" denotes using the joint similarity for person ReID.

Table 1 shows that, supervised learning on the target domain achieves promising performance. However, directly transferring the supervised model to different domains leads to substantial performance drop, *e.g.*, the rank1 accuracy drops to 44.2% on Market1501 and 48.4% on DukeMTMC-reID after direct transfer. The performance drop is mainly caused by the domain bias between datasets.

It is also clear that, SAC consistently outperforms direct transfer by large margins. For instance, SAC improves the rank1 accuracy from 42.1% to 64.5% and 31.8% to 59.4% on Market-1501 and DukeMTMC-reID, respectively. This shows that, although SAC is efficient to compute, it effectively boosts the ReID performance on target domain. Compared with the baseline, MTC uses joint similarity for label prediction. Table 1 shows that, MTC performs better than the baseline, *e.g.*, outperforms baseline by 9.8% and 5.2% in mAP on Market1501 and DukeMTMC-reID, respectively. This performance gain clearly indicates the robustness of our joint similarity.

After combining SAC and MTC, JVTC achieves more substantial performance gains on two datasets. For instance, JVTC achieves mAP of 61.1% on Market1501, much better than the 46.6% of baseline. "JVTC+" further uses joint similarity to compute the query-gallery similarity. It achieves the best performance, and outperforms the supervised training on target domain. We hence could conclude that, each component in our method is important for performance boost, and their combination achieves the best performance.

Hyper-parameter Analysis: This section investigates some important hyper-parameters in our method, including the temperature factors β_1, β_2, and the loss weights w_1 and w_2, respectively. To make the evaluation possible, each experiment varies the value of one hyper-parameter while keeping others fixed. All experiments are conducted with unsupervised domain adaptive ReID setting on both Market-1501 and DukeMTMC-reID.

Figure (3)(a) shows the effects of temperature factors β_1 and β_2 in Eq. (8) and Eq. (14). We can see that, a small temperature factor usually leads to better ReID performance. This is because that smaller temperature factor leads to

Table 2. Comparison with unsupervised, domain adaptive, and semi-supervised ReID methods on Market1501 and DukeMTMC-reID.

Dataset	Market1501						DukeMTMC					
Method	Source	mAP	r1	r5	r10	r20	Source	mAP	r1	r5	r10	r20
Supervised	Market	69.7	86.3	94.3	96.5	97.6	Duke	61.0	80.2	89.1	91.9	94.2
Direct transfer	Duke	18.2	42.1	60.7	67.9	74.8	Market	16.6	31.8	48.4	55.0	61.7
LOMO [20]	None	8.0	27.2	41.6	49.1	–	None	4.8	12.3	21.3	26.6	–
BOW [36]	None	14.8	35.8	52.4	60.3	–	None	8.3	17.1	28.8	34.9	–
BUC [21]	None	38.3	66.2	79.6	84.5	–	None	27.5	47.4	62.6	68.4	–
DBC [3]	None	41.3	69.2	83.0	87.8	–	None	30.0	51.5	64.6	70.1	–
JVTC	None	41.8	72.9	84.2	88.7	92.0	None	42.2	67.6	78.0	81.6	84.5
JVTC+	None	**47.5**	**79.5**	**89.2**	**91.9**	**94.0**	None	**50.7**	**74.6**	**82.9**	**85.3**	**87.2**
PTGAN [31]	Duke	–	38.6	–	66.1	–	Market	–	27.4	–	50.7	–
CamStyle [41]	Duke	27.4	58.8	78.2	84.3	88.8	Market	25.1	48.4	62.5	68.9	74.4
T-Fusion [22]	CUHK01	–	60.8	74.4	79.3	–	–	–	–	–	–	–
ARN [19]	Duke	39.4	70.3	80.4	86.3	93.6	Market	33.4	60.2	73.9	79.5	82.5
MAR [34]	MSMT17	40.0	67.7	81.9	87.3	–	MSMT17	48.0	67.1	79.8	84.2	–
ECN [39]	Duke	43.0	75.1	87.6	91.6	–	Market	40.4	63.3	75.8	80.4	–
PDA-Net [18]	Duke	47.6	75.2	86.3	90.2	–	Market	45.1	63.2	77.0	82.5	–
PAST [35]	Duke	54.6	78.4	–	–	–	Market	54.3	72.4	–	–	–
CAL-CCE [25]	Duke	49.6	73.7	–	–	–	Market	45.6	64.0	–	–	–
CR-GAN [1]	Duke	54.0	77.7	89.7	92.7	–	Market	48.6	68.9	80.2	84.7	–
SSG [5]	Duke	58.3	80.0	90.0	92.4	–	Market	53.4	73.0	80.6	83.2	–
TAUDL [15]	Tracklet	41.2	63.7	–	–	–	Tracklet	43.5	61.7	–	–	–
UTAL [16]	Tracklet	46.2	69.2	–	–	–	Tracklet	43.5	62.3	–	–	–
SSG+ [5]	Duke	62.5	81.4	91.6	93.8	–	Market	56.7	74.2	83.5	86.7	–
SSG++ [5]	Duke	**68.7**	86.2	94.6	96.5	–	Market	60.3	76.0	85.8	89.3	–
JVTC	Duke	61.1	83.8	93.0	95.2	96.9	Market	56.2	75.0	85.1	88.2	90.4
JVTC+	Duke	67.2	**86.8**	**95.2**	**97.1**	**98.1**	Market	**66.5**	**80.4**	**89.9**	**92.2**	**93.7**

a smaller entropy in the classification score, which is commonly beneficial for classification loss computation. However, too small temperature factor makes the training hard to converge. According to Fig. 3(a), we set $\beta_1 = 0.1$, $\beta_2 = 0.05$.

Figure 3(b) shows effects of loss weight w_1 and w_2 in network training. We vary the loss weight w_1 and w_2 from 0 to 5. $w_1(w_2) = 0$ means we don't consider the corresponding loss during training. It is clear that, a positive loss weight is beneficial for the ReID performance on both datasets. As we increase the loss weights, the ReID performance starts to increase. The best performance is achieved with $w_1 = 1$ and $w_2 = 0.2$ on two datasets. Further increasing the loss weights substantially drops the ReID performance. This is because increasing w_1 and w_2 decreases the weight of $\mathcal{L}_{src}$, which is still important. Based on this observation, we set $w_1 = 1$ and $w_2 = 0.2$ in following experiments.

4.4 Comparison with State-of-the-Art Methods

This section compares our method against state-of-the-art unsupervised, unsupervised domain adaptive, and semi-supervised methods on three datasets. Comparisons on Market1501 and DukeMTMC-reID are summarized in Table 2. Comparisons on MSMT17 are summarized in Table 3. In those tables, "Source" refers to the labeled source dataset, which is used for training in unsupervised domain adaptive ReID. "None" denotes unsupervised ReID.

Comparison on Market1501 and DukeMTMC-reID: We first compare our method with unsupervised learning methods. Compared methods include hand-crafted features LOMO [20] and BOW [36], and deep learning methods DBC [3] and BUC [21]. It can be observed from Table 2 that, hand-crafted features LOMO and BOW show unsatisfactory performance, even worse than directly transfer. Using unlabeled training dataset for training, deep learning based methods outperform hand-crafted features. BUC and DBC first treat each image as a single cluster, then merge clusters to seek pseudo labels for training. Our method outperforms them by large margins, *e.g.*, our rank1 accuracy on Market1501 achieves 72.9% *vs.* their 66.2% and 69.2%, respectively. The reasons could be because our method considers both visual similarity and temporal consistency to predict labels. Moreover, our method further computes classification loss in each training batch with SAC. By further considering temporal consistency during testing, JVTC+ gets further performance promotions on both datasets, even outperforms several unsupervised domain adaptive methods.

We further compare our method with unsupervised domain adaptive methods including PTGAN [31], CamStyle [41], T-Fusion [22], ARN [19], MAR [34], ECN [39], PDA-Net [18], PAST [35], CAL-CCE [25], CR-GAN [1] and SSG [5], and semi-supervised methods including TAUDL [15], UTAL [16], SSG+ [5], and SSG++ [5]. Under the unsupervised domain adaptive training setting, our method achieves the best performance on both Market1501 and DukeMTMC-reID in Table 2. For example, our method achieves 83.8% rank1 accuracy on Market1501 and gets 75.0% rank1 accuracy on DukeMTMC-reID. T-Fusion [22] also use temporal cues for unsupervised ReID, but achieves unsatisfactory performance, *e.g.*, 60.8% rank1 accuracy on Market1501 dataset. The reason may because that T-Fusion directly multiplies the visual and temporal probabilities, while our method fuses the visual and temporal similarities through more reasonable smooth fusion to boost the robustness. Our method also consistently outperforms the recent SSG [5] on those two datasets. SSG clusters multiple visual features and needs to train 2100 epoches before convergence. Differently, our method only uses global feature and could be well-trained in 100 epoches. We hence could conclude that, our method is also more efficient than SSG. By further considering temporal consistency during testing, JVTC+ outperforms semi-supervised method SSG++ [5] and supervised training on target domain.

Comparison on MSMT17: MSMT17 is more challenging than Market1501 and DukeMTMC-reID because of more complex lighting and scene variations. Some works have reported performance on MSMT17, including unsupervised

Table 3. Comparison with unsupervised and domain adaptive methods on MSMT17.

Method	Source	mAP	r1	r5	r10	r20
Supervised	MSMT17	35.9	63.3	77.7	82.4	85.9
JVTC	None	15.1	39.0	50.9	56.8	61.9
JVTC+	None	17.3	43.1	53.8	59.4	64.7
PTGAN [31]	Market1501	2.9	10.2	24.4	–	
ECN [39]		8.5	25.3	36.3	42.1	–
SSG [5]		13.2	31.6	49.6	–	–
SSG++ [5]		16.6	37.6	57.2	–	–
JVTC		19.0	42.1	53.4	58.9	64.3
JVTC+		**25.1**	**48.6**	**65.3**	**68.2**	**75.2**
PTGAN [31]	DukeMTMC	3.3	11.8	27.4	–	–
ECN [39]		10.2	30.2	41.5	46.8	–
SSG [5]		13.3	32.2	51.2	–	–
SSG++ [5]		18.3	41.6	62.2	–	–
JVTC		20.3	45.4	58.4	64.3	69.7
JVTC+		**27.5**	**52.9**	**70.5**	**75.9**	**81.2**

domain adaptive methods PTGAN [31], ECN [39] and SSG [5], and semi-supervised method SSG++ [5], respectively. The comparison on MSMT17 are summarized in Table 3. As shown in the table, our method outperforms existing methods by large margins. For example, our method achieves 45.4% rank1 accuracy when using DukeMTMC-reID as the source dataset, which outperforms the unsupervised domain adaptive method SSG [5] and semi-supervised method SSG++ [5] by 13.2% and 3.8%, respectively. We further achieves 52.9% rank1 accuracy after applying the joint similarity during ReID. This outperforms the semi-supervised method SSG++ [5] by 11.3%. The above experiments on three datasets demonstrate the promising performance of our JVTC.

5 Conclusion

This paper tackles unsupervised domain adaptive person ReID through jointly enforcing visual and temporal consistency in the combination of local one-hot classification and global multi-class classification. Those two classification tasks are implemented by SAC and MTC, respectively. SAC assigns images in the training batch with distinct person ID labels, then adopts a self-adaptive classier to classify them. MTC predicts multi-class labels by considering both visual similarity and temporal consistency to ensure the quality of label prediction. The two classification models are combined in a unified framework for discriminative feature learning on target domain. Experimental results on three datasets demonstrate the superiority of the proposed method over state-of-the-art unsupervised and domain adaptive ReID methods.

Acknowledgments. This work is supported in part by Peng Cheng Laboratory, The National Key Research and Development Program of China under Grant No. 2018YFE0118400, in part by Beijing Natural Science Foundation under Grant No. JQ18012, in part by Natural Science Foundation of China under Grant No. 61936011, 61620106009, 61425025, 61572050, 91538111.

References

1. Chen, Y., Zhu, X., Gong, S.: Instance-guided context rendering for cross-domain person re-identification. In: ICCV (2019)
2. Deng, J., Dong, W., Socher, R., Li, L.J., Li, K., Fei-Fei, L.: ImageNet: a large-scale hierarchical image database. In: CVPR (2009)
3. Ding, G., Khan, S., Yin, Q., Tang, Z.: Dispersion based clustering for unsupervised person re-identification. In: BMVC (2019)
4. Ester, M., Kriegel, H.P., Sander, J., Xu, X.: Density-based spatial clustering of applications with noise. In: KDD (1996)
5. Fu, Y., Wei, Y., Wang, G., Zhou, Y., Shi, H., Huang, T.S.: Self-similarity grouping: a simple unsupervised cross domain adaptation approach for person re-identification. In: ICCV (2019)
6. Ganin, Y., Lempitsky, V.: Unsupervised domain adaptation by backpropagation. arXiv preprint arXiv:1409.7495 (2014)
7. Ghifary, M., Kleijn, W.B., Zhang, M., Balduzzi, D., Li, W.: Deep reconstruction-classification networks for unsupervised domain adaptation. In: Leibe, B., Matas, J., Sebe, N., Welling, M. (eds.) ECCV 2016. LNCS, vol. 9908, pp. 597–613. Springer, Cham (2016). https://doi.org/10.1007/978-3-319-46493-0_36
8. Gray, D., Tao, H.: Viewpoint invariant pedestrian recognition with an ensemble of localized features. In: Forsyth, D., Torr, P., Zisserman, A. (eds.) ECCV 2008. LNCS, vol. 5302, pp. 262–275. Springer, Heidelberg (2008). https://doi.org/10.1007/978-3-540-88682-2_21
9. Gretton, A., Borgwardt, K., Rasch, M., Schölkopf, B., Smola, A.J.: A kernel method for the two-sample-problem. In: NeurIPS (2007)
10. He, K., Zhang, X., Ren, S., Sun, J.: Deep residual learning for image recognition. In: CVPR (2016)
11. Li, J., Wang, J., Tian, Q., Gao, W., Zhang, S.: Global-local temporal representations for video person re-identification. In: ICCV (2019)
12. Li, J., Zhang, S., Huang, T.: Multi-scale 3D convolution network for video based person re-identification. In: AAAI (2019)
13. Li, J., Zhang, S., Huang, T.: Multi-scale temporal cues learning for video person re-identification. IEEE Trans. Image Process. **29**, 4461–4473 (2020)
14. Li, J., Zhang, S., Tian, Q., Wang, M., Gao, W.: Pose-guided representation learning for person re-identification. IEEE Trans. Pattern Anal. Mach. Intell. (2019)
15. Li, M., Zhu, X., Gong, S.: Unsupervised person re-identification by deep learning tracklet association. In: Ferrari, V., Hebert, M., Sminchisescu, C., Weiss, Y. (eds.) ECCV 2018. LNCS, vol. 11208, pp. 772–788. Springer, Cham (2018). https://doi.org/10.1007/978-3-030-01225-0_45
16. Li, M., Zhu, X., Gong, S.: Unsupervised tracklet person re-identification. IEEE Trans. Pattern Anal. Mach. Intell. **42**, 1770–1778 (2019)
17. Li, W., Zhu, X., Gong, S.: Harmonious attention network for person re-identification. In: CVPR (2018)

18. Li, Y.J., Lin, C.S., Lin, Y.B., Wang, Y.C.F.: Cross-dataset person re-identification via unsupervised pose disentanglement and adaptation. arXiv preprint arXiv:1909.09675 (2019)
19. Li, Y.J., Yang, F.E., Liu, Y.C., Yeh, Y.Y., Du, X., Frank Wang, Y.C.: Adaptation and re-identification network: an unsupervised deep transfer learning approach to person re-identification. In: CVPR Workshops (2018)
20. Liao, S., Hu, Y., Zhu, X., Li, S.Z.: Person re-identification by local maximal occurrence representation and metric learning. In: CVPR (2015)
21. Lin, Y., Dong, X., Zheng, L., Yan, Y., Yang, Y.: A bottom-up clustering approach to unsupervised person re-identification. In: AAAI (2019)
22. Lv, J., Chen, W., Li, Q., Yang, C.: Unsupervised cross-dataset person re-identification by transfer learning of spatial-temporal patterns. In: CVPR (2018)
23. Mao, S., Zhang, S., Yang, M.: Resolution-invariant person re-identification. In: IJCAI (2019)
24. Pan, Y., Yao, T., Li, Y., Wang, Y., Ngo, C.W., Mei, T.: Transferrable prototypical networks for unsupervised domain adaptation. In: CVPR (2019)
25. Qi, L., Wang, L., Huo, J., Zhou, L., Shi, Y., Gao, Y.: A novel unsupervised camera-aware domain adaptation framework for person re-identification. arXiv preprint arXiv:1904.03425 (2019)
26. Ristani, E., Solera, F., Zou, R., Cucchiara, R., Tomasi, C.: Performance measures and a data set for multi-target, multi-camera tracking. In: Hua, G., Jégou, H. (eds.) ECCV 2016. LNCS, vol. 9914, pp. 17–35. Springer, Cham (2016). https://doi.org/10.1007/978-3-319-48881-3_2
27. Su, C., Li, J., Zhang, S., Xing, J., Gao, W., Tian, Q.: Pose-driven deep convolutional model for person re-identification. In: ICCV (2017)
28. Sun, B., Feng, J., Saenko, K.: Return of frustratingly easy domain adaptation. In: AAAI (2016)
29. Wang, D., Zhang, S.: Unsupervised person re-identification via multi-label classification. In: CVPR (2020)
30. Wang, G., Lai, J., Huang, P., Xie, X.: Spatial-temporal person re-identification. In: AAAI (2019)
31. Wei, L., Zhang, S., Gao, W., Tian, Q.: Person transfer GAN to bridge domain gap for person re-identification. In: CVPR (2018)
32. Wei, L., Zhang, S., Yao, H., Gao, W., Tian, Q.: Glad: Global-local-alignment descriptor for pedestrian retrieval. In: ACM MM (2017)
33. Wu, Z., Xiong, Y., Yu, S.X., Lin, D.: Unsupervised feature learning via non-parametric instance discrimination. In: CVPR (2018)
34. Yu, H.X., Zheng, W.S., Wu, A., Guo, X., Gong, S., Lai, J.H.: Unsupervised person re-identification by soft multilabel learning. In: CVPR (2019)
35. Zhang, X., Cao, J., Shen, C., You, M.: Self-training with progressive augmentation for unsupervised cross-domain person re-identification. In: ICCV (2019)
36. Zheng, L., Shen, L., Tian, L., Wang, S., Wang, J., Tian, Q.: Scalable person re-identification: a benchmark. In: ICCV (2015)
37. Zheng, Z., Zheng, L., Yang, Y.: Unlabeled samples generated by GAN improve the person re-identification baseline in vitro. In: ICCV (2017)
38. Zhong, Y., Wang, X., Zhang, S.: Robust partial matching for person search in the wild. In: CVPR (2020)
39. Zhong, Z., Zheng, L., Luo, Z., Li, S., Yang, Y.: Invariance matters: exemplar memory for domain adaptive person re-identification. In: CVPR (2019)
40. Zhong, Z., Zheng, L., Zheng, Z., Li, S., Yang, Y.: Camera style adaptation for person re-identification. In: CVPR (2018)

41. Zhong, Z., Zheng, L., Zheng, Z., Li, S., Yang, Y.: CamStyle: a novel data augmentation method for person re-identification. IEEE Trans. Image Process. **28**(3), 1176–1190 (2018)
42. Zhu, J.Y., Park, T., Isola, P., Efros, A.A.: Unpaired image-to-image translation using cycle-consistent adversarial networks. In: ICCV (2017)

Autoencoder-Based Graph Construction for Semi-supervised Learning

Mingeun Kang[1], Kiwon Lee[1], Yong H. Lee[2], and Changho Suh[1](✉)

[1] KAIST, Daejeon, South Korea
{minkang23,kaiser5072,chsuh}@kaist.ac.kr
[2] UNIST, Ulsan, South Korea
yohlee@unist.ac.kr

Abstract. We consider graph-based semi-supervised learning that leverages a similarity graph across data points to better exploit data structure exposed in unlabeled data. One challenge that arises in this problem context is that conventional matrix completion which can serve to construct a similarity graph entails heavy computational overhead, since it re-trains the graph independently whenever model parameters of an interested classifier are updated. In this paper, we propose a holistic approach that employs a parameterized neural-net-based autoencoder for matrix completion, thereby enabling simultaneous training between models of the classifier and matrix completion. We find that this approach not only speeds up training time (around a three-fold improvement over a prior approach), but also offers a higher prediction accuracy via a more accurate graph estimate. We demonstrate that our algorithm obtains state-of-the-art performances by respectful margins on benchmark datasets: Achieving the error rates of 0.57% on MNIST with 100 labels; 3.48% on SVHN with 1000 labels; and 6.87% on CIFAR-10 with 4000 labels.

Keywords: Semi-supervised learning · Matrix completion · Autoencoders

1 Introduction

While deep neural networks can achieve human-level performance on a widening array of supervised learning problems, they come at a cost: Requiring a large collection of *labeled* data and thus relying on a huge amount of human effort to manually label examples. Semi-supervised learning (SSL) serves as a powerful framework that leverages *unlabeled* data to address the lack of labeled data.

M. Kang and K. Lee—Equal contribution.

Electronic supplementary material The online version of this chapter (https://doi.org/10.1007/978-3-030-58586-0_30) contains supplementary material, which is available to authorized users.

A. Vedaldi et al. (Eds.): ECCV 2020, LNCS 12369, pp. 500–517, 2020.
https://doi.org/10.1007/978-3-030-58586-0_30

One popular SSL paradigm is to employ *consistency loss* which captures a similarity between prediction results of the same data points yet with different small perturbations. This methodology is inspired by the key smoothness assumption that nearby data points are likely to have the same class, and a variety of algorithms have been developed inspired by this assumption [18,27,29,37]. However, it comes with one limitation: Taking into account only the same single data point (yet with perturbations) while not exploiting any relational structure across distinct data points.

This has naturally motivated another prominent SSL framework, named *graph-based SSL* [24,36,43]. The main idea is to employ a *similarity graph*, which represents the relationship among a pair of data points, to form another loss term, called *feature matching loss*, and then incorporate it as a regularization term into a considered optimization. This way, one can expect that similar data points would be embedded tighter in a low-dimensional space, thus yielding a higher classification performance [5] (also see Fig. 2 for visualization).

One important question that arises in graph-based SSL is: How to construct such similarity graph? Most prior works rely on features in the input space [2,43] or given (and/or predicted) labels in the output space [24]. A recent approach is to incorporate a matrix completion approach [36]. The approach exploits both features and available labels to form an augmented matrix having fully-populated features yet with very sparse labels. Leveraging the low-rank structure of such augmented matrix, it resorts to a well-known algorithm (nuclear norm minimization [8]), thereby estimating a similarity graph. However, we face one challenge in this approach. The challenge is that the graph estimation is done *independently* of the model parameter update of an interested classifier, which in turn incurs a significant computational complexity. Notice that the graph update should be done whenever changes are made on classifier model parameters.

Our main contribution lies in developing a novel integrated framework that holistically combines the classifier parameter update with the matrix-completion-based graph update. We introduce a parameterized neural-net-based autoencoder for matrix completion and define a new loss, which we name *autoencoder loss*, to reflect the quality of the graph estimation. We then integrate the autoencoder loss with the original supervised loss (computed only via few available labels) together with two additional losses: (i) consistency loss (guiding the same data points with small perturbations to yield the same class); (ii) feature matching loss (encouraging distinct yet similar data points, reflected in the estimated graph, to have the same class). As an autoencoder, we develop a novel structure inspired by the recent developments tailored for matrix completion [10,33]; see Remark 1 for details.

We emphasize two key aspects of our holistic approach. First, model parameters of the classifier (usually CNN) and the matrix completion block (autoencoder) are *simultaneously* trained, thus exhibiting a significant improvement in computational complexity (around a three-fold gain in various real-data experiments), relative to the prior approach [36] which performs the two procedures *separately in an alternate manner*. Second, our approach exploits the smoothness property of *both* same-yet-perturbed data points and distinct-yet-similar data points. This together with an accurate graph estimate due to our autoencoder-based matrix completion turns out to offer greater error rate performances on various benchmark datasets (MNIST, SVHN and CIFAR-10) with respectful margins over many other state of the arts [18,22,24,32,36–38]. The improvements are more significant when available labels are fewer. See Tables 1 and 2 in Sect. 5 for details.

2 Related Work

2.1 Semi-supervised Learning

There has been a proliferation of SSL algorithms. One major stream of the algorithms is along the idea of adversarial training [13]. While the literature based on such methodology is vast, we list a few recent advances with deep learning [9,22,32,35]. Springenberg [35] has proposed a categorical GAN (Cat-GAN) that learns a discriminative classifier so as to maximize mutual information between inputs and predicted labels while being robust to bogus examples produced by an adversarial generative model. Salimans et al. [32] propose a new loss, called feature matching loss, in an effort to stabilize GAN training, also demonstrating the effectiveness of the technique for SSL tasks. Subsequently Li et al. [22] introduce another third player on top of the two players in GANs to exploit label information, thus improving SSL performances. Dai et al. [9] provide a deeper understanding on the role of generator for SSL tasks, proposing another framework. While the prior GAN approaches yield noticeable classification performances, they face one common challenge: Suffering from training instability as well as high computational complexity.

Another prominent stream is not based on the GAN framework, instead employing *consistency loss* that penalizes the distinction of prediction results for the same data points having small perturbations [18,27,29,30,37]. Depending on how to construct the consistency loss, there are a variety of algorithms including: (i) ladder networks [30]; (ii) TempEns [18]; (iii) Mean Teacher (MT) [37]; (iv) Virtual Adversarial Training (VAT) [27]; (v) Virtual Adversarial Dropout (VAdD) [29]. The recent advances include [1,3,4,34,38,42,44,46]. In particular, [3,4,34,38,42,46] propose an interesting idea of mixing (interpolating) data points and/or predictions, to better exploit the consistency loss, thus achieving promising results.

Most recently, it has been shown in [48] that consistency loss can be better represented with the help of self-supervised learning, which can be interpreted as a feature representation approach via pretext tasks (such as predicting context or image rotation) [17].

2.2 Graph-Based SSL

In an effort to capture the *relational* structure across *distinct* data points which are not reflected by the prior approaches, a more advanced technique has been introduced: graph-based SSL. The key idea is to exploit such relational structure via a *similarity graph* that represents the closeness between data points. The similarity graph is desired to be constructed so as to well propagate information w.r.t. few labeled data into the unlabeled counterpart. Numerous algorithms have been developed depending on how to construct the similarity graph [2,6,12, 15,24,36,40,41,43,45,50,51]. Recent developments include [40,51] that exploit graph convolutional networks (GCNs) to optimize the similarity graph and a classifier.

Among them, two most related works are: (i) Smooth Neighbors on Teach Graphs (SNTG) [24]; (ii) GSCNN [36]. While the SNTG [24] designs the graph based solely on *labels* (given and/or predicted) in the output space, the GSCNN [36] incorporates also features in the input space to better estimate the similarity graph with the aid of the *matrix completion* idea. However, the employed matrix completion takes a conventional approach based on nuclear norm minimization [8], which operates independently of a classifier's parameter update and therefore requires two separate updates. This is where our contribution lies in. We employ a neural-net-based matrix completion so that it can be trained simultaneously with the interested classifier, thereby addressing the computational complexity issue.

2.3 Matrix Completion

Most traditional algorithms are based on rank minimization. Although the rank minimization is NP-hard, Candes and Recht [8] proved that for some enough number of observed matrix entries, one can perfectly recover an unknown matrix via nuclear norm minimization (NNM). This NNM algorithm formed the basis of the GSCNN [36]. Instead we employ a deep learning based approach for matrix completion. Our approach employs a *parameterized* neural-net-based autoencoder [10,23,33] and therefore the model update can be gracefully merged with that of the classifier. The parameterization that leads to simultaneous training is the key to speeding up training time.

3 Problem Formulation

Let $\mathcal{L} = \{(x_i, y_i)\}_{i=1}^{m}$ and $\mathcal{U} = \{x_i\}_{i=m+1}^{n}$ be labeled and unlabeled datasets respectively, wherein $x_i \in \mathcal{X}$ indicates the ith data (observation) and $y_i \in \mathcal{Y} = \{1, 2, \ldots, c\}$ denotes the corresponding label. Here c is the number of classes. Usually available labels are limited, i.e., $m \ll n$. The task of SSL is to design a classifier f (parameterized with θ) so that it can well predict labels for *unseen* examples. With regard to a loss function, we employ one conventional approach that takes into account the two major terms: (i) supervised loss (quantifying prediction accuracy w.r.t. labeled data); (ii) regularization terms (reflecting the data structure exposed in unlabeled data).

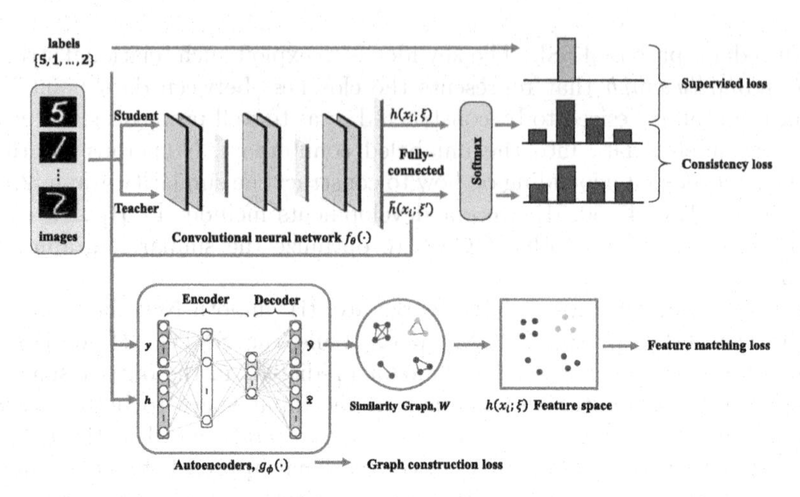

Fig. 1. A graph-based SSL architecture with a CNN classifier (parameterized with θ) and an autoencoder for matrix completion (parameterized with ϕ).

Supervised Loss. We consider:

$$L_s(\theta) = \sum_{i=1}^{m} \ell_s(f(x_i;\theta), y_i) \tag{1}$$

where $\ell_s(\cdot,\cdot)$ denotes cross entropy loss and $f(x_i;\theta)$ is the classifier softmax output that aims to represent the ground-truth conditional distribution $p(y_i|x_i;\theta)$.

As regularization terms, we employ two unsupervised losses: (i) consistency loss (that captures the distinction of prediction results of the same data points with different yet small perturbations); (ii) feature matching loss (that quantifies the difference between distinct data points via a similarity graph together with some features). Below are detailed formulas that we would take.

Consistency Loss. We use one conventional formula as in [18,24,27,29,37]:

$$L_c(\theta) = \sum_{i=1}^{n} \ell_c(f(x_i;\theta,\xi), \tilde{f}(x_i;\theta',\xi')) \tag{2}$$

where $f(x_i;\theta,\xi)$ denotes the prediction of one model, say student model, with parameter θ and random perturbation ξ (imposed to the input x_i); $\tilde{f}(x_i;\theta',\xi')$ indicates that of another model, say teacher model, with parameter θ' and different perturbation ξ' yet w.r.t. the same input x_i; and $\ell_c(\cdot,\cdot)$ is a distance measure between the outputs of the two models (for instance, the Euclidean distance or KL divergence). Here we exploit both labeled and unlabeled datasets. Notice that this loss penalizes the difference between the predictions of the same data point x_i yet with different perturbations, reflected in ξ and ξ'. We consider a simple setting where $\theta' = \theta$, although these can be different in general [18,27,29,37].

Feature Matching Loss. Another unsupervised loss that we will use to exploit the relational structure of different data points is feature matching loss:

$$L_g(\theta, \mathbf{W}) = \sum_{x_i, x_j \in \mathcal{L} \cup \mathcal{U}} \ell_g(h(x_i; \theta), h(x_j; \theta), W_{ij}) \tag{3}$$

where $h : \mathcal{X} \to \mathbb{R}^p$ is a mapping from the input space to a low dimensional feature space; W_{ij} denotes the (i, j) entry of a similarity graph matrix $\mathbf{W}$ (taking 1 if x_i and x_j are of the same class, 0 otherwise); and $\ell_g(\cdot, \cdot)$ is another distance measure between the ith and jth features which varies depending on W_{ij}. Here the function ℓ_g is subject to our design choice [24, 36], and we use the contrastive Siamese networks [7] as in [24]:

$$\ell_g = \begin{cases} \|h(x_i) - h(x_j)\|^2 & \text{if } W_{ij} = 1 \\ \max(0, \mathsf{margin} - \|h(x_i) - h(x_j)\|)^2 & \text{if } W_{ij} = 0 \end{cases} \tag{4}$$

where margin denotes a pre-defined positive value which serves as a threshold in feature difference for declaring distinct classes, and $\| \cdot \|^2$ is the Euclidean distance.

Various methods have been developed for construction of $\mathbf{W}$. One recent work is the SNTG [24] which uses the predictions of the teacher model for the construction. Another recent work [36] takes a *matrix completion* approach which additionally exploits features on top of given labels. Here the predicted labels due to matrix completion serve to construct the graph.

SSL Framework. Here is the unified loss function of our consideration:

$$L_s(\theta) + \lambda_c L_c(\theta) + \lambda_g L_g(\theta, \mathbf{W}) \tag{5}$$

where λ_c and λ_g are regularization factors that serve to balance across three different loss terms.

In (5), we face one challenge when employing a recent advance [36] that relies on conventional matrix completion for $\mathbf{W}$. The GSCNN [36] solves nuclear norm minimization [8] for matrix completion via a soft impute algorithm [26]. Here an issue arises: The algorithm [26] has to *retrain the graph* $\mathbf{W}$ *for every iteration whenever* features of the classifier, affected by θ, are updated. In other words, matrix completion is done in an *alternate* manner with the classifier update. For each iteration, say t, the graph is estimate to output, say $\mathbf{W}^{(t)}$, from the current classifier parameter, say $\theta^{(t)}$, and this updated $\mathbf{W}^{(t)}$ yields the next parameter estimate $\theta^{(t+1)}$ as per (5), and this process is repeated. Here the challenge is that this alternating update incurs significant computational complexity.

4 Our Approach

To address such challenge w.r.t. computational complexity, we invoke a parameterization trick. The idea is to parameterize a matrix completion block with a neural network, and to *simultaneously* train both parameters, one for classifier and the other for matrix completion. Specifically we employ an *autoencoder*-type neural network with parameter, say ϕ, for matrix completion; introduce another loss that we call *autoencoder loss*, which captures the quality of matrix completion and therefore the graph estimate; integrate the new loss with the other three loss terms in (5); and then simultaneously update both θ (classifier) and ϕ (autoencoder) guided by the integrated loss. We will detail the idea in the next sections.

4.1 Learning the Graph with Autoencoder

For a classifier, we employ a CNN model. Let $h(x_1), \ldots, h(x_n) \in \mathbb{R}^d$ be the feature vectors w.r.t. n examples prior to the softmax layer. We first construct a feature matrix that stacks all of the vectors: $\mathbf{X} = [h(x_1), \ldots, h(x_n)] \in \mathbb{R}^{d \times n}$. Let $u(y_1), \ldots, u(y_m) \in \mathbb{R}^c$ be the one-hot-coded label vectors. We then construct a label matrix stacking all of them: $\mathbf{Y} = [u(y_1), \ldots, u(y_m), \underbrace{\mathbf{0}_c, \ldots, \mathbf{0}_c}_{(n-m)\ vectors}] \in \mathbb{R}^{c \times n}$ where $\mathbf{0}_c$ is the all-zero vector of size c. Notice that $\mathbf{X}$ is fully populated while $\mathbf{Y}$ is very sparse due to many missing labels. Let $\Omega_{\mathbf{Y}}$ be the set of indices for the observed entries in $\mathbf{Y}$: $(i, j) \in \Omega_{\mathbf{Y}}$ when y_j is an observed label. Now we construct an augmented matrix that stacks $\mathbf{X}$ and $\mathbf{Y}$ in the row wise:

$$\mathbf{Z} = \begin{bmatrix} \mathbf{Y} \\ \mathbf{X} \end{bmatrix}. \tag{6}$$

This is an interested matrix for completion. Notice that completing $\mathbf{Z}$ enables predicting all the missing labels, thus leading to a graph estimate.

For matrix completion, we employ an autoencoder-type neural network which outputs a completed matrix fed by $\mathbf{Z}$. We also adopt the idea of input dropout [39]: Probabilistically replacing some of the label-presence column vectors with zero vector $\mathbf{0}_c$. For instance, when the jth column vector $\mathbf{z}_j = \begin{bmatrix} u(y_j) \\ h(x_j) \end{bmatrix}$ is dropped out, the output reads: $\bar{\mathbf{z}}_j = \begin{bmatrix} \mathbf{0}_c \\ h(x_j) \end{bmatrix}$. With the input dropout, we get:

$$\bar{\mathbf{Z}} = \begin{bmatrix} \bar{\mathbf{Y}} \\ \mathbf{X} \end{bmatrix}. \tag{7}$$

For training our autoencoder function g parameterized by ϕ, we employ the following loss (autoencoder loss) that reflects the reconstruction quality of the interested matrix:

$$L_{AE}(\theta, \phi) = \sum_{(i,j) \in \Omega_{\mathbf{Y}}} (\mathbf{Y}_{ij} - \hat{\mathbf{Y}}_{ij})^2 + \mu(t) \|\mathbf{X} - \hat{\mathbf{X}}\|_F^2 \tag{8}$$

where $(\hat{\mathbf{Y}}, \hat{\mathbf{X}})$ indicates the output of the ϕ-parameterized autoencoder $g_\phi(\bar{\mathbf{Z}})$; $\| \cdot \|_F$ denotes the Frobenius norm; and $\mu(t)$ is a hyperparameter that balances the reconstruction loss of observed labels (reflected in the first term) against that of the features (reflected in the second term). Since the quality of the feature matrix estimation improves with the learning progress (relative to the fixed label counterpart), we use a monotonically increasing ramp-up function $\mu(t)$ [18].

Let $\tilde{u}(y_i)$ be the predicted label vector for the ith example, obtained by the autoencoder $g_\phi(\bar{\mathbf{Z}})$. Let $\tilde{y}_i = \text{argmax}_k [\tilde{u}(y_i)]_k$ where $[\cdot]_k$ is the kth component of the vector (the estimated probability that the ith example belongs to class k). This then enables us to construct a similarity graph:

$$W_{ij} = \begin{cases} 1 & \text{if } \tilde{y}_i = \tilde{y}_j; \\ 0 & \text{if } \tilde{y}_i \neq \tilde{y}_j. \end{cases} \tag{9}$$

We consider a simple binary case where $W_{ij} \in \{0, 1\}$ as in [24]. One may use a weighted graph with soft-decision values. The constructed graph is then applied to the feature matching loss (3) to train the classifier model (CNN).

Remark 1. (Our choice for autoencoder structure): The autoencoder that we design is of a special structure: A nonlinear (multi-layered) encoder followed by a linear (single-layer) decoder. We name this the *basis learning autoencoder* [20, 21]. Note that the structure helps learning the basis via a linear decoder, as it is guided to linearly combine the vectors generated by a non-linear encoder. The rationale behind this choice is that any matrix can be represented as a linear combination of the basis vectors of its row or column space, and *the basis vectors serve as the most efficient features of a matrix.* The number of nodes in the last layer of the encoder can serve as an effective *rank* of the matrix (the number of basis vectors extracted from the encoder). We observe through experiments that matrix completion performs well when the effective rank is greater than or equal to the number of classes. Also we have empirically verified that the bases of the data are indeed learned by the autoencoder which has a linear decoder. We leave the detailed empirical analysis in the supplementary. ■

4.2 Simultaneous Training

We aim to update autoencoder parameter ϕ *simultaneously* with classifier parameter θ, unlike the prior approach [36] that takes an *alternating* update. To this end, we integrate the autoencoder loss (8) with the prior three loss terms (5):

$$L_s(\theta) + w(t) \cdot (L_c(\theta) + \lambda \cdot L_g(\theta, \mathbf{W})) + L_{AE}(\theta, \phi) \tag{10}$$

where $L_s(\theta)$, $L_c(\theta)$, $L_g(\theta, \mathbf{W})$ and $L_{AE}(\theta, \phi)$ indicate supervised loss (1), consistency loss (2), feature matching loss (3), and autoencoder loss (8), respectively. Here we employ three hyperparameters: λ controls the ratio between $L_c(\theta)$ and $L_g(\theta, \mathbf{W})$; $w(t)$ is a ramp-up function that increases with epoch; and $\mu(t)$ is another ramp-up function placed inside $L_{AE}(\theta, \phi)$ (see (8)). We use the same ramp-up function as in [18,24], since features are not reliable in the initial phase of the training. For simplicity, we apply the same weight between $L_{AE}(\theta, \phi)$ and $L_s(\theta)$.

Algorithm 1. Autoencoder-based Graph Construction

1: **Input:** Labeled dataset $\mathcal{L} = \{(x_i, y_i)\}_{i=1}^{m}$ and unlabeled dataset $\mathcal{U} = \{x_i\}_{i=m+1}^{n}$, hyperparameters $\lambda, \mu(t), w(t)$ and the number of training epochs T
2: **Initialize:** $f_\theta(x)$ (student model), $\tilde{f}_\theta(x)$ (teacher model), and $g_\phi(x)$ (ϕ-parameterized autoencoder)
3: **for** $t = 1$ to T **do**
4: Randomly sample mini-batches from $\mathcal{L}$ and $\mathcal{U}$
5: **for** each mini-batch $\mathcal{B}$ **do**
6: $(z_i, h_i) \leftarrow f_\theta(x_{i\in\mathcal{B}})$ obtain outputs & features
7: $(\tilde{z}_i, \tilde{h}_i) \leftarrow \tilde{f}_\theta(x_{i\in\mathcal{B}})$ (for teacher model)
8: $\hat{y} \leftarrow g_\phi([\tilde{h}_i, y_{i\in\mathcal{B}}])$ predict labels via AE
9: **for** $(\hat{y}_i, \hat{y}_j)$ **do**
10: Compute W_{ij} from $\hat{y}$ as per (9)
11: **end for**
12: update θ and ϕ by minimizing (10)
13: **end for**
14: **end for**

Our proposed algorithm is summarized in Algorithm 1. In Step 4, labeled and unlabeled data are randomly sampled while maintaining the constant ratio between them for every mini-batch. Steps 6–8 indicate the feed-forward process of student model, teacher model, and autoencoder, respectively. Step 10 constructs a similarity graph by comparing pairs of the labels predicted by autoencoder. We then update all parameters (θ, ϕ) simultaneously as per the single integrated loss function that we design in (10). This process is repeated until the maximum epoch is reached.

Tables 1 and 2 show performance comparisons with several other SSLs. Among consistency-loss-based approaches, we consider Π-model [18], Mean-Teacher [37] and ICT [38]. The performance results reported for baselines come from corresponding papers.

5 Experiments

We conduct real-data experiments on three benchmark datasets: MNIST, SVHN, and CIFAR-10. In all of the datasets, only a small fraction of the training data is used as labeled data, while the remaining being considered as unlabeled data. The MNIST consists of 70,000 images of size 28×28 with 60,000 for training and 10,000 for testing. The SVHN (an image of size 32×32 representing a close-up house number) consists of 73,257 color images for training, and 26,032 for testing. The CIFAR-10 contains 50,000 color images (each of size 32×32) for training and 10,000 for testing.

We adopt the same preprocessing technique as in [18,24,37,38]: Whitening with zero mean and unit variance for MNIST and SVHN; and Zero-phase Component Analysis (ZCA) whitening for CIFAR-10. We apply the translation augmentation to SVHN, and the mirror and translation augmentation to

CIFAR-10. For a classifier model, we employ CNN-13, the standard benchmark architecture used in the prior works [18,24,27,29,37,38]. We leave the detailed network architecture in the supplementary. We mostly follow hyperparameter search from [24,37,38]. See the supplementary for details. As for the choice of labeled data, we follow the common practice in [24,30,32]: Randomly sampling 100, 1000, 4000 labels for MNIST, SVHN and CIFAR-10, respectively. We also include experiments in which fewer labels are available. The results (to be reported) are the ones averaged over 10 trials with different random seeds for data splitting.

Very recently, graph-based SSL [51] has been proposed that employs a GCN-based classifier. Since the framework enforces the input of graph learning layer to be always fixed (such as pixel values or features from a CNN descriptor), the performance is limited by the quality of the input feature. Hence, no direct comparison has been made in our paper.

5.1 Performance Evaluations

Tables 1, 2, and 3 demonstrate the error rates respectively on: MNIST with 100/50/20 labels; SVHN with 1000/500/250 labels; and CIFAR-10 with 4000/2000/1000 labels. The underlines indicate the state of the arts among baselines, e.g., SNTG [24] in MNIST and SVHN; and ICT [38] in CIFAR-10. The number in parenthesis denotes the relative performance gain over the best baseline. We see noticeable performance improvements especially on MNIST dataset with 100 labels. It could seems that the absolute performance gain is not dramatic. In [24], however, this minor-looking absolute gain is considered to be significant, and it is well reflected in relative performance gain, which is 13.6% over the best

Table 1. Error rates (%) on MNIST, averaged over 10 trials. The boldface, underline and the number in parenthesis indicate the best results, the best among baselines, and performance gain over the best baseline, respectively. The results that we reproduced are marked as *.

Models	MNIST		
	100 labels	50 labels	20 labels
ImprovedGAN [32]	0.98 ± 0.065	2.21 ± 1.36	16.77 ± 4.52
TripleGAN [22]	0.91 ± 0.58	1.56 ± 0.72	4.81 ± 4.95
Π-model [18]	0.89 ± 0.15	1.02 ± 0.37	6.32 ± 6.90
TempEns [18]	$1.55 \pm 0.49^*$	$3.33 \pm 1.42^*$	$17.52 \pm 6.56^*$
MT [37]	$1.00 \pm 0.54^*$	$1.36 \pm 0.59^*$	$6.08 \pm 5.19^*$
VAT [27]	$0.88 \pm 0.38^*$	$1.34 \pm 0.60^*$	$5.15 \pm 4.77^*$
SNTG [24]	$\underline{0.66 \pm 0.07}$	$\underline{0.94 \pm 0.42}$	$\underline{1.36 \pm 0.78}$
ICT [38]	$0.95 \pm 0.29^*$	1.29 ± 0.34	3.83 ± 2.67
Our model	$\mathbf{0.57 \pm 0.06}$ (13.6%)	$\mathbf{0.64 \pm 0.14}$ (31.9%)	$\mathbf{0.85 \pm 0.21}$ (37.6%)

Table 2. Error rates (%) on SVHN, averaged over 10 trials. The boldface indicates the best results and the underline indicates the best among baselines. The number in parenthesis is the performance gain over the best baseline. The results that we reproduced are marked as *.

Models	SVHN		
	1000 labels	500 labels	250 labels
Supervised-only	12.83 ± 0.47	22.93 ± 0.67	40.62 ± 0.95
Π-model [18]	4.82 ± 0.17	6.65 ± 0.53	9.93 ± 1.15
TempEns [18]	4.42 ± 0.16	5.12 ± 0.13	12.62 ± 2.91
MT [37]	3.95 ± 0.19	4.18 ± 0.27	4.35 ± 0.50
VAT [27]	$3.94 \pm 0.12^*$	$4.71 \pm 0.29^*$	$5.49 \pm 0.34^*$
SNTG [24]	$\underline{3.82 \pm 0.25}$	$\underline{3.99 \pm 0.24}$	$\underline{4.29 \pm 0.23}$
ICT [38]	3.89 ± 0.04	4.23 ± 0.15	4.78 ± 0.68
Our model	$\mathbf{3.48 \pm 0.13}$ (8.90%)	$\mathbf{3.64 \pm 0.15}$ (9.02%)	$\mathbf{3.97 \pm 0.20}$ (7.46%)

Table 3. Error rates (%) on CIFAR-10, averaged over 10 trials. The boldface indicates the best results and the underline indicates the best among baselines. The number in parenthesis is the performance gain over the best baseline. The results that we reproduced are marked as *.

Models	CIFAR-10		
	4000 labels	2000 labels	1000 labels
Supervised-only	20.26 ± 0.38	31.16 ± 0.66	39.95 ± 0.75
Π-model [18]	12.36 ± 0.31	31.65 ± 1.20	17.57 ± 0.44
TempEns [18]	12.16 ± 0.24	15.64 ± 0.39	23.31 ± 1.01
MT [37]	12.31 ± 0.28	15.73 ± 0.31	21.55 ± 1.48
VAT [27]	$11.23 \pm 0.21^*$	$14.07 \pm 0.38^*$	$19.21 \pm 0.76^*$
SNTG [24]	9.89 ± 0.34	13.64 ± 0.32	18.41 ± 0.52
GSCNN [36]	15.49 ± 0.64	18.98 ± 0.62	16.82 ± 0.47
ICT [38]	$\underline{7.29 \pm 0.02}$	$\underline{9.26 \pm 0.09}$	$\underline{15.48 \pm 0.78}$
Our model	$\mathbf{6.87 \pm 0.19}$ (5.76%)	$\mathbf{8.13 \pm 0.22}$ (12.20%)	$\mathbf{12.50 \pm 0.58}$ (19.25%)

baseline. Observe more significant improvements with a decrease in the number of available labels. We expect this is because our well-constructed graph plays a more crucial role in challenging scenarios with fewer labels.

5.2 Comparison to Graph-Based SSL [18,24,36]

We put a particular emphasis on performance comparisons with the most related graph-based approaches [18,24,36], both in view of error rate and computational

Table 4. Comparison to the other graph-based SSLs on MNIST with 100 labels without augmentation.

Models	Error rate (%)	Running time (s)	# of parameters
Π-model [18]	0.89 ± 0.15	$18,343.81$	$3,119,508$
SNTG [24]	0.66 ± 0.07	$19,009,74$	$3,119,508$
GSCNN [36] (see Footnote 1)	0.60 ± 0.13	$62,240.97$	$3,119,508$
Our model	0.57 ± 0.06	$20,108.16$	$3,255,898$

complexity. The training time is measured w.r.t. MNIST with 100 labels and without data augmentation. Our implementation is done via TensorFlow on Xeon E5-2650 v4 CPU and TITAN V GPU.

Table 4 shows both error rate and training (running) time. Notice that our matrix completion yields a higher accuracy than SNTG [24] which only exploits the softmax values in the output space. This implies that using *both* features and predictions is more effective in better estimating a similarity graph, although it requires slightly increased model parameters and therefore slows down running time yet by a small margin.

In comparison to another matrix completion approach (GSCNN [36][1]), our framework offers much faster training time (around **3.1** times faster) in addition to the performance improvement in error rate. This is because our integrated approach enables *simultaneous training* between the models of the classifier and matrix completion. While our algorithm introduces a *parameterized autoencoder* and thus requires more model parameters, this addition is negligible relative to the complexity of the CNN classifier model employed.

Under graph-based approaches, even a slightly inaccurate graph may yield non-negligible performance degradation, as the error can be propagated through iterations. An inaccurate graph is likely to appear in the initial phase of training. To overcome this, we adjusted the reliability of the estimated graph using $w(t)$ monotonically increasing function in feature matching loss as in [24]. [24] already showed stable convergence when obtaining a graph from the teacher model classifier. Empirically, the graph accuracy obtained from our autoencoder is always higher than that of [24]. We leave the empirical result in the supplementary. As a result, we did not observe any divergence in all our experiments.

5.3 Ablation Study

In an effort to quantify the impact of every individual component employed in our framework, we also conduct an ablation study. This is done for CIFAR-10 with 4000 labels. We use supervised loss by default. We sequentially incorporate consistency loss, feature matching loss and graph construction via our autoencoder,

[1] For GSCNN, we use the same CNN structure as in this paper, and incorporate a consistency loss for a fair comparison.

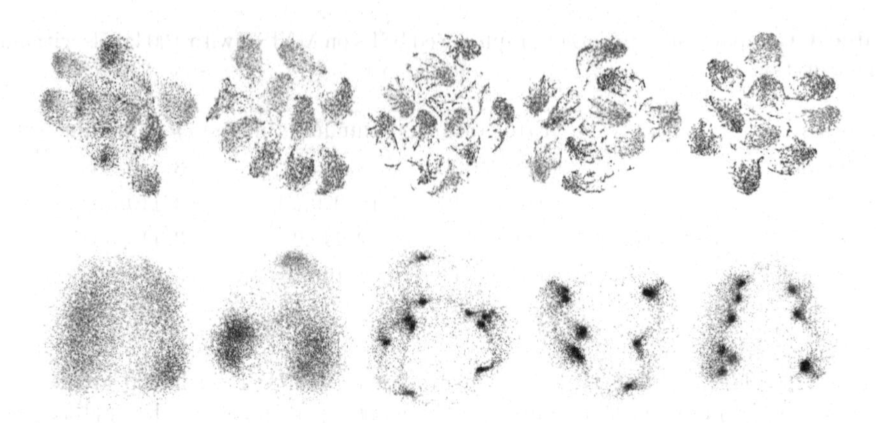

Fig. 2. Embeddings of CIFAR-10 test data with 10,000 images. These features are projected onto a two-dimensional space using t-SNE [25] (a–e) and PCA (f–j). (a) Supervised-only. (b) Supervised + Consistency. (c) Supervised + Feature matching. (d) Supervised + Consistency + Feature matching. (e) Supervised + Consistency + Feature matching with our Autoencoder.

as illustrated in Table 5. We also consider three representative methods (*Π*-model [18], Mean-Teacher [37], and ICT [38]) to see the effects of our method under various consistency losses.

Table 5 shows the results on CIFAR-10 with 4000 labels. Note that ICT [38] offers the most powerful consistency loss relative to *Π*-model [18] and MT [37]. Feature matching loss indeed plays a role and the effect is more significant with our autoencoder approach. This suggests that the precise graph due to our autoencoder maximizes the benefit of feature matching.

Moreover, we visualize 2D embeddings on CIFAR-10 test data using Embedding projector in Tensorflow. The feature vectors $h(x_i) \in \mathbb{R}^{128}$, extracted from the CNN, are projected onto a two-dimensional space using t-SNE [25]. We set perplexity to 25 and the learning rate to 10 as hyperparameters for t-SNE. In Fig. 2, we see the impact of each loss term upon clustering structures in an embedding domain. Here ICT is employed for consistency loss. From Fig. 2(c), we can see a clearer clustering structure (relative to (b)), suggesting that feature matching loss helps embedded data to move away from the decision boundary. Taking both losses (Figures (d) and (e)), we observe tighter clusters particularly with our autoencoder approach.

5.4 Wide ResNet Results

Recent works [4,28] employ a 28-layer Wide ResNet [47] architecture, instead of the standard CNN-13 that we have used for the above experiments. In an effort to investigate whether our framework can be gracefully merged with another, we also consider this architecture to conduct more experiments. We followed the simulation settings given in [4]. Table 6 demonstrates the error rates on CIFAR-10 with 4000/2000/1000/500/250 labels. The results show that our model is

Table 5. Ablation study on CIFAR-10 with 4000 labels. The number in parenthesis is the performance gain that we can obtain via the similarity graph due to our autoencoder.

	Error rate (%)
Supervised loss	20.26 ± 0.38
Supervised + Feature matching	13.99 ± 0.20
Supervised + Feature matching with AE	13.37 ± 0.17 (+4.43%)
***Π*-model**	
Supervised + Consistency	12.36 ± 0.31
Supervised + Consistency + Feature matching	11.00 ± 0.13
Supervised + Consistency + Feature matching with AE	10.81 ± 0.13 (+1.72%)
MT	
Supervised + Consistency	12.31 ± 0.28
Supervised + Consistency + Feature matching	12.12 ± 0.14
Supervised + Consistency + Feature matching with AE	11.54 ± 0.34 (+3.35%)
ICT	
Supervised + Consistency	7.25 ± 0.20[a]
Supervised + Consistency + Feature matching	7.18 ± 0.13
Supervised + Consistency + Feature matching with AE	6.87 ± 0.19 (+3.94%)

[a]In Table 5, the original paper reported 7.29±0.02 as the average and standard deviation for 3 runs. We replaced them with those with 10 runs.

Table 6. Error rate (%) with a 28-layer Wide ResNet on CIFAR-10, averaged over 5 trials.

Models	CIFAR-10				
	250 labels	500 labels	1000 labels	2000 labels	4000 labels
Π-model [18]	53.02 ± 2.05	41.82 ± 1.52	31.53 ± 0.98	23.07 ± 0.66	17.41 ± 0.37
PseudoLabel [19]	49.98 ± 1.17	40.55 ± 1.70	30.91 ± 1.73	21.96 ± 0.42	16.21 ± 0.11
Mixup [49]	47.43 ± 0.92	36.17 ± 1.36	25.72 ± 0.66	18.14 ± 1.06	13.15 ± 0.20
VAT [27]	36.03 ± 2.82	26.11 ± 1.52	18.68 ± 0.40	14.40 ± 0.15	11.05 ± 0.31
MT [37]	47.32 ± 4.71	42.01 ± 5.86	17.32 ± 4.00	12.17 ± 0.22	10.36 ± 0.25
MixMatch [4]	11.08 ± 0.87	9.65 ± 0.94	7.75 ± 0.32	7.03 ± 0.15	6.24 ± 0.06
Our model	10.87 ± 0.40	9.53 ± 0.85	7.69 ± 0.17	7.08 ± 0.14	6.24 ± 0.06

well merged with the Wide ResNet architecture, while offering slightly better performance relative to the state-of-the-art algorithm [4]. We do expect other techniques [3,34,46] can be gracefully merged with our approach (that additionally employs feature matching loss), yielding further improvements, as also demonstrated in Table 6 w.r.t. other technique [4].

6 Conclusion

We proposed a holistic training approach to graph-based SSL that is computationally efficient and offers the state-of-the-art error rate performances on benchmark datasets. The key idea is to employ an autoencoder-based matrix completion for similarity graph which enables simultaneous training between model parameters of an interested classifier and matrix completion. Experiments on three benchmark datasets demonstrate that our model outperforms the previous state of the arts. In particular, the performance gain is more distinct with a decrease in the number of available labels. Ablation study emphasizes the role of our key distinctive component: Autoencoder for graph construction. Future works of interest include: (a) a graceful merge with the state-of-the-art consistency loss-based approach [3,34,46,48]; and (b) evaluations of our model on various large-scale datasets such as CIFAR-100 and ImageNet.

Acknowledgments. This work was supported by the ICT R&D program of MSIP/IITP (2016-0-00563, Research on Adaptive Machine Learning Technology Development for Intelligent Autonomous Digital Companion), and Institute of Information & Communications Technology Planning & Evaluation (IITP) grant funded by the Korea government (MSIT) (2020-0-00626, Ensuring high AI learning performance with only a small amount of training data).

References

1. Athiwaratkun, B., Finzi, M., Izmailov, P., Wilson, A.G.: There are many consistent explanations of unlabeled data: why you should average. In: Proceedings of the International Conference on Learning Representation (ICLR) (2019)
2. Belkin, M., Niyogi, P., Sindwani, V.: Manifold regularization: a geometric framework for learning from labeled and unlabeled examples. J. Mach. Learn. Res. **7**, 2399–2434 (2006)
3. Berthelot, D., et al.: RemixMatch: semi-supervised learning with distribution alignment and augmentation anchoring. In: Proceedings of the International Conference on Representation Learning (ICLR) (2020)
4. Berthelot, D., Carlini, N., Goodfellow, I., Papernot, N., Oliver, A., Raffel, C.: MixMatch: a holistic approach to semi-supervised learning. In: Advances in Neural Information Processing Systems (NIPS), December 2019
5. Berton, L., Andrade Lopes, A.D.: Graph construction for semi-supervised learning. In: Proceedings of the Twenty-Fourth International Joint Conference on Artificial Intelligence, pp. 4343–4344 (2015)
6. Blum, A., Chawla, S.: Learning from labeled and unlabeled data using graph mincuts. In: Proceedings of the International Conference on Machine Learning, pp. 19–26, June 2001
7. Bromley, J., Guyon, I., LeCun, Y., Sickinger, E., Shah, R.: Signature verification using a "Siamese" time delay neural network. In: Advances in Neural Information Processing Systems, pp. 737–744 (1994)
8. Candés, E.J., Recht, B.: Exact matrix completion via convex optimization. Found. Comput. Math. **9**(6), 717–772 (2009)

9. Dai, X., Yang, Z., Yang, F., Cohen, W.W., Salakhutdinov, R.: Good semi-supervised learning that requires a bad GAN. In: Advances in Neural Information Processing Systems (NIPS), pp. 6510–6520, December 2017
10. Dong, X., Yu, L., Wu, Z., Sun, Y., Yuan, L., Zhang, F.: A hybrid collaborative filtering model with deep structure for recommender systems. In: Proceedings of the 31st AAAI Conference on Artificial Intelligence, pp. 1309–1315 (2017)
11. Glorot, X., Bengio, Y.: Understanding the difficulty of training deep feedforward neural networks. In: International Conference on Artificial Intelligence and Statistics (2010)
12. Gong, C., Liu, T., Tao, D., Fu, K., Tu, E., Yang, J.: Deformed graph Laplacian for semisupervised learning. IEEE Trans. Neural Netw. Learn. Syst. **26**(10), 717–772 (2015)
13. Goodfellow, I.J., et al.: Generative adversarial nets. In: Advances in Neural Information Processing Systems (NIPS), December 2014
14. He, K., Zhang, X., Ren, S., Sun, J.: Delving deep into rectifiers: surpassing human-level performance on imagenet classification. In: Proceedings of the IEEE International Conference on Computer Vision (ICCV) (2015)
15. Iscen, A., Tolias, G., Avrithis, Y., Chum, O.: Label propagation for deep semi-supervised learning. In: Proceedings of the IEEE Conference on Computer Vision and Pattern Recognition. pp. 5070–5079, June 2019
16. Kingma, D.P., Ba, J.: Adam: a method for stochastic optimization. In: 3rd International Conference for Learning Representations (2015)
17. Kolesnikov, A., Zhai, X., Beyer, L.: Revisiting self-supervised visual representation learning. In: Proceedings of IEEE Conference on Computer Vision and Pattern Recognition (CVPR) (2019)
18. Laine, S., Aila, T.: Temporal ensembling for semi-supervised learning. In: Proceedings of the International Conference on Representation Learning (ICLR), April 2017
19. Lee, D.H.: Pseudo-label: the simple and efficient semi-supervised learning method for deep neural networks. In: Workshop on Challenges in Representation Learning, ICML, vol. 3, p. 2 (2013)
20. Lee, K., Jo, H., Kim, H., Lee, Y.H.: Basis learning autoencoders for hybrid collaborative filtering in cold start setting. In: Proceedings of the 29th International Workshop on Machine Learning for Signal Processing (MLSP) (2019)
21. Lee, K., Lee, Y.H., Suh, C.: Alternating autoencoders for matrix completion. In: Proceedings of the IEEE Data Science Workshop (DSW) (2018)
22. Li, C., Xu, K., Zhu, J., Zhang, B.: Triple generative adversarial nets. In: Advances in Neural Information Processing Systems (NIPS), pp. 4088–4098, December 2017
23. Li, S., Kawale, J., Fu, Y.: Deep collaborative filtering via marginalized denoising auto-encoder. In: Proceedings of the 24th ACM International Conference on Information and Knowledge Management, pp. 811–820. ACM (2015)
24. Luo, Y., Zhu, J., Li, M., Ren, Y., Zhang, B.: Smooth neighbors on teacher graphs for semi-supervised learning. In: Proceedings of IEEE Conference on Computer Vision and Pattern Recognition (CVPR), pp. 8896–8905, June 2018
25. Maaten, L., Hinton, G., Johnson, I.: Visualizing data using t-SNE. J. Mach. Learn. Res. **9**, 2579–2605 (2008)
26. Mazumder, R., Hastie, T., Tibshirani, R.: Spectral regularization algorithms for learning large incomplete matrices. J. Mach. Learn. Res. **11**, 2287–2322 (2010)
27. Miyato, T., Ichi Maeda, S., Koyama, M., Ishii, S.: Virtual adversarial training: a regularization method for supervised and semi-supervised learning. In: Proceedings of the International Conference on Representation Learning (ICLR), April 2017

28. Oliver, A., Odena, A., Raffel, C., Cubuk, E.D., Goodfellow, I.J.: Realistic evaluation of semi-supervised learning algorithms. In: Advances in Neural Information Processing Systems (NIPS), pp. 3235–3246, December 2018
29. Park, S., Park, J., Shin, S.J., Moon, I.C.: Adversarial dropout for supervised and semi-supervised learning. In: Proceedings of the Thirty-Second AAAI Conference on Artificial Intelligence, April 2018
30. Rasmus, A., Valpola, H., Honkala, M., Berglund, M., Raiko, T.: Semi-supervised learning with ladder networks. In: Advances in Neural Information Processing Systems (NIPS), pp. 3546–3554, December 2015
31. Salimans, T., Kingma, D.: Weight normalization: a simple reparameterization to accelerate training of deep neural networks. In: Advances in Neural Information Processing Systems (NIPS) (2016)
32. Salimans, T., Goodfellow, I., Zaremba, W., Cheung, V., Radford, A., Chen, X.: Improved techniques for training GANs. In: Advances in Neural Information Processing Systems (NIPS), pp. 2234–2242, December 2016
33. Sedhain, S., Krishna, M., Scanner, S., Xie, L.: AutoRec: autoencoders meet collaborative filtering. In: Proceedings of the 24th International Conference on World Wide Web, pp. 111–112 (2015)
34. Sohn, K., et al.: Fixmatch: simplifying semi-supervised learning with consistency and confidence. arXiv preprint arXiv:2001.07685 (2020)
35. Springenberg, J.T.: Unsupervised and semi-supervised learning with categorical generative adversarial networks. In: Proceedings of the International Conference on Learning Representation (ICLR), May 2016
36. Taherkhani, F., Kazemi, H., Nasrabadi, N.M.: Matrix completion for graph-based deep semi-supervised learning. In: Proceedings of the 33rd AAAI Conference on Artificial Intelligence. pp. 8896–8905, January 2019
37. Tarvainen, A., Valpola, H.: Mean teachers are better role models: weight-averaged consistency targets improve semi-supervised deep learning results. In: Advances in Neural Information Processing Systems (NIPS), pp. 1195–1204, December 2017
38. Verma, V., Lamb, A., Kannala, J., Bengio, Y., Lopez-Paz, D.: Interpolation consistency training for semi-supervised learning. In: International Joint Conference on Artificial Intelligence, pp. 3635–3641, August 2019
39. Volkovs, M., Yu, G., Poutanen, T.: DropOutNet: addressing cold start in recommender systems. In: Advances in Neural Information Processing Systems (NIPS), December 2015
40. Wan, S., Gong, C., Zhong, P., Du, B., Zhang, L., Yang, J.: Multi-scale dynamic graph convolutional network for hyperspectral image classification. IEEE Trans. Geosci. Remote Sens. 1–16 (2019)
41. Wang, B., Tu, Z., Tsotsos, J.K.: Dynamic label propagation for semi-supervised multi-class multi-label classification. In: Proceedings of the IEEE International Conference on Computer Vision, pp. 425–432 (2013)
42. Wang, Q., Li, W., Gool, L.V.: Semi-supervised learning by augmented distribution alignment. In: Proceedings of the IEEE International Conference on Computer Vision, pp. 1466–1475 (2019)
43. Weston, J., Ratle, F., Collobert, R.: Deep learning via semi-supervised embedding. In: Proceedings of the 25th International Conference on Machine Learning, pp. 1168–1175, July 2008
44. Wu, S., Li, J., Liu, C., Yu, Z., Wong, H.S.: Mutual learning of complementary networks via residual correction for improving semi-supervised classification. In: Proceedings of IEEE Conference on Computer Vision and Pattern Recognition (CVPR), pp. 6500–6509 (2019)

45. Wu, X., Zhao, L., Akoglu, L.: A quest for structure: jointly learning the graph structure and semi-supervised classification. In: In Proceedings of the 27th ACM International Conference on Information and Knowledge Management (2018)
46. Xie, Q., Dai, Z., Hovy, E., Luong, M.T., Le, Q.V.: Unsupervised data augmentation for consistency training. arXiv preprint arXiv:1904.12848 (2019)
47. Zagoruyko, S., Komodakis, N.: Wide residual networks. In: Richard C. Wilson, E.R.H., Smith, W.A.P. (eds.) Proceedings of the British Machine Vision Conference (BMVC), pp. 87.1–87.12. BMVA Press, September 2016. https://doi.org/10.5244/C.30.87
48. Zhai, X., Oliver, A., Kolesnikov, A., Beyer, L.: S4l: self-supervised semi-supervised learning. In: Proceedings of the IEEE International Conference on Computer Vision (ICCV), November 2019
49. Zhang, H., Cisse, M., Dauphin, Y.N., Lopez-Paz, D.: Mixup: beyond empirical risk minimization. In: Proceedings of the International Conference on Learning Representation (2018)
50. Zhul, X., Ghahramani, Z.: Learning from labeled and unlabeled data with label propagation. Technical report (2002)
51. Ziang, B., Zhang, Z., Lin, D., Tang, J., Luo, B.: Semi-supervised learning with graph learning-convolutional networks. In: Proceedings of IEEE Conference on Computer Vision and Pattern Recognition (CVPR) (2019)

Virtual Multi-view Fusion for 3D Semantic Segmentation

Abhijit Kundu(✉), Xiaoqi Yin, Alireza Fathi, David Ross, Brian Brewington, Thomas Funkhouser, and Caroline Pantofaru

Google Research, San Francisco, USA
abhijitkundu@google.com

Abstract. Semantic segmentation of 3D meshes is an important problem for 3D scene understanding. In this paper we revisit the classic multiview representation of 3D meshes and study several techniques that make them effective for 3D semantic segmentation of meshes. Given a 3D mesh reconstructed from RGBD sensors, our method effectively chooses different virtual views of the 3D mesh and renders multiple 2D channels for training an effective 2D semantic segmentation model. Features from multiple per view predictions are finally fused on 3D mesh vertices to predict mesh semantic segmentation labels. Using the large scale indoor 3D semantic segmentation benchmark of ScanNet, we show that our virtual views enable more effective training of 2D semantic segmentation networks than previous multiview approaches. When the 2D per pixel predictions are aggregated on 3D surfaces, our virtual multiview fusion method is able to achieve significantly better 3D semantic segmentation results compared to all prior multiview approaches and recent 3D convolution approaches.

Keywords: 3D semantic segmentation · Scene understanding

1 Introduction

Semantic segmentation of 3D scenes is a fundamental problem in computer vision. Given a 3D representation of a scene (e.g., a textured mesh of an indoor environment), the goal is to output a semantic label for every surface point. The output could be used for semantic mapping, site monitoring, training autonomous navigation, and several other applications.

State-of-the-art (SOTA) methods for 3D semantic segmentation currently use 3D sparse voxel convolution operators for processing input data. For example, MinkowskiNet [7] and SparseConvNet [11] each load the input data into a sparse 3D voxel grid and extract features with sparse 3D convolutions. These "place-centric" methods are designed to recognize 3D patterns and thus work well for

Electronic supplementary material The online version of this chapter (https://doi.org/10.1007/978-3-030-58586-0_31) contains supplementary material, which is available to authorized users.

A. Vedaldi et al. (Eds.): ECCV 2020, LNCS 12369, pp. 518–535, 2020.
https://doi.org/10.1007/978-3-030-58586-0_31

types of objects with distinctive 3D shapes (e.g., chairs), and not so well for others (e.g., wall pictures). They also take a considerable amount of memory, which limits spatial resolutions and/or batch sizes.

Alternatively, when posed RGB-D images are available, several researchers have tried using 2D networks designed for processing photographic RGB images to predict dense features and/or semantic labels and then aggregate them on visible 3D surfaces [15,41], and others project features onto visible surfaces and convolve them further in 3D [10,18,19,40]. Although these "view-centric" methods utilize massive image processing networks pretrained on large RGB image datasets, they do not achieve SOTA performance on standard 3D segmentation benchmarks due to the difficulties of occlusion, lighting variation, and camera pose misalignment in RGB-D scanning datasets. None of the view-based methods is currently in the top half of the current leaderboard for the 3D Semantic Label Challenge of the ScanNet benchmark.

In this paper, we propose a new view-based approach to 3D semantic segmentation that overcomes the problems with previous methods. The key idea is to use synthetic images rendered from "virtual views" of the 3D scene rather than restricting processing to the original photographic images acquired by a physical camera. This approach has several advantages that address the key problems encountered by previous view-centric method [3,21]. First, we select camera intrinsics for virtual views with unnaturally wide field-of-view to increase the context observed in each rendered image. Second, we select virtual viewpoints at locations with small variation in distances/angles to scene surfaces, relatively few occlusions between objects, and large surface coverage redundancy. Third, we render non-photorealistic images without view-dependent lighting efffects and occlusions by backfacing surfaces – i.e., virtual views can look into a scene from behind the walls, floors, and ceilings to provide views with relatively large context and little occlusion. Fourth, we aggregate pixel-wise predictions onto 3D surfaces according to exactly known camera parameters of virtual views, and thus do not encounter "bleeding" of semantic labels across occluding contours. Fifth, virtual views during training and inference can mimic multi-scale training and testing and avoid scale in-variance issues of 2D CNNs. We can generate as many virtual views as we want during both training and testing. During training, more virtual views provides robustness due to data augmentation. During testing, more views provides robustness due to vote redundancy. Finally, the 2D segmentation model in our multiview fusion approach can benefit from large image pre-training data like ImageNet and COCO, which are unavailable for pure 3D convolution approaches.

We have investigated the idea of using virtual views for semantic segmentation of 3D surfaces using a variety of ablation studies. We find that the broader design space of view selection enabled by virtual cameras can significantly boost the performance of multiview fusion as it allows us to include physically impossible but useful views (e.g., behind walls). For example, using virtual views with original camera parameters improves 3D mIoU by 3.1% compared with using original photographic images, using additional normal and coordinates channels and higher field of view can further boost mIoU by 5.7%, and an additional gain of 2.1% can be achieved by carefully selecting virtual camera poses to best capture the 3D information in the scenes and optimize for training 2D CNNs.

Overall, our simple system is able to achieve state-of-the-art results on both 2D and 3D semantic labeling tasks in ScanNet Benchmark [9], and is significantly better than the best performing previous multi-view methods and very competitive with recent 3D methods based on convolutions of 3D point sets and meshes. In addition, we show that our proposed approach consistently outperforms 3D convolution and real multi-view fusion approaches when there are fewer scenes for training. Finally, we show that similar performance can be obtained with significantly fewer views in the inference stage. For example, multi-view fusion with ∼12 virtual views per scene will outperform that with all ∼1700 original views per scene.

The rest of the paper is organized as follows. We introduce the research landscape and related work in Sect. 2. We describe the proposed virtual multi-view fusion approach in detail in Sect. 3–Sect. 5. Experiment results and ablation studies of our proposed approach are presented in Sect. 6. Finally we conclude the paper with discussions of future directions in Sect. 7.

2 Related Work

There has been a large amount of previous work on semantic segmentation of 3D scenes. The following reviews only the most related work.

Multi-view Labeling. Motivated by the success of view-based methods for object classification [35], early work on semantic segmentation of RGB-D surface reconstructions relied on 2D networks trained to predict dense semantic labels for RGB images. Pixel-wise semantic labels were backprojected and aggregated onto 3D reconstructed surfaces via weighted averaging [15,41], CRFs [25], Bayesian fusion [24,41,46], or 3D convolutions [10,18,19]. These methods performed multiview aggregation only for the originally captured RGB-D photographic images, which suffer from limited fields-of-view, restricted viewpoint ranges, view-dependent lighting effects, and misalignments with reconstructed surface geometry, all of which reduce semantic segmentation performance. To overcome these problems, some recent work has proposed using synthetic images of real data in a multiview labeling pipeline [3,12,21], but they still use camera parameters typical of real images (e.g., small field of view), propose methods suitable only for outdoor environments (lidar point clouds of cities), and do not currently achieve state-of-the-art results.

3D Convolution. Recent work on 3D semantic segmentation has focused on methods that extract and classify features directly with 3D convolutions. Network architectures have been proposed to extract features from 3D point clouds [16,29–31,33,38], surface meshes [14,17], voxel grids [34], and octrees [32]. Current state-of-the-art methods are based on sparse 3D voxel convolutions [7,8,11], where submanifold sparse convolution operations are used to compute features on sparse voxel grids. These methods utilize memory more efficiently than dense voxel grids, but are still limited in spatial resolution in comparison to 2D images and can train with supervision only on 3D datasets, which generally are very small in comparison to 2D image datasets.

Synthetic Data. Other work has investigated training 2D semantic segmentation networks using computer graphics renderings of 3D synthetic data [47]. The main advantage of this approach is that image datasets can be created with unlimited size by rendering novel views of a 3D scene [22,26]. However, the challenge is generally domain adaptation – networks trained on synthetic data and tested on real data usually do not perform well. Our method avoids this problem by training and testing on synthetic images rendered with the same process.

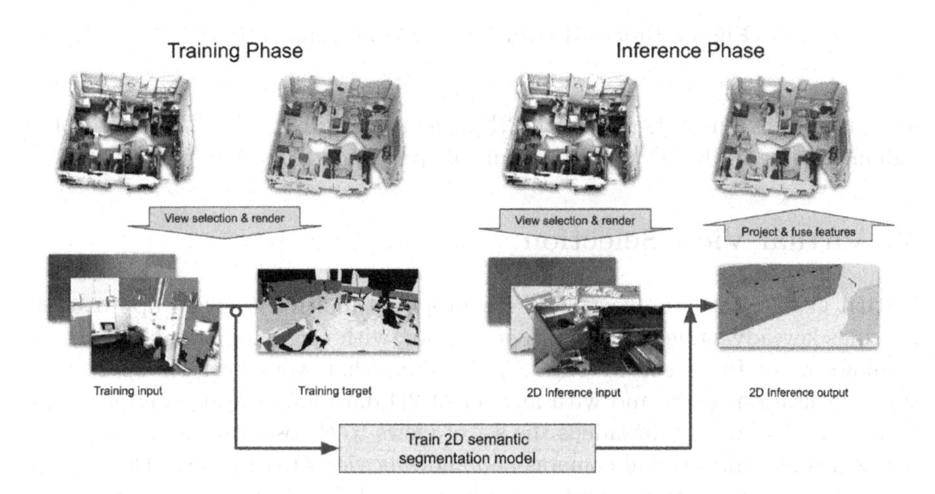

Fig. 1. Virtual multi-view fusion system overview.

3 Method Overview

The proposed multiview fusion approach is illustrated in Fig. 1. At a high level, it consists of the following steps.

Training Stage. During the training stage, we first select virtual views for each 3D scene, where for each virtual view we select camera intrinsics, camera extrinsics, which channels to render, and rendering parameters (e.g., depth range, backface culling). We then generate training data by rendering the selected virtual views for the selected channels and ground truth semantic labels. We train 2D semantic segmentation models using the rendered training data and use the model in the inference stage.

Inference Stage. At inference stage, we select and render virtual views using a similar approach as in the training stage, but without the ground truth semantic labels. We conduct 2D semantic segmentation on the rendered virtual views

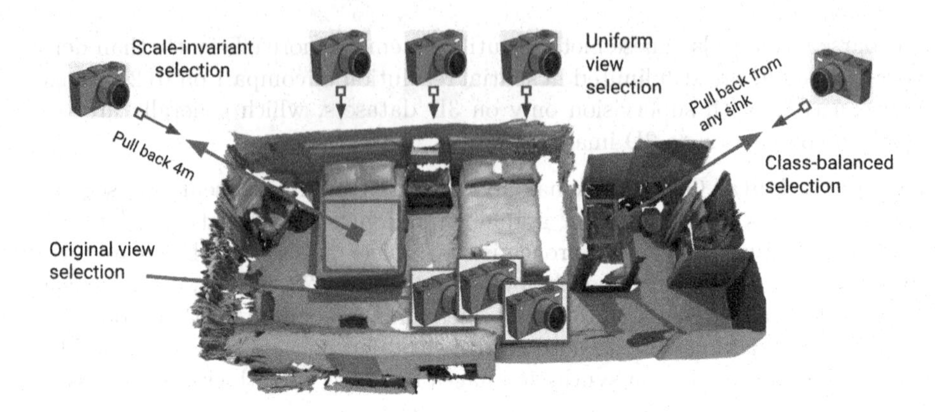

Fig. 2. Proposed virtual view selection approaches.

using the trained model, project the 2D semantic features to 3D, then derive the semantic category in 3D by fusing multiple projected 2D semantic features.

4 Virtual View Selection

Virtual view selection is central to the proposed multiview fusion approach as it brings key advantages over multiview fusion with original image views. First, it allows us to freely select camera parameters that work best for 2D semantic segmentation tasks, and with any set of 2D data augmentation approaches. Second, it significantly broadens the set of views to choose from by relaxing the physical constraints of real cameras and allowing views from unrealistic but useful camera positions that significantly boost model performance, e.g. behind a wall. Third, it allows 2D views to capture additional channels that are difficult to capture with real cameras, e.g., normals and coordinates. Finally, by selecting and rendering virtual views, we have essentially eliminated any errors in the camera calibration and pose estimation, which are common in the 3D reconstruction process. Lastly, sampling views consistently at different scales resolves scale in-variance issues of traditional 2D CNNs.

Camera Intrinsics. A significant constraint of original image views is the FOV - images may have been taken very close to objects or walls, say, and lack the object features and context necessary for accurate classification. Instead, we use a pinhole camera model with significantly higher field of view (FOV) than the original cameras, providing larger context that leads to more accurate 2D semantic segmentation [27]. Figure 3 shows an example of original views compared with virtual views with high FOV.

Camera Extrinsics. We use a mixture of the following sampling strategies to select camera extrinsics as shown in Fig. 2 and Fig. 4.

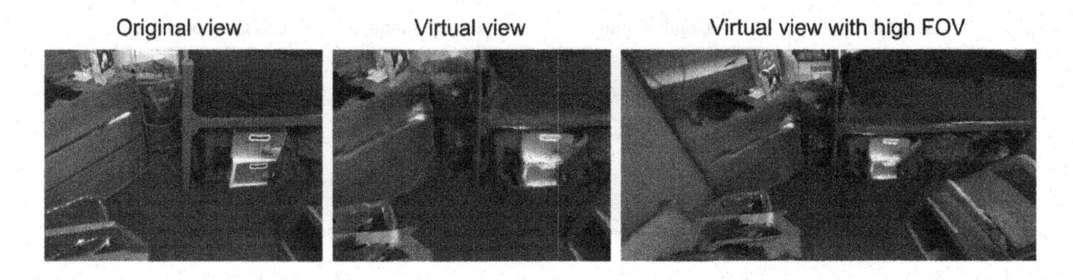

Fig. 3. Original views vs. virtual views. High FOV provides larger context of the scene which helps 2D perception, e.g., the chair in the bottom right corner is partially represented in the original view but can easily segmented in the high FOV virtual view.

Fig. 4. Example virtual view selection on two ScanNet scenes. Green curve is the trajectory of the original camera poses; Blue cameras are the selected views with the proposed approach. Note that we only show a random subset of all selected views for illustration purposes. (Color figure online)

- Uniform sampling. We want to uniformly sample camera extrinsics to generate many novel views, independently from the specific structure of the 3D scene. Specifically, we use top-down views from uniformly sampled positions at the top of the 3D scene, as well as views that look through the center of the scene but with uniformly sampled positions in the 3D scene.
- Scale-invariant sampling. As 2D convolutional neural networks are generally not scale invariant, the model performance may suffer if the scales of views do not match the 3D scene. To overcome this limitation, we propose sampling views at a range of scales with respect to segments in the 3D scene. Specifically, we do an over-segmentation of the 3D scene, and for each segment, we position the cameras to look at the segment by pulling back to a certain range of distances along the normal direction. We do a depth check to avoid occlusions by foreground objects. If backface culling is disabled in the rendering stage (discussed in more detail below), we do a ray tracing and drop

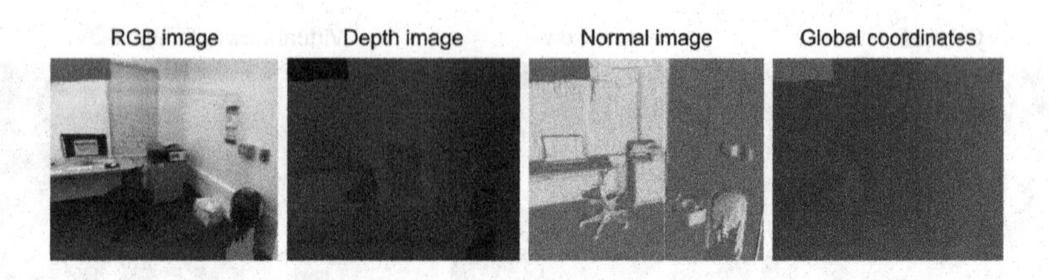

Fig. 5. Example virtual rendering of selected channels.

any views blocked by the backfaces. Note the over-segmentation of the 3D scene is unsupervised and does not use the ground truth semantic labels, so the scale-invariant sampling can be applied both in the training and inference stages.

- Class-balanced sampling. Class balancing has been extensively used as data augmentation approaches for 2D semantic segmentation. We conduct class balancing by selecting views that look at mesh segments of under-represented semantic categories, similar to the scale-invariant sampling approach. Note this sampling approach only applies to the training stage when the ground truth semantic labels are available.
- Original views sampling. We also sample from the original camera views as they represent how a human would choose camera views in the real 3D scene with real physical constraints. Also, the 3D scene is reconstructed from the original views, so including them can make sure we cover corner cases that would otherwise be difficult as random virtual views.

Channels for Rendering. To exploit all the 3D information available in the scene, we render the following channels: RGB color, normal, normalized global XYZ coordinates. The additional channels allow us to go beyond the limitations of the existing RGB-D sensors. While depth image also contains the same information, we think normalized global coordinate image makes the learning problem simpler as now just like the normal and color channel, coordinate values of the same 3D point is view invariant. Figure 5 shows example rendered views of the selected channels.

Rendering Parameters. We turn on backface culling in the rendering so that the backfaces do not block the camera views, further relaxing the physical constraints of the 3D scene and expanding the design space of the view selection. For example, as shown in Fig. 6, in an indoor scenario, we can select views from outside a room which typically include more context of the room and can potentially improve model performance; On the other hand, with backface culling turned off, we either are constrained ourselves to views inside the room therefore limited context, or suffer from high occlusion by the backfaces of the walls.

Training vs. Inference Stage. We want to use similar view selection approaches for the training and inference stages to avoid creating a domain gap, e.g., if we sampled many top-down views in the training stage but used

Fig. 6. Effect of backface culling. Backface culling allows the virtual camera to see more context from views that are not physically possible with real cameras.

lots of horizontal views in the inference stage. The main difference between the view selection strategies between the two stages is the class-balancing which can only be done in the training stage. Also, while the inference cost may matter in real-world applications, in this paper we consider offline 3D segmentation tasks and do not optimize the computation cost in either stage, so we can use as many virtual views as needed in either stage.

5 Multiview Fusion

5.1 2D Semantic Segmentation Model

With rendered virtual views as training data, we are now ready to train a 2D semantic segmentation models. We use a xcpetion65 [6] feature extractor and DeeplabV3+ [4] decoder. We initialize our model from pre-trained classification model checkpoints trained on ImageNet. When training a model with additional input channels like normal image and co-ordinate image we modify the first layer of the pre-training checkpoints by tiling the weights across the additional channels and normalize them across each spatial position such that the sum of weights along the channel dimension remains the same.

5.2 3D Fusion of 2D Semantic Features

During inference, we run the 2D semantic segmentation model on virtual views and obtain image features (e.g., unary probabilities for each pixel). To project the 2D image features to 3D, we use the following approach: We render a depth channel on the virtual views; For each 3D point, we project it back to each of the virtual views, and accumulate the image feature of the projected pixel only if the depth of the pixel matches the point-to-camera distance. This approach achieves better computational efficiency than the alternative approach of casting rays from each pixel to find the 3D point to aggregate. First, the number of 3D points in a scene are much less than the total number of pixels in all rendered images of the scene. Secondly, projecting a 3D point with a depth check is faster than operations involving ray casting.

Formally, let $\mathbf{X}_k \in \mathbb{R}^3$ be the 3D position of the kth point, $\mathbf{x}_{k,i} \in \mathbb{R}^2$ be the pixel coordinates by projecting the kth 3D point to virtual view $i \in \mathcal{I}$, $\mathbf{K}_i$ be its instrinsics matrix while $\mathbf{R}_i$ be the rotation, $\mathbf{t}_i$ the translation in the extrinsics, $\mathcal{A}_i$ be the set of valid pixel coordinates. Let $c_{k,i}$ be the distance between the position of camera i and kth 3D point. We have:

$$\mathbf{x}_{k,i} = \mathbf{K}_i(\mathbf{R}_i\mathbf{X}_k + \mathbf{t}_i) \quad (1)$$

$$c_{k,i} = \left\|\mathbf{X}_k - \mathbf{R}_i^{-1}\mathbf{t}_i\right\|_2 \quad (2)$$

Let $\mathcal{F}_k$ be the set of image features projected to the kth 3D point, $\mathbf{f}_i(\cdot)$ be the mapping from pixel coordinates in virtual image i to the image feature vector, $d_i(\cdot)$ be the mapping from pixel coordinates to the depth since we render depth channel. Then:

$$\mathcal{F}_k = \{\mathbf{f}_i(\mathbf{x}_{k,i}) \mid \mathbf{x}_{k,i} \in \mathcal{A}_i, |d_i(\mathbf{x}_{k,i}) - c_{k,i}| < \delta, \forall i \in \mathcal{I}\} \quad (3)$$

where $\delta > 0$ is the threshold for depth matching.

To fuse projected features $\mathcal{F}_k$ for 3D point k, we simply take the average of all features in $\mathcal{F}_k$ and obtain the fused feature. There simple fusion function was better than other alternatives like picking the category with maximum probability across all projected features.

6 Experiments

We ran a series of experiments to evaluate how well our proposed method for 3D semantic segmentation of RGB-D scans works compared to alternative approaches and to study how each component of our algorithm affects the results.

6.1 Evaluation on ScanNet Dataset

We evaluate our approach on ScanNet dataset [9], on the hidden test set for the task of both 3D mesh semantic segmentation and 2D image semantic segmentation. We also perform a detailed ablation study on the validation set of ScanNet in Sect. 6.3. Unlike our ablation studies, we use xception101 [6] as the 2D backbone and we additionally use ADE20K [48] for pre-training the 2D segmentation model. We compare our virtual multiview-fusion approach against state-of-the-art methods for 3D semantic segmentation, most of which utilize 3D convolutions of sparse voxels or point clouds. We also compare our 2D image segmentation results obtained by projecting back 3D labels obtained by our multiview fusion approach. Results are available in Table 1.

From these results, we see that our approach outperforms previous approaches based on convolutions of 3D point sets [16,30,38,43,44], and it achieves results comparable to the SOTA methods based on sparse voxel convolutions [7,11,13]. Our method achieves the best 2D segmentation results (74.5%). In Sect. 6.3, we also demonstrate improvement in single frame 2D semantic segmentation.

Table 1. Semantic segmentation results on ScanNet validation and test splits.

Method	3D mIoU (val split)	3D mIoU (test split)	2D mIoU (test split)
PointNet [30]	53.5	55.7	-
3DMV [10]	-	48.4	49.8
SparseConvNet [11]	69.3	72.5	-
PanopticFusion [28]	-	52.9	-
PointConv [43]	61.0	66.6	-
JointPointBased [5]	69.2	63.4	-
SSMA [39]	-	-	57.7
KPConv [38]	69.2	68.4	-
MinkowskiNet [7]	72.2	73.6	-
PointASNL [44]	63.5	66.6	-
OccuSeg [13]	-	**76.4**	-
JSENet [16]	-	69.9	-
Ours	**76.4**	74.6	**74.5**

Our approach performs significantly better than any previous multiview fusion methods [10,28] on ScanNet semantic labeling benchmark. The mean IoU of the previously best performing multiview method on the ScanNet test set is 52.9% [28], which is significantly less than our results of 74.6%. By using our virtual views, we are able to learn 2D semantic segmentation networks that provide more accurate and more consistent semantic labels when aggregated on 3D surfaces. The result is semantic segmentations of high accuracy and sharp boundaries, as shown in Fig. 7 (Table 2).

Table 2. Results on the Stanford 3D Indoor Spaces (S3DIS) dataset [1]. Following previous works we use Fold-1 split with Area5 as the test set.

Method	mIOU	ceiling	floor	wall	beam	column	window	door	chair	table	bookcase	sofa	board	clutter
PointNet [30]	41.09	88.8	97.3	69.8	0.1	3.9	46.3	10.8	52.6	58.9	40.3	5.9	26.4	33.2
SegCloud [37]	48.92	90.1	96.1	69.9	0.0	18.4	38.4	23.1	75.9	70.4	58.4	40.9	13.0	41.6
TangentConv [36]	52.80	90.5	97.7	74.0	0.0	20.7	39.0	31.3	77.5	69.4	57.3	38.5	48.8	39.8
3D RNN [45]	53.40	95.2	98.6	77.4	0.8	9.8	52.7	27.9	76.8	78.3	58.6	27.4	39.1	51.0
PointCNN [23]	57.26	92.3	98.2	79.4	0.0	17.6	22.7	62.1	80.6	74.4	66.7	31.7	62.1	56.7
SuperpointGraph [20]	58.04	89.4	96.9	78.1	0.0	42.8	48.9	61.6	84.7	75.4	69.8	52.6	2.1	52.2
PCCN [42]	58.27	90.3	96.2	75.9	0.3	6.0	69.5	63.5	66.9	65.6	47.3	68.9	59.1	46.2
PointASNL [44]	62.60	94.3	98.4	79.1	0.0	26.7	55.2	66.2	83.3	86.8	47.6	68.3	56.4	52.1
MinkowskiNet [7]	65.35	91.8	98.7	86.2	0.0	34.1	48.9	62.4	89.8	81.6	74.9	47.2	74.4	58.6
Ours	**65.38**	92.9	96.9	85.5	0.8	23.3	65.1	45.7	85.8	76.9	74.6	63.1	82.1	57.0

6.2 Evaluation on Stanford 3D Indoor Spaces (S3DIS)

We also evaluated our method on the Stanford Large-Scale 3D Indoor Spaces Dataset (S3DIS) [1,2] for the task of semantic 3D segmentation. The proposed

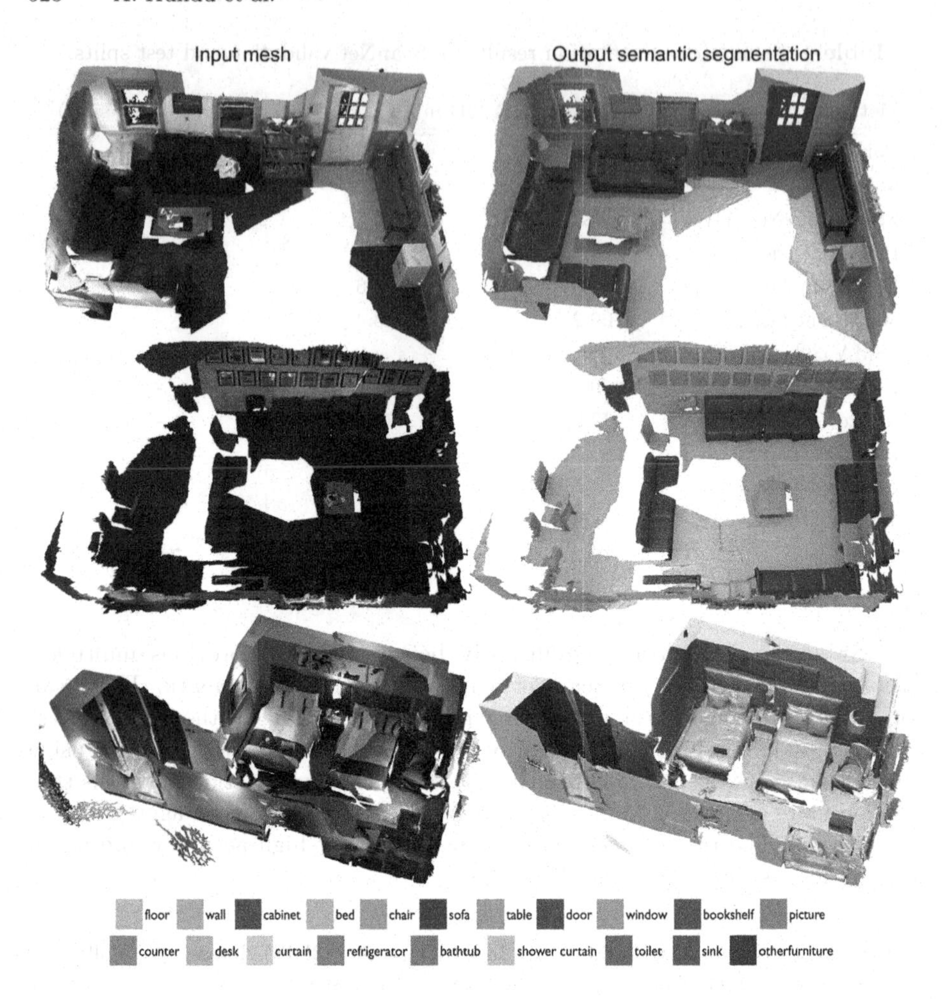

Fig. 7. Qualitiative 3D semantic segmentation results on ScanNet test set.

virtual multi-view fusion approach achieves 65.4% 3D mIoU, outperforming recent SOTA methods MinkowskiNet [7] (65.35%) and PointASNL [44] (62.60%). See Table 1 for quantitative evaluation. Figure 8 shows the output of our approach on Area5 scene from S3DIS dataset.

6.3 Ablation Studies

We investigate which aspects of our proposed method make the most difference we performed ablation study on the ScanNet [9]. To perform this experiment, we started with a baseline method that trains a model to compute 2D semantic segmentation for the original photographic images, uses it to predict semantics for all the original views in the validation set, and then aggregates the class probabilities on backprojected 3D surfaces using the simple averaging method

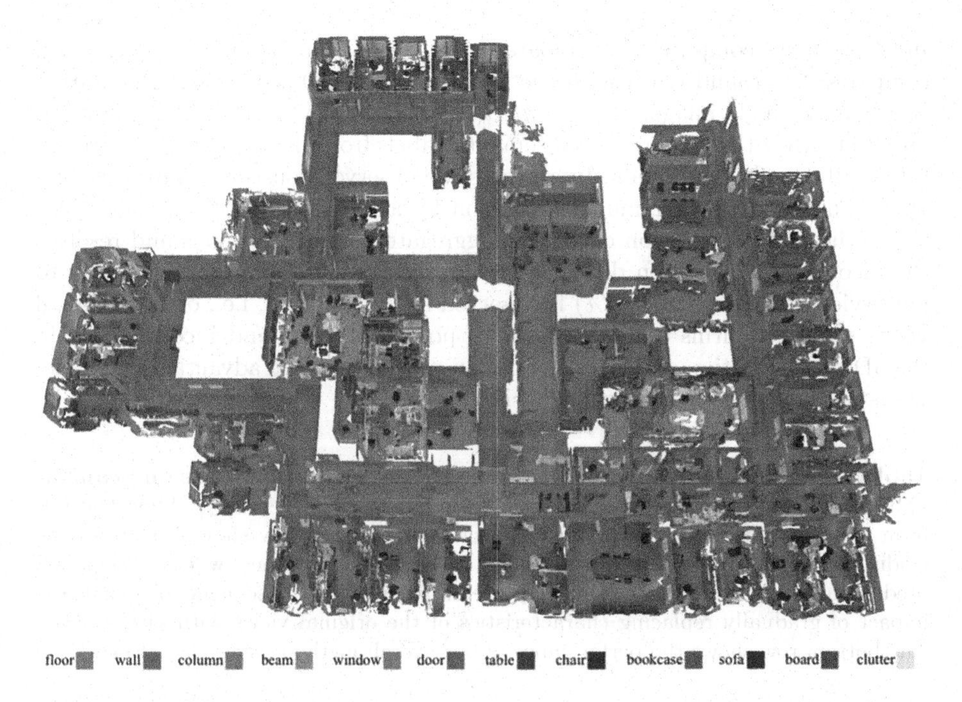

Fig. 8. Qualitiative 3D semantic segmentation results on Area5 of Stanford 3D Indoor Spaces (S3DIS) Dataset. Semantic label colors are overlayed on the textured mesh. *Ceiling* not shown for clarity.

described in Sect. 3. This mean class IoU of this baseline result is shown in the top row of Table 3. We then performed a series of tests where we included features of our virtual view algorithm one-by-one and measured the impact on performance. The second row shows the impact of using rendered images rather than photographic ones; the third shows the impact of adding additional image normal and coordinate channels captured during rendering; the fourth row shows the impact of rendering images with two times larger field-of-view; the fifth row shows the impact of our virtual viewpoints selection algorithm. We find that each of these ideas improves the 3D segmentation IoU performance significantly.

Specifically, with fixed camera extrinsics matching the original views, we compare the effect of virtual view renderings versus the original photographic images: using virtual views leads to 3.1% increase of 3D mIoU as it removes any potential errors in the 3D reconstruction and pose estimation process. Using additional channels of normal and global coordinates achieves another 2.9% performance boost in 3D mIoU as it allows the 2D semantic segmentation model to exploit the 3D information in the scene other than RGB. Increasing the FOV further improves the 3D mIoU by 1.8% since it allows the 2D model to use more context. Lastly, view sampling with backface culling achieves the best performance and an 2.2% improvement compared to the original views, showing that the camera poses can significantly affect the perception of 3D scenes.

In addition, we compute and compare a) the *single-view* 2D image mIoU, which compares 2D ground truth with the prediction of a 2D semantic segmentation model from single image, and b) *multi-view* 2D image mIoU, which compares ground truth with the reprojected semantic labels from the 3D semantic segmentation after multiview fusion. In all cases, we observed consistent improvements of 2D image mIoU after multiview fusion by a margin of 5.3% to 8.4%. This shows the multiview fusion effectively aggregates the observations and resolves the inconsistency between different views. Note that the largest single-view to multi-view improvement (8.4%) is observed in the first row, i.e., on the original views, which confirms our hypothesis of potential errors and inconsistency in the 3D reconstruction and pose estimation process and the advantage of virtual views on removing these inconsistencies.

Table 3. Evaluation on 2D and 3D Semantic segmentation tasks on ScanNet validation set. Ablation study evaluating the impact of sequentially adding features from our proposed virtual view fusion algorithm. The top row shows results of the traditional semantic segmentation approach with multiview fusion – where all semantic predictions are made on the original captured input images. Subsequent rows show the impact of gradually replacing characteristics of the original views with virtual ones. The bottom row shows the performance of our overall method using virtual views.

Image Type	Input Image Channels	Intrinsics	Extrinsics	2D Image IoU (Single View)	3D Mesh IoU (Multiview)	2D Image IoU (Multiview)
Real	RGB	Original	Original	60.1	60.1	68.5
Virtual	RGB	Original	Original	64.4	63.2	69.8
Virtual	RGB + Normal + Coordinates	Original	Original	66.1	66.1	70.8
Virtual	RGB + Normal + Coordinates	High FOV	Original	66.9	67.9	72.2
Virtual	RGB + Normal + Coordinates	High FOV	View sampling	67.0	70.1	74.9

Effect of Training Set Size. Our next experiment investigates the impact of the training set size on our algorithm. We hypothesize that generating large numbers of virtual views provides a form of data augmentation that improves generalizability of small training sets. To test this idea, we randomly sampled different numbers of scenes from the training set and trained our algorithm only on them. We compare performance of multiview fusion using a 2D model trained from virtual views rendered from those scenes versus from the original photographic images, as well as a 3D convolution method SparseConv (Fig. 9a). Note that we conduct the experiments on ScanNet low resolution meshes while for others we use high resolution ones. For virtual/real multiview fusion approaches, we use the same set of views for each scene across different experiments. We find that the virtual multiview fusion approach consistently outperforms 3D SparseConv and real multiview fusion even with a small number of scenes.

Effect of Number of Views at Inference. Next we investigate the impact of number of virtual views used in the inference stage on our algorithm. We run

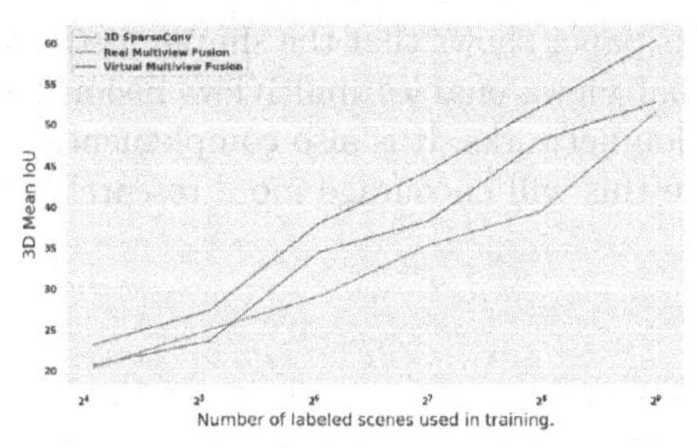

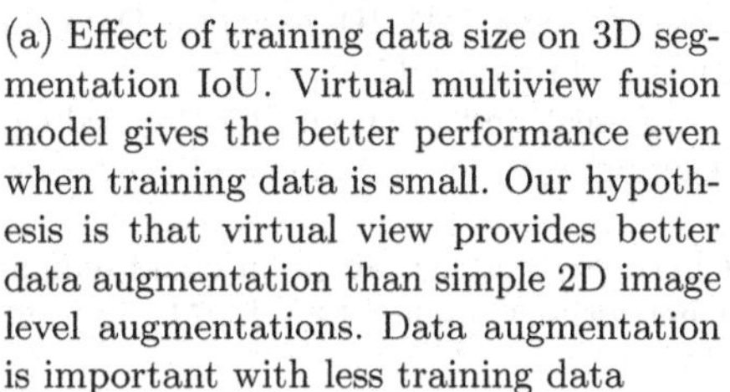

(a) Effect of training data size on 3D segmentation IoU. Virtual multiview fusion model gives the better performance even when training data is small. Our hypothesis is that virtual view provides better data augmentation than simple 2D image level augmentations. Data augmentation is important with less training data

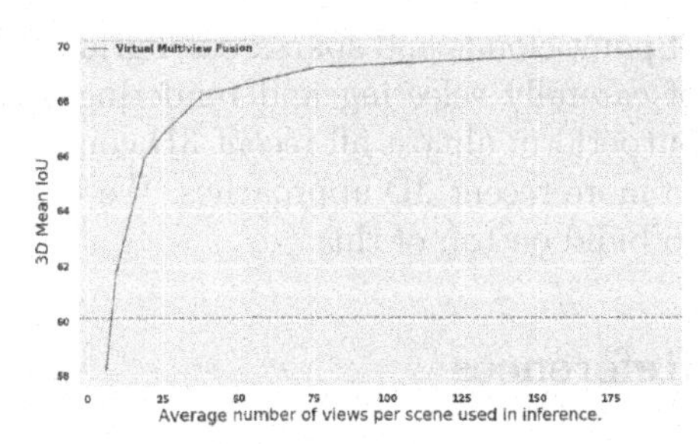

(b) Effect of number of views used at inference time on 3D segmentation. The dotted green line shows the best mIoU (60.1) obtained with multi-view fusion using all original views ($\approx$ 1700 views per scene). Our virtual multiview fusion model achieves the same accuracy with just $\approx$ 12 views per scene.

Fig. 9. Impact of data size (number of views) during training and inference.

our virtual view selection algorithms on the ScanNet validation dataset, run a 2D model on them, and then do multiview fusion using only a random subset of the virtual views. As shown in Fig. 9b, the 3D mIoU increases with the number of virtual views with diminishing returns. The virtual multiview fusion approach is able to achieve good performance even with a significantly smaller inference set. For example, while we achieve 70.1% 3D mIoU with all virtual views ($\sim$2000 views per scene), we can reach 61.7% mIoU even with $\sim$10 views per scene, and 68.2% with $\sim$40 views per scene. In addition, the result shows that using more views selected with the same approach as for training views does not negatively affect the multiview fusion performance, which is not obvious as the confident but wrong prediction of one single view can harm the overall performance.

7 Conclusion

In this paper, we propose a virtual multiview fusion approach to 3D semantic segmentation of textured meshes. This approach builds off a long history of representing and labeling meshes with images, but introduces several new ideas that significantly improve labeling performance: virtual views with additional channels, back-face culling, wide field-of-view, multiscale aware view sampling. As a result, it overcomes the 2D-3D misalignment, occlusion, narrow view, and scale invariance issues that have vexed most previous multiview fusion approaches.

The surprising conclusion from this paper is that multiview fusion algorithms are a viable alternative to 3D convolution for semantic segmentation of 3D textured meshes. Although early work on this task considered multiview fusion, the general approach has been abandoned in recent years in favor of 3D convolutions of point clouds and sparse voxel grids. This paper shows that the simple method of carefully selecting and rendering virtual views enables multiview fusion to outperform almost all recent 3D convolution networks. It is also complementary to more recent 3D approaches. We believe this will encourage more researchers to build on top of this.

References

1. Armeni, I., Sax, A., Zamir, A.R., Savarese, S.: Joint 2D–3D-Semantic Data for Indoor Scene Understanding. ArXiv e-prints, February 2017
2. Armeni, I., et al.: 3D semantic parsing of large-scale indoor spaces. In: Proceedings of the IEEE International Conference on Computer Vision and Pattern Recognition (2016)
3. Boulch, A., Guerry, J., Le Saux, B., Audebert, N.: Snapnet: 3D point cloud semantic labeling with 2D deep segmentation networks. Comput. Graph. **71**, 189–198 (2018)
4. Chen, L.-C., Zhu, Y., Papandreou, G., Schroff, F., Adam, H.: Encoder-decoder with atrous separable convolution for semantic image segmentation. In: Ferrari, V., Hebert, M., Sminchisescu, C., Weiss, Y. (eds.) ECCV 2018. LNCS, vol. 11211, pp. 833–851. Springer, Cham (2018). https://doi.org/10.1007/978-3-030-01234-2_49
5. Chiang, H., Lin, Y., Liu, Y., Hsu, W.H.: A unified point-based framework for 3D segmentation. In: 2019 International Conference on 3D Vision (3DV), pp. 155–163, September 2019
6. Chollet, F.: Xception: deep learning with depthwise separable convolutions. In: The IEEE Conference on Computer Vision and Pattern Recognition (CVPR) (2017)
7. Choy, C., Gwak, J., Savarese, S.: 4D spatio-temporal convnets: Minkowski convolutional neural networks. In: Proceedings of the IEEE Conference on Computer Vision and Pattern Recognition, pp. 3075–3084 (2019)
8. Choy, C., Park, J., Koltun, V.: Fully convolutional geometric features. In: Proceedings of the IEEE International Conference on Computer Vision, pp. 8958–8966 (2019)
9. Dai, A., Chang, A.X., Savva, M., Halber, M., Funkhouser, T., Nießner, M.: Scannet: richly-annotated 3D reconstructions of indoor scenes. In: Proceedings of the IEEE Conference on Computer Vision and Pattern Recognition, pp. 5828–5839 (2017)
10. Dai, A., Nießner, M.: 3DMV: joint 3D-multi-view prediction for 3D semantic scene segmentation. In: Ferrari, V., Hebert, M., Sminchisescu, C., Weiss, Y. (eds.) ECCV 2018. LNCS, vol. 11214, pp. 458–474. Springer, Cham (2018). https://doi.org/10.1007/978-3-030-01249-6_28
11. Graham, B., Engelcke, M., van der Maaten, L.: 3D semantic segmentation with submanifold sparse convolutional networks. In: Proceedings of the IEEE Conference on Computer Vision and Pattern Recognition, pp. 9224–9232 (2018)
12. Guerry, J., Boulch, A., Le Saux, B., Moras, J., Plyer, A., Filliat, D.: Snapnet-R: consistent 3D multi-view semantic labeling for robotics. In: Proceedings of the IEEE International Conference on Computer Vision Workshops, pp. 669–678 (2017)
13. Han, L., Zheng, T., Xu, L., Fang, L.: Occuseg: occupancy-aware 3D instance segmentation. In: Proceedings of the IEEE/CVF Conference on Computer Vision and Pattern Recognition, pp. 2940–2949 (2020)

14. Hanocka, R., Hertz, A., Fish, N., Giryes, R., Fleishman, S., Cohen-Or, D.: MeshCNN: a network with an edge. ACM Trans. Graph. (TOG) **38**(4), 1–12 (2019)
15. Hermans, A., Floros, G., Leibe, B.: Dense 3D semantic mapping of indoor scenes from RGB-D images. In: 2014 IEEE International Conference on Robotics and Automation (ICRA), pp. 2631–2638. IEEE (2014)
16. Hu, Z., Zhen, M., Bai, X., Fu, H., Tai, C.l.: JSENet: joint semantic segmentation and edge detection network for 3D point clouds. In: ECCV (2020)
17. Huang, J., Zhang, H., Yi, L., Funkhouser, T., Nießner, M., Guibas, L.J.: Texturenet: consistent local parametrizations for learning from high-resolution signals on meshes. In: Proceedings of the IEEE Conference on Computer Vision and Pattern Recognition, pp. 4440–4449 (2019)
18. Jaritz, M., Gu, J., Su, H.: Multi-view pointnet for 3D scene understanding. In: Proceedings of the IEEE International Conference on Computer Vision Workshops (2019)
19. Lai, K., Bo, L., Fox, D.: Unsupervised feature learning for 3D scene labeling. In: 2014 IEEE International Conference on Robotics and Automation (ICRA), pp. 3050–3057. IEEE (2014)
20. Landrieu, L., Simonovsky, M.: Large-scale point cloud semantic segmentation with superpoint graphs. In: The IEEE Conference on Computer Vision and Pattern Recognition (CVPR), June 2018
21. Lawin, F.J., Danelljan, M., Tosteberg, P., Bhat, G., Khan, F.S., Felsberg, M.: Deep projective 3D semantic segmentation. In: Felsberg, M., Heyden, A., Krüger, N. (eds.) CAIP 2017. LNCS, vol. 10424, pp. 95–107. Springer, Cham (2017). https://doi.org/10.1007/978-3-319-64689-3_8
22. Li, W., et al.: Interiornet: mega-scale multi-sensor photo-realistic indoor scenes dataset. arXiv preprint arXiv:1809.00716 (2018)
23. Li, Y., Bu, R., Sun, M., Wu, W., Di, X., Chen, B.: PointCNN: convolution on x-transformed points. In: Advances in Neural Information Processing Systems, pp. 820–830 (2018)
24. Ma, L., Stückler, J., Kerl, C., Cremers, D.: Multi-view deep learning for consistent semantic mapping with RGB-D cameras. In: 2017 IEEE/RSJ International Conference on Intelligent Robots and Systems (IROS), pp. 598–605. IEEE (2017)
25. McCormac, J., Handa, A., Davison, A., Leutenegger, S.: Semanticfusion: dense 3D semantic mapping with convolutional neural networks. In: 2017 IEEE International Conference on Robotics and automation (ICRA), pp. 4628–4635. IEEE (2017)
26. McCormac, J., Handa, A., Leutenegger, S., Davison, A.J.: Scenenet RGB-D: can 5m synthetic images beat generic imagenet pre-training on indoor segmentation? In: Proceedings of the IEEE International Conference on Computer Vision, pp. 2678–2687 (2017)
27. Mottaghi, R., et al.: The role of context for object detection and semantic segmentation in the wild. In: The IEEE Conference on Computer Vision and Pattern Recognition (CVPR), June 2014
28. Narita, G., Seno, T., Ishikawa, T., Kaji, Y.: Panopticfusion: online volumetric semantic mapping at the level of stuff and things. arXiv preprint arXiv:1903.01177 (2019)
29. Pham, Q.H., Nguyen, T., Hua, B.S., Roig, G., Yeung, S.K.: JSIS3D: joint semantic-instance segmentation of 3D point clouds with multi-task pointwise networks and multi-value conditional random fields. In: Proceedings of the IEEE Conference on Computer Vision and Pattern Recognition, pp. 8827–8836 (2019)

30. Qi, C.R., Su, H., Mo, K., Guibas, L.J.: Pointnet: deep learning on point sets for 3D classification and segmentation. In: Proceedings of the IEEE Conference on Computer Vision and Pattern Recognition, pp. 652–660 (2017)
31. Qi, C.R., Yi, L., Su, H., Guibas, L.J.: Pointnet++: deep hierarchical feature learning on point sets in a metric space. In: Advances in Neural Information Processing Systems, pp. 5099–5108 (2017)
32. Riegler, G., Osman Ulusoy, A., Geiger, A.: OctNet: Learning deep 3D representations at high resolutions. In: Proceedings of the IEEE Conference on Computer Vision and Pattern Recognition, pp. 3577–3586 (2017)
33. Shi, S., et al.: PV-RCNN: point-voxel feature set abstraction for 3D object detection. arXiv preprint arXiv:1912.13192 (2019)
34. Song, S., Yu, F., Zeng, A., Chang, A.X., Savva, M., Funkhouser, T.: Semantic scene completion from a single depth image. In: Proceedings of the IEEE Conference on Computer Vision and Pattern Recognition, pp. 1746–1754 (2017)
35. Su, H., Maji, S., Kalogerakis, E., Learned-Miller, E.: Multi-view convolutional neural networks for 3D shape recognition. In: Proceedings of the IEEE International Conference on Computer Vision, pp. 945–953 (2015)
36. Tatarchenko, M., Park, J., Koltun, V., Zhou, Q.Y.: Tangent convolutions for dense prediction in 3D. In: The IEEE Conference on Computer Vision and Pattern Recognition (CVPR), June 2018
37. Tchapmi, L., Choy, C., Armeni, I., Gwak, J., Savarese, S.: Segcloud: semantic segmentation of 3D point clouds. In: 2017 International Conference on 3D Vision (3DV), pp. 537–547. IEEE (2017)
38. Thomas, H., Qi, C.R., Deschaud, J.E., Marcotegui, B., Goulette, F., Guibas, L.J.: KPConv: flexible and deformable convolution for point clouds. In: Proceedings of the IEEE International Conference on Computer Vision, pp. 6411–6420 (2019)
39. Valada, A., Mohan, R., Burgard, W.: Self-supervised model adaptation for multimodal semantic segmentation. Int. J. Comput. Vis. 1–47 (2019)
40. Valentin, J., et al.: SemanticPaint: Interactive 3D labeling and learning at your fingertips. In: ACM Transactions on Graphics. ACM (2015)
41. Vineet, V., et al.: Incremental dense semantic stereo fusion for large-scale semantic scene reconstruction. In: 2015 IEEE International Conference on Robotics and Automation (ICRA), pp. 75–82. IEEE (2015)
42. Wang, S., Suo, S., Ma, W.C., Pokrovsky, A., Urtasun, R.: Deep parametric continuous convolutional neural networks. In: The IEEE Conference on Computer Vision and Pattern Recognition (CVPR), June 2018
43. Wu, W., Qi, Z., Fuxin, L.: PointConv: deep convolutional networks on 3D point clouds. In: Proceedings of the IEEE Conference on Computer Vision and Pattern Recognition, pp. 9621–9630 (2019)
44. Yan, X., Zheng, C., Li, Z., Wang, S., Cui, S.: PointASNL: robust point clouds processing using nonlocal neural networks with adaptive sampling. arXiv preprint arXiv:2003.00492 (2020)
45. Ye, X., Li, J., Huang, H., Du, L., Zhang, X.: 3D recurrent neural networks with context fusion for point cloud semantic segmentation. In: Ferrari, V., Hebert, M., Sminchisescu, C., Weiss, Y. (eds.) ECCV 2018. LNCS, vol. 11211, pp. 415–430. Springer, Cham (2018). https://doi.org/10.1007/978-3-030-01234-2_25
46. Zhang, C., Liu, Z., Liu, G., Huang, D.: Large-scale 3D semantic mapping using monocular vision. In: 2019 IEEE 4th International Conference on Image, Vision and Computing (ICIVC), pp. 71–76. IEEE (2019)

47. Zhang, Y., et al.: Physically-based rendering for indoor scene understanding using convolutional neural networks. In: Proceedings of the IEEE Conference on Computer Vision and Pattern Recognition, pp. 5287–5295 (2017)
48. Zhou, B., Zhao, H., Puig, X., Fidler, S., Barriuso, A., Torralba, A.: Scene parsing through ADE20K dataset. In: Proceedings of the IEEE Conference on Computer Vision and Pattern Recognition, pp. 633–641 (2017)

Decoupling GCN with DropGraph Module for Skeleton-Based Action Recognition

Ke Cheng[1,2], Yifan Zhang[1,2](✉), Congqi Cao[4], Lei Shi[1,2], Jian Cheng[1,2,3], and Hanqing Lu[1,2]

[1] NLPR & AIRIA, Institute of Automation, Chinese Academy of Sciences, Beijing, China
chengke2017@ia.ac.cn, {yfzhang,lei.shi,jcheng,luhq}@nlpr.ia.ac.cn
[2] School of Artificial Intelligence, University of Chinese Academy of Sciences, Beijing, China
[3] CAS Center for Excellence in Brain Science and Intelligence Technology, Beijing, China
[4] School of Computer Science, Northwestern Polytechnical University, Xi'an, China
congqi.cao@nwpu.edu.cn

Abstract. In skeleton-based action recognition, graph convolutional networks (GCNs) have achieved remarkable success. Nevertheless, how to efficiently model the spatial-temporal skeleton graph without introducing extra computation burden is a challenging problem for industrial deployment. In this paper, we rethink the spatial aggregation in existing GCN-based skeleton action recognition methods and discover that they are limited by coupling aggregation mechanism. Inspired by the decoupling aggregation mechanism in CNNs, we propose decoupling GCN to boost the graph modeling ability with no extra computation, no extra latency, no extra GPU memory cost, and less than 10% extra parameters. Another prevalent problem of GCNs is over-fitting. Although dropout is a widely used regularization technique, it is not effective for GCNs, due to the fact that activation units are correlated between neighbor nodes. We propose DropGraph to discard features in correlated nodes, which is particularly effective on GCNs. Moreover, we introduce an attention-guided drop mechanism to enhance the regularization effect. All our contributions introduce zero extra computation burden at deployment. We conduct experiments on three datasets (NTU-RGBD, NTU-RGBD-120, and Northwestern-UCLA) and exceed the state-of-the-art performance with less computation cost.

Keywords: Skeleton-based action recognition · Decoupling GCN · DropGraph

Electronic supplementary material The online version of this chapter (https://doi.org/10.1007/978-3-030-58586-0_32) contains supplementary material, which is available to authorized users.

A. Vedaldi et al. (Eds.): ECCV 2020, LNCS 12369, pp. 536–553, 2020.
https://doi.org/10.1007/978-3-030-58586-0_32

1 Introduction

Human action recognition, which plays an essential role in video understanding and human-computer interaction, attracts more attention in recent years [39,43,46]. Compared with action recognition with RGB video, skeleton-based action recognition is robust to circumstance changes and illumination variations [3,17,25,29,33,34,36,40,46,48,50].

Traditional methods mainly focus on designing hand-crafted features [4,39]. However, the performance of these handcrafted-features-based methods is barely satisfactory. Deep learning methods usually rearrange a skeleton sequence as a pseudo-image [9,14,21] or a series of joint coordinates [25,32,50], then use CNNs or RNNs to predict action labels. Recently, graph convolutional networks (GCNs), which generalize CNNs from image to graph, have been successfully adopted to model skeleton data [46]. The key component of GCN is spatial aggregation, which aggregates features of different body joints. To increase the flexibility of the skeleton graph construction, researchers propose various modules to enhance the spatial aggregation ability for GCNs [13,18,34,35,44].

In this paper, we rethink the spatial aggregation of GCNs, which is derived from CNNs. We discover that existing GCN-based skeleton action recognition methods neglect an important mechanism in CNNs: *decoupling aggregation*. Concretely, every channel has an independent spatial aggregation kernel in CNNs, capturing different spatial information in different frequencies, orientations and colors, which is crucial for the success of CNNs. However, all the channels in a graph convolution share one spatial aggregation kernel: the adjacent matrix. Although some researchers partition one adjacent matrix into multiple adjacent matrices and ensemble multiple graph convolution results of these adjacent matrices [13,17,33–35,44,46], the number of adjacent matrices is typically less than 3, which limits the expressiveness of spatial aggregation. Increasing the number of adjacent matrices will cause multiplying growth of computation cost and reduce efficiency.

Inspired by the decoupling aggregation in CNNs, we propose DeCoupling Graph Convolutional Networks (DC-GCN) to address the above dilemma with no extra FLOPs, latency, and GPU memory. DC-GCN split channels into g decoupling groups, and each group has a trainable adjacent matrix, which largely increases the expressiveness of spatial aggregation. Note that the FLOPs of decoupling graph convolution is exactly the same with conventional graph convolution. More importantly, DC-GCN is hardware-friendly and increases no extra time and GPU memory, the two most determining factors in industrial deployment. Besides, DC-GCN only cost 5%–10% extra parameters.

Another prevalent problem in graph convolution is over-fitting. Although dropout [37] is widely used in GCNs, we discover that the performance does not increase obviously with dropout layer. Because graph convolution is actually a special form of Laplacian smoothing [19], activation units are correlated between neighbor nodes. Even one node in a graph is dropped, information about this node can still be obtained from its neighbor nodes, which causes over-fitting. To relieve the over-fitting problem, we propose DropGraph, a particularly effective regularization method for graph convolutional networks. The key idea is: when we drop one node, we drop its neighbor node together. In addition, we propose an attention-guided drop mechanism to enhance the regularization effect.

The main contributions of this work are summarized as follows: 1) We propose DC-GCN, which efficiently enhances the expressiveness of graph convolution with zero extra computation cost. 2) We propose ADG to effectively relieve the crucial over-fitting problem in GCNs. 3) Our approach exceeds the state-of-the-art method with less computation cost. Code will be available at https://github.com/kchengiva/DecoupleGCN-DropGraph.

2 Background

Human Skeleton Graph Convolution. The skeleton data represents a human action as multiple skeleton frames. Every skeleton frame is represented as a graph $\mathcal{G}(V, E)$, where V is the set of n body joints and E is a set of m bones. For example, 3D joints skeleton positions across T frames can be represented as $\mathcal{X} \in \mathbb{R}^{n \times 3 \times T}$, and the 3D joints skeleton frames in the t-th frame is denoted as $\mathbf{X}_t = \mathcal{X}_{:,:,t} \in \mathbb{R}^{n \times 3}$. GCN-based action recognition models [13,18,33–35,44,46] are composed of several spatial-temporal GCN blocks, where spatial graph convolution is the key component.

Let $\mathbf{X} \in \mathbb{R}^{n \times C}$ be the input features in one frame, and $\mathbf{X}' \in \mathbb{R}^{n \times C'}$ be the output features of these joints, where C and C' are the input and output feature dimension respectively. The spatial graph convolution is

$$\mathbf{X}' = \sum_{p \in \mathcal{P}} \mathbf{X}'^{(p)} = \sum_{p \in \mathcal{P}} \widetilde{\mathbf{A}^{(p)}} \mathbf{X} \mathbf{W}^{(p)}, \tag{1}$$

where $\mathcal{P} = \{\text{root, centripetal, centrifugal}\}$ denotes the partition subsets [46]. $\widetilde{\mathbf{A}^{(p)}}$ is initialized as $\mathbf{D}^{(p)^{-\frac{1}{2}}} \mathbf{A}^{(p)} \mathbf{D}^{(p)^{-\frac{1}{2}}}$, where $\mathbf{D}_{ii}^{(p)} = \sum_j (\mathbf{A}_{ij}^{(p)}) + \varepsilon$. Here ε is set to 0.001 to avoid empty rows. Recent works let both $\widetilde{\mathbf{A}^{(p)}} \in \mathbb{R}^{n \times n}$ and $\mathbf{W}^{(p)} \in \mathbb{R}^{C \times C'}$ trainable [33,34].

Regularization Method
Over-fitting is a crucial problem in deep neural networks, including GCNs. Dropout [37] is a common regularization method. Although dropout is very effective at regularizing fully-connected layers, it is not powerful when used in convolutional layers. [2] proposed Cutout for regularizing CNNs, which randomly removes contiguous region in the input images. [5] proposed DropBlock, which applying Cutout at every feature map in CNNs. The reason why Cutout and DropBlock are efficient is that features are spatially correlated in CNNs. GCNs have the similar problems with CNNs, where common dropout is not effective. Inspired from DropBlock [5] in CNNs, we proposed DropGraph to effectively regularize GCNs.

3 Approach

In this section, we analyze the limitation of the human skeleton graph convolutional networks and propose DeCoupling Graph Convolutional Network (DC-GCN). In addition, we propose an attention-guided DropGraph (ADG) to relieve the prevalent overfitting problem in GCNs.

3.1 Decoupling Graph Convolutional Network

For clarity, we first discuss the case of graph convolution with a single partition set, then naturally extend to the multiple partition case.

Motivation. Graph convolution contains two matrix multiplication processes: $\widetilde{\mathbf{A}}\mathbf{X}$ and $\mathbf{XW}$. $\widetilde{\mathbf{A}}\mathbf{X}$ computes the aggregation information between different skeletons, so we call it *spatial aggregation*. $\mathbf{XW}$ compute the correlate information between different channels, so we call it *channel correlation*.

As shown in Fig. 1 (a), the spatial aggregation ($\widetilde{\mathbf{A}}\mathbf{X}$) can be decomposed into computing the aggregation on every channel respectively. Note that all the channels of feature $\mathbf{X}$ share one adjacency matrix $\mathbf{A}$ (drawn in the same color), which means all the channels share the same aggregation kernel. We call it *coupling aggregation*. All existing GCN-based skeleton action recognition methods adopt the *coupling aggregation*, such as ST-GCN [46], Nonlocal adaptive GCN [34], AS-GCN [18], Directed-GNN [33]. We collectively call them *coupling graph convolution*.

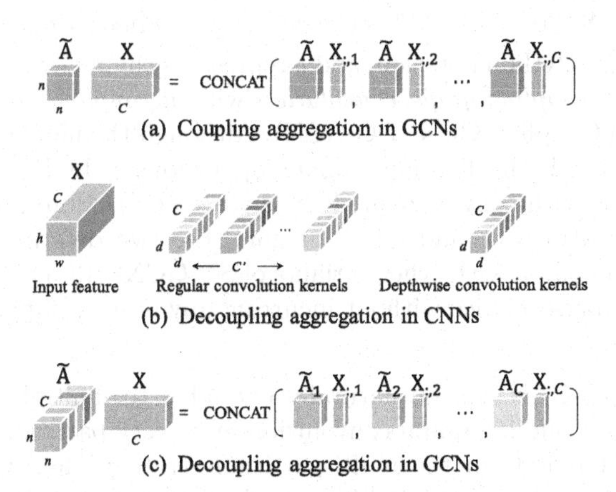

Fig. 1. Conventional GCNs (a) employ coupling aggregation, while CNNs (b) employ decoupling aggregation. We introduce the decoupling aggregation mechanism into GCNs and propose decoupling GCN (c).

However, CNNs, the source of inspiration for GCNs, do not adopt the *coupling aggregation*. As shown in Fig. 1 (b), different channels have independent spatial aggregation kernels, shown in different color. We call this mechanism *decoupling aggregation*. Decoupling aggregation mechanism can largely increase the spatial aggregation ability, which is essential for the success of CNNs.

DeCoupling GCN. Conventional graph convolution limited by coupling aggregation can be analogous to a "degenerate depthwise convolution" whose convolution kernels are shared between channels. The expressiveness of the "degenerate depthwise convolution" is notably weaker than a standard depthwise convolution. Therefore, we deduce that existing GCN-based skeleton action recognition models [18,33–35,45,46] lack the *decoupling aggregation* mechanism.

In this paper, we propose *decoupling graph convolution* for skeleton action recognition, where different channel has independent trainable adjacent matrix, shown in Fig. 1 (c). Decoupling graph convolution largely increases the variety of adjacent matrix. Similar to the redundancy of CNN kernels [30], decoupling graph convolution may introduce redundant adjacent matrix. Hence we split channels into g groups. Channels in a group share one trainable adjacent matrix. When $g = C$, every channel has its own spatial aggregation kernel which causes large number of redundant parameters; when $g = 1$, decoupling graph convolution degenerates into coupling graph convolution. Interestingly, experiments show that 8–16 groups are enough. In this case, we only increase 5%–10% extra parameters. The equation of decoupling graph convolution is shown as below:

$$\mathbf{X}' = \widetilde{\mathbf{A}^d}_{:,:,1}\mathbf{X^w}_{:,:\lfloor\frac{\mathbf{C}}{\mathbf{g}}\rfloor} \| \widetilde{\mathbf{A}^d}_{:,:,2}\mathbf{X^w}_{:,\lfloor\frac{\mathbf{C}}{\mathbf{g}}\rfloor:\lfloor\frac{\mathbf{2C}}{\mathbf{g}}\rfloor} \| \cdots \| \widetilde{\mathbf{A}^d}_{:,:,g}\mathbf{X^w}_{:,\lfloor\frac{\mathbf{(g-1)C}}{\mathbf{g}}\rfloor:} \tag{2}$$

where $\mathbf{X^w} = \mathbf{XW}$, $\widetilde{\mathbf{A}^d} \in \mathbb{R}^{n \times n \times g}$ is the decoupling adjacent matrices. Indexes of $\widetilde{\mathbf{A}^d}$ and $\mathbf{X^w}$ are in Python notation, and $\|$ represents channel-wise concatenation.

By replacing *coupling graph convolution* with *decoupling graph convolution*, we construct DeCoupling GCN (DC-GCN). Although the number of parameters is slightly increased, the floating-number operations (FLOPs) of DC-GCN is exactly the same with conventional GCN ($n^2C + nC^2$). More importantly, DC-GCN costs no extra time and GPU memory, the two determining factors for deployment. Compared with other variants of ST-GCNs [13,33,34,44], DC-GCN achieves higher performance without incurring any extra computations.

Discussion. DC-GCN can be naturally extended to multiple partition cases by introducing decoupling graph convolution into every partition. Note that our DC-GCN is different from the multi-partition strategy [46], which ensembles multiple graph convolutions with different adjacent matrices. The FLOPs of the multi-partition strategy is proportional to the number of adjacency matrices, while DC-GCN introduces various adjacency matrices with no extra computation. Besides, all our experiments use 3-partition ST-GCN as baseline, which shows the complementarity between multi-partition strategy and DC-GCN.

DC-GCN is different from SemGCN [49] in many aspects: 1) SemGCN focus on pose regression, while we focus on action recognition. 2) The receptive field of SemGCN is localized, and a heavy non-local module is inserted for non-local modeling. Our DC-GCN has non-local receptive fields and increases no extra FLOPs. 3) The parameter cost of SemGCN is nearly double of baseline. Our DC-GCN only increases 5%–10% extra parameters.

3.2 Attention-Guided DropGraph

Motivation. Although dropout is a widely used regularization method, the performance of GCNs does not increase obviously with dropout layer. A possible reason is that graph features are correlated between nearby neighbors. As shown in [19], graph convolution is a special form of Laplacian smoothing, which mixes the features of a node and its neighbors. Even one node is dropped, information about this node can still be obtained from its neighbor node, leading to over-fitting. We propose DropGraph to effectively regularize GCNs, and design an attention-guided drop mechanism to further enhance the regularization effect.

DropGraph. The main idea of DropGraph is: when we drop one node, we drop its neighbor node set together. DropGraph has two main parameters: γ and K. γ controls the sample probability, and K controls the size of the neighbor set to be dropped. On an input feature map, we first sample root nodes v_{root} with the Bernoulli distribution with probability γ, then drop the activation of v_{root} and the nodes that are at maximum K steps away from v_{root}. DropGraph can be implemented as Algorithm 1.

Algorithm 1. DropGraph

Input: a GCN feature $\mathbf{X} \in \mathbb{R}^{n \times C}$, adjacent matrix $\mathbf{A}$, γ, K, *mode*
1: **if** $mode == Inference$ **then**
2: return $\mathbf{X}$
3: **else**
4: Randomly sample $\mathbf{V}_{root} \in \mathbb{R}^n$, every element in $\mathbf{V}_{root}$ is in Bernoulli distribution with probability γ.
5: Compute the drop mask $\mathbf{M} \in \mathbb{R}^n$ to mask the nodes that are at maximum K steps away from $\mathbf{V}_{root}$:
$\mathbf{M} = 1 - \mathbf{Bool}((\mathbf{A}+\mathbf{I})^K \mathbf{V}_{root}^\top)$, where **Bool** is function setting non-zero element to 1.
6: Apply the mask: $\mathbf{X} = \mathbf{X} \times \mathbf{M}$
7: Normalize the feature:
$\mathbf{X} = \mathbf{X} \times count(\mathbf{M}) / count\_ones(\mathbf{M})$
8: **end if**

Let *keep_prob* denote the probability of an activation unit to be kept. For conventional dropout, $keep\_prob = 1-\gamma$. But for DropGraph, every zero entry on v_{root} is expanded to its $1^{st}, 2^{ed}, \cdots, K^{th}$-order neighborhood. Thus, *keep_prob* depends on both γ and K. In a graph with n nodes and e edges, we define the average degree of each node as $d_{ave} = 2e/n$. The expectation number of nodes in the ith-order neighborhood of a random sampled node can be estimated as:

$$B_i = d_{ave} \times (d_{ave} - 1)^{i-1} \tag{3}$$

The average expanded drop size is estimated as:

$$drop\_size = 1 + \sum_{i=1}^{K} B_i \tag{4}$$

If we want to keep activation units with the probability of *keep_prob*, we set:

$$\gamma = \frac{1 - keep\_prob}{drop\_size} \tag{5}$$

Note that there might be some overlaps between drop areas, so this equation is only an approximation. In our experiments, we first estimate the *keep_prob* to use (between 0.75–0.95), and then compute γ as Eq. 5.

Attention-Guided Drop Mechanism. To enhance the regularization effect, we let the attention area have higher probability to sample v_{root}. Let v be a node, γ_v denote the probability of sampling the node v as v_{root}. We modify Eq. 5 as:

$$\gamma_v = \widetilde{\alpha_v} \frac{1 - keep\_prob}{drop\_size} = \alpha_v \frac{count(\alpha)}{\sum \alpha} \frac{1 - keep\_prob}{drop\_size} \tag{6}$$

where α is the attention map, $\widetilde{\alpha}$ is the normalized attention map, $count(\alpha)$ is the number of elements in α. To assess the distribution of attention area, a common implicit assumption is that the absolute value of an activation is an indication about the importance of one unit [47]. We follow this assumption and generate α by averaging the absolute value across the channel dimension.

Spatial-Temporal ADG. In skeleton action recognition, the input of attention-guided DropGraph (ADG) is a spatiotemporal feature $\mathcal{X} \in \mathbb{R}^{n \times C \times T}$. As shown in Fig. 2, we apply ADG to spatial graph and temporal graph respectively.

The spatial aspect of the graph is the human physical structure with the number of nodes n. We generate spatial attention on every skeleton $\alpha_S \in \mathbb{R}^n$ by compressing the absolute value of $\mathcal{X}$ using average pooling on channel dimension and temporal dimension. After sampling v_{root}, we expand the drop area to its spatial neighbors. Then we broadcast the drop area to all temporal frames.

The temporal aspect of the graph is constructed by connecting consecutive frames on temporal dimension, with the number of nodes T. We generate temporal attention on every frame $\alpha_T \in \mathbb{R}^T$ by compressing the absolute value of $\mathcal{X}$ using average pooling on channel dimension and skeleton dimension. After sampling v_{root}, we expand the drop area to its temporal neighbors. Then we broadcast the drop area to all body joints.

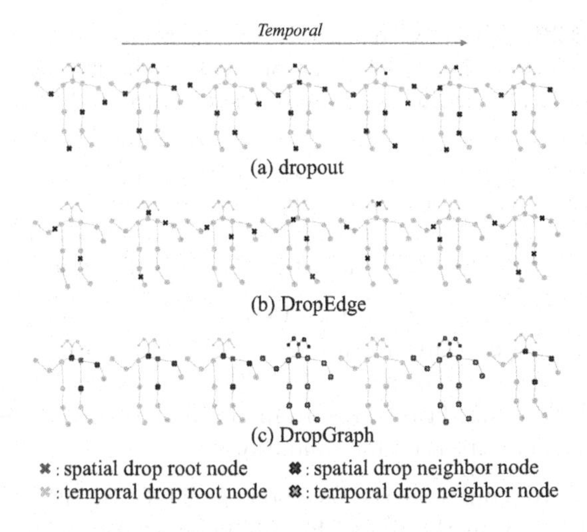

Fig. 2. Spatial-temporal DropGraph.

We cascade spatial ADG and temporal ADG to construct spatiotemporal ADG. We apply ADG on both GCN branch and skip connection branch. We adopt linear scheme [5] to decreasing *keep_prob* over time from 1 to target value.

Comparison with Other Regularization Methods. We compare DropGraph with other two regularization methods for GCNs: (a) dropout [37], which randomly drops the nodes with a certain probability; (b) DropEdge [31], which randomly drop the edges in a graph with a certain probability. As shown in Fig. 2, the drop area of both dropout and DropEdge are isolated, which can not effectively remove related information of the dropped node. For dropout, even if one node is dropped, information about this node can still be obtained from its neighbor node. For DropEdge, even if one edge is dropped, related information can still reach this node through other edges. DropGraph addresses their drawbacks and achieve notably better performance (details in Sect. 4.2).

4 Experiments

4.1 Datasets and Model Configuration

NTU-RGBD. NTU-RGBD is the most widely used 3D joint coordinates dataset. It contains 56,880 action samples in 60 action classes. These samples are performed by 40 distinct subjects. The 3D skeleton data is captured by Kinect V2. Each action is captured by 3 cameras from different horizontal angles: $-45°$, $0°$, $45°$. The original paper [32] recommends two protocols. 1) Cross-Subject (X-sub): training data comes from 20 subjects, and the remaining 20 subjects are used for validation. 2) Cross-View (X-view): training data comes from the camera $0°$ and $45°$, and validation data comes from camera $-45°$.

NTU-RGBD-120. NTU-RGBD-120 is the extended version of the NTU-RGBD dataset. It contains 114,480 action samples in 120 action classes, performed by 106 distinct subjects. This dataset contains 32 setups, and every different setup has a specific location and background. The original paper [22] recommends two evaluation protocols. 1). Cross-Subject (X-sub): training data comes from 53 subjects, and the remaining 53 subjects are used for validation. 2). Cross-Setup (X-setup): picking all the samples with even setup IDs for training, and the remaining samples with odd setup IDs for validation.

Northwestern-UCLA. Northwestern-UCLA (NW-UCLA) dataset [42] contains 1494 video clips covering 10 categories, which is captured by three Kinect cameras. Each action is performed by 10 different subjects. We adopt the same protocol as [42]: training data comes from the first two cameras, and samples from the other camera are used for validation.

Model Setting. We construct the backbone as ST-GCN [46]. The batch size is 64. We use SGD to train the model for 100 epochs. We use momentum of 0.9 and weight decay of 1e–4. The learning rate is set as 0.1 and is divided by 10 at epoch 60 and 80. For NTU-RGBD and NTU-RGBD-120, we use the same data preprocess as [34]. For NW-UCLA, we use the same data preprocess as [35].

4.2 Ablation Study

Decoupling Graph Convolution. In this subsection, we demonstrate the effectiveness and efficiency of DC-GCN.

(1) Efficacy of DC-GCN. We perform ablation study on different decoupling groups, shown in Fig. 3. Our baseline is ST-GCN [46]. We also compare the performance with non-local adaptive graph module (CVPR 2019) [34] and SE module [8]. From Fig. 3, we can draw the following conclusions:

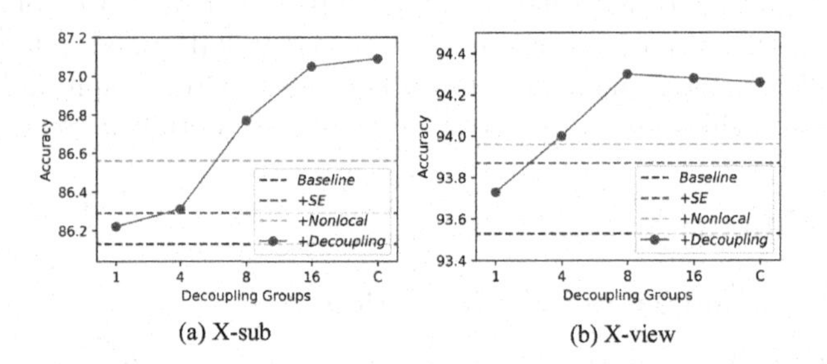

(a) X-sub (b) X-view

Fig. 3. Decoupling GCN on NTU-RGBD dataset.

- DC-GCN outperforms the baseline at 1.0% on NTU X-sub task and 0.8% on NTU X-view task. Compared with non-local adaptive graph and SE module, DC-GCN achieves higher performance at no extra computation.

- Compared to coupling graph convolution network (group = 1), decoupling graph convolution network achieves higher performance. We do not need to decouple the adjacent matrix of every channel. 8 groups are enough for NTU-RGBD X-view task and 16 groups are enough for NTU-RGBD X-sub task, which is used as our default setting in the following discussion.
- Non-local adaptive GCN [34] uses a non-local module to predict the data-dependent graph for each sample. Compared to it, our DC-GCN employs several static graphs, but get even higher performance with less FLOPs.

Table 1. FLOPs, speed and GPU memory cost. The FLOPs is for one sample on NTU-RGBD dataset. The speed and memory is measured on 1 NVIDIA Tesla K80 GPU with batch size = 64 in PyTorch evaluation mode. The time is network time, without the data loading time.

Model	GFLOPs	Network time (ms/batch)	Memory (G)
Baseline	16.2	12	5.014
+SE	16.2	16	5.494
+Nonlocal	17.9	23	5.603
+Coupling $g = 1$	16.2	12	5.014
+Decoupling $g = 4, 8, 16, C$	16.2	12	5.014

(2) FLOPs, speed and GPU memory. DC-GCN is not only theoretically efficient but also has high throughput and efficient GPU memory cost, as shown in Table 1. We can conclude that:

- Decoupling GCN ($g = 4, 8, 16, C$) introduces no extra FLOPs/latency/GPU memory compared to coupling GCN ($g = 1$). In addition, the latency and GPU memory cost of DC-GCN are almost the same as ST-GCN baseline.
- DC-GCN is more efficient than non-local adaptive GCN. Non-local adaptive GCN increases 10% extra FLOPs and 589M extra GPU memory, and cost 92 % extra time compared to DC-GCN.
- Although SE module is efficient at FLOPs, it costs 33% extra time and 480M extra GPU memory compared with baseline. This is, the theoretical efficiency is not equivalent to fast speed and efficient GPU memory cost. Compared with SE module, our DC-GCN is a hardware-friendly approach.

(3) Parameter cost. Decoupling GCN introduces extra parameters to baseline. Because our group decoupling mechanism, DC-GCN only introduces 5%–10% extra parameters when g = 8–16. If we increase the number of channels of baseline to the same parameter cost, we do not get notable improvement.

(4) Visualization of the learned adjacent matrices. We visualize the learned adjacent matrices of coupling GCN (group = 1) and decoupling GCN

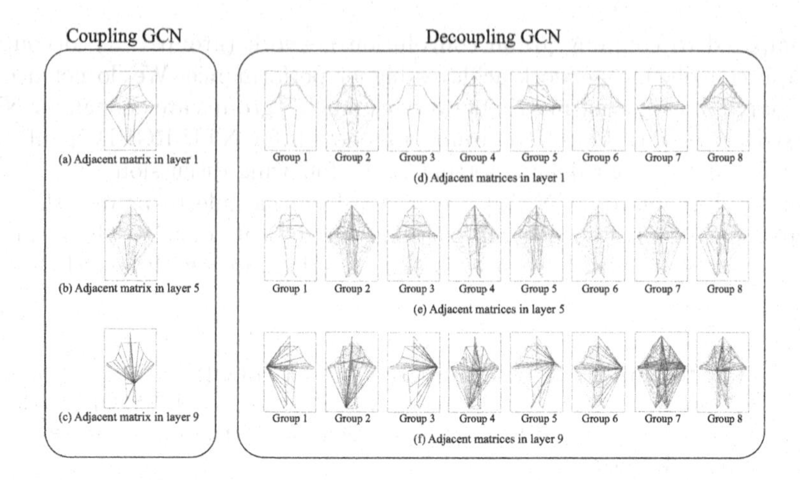

Fig. 4. Visualization of the learned adjacent matrices. The green lines show the body physical connections. The thickness of red lines shows the connection strength of the learned adjacent matrices. (Color figure online)

(group = 8), shown in Fig. 4. Compared with coupling GCN where every channel shares one adjacent matrix, decoupling GCN (group = 8) largely increases the variety of spatial aggregation. In this way, decoupling GCN can model diverse relations among joints.

In shallow layers (e.g., layer 1), the skeleton connections in decoupling GCN tend to be local, as shown in Fig. 4 (d). For example, some adjacent matrices have strong connections between head and hand (e.g., Group 4 and Group 8 in Fig. 4 (d)), which are helpful for recognizing "wipe face" and "brush teeth"; some adjacent matrices have strong connections between head and neck (e.g., Group 1 in Fig. 4 (d)), which are helpful for recognizing "nod head" and "shake head"; some adjacent matrices have strong connections between hand and wrist (e.g., Group 3 in Fig. 4 (d)), which are helpful for recognizing "write" and "count money" ; some adjacent matrices have strong connections between two hands (e.g., Group 2, Group 5 and Group 7 in Fig. 4 (d)), which are helpful for recognizing "clap" and "rub two hands". This characteristic makes decoupling GCN work well in action recognition tasks.

In deep layers (e.g., layer 9), the skeleton connections in decoupling GCN tend to be global, as shown in Fig. 4 (f). These adjacent matrices tend to gather the global feature to one joint. In this way, the deep layers can integrate global information (the whole human body) with local information (each single joint), which helps predict the final classification score.

Attention-Guided DropGraph. In this subsection, we demonstrate the effectiveness of attention-guided DropGraph (ADG).

(1) Comparison with other regularization methods. We compare with three regularization methods: dropout [37], label smoothing [38] and DropEdge

[31]. For our proposed DropGraph, we set $K = 1$ for spatial DropGraph and $K = 20$ for temporal DropGraph respectively. The detail ablation study on K is provided in the supplement material. Note that when $K = 0$, DropGraph degenerates to dropout [37]. As shown in Table 2, dropout is not powerful in GCN. DropGraph notably exceeds the other regularization methods. With the attention-guide drop mechanism, the regularization effect is further enhanced.

Table 2. Compare with other regularization methods. The top-1 accuracy (%) is evaluated on NTU-RGBD. Δ shows the improvement of accuracy.

Model	Regularization method	X-sub	Δ	X-view	Δ
DC-GCN	–	87.1	0	94.3	0
	Dropout [37]	87.2	+0.1	94.4	+0.1
	Label smoothing [38]	87.1	+0.0	94.4	+0.1
	DropEdge [31]	87.6	+0.5	94.7	+0.4
	DropGraph (ours)	88.0	+0.9	95.0	+0.7
	Attention-guided DropGraph (ours)	**88.2**	**+1.1**	**95.2**	**+0.9**

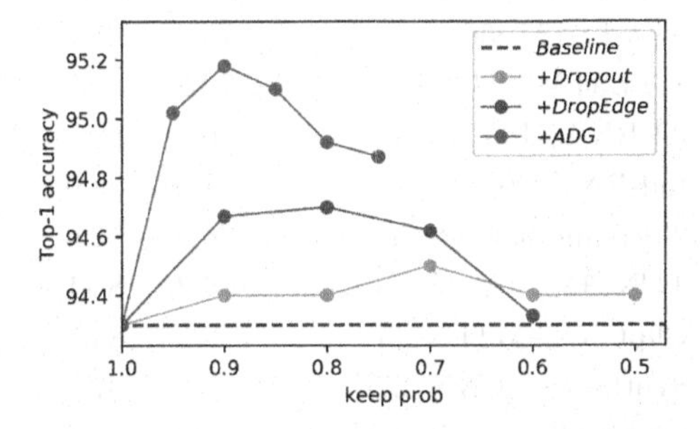

Fig. 5. Compare dropout, DropEdge and our ADG at different *keep_prob*.

(2) The setting of ***keep_prob*****.** We discuss the setting of *keep_prob* on dropout, DropEdge and our proposed ADG. As shown in Fig. 5, ADG provides efficient regularization when *keep_prob* =0.85–0.9. ADG has a notable improvement compared to the best result of dropout and DropEdge.

Ablation Studies on NW-UCLA and NTU-RGBD-120. Besides the above ablation study on NTU-RGBD dataset, we also perform ablation studies on NW-UCLA and NTU-RGBD-120 datasets, shown in Table 3.

Table 3. Ablation study on NW-UCLA and NTU-RGBD-120. DC refers to decoupling.

Backbone	+DC	+ADG	NW-UCLA (%)	NTU120 X-sub (%)	NTU120 X-setup (%)
ST-GCN			89.8	79.7	81.3
ST-GCN	✓		91.6	81.3	82.7
ST-GCN	✓	✓	93.8	82.4	84.3

4.3 Comparisons to the State-of-the-Art

Multi-stream strategy is commonly employed in previous state-of-the-art approaches [33–35,43,46]. We adopt the same multi-stream ensemble strategy with [33], which ensembles 4 streams: joint, bone, motion, and bone motion. The joint stream uses the original joint position as input; the bone stream uses the difference between adjacent joints as input; the motion stream uses the difference between adjacent frames as input.

Table 4. Comparisions of the top-1 accuracy (%) with the state-of-the-art methods on the NTU-RGBD dataset.

Methods	X-sub	X-view
Lie Group [39]	50.1	52.8
STA-LSTM [36]	73.4	81.2
VA-LSTM [48]	79.2	87.7
ARRN-LSTM [16]	80.7	88.8
Ind-RNN [20]	81.8	88.0
2-Stream 3DCNN [21]	66.8	72.6
TCN [11]	74.3	83.1
ClipCNN+MTLN [9]	79.6	84.8
Synthesized CNN [27]	80.0	87.2
CNN+Motion+Trans [15]	83.2	88.8
ST-GCN [46]	81.5	88.3
Motif+VTDB [44]	84.2	90.2
STGR-GCN [13]	86.9	92.3
AS-GCN [18]	86.8	94.2
Non-local adaptive GCN [34]	88.5	95.1
AGC-LSTM [35]	89.2	95.0
Directed-GNN [33]	89.9	96.1
DC-GCN+ADG (ours)	**90.8**	**96.6**

We conduct extensive experiments on three datasets: NTU-RGBD dataset, NW-UCLA dataset, and the recently proposed NTU-RGBD-120 dataset, shown

Table 5. Comparisons of the accuracy (%) with the state-of-the-art methods on the NW-UCLA dataset.

Methods	Year	Top-1
Lie Group [39]	2014	74.2
Actionlet ensemble [41]	2014	76.0
Visualization CNN [28]	2017	86.1
Ensemble TS-LSTM [12]	2017	89.2
2s AGC-LSTM [35]	2019	93.3
DC-GCN+ADG (ours)	–	**95.3**

Table 6. Comparisions of the top-1 accuracy (%) with the state-of-the-art methods on the NTU-RGBD-120 dataset.

Methods	X-sub	X-setup
Part-Aware LSTM [32]	25.5	26.3
Soft RNN [7]	36.3	44.9
Dynamic Skeleton [6]	50.8	54.7
Spatio-Temporal LSTM [25]	55.7	57.9
Internal Feature Fusion [24]	58.2	60.9
GCA-LSTM [26]	58.3	59.2
Multi-Task Learning Network [9]	58.4	57.9
FSNet [23]	59.9	62.4
Multi CNN + RotClips [10]	62.2	61.8
Pose Evolution Map [29]	64.6	66.9
SkeleMotion [1]	67.7	66.9
DC-GCN+ADG (ours)	**86.5**	**88.1**

in Table 4, Table 5, and Table 6 respectively. Our approach exceeds all the previous methods with a notable margin.

Note that the comparison with Directed-GNN is unfair because of the computational cost disparity. Directed-GNN doubles the number of channels in temporal convolution and introduces extra directed graph modules, whose computational cost (127G FLOPs) is nearly double of ours (65G FLOPs)[1]. Nevertheless, we outperform the current state-of-the-art method Directed-GNN at 0.9% on NTU-RGBD X-sub task. On NW-UCLA, we outperform the current state-of-the-art method AGC-LSTM at 2.0%. On NTU-120 RGB+D dataset, we obviously exceed all previously reported performance.

[1] Details about the computational complexity are provided in supplement material.

5 Conclusion

In this work, we propose decoupling GCN to boost the graph modeling ability for skeleton-based action recognition. In addition, we propose an attention-guided DropGraph module to effectively relieve the crucial over-fitting problem in GCNs. Both these two contributions introduce zero extra computation, zero extra latency, and zero extra GPU memory cost at deployment. Hence, our approach is not only theoretically efficient but also has well practicality and application prospects. Our approach exceeds the current state-of-the-art method on three datasets: NTU-RGBD, NTU-RGBD-120, and NW-UCLA with even less computation. Since enhancing the effectiveness of the graph modeling and reducing the over-fitting risk are two prevalent problems in GCNs, our approach has potential application value for other GCN tasks, such as recommender systems, traffic analysis, natural language processing, computational chemistry.

Acknowledgement. This work was supported in part by the National Natural Science Foundation of China under Grant 61876182 and 61872364, in part by the Jiangsu Leading Technology Basic Research Project BK20192004. This work was partly supported by the Open Projects Program of National Laboratory of Pattern Recognition.

References

1. Caetano, C., Sena, J., Brémond, F., Santos, J.A.d., Schwartz, W.R.: Skelemotion: a new representation of skeleton joint sequences based on motion information for 3D action recognition. arXiv preprint arXiv:1907.13025 (2019)
2. DeVries, T., Taylor, G.W.: Improved regularization of convolutional neural networks with cutout. arXiv preprint arXiv:1708.04552 (2017)
3. Du, Y., Wang, W., Wang, L.: Hierarchical recurrent neural network for skeleton based action recognition. In: Proceedings of the IEEE Conference on Computer Vision and Pattern Recognition, pp. 1110–1118 (2015)
4. Fernando, B., Gavves, E., Oramas, J.M., Ghodrati, A., Tuytelaars, T.: Modeling video evolution for action recognition. In: Proceedings of the IEEE Conference on Computer Vision and Pattern Recognition, pp. 5378–5387 (2015)
5. Ghiasi, G., Lin, T.Y., Le, Q.V.: Dropblock: a regularization method for convolutional networks. In: Advances in Neural Information Processing Systems, pp. 10727–10737 (2018)
6. Hu, J.F., Zheng, W.S., Lai, J., Zhang, J.: Jointly learning heterogeneous features for RGB-D activity recognition. In: Proceedings of the IEEE Conference on Computer Vision and Pattern Recognition, pp. 5344–5352 (2015)
7. Hu, J.F., Zheng, W.S., Ma, L., Wang, G., Lai, J.H., Zhang, J.: Early action prediction by soft regression. IEEE Trans. Pattern Anal. Mach. Intell. **41**, 2568–2583 (2018)
8. Hu, J., Shen, L., Sun, G.: Squeeze-and-excitation networks. In: Proceedings of the IEEE Conference on Computer Vision and Pattern Recognition, pp. 7132–7141 (2018)
9. Ke, Q., Bennamoun, M., An, S., Sohel, F., Boussaid, F.: A new representation of skeleton sequences for 3D action recognition. In: Proceedings of the IEEE Conference on Computer Vision and Pattern Recognition, pp. 3288–3297 (2017)

10. Ke, Q., Bennamoun, M., An, S., Sohel, F., Boussaid, F.: Learning clip representations for skeleton-based 3D action recognition. IEEE Trans. Image Process. **27**(6), 2842–2855 (2018)
11. Kim, T.S., Reiter, A.: Interpretable 3D human action analysis with temporal convolutional networks. In: 2017 IEEE Conference on Computer Vision and Pattern Recognition Workshops (CVPRW), pp. 1623–1631. IEEE (2017)
12. Lee, I., Kim, D., Kang, S., Lee, S.: Ensemble deep learning for skeleton-based action recognition using temporal sliding LSTM networks. In: Proceedings of the IEEE International Conference on Computer Vision, pp. 1012–1020 (2017)
13. Li, B., Li, X., Zhang, Z., Wu, F.: Spatio-Temporal Graph Routing For Skeleton-based Action Recognition (2019)
14. Li, B., Dai, Y., Cheng, X., Chen, H., Lin, Y., He, M.: Skeleton based action recognition using translation-scale invariant image mapping and multi-scale deep CNN. In: 2017 IEEE International Conference on Multimedia & Expo Workshops (ICMEW), pp. 601–604. IEEE (2017)
15. Li, C., Zhong, Q., Xie, D., Pu, S.: Skeleton-based action recognition with convolutional neural networks. In: 2017 IEEE International Conference on Multimedia & Expo Workshops, ICME Workshops, Hong Kong, China, 10–14 July 2017, pp. 597–600 (2017). https://doi.org/10.1109/ICMEW.2017.8026285
16. Li, L., Zheng, W., Zhang, Z., Huang, Y., Wang, L.: Skeleton-based relational modeling for action recognition. CoRR abs/1805.02556 (2018). http://arxiv.org/abs/1805.02556
17. Li, M., Chen, S., Chen, X., Zhang, Y., Wang, Y., Tian, Q.: Actional-structural graph convolutional networks for skeleton-based action recognition. In: The IEEE Conference on Computer Vision and Pattern Recognition (CVPR), June 2019
18. Li, M., Chen, S., Chen, X., Zhang, Y., Wang, Y., Tian, Q.: Actional-structural graph convolutional networks for skeleton-based action recognition. In: Proceedings of the IEEE Conference on Computer Vision and Pattern Recognition, pp. 3595–3603 (2019)
19. Li, Q., Han, Z., Wu, X.M.: Deeper insights into graph convolutional networks for semi-supervised learning. In: Thirty-Second AAAI Conference on Artificial Intelligence (2018)
20. Li, S., Li, W., Cook, C., Zhu, C., Gao, Y.: Independently recurrent neural network (indrnn): Building a longer and deeper RNN. In: 2018 IEEE Conference on Computer Vision and Pattern Recognition, CVPR 2018, Salt Lake City, UT, USA, 18–22 June 2018, pp. 5457–5466 (2018). https://doi.org/10.1109/CVPR.2018.00572, http://openaccess.thecvf.com/content_cvpr_2018/html/Li_Independently_Recurrent_Neural_CVPR_2018_paper.html
21. Liu, H., Tu, J., Liu, M.: Two-stream 3D convolutional neural network for skeleton-based action recognition. arXiv preprint arXiv:1705.08106 (2017)
22. Liu, J., Shahroudy, A., Perez, M., Wang, G., Duan, L., Kot, A.C.: NTU RGB+D 120: a large-scale benchmark for 3D human activity understanding. CoRR abs/1905.04757 (2019). http://arxiv.org/abs/1905.04757
23. Liu, J., Shahroudy, A., Wang, G., Duan, L.Y., Chichung, A.K.: Skeleton-based online action prediction using scale selection network. IEEE Trans. Pattern Anal. Mach. Intell. **42**, 1453–1467 (2019)
24. Liu, J., Shahroudy, A., Xu, D., Kot, A.C., Wang, G.: Skeleton-based action recognition using spatio-temporal LSTM network with trust gates. IEEE Trans. Pattern Anal. Mach. Intell. **40**(12), 3007–3021 (2017)

25. Liu, J., Shahroudy, A., Xu, D., Wang, G.: Spatio-temporal LSTM with trust gates for 3D human action recognition. In: Leibe, B., Matas, J., Sebe, N., Welling, M. (eds.) ECCV 2016. LNCS, vol. 9907, pp. 816–833. Springer, Cham (2016). https://doi.org/10.1007/978-3-319-46487-9_50
26. Liu, J., Wang, G., Hu, P., Duan, L.Y., Kot, A.C.: Global context-aware attention LSTM networks for 3D action recognition. In: Proceedings of the IEEE Conference on Computer Vision and Pattern Recognition, pp. 1647–1656 (2017)
27. Liu, M., Liu, H., Chen, C.: Enhanced skeleton visualization for view invariant human action recognition. Pattern Recogn. **68**, 346–362 (2017). https://doi.org/10.1016/j.patcog.2017.02.030
28. Liu, M., Liu, H., Chen, C.: Enhanced skeleton visualization for view invariant human action recognition. Pattern Recogn. **68**, 346–362 (2017)
29. Liu, M., Yuan, J.: Recognizing human actions as the evolution of pose estimation maps. In: Proceedings of the IEEE Conference on Computer Vision and Pattern Recognition, pp. 1159–1168 (2018)
30. Molchanov, P., Tyree, S., Karras, T., Aila, T., Kautz, J.: Pruning convolutional neural networks for resource efficient inference. arXiv preprint arXiv:1611.06440 (2016)
31. Rong, Y., Huang, W., Xu, T., Huang, J.: Dropedge: towards deep graph convolutional networks on node classification. In: International Conference on Learning Representations (2020)
32. Shahroudy, A., Liu, J., Ng, T.T., Wang, G.: NTU RGB+D: a large scale dataset for 3D human activity analysis. In: Proceedings of the IEEE Conference on Computer Vision and Pattern Recognition, pp. 1010–1019 (2016)
33. Shi, L., Zhang, Y., Cheng, J., Lu, H.: Skeleton-based action recognition with directed graph neural networks. In: The IEEE Conference on Computer Vision and Pattern Recognition (CVPR), June 2019
34. Shi, L., Zhang, Y., Cheng, J., Lu, H.: Two-stream adaptive graph convolutional networks for skeleton-based action recognition. In: The IEEE Conference on Computer Vision and Pattern Recognition (CVPR), June 2019
35. Si, C., Chen, W., Wang, W., Wang, L., Tan, T.: An attention enhanced graph convolutional LSTM network for skeleton-based action recognition. In: The IEEE Conference on Computer Vision and Pattern Recognition (CVPR), June 2019
36. Song, S., Lan, C., Xing, J., Zeng, W., Liu, J.: An end-to-end spatio-temporal attention model for human action recognition from skeleton data. In: Thirty-First AAAI Conference on Artificial Intelligence (2017)
37. Srivastava, N., Hinton, G., Krizhevsky, A., Sutskever, I., Salakhutdinov, R.: Dropout: a simple way to prevent neural networks from overfitting. J. Mach. Learn. Res. **15**(1), 1929–1958 (2014)
38. Szegedy, C., Vanhoucke, V., Ioffe, S., Shlens, J., Wojna, Z.: Rethinking the inception architecture for computer vision. In: Proceedings of the IEEE Conference on Computer Vision and Pattern Recognition, pp. 2818–2826 (2016)
39. Veeriah, V., Zhuang, N., Qi, G.J.: Differential recurrent neural networks for action recognition. In: Proceedings of the IEEE International Conference on Computer Vision, pp. 4041–4049 (2015)
40. Vemulapalli, R., Arrate, F., Chellappa, R.: Human action recognition by representing 3D skeletons as points in a lie group. In: 2014 IEEE Conference on Computer Vision and Pattern Recognition, CVPR 2014, Columbus, OH, USA, 23–28 June 2014, pp. 588–595 (2014). https://doi.org/10.1109/CVPR.2014.82
41. Wang, J., Liu, Z., Wu, Y., Yuan, J.: Learning actionlet ensemble for 3D human action recognition. IEEE Trans. Pattern Anal. Mach. Intell. **36**(5), 914–927 (2013)

42. Wang, J., Nie, X., Xia, Y., Wu, Y., Zhu, S.C.: Cross-view action modeling, learning and recognition. In: Proceedings of the IEEE Conference on Computer Vision and Pattern Recognition, pp. 2649–2656 (2014)
43. Wang, L., et al.: Temporal segment networks: towards good practices for deep action recognition. In: Leibe, B., Matas, J., Sebe, N., Welling, M. (eds.) ECCV 2016. LNCS, vol. 9912, pp. 20–36. Springer, Cham (2016). https://doi.org/10.1007/978-3-319-46484-8_2
44. Wen, Y.H., Gao, L., Fu, H., Zhang, F.L., Xia, S.: Graph CNNs with motif and variable temporal block for skeleton-based action recognition. In: Proceedings of the AAAI Conference on Artificial Intelligence, vol. 33, pp. 8989–8996 (2019)
45. Wen, Y., Gao, L., Fu, H., Zhang, F., Xia, S.: Graph CNNs with motif and variable temporal block for skeleton-based action recognition. In: The Thirty-Third AAAI Conference on Artificial Intelligence, AAAI 2019, The Thirty-First Innovative Applications of Artificial Intelligence Conference, IAAI 2019, The Ninth AAAI Symposium on Educational Advances in Artificial Intelligence, EAAI 2019, Honolulu, Hawaii, USA, 27 January–1 February 2019, pp. 8989–8996 (2019). https://aaai.org/ojs/index.php/AAAI/article/view/4929
46. Yan, S., Xiong, Y., Lin, D.: Spatial temporal graph convolutional networks for skeleton-based action recognition. In: Thirty-Second AAAI Conference on Artificial Intelligence (2018)
47. Zagoruyko, S., Komodakis, N.: Paying more attention to attention: improving the performance of convolutional neural networks via attention transfer. arXiv preprint arXiv:1612.03928 (2016)
48. Zhang, P., Lan, C., Xing, J., Zeng, W., Xue, J., Zheng, N.: View adaptive recurrent neural networks for high performance human action recognition from skeleton data. In: Proceedings of the IEEE International Conference on Computer Vision, pp. 2117–2126 (2017)
49. Zhao, L., Peng, X., Tian, Y., Kapadia, M., Metaxas, D.N.: Semantic graph convolutional networks for 3D human pose regression. In: IEEE Conference on Computer Vision and Pattern Recognition, CVPR 2019, Long Beach, CA, USA, 16–20 June 2019, pp. 3425–3435. Computer Vision Foundation/IEEE (2019)
50. Zheng, W., Li, L., Zhang, Z., Huang, Y., Wang, L.: Skeleton-based relational modeling for action recognition. arXiv preprint arXiv:1805.02556 (2018)

Deep Shape from Polarization

Yunhao Ba[1], Alex Gilbert[1], Franklin Wang[1], Jinfa Yang[2], Rui Chen[2], Yiqin Wang[1], Lei Yan[2], Boxin Shi[2(✉)], and Achuta Kadambi[1(✉)]

[1] University of California, Los Angeles, USA
{yhba,alexrgilbert}@ucla.edu, franklinxzw@gmail.com, achuta@ee.ucla.edu
[2] Peking University, Beijing, China
{jinfayang,shiboxin}@pku.edu.cn

Abstract. This paper makes a first attempt to bring the Shape from Polarization (SfP) problem to the realm of deep learning. The previous state-of-the-art methods for SfP have been purely physics-based. We see value in these principled models, and blend these physical models as priors into a neural network architecture. This proposed approach achieves results that exceed the previous state-of-the-art on a challenging dataset we introduce. This dataset consists of polarization images taken over a range of object textures, paints, and lighting conditions. We report that our proposed method achieves the lowest test error on each tested condition in our dataset, showing the value of blending data-driven and physics-driven approaches.

Keywords: Shape from Polarization · 3D reconstruction · Physics-based deep learning

1 Introduction

While deep learning has revolutionized many areas of computer vision, the deep learning revolution has not yet been studied in context of Shape from Polarization (SfP). The SfP problem is fascinating because, if successful, shape could be obtained in completely passive lighting conditions without estimating lighting direction. Recent progress in CMOS sensors has spawned machine vision cameras that capture the required polarization information in a single shot [42], making the capture process more relaxed than photometric stereo.

This SfP problem can be stated simply: light that reflects off an object has a polarization state that corresponds to shape. In reality, the underlying physics is among the most optically complex of all computer vision problems. For this reason, previous SfP methods have high error rates (in context of mean angular

A. Gilbert and F. Wang—Equal contribution.
Project page: https://visual.ee.ucla.edu/deepsfp.htm.

Electronic supplementary material The online version of this chapter (https://doi.org/10.1007/978-3-030-58586-0_33) contains supplementary material, which is available to authorized users.

A. Vedaldi et al. (Eds.): ECCV 2020, LNCS 12369, pp. 554–571, 2020.
https://doi.org/10.1007/978-3-030-58586-0_33

error (MAE) of surface normal estimation), and limited generalization to mixed materials and lighting conditions.

The physics of SfP are based on the Fresnel Equations. These equations lead to an underdetermined system—the so-called *ambiguity problem*. This problem arises because a linear polarizer cannot distinguish between polarized light that is rotated by π radians. This results in two confounding estimates for azimuth angle at each pixel. Previous work in SfP has used additional information to constrain the ambiguity problem. For instance, Smith *et al.* [51] use both polarization and shading constraints as linear equations when solving object depth, and Mahmoud *et al.* [33] use shape from shading constraints to correct the ambiguities. Other authors assume surface convexity to constrain the azimuth angle [4,37] or use a coarse depth map to constrain the ambiguity [21,22]. There are also additional binary ambiguities based on reflection type, as discussed in [4,33]. Table 1 compares our proposed technique with prior work.

Table 1. Deep SfP vs Previous Methods. We compare the input constraints and result quality of the proposed hybrid of physics and learning compared to previous, physics-based SfP methods.

Method	Inputs	Mean Angular Error	Robustness to Texture-Copy	Lighting Invariance
Miyazaki [37]	Polarization Images	High	Strong	Moderate
Mahmoud [33]	Polarization Images	High	Not Observed	Moderate
Smith [52]	Polarization Images Lighting Estimate	Moderate	Strong	Moderate
Proposed	Polarization Images	Lowest	Strong	Strong

Another contributing factor to the underdetermined nature of SfP is the *refractive problem*. SfP needs knowledge of per-pixel refractive indices. Previous work has used hard-coded values to estimate the refractive index of scenes [37]. This leads to a relative shape recovered with refractive distortion.

Yet another limitation of the physical model is particular susceptibility to *noise*. The polarization signal is very subtle for fronto-parallel geometries so it is important that the input images are relatively noise-free. Unfortunately, a polarizing filter reduces the captured light intensity by 50%, worsening the effects of Poisson shot noise, encouraging a noise tolerant SfP algorithm.[1]

In this paper, we address these SfP pitfalls by moving away from a physics-only solution, toward the realm of data-driven techniques. While it is tempting to apply traditional deep learning models to the SfP problem, we find this approach does not maximize performance. Instead, we propose a physics-based learning algorithm that not only outperforms traditional deep learning, but also outperforms three baseline comparisons to physics-based SfP. We summarize our contributions as follows:

[1] For a detailed discussion of other sources of noise please refer to Schechner [47].

- a first attempt to apply deep learning techniques to solve the SfP problem;
- incorporation of the existing physical model into the deep learning approach;
- demonstration of significant error reduction; and
- introduction of the first polarization image dataset with ground truth shape, laying a foundation for future data-driven methods.

Limitations: As a physics-based learning approach, our technique still relies on computing the physical priors for every test example. This means that the per-frame runtime would be the sum of the compute time for the forward pass and that of the physics-based prior. Our runtime details are in the supplement. Future work could parallelize compute of the physical prior. Another limitation pertains to the accuracy inherent to SfP. Our average MAE on the test set is 18.5°. While this is the best SfP performer on our challenging dataset, the error is higher than with a more controlled technique like photometric stereo.

2 Related Work

Polarization cues have been employed for various tasks, such as reflectometry estimation [12], radiometric calibration [58], facial geometry reconstruction [13], dynamic interferometry [32], polarimetric spatially varying surface reflectance functions (SVBRDF) recovery [5], and object shape acquisition [14,31,43,64]. This paper is at the seamline of deep learning and SfP, offering unique performance tradeoffs from prior work. Refer to Table 1 for an overview.

Shape from Polarization infers the shape (usually represented in surface normals) of a surface by observing the correlated changes of image intensity with the polarization information. Changes of polarization information could be captured by rotating a linear polarizer in front of an ordinary camera [2,60] or polarization cameras using a single shot in real time (e.g., PolarM [42] in [62]). Conventional SfP decodes such information to recover the surface normal up to some ambiguity. If only images with different polarization information are available, heuristic priors such as the surface normals along the boundary and convexity of the objects are employed to remove the ambiguity [4,37]. Photometric constraints from shape from shading [33] and photometric stereo [1,11,39] complements polarization constraints to make the normal estimates unique. If multi-spectral measurements are available, surface normal and its refractive index could be estimated at the same time [16,17]. More recently, a joint formulation of shape from shading and SfP in a linear manner is shown to be able to directly estimate the depth of the surface [51,52,59]. This paper is the first attempt at combining deep learning and SfP.

Polarized 3D involves stronger assumptions than SfP and has different inputs and outputs. Recognizing that SfP alone is a limited technique, the Polarized 3D class of methods integrate SfP with a low resolution depth estimate. This additional constraint allows not just recovery of shape but also a high-quality 3D model. The low resolution depth could be achieved by employing two-view

[3,6,35], three-view [8], multi-view [9,36] stereo, or even in real time by using a SLAM system [62]. These depth estimates from geometric methods are not reliable in textureless regions where finding correspondence for triangulation is difficult. Polarimetric cues could be jointly used to improve such unreliable depth estimates to obtain a more complete shape estimation. A depth sensor such as the Kinect can also provide coarse depth prior to disambiguate the ambiguous normal estimates given by SfP [21,22]. The key step that characterizes Polarized 3D is a holistic approach that rethinks both SfP and the depth-normal fusion process. The main limitation of Polarized 3D is the strong requirement of a coarse depth map, which is not true for our proposed technique.

Data-driven computational imaging approaches draw much attention in recent years thanks to the powerful modeling ability of deep neural networks. Various types of convolutional neural networks (CNNs) are designed to enable 3D imaging for many types of sensors and measurements. From single photon sensor measurements, a multi-scale denoising and upsampling CNN is proposed to refine depth estimates [28]. CNNs also show advantage in solving phase unwrapping, multipath interference, and denoising jointly from raw time-of-flight measurements [34,54]. From multi-directional lighting measurements, a fully-connected network is proposed to solve photometric stereo for general reflectance with a pre-defined set of light directions [45]. Then the fully convolutional network with an order-agnostic max-pooling operation [7] and the observation map invariant to the number and permutation of the images [18] are concurrently proposed to deal with an arbitrary set of light directions. Normal estimates from photometric stereo can also be learned in an unsupervised manner by minimizing reconstruction loss [57]. Other than 3D imaging, deep learning has helped solve several inverse problems in the field of computational imaging [30,46,55,56]. Separation of shape, reflectance and illuminance maps for wild facial images can be achieved with the CNNs as well [48]. CNNs also exhibit potential for modeling SVBRDF of a near-planar surface [10,25,26,63], and more complex objects [27]. The challenge with existing deep learning frameworks is that they do not leverage the unique physics of polarization.

3 Proposed Method

In this section, we first introduce basic knowledge of SfP, and then present our physics-based CNN. Blending physics and deep learning improves the performance and generalizability of the method.

3.1 Image Formation and Physical Solution

Our objective is to reconstruct surface normals $\hat{\boldsymbol{N}}$ from a set of polarization images $\{\boldsymbol{I}_{\phi_1}, \boldsymbol{I}_{\phi_2}, ..., \boldsymbol{I}_{\phi_M}\}$ with different polarization angles. For a specific polarization angle ϕ_{pol}, the intensity at a pixel of a captured image follows a sinusoidal variation under unpolarized illumination:

$$I(\phi_{pol}) = \frac{I_{max} + I_{min}}{2} + \frac{I_{max} - I_{min}}{2} \cos(2(\phi_{pol} - \phi)), \tag{1}$$

where ϕ denotes the phase angle, and I_{min} and I_{max} are lower and upper bounds for the observed intensity. Equation (1) has a π-***ambiguity*** in context of ϕ: two phase angles, with a π shift, will result in the same intensity in the captured images. Based on the phase angle ϕ, the azimuth angle φ can be retrieved with $\frac{\pi}{2}$-***ambiguity*** as follows [9]:

$$\phi = \begin{cases} \varphi, & \text{if diffuse reflection dominates} \\ \varphi - \frac{\pi}{2}, & \text{if specular reflection dominates} \end{cases}. \tag{2}$$

The zenith angle θ is related to the degree of polarization ρ, which can be written as:

$$\rho = \frac{I_{max} - I_{min}}{I_{max} + I_{min}}. \tag{3}$$

When diffuse reflection is dominant, the degree of polarization can be expressed with the zenith angle θ and the refractive index n as follows [4]:

$$\rho_d = \frac{(n - \frac{1}{n})^2 \sin^2\theta}{2 + 2n^2 - (n + \frac{1}{n})^2 \sin^2\theta + 4\cos\theta\sqrt{n^2 - \sin^2\theta}}. \tag{4}$$

The dependency of ρ_d on n is weak [4], and we assume $n = 1.5$ throughout the rest of this paper. With this known n, Eq. (4) can be rearranged to obtain a close-form estimation of the zenith angle for the diffuse dominant case.

When specular reflection is dominant, the degree of polarization can be written as [4]:

$$\rho_s = \frac{2\sin^2\theta\cos\theta\sqrt{n^2 - \sin^2\theta}}{n^2 - \sin^2\theta - n^2\sin^2\theta + 2\sin^4\theta}. \tag{5}$$

Equation (5) can not be inverted analytically, and solving the zenith angle with numerical interpolation will produce two solutions if there are no additional constraints. For real world objects, specular reflection and diffuse reflection are mixed depending on the surface material of the object. As shown in Fig. 1, the ambiguity in the azimuth angle and uncertainty in the zenith angle are fundamental limitations of SfP. Overcoming these limitations through physics-based neural networks is the primary focus of this paper.

3.2 Learning with Physics

A straightforward approach to estimating the normals, from polarization would be to simply take the set of polarization images as input, encode it into a feature map using a CNN, and feed the feature map into a normal-regression sub-network. Unsurprisingly, we find this results in normal reconstructions with higher MAE and undesirable lighting artifacts (see Fig. 7). To guide the network towards more optimal solutions from the polarization information, one possible method is to force our learned solutions to adhere to the polarization equations described in Sect. 3.1, similar to the method used in [23]. However, it is difficult to use these physical solutions for SfP tasks due to the following reasons:

1. Normals derived from the equations will inherently have ambiguous azimuth angles. **2.** Specular reflection and diffuse reflection coexist simultaneously, and determining the proportion of each type is complicated. **3.** Polarization images are usually noisy, causing error in the ambiguous normals, especially when the degree of polarization is low. Shifting the azimuth angles by π or $\frac{\pi}{2}$ could not reconstruct the surface normals properly for noisy images.

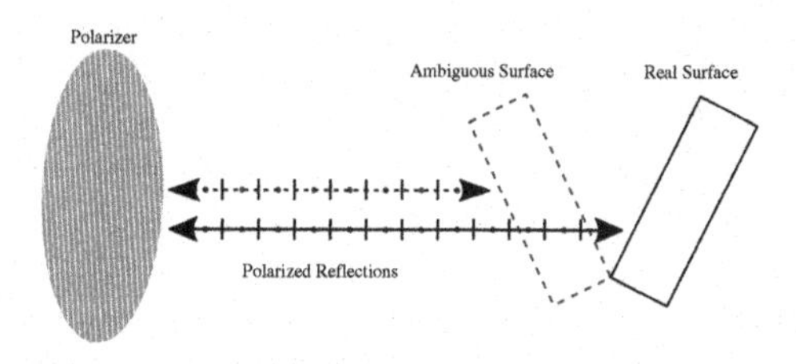

Fig. 1. SfP is underdetermined and one causal factor is the *ambiguity problem*. Here, two different surface orientations could result in exactly the same polarization signal, represented by dots and hashes. The dots represent polarization out of the plane of the paper and the hashes represent polarization within the plane of the board. Based on the measured data, it is unclear which orientation is correct. Ambiguities can also arise due to specular and diffuse reflections (which change the phase of light). For this reason, our network uses multiple physical priors.

Therefore, we propose directly feeding both the polarization images and ambiguous normal maps into the network, and leave the network to learn how to combine both of these inputs effectively from training data. The estimated surface normals can be structured as following:

$$\hat{\boldsymbol{N}} = f(\boldsymbol{I}_{\phi_1}, \boldsymbol{I}_{\phi_2}, ..., \boldsymbol{I}_{\phi_M}, \boldsymbol{N}_{diff}, \boldsymbol{N}_{spec1}, \boldsymbol{N}_{spec2}), \tag{6}$$

where $f(\cdot)$ is the proposed prediction model, $\{\boldsymbol{I}_{\phi_1}, \boldsymbol{I}_{\phi_2}, ..., \boldsymbol{I}_{\phi_M}\}$ is a set of polarization images, and $\hat{\boldsymbol{N}}$ is the estimated surface normals. We use the diffuse model in Sect. 3.1 to calculate $\boldsymbol{N}_{diff}$, and $\boldsymbol{N}_{spec1}, \boldsymbol{N}_{spec2}$ are the two solutions from the specular model. These ambiguous normals can implicitly direct the proposed network to learn the surface normal information from the polarization.

Our network structure is illustrated in Fig. 2. It consists of a fully convolutional encoder to extract and combine high-level features from the ambiguous physical solutions and the polarization images, and a decoder to output the estimated normals, $\hat{\boldsymbol{N}}$. Although three polarization images are sufficient to capture the polarization information, we use images with a polarizer at $\phi_{pol} \in \{0°, 45°, 90°, 135°\}$. These images are concatenated channelwise with the ambiguous normal solutions as the model input.

Note that the fixed nature of our network input is not arbitrary, but based on the output of standard polarization cameras. Such cameras utilize a layer of polarizers above the photodiodes to capture these four polarization images in a single shot. Our network design is intended to enable applications using this current single-shot capture technology. Single-shot capture is a clear advantage of our method over alternative reconstruction approaches, such as photometric stereo, since it allows images to be captured in a less constrained setting.

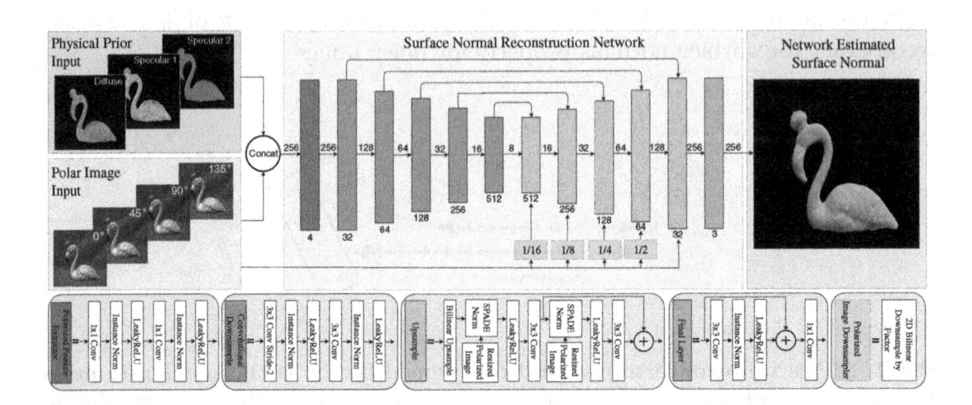

Fig. 2. Overview of our proposed physics-based neural network. The network is designed according to the encoder-decoder architecture in a fully convolutional manner. The blocks comprising the network are shown below the high-level diagram of our network pipeline. We use a block based on spatially-adaptive normalization as previously implemented in [40]. The numbers below the blocks refer to the number of output channels and the numbers next to the arrows refer to the spatial dimension.

After polarization feature extraction, there are five encoder blocks to encode the input to a $B \times 512 \times 8 \times 8$ tensor, where B is the minibatch size. The encoded tensor is then decoded by the same number of decoder blocks, with skip connections between blocks at the same hierarchical level as proposed in U-Net [44]. It has been noted that such deep architectures may wash away some necessary information from the input [15,53], so we apply spatially-adaptive normalization (SPADE) [40] to address this problem. Motivated by their architecture, we replace the modulation parameters of batch normalization layers [19] in each decoder block with parameters learned from downsampled polarization images using simple, two-layer convolutional sub-networks. The details of our adaptations to the SPADE module are depicted in Fig. 3. Lastly, we normalize the output estimated normal vectors to unit length, and apply the cosine similarity loss function:

$$L_{cosine} = \frac{1}{W \times H} \sum_{i}^{W} \sum_{j}^{H} (1 - \langle \hat{\boldsymbol{N}}_{ij}, \boldsymbol{N}_{ij} \rangle), \tag{7}$$

where $\langle \cdot, \cdot \rangle$ denotes the dot product, $\hat{\boldsymbol{N}}_{ij}$ is the estimated surface normal at pixel location (i, j), and $\boldsymbol{N}_{ij}$ is the corresponding ground truth surface normal. This loss is minimized when $\hat{\boldsymbol{N}}_{ij}$ and $\boldsymbol{N}_{ij}$ have identical orientation.

4 Dataset and Implementation Details

In what follows, we describe the dataset capture and organization as well as software implementation details. This is the first real-world dataset of its kind in the SfP domain, containing polarization images and corresponding ground truth surface normals for a variety of objects, under multiple different lighting conditions. The Deep Shape from Polarization dataset can thus provide a baseline for future attempts at applying learning to the SfP problem.

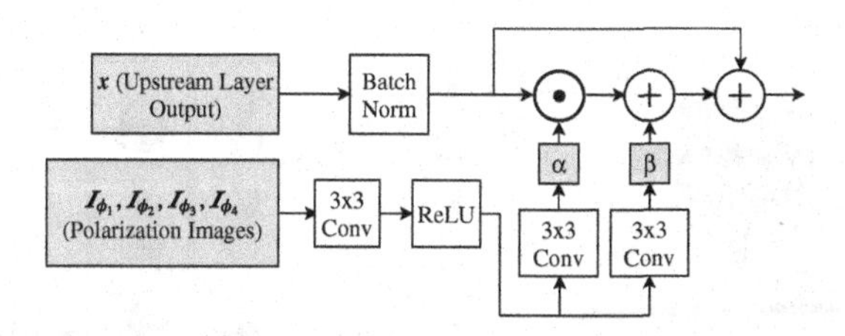

Fig. 3. Diagram of SPADE normalization block. We use the polarization images to hierarchically inject back information in upsampling. The SPADE block, which takes a feature map $\boldsymbol{x}$ and a set of downsampled polarization images $\{\boldsymbol{I}_{\phi_1}, \boldsymbol{I}_{\phi_2}, \boldsymbol{I}_{\phi_3}, \boldsymbol{I}_{\phi_4}\}$ as the input, learns affine modulation parameters $\boldsymbol{\alpha}$ and $\boldsymbol{\beta}$. The circle dot sign represents elementwise multiplication, and the circle plus sign represents elementwise addition.

4.1 Dataset

A polarization camera [29] with a layer of polarizers above the photodiodes (as described in Sect. 3.2) is used to capture four polarization images at angles $0°, 45°, 90°$ and $135°$ in a single shot. Then a structured light based 3D scanner [50] (with single shot accuracy no more than 0.1 mm, point distance from 0.17 mm to 0.2 mm, and a synchronized turntable for automatically registering scanning from multiple viewpoints) is used to obtain high-quality 3D shapes. Our real data capture setup is shown in Fig. 4. The scanned 3D shapes are aligned from the scanner's coordinate system to the image coordinate system of the polarization camera by using the shape-to-image alignment method adopted in [49]. Finally, we compute the surface normals of the aligned shapes by using the Mitsuba renderer [20]. Our introduced dataset consists of 25 different objects, each object with 4 different orientations for a total of 100 object-orientation combinations. For each object-orientation combination, we capture images in 3 lighting conditions: indoors, outdoors on an overcast day, and outdoors on a sunny day. In total, we capture 300 images for this dataset, each with 4 polarization angles.[2]

[2] The dataset is available at: https://visual.ee.ucla.edu/deepsfp.htm.

4.2 Software Implementation

Our model was implemented in PyTorch [41], and trained for 500 epochs with a batch size of 4. It took around 8 h for the network to converge with a single NVIDIA GeForce RTX 2070. We used the Adam optimizer [24] with default parameters with a base learning rate of 0.01. We train our model on randomly cropped 256 × 256 image patches, which is relatively common in shape estimation tasks [38,61] as a form of data augmentation. Further implementation details are in the supplement.

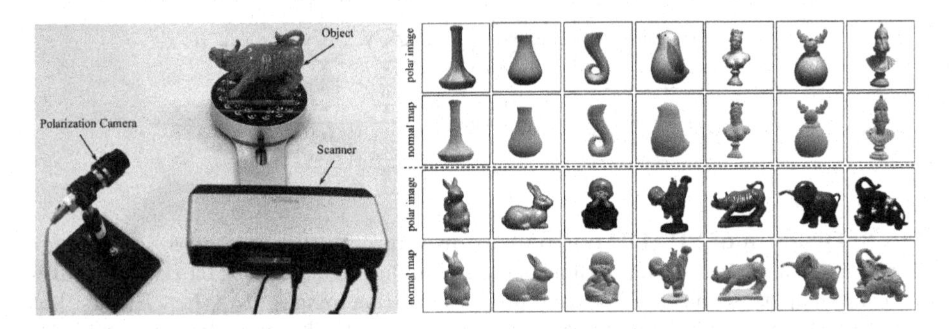

Fig. 4. This is the first dataset of its kind for the SfP problem. The capture setup and several example objects are shown above. We use a polarization camera to capture four gray-scale images of an object with four polarization angles in a single shot. The scanner is put next to the camera for obtaining the 3D shape of the object. The polarization images shown have a polarizer angle of 0°. The corresponding normal maps are aligned below. For each object, the capture process was repeated for 4 different orientations (front, back, left, right) and under 3 different lighting conditions (indoor lighting, outdoor overcast, and outdoor sunlight).

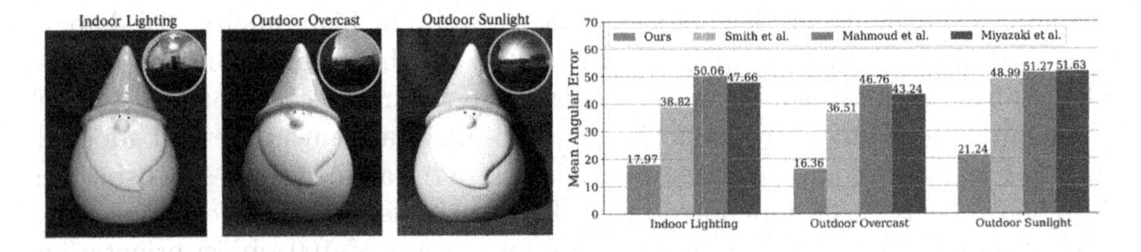

Fig. 5. The proposed method handles objects under varied lighting conditions. Note that our method has very similar mean angular error among all test objects across the three lighting conditions (bottom row). Please see supplement for further comparisons of lighting invariance.

5 Experimental Results

In this section, we evaluate our model with the presented challenging real-world scene benchmark, and compare it against three physics-only methods for SfP. All neural networks were trained on the same training data as discussed in Sect. 4.1. To quantify shape accuracy, we compute the widely used mean angular error (MAE) score on the surface normals.

5.1 Comparisons to Physics-Based SfP

We used a test dataset consisting of scenes that include BALL, HORSE, VASE, CHRISTMAS, FLAMINGO, DRAGON. On this test set, we implement three physics-based methods for SfP as a baseline: **1.** Smith *et al.* [52]. **2.** Mahmoud *et al.* [33]. **3.** Miyazaki *et al.* [37]. The first method recovers the depth map directly, and we only use the diffuse model due to the lack of specular reflection masks. The surface normals are obtained from the estimated depth with bicubic fit. Both the first and the second methods require lighting input, and we use the estimated lighting from the first method during comparison. The second method also requires known albedo, and following convention, we assume a uniform albedo of 1. Note the method proposed in [37] is the same as that presented in [4]. We omit comparison with Tozza *et al.* [59], as it requires two unpolarized intensity images, with two different light source directions. To motivate a fair comparison, we obtained the comparison codes directly from Smith *et al.* [52].[3]

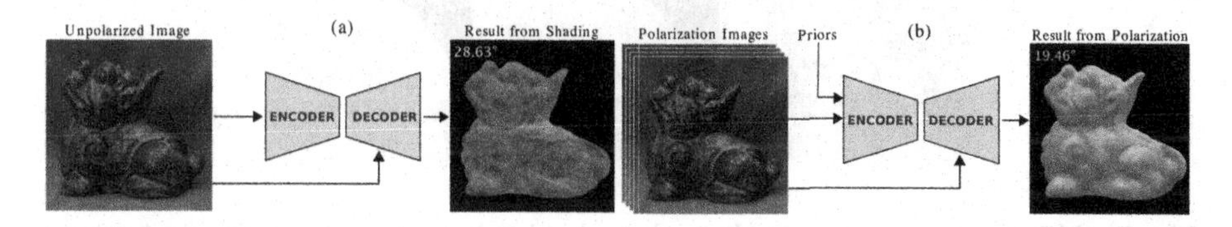

Fig. 6. Our network is learning from polarization cues, not just shading cues. An ablation study conducted on the DRAGON scene. In (a) the network does not have access to polarization inputs. In (b) the network can learn from polarization inputs and polarization physics. Please refer to Fig. 8, row c, for the ground truth shape of the DRAGON.

5.2 Robustness to Lighting Variations

Figure 5 shows the robustness of the method to various lighting conditions. Our dataset includes lighting in three broad categories: (a) indoor lighting; (b) outdoor overcast; and (c) outdoor sunlight. Our method has the lowest MAE, over the three lighting conditions. Furthermore, our method is consistent across conditions, with only slight differences in MAE for each object between lightings.

---

[3] https://github.com/waps101/depth-from-polarisation.

5.3 Importance of Polarization

An interesting question is how much of the shape information is learned from polarization cues as compared to shading cues. Figure 6 explores the benefit of polarization by ablating network inputs. We compare two cases. Figure 6(a) shows the resulting shape reconstruction when using a network architecture optimized for an unpolarized image input. The shape has texture copy and a high MAE of 28.63°. In contrast, Fig. 6(b) shows shape reconstruction from our proposed method of learning from four polarization images and a model of polarization physics. We observe that shape reconstruction using polarization cues is more robust to texture copy artifacts, and has a lower MAE of only 19.46°. Although only one image is used in the shading network (as is typical for shape from shading), this image is computed using an average of the four polarization images. Thus the distinction between the two cases in Fig. 6(a) and 6(b) is the polarization diversity, rather than improvements in photon noise.

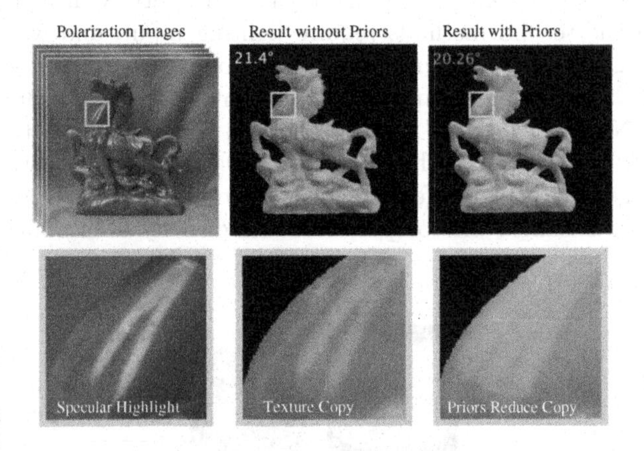

Fig. 7. Ablation test shows that the physics-based prior reduces texture copy artifacts. We see that the specular highlight in the input polarization image is directly copied into the normal reconstruction without priors. Note that our prior-based method shows stronger suppression of the copy artifact. Please see supplement for further examples of the effects of priors on texture copy.

5.4 Importance of Physics Revealed by Ablating Priors

Figure 7 highlights the importance of physics-based learning, as compared to traditional machine learning. Here, we refer to "traditional machine learning" as learning shape using only the polarization images as input. These results are shown in the middle column of Fig. 7. Shape reconstructions based on traditional machine learning exhibit image-based artifacts, because the polarization images contain brightness variations that are not due to geometry, but due to specular highlights (e.g., the HORSE is shiny). Learning from just the polarization images alone causes these image-based variations to masquerade as shape variations, as shown in the zoomed inset of Fig. 7. A term used for this is *texture copy*, where image texture is undesirably copied onto the geometry [21]. In contrast, the proposed results with physics priors are shown in the rightmost inset of Fig. 7, showing less dependence on image-based texture (because we also input the geometry-based physics model).

5.5 Quantitative Evaluation on Our Test Set

We use MAE[4] to make a quantitative comparison between our method and the previous physics-based approaches. Table 2 shows that the proposed method has the lowest MAE on each object, as well as the overall test set. The two most challenging scenes in the test set are the HORSE and the DRAGON. The former has intricate detail and specularities, while the latter has a mixed material surface. The physics-based methods struggle on these challenging scenes as all scenes have over 49° of mean angular error. The method from Smith *et al.* [52] has the second-lowest error on the DRAGON scene, but the method from Miyazaki *et al.* [37] has the second-lowest error on the HORSE scene. On the overall test set, the physics-based methods are all clustered between 41.4 and 49.0°, while the physics-based deep learning approach we propose achieves over a two-fold reduction in error to 18.5°.

Table 2. Our method outperforms previous methods for each object in the test set. Numbers represent the MAE averaged across the three lighting conditions for each object. The best model is marked in bold and the second-best is in italic.

Scene	**Proposed**	Smith [52]	Mahmoud [33]	Miyazaki [37]
BOX	**23.31°**	*31.00°*	41.51°	45.47°
DRAGON	**21.55°**	*49.16°*	70.72°	57.72°
FATHER CHRISTMAS	**13.50°**	39.68°	*39.20°*	41.50°
FLAMINGO	**20.19°**	*36.05°*	47.98°	45.58°
HORSE	**22.27°**	55.87°	*50.55°*	51.34°
VASE	**10.32°**	*36.88°*	44.23°	43.47°
WHOLE SET	**18.52°**	*41.44°*	49.03°	47.51°

[4] MAE is the most commonly reported measure for surface normal reconstruction, but in many cases it is a deceptive metric. We find that a few outliers in high-frequency regions can skew the MAE for entire reconstructions. Accordingly, we emphasize the qualitative comparisons of the proposed method to its physics-based counterparts.

The reader may wonder why the physics-based methods perform poorly on tested scenes. The result from Smith *et al.* [52] assumes a reflection model and combinatorial lighting estimation, which do not appear to scale to unconstrained, real world environments, resulting in a normal map with a larger error. Mahmoud *et al.* [33] uses shading constraints that assume a distant light source, which is not the case for some of the tested scenes, especially the indoor ones. Finally, the large region-wise anomalies on many of the results from Miyazaki *et al.* [37] are due to the sensitive nature of their histogram normalization method.

5.6 Qualitative Evaluation on Our Test Set

Figure 8 shows qualitative and quantitative data for various objects in our test set. The RGB images in (row a) are not used as input, but are shown in the top row of the figure for context about material properties. The input to all the methods shown is four polarization images, shown in (row b) of Fig. 8. The ground truth shape is shown in (row c), and corresponding shape reconstructions for the proposed method are shown in (row d). Comparison methods are shown in (row e) through (row g). It is worth noting that the physics-based methods particularly struggle with *texture copy* artifacts, where color variations masquerade as geometric variations. This can be seen in Fig. 8, (row f), where the physics-based reconstruction of Mahmoud [33] confuses the color variation in the beak of the FLAMINGO with a geometric variation. In contrast, our proposed method, shown in (row d), recovers the beak more accurately. Beyond texture copy, another limitation of physics-based methods lies in the difficulty of solving the *ambiguity problem*, discussed earlier in this paper. In row g, the physics-based approach from Miyazaki *et al.* [37] has significant ambiguity errors. This can be seen as the fixed variations in color of normal maps, which are not due to random noise. Although less drastic, the physics-based method of Smith *et al.* [52] also shows such fixed pattern artifacts, due to the underdetermined nature of the problem. Our proposed method is fairly robust to fixed pattern error, and our deviation from ground truth is largely in areas with high-frequency detail. Although the focus of Fig. 8 is to highlight qualitative comparisons, it is worth noting that the MAE in of the proposed method is the lowest for all these scenes (lowest MAE is highlighted in green font).

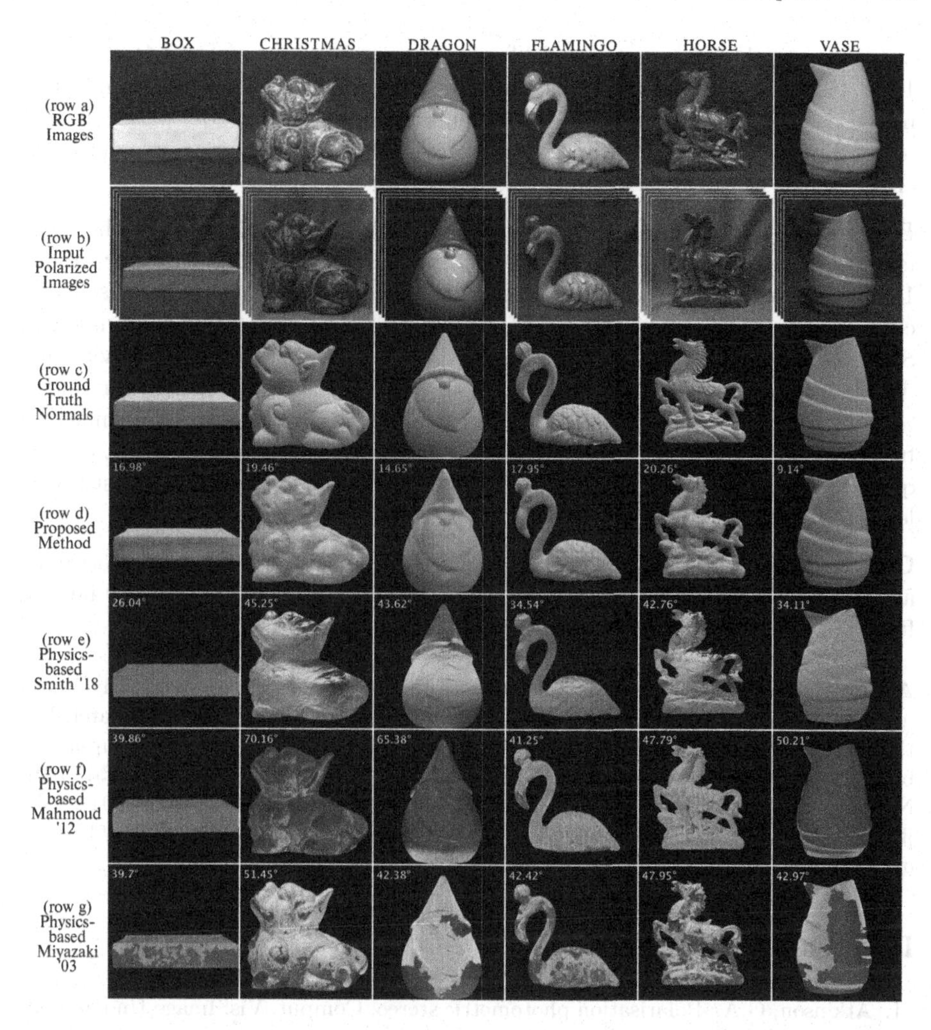

Fig. 8. The proposed method shows qualitative and quantitative improvements in shape recovery our test dataset. (row a) Shows the RGB scene photographs for context - these are not used as the input to any of the methods. (row b) The input to all methods are a stack of four polarization photographs at angles of 0°, 45°, 90°, and 135° (row c). The ground truth normals, obtained experimentally. (row d) The proposed approach for shape recovery. (row e-g) We compare with physics-based SfP methods by Smith *et al.* [51], Mahmoud *et al.* [33] and Miyazaki *et al.* [37]. (We omit the results from Atkinson *et al.* [4], which uses a similar method as [37]). Please see supplement for further comparisons. (Color figure online)

6 Discussion

In summary, we presented a first attempt re-examining SfP through the lens of deep learning, and specifically, physics-based deep learning. Table 2 shows that our network achieves over a two-fold reduction in shape error, from 41.4° [52] to 18.5°. An ablation test verifies the importance of using the physics-based prior in the deep learning model. In experiments, the proposed model performs well under varied lighting conditions, while previous physics-based approaches have either higher error or variation across lighting.

Future Work. The framerate of our technique is limited both by the feed-forward pass, as well as the time required to calculate the physical prior (about 1 s per frame). Future work could explore parallelizing the physics-based calculations or using approximations for more efficient compute. As discussed in Sect. 5.5, the high MAE is largely due to a few regions with extremely fine detail. Finding ways to effectively weight these areas more heavily or add a refinement stage focused on these challenging regions, are promising avenues for future exploration. Moreover, identifying a metric better able to capture the quality of reconstructions than MAE would be valuable for continued study of learning-based SfP.

Conclusion. We hope the results of this study encourage future explorations at the seamline of deep learning and polarization as well as the broader field of fusion of data-driven and physics-driven techniques.

Acknowledgements. The work of UCLA authors was supported by a Sony Imaging Young Faculty Award, Google Faculty Award, and the NSF CRII Research Initiation Award (IIS 1849941). The work of Peking University authors was supported by National Natural Science Foundation of China (61872012, 41842048, 41571432), National Key R&D Program of China (2019YFF0302902, 2017YFB0503004), Beijing Academy of Artificial Intelligence (BAAI), and Education Department Project of Guizhou Province.

References

1. Atkinson, G.A.: Polarisation photometric stereo. Comput. Vis. Image Understand. **160**, 158–167 (2017)
2. Atkinson, G.A., Ernst, J.D.: High-sensitivity analysis of polarization by surface reflection. Mach. Vis. Appl. **29**, 1171–1189 (2018)
3. Atkinson, G.A., Hancock, E.R.: Multi-view surface reconstruction using polarization. In: ICCV (2005)
4. Atkinson, G.A., Hancock, E.R.: Recovery of surface orientation from diffuse polarization. In: IEEE TIP (2006)
5. Baek, S.H., Jeon, D.S., Tong, X., Kim, M.H.: Simultaneous acquisition of polarimetric SVBRDF and normals. In: ACM SIGGRAPH (TOG) (2018)
6. Berger, K., Voorhies, R., Matthies, L.H.: Depth from stereo polarization in specular scenes for urban robotics. In: ICRA (2017)
7. Chen, G., Han, K., Wong, K.-Y.K.: PS-FCN: a flexible learning framework for photometric stereo. In: Ferrari, V., Hebert, M., Sminchisescu, C., Weiss, Y. (eds.) ECCV 2018. LNCS, vol. 11213, pp. 3–19. Springer, Cham (2018). https://doi.org/10.1007/978-3-030-01240-3_1

8. Chen, L., Zheng, Y., Subpa-asa, A., Sato, I.: Polarimetric three-view geometry. In: Ferrari, V., Hebert, M., Sminchisescu, C., Weiss, Y. (eds.) ECCV 2018. LNCS, vol. 11220, pp. 21–37. Springer, Cham (2018). https://doi.org/10.1007/978-3-030-01270-0_2
9. Cui, Z., Gu, J., Shi, B., Tan, P., Kautz, J.: Polarimetric multi-view stereo. In: CVPR (2017)
10. Deschaintre, V., Aittala, M., Durand, F., Drettakis, G., Bousseau, A.: Single-image SVBRDF capture with a rendering-aware deep network. In: ACM SIGGRAPH (TOG) (2018)
11. Drbohlav, O., Sara, R.: Unambiguous determination of shape from photometric stereo with unknown light sources. In: ICCV (2001)
12. Ghosh, A., Chen, T., Peers, P., Wilson, C.A., Debevec, P.: Circularly polarized spherical illumination reflectometry. In: ACM SIGGRAPH (TOG) (2010)
13. Ghosh, A., Fyffe, G., Tunwattanapong, B., Busch, J., Yu, X., Debevec, P.: Multiview face capture using polarized spherical gradient illumination. In: ACM SIGGRAPH (TOG) (2011)
14. Guarnera, G.C., Peers, P., Debevec, P., Ghosh, A.: Estimating surface normals from spherical stokes reflectance fields. In: Fusiello, A., Murino, V., Cucchiara, R. (eds.) ECCV 2012. LNCS, vol. 7584, pp. 340–349. Springer, Heidelberg (2012). https://doi.org/10.1007/978-3-642-33868-7_34
15. Huang, G., Sun, Y., Liu, Z., Sedra, D., Weinberger, K.: Deep networks with stochastic depth. CoRR (2016)
16. Huynh, C.P., Robles-Kelly, A., Hancock, E.R.: Shape and refractive index recovery from single-view polarisation images. In: CVPR (2010)
17. Huynh, C.P., Robles-Kelly, A., Hancock, E.R.: Shape and refractive index from single-view spectro-polarimetric images. IJCV **101**, 64–94 (2013). https://doi.org/10.1007/s11263-012-0546-3
18. Ikehata, S.: CNN-PS: CNN-based photometric stereo for general non-convex surfaces. In: Ferrari, V., Hebert, M., Sminchisescu, C., Weiss, Y. (eds.) ECCV 2018. LNCS, vol. 11219, pp. 3–19. Springer, Cham (2018). https://doi.org/10.1007/978-3-030-01267-0_1
19. Ioffe, S., Szegedy, C.: Batch normalization: accelerating deep network training by reducing internal covariate shift. arXiv preprint arXiv:1502.03167 (2015)
20. Jakob, W.: Mitsuba renderer (2010). http://www.mitsuba-renderer.org
21. Kadambi, A., Taamazyan, V., Shi, B., Raskar, R.: Polarized 3D: high-quality depth sensing with polarization cues. In: ICCV (2015)
22. Kadambi, A., Taamazyan, V., Shi, B., Raskar, R.: Depth sensing using geometrically constrained polarization normals. IJCV **125**, 34–51 (2017). https://doi.org/10.1007/s11263-017-1025-7
23. Karpatne, A., Watkins, W., Read, J., Kumar, V.: Physics-guided neural networks (PGNN): an application in lake temperature modeling. CoRR (2017)
24. Kingma, D.P., Ba, J.: Adam: a method for stochastic optimization. arXiv preprint arXiv:1412.6980 (2014)
25. Li, X., Dong, Y., Peers, P., Tong, X.: Modeling surface appearance from a single photograph using self-augmented convolutional neural networks. In: ACM SIGGRAPH (TOG) (2017)
26. Li, Z., Sunkavalli, K., Chandraker, M.: Materials for masses: SVBRDF acquisition with a single mobile phone image. In: Ferrari, V., Hebert, M., Sminchisescu, C., Weiss, Y. (eds.) ECCV 2018. LNCS, vol. 11207, pp. 74–90. Springer, Cham (2018). https://doi.org/10.1007/978-3-030-01219-9_5

27. Li, Z., Xu, Z., Ramamoorthi, R., Sunkavalli, K., Chandraker, M.: Learning to reconstruct shape and spatially-varying reflectance from a single image. In: ACM SIGGRAPH Asia (TOG) (2018)
28. Lindell, D.B., O'Toole, M., Wetzstein, G.: Single-photon 3D imaging with deep sensor fusion. In: ACM SIGGRAPH (TOG) (2018)
29. Lucid Vision Phoenix polarization camera (2018). https://thinklucid.com/product/phoenix-5-0-mp-polarized-model/
30. Lyu, Y., Cui, Z., Li, S., Pollefeys, M., Shi, B.: Reflection separation using a pair of unpolarized and polarized images. In: Advances in Neural Information Processing Systems, pp. 14559–14569 (2019)
31. Ma, W.C., Hawkins, T., Peers, P., Chabert, C.F., Weiss, M., Debevec, P.: Rapid acquisition of specular and diffuse normal maps from polarized spherical gradient illumination. In: Eurographics Conference on Rendering Techniques (2007)
32. Maeda, T., Kadambi, A., Schechner, Y.Y., Raskar, R.: Dynamic heterodyne interferometry. In: ICCP (2018)
33. Mahmoud, A.H., El-Melegy, M.T., Farag, A.A.: Direct method for shape recovery from polarization and shading. In: ICIP (2012)
34. Marco, J., et al.: Deeptof: off-the-shelf real-time correction of multipath interference in time-of-flight imaging. In: ACM SIGGRAPH (TOG) (2017)
35. Miyazaki, D., Kagesawa, M., Ikeuchi, K.: Transparent surface modeling from a pair of polarization images. In: PAMI (2004)
36. Miyazaki, D., Shigetomi, T., Baba, M., Furukawa, R., Hiura, S., Asada, N.: Surface normal estimation of black specular objects from multiview polarization images. Int. Soc. Opt. Photon. Opt. Eng. (2016)
37. Miyazaki, D., Tan, R.T., Hara, K., Ikeuchi, K.: Polarization-based inverse rendering from a single view. In: ICCV (2003)
38. Mo, Z., Shi, B., Lu, F., Yeung, S.K., Matsushita, Y.: Uncalibrated photometric stereo under natural illumination. In: Proceedings of the IEEE Conference on Computer Vision and Pattern Recognition, pp. 2936–2945 (2018)
39. Ngo, T.T., Nagahara, H., Taniguchi, R.: Shape and light directions from shading and polarization. In: CVPR (2015)
40. Park, T., Liu, M.Y., Wang, T.C., Zhu, J.Y.: Semantic image synthesis with spatially-adaptive normalization. In: CVPR (2019)
41. Paszke, A., et al.: Automatic differentiation in pytorch. In: NIPS-W (2017)
42. PolarM polarization camera (2017). http://www.4dtechnology.com/products/polarimeters/polarcam/
43. Riviere, J., Reshetouski, I., Filipi, L., Ghosh, A.: Polarization imaging reflectometry in the wild. In: ACM SIGGRAPH (TOG) (2017)
44. Ronneberger, O., Fischer, P., Brox, T.: U-Net: convolutional networks for biomedical image segmentation. In: Navab, N., Hornegger, J., Wells, W.M., Frangi, A.F. (eds.) MICCAI 2015. LNCS, vol. 9351, pp. 234–241. Springer, Cham (2015). https://doi.org/10.1007/978-3-319-24574-4_28
45. Santo, H., Samejima, M., Sugano, Y., Shi, B., Matsushita, Y.: Deep photometric stereo network. In: ICCV Workshops (2017)
46. Satat, G., Tancik, M., Gupta, O., Heshmat, B., Raskar, R.: Object classification through scattering media with deep learning on time resolved measurement. OSA Opt. Exp. **25**, 17466–17479 (2017)
47. Schechner, Y.Y.: Self-calibrating imaging polarimetry. In: ICCP (2015)
48. Sengupta, S., Kanazawa, A., Castillo, C.D., Jacobs, D.W.: SfSnet: learning shape, reflectance and illuminance of faces in the wild. In: CVPR (2018)

49. Shi, B., Mo, Z., Wu, Z., Duan, D., Yeung, S.K., Tan, P.: A benchmark dataset and evaluation for non-Lambertian and uncalibrated photometric stereo. In: PAMI (2019)
50. SHINING 3D scanner (2018). https://www.einscan.com/einscan-se-sp
51. Smith, W.A.P., Ramamoorthi, R., Tozza, S.: Linear depth estimation from an uncalibrated, monocular polarisation image. In: Leibe, B., Matas, J., Sebe, N., Welling, M. (eds.) ECCV 2016. LNCS, vol. 9912, pp. 109–125. Springer, Cham (2016). https://doi.org/10.1007/978-3-319-46484-8_7
52. Smith, W.A.P., Ramamoorthi, R., Tozza, S.: Height-from-polarisation with unknown lighting or albedo. In: PAMI (2018)
53. Srivastava, R.K., Greff, K., Schmidhuber, J.: Highway networks. CoRR (2015)
54. Su, S., Heide, F., Wetzstein, G., Heidrich, W.: Deep end-to-end time-of-flight imaging. In: CVPR (2018)
55. Tancik, M., Satat, G., Raskar, R.: Flash photography for data-driven hidden scene recovery. arXiv preprint arXiv:1810.11710 (2018)
56. Tancik, M., Swedish, T., Satat, G., Raskar, R.: Data-driven non-line-of-sight imaging with a traditional camera. In: OSA Imaging and Applied Optics (2018)
57. Taniai, T., Maehara, T.: Neural inverse rendering for general reflectance photometric stereo. In: ICML (2018)
58. Teo, D., Shi, B., Zheng, Y., Yeung, S.K.: Self-calibrating polarising radiometric calibration. In: CVPR (2018)
59. Tozza, S., Smith, W.A.P., Zhu, D., Ramamoorthi, R., Hancock, E.R.: Linear differential constraints for photo-polarimetric height estimation. In: ICCV (2017)
60. Wolff, L.B.: Polarization vision: a new sensory approach to image understanding. Image Vis. Comput. **15**, 81–93 (1997)
61. Xiong, Y., Chakrabarti, A., Basri, R., Gortler, S.J., Jacobs, D.W., Zickler, T.: From shading to local shape. IEEE Trans. Pattern Anal. Mach. Intell. **37**(1), 67–79 (2014)
62. Yang, L., Tan, F., Li, A., Cui, Z., Furukawa, Y., Tan, P.: Polarimetric dense monocular SLAM. In: CVPR (2018)
63. Ye, W., Li, X., Dong, Y., Peers, P., Tong, X.: Single image surface appearance modeling with self-augmented CNNs and inexact supervision. Wiley Online Library Computer Graphics Forum (2018)
64. Zhu, D., Smith, W.A.P.: Depth from a polarisation + RGB stereo pair. In: CVPR (2019)

A Boundary Based Out-of-Distribution Classifier for Generalized Zero-Shot Learning

Xingyu Chen[1], Xuguang Lan[1(✉)], Fuchun Sun[2], and Nanning Zheng[1]

[1] Xi'an Jiaotong University, Xi'an, China
xingyuchen1990@gmail.com
{xglan,nnzheng}@mail.xjtu.edu.cn
[2] Tsinghua University, Beijing, China
fcsun@tsinghua.edu.cn

Abstract. Generalized Zero-Shot Learning (GZSL) is a challenging topic that has promising prospects in many realistic scenarios. Using a gating mechanism that discriminates the unseen samples from the seen samples can decompose the GZSL problem to a conventional Zero-Shot Learning (ZSL) problem and a supervised classification problem. However, training the gate is usually challenging due to the lack of data in the unseen domain. To resolve this problem, in this paper, we propose a boundary based Out-of-Distribution (OOD) classifier which classifies the unseen and seen domains by only using seen samples for training. First, we learn a shared latent space on a unit hyper-sphere where the latent distributions of visual features and semantic attributes are aligned class-wisely. Then we find the boundary and the center of the manifold for each class. By leveraging the class centers and boundaries, the unseen samples can be separated from the seen samples. After that, we use two experts to classify the seen and unseen samples separately. We extensively validate our approach on five popular benchmark datasets including AWA1, AWA2, CUB, FLO and SUN. The experimental results show that our approach surpasses state-of-the-art approaches by a significant margin.

Keywords: Generalized zero-shot learning · Boundary based out-of-distribution classifier

1 Introduction

Zero-Shot Learning (ZSL) is an important topic in the computer vision community which has been widely adopted to solve challenges in real-world tasks. In the conventional setting, ZSL aims at recognizing the instances drawn from the unseen domain, for which the training data are lacked and only the semantic auxiliary information is available. However, in real-world scenarios, the instances are drawn from either unseen or seen domains, which is a more challenging task called Generalized Zero-Shot Learning (GZSL).

A. Vedaldi et al. (Eds.): ECCV 2020, LNCS 12369, pp. 572–588, 2020.
https://doi.org/10.1007/978-3-030-58586-0_34

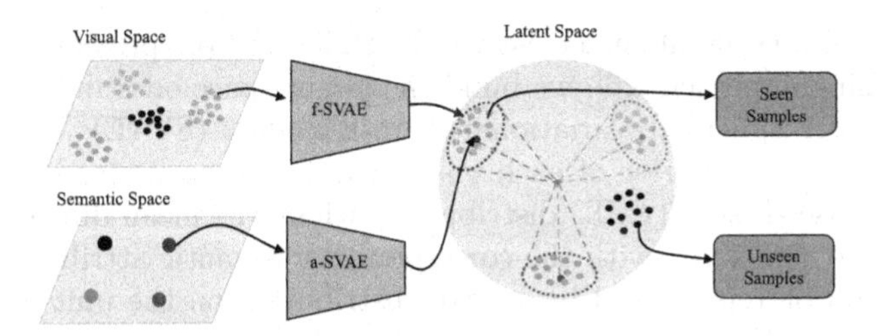

Fig. 1. The boundary based OOD classifier learns a bounded manifold for each seen class on a unit hyper-sphere (latent space). By using the manifold boundaries (dotted circles) and the centers (dark-colored dots), the unseen samples (black dots) can be separated from the seen samples (colored dots).

Previous GZSL algorithms can be grouped into three lines: (1) Embedding methods [1–3,5,10,14,18,25,27,28,33] which aim at learning embeddings that unify the visual features and semantic attributes for similarity measurement. However, due to the bias problem [33], the projected feature anchors of unseen classes may be distributed too near to that of seen classes in the embedding space. Consequently, the unseen samples are easily classified into nearby seen classes. (2) Generative methods [6,9,15,21,26,31] which focus on generating synthetic features for unseen classes by using generative models such as GAN [11] or VAE [13]. By leveraging the synthetic data, the GZSL problem can be converted to a supervised problem. Although the generative methods substantially improve the GZSL performance, they are still bothered by the feature confusion problem [17]. Specifically, the synthetic unseen features may be tangled with the seen features. As a result, a classifier will be confused by the features which have strong similarities but different labels. An intuitive phenomenon is that previous methods usually make trade-offs between the accuracy of seen classes and unseen classes to get higher Harmonic Mean values. (3) Gating methods [4,27] which usually incorporates a gating mechanism with two experts to handle the unseen and seen domains separately. Ideally, if the binary classifier is reliable enough, the GZSL can be decomposed to a ZSL problem and a supervised classification problem, which does not suffer from the feature confusion or bias problem. Unfortunately, it is usually difficult to learn such a classifier because unseen samples are not available during training.

To resolve the main challenge in the gating methods, we propose a boundary based Out-of-Distribution (OOD) classifier for GZSL in this paper. As illustrated in Fig. 1, the key idea of our approach is to learn a bounded manifold for each seen class in the latent space. A datum that can be projected into the bounded manifold will be regarded as a seen sample. Otherwise, we believe it is an unseen sample. In this way, we can easily separate unseen classes from seen classes even we do not use any unseen samples for training.

To learn a bounded manifold for each seen class, the proposed OOD classifier learns a shared latent space for both visual features and the semantic attributes. In the latent space, the distributions of visual features and semantic attributes are aligned class-wisely. Different from previous latent distribution aligning approach [26], we build the latent space on a unit hyper-sphere by using Hyper-Spherical Variational Auto-Encoders (SVAE) [8]. Specifically, the approximated posterior of each visual feature is encouraged to be aligned with a von Mises-Fisher (vMF) distribution, where the mean direction and concentration are associated with the corresponding semantic attribute. Therefore, each class can be represented by a vMF distribution on the unit hyper-sphere, which is easy to find the manifold boundary. In addition, the mean direction predicted by semantic attribute can be regarded as the class center. By leveraging the boundary and the class center, we can determine if a datum is projected into the manifold. In this way, the unseen features can be separated from the seen features. After that, we apply two experts to classify the seen features and unseen features separately.

The proposed classifier can incorporate with any state-of-the-art ZSL method. The core idea is very straightforward and easy to implement. We evaluate our approach on five popular benchmark datasets, i.e. AWA1, AWA2, CUB, FLO and SUN for generalized zero-shot learning. The experimental results show that our approach surpasses the state-of-the-art approaches by a significant margin.

2 Related Work

Embedding Methods. To solve GZSL, the embedding methods [1–3,5,10,14, 18,19,25,27,28,32–35] usually learn a mapping to unify the visual features and semantic attributes for similarity measurement. For example, Zhang et al. [35] embed features and attributes into a common space where each point denotes a mixture of seen class proportions. Other than introducing common space, Kodirov et al. [14] propose a semantic auto-encoder which aims to embed visual feature vector into the semantic space while constrain the projection must be able to reconstruct the original visual feature. On the contrary, Long et al. [19] learn embedding from semantic space into visual space. However, due to the bias problem, previous embedding methods usually misclassify the unseen classes into seen classes. To alleviate the bias problem, Zhang et al. [33] propose a co-representation network which adopts a single-layer cooperation module with parallel structure to learn a more uniform embedding space with better representation.

Generative Methods. The generative methods [6,9,15,21,26,31] treat GZSL as a case of missing data and try to generate synthetic samples of unseen classes from semantic information. By leveraging the synthetic data, the GZSL problem can be converted to a supervised classification problem. Therefore, These methods usually rely on generative models such as GAN [11] and VAE [13]. For example, Xian et al. [31] directly generate image features by pairing a conditional WGAN with a classification loss. Mishara et al. [21] utilize a VAE to generate image features conditional on the class embedding vector. Felix et al.

[9] propose a multi-modal cycle-consistent GAN to improve the quality of the synthetic features. Compared to the embedding methods, the generative methods significantly improve the GZSL performance. However, Li et al. [17] find that the generative methods are bothered by the feature confusion problem. To alleviate this problem, they present a boundary loss which maximizes the decision boundary of seen categories and unseen ones while training the generative model.

Gating Methods. There are a few works using a gating based mechanism to separate the unseen samples from the seen samples for GZSL. The gate usually incorporates two experts to handle seen and unseen domains separately. For example, Socher et al. [27] propose a hard gating model to assign test samples to each expert. Only the selected expert is used for prediction, ignoring the other expert. Recently, Atzmon et al. [4] propose a soft gating model which makes soft decisions if a sample is from a seen class. The key to the soft gating is to pass information between three classifiers to improve each one's accuracy. Different from the embedding methods and the generative methods, the gating methods do not suffer from the bias problem or the feature confusion problem. However, a key difficulty in gating methods is to train a binary classifier by only using seen samples. In this work, we propose a boundary based OOD classifier by only using seen samples for training. The proposed classifier is a hard gating model. Compared to previous gating methods, it provides much more accurate classification results.

3 Revisit Spherical Variational Auto-Encoders

The training objective of a general variational auto-encoder is to maximize $\log \int p_\phi(x, z)dz$, the log-likelihood of the observed data, where x is the training data, z is the latent variable and $p_\phi(x, z)$ is a parameterized model representing the joint distribution of x and z. However, computing the marginal distribution over the latent variable z is generally intractable. In practice, it is implemented to maximize the Evidence Lower Bound (ELBO).

$$\log \int p_\phi(x, z)dz \geq \mathbb{E}_{q(z))} [\log p_\phi(x|z)] - KL(q(z)||p(z)), \tag{1}$$

where $q(z)$ approximates the true posterior distribution and $p(z)$ is the prior distribution. $p_\phi(x|z)$ is to map a latent variable to a data point x which is parameterized by a decoder network. $KL(q(z)||p(z))$ is the Kullback-Leibler divergence which encourages $q(z)$ to match the prior distribution. The main difference for various variational auto-encoders is in the adopted distributions.

For SVAE [8], both of the prior and posterior distributions are based on von Mises-Fisher(vMF) distributions. A vMF distribution can be regarded as a Normal distribution on a hyper-sphere, which is defined as:

$$q(z|\mu, \kappa) = C_m(\kappa) \exp(\kappa \mu^T z) \tag{2}$$

$$C_m(\kappa) = \frac{\kappa^m/2 - 1}{(2\pi)^{m/2} I_{m/2-1}(\kappa)} \quad (3)$$

where $\mu \in \mathbb{R}^m, ||\mu||_2 = 1$ represents the direction on the sphere and $\kappa \in \mathbb{R}_{\geq 0}$ represents the concentration around μ. $C_m(\kappa)$ is the normalizing constant, I_v is the modified Bessel function of the first kind at order v.

Theoretically, $q(z)$ should be optimized over all data points, which is not tractable for large dataset. Therefore it uses $q_\theta(z|x) = q(z|\mu(x), \kappa(x))$ which is parameterized by an encoder network to do stochastic gradient descent over the dataset. The final training objective is defined as:

$$L_{\text{SVAE}}(\theta, \phi; x) = \mathbb{E}_{q_\theta(z|x)}\left[\log p_\phi(x|z)\right] - KL(q_\theta(z|x)||p(z)). \quad (4)$$

4 Proposed Approach

4.1 Problem Formulation

We first introduce the definitions of OOD classification and GZSL. We are given a set of training samples of seen classes $\mathcal{S} = \{(x, y, a)|x \in \mathcal{X}, y \in \mathcal{Y}_s, a \in \mathcal{A}_s\}$ where x represents the feature of an image extracted by a CNN, y represents the class label in $\mathcal{Y}_s = \{y_s^1, y_s^2, ..., y_s^N\}$ consisting of N seen classes and a represents corresponding class-level semantic attribute which is usually hand-annotated or a Word2Vec feature [20]. We are also given a set $\mathcal{U} = \{(y, a)|y \in \mathcal{Y}_u, a \in \mathcal{A}_u\}$ of unseen classes $\mathcal{Y}_u = \{y_u^1, y_u^2, ..., y_u^M\}$. The zero shot recognition states that $\mathcal{Y}_s \cap \mathcal{Y}_u = \varnothing$. Given $\mathcal{S}$ and $\mathcal{U}$, the OOD classifier aims at learning a binary classifier $f_{OOD} : \mathcal{X} \to \{0, 1\}$ that distinguishes if a datum belongs to $\mathcal{S}$ or $\mathcal{U}$. The task of GZSL aims at learning a classifier $f_{gzsl} : \mathcal{X} \to \mathcal{Y}_s \cup \mathcal{Y}_u$.

4.2 Boundary Based Out-of-Distribution Classifier

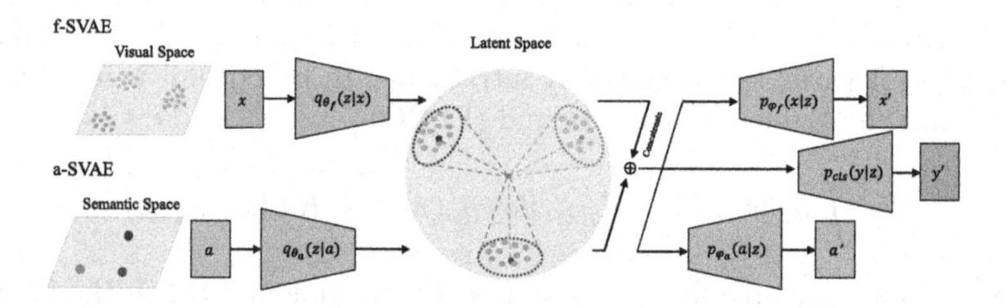

Fig. 2. Our model consists of two SVAEs, one for visual features and another for semantic attributes. By combining the objective functions of the two SVAEs with a cross-reconstruction loss and a classification loss, we train our model to align the latent distributions of visual features and semantic attributes class-wisely. In this way, each class can be represented by a vMF distribution whose boundary is easy to find.

The proposed OOD classifier aims to classify the unseen and seen domains by only using seen samples for training. The core idea of our approach is quite straightforward. First, we build a latent space for visual features and semantic attributes. Then we learn a bounded manifold for each seen class. Next we find the boundaries of the learned manifolds. By leveraging the boundaries, we can determine if a test sample is projected into the manifolds. For the samples which can be projected into the manifolds, we believe they belong to the seen domain and assign them to a seen expert. Otherwise, we assign them to an unseen expert.

Build the Latent Space on a Unit Hyper-sphere. Different from previous works, we build the latent space on a unit hyper-sphere by using hyper-spherical variational auto-encoders. In the latent space, each class is approximately represented by a vMF distribution of which the mean direction can be regarded as the class center. Using the spherical representation has two advantages. First, we can naturally use cosine similarity as the distance metric since all latent variables and class centers are located on the unit hyper-sphere. Second, for each seen class, it is easy to find the manifold boundary. Specifically, we can find a threshold based on the cosine similarities between the latent variables and the class center. According to the class center and the corresponding boundary, we can determine if a visual feature is projected into the manifold.

Learn a Bounded Manifold for Each Class. To learn a bounded manifold for each class, inspired by [26], we encourage the latent distributions of visual features and the corresponding semantic attribute to be aligned with each other in the latent space. As illustrated in Fig. 2, our model consists of two SVAEs correspond to two data modalities, one for visual features and another for semantic attributes, denoted as f-SVAE and a-SVAE, respectively. Given an attribute $a \in \mathcal{A}_s$, the encoder of a-SVAE predicts a vMF distribution $q_{\theta_a}(z|a) = q(z|\mu(a), \kappa(a))$. Meanwhile, given the corresponding visual feature x, the encoder of f-SVAE predicts a vMF distribution $q_{\theta_f}(z|x) = q(z|\mu(x), \kappa(x))$. Each SVAE regards the distribution predicted by another SVAE as the prior distribution. Therefore, we can align the two distributions by optimizing the objective functions of f-SVAE and a-SVAE simultaneously. We further adopt a cross-reconstruction loss and a classification loss to ensure the latent representations capture the modality invariant information while preserving discrimination. Therefore, the training objective consists four parts.

f-SVAE: For the f-SVAE, we expect to maximize the log-likelihood and minimize the discrepancy between the approximated posterior $q_{\theta_f}(z|x)$ and the prior distribution $q_{\theta_a}(z|a)$. Therefore, the training objective is defined as:

$$L_{f-SVAE} = \mathbb{E}_{p(x,a)}[\mathbb{E}_{q_{\theta_f}(z|x)}[\log p_{\phi_f}(x|z)] - \lambda_f D_z(q_{\theta_f}(z|x) \parallel q_{\theta_a}(z|a))], \quad (5)$$

where $\mathbb{E}_{q_{\theta_f}(z|x)}[\log p_{\phi_f}(x|z)]$ represents the expectation of log-likelihood over latent variable z. In practice, we use the negative reconstruction error of visual feature x instead. $p_{\phi_f}(x|z)$ is the decoder network of f-SVAE. $D_z(q_{\theta_f}(z|x) \parallel q_{\theta_a}(z|a))$ represents the discrepancy between the two vMF distributions. λ_f is a hyper-parameter to weight the discrepancy term. It worth noting that

$D_z(q_{\theta_f}(z|x) \parallel q_{\theta_a}(z|a))$ is the Earth Mover's Distance (EMD) between the two distributions which is defined as:

$$D_z(q_{\theta_f}(z|x) \parallel q_{\theta_a}(z|a)) = \inf_{\Omega \in \prod(q_{\theta_f}, q_{\theta_a})} \mathbb{E}_{(z_1,z_2)\sim\Omega}[\parallel z_1 - z_2 \parallel]. \tag{6}$$

The reason we use EMD instead of the KL-divergence is that the KL-divergence may fail when the support regions of the two distributions $q_{\theta_f}(z|x)$ and $q_{\theta_a}(z|a)$ do not completely coincide. To calculate the EMD, we utilize the Sinkhorn iteration algorithm in [7].

a-SVAE: Similarly, for the a-SVAE, $q_{\theta_f}(z|x)$ is regarded as the prior distribution. The objective function is defined as:

$$L_{a-SVAE} = \mathbb{E}_{p(x,a)}[\mathbb{E}_{q_{\theta_a}(z|a)}[\log p_{\phi_a}(a|z)] - \lambda_a D_z(q_{\theta_a}(z|a) \parallel q_{\theta_f}(z|x))], \tag{7}$$

where $\mathbb{E}_{q_{\theta_a}(z|a)}[\log p_{\phi_a}(a|z)]$ represents the negative reconstruction error of semantic attribute a. $D_z(q_{\theta_a}(z|a) \parallel q_{\theta_f}(z|x))$ is the discrepancy between the two vMF distributions. As EMD is symmetrical, $D_z(q_{\theta_a}(z|a) \parallel q_{\theta_f}(z|x))$ equals to $D_z(q_{\theta_f}(z|x) \parallel q_{\theta_a}(z|a))$, weighted by hyper-parameter λ_a.

Cross-reconstruction Loss: Since we learn a shared latent space for the two different modalities, the latent representations should capture the modality invariant information. For this purpose, we also adopt a cross-reconstruction regularizer:

$$L_{cr} = \mathbb{E}_{p(x,a)}[\mathbb{E}_{q_{\theta_a}(z|a)}[\log p_{\phi_f}(x|z)] + \mathbb{E}_{q_{\theta_f}(z|x)}[\log p_{\phi_a}(a|z)]], \tag{8}$$

where $\mathbb{E}_{q_{\theta_a}(z|a)}[\log p_{\phi_f}(x|z)]$ and $\mathbb{E}_{q_{\theta_f}(z|x)}[\log p_{\phi_a}(a|z)]$ also represent negative reconstruction errors.

Classification Loss: To make the latent variables more discriminate, we introduce the following classification loss:

$$L_{cls} = \mathbb{E}_{p(x,y,a)}[\mathbb{E}_{q_{\theta_a}(z|a)}[\log p_{\phi_{cls}}(y|z)] + \mathbb{E}_{q_{\theta_f}(z|x)}[\log p_{\phi_{cls}}(y|z)]], \tag{9}$$

where ϕ_{cls} represents the parameters of a linear softmax classifier. Although the classification loss may hurt the inter-class association between seen and unseen classes, it also reduces the risk for unseen features being projected into the manifolds of seen classes, which benefits to the binary classification. The reason is that our OOD classifier only cares about separating unseen features from the seen features, but not cares about which class the unseen features belong to.

Overall Objective: Finally, we train our model by maximizing the following objective:

$$L_{overall} = L_{f-SVAE} + L_{a-SVAE} + \alpha L_{cr} + \beta L_{cls}, \tag{10}$$

where α, β are the hyper-parameters used to weight the two terms.

Find the Boundaries for OOD Classification. When the proposed model is trained to convergence, the visual features and the semantic attributes are aligned class-wisely in the latent space. Each class is represented by a vMF distribution. Therefore, the manifold of each class can be approximately represented by a circle on the unit hyper-sphere. By leveraging the center and the boundary, we can determine whether a latent variable locates in the manifold.

For class $y^i \in \mathcal{Y}_s$, the class center can be found by using its semantic attribute. Given $a^i \in \mathcal{A}_s$, a-SVAE predicts a vMF distribution $q(z|\mu(a^i), \kappa(a^i))$ of which $\mu(a^i)$ is regarded as the class center.

There could be many ways to find the boundaries. In this paper, we present a simple yet effective one. We first encode all training samples of seen classes to latent variables. After that we calculate the cosine similarity $S(z^i, \mu(a^i))$ between each latent variable z^i and the corresponding class center $\mu(a^i)$. Then we search a threshold η which is smaller than $\gamma \in (0, 100\%)$ and larger than $1 - \gamma$ of the cosine similarities. We adopt η for all seen classes to represent the boundaries. Here, γ can be viewed as the OOD classification accuracy on training samples. Given a γ, we can find the corresponding threshold η.

Given a test sample x which may come from a seen class or an unseen class, we first encode it to latent variable z. Then we compute the cosine similarities between it to all seen class centers and find the maximum. By leveraging the threshold η, we determine the test sample belongs to unseen class or seen class using Eq. 11,

$$y^{OOD} = \begin{cases} 0, & \text{if} \quad \max\{S(z, \mu(a^i)) | \forall a^i \in \mathcal{A}_s\} < \eta \\ 1, & \text{if} \quad \max\{S(z, \mu(a^i)) | \forall a^i \in \mathcal{A}_s\} \geq \eta \end{cases} \tag{11}$$

where 0 stands for unseen class and 1 for seen class.

Generalized Zero-Shot Classification. For the GZSL task, we incorporate the proposed OOD classifier with two domain experts. Given a test sample, the OOD classifier determines if it comes from a seen class. Then, according to the predicted label, the test sample is assigned to a seen expert or an unseen expert for classification.

4.3 Implementation Details

OOD Classifier. For the f-SVAE, we use two 2-layer Fully Connected (FC) network for the encoder and decoder networks. The first FC layer in the encoder has 512 neurons with ReLU followed. The output is then fed to two FC layers to produce the mean direction and the concentration for the reparameterize trick. The mean direction layer has 64 neurons and the concentration layer only has 1 neuron. The output of mean direction layer is normalized by its norm such that it lies on the unit hyper-sphere. The concentration layer is followed by a Softplus activation to ensure its output larger than 0. The decoder consists of two FC layers. The first layer has 512 neurons with ReLU followed. The second layer has 2048 neurons.

The structure of a-SVAE is similar to f-SVAE except for the input dimension and the neuron number of the last FC layer equal to the dimension of the semantic attributes. We use a Linear Softmax classifier which takes the latent variables as input for calculating the classification loss. The structure is same as in [31].

We train our model by the Adam optimizer with learning rate 0.001. The batch size is set to 128. The hyper-parameter λ_f, λ_a, α, β are set to 0.1, 0.1, 1.0 and 1.0, respectively.

Unseen and Seen Experts. For the unseen samples, we use the f-CLSWGAN [31] with the code provided by the authors. For the seen samples, we directly combine the encoder of f-SVAE and the linear softmax classifier for classification.

5 Experiments

The proposed approach is evaluated on five benchmark datasets, where plenty of recent state-of-the-art methods are compared. Moreover, the features and settings used in experiments follow the paper [30] for fair comparison.

5.1 Datasets, Evaluation and Baselines

Datasets. The five benchmark datasets include Animals With Attributes 1 (AWA1)[16], Animal With Attributes 2 (AWA2) [30], Caltech-UCSD-Birds (CUB) [29], FLOWER (FLO) [22] and SUN attributes (SUN) [23]. Specifically, AWA1 contains 30,475 images and 85 kinds of properties, where 40 out of 50 classes are obtained for training. In AWA2, 37,322 images in the same classes are re-collected because original images in AWA1 are not publicly available. CUB has 11,788 images from 200 different types of birds annotated with 312 properties, where 150 classes are seen and the others are unseen during training. FLO consists of 8,189 images which come from 102 flower categories, where 82/20 classes are used for training and testing. For this dataset, we use the same semantic descriptions provided by [24]. SUN has 14,340 images of 717 scenes annotated with 102 attributes, where 645 classes are regarded as seen classes and the rest are unseen classes.

Evaluation. For OOD classification, the in-distribution samples are regarded as the seen samples and the out-of-distribution samples are regarded as unseen samples. The True-Positive-Rate (**TPR**) indicates the classification accuracy of seen classes and the False-Positive-Rate (**FPR**) indicates the accuracy of unseen classes. We also measure the Area-Under-Curve (**AUC**) by sweeping over classification threshold.

For GZSL, the average of per-class precision (AP) is measured. The "**ts**" and "**tr**" denote the Average Precision (AP) of images from unseen and seen classes, respectively. "**H**" is the harmonic mean which is defined as: $H = 2 * tr * ts/(tr + ts)$. The harmonic mean reflects the ability of method that recognizes seen and unseen images simultaneously.

GZSL Baselines. We compare our approach with three lines of previous works in the experiments. (1) Embedding methods which focus on learning embeddings that unify the visual features and semantic attributes for similarity measurement. We include the recent competitive baselines: SJE [2], ALE [1], PSR [3], SAE [14], EZSL [25], LESAE [18], ReViSE [28], CMT [27], SYNC [5], DeViSE [10] and CRnet [33]. (2) Generative methods which focus on generating synthetic features or images for unseen classes using GAN or VAE. We also compare our approach with the recent state-of-the-arts such as CVAE [21], SP-AEN [6], f-CLSWGAN [31], CADA-VAE [26], cycle-(U)WGAN [9], SE [15] and AFC-GAN [17]. (3) Gating methods which aim at learning a classifier to distinguish the unseen features from the seen features. We compare our approach with the recent state-of-the-art COSMO [4].

5.2 Out-of-Distribution Classification

In this experiment, We conduct OOD classification experiments on the five benchmark datasets.

Table 1. Comparison with various gating models on validation set. **AUC** denotes Area-Under-Curve when sweeping over detection threshold. **FPR** denotes False-Positive-Rate on the threshold that yields 95% True Positive Rate for detecting in-distribution samples. The best results are highlighted with bold numbers.

Method	AWA1			CUB			SUN		
	H	AUC	FPR	H	AUC	FPR	H	AUC	FPR
MAX-SOFTMAX-3 [12]	53.1	88.6	56.8	43.6	73.4	79.6	38.4	61.0	92.3
CB-GATING-3 [4]	56.8	92.5	45.5	44.8	82.0	72.0	40.1	77.7	77.5
Ours	**70.1**	**95.0**	**12.5**	**67.7**	**99.4**	**2.5**	**71.0**	**99.5**	**1.6**

Table 2. OOD classification results of our approach by selecting different thresholds using γ.

	AWA1		AWA2		CUB		FLO		SUN	
	TPR	FPR	TPR	FPR	TPR	FPR	TPR	FPR	TPR	FPR
$\gamma = 0.85$	85.0	5.3	85.2	6.8	84.2	0.7	85.3	0.4	85.4	0.2
$\gamma = 0.90$	90.1	6.3	89.8	8.2	89.5	0.9	88.2	0.6	90.6	0.2
$\gamma = 0.95$	95.4	7.9	95.2	10.6	94.9	1.1	94.4	0.8	95.1	0.4

We first compare the boundary based OOD classifier with two state-of-the-art gating-based methods: (1) MAX-SOFTMAX-3 is a baseline gating model of [12]. (2)CB-GATING-3 is the best confidence-based gating model in [4].

For a fair comparison, we use the same dataset splitting as in [4]. Table 1 shows the classification results of the proposed OOD classifier compared to the two baseline methods. It worth noting that the FPR scores are reported on the threshold that yields 95% TPR for detecting in-distribution samples. It can be seen that the two baseline methods have much higher FPR values. For example, the FPR of CB-GATING-3 is 45.5% on AWA1, 72.0% on CUB and 77.5% on SUN. It indicates that most of the unseen samples are misclassified to the seen samples. However, the FPR of our approach is reduced to 12.5% on AWA1, 2.5% on CUB and 1.6% on SUN, which significantly outperforms the baselines methods. Therefore, we achieve the best harmonic mean and AUC scores. Our approach can be categorized as a hard-gating approach. Compared to the soft-gating method in [4], our approach is more straightforward and more effective.

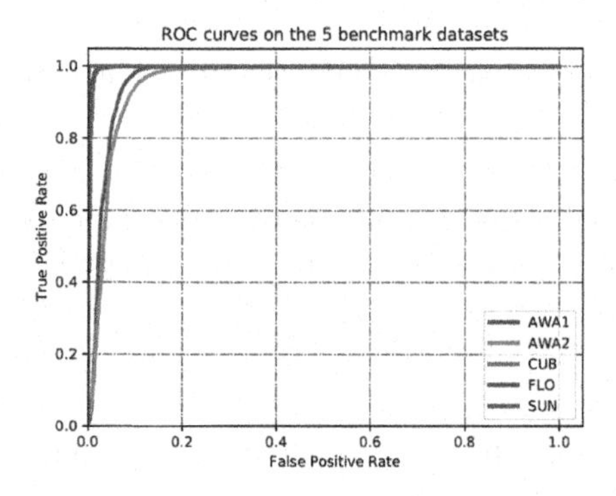

Fig. 3. The ROC curves on the five benchmark datasets.

We also present the OOD classification results on the test sets of the five benchmark datasets in Table 2. It can be seen that the proposed OOD classifier shows stable performance when we sweep the threshold. The ROC curves are shown in Fig. 3, where the AUC is 96.8% on AWA1, 95.7% on AWA2, 99.6% on CUB, 99.8% on FLO and 99.9% on SUN.

5.3 Comparison with State-of-the-Arts

We further evaluate our approach on the five benchmark datasets under the GZSL setting. We report the top-1 accuracy and harmonic mean of each method in Table 3 where "-" indicates that the result is not reported.

We see that most of the embedding methods suffer from the bias problem. For example, the ts values of baseline methods [1–3,5,10,14,18,25,27,28] are much lower than the tr values, which leads to poor harmonic results. Compared to the embedding methods, the generative methods [6,9,15,21,26,31] show much higher harmonic mean results. However, due to the feature confusion problem, these methods have to make trade-offs between ts and tr values to get higher harmonic

Table 3. Generalized Zero-Shot Learning results on AWA1, AWA2, CUB, FLO and SUN. We measure the AP of Top-1 accuracy in %. The best results are highlighted with bold numbers.

Method	AWA1			AWA2			CUB			FLO			SUN		
	ts	tr	H	ts	tr	H	ts	tr	H	ts	tr	H	ts	tr	H
SJE [2]	11.3	74.6	19.6	8.0	73.9	14.4	23.5	59.2	33.6	13.9	47.6	21.5	14.7	30.5	19.8
ALE [1]	16.8	76.1	27.5	14.0	81.8	23.9	23.7	62.8	34.4	13.3	61.6	21.9	21.8	33.1	26.3
PSR [3]	–	–	–	20.7	73.8	32.3	24.6	54.3	33.9	–	–	–	20.8	37.2	26.7
SAE [14]	16.7	82.5	27.8	8.0	73.9	14.4	18.8	58.5	29.0	–	–	–	8.8	18.0	11.8
ESZSL [25]	6.6	75.6	12.1	5.9	77.8	11.0	12.6	63.8	21.0	11.4	56.8	19.0	11.0	27.9	15.8
LESAE [18]	19.1	70.2	30.0	21.8	70.6	33.3	24.3	53.0	33.3	–	–	–	21.9	34.7	26.9
ReViSE [28]	46.1	37.1	41.1	46.4	39.7	42.8	37.6	28.3	32.3	–	–	–	24.3	20.1	22.0
CMT [27]	0.9	87.6	1.8	0.5	90.0	1.0	7.2	49.8	12.6	–	–	–	8.1	21.8	11.8
SYNC [5]	8.9	87.3	16.2	10.0	90.5	18.0	11.5	70.9	19.8	–	–	–	7.9	43.3	13.4
DeViSE [10]	13.4	68.7	22.4	17.1	74.7	27.8	23.8	53.0	32.8	9.9	44.2	16.2	16.9	27.4	20.9
CRnet [33]	58.1	74.7	65.4	52.6	78.8	63.1	45.5	56.8	50.5	–	–	–	34.1	36.5	35.3
CVAE [21]	–	–	47.2	–	–	51.2	–	–	34.5	–	–	–	–	–	26.7
SP-AEN [6]	–	–	–	23.3	90.9	37.1	34.7	70.6	46.6	–	–	–	24.9	38.6	30.3
f-CLSWGAN [31]	57.9	61.4	59.6	52.1	68.9	59.4	43.7	57.7	49.7	59.0	73.8	65.6	42.6	36.6	39.4
cycle-(U)WGAN [9]	59.6	63.4	59.8	–	–	–	47.9	59.3	53.0	61.6	69.2	65.2	47.2	33.8	39.4
SE [15]	56.3	67.8	61.5	58.3	68.1	62.8	41.5	53.3	46.7	–	–	–	40.9	30.5	34.9
CADA-VAE [26]	57.3	72.8	64.1	55.8	75.0	63.9	51.6	53.5	52.4	–	–	–	47.2	35.7	40.6
AFC-GAN [17]	–	–	–	58.2	66.8	62.2	53.5	59.7	56.4	60.2	80.0	68.7	49.1	36.1	41.6
COSMO + fCLSWGAN [4]	64.8	51.7	57.5	–	–	–	41.0	60.5	48.9	59.6	81.4	68.8	35.3	40.2	37.6
COSMO + LAGO [4]	52.8	80.0	63.6	–	–	–	44.4	57.8	50.2	–	–	–	44.9	37.7	41.0
Ours($\gamma = 0.95$)	59.0	94.3	**72.6**	55.9	94.9	**70.3**	53.8	94.6	**68.6**	61.9	91.7	**73.9**	57.8	95.1	**71.9**

mean results. For example, the ts values of f-CLSWGAN, cycle-(U)WGAN, SE, CADA-VAE and AFC-GAN are higher than the tr values on the SUN dataset, which means the accuracy of seen classes are even worse than the unseen classes. The gating based method [4] is not good enough to classify the unseen and seen domains. Therefore the performance does not show obvious improvement compared to generative methods.

It can be seen that our approach achieves superior performance compared to the previous methods on all datasets, e.g. we achieve 72.6% harmonic mean on AWA1, 70.3% on AWA2, 68.6% on CUB, 73.9% on FLO and 71.9% on SUN, which significantly outperforms the baseline methods. In our experiments, we incorporate the proposed OOD classifier with the ZSL classifier of f-CLSWGAN. By using the proposed classifier, the ts values are improved compared to the original approach. Moreover, in our approach the tr values are significantly higher. Compared to the gating based method COSMO + fCLSWGAN which also uses the ZSL classifier of f-CLSWGAN, our approach also has much higher harmonic mean results. It indicates that the proposed OOD classifier is more reliable.

Obviously, the GZSL performance of our approach mainly depends on the OOD classifier and the ZSL classifier. As our OOD classifier is reliable enough to separate the unseen features from the seen features, the GZSL problem can be substantially simplified. Therefore, our approach does not suffer the bias problem or feature confusion problem. In practice, we can replace the ZSL classifier by any state-of-the-art models. Consequently, the Harmonic mean of our approach could be further improved by using more powerful ZSL models.

5.4 Model Analysis

In this section, to give a deep insight into our approach, we analyze our model under different settings.

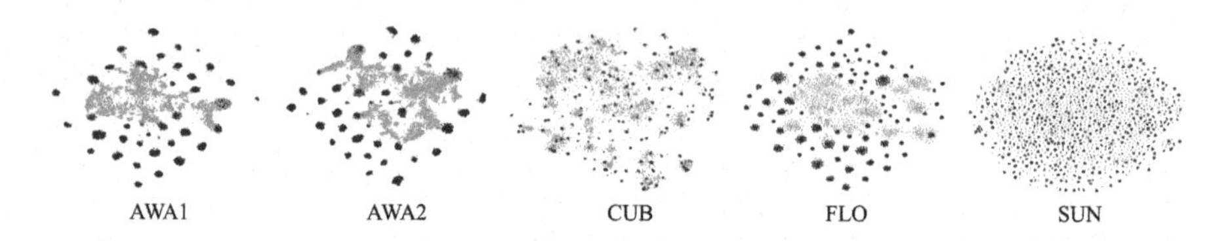

Fig. 4. The t-SNE visualization results for the learned latent space on the test sets of AWA1, AWA2, CUB, FLO and SUN. The blue dots represent the variables encoded from seen classes. The orange dots represent the variables encoded from unseen classes. (Color figure online)

Latent Space Visualization. To demonstrate the learned latent space, we visualize the latent variables of seen features and unseen features to the 2D-plane by using t-SNE. The visualization results of five datasets are shown in Fig. 4 where the blue dots represent the seen variables and the orange dots represent the unseen variables. It can be seen that the features in each seen class are clustered together in the latent space so that they can be easily classified. The unseen features are encoded to the latent variables chaotically scattered across the latent space. We see that most of the unseen variables locate out of the manifolds of seen classes. Although the inter-class association between seen and unseen classes is broken, the unseen variables can be easily separated from the seen variables.

Ablation Study. As defined in Eq. 10, the overall objective of our model consists of L_{f-SVAE}, L_{a-SVAE}, L_{cr} and L_{cls}. In this experiment, we analyze the impact of each term on AWA1 and CUB datasets. We report the AUC and the FPR scores on the threshold corresponding to $\gamma = 0.95$ for four objective functions in Table 4, where "+" stands for the combination of different terms. When there lacks of L_{cr} and L_{cls}, we observe that the first objective function only achieves 62.5% AUC score on AWA1, and 56.1% on CUB. The FPR score are 93.3% and 88.5%, respectively. It can be seen that the unseen samples can hardly be separated from the seem samples. When we further add L_{cr}, the AUC score increases to 89.3% and the FPR decreases to 44.2% on AWA1. However, the results only have small improvements on CUB dataset. We find that learning the modality invariant information helps to improve the OOD classification. But the improvement is influenced by the number of classes. When we add L_{cls} to the first objective function, the AUC score is improved to 94.9% on AWA1 and 98.2% on CUB. It can be seen that the classification loss heavily affects the binary classification. When we combine both L_{cr} and L_{cls}, the overall objective achieves the best OOD classification results.

Table 4. Binary classification results of different training objective functions. We report the AUC and the FPR corresponding to $\gamma = 0.95$.

Objective function	AWA1		CUB	
	AUC	FPR	AUC	FPR
$L_{f-SVAE} + L_{a-SVAE}$	62.5	93.3	56.1	88.5
$L_{f-SVAE} + L_{a-SVAE} + L_{cr}$	89.3	44.2	60.6	86.7
$L_{f-SVAE} + L_{a-SVAE} + L_{cls}$	94.9	15.7	98.2	9.2
$L_{overall}$	96.8	7.9	99.6	1.1

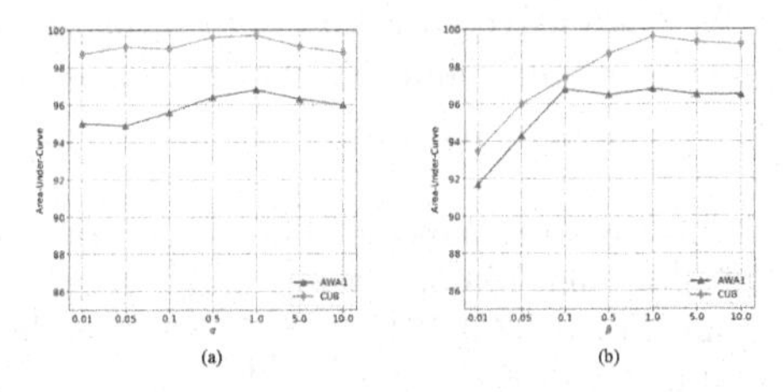

(a) (b)

Fig. 5. Parameter sensitivity on AWA1 and CUB datasets.

Parameters Sensitivity. The hyper-parameters in our approach are tuned by cross-validation. Fixing λ_f and λ_a to 0.1, we mainly tune α and β for our approach. Figure 5 shows the AUC scores influenced by each hyper-parameter on AWA1 and CUB datasets, where each hyper-parameter is varied with the others are fixed. It can be seen that our method can work stably with different parameters.

Table 5. The influence of latent space dimension on the AUC score for AWA1 and CUB datasets.

Dimension	16	32	64	128	256
AWA1	95.2	95.7	96.8	90.5	86.2
CUB	95.8	96.5	99.6	97.7	95.1

Dimension of Latent Space. In this analysis, we explore the robustness of our OOD classifier to the dimension of latent space. We report the AUC score in Table 5 with respect to different dimensions on AWA1 and CUB, ranging from 16, 32, 64, 128, and 256. We observe that the AUC score increases while we increase the latent space dimension and reaches the peak for both datasets at 64. When we continue to increase the dimension, the AUC score begins to decline, which indicates that increasing the dimension also may increase the risk of overfitting. For general consideration, we set the dimension to 64 for all datasets.

6 Conclusions

In this paper, we present an Out-of-Distribution classifier for the Generalized Zero-Shot learning problem. The proposed classifier is based on multi-modal hyper-spherical variational auto-encoders which learns a bounded manifold for each seen class in the latent space. By using the boundaries, we can separate the unseen samples from the seen samples. After that, we use two experts to classify the unseen samples and the seen samples separately. In this way, the GZSL problem is simplified to a ZSL problem and a conventional supervised classification problem. We extensively evaluate our approach on five benchmark datasets. The experimental results show that our approach surpasses state-of-the-art approaches by a significant margin.

Acknowledgements. This work was supported in part by Trico-Robot plan of NSFC under grant No.91748208, National Major Project under grant No.2018ZX01028-101, Shaanxi Project under grant No.2018ZDCXLGY0607, NSFC under grant No.61973246, and the program of the Ministry of Education for the university.

References

1. Akata, Z., Perronnin, F., Harchaoui, Z., Schmid, C.: Label-embedding for image classification. IEEE Trans. Pattern Anal. Mach. Intell. **38**(7), 1425–1438 (2016)
2. Akata, Z., Reed, S., Walter, D., Lee, H., Schiele, B.: Evaluation of output embeddings for fine-grained image classification. In: Proceedings of the IEEE Conference on Computer Vision and Pattern Recognition, pp. 2927–2936 (2015)
3. Annadani, Y., Biswas, S.: Preserving semantic relations for zero-shot learning. In: Proceedings of the IEEE Conference on Computer Vision and Pattern Recognition, pp. 7603–7612 (2018)
4. Atzmon, Y., Chechik, G.: Adaptive confidence smoothing for generalized zero-shot learning. In: Proceedings of the IEEE Conference on Computer Vision and Pattern Recognition, pp. 11671–11680 (2019)
5. Changpinyo, S., Chao, W.L., Gong, B., Sha, F.: Synthesized classifiers for zero-shot learning. In: Proceedings of the IEEE Conference on Computer Vision and Pattern Recognition, pp. 5327–5336 (2016)
6. Chen, L., Zhang, H., Xiao, J., Liu, W., Chang, S.F.: Zero-shot visual recognition using semantics-preserving adversarial embedding networks. In: Proceedings of the IEEE Conference on Computer Vision and Pattern Recognition, pp. 1043–1052 (2018)

7. Cuturi, M.: Sinkhorn distances: lightspeed computation of optimal transport. In: Advances in Neural Information Processing Systems, pp. 2292–2300 (2013)
8. Davidson, T.R., Falorsi, L., De Cao, N., Kipf, T., Tomczak, J.M.: Hyperspherical variational auto-encoders. In: 34th Conference on Uncertainty in Artificial Intelligence (UAI 2018) (2018)
9. Felix, R., Kumar, V.B., Reid, I., Carneiro, G.: Multi-modal cycle-consistent generalized zero-shot learning. In: Proceedings of the European Conference on Computer Vision (ECCV), pp. 21–37 (2018)
10. Frome, A., Corrado, G.S., Shlens, J., Bengio, S., Dean, J., Mikolov, T., et al.: Devise: A deep visual-semantic embedding model. In: Advances in Neural Information Processing Systems, pp. 2121–2129 (2013)
11. Goodfellow, I., et al.: Generative adversarial nets. In: Advances in Neural Information Processing Systems, pp. 2672–2680 (2014)
12. Hendrycks, D., Gimpel, K.: A baseline for detecting misclassified and out-of-distribution examples in neural networks. In: Proceedings of International Conference on Learning Representations (2017)
13. Kingma, D.P., Welling, M.: Auto-encoding variational bayes. arXiv preprint arXiv:1312.6114 (2013)
14. Kodirov, E., Xiang, T., Gong, S.: Semantic autoencoder for zero-shot learning. arXiv preprint arXiv:1704.08345 (2017)
15. Kumar Verma, V., Arora, G., Mishra, A., Rai, P.: Generalized zero-shot learning via synthesized examples. In: Proceedings of the IEEE Conference on Computer Vision and Pattern Recognition, pp. 4281–4289 (2018)
16. Lampert, C.H., Nickisch, H., Harmeling, S.: Attribute-based classification for zero-shot visual object categorization. IEEE Trans. Pattern Anal. Mach. Intell. **36**(3), 453–465 (2014)
17. Li, J., Jing, M., Lu, K., Zhu, L., Yang, Y., Huang, Z.: Alleviating feature confusion for generative zero-shot learning. In: Proceedings of the 27th ACM International Conference on Multimedia, pp. 1587–1595. ACM (2019)
18. Liu, Y., Gao, Q., Li, J., Han, J., Shao, L.: Zero shot learning via low-rank embedded semantic autoencoder. In: IJCAI, pp. 2490–2496 (2018)
19. Long, Y., Liu, L., Shao, L., Shen, F., Ding, G., Han, J.: From zero-shot learning to conventional supervised classification: unseen visual data synthesis. In: CVPR, pp. 1627–1636 (2017)
20. Mikolov, T., Sutskever, I., Chen, K., Corrado, G.S., Dean, J.: Distributed representations of words and phrases and their compositionality. In: Advances in Neural Information Processing Systems, pp. 3111–3119 (2013)
21. Mishra, A., Krishna Reddy, S., Mittal, A., Murthy, H.A.: A generative model for zero shot learning using conditional variational autoencoders. In: Proceedings of the IEEE Conference on Computer Vision and Pattern Recognition Workshops, pp. 2188–2196 (2018)
22. Nilsback, M.E., Zisserman, A.: Automated flower classification over a large number of classes. In: 2008 Sixth Indian Conference on Computer Vision, Graphics & Image Processing, pp. 722–729. IEEE (2008)
23. Patterson, G., Xu, C., Su, H., Hays, J.: The sun attribute database: beyond categories for deeper scene understanding. Int. J. Comput. Vis. **108**(1–2), 59–81 (2014). https://doi.org/10.1007/s11263-013-0695-z
24. Reed, S., Akata, Z., Lee, H., Schiele, B.: Learning deep representations of fine-grained visual descriptions. In: Proceedings of the IEEE Conference on Computer Vision and Pattern Recognition, pp. 49–58 (2016)

25. Romera-Paredes, B., Torr, P.: An embarrassingly simple approach to zero-shot learning. In: International Conference on Machine Learning, pp. 2152–2161 (2015)
26. Schonfeld, E., Ebrahimi, S., Sinha, S., Darrell, T., Akata, Z.: Generalized zero-and few-shot learning via aligned variational autoencoders. In: Proceedings of the IEEE Conference on Computer Vision and Pattern Recognition, pp. 8247–8255 (2019)
27. Socher, R., Ganjoo, M., Manning, C.D., Ng, A.: Zero-shot learning through cross-modal transfer. In: Advances in Neural Information Processing Systems, pp. 935–943 (2013)
28. Tsai, Y.H.H., Huang, L.K., Salakhutdinov, R.: Learning robust visual-semantic embeddings. In: 2017 IEEE International Conference on Computer Vision (ICCV), pp. 3591–3600. IEEE (2017)
29. Wah, C., Branson, S., Welinder, P., Perona, P., Belongie, S.: The caltech-ucsd birds-200-2011 dataset (2011)
30. Xian, Y., Lampert, C.H., Schiele, B., Akata, Z.: Zero-shot learning-acomprehensive evaluation of the good, the bad and the ugly. IEEE Trans. Pattern Anal. Mach. Intell. **41**, 2251–2265 (2018)
31. Xian, Y., Lorenz, T., Schiele, B., Akata, Z.: Feature generating networks for zero-shot learning. In: Proceedings of the IEEE Conference on Computer Vision and Pattern Recognition, pp. 5542–5551 (2018)
32. Xu, X., Shen, F., Yang, Y., Zhang, D., Shen, H.T., Song, J.: Matrix tri-factorization with manifold regularizations for zero-shot learning. In: CVPR, pp. 3798–3807 (2017)
33. Zhang, F., Shi, G.: Co-representation network for generalized zero-shot learning. In: International Conference on Machine Learning, pp. 7434–7443 (2019)
34. Zhang, H., Koniusz, P.: Zero-shot kernel learning. In: CVPR, pp. 7670–7679 (2018)
35. Zhang, Z., Saligrama, V.: Zero-shot learning via semantic similarity embedding. In: Proceedings of the IEEE International Conference on Computer Vision, pp. 4166–4174 (2015)

Mind the Discriminability: Asymmetric Adversarial Domain Adaptation

Jianfei Yang[1], Han Zou[2], Yuxun Zhou[2], Zhaoyang Zeng[3], and Lihua Xie[1](✉)

[1] Nanyang Technological University, Singapore, Singapore
{yang0478,elhxie}@ntu.edu.sg
[2] University of California, Berkeley, USA
[3] Sun Yat-sen University, Guangzhou, China

Abstract. Adversarial domain adaptation has made tremendous success by learning domain-invariant feature representations. However, conventional adversarial training pushes two domains together and brings uncertainty to feature learning, which deteriorates the discriminability in the target domain. In this paper, we tackle this problem by designing a simple yet effective scheme, namely Asymmetric Adversarial Domain Adaptation (AADA). We notice that source features preserve great feature discriminability due to full supervision, and therefore a novel asymmetric training scheme is designed to keep the source features fixed and encourage the target features approaching to the source features, which best preserves the feature discriminability learned from source labeled data. This is achieved by an autoencoder-based domain discriminator that only embeds the source domain, while the feature extractor learns to deceive the autoencoder by embedding the target domain. Theoretical justifications corroborate that our method minimizes the domain discrepancy and spectral analysis is employed to quantize the improved feature discriminability. Extensive experiments on several benchmarks validate that our method outperforms existing adversarial domain adaptation methods significantly and demonstrates robustness with respect to hyper-parameter sensitivity.

Keywords: Adversarial domain adaptation · Asymmetric training

1 Introduction

Learning robust representations from large-scale labeled datasets, deep neural networks have achieved huge success in kinds of applications, such as visual recognition and neural language processing [7,16]. Nevertheless, well-trained deep models are sensitive to cross-domain distribution shift (*domain shift*) that exists when applying them to a new domain, which usually requires tremendous

Electronic supplementary material The online version of this chapter (https://doi.org/10.1007/978-3-030-58586-0_35) contains supplementary material, which is available to authorized users.

A. Vedaldi et al. (Eds.): ECCV 2020, LNCS 12369, pp. 589–606, 2020.
https://doi.org/10.1007/978-3-030-58586-0_35

efforts on annotating new labels. To render data-hungry model strong representation ability like data-adequate model, domain adaptation is proposed to transfer the knowledge from a label-rich domain (*source domain*) to a label-scarce or unlabeled domain (*target domain*) [26]. It alleviates the negative effect of *domain shift* in transfer learning and reduces the manual overhead for labeling.

Prevailing domain adaptation [25] tackles the problem of *domain shift* by enhancing the transferability of feature learning, *i.e.* aligning the marginal distributions across domains. For deep neural networks, aligning the deep features mainly relies on two categories of approaches. One category reduces the distribution discrepancy by measuring the statistics [33,36], which is simple to implement and usually possesses stable convergence. Another category is adversarial domain adaptation, inspired by the Generative Adversarial Network (GAN) [12]. A binary domain discriminator is introduced to distinguish the domain labels while the feature extractor learns to fool the discriminator [9]. Adversarial domain adaptation approaches have achieved prominent performance on many challenging tasks including semantic segmentation [38], 3D estimation [43], sentiment classification [18] and wireless sensing [41,44].

Though adversarial domain adaptation methods yield superior performance, they bring "uncertainty" to feature learning [5]. Such uncertainty is due to the side effect of domain adversarial training (DAT). Specifically, any features that can deceive the domain discriminator and perform well in the source domain conform with the goal of DAT. The worst consequence could be overfitting to the source domain and generating meaningless features in the target domain as long as they can fool the domain discriminator. Therefore, this uncertainty in adversarial training is severely detrimental to learning discriminative features in the target domain. This explains why many subsequent works aim to preserve semantic information [39] or adjust the boundary of classifier during DAT [29,32]. These solutions ameliorate the traditional DAT yet by adding more learning steps, which either increases the computational overhead or requires a sophisticated hyper-parameter tuning process for multiple objectives.

In this paper, we address the uncertainty problem by proposing an Asymmetric Adversarial Domain Adaptation (AADA). The key problem of DAT consists in the symmetric objective which is to equally push two domains as close as possible, which deteriorates feature learning and neglects the decision boundary. Inspired by the fact that the source domain has great discriminability due to full supervision, AADA aims to fix the source domain and only adapts the target domain. To this end, we design an autoencoder serving as a domain discriminator to embed the source features, while a feature extractor is trained to deceive it—to embed the target features. Such adversarial process can be realized by a minimax game with a margin loss. As shown in Fig. 1, DAT employs a binary domain discriminator to align two domains together, while we only force the target domain to approach to the source domain that possesses good discriminability. Furthermore, it is acknowledged that the autoencoder is an energy function that learns to map the observed sample to the low-energy space [17]. Therefore, energy function can cluster similar data to form a high density manifold, which helps to preserve more semantic information. We leverage the energy function to fix the source domain by associating lower energies to it while pushing the target domain to the low-energy space via adversarial training in an innovative manner. *The proposed method is a novel and new fundamental domain alignment technique which can be easily integrated with other domain adaptation approaches.* Our contribution in this paper is threefold:

- We propose a novel asymmetric adversarial scheme that replaces the conventional domain classifier with an autoencoder, which incorporates only the target domain into the adversarial feature training, circumventing the loss of discriminability in traditional domain adversarial training.
- The autoencoder is an energy function that maps the two domains to the low-energy space, which encourages the feature clusters to be tight and thus further benefits the classification task in an unsupervised manner.
- AADA is a generic domain alignment approach that can be used as an ingredient in existing domain adaptation approaches. The experiment validates that AADA outperforms other domain alignment approaches significantly. We further demonstrate the boosted discriminability by spectral analysis.

The paper is organized as follows. We firstly revisit adversarial domain adaptation while highlighting its limitations in Sect. 2. Then the AADA method is detailed in Sect. 3. Section 4 demonstrates the effectiveness and superiority of AADA. Section 5 compares AADA with other relevant approaches and Sect. 6 concludes the paper.

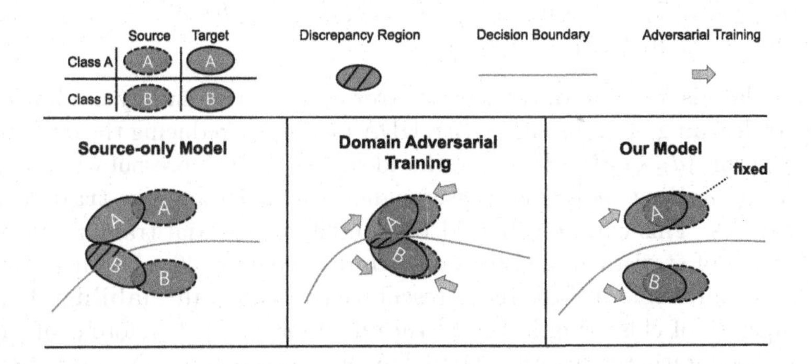

Fig. 1. Comparison between previous Domain Adversarial Training (DAT) method and ours. **Left:** The discrepancy region is large before adaptation. **Middle:** DAT based on a domain classifier aligns two domains together and pushes them as close as possible, which hurts the feature discriminability in the target domain. **Right:** AADA fixes the source domain that is regarded as the low-energy space, and pushes the target domain to approach the source one, which makes use of the well-trained classifier in the labeled source domain.

2 Revisiting Transferability and Discriminability in Symmetric Adversarial Domain Adaptation

We first revisit the learning theory of unsupervised domain adaptation (UDA) [1] where we analyze how the target expected error $\epsilon_{\mathcal{T}}(h)$ is connected to transferability and discriminability in representation learning. Then we highlight the problem of forfeiting discriminability due to symmetric adversarial training, and excogitate the insights of our approach that conceals behind.

2.1 The Theory of Domain Adaptation

In unsupervised domain adaptation, we have access to N_s labeled examples from a source domain $\mathcal{D}_S = \{\mathbf{x}_i^s, y_i^s | \mathbf{x}_i^s \in \mathbf{X}_s, y_i^s \in Y_s\}$ and N_t unlabeled examples from a target domain $\mathcal{D}_T = \{x_i^t | \mathbf{x}_i^t \in \mathbf{X}_t\}$, which are sampled from distinct distributions $\mathbb{P}$ and $\mathbb{Q}$, respectively. The objective of UDA is to learn a model that performs well for the target domain. The learning theory of UDA was proposed by Ben-David [1].

Theorem 1. *Let $\mathcal{H}$ be the common hypothesis class for source and target. Let ϵ_s and ϵ_t be the source and target generalization error functions, respectively. The expected error for the target domain is upper bounded as*

$$\epsilon_t(h) \leq \epsilon_s(h) + \frac{1}{2} d_{\mathcal{H}\Delta\mathcal{H}}(\mathcal{D}_S, \mathcal{D}_T) + \lambda, \forall h \in \mathcal{H}, \quad (1)$$

where $d_{\mathcal{H}\Delta\mathcal{H}}(\mathcal{D}_S, \mathcal{D}_T) = 2 \sup_{h_1, h_2 \in \mathcal{H}} |\Pr_{x \sim \mathcal{D}_S}[h_1(x) \neq h_2(x)] - \Pr_{x \sim \mathcal{D}_T}[h_1(x) \neq h_2(x)]|$ and $\lambda = \min_h [\epsilon_s(h^*) + \epsilon_t(h^*)]$.

As source data is annotated, the source error $\epsilon_s(h)$ can be simply minimized via supervised learning. To minimize $\epsilon_t(h)$, UDA focuses on reducing the domain discrepancy term $\frac{1}{2} d_{\mathcal{H}\Delta\mathcal{H}}(\mathcal{D}_S, \mathcal{D}_T)$ and the ideal risk λ. In representation learning for UDA, minimizing the domain discrepancy is able to improve **transferability** of features. This can be achieved by domain adversarial training [9] or the minimization of statistical measures of such discrepancy [36]. Another criterion that plays a vital role in feature representations is **discriminability**. It refers to the capacity of clustering in the feature manifold, and therefore controls the easiness of separating categories. For UDA, we pursue good discriminability in both source and target domains simultaneously. As we typically use a shared feature extractor for two domains, enhancing discriminability is equivalent to seeking for a better ideal joint hypothesis $h^* = \min_h [\epsilon_s(h) + \epsilon_t(h)]$ [5].

2.2 Limitations and Insights

From the analyses above, the feature learning of UDA should guarantee both transferability and discriminability, which inspires us to investigate the existing domain alignment methods from these two perspectives. Adversarial domain adaptation methods have shown prominent performance and become increasingly popular. Ganin *et al.* [9] pioneered it by proposing Domain Adversarial Neural Network (**DANN**) that learns domain-invariant features using the reverse gradients from a domain classifier. Typically, adversarial UDA approaches consist of a shared feature extractor $\mathbf{f} = G_f(x)$, a label predictor $\mathbf{y} = G_y(x)$ and a domain discriminator $\mathbf{d} = G_d(x)$. Apart from standard supervised learning on the source domain, a minimax game between $\mathbf{f}$ and $\mathbf{d}$ is designed. The domain discriminator $\mathbf{d}$ is trained to distinguish the domain label between the source domain and the target domain, while the feature extractor $\mathbf{f}$ learns to deceive the domain classifier $\mathbf{d}$. In this manner, the domain adversarial training enables the model to learn transferable features across domains when the Nash Equilibrium is achieved. The whole process can be formulated as

$$\min_{G_f, G_y} \mathcal{L}_c(\mathbf{X}_s, Y_s) - \gamma \mathcal{L}_c(\mathbf{X}_s, \mathbf{X}_t), \tag{2}$$

$$\min_{G_d} \mathcal{L}_c(\mathbf{X}_s, \mathbf{X}_t), \tag{3}$$

where $\mathcal{L}_c$ is the classification loss such as cross-entropy loss, and the hyper-parameter γ decides the importance of transferability in feature learning.

Symmetric DAT does improve the transferability across domains but sacrifices the discriminability in the target domain. Let us analyze the objectives of training feature extractor $\mathbf{f}$ that are two-folds: (1) good discriminability in the source domain and (2) learning representations that are indistinguishable to the domain discriminator. There is no constraint on the discriminability in the target domain. As depicted in Fig. 1, DAT is symmetric and makes two domains as close as possible. Theoretically, the worst case is that the feature extractor generates meaningless representations on the target domain, as long as they can deceive the domain classifier. Hence, a good decision boundary on the source domain cannot perform well on the target domain. Previous works quantified the discriminability by spectral analysis and drew a similar conclusion [5,20].

We believe that *the decreasing discriminability is caused by symmetric adversarial training that involves both source domain and target domain in adversarial feature learning* as shown in the second term of Eq. (2). Specifically, to deceive the binary domain discriminator, DAT aims to push two domains close. In this process, DAT cannot control how the domains are aligned and cannot guarantee whether the decision boundary separates the categorical clusters in the target domain. This motivates us to fix the source domain and only render the target domain to approach to the source, as depicted in Fig. 1. In this fashion, the feature discriminability is preserved and a good classifier is easily obtained. To achieve symmetric adversarial training, we innovatively propose to leverage the autoencoder as the domain classifier.

3 Asymmetric Adversarial Domain Adaptation

Maintaining the source manifolds during DAT is not a trivial task with the consideration of the complexity of network architecture. This requires us to design a simple yet effective asymmetric adversarial mechanism. In this section, Asymmetric Adversarial Domain Adaptation (AADA) is detailed and we theoretically justify how it reduces domain discrepancy.

3.1 The Learning Framework

As shown in Fig. 2, our model consists of a shared feature extractor G_f parameterized by θ_f, a label predictor G_y parameterized by θ_y and an autoencoder G_a parameterized by θ_a. The feature extractor G_f, typically composed of multiple convolutional layers, embeds an input sample to a feature embedding z, and then the label predictor G_y that usually consists of several fully connected layers maps the feature embedding to the predicted label $\hat{y}$. The autoencoder G_a reconstructs an embedding z to $\hat{z}$.

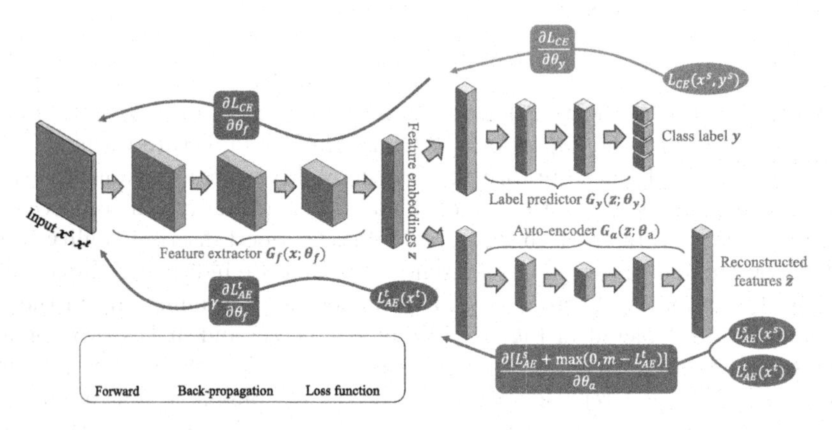

Fig. 2. AADA model constitutes a shared feature extract G_f, a classifier G_y and an autoencoder G_a. Except for supervised learning on the source domain, the autoencoder plays a domain discriminator role that learns to embed the source features and push the target features away, while the feature extractor learns to generate target features that can deceive the autoencoder. Such process is an asymmetric adversarial game that pushes the target domain to the source domain in the feature space.

In the learning phase, the first objective of the model is to learn feature discriminability in the source domain. As we have access to the labeled source data $(\mathbf{X}_s, Y_s)$, it is simply achieved by minimizing the cross-entropy loss via back-propagation:

$$\begin{aligned} &\min_{G_f, G_y} \mathcal{L}_{CE}(\mathbf{X}_s, Y_s) \\ &= -\mathbb{E}_{(\mathbf{x}_s, y_s)\sim(\mathbf{X}_s, Y_s)} \sum_{n=1}^{N_s} [\mathbb{I}_{[l=y_s]} \log G_y(G_f(\mathbf{x}_s))]. \end{aligned} \tag{4}$$

With robust feature learning in the source domain, the next objective is to learn transferable representations using the unlabeled data $(\mathbf{X}_t)$ in the target domain. To this end, we propose an asymmetric adversarial training scheme that involves an autoencoder G_a with a margin Mean Squared Error (MSE) loss. The autoencoder G_a, which plays a domain discriminator role, only learns to embed features from the source domain, but not to embed features from the target domain. The objective of the autoencoder-based domain discriminator is formulated as:

$$\min_{G_a} \mathcal{L}_{AE}(\mathbf{X_s}) + \max(0, m - \mathcal{L}_{AE}(\mathbf{X_t})), \tag{5}$$

where m is the margin between two domains in the feature space. Here, the MSE loss of the autoencoder is defined as:

$$\mathcal{L}_{AE}(\mathbf{x}_i) = ||G_a(G_f(\mathbf{x}; \theta_f); \theta_a) - \mathbf{x}_i||_2^2, \tag{6}$$

where $|| \cdot ||_2^2$ denotes the squared L_2-norm. Such unsupervised loss introduces a cycle-consistent constraint that improves feature discriminability.

To play the adversarial game, the feature extractor G_f learns to fool the autoencoder G_a by generating source-like features for the samples in the target domain. When the feature extractor succeeds, the representations of the target domain can inherit good discriminability from the source domain, and the label predictor G_y trained in the source domain applies equally. The adversarial training of the feature extractor G_f is formulated by:

$$\min_{G_f} \mathcal{L}_{AE}(\mathbf{X_t}). \tag{7}$$

The overall optimization of the proposed AADA model is formally defined by:

$$\begin{aligned} &\min_{G_f, G_y} \mathcal{L}_{CE}(\mathbf{X}_s, Y_s) + \gamma \mathcal{L}_{AE}(\mathbf{X_t}), \\ &\min_{G_a} \mathcal{L}_{AE}(\mathbf{X_s}) + \max(0, m - \mathcal{L}_{AE}(\mathbf{X_t})), \end{aligned} \tag{8}$$

where γ is a hyper-parameter that controls the importance of transferability. In this fashion, our approach only incorporates the $\mathcal{L}_{AE}(\mathbf{X_t})$ term into the training of G_f, which pushes the target domain to the source domain. Oppositely, the objective of G_a serves as a domain discriminator that pushes two domains away from a margin m.

3.2 Discussions and Theories

The autoencoder used in our approach is an energy function, which associates lower energies (*i.e.* MSE) to the observed samples in a binary classification problem [17]. For UDA, the autoencoder in our model associates low energies to the source features, and AADA compels the target features to approach to low-energy space. The design of such adversarial scheme is inspired by the Energy-based GAN which theoretically proves that using an energy function in GAN, the true distribution can be simulated by the generator at *Nash Equilibrium* [42]. Similarly, in AADA, the feature extractor G_f can mimic the source distribution for the samples in the target domain when the model achieves convergence.

As theoretically justified in [9], domain adversarial training using a domain classifier effectively reduces the domain discrepancy term $d_{\mathcal{H}\Delta\mathcal{H}}(\mathcal{D}_S, \mathcal{D}_T)$ in Theorem 1. In our approach, we notice that the autoencoder can be treated as a form of a domain classifier with a margin. The MSE loss of the autoencoder in our framework, *i.e.* $\mathcal{L}_{AE}(\cdot)$, works as the same way of the domain classifier. The training objective of $\mathcal{L}_{AE}(\cdot)$ is $\mathcal{L}_{AE}(\mathbf{X_s}) = 0$ and $\mathcal{L}_{AE}(\mathbf{X_t}) = m$, which is equivalent to the functionality of the standard domain classifier. Therefore, following the theorem proved in [9], the proposed autoencoder can maximize the domain divergence while the adversarial feature learning minimizes the divergence by deceiving the autoencoder. Moreover, as the autoencoder memorizes more domain information than a binary classifier, it transfers more knowledge during asymmetric adversarial training.

4 Experiments

We evaluate the proposed AADA on three UDA benchmarks and compare our model with the prevailing approaches. Then we validate the improved feature discriminability by spectral analysis. In the discussions, we verify our motivations by quantitatively analyzing the transferability and discriminability.

4.1 Experimental Setup

Digits [10]. We use five digits datasets *MNIST, MNIST-M, USPS, SVHN* and *Synthetic Digits* (*SYN-DIGIT*)), all of which consist of 32×32 images. We assess five types of adaptation scenarios with distinct levels of domain shift. We follow the experimental settings of **DANN** [9] that uses the official training splits in two domains for training and evaluates the model on the testing split.

Image-CLEF[1] is a domain adaptation dataset for ImageCLEF Challenge. It constitutes three domains: *Caltech-256* (**C**), *ImageNet ILSVRC 2012* (**I**) and *Pascal VOC 2012* (**P**), which form six transfer tasks.

Office-Home [37] consists of 65 categories in office and home settings, and has more than 15500 images. It is a very challenging dataset with four extremely distinct domains: *Artistic images* (**Ar**), *Clip Art* (**Cl**), *Product images* (**Pr**) and *Real-World images* (**Rw**), which forms 12 transfer tasks. For Image-CLEF and Office-Home, we employ the full training protocol in [22] that employs all images from the source domain and the target domain for training.

Baselines. We compare our AADA with the state-of-the-art UDA methods: **MMD** [36], Deep Adaptation Network (**DAN**) [22], Deep **CORAL** [33], **DANN** [9], Self-ensembling **SE** [8], Deep Reconstruction Classification Network (**DRCN**) [11], Domain Separate Network (**DSN**) [3], **ADDA** [35], **CoGAN** [21], **CyCADA** [14], Maximum Classifier Discrepancy (**MCD**) [29], Conditional Domain Adversarial Network (**CDAN**) [23], Batch Spectral Penalization (**BSP**) [5], **CCN** [15], **GTA** [31] and **MCS** [19]. To further prove that our method can

[1] http://imageclef.org/2014/adaptation.

be integrated with other UDA methods to achieve better adaptation, we integrate it with the Constrained Clustering Network (**CCN**) [15] that employs self-training to improve feature transferability.

Implementation Details. For the task with 32×32 images, we use the same network architecture as DANN, while for Image-CLEF and Office-Home, we use **ResNet-50** with pretrained parameters on ImageNet [6]. In AADA, we use an autoencoder with only fully connected layers, and the detailed network architectures are in the appendix. We use Adam optimizer with the constant learning rate $\mu = 5e^{-4}$ for digit adaptation, and SGD with the decaying learning rate in DANN for object recognition. For hyperparameter m and γ, we set $m = 0.5, \gamma = 1e^{-2}$ for 32×32 images, and $m = 1, \gamma = 1e^{-1}$ for object recognition datasets, which are empirically obtained by cross-validation on MNIST→USPS. The whole experiment is implemented by **PyTorch** framework.

Table 1. Accuracy (%) of domain adaptation tasks on *Digit*.

Source Target	MNIST USPS	USPS MNIST	SVHN MNIST	SYN-DIGIT SVHN	MNIST MNIST-M
Source-only	78.2	63.4	54.9	86.7	56.3
Domain Alignment Methods					
MMD [36]	81.1	–	71.1	88.0	76.9
CORAL [33]	80.7	–	63.1	85.2	57.7
Adversarial Training based Methods					
DANN [9]	85.1	73.0	74.7	90.3	76.8
CoGAN [21]	91.2	89.1	–	–	62.0
ADDA [35]	89.4	90.1	76.0	–	–
CDAN [23]	93.9	96.9	88.5	–	–
GTA [31]	95.3	90.8	92.4	–	–
CyCADA [14]	95.6	96.5	90.4	–	–
Other State-of-the-Art Methods					
DRCN [11]	91.8	73.7	82.0	87.5	68.3
DSN [3]	91.3	–	82.7	91.2	83.2
MCD [29]	96.5	94.1	96.2	–	–
BSP + DANN [5]	94.5	97.7	89.4	–	–
BSP + CDAN [5]	95.0	98.1	92.1	–	–
MCS + GTA [19]	97.8	98.2	91.7	–	–
AADA_{opt}	95.6 ± 0.3	92.7 ± 0.5	74.8 ± 0.8	88.8 ± 0.2	47.1 ± 1.2
AADA (Ours)	**98.4 ± 0.3**	**98.6 ± 0.3**	**98.1 ± 0.5**	**92.2 ± 0.4**	**95.5 ± 0.2**

4.2 Overall Results

We first compare our methods with MMD, CORAL, DANN, DAN and JAN that are only based on domain alignment. In Table 1, our approach shows significant improvement against the standard DAT (DANN), even outperforming the state-of-the-art methods that require much higher computation overhead such as CyCADA and GTA due to the training of cyclic or generative networks. For pixel-level domain shift in MNIST→MNIST-M, the traditional domain alignment methods such as MMD and DANN cannot effectively deal with them, but AADA achieves 95.5% accuracy. Moreover, for SVHN→MNIST with larger domain shift, AADA surpasses DANN by 23.4% and BSP+DANN by 8.7%. For more challenging tasks in Table 2 and 3, the proposed AADA surpasses all the methods that are purely based on domain alignment. AADA outperforms DANN by 1.5% on Image-CLEF and 7.7% on OfficeHome. Since the domain shift is small in Image-CLEF, the improvement margin is not large.

Table 2. Classification accuracy (%) on *Image-CLEF* (ResNet-50).

Method	I→P	P→I	I→C	C→I	C→P	P→C	Avg
ResNet-50 [13]	74.8	83.9	91.5	78.0	65.5	91.2	80.7
DAN [22]	74.5	82.2	92.8	86.3	69.2	89.8	82.5
DANN [9]	75.0	86.0	96.2	87.0	74.3	91.5	85.0
CCN [15]	77.1	87.5	94.0	86.0	74.5	91.7	85.1
CDAN [23]	76.7	90.6	**97.0**	90.5	74.5	93.5	87.1
AADA	78.0	90.3	94.0	87.8	75.2	93.5	86.5
AADA + CCN	**79.2**	**92.5**	96.2	**91.4**	**76.1**	**94.7**	**88.4**

AADA is a generic domain alignment approach that can be integrated to other novel UDA frameworks, achieving better performance. We integrate AADA with CCN [15] and evaluate it on challenging tasks. As shown in Table 2 and 3, the proposed AADA+CCN achieves the state-of-the-art accuracies, outperforming CDAN and BSP that also aim to improve discriminability. We can see that AADA improves CCN by 2.9% on Image-CLEF and 7.3% on Office-Home, which implies that AADA can bring much improvement to existing UDA methods by preserving more discriminability in domain alignment.

Table 3. Classification accuracy (%) on *Office-Home* (ResNet-50).

Method	Ar→Cl	Ar→Pr	Ar→Rw	Cl→Ar	Cl→Pr	Cl→Rw	Pr→Ar	Pr→Cl	Pr→Rw	Rw→Ar	Rw→Cl	Rw→Pr	Avg
ResNet-50 [13]	34.9	50.0	58.0	37.4	41.9	46.2	38.5	31.2	60.4	53.9	41.2	59.9	46.1
DAN [22]	43.6	57.0	67.9	45.8	56.5	60.4	44.0	43.6	67.7	63.1	51.5	74.3	56.3
DANN [9]	45.6	59.3	70.1	47.0	58.5	60.9	46.1	43.7	68.5	63.2	51.8	76.8	57.6
CCN [15]	47.3	65.2	70.1	51.3	60.5	60.9	48.1	45.5	71.3	65.1	53.5	77.0	59.7
SE [8]	48.8	61.8	72.8	54.1	63.2	65.1	50.6	49.2	72.3	66.1	55.9	78.7	61.5
CDAN [23]	49.0	69.3	74.5	54.4	66.0	68.4	55.6	48.3	75.9	68.4	55.4	80.5	63.8
BSP+DANN [5]	51.4	68.3	75.9	56.0	67.8	68.8	57.0	49.6	75.8	70.4	57.1	80.6	64.9
AADA	52.3	69.5	76.3	59.7	68.2	70.2	58.2	48.9	75.9	69.1	54.3	80.5	65.3
AADA+CCN	**54.0**	**71.3**	**77.5**	**60.8**	**70.8**	**71.2**	**59.1**	**51.8**	76.9	**71.0**	**57.4**	**81.8**	**67.0**

4.3 Spectral Analysis Using SVD

The intuition and the theoretical analysis validate that the proposed model achieves a good trade-off between transferability and discriminability. Here we employ the quantitative method to further demonstrate it. The prior research proposes to apply Singular Value Decomposition (SVD) to the representation z, and then infer transferability and discriminability by the Singular Values (SV) and the Corresponding Angles (CA) of eigenvectors, respectively [5]. Motivated by this, we conducted an experiment on a digit adaptation task SVHN→MNIST. Using the feature extractor G_f, we obtained the target feature matrix $\mathbf{F_t} = [\mathbf{f_t^1} \dots \mathbf{f_t^b}]$ where b is the batch size. Then we apply SVD to the target feature matrix as follows:

$$\mathbf{F_t} = \mathbf{U_t} \mathbf{\Sigma_t} \mathbf{V_t^T}, \tag{9}$$

where $\mathbf{\Sigma_t}$ denotes the eigenvalue, $\mathbf{U_t}$ denotes the eigenvector and $\mathbf{V_t}$ is an unitary matrix. SVD is also applied to the source feature matrix to obtain $\mathbf{\Sigma_s}$ and $\mathbf{U_s}$. In Fig. 3(a), we plot the top-20 normalized singular values $\mathbf{\Sigma_t}$ *w.r.t* three models including the source-only model, DANN and our AADA model. It is observed that the largest singular value of the DANN target feature matrix is greater than the other values, which impairs the semantic information included in other smaller singular values. In comparison, the distribution of singular values for AADA is similar to that for the source-only model that preserves more discriminability in feature learning. In Fig. 3(b), we show the normalized corresponding angles of singular values. The corresponding angle depicts the commonality between the source eigenvectors $\mathbf{U_s}$ and the target eigenvectors $\mathbf{U_t}$, which indicates the transferability of the features. For DANN, the sharp distribution of the angles indicates that DANN only utilizes several peak transferable features. This deteriorates the informative representation in the target domain. Whereas, AADA obtains more transferable features that also show better discriminability, which has good repercussion for learning a common decision boundary.

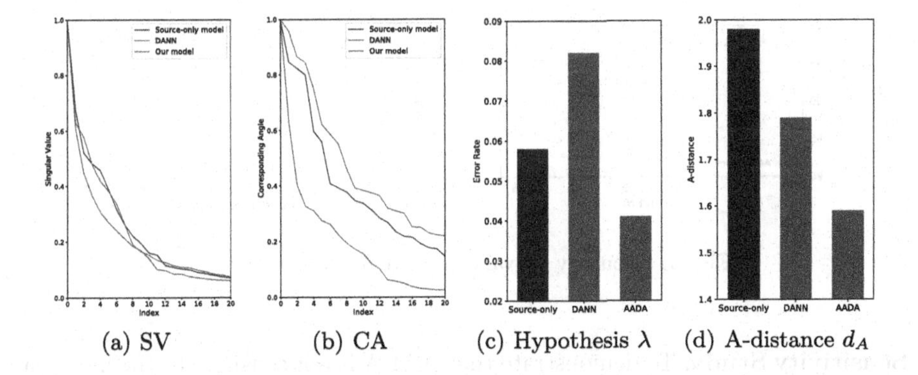

(a) SV (b) CA (c) Hypothesis λ (d) A-distance d_A

Fig. 3. Measures of discriminability and transferability.

4.4 Analytics and Discussions

Opposite Direction of AADA. The proposed AADA fixes the source domain and then learns to force the target domain to approach to it, which utilizes the good classifier of the source domain. What if we fix the target domain and force the source domain to approach to it? We denote this situation as the opposite form of AADA, namely AADA_{opt}. The optimization procedure of AADA_{opt} is written as:

$$\begin{aligned} &\min_{G_f, G_y} \mathcal{L}_{CE}(\mathbf{X}_s, Y_s) + \gamma \mathcal{L}_{AE}(\mathbf{X_s}), \\ &\min_{G_a} \mathcal{L}_{AE}(\mathbf{X_t}) + \max(0, m - \mathcal{L}_{AE}(\mathbf{X_s})). \end{aligned} \tag{10}$$

Intuitively, as the decision boundary of the target domain keeps changing during training, this may weaken the discriminability in domain adversarial training, which is similar to DANN. The result in Table 1 proves our analysis. AADA_{opt} only produces similar results as DANN, which implies that it is infeasible to align two domains and the boundaries of classifier simultaneously.

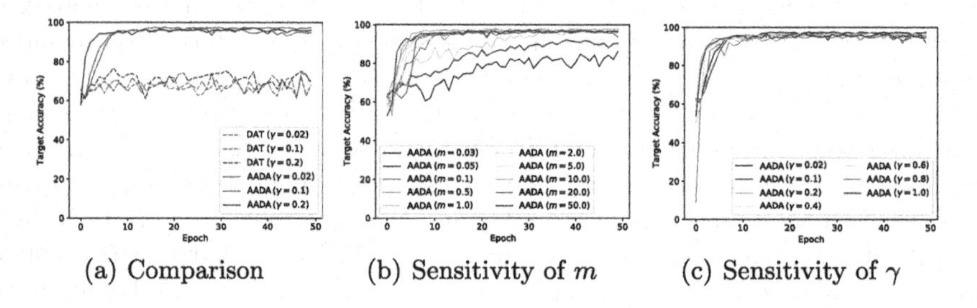

(a) Comparison (b) Sensitivity of m (c) Sensitivity of γ

Fig. 4. The training procedures *w.r.t.* the hyper-parameters m and γ.

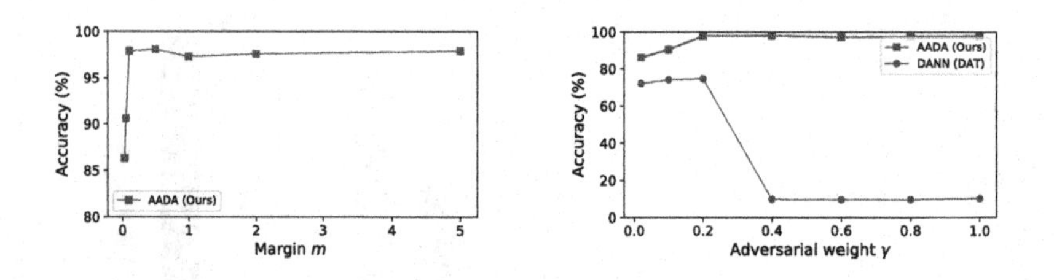

Fig. 5. Accuracy by varying m (left) and γ (right).

Sensitivity Study. To demonstrate that AADA is not sensitive to the hyperparameters, we conduct the experiments on SVHN→MNIST across multiple m and γ. Fig. 5 illustrates the sensitivity results in terms of the margin m. As the margin increases, the accuracy increases until 0.5, which conforms with our intuition.

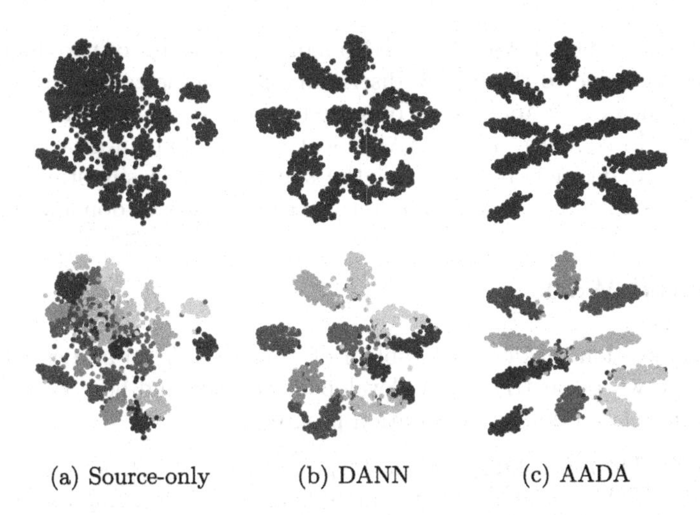

Fig. 6. The t-SNE visualization of the embeddings z on task **SVHN→MNIST**. The top figures are visualization with domain labels (blue: source, red: target). The bottom figures are visualization with category labels (10 classes). (Color figure online)

When the margin is small, the degree of transferability is limited. If the margin is greater than 0.1, we observe that the results are very stable in Fig. 4(b). As to γ, we compare our approach with Domain Adversarial Training (DAT) that is the reproduced DANN where γ is the weight of the adversarial loss in Eq. 2. In Fig. 5 and Fig. 4(c), the results of AADA are robust given large γ but DANN cannot converge with large γ. In Fig. 4(a), we can observe that the target accuracies of DAT fluctuate a lot during training while AADA provides more stable training procedure. The sensitivity study is consistent with the insights provided earlier, and moreover the results show more robustness *w.r.t* training procedure, which makes it easier to apply our method in an unsupervised manner.

Ideal Joint Hypothesis. We estimate the ideal joint hypothesis λ in Theorem 1 to show the **discriminability** of feature embeddings. To this end, we train an MLP classifier on all source and target data with labels. As shown in Fig. 3(c), AADA has the lowest λ in the feature space on task SVHN→MNIST. This demonstrates that AADA preserves more feature discriminability in two domains by asymmetric and cycle-consistent objectives. It helps learn a good decision boundary that separates the data from two domains.

Distribution Discrepancy. As proposed in the theory of domain adaptation [1], the A-distance is a measure of domain discrepancy that quantifies the **transferability** of feature embeddings. It is defined as $d_A = 2(1 - 2\epsilon)$ where ϵ is the error of a domain classifier. We train an MLP classifier to discriminate source and target domains on task SVHN→MNIST. Results are shown in Fig. 3(d), and it is observed that AADA has better transferability than DANN.

Visualization Representation. We visualize learned features on the *Digit* task SVHN→MNIST via t-SNE [24] and present it in Fig. 6. The visualization validates the insights of AADA. DANN aligns the features of two domains, but we can see that due to the lack of discriminability, some categories of data are confused. Hence many approaches proposed to adjust the boundary for DANN features [29]. In comparison, the AADA features are domain-invariant and preserve the good decision boundary of the source domain simultaneously. This further proves that AADA learns better discriminability in domain alignment.

5 Related Work

Domain adaptation tackles the problem of *domain shift* in statistical learning [26]. Massive works on UDA were developed recently, and here we discuss and compare the related domain adaptation progress.

Adversarial Domain Adaptation. Inspired by GAN [12], adversarial domain adaptation (ADA) methods yield remarkable results by learning representations that cannot be distinguished by a domain discriminator [9,32,34,35]. ADA can act on both feature-level and pixel-level alignment [2,14]. Adversarial training can also maximize classifier discrepancy to learn adapted classifier [29]. ADA is generalized to new scenarios including partial adversarial domain adaptation [4] and open set domain adaptation [30]. As such, ADA becomes a necessary ingredient for many subsequent UDA approaches.

Enhancing Discriminability in UDA. Though ADA methods contribute to learning domain-invariant features, adversarial learning hinders the feature discriminability in the target domain [5]. Conditional adversarial domain adaptation captures the cross-covariance between features and predictions to improve the discriminability [23]. More methods propose to learn semantic features by clustering and self-training [39]. BSP penalizes the largest singular values of features [5], and Xu *et al.* [40] propose that larger feature norm boosts the discriminability. These methods effectively increase the discriminability by adding extra regularization terms or building self-training algorithms, which can lead to much complexity or more difficult hyper-parameter tuning process.

Asymmetric Training. Asymmetric training means an unequal training process for multiple networks or parts of networks. Satio *et al.* pioneered an asymmetric tri-training for domain adaptation, which leverages two networks for generating pseudo labels and one network to learn target representations [28]. Asymmetric training between two feature extractors is developed in ADDA with untied sharing weights [35,45]. Our approach proposes an asymmetric training between the two players of adversarial game.

Autoencoder in Domain Adaptation. Autoencoder can learn representations in an unsupervised manner, and it has been directly utilized for learning target domain in DRCN [11] and DSN [3]. It also enables cyclic methods such as CyCADA [14]. All these methods use auto-encoder to learn target domain features in a straightforward way. In AADA, we employ autoencoder as a domain classifier which empowers asymmetric adversarial training and hence improve the discriminability. From the perspective of energy-based model [17], autoencoder is an energy function [27,42] that maps the correct variables to low energies. In this paper, it is designed to assign low energies to the source domain, and encourage the target domain to approach to the low-energy space.

6 Conclusion

In this paper, we propose a novel asymmetric adversarial regime for unsupervised domain adaptation. As the conventional adversarial UDA methods affect the discriminability while improving the transferability of feature representations, our method aims to preserve the discriminability by encouraging the target domain to approach to the source domain in the feature space, which is achieved by an autoencoder with an asymmetric adversarial training scheme. Spectral analysis is utilized to justify the improved discriminability and transferability. The experimental results demonstrate its robustness and superiority on several public UDA datasets.

References

1. Ben-David, S., Blitzer, J., Crammer, K., Kulesza, A., Pereira, F., Vaughan, J.W.: A theory of learning from different domains. Mach. Learn. **79**(1–2), 151–175 (2010). https://doi.org/10.1007/s10994-009-5152-4
2. Bousmalis, K., Silberman, N., Dohan, D., Erhan, D., Krishnan, D.: Unsupervised pixel-level domain adaptation with generative adversarial networks. In: Proceedings of the IEEE Conference on Computer Vision and Pattern Recognition, pp. 3722–3731 (2017)
3. Bousmalis, K., Trigeorgis, G., Silberman, N., Krishnan, D., Erhan, D.: Domain separation networks. In: Advances in Neural Information Processing Systems, pp. 343–351 (2016)
4. Cao, Z., Ma, L., Long, M., Wang, J.: Partial adversarial domain adaptation. In: Proceedings of the European Conference on Computer Vision (ECCV), pp. 135–150 (2018)
5. Chen, X., Wang, S., Long, M., Wang, J.: Transferability vs. discriminability: Batch spectral penalization for adversarial domain adaptation. In: Chaudhuri, K., Salakhutdinov, R. (eds.) Proceedings of the 36th International Conference on Machine Learning. Proceedings of Machine Learning Research, 09–15 June 2019, Long Beach, California, USA, vol. 97, pp. 1081–1090. PMLR (2019)
6. Deng, J., Dong, W., Socher, R., Li, L.J., Li, K., Fei-Fei, L.: Imagenet: a large-scale hierarchical image database. In: 2009 IEEE Conference on Computer Vision and Pattern Recognition, pp. 248–255. IEEE (2009)
7. Devlin, J., Chang, M.W., Lee, K., Toutanova, K.: Bert: Pre-training of deep bidirectional transformers for language understanding. arXiv preprint arXiv:1810.04805 (2018)

8. French, G., Mackiewicz, M., Fisher, M.: Self-ensembling for visual domain adaptation. In: International Conference on Learning Representations, no. 6 (2018)
9. Ganin, Y., Lempitsky, V.: Unsupervised domain adaptation by backpropagation. In: Bach, F., Blei, D. (eds.) Proceedings of the 32nd International Conference on Machine Learning. Proceedings of Machine Learning Research, Lille, France, 07–09 July 2015, vol. 37, pp. 1180–1189. PMLR (2015)
10. Ganin, Y.: Domain-adversarial training of neural networks. J. Mach. Learn. Res. **17**(1), 2030–2096 (2016)
11. Ghifary, M., Kleijn, W.B., Zhang, M., Balduzzi, D., Li, W.: Deep reconstruction-classification networks for unsupervised domain adaptation. In: Leibe, B., Matas, J., Sebe, N., Welling, M. (eds.) ECCV 2016. LNCS, vol. 9908, pp. 597–613. Springer, Cham (2016). https://doi.org/10.1007/978-3-319-46493-0_36
12. Goodfellow, I., et al.: Generative adversarial nets. In: Advances in Neural Information Processing Systems, pp. 2672–2680 (2014)
13. He, K., Zhang, X., Ren, S., Sun, J.: Deep residual learning for image recognition. In: Proceedings of the IEEE Conference on Computer Vision and Pattern Recognition, pp. 770–778 (2016)
14. Hoffman, J., et al.: CyCADA: cycle-consistent adversarial domain adaptation. In: Proceedings of the 35th International Conference on Machine Learning, pp. 1989–1998 (2018)
15. Hsu, Y.C., Lv, Z., Kira, Z.: Learning to cluster in order to transfer across domains and tasks. In: International Conference on Learning Representations (ICLR) (2018)
16. LeCun, Y., Bengio, Y., Hinton, G.: Deep learning. Nature **521**(7553), 436 (2015)
17. LeCun, Y., Chopra, S., Hadsell, R., Ranzato, M., Huang, F.: A tutorial onenergy-based learning. Predicting Struct. Data **1** (2006)
18. Li, Z., Zhang, Y., Wei, Y., Wu, Y., Yang, Q.: End-to-end adversarial memory network for cross-domain sentiment classification. In: IJCAI, pp. 2237–2243 (2017)
19. Liang, J., He, R., Sun, Z., Tan, T.: Distant supervised centroid shift: a simple and efficient approach to visual domain adaptation. In: Proceedings of the IEEE Conference on Computer Vision and Pattern Recognition, pp. 2975–2984 (2019)
20. Liu, H., Long, M., Wang, J., Jordan, M.: Transferable adversarial training: a general approach to adapting deep classifiers. In: International Conference on Machine Learning, pp. 4013–4022 (2019)
21. Liu, M.Y., Tuzel, O.: Coupled generative adversarial networks. In: Advances in Neural Information Processing Systems, pp. 469–477 (2016)
22. Long, M., Cao, Y., Wang, J., Jordan, M.I.: Learning transferable features with deep adaptation networks. In: Proceedings of the 32nd International Conference on International Conference on Machine Learning - Volume 37, pp. 97–105. ICML 2015 (2015)
23. Long, M., Cao, Z., Wang, J., Jordan, M.I.: Conditional adversarial domain adaptation. In: Advances in Neural Information Processing Systems (2018)
24. Maaten, L.V.D., Hinton, G.: Visualizing data using t-SNE. J. Mach. Learn. Res. **9**, 2579–2605 (2008)
25. Pan, S.J., Tsang, I.W., Kwok, J.T., Yang, Q.: Domain adaptation via transfer component analysis. IEEE Trans. Neural Netw. **22**(2), 199–210 (2011)
26. Pan, S.J., Yang, Q., et al.: A survey on transfer learning. IEEE Trans. Knowl. Data Eng. **22**(10), 1345–1359 (2010)
27. Ranzato, M., Boureau, Y.L., Chopra, S., LeCun, Y.: A unified energy-based framework for unsupervised learning. In: Artificial Intelligence and Statistics, pp. 371–379 (2007)

28. Saito, K., Ushiku, Y., Harada, T.: Asymmetric tri-training for unsupervised domain adaptation. In: Proceedings of the 34th International Conference on Machine Learning-Volume 70, pp. 2988–2997. JMLR. org (2017)
29. Saito, K., Watanabe, K., Ushiku, Y., Harada, T.: Maximum classifier discrepancy for unsupervised domain adaptation. In: Proceedings of the IEEE Conference on Computer Vision and Pattern Recognition, pp. 3723–3732 (2018)
30. Saito, K., Yamamoto, S., Ushiku, Y., Harada, T.: Open set domain adaptation by backpropagation. In: The European Conference on Computer Vision, ECCV, September 2018
31. Sankaranarayanan, S., Balaji, Y., Castillo, C.D., Chellappa, R.: Generate to adapt: aligning domains using generative adversarial networks. In: Proceedings of the IEEE Conference on Computer Vision and Pattern Recognition, pp. 8503–8512 (2018)
32. Shu, R., Bui, H.H., Narui, H., Ermon, S.: A dirt-t approach to unsupervised domain adaptation. In: Proceedings 6th International Conference on Learning Representations (2018)
33. Sun, B., Saenko, K.: Deep CORAL: correlation alignment for deep domain adaptation. In: Hua, G., Jégou, H. (eds.) ECCV 2016. LNCS, vol. 9915, pp. 443–450. Springer, Cham (2016). https://doi.org/10.1007/978-3-319-49409-8_35
34. Tzeng, E., Hoffman, J., Darrell, T., Saenko, K.: Simultaneous deep transfer across domains and tasks. In: Proceedings of the IEEE International Conference on Computer Vision, pp. 4068–4076 (2015)
35. Tzeng, E., Hoffman, J., Saenko, K., Darrell, T.: Adversarial discriminative domain adaptation. In: Computer Vision and Pattern Recognition (CVPR), vol. 1, p. 4 (2017)
36. Tzeng, E., Hoffman, J., Zhang, N., Saenko, K., Darrell, T.: Deep domain confusion: maximizing for domain invariance. CoRR abs/1412.3474 http://arxiv.org/abs/1412.3474 (2014)
37. Venkateswara, H., Eusebio, J., Chakraborty, S., Panchanathan, S.: Deep hashing network for unsupervised domain adaptation. In: Proceedings CVPR, pp. 5018–5027 (2017)
38. Vu, T.H., Jain, H., Bucher, M., Cord, M., Perez, P.: Advent: adversarial entropy minimization for domain adaptation in semantic segmentation. In: The IEEE Conference on Computer Vision and Pattern RecognitionCVPR, June 2019
39. Xie, S., Zheng, Z., Chen, L., Chen, C.: Learning semantic representations for unsupervised domain adaptation. In: Dy, J., Krause, A. (eds.) Proceedings of the 35th International Conference on Machine Learning. Proceedings of Machine Learning Research, Stockholmsmässan, Stockholm Sweden, 10–15 July 2018, vol. 80, pp. 5423–5432. PMLR (2018)
40. Xu, R., Li, G., Yang, J., Lin, L.: Larger norm more transferable: an adaptive feature norm approach for unsupervised domain adaptation. In: Proceedings of the IEEE International Conference on Computer Vision, pp. 1426–1435 (2019)
41. Yang, J., Zou, H., Cao, S., Chen, Z., Xie, L.: Mobileda: towards edge domain adaptation. IEEE Internet Things J. **7**, 6909–6918 (2020)
42. Zhao, J., Mathieu, M., LeCun, Y.: Energy-based generative adversarial network. In: Proceedings 5th International Conference on Learning Representations (2017)
43. Zhou, X., Karpur, A., Gan, C., Luo, L., Huang, Q.: Unsupervised domain adaptation for 3D keypoint estimation via view consistency. In: ECCV, pp. 137–153 (2018)

44. Zou, H., Yang, J., Zhou, Y., Xie, L., Spanos, C.J.: Robust wifi-enabled device-free gesture recognition via unsupervised adversarial domain adaptation. In: 2018 27th International Conference on Computer Communication and Networks (ICCCN), pp. 1–8. IEEE (2018)
45. Zou, H., Zhou, Y., Yang, J., Liu, H., Das, H.P., Spanos, C.J.: Consensus adversarial domain adaptation. In: Proceedings of the AAAI Conference on Artificial Intelligence, vol. 33, pp. 5997–6004 (2019)

SeqXY2SeqZ: Structure Learning for 3D Shapes by Sequentially Predicting 1D Occupancy Segments from 2D Coordinates

Zhizhong Han[1,2], Guanhui Qiao[1], Yu-Shen Liu[1(✉)], and Matthias Zwicker[2]

[1] School of Software, BNRist, Tsinghua University, Beijing, People's Republic of China
qiaogh18@mails.tsinghua.edu.cn, liuyushen@tsinghua.edu.cn

[2] Department of Computer Science, University of Maryland, College Park, USA
h312h@umd.edu, zwicker@cs.umd.edu

Abstract. Structure learning for 3D shapes is vital for 3D computer vision. State-of-the-art methods show promising results by representing shapes using implicit functions in 3D that are learned using discriminative neural networks. However, learning implicit functions requires dense and irregular sampling in 3D space, which also makes the sampling methods affect the accuracy of shape reconstruction during test. To avoid dense and irregular sampling in 3D, we propose to represent shapes using 2D functions, where the output of the function at each 2D location is a sequence of line segments inside the shape. Our approach leverages the power of functional representations, but without the disadvantage of 3D sampling. Specifically, we use a voxel tubelization to represent a voxel grid as a set of tubes along any one of the X, Y, or Z axes. Each tube can be indexed by its 2D coordinates on the plane spanned by the other two axes. We further simplify each tube into a sequence of occupancy segments. Each occupancy segment consists of successive voxels occupied by the shape, which leads to a simple representation of its 1D start and end location. Given the 2D coordinates of the tube and a shape feature as condition, this representation enables us to learn 3D shape structures by sequentially predicting the start and end locations of each occupancy segment in the tube. We implement this approach using a Seq2Seq model with attention, called SeqXY2SeqZ, which learns the mapping from a sequence of 2D coordinates along two arbitrary axes to a sequence of 1D locations along the third axis. SeqXY2SeqZ not only benefits from the regularity of voxel grids in training and testing,

This work was supported by National Key R&D Program of China (2020YFF0304100, 2018YFB0505400), the National Natural Science Foundation of China (62072268), and NSF (award 1813583).

Electronic supplementary material The online version of this chapter (https://doi.org/10.1007/978-3-030-58586-0_36) contains supplementary material, which is available to authorized users.

A. Vedaldi et al. (Eds.): ECCV 2020, LNCS 12369, pp. 607–625, 2020.
https://doi.org/10.1007/978-3-030-58586-0_36

but also achieves high memory efficiency. Our experiments show that SeqXY2SeqZ outperforms the state-of-the-art methods under the widely used benchmarks.

Keywords: 3D reconstruction · Voxel grids · Implicit function · RNN · Attention

1 Introduction

3D voxel grids are an attractive representation for 3D structure learning because they can represent shapes with arbitrary topology and they are well suited to convolutional neural network architectures. However, these advantages are dramatically diminished by the disadvantage of cubic storage and computation complexity, which significantly affects the structure learning efficiency and accuracy of deep learning models.

Recently, implicit functions have been drawing research attention as a promising 3D representation to resolve this issue. By representing a 3D shape as a function, discriminative neural networks can be trained to learn the mapping from a 3D location to a label, which can either indicate the inside or outside of the shape [4,42,51] or a signed distance to the surface [47,60]. As a consequence, shape reconstruction requires sampling the function in 3D, where the 3D locations are required to be sampled near the 3D surface for training. Recent approaches based on implicit functions have shown superiority over point clouds in terms of geometry details, and advantages over meshes in terms of being able to represent arbitrary topologies. Although it is very memory efficient to learn implicit functions using discriminative models, these approaches require sampling dense 3D locations in a highly irregular manner during training, which also makes the sampling methods affect the accuracy of shape reconstruction during test.

To resolve this issue, we propose a method for 3D shape structure learning by leveraging the advantages of learning shape representations based on continuous functions without requiring sampling in 3D. Rather than regarding a voxel grid as a set of individual 3D voxels, which suffers from cubic complexity in learning, we represent voxel grids as functions over a 2D domain that map 2D locations to 1D voxel tubes. This voxel tubelization regards a voxel grid as a set of tubes along any one of three dimensions, for example Z, and indexes each tube by its 2D location on the plane spanned by the other two dimensions, i.e., X and Y. In addition, each tube is represented as a sequence of occupancy segments, where each segment consists of successive occupied voxels given by two 1D locations indicating the start and end points. Given a shape feature as a condition, this voxel tubelization enables us to propose a Seq2Seq model with attention as a discriminative model to predict each tube from its 2D location. Specifically, we leverage an RNN encoder to encode the 2D coordinates of a tube with a shape condition, and leverage an RNN decoder to sequentially predict the start and end locations of each occupancy segment in the tube. Because our approach essentially maps a coordinate sequence to another coordinate sequence, we call our method *SeqXY2SeqZ*. Given the 2D coordinates of a tube, SeqXY2SeqZ produces the 1D coordinates of the occupancy segments along the third dimension. Not only can SeqXY2SeqZ be evaluated with a number of RNN steps that is quadratic in the grid resolution during test, but it is also memory efficient enough to learn high resolution shape representations. Experimental results show that SeqXY2SeqZ outperforms the state-of-the-art methods.

Our contributions are as follows. First, we propose a novel shape representation based on 2D functions that map 2D locations to sequences of 1D voxel tubes, avoiding the cubic complexity of voxel grids. Our representation enables 3D structure learning of voxel grids in a tube-by-tube manner via discriminative neural networks. Second, we propose SeqXY2SeqZ, an RNN-based Seq2Seq model with attention, to implement the mapping from 2D locations to 1D sequences. Given a 2D coordinate and a shape condition, SeqXY2SeqZ sequentially predicts occupancy segments in a 1D tube. It requires a number of RNN steps that grows only quadratically with resolution, and achieves high resolutions due to its memory efficiency. Third, SeqXY2SeqZ demonstrates the feasibility of generating 3D voxel grids using discriminative neural networks in a more efficient way, and achieves state-of-the-art results in shape reconstruction.

2 Related Work

Deep learning models have made big progress in 3D shape understanding tasks [13–20,22–25,27,28,40,41,49,62,63]. Recent 3D structure learning methods are also mainly based on deep learning models.

Voxel-Based Models. Because of their regularity, many previous studies learned 3D structures from voxel grids with 3D supervision [6,50] or 2D supervision with the help of differentiable renderers [8,9,56,57,66,67]. Due to the cubic complexity of voxel grids, these generative models are limited to relatively low resolution, such as 32^3. Recent studies [6,65,70] employed shallow 3D convolutional networks to reconstruct voxel grids in higher resolutions of 128^3, however, the computational cost is still very large. To remedy this issue, some methods employed a multi-resolution strategy [26,54]. However, these methods are very complicated to implement and additionally require multiple passes over the input. Another alternative was introduced to represent 3D shapes using multiple depth images [50]. However, it is hard to obtain consistency across multiple generated depth images during inference.

Different from these generative neural networks, we provide a novel perspective to benefit from the regularity of voxel grids but avoid their cubic complexity by leveraging discriminative neural networks in shape generation. Moreover, our representation is different from multi-layer depth maps [52] and scanline [64], since we do not require additional support, such as binary masks [52] or edge end point determination.

Point Cloud-Based Models. As pioneers, PointNet [48] and PointNet++ [49] enabled the learning of 3D structure from point clouds. Later, different variations were proposed to improve the learning of 3D structures from 3D point clouds [7,24,40] or 2D images with various differentiable renderers [11,29,32,33,44,68]. Although point clouds are a compact and memory efficient 3D representation, they cannot express geometry details without additional non-trivial post-processing steps to generate meshes.

Mesh-Based Models. 3D meshes are also attractive in deep learning [3,10,31,34,35,38,58,61]. Supervised methods employed 3D meshes as supervision to train networks by minimizing the location error of vertices with geometry constraints [10,58,61], while unsupervised methods relied on differentiable renderers to reconstruct meshes from multiple views [3,31,34,35,38]. However, these methods cannot generate arbitrary vertex topology but inherit the connectivity of the template mesh.

Implicit Function-Based Models. Recently, implicit functions have become a promising 3D representation in deep learning models [4,42,43,46,47,51,60]. By representing a 3D shape as a 3D function, these methods employ discriminative neural networks to learn the function from a 3D location to an indicator labelling inside or outside of the shape [4,42,51] or a signed distance to the surface [47,60]. However, these methods required to sample points near 3D surfaces during training. To learn implicit functions without 3D supervision, different differentiable renderers were proposed to back propagate the loss calculated on 2D images [12,30,36,39,45,53,69]. Although it is very memory efficient to learn 3D implicit functions using discriminative models in a point-by-point manner, supervised method require sampling dense and irregular 3D locations during training, which also makes the sampling methods affect the accuracy of shape reconstruction during test.

Although our method is also a discriminative network for 3D structure learning, it can benefit from the regularity of voxel grids by learning a 2D function. It is memory efficient and avoids the dense and irregular sampling during training.

3 Overview

The core idea of SeqXY2SeqZ is to represent shapes as 2D functions that map each 2D location to a sequence of 1D occupancy segments. More specifically, we interpret each 3D shape $\boldsymbol{M}$ as a set of 1D tubes $\boldsymbol{t}_i$, where each tube $\boldsymbol{t}_i$ is indexed by its 2D coordinate $\boldsymbol{c}_i$. Tube $\boldsymbol{t}_i$ consists of a sequence of occupancy segments, where we represent each segment o_j by its 1D start and end locations s_j and e_j. To generate $\boldsymbol{M}$, SeqXY2SeqZ learns a 2D function to predict each tube $\boldsymbol{t}_i$ from its coordinate $\boldsymbol{c}_i$ and a shape condition by generating the start and end locations s_j and e_j of each occupancy segment o_j in $\boldsymbol{t}_i$.

Figure 1 illustrates how SeqXY2SeqZ generates a tube along the Z axis from its 2D coordinates on the X-Y plane. Specifically, we input the 2D coordinate $X = 5$ and $Y = 5$ sequentially into an encoder, and a decoder sequentially predicts the start and end locations of two occupancy segments along the Z axis. In the figure, there is one occupancy segment with only one voxel starting at $Z = 1$ and ending at $Z = 1$, and a second segment starting at $Z = 3$ and ending at $Z = 5$. Therefore, the decoder sequentially predicts $Z = 1$, $Z = 1$, $Z = 3$, $Z = 5$ to reconstruct the tube at $X = 5$ and $Y = 5$. In addition, the decoder outputs a binary flag to indicate whether there is any occupancy segment in this tube at all. The encoder also requires a shape condition from an image or a learned feature as input to provide information about the reconstructed shape.

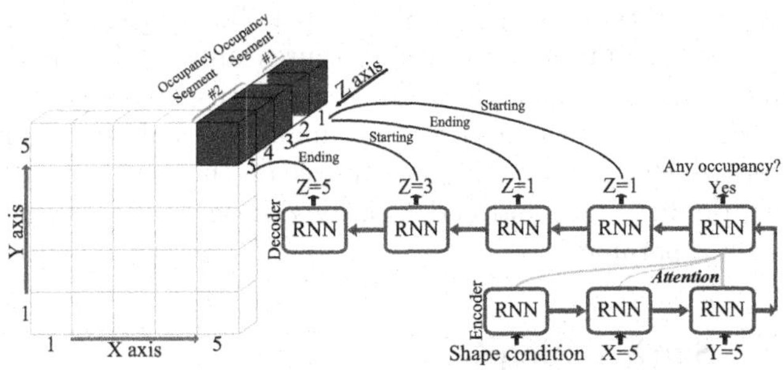

Fig. 1. The overview of SeqXY2SeqZ.

4 Voxel Tubelization

To train the SeqXY2SeqZ model, we first need to convert each 3D voxel grid into a tubelized representation consisting of sets of 1D voxel tubes over a 2D plane. For a 3D shape $\boldsymbol{M}$ represented by a grid with a resolution of R^3, voxel tubelization re-organizes these R^3 voxels into a set of $R \times R$ tubes $\boldsymbol{t}_i$ along one of the three axes. Each tube $\boldsymbol{t}_i$ can then be indexed by its location on the plane spanned by the other two dimensions using a 2D coordinate $\boldsymbol{c}_i$, such that $\boldsymbol{M} = \{(\boldsymbol{c}_i, \boldsymbol{t}_i) | i \in [1, R^2]\}$. We further represent each tube $\boldsymbol{t}_i$ using run-length encoding of its J_i occupancy segments o_j, where $j \in [1, J_i]$ and $J_i \in [1, R]$. An occupancy segment is a set of consecutive voxels that are occupied by the shape, which we encode as a sequence of start and end locations s_j and e_j. Note that s_j and e_j are discrete 1D indices, which we will predict using a discriminative approach. We denote the tubes consisting of occupancy segments as $\boldsymbol{t}_i = [s_1, e_1, ..., s_j, e_j, ..., s_{J_i}, e_{J_i}]$. In our experimental section we show that this representation is effective irrespective of the axis that is leveraged for the tubelization.

Our approach takes advantage of the following properties of voxel tubelization and run-length encoding of occupancy segments. First, run-length encoding of occupancy segments significantly reduces the memory complexity of 3D grids, since only two indices are needed to encode each segment, irrespective of its length. Second, our approach allows us to represent shapes as 2D functions that map 2D locations to sequences of 1D occupancy segments, which we will implement using discriminative neural networks. This is similar to shape representations based on 3D implicit functions implemented by discriminative networks, but our approach requires only $\mathcal{O}(R^2)$ RNN evaluation steps during shape reconstruction. Third, networks that predict voxel occupancy using a scalar probability require an occupancy probability threshold as a hyperparameter, which can have a large influence on the reconstruction accuracy. In contrast, we predict start and end locations of occupancy segments and do not require such a parameter.

5 SeqXY2SeqZ

SeqXY2SeqZ generates each tube $\boldsymbol{t}_i$ from its coordinate $\boldsymbol{c}_i$ and a shape condition. We use an RNN encoder to encode the coordinate $\boldsymbol{c}_i$ and the shape condition, while an RNN decoder produces the start and end locations of the occupancy segments o_j in $\boldsymbol{t}_i$.

RNN Encoder. We condition the RNN encoder on a global shape feature $\boldsymbol{f} \in \mathbb{R}^{1\times D}$ that represents the unique 3D structure of each object. For example, in 3D shape reconstruction from a single image, $\boldsymbol{f}$ could be a feature vector extracted from an image to guide the 3D shape reconstruction. In a 3D shape to 3D shape translation application, $\boldsymbol{f}$ could be a feature vector that can be jointly learned with other parameters in the networks, such as shape memories [21] or codes [47].

As shown in Fig. 2(a), the RNN encoder aggregates the shape condition $\boldsymbol{f}$ and a 2D coordinate $\boldsymbol{c}_i = [c_i^1, c_i^2]$ into a hidden state $\boldsymbol{h}_i$, which is subsequently leveraged by the RNN decoder to generate the corresponding tube $\boldsymbol{t}_i$. Rather than directly employing a location c_i^1 or c_i^2 as a discrete integer, we leverage the location as a location embedding $\boldsymbol{x}_i^1$ or $\boldsymbol{x}_i^2$, which makes locations meaningful in feature space. In this way, we have a location embedding matrix along each axis, i.e., $\mathbf{F}_X$, $\mathbf{F}_Y$ and $\mathbf{F}_Z$. Each matrix holds the location embedding of all R locations along an axis as R rows, i.e., $\mathbf{F}_X \in \mathbb{R}^{R\times D}$, $\mathbf{F}_Y \in \mathbb{R}^{R\times D}$ and $\mathbf{F}_Z \in \mathbb{R}^{R\times D}$, so that we can get an embedding for a specific location by looking up the location embedding matrix. In the case of tubelizing along the Z axis demonstrated in Fig. 1, the RNN encoder would employ the location embeddings along the X and Y axes, that is $\boldsymbol{x}_i^1 = \mathbf{F}_X(c_i^1)$ and $\boldsymbol{x}_i^2 = \mathbf{F}_Y(c_i^2)$.

We employ Gated Recurrent Units (GRU) [5] as the RNN cells in SeqXY2SeqZ. At each step, a hidden state is produced, and the hidden state $\boldsymbol{h}_i$ at the last step is leveraged by the RNN decoder to predict a tube $\boldsymbol{t}_i$ for the reconstruction of a shape conditioned on $\boldsymbol{f}$, where $\boldsymbol{h}_i \in \mathbb{R}^{1\times H}$. The encoding process is detailed in our supplemental material.

Location Embedding. Although we could employ three different location embedding matrices to hold embeddings for locations along the X, Y, and Z axes separately, we use $\mathbf{F}_X$, $\mathbf{F}_Y$ and $\mathbf{F}_Z$ in a shareable manner. For example, we can employ the same location embedding matrix on the plane used for indexing the 1D tubes, such as $\mathbf{F}_X = \mathbf{F}_Y$ in the case shown in Fig. 1. In our experiments, we justify that we can even employ only one location embedding matrix along all three axes, that is $\mathbf{F}_X = \mathbf{F}_Y = \mathbf{F}_Z$. The shareable location embeddings significantly increase the memory efficiency of SeqXY2SeqZ.

RNN Decoder. With the hidden state $\boldsymbol{h}_i$ from the RNN encoder, the RNN decoder needs to generate a tube $\boldsymbol{t}_i$ for the shape indicated by condition $\boldsymbol{f}$ via sequentially predicting the start and end locations of each occupancy segment o_j. To interpret the prediction of tubes with no occupancy segments, we include an additional global occupancy indicator b that the decoder predicts first, where $b = 1$ indicates that there are occupancy segments in the current tube.

We denote $\boldsymbol{w}_i$ as the concatenation of b and $\boldsymbol{t}_i$, such that $\boldsymbol{w}_i = [b, s_1, e_1, ..., s_{J_i}, e_{J_i}]$, where each element in $\boldsymbol{w}_i$ is uniformly denoted as w_i^k and $k \in [1, 2 \times J_i + 1]$. Note that the start and end points s_j and e_j are discrete voxel locations, which we interpret as class labels. In each step, the RNN decoder selects a discrete label to determine either start or end location. Therefore, we leverage the following cross entropy classification loss to push the decoder to predict the correct label sequence $\boldsymbol{w}_i$ as accurately as possible under the training set,

$$L = - \sum_{k \in [1, 2 \times J_i + 1]} \log p(w_i^k | w_i^{<k}, \boldsymbol{h}_i), \tag{1}$$

where w_i^k is the k-th element in the sequence $\boldsymbol{w}_i$, $w_i^{<k}$ represents the elements in front of w_i^k, $p(w_i^k | w_i^{<k}, \boldsymbol{h}_i)$ is the probability of correctly predicting the k-th element according to the previous elements $w_i^{<k}$ and the hidden state $\boldsymbol{h}_i$ from the encoder. Finally, our objective function is given as

$$\mathbf{F}_X^*, \mathbf{F}_Y^*, \mathbf{F}_Z^*, \boldsymbol{\theta}^*, \boldsymbol{f}^* = \underset{\mathbf{F}_X, \mathbf{F}_Y, \mathbf{F}_Z, \boldsymbol{\theta}, \boldsymbol{f}}{\arg\min} \quad L, \tag{2}$$

where $\boldsymbol{\theta}$ denotes the parameters of the RNN encoder and decoder, $\boldsymbol{f}$ is the shape condition, which is fixed or trainable depending on the application, and the location embedding matrices $\mathbf{F}_X, \mathbf{F}_Y, \mathbf{F}_Z$ could be shareable.

Training progress in a step by step manner is shown in Fig. 2(b). At the k-th step, element w_i^k in sequence $\boldsymbol{w}_i$ is predicted through a softmax layer. For example, w_i^1 is either true or false for the global occupancy indicator b, and w_i^2 and w_i^3 are the start and end locations s_1 and e_1 of the occupancy segment o_1 in the range of $[1, R]$, etc. In addition, for each w_i^k we look up its location embedding $\boldsymbol{y}_i^k$ from the location embedding matrix of the coordinate axis corresponding to the tube direction. The embedding $\boldsymbol{y}_i^k$ is then used in the prediction of w_i^{k+1} at the next step. For example, in the tubelization along the Z axis demonstrated in Fig. 1, $\boldsymbol{y}_i^k$ is looked up in $\boldsymbol{F}_Z$, such that $\boldsymbol{y}_i^k = \boldsymbol{F}_Z(w_i^k)$, where each row of $\boldsymbol{F}_Z$ represents an embedding for a location, and two additional rows for a true or false of b.

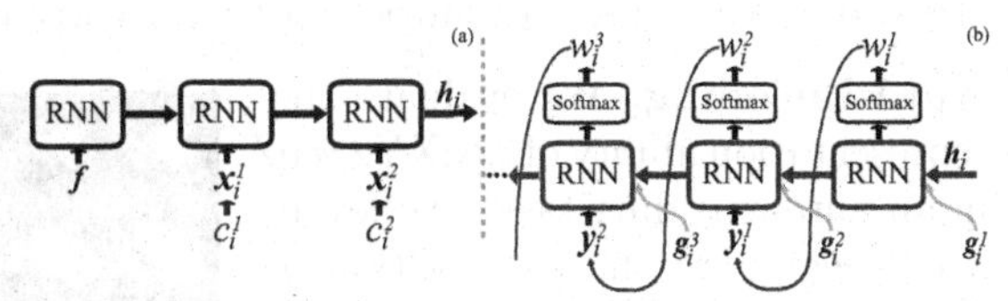

Fig. 2. The illustration of RNN encoder (a) and RNN decoder (b).

Attention. Finally, we leverage a state-of-the-art attention mechanism [1] to increase the prediction accuracy of the predicted locations. We employ a context vector $\boldsymbol{g}_i^k$ for the prediction of w_i^k, where $\boldsymbol{g}_i^k$ summarizes how well each step of the encoder matches the prediction of w_i^k. The decoding with attention is detailed in supplemental material.

6 Experiments and Analysis

We employ tubelization along the Y axis in all our experiments and learn only two location embedding matrices. We share the location embedding matrices along the X and Z axes providing the 2D coordinates of tubes, such that $\mathbf{F}_X = \mathbf{F}_Z$, while we use a separate matrix along the Y axis. The location embedding is $D = 512$-dimensional, and the hidden state of the RNNs is also $H = 512$-dimensional, where the RNN encoder is bidirectional.

We train SeqXY2SeqZ using the Adam optimizer with $\epsilon = 8 \times 10^{-6}$, with a batch size of 64 and a learning rate of 1×10^{-3} in all experiments. The maximum number of steps in the encoder and decoder are 4 and 30, respectively. We employ volumetric IoU to evaluate the accuracy of the reconstructed shapes, and all reported IoU values are multiplied by 10^2.

Table 1. Reconstruction (64^3) comparison in terms of IoU.

Methods	Plane	Car	Chair	Rifle	Table
IM-AE [4]	78.77	89.36	65.65	72.88	71.44
CNN-AE [4]	86.07	90.73	74.22	78.37	84.67
OccNet(Train) [42]	–	–	89.00	–	–
Our(512–512)	**90.35**	**91.18**	74.32	**84.46**	**86.21**
Our(1024–2048)	–	–	**93.10**	–	–

6.1 Representation Ability

Dataset. For fair comparison, we leverage five widely used categories from ShapeNetCore [2] in this subsection, including airplane, car, chair, rifle, and table, and keep the same train and test splitting as [4]. The ground truth shapes are also voxelized at a resolution of 64^3, such that $R = 64$.

Auto-Encoding. We evaluate the representation ability of SeqXY2SeqZ in an auto-encoding task. We leverage a learnable shape condition $\boldsymbol{f}$ to represent each shape. Specifically, shape features $\boldsymbol{f}$ are learned together with the other parameters in the RNN during training. During testing, we keep updating the shape features while fixing the parameters in the RNN including the location embedding matrices, which is similar as introduced by shape memories [21] or codes [47]. Note that $\boldsymbol{f}$ are also $D = 512$-dimensional vectors, similar as the location embeddings.

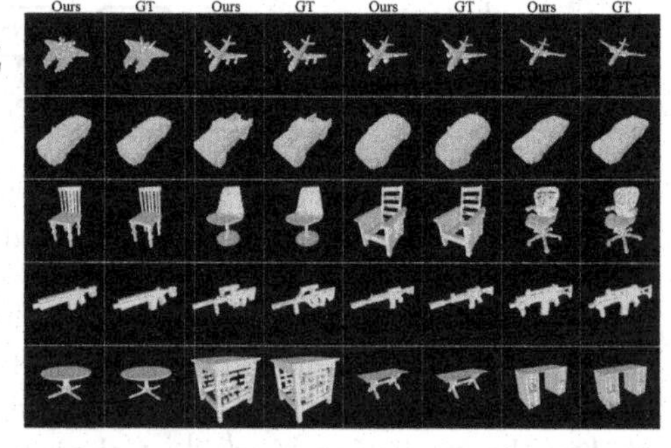

Fig. 3. Auto-decoded shapes by learned features.

In this task, we compare SeqXY2SeqZ with results from the implicit decoder (IM) [4] and occupancy network (OccNet) [42]. We show the comparison in Table 1, where the mean IoU over the first 100 shapes in the test set of each category is reported by IM while OccNet only reported its results on the training set of chair at a resolution of 256.

As shown by "Our(512-512)" in Table 1, our results with $D = 512$-dimensional location embeddings and $H = 512$-dimensional hidden states are the best among all compared methods under all shape categories. If we increase the learning ability of SeqXY2SeqZ by using location embeddings and hidden states with higher dimensions, such as $D = 2048$ and $H = 1024$ shown by "Our(2048-1024)", we achieve an even higher IoU of 93.10 under the challenging chair class.

Table 2. Tubelization direction comparison.

	Along Y	Along Z	Along X
IoU	**90.35**	89.96	90.21

In Fig. 3, we visualize the reconstructed shapes in the test set of each category with our best results in Table. 1. The reconstructed shapes with high fidelity demonstrate that SeqXY2SeqZ is capable of learning very complex structures of 3D shapes, such as the ones on chairs and tables.

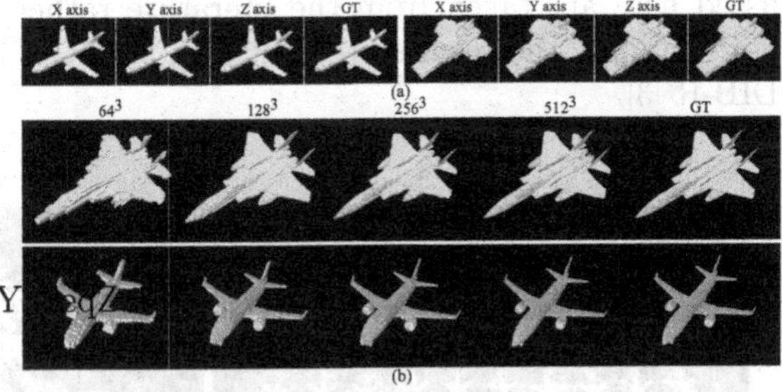

Fig. 4. (a) Qualitative comparison of reconstructions with tubelization along different axes. (b) Auto-encoded shapes in different resolutions.

Tubelization Direction. We can tubelize a voxel grid along any one of the X, Y or Z axes, which should be kept consistent in training and testing. Although the tubelization direction may lead to different ways of 3D structure learning, SeqXY2SeqZ does not exhibit any bias on the tubelization direction. We demonstrate this by training SeqXY2SeqZ using voxel grids tubelized under the X, Y and Z axis, respectively. Table 2 shows that we achieve comparable results along the three tubelization directions under the airplane class. Visual comparisons are shown in Fig. 4(a).

High Resolutions. Thanks to the 2D functions and the shareable location embedding matrices, SeqXY2SeqZ is memory efficient enough to reconstruct shapes in high resolutions. We show auto-encoded airplanes in different resolutions in Fig. 4(b). The high fidelity shapes justify our capabilities of high resolution reconstruction.

6.2 Single Image 3D Reconstruction

Dataset. We employ the dataset released from [6], which contains 3D shapes from 13 categories in the ShapeNetCore [2]. We also use the same train and test splitting, where each shape is represented as a voxel grid with a resolution of 32^3 accompanying 24 rendered images. While many 3D reconstruction techniques (including ours, see Table 1 and Fig. 4) support higher resolutions, we follow previous works [37,50,55] and choose ground truth voxel grids in the benchmark to provide a comparison to a broad range of competing approaches.

Single Image Reconstruction. We leverage a CNN encoder from [38] to extract a 512 dimensional feature from a rendered image as a shape condition in this experiment. We compare with the state-of-the-art supervised and unsupervised methods in Table 3. Among these methods, "DISN-V" is a network formed by a DISN [60] encoder and a 3D CNN decoder, "DISN-C" is DISN [60] working with the estimated camera poses which is required in the reconstruction, "PTN-R" is the result using retrieval from PTN [67]. Besides the voxel-based methods including R2N2 [6], PTN [67] and Matryoshka [50], all the other methods represent 3D shapes as triangle meshes, where IM [4], OccNet [42], and DISN [60] are based on learning 3D implicit functions. For fair comparison, all the results listed here are taken from the literature rather than being reproduced by us. For example, the results of NMR [31], SoftRas [37] and DIB-R [3] are all from DIB-R [3].

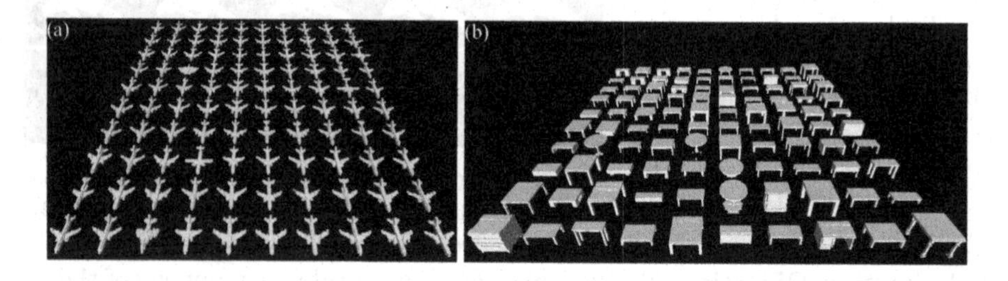

Fig. 5. Single image reconstruction for airplanes (a) and tables (b).

Table 3 demonstrates the performance of our method, showing that in terms of the mean IoU we improve by 6.3 over the best 3D implicit function based method (DISN) and by 2.1 over the best unsupervised method (DIB-R). We achieve the best IoU in 7 out of 13 categories among all supervised methods, and in 8 out of 13 categories among all unsupervised methods. Matryoshka [50] comes closest to our performance, but it employs non-standard augmentation on training images, which we omit. Figure 6 shows a visual comparison, where the shapes are reconstructed from the same input images using the trained network parameters released by different methods. Although we trained our method at a resolution of 32^3, the high accuracy enables us to reveal complex geometry that other methods cannot handle, which makes our results comparable to the meshes reconstructed by other methods. Figure 5 shows additional airplanes and tables reconstructed by our method.

Table 3. Quantitative comparison of single image 3D shape reconstruction in terms of IoU.

	Method	Modality	Plane	Bench	Cabinet	Car	Chair	Display	Lamp	Speaker	Rifle	Sofa	Table	Phone	Boat	Mean
Supervised	AtlasNet [10]	Mesh	39.2	34.2	20.7	22.0	25.7	36.4	21.3	23.2	45.3	27.9	23.3	42.5	28.1	30.0
	Pixel2mesh [59]		51.5	40.7	43.4	50.1	40.2	55.9	29.1	52.3	50.9	60.0	31.2	69.4	40.1	47.3
	3DN [60]		54.3	39.8	49.4	59.4	34.4	47.2	35.4	45.3	57.6	60.7	31.3	71.4	46.4	48.7
	R2N2 [6]	Voxel	51.3	42.1	71.6	79.8	46.6	46.8	38.1	66.2	54.4	62.8	51.3	66.1	51.3	56.0
	Matryoshka [51]		64.7	57.7	**77.6**	**85.0**	**54.7**	53.2	40.8	**70.1**	61.6	68.1	**57.3**	75.6	**59.1**	63.5
	IM [4]	3D Implicit	55.4	49.5	51.5	74.5	52.2	56.2	29.6	52.6	52.3	64.1	45.0	70.9	56.6	54.6
	OccNet [43]		54.7	45.2	73.2	73.1	50.2	47.9	37.0	65.3	45.8	67.1	50.6	70.9	52.1	56.4
	DISN-V [61]		50.6	44.3	52.3	76.9	52.6	51.5	36.2	58.0	50.5	67.2	50.3	70.9	57.4	55.3
	DISN-C [61]		57.5	52.9	52.3	74.3	54.3	56.4	34.7	54.9	59.2	65.9	47.9	72.9	55.9	57.0
	Ours	2D Implicit	**73.2**	**58.5**	71.0	78.1	50.3	**60.0**	**44.7**	62.2	**66.7**	**68.4**	55.0	**80.2**	58.4	**63.6**
Unsupervised	NMR [31]	Mesh	58.5	45.7	74.1	71.3	41.4	55.5	36.7	67.4	55.7	60.2	39.1	76.2	59.4	57.0
	SoftRas [37]		58.4	44.9	73.6	77.1	49.7	54.7	39.1	68.4	62.0	63.6	45.3	75.5	58.9	59.3
	DIB-R [3]		57.0	49.8	**76.3**	**78.8**	52.7	58.8	40.3	**72.6**	56.1	67.7	50.8	74.3	**60.9**	61.2
	PTN-R [68]	Voxel	55.6	48.8	57.1	65.2	35.1	39.6	29.1	46.0	51.3	53.1	31.0	67.0	40.8	47.7
	PTN [68]		55.6	49.2	68.2	71.2	44.9	54.0	42.2	58.7	59.9	62.2	49.4	75.0	55.1	57.4
	IMRender [40]	3D Implicit	65.1	53.6	-	78.2	**54.8**	-	-	-	-	-	51.5	-	60.8	60.7
	Ours	2D Implicit	**73.2**	**58.5**	71.0	78.1	50.3	**60.0**	**44.7**	62.2	**66.7**	**68.4**	**55.0**	**80.2**	58.4	**63.6**

6.3 Ablation Studies and Analysis

Ablation studies. We highlight some elements in our method by ablation studies in single image reconstruction under the chair class in Table 4. We compare our result with the ones without attention ("NoAtt"), the ones with LSTM RNN cells ("LSTM"), and the ones with single direction RNN encoder ("SingleDir"). We find that GRU performs better than LSTM, and both attention mechanism and bidirectional RNN encoder contribute to the performance.

Table 4. Ablation studies under chair class.

	NoAtt	LSTM	SingleDir	ShareableXYZ	Our(GRU)
IoU	47.5	49.8	48.8	49.1	**50.3**

Shareable Location Embedding Matrix. The memory efficiency is one advantage of SeqXY2SeqZ. We achieve this not only by avoiding the direct involvement of 3D voxel grids, but also by sharing the location embedding matrices. The above experiments have shown the effectiveness of shared location embedding matrices for the X and Z axes to define the plane indexing the tubes. In this experiment, we step further by employing only one location embedding matrix for all three axes. We also tubelize the voxel grids along the Y axis, and train SeqXY2SeqZ under the chair class in sigle image 3D reconstruction. In Table 4, "ShareableXYZ" still achieves the comparable result with "Our(GRU)".

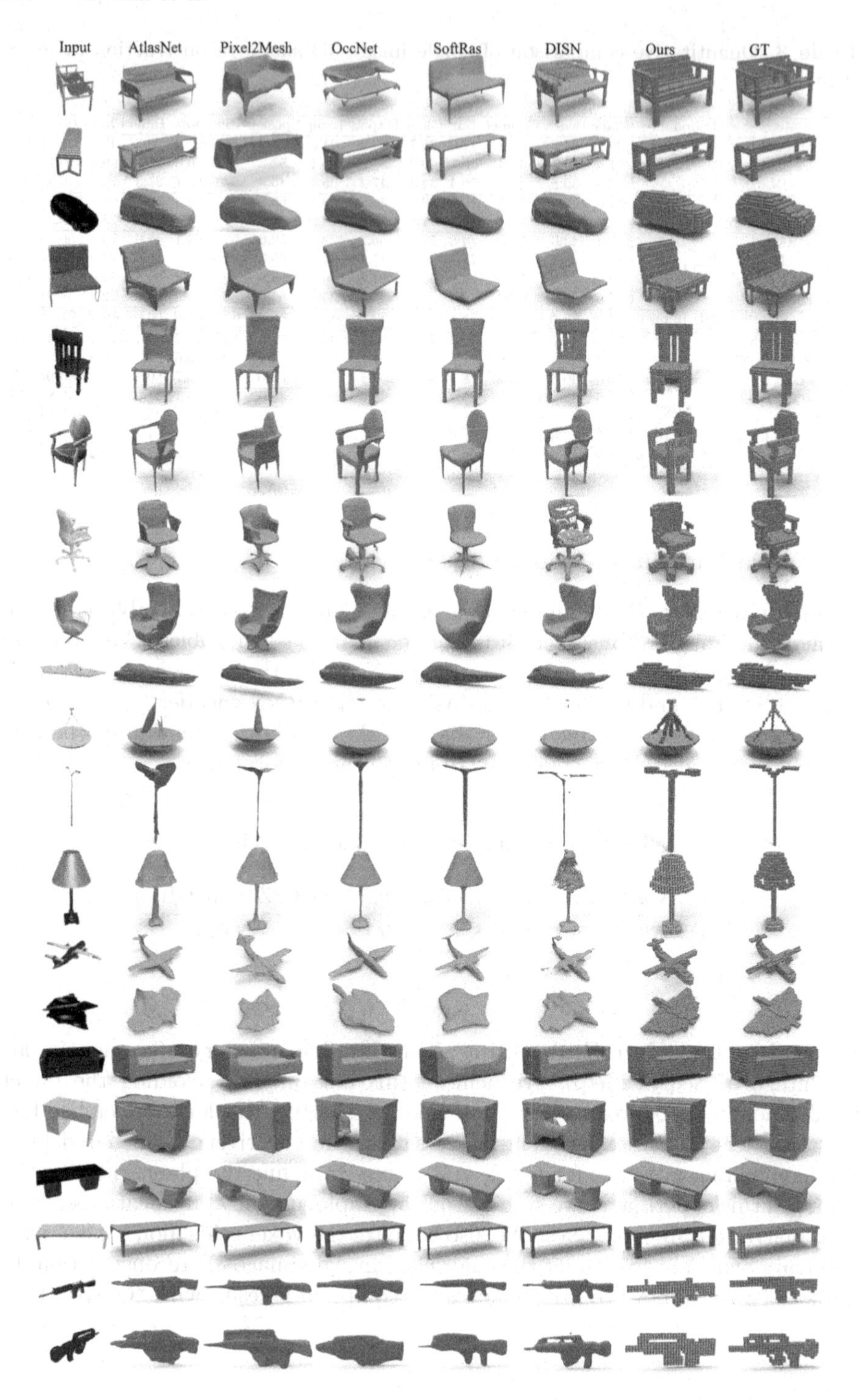

Fig. 6. Qualitative comparison with the state-of-the-art supervised and unsupervised methods.

Location Embedding Visualization. We visualize the location embeddings learned in auto-encoding of Table 1 in Fig. 7(a), where each class leverages two sets of location embeddings including one shared by the X and Z axes, and the other along the Y axis. We visualize each set of location embeddings using a cosine distance matrix whose element is the pairwise cosine distance between arbitrary two location embeddings. The structure of a shape category is demonstrated by the distinctive patterns on the cosine distance matrix in different shape categories, which demonstrates the effectiveness of the learned location embeddings. In each similarity matrix, blue means more similar between two location embeddings while yellow means more different. The similarity indicates whether the two corresponding locations show similar occupancy surrounding. For a class containing shapes with similar structures, like cars, the patterns are more obvious, while a class containing shapes with large structure variations, like chairs, the patterns are less obvious. In addition, we visualize the location embeddings learned in single image reconstruction of Table 3 in Fig. 7(b), where we also observe the different patterns on the cosine distance matrix in different shape categories. Note that we show the 64^2 dimensional distance matrix in Fig. 7(a) and the 32^2 dimensional distance matrix in Fig. 7(b) in the same size.

Attention Visualization. We further visualize the attention learned in auto-encoding of Table 1. At each 2D coordinate, an attention vector $\boldsymbol{a}$ is learned at each decoder step for all encoder steps. For each decoder step, we leverage entropy $(-\boldsymbol{a}.*log_2\boldsymbol{a})$ to visualize $\boldsymbol{a}$ at all 2D coordinates (if there is no output at this decoder step, we encode -1 at this 2D coordinate) into an attention image, and we normalize the whole attention image using the maximal entropy. We show five attention images at the first five decoder steps for each shape in Fig. 8(a). In each image, the higher entropy (above 0, the lighter color) indicates this decoder step is paying attention more equally on all encoder steps to generate more complex structure, such as chairs, while the lower entropy (above 0, the darker color) indicates this decoder step is focusing on a specific encoder step to generate relatively simple structure, such as cars. Similarly, we visualize the attention learned in single image reconstruction of Table 3 in Fig. 8(b), where the chair can be reconstructed by only one occupancy segment at all 2D coordinates, which makes the attention much simpler than the one for the chair in Fig. 8(a). Note that we show the 64^2 dimensional attention images in Fig. 8(a) and the 32^2 dimensional attention images in Fig. 8(b) in the same size.

Memory and Computation Time. We compare the memory and computation time requirements with methods based on learning 3D implicit functions in Table 5, including OccNet [42] and DISN [60]. To reconstruct a 3D shape at a resolution of R^3 from a single image during test, OccNet [42] requires to get occupancy values for about $3.8 * R^3$ sampled points with sub additional steps of subdivision, while DISN requires to get SDF values for R^3 sampled points, both of which are higher complexity than our $\mathcal{O}(R^2)$ RNN steps. Since DISN cannot run on a single GPU as OccNet and SeqXY2SeqZ, we report a fair comparison in terms of the CPU run time and RAM space with $R = 64$ and $sub = 2$

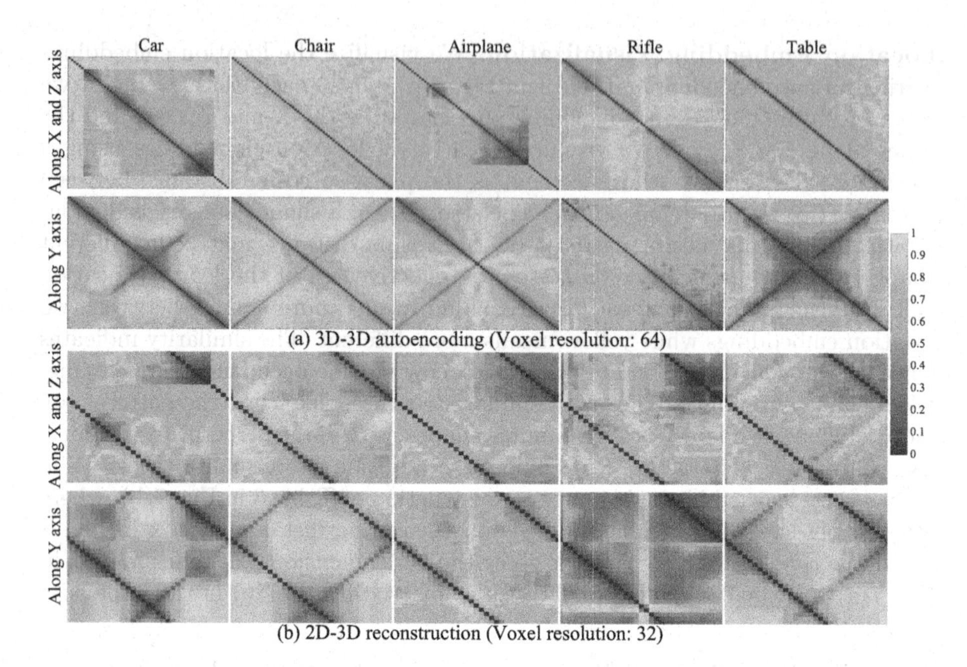

Fig. 7. Pairwise cosine distances of location embeddings learned in auto-encoding (a) and single image reconstruction (b). In each similarity matrix, blue means two locations indicated by their embeddings show similar occupancy surrounding while yellow means more different.

Table 5. Complexity comparison with 3D implicit functions.

	OccNet [42]	DISN [60]	Ours
Network evaluations	$\mathcal{O}(3.8 * R^2)$	$\mathcal{O}(R^3)$	$\mathcal{O}(R^2)$
Time (CPU)	55.80 s	14.68 s	**8.79** s
Space	1175 MB	>11 GB	**286** MB

for reconstructing one shape from a single image. Benefiting from learning 2D functions that predict sparse representations of 1D voxel tubes, SeqXY2SeqZ achieves both the lowest time and memory requirements by a large margin.

More Analysis. More analysis on the efficiency of our voxel tubelization and the feature space learned by our method can be found in our supplemental material.

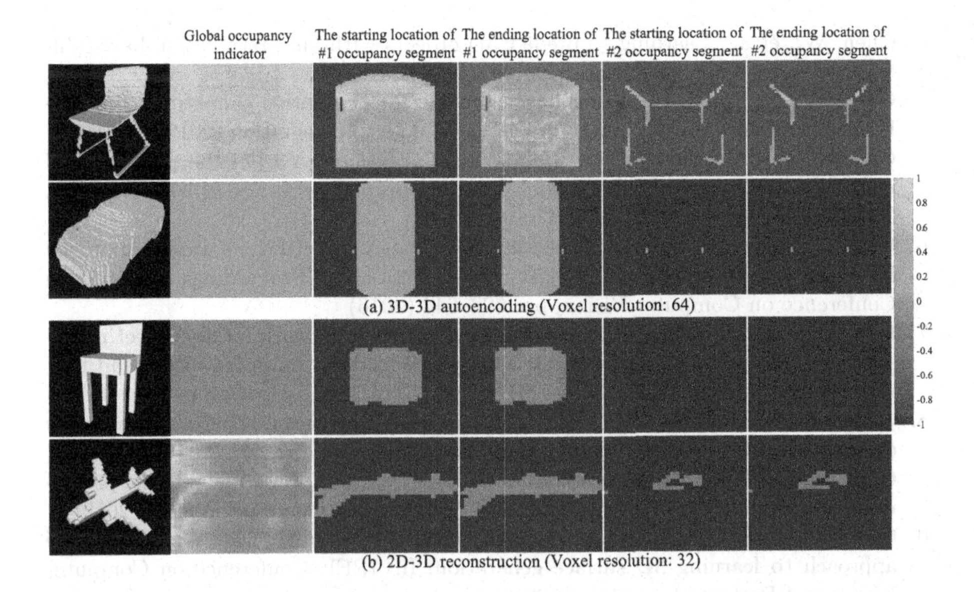

Fig. 8. The visualization of attention learned in auto-encoding. We visualize the attention weights learned at the first five steps of the decoder. The attention at each step for all 2D coordinates is shown as an image, where attention weights on the encoder at each 2D coordinate are encoded as entropy shown by color.

7 Conclusion

We propose SeqXY2SeqZ to learn the structure of 3D shapes using a discriminative neural network not only benefiting from the regularity inherent in voxel grids during both training and testing, but also avoiding cubic complexity for high memory efficiency. SeqXY2SeqZ successfully resolves the issue of dense and irregular sampling during structure learning or inference required by 3D implicit function-based methods, which leads to higher inference times compared to our approach. This is achieved based on the encoding of voxel grids by our 1D voxel tubelization, which effectively represents a voxel grid as a mapping from discrete 2D coordinates to sequences of discrete 1D locations. This mapping further enables SeqXY2SeqZ to effectively learn the 3D structures as 2D functions. We demonstrate that SeqXY2SeqZ outperforms the state-of-the-art methods under widely used benchmarks.

References

1. Bahdanau, D., Cho, K., Bengio, Y.: Neural machine translation by jointly learning to align and translate. CoRR abs/1409.0473 (2014)
2. Chang, A.X., et al.: ShapeNet: an information-rich 3D model repository. CoRR abs/1512.03012 (2015)

3. Chen, W., et al.: Learning to predict 3D objects with an interpolation-based differentiable renderer. CoRR abs/1908.01210 (2019)
4. Chen, Z., Zhang, H.: Learning implicit fields for generative shape modeling. In: IEEE Conference on Computer Vision and Pattern Recognition (2019)
5. Cho, K., van Merrienboer, B., Bahdanau, D., Bengio, Y.: On the properties of neural machine translation: encoder-decoder approaches. In: SSST@EMNLP, pp. 103–111 (2014)
6. Choy, C.B., Xu, D., Gwak, J., Chen, K., Savarese, S.: 3D–R2N2: a unified approach for single and multi-view 3D object reconstruction. In: Proceedings of European Conference on Computer Vision, pp. 628–644 (2016)
7. Fan, H., Su, H., Guibas, L.J.: A point set generation network for 3D object reconstruction from a single image. In: IEEE Conference on Computer Vision and Pattern Recognition, pp. 2463–2471 (2017)
8. Gadelha, M., Maji, S., Wang, R.: 3D shape induction from 2D views of multiple objects. In: International Conference on 3D Vision, pp. 402–411 (2017)
9. Gadelha, M., Wang, R., Maji, S.: Shape reconstruction using differentiable projections and deep priors. In: International Conference on Computer Vision (2019)
10. Groueix, T., Fisher, M., Kim, V.G., Russell, B.C., Aubry, M.: A papier-mâché approach to learning 3D surface generation. In: IEEE Conference on Computer Vision and Pattern Recognition (2018)
11. Han, Z., Chen, C., Liu, Y.S., Zwicker, M.: DRWR: a differentiable renderer without rendering for unsupervised 3D structure learning from silhouette images. In: ICML (2020)
12. Han, Z., Chen, C., Liu, Y.S., Zwicker, M.: DRWR: a differentiable renderer without rendering for unsupervised 3D structure learning from silhouette images. In: International Conference on Machine Learning (2020)
13. Han, Z., Chen, C., Liu, Y.S., Zwicker, M.: ShapeCaptioner: generative caption network for 3D shapes by learning a mapping from parts detected in multiple views to sentences. In: ACM International Conference on Multimedia (2020)
14. Han, Z., Liu, X., Liu, Y.S., Zwicker, M.: Parts4Feature: learning 3D global features from generally semantic parts in multiple views. In: IJCAI (2019)
15. Han, Z., Liu, Z., Han, J., Vong, C.M., Bu, S., Chen, C.: Unsupervised learning of 3D local features from raw voxels based on a novel permutation voxelization strategy. IEEE Trans. Cybern. **49**(2), 481–494 (2019)
16. Han, Z., Liu, Z., Han, J., Vong, C.M., Bu, S., Chen, C.: Mesh convolutional restricted Boltzmann machines for unsupervised learning of features with structure preservation on 3D meshes. IEEE Trans. Neural Netw. Learn. Syst. **28**(10), 2268–2281 (2017)
17. Han, Z., Liu, Z., Han, J., Vong, C.M., Bu, S., Li, X.: Unsupervised 3D local feature learning by circle convolutional restricted Boltzmann machine. IEEE Trans. Image Process. **25**(11), 5331–5344 (2016)
18. Han, Z., et al.: Deep spatiality: unsupervised learning of spatially-enhanced global and local 3D features by deep neural network with coupled softmax. IEEE Trans. Image Process. **27**(6), 3049–3063 (2018)
19. Han, Z., et al.: BoSCC: bag of spatial context correlations for spatially enhanced 3D shape representation. IEEE Trans. Image Process. **26**(8), 3707–3720 (2017)
20. Han, Z., et al.: 3D2SeqViews: aggregating sequential views for 3D global feature learning by CNN with hierarchical attention aggregation. IEEE Trans. Image Process. **28**(8), 3986–3999 (2019)

21. Han, Z., Shang, M., Liu, Y.S., Zwicker, M.: View inter-prediction GAN: unsupervised representation learning for 3D shapes by learning global shape memories to support local view predictions. In: AAAI, pp. 8376–8384 (2019)
22. Han, Z., et al.: SeqViews2SeqLabels: learning 3D global features via aggregating sequential views by rnn with attention. IEEE Trans. Image Process. **28**(2), 658–672 (2019)
23. Han, Z., Shang, M., Wang, X., Liu, Y.S., Zwicker, M.: Y2Seq2Seq: cross-modal representation learning for 3D shape and text by joint reconstruction and prediction of view and word sequences. In: AAAI, pp. 126–133 (2019)
24. Han, Z., Wang, X., Liu, Y.S., Zwicker, M.: Multi-angle point cloud-vae: unsupervised feature learning for 3D point clouds from multiple angles by joint self-reconstruction and half-to-half prediction. In: ICCV (2019)
25. Han, Z., Wang, X., Vong, C.M., Liu, Y.S., Zwicker, M., Chen, C.P.: 3DViewGraph: learning global features for 3D shapes from a graph of unordered views with attention. In: IJCAI (2019)
26. Hane, C., Tulsiani, S., Malik, J.: Hierarchical surface prediction for 3D object reconstruction. In: International Conference on 3D Vision, pp. 412–420 (2017)
27. Hu, T., Han, Z., Shrivastava, A., Zwicker, M.: Render4Completion: synthesizing multi-view depth maps for 3D shape completion. ArXiv abs/1904.08366 (2019)
28. Hu, T., Han, Z., Zwicker, M.: 3D shape completion with multi-view consistent inference. In: AAAI (2020)
29. Insafutdinov, E., Dosovitskiy, A.: Unsupervised learning of shape and pose with differentiable point clouds. In: Advances in Neural Information Processing Systems, pp. 2807–2817 (2018)
30. Jiang, Y., Ji, D., Han, Z., Zwicker, M.: SDFDiff: differentiable rendering of signed distance fields for 3D shape optimization. In: IEEE Conference on Computer Vision and Pattern Recognition (2020)
31. Kato, H., Ushiku, Y., Harada, T.: Neural 3D mesh renderer. In: IEEE Conference on Computer Vision and Pattern Recognition, pp. 3907–3916 (2018)
32. L., N.K., Mandikal, P., Agarwal, M., Babu, R.V.: Capnet: continuous approximation projection for 3D point cloud reconstruction using 2D supervision. In: AAAI (2019)
33. Lin, C.H., Kong, C., Lucey, S.: Learning efficient point cloud generation for dense 3D object reconstruction. In: AAAI Conference on Artificial Intelligence (2018)
34. Liu, H.T.D., Tao, M., Jacobson, A.: Paparazzi: surface editing by way of multi-view image processing. ACM Trans. Graph. **37**, 221 (2018)
35. Liu, H.T.D., Tao, M., Li, C.L., Nowrouzezahrai, D., Jacobson, A.: Beyond pixel norm-balls: Parametric adversaries using an analytically differentiable renderer. In: International Conference on Learning Representations (2019)
36. Liu, S., Zhang, Y., Peng, S., Shi, B., Pollefeys, M., Cui, Z.: DIST: rendering deep implicit signed distance function with differentiable sphere tracing. In: IEEE Conference on Computer Vision and Pattern Recognition (2020)
37. Liu, S., Chen, W., Li, T., Li, H.: Soft rasterizer: differentiable rendering for unsupervised single-view mesh reconstruction. CoRR abs/1901.05567 (2019)
38. Liu, S., Li, T., Chen, W., Li, H.: Soft rasterizer: a differentiable renderer for image-based 3D reasoning. In: The IEEE International Conference on Computer Vision (2019)
39. Liu, S., Saito, S., Chen, W., Li, H.: Learning to infer implicit surfaces without 3D supervision. In: Advances in Neural Information Processing Systems (2019)

40. Liu, X., Han, Z., Liu, Y.S., Zwicker, M.: Point2Sequence: learning the shape representation of 3D point clouds with an attention-based sequence to sequence network. In: AAAI, pp. 8778–8785 (2019)
41. Liu, X., Han, Z., Xin, W., Liu, Y.S., Zwicker, M.: L2G auto-encoder: understanding point clouds by local-to-global reconstruction with hierarchical self-attention. In: ACMMM (2019)
42. Mescheder, L., Oechsle, M., Niemeyer, M., Nowozin, S., Geiger, A.: Occupancy networks: learning 3D reconstruction in function space. In: IEEE Conference on Computer Vision and Pattern Recognition (2019)
43. Michalkiewicz, M., Pontes, J.K., Jack, D., Baktashmotlagh, M., Eriksson, A.P.: Deep level sets: implicit surface representations for 3D shape inference. CoRR abs/1901.06802 (2019)
44. Navaneet, K.L., Mandikal, P., Jampani, V., Babu, R.V.: DIFFER: moving beyond 3D reconstruction with differentiable feature rendering. In: CVPR Workshops (2019)
45. Niemeyer, M., Mescheder, L., Oechsle, M., Geiger, A.: Differentiable volumetric rendering: learning implicit 3D representations without 3D supervision. In: IEEE Conference on Computer Vision and Pattern Recognition (2020)
46. Oechsle, M., Mescheder, L., Niemeyer, M., Strauss, T., Geiger, A.: Texture fields: learning texture representations in function space (2019)
47. Park, J.J., Florence, P., Straub, J., Newcombe, R., Lovegrove, S.: DeepSDF: learning continuous signed distance functions for shape representation. In: IEEE Conference on Computer Vision and Pattern Recognition (2019)
48. Qi, C.R., Su, H., Mo, K., Guibas, L.J.: PointNet: deep learning on point sets for 3D classification and segmentation. In: IEEE Conference on Computer Vision and Pattern Recognition (2017)
49. Qi, C.R., Yi, L., Su, H., Guibas, L.J.: PointNet++: deep hierarchical feature learning on point sets in a metric space. In: Advances in Neural Information Processing Systems, pp. 5105–5114 (2017)
50. Richter, S.R., Roth, S.: Matryoshka networks: predicting 3D geometry via nested shape layers. In: CVPR, pp. 1936–1944 (2018)
51. Saito, S., Huang, Z., Natsume, R., Morishima, S., Kanazawa, A., Li, H.: PIFu: pixel-aligned implicit function for high-resolution clothed human digitization. In: IEEE International Conference on Computer Vision (2019)
52. Shin, D., Ren, Z., Sudderth, E.B., Fowlkes, C.C.: 3D scene reconstruction with multi-layer depth and epipolar transformers. In: IEEE International Conference on Computer Vision (2019)
53. Sitzmann, V., Zollhöfer, M., Wetzstein, G.: Scene representation networks: continuous 3D-structure-aware neural scene representations. In: Advances in Neural Information Processing Systems (2019)
54. Tatarchenko, M., Dosovitskiy, A., Brox, T.: Octree generating networks: efficient convolutional architectures for high-resolution 3D outputs. In: IEEE International Conference on Computer Vision, pp. 2107–2115 (2017)
55. Tatarchenko, M., Richter, S.R., Ranftl, R., Li, Z., Koltun, V., Brox, T.: What do single-view 3D reconstruction networks learn? In: The IEEE Conference on Computer Vision and Pattern Recognition (2019)
56. Tulsiani, S., Efros, A.A., Malik, J.: Multi-view consistency as supervisory signal for learning shape and pose prediction. In: Computer Vision and Pattern Recognition (2018)

57. Tulsiani, S., Zhou, T., Efros, A.A., Malik, J.: Multi-view supervision for single-view reconstruction via differentiable ray consistency. In: IEEE Conference on Computer Vision and Pattern Recognition. pp. 209–217 (2017)
58. Wang, N., Zhang, Y., Li, Z., Fu, Y., Liu, W., Jiang, Y.: Pixel2mesh: generating 3D mesh models from single RGB images. In: European Conference on Computer Vision, pp. 55–71 (2018)
59. Wang, W., Ceylan, D., Mech, R., Neumann, U.: 3DN: 3D deformation network. In: CVPR (2019)
60. Wang, W., Xu, Q., Ceylan, D., Mech, R., Neumann, U.: DISN: deep implicit surface network for high-quality single-view 3D reconstruction. In: NeurIPS (2019)
61. Wen, C., Zhang, Y., Li, Z., Fu, Y.: Pixel2Mesh++: multi-view 3D mesh generation via deformation. In: IEEE International Conference on Computer Vision (2019)
62. Wen, X., Han, Z., Youk, G., Liu, Y.S.: CF-SIS: semantic-instance segmentation of 3D point clouds by context fusion with self-attention. In: ACM International Conference on Multimedia (2020)
63. Wen, X., Li, T., Han, Z., Liu, Y.S.: Point cloud completion by skip-attention network with hierarchical folding. In: The IEEE Conference on Computer Vision and Pattern Recognition (2020)
64. Whitted, T.: A scan line algorithm for computer display of curved surfaces. In: The 5th Annual Conference on Computer Graphics and Interactive Techniques SIGGRAPH, p. 26 (1978)
65. Wu, J., Wang, Y., Xue, T., Sun, X., Freeman, B., Tenenbaum, J.: MarrNet: 3D shape reconstruction via 2.5D sketches. In: Advances in Neural Information Processing Systems, pp. 540–550 (2017)
66. Wu, J., Zhang, C., Xue, T., Freeman, B., Tenenbaum, J.: Learning a probabilistic latent space of object shapes via 3D generative-adversarial modeling. In: Advances in Neural Information Processing Systems, pp. 82–90 (2016)
67. Yan, X., Yang, J., Yumer, E., Guo, Y., Lee, H.: Perspective transformer nets: learning single-view 3D object reconstruction without 3D supervision. In: Advances in Neural Information Processing Systems, pp. 1696–1704 (2016)
68. Yifan, W., Serena, F., Wu, S., Öztireli, C., Sorkine-Hornung, O.: Differentiable surface splatting for point-based geometry processing. ACM Trans. Graph. **38**(6), 1–14 (2019)
69. Zakharov, S., Kehl, W., Bhargava, A., Gaidon, A.: Autolabeling 3D objects with differentiable rendering of SDF shape priors. In: IEEE Conference on Computer Vision and Pattern Recognition (2020)
70. Zhang, X., Zhang, Z., Zhang, C., Tenenbaum, J., Freeman, B., Wu, J.: Learning to reconstruct shapes from unseen classes. In: Advances in Neural Information Processing Systems, pp. 2257–2268 (2018)

Simultaneous Detection and Tracking with Motion Modelling for Multiple Object Tracking

Shijie Sun[1], Naveed Akhtar[2], Xiangyu Song[3], Huansheng Song[1(✉)], Ajmal Mian[2], and Mubarak Shah[4]

[1] Chang'an University, Xi'an, Shaanxi, China
{shijiesun,hshsong}@chd.edu.cn
[2] University of Western Australia, 35 Stirling Highway, Crawley, WA, Australia
{naveed.akhtar,ajmal.mian}@uwa.edu.au
[3] Deakin University, RWaurn Ponds, Victoria 3216, Melbourne, Australia
xiangyu.song@deakin.edu.au
[4] University of Central Florida, Orlando, FL, USA
shah@crcv.ucf.edu

Abstract. Deep learning based Multiple Object Tracking (MOT) currently relies on off-the-shelf detectors for tracking-by-detection. This results in deep models that are detector biased and evaluations that are detector influenced. To resolve this issue, we introduce Deep Motion Modeling Network (DMM-Net) that can estimate multiple objects' motion parameters to perform joint detection and association in an end-to-end manner. DMM-Net models object features over multiple frames and simultaneously infers object classes, visibility and their motion parameters. These outputs are readily used to update the tracklets for efficient MOT. DMM-Net achieves PR-MOTA score of 12.80 @ 120+ fps for the popular UA-DETRAC challenge - which is better performance and orders of magnitude faster. We also contribute a synthetic large-scale public dataset Omni-MOT for vehicle tracking that provides precise ground-truth annotations to eliminate the detector influence in MOT evaluation. This 14M+ frames dataset is extendable with our public script (Code at Dataset, Dataset Recorder, Omni-MOT Source). We demonstrate the suitability of Omni-MOT for deep learning with DMM-Net, and also make the source code of our network public.

Keywords: Multiple object tracking · Tracking-by-detection · Deep learning · Simultaneous detection and tracking.

1 Introduction

Multiple Object Tracking (MOT) [10,25,42,44] is a longstanding problem in Computer Vision [27]. Contemporary deep learning based MOT has widely

Electronic supplementary material The online version of this chapter (https://doi.org/10.1007/978-3-030-58586-0_37) contains supplementary material, which is available to authorized users.

A. Vedaldi et al. (Eds.): ECCV 2020, LNCS 12369, pp. 626–643, 2020.
https://doi.org/10.1007/978-3-030-58586-0_37

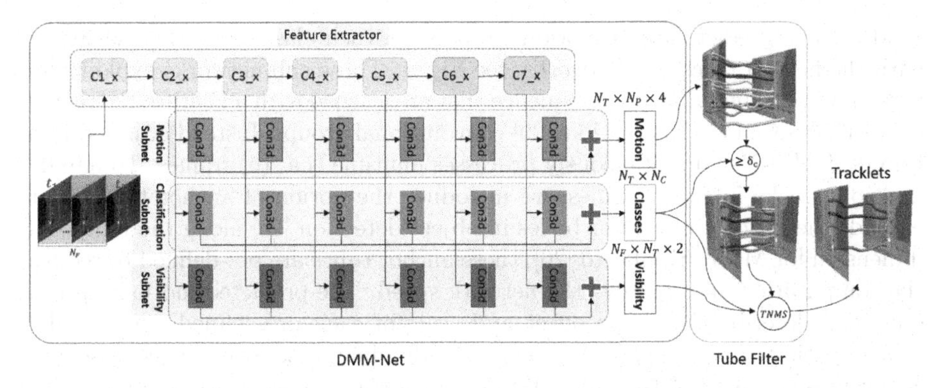

Fig. 1. Schematics of the proposed end-to-end trainable DMM-Net: N_F frames from time stamp $t_1 : t_2$ are input to the network. The frame sequence is first processed with a *Feature Extractor* comprising 3D ResNet-like [20] convolutional groups. Outputs of the last groups 2 to 7 are processed by Motion Subnet, Classifier Subnet, and Visibility Subnet. Each sub-network uses 3D convolutions to learn features that are concatenated to predict anchor tubes' motion parameters ($\boldsymbol{O}_M \in \mathbb{R}^{N_T \times N_P \times 4}$), object classes ($\boldsymbol{O}_C \in \mathbb{R}^{N_T \times N_C}$), and visibility ($\boldsymbol{O}_V \in \mathbb{R}^{N_F \times N_T \times 2}$), where N_T, N_P and N_C denote the number of anchor tubes, motion parameters and object classes (including background). We explain anchor tubes and motion parameters in Sect. 4. DMM-Net is trained with its specialized loss. For deployment, the anchor tubes predicted by the DMM-Net are filtered to compute tracklets defined over multiple frames. These tracklets are later combined to form a complete track.

adopted the tracking-by-detection paradigm [43], that capitalizes on the natural division of *detection* and *data association* tasks for the problem. In standard MOT evaluation protocol, the object detections are assumed to be known and public detections are provided on evaluation sequences, and MOT algorithms are expected to output object tracks by solving the data association problem.

Although attractive, using off-the-shelf detectors for MOT also has undesired ramifications. For instance, a deep model employed for the subsequent data association task (or a constituent sub-task) gets biased to the detector. The detector performance can also become a bottle-neck for the overall tracker. Additionally, composite techniques resulting from independent detectors fail to harness the true representation power of deep learning by sacrificing end-to-end training etc. It also seems unnatural that a MOT tracker must be evaluated on different detectors to interpret its tracking performance. All these problems are potentially solvable if trackers can implicitly detect the target objects, and detector bias can be removed from the ground-truth labeling of the tracks. This work makes strides towards these solutions.

We make two major contributions towards setting the tracking-by-detection paradigm free from off-the-shelf detectors in deep learning settings. Our first contribution comes as the first-of-its-kind deep network that performs object detections and data association by estimating multiple object motion parameters in an end-to-end manner. Our network, called Deep Motion Modeling Network

(DMM-Net), predicts object motion parameters, their classes and their visibility with the help of three sub-networks, see Fig. 1. These sub-networks exploit feature maps of frames in a video sequence that are learned with a Feature Extractor comprising seven 3D ResNet-like [20] convolutional groups. Instead of individual frames, DMM-Net simultaneously processes multiple (i.e. 16) frames. To handle multiple tracks in those frames, we introduce the notion of anchor tubes that extends the concept of anchor boxes in object detection [26] along the temporal dimension for MOT. Similar to [26], these anchor tubes are pre-defined to reduce the computation and improve the network speed. The predicted motion parameters can describe the shape offset and scale of each pre-defined anchor tube in the spatio-temporal space. We propose individual losses over the comprehensive representations of the sub-networks to predict object motion parameters, object classes and visibility. The DMM-Net output is readily usable to update the tracks. As our second major contribution, we propose a realistic large-scale dataset with accurate and extensive ground-truth annotations. The proposed dataset, termed Omni-MOT for its comprehensive coverage of the MOT scenarios, is generated with CARLA simulator [13] for vehicle tracking. The dataset comprises 14M+ frames, 250K tracks, 110 million bounding boxes, three weather conditions, three crowd levels and three camera views in five simulated towns. By eliminating the use of off-the-shelf detectors in ground-truth labeling, it removes any detector bias in evaluating the techniques.

The Omni-MOT dataset and DMM-Net source code are both publicly available for the broader research community. For the former, we also provide software tools to freely extend the dataset. We demonstrate the suitability of the Omni-MOT for deep models by training and evaluating DMM-Net on its videos. We also augment DMM-Net with Omni-MOT and evaluate our technique on the popular UA-DETRAC challenge [45]. The remote server computed results show that DMM-Net is able to achieve a very competitive 12.80 score for the comprehensive PR-MOTA metric with the overall speed of 123 fps. The orders of magnitude increase in the speed is directly attributed to the intrinsic detections in our tracking-by-detection technique.

2 Related Work

Multiple Object Tracking (MOT) is a fundamental problem in Computer Vision that has attracted significant interest of researchers in recent years. For a general review of the related literature, we refer to Luo et al. [27] and Emami et al. [14]. Here, we focus on the key contributions that relate to this work more closely. With the recent advances in object detectors, tracking-by-detection is fast becoming the common contemporary paradigm for MOT [38,39,43,46]. In this scheme, objects are first detected frame-wise and later associated with each other across multiple frames. Relying on off-the-shelf detectors, techniques following this paradigm mainly focus on object association, which can make them inherently detector biased. These methods can be broadly categorized as local [34,41] and global [10,37,48] approaches. The former use two frames for data association, while the latter associate objects over a larger number of frames.

More recent global techniques cast data association into a network flow problem [4,7,33,40]. For instance, Berclaz et al. [4] solved a constrained flow optimization problem for MOT and used the k-shortest paths algorithm for associating the tracks. Chari et al. [8] added a pairwise cost to the min-cost network flow framework and proposed a convex relaxation of the problem. However, such methods rely on object detectors rather strongly, which makes them less attractive in the presence of occlusions and misdetections. To mitigate the problems resulting from occlusions, Milan et al. [29] employed a continuous energy minimization framework for MOT that incorporates explicit occlusion reasoning and appearance modeling. Wen et al. [47] also proposed a data association technique based on undirected hierarchical relation hyper-graph.

Sun et al. [42] proposed a deep affinity network to model features of pre-detected objects and compute object affinities by the same network. Bea et al. [3] modified the Siameses Network to learn deep representations for MOT with object association. There are few instances of deep learning techniques that aim at removing the reliance of tracking on independent detectors. For instance, Feichtenhofer et al. [15] proposed an R-FCN based network [9] that performs object detection and jointly builds an object model to be used for data association. However, their method is limited to frame-wise detections. Consequently, it only allows frame-by-frame association, requiring manual adjustment of temporal stride. Our technique is inherently different, as it directly computes tracklets over multiple frames by motion modeling, enabling realtime solutions while considering all the frames.

Besides the development of novel techniques, the role of datasets is central to the advancement of deep learning based MOT. Currently, a few large datasets for this task are available, e.g. PETS2009 [17], KITTI [18], DukeMTMC [36], PoseTrack [23], and MOT Challenge datasets [25,28]. These datasets are recorded in the real world with pedestrians and vehicles as the objects of interest. We refer to the original works for more details on these datasets. Below, we briefly discuss UA-DETRAC [45] for its high relevance to our contribution.

UA-DETRAC [45] is a large dataset for traffic scenes MOT. It provides object bounding boxes, their IDs and information on the overlapped ratio of the objects. However, the provided detections are individually generated by detectors and hence are prone to errors. This results in an undesired detector-bias in tracker evaluation. Besides, different pre-processing procedures of the detectors employed by the dataset also cause problems in fair evaluation. Although UA-DETRAC has served a great purpose in advancing the state-of-the-art in MOT for vehicles, the aforementioned issues call for a more transparent dataset that does not rely on off-the-shelf detectors for evaluating trackers. This work provides such a dataset with realistic settings and complete control over the environment conditions and camera viewpoints.

3 Omni-MOT Dataset

We term the proposed dataset as **Omni-MOT** (OMOT) dataset for its comprehensive coverage of the conditions and scenarios possible in MOT. The dataset is publicly available for download. Moreover, we also make the recording script for the dataset public that will enable the community to further extend the data The provided script has the ability to generate multi-camera videos. The dataset is recorded using virtual cameras in the CARLA simulator [13].

To the best of our knowledge, Omni-MOT is the first realistic large dataset for tracking vehicles that completely relinquishes off-the-shelf detectors in ground truth generation, and provides comprehensive annotations. Moreover, with the provided scripts, the dataset can easily be extended for future research. To put the scale of Omni-MOT into perspective, the provided number of frames is almost 1,200 times larger than MOT17. The number of provided tracks and boxes respectively are 210 and 30 times larger than UA-DETRAC. Not to mention, all the boxes and tracks are simulator generated that avoids any labeling error. Considering that OMOT can also be used for data augmentation, we make the ground truth for the test videos public as well. Please see the supplementary material of the paper for complete details of the dataset.

4 Deep Motion Modeling Network

To absolve deep learning based tracking-by-detection from independently pre-trained off-the-shelf detectors, we propose **D**eep **M**otion **M**odeling **Net**work (DMM-Net) for online MOT (see Fig. 1). Our network enables MOT by jointly performing object detection, tracking, and classification across multiple video frames without requiring pre-detections and subsequent data association. For a given input video, it outputs objects' motion parameters, classes, and their visibility across the input frames. We provide a detailed discussion of our network below. However, we first introduce the used notations and conventions.

- N_F, N_C, N_P, N_T denote the number of input frames, object classes (0 for 'background'), time-related motion parameters, and anchor tubes.
- W, H are the frame width, and frame height.
- $\boldsymbol{I}_t$ denotes the video frame at time t. Subsequently, a 4-D tensor $\boldsymbol{I}_{t_1:t_2:N_F} \in \mathbb{R}^{3\times N_F\times W\times H}$ denotes N_F video frames from time t_1 to t_2-1. For simplicity, we often ignore the subscript ": N_F".
- $\boldsymbol{B}_{t_1:t_2:N_F}, \boldsymbol{C}_{t_1:t_2:N_F}, \boldsymbol{V}_{t_1:t_2:N_F}$ respectively denote the ground truth boxes, classes and visibilities in the selected N_F video frames from time t_1 to t_2-1. The text also ignores ": N_F" for these notations.
- $\boldsymbol{O}_{M,t_1:t_2:N_F}, \boldsymbol{O}_{C,t_1:t_2:N_F}, \boldsymbol{O}_{V,t_1:t_2:N_F}$ denote the estimated motion parameters, classes, and visiblities. With time stamps and frames clear from the context, we simplify these notations as O_M, O_C, O_V. In Fig. 2, we illustrate object visibility, classes and motion parameters.

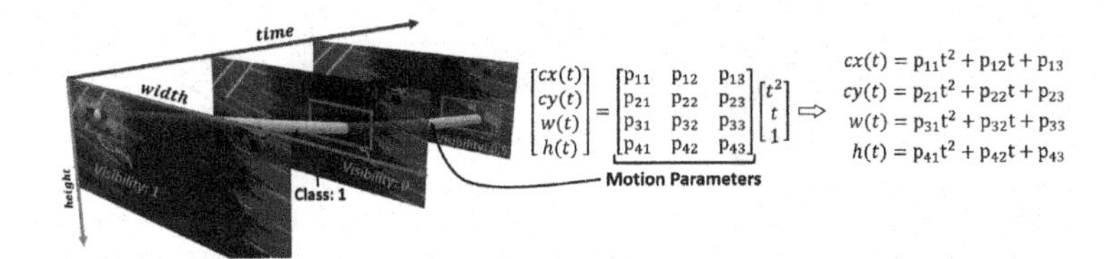

Fig. 2. (Best viewed in color) Illustration of motion visibility, class, and motion parameters: The visibility of the vehicle goes from 1 (fully visible) to 0 (invisible) and back to 0.5 (partially visible). Classes are predefined as 1 for vehicle and 0 for everything else. Motion parameters $\{p_{11}, \cdots, p_{43}\}$ are used to locate the center x (cx), center y (cy), width (w) and height (h) of the tube/tracklet at anytime. We employ a quadratic motion model leveraging 4×3 matrices. (Color figure online)

Table 1. Inputs for training and testing DMM-Net. The tensor dimensions are given as Channels×Duration×Width×Height. $n_{t_1:t_2}$ is the number of tracks in $I_{t_1:t_2}$.

Input	Dimensions/Size	Train	Test
$\boldsymbol{I}_{t_1:t_2}$	$3 \times N_F \times W \times H$	✓	✓
$\boldsymbol{t}_{t_1:t_2}$	N_F	✓	✓
$\boldsymbol{B}_{t_1:t_2}$	$N_F \times n_{t_1:t_2} \times 4$	✓	×
$\boldsymbol{C}_{t_1:t_2}$	$n_{t_1:t_2}$	✓	×
$\boldsymbol{V}_{t_1:t_2}$	$N_F \times n_{t_1:t_2}$	✓	×

Conventions: The shape of a network output tensor is considered to be Batch × Channels × Duration × Width × Height, where Duration accounts for the number of frames. For brevity, we often ignore the Batch dimension.

4.1 Data Preparation

Appropriate data preparation is important for training (and later testing) our network. In this process, we also avail the opportunity of augmenting the data to induce robustness in our model against camera photometric distortions, background scene variations, and other practical factors. The inputs expected by our network are summarized in Table 1. For training a robust network, we perform the following data pre-processing:

1. *Stochastic frame sequence* is introduced by randomly choosing N_F frames from $2N_F$ frames.
2. *Static scene emulation* is done by duplicating selected frames N_F times.
3. *Photometric camera distortions* are introduced in the frames by scaling each pixel by a random value in the range [0.7, 1.5]. The resulting frames are converted to HSV format, and their saturation channel is again scaled by a random value in [0.7, 1.5]. A frame is then converted back to RGB format and re-scaled by a random value in the same range. This process of photometric distortion is inspired by [42].

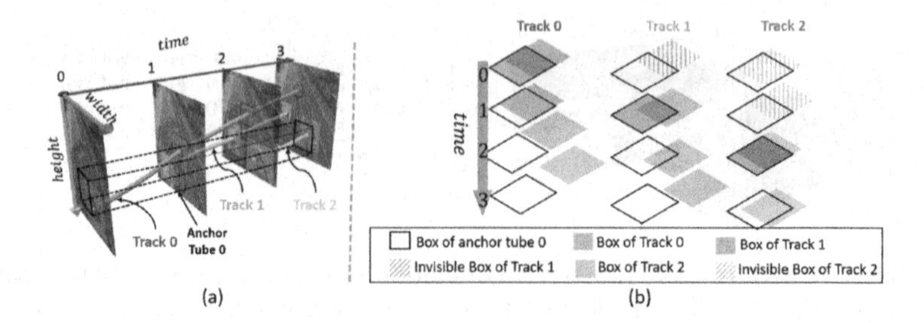

Fig. 3. (a) One anchor tube and three tracks are shown to illustrate the search for the best track of the anchor tube. (b) is the simplified demonstration of (a). Based on the largest overlap between the first tube box and the first *visible* track box, Track 2 is selected as the best match. (Color figure online)

4. *Frame expansion* is performed by enlarging video frames by a random factor in the range [1, 1.2]. To that end, we pad the original frames with extra pixels whose value is set to the mean pixel value of the training data.

In our pre-processing, the above-mentioned steps 2–4 are applied to the frames with probability 0.01, 0.5, 0.5, respectively. The resulting frame are then resized to the fixed dimension $H \times W \times 3$, and horizontally flipped with a probability of 0.5. We simultaneously process the selected N_F frames by applying the above mentioned transformations. These RGB video frames are then arranged as 4D-tensors $\boldsymbol{I}_{t_1:t_2} \in \mathbb{R}^{3 \times N_F \times W \times H}$. We fill the ground truth data matrices $\boldsymbol{B}_{t_1:t_2}$, $\boldsymbol{C}_{t_1:t_2}$, $\boldsymbol{V}_{t_1:t_2}$ with the detected boxes, and visibilities in the video frames from t_1 to $t_2 - 1$. We set the labeled boxes, classes, and visibilities of fully occluded boxes to 0. We also remove the tracks whose ratio of fully occluded boxes are greater than δ_v. We let $N_F = 16$ in our experiments.

4.2 Anchor Tubes

Inspired by the effective Single Shot Detector (SSD) [26] for object detection, we extend the core concept of this technique for MOT. Analogous to the anchor boxes of SSD, we introduce *anchor tubes* for DMM-Net. Here, an anchor tube is a set of anchor boxes (and associated object class and visibility), that share the same location in multiple frames along the temporal dimension (see supplementary material for further visualization). The information of anchor tubes is encoded with the tensors $\boldsymbol{B}_{t_1:t_2}$, $\boldsymbol{C}_{t_1:t_2}$, and $\boldsymbol{V}_{t_1:t_2}$. The DMM-Net is designed to predict the tube shape offset parameters along the temporal dimension, the confidence for each object class, and the visibility of each box in the tube.

Computing anchor tubes for network training can also be interpreted as further data preparation for DMM-Net. We first employ a search strategy to find the best-matched track for each anchor tube, and subsequently encode the tubes by their best-matched tracks. For the former, we specify the overlap between an anchor tube and a track as the overlap ratio between the first visible box of the track and the corresponding box in the anchor tube. A simplified illustration of

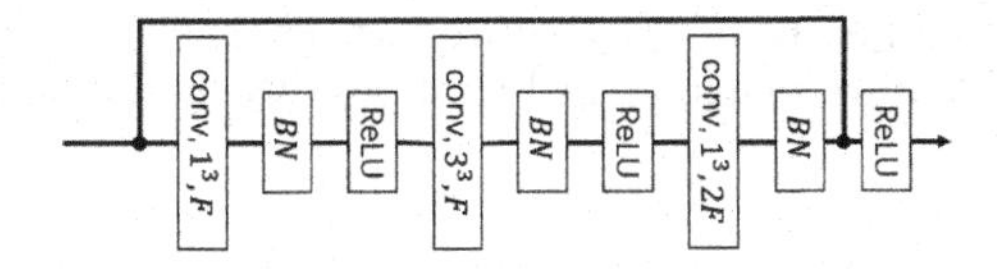

Fig. 4. Used ResNeXt block: (conv, x^3, F) denote 'F' convolutional kernels of size $x \times x \times x$. *BN* is Batch Normalization [22] and shortcut connection is summation.

this concept is presented in Fig. 3 with one anchor tube and three tracks. The anchor tube 0 iterates over all the tracks to find the largest overlap ratio. The first box of Track 1 and the first two boxes of Track 2 are occluded. Therefore, the overlap ratio of the anchor box filled red is used for selecting the best-matched track, i.e. Track 2 for the largest overlap.

To encode the anchor tubes, we employ their best-matched tracks ($\boldsymbol{B}_{t_1:t_2}$) along with the classes ($\boldsymbol{C}_{t_1:t_2}$) and visibility ($\boldsymbol{V}_{t_1:t_2}$). We denote a box of the ith anchor tube at the tth frame as $a_{i,t} = (a^{cx}_{i,t}, a^{cy}_{i,t}, a^{w}_{i,t}, a^{h}_{i,t})$, where the superscripts cx and cy indicate the (x, y) location of the center, and w, h denote the width and height of the box. We use the same notation in the following text as well. Each anchor box is encoded by the box of its best-matched track as follows:

$$\begin{cases} g^{w}_{i,t} = log(b^{w}_{i,t}/a^{w}_{i,t}) \\ g^{h}_{i,t} = log(b^{h}_{i,t}/a^{h}_{i,t}) \\ g^{cx}_{i,t} = (b^{cx}_{i,t} - a^{cx}_{i,t})/a^{w}_{i,t} \\ g^{cy}_{i,t} = (b^{cy}_{i,t} - a^{cy}_{i,t})/a^{h}_{i,t}, \end{cases} \tag{1}$$

where $(b^{cx}_{i,t}, b^{cy}_{i,t}, b^{w}_{i,t}, b^{h}_{i,t})$ describe the box of the best-matched track at the tth frame, and $(g^{cx}_{i,t}, g^{cy}_{i,t}, g^{w}_{i,t}, g^{h}_{i,t})$ represent the resulting encoded box for the best-matched track. Following this encoding, we replace each original box, class, and visibility by its newly encoded counterpart. In the above-mentioned encoding, an anchor tube that has its best-matched track overlap ratio less than δ_o, is identified as the 'background' class.

4.3 Motion Model

For motion modeling, we leverage a quadratic model in time. One of the outputs of DMM-Net is a tensor of motion parameters $\boldsymbol{O}_M \in \mathbb{R}^{N_T \times 4 \times N_P}$, where $N_P = 3$ in our experiments and N_T indicates the number of anchor tubes. We estimate an encoded anchor tube for a track as:

$$\begin{cases} \hat{g}^{w}_{i,t} = p_{11}t^2 + p_{12}t + p_{13} \\ \hat{g}^{h}_{i,t} = p_{21}t^2 + p_{22}t + p_{23} \\ \hat{g}^{cx}_{i,t} = p_{31}t^2 + p_{32}t + p_{33} \\ \hat{g}^{cy}_{i,t} = p_{41}t^2 + p_{42}t + p_{43}, \end{cases} \tag{2}$$

where $\hat{g}_{i,t}$ indicates an estimated box descriptor of the ith encoded anchor tube at the tth frame, the superscripts cx, cy, w and h indicate the center x, center y,

Table 2. Convolutional groups of the Feature Extractor. Conv is abbreviated as 'C', 'F' is the number of feature maps, as in Fig. 4, and 'N' is the number of bottleneck blocks in each layer.

C1_x	C2_x		C3_x		C4_x		C5_x		C6_x		C7_x	
conv, 7^3, 64	F	N	F	N	F	N	F	N	F	N	F	N
	64	3	128	4	256	23	512	3	32	3	32	3

height, width of the box respectively, and $\{p_{11}, \cdots, p_{43}\}$ are the motion parameters . Each encoded ground-truth anchor tube can be decoded into a ground truth track by Eq. (1). We further elaborate on the relationship between the motion parameters and the ground truth tracks in the supplementary material. Note that the used motion function is replaceable by any differential function in our technique. We choose quadratic motion modeling considering vehicle tracking as our primary objective.

4.4 Architecture

As shown in Fig. 1, our network comprises a Feature Extractor and three subnetwork. The network simultaneously processes N_F video frames for which features are first extracted. The feature maps of six intermediate layers of the Feature Extractor are fed to the sub-networks. These networks predict tensors containing motion parameters ($\boldsymbol{O}_M \in \mathbb{R}^{N_T \times N_P \times 4}$), object classes ($\boldsymbol{O}_C \in \mathbb{R}^{N_T \times N_C}$), and visibility information ($\boldsymbol{O}_V \in \mathbb{R}^{N_F \times N_T \times 2}$).

Feature Extractor: We construct our Feature Extractor based on the ResNeXt blocks of the 3D ResNet [20][1]. The architectural details of the blocks are provided in Fig. 4. The blocks accept 'F' channel input that allows us to simultaneously model spatio-temporal features in N_F frames. We enhance the 3D ResNet architecture by removing the *fully-connected* and *softmax* layers of the network and appending two extra convolutional groups denoted as Conv6_x and Conv7_x. We perform this enhancement because the additional convolutional layers are still able to encode further higher level spatio-temporal features of the frames. Details on the convolutional groups used by the Feature Extractor are provided in Table 2. Here, Conv1_x (abbreviated as C1_x) contains 64 convolutional kernels of size $7 \times 7 \times 7$, stride $1 \times 2 \times 2$. A $3 \times 3 \times 3$ max-pooling layer with stride 2 is inserted before Conv2_x for down sampling. Each convolutional layer is followed by batch normalization [21] and ReLU [31]. Spatio-temporal down-sampling is performed by Conv3_1, Conv4_1 and Conv5_1 with a stride of 2.

Sub-networks: The DMM-Net has three sub-networks to compute motion parameters, object classes, and their visibility. These networks use the outputs of Conv2_x to Conv7_x groups of the Feature Extractor. Each sub-network processes the input with six convolutional layers. The architectural details of

[1] Notation are adopted from the original work.

Table 3. Details of the Motion Subnet (M.S.), Classifier Subnet (C.S.), and Visibility Subnet (V.S.): K.S., P.S., S.S., and N.K. respectively denote the kernel shape, padding size, stride size, and the number of kernels. The row 'Output' reports output tensor shape. $\mathcal{K} = \{10, 8, 8, 5, 4, 4\}$ is the number of anchor tubes for each of the six input feature maps for each sub-network.

	M.S	C.S	V.S
K.S.	$\{8, 4, 2, 1, 1, 1\} \times 3 \times 3$		
P.S.	$0 \times 1 \times 1$		
S.S.	$1 \times 1 \times 1$		
N.K.	$4\mathcal{K}N_P$	$\mathcal{K}N_C$	$2\mathcal{K}$
Output	$N_T \times N_P \times 4$	$N_T \times N_C$	$N_F \times N_T \times 2$

all three sub-networks are summarized in Table 3. The kernel shape, padding size and stride are the same for each network, whereas the employed numbers of kernels are different. We fix the temporal kernel size for each network to $\{8, 4, 2, 1, 1, 1\}$. Denoting the numbers of anchor tubes defined for the six input feature maps of a sub-network by $\mathcal{K}$, we let $\mathcal{K} = \{10, 8, 8, 5, 4, 4\}$ in our experiments. For each sub-network, we concatenate the output feature maps of each convolutional layer. We reshape the concatenated feature according to the dimensions mentioned in the last row of the table. These outputs are subsequently used to compute the network loss.

Network Loss: Based on the architecture, the overall loss of DMM-Net is defined as a weighted sum of three sub-losses, including the motion loss ($\mathcal{L}_M$), the classification loss ($\mathcal{L}_C$) and the visibility loss ($\mathcal{L}_V$), given as:

$$\mathcal{L} = \frac{1}{N}(\alpha\mathcal{L}_M + \beta\mathcal{L}_C + \mathcal{L}_V), \tag{3}$$

where N is the number of *positive* anchor tubes, and α, β are both hyper-parameters. The positive anchor tubes exclude the background tubes.

The motion loss $\mathcal{L}_M$ is defined as the sum of *Smooth*-L1 losses between the ground truth encoded tracks $g_{i,t}$ and their predicted encoded tracks $\hat{g}_{i,t}$, formally:

$$\mathcal{L}_M = \sum_{t=0}^{N_F-1} \sum_{i \in Pos} \sum_{m \in \{cx, cy, h, w\}} ||g_{i,t}^m - \hat{g}_{i,t}^m||_1, \tag{4}$$

where Pos is the set of positive encoded anchor tube indices. We compute $\hat{g}_{i,t}^m$ using Eq. (2) discussed in Sect. 4.3.

For the classification loss $\mathcal{L}_C$, we employ the hard negative mining strategy [26] to balance the positive and negative anchor tubes, where the negative tubes correspond to the background. We denote $x_{i,j,t}^p \in \{1, 0\}$ to be an indicator for matching the ith anchor tube and the jth ground-truth track of class 'p' at the tth frame. Each anchor tube at least has one best-matched ground-truth

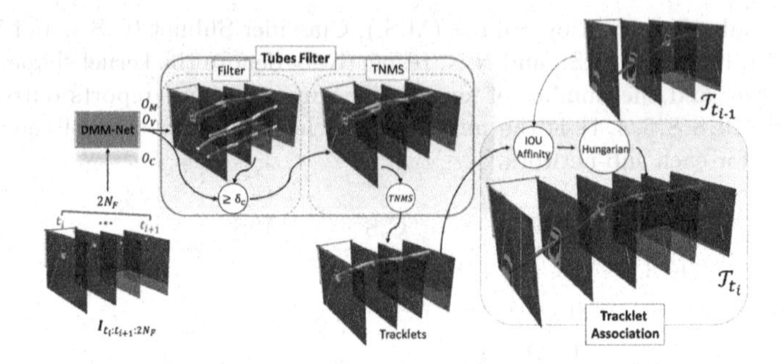

Fig. 5. Deployment of DMM-Net: For the $2N_F$ frames $\boldsymbol{I}_{t_i:t_{i+1}:2N_F}$ from t_i to t_{i+1}, the trained DMM-Net selects N_F frames as its input and outputs predicted tubes encoded by object motion parameter matrix ($\boldsymbol{O}_M$), classification matrix ($\boldsymbol{O}_C$) and visibility matrix ($\boldsymbol{O}_V$). These matrices are filtered and the track set $\boldsymbol{\mathcal{T}}_{t_i}$ is updated by associating the filtered tracklets by their IOU with the previous track set $\boldsymbol{\mathcal{T}}_{t_{i-1}}$.

track, implying $\sum_i x^p_{i,j,t} \geq 1$ for any t. The classification loss of the DMM-Net is defined as:

$$\mathcal{L}_C = \sum_{t=0}^{N_F} \left(- \sum_{i \in Pos} x^p_{i,j,t} \, log(\hat{c}^p_{i,t}) - \sum_{i \in Neg} log(\hat{c}^0_{i,t}) \right), \tag{5}$$

where $\hat{c}^p_{i,t} = \frac{exp\, c^p_{i,t}}{\sum_p exp\, c^p_{i,t}}$ such that $c^p_{i,t}$ is the predicted confidence of the object being for class $p : p \in \{1, \cdots, N_C\}$, Neg is the set of negative encoded anchor tube indices. We also consider the visibility estimation task fro classification viewpoint and classify each positive box into invisible '0' or visible '1' box. Based on that, the loss is computed as:

$$\mathcal{L}_V = \sum_{t=0}^{N_F} \left(- \sum_{i \in Pos} y^q_{i,j,t} \, log(\hat{v}^q_{i,t}) \right), \tag{6}$$

where $\hat{v}^q_{i,t} = \frac{exp\, v^q_{i,t}}{\sum_q exp\, v^q_{i,t}}$, and $y^q_{i,j,t}$ is an indicator for matching the ith anchor tube and the jth ground-truth track of visibility 'q' at the tth frame, such that $q \in \{0, 1\}$.

4.5 Deployment

The trained DMM-Net is readily deployable for tracking (Fig. 5). The overall tracker processes $2N_F$ frames $\boldsymbol{I}_{t_i:t_{i+1}:2N_F}$, where the DMM-Net first selects N_F consecutive frames and outputs the predicted tubes encoded with object motion parameters, classes and visibility using Eq. (2). The tubes are decoded with Eq. (1). A filtration is then done to remove the undesired tubes and we get

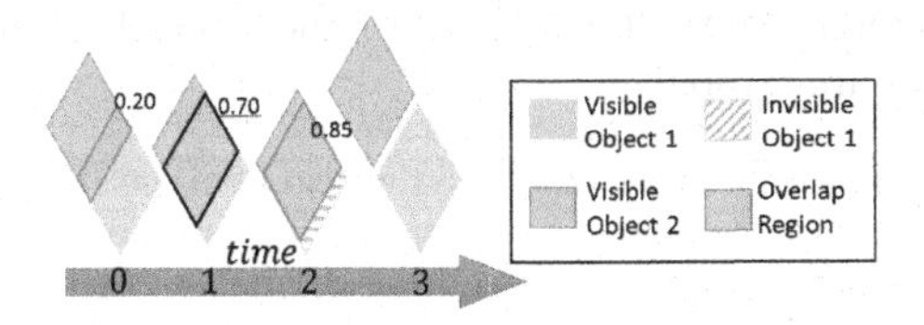

Fig. 6. Example of calculating Tube IOU. There are two tracklets (yellow, cyan). Their intersection is green. The tracklet IOU is the maximum IOU of the visible box pair. Although IOU at $t = 2$ is the largest, the lined yellow object is invisible, hence, we select the second largest IOU at $t = 1$ as the tracklet IOU.

tracklets (details to follow). We compute updated trajectory set $\mathcal{T}_{t_i}$ by associating the tracklets with the previous trajectory set $\mathcal{T}_{t_{i-1}}$ using their IOUs.

To filter, we first remove tubes with confidence lower than a threshold δ_c. We subsequently perform a Tube None Maximum Suppression (TNMS). To that end, we cluster the detected tubes of the same category into multiple groups by their IOUs with a threshold δ_{nms}. Then, we only keep one tube for each group that has the maximum confidence of being positive. The kept tubes, namely "tracklets", are employed to update trajectory set.

We initialize our track set $\mathcal{T}_{t_i}$ with as many trajectories as the number of tracklets. The track set is updated from t_i^{th} to t_{i+1}^{th} stamp using the Hungarian algorithm [30] applied to an IOU matrix $\boldsymbol{\Psi} \in \mathbb{R}^{n'_{t_{i-1}} \times n_{t_i}}$, where $n'_{t_{i-1}}$ is the number of element in the previous track set, and n_{t_i} is the number of tracklets. Notice that we perform association of tracklets and not the individual objects. The object association remains implicit and is done inside the DMM-Net which lets us define the tubes. The association of tracklets, defined across multiple frames, leads to significant computational gain. This is one of the core reasons of the orders of magnitude gain in the tracking speed of our technique over existing methods, as will be clear in Sect. 5.

To form the matrix $\boldsymbol{\Psi}$, we must compute IOU between the existing tracklets and the new tracklets. Figure 6 shows the procedure adopted for calculating the IOU between two tracklets with a simplified example. Overall, our tracker is an on-line technique in the sense that it does not use future frames to predict object trajectories. Concurrently, DMM-Net can perform tracking in real-time that makes it a highly desirable technique for practical applications.

5 Experiments

We evaluate our technique using the proposed Omni-MOT dataset and the popular UA-DETRAC challenge [45]. The former is to demonstrate both the effectiveness of our network for MOT and trainability of deep models for MOT with Omni-MOT. We mainly focus on 'vehicle' tracking in this work. The proposed dataset is also for the vehicle tracking problem. Whereas DMM-Net is trained here for vehicle tracking, it is possible to train it for e.g.. pedestrian tracking.

Table 4. Results on Omni-MOT: The symbol ↑ indicates higher values are better, and ↓ implies lower values are favored.

Type	Camera	IDF1↑	IDP↑	IDR↑	Rcll↑	Prcn↑	GT↑	MT↑	PT↑	ML↓	FP↓	FN↓	IDs↓	FM↓	MOTA↑	MOTP↑
Test	Camera_0	72.2%	69.3%	75.4%	90.6%	83.2%	41	38	3	0	1762	911	18	98	72.1%	79.3%
	Camera_1	59.3%	56.2%	62.7%	81.9%	73.4%	35	19	15	1	1199	730	12	61	52.0%	75.5%
	Camera_5	40.2%	61.3%	29.9%	34.7%	71.1%	44	9	20	15	2779	12855	38	80	20.4%	76.1%
	Camera_7	77.8%	79.6%	76.1%	88.3%	92.4%	37	29	8	0	1518	2457	20	99	80.9%	80.3%
	Camera_8	30.5%	47.3%	22.5%	33.5%	70.7%	38	0	28	10	1789	8548	36	110	19.3%	70.3%
Train	Camera_0	59.2%	58.6%	59.7%	87.7%	86.0%	48	40	8	0	2517	2174	43	172	73.2%	79.5%
	Camera_1	47.6%	44.2%	51.6%	80.0%	68.6%	46	29	16	1	3681	2005	50	137	42.9%	76.9%
	Camera_5	45.0%	52.7%	39.2%	57.0%	76.7%	49	6	37	6	4844	12000	122	234	39.2%	77.3%
	Camera_7	74.7%	71.9%	77.8%	90.1%	83.3%	43	31	12	0	1804	985	16	91	71.8%	79.7%
	Camera_8	41.7%	52.2%	34.7%	44.1%	66.3%	43	0	41	2	1884	4705	18	117	21.4%	68.3%
Average		55.5%	61.1%	50.9%	66.5%	79.8%	424	201	188	35	23777	47370	373	1199	49.4%	77.8%

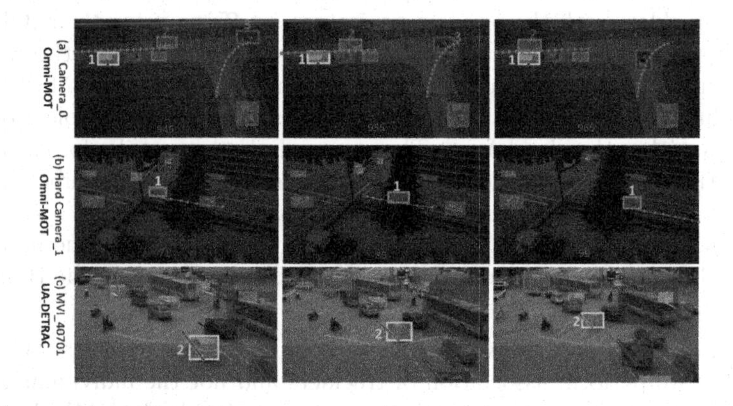

Fig. 7. Tracking illustration of DMM-Net on Omni-MOT (first and second row) and UA-DETRAC (third row).The mentioned IDs are for reference in the text only.

Implementation Details: We implement the DMM-Net using Pytorch [32] and train it using NVIDIA GeForce GTX Titan GPU. The hyper-parameter values are selected with cross-validation to maximize the MOTA metric on the validation set of UA-DETRAC. The chosen values of the hyperparameters are as follows. Batch size $B = 8$, number of training epochs per model = 20, number of input frames = 16, input frame size = 168×168, We use Adam Optimizer [24] for training. Other hyper-parameter values are $\delta_c = 0.4$, $\delta_{nms} = 0.3$, and $\delta_o = 0.5$.

Omni-MOT Dataset: To demonstrated the efficacy of DMM-Net and trainability of deep models on our dataset, we select five scenes from the Omni-MOT and perform tracking. The scenes are chosen with Easy camera viewpoint with clear weather conditions. The selected scenes are from Town 02 (with 50 vehicles) that are indexed 0, 1, 5, 7 and 8 in the dataset. In this experiment, we train and test DMM-Net using the train and test partitions of the selected scene, as provided by Omni-MOT. Our training required 59.2 h for 22 epochs.

We use both CLEAR MOT [5], and MT/ML [35] metrics and summarize the results in Table 4. For the definitions of metrics we refer to the original

Table 5. Results on UA-DETRAC: 'T.S.' is tracker speed (fps), 'D.S.' is detector speed (fps), and 'A.S.' is the overall speed (fps). The DMM-Net does not need explicit detector. DMM-Net+ uses a subset of Omni-MOT for data augmentation.

Tracker	Detector	PR-MOTA↑	PR-MT↑	PR-ML↓	PR-IDS↓	PR-FRAG↓	PR-FP↓	PR-FN↓	T.S.↑	D.S.↑	A.S.↑
CEM [1]	ACF [12]	4.50%	2.90%	37.10%	265.4	366.0	15180.3	270643.2	3.74	0.67	0.57
CMOT [49]	ACF [12]	7.80%	14.30%	20.70%	418.3	2161.7	81401.4	183400.2	3.12	0.67	0.55
DCT [2]	ACF [12]	7.90%	4.80%	34.40%	108.1	101.4	13059.7	251166.4	1.29	0.67	0.44
GOG [33]	ACF [12]	10.80%	12.20%	22.30%	3950.8	3987.3	45201.5	197094.2	319.29	0.67	0.67
H2T [47]	ACF [12]	8.20%	13.10%	21.30%	1122.8	1445.8	71567.4	189649.1	1.08	0.67	0.41
IHTLS [11]	ACF [12]	6.60%	11.50%	22.40%	1243.1	4723.0	72757.5	198673.5	5.09	0.67	0.59
CEM [1]	DPM [16]	3.30%	1.30%	37.80%	265	317.1	13888.7	270718.5	4.49	0.17	0.16
DCT [2]	DPM [16]	2.70%	0.50%	42.70%	72.2	**68.8**	7785.8	280762.2	2.85	0.17	0.16
GOG [33]	DPM [16]	5.50%	4.10%	27.70%	1873.9	1988.5	38957.6	230126.6	476.52	0.17	0.17
IOUT [6]	DPM [16]	1.92%	0.84%	43.70%	**61.4**	106.0	**3111.5**	290412.2	100842	0.17	0.17
CEM [1]	R-CNN [19]	2.70%	2.30%	34.10%	778.9	1080.4	34768.9	269043.8	5.4	0.1	0.10
DCT [2]	R-CNN [19]	11.70%	10.10%	22.80%	758.7	742.9	336561.2	210855.6	0.71	0.1	0.09
CMOT [49]	R-CNN [19]	11.00%	**15.70%**	19.00%	506.2	22551.1	74253.6	**177532.6**	3.59	0.1	0.10
GOG [33]	R-CNN [19]	10.00%	13.50%	20.10%	7834.5	7401.0	58378.5	192302.7	352.8	0.1	0.10
H2T [47]	R-CNN [19]	11.10%	14.60%	19.80%	1481.9	1820.8	66137.2	184358.2	2.78	0.1	0.10
IHTLS [11]	R-CNN [19]	8.30%	12.00%	21.40%	1536.4	5954.9	68662.6	199268.8	11.96	0.1	0.10
DMM-Net	-	**11.80%**	10.30%	**15.20%**	230.3	658.0	36238.8	194886.4	-	-	**123.25**
DMM-Net+	-	**12.20%**	10.80%	**14.90%**	228.2	674.1	36355.8	192289.6	-	-	**123.25**

works. We refer to [25] for the details on the comprehensive metrics MOTA and MOTAP. The table provides metric values for both training and test partitions. The variation between these values is a clear indication that despite the Easy camera view, the dataset provides reasonable challenges for the MOT task. We provide results with additional view points in the supplementary material to further put the reported values into better perspective.

An illustration of tracking for Camera_0 is also provided in Fig. 7 (top). Our technique is able to easily track vehicles that are stationary, (e.g. 1), moving along a straight path (e.g. 2), and moving along a curved path (e.g. 3), which justifies the selection of our motion model. Additionally, our tracker has the power to deal with even full occlusions between the frames. We show such a case for a hard camera view point in the middle row of Fig. 7. In this case, vehicle-1 is totally occluded at the 86th frame which can be detected and tracked by our tracker. At the 99th frame, vehicle-1 is partially occluded, but the tracklets association performs correctly. Note that, our tracker does not need to be evaluated with (detector-based) UA-DETRAC metrics [45] as the Omni-MOT dataset provides ground truth detections without using off-the-shelf detectors.

UA-DETRAC: The UA-DETRAC challenge [45] is arguably the most widely used large-scale benchmark for MOT of vehicles. It comprises 100 videos @ 25 fps, recorded in 24 locations with frame size 540 × 960. For this challenge, results are computed by a remote server using CLEAR MOT and MT/ML metrics with Precision-Recall curve of the detection. We refer to [45] for further details on the metrics. In Table 5, we show our results on the UA-DETRAC challenge. The pre-fix PR for the metrics indicates the use of Precision-Recall curve. It can be seen that DMM-Net achieves excellent MOTA score, which is widely considered as the most comprehensive metric for MOT. We also provide results for DMM-Net+ which augments the UA-DETRAC training set with a subset of Omni-MOT. This subset contained 8 random videos from Omni-MOT. We can see an overall performance gain with this augmentation, exemplifying the utility of Omni-MOT in data augmentation.

Notice that our technique does not require an external 'detector' and achieves highly promising PR-MOTA score for UA-DETRAC @ 120+ fps - orders of magnitude faster than the existing methods. In the third row of Fig. 7, we show an example of tracking result. The box color indicates the predicted object identity. Our tracker leverages information from previous frames to detect partially occluded objects, e.g.. vehicle-3 in frame 518. The figure also shows the generalization power of our in-built detection mechanism. For instance, vehicle-1 is a very rare vehicle that is consistently assigned the correct identity. In the supplementary material, we provide further example tracking videos that clearly demonstrate successful tracking by DMM-Net for UA-DETRAC.

6 Conclusion

In the context of tracking-by-detection, we proposed a deep network DMM-Net that removes the need for an explicit external detector and performs tracking at 120+ fps to achieve 12.80% PR-MOTA value on the UA-DETRAC challenge. Our network meticulously models object motions, classes and their visibility that are subsequently used for efficiently associating the object tracks. The proposed network provides an end-to-end solution for detection and tracklet generation across multiple video frames. We also propose a realistic CARLA simulator based large-scale dataset with over 14M frames for vehicle tracking. The dataset provides precise and comprehensive ground truth with full control over data parameters, which allows for the much needed transparency in evaluation. We also provide scripts to generate more data under our framework and we make the code of DMM-Net public.

References

1. Andriyenko, A., Schindler, K.: Multi-target tracking by continuous energy minimization. In: Proceedings of the IEEE Computer Society Conference on Computer Vision and Pattern Recognition, pp. 1265–1272 (2011). https://doi.org/10.1109/CVPR.2011.5995311
2. Andriyenko, A., Schindler, K., Roth, S.: Discrete-continuous optimization for multi-target tracking. In: Proceedings of the IEEE Computer Society Conference on Computer Vision and Pattern Recognition, pp. 1926–1933 (2012). https://doi.org/10.1109/CVPR.2012.6247893
3. Bae, S.H., Yoon, K.J.: Confidence-based data association and discriminative deep appearance learning for robust online multi-object tracking. IEEE Trans. Pattern Anal. Mach. Intell. **40**(3), 595–610 (2018). https://doi.org/10.1109/TPAMI.2017.2691769

4. Berclaz, J., Fleuret, F., Türetken, E., Fua, P.: Multiple object tracking using k-shortest paths optimization. IEEE TPAMI **33**(9), 1806–1819 (2011). https://doi.org/10.1109/TPAMI.2011.21
5. Bernardin, K., Stiefelhagen, R.: Evaluating multiple object tracking performance: the CLEAR MOT metrics. Eurasip J. Image Video Process. **2008** (2008). https://doi.org/10.1155/2008/246309
6. Bochinski, E., Eiselein, V., Sikora, T.: High-Speed tracking-by-detection without using image information. In: 2017 14th IEEE International Conference on Advanced Video and Signal Based Surveillance, AVSS 2017 (2017). https://doi.org/10.1109/AVSS.2017.8078516
7. Butt, A.A., Collins, R.T.: Multi-target tracking by Lagrangian relaxation to min-cost network flow. In: Proceedings of CVPR, pp. 1846–1853 (2013)
8. Chari, V., Lacoste-Julien, S., Laptev, I., Sivic, J.: On pairwise costs for network flow multi-object tracking. In: Proceedings of CVPR, 07–12 June, pp. 5537–5545 (2015). https://doi.org/10.1109/CVPR.2015.7299193
9. Dai, J., Li, Y., He, K., Sun, J.: R-FCN: object detection via region-based fully convolutional networks. In: NIPS 2016 Proceedings of the 30th International Conference on Neural Information Processing Systems, pp. 379–387 (2016). https://academic.microsoft.com/paper/2407521645
10. Dehghan, A., Modiri Assari, S., Shah, M.: GMMCP tracker: globally optimal generalized maximum multi clique problem for multiple object tracking. In: Proceedings of CVPR, pp. 4091–4099 (2015)
11. Dicle, C., Camps, O.I., Sznaier, M.: The way they move: Tracking multiple targets with similar appearance. In: Proceedings of the IEEE International Conference on Computer Vision, pp. 2304–2311 (2013). https://doi.org/10.1109/ICCV.2013.286
12. Dollar, P., Appel, R., Belongie, S., Perona, P.: Fast feature pyramids for object detection. IEEE Trans. Pattern Anal. Mach. Intell. (2014). https://doi.org/10.1109/TPAMI.2014.2300479
13. Dosovitskiy, A., Ros, G., Codevilla, F., Lopez, A., Koltun, V.: CARLA: an Open Urban Driving Simulator. In: Proceedings of the 1st Annual Conference on Robot Learning, pp. 1–16 (2017). http://arxiv.org/abs/1711.03938
14. Emami, P., Pardalos, P.M., Elefteriadou, L., Ranka, S.: Machine learning methods for solving assignment problems in multi-target tracking. arXiv:1802.06897 **1**(1), 1–35 (2018)
15. Feichtenhofer, C., Pinz, A., Zisserman, A.: Detect to track and track to detect. In: Proceedings of the IEEE International Conference on Computer Vision 2017-October, pp. 3057–3065 (2017). https://doi.org/10.1109/ICCV.2017.330
16. Felzenszwalb, P.F., Girshick, R.B., McAllester, D., Ramanan, D.: Object detection with discriminatively trained part-based models. IEEE TPAMI **32**(9), 1627–1645 (2010)
17. Ferryman, J., Shahrokni, A.: PETS2009: Dataset and challenge. In: Proceedings of the 12th IEEE International Workshop on Performance Evaluation of Tracking and Surveillance, PETS-Winter 2009 (2009). https://doi.org/10.1109/PETS-WINTER.2009.5399556
18. Geiger, A., Lenz, P., Urtasun, R.: Are we ready for autonomous driving? the KITTI vision benchmark suite. In: Proceedings of the IEEE Computer Society Conference on Computer Vision and Pattern Recognition, pp. 3354–3361 (2012). https://doi.org/10.1109/CVPR.2012.6248074

19. Girshick, R., Donahue, J., Darrell, T., Malik, J.: Rich feature hierarchies for accurate object detection and semantic segmentation. In: Proceedings of the IEEE Computer Society Conference on Computer Vision and Pattern Recognition (2014). https://doi.org/10.1109/CVPR.2014.81
20. Hara, K., Kataoka, H., Satoh, Y.: Can Spatiotemporal 3D CNNs Retrace the History of 2D CNNs and ImageNet? In: CVPR, pp. 6546–6555 (2018). https://doi.org/10.1109/CVPR.2018.00685, http://arxiv.org/abs/1711.09577
21. He, K., Zhang, X., Ren, S., Sun, J.: Deep residual learning for image recognition. In: Proceedings of the IEEE Computer Society Conference on Computer Vision and Pattern Recognition (2016). https://doi.org/10.1109/CVPR.2016.90
22. Ioffe, S., Szegedy, C.: Batch normalization: accelerating deep network training by reducing internal covariate shift. arXiv (2015)
23. Iqbal, U., Milan, A., Gall, J.: PoseTrack: joint multi-person pose estimation and tracking. In: Proceedings - 30th IEEE Conference on Computer Vision and Pattern Recognition, CVPR 2017 (2017). https://doi.org/10.1109/CVPR.2017.495
24. Kingma, D.P., Ba, J.L.: Adam: a Method for Stochastic Optimization. In: ICLR 2015 : International Conference on Learning Representations 2015 (2015). https://academic.microsoft.com/paper/2964121744
25. Leal-Taixé, L., Milan, A., Reid, I., Roth, S., Schindler, K.: MOTChallenge 2015: towards a benchmark for multi-target tracking. arXiv:1504.01942 [cs] pp. 1–15 (2015). http://arxiv.org/abs/1504.01942
26. Liu, W., et al.: SSD: single shot multibox detector. In: Leibe, B., Matas, J., Sebe, N., Welling, M. (eds.) ECCV 2016. LNCS, vol. 9905, pp. 21–37. Springer, Cham (2016). https://doi.org/10.1007/978-3-319-46448-0_2
27. Luo, W., et al.: Multiple object tracking: a literature review. arXiv:1409.7618v4, pp. 1–18 (2017). https://doi.org/10.1145/0000000.0000000
28. Milan, A., Leal-Taixe, L., Reid, I., Roth, S., Schindler, K.: MOT16: a benchmark for multi-object tracking. CoRR abs/1603.0 (2016). http://arxiv.org/abs/1603.00831
29. Milan, A., Roth, S., Schindler, K.: Continuous energy minimization for multitarget tracking. IEEE Trans. Pattern Anal. Mach. Intell. **36**(1), 58–72 (2014). https://doi.org/10.1109/TPAMI.2013.103
30. Munkres, J.: Algorithms for the assignment and transportation problems. J. Soc. Ind. Appl. Math. **5**(1), 32–38 (1957). https://doi.org/10.1137/0105003
31. Nair, V., Hinton, G.: Rectified linear units improve restricted Boltzmann machines. In: Proceedings of the 27th International Conference on Machine Learning (2010)
32. Paszke, A., et al.: Automatic differentiation in PyTorch. Adv. Neural Inf. Process. Syst. **30**(Nips), 1–4 (2017)
33. Pirsiavash, H., Ramanan, D., Fowlkes, C.C.: Globally-optimal greedy algorithms for tracking a variable number of objects. Proceedings of the IEEE Computer Society Conference on Computer Vision and Pattern Recognition, pp. 1201–1208 (2011). https://doi.org/10.1109/CVPR.2011.5995604
34. Reid, D., et al.: An algorithm for tracking multiple targets. IEEE Trans. Autom. Control **24**(6), 843–854 (1979)
35. Ristani, E., Solera, F., Zou, R., Cucchiara, R., Tomasi, C.: Performance measures and a data set for multi-target, multi-camera tracking. In: Hua, G., Jégou, H. (eds.) ECCV 2016. LNCS, vol. 9914, pp. 17–35. Springer, Cham (2016). https://doi.org/10.1007/978-3-319-48881-3_2
36. Ristani, E., Tomasi, C.: Tracking multiple people online and in real time. In: Cremers, D., Reid, I., Saito, H., Yang, M.-H. (eds.) ACCV 2014. LNCS, vol. 9007, pp. 444–459. Springer, Cham (2015). https://doi.org/10.1007/978-3-319-16814-2_29

37. Roshan Zamir, A., Dehghan, A., Shah, M.: GMCP-tracker: global multi-object tracking using generalized minimum clique graphs. In: Fitzgibbon, A., Lazebnik, S., Perona, P., Sato, Y., Schmid, C. (eds.) ECCV 2012. LNCS, vol. 7573, pp. 343–356. Springer, Heidelberg (2012). https://doi.org/10.1007/978-3-642-33709-3_25
38. Shafique, K., Shah, M.: A noniterative greedy algorithm for multiframe point correspondence. IEEE Trans. Pattern Anal. Mach. Intell. **27**(1), 51–65 (2005)
39. Sheng, H., Zhang, Y., Chen, J., Xiong, Z., Zhang, J.: Heterogeneous association graph fusion for target association in multiple object tracking. IEEE Trans. Circ. Syst. Video Technol. **29**, 3269–3280 (2018)
40. Shitrit, H.B., Berclaz, J., Fleuret, F., Fua, P.: Multi-commodity network flow for tracking multiple people. IEEE TPAMI **36**(8), 1614–1627 (2014)
41. Shu, G., Dehghan, A., Oreifej, O., Hand, E., Shah, M.: Part-based multiple-person tracking with partial occlusion handling. In: Proceedings of CVPR, pp. 1815–1821. IEEE (2012)
42. Sun, S., Akhtar, N., Song, H., Mian, A., Shah, M.: Deep affinity network for multiple object tracking **13**(9), 1–15 (2018). http://arxiv.org/abs/1810.11780
43. Tian, Y., Dehghan, A., Shah, M.: On detection, data association and segmentation for multi-target tracking. IEEE Trans. Pattern Anal. Mach. Intell. **41**, 2146–2160 (2018)
44. Voigtlaender, P., et al.: Mots: multi-object tracking and segmentation. In: Proceedings of CVPR, pp. 7942–7951 (2019)
45. Wen, L., et al.: UA-DETRAC: a new benchmark and protocol for multi-object detection and tracking (2015). http://arxiv.org/abs/1511.04136
46. Wen, L., Du, D., Li, S., Bian, X., Lyu, S.: Learning non-uniform hypergraph for multi-object tracking. In: Proceedings of the AAAI Conference on Artificial Intelligence, vol. 33, pp. 8981–8988 (2019)
47. Wen, L., Li, W., Yan, J., Lei, Z., Yi, D., Li, S.Z.: Multiple target tracking based on undirected hierarchical relation hypergraph. In: Proceedings of the IEEE Computer Society Conference on Computer Vision and Pattern Recognition, vol. 1, pp. 1282–1289 (2014). https://doi.org/10.1109/CVPR.2014.167
48. Wu, B., Nevatia, R.: Detection and tracking of multiple, partially occluded humans by Bayesian combination of edgelet based part detectors. Int. J. Comput. Vis. **75**(2), 247–266 (2007)
49. Zhu, J., Yang, H., Liu, N., Kim, M.: Online Multi-Object Tracking with Dual Matching Attention Networks, pp. 1–17 (2019)

Deep FusionNet for Point Cloud Semantic Segmentation

Feihu Zhang[1(✉)], Jin Fang[2], Benjamin Wah[3], and Philip Torr[1]

[1] University of Oxford, Oxford, UK
feihu.zhang@eng.ox.ac.uk
[2] Baidu Research, Beijing, China
[3] Chinese University of Hong Kong, Sha Tin, Hong Kong S.A.R.

Abstract. Many point cloud segmentation methods rely on transferring irregular points into a voxel-based regular representation. Although voxel-based convolutions are useful for feature aggregation, they produce ambiguous or wrong predictions if a voxel contains points from different classes. Other approaches (such as PointNets and point-wise convolutions) can take irregular points for feature learning. But their high memory and computational costs (such as for neighborhood search and ball-querying) limit their ability and accuracy for large-scale point cloud processing. To address these issues, we propose a deep fusion network architecture (FusionNet) with a unique voxel-based "mini-PointNet" point cloud representation and a new feature aggregation module (fusion module) for large-scale 3D semantic segmentation. Our FusionNet can learn more accurate point-wise predictions when compared to voxel-based convolutional networks. It can realize more effective feature aggregations with lower memory and computational complexity for large-scale point cloud segmentation when compared to the popular point-wise convolutions. Our experimental results show that FusionNet can take more than one million points on one GPU for training to achieve state-of-the-art accuracy on large-scale Semantic KITTI benchmark.The code will be available at https://github.com/feihuzhang/LiDARSeg.

1 Introduction

Semantic segmentation of 3D Point cloud is widely applicable in many scenarios, including remote sensing, AR/VR, robotics, and self-driving cars. Many deep neural network models have been proposed for this important task.

Approaches typically rely on a voxel-based regular representation that converts unordered points to regular 3D voxel/grids before using 3D/2D convolutions for feature learning [11,23,29,36,36,43,54,54]. Inspired by the success in image-based segmentation, these volumetric networks can be efficiently designed and trained for semantic segmentation of 3D point clouds, using regular convolutions which have been shown useful for coarse-grain feature aggregation/learning

Electronic supplementary material The online version of this chapter (https://doi.org/10.1007/978-3-030-58586-0_38) contains supplementary material, which is available to authorized users.

A. Vedaldi et al. (Eds.): ECCV 2020, LNCS 12369, pp. 644–663, 2020.
https://doi.org/10.1007/978-3-030-58586-0_38

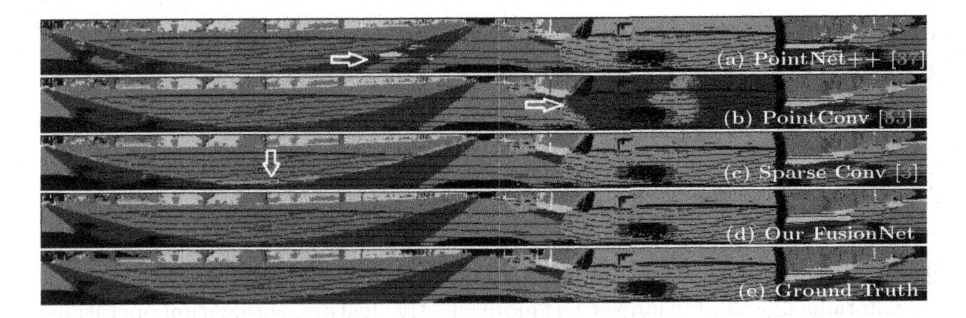

Fig. 1. Problem illustrations with large-scale point clouds (Semantic KITTI [2]). The LiDAR frame contains more than 120K points in a large 3D space of 160 m × 160 m × 20 m. Results are projected into 2D cylindrical images for visual comparisons. *Top rows (a) and (b):* State-of-the-art PointNet++ [37] and point-wise convolutions [53] fail in large objects/areas (*e.g.* sidewalk, car) due to insufficient feature propagations/aggregations for large-scale point clouds. *Third row (c):* State-of-the-art sparse convolutions [3] predict wrong labels at the border between different classes due to their ambiguous and coarse voxel-level learning. *Fourth row (d):* Our FusionNet gives accurate segmentation labels for the large-scale point cloud. *Last row (e):* ground truth.

[3,7,61]. However, they can only give voxel-level predictions. Moreover, their voxelization process transfers many raw points to one single voxel that will produce ambiguous or wrong predictions at object borders when voxels consist of points from different classes (as illustrated in Fig. 1(c)).

When compared to regular convolution operations, PointNets [35,37] can take irregular points for feature learning. However, the Multi-Layer-Perceptron (MLP) [35] in PointNets keeps only the most significant activation and may lose some useful detailed information for segmentation. Also, these models are limited for training deep and robust networks for large-scale point clouds (Fig. 1(a)) because they have high memory and computational complexity in their neighborhood search, sampling, and ball-querying operations [35,37,50,57].

There is also plenty of work on point-based convolutions that explores the idea of convolutions directly on irregular points [8,25,26,46,47,55]. They learn to approximate a weight function or interpolate convolutional weights [10,25,28,32, 50,53,55]. Compared to volumetric convolutions that can directly index a kernel with fixed relative positions of the neighborhoods, the positions of neighbors become unpredictable in these point-wise convolutions as the points are scattered irregularly. Hence, the kernel for neighboring points must be calculated on the fly. The additional memory costs and matrix multiplications will limit the training of effective networks for large-scale point clouds and may produce wrong predictions in some objects due to insufficient feature aggregations (Fig. 1(b)).

To address the challenges of designing effective network architectures for accurate point-wise segmentation of large-scale point clouds, we develop a deep fusion network architecture with a unique voxel-based "mini-PointNet" structure for point cloud representation and a novel fusion module (Fig. 2) for feature aggregation. The proposed FusionNet realizes both the voxel aggregation and the fine-grain point-wise feature learning. It inherits all the advantages (in terms of effectiveness and efficiency) of volumetric convolutional networks (*e.g.*, 3D UNet [41]) while being able to learn point-wise features for accurate label predictions.

FusionNet possesses many advantages over existing models in large-scale point cloud segmentation:

1) When compared to existing voxel networks [3,7,61], FusionNet can predict point-wise labels and avoid those ambiguous/wrong predictions when a voxel has points from different classes.
2) When compared to the popular PointNets [35,37,50,57] and point-based convolutions [25,53], FusionNet has more effective feature aggregation operations (including the efficient neighborhood-voxel aggregations and the fine-grain inner-voxel point-level aggregations). These operations help produce better accuracy for large-scale point cloud segmentation.
3) FusionNet takes full advantage of the sparsity property to reduce its memory footprint. For instance, it can take more than one million points in training and use only one GPU to achieve state-of-the-art accuracy.

With an effective feature aggregation module and lower memory and computation costs, we can train our FusionNet to learn more effective deep features for accurate large-scale point cloud segmentation (as illustrated in Fig. 1d). In our experiments, FusionNet achieves state-of-the-art performance on large-scale point cloud datasets. Especially, it outperforms state-of-the-art PointNets [37] (by 40%), point-wise convolutions [53] (by 10%) and sparse CNN [3] (by 7%) in mean IoU evaluations on the challenging Semantic KITTI benchmark.

2 Related Work

Existing approaches for 3D point cloud segmentation can be roughly categorized into two types: regular voxel-based networks and irregular point-based networks.

2.1 Voxel-Based Networks

There is substantial previous work on 3D CNN to convert point clouds to 2D or 3D volumetric grids/voxels (or similar slices/lattices) [29,36,54,61]. For example, 3D point clouds or shapes have been projected into several 2D image-like grids with 2D convolutions for shape classification and retrieval tasks [36,43]. Other studies [11,23,29,36,54] voxelize point clouds into 3D volumetric voxels and apply 3D convolution networks for point cloud processing and understanding.

Among these studies, VV-Net [30] uses a kernel-based interpolated variational auto encoder (VAE) on regular voxel grids; instance segmentation models have been proposed on dense 3D voxels/2D grids [19,61]; panoptic labels can be predicted using a spatially hashed volumetric map [31]; high-resolution RGB inputs have been leveraged by associating 2D images with the volumetric grid [11]; efficient feature aggregations have been explored in PVCNN [27] through a voxel-based regular CNN in low resolution to avoid random memory accesses.

To extend voxel-based networks to high-resolution scene segmentation tasks [2,4], a set of unbalanced octrees have been used to improve the resolution [40]. Sparse Submanifold CNN [3,7] computes the convolutions only at activated points to design high-resolution volumetric inputs. However, higher resolution inputs mean more sparse inputs, making convolutional layers less efficient for feature aggregation. Also, it will increase memory and computation costs.

2.2 Point-Based Networks

Recent work takes raw points as input for feature learning and label predictions.

PointNets: These methods aggregate features from different points by shared multi-layer perceptrons (MLP) [35]. PointNet++ improves its network by adding a hierarchical structure [37]. Wang *et al.* associate instances and semantics segmentation [49] with feature encoder of stacked PointNet layers. He *et al.* learn deep geodesic-aware representations for PointNet/PointNet++ [9]. The issue with the MLP module is that they keep only the most significant activation on features, leading to the loss of some useful detailed information for segmentation.

Point-Wise Convolutions: Other recent work explores the extension of convolutions on irregular and unordered point-cloud inputs [8,25,26,46,47,55].

For unordered points, PointCNN performs convolutions on transformed points [25]. Zhang *et al.* use statistics from concentric spherical shells to resolve point-order ambiguities [62]. FeaStNet generalizes conventional convolution layers by adding a soft-assignment matrix [47]. KPConv learns flexible and deformable point convolutions [46]. Point-wise Conv [13,53], GeoCNN [20] and annular convolution [18] query the nearest neighbors for convolutions with different kernels.

Some work explores contiguous convolutions [32,50,53] for point clouds. For instance, PointConv [53] treats convolution kernels as nonlinear functions of the coordinates and uses (MLP) modules to learn continuous weight functions for point-wise convolutions. Hermosilla *et al.* develop Monte Carlo convolutions for learning non-uniformly sampled point clouds [10]. Similarly, Mao *et al.* propose to interpolate discrete convolutional weights for convolutions on points [28].

Other work explores the ideas of convolutions on unique surfaces. Rao *et al.* map 3D points onto a discretized sphere for convolution definition [38]. Huang *et al.* utilize a 4-rotational symmetric field to define a domain for convolution on a surface [14]. Tangent convolution projects local surface geometry on a tangent plane around every point, producing planar convolution-able tangent images [44].

In the original regular 2D/3D convolutions, the relative positions of the neighborhoods are fixed, and the convolutional kernel can be directly indexed. However, for point-based convolutions, the points are irregularly scattered over a 3D space, making the relative positions of neighbors unpredictable. Hence, the kernel for each neighboring point must be calculated on the fly using additional matrix multiplications.

There are also graph or tree-based convolutions for point clouds, including local spectral graph convolutions [48], graph attention convolutions [49], regular-

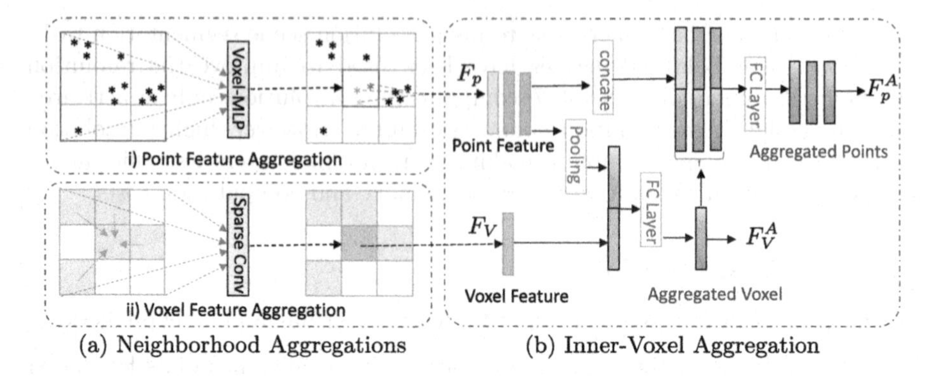

(a) Neighborhood Aggregations (b) Inner-Voxel Aggregation

Fig. 2. Illustartion of the FusionNet module. (a) Neighborhood aggregation, including i) point-level feature aggregation by the proposed Voxel-MLP and the ii) voxel-level feature aggregation with sparse convolutions. (b) Inner-voxel aggregation of each voxel in a feedback fashion. Features of all the points in the "mini-PointNet" are fused into voxel-level features by average-pooling and concatenation. The voxel-level features are then fed back to each point (with concatenation and refinement).

ized graph CNN (RGCNN) [45], and octree guided CNN with spherical kernels [24]. The unique feature of them is that they often rely on a graph construction and a learned kernel function for convolutions on unordered points. These are realized by additional modules/layers with extra computations and memory overhead which limit their abilities for designing effective deep architectures in large-scale point cloud processing.

Other Point Networks: There are also special networks developed for various applications to directly take raw points as input for sampling [58,60], semantic [17,39,51,59,63,64], and instance segmentation [33,57]. Some employ new and special data structures for point cloud processing, including 3D point capsule encoder and decoder networks [65], over segmentation on graph structure [21], basis point sets [34], multi-resolution tree-structured networks [5], bilateral convolutions on a sparse lattice of the point cloud [42], recurrent neural networks (RNN) on slices of a point cloud [15], and the superpoint graph-based semantic segmentation [22].

3 FusionNet

Many previous models are either limited by the ambiguous voxel-level predictions [7,29,61] or have insufficient feature aggregations that use high memory or computational costs [25,27,37,61] in large-scale point-cloud segmentation. This section describes our deep FusionNet model that can learn accurate point-wise features and realize more effective and memory-efficient feature aggregations in large-scale processing. It utilizes a unique voxel-based "mini-PointNet" for sparse representation and the novel FusionNet modules for feature aggregations.

Figure 2 illustrates the FusionNet module that includes both **neighborhood voxel aggregation** (Fig. 2(a)) for voxel-level learning and **inner-voxel aggregation** for fine-grain point-level learning (Fig. 2(b)). The rest of this section is organized as follows. Section 3.1 describes the unique point cloud representation. Section 3.2 and 3.3 present the design of the FusionNet module (Fig. 2). Section 3.4 shows the up- and down-sampling layers. Finally, Sect. 3.5 describes the architecture and its sparse implementation.

3.1 Point Cloud Representation

Our point cloud representation is based on a unique voxel-based "mini-PointNet" that has two steps: voxelization and "mini-PointNet" construction.

Voxelization: Given point cloud P in a range of $L_x \times L_y \times L_z$ as input, we transfer the irregular points into regular 3D voxels with a resolution of $H \times W \times D$. The resolution is controlled by the voxel parameter $s = (s_x, s_y, s_z)$ (length/width/height of the each voxel). Here, $(H, W, D) = (L_x/s_x, L_y/s_y, L_z/s_z)$.

Mini-PointNet: Each point p can be represented as $p = \{(p_x, p_y, p_z), F_p\}$ with (p_x, p_y, p_z) as its 3D space location and F_p as its point features (*e.g.* color, intensity). Since many voxels may have more than one point, voxel V with m points can be represented as a "mini-PointNet": $V = \{(V_x, V_y, V_z), F_V, \{p_1, p_2, ...p_m\}\}$, where (V_x, V_y, V_z) is the coordinate of the voxel that contains m points $\{p_1, p_2...p_m\}$, and F_V is the voxel features that can be learned from those points in each voxel.

The voxel-based "mini-PointNets" are stored in sparse data structures to reduce their memory requirement. Empty or invalid voxels are not be stored or processed during feature aggregations.

After transferring the point cloud into the voxel-based "mini-PointNet" representation, our FusionNets use the proposed feature aggregation modules to learn deep feature representation for segmentation.

3.2 Neighborhood Aggregations

As illustrated in Fig. 2(a), there are two types of feature aggregations can propagate information from neighborhood voxels to the current voxel: voxel feature aggregation and point feature aggregation.

Voxel Feature Aggregation: We utilize regular convolutions for voxel feature aggregation, as the convolutional layer has been proven to be successful in 2D-image semantic segmentation. The knowledge can be easily transferred to the design and training of deep network models for 3D point clouds. Regular voxel-based convolutions are also more efficient than many existing point-wise convolution models [10,25,28,32,50,53,55]. These models requires extra memory and computations to learn or interpolate the weight kernels of the point

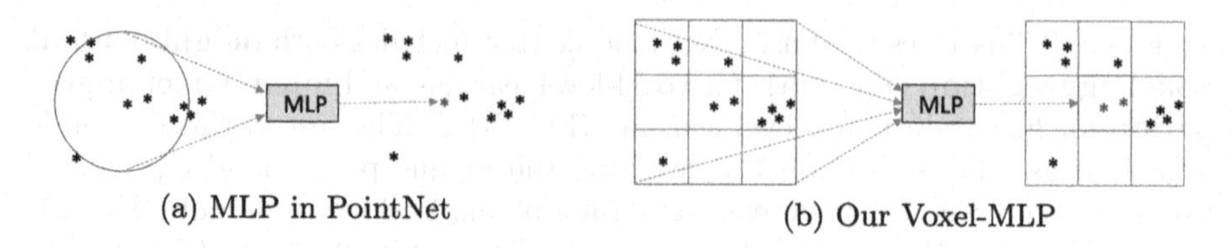

Fig. 3. (a) The original MLP module widely used in PointNets [35,37] and point-wise convolutions [13,53]. It relies on neighborhood search/ball querying for each pixel with high time complexity ($O(n^2)$ or $O(n\log(n))$ with k-d tree). (b) Our voxel-level MLP (*e.g.* kernel size = 3) directly aggregates features from all the points in the neighborhood voxels to points in the target voxel, allowing neighborhoods to be easily visited in regular voxels. The Voxel-MLP has a lower $O(n)$ time complexity.

Algorithm 1: Implementation of the Voxel-MLP

---

for *each voxel V:* **do**
  for *all points p ($p \in N(V)$) in neighboor voxels $N(V)$ of V* **do**
    Spatial encoding:
      i) spatial coordinate shifts $\Delta p \leftarrow (p_x, p_y, p_z) - (V_x, V_y, V_z)$;
      ii) concatenation and FC layer: $\mathbf{F'}_p \leftarrow \text{fc}(\text{cat}\{\mathbf{F}_p, \Delta p\})$;
    Point feature max-pooling: $\mathbf{F}_{max} \leftarrow \max\{\mathbf{F'}_p, p \in N(V)\}$;
  end
  for *each point p in current voxel V* **do**
    Concatenate $\mathbf{F'}_p$ and $\mathbf{F}_{max}$: $\mathbf{F''}_p \leftarrow \text{cat}(\mathbf{F'}_p, \mathbf{F}_{max})$;
    Refinement with point-wise fully-connected layer: $\mathbf{F}_p^A \leftarrow \text{fc}(\mathbf{F''}_p)$;
  end
end
Result: Aggregated point feature $\mathbf{F}_p^A$ ($p \in V$)

---

convolutions, as well as to locate the neighborhood points from irregular point cloud by KNN search or ball querying.

To reduce the memory footprint and to develop deep neural networks for the segmentation of large-scale point clouds, we use the Sparse Submanifold Convolution layer [3,7] for voxel-based aggregation to compute the convolution only at activated voxels (as illustrated in step ii) of Fig. 2(a)). This approach helps minimize the memory needed.

Point Feature Aggregation: Figure 3(b) presents a more efficient voxel-based multi-layer perceptron (Voxel-MLP) for neighborhood feature aggregation.

For the m points $\{p_1, p_2, ...p_m\}$ in the current voxel V, their point feature can be aggregated from all the points in the neighborhood voxels $N(V)$ (*e.g.* 27 neighborhood voxels if the kernel size is 3) by our Voxel-MLP as follows:

$$\text{mlp}(p_1, p_2, ...p_m) = \gamma\left(\max_{p_k \in N(V)} \{h(p_k)\}, \{p_1, p_2, ...p_m\}\right). \quad (1)$$

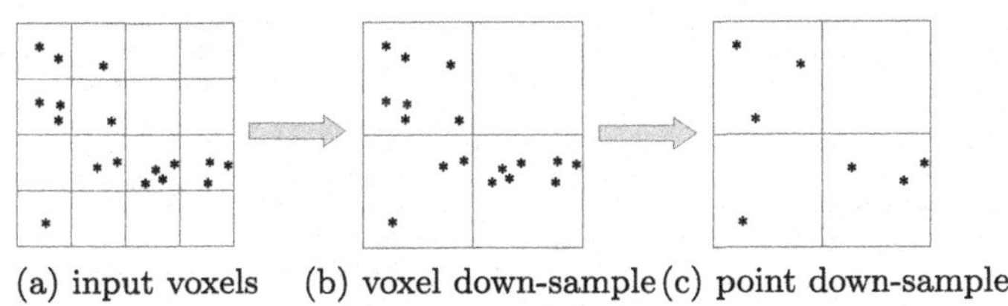

Fig. 4. Illustration of feature down-sampling. (a) Voxels and "mini-PointNets" inputs. (b) After down-sampling with stirde of 2, the size of the voxels is doubled to get a lower resolution. (c) The number of points in each voxel is also reduced to realize the "mini-PointNet" down-sampling.

The implementation details are presented in Algorithm 1. The response of h can be interpreted as the spatial encoding [35] of a point. The feature aggregation γ can be realized by concatenations and fully connected layers.

Our Voxel-MLP (Fig. 3(b)) is similar to the original MLP module (Fig. 3(a)) widely used in PointNets [35,37] and point-wise convolutions [13,53] for segmentation. However, the ball querying, sampling, or KNN neighborhood search in these existing models need $O(n^2)$ time complexity for points selection in irregular point clouds. Although k-d trees can be used to accelerate the search with $O(n \log(n))$ complexity, building a k-d tree for each layer will also cost extra memory and computations. In contrast, in our Voxel-MLP, all the neighborhood points can be directly visited more efficiently from neighborhood voxels (by hash mapping with $O(1)$ time complexity for each point–in total $O(n)$).

The receptive field of the Voxel-MLP is controlled by the kernel size that is the same as our convolutional aggregation. For example, when setting the kernel size to 3, neighborhood points will be selected from 27 neighborhood voxels.

3.3 Inner-Voxel Aggregation

Voxel-level aggregations only learn coarse-grain voxel features that usually produce ambiguous/wrong predictions at the object borders when voxels consist of points from different classes. To address this issue, we design the inner-voxel aggregation for fine-grain point-level feature aggregations and label predictions.

Figure 2(b) illustrates the inner-voxel aggregation realized by a feedback design of the feature fusion. The point-wise feature F_p and voxel feature F_V are from the outputs of the neighborhood aggregation block (Fig. 2(a)). They are sent to the inner-voxel block for fusion and aggregation. We first use point-wise average pooling to achieve the pooled feature vector from m points in the "mini-PointNet." The feature vector is then fused into the voxel feature F_V by concatenation. After refinement by one 1×1 convolution/FC layer, the aggregated voxel-level features F_V^A are then fed back to each point in the "mini-PointNet" by concatenations. Finally, a point-wise fully connected layer is employed to get the aggregated point feature F_p^A.

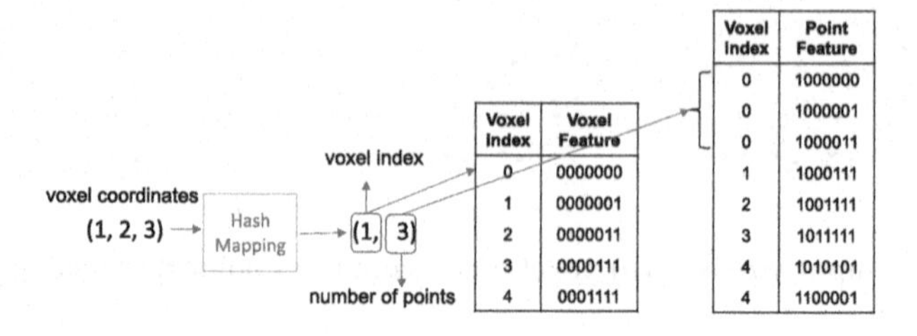

Fig. 5. Illustration of the sparse data structure and the voxel/point visiting by hash mapping.

Namely, for each valid voxel V with m points $(\{p_1, p_k...p_m\} \in V)$ in the "mini-PointNet", the inner-voxel aggregation can be represented as:

$$\begin{aligned} F_V^A &= \gamma_1 \left(\underset{p \in V}{\operatorname{avg}} \left\{ h(F_p, p)) \right\}, F_V \right) \\ F_p^A &= \gamma_2 \left(h(F_p, p), F_V^A \right), \ p \in V. \end{aligned} \tag{2}$$

Where the response of h can be interpreted as the spatial encoding of the point p. It can be learned by fully-connected layers with the point feature F_p and its spatial shifts to the voxel center $(p_x - V_x, p_y - V_y, p_z - V_z)$ as inputs (which is similar to [35] and the Voxel-MLP). γ_1 and γ_2 are the feature fusion functions which are realized by concatenations and fully-connected layer.

After the circulation of the inner-voxel aggregation, the voxel-level features and the point features are deeply fused. They are then sent to the next FusionNet module/layer for further aggregations.

3.4 Down/Up-Sampling

Down- and up-sampling are essential in neural networks for segmentation since they help capture pyramidal features in different resolutions and receptive fields.

For regular voxels, we use convolutions and transposed convolutions with strides (*e.g.* 2) for down- and up-sampling. These are similar to existing image segmentation models (*e.g.* UNet [41]) that have been found to be effective.

Figure 4 illustrates our point feature down-sampling that strictly follows voxel-level sampling. After voxel-level down-sampling, the size of a voxel is expanded to get a lower-resolution representation. To realize point-level down-sampling, we reduce the number of points in each voxel by re-sampling to get the new "mini-PointNet". For example, we reduce the number of points of each voxel to half while ensuring at least one valid point to avoid new empty voxels (Fig. 4(c)).

When compared to the slow iterative farthest-sampling strategy [37] (which sometimes takes 1–2 s for large-scale point cloud), the point sampling in FusionNet is realized in each voxel independently and can be computed in parallel by GPU in super fast speed (10–100 times faster than the farthest-sampling strategy). Moreover, voxel-based sampling guarantees each valid voxel to have at least one point and avoids the appearance of new empty voxels. This is especially important for sparse point cloud or regions (such as LiDAR point cloud where points are very sparse in the distance). Our voxel-based down-sampling can keep the consistency between point-level and voxel-level networks that cannot be realized by random sampling and the farthest-sampling strategy.

For point up-sampling, we use point interpolations proposed in PointNet [35] to recover the high-resolution point representations.

3.5 Architecture and Sparse Implementation

Architecture:[1] We use the 3D UNet backbone which is the same as that in [3,7]. At each pyramid level, one FusionNet module is employed to replace the convolutional layer. Both point-level and voxel-level features are densely connected to the previous layers by concatenations.

Sparse Implementation: Figure 5 illustrates the storage of the points and voxels in a sparse data structure. Each voxel and its points can be visited quickly by hash mapping. The coordinates (including the batch ID) of the voxels are used as the keys to the hash map, and a querying coordinate directly points to the location of each voxel and the "mini-PointNet".

When compared to dense regular convolution nets [46,61], FusionNet takes full advantage of sparsity to minimize its memory requirement. It differs from PointNets [35,37] and point-wise convolutional nets [13,53] that need high computational complexity for neighborhood search. As FusionNet utilizes a regular-voxel representation and hash mapping to realize the $O(1)$ point/voxel indexing, it is more suitable for large-scale point cloud segmentation tasks.

4 Experiments

In our experiments, we train our model with conventional cross-entropy loss for 40k iterations on one GPU. We use Momentum SGD with the Poly scheduler to train networks from learning rate 1e-1 and apply data augmentation including rotation around the gravity axis, scaling, spatial translation and spatial elastic distortion. For evaluations, we use the standard mean Intersection over Union (mIoU) and mean Accuracy (mAcc) as metrics following the previous works.

4.1 Datasets

We use three point cloud datasets for evaluations. Two of them are collected from the indoor scenes, and the rest one is from the large-scale driving scenes.

S3DIS: Stanford 3D Indoor Spaces (S3DIS) dataset [1] contains scans of six floors of three buildings. We use the Fold #1 split following many previous works.

[1] Architecture details are in the supplementary materials: www.feihuzhang.com.

ScanNet: The ScanNet [4] 3D semantic segmentation benchmark consists of 3D reconstructions of real rooms which has 1.5k rooms and 20 classes for semantic segmentation. Each point cloud scan contains 7–550k points.

Semantic KITTI: The Semantic KITTI dataset is a new large-scale LiDAR point cloud dataset in driving scenes. It has 22 sequences with 19 valid classes, and each scan contains 10–13k points in a large space of $160\,\text{m} \times 160\,\text{m} \times 20\,\text{m}$. We use Sequences 0–7 and 9–10 as the training set (in total 19k frames), Sequence 8 (4k frames) as the validation set, and the remaining 11 sequences (20k frames) as the test set. Different from other point cloud datasets, LiDAR points are distributed irregularly in a large 3D space. There are many small objects with only a few points and the points are very sparse in the distance. All these make it challenging for semantic segmentation of the large-scale LiDAR point clouds.

4.2 Ablation Study

We test the performance of the models with different feature aggregation settings: 1) only using the regular convolutional aggregation like that in [3,7]; 2) with both the convolutional voxel-feature aggregation and the neighborhood-voxel point feature aggregation; and 3) with all three types of feature aggregations (our FusionNet).

Table 1 shows that only with the regular voxel-based convolutional aggregation, it will become a sparse convolutional nets [3,7] and achieves 54.8% in the mean IoU evaluation. By introducing the neighborhood point feature aggregation, the results can be improved by 4%. On the contrary, the FusionNet will degenerate to pointnets [37] if we only use the point feature aggregation. Finally, the full settings of the FusionNet get the best mean IoU of 63.7%.

4.3 Benchmark Evaluations

1) Semantic KITTI: We use a voxel size of 5 cm for building our voxel-based "mini-PointNet" representation. Each valid "mini-PointNet" contains 1–60 points. The model is trained on the training set, and the results are submitted to the online benchmark for evaluation using the 20k test set.

The results are presented in Table 2 and visualized in Fig. 6 for comparisons. Our FusionNet can achieve a new state-of-the-art performance in both the mean IoU and the accuracy evaluations. It outperforms the PointNet++ [37] (by 40%), state-of-the-art point-wise convolutions [53] (by 10%) and the sparse convolution [3] (by 7%) in mean IoU evaluations. Moreover, It achieves the best IoUs in 14 out of the 19 classes. For many small objects (*e.g.* person, bicycle, motorcycle), it outperforms other existing models by at least 10–25%.

FusionNet has many advantages for the large-scale LiDAR point cloud segmentation. Compared to state-of-the-art voxel-based networks [3,7], FusionNet can predict point-wise labels and avoid those ambiguous/wrong predictions at object boundaries when a voxel has points from different classes. When compared to state-of-the-art point-wise convolutions (*e.g.* [50]), our FusionNet gets

Table 1. Ablation study on large-scale Semantic KITTI dataset.

Neighborhood-voxel aggregation		Inner-voxel aggregation	Mean accuracy (%)	Mean IoU
Voxel feature agg	Point feature agg			
√			89.6	54.8
	√	√	86.2	41.6
√	√		90.7	58.9
√	√	√	91.8	63.7

Table 2. Single scan results (19 classes) on Semantic KITTI benchmarks. All models are trained on the training set and evaluated on the online benchmark.

Approach	**mIoU**	**mAcc**	road	sidewalk	parking	other-ground	building	car	truck	bicycle	motorcycle	other-vehicle	vegetation	trunk	terrain	person	bicyclist	motorcyclist	fence	pole	traffic sign
PointNet [35]	14.6	-	61.6	35.7	15.8	1.4	41.4	46.3	0.1	1.3	0.3	0.8	31.0	4.6	17.6	0.2	0.2	0.0	12.9	2.4	3.7
SPGraph [22]	17.4	-	45.0	28.5	0.6	0.6	64.3	49.3	0.1	0.2	0.2	0.8	48.9	27.2	24.6	0.3	2.7	0.1	20.8	15.9	0.8
SPLATNet [42]	18.4	-	64.6	39.1	0.4	0.0	58.3	58.2	0.0	0.0	0.0	0.0	71.1	9.9	19.3	0.0	0.0	0.0	23.1	5.6	0.0
PointNet++ [37]	20.1	-	72.0	41.8	18.7	5.6	62.3	53.7	0.9	1.9	0.2	0.2	46.5	13.8	30.0	0.9	1.0	0.0	16.9	6.0	8.9
SqzeSegV2[52]	39.7	-	88.6	67.6	45.8	17.7	73.7	81.8	13.4	18.5	17.9	14.0	71.8	35.8	60.2	20.1	25.1	3.9	41.1	20.2	36.3
TanConv[44]	40.9	-	83.9	63.9	33.4	15.4	83.4	90.8	15.2	2.7	16.5	12.1	79.5	49.3	58.1	23.0	28.4	8.1	49.0	35.8	28.5
PointASNL[56]	46.8	-	87.4	74.3	24.3	1.8	83.1	87.9	39.0	0.0	25.1	29.2	84.1	52.2	**70.6**	34.2	**57.6**	0.0	43.9	57.8	36.9
DarkNet53[4]	49.9	87,8	**91.8**	74.6	64.8	27.9	84.1	86.4	25.5	24.5	32.7	22.6	78.3	50.1	64.0	36.2	33.6	4.7	55.0	38.9	52.2
RandLANet[12]	50.3	85.9	88.0	67.9	56.9	15.5	81.1	94.0	**42.7**	19.8	21.4	**38.7**	78.3	60.3	59.0	47.5	48.8	4.6	49.7	44.2	38.1
PointConv[53]	51.2	87.1	88.9	68.4	58.9	19.7	84.6	93.1	37.8	20.7	22.9	38.1	79.9	62.8	60.7	46.2	39.1	5.5	52.0	48.1	44.7
MinkNet42[3]	54.3	89.5	91.1	69.7	63.8	29.3	**92.7**	94.3	26.1	23.1	26.2	36.7	83.7	68.4	64.7	43.1	36.4	7.9	57.1	57.3	60.1
FusionNet	**61.3**	**91.2**	**91.8**	**77.1**	**68.8**	**30.8**	92.5	**95.3**	41.8	**47.5**	**37.7**	34.5	**84.5**	**69.8**	68.5	**59.5**	56.8	**11.9**	**69.4**	**60.4**	**66.5**

much better segmentation accuracy in the large-scale LiDAR dataset. This is because our FusionNet is realized with more effective feature aggregation operations (including the effective voxel-level neighborhood aggregations and the fine-grain inner-voxel point-level aggregations).

2) 3DSIS and ScanNet: We also evaluate our FusionNet on two indoor-scene datasets (3DSIS [1] and ScanNet[4]). The results are presented in Table 3 and Table 4. The voxel size is set to 5cm for our voxel-based "mini-PointNet" representation. When compared with the sparse convolutional nets [3] which use the same resolution as input, our FusionNet model can learn fine-grain point-wise features and give point-wise predictions. Therefore, it can get better mean IoUs (68.8% on ScanNet benchmark and 67.2% on S3DIS test set). Also, our FusionNet outperforms state-of-the-art point-wise convolutions (*e.g.* PointCNN [25], PointConv [53]) by 2–10% in the mean IoU evaluations. This is because our FusionNet is realized with a more powerful feature aggregation module for learning more effective features in the segmentation.

4.4 Analysis of the Voxel-Based "Mini-PointNet"

Our FusionNet is realized with the unique voxel-based "mini-PointNet" representation. The voxel size is the key parameter that decides the resolution of our "mini-PointNet" representation and will influence the accuracy of the segmentation results. We evaluate the performance by using different voxel sizes for

Table 3. Stanford Area 5 Test (Fold #1) (S3DIS) [1]

Methods	mIoU	mAcc
PointNet [35]	41.09	48.98
SparseUNet [6]	41.72	64.62
PVConv [27]	52.3	–
TangentConv [44]	52.80	60.70
3D RNN [59]	53.40	71.30
PointCNN[25]	57.26	63.86
SuperpointGraph [22]	58.04	66.50
MinkNet32 [3]	65.35	71.71
KPConv[46]	67.1	**72.8**
Our FusionNet	**67.2**	72.3

Table 4. 3D Semantic Label Benchmark on ScanNet [3].

Methods	mIoU
ScanNet [4]	30.6
PointNet++ [37]	33.9
TangetConv [44]	43.8
SurfaceConv [32]	44.2
MVPNet [16]	64.1
PointConv [53]	66.6
PointASNL [56]	66.6
MinkNet42 (5cm) [3]	67.9
KPConv [46]	68.4
FusionNet (5cm)	**68.8**

building the networks. Similar networks (PVCNN [27] and sparse convolutions networks [3]) are used for comparisons. They use either the dense voxel representation [27] or the sparse voxel representation [3]. Models are tested on the Semantic KITTI validation set. We fixed the number of points to 125k for the memory test (in the training phase).

As shown in Fig. 7, higher resolution can help get better accuracy. But, the memory and computational costs significantly increase (almost cubically) which is a big limitation for learning on large-scale point clouds. State-of-the-art voxel-based convolutional nets [3,7] require an extremely high resolution to achieve state-of-the-art accuracy. By using the same resolution setting, our FusionNet can get far better accuracy. The FusionNet with a lower resolution (5 cm) outperforms the high-resolution (3 cm) sparse convolutional network [3]. This is benefited from our unique voxel-based "mini-PointNet" representation that can help learn point-wise predictions with more effective feature aggregations.

4.5 Efficiency for Large-Scale Point Cloud Processing

In this section, we systematically evaluate the efficiency and abilities of our FusionNet on processing real-world large-scale point clouds. The Semantic KITTI dataset [2] is used for evaluations. We compare our FusionNet with the recent work that can also produce point-wise predictions. For fair comparisons, we fixed the number of points (to 120k) for each LiDAR scan during the speed and memory test. In addition, we also measure the maximum number of 3D points each network can take in a single pass to infer per-point semantics. All experiments are conducted on the same device (RTX TITAN GPU).

Due to the high memory and computational costs and the limitation of the architecture design, many state-of-the-art point-wise segmentation models [12, 35,37,53] require re-sampling to reduce and fix the number of points in each point

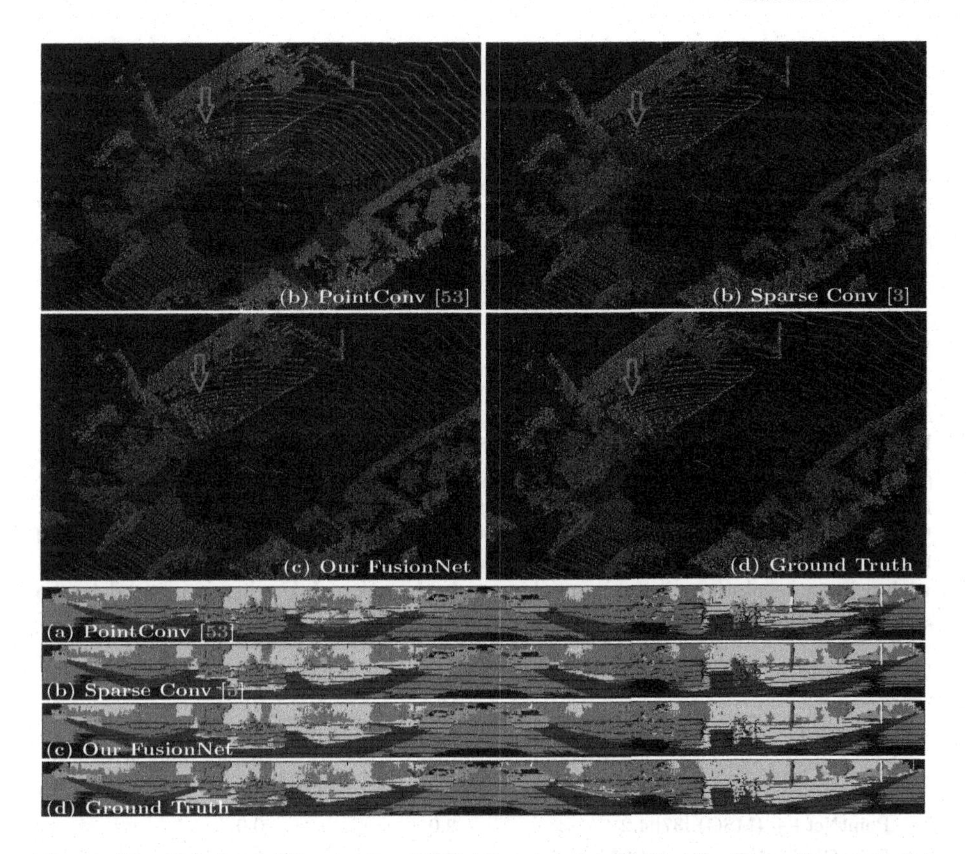

Fig. 6. Visualization of the results of LiDAR point clouds. Top (a-d) are the raw point cloud results. Bottom (a-d) are the visual comparisons by projecting the point clouds to cylindrical images. (a) State-of-the-art point-wise convolutions [53], (b) state-of-the-art sparse convolutions [3], (c) our FusionNet, (d) ground truths. The differences are as illustrated by red arrows (where bicycles are predicted as other objects by [3,53]).

cloud frame for training and testing. This need to split the point cloud. But, in LiDAR point clouds, points are very sparse in the distance, down-sampling will further increase the sparsity and lead to worse segmentation accuracy due to the insufficient feature aggregation from neighborhoods.

As a comparison, our FusionNet can directly take the entire point cloud as input for training and testing that is more flexible in real applications. As shown in Table 5, it can take up to 2.5 million points in a single pass to infer the point-wise segmentation which is four times as many as that of PointNet++.

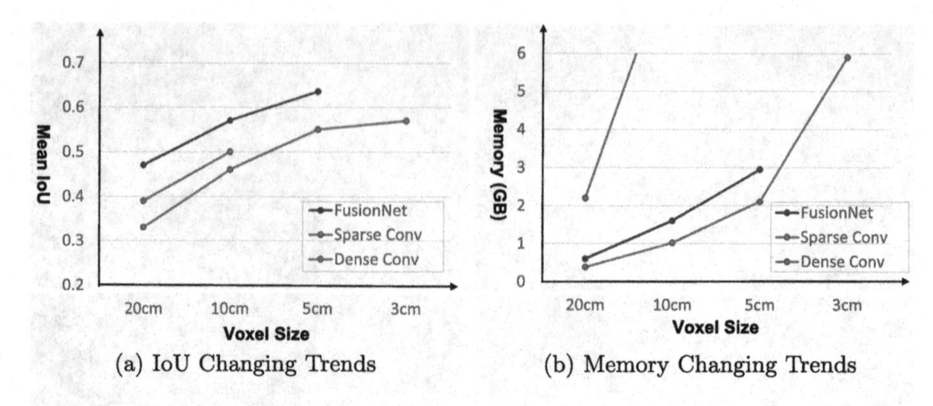

(a) IoU Changing Trends (b) Memory Changing Trends

Fig. 7. Performance illustration with different voxel sizes. (a) The mean IoU changes along with the voxel sizes. PVCNN [27] and sparse convolutions [3]) are used for comparisons. Higher resolution can help get a better mean IoU. (b) The memory consumption changes along with the voxel sizes (training phase using LiDAR frame [2]). The memory consumption significantly increases as the resolution goes up. Our FusionNet with a lower resolution (5 cm) can achieve a better mean IoU when compared to the sparse convolutions [3] (with a high resolution of 3 cm).

Table 5. Efficiency comparisons for large-scale point cloud processing.

Approach	Memory consumption (GB)	Elapsed time (per scan)	Max inference points (million)
PointNet [35]	3.1	0.2	0.8
PointNet++ (MSG) [37]	4.2	2.0	0.6
PointCNN [25]	13.4	3.0	0.2
PointConv [53]	3.0	3.0	0.8
Our FusionNet	1.0	0.9	2.3

5 Conclusions and Future Work

In this paper, we propose a deep fusion network architecture (FusionNet) with a new feature aggregation module for large-scale 3D semantic segmentation. The proposed FusionNet utilizes a unique voxel-based "mini-PointNet" for point cloud representation and learning. It can realize both the neighborhood voxel-level feature aggregation and the fine-grain point-wise feature learning. Therefore, FusionNet can produce more accurate point-wise predictions when compared to voxel-based convolutional networks. Also, when compared to popular point-wise convolutions, it realizes more effective feature aggregations with lower memory and computational costs.

Currently, part of our system (the hash mapping) is implemented on the CPU that limits the running speed. We will try the GPU hash mapping in the future that will significantly accelerate our FusionNet for real-time applications.

Acknowledgement. Research is supported by Baidu, the ERC grant ERC-2012-AdG 321162-HELIOS, EPSRC grant Seebibyte EP/M013774/1 and EPSRC/MURI grant EP/N019474/1. We would also like to acknowledge the Royal Academy of Engineering and FiveAI.

References

1. Armeni, I., et al.: 3D semantic parsing of large-scale indoor spaces. In: Proceedings of the IEEE Conference on Computer Vision and Pattern Recognition (CVPR), pp. 1534–1543 (2016)
2. Behley, J., et al.: Semantickitti: a dataset for semantic scene understanding of lidar sequences. In: Proceedings of the IEEE International Conference on Computer Vision (ICCV), pp. 9297–9307 (2019)
3. Choy, C., Gwak, J., Savarese, S.: 4D spatio-temporal convnets: minkowski convolutional neural networks. In: Proceedings of the IEEE Conference on Computer Vision and Pattern Recognition (CVPR), pp. 3075–3084 (2019)
4. Dai, A., Chang, A.X., Savva, M., Halber, M., Funkhouser, T., Nießner, M.: Scannet: richly-annotated 3D reconstructions of indoor scenes. In: Proceedings of the IEEE Conference on Computer Vision and Pattern Recognition (CVPR), pp. 5828–5839 (2017)
5. Gadelha, M., Wang, R., Maji, S.: Multiresolution tree networks for 3D point cloud processing. In: Ferrari, V., Hebert, M., Sminchisescu, C., Weiss, Y. (eds.) ECCV 2018. LNCS, vol. 11211, pp. 105–122. Springer, Cham (2018). https://doi.org/10.1007/978-3-030-01234-2_7
6. Graham, B.: Sparse 3D convolutional neural networks. arXiv preprint arXiv:1505.02890 (2015)
7. Graham, B., Engelcke, M., van der Maaten, L.: 3D semantic segmentation with submanifold sparse convolutional networks. In: Proceedings of the IEEE Conference on Computer Vision and Pattern Recognition (CVPR) (2018)
8. Groh, F., Wieschollek, P., Lensch, H.P.A.: Flex-convolution. In: Jawahar, C.V., Li, H., Mori, G., Schindler, K. (eds.) ACCV 2018. LNCS, vol. 11361, pp. 105–122. Springer, Cham (2019). https://doi.org/10.1007/978-3-030-20887-5_7
9. He, T., Huang, H., Yi, L., Zhou, Y., Wu, C., Wang, J., Soatto, S.: Geonet: deep geodesic networks for point cloud analysis. In: Proceedings of the IEEE Conference on Computer Vision and Pattern Recognition (CVPR), June 2019
10. Hermosilla, P., Ritschel, T., Vázquez, P.P., Vinacua, À., Ropinski, T.: Monte Carlo convolution for learning on non-uniformly sampled point clouds. ACM Trans. Graph. (TOG) **37**(6), 1–12 (2018)
11. Hou, J., Dai, A., Nießner, M.: 3D-SIS: 3D semantic instance segmentation of RGB-D scans. In: Proceedings of the IEEE Conference on Computer Vision and Pattern Recognition (CVPR), pp. 4421–4430 (2019)
12. Hu, Q., et al.: Randla-net: efficient semantic segmentation of large-scale point clouds. arXiv preprint arXiv:1911.11236 (2019)
13. Hua, B.S., Tran, M.K., Yeung, S.K.: Pointwise convolutional neural networks. In: Proceedings of the IEEE Conference on Computer Vision and Pattern Recognition (CVPR), pp. 984–993 (2018)
14. Huang, J., Zhang, H., Yi, L., Funkhouser, T., Nießner, M., Guibas, L.J.: Texturenet: consistent local parametrizations for learning from high-resolution signals on meshes. In: Proceedings of the IEEE Conference on Computer Vision and Pattern Recognition (CVPR), pp. 4440–4449 (2019)

15. Huang, Q., Wang, W., Neumann, U.: Recurrent slice networks for 3D segmentation of point clouds. In: Proceedings of the IEEE Conference on Computer Vision and Pattern Recognition (CVPR), June 2018
16. Jaritz, M., Gu, J., Su, H.: Multi-view pointnet for 3D scene understanding. In: Proceedings of the IEEE International Conference on Computer Vision Workshops (2019)
17. Jiang, L., Zhao, H., Liu, S., Shen, X., Fu, C.W., Jia, J.: Hierarchical point-edge interaction network for point cloud semantic segmentation. In: Proceedings of the IEEE International Conference on Computer Vision (ICCV), October 2019
18. Komarichev, A., Zhong, Z., Hua, J.: A-CNN: annularly convolutional neural networks on point clouds. In: Proceedings of the IEEE Conference on Computer Vision and Pattern Recognition (CVPR), June 2019
19. Lahoud, J., Ghanem, B., Pollefeys, M., Oswald, M.R.: 3D instance segmentation via multi-task metric learning. In: Proceedings of the IEEE International Conference on Computer Vision (ICCV), pp. 9256–9266 (2019)
20. Lan, S., Yu, R., Yu, G., Davis, L.S.: Modeling local geometric structure of 3D point clouds using geo-CNN. In: Proceedings of the IEEE Conference on Computer Vision and Pattern Recognition (CVPR), June 2019
21. Landrieu, L., Boussaha, M.: Point cloud oversegmentation with graph-structured deep metric learning. In: Proceedings of the IEEE Conference on Computer Vision and Pattern Recognition (CVPR), June 2019
22. Landrieu, L., Simonovsky, M.: Large-scale point cloud semantic segmentation with superpoint graphs. In: Proceedings of the IEEE Conference on Computer Vision and Pattern Recognition (CVPR), June 2018
23. Le, T., Duan, Y.: Pointgrid: a deep network for 3D shape understanding. In: Proceedings of the IEEE Conference on Computer Vision and Pattern Recognition (CVPR), June 2018
24. Lei, H., Akhtar, N., Mian, A.: Octree guided CNN with spherical kernels for 3D point clouds. In: Proceedings of the IEEE Conference on Computer Vision and Pattern Recognition (CVPR), June 2019
25. Li, Y., Bu, R., Sun, M., Wu, W., Di, X., Chen, B.: PointCNN: convolution on x-transformed points. In: Advances in Neural Information Processing Systems (NeurIPS), pp. 820–830 (2018)
26. Liu, Y., Fan, B., Meng, G., Lu, J., Xiang, S., Pan, C.: Densepoint: learning densely contextual representation for efficient point cloud processing. In: Proceedings of the IEEE International Conference on Computer Vision (ICCV), October 2019
27. Liu, Z., Tang, H., Lin, Y., Han, S.: Point-voxel CNN for efficient 3D deep learning. In: Advances in Neural Information Processing Systems (NeurIPS), pp. 963–973 (2019)
28. Mao, J., Wang, X., Li, H.: Interpolated convolutional networks for 3D point cloud understanding. In: Proceedings of the IEEE International Conference on Computer Vision (ICCV), October 2019
29. Maturana, D., Scherer, S.: Voxnet: a 3D convolutional neural network for real-time object recognition. In: 2015 IEEE/RSJ International Conference on Intelligent Robots and Systems (IROS), pp. 922–928. IEEE (2015)
30. Meng, H.Y., Gao, L., Lai, Y.K., Manocha, D.: VV-Net: voxel VAE net with group convolutions for point cloud segmentation. In: Proceedings of the IEEE International Conference on Computer Vision (ICCV), October 2019
31. Narita, G., Seno, T., Ishikawa, T., Kaji, Y.: Panopticfusion: online volumetric semantic mapping at the level of stuff and things. arXiv preprint arXiv:1903.01177 (2019)

32. Pan, H., Liu, S., Liu, Y., Tong, X.: Convolutional neural networks on 3D surfaces using parallel frames. arXiv preprint arXiv:1808.04952 (2018)
33. Pham, Q.H., Nguyen, T., Hua, B.S., Roig, G., Yeung, S.K.: JSIS3D: joint semantic-instance segmentation of 3D point clouds with multi-task pointwise networks and multi-value conditional random fields. In: Proceedings of the IEEE Conference on Computer Vision and Pattern Recognition (CVPR), June 2019
34. Prokudin, S., Lassner, C., Romero, J.: Efficient learning on point clouds with basis point sets. In: Proceedings of the IEEE International Conference on Computer Vision (ICCV), October 2019
35. Qi, C.R., Su, H., Mo, K., Guibas, L.J.: Pointnet: deep learning on point sets for 3D classification and segmentation. In: Proceedings of the IEEE Conference on Computer Vision and Pattern Recognition (CVPR), pp. 652–660 (2017)
36. Qi, C.R., Su, H., Nießner, M., Dai, A., Yan, M., Guibas, L.J.: Volumetric and multi-view CNNs for object classification on 3D data. In: Proceedings of the IEEE Conference on Computer Vision and Pattern Recognition (CVPR), pp. 5648–5656 (2016)
37. Qi, C.R., Yi, L., Su, H., Guibas, L.J.: Pointnet++: deep hierarchical feature learning on point sets in a metric space. In: Advances in Neural Information Processing Systems (NeurIPS), pp. 5099–5108 (2017)
38. Rao, Y., Lu, J., Zhou, J.: Spherical fractal convolutional neural networks for point cloud recognition. In: Proceedings of the IEEE Conference on Computer Vision and Pattern Recognition (CVPR), June 2019
39. Rethage, D., Wald, J., Sturm, J., Navab, N., Tombari, F.: Fully-convolutional point networks for large-scale point clouds. In: Ferrari, V., Hebert, M., Sminchisescu, C., Weiss, Y. (eds.) ECCV 2018. LNCS, vol. 11208, pp. 625–640. Springer, Cham (2018). https://doi.org/10.1007/978-3-030-01225-0_37
40. Riegler, G., Osman Ulusoy, A., Geiger, A.: OctNet: learning deep 3D representations at high resolutions. In: Proceedings of the IEEE Conference on Computer Vision and Pattern Recognition (CVPR), pp. 3577–3586 (2017)
41. Ronneberger, O., Fischer, P., Brox, T.: U-Net: convolutional networks for biomedical image segmentation. In: Navab, N., Hornegger, J., Wells, W.M., Frangi, A.F. (eds.) MICCAI 2015. LNCS, vol. 9351, pp. 234–241. Springer, Cham (2015). https://doi.org/10.1007/978-3-319-24574-4_28
42. Su, H., et al.: Splatnet: sparse lattice networks for point cloud processing. In: Proceedings of the IEEE Conference on Computer Vision and Pattern Recognition (CVPR), June 2018
43. Su, H., Maji, S., Kalogerakis, E., Learned-Miller, E.: Multi-view convolutional neural networks for 3D shape recognition. In: Proceedings of the IEEE International Conference on Computer Vision (ICCV), pp. 945–953 (2015)
44. Tatarchenko, M., Park, J., Koltun, V., Zhou, Q.Y.: Tangent convolutions for dense prediction in 3D. In: Proceedings of the IEEE Conference on Computer Vision and Pattern Recognition (CVPR), pp. 3887–3896 (2018)
45. Te, G., Hu, W., Zheng, A., Guo, Z.: RGCNN: regularized graph CNN for point cloud segmentation. In: Proceedings of the 26th ACM International Conference on Multimedia, pp. 746–754 (2018)
46. Thomas, H., Qi, C.R., Deschaud, J.E., Marcotegui, B., Goulette, F., Guibas, L.J.: KPConv: flexible and deformable convolution for point clouds. In: Proceedings of the IEEE International Conference on Computer Vision (ICCV), pp. 6411–6420 (2019)

47. Verma, N., Boyer, E., Verbeek, J.: Feastnet: feature-steered graph convolutions for 3D shape analysis. In: Proceedings of the IEEE Conference on Computer Vision and Pattern Recognition (CVPR), pp. 2598–2606 (2018)
48. Wang, C., Samari, B., Siddiqi, K.: Local spectral graph convolution for point set feature learning. In: Ferrari, V., Hebert, M., Sminchisescu, C., Weiss, Y. (eds.) ECCV 2018. LNCS, vol. 11208, pp. 56–71. Springer, Cham (2018). https://doi.org/10.1007/978-3-030-01225-0_4
49. Wang, L., Huang, Y., Hou, Y., Zhang, S., Shan, J.: Graph attention convolution for point cloud semantic segmentation. In: Proceedings of the IEEE Conference on Computer Vision and Pattern Recognition (CVPR), June 2019
50. Wang, S., Suo, S., Ma, W.C., Pokrovsky, A., Urtasun, R.: Deep parametric continuous convolutional neural networks. In: Proceedings of the IEEE Conference on Computer Vision and Pattern Recognition (CVPR), June 2018
51. Wang, X., He, J., Ma, L.: Exploiting local and global structure for point cloud semantic segmentation with contextual point representations. In: Advances in Neural Information Processing Systems (NeurIPS), pp. 4573–4583 (2019)
52. Wu, B., Zhou, X., Zhao, S., Yue, X., Keutzer, K.: SqueezeSegV2: improved model structure and unsupervised domain adaptation for road-object segmentation from a lidar point cloud. In: 2019 International Conference on Robotics and Automation (ICRA), pp. 4376–4382. IEEE (2019)
53. Wu, W., Qi, Z., Fuxin, L.: PointConv: deep convolutional networks on 3D point clouds. In: Proceedings of the IEEE Conference on Computer Vision and Pattern Recognition (CVPR), June 2019
54. Wu, Z., et al.: 3D shapenets: a deep representation for volumetric shapes. In: Proceedings of the IEEE Conference on Computer Vision and Pattern Recognition (CVPR), pp. 1912–1920 (2015)
55. Xu, Y., Fan, T., Xu, M., Zeng, L., Qiao, Yu.: SpiderCNN: deep learning on point sets with parameterized convolutional filters. In: Ferrari, V., Hebert, M., Sminchisescu, C., Weiss, Y. (eds.) ECCV 2018. LNCS, vol. 11212, pp. 90–105. Springer, Cham (2018). https://doi.org/10.1007/978-3-030-01237-3_6
56. Yan, X., Zheng, C., Li, Z., Wang, S., Cui, S.: PointASNL: robust point clouds processing using nonlocal neural networks with adaptive sampling. In: Proceedings of the IEEE Conference on Computer Vision and Pattern Recognition (CVPR), June 2020
57. Yang, B., et al.: Learning object bounding boxes for 3D instance segmentation on point clouds. In: Advances in Neural Information Processing Systems (NeurIPS), pp. 6737–6746 (2019)
58. Yang, J., et al.: Modeling point clouds with self-attention and gumbel subset sampling. In: Proceedings of the IEEE Conference on Computer Vision and Pattern Recognition (CVPR), June 2019
59. Ye, X., Li, J., Huang, H., Du, L., Zhang, X.: 3D recurrent neural networks with context fusion for point cloud semantic segmentation. In: Ferrari, V., Hebert, M., Sminchisescu, C., Weiss, Y. (eds.) ECCV 2018. LNCS, vol. 11211, pp. 415–430. Springer, Cham (2018). https://doi.org/10.1007/978-3-030-01234-2_25
60. Yifan, W., Wu, S., Huang, H., Cohen-Or, D., Sorkine-Hornung, O.: Patch-based progressive 3D point set upsampling. In: Proceedings of the IEEE Conference on Computer Vision and Pattern Recognition (CVPR), June 2019
61. Zhang, F., et al.: Instance segmentation of lidar point clouds. In: International Conference on Robotics and Automation (ICRA) (2020)

62. Zhang, Z., Hua, B.S., Yeung, S.K.: Shellnet: efficient point cloud convolutional neural networks using concentric shells statistics. In: Proceedings of the IEEE International Conference on Computer Vision (ICCV), October 2019
63. Zhao, H., Jiang, L., Fu, C.W., Jia, J.: Pointweb: enhancing local neighborhood features for point cloud processing. In: Proceedings of the IEEE Conference on Computer Vision and Pattern Recognition (CVPR), pp. 5565–5573 (2019)
64. Zhao, H., et al.: PSANet: point-wise spatial attention network for scene parsing. In: Ferrari, V., Hebert, M., Sminchisescu, C., Weiss, Y. (eds.) ECCV 2018. LNCS, vol. 11213, pp. 270–286. Springer, Cham (2018). https://doi.org/10.1007/978-3-030-01240-3_17
65. Zhao, Y., Birdal, T., Deng, H., Tombari, F.: 3D point capsule networks. In: Proceedings of the IEEE Conference on Computer Vision and Pattern Recognition (CVPR), June 2019

Deep Material Recognition in Light-Fields via Disentanglement of Spatial and Angular Information

Bichuan Guo[1], Jiangtao Wen[1], and Yuxing Han[2](✉)

[1] Tsinghua University, Beijing, China
gbc16@mails.tsinghua.edu.cn
[2] Research Institute of Tsinghua University in Shenzhen, Shenzhen, China
yuxinghan@tsinghua-sz.org

Abstract. Light-field cameras capture sub-views from multiple perspectives simultaneously, with possibly reflectance variations that can be used to augment material recognition in remote sensing, autonomous driving, etc. Existing approaches for light-field based material recognition suffer from the entanglement between angular and spatial domains, leading to inefficient training which in turn limits their performances. In this paper, we propose an approach that achieves decoupling of angular and spatial information by establishing correspondences in the angular domain, then employs regularization to enforce a rotational invariance. As opposed to relying on the Lambertian surface assumption, we align the angular domain by estimating sub-pixel displacements using the Fourier transform. The network takes sparse inputs, i.e. sub-views along particular directions, to gain structural information about the angular domain. A novel regularization technique further improves generalization by weight sharing and max-pooling among different directions. The proposed approach outperforms any previously reported method on multiple datasets. The accuracy gain over 2D images is improved by a factor of 1.5. Ablation studies are conducted to demonstrate the significance of each component.

Keywords: Light field · Material recognition · Angular registration

1 Introduction

Material recognition is a fundamental problem in vision and plays an important role in many industrial applications. For example, drones are used in topography [21] for recognizing ground surfaces, and mowers can operate autonomously by inspecting surrounding surface materials. As a result, material recognition has been an active area of research, which is inherently challenging because the appearances of materials depend on various factors such as shape, lighting and

Electronic supplementary material The online version of this chapter (https://doi.org/10.1007/978-3-030-58586-0_39) contains supplementary material, which is available to authorized users.

A. Vedaldi et al. (Eds.): ECCV 2020, LNCS 12369, pp. 664–679, 2020.
https://doi.org/10.1007/978-3-030-58586-0_39

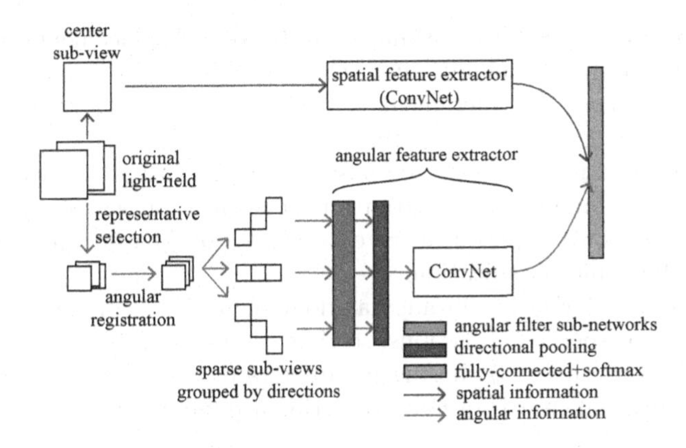

Fig. 1. Overview of the proposed framework. The light-field is represented by a spatial component (up), as well as an angular component (down) which is cropped to avoid occlusion and to facilitate alignment. Angular features are extracted from sparse sub-views grouped by perspective directions; all directions share filter weights and are aggregated via max-pooling. Spatial and angular features are combined at the end to produce class probabilities

exhibit strong visual variations. A variety of clues, such as texture, reflectance and scene context, need be taken into account to achieve good performance.

The recent introduction of commercially available and compact light-field cameras (e.g. Lytro Illum) provides an efficient way to improve image based recognition, making it possible to significantly boost the performance of the above-mentioned applications with a simple optical upgrade. These cameras can capture sub-views from multiple viewpoints on a regular grid (i.e. the angular domain) in a single shot. Such rich information can be used to obtain critical cues such as depth information and intensity variation in different perspectives, which are not directly available from 2D images. Existing research have investigated using light-field cameras for recognizing not only materials [25], but also objects [14] and reflectance [13]. They confirmed that significant accuracy boost can be achieved by using light-field images over 2D images.

Several key aspects of light-field images were not addressed by existing methods. First, due to light-field camera optics, the angular domain is intertwined with the spatial domain to express reflectance variation. As a result, filters directly applied to each domain separately cannot disentangle reflectance variation efficiently. Moreover, occlusion in the scene can cause discontinuity of information distribution and misalignment, where same positions in sub-views correspond to different objects and materials. Third, the angular domain is highly redundant as sub-views exhibit strong similarities. Without regularization, it is hard to learn structural information and can easily lead to overfitting.

We propose a novel framework to tackle these problems. An overview of our methods is shown in Fig. 1. Our contributions are summarized below:

- Sub-views are aligned by estimating sub-pixel displacements using their Fourier transforms. A simple algorithm is deduced by leveraging angular domain geometry. This alignment procedure disentangles reflectance variation in angular domain from spatial domain.
- A representative region is selected based on inferred depth to avoid occlusion and to ease alignment. The spatial information is preserved by a separate, full sub-view. Two expert feature extractors (i.e. spatial and angular) are combined for joint classification.
- We employ regularization in angular domain by (1) using sparse sub-views grouped by perspective directions, so that the network is aware of the angular domain structure; (2) sharing parameters in all directions and enforcing rotational invariance with a novel directional pooling layer.

2 Related Work

Material Recognition. Two distinct approaches have been taken in the literature for image based material recognition. One relies on information beyond object appearance, such as reflectance disks [33], scene depth [34], 3D surface geometry [7] and spatial thermal textures [5]. They incorporate special measurements and properties that are closely related to material characteristics. The other approach relies on 2D images alone, usually by efficient use of context information including object, texture and background. Schwartz and Nishino [17] proposed to separate local material from object/scene context and combine them later. Cimpoi et al. [6] introduced a new texture descriptor, which was used along with object descriptors. These 2D image based methods are not directly applicable to 4D light-field data studied in this paper; however, we are inspired by the idea of global-local decomposition in that we use separate expert models to extract spatial and angular features.

Qi et al. [15] explored the transform invariance of co-occurrence features, and introduced a pairwise rotation invariant feature for 2D images. We instead design a ConvNet architecture to enforce rotational invariance in angular domain for light-fields. Xue et al. [32] used a stereo pair to perform material recognition, by feeding their difference and one view to a neural network. Wang et al. [25] proposed multiple ConvNet architectures for recognizing materials using light-field images. Our work further explores this idea of using reflectance variation to improve material recognition, which requires algorithmic adaptations to several key characteristics of light-field images.

Computer Vision for Light-Fields. Light-fields received increasing popularity in the vision community over the years. One direction is to enhance the quality of light-fields, as current acquisition techniques are still rudimentary. This includes super-resolution in spatial domain [27] or angular domain [31], denoising [4] and light-field video interpolation [26]. Another line of research is to tackle existing computer vision problems using light-field images. Many depth estimation algorithms [11,12,24,28] that rely on efficient use of angular information were proposed. Similar methodologies were also adopted in other tasks. Lu et al. [14] used

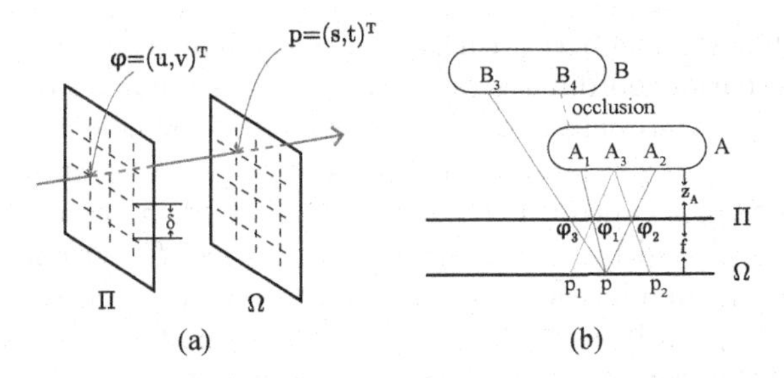

Fig. 2. Light-field geometry. (a) The 4D light-field parametrized by two parallel planes: the lens plane Π (angular domain) and the sensor plane Ω (spatial domain). (b) Green lines: Pixels with same spatial coordinates $\boldsymbol{p}$ correspond to different objects. Red lines: Disparity $\boldsymbol{p}_1 - \boldsymbol{p}_2$ is linear to lens translation $\varphi_1 - \varphi_2$ (Color figure online)

domain-interleaved filters to simultaneously extract spatial and angular features for object classification. Alperovich et al. [1] used an autoencoder to find compact representations for epipolar plane images [2] (EPI), which can be decoded to yield diffuse/specular intrinsic components. Chen et al. [3] used the surface camera to model reflectance variation of 3D points, leading to better stereo matching. We borrowed this idea in that our approach constructs the surface camera via alignment to disentangle angular and spatial variations.

Neural Networks for Light-Fields. Many papers employed neural networks to analyze light-fields. To recognize materials or bidirectional reflectance distribution functions (BRDF), [13,25] proposed to stack sub-views and take convolutions in spatial domain, or instead apply filters in angular domain. We extend the above ideas by enforcing sparsity in both domains in order to avoid occlusion and reduce overfitting. Heber et al. [9] introduced stacking in EPI domain for shape inference. [13,25,30] used alternating spatial and angular domain convolutions to mimic a full 4D filter while keeping computational costs low. Shin et al. [19] proposed to only use horizontal or crosshair sub-views to increase computational speed. Xue et al. [32] proposed the concept of angular gradients which enables the network to be aware of angular structure. In this work we further explore these ideas and find that, combined with weight sharing and pooling, such sparsity in angular domain improves generalization by enforcing rotational invariance, and improves the efficiency of extracting angular gradients.

3 Light-Field Imaging Model

3.1 Light-Field Geometry

The 4D light-field is usually parametrized using two parallel planes: a lens plane Π, representing the angular domain, and a sensor plane Ω, representing the spatial domain, as shown in Fig. 2(a). A ray of light passes through a micro-lens on Π and is captured by a pixel on Ω. The light-field can be viewed as a function

$$L : \Pi \times \Omega \to \mathbb{R}^C, (\boldsymbol{\varphi}, \boldsymbol{p}) \mapsto L(\boldsymbol{\varphi}, \boldsymbol{p}), \tag{1}$$

where φ is a micro-lens on Π with angular coordinates $(u, v)^\mathsf{T}$, $\boldsymbol{p}$ is a pixel on sensor Ω with spatial coordinates $(s, t)^\mathsf{T}$, and C is the number of color channels. The behavior of a micro-lens can be analyzed with a pinhole camera model. We first fix the pixel $\boldsymbol{p}$ in spatial domain and vary the lenses φ in angular domain. In Fig. 2(b), φ_1, φ_2 capture 3D points A_1, A_2, and φ_3 captures a 3D point B_3. Here B is occluded by A since φ_1 would capture B_4 without A. It is clear that variations in angular domain may correspond to multiple 3D points, or even different objects in case of occlusion. However, for material recognition, the key is to obtain reflectance variation of individual surface points in different perspectives. This can be done by fixing the 3D point and correcting for disparity in spatial domain. The light rays emitted from a 3D point A_3 with depth z_A reach two lenses φ_1 and φ_2, and are captured by sensors at $\boldsymbol{p}_1$ and $\boldsymbol{p}_2$. The disparity between $\boldsymbol{p}_1$ and $\boldsymbol{p}_2$ is related to lens translation linearly:

$$\boldsymbol{p}_1 - \boldsymbol{p}_2 = \left(1 + \frac{f}{z_A}\right)(\varphi_1 - \varphi_2). \tag{2}$$

If scene depth z_A and focal length f are known, one can use (2) to group pixels in different sub-views with same originations. This is often called the angular sampling image (ASI) [18], describing the reflectance variation of individual 3D points. Operating on the ASI has the advantage that reflectance information is decoupled from spatial variations, which allows specialized expert models to be applied to each aspect, and later combined to form a joint decision.

3.2 Analysis of Baseline Methods

We now analyze several baseline methods (in *italic*) proposed in [25]. *2D-average* and *viewpool* feed each sub-view to a ConvNet and perform averaging or max-pooling for aggregation. They fail to exploit between-view correlation and perform poorly. *Stack* stacks all sub-views before a ConvNet, the first layer of which takes convolutions in spatial and angular domains simultaneously. We see previously that these two domains are intertwined; this complex coupling effect causes the model to perform only slightly better than 2D models. *EPI* is similar to *stack* as it stacks in EPI domain instead, achieving similar performance.

Two winning models from [25], namely *ang-filter* (referred to as *angular-filter* in the original paper) and *4D-filter*, provide significant performance boosts over 2D models. Our **first motivation** is due to the baseline method *ang-filter* outperforms *stack*. These two methods both apply convolutional filters in angular domain, the difference is that "ang-filter" has 1x1 spatial filters while "stack" has 3x3 spatial filters. This means *ang-filter* is only convolving in the angular domain while *stack* is convolving in spatial and angular domains at the same time, which suggests us to decouple these two domains. The **second motivation** can be considered as a step further from the first motivation. According to the analysis in Sect. 3.1, the angular domain itself is both affected by spatial and reflectance variations. This suggests us to align pixels into ASIs, which will decouple reflectance variations from spatial variations.

4 Methods

In our proposed framework, we first decompose the light-field into a spatial component and an angular component; features from both components are extracted separately and combined at the end to produce class probabilities, as shown in Fig. 1. The spatial component is the center sub-view, containing overall appearance, object and scene context. The angular component is then aligned into ASIs through *representative selection* and *angular registration*.

4.1 Representative Selection

ASIs can be formed by collecting pixels from sub-views according to (2). In practice, the dense depth field is usually not available and needs to be estimated. However, most depth estimation algorithms [10,18,24] rely on the assumption that objects are made of Lambertian materials with uniform reflectance across all viewing angles. This is clearly an oversimplification as our key to the problem is reflectance variation. Even for methods that are robust to non-Lambertian surfaces, dense depth estimation is still error-prone in case of occlusion [24].

We notice that by selecting a region with constant depth, the situation is greatly simplified. By the disparity linearity (2), the disparity between two sub-views becomes a constant for any 3D point in this region, which implies that we can form ASIs by simply translating sub-views. Also, a region with constant depth is mostly occlusion-free. Therefore, we crop the angular component to a representative region that has approximately constant depth. As the spatial component remains full resolution, we will not lose much spatial information by cropping the angular component.

The dense depth map $D(\Omega)$ estimated from [11] is used for region selection, which is less error-prone than using it to directly warp pixels (we verify this experimentally in Sect. 5.3). The spatial domain is partitioned into an $N \times N$ grid, and denote the depth values in each block by

$$\forall 1 \leq i,j \leq N, D_{ij} = \left\{ D(s,t) : \frac{(i-1)S}{N} < s \leq \frac{iS}{N}, \frac{(j-1)T}{N} < t \leq \frac{jT}{N} \right\}, \quad (3)$$

where $S \times T$ is the spatial resolution. The representative region is selected as the one that minimizes the following loss

$$(i^*, j^*) = \underset{i,j}{\operatorname{argmin}}\, \sigma[D_{ij}] + \lambda \big| \mu[D_{ij}] - \operatorname{med}[D(\Omega)] \big|, \quad (4)$$

where $\sigma[\cdot]$, $\mu[\cdot]$ and $\operatorname{med}[\cdot]$ denote the standard deviation, mean and median of a set. The first term penalizes the amount of depth variation as we wish it to be approximately constant, and the second term penalizes outlier regions that significantly deviate from the overall median depth, which usually correspond to backgrounds. The Lagrangian coefficient λ is empirically set to 1.0 to provide a good trade-off between these two terms.

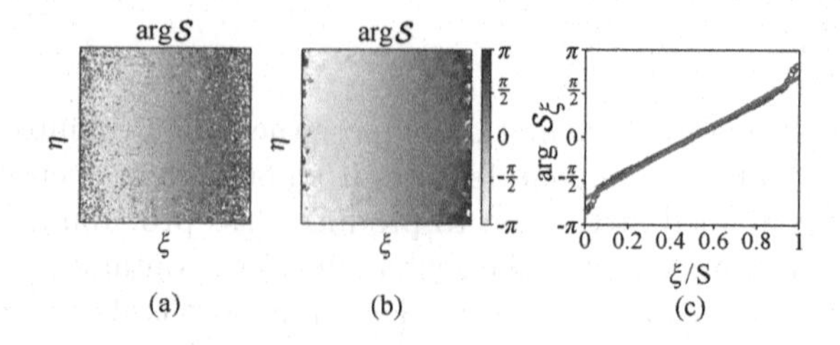

Fig. 3. Displacement estimation in Fourier domain. (a) If $\Delta t = 0$, $\arg \mathcal{S}$ is constant in η. There is a large amount of noise due to aliasing. (b) Median filtering on $\arg \mathcal{S}$ to remove isolated noise. (c) Median values of $\arg \mathcal{S}$ at each ξ and its linear regression

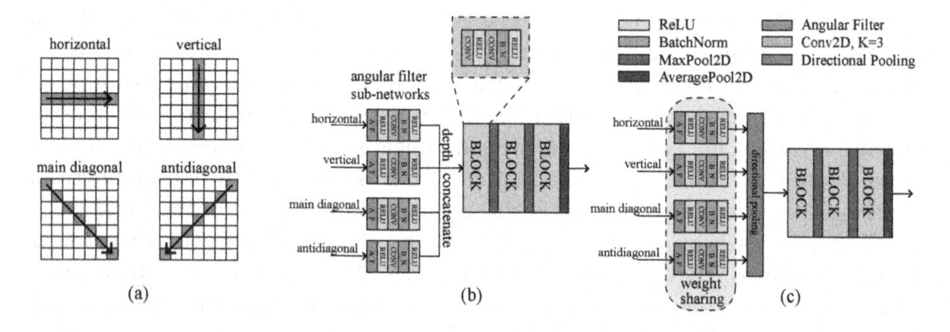

Fig. 4. Network architecture of the angular feature extractor. (a) Sub-views along particular directions are stacked and fed to the network. (b) *Angular-4*. Stacks of each direction have their own angular filter sub-networks. They are concatenated depth-wise before shared convolutional layers, which consist of three basic building blocks, each basic block is a sequence of "Conv-ReLU-Conv-BN-ReLU" layers. (c) *Angular-S*. Comparing to *angular-4*, the angular filter sub-networks have shared weights. The directional pooling layer is a MaxPool3D layer pooling among four directions. In (b) and (c), all "AF" layers are Conv2D layers with kernel size $K_{\mathrm{af}} \in \{1, 3\}$, SAME padding, containing $\#\mathrm{af} \in \{16, 32, 64\}$ different angular filters.

4.2 Angular Registration

As the representative region has approximately constant depth, we can align (also called *register* in literature) any sub-view with the center sub-view via translation. By constraining disparity to be shared across spatial domain, the pixel level Lambertian surface assumption is relaxed to sub-view level translation, which holds as long as changing viewpoints does not cause drastic overall changes. Jeon et al. [11] used the Fourier domain to perform sub-pixel displacement. Here we instead use the Fourier domain to estimate the displacement itself. The Fourier shift theorem [16] states that if two images I_1 and I_2 of same size $S \times T$ are related by a translation $\Delta \boldsymbol{p} = (\Delta s, \Delta t)$, $I_2(s,t) = I_1(s + \Delta s, t + \Delta t)$, then their discrete Fourier transforms $\mathcal{F}\{I_1\}$, $\mathcal{F}\{I_2\}$ are related by

$$\mathcal{F}\{I_2\}(\xi, \eta) = \mathcal{F}\{I_1\}(\xi, \eta) \cdot e^{2\pi i (\xi \Delta s / S + \eta \Delta t / T)}, \tag{5}$$

therefore we can perform sub-pixel translation $\Delta \boldsymbol{p}$ to I_1 by

$$I_2 = \mathcal{F}^{-1}\{\mathcal{F}\{I_1\}(\boldsymbol{\omega}) \exp(2\pi i \boldsymbol{\omega} \cdot (\Delta s/S, \Delta t/T)\}. \tag{6}$$

To estimate the translation between I_1 and I_2, take the inverse Fourier transform of the cross-power spectrum

$$\mathcal{S} = \frac{\mathcal{F}\{I_1\}\mathcal{F}^*\{I_2\}}{|\mathcal{F}\{I_1\}\mathcal{F}\{I_2\}|} = \exp[-2\pi i (\frac{\xi}{S}\Delta s + \frac{\eta}{T}\Delta t)], \tag{7}$$

to arrive at an impulse function at $(\Delta s, \Delta t)$, where F^* denotes the complex conjugate of F. However, this method can only measure integral translation, and its performance is not robust to aliasing, i.e. imperfect correspondence which is prevalent in case of non-Lambertian materials. Stone et al. [22] proposed a generic sub-pixel registration algorithm that deals with aliasing. Here we deduce a simplified approach by leveraging angular domain geometry.

By (2), $\Delta \boldsymbol{p}$ is linear to the lens translation $\Delta \boldsymbol{\varphi} = (\Delta u, \Delta v)$. If we choose two sub-views with $\Delta v = 0$, Δt will also be zero, and the phase of the cross-power spectrum $\arg \mathcal{S} = -2\pi \Delta s \cdot \xi/S$ will be constant in η, as shown in Fig. 3(a). We first apply a median filter to $\arg \mathcal{S}$ to remove isolated noise due to aliasing, as shown in Fig. 3(b). Since we are dealing with periodic phase data, we take circular medians [23] instead of ordinary medians. To reduce clustered noise along the η axis, the median value $\arg \mathcal{S}_\xi = \text{med}[\mathcal{S}(\xi, :)]$ is taken along η. Finally, we regress $\arg \mathcal{S}_\xi$ on ξ/S using ordinary least squares to further reduce noise along the ξ axis, as shown in Fig. 3(c). The estimated slope is then -2π times the estimated displacement $\Delta \hat{s}$. Note that if Δs is larger than one pixel, $\arg \mathcal{S}_\xi$ will wrap around $\pi/-\pi$. Therefore, we first locate the impulse $\mathcal{F}^{-1}\{\mathcal{S}\}$ to register I_1 and I_2 to the nearest integral pixel, then proceed with sub-pixel refinement.

The micro-lens array of a light-field camera can be modeled with a square grid, with constant translation δ between adjacent lenses [35], as shown in Fig. 2(a). Therefore, the disparity Δ between adjacent sub-views is also a constant. We estimate Δs from two sub-views at both ends of the center row, and Δt from two sub-views at both ends of the center column. Δ is estimated by averaging along two axes:

$$\Delta = \frac{1}{2}(\frac{\Delta s}{U-1} + \frac{\Delta t}{V-1}), \tag{8}$$

where U and V are the numbers of sub-views along axes u and v. With Δ, we register all sub-views to the center sub-view using (6), so that the angular domain of each spatial pixel forms an ASI. We provide some visual examples in the supplementary material.

4.3 Network Architecture

We extract spatial features from the spatial component with a standard 2D image feature extractor (e.g. ResNet [8]). Angular features are extracted from the cropped and registered angular component using an angular feature extractor, shown in Fig. 4. We employ a multi-stream network [19] that reads sub-views in some certain directions, i.e. center horizontal, center vertical, main diagonal and antidiagonal, as shown in Fig. 4(a). The angular feature extractor *angular-4* (4 directions) is shown in Fig. 4(b). Sub-views in each direction are stacked and encoded separately by angular filter sub-networks at the beginning to produce meaningful representations. This not only explicitly supplies the network with structural information of angular domain, but also reduces angular redundancy. The first layer "AF" is a convolutional layer consisting of multiple filters that convolve across the angular domain. *Angular-4* is a modification of the multi-stream architecture in [19]. Convolutional kernel sizes and network depth are altered; the last convolutional layer is replaced by global average pooling to produce feature vectors. See more details in the caption of Fig. 4 and the supplementary material.

The multi-stream architecture was originally proposed for depth estimation. However, there is a major difference between depth estimation and material recognition: depth is a 2D function with axes orientations, but material is invariant to camera and object rotations. These actions cause angular domain axes to rotate. We can incorporate this prior knowledge by employing an invariance mechanism, which we call *angular rotational invariance.* Our final architecture *angular-S* (shared) augments *angular-4* with this invariance, as shown in Fig. 4(c). All four angular filter sub-networks have shared weights, which is implemented by passing all sub-view stacks to a single sub-network separately to produce four feature maps. These feature maps are concatenated along a new "direction" dimension, and a max-pooling layer is applied to this new dimension, which we call *directional pooling.* We can verify this architecture indeed enforces angular rotational invariance by observing that, if angular domain axes rotate by multiples of $\pi/4$, it results in a permutation of sub-view stacks, and the output of the directional pooling layer is invariant.

5 Experimental Results

5.1 Data and Training Procedure

We report results on multiple light-field material datasets to demonstrate the robustness of our methods. The first dataset is LFMR [25], which contains real light-field images captured by Lytro with spatial resolution 376×541 and angular resolution 7×7. The second dataset is the rendered light-field images with same resolutions from the BTF dataset [29], which was also used in [25] for evaluation. Our data preprocessing procedures follow [25]: square sample patches are extracted from whole light-field images; their centers are separated by at least half the patch size, and more than 50% of the pixels correspond to the target material. Unless stated otherwise, most of our experiments use 128×128 patches. Each dataset is randomly split into a training set and a test set by 7:3, 1/7 of the training set is used as a validation set for hyper-parameter selection.

Patches from the same light-field image only appear in one set to avoid strong correlation between the train/val/test sets.

Table 1. Classification accuracy (in percentage) comparison on test sets. "2D": use spatial features for classification. "model-4/S": *angular-4/S* is used as the angular feature extractor. "gain$_{2D}$": accuracy improvement over "2D"

Dataset	LFMR [25]		BTF [29]	
Method	Accuracy	gain$_{2D}$	Accuracy	gain$_{2D}$
2D	70.45 ± 0.23	–	67.35 ± 0.33	–
StackNet [13]	72.67 ± 2.39	2.22	70.19 ± 1.87	2.84
AngConvNet [13]	73.23 ± 2.29	2.78	72.16 ± 1.08	4.81
Lu et al. [14]	76.48 ± 1.23	6.03	$\mathbf{73.84} \pm 1.31$	**6.49**
4D-filter [25]	77.29 ± 1.05	6.84	71.81 ± 1.93	4.46
ang-filter [25]	$\mathbf{77.83} \pm 0.89$	**7.38**	73.27 ± 0.85	5.92
MDAIN [32]	75.73 ± 1.88	5.28	69.43 ± 1.52	2.08
model-4 (ours)	80.38 ± 0.53	9.93	75.14 ± 0.66	7.79
model-S (ours)	$\mathbf{81.75} \pm 0.61$	**11.30**	$\mathbf{77.52} \pm 0.58$	**10.17**

Data augmentation is carried out in spatial domain, including random horizontal flipping, randomly cropping the spatial resolution to the largest factor of 224 (e.g. if the patch size is 128×128, we take 112×112 random crops), and then upsampling to 224×224. We also normalize all sub-views by subtracting half the max intensity (e.g. 128 for 8-bit images) uniformly in all color channels. At test time, we perform centered cropping instead of random cropping, followed by normalization. The angular feature extractor is trained from scratch, and the spatial feature extractor is ResNet-18 (except the last fully-connected layer), since it provides fast training and good generalization. All models are trained with stochastic gradient descent and cross entropy loss for 200 epochs, with a base learning rate of 10^{-4}, momentum 0.9 and batch size 128. The angular feature extractor and the fully connected layer use $10\times$ the base learning rate. All experiments use the top-1 accuracy as their performance measures.

5.2 Overall Performance

Table 1 shows the results of our proposed framework on the test sets, and compares with winning baseline methods from [25]. We also compare with architectures designed for other closely related tasks. StackNet and AngConvNet were proposed in [13] for surface BRDF recognition using light-field images. MDAIN [32] uses multiple stereo pairs for material recognition. The authors tested on light-field datasets by selecting 4 sub-view pairs of entire light-field images. We replicate their test conditions except that patches are classified rather than the whole light-field. Lu et al. [14] proposed a domain-interleaved architecture that resembles *4D-filter* for light-field object classification. Since it requires 8×8 angular resolution, we pad angular domain by replication and use it for material classification. The classification accuracy is reported by averaging 5 random train/test splits. More details are given in the table caption.

For fair comparison, all previous methods also use ResNet-18 as their backbone networks. Note that [25] originally used VGG-16 [20] as the backbone architecture and achieved 7% gain on both *ang-filter* and *4D-filter*. By switching to ResNet-18 we reproduce similar gains but with much less memory and shorter runtime. We see that comparing to the best baseline method, our *model-S* achieves 11.30%/7.38% = 1.53 times gain on the LFMR dataset, and 10.17%/6.49% = 1.57 times gain on the BTF dataset. This result shows that our proposed framework significantly outperforms previously reported methods. While previous methods achieve gains over 2D by *utilizing additional data* which is arguably expected, our method outperforms these methods by *more efficient usage of these additional data.*

5.3 Ablation Studies

We conduct extensive ablation experiments using the LFMR dataset, including hyper-parameter choices, contribution of each technical component and robustness tests.

Table 2. Left: classification accuracy on validation set. Representative selection is enabled. Right: classification accuracy with different N for representative selection. $N = 1$ means no cropping. We report results using both angular and spatial features ("angular+spatial"), and only using angular features ("angular") for classification. In both tables, *angular-S* is the angular feature extractor, angular registration is enabled

K_{af}	#af	mean acc. (%)	std. (%)
1	16	81.35	0.57
1	32	**81.92**	0.56
1	64	80.93	0.60
3	32	81.16	0.60

feature	N	mean acc. (%)	std. (%)
angular+spatial	1	81.45	0.42
angular+spatial	2	81.56	0.69
angular+spatial	3	**81.92**	0.56
angular+spatial	4	81.28	0.53
angular	1	**66.34**	1.22
angular	2	60.15	1.04
angular	3	55.31	1.41
angular	4	51.93	1.53

Hyper-parameters. Table 2 (left) compares different angular filter kernel sizes K_{af} and numbers of angular filters #af. The classification accuracy on the validation set is reported by averaging 5 random train/validation splits. This agrees with the observation in [25] that a medium #af offers the best performance, while a low #af reduces representation power, and a large #af leads to overfitting. We also observe that $K_{af} = 1$ outperforms $K_{af} = 3$. $K_{af} = 1$ corresponds to taking convolutions only in angular domain, i.e. aligned ASIs, while $K_{af} = 3$ corresponds to taking convolutions both in the ASI and spatial domain. This result confirms that decoupling angular and spatial information is beneficial.

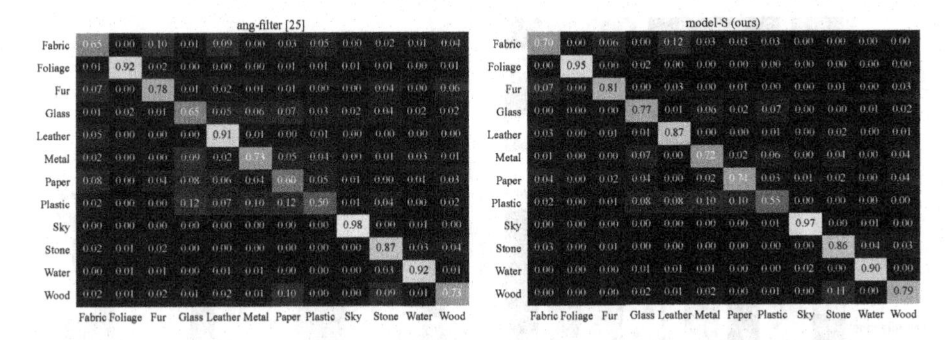

Fig. 5. Confusion matrices of *ang-filter* [25] and *model-S* on the LFMR dataset. Paper, glass, woods and plastic are the most improved categories

Table 2 (right) compares different N used for representative selection. $N = 1$ means the entire patch is used and selection is disabled. Besides the proposed framework ("angular+spatial"), we also report results where only angular features are used for classification ("angular"). For "angular+spatial", a medium sized region offers the best performance, as a large region may violate the constant depth assumption, leading to potential occlusion and bad registration; while a small region carries little information. In contrast, as the spatial component is missing in "angular", the best choice is to use the entire patch, which contains the most spatial information. This result shows that by decomposing light-fields into two components, we can conveniently represent local properties with a small region without worrying about losing much spatial information.

Table 3. Left: ablation study on the test set of LFMR. "model-4/S": *angular-4*/*angular-S* is used as the angular feature extractor. "select": use representative selection. "register": use angular registration. Right: classification accuracy (in percentage) comparison using different light-field patch sizes

id	method	select	register	mean acc.	std.
1	ang-filter	-	-	77.83	0.89
2	model-4	✓	✓	**80.38**	0.53
3		✓	✗	79.85	0.45
4		✗	✓	79.36	1.05
5		✗	✗	78.91	0.75
6	model-S	✓	✓	**81.75**	0.61
7		✓	✗	81.00	0.56
8		✗	✓	81.45	0.42
9		✗	✗	80.74	0.52
10		random	✓	81.41	0.73
11		✗	warp	80.35	0.55

data	2D	ang-filter	model-S
size=32	49.18	58.79	**63.29**
size=64	58.73	66.38	**71.73**
size=128	70.45	77.83	**81.75**
size=256	78.80	82.37	**85.69**

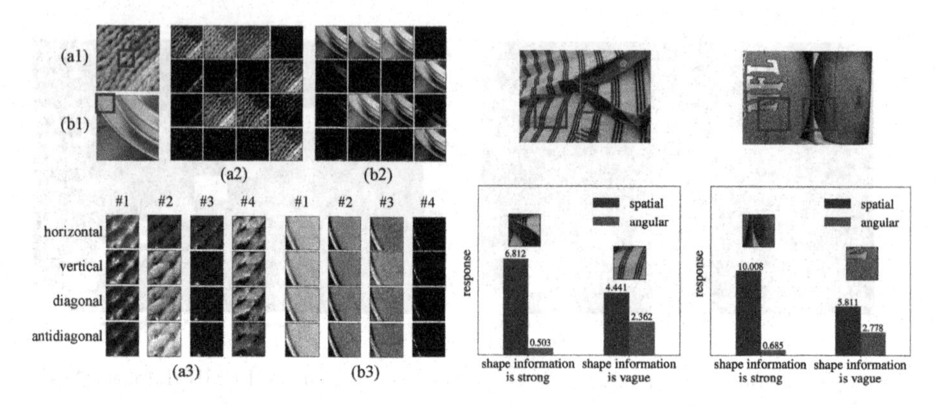

Fig. 6. Left: angular filter activations of *ang-filter* and *angular-S*. (a1) and (b1) are the original light-field patches, representative regions are outlined in red. (a2) and (b2) are the first 16 feature maps from *ang-filter*. Columns in (a3) and (b3) correspond to the first 4 feature maps from *angular-S*, each row corresponds to a stack of sub-views. Right: angular/spatial responses of different patches in the same light-field image. Top: original light-field images, selected patches are outlined in red. Bottom: corresponding patches are displayed above bar graphs (Color figure online)

Components. Table 3 (left) provides a breakdown of each component's contribution. Compare row 1 and 5, we see that using sparse inputs and explicit directional information provides 1.1% gain. Compare row 5 and 9, we see that enforcing angular rotational invariance provides 1.8% further gain. Within the *model-4* group, angular registration offers 0.5% gain, and representative selection offers 0.9% gain. When both are used, the combined gain 1.5% is nearly additive. This gain becomes less significant (1.0%) for *model-S*, implying that angular rotational invariance increases the network's robustness to misaligned and redundant data (Fig. 5).

We analyze representative selection and angular registration in more detail with two more ablation test cases. Row 10 selects representative regions randomly rather than using (4) to select the best candidate. Its performance drops since it is susceptible to occlusion and background. Row 11 uses the estimated depth map to directly warp pixels, rather than to select representative regions. We observe a performance degradation comparing to row 8 as it imposes the Lambertian assumption to all materials and alters ASIs. This further proves the necessity and significance of our proposed methods.

Robustness. We verify the robustness of our methods by varying the light-field patch sizes. Table 3 (right) compares performances of the 2D model using spatial features ("2D"), the best baseline method *ang-filter* and our best method *model-S*. It can be seen that under various input sizes, our method consistently outperforms *ang-filter* and significantly improves the gain over "2D".

5.4 Visualization

Angular Filter Output. Both *ang-filter* and our angular feature extractors use angular filters to perform convolutions in angular domain before subsequent layers. Figure 6 (left) compares their activations in *ang-filter* and *angular-S*, both using 1×1 kernels. We observe that in (a3), the activations of different sub-view stacks differ, indicating the material (fabric) has strong reflection variation. Because all angular filter sub-networks have shared weights, difference of activations can only be caused by difference of inputs. Meanwhile in (b3), the material (plastic) has homogeneous reflection, and its activations are similar across directions. In contrast, this pattern is not present in (a2) and (b2), where directional information is hard to visualize.

Angular v.s. Spatial Responses. By combining angular and spatial features at the end, we can use the network to evaluate their relative strengths. For the neuron i corresponding to the true class in the softmax layer, its pre-activation is a linear transform of angular and spatial feature vectors. Define the angular and spatial responses as the absolute values of $W_i^a \phi_a$ and $W_i^s \phi_s$, so that the pre-activation y_i of the true class neuron i has the decomposition $y_i = W_i^a \phi_a + W_i^s \phi_s$, where W_i^s, W_i^a are weights of the fully-connected layer connected to neuron i, ϕ_a and ϕ_s are angular and spatial feature vectors, respectively. Figure 6 (right) compares angular and spatial responses for two different patches in the same light-field image. If the patch contains object or shape information so that the material can be easily inferred, the spatial response is much higher than angular response. Conversely, if context information is vague, then material has to be inferred from local properties such as texture and reflection, the angular response becomes more significant.

6 Conclusion

In this paper, we propose a novel framework for material recognition using light-fields that can potentially boost many industrial applications with a simple optical upgrade. The light-field is decomposed into the center sub-view and a representative crop, responsible for spatial and angular feature extraction, respectively. Thanks to the spatial-angular decomposition, we can keep most spatial information intact while cropping the angular component to avoid occlusion and for better registration. The angular feature extractor employs directional regularization by weight sharing in angular filter sub-networks and directional pooling; they together enforce rotational invariance in angular domain. Our methodology is verified by thorough ablation and robustness studies. It also casts light on how to efficiently learn from data with intertwined dimensions.

Acknowledgement. Yuxing Han is the corresponding author. This work was supported by the Natural Science Foundation of China (Project Number 61521002) and Shenzhen International Collaborative Research Project (Grant GJHZ2018092915 1604875).

References

1. Alperovich, A., Johannsen, O., Strecke, M., Goldluecke, B.: Light field intrinsics with a deep encoder-decoder network. In: CVPR (2018)
2. Bolles, R.C., Baker, H.H., Marimont, D.H.: Epipolar-plane image analysis: an approach to determining structure from motion. IJCV **1**(1), 7–55 (1987). https://doi.org/10.1007/BF00128525
3. Chen, C., Lin, H., Yu, Z., Bing Kang, S., Yu, J.: Light field stereo matching using bilateral statistics of surface cameras. In: CVPR (2014)
4. Chen, J., Hou, J., Chau, L.P.: Light field denoising via anisotropic parallax analysis in a CNN framework. IEEE Signal Process. Lett. **25**(9), 1403–1407 (2018)
5. Cho, Y., Bianchi-Berthouze, N., Marquardt, N., Julier, S.J.: Deep thermal imaging: proximate material type recognition in the wild through deep learning of spatial surface temperature patterns. In: Proceedings of the 2018 CHI Conference on Human Factors in Computing Systems (2018)
6. Cimpoi, M., Maji, S., Vedaldi, A.: Deep filter banks for texture recognition and segmentation. In: CVPR (2015)
7. DeGol, J., Golparvar-Fard, M., Hoiem, D.: Geometry-informed material recognition. In: CVPR (2016)
8. He, K., Zhang, X., Ren, S., Sun, J.: Deep residual learning for image recognition. In: CVPR (2016)
9. Heber, S., Yu, W., Pock, T.: Neural EPI-volume networks for shape from light field. In: ICCV (2017)
10. Honauer, K., Johannsen, O., Kondermann, D., Goldluecke, B.: A dataset and evaluation methodology for depth estimation on 4D light fields. In: Asian Conference on Computer Vision (2016)
11. Jeon, H.G., et al.: Accurate depth map estimation from a lenslet light field camera. In: CVPR (2015)
12. Johannsen, O., Sulc, A., Goldluecke, B.: What sparse light field coding reveals about scene structure. In: CVPR (2016)
13. Lu, F., He, L., You, S., Chen, X., Hao, Z.: Identifying surface BRDF from a single 4-D light field image via deep neural network. IEEE J. Sel. Top. Signal Process. **11**(7), 1047–1057 (2017)
14. Lu, Z., Yeung, H.W.F., Qu, Q., Chung, Y.Y., Chen, X., Chen, Z.: Improved image classification with 4D light-field and interleaved convolutional neural network. Multimedia Tools Appl. **78**(20), 29211–29227 (2018). https://doi.org/10.1007/s11042-018-6597-x
15. Qi, X., Xiao, R., Li, C.G., Qiao, Y., Guo, J., Tang, X.: Pairwise rotation invariant co-occurrence local binary pattern. IEEE TPAMI **36**(11), 2199–2213 (2014)
16. Reddy, B.S., Chatterji, B.N.: An FFT-based technique for translation, rotation, and scale-invariant image registration. IEEE TIP **5**(8), 1266–1271 (1996)
17. Schwartz, G., Nishino, K.: Material recognition from local appearance in global context. arXiv preprint arXiv:1611.09394 (2016)
18. Sheng, H., Zhang, S., Cao, X., Fang, Y., Xiong, Z.: Geometric occlusion analysis in depth estimation using integral guided filter for light-field image. IEEE TIP **26**(12), 5758–5771 (2017)
19. Shin, C., Jeon, H.G., Yoon, Y., So Kweon, I., Joo Kim, S.: EPINET: a fully-convolutional neural network using epipolar geometry for depth from light field images. In: CVPR (2018)

20. Simonyan, K., Zisserman, A.: Very deep convolutional networks for large-scale image recognition. In: ICLR (2015)
21. Sonnemann, T., Ulloa Hung, J., Hofman, C.: Mapping indigenous settlement topography in the Caribbean using drones. Remote Sens. **8**(10), 791 (2016)
22. Stone, H.S., Orchard, M.T., Chang, E.C., Martucci, S.A.: A fast direct Fourier-based algorithm for subpixel registration of images. IEEE Trans. Geosci. Remote Sens. **39**(10), 2235–2243 (2001)
23. Storath, M., Weinmann, A.: Fast median filtering for phase or orientation data. IEEE TPAMI **40**(3), 639–652 (2018)
24. Wang, T.C., Efros, A.A., Ramamoorthi, R.: Occlusion-aware depth estimation using light-field cameras. In: ICCV (2015)
25. Wang, T.C., Zhu, J.Y., Hiroaki, E., Chandraker, M., Efros, A.A., Ramamoorthi, R.: A 4D light-field dataset and CNN architectures for material recognition. In: ECCV (2016)
26. Wang, T.C., Zhu, J.Y., Kalantari, N.K., Efros, A.A., Ramamoorthi, R.: Light field video capture using a learning-based hybrid imaging system. ACM TOG **36**(4), 1–13 (2017)
27. Wang, Y., Liu, F., Zhang, K., Hou, G., Sun, Z., Tan, T.: LFNet: a novel bidirectional recurrent convolutional neural network for light-field image super-resolution. IEEE TIP **27**(9), 4274–4286 (2018)
28. Wanner, S., Goldluecke, B.: Reconstructing reflective and transparent surfaces from epipolar plane images. In: German Conference on Pattern Recognition (2013)
29. Weinmann, M., Gall, J., Klein, R.: Material classification based on training data synthesized using a BTF database. In: ECCV (2014)
30. Wing Fung Yeung, H., Hou, J., Chen, J., Ying Chung, Y., Chen, X.: Fast light field reconstruction with deep coarse-to-fine modeling of spatial-angular clues. In: ECCV (2018)
31. Wu, G., Zhao, M., Wang, L., Dai, Q., Chai, T., Liu, Y.: Light field reconstruction using deep convolutional network on EPI. In: CVPR (2017)
32. Xue, J., Zhang, H., Dana, K., Nishino, K.: Differential angular imaging for material recognition. In: CVPR (2017)
33. Zhang, H., Dana, K., Nishino, K.: Reflectance hashing for material recognition. In: CVPR (2015)
34. Zhao, C., Sun, L., Stolkin, R.: A fully end-to-end deep learning approach for real-time simultaneous 3D reconstruction and material recognition. In: 2017 18th International Conference on Advanced Robotics (2017)
35. Zhao, S., Chen, Z.: Light field image coding via linear approximation prior. In: ICIP (2017)

Dual Adversarial Network for Deep Active Learning

Shuo Wang[1,2], Yuexiang Li[2(✉)], Kai Ma[2], Ruhui Ma[1(✉)], Haibing Guan[1], and Yefeng Zheng[2]

[1] School of Electronic Information and Electrical Engineering, Shanghai Jiao Tong University, Shanghai 200240, China
ruhuima@sjtu.edu.cn

[2] Tencent Jarvis Lab, Shenzhen, China
vicyxli@tencent.com

Abstract. Active learning, reducing the cost and workload of annotations, attracts increasing attentions from the community. Current active learning approaches commonly adopted uncertainty-based acquisition functions for the data selection due to their effectiveness. However, data selection based on uncertainty suffers from the overlapping problem, i.e., the top-K samples ranked by the uncertainty are similar. In this paper, we investigate the overlapping problem of recent uncertainty-based approaches and propose to alleviate the issue by taking representativeness into consideration. In particular, we propose a dual adversarial network, namely DAAL, for this purpose. Different from previous hybrid active learning methods requiring multi-stage data selections i.e., step-by-step evaluating the uncertainty and representativeness using different acquisition functions, our DAAL learns to select the most uncertain and representative data points in one-stage. Extensive experiments conducted on three publicly available datasets, i.e., CIFAR10/100 and Cityscapes, demonstrate the effectiveness of our method—a new state-of-the-art accuracy is achieved.

Keywords: Active learning · Generative adversarial network · Unsupervised video summarization · Deep learning

1 Introduction

Benefiting from large-scale annotated datasets, deep learning has shown its great success in various computer vision tasks such as image classification, object detection, and semantic segmentation. Yet, the annotation of large-scale datasets is extremely laborious and costly to obtain, especially for the dense pixel-level annotation and the one requiring experienced annotators to tackle (e.g., medical images). For this reason, semi-supervised learning methods [3,18,25,29,30] and

S. Wang—Intern at Tencent Jarvis Lab.

A. Vedaldi et al. (Eds.): ECCV 2020, LNCS 12369, pp. 680–696, 2020.
https://doi.org/10.1007/978-3-030-58586-0_40

unsupervised learning methods [1,6,28,44] attract increasing attention. However, given a fixed amount of data, their performance is still bound to that of fully-supervised learning.

Active learning (AL) that incrementally queries the most informative samples from the data pool to reduce the overall annotation effort has thus emerged as a promising research avenue for the use of deep learning [9]. Among recent AL-related works, pool-based AL methods [10,32,36,41], which iteratively select data points from a large unlabeled data pool for annotation according to the acquisition function, are the most successful. Accurate estimation of data informativeness is the core of pool-based AL. Many researches focused on exploring effective acquisition functions to achieve this goal, which can be classified to two categories—uncertainty-based and representation-based. The AL using uncertainty-based acquisition functions [36,41] prefer to select samples confusing the classifier (i.e., high uncertainty), while the representation-based ones [39,45] select samples best representing the unlabeled pool. The estimation of data informativeness is not comprehensive in either term of uncertainty or representativeness. Therefore, some studies [39,45] proposed the hybrid strategies. However, their inferior performance compared to the uncertainty-based approaches [36,41] illustrates that the benefit of representativeness is not actually exploited.

Recent years witnessed the success of adversarial networks and several studies [11,36] tried to apply the adversarial learning for more accurate estimation of data informativeness. For example, Sinha et al. [36] proposed the VAAL by using a variational autoencoder (VAE) [20] and a discriminator, where the VAE embedded the labeled and unlabeled images to a latent space and the discriminator was utilized as a binary classifier to measure the uncertainty of the input samples. However, the selected samples may not be the most informative ones for the task model performing the target task (e.g., image classification and semantic segmentation) since the VAAL is fully task-agnostic.

In contrast, Yoo et al. [41] employed the loss of task model as the criterion to estimate the contribution of data made to the target task—a loss prediction module was proposed to estimate the loss of task model for the unlabeled data. Since the loss prediction module directly utilizes features from the task model, it gives more accurate estimation of the informativeness of current input to the task model. However, this loss prediction module suffers from the overlapping problem [32,41]—the information contained in the top K selected samples is similar. The solution [32,41] to this problem is performed in a two-stage manner—a random subset with a certain size from the unlabeled pool is firstly created to ensure the diversity of data, and then apply uncertainty-based acquisition function to that subset for data point selection. Although this simple random subset selection (RSS) strategy can alleviate the overlapping problem, its effectiveness to other uncertainty-based AL methods [36] has not been explored. Current multi-stage AL frameworks, i.e., step-by-step evaluating the uncertainty and representativeness using different methods, are another potential solutions to the overlapping problem. However, they are time-consuming for data selection and difficult to provide proper estimation of data informativeness due to the trade-off between two evaluation methods. Therefore, a one-stage AL method without suffering from the overlapping problem is worthwhile to develop.

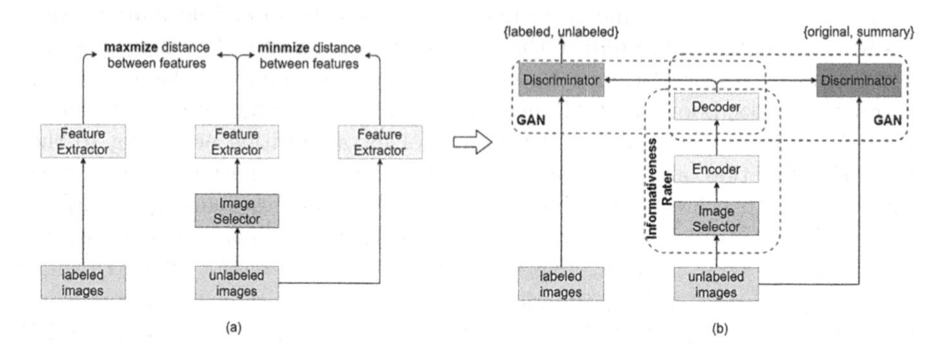

Fig. 1. Overview: (a) Our goal is to simultaneously select the most uncertain and representative images for the task model from the unlabeled pool. The image selector is required to select the samples maximizing the distance to labeled images (uncertainty), while minimizing the distance to unlabeled ones (representativeness). (b) The generative adversarial framework is proposed to assist image selector to properly measure the distances between deep features.

To this end, we propose a one-stage dual adversarial network for active learning, namely DAAL, to accurately select the most informative data points from the unlabeled pool for the task model by simultaneously considering the uncertainty and representativeness. Inspired by the unsupervised video summarization [15,26] in which a frame selector was proposed to learn a representative summary[1] of the original video, we propose to use an image selector as the acquisition function to find a sparse and representative subset from the unlabeled pool, which is simultaneously with high uncertainty for the task model.

The overview of our approach is illustrated in Fig. 1 (a). The image selector is required to find samples with larger feature distance to the labeled samples (i.e., high uncertainty to the classifier) and smaller distance to the unlabeled data (i.e., good representativeness to the unlabeled data). However, specifying a suitable distance of deep features is difficult [23]. Therefore, we use a generative adversarial (GAN) [13] framework to assist the image selector in the distance measurement between deep feature representations. As shown in Fig. 1 (b), samples selected by the image selector are sent to a VAE, which embeds the features of selected samples into the same latent space and then reconstructs them. The reconstructed features are fed to two discriminators, which encourage the image selector to simultaneously take uncertainty (labeled/unlabeled) and representativeness (original/summary) into consideration during data selection. The informativeness rater (consisting of image selector and VAE) and discriminators are

[1] Summary is a sparse subset of video frames which optimally represent the input video.

framed as a multi-player competition, similar to GAN [13]. The informativeness rater is trained to trick the discriminators. The performances of informativeness rater and discriminators are improved by iteratively optimization.

In summary, our contributions are manifold. First, we propose a one-stage dual adversarial network for active learning, namely DAAL, which can simultaneously learn to select the most uncertain and representative data points from the unlabeled pool. Second, our designed DAAL is more effective to alleviate the overlapping problem, compared to the approaches using random subset selection and other hybrid methods. Last but not least, extensive experiments conducted on three publicly available datasets (CIFAR10/100 and Cityscapes) show that our approach outperforms the benchmarking methods and achieves a new state-of-the-art.

2 Related Work

Active Learning. The methods in the area of active learning can be roughly classified to two categories—uncertainty-based and representation-based methods.

Uncertainty-Based Methods. The core idea is to find the samples, which are difficult for the classifier to correctly classify (i.e., with high uncertainty to the classifier). These methods can be further categorized to Bayesian and non-Bayesian frameworks. For Bayesian active learning methods, probabilistic models such as Bayesian neural networks [7] and Gaussian processes [19] are used to estimate the uncertainty of samples. Houlsby et al. [16] proposed a Bayesian active learning, which used the mutual information of the training examples as a proxy uncertainty measurement for sample selection. For non-Bayesian active learning methods, the sample uncertainty can be measured in various ways such as the distance between samples and the decision boundary [4], information entropy [17] and risk expectation [38]. In more recent works, Gal et al. [9] proposed to utilize dropout layers to estimate the uncertainty of the prediction yielded by a neural network for sample query. Yoo et al. [41] proposed to use an auxiliary loss prediction module to learn the target loss of inputs jointly with the training phase and samples with high predicted losses are selected. Sinha et al. [36] proposed a framework (namely VAAL) for active learning, consisting of a variational autoencoder (VAE) and generative adversarial network (GAN). The probability of discriminator is seen as the uncertainty estimation for sample selection.

Representation-Based Methods. This kind of approaches aims to constitute a set of diverse samples, which are the most representative of the entire dataset. Sener et al. [31] proposed a core-set selection method, which selected the samples minimizing the Euclidean distance between the selected data and the unlabeled data pool in the feature space. There are also some hybrid methods [8,40] taking both uncertainty and diversity into account.

Active Learning for Semantic Segmentation. Semantic segmentation is one of the most prevailing tasks for active learning due to its expensive annotation, which has been broadly investigated in recent studies [24,34,36,39]. Yang et al. [39] proposed a hybrid framework, namely suggestive annotation (SA), combining the measurements of uncertainty and representativeness. This framework estimated the uncertainty of data points using an ensemble of models and measured the representativeness using the core-set approach [31].

Variational Autoencoder. Autoencoders are commonly used to effectively learn a feature representation for various tasks [2]. Variational autoencoder is a variant of autoencoder, which defines a posterior distribution over the observed data, given an unobserved latent variable. A VAE is used to embed the labeled and unlabeled images into the same latent space by VAAL [36]. Given $e \sim p_e(e)$ as a priori over the unobserved latent variable, we can formulate the objective function of VAE with observed data x as:

$$\mathcal{L}_{VAE} = -\log \frac{p(x|e)p(e)}{q(e|x)} = \underbrace{-\log(p(x|e))}_{\mathcal{L}_{recon}} + \underbrace{D_{KL}(q(e|x)||p(e))}_{\mathcal{L}_{prior}} \tag{1}$$

where D_{KL} is the Kullback-Leibler divergence; $q(e|x)$ is the probability of observing e given x; $p_e(e)$ is the standard normal distribution; and $p(x|e)$ is the conditional generative distribution for x.

Generative Adversarial Network. The typical generative adversarial network (GAN) [13] consists of a generator network and a discriminator. The generator generates data simulating an unknown distribution and the discriminator network aims to distinguish the generated/fake samples from the real ones. The generator and the discriminator are alternately trained to force the generator fitting the real data distribution while maximizing the probability of the discriminator making a mistake. Given x as the true data, $e \sim p_e(e)$ as the prior input noise, and $\hat{x} = G(e)$ as the generated sample, the objective function of a typical GAN can be formulated as:

$$\min_G \max_D \left[\mathbb{E}_x[\log D(x)] + \mathbb{E}_e[\log(1 - D(\hat{x}))]\right] \tag{2}$$

where the discriminator D is trained to maximize the probability of real/fake classification and the generator G is trained to minimize $\log(1 - D(\hat{x}))$.

3 Method

In this section, we introduce the proposed DAAL in details. Specific descriptions of network architecture are introduced in Sect. 3.1. In Sect. 3.2, the detailed training procedure of the DAAL is illustrated.

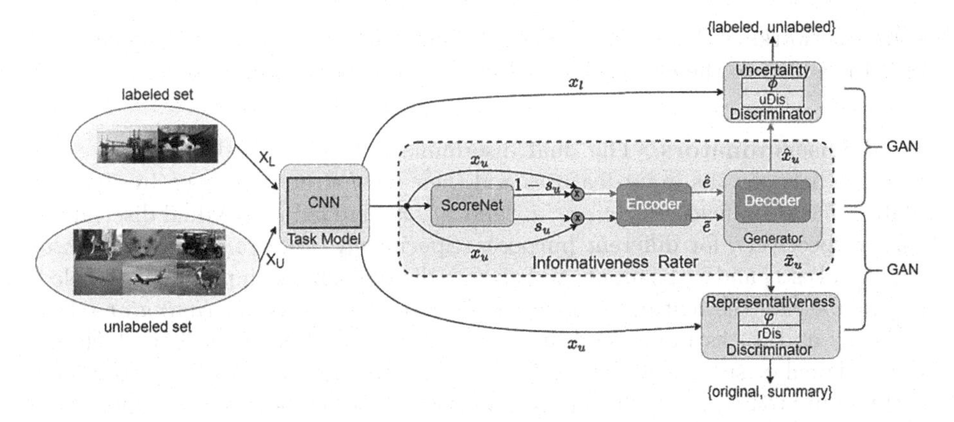

Fig. 2. Major components of our approach. The informativeness rater consists of a ScoreNet (image selector) and a VAE (encoder-decoder architecture). Given the features encoded by the task model for labeled (x_l) and unlabeled images (x_u), the ScoreNet assigns importance score s_u to x_u. The encoder encodes the sample with a pair of features ($\hat{e}$, $\tilde{e}$), which are fed to the decoder for reconstruction ($\hat{x}_u$, $\tilde{x}_u$). The reconstructed features are sent to the dual discriminators, respectively. The uDis is required to classify whether the $\hat{x}_u$ belongs to labeled pool or not. The rDis aims to identify the reconstructed feature $\tilde{x}_u$ from the original one x_u. The decoder/generator and dual discriminators are adversarially trained until the uDis cannot discriminate between labeled and unlabeled data points and the rDis is not able to distinguish between the summary and original datasets.

3.1 Dual Adversarial Network for Deep Active Learning

The detailed information of our DAAL including the information flow is illustrated in Fig. 2. The ScoreNet and VAE form an independent function unit, namely informativeness rater, which cooperates with two different discriminators to construct the dual adversarial network. To accurately measure the informativeness of input samples for the task model, the features of labeled and unlabeled images encoded by the task model are adopted as input of our DAAL.

Informativeness Rater. Given the deep features of images from the unlabeled pool ($X_U = \{x_u : u = 1, ..., N\}$) generated by the task model, the ScoreNet assigns a relative importance score ($s = \{s_u : s_u \in [0, 1], u = 1, ..., N\}$) to each of them. Original input features x_u are weighted using these scores. Note that we use $1 - s_u$ and s_u as the weights for uncertainty and representativeness branches, respectively, to ensure the optimization of these two terms in the same direction. These weighted deep features are sent to a VAE which consists of an encoder and a decoder/generator. The encoder maps the inputs to the features ($e = \{\hat{e}, \tilde{e}\}$) in the same latent space, while the decoder/generator reconstructs the embedded features ($x_{\text{recon}} = \{\hat{x}_u, \tilde{x}_u\}$). The ScoreNet adopts a simple architecture, consisting of a 5-layer multi-layer perceptron (MLP) with Xavier initialization [12], to map the input feature to a 1×1 score vector. Both the encoder and decoder are neural networks with 7-layer MLP. A dropout layer [37] with the dropout ratio of 0.4 is added to the end of each MLP layer to avoid overfitting.

Dual Discriminators. The dual discriminators are utilized to measure the distance between the input features and their reconstructions given by the generator. The deep features, i.e., x_l, $\hat{x}_u$ and x_u, $\tilde{x}_u$, are fed to the dual discriminators, respectively, for different purposes. Specifically, *uncertainty discriminator* (uDis) takes x_l and $\hat{x}_u$ as input, and aims to distinguish which pool (i.e., 'labeled' or 'unlabeled') the features belong to. *Representativeness discriminator* (rDis) tasks x_u and $\tilde{x}_u$ as input, and aims to classify them into two distinct classes (i.e., original or summary). The dual discriminators in the GAN adopt the same architecture to ScoreNet without dropout layers. The class 'summary' represents the reconstruction of weighted deep features of the input batch. If the discriminator cannot distinguish the summary batch from the original one, the images with high scores are seen to have good representativeness to the small unlabeled pool.

Training Strategy: Multi-player Competition. Our DAAL involves two adversarial games between the VAE and dual discriminators, which iteratively optimize the ScoreNet for accurate data selection. Due to the score $s \in [0, 1]$, the features with scores closed to 1 are easier for decoder/generator to reconstruct. For the uDis, if a well-reconstructed $\hat{x}_u$ fools it, the lower score ($1 - s_u$ as the weights for $\hat{x}_u$) should be maintained. The VAE and uDis optimize the ScoreNet via adversarial training. Simultaneously, as aforementioned, the adversarial learning between VAE and rDis encourage the ScoreNet to assign larger scores to the samples representative to the unlabeled pool. Therefore, the importance scores yielded by our ScoreNet can simultaneously evaluate the uncertainty and representativeness of data points—a higher score intrinsically represents the image with both larger uncertainty and representativeness. During the sample selection of active learning, samples with top k largest scores ranked by the ScoreNet are selected from the unlabeled pool for annotation.

3.2 Training Procedure of DAAL

Denote the network weights of ScoreNet, encoder and decoder of VAE and the dual discriminator (uDis and rDis) as $\{w_s; w_e, w_d; w_u, w_r\}$. The training procedure of DAAL is summarized in Algorithm 1. The proposed DAAL is supervised by four loss functions, which are 1) prior loss $\mathcal{L}_{prior}$ (as defined in Eq. 1) for the encoder of VAE, 2) reconstruction loss $\mathcal{L}_{recon}$ for VAE, 3) GAN loss $\mathcal{L}_{GAN}$, and 4) sparsity loss $\mathcal{L}_{\text{sparsity}}$ for the ScoreNet.

Reconstruction Loss $\mathcal{L}_{\text{recon}}$. Instead of using the standard reconstruction loss for autoencoder networks, i.e., $||x - \hat{x}||_2$ where x and $\hat{x}$ are the input and corresponding reconstruction, respectively, we follow the practice in [23] to use the last output layer of the discriminators for the calculation of $\mathcal{L}_{\text{recon}}$. Denote the

Algorithm 1. Training dual adversarial network

1: **Input:** Features (X_L and X_U) encoded by the task model for labeled and unlabeled images, respectively.
2: **Output:** Learned parameters $\{w_s, w_e, w_d, w_u, w_r\}$.
3: **Function:**
4: $f(x; w)$: forward the input x through neural network (w).
5: *update*(.): backward to update the neural network weights.
6: $\mathcal{L}(.)$: loss function.
7: **Procedure:**
8: Initialize all parameters $\{w_s, w_e, w_d, w_u, w_r\}$
9: **for** batch (x_l, x_u) from X_L and X_U **do**
10: $s_u \leftarrow f(x_u; w_s)$ // *select images*
11: $(\hat{e}, \tilde{e}) \leftarrow f((x_u, s_u); w_e)$ // *encoding*
12: $(\hat{x}_u, \tilde{x}_u) \leftarrow f((\hat{e}, \tilde{e}); w_r)$ // *reconstruction*
13: // *Updates using stochastic gradient:*
14: $\{w_s, w_e\} \leftarrow update(\mathcal{L}_{recon}(x_u, \hat{e}, \tilde{e}) + \mathcal{L}_{prior}(x_u, \hat{e}, \tilde{e}) + \mathcal{L}_{sparsity}(x_u, s_u))$
15: $\{w_d\} \leftarrow update(\mathcal{L}_{\text{recon}}(x_u, \hat{e}, \tilde{e}) + \mathcal{L}_{\text{GAN}}(x_u, x_l, \hat{x}_u, \tilde{x}_u))$
16: $\{w_u, w_r\} \leftarrow update(\mathcal{L}_{\text{GAN}}(x_u, x_l, \hat{x}_u, \tilde{x}_u))$ // *maximization update*
17: **end for**

output of the last hidden layer of uDis and rDis as $\phi(x_u)$ and $\varphi(x_u)$, for input x_u, respectively. Given embedded features $\hat{e}$ and $\tilde{e}$ of input x_u, $\mathcal{L}_{\text{recon}}$ can be formulated as:

$$\mathcal{L}_{\text{recon}} = \mathbb{E}[-\log p(\phi(x_u)|\hat{e})] + \mathbb{E}[-\log p(\varphi(x_u)|\tilde{e})] \tag{3}$$

where expectation $\mathbb{E}$ is approximated as the empirical mean of the training samples.

GAN loss $\mathcal{L}_{\text{GAN}}$. The adversarial learning between generator and dual discriminators is supervised by the GAN loss, which can be formulated as:

$$\begin{aligned}\mathcal{L}_{\text{GAN}} = &\log(\text{uDis}(x_l)) + \log(1 - \text{uDis}(\hat{x}_u)) \\ &+ \log(\text{rDis}(x_u)) + \log(1 - \text{rDis}(\tilde{x}_u)).\end{aligned} \tag{4}$$

where uDis(.) and rDis(.) represent the model functions of uncertainty discriminator and representativeness discriminator, respectively.

Sparsity Loss. The sparsity loss is a regularization term for the ScoreNet to prevent it from assigning equal importance to all data points. The sparsity loss consists of a length regularizer loss $\mathcal{L}_{LR}$ and a determinantal point process (DPP) loss $\mathcal{L}_{DPP}$ [22]. The $\mathcal{L}_{LR}$ limits the number of elements selected by the ScoreNet, while the $\mathcal{L}_{DPP}$ ensures the diversity of selected data points. The overall sparsity loss [35] can be defined as:

$$\mathcal{L}_{\text{sparsity}} = \mathcal{L}_{\text{LR}} + \mathcal{L}_{\text{DPP}} \tag{5}$$

where

$$\mathcal{L}_{\text{LR}} = \left\| \sigma - \frac{1}{n}\sum_{t=1}^{n} s_t \right\|_2, \quad \mathcal{L}_{\text{DPP}} = -\log(P(s_{x'})) \tag{6}$$

where σ represents the percentage of images for subset selection; $s_{x'}$ is the importance scores for a subset $x' \subset X_U$. The probability function P in $\mathcal{L}_{\text{DPP}}$ can be written as:

$$P(s_{x'}; D) = \frac{|D(s_{x'})|}{|D + I|} \tag{7}$$

where $D \in R^{n \times n}$ with $D_{i,j} = s_i s_j x_i^T x_j$ and $D(s_{x'}) \in R^{\sigma n \times \sigma n}$ (i.e., a submatrix of D given $s_{x'}$); $|.|$ denotes determinant and I is the identity matrix.

4 Experiments

In this section, we evaluate the effectiveness of our DAAL on three publicly available datasets. The evaluation results are presented in Sects. 4.3 and 4.4. Furthermore, we conduct an in-depth investigation on the drawback of uncertainty-only AL approaches. The related results can be found in Sect. 4.5. Finally, we analyze the importance of each component in the DAAL network in Sect. 4.6.

4.1 Datasets

CIFAR10 [21] **and CIFAR100** [21]**.** We evaluate the proposed DAAL on CIFAR10 and CIFAR100 datasets for image classification. Both datasets contain 60,000 images with a uniform size of 32 × 32 pixels. CIFAR10 and CIFAR100 have 10 and 100 categories, respectively. The training set and test set consist of 50,000 images and 10,000 images, respectively. The average classification accuracy is adopted as the metric for performance evaluation in this task.

Cityscapes [5]**.** Our DAAL is further evaluated on the Cityscapes dataset for semantic segmentation. Cityscapes is a large scale driving video dataset which contains 3,475 frames with instance segmentation annotations of 19 classes. The images have a uniform size of 2048 × 1024 pixels. The Cityscapes dataset is separated to the public training set and test set. For fair comparison, we adopt the public dataset partition in our experiments. The mean IoU is utilized to evaluate the performance of semantic segmentation.

4.2 Experimental Settings

Consistent to the existing approaches [36,41], we randomly select 10% samples from the training set for labeling and use them as the initial labeled pool at the beginning of the experiments. The rest of the training data set is treated as the unlabeled pool. In each iteration of active learning, we augment the labeled pool with 5% of the whole training set by selecting samples from the unlabeled pool for oracles to annotate using the acquisition function.

Baseline Methods. We involve various previous AL approaches as baselines for comparison, including common AL approaches for both image classification and semantic segmentation, e.g., Entropy [33], Learning Loss (LL) [41], Core-set [32], and VAAL [36], and task-specific ones for semantic segmentation only, e.g., MC Dropout [14] and Suggestive Annotation [32]. The results using *random sampling* is also reported for comparison. We notice that the idea of SA is closed to our DAAL, which both considers the uncertainty and representativeness for data point selection. However, SA is a two-step hybrid ensemble method, using the bootstrapping and Core-set for uncertainty and representativeness estimation, respectively. In contrast, our DAAL can select the samples in both terms of uncertainty and representativeness in one-step.

Implementation Details. For the image classification task, we use Wide-Resnet-Network [43] with depth = 28, width = 2 (WRN-28-2) as the backbone of the task model, while the dilated residual network [42] is used for the semantic segmentation task. The dual adversarial network is implemented using the PyTorch toolbox. For fair comparison, the baselines adopt the same training protocol. To alleviate the influence caused by the random nature of a neural network and the random initial labeled pool, all experimental results reported are the average of three repeated experiments. Note that we evaluate all methods without any data augmentation. One reason is that we expect to select the most representative subset of the raw dataset rather than the enlarged dataset by data augmentation. The other is that as reported in [27], the performance of AL methods with/without using data augmentation differs drastically, which is difficult for a fair comparison.

Image Classification. The task model and our DAAL are trained for 150 epochs. The stochastic gradient descent (SGD) optimizer is adopted to supervise the training of the task model. The initial learning rate is set to 0.1 and decreases to 0.01 after 80 epochs and 0.001 after 120 epochs, respectively. For the training of our dual adversarial network, the Adam optimizer is used with the learning rate of 5×10^{-4}. The batch size during adversarial learning is set to 128 and σ of Eq. 6 is set to 0.2.

Semantic Segmentation. The task model and our DAAL are trained for 50 epochs and 100 epochs, respectively, using the Adam optimizer. The learning rate is set to 5×10^{-4} and σ of Eq. 6 is set to 0.2.

4.3 Image Classification on CIFAR10/100

We compare the proposed DAAL with the baselines on CIFAR10/100. The evaluation results are presented in Fig. 3. On CIFAR10, the WRN-28-2 trained with the samples selected by our DAAL achieves the highest average classification accuracy, e.g., $84.17 \pm 0.24\%$ with 40% data from the training dataset, which is close to the accuracy yielded by training on the entire dataset (i.e.,

88.34 ± 0.43%). As shown in Fig. 3, the average classification accuracy between different AL approaches is similar with an extremely small labeled pool (i.e., 15%). The underlying reason is that the task model is not well trained and samples in the unlabeled pool may contain similar informativeness for the task model. As the size of labeled pool increases, the DAAL shows its advantage on data selection and surpasses the runner-up (i.e., Learning Loss [41]). The random sampling strategy yields the lowest classification accuracy under most settings of labeled data amount.

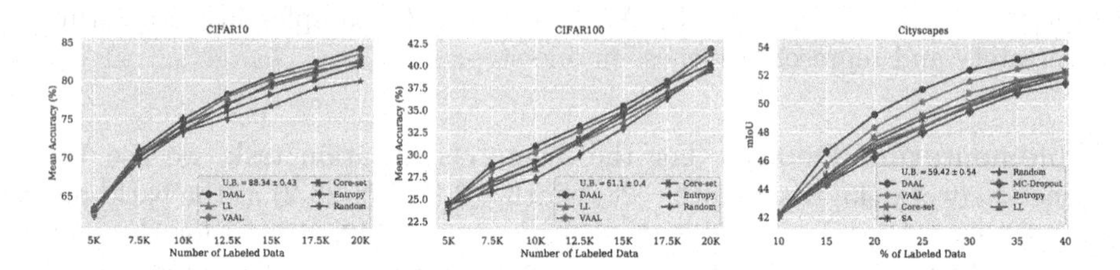

Fig. 3. Performance comparison of different AL approaches, including VAAL [36], Learning Loss (LL) [41], Core-set [32], Entropy [33], random sampling and our DAAL on CIFAR10/100. For semantic segmentation, the additional approaches, MC-Dropout [14] and Suggestive Annotation [32] are compared. The U.B. denotes the upper bound performance given by the task model trained on the entire data set.

Similar trends of improvement are observed on CIFAR100. Our DAAL achieves an average classification accuracy of 41.92 ± 0.31% with 40% training data, which is the highest record among the listed AL approaches. It is worthwhile to mention that only our DAAL provides consistent improvements on both CIFAR10 and CIFAR100. The runner-up approach (i.e., Learning Loss) on CIFAR10 achieves an average classification accuracy of 39.15 ± 0.41% on CIFAR100 with 40% training data, which ranks the fourth place—even lower than the random sampling strategy. The experimental results demonstrate that our DAAL can comprehensively estimate the informativeness of data points and select the samples with larger contributions to the optimization of model for the image classification task.

4.4 Semantic Segmentation on Cityscapes

We illustrate the performance of DAAL and the baseline methods on Cityscapes in Fig. 3. Our DAAL achieves an mIoU of 53.8 ± 0.24% by using only 40% labeled data, which is comparable to the performance of training on the entire dataset (i.e., 59.42 ± 0.29%). As the proposed DAAL takes both the uncertainty and representativeness into consideration, it outperforms the uncertainty-only AL approach (e.g., VAAL and MC Dropout). Although the SA selects samples based on a combination term of uncertainty and representativeness, its performance is even lower than the uncertainty-only VAAL, which demonstrates the benefit of the two terms (i.e., uncertainty and representativeness) is not fully exploited. The state-of-the-art AL approach, i.e., LL, which predicts the loss of neural network as its uncertainty estimation, yields a similar accuracy to random sampling on the Cityscapes test set. The reason for this phenomenon is that LL can provide excellent uncertainty estimation only for datasets containing fewer classes (e.g., CIFAR10). As the task difficulty increases, such as more classes (i.e., CIFAR100) or complicated targets (i.e., pixel-wise prediction on Cityscapes), the LL often fails to accurately predict the loss of neural network and consequently selects the less informative samples.

4.5 Performance Analysis

To further evaluate the effectiveness of our DAAL, we conduct experiments to compare the state-of-the-art AL approaches (Learning Loss [41] and VAAL [36]) with the proposed DAAL.

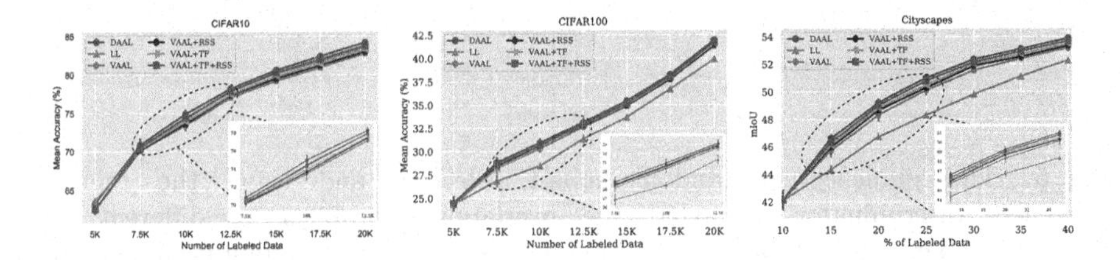

Fig. 4. Performance comparison of DAAL, original VAAL, VAAL using the deep features of the task model, and LL on CIFAR10/100. TF denotes the deep features of the task model and RSS represents the random subset selection.

Influence of Feature Representation in VAAL. We have analyzed the shortcomings of LL and VAAL on CIAFR10 and CIAFR100 and presented some explanations in previous section. As the VAAL does not involve the information of task model for data selection, the comparison between VAAL and our DAAL may be unfair. In this regard, we build a VAAL taking the features encoded by the task model as input (denoted as VAAL + TF), instead of the image data. The evaluation results on the three datasets are presented in Fig. 4. It can be observed that the accuracy of VAAL increases on CIFAR10 by using the features encoded by the task model, which achieves a comparable accuracy to LL. Furthermore, LL [41] constitutes a random subset with the size of 10,000 samples from unlabeled pool during each active learning iteration to ensure the diversity of selected samples, where the K-most uncertain samples are chosen. We evaluate this random subset strategy with VAAL. The variants are denoted as VAAL + RSS and VAAL + TF + RSS, respectively. As illustrated in Fig. 4, the performances of VAAL and VAAL + TF are both improved by using the random subset strategy. However, due to the lack of consideration of representativeness, the performances of VAAL variants are still inferior to our DAAL.

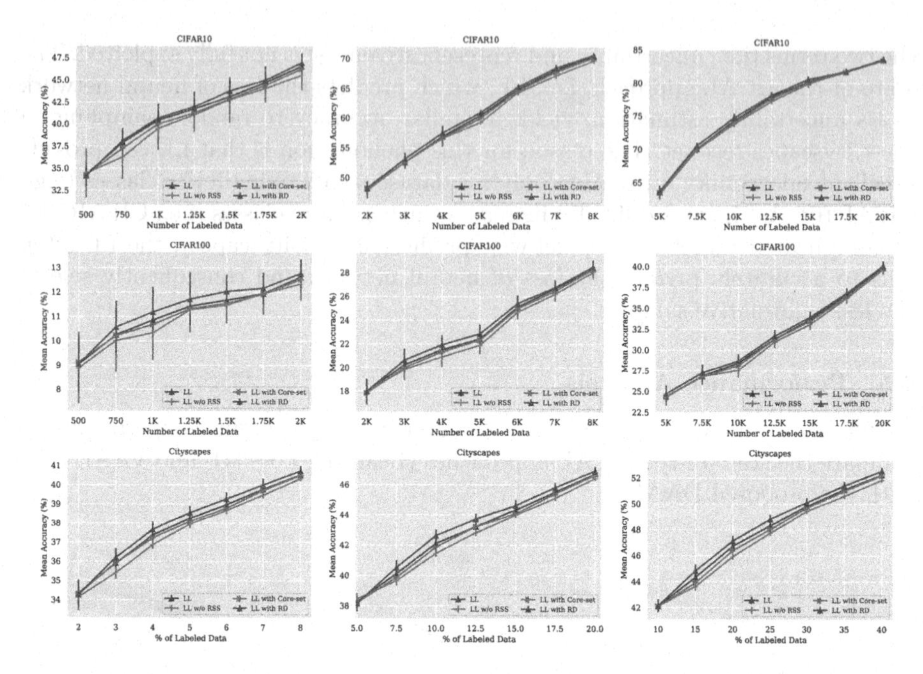

Fig. 5. The influence of random subset selection and using the representativeness discriminator to avoid the overlapping problem in different budget sizes (i.e., the number of samples selected for annotation in each iteration) on CIFAR10 (top), CIFAR100 (middle), and Cityscapes (bottom). The budget sizes are (250, 1000, 2500), (250, 1000, 2500) and (30, 75, 150) for the CIFAR10, CIFAR100, and Cityscapes dataset, respectively. RSS denotes random subset strategy and RD denotes the representativeness discriminator.

Overlapping Problem Occurring in Learning Loss. Several studies [32,41] stated that there was an overlapping problem occurring in the current uncertainty-only AL approaches. In particular, if the uncertainty-only AL approaches are asked to select K samples from the unlabeled pool, the information contained in the K selected samples may be similar, due to the single criterion (i.e., uncertainty to the classifier). When the budget for oracle annotation is small, this overlapping problem tends to be more severe. To verify this intuition, we further conduct experiments evaluating the performance of LL with different budget sizes, which are (250, 1000, 2500), (250, 1000, 2500), and (30, 75, 150) for CIFAR10, CIFAR100, and Cityscapes, respectively. The evaluation results are presented in Fig. 5. It can be observed that the influence caused by the overlapping problem decreases as the budget size increases—the LL and LL without random subset strategy (LL w/o RRS) achieve similar performance with the largest budget size.

To evaluate the benefit generated by integrating the additional criteria (e.g., representativeness) into the process of data selection, our representativeness discriminator is added to LL without RSS, which forms an one-stage framework, denoted as LL + RD. To further evaluate the drawback of multi-stage hybrid approaches, the representative-based Core-set is integrated to LL, denoted as LL + Core-set, which first selects a representative subset by Core-set and then chooses the K-most uncertain samples from them by LL. It can be observed from Fig. 5 that without using RSS, the performance of LL significantly drops due to the overlapping problem. The use of Core-set can alleviate the overlapping problem, which achieves similar accuracies to LL + RSS. Oppositely, the representativeness discriminator remarkably boosts the accuracy of LL w/o RSS, which consistently surpasses the multi-stage approaches (LL + RSS and LL + Core-set) under different settings (e.g., budget size). The experimental results demonstrate the superiority of representativeness tackling the overlapping problem and the one-stage framework, which fully exploits the benefit of representativeness.

4.6 Ablation Study

The contribution of each component of DAAL on CIFAR10/100 and Cityscapes is illustrated in Fig. 6. The performance of DAAL degrades by removing either of the discriminators. The uDis-only DAAL yields similar accuracies to LL, while the rDis-only DAAL can only surpass the random sampling. Hence, the estimation of data informativeness is not comprehensive in either term of uncertainty and representativeness. Our one-stage DAAL is a more proper solution for the data selection during active learning.

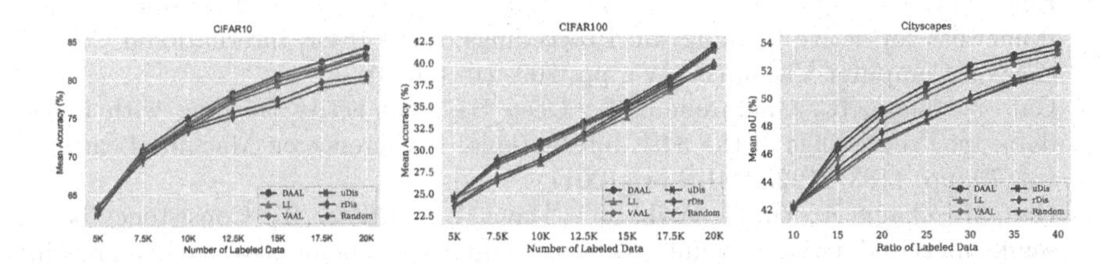

Fig. 6. Impact of the dual discriminators of the proposed DAAL. uDis and rDis represent using uncertainty discriminator or representativeness discriminator only to train the informativeness rater, respectively.

5 Conclusion

In this paper, we proposed a novel one-stage pool-based active learning approach, namely DAAL, which learns to select samples with high uncertainty to the classifier and good representativeness to the unlabeled samples. The proposed AL framework involves a informativeness rater and dual discriminators. Through the adversarial learning between the informativeness rater and discriminators, our

DAAL framework is able to comprehensively estimate the data informativeness to the optimization of the task model. Extensive experiments were conducted on three publicly available datasets (i.e., CIFAR10/100 and Cityscapes). The experimental results showed that our DAAL surpassed the state-of-the-art AL approaches.

References

1. Agrawal, P., Carreira, J., Malik, J.: Learning to see by moving. In: Proceedings of the IEEE International Conference on Computer Vision (ICCV), pp. 37–45 (2015)
2. Bengio, Y., Courville, A., Vincent, P.: Representation learning: a review and new perspectives. IEEE Trans. Pattern Anal. Mach. Intell. **35**(8), 1798–1828 (2013)
3. Berthelot, D., Carlini, N., Goodfellow, I., Papernot, N., Oliver, A., Raffel, C.A.: Mixmatch: a holistic approach to semi-supervised learning. In: Advances in Neural Information Processing Systems (NeurIPS), pp. 5050–5060 (2019)
4. Brinker, K.: Incorporating diversity in active learning with support vector machines. In: Proceedings of the 20th International Conference on Machine Learning (ICML 2003), pp. 59–66 (2003)
5. Cordts, M., et al.: The cityscapes dataset for semantic urban scene understanding. In: Proceedings of the IEEE Conference on Computer Vision and Pattern Recognition, pp. 3213–3223 (2016)
6. Doersch, C., Gupta, A., Efros, A.A.: Unsupervised visual representation learning by context prediction. In: Proceedings of the IEEE International Conference on Computer Vision (ICCV), pp. 1422–1430 (2015)
7. Ebrahimi, S., Rohrbach, A., Darrell, T.: Gradient-free policy architecture search and adaptation. arXiv preprint arXiv:1710.05958 (2017)
8. Elhamifar, E., Sapiro, G., Yang, A., Shankar Sasrty, S.: A convex optimization framework for active learning. In: Proceedings of the IEEE International Conference on Computer Vision (ICCV), pp. 209–216 (2013)
9. Gal, Y., Islam, R., Ghahramani, Z.: Deep Bayesian active learning with image data. In: Proceedings of the 34th International Conference on Machine Learning, vol. 70, pp. 1183–1192. JMLR.org (2017)
10. Gao, M., Zhang, Z., Yu, G., Arik, S.O., Davis, L.S., Pfister, T.: Consistency-based semi-supervised active learning: towards minimizing labeling cost. arXiv preprint arXiv:1910.07153 (2019)
11. Gissin, D., Shalev-Shwartz, S.: Discriminative active learning. arXiv preprint arXiv:1907.06347 (2019)
12. Glorot, X., Bengio, Y.: Understanding the difficulty of training deep feedforward neural networks. In: Proceedings of the Thirteenth International Conference on Artificial Intelligence and Statistics, pp. 249–256 (2010)
13. Goodfellow, I., et al.: Generative adversarial nets. In: Advances in Neural Information Processing Systems (NeurIPS), pp. 2672–2680 (2014)
14. Gorriz, M., Carlier, A., Faure, E., Giro-i Nieto, X.: Cost-effective active learning for melanoma segmentation. arXiv preprint arXiv:1711.09168 (2017)
15. He, X., et al.: Unsupervised video summarization with attentive conditional generative adversarial networks. In: Proceedings of the 27th ACM International Conference on Multimedia, pp. 2296–2304 (2019)
16. Houlsby, N., Huszár, F., Ghahramani, Z., Lengyel, M.: Bayesian active learning for classification and preference learning. arXiv preprint arXiv:1112.5745 (2011)

17. Joshi, A.J., Porikli, F., Papanikolopoulos, N.: Multi-class active learning for image classification. In: 2009 IEEE Conference on Computer Vision and Pattern Recognition (CVPR), pp. 2372–2379. IEEE (2009)
18. Joulin, A., van der Maaten, L., Jabri, A., Vasilache, N.: Learning visual features from large weakly supervised data. In: Leibe, B., Matas, J., Sebe, N., Welling, M. (eds.) ECCV 2016. LNCS, vol. 9911, pp. 67–84. Springer, Cham (2016). https://doi.org/10.1007/978-3-319-46478-7_5
19. Kapoor, A., Grauman, K., Urtasun, R., Darrell, T.: Active learning with gaussian processes for object categorization. In: 2007 IEEE 11th International Conference on Computer Vision (ICCV), pp. 1–8. IEEE (2007)
20. Kingma, D.P., Welling, M.: Auto-encoding variational bayes. arXiv preprint arXiv:1312.6114 (2013)
21. Krizhevsky, A., Hinton, G., et al.: Learning multiple layers of features from tiny images. Tech report (2009)
22. Kulesza, A., Taskar, B., et al.: Determinantal point processes for machine learning. Found. Trends® Mach. Learn. **5**(2–3), 123–286 (2012)
23. Larsen, A.B.L., Sønderby, S.K., Larochelle, H., Winther, O.: Autoencoding beyond pixels using a learned similarity metric. arXiv preprint arXiv:1512.09300 (2015)
24. Mackowiak, R., Lenz, P., Ghori, O., Diego, F., Lange, O., Rother, C.: Cereals-cost-effective region-based active learning for semantic segmentation. arXiv preprint arXiv:1810.09726 (2018)
25. Mahajan, D., et al.: Exploring the limits of weakly supervised pretraining. In: Proceedings of the European Conference on Computer Vision (ECCV), pp. 181–196 (2018)
26. Mahasseni, B., Lam, M., Todorovic, S.: Unsupervised video summarization with adversarial LSTM networks. In: Proceedings of the IEEE Conference on Computer Vision and Pattern Recognition (CVPR), pp. 202–211 (2017)
27. Mittal, S., Tatarchenko, M., Çiçek, Ö., Brox, T.: Parting with illusions about deep active learning. arXiv preprint arXiv:1912.05361 (2019)
28. Noroozi, M., Pirsiavash, H., Favaro, P.: Representation learning by learning to count. In: Proceedings of the IEEE International Conference on Computer Vision (ICCV), pp. 5898–5906 (2017)
29. Papandreou, G., Chen, L.C., Murphy, K.P., Yuille, A.L.: Weakly-and semi-supervised learning of a deep convolutional network for semantic image segmentation. In: Proceedings of the IEEE International Conference on Computer Vision (CVPR), pp. 1742–1750 (2015)
30. Rasmus, A., Berglund, M., Honkala, M., Valpola, H., Raiko, T.: Semi-supervised learning with ladder networks. In: Advances in Neural Information Processing Systems (NeurIPS), pp. 3546–3554 (2015)
31. Sener, O., Savarese, S.: Active learning for convolutional neural networks: a core-set approach. arXiv preprint arXiv:1708.00489 (2017)
32. Sener, O., Savarese, S.: Active learning for convolutional neural networks: a core-set approach. In: International Conference on Learning Representations (2018)
33. Shannon, C.E.: A mathematical theory of communication. Bell Syst. Tech. J. **27**(3), 379–423 (1948)
34. Siddiqui, Y., Valentin, J., Nießner, M.: Viewal: active learning with viewpoint entropy for semantic segmentation. arXiv preprint arXiv:1911.11789 (2019)
35. Singh, A., Virmani, L., Subramanyam, A.: Image corpus representative summarization. In: 2019 IEEE Fifth International Conference on Multimedia Big Data (BigMM), pp. 21–29. IEEE (2019)

36. Sinha, S., Ebrahimi, S., Darrell, T.: Variational adversarial active learning. In: Proceedings of the IEEE International Conference on Computer Vision (ICCV), pp. 5972–5981 (2019)
37. Srivastava, N., Hinton, G., Krizhevsky, A., Sutskever, I., Salakhutdinov, R.: Dropout: a simple way to prevent neural networks from overfitting. J. Mach. Learn. Res. **15**(1), 1929–1958 (2014)
38. Tong, S., Koller, D.: Support vector machine active learning with applications to text classification. J. Mach. Learn. Res. **2**, 45–66 (2001)
39. Yang, L., Zhang, Y., Chen, J., Zhang, S., Chen, D.Z.: Suggestive annotation: a deep active learning framework for biomedical image segmentation. In: Descoteaux, M., Maier-Hein, L., Franz, A., Jannin, P., Collins, D.L., Duchesne, S. (eds.) MICCAI 2017. LNCS, vol. 10435, pp. 399–407. Springer, Cham (2017). https://doi.org/10.1007/978-3-319-66179-7_46
40. Yang, Y., Ma, Z., Nie, F., Chang, X., Hauptmann, A.G.: Multi-class active learning by uncertainty sampling with diversity maximization. Int. J. Comput. Vision **113**(2), 113–127 (2015)
41. Yoo, D., Kweon, I.S.: Learning loss for active learning. In: Proceedings of the IEEE Conference on Computer Vision and Pattern Recognition (CVPR), pp. 93–102 (2019)
42. Yu, F., Koltun, V., Funkhouser, T.: Dilated residual networks. In: Proceedings of the IEEE Conference on Computer Vision and Pattern Recognition (CVPR), pp. 472–480 (2017)
43. Zagoruyko, S., Komodakis, N.: Wide residual networks. arXiv preprint arXiv:1605.07146 (2016)
44. Zhang, R., Isola, P., Efros, A.A.: Split-brain autoencoders: unsupervised learning by cross-channel prediction. In: Proceedings of the IEEE Conference on Computer Vision and Pattern Recognition (CVPR), pp. 1058–1067 (2017)
45. Zheng, H., et al.: Biomedical image segmentation via representative annotation. In: Proceedings of the AAAI Conference on Artificial Intelligence, vol. 33, pp. 5901–5908 (2019)

Fully Convolutional Networks for Continuous Sign Language Recognition

Ka Leong Cheng[1], Zhaoyang Yang[2], Qifeng Chen[1], and Yu-Wing Tai[1,3](✉)

[1] The Hong Kong University of Science and Technology, Clear Water Bay, Hong Kong
{klchengad,cqf}@ust.hk

[2] Tencent, Shenzhen, China
yangzhaoyang6@126.com

[3] Kwai Inc., Beijing, China
yuwing@gmail.com

Abstract. Continuous sign language recognition (SLR) is a challenging task that requires learning on both spatial and temporal dimensions of signing frame sequences. Most recent work accomplishes this by using CNN and RNN hybrid networks. However, training these networks is generally non-trivial, and most of them fail in learning unseen sequence patterns, causing an unsatisfactory performance for online recognition. In this paper, we propose a fully convolutional network (FCN) for online SLR to concurrently learn spatial and temporal features from weakly annotated video sequences with only sentence-level annotations given. A gloss feature enhancement (GFE) module is introduced in the proposed network to enforce better sequence alignment learning. The proposed network is end-to-end trainable without any pre-training. We conduct experiments on two large scale SLR datasets. Experiments show that our method for continuous SLR is effective and performs well in online recognition.

Keywords: Continuous sign language recognition · Fully convolutional network · Joint training · Online recognition

1 Introduction

Sign language is a common communication method for people with disabled hearing. It composes of a variety range of gestures, actions, and even facial emotions. In linguistic terms, a gloss is regarded as the unit of the sign language [27]. To sign a gloss, one may have to complete one or a series of gestures and actions. However, many glosses have very similar gestures and movements because of the richness of the vocabulary in a sign language. Also, because different people have

Electronic supplementary material The online version of this chapter (https://doi.org/10.1007/978-3-030-58586-0_41) contains supplementary material, which is available to authorized users.

A. Vedaldi et al. (Eds.): ECCV 2020, LNCS 12369, pp. 697–714, 2020.
https://doi.org/10.1007/978-3-030-58586-0_41

different action speeds, a same signing gloss may have different lengths. Not to mention that different from spoken languages, sign language like ASL [22] usually does not have a standard structured grammar. These facts place additional difficulties in solving continuous SLR because it requires the model to be highly capable of learning spatial and temporal information in the signing sequences.

Early work on continuous SLR [6,18,34] utilizes hand-crafted features followed by Hidden Markov Models (HMMs) [43,48] or Dynamic Time Warping (DTW) [47] as common practices. More recent approaches achieve state-of-the-art results using CNN and RNN hybrid models [4,14,44]. However, we observe that these hybrid models tend to focus on the sequential order of seen signing sequences in the training data but not the glosses, due to the existence of RNN. So, it is sometimes hard for these trained networks to recognize unseen signing sequences with different sequential patterns. Also, training of these models is generally non-trivial, as most of them require pre-training and incorporate iterative training strategy [4], which greatly lengthens the training process. Furthermore, the robustness of previous models is limited to sentence recognition only; most of the methods fail when the test cases are signing videos of a phrase (sentence fragment) or a paragraph (several sentences). Online recognition requires good recognition responses for partial sentences, but these models usually cannot give correction recognition until the signer finishes all the signing glosses in a sentence. Such limitation in robustness makes online recognition almost impossible for CNN and RNN hybrid models.

In this paper, we propose a fully convolutional network [24] for continuous SLR to address these challenges. The proposed network can be trained end-to-end without any pre-training. On top of this, we introduce a GFE module to enhance the representativeness of features. The FCN design enables the proposed network to recognize new unseen signing sentences, or even unseen phrases and paragraphs. We conduct different sets of experiments on two public continuous SLR datasets. The major contribution of this work can be summarized:

1. We are the first to propose a fully convolutional end-to-end trainable network for continuous SLR. The proposed FCN method models the semantic structure of sign language as glosses instead of sentences. Results show that the proposed network achieves state-of-the-art accuracy on both datasets, compared with other RGB-based methods.
2. The proposed GFE module enforces additional rectified supervision and is jointly trained along with the main stream network. Compared with iterative training, joint training with the GFE module fastens the training process because joint training does not require additional fine-tuning stages.
3. The FCN architecture achieves better adaptability in more complex real-world recognition scenarios, where previous LSTM based methods would almost fail. This attribute makes the proposed network able to do online recognition and is very suitable for real-world deployment applications.

2 Related Work

There are mainly two scenarios in SLR: isolated SLR and continuous SLR. Isolated SLR mainly focuses on the scenario where glosses have been well segmented temporally. Work in the field generally solves the task with methods such as Hidden Markov Models (HMMs) [10,12,13,29,35,42], Dynamic Time Warping (DTW) [36], and Conditional Random Field (CRF) [40,41]. As for continuous SLR, the task becomes more difficult as it aims to recognize glosses in the scenarios where no gloss segmentation is available but only sentence-level annotations as a whole. Learning separated individual glosses becomes more difficult in the weakly supervised setting. Many approaches propose to estimate the temporal boundary of different glosses first and then apply isolated SLR techniques and sequence to sequence methods [7,16] to construct the sentence.

Concerning temporal boundary estimation, Cooper and Bowden [3] develop a method to extract similar video regions for inferring alignments in videos by using data mining and head and hand tracking. Farhadi and Forsyth [8] also come up with a method that utilizes HMMs to build a discriminative model for estimating the start and end frames of the glosses in video streams with a voting method. Yin et al. [46] make further improvements by introducing a weakly supervised metric learning framework to address the inter-signer variation problem in real applications of SLR.

As for sequence to sequence methods, much work follows the framework used in the topic of speech recognition [25,33], handwriting recognition [23,32], and video captioning [39]. Specifically, an encoder module is responsible for extracting features in the input video frame sequences, and a CTC module acts as a cost function to learn the ground truth sequences. This framework also shows good performance on continuous SLR, and more recent work applies CNN and RNN hybrid models to infer gloss alignments implicitly [2,14,26,30]. However, RNNs are sometimes more sensitive to the sequential order than the spatial features. As a result, these models tend to learn much about the sequential signing patterns but little about the glosses (words), causing the failure of the recognition for unseen phrases and paragraphs.

3 Method

Formally, the proposed network aims to learn a mapping $H : \mathcal{X} \mapsto \mathcal{Y}$ that can transform an input video frame sequence $\mathcal{X}$ to a target sequence $\mathcal{Y}$. The feature extraction contains two main steps: a frame feature encoder and a two-level gloss feature encoder. On top of them, a gloss feature enhancement (GFE) module is introduced to enhance the feature learning. An overview of the proposed network is shown in Fig. 1.

3.1 Main Stream Design

Frame Feature Encoder. The proposed network first encodes spatial features of the input RGB frames. The frame feature encoder S composes of a convolutional backbone S_{cnn} to extract features in the frames and a global average

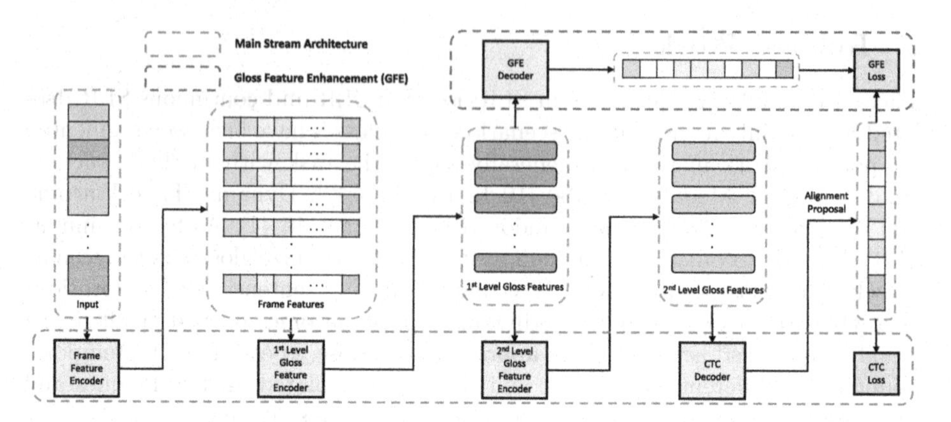

Fig. 1. Overview of the proposed network. The network is fully convolutional and divides the feature encoding process into two main steps. A GFE module is introduced to enhance the feature learning

pooling layer S_{gap} to compress the spatial features into feature vectors. Formally, each signing sequence is a tensor with shape (t, c, h, w), where t denotes the length of the sequence, c denotes the number of channels, and h, w denotes the height and width of the frames. The process of encoder S can be described as:

$$\{s\}^{t\times f_s} = S(\{x\}^{t\times c\times h\times w}) = S_{gap}(S_{cnn}(\{x\}^{t\times c\times h\times w})). \tag{1}$$

The output is of shape $\{s\}^{t\times f_s}$. Note that frame feature encoder treats each frame independently for the frame (spatial) feature learning.

Two-Level Gloss Feature Encoder. The two-level gloss feature encoder G follows S immediately and aims to encode gloss features. Instead of using LSTM layers, a common practice in temporal feature encoding, we achieve this by using 1D convolutional layers over time dimension. Precisely, the first level encoder G_1 consists of 1D-CNNs with a relatively larger filter size. Pooling layers can be used between convolutional layers to increase the window size when needed. Differently, the filter size is relatively smaller for the 1D-CNNs in the second level encoder G_2, with no pooling layers used in G_2. So, G_2 does not change the temporal dimension but only reconsider the contextual information between glosses by taking into account the neighboring glosses.

The overall convolutional process of G can be interpreted as a sliding window on the frame feature vector $\{s\}^{t\times f_s}$ along the time dimension. The sliding window size l and the stride δ are determined by the accumulated receptive field size and the accumulated stride of 1D-CNNs in G_1. Let $\{g\}^{k\times f_g}$ and $\{g'\}^{k\times f_{g'}}$ be the output tensor of gloss feature encoder G_1 and G_2, respectively. The operation of the encoder G can be formulated as:

$$\{g'\}^{k\times f_{g'}} = G(\{s\}^{t\times f_s}) = G_2(G_1(\{s\}^{t\times f_s})) = G_2(\{g\}^{k\times f_g}), \tag{2}$$

where k is the number of encoded gloss features and can be calculated with:

$$k = \lfloor \frac{t - l}{\delta} \rfloor + 1. \quad (3)$$

The two-level gloss feature encoder takes into account only multiple frames at a time. The window size l should be designed to be around the average length of the signing glosses to ensure good performance during the gloss feature extraction. With a proper window size design, G_1 can better model the semantic information of a "gloss" in sign language. G_2 further considers the gloss neighborhood information to achieve better prediction.

One benefit of our FCN design over previous LSTM design is that it greatly increases the adaptability of recognition, especially for online applications. Our proposed network can provide high-quality recognition on sequences with various length, which is essential in real-world recognition scenarios. We will further discuss the advantages of the FCN design in Sect. 4.4.

CTC Decoder. The Connectionist Temporal Classification (CTC) [9] is used as the network decoder. The CTC decoder D aims to decode the encoded gloss feature $\{g'\}^{k\times f_{g'}}$. CTC is an objective function that considers all possible underlying alignments between the input and target sequence. An extra "blank" label is added in the prediction space to match the output sequence with the target sequence in temporal dimension. Specifically, we employ a fully connected layer D_{fc} after G to cast the gloss feature dimension from $(k,\ f_{g'})$ to $(k,\ u)$ and a Softmax activation to finally transform the gloss feature to the prediction space $\{z\}^{k\times u}$:

$$\{z\}^{k\times u} = D(\{g'\}^{k\times f_{g'}}) = softmax(D_{fc}(\{g'\}^{k\times f_{g'}})), \quad (4)$$

where v is the vocabulary size and $u = v + 1$ is the size of each output with the extra "blank" added.

With normalized probabilities $\{z\}^{k\times u}$, the output alignment $\boldsymbol{\pi}$ can then be generated by taking the label with maximum likelihood at every decoding step. The final recognition result $\boldsymbol{y}$ is obtained by using the many-to-one function $\mathcal{B}$ introduced in CTC to remove repeated and blank predictions in $\boldsymbol{\pi}$. The CTC objective function is defined as the negative log-likelihood of all possible alignments matched to the ground truth:

$$\mathcal{L}_{ctc}(\boldsymbol{x}, \boldsymbol{y}) = -log\ p(\boldsymbol{y}|\boldsymbol{x}). \quad (5)$$

With the additional l_2 regularizer $\mathcal{L}_{reg}$ on the network parameters $\boldsymbol{W}$, the objective function of the main stream of the network $\mathcal{L}_{main}$ is defined as:

$$\begin{aligned}\mathcal{L}_{main} &= \mathcal{L}_{ctc} + \lambda_1 \mathcal{L}_{reg} \\ &= \frac{1}{|\mathcal{S}|} \sum_{(\boldsymbol{x},\boldsymbol{y})\in\mathcal{S}} \mathcal{L}_{ctc}(\boldsymbol{x}, \boldsymbol{y}) + \lambda_1 ||\boldsymbol{W}||^2,\end{aligned} \quad (6)$$

where $\mathcal{S}$ is the sample space, and λ_1 is the weight factor of the regularizer.

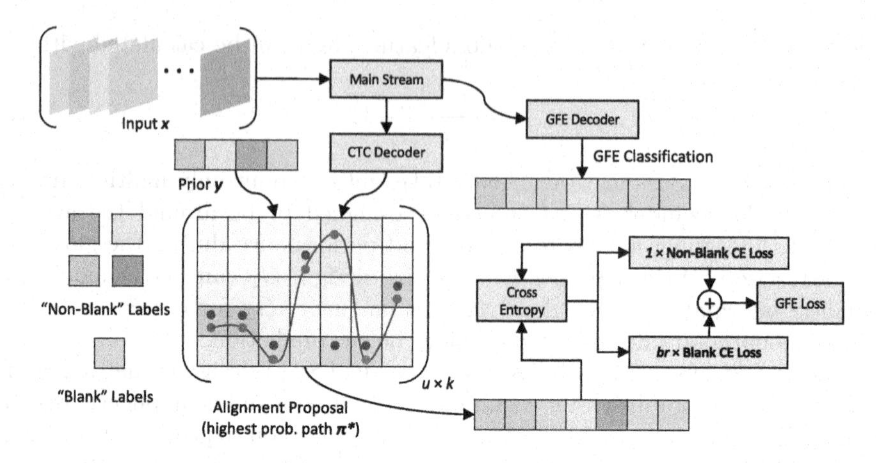

Fig. 2. The GFE module. The red dots in the prediction map are network outputs, while the blue ones are alignment proposals. The proposal rectifies the false predictions in the output to match the ground truth (Color figure online)

3.2 Gloss Feature Enhancement

The main stream of the network has mainly two tasks: (1) alignment inference and (2) gloss prediction. The performance highly depends on how well the network can generalize on glosses, as they are the unit of sign language. Therefore, it is essential to improve the quality of gloss features. Previous methods generally achieve this by incorporating iterative training strategies [4,5,20,31]. They first break training into several stages and then gradually refine feature extraction as each stage processed. However, with this strategy, whenever the training is switched to another stage, the network needs to first gradually adapt to a different objective, which greatly lengthens the number of training epochs and reduces the training efficiency. Moreover, the supervision used in some methods is generally the output of the network, which may contain some false predictions that can further reduce learning efficiency and limit the effectiveness of the refinement.

To remedy these problems, we propose the GFE module. The GFE module uses rectified supervision and is jointly trained with the main stream of the network, so it can improve the main stream network performance on the line. We illustrate the idea of the GFE module in Fig. 2.

Alignment Proposal. High-quality supervision can significantly improve the effectiveness of feature enhancement. Similar to [5], we make use of the network prediction map to find a better alignment proposal as the supervision. Specifically, given an input sequence, the CTC decoder D generates a prediction map $\{z\}^{k \times u}$, which is the probability of emissions in each decoding step. Let π^* denote the element in the alignment proposal. The alignment proposal $\boldsymbol{\pi}^* = \{\pi^*\}^k$ used in the GFE module can then be generated by searching the alignment proposal with the highest probability that can be matched to the ground truth sequence:

$$\boldsymbol{\pi}^* = \arg\max_{\boldsymbol{\pi} \in \mathcal{B}^{-1}(\boldsymbol{y})} p(\boldsymbol{\pi}|\boldsymbol{x}), \tag{7}$$

where $\mathcal{B}^{-1}$ is the inverse function of $\mathcal{B}$. Hence, the alignment proposal is guaranteed to be a matched alignment of the ground truth sequence. Each π^* in $\boldsymbol{\pi}^*$ can be paired with a first level gloss feature vector g at the corresponding time step in $\boldsymbol{g}$, which gives a pair of learning sample $(g, \pi^*) \in \mathcal{V}$.

Joint Training with Weighted Cross-entropy. To use the learning pairs in $\mathcal{V}$ as enhancement supervision, we add a fully connected layer F_{fc} followed by a Softmax activation after G_1. When joint training with the GFE module, gradients along this addition branch only propagate back to F and G_1 to enhance the frame and gloss feature learning. Formally, the GFE module contains a GFE decoder F, that takes $\boldsymbol{g} = \{g\}^{k \times f_g}$, the output vector of G_1, as inputs, and outputs the predicted gloss sequence in prediction space $\hat{\boldsymbol{\pi}}^* = \{\hat{\pi}^*\}^{k \times u}$:

$$\{\hat{\pi}^*\}^{k \times u} = F(\{g\}^{k \times f_g}) = softmax(F_{fc}(\{g\}^{k \times f_g})). \tag{8}$$

It is intuitive to train the gloss feature enhancement branch with cross-entropy loss. However, it is common that most of the label π^* in $\boldsymbol{\pi}^*$ is "blank" as k is generally much bigger than the number of glosses in the ground truth, causing the imbalance of samples in $\mathcal{V}$. The sample imbalance may limit the effectiveness of training for the GFE module. Therefore, we introduce a balance ratio to decrease the loss from "blank" labels. The balance ratio is defined as the proportion of "non-blank" labels in the given proposal $\boldsymbol{\pi}^*$:

$$br = \frac{\#non\text{-}blank}{\#total}. \tag{9}$$

For every $(g, \pi^*) \in \mathcal{V}$, we re-scale the cross-entropy loss to obtain the GFE loss, where the scaling factor w_i equals to *br* if it is blank label $(i = u)$, otherwise w_i equals to 1:

$$\mathcal{L}_{gfe}(g, \pi^*) = -\frac{1}{u}\sum_{i=1}^{u} w_i log\ p(\pi = \pi_i^*|g). \tag{10}$$

With the GFE module, the overall objective of the proposed network $\mathcal{L}$ becomes:

$$\begin{aligned} \mathcal{L} =& \mathcal{L}_{main} + \lambda_2 \mathcal{L}_{gfe} \\ =& \frac{1}{|\mathcal{S}|} \sum_{(\boldsymbol{x},\boldsymbol{y}) \in \mathcal{S}} \mathcal{L}_{ctc}(\boldsymbol{x}, \boldsymbol{y}) \\ &+ \frac{\lambda_2}{|\mathcal{V}|} \sum_{(g,\pi^*) \in \mathcal{V}} \mathcal{L}_{gfe}(g, \pi^*) + \lambda_1 ||\boldsymbol{W}||^2, \end{aligned} \tag{11}$$

where λ_2 is the weight factor for the GFE module. Note that the network objective is unified with joint training, so the training process is more efficient.

4 Experiments

We conduct experiments on the Chinese Sign Language (CSL) dataset [14] and the RWTH-PHOENIX-Weather-2014 (RWTH) dataset [18]. We detail the experimental setup, results, ablation studies, and online recognition in this section.

4.1 Experimental Setup

Dataset. The RWTH-PHOENIX-Weather-2014 (RWTH) dataset is recorded from a public weather broadcast television station in Germany. All signers wear dark clothes and perform sign languages in front of a clean background. There are 6,841 different sentences signed by 9 different signers (around 80,000 glosses with a vocabulary of size 1,232). All videos are pre-processed to a resolution of 210×260 and 25 frames per second (FPS). The dataset is officially split with 5,672 training samples, 540 validation samples, and 629 testing samples.

The Chinese Sign Language (CSL) dataset contains 100 sentences, each being signed for 5 times by 50 signers (in total 25,000 videos). Videos are shoot using a Microsoft Kinect camera with a resolution of 1280×720 and a frame rate of 30 FPS. The vocabulary size is relatively small (178); however, the dataset is richer in performance diversity, since signers wear different clothes and sign with different speeds and action ranges. With no official split given, we divide the dataset into a training set of 20,000 samples and a testing set of 5,000 samples and ensure that the sentences in the training and testing sets are the same, but the signers are different.

Evaluation Metric. We use word error rate (WER), which is the metric commonly used in continuous SLR, to evaluate the performance of recognition:

$$WER(H(\boldsymbol{x}), \boldsymbol{y})) = \frac{\#ins + \#del + \#sub}{\#labels\ in\ \boldsymbol{y}}. \tag{12}$$

We treat a Chinese character as a word during evaluation for the CSL dataset.

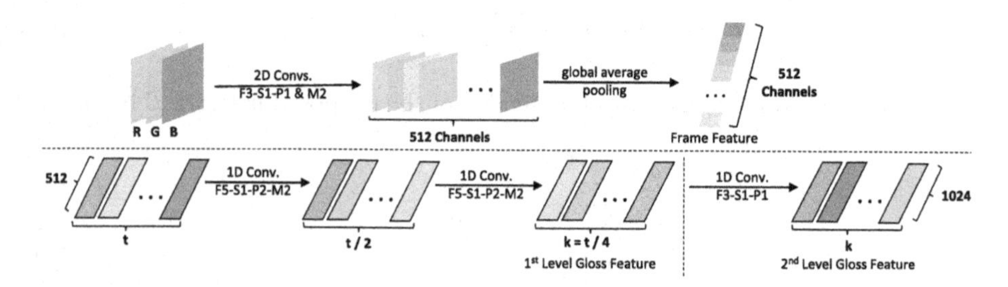

Fig. 3. The setting of the proposed main stream network. F, M refer to the filter size of convolution and max-pooling, respectively. S, P refer to the stride and padding size of convolution, respectively. Numbers aside are their actual size

Implementation Details. The main stream network setting used in our experiments is shown in Fig. 3. For S_{cnn}, the channel number gradually increases in this pattern: 3-32-64-64-128-128-256-256-512-512. F3-S1-P1 is used in each layer, and an additional M2 is added if the channel number increases. S_{gap} does global average pooling on each channel, so each frame is encoded as an array with a length of 512. For encoder G_1, two F5-S1-P2-M2 layers are used, and the channel number remains unchanged as 512. For encoder G_2, one F3-S1-P1 layer is used, and the channel number increases to 1024. Both fully connected layers D_{fc} and F_{fc} in the main stream and the GFE module cast the input channel number to u, the number of vocabulary size plus one blank label.

Batch normalization [15] is added after every convolutional layer to accelerate training. The input resolution of the network is 224×224. The window size in the first level gloss feature encoder is set to be 16 (about 0.5–0.6 s), which is the average time needed for completing a gloss, and the stride of the window is set to be 4. The second level gloss feature encoder further considers 3 adjacent gloss features for better prediction.

We use Adam [17] optimizer for training. We set the initial learning rate to be 10^{-4}. The weight factor λ_1 and λ_2 are empirically set to be 10^{-4} and 0.05, respectively. For the RWTH dataset, we train the proposed network for 80 epochs and halve learning rate at epoch 40 and 60. For the CSL dataset, the network is trained for 60 epochs, with the learning rate reduced by half at epoch 30 and 45. For data augmentation, all frames are first resized to 256×256 and then randomly cropped to fit the input shape. We also do temporal augmentation by first scaling up the sequence by +20% and then by −20%. Joint training with the GFE module is activated after epoch 15 for RWTH and epoch 10 for CSL, which are chosen through experiments to avoid unreliable alignment proposal at the initial optimization stage. The alignment proposals in the GFE module are updated every 10 epochs. When updating the proposal, temporal augmentation is disabled.

4.2 Results

We give a thorough comparison between the proposed network and previous RGB-based methods on both datasets. The results of previous methods are collected from their original papers. Please note that we mainly focus on online recognition in SLR, where the inputs are usually RGB video frames. Hence, we only compare our results with previous methods that use solely RGB modality.

Results on the CSL dataset and the RWTH dataset are shown in Table 1 and Table 2, respectively. We see that the proposed network achieves state-of-the-art performance on both datasets for RGB-based methods. The best result achieves 3.0% for the CSL dataset. For the RWTH dataset, our model reports 23.7% for the development set and 23.9% for the testing set. We also test the performance on the recognition partition (without translation) of the RWTH-PHOENIX Weather 2014 **T** dataset [1]. Our WER on the development set and testing set are 23.3% and 25.1%, respectively. These results illustrate the effectiveness of the proposed network.

Table 1. Result comparison on CSL

Methods	WER
DTW-HMM [47]	28.4
LSTM [38]	26.4
S2VT [37]	25.5
LSTM-A [45]	24.3
LSTM-E [28]	23.2
HAN [43]	20.7
LS-HAN [14]	17.3
SubUNet [2]	11.0
HLSTM [11]	7.6
HLSTM-attn [11]	7.1
Align-iOpt [31]	6.1
SF-Net [44]	3.8
Ours	**3.0**

Table 2. Result comparison on RWTH

Methods	WER	
	Dev	Test
Koller et al. [18]	57.3	55.6
Deep Hand [19]	47.1	45.1
Deep Sign [21]	38.3	38.8
SubUNet [2]	40.8	40.7
Cui et al. [4]	39.4	38.7
LS-HAN [14]	-	38.3
Align-iOpt [31]	37.1	36.7
SF-Net [44]	38.0	38.1
Re-Sign [20]	27.1	26.8
STMC (RGB) [49]	25.0	-
Cui et al. (RGB) [5]	23.8	24.4
Ours	**23.7**	**23.9**

To further demonstrate the effectiveness of our methods, we show some sample outputs in Fig. 4 to compare the recognition quality of different network settings. We observe that in the LSTM setting, models recognize the identical glosses (such as "OST" in sample 1 and "HIER" in sample 3) in a sentence as different words, and that errors usually occur adjacently in the sentences. In contrast, the proposed network produces consistent results for the identical glosses in a sentence, and errors are usually isolated. The observations infer that the LSTM based methods tend to learn robust sequential information, while the proposed network focuses on learning strong gloss features. So, we claim that the proposed network has a better generalization capability, because identical glosses are consistently classified, and errors do not have significant effects on neighboring glosses. On top of that, the GFE module further improves the performance by rectifying wrong recognition (such as "IN-KOMMEND" in sample 3), finding missing recognition (such as "AUCH" in sample 2), and adjusting alignments. More qualitative comparison can be found in the supplementary material.

4.3 Ablation Studies

In this section, we present further ablation studies to demonstrate the effectiveness of our method.

Temporal Feature Encoder. We first conduct a set of experiments to compare network performance with different temporal feature encoder designs. We test 6 different design combinations for the temporal feature encoder. For notation, **None** means no architecture; **LSTM** or **BiLSTM** means 1 LSTM or BiLSTM

Fig. 4. Sample outputs for different network settings. Wrong recognitions (except deletion errors) are highlighted in red. Ground truths are manually aligned (Color figure online)

Table 3. Network performance with different temporal feature encoder design

1st level	2nd level	WER	
		RWTH	CSL
None	1D-CNN	60.5	23.3
1D-CNN	None	42.1	10.4
LSTM	1D-CNN	32.1	10.8
LSTM	BiLSTM	30.8	3.6
1D-CNN	BiLSTM	26.5	**3.4**
1D-CNN	1D-CNN	**26.0**	8.2

Table 4. Network performance in different GFE module settings, br refers to the balance ratio

	GFE	br	WER
RWTH	✗	✗	26.0
	✓	✗	25.4
	✓	✓	**23.9**
CSL	✗	✗	8.2
	✓	✗	4.5
	✓	✓	**3.0**

layer with 512 hidden states, respectively; **1D-CNN** means two F5-S1-P2-M2 1DConvs for the 1st level or one F3-S1-P1 1DConv for the 2nd level. We show results on the testing set of the RWTH and CSL datasets in Table 3. Note that in this set of experiments, the GFE module is not activated.

We see that both levels of gloss feature encoder are essential for recognition as WER raises significantly when either of them is absent. CNN-based designs consistently outperform their LSTM counterparts for the RWTH dataset, but networks with BiLSTM give the best results for the CSL dataset. We should be reminded of the differences between the two datasets. The CSL dataset has richer diversity in the spatial dimension (such as different cloth colors) than in the temporal term. In other words, it is easier for the network to learn the temporal features than the spatial features in the CSL dataset. Also, different from that all the testing sentences are unseen in the RWTH training set, all the testing sentences in the CSL dataset are already seen in the training set, but just signed by different people. The working nature of 1D-CNN and LSTM is also different. The LSTM based method has direct access to sentence-level information, since LSTM layers have access to all the frame information. While the FCN method has less sentence-level supervision (only indirect access through the CTC decoding function), as the FCN model uses only a fixed number of frames to predict a gloss at each time step with 1D-CNNs.

Therefore, we claim that LSTM based methods tend to "remember" all the signing sequences in the training set instead of trying to learn the glosses independently. When testing the sentences in the CSL dataset, whose sequential sentences are the same as the training samples, the strong sequential information learned by LSTM based methods is significant and helpful, causing a relatively low reported WER. While for our FCN method, it is hard to fully extract the spatial and temporal features with weak supervision without substantial sentence-level information. Thus, it is essential to have a GFE module for feature enhancement, and the full version of our proposed network outperforms all the LSTM based methods for both datasets.

GFE Module. We conduct a set of experiments to investigate the effectiveness of the GFE module. We test different settings on both the RWTH and CSL datasets, including learning without the GFE module, with the GFE module but without balance ratio, and with the GFE module and balance ratio. Results on the testing sets are shown in Table 4.

We see that when the balance ratio is used, the GFE module can significantly fine-tune the features and improve the performances accordingly for both datasets. The GFE module with balance ratio improves the testing WER for the RWTH dataset by 2.1%; the improvement is more prominent for the CSL dataset by 5.2%. The difference of improvement is because the CSL dataset has much richer diversity in the spatial dimension than the RWTH dataset, making the spatial features in the CSL dataset more difficult to be learned without the GFE module.

4.4 Online Recognition

We mention in Sect. 4.3 that the proposed network has weaker supervision on sentence information. The FCN method focuses more on glosses rather than sentences, which directs us for some interesting experiments with different setups.

Simulating Experiments. For recognition in the real world, it is natural to consider the following three scenarios: (1) Signers sign several sentences at a time. (2) Signers sign only a phrase at a time. (3) Signers may pause for some while (stutters of actions) in the middle of signing. Accordingly, as shown in Fig. 5, we design three types of experiments which are conducted on the RWTH dataset to simulate online recognition: (1) concatenate multiple samples for a new sample (numbers after **Concat** indicate the number of samples being concatenated);

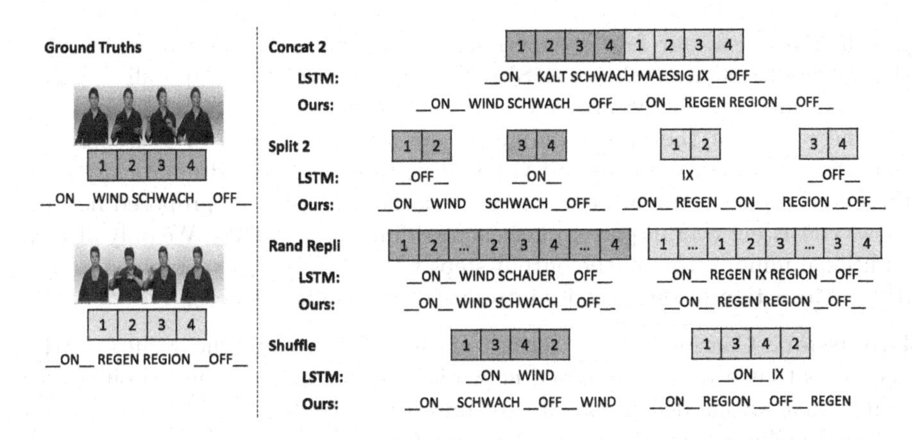

Fig. 5. Two samples are chosen for illustrating four different simulating scenarios for real-world recognition. Given that both the LSTM based method and the proposed network can recognize the original samples correctly, in all the simulating scenarios, the LSTM based method provides many false predictions while ours can preserve its accuracy. Errors are marked in red (Color figure online)

Table 5. Network performances in different real-world simulating scenarios on RWTH

Setups	Dev		Test	
	LSTM	Ours	LSTM	Ours
Original	31.7	23.7	30.8	23.9
Split 2	45.6	26.6	42.3	26.4
Split 3	50.2	28.0	45.3	27.6
Concat 2	40.5	24.3	41.1	24.9
Concat 3	46.0	24.8	45.7	25.0
Concat All	–	25.5	–	25.3
Rand Repli	39.1	24.6	39.5	24.6
Shuffle	58.9	27.3	55.2	28.1

(2) evenly split a sample into multiple samples (numbers after **Split** indicate the number of equal segments); (3) randomly select 5 frames in each sample and replace them with 12 replications in place (**Rand Repli** means random replication).

The LSTM based method may have too strong supervision of the sequential order, making it sensitive to gloss order. But one advantage of the proposed network is that the FCN model learns the glosses independently, so it is more robust to order-independent representations. To show this advantage, we may experimentally by shuffling the glosses in the testing samples. However, given no isolated annotations for individual glosses, we cannot manually construct "new" sentences with different signing orders by random shuffling. To mimic the shuffling idea, our fourth experiment is an imperfect but reasonable gloss shuffling experiment, as shown as **Shuffle** in Fig. 5. We first temporally segment the input into two equal parts and insert one into another.

The results of the four experiments are shown in Table 5. It is observed that the performance of the LSTM based network degrades dramatically in all these four types of simulating scenarios. On the contrary, the proposed network shows consistent overall performance across different scenarios. We only observe additional minor errors in the output steps where samples are combined or split due to the action inconsistencies introduced in boundary places.

Discussion. Considering the nature of the FCN design, the results further inspire us that even the proposed network is continuously being fed only a few frames that are needed (adequate) to infer the output, it can still combine all the intermediate outputs to give the same final recognition result. We use this technique to test for the **Concat All** scenario, where all testing samples are concatenated together as one large sample. Unfortunately, the LSTM based model fails to provide any result for the **Concat All** scenario, as the memory capacity limits the network to take such a large sample as an input.

All the results Table 5 indicate that the proposed network has more generalization capability and better flexibility for recognition in complex real-world recognition scenarios. The FCN design enables the proposed network to significantly reduce the memory usage for recognition. Meanwhile, indicated by the results in the **Split** and **Concat** scenarios, besides recognizing signing sentences, our method gives accurate recognition results for signing phrases and paragraphs. We can further conclude from the results in the **Split** scenarios that there is no need to wait for the arrival of all the signing glosses during the recognition process, as accurate intermediate (partial) recognition results can be given whenever adequate frames are available to the proposed network. With this great property, our method can provide intermediate results word by word along time, which is very friendly from a human-computer interaction perspective for SLR users. These properties make the proposed network have a promising application prospect for online recognition. A visual demonstration is shown in the supplementary demo video.

5 Conclusions

In this paper, we are the first to propose a fully convolutional network that can be trained end-to-end without pre-training for continuous SLR. A jointly trained GFE module is introduced to enhance the representativeness of features. Experimental results show that the proposed network achieves state-of-the-art performance on benchmark datasets with RGB-based methods. For recognition in real-world scenarios where the LSTM based network mostly fails, the proposed network reaches consistent performance and shows many great advantageous properties. These advantages make our proposed network robust enough to do online recognition.

One possible future research direction for continuous SLR is to strengthen the supervision by using the fact that some glosses are combinations of letter signs; however, this may require additional labeling pre-processing and professional knowledge in sign language. Also, the better gloss recognition accuracy obtained by the proposed network may have a good research prospect in sign language translation (SLT). Furthermore, we hope the proposed network can inspire future studies on sequence recognition tasks to investigate FCN architecture as an alternative to LSTM based models, especially for those tasks with limited data for training.

References

1. Camgoz, N., Hadfield, S., Koller, O., Ney, H., Bowden, R.: Neural sign language translation. In: Proceedings of IEEE Conference on Computer Vision and Pattern Recognition, pp. 7784–7793 (2018)
2. Camgoz, N.C., Hadfield, S., Koller, O., Bowden, R.: Subunets: end-to-end hand shape and continuous sign language recognition. In: Proceedings of IEEE International Conference on Computer Vision, pp. 3075–3084 (2017)
3. Cooper, H., Bowden, R.: Learning signs from subtitles: a weakly supervised approach to sign language recognition. In: Proceedings of IEEE Conference on Computer Vision and Pattern Recognition, pp. 2568–2574 (2009)
4. Cui, R., Liu, H., Zhang, C.: Recurrent convolutional neural networks for continuous sign language recognition by staged optimization. In: Proceedings of IEEE Conference on Computer Vision and Pattern Recognition, pp. 1610–1618 (2017)
5. Cui, R., Liu, H., Zhang, C.: A deep neural framework for continuous sign language recognition by iterative training. IEEE Trans. Multimedia **21**, 1880–1891 (2019)
6. Evangelidis, G.D., Singh, G., Horaud, R.: Continuous gesture recognition from articulated poses. In: Proceedings of European Conference on Computer Vision, pp. 595–607 (2015)
7. Fang, G., Gao, W.: A SRN/HMM system for signer-independent continuous sign language recognition. In: Proceedings of IEEE International Conference on Automatic Face Gesture Recognition, pp. 312–317 (2002)
8. Farhadi, A., Forsyth, D.: Aligning ASL for statistical translation using a discriminative word model. In: Proceedings of IEEE Conference on Computer Vision and Pattern Recognition, pp. 1471–1476 (2006)
9. Graves, A., Fernández, S., Gomez, F., Schmidhuber, J.: Connectionist temporal classification: Labelling unsegmented sequence data with recurrent neural networks. In: Proceedings of International Conference on Machine Learning, pp. 369–376 (2006)
10. Guo, D., Zhou, W., Li, H., Wang, M.: Online early-late fusion based on adaptive HMM for sign language recognition. ACM Trans. Multimedia Comput. Communi. Appl. **14**, 1–18 (2017)
11. Guo, D., Zhou, W., Li, H., Wang, M.: Hierarchical LSTM for sign language translation. In: Proceedings of AAAI Conference on Artificial Intelligence, pp. 6845–6852 (2018)
12. Guo, D., Zhou, W., Wang, M., Li, H.: Sign language recognition based on adaptive HMMs with data augmentation. In: Proceedings of IEEE International Conference on Image Processing, pp. 2876–2880 (2016)

13. Han, J., Awad, G., Sutherland, A.: Modelling and segmenting subunits for sign language recognition based on hand motion analysis. Pattern Recogn. Lett. **30**, 623–633 (2009)
14. Huang, J., Zhou, W., Zhang, Q., Li, H., Li, W.: Video-based sign language recognition without temporal segmentation. In: Proceedings of AAAI Conference on Artificial Intelligence, pp. 2257–2264 (2018)
15. Ioffe, S., Szegedy, C.: Batch normalization: accelerating deep network training by reducing internal covariate shift. In: Proceedings of International Conference on Machine Learning, pp. 448–456 (2015)
16. Kelly, D., McDonald, J., Markham, C.: Recognizing spatiotemporal gestures and movement epenthesis in sign language. In: Proceedings of IEEE International Conference on Image Processing and Machine Vision, pp. 145–150 (2009)
17. Kingma, D.P., Ba, J.: Adam: a method for stochastic optimization. CoRR preprint CoRR:1412.6980 (2014)
18. Koller, O., Forster, J., Ney, H.: Continuous sign language recognition: towards large vocabulary statistical recognition systems handling multiple signers. Comput. Vis. Image Underst. **141**, 108–125 (2015)
19. Koller, O., Ney, H., Bowden, R.: Deep hand: how to train a CNN on 1 million hand images when your data is continuous and weakly labelled. In: Proceedings of IEEE Conference on Computer Vision and Pattern Recognition, pp. 3793–3802 (2016)
20. Koller, O., Zargaran, S., Ney, H.: Re-sign: re-aligned end-to-end sequence modelling with deep recurrent CNN-HMMs. In: Proceedings of the IEEE Conference on Computer Vision and Pattern Recognition, pp. 3416–3424 (2017)
21. Koller, O., Zargaran, S., Ney, H., Bowden, R.: Deep sign: hybrid CNN-HMM for continuous sign language recognition. In: Proceedings of British Machine Vision Conference, pp. 136.1–136.12 (2016)
22. Liddell, S.K.: Grammar, Gestures, and Meaning in American Sign Language, pp. 52–53. Cambridge University Press, Cambridge (2003)
23. Liwicki, M., Graves, A., Bunke, H., Schmidhuber, J.: A novel approach to on-line handwriting recognition based on bidirectional long short-term memory networks. In: Proceedings of International Conference on Document Analysis and Recognition, pp. 367–371 (2007)
24. Long, J., Shelhamer, E., Darrell, T.: Fully convolutional networks for semantic segmentation. In: Proceedings of the IEEE Conference on Computer Vision and Pattern Recognition, pp. 3431–3440 (2015)
25. Miao, Y., Gowayyed, M., Metze, F.: Eesen: end-to-end speech recognition using deep RNN models and WFST-based decoding. In: IEEE Conference on Automatic Speech Recognition and Understanding Workshops, pp. 167–174 (2015)
26. Molchanov, P., Yang, X., Gupta, S., Kim, K., Tyree, S., Kautz, J.: Online detection and classification of dynamic hand gestures with recurrent 3D convolutional neural networks. In: Proceedings of IEEE Conference on Computer Vision and Pattern Recognition, pp. 4207–4215 (2016)
27. Ong, S., Ranganath, S.: Automatic sign language analysis: a survey and the future beyond lexical meaning. IEEE Trans. Pattern Anal. Mach. Intell. **27**, 873–91 (2005)
28. Pan, Y., Mei, T., Yao, T., Li, H., Rui, Y.: Jointly modeling embedding and translation to bridge video and language. In: Proceedings of IEEE Conference on Computer Vision and Pattern Recognition, pp. 4594–4602 (2015)
29. Pitsikalis, V., Theodorakis, S., Vogler, C., Maragos, P.: Advances in phonetics-based sub-unit modeling for transcription alignment and sign language recognition. In: IEEE Conference on Computer Vision and Pattern Recognition Workshops, pp. 1–6 (2011)

30. Pu, J., Zhou, W., Li, H.: Dilated convolutional network with iterative optimization for continuous sign language recognition. In: Proceedings of International Joint Conference on Artificial Intelligence, pp. 885–891 (2018)
31. Pu, J., Zhou, W., Li, H.: Iterative alignment network for continuous sign language recognition. In: Proceedings of IEEE Conference on Computer Vision and Pattern Recognition, pp. 4165–4174 (2019)
32. Puigcerver, J.: Are multidimensional recurrent layers really necessary for handwritten text recognition? In: Proceedings of International Conference on Document Analysis and Recognition, pp. 67–72 (2017)
33. Sak, H., Senior, A., Rao, K., İrsoy, O., Graves, A., Beaufays, F., Schalkwyk, J.: Learning acoustic frame labeling for speech recognition with recurrent neural networks. In: Proceedings of IEEE International Conference on Acoustics, Speech and Signal Processing, pp. 4280–4284 (2015)
34. Sun, C., Zhang, T., Bao, B.K., Xu, C., Mei, T.: Discriminative exemplar coding for sign language recognition with kinect. IEEE Trans. Cybern. **43**, 1418–1428 (2013)
35. Theodorakis, S., Katsamanis, A., Maragos, P.: Product-HMMs for automatic sign language recognition. In: Proceedings of IEEE International Conference on Acoustics, Speech and Signal Processing, pp. 1601–1604 (2009)
36. Vela, A.H., et al.: Probability-based dynamic time warping and bag-of-visual-and-depth-words for human gesture recognition in RGB-D. Pattern Recogn. Lett. **50**, 112–121 (2014)
37. Venugopalan, S., Rohrbach, M., Donahue, J., Mooney, R., Darrell, T., Saenko, K.: Sequence to sequence - video to text. In: Proceedings of IEEE International Conference on Computer Vision, pp. 4534–4542 (2015)
38. Venugopalan, S., Xu, H., Donahue, J., Rohrbach, M., Mooney, R., Saenko, K.: Translating videos to natural language using deep recurrent neural networks. In: Proceedings of the Conference of the North American Chapter of the Association for Computational Linguistics: Human Language Technologies, pp. 1494–1504 (2015)
39. Wang, B., Ma, L., Zhang, W., Liu, W.: Reconstruction network for video captioning. In: Proceedings of IEEE Conference on Computer Vision and Pattern Recognition, pp. 7622–7631 (2018)
40. Yang, H.D., Lee, S.W.: Robust sign language recognition with hierarchical conditional random fields. In: Proceedings of IEEE International Conference on Pattern Recognition, pp. 2202–2205 (2010)
41. Yang, R., Sarkar, S.: Detecting coarticulation in sign language using conditional random fields. In: Proceedings of IEEE International Conference on Pattern Recognition, pp. 108–112 (2006)
42. Yang, R., Sarkar, S.: Gesture recognition using hidden Markov models from fragmented observations. In: Proceedings of IEEE Computer Society Conference on Computer Vision and Pattern Recognition, pp. 766–773 (2006)
43. Yang, W., Tao, J., Ye, Z.: Continuous sign language recognition using level building based on fast hidden Markov model. Pattern Recogn. Lett. **78**, 28–35 (2016)
44. Yang, Z., Shi, Z., Shen, X., Tai, Y.W.: SF-net: structured feature network for continuous sign language recognition. arXiv preprint arXiv:1908.01341 (2019)
45. Yao, L., et al.: Describing videos by exploiting temporal structure. In: Proceedings of IEEE International Conference on Computer Vision, pp. 4507–4515 (2015)
46. Yin, F., Chai, X., Zhou, Y., Chen, X.: Weakly supervised metric learning towards signer adaptation for sign language recognition. In: Proceedings of British Machine Vision Conference, pp. 35.1–35.12 (2015)

47. Zhang, J., Zhou, W., Li, H.: A threshold-based HMM-DTW approach for continuous sign language recognition. In: Proceedings of International Conference on Internet Multimedia Computing and Service, pp. 237–240 (2014)
48. Zhang, J., Zhou, W., Xie, C., Pu, J., Li, H.: Chinese sign language recognition with adaptive HMM. In: Proceedings of IEEE International Conference on Multimedia and Expo, pp. 1–6 (2016)
49. Zhou, H., Zhou, W., Zhou, Y., Li, H.: Spatial-temporal multi-cue network for continuous sign language recognition. In: Proceedings of AAAI Conference on Artificial Intelligence, pp. 13009–13016 (2020)

Self-adapting Confidence Estimation for Stereo

Matteo Poggi(✉), Filippo Aleotti, Fabio Tosi, Giulio Zaccaroni, and Stefano Mattoccia

University of Bologna, Viale del Risorgimento 2, Bologna, Italy
m.poggi@unibo.it

Abstract. Estimating the confidence of disparity maps inferred by a stereo algorithm has become a very relevant task in the years, due to the increasing number of applications leveraging such cue. Although self-supervised learning has recently spread across many computer vision tasks, it has been barely considered in the field of confidence estimation. In this paper, we propose a flexible and lightweight solution enabling self-adapting confidence estimation agnostic to the stereo algorithm or network. Our approach relies on the minimum information available in any stereo setup (i.e., the input stereo pair and the output disparity map) to learn an effective confidence measure. This strategy allows us not only a seamless integration with any stereo system, including consumer and industrial devices equipped with undisclosed stereo perception methods, but also, due to its self-adapting capability, for its out-of-the-box deployment in the field. Exhaustive experimental results with different standard datasets support our claims, showing how our solution is the first-ever enabling online learning of accurate confidence estimation for any stereo system and without any requirement for the end-user.

Keywords: Stereo matching · Confidence · Online adaptation

1 Introduction

Stereo is one of the most popular strategies to accurately perceive the 3D structure of the scene through two synchronized cameras and several algorithms, either hand-designed or based on deep neural networks, exist. In many practical applications, alongside with disparity inference, confidence estimation is often performed as well. Purposely, a wide range of methods based either on hand-crafted measures [18] or *learning-based* strategies [42] have been proposed. Recent works [13,21,60] showed how state-of-the-art networks processing cues available from any stereo setup, i.e. the input stereo pair and the output disparity map, are substantially equivalent to those processing the entire cost volume [21], further supporting the evidence that the disparity map itself contains sufficient clues to identify outliers as initially proposed in [37,38]. Such a feature is

Electronic supplementary material The online version of this chapter (https://doi.org/10.1007/978-3-030-58586-0_42) contains supplementary material, which is available to authorized users.

A. Vedaldi et al. (Eds.): ECCV 2020, LNCS 12369, pp. 715–733, 2020.
https://doi.org/10.1007/978-3-030-58586-0_42

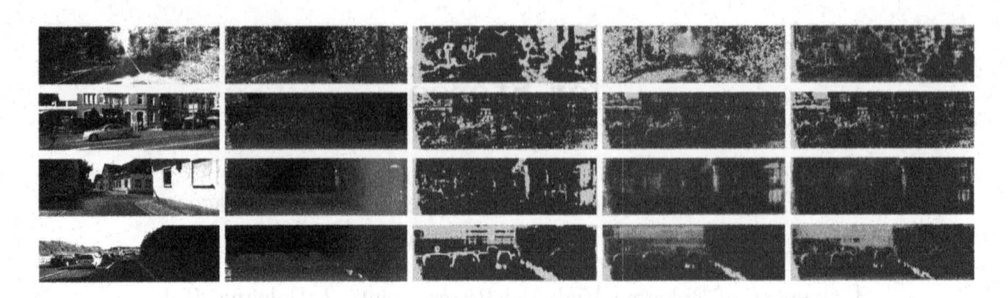

Fig. 1. Self-supervised confidence estimation. From left, reference image, disparity from various algorithms and confidence estimated by self-supervised frameworks [31,61] and ours. From top to bottom: Census-CBCA, MCCNN-fst-CBCA, Census-SGM and MCCNN-fst-SGM. Color encoding details in the supplementary material.

highly desirable since it potentially paves the way for learning confidence estimation for any stereo camera even without any knowledge about the stereo algorithm/network deployed. This fact is very appealing since it frequently occurs with most industrial/off-the-shelf (e.g. Stereolabs ZED 2) or consumer devices (e.g. smartphones). Nonetheless, this opportunity was investigated only partially in the literature. Moreover, all these methods are strongly constrained to the need for ground truth depth labels acquired in the target domain. However, since achieving such labels is cumbersome and time-consuming, two self-supervised paradigms have been proposed in the literature [31,61]. Although these methods proved that confidence estimation could be learned without needing active sensors, they have severe constraints. Individually, [31] requires static stereo sequences while [61] needs access to the *cost volume*, rarely exposed in the case of off-the-shelf stereo sensors or not defined at all in most modern neural networks [25,29,59]. As a consequence, both are not thought to handle *adaptation*, required to soften domain-shift issues. Thus, a solution for out-of-the-box deployment of *self-adapting* confidence estimation would be highly desirable for many practical applications. A notable example concerns smartphone (e.g. Apple iPhone) nowadays equipped with multiple cameras and undisclosed stereo algorithms/networks deployed for augmented reality or other applications in unpredictable environments.

Therefore in this paper, inspired by recent works performing continuous learning [3,59] for depth estimation, we propose the first-ever solution for self-adaptation of a confidence measure unconstrained to the target stereo system. For this purpose, we deploy a novel loss function built upon cues available from the input stereo pair and the output disparity only, needing no additional information to learn/adapt to the sensed environment. Our solution is comparable, and often better, w.r.t known strategies requiring full access to the cost volume [61] or static scenes for training [31], as shown in Fig. 1 on a variety of algorithms.

Extensive experimental results on KITTI, Middlebury 2014, ETH3D and DrivingStereo datasets support the following main claims of our novel confidence estimation paradigm: 1) competitive (often, better) with state-of-the-art when trained in a conventional, offline manner and tested on KITTI; 2) superior generalization capability on other datasets (e.g., Middlebury and ETH3D) compared to known self-supervised methods; 3) capable of online adaptation, outperforming competitors in unseen environments (e.g., DrivingStereo).

2 Related Work

In this section, we review the literature concerning confidence measures and recent trends in stereo matching.

Confidence Measures for Stereo. Confidence measures have been, at first, reviewed and evaluated in [18] and, more recently, in [42] highlighting that two broad categories exist: *hand-made* and *learned* measures. The former class consists of conventional method computed typically from cost volume analysis such as the ratio between two minima, as in PKR [16], or, as more recently proposed, determining local properties of the disparity map like the number of pixels with the same disparity hypothesis (DA [37]). Concerning learned measures [42], hand-made cues are usually combined and fed as input to a random forest classifier [14,22,32,33,37,43,53] or to a CNN [8,13,21,38,39,41,51,60] appropriately trained deploying depth labels. Learned methods may require 1) full access to the cost volume to extract hand-made features [14,22,32,33,39,41,53] or process the volume itself [13,21,23,24], 2) disparity maps for both left and right viewpoint [51] or 3) only the input image and its corresponding disparity map [8,37,38,43,60]. These three requirements translate into harder to softer constraints at deployment, most of them usually not met by off-the-shelf stereo cameras since exposing only the input stereo pair and the output disparity map to the user. Latest works [13,21] showed that, although a CNN with access to the full cost volume can perform better than networks processing disparity and reference image only, the margin between the two approaches is small and in most cases negligible, at the cost of a much minor versatility of the former.

Applications of Confidence Measures. In addition to the traditional outliers filtering task, many higher-level applications exploit such cue for different purposes. Again, two main categories exist, acting inside a stereo algorithm or outside it. Belonging to the former, Spyropoulos and Mordohai [53,55] estimate confidence and detect *ground control points* to improve global optimization. Park and Yoon [32,33] proposed a confidence-based modulation of the cost volume applied before SGM optimization, Poggi and Mattoccia [37,43] reduced the streaking effects of the SGM [15] stereo algorithm by using a weighted sum of the scanlines according to a confidence measure. Schonberger et al. [49] act similarly, fusing multiple scanlines of SGM using a random forest classifier. Seki and Pollefeys [51] changed P1 and P2 penalties of SGM dynamically according to the estimated confidence. Methods acting outside the stereo algorithms have been proposed for stereo algorithm fusion [36,54], sensor fusion [27,34], and unsupervised adaptation of deep models for stereo matching [56,57].

Self-supervised Confidence Estimation. Self-supervised learning has been barely investigated for confidence estimation. Mostegel et al. [31] leverage stereo videos looking at consistencies and contradictions between the different viewpoints of a static scene in order to obtain correct and wrong candidates from a given stereo algorithm. Tosi et al. [61] instead rely on traditional confidence measures to obtain these two sets according to a consensus among them.

Deep Stereo and Self-adaptation. At first, CNNs have replaced single steps in the stereo pipeline [47], such as cost computation [5,26,67], rapidly converging towards end-to-end solutions estimating dense disparity maps by means of 2D [19,25,29,52,65,66,66] or 3D networks [4,7,20,40,68]. The latest trend consists of training stereo networks from scratch using proxy labels [1,63] or casting disparity estimation as a continuous learning problem. A first work in this latter direction is [71], while more recent ones further moved in the direction of real-time continuous adaptation [59] to new environments or meta-learning [58].

3 Learning a Confidence Measure Out-of-the-Box

This work aims at proposing a self-supervised paradigm suited for learning a confidence measure, unconstrained from the specific stereo method deployed and capable of self-adaptation. We first classify stereo systems into different categories according to the data they make available, and then we introduce a novel strategy compatible with all of them.

3.1 Taxonomy of Stereo Matching Systems

In this section, we define three main broad categories of stereo matching solutions, each one characterized by different data made available during deployment. From now on, we will refer to a generic rectified stereo pair as ($\mathcal{I}_L$, $\mathcal{I}_R$), respectively made of left and right images, and to a generic stereo algorithm or deep network as $\mathcal{S}$. In the remainder, to simplify notation, we omit (x, y) coordinates if not strictly necessary.

Black-Box Models. Given any stereo algorithm processing a stereo pair ($\mathcal{I}_L$, $\mathcal{I}_R$), we define the output disparity map, computed assuming $\mathcal{I}_L$ as the **reference** image, as $\mathcal{D}_L = \mathcal{S}(\mathcal{I}_L, \mathcal{I}_R)$. This image triplet is the minimum amount of data available out of any stereo method, and we define as **black-box** all the systems making available only these cues. Such systems are highly representative of off-the-shelf stereo cameras (e.g., Stereolabs ZED 2) or stereo methods implemented in consumer devices (e.g., Apple iPhones). They neither allow end-users to access the implementation nor provide explicit ways (APIs) to call for it. For each ($\mathcal{I}_L$, $\mathcal{I}_R$) acquired in the field by the device, they provide the corresponding disparity map typically with undisclosed approaches based either on conventional stereo algorithms or deep networks. Hence, learning confidence measures for these systems is particularly challenging, yet appealing.

Gray-Box Models. Although black-box systems provide cues available in any stereo system, when explicit calls to the algorithm APIs are exposed, additional cues can be retrieved. Hence, we define a second family of systems for which, although it is given no access to the algorithm implementation or its intermediate data, explicit calls to the method itself are possible (e.g. stereo algorithms provided by pre-compiled libraries). Most deep stereo networks prevent the deployment of their internal representation since too abstract and substantially unintelligible, e.g. 2D architectures [19,25,29,52,65,66]. We define systems belonging to this class as **gray-box**, since multiple calls to $\mathcal{S}$ allow for retrieving additional cues. For instance, it is straightforward to compute the Left to Right Consistency (LRC) of the disparity maps, a popular strategy to obtain a confidence estimator, even if not explicitly provided by $\mathcal{S}$ itself in its original implementation. Given the possibility to call $\mathcal{S}$ two times, consistency checking can be performed analyzing $\mathcal{D}_L$ and a second disparity map, namely $\mathcal{D}_R$ obtained by assuming $\mathcal{I}_R$ as the reference images. Defining $\leftarrow$ the horizontal flipping operator, $\mathcal{D}_R$ is obtained as follows:

$$\mathcal{D}_R = \overleftarrow{\mathcal{S}(\overleftarrow{\mathcal{I}_R}, \overleftarrow{\mathcal{I}_L})} \tag{1}$$

Once obtained $\mathcal{D}_R$, the consistency between the two can be checked as

$$\text{LRC} = |\mathcal{D}_L - \pi(\mathcal{D}_L, \mathcal{D}_R)| < \delta \tag{2}$$

with $\pi(a, b)$ a sampling operator, collecting values at coordinate a from b, and δ a threshold value (usually 1) above which $\mathcal{D}_L$ and $\mathcal{D}_R$ are considered inconsistent. Although less effective than other measures [18], it comes at a lower price.

White-Box Models. Finally, if the implementation of $\mathcal{S}$ is accessible, additional cues can be sourced by processing intermediate data structures, if meaningful. The preferred one is the cost volume $\mathcal{V}$, containing matching costs $\mathcal{V}(x, y, d)$ for pixels at coordinates (x, y) and any disparity hypothesis $d \in [0, d_{max}]$. This class of systems, referred to as **white-boxes**, enables computation of any confidence measure, either conventional [18] or learning-based [13,21,42]. Popular traditional confidence measures obtained from $\mathcal{V}$ are the Peak-Ratio (PKR) and Left-Right Difference (LRD) defined, respectively, as

$$\text{PKR} = \frac{\mathcal{V}(d_{2m})}{\mathcal{V}(d_1)} \quad \text{and} \quad \text{LRD} = \frac{\mathcal{V}(d_2) - \mathcal{V}(d_1)}{\mathcal{V}(d_1) - \min_d \mathcal{V}_R(x - d_1, y, d)} \tag{3}$$

with d_1, d_2 and d_{2m}, respectively, the disparity hypotheses corresponding to the minimum cost, the second minimum and the second local minima [18]. Regarding LRD, given the cost volume $\mathcal{V}_R$ computed assuming $\mathcal{I}_R$ as the reference image, for any pixel (x, y) costs are sampled at $(x - d_1, y)$.

Motivations and Challenges. Indeed, for the reasons outlined so far, black-box models represent the most challenging, yet general and appealing target when dealing with confidence estimation since their constraints prevent the deployment of most state-of-the-art measures [13,21], as well as self-supervised strategy existing in the literature [31,61]. Hence, first and foremost, we aim at devising a general-purpose strategy enabling self-supervised confidence estimation in such constrained settings. As a notable consequence, this fact paves the way to tackle the same task even for state-of-the-art CNNs. Finally, having achieved this goal, out-of-the-box learning of confidence estimation with any stereo setup and self-adaptation in any environment is at hand.

3.2 Self-supervision Cues for Black-Box Models

In order to develop a self-supervised strategy suited for any stereo system, it is crucial to identify cues that are effective to source a robust supervision signal. According to the previous discussion, in the case of black-box models, we can rely on $(\mathcal{I}_L, \mathcal{I}_R)$ and $\mathcal{D}_L$ only. In this circumstance, although relevant information is not available compared to other models, we introduce three terms to obtain the desired self-supervised signal from the meagre cues available.

Image Reprojection Error. In recent literature, several works proved how the reprojection across the two viewpoints available in a rectified stereo pair could be a powerful source of supervision, either for monocular [11,12,35] or stereo [59,70] depth estimation. Specifically, we can reproject $\mathcal{I}_R$ on the reference image coordinates as $\tilde{\mathcal{I}_R} = \pi(\mathcal{D}_L, \mathcal{I}_R)$ Then, the difference between $\mathcal{I}_L$ and warped right view $\tilde{\mathcal{I}_R}$ appearance encodes how correct the reprojection is. To this aim, the most popular choice is a weighted sum between two terms, respectively SSIM [62] and absolute difference.

$$\Delta_{(\mathcal{I}_L, \tilde{\mathcal{I}_R})} = \alpha \cdot (1 - \text{SSIM}(\mathcal{I}_L, \tilde{\mathcal{I}_R})) + (1 - \alpha)|\mathcal{I}_L - \tilde{\mathcal{I}_R}| \quad (4)$$

with α usually tuned to 0.85 [11]. The higher it is, the more likely $\mathcal{D}_L$ is wrong. By definition, matching pixels is particularly challenging in ambiguous regions, such as textureless portions of the image. To this aim, we first aim at detecting regions with rich texture, being more likely to be correctly estimated by $\mathcal{S}$, by comparing Δ computed between $(\mathcal{I}_L, \mathcal{I}_R)$ with the one after reprojection as $\mathcal{T} = \Delta_{(\mathcal{I}_L, \mathcal{I}_R)} > \Delta_{(\mathcal{I}_L, \tilde{\mathcal{I}_R})}$. In large ambiguous regions, $\Delta_{(\mathcal{I}_L, \mathcal{I}_R)}$ will result equal (or even minor) than the reprojection error [12], thus identifying pixels on which stereo is prone to errors.

Agreement Among Neighboring Matches. Since most regions of a disparity map should be smooth, variations in nearby pixels should be small except at depth boundaries. As highlighted in [37,43], $\mathcal{D}_L$ itself allows for the extraction of meaningful cues to assess the quality of disparity assignments. Purposely, we rely on the **disparity agreement** between neighbouring pixels, defined as

$$\text{DA} = \frac{\mathcal{H}_{N \times N}(d_1)}{N \times N} \quad (5)$$

$\mathcal{H}_{N \times N}$ is a histogram encoding, for each pixel (x, y), the number of neighbours in a $N \times N$ window having the same disparity d (in case of subpixel precision, within 1 pixel). In the absence of depth discontinuities, the majority of pixels in the neighbourhood should share the same, or very similar, disparity hypothesis. Hence, we define a second criterion to identify reliable stereo correspondences as $\mathcal{A} = \text{DA} > 0.5$, assuming that more than half of the pixels in the neighbourhood share the same disparity. It is worth noting that this criterion is often not met in the presence of depth boundaries, even in case of correct disparities.

Uniqueness Constraint. In an ideal frontal-parallel scene observed by a stereo camera in standard form, for each pixel in $\mathcal{I}_L$ exists at most one match in $\mathcal{I}_R$ and

Fig. 2. Effects of different criteria. Given the highlighted region, we show inliers (green) and outliers (red) guesses by using the following cues in multi-modal binary cross-entropy: a) $\mathcal{T}^p,\mathcal{T}^q$ b) $\mathcal{A}^p,\mathcal{A}^q$ c) $\mathcal{U}^p,\mathcal{U}^q$ d) $\mathcal{T}^p,\mathcal{A}^p,\mathcal{U}^p,\mathcal{T}^q$ e) $\mathcal{T}^p,\mathcal{A}^p,\mathcal{U}^p,\mathcal{T}^q,\mathcal{A}^q,\mathcal{U}^q$. For black pixels, the considered configuration gives no guesses. (Color figure online)

vice-versa. Leveraging this property, known as uniqueness, is particularly useful [6] to detect outliers in occluded regions and represents a reliable alternative to LRC and LRD measures, not usable when dealing with black-box models. Uniqueness Constraint (UC) is encoded as

$$\text{UC} = [x - \mathcal{D}_L(x,y)] \notin \bigcup_k [(x+k) - \mathcal{D}_L(x+k,y)] \tag{6}$$

with $k \in [-d^*_{max}, -1] \cup [1, d^*_{max}]$ and $d^*_{max} = d_{max} - \mathcal{D}_L(x,y)$. In other words, the uniqueness for any pixel in $\mathcal{I}_L$ holds if it does not collide in the target image with any other pixel, i.e., not matching the same pixel in $\mathcal{I}_R$ matched by any other. We exploit this property to define our third criterion as $\mathcal{U} = \text{UC}$. We conclude observing that, although effective at detecting mostly occlusions, the uniqueness constraint is often violated in the presence of slanted surfaces.

3.3 Multi-modal Binary Cross Entropy

Given the three criteria outlined above, we revise the traditional binary cross entropy loss to take into account multiple label hypotheses. We refer to this variant as **Multi-modal Binary Cross Entropy** (MBCE), defined as

$$\mathcal{L}_{\text{MBCE}} = -\left[\left(\prod_{p \in \mathcal{P}} p\right) \cdot log(o) + \left(\prod_{q \in \mathcal{Q}} q\right) \cdot log(1-o)\right] \tag{7}$$

with o the output of the neural network $\in [0,1]$, i.e. passed through a sigmoid activation, $\mathcal{P}$ and $\mathcal{Q}$ two sets of **proxy labels** derived respectively by a criterion being met or not. For instance, pixels satisfying the first criterion on image reprojection will have labels $\mathcal{T}^p = 1$, $\mathcal{T}^q = 0$ and vice versa when they do not. Unlike traditional binary cross entropy, where a single label y and its counterpart $(1-y)$ are used, we define disjoint sets of proxies allowing for a flexible configuration of the loss function according to the three criteria described so far. For instance, by setting $\mathcal{P} = [\mathcal{T}^p, \mathcal{A}^p]$ and $\mathcal{Q} = [\mathcal{T}^q]$ we will train the network to detect good matches using image reprojection plus agreement and outliers using the former only. Adding elements to the sets $\mathcal{P}$ and $\mathcal{Q}$ reduces progressively the number of pixels considered correct or wrong, respectively. Figure 2 shows this, highlighting how combining multiple guesses as in d) and e) for some pixels no supervision is given when criteria do not match. We will report the impact of this and the different configurations in a thorough ablation study.

4 Experimental Results

In this section, we report an exhaustive evaluation to assess the effectiveness of our strategy, referred to as *Out-of-The-Box* (OTB), by conducting three main experiments, respectively: 1) ablation study on the MBCE loss, 2) comparison with self-supervised approaches [31,61] in a conventional offline training and 3) an evaluation concerning online adaptation of OTB. Code for this latter experiment is available at http://github.com/mattpoggi/self-adapting-confidence.

4.1 Implementation Details

We now report all the details to understand and reproduce our experiments fully.

Evaluation Protocol. To measure the effectiveness of the learned confidence measures, we compute the Area Under Curve (AUC) of the sparsification plots [18,21,42,60]. Given a disparity map, pixels are sorted in increasing order of confidence and gradually removed (e.g., 5% each time) from the disparity map. At each iteration, the error rate is computed over the sparse disparity map as the percentage of pixels having absolute error larger than τ. Plotting the error rate results in a sparsification curve, whose AUC quantitatively assesses the confidence effectiveness (the lower, the better). Optimal AUC is obtained by sampling the pixels in decreasing order of absolute error or as in [18].

Confidence Networks. Since the goal of this work is to define an effective self-supervised strategy suited for online learning rather than proposing a novel architecture, in our experiments, we test our proposal to train existing networks. Purposely, we consider three architectures: CCNN [38], ConfNet and LGC [60] to carry out our experiments because 1) they process only disparity map and reference image, thus are suited to all methods from white-box to black-box, 2) according to recent works [13,21], the most accurate one (LGC) is on par with state-of-the-art networks processing the cost volume and 3) the source code is fully available, conversely to [13,21]. Moreover, in ConfNet we replaced deconvolutions with bilinear upsampling followed by 3×3 convolutions and process $\mathcal{D}_L$ only, significantly improving its performance and thus filling most of the gap with CCNN and LGC. We defined a training schedule for each network, kept constant in all experiments. For CCNN, we use batches of 128 patches for 1M iterations, for ConfNet batches of single, 320×1216 crops 25K iterations, finally for LGC batches of 128 patches 300K iterations, starting from pre-trained CCNN and ConfNet models. We trained all networks with SGD optimizer and a constant learning rate of 0.001. For patch-based methods, proxy signals are computed offline on the full resolution image.

Datasets. We consider five standard datasets: KITTI 2012 [10], KITTI 2015 [30], Middlebury 2014[1] [46], ETH3D [50] and DrivingStereo [64], setting τ respectively to 3, 3, 1, 1 and 3. Being ground truth required to assess performance, we refer to the training set of such datasets. To train confidence estimation networks, we select the first 20 images from KITTI 2012 as in [42,60] for supervised training and the 400 images from the first 20 sequences of the KITTI 2012 multiview extension used in [31,61] for self-supervised ones. To evaluate the trained confidence networks, we use the remaining 174 images from KITTI 2012 as the validation set and the totality of images available from KITTI 2015 for experiments on environments similar to the training set. Moreover, we also assess their generalization performance on the whole Middlebury 2014 and ETH3D datasets. In these experiments, only the KITTI 2012 images listed above are used for training, thus the networks are transferred without any fine-tuning. Finally, to test self-adaptation peculiar of OTB we use a sequence from the DrivingStereo dataset, namely 2018-10-25-07-37, made of 7K frames.

Stereo Algorithms. Following the recent literature [21,42,60], we evaluate the effectiveness of our strategy on a variety of stereo algorithms with different degrees of accuracy, in order to highlight how strong is our self-supervised paradigm in the presence of heterogenous disparity maps. We consider four main stereo algorithms deploying the code provided Zbontar and LeCun [67] under different settings. Specifically: Census-CBCA, Census-SGM, MCCNN-fst-CBCA and MCCNN-fst-SGM. The first two rely on a census-based matching cost computation, respectively, optimized by a Cross Based Cost Aggregation (CBCA) strategy [69] and SGM [15]. The latter two replace the census-based matching costs with predictions obtained by MCCNN-fst, for which we use pre-trained weights on KITTI 2012, 2015 and Middlebury provided by the authors and tested on the same datasets. For ETH3D, Middlebury weights have been used. No post-processing is applied to any output. Furthermore, to evaluate the impact of self-adaptation made possible by OTB with a real black-box method, we also consider two recent deep stereo network. We choose MADNet [59] and GANet [68], both trained on synthetic images [29] and then fine-tuned with ground truth on KITTI 2015, because of the availability of trained model and its accuracy-speed trade-off. Since fine-tuned on KITTI, we conduct experiments with MADNet and GANet on DrivingStereo only.

Competitors. We compare the proposed OTB strategy with existing methods proposed by Mostegel et al. [31] (named SELF) and by Tosi et al. [61] (named WILD). The former reasons about contradictions on observations from multiple viewpoints: given a stereo sequence framing a static scene with a moving camera, $\mathcal{D}_L$ and $\mathcal{D}_R$ are computed for each pair, registered and checked for inconsistencies. Since it requires both $\mathcal{D}_L$ and $\mathcal{D}_R$ disparity maps, SELF is suited only for systems belonging to gray-box and white-box categories. Concerning WILD, it requires a pool of six confidence measures extracted from the cost volume to identify inliers and outliers according to heuristic thresholding on the mea-

[1] We use the quarter resolution split as in previous works [21,42,60].

Table 1. Ablation study on the proposed multi-modal binary cross entropy. We report AUC scores for networks trained on KITTI 2012 (20 or 400 images) and tested on KITTI 2012 (174 images, top) and Middlebury (15 images, bottom).

KITTI 2012

Match cost						Census		MCCNN-fst		Census		MCCNN-fst		Census		MCCNN-fst	
Aggregation						CBCA	SGM	CBCA	SGM	CBCA	SGM	CBCA	SGM	CBCA	SGM	CBCA	SGM
$\Delta_{\mathcal{I}}(\mathcal{I}_L, \tilde{\mathcal{I}}_R)$						0.210	0.086	0.165	0.044	0.210	0.086	0.165	0.044	0.210	0.086	0.165	0.044
DA						0.112	0.047	0.063	0.023	0.112	0.047	0.063	0.023	0.112	0.047	0.063	0.023
UC						0.165	0.063	0.123	0.034	0.165	0.063	0.123	0.034	0.165	0.063	0.123	0.034
$\mathcal{T}^p$	$\mathcal{A}^p$	$\mathcal{U}^p$	$\mathcal{T}^q$	$\mathcal{A}^q$	$\mathcal{U}^q$	CCNN				ConfNet				LGC			
✓			✓			0.080	0.045	0.047	0.018	0.077	0.033	0.045	0.014	0.082	0.058	0.046	0.026
	✓			✓		0.105	0.045	0.073	0.023	0.087	0.035	0.049	0.017	0.110	0.040	0.074	0.022
		✓			✓	0.111	0.035	0.087	0.022	0.101	0.038	0.065	0.020	0.114	0.035	0.077	0.020
✓	✓		✓			0.078	0.033	0.050	0.019	0.072	0.030	0.038	0.014	0.075	0.034	0.049	0.023
✓	✓		✓	✓		0.089	0.035	0.059	0.023	0.071	0.029	0.038	0.014	0.082	0.031	0.066	0.020
✓		✓	✓			0.072	0.038	0.053	0.019	0.074	0.029	0.040	0.013	0.070	0.036	0.042	0.016
✓		✓	✓		✓	0.088	0.032	0.075	0.020	0.076	0.030	0.041	0.013	0.084	0.031	0.071	0.017
✓	✓	✓	✓			0.068	0.034	0.046	0.018	0.070	0.029	0.037	0.013	0.068	0.032	0.041	0.016
✓	✓	✓	✓	✓	✓	0.085	0.029	0.057	0.017	0.071	0.028	0.038	0.012	0.081	0.028	0.050	0.015

Middlebury

Match cost						Census		MCCNN-fst		Census		MCCNN-fst		Census		MCCNN-fst	
Aggregation						CBCA	SGM	CBCA	SGM	CBCA	SGM	CBCA	SGM	CBCA	SGM	CBCA	SGM
$\Delta_I(\mathcal{I}_L, \mathcal{I}_R)$						0.190	0.180	0.179	0.134	0.190	0.180	0.179	0.134	0.190	0.180	0.179	0.134
DA						0.161	0.168	0.099	0.087	0.161	0.168	0.099	0.087	0.161	0.168	0.099	0.087
$\mathcal{U}$						0.193	0.188	0.192	0.145	0.193	0.188	0.192	0.145	0.193	0.188	0.192	0.145
$\mathcal{T}^p$	$\mathcal{A}^p$	$\mathcal{U}^p$	$\mathcal{T}^q$	$\mathcal{A}^q$	$\mathcal{U}^q$	CCNN				ConfNet				LGC			
✓	✓	✓	✓			0.116	0.123	0.087	0.077	0.133	0.112	0.087	0.067	0.127	0.111	0.090	0.064
✓	✓	✓	✓	✓	✓	0.153	0.146	0.095	0.081	0.134	0.122	0.095	0.069	0.138	0.142	0.099	0.080

sures. Since it requires access to the cost volume, WILD is suited for white-box algorithms only. In contrast, among other advantages discussed next, it worth stressing that our OTB approach is suited for black-box systems and agnostic to the scene content, in contrast to SELF that requires static scenes.

4.2 Ablation Study

At first, we study the impact of the different terms in the proposed self-supervised loss function. To this aim, on KITTI 2012 and as for other experiments, we train 9 variants of each network for each of the four stereo algorithms. Then, we evaluate confidences on the KITTI 2012 dataset and, without retraining, on Middlebury 2014. Table 1 collects the outcome of this evaluation, reporting on top results on KITTI 2012 and, at the bottom, on Middlebury. We report as baselines the performance of $\Delta_{(\mathcal{I}_L, \tilde{\mathcal{I}}_R)}$, DA and UC. DA is computed on 5×5 windows.

On KITTI (top of the table), we first report the results achieved by training the three networks selecting only one of the three cues used to distinguish between correct and wrong matches, i.e. $[\mathcal{T}^p, \mathcal{T}^q]$, $[\mathcal{A}^p, \mathcal{A}^q]$ and $[\mathcal{U}^p, \mathcal{U}^q]$ configurations. We can notice that each of them outperforms the performance of the corresponding baseline used for supervision. This trend occurs on all the algorithms and for each network, showing the surprisingly robust capacity of the

Table 2. Evaluation on KITTI. We report AUC scores for networks trained on KITTI 2012 (20 or 400 images) and tested on 2012 (174 images) and 2015 (200 images).

	Match cost	Census				MCCNN-fst			
	Aggregation	CBCA		SGM		CBCA		SGM	
	KITTI split	2012	2015	2012	2015	2012	2015	2012	2015
	Bad$\mathcal{T}$ %	27.193	22.281	10.330	8.998	18.875	16.926	6.084	6.028
Traditional	LRD	0.096	0.080	0.033	0.032	0.080	0.077	0.017	0.023
	PKR	0.106	0.089	0.028	0.029	0.065	0.062	0.010	0.017
	LRC	0.142	0.113	0.062	0.056	0.103	0.092	0.036	0.041
	$\Delta_{(\mathcal{I}_L,\mathcal{I}_R)}$	0.210	0.175	0.086	0.079	0.165	0.150	0.044	0.041
	DA	0.112	0.090	0.047	0.046	0.063	0.059	0.023	0.028
	UC	0.165	0.131	0.063	0.058	0.123	0.111	0.034	0.037
CCNN	Supervised	0.059	0.046	0.018	0.017	0.031	0.032	0.009	0.012
	WILD [61]	0.076	0.065	**0.026**	**0.026**	0.052	0.047	**0.012**	**0.017**
	SELF [31]	0.076	0.065	0.047	0.046	**0.038**	**0.041**	**0.012**	0.018
	OTB (Ours)	0.068	0.055	0.029	0.031	0.046	0.048	0.017	0.022
ConfNet	Supervised	0.061	0.049	0.017	0.016	0.033	0.034	0.006	0.010
	WILD [61]	0.089	0.067	0.024	0.020	0.054	0.050	0.010	0.016
	SELF [31]	0.075	0.066	0.024	0.024	0.041	0.044	0.014	0.016
	OTB (Ours)	**0.070**	**0.058**	0.028	0.032	0.037	0.040	0.012	0.017
LGC	Supervised	0.056	0.044	0.016	0.016	0.029	0.030	0.007	0.010
	WILD [61]	0.089	0.065	**0.026**	**0.025**	0.049	0.045	**0.011**	**0.017**
	SELF [31]	0.089	0.081	**0.026**	0.026	0.056	0.057	0.020	0.021
	OTB (Ours)	0.068	0.055	0.028	0.032	**0.041**	**0.044**	0.015	0.019
	Optimal	0.047	0.034	0.008	0.008	0.024	0.022	0.003	0.005

networks to learn how to estimate confidence better than a noisy supervision signal used for training. In general, the models trained on $[\mathcal{T}^p, \mathcal{T}^q]$ outperforms the others, except rare cases (i.e. CCNN and LGC on Census-SGM, outperformed by $[\mathcal{U}^p, \mathcal{U}^q]$ setting). Although effective at detecting textureless and ambiguous regions, the reprojection fails at filtering outliers due to slanted surfaces and occlusions. Thus, we incrementally add a single criterion, i.e. $\mathcal{A}^p$ or $\mathcal{U}^p$ to filter out false positives obtained by $[\mathcal{T}^p, \mathcal{T}^q]$ configuration. We incrementally add, on another configuration, the corresponding negative criterion to remove pixels wrongly categorized as outliers by $\mathcal{T}^q$. In most cases, adding a single criterion to $\mathcal{P}$ is beneficial, while we can notice how introducing negative criteria degrades the performance on CBCA algorithms. This occurs because adding $\mathcal{A}^q$ or $\mathcal{U}^q$ makes textureless regions no longer labelled as outliers, as shown in Fig. 2 left comparing patches d) and e). Finally, adding both $\mathcal{A}^p$ and $\mathcal{U}^p$ produces the best overall results for CBCA methods. By introducing $\mathcal{A}^q$ and $\mathcal{U}^q$ too we obtain better results only on SGM methods, since much more accurate than CBCA ones and thus more false outliers are introduced if $\mathcal{A}^q$ and $\mathcal{U}^q$ are not used, as shown in Fig. 2 right, comparing d) and e). On the other hand, by testing the best configurations on Middlebury 2014, enabling all the positive criteria and only $\mathcal{T}^q$ for negative allows for better generalization to unseen environments.

4.3 Comparison with Offline Methods

Having found the best configuration for the $\mathcal{L}_{\text{MBCE}}$ loss, we compare our supervision paradigm with known self-supervised approaches [31,61]. In our experiments, we obtain proxy labels for SELF and WILD using the code provided by

Table 3. Generalization on Middlebury and ETH3D. We report AUC scores for networks trained on KITTI 2012 (20 or 400 images) and tested on Middlebury (15 images) and ETH3D (27 images) without retraining or adaptation.

	Match cost	Census				MCCNN-fst			
	Aggregation	CBCA		SGM		CBCA		SGM	
	Dataset	Midd	ETH	Midd	ETH	Midd	ETH	Midd	ETH
	Bad1 %	28.701	21.270	26.682	15.471	29.799	34.279	21.799	12.594
Traditional	LRD	0.117	0.082	0.113	0.059	0.107	0.185	0.075	0.051
	PKR	0.124	0.086	0.112	0.056	0.095	0.181	0.059	0.042
	LRC	0.189	0.135	0.197	0.114	0.188	0.239	0.149	0.091
	$\Delta_{(\mathcal{I}_L,\mathcal{I}_R)}$	0.190	0.162	0.180	0.119	0.179	0.257	0.134	0.097
	DA	0.161	0.119	0.168	0.093	0.099	0.159	0.087	0.047
	UC	0.193	0.148	0.188	0.114	0.192	0.264	0.145	0.096
CCNN	Supervised	0.110	0.096	0.118	0.076	0.079	0.138	0.068	0.046
	WILD [61]	0.136	0.114	0.140	0.086	0.095	0.154	0.081	0.046
	SELF [31]	0.163	0.174	0.217	0.174	0.090	0.147	0.081	0.076
	OTB (Ours)	0.116	0.084	**0.123**	**0.070**	0.087	0.137	**0.077**	**0.042**
ConfNet	Supervised	0.121	0.086	0.104	0.063	0.086	0.138	0.062	0.036
	WILD [61]	**0.122**	0.101	0.117	**0.063**	0.091	0.160	0.073	0.037
	SELF [31]	0.154	0.120	0.121	0.067	0.096	0.172	0.084	0.048
	OTB (Ours)	0.133	**0.093**	**0.112**	0.067	0.087	**0.138**	**0.067**	0.035
LGC	Supervised	0.111	0.080	0.111	0.061	0.083	0.136	0.065	0.040
	WILD [61]	0.136	0.104	0.133	0.082	0.098	0.156	0.084	0.050
	SELF [31]	0.128	0.105	0.117	0.066	0.091	0.154	0.086	0.060
	OTB (Ours)	**0.127**	0.084	0.111	0.056	**0.090**	**0.139**	0.064	0.035
	Optimal	0.053	0.041	0.046	0.022	0.057	0.103	0.030	0.014

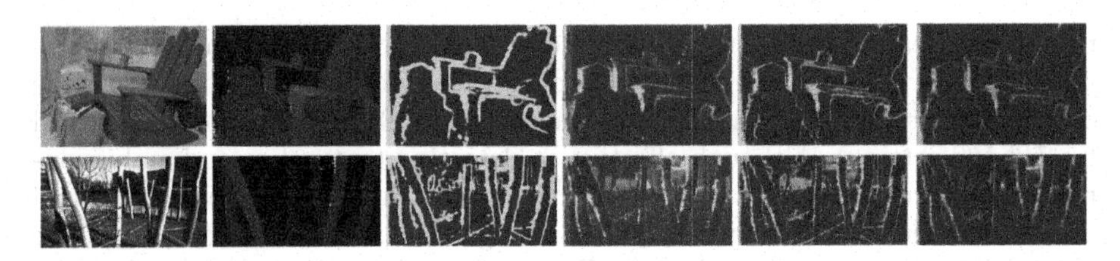

Fig. 3. Qualitative results for generalization. From left: reference image, disparity by MCCNN-fst-SGM, ConfNet trained with [31,61], our method and ground-truth. On top, Adirondack (Middlebury), at the bottom, Playground_3l (ETH3D).

the respective authors. We collect the outcome of these experiments in Tables 2 and 3. We label with different colors methods ranging from stronger constraints (need for ground truth) to weaker (ours). For each architecture, stereo algorithm and evaluation set triplet we label in **bold** the best self-supervision approach, while in red the couple architecture/self-supervision on an entire evaluation set.

KITTI Datasets. Table 2 collects evaluations on the KITTI 2012 and 2015 datasets, respectively, using the 174 validation set from 2012 and the full 2015 set. We point out that all self-supervised strategies outperform traditional measures, reported on top as baselines, such as LRD, PKR, LRC and the cues used in our $\mathcal{L}_{\text{MBCE}}$ loss, struggling only when dealing with the very accurate MCCNN-fst-SGM algorithm. Comparing the different architectures, we can notice how the self-supervised paradigms break the hierarchy (i.e., self-supervised LGC is often outperformed by ConfNet). On Census-CBCA, our strategy always outperforms

Table 4. Self-adaptation. We report AUC scores for networks trained on KITTI 2012 (20 or 400 images) and tested on a DrivingStereo sequence (6905 frames).

	Traditional			Supervision						
Algorithm	$\Delta_{(\mathcal{I}_L,\tilde{\mathcal{I}}_R)}$	DA	UC	Supervised	WILD [61]	SELF [31]	OTB	**OTB (online)**	Opt.	Badτ %
Census-SGM	0.179	0.106	0.162	0.061	0.067	0.074	0.072	0.064	0.029	21.007
MADNet [59]	0.133	0.147	0.155	0.116	-	0.135	0.139	0.126	0.021	16.226
GANet [68]	0.019	0.018	0.025	0.017	-	0.021	0.019	0.015	0.001	2.897

Fig. 4. Qualitative results on DrivingStereo. From left: reference image, disparity by Census-SGM, ConfNet trained with [31,61], OTB and online-adapted OTB.

SELF and WILD when used to train any architecture. The same behaviour is confirmed on MCCNN-fst-CBCA, except for CCNN resulting better with SELF but with the best performance achieved by ConfNet trained with OTB. This outcome highlights the outstanding performance of OTB with noisy algorithms (about 27 and 19% error rates on the validation set), close to full supervision. On SGM algorithms, OTB results comparable with SELF and WILD, although sourcing supervision only from images and $\mathcal{D}_L$, thus in a much weaker form compared to the competitors. On three out of four algorithms, ConfNet results to be the most effective architecture when trained in a self-supervised manner.

Generalization on Middlebury and ETH3D. Table 3 reports results on the Middlebury 2014 and ETH3D datasets. The same networks evaluated so far (trained on KITTI 2012 images) are transferred here without retraining or adaptation, enabling to assess the generalization properties of each network/supervision configuration. Not surprisingly, the margin between learned and traditional measures is much smaller because of the domain shift. Nonetheless, in many cases the performance is still in favor of learned approaches, with some exceptions. We point out that networks trained with OTB self-supervision always outperform SELF and WILD, except for ConfNet in two cases. Moreover, networks trained with OTB generalize better than their fully supervised counterparts in some cases, mostly on the ETH3D dataset (e.g., CCNN with all algorithms, ConfNet with MCCNN-fst-SGM and LGC with both SGM methods). Figure 3 shows qualitative examples of this test.

4.4 Self-adapting In-the-wild

Finally, we conduct experiments aimed at assessing how effective our strategy is for self-adaptation of a confidence measure in unseen environments. Purposely, we simulate deployment in an autonomous driving scenario, selecting a sequence from the DrivingStereo dataset [64]. We use sequence 2018-10-25-07-37, containing 6905 stereo pairs acquired in unconstrained (i.e., dynamic) environment.

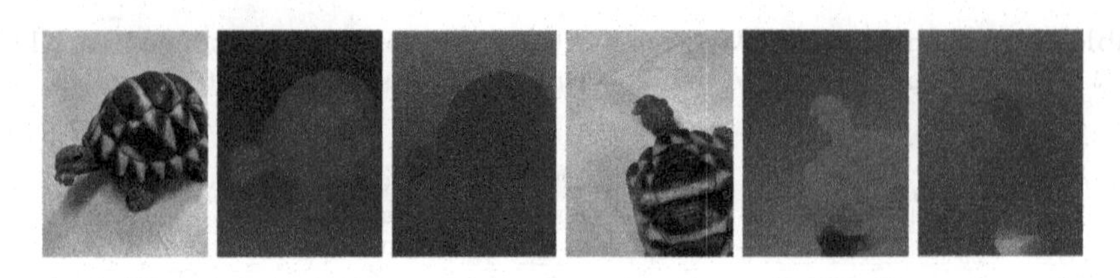

Fig. 5. Qualitative results with Apple iPhone XS. We show two examples of reference image and disparity map acquired with the iPhone XS, followed by estimated confidence map after few iterations of on-the-fly learning.

For this evaluation, we choose Census-SGM, MADNet and GANet. The former because it represents the preferred choice for hardware implementation on custom stereo cameras [2,9,17,28,44,45,48]. The remaining two because well representing modern end-to-end CNNs that are fast (MADNet) or yield state-of-the-art accuracy (GANet). For confidence networks we select ConfNet, because effective with accurate algorithms and well-suited for online adaptation.

In this experiment, we assume to have pre-trained versions of ConfNet with the different self-supervision paradigms, again on KITTI 2012. For OTB, we use $[\mathcal{T}^p, \mathcal{A}^p, \mathcal{U}^p, \mathcal{T}^q, \mathcal{A}^q, \mathcal{U}^q]$ for offline training and $[\mathcal{T}^p, \mathcal{A}^p, \mathcal{U}^p, \mathcal{T}^q]$ during adaptation. When performing online adaptation (**online** entry), for each stereo pair the confidence is estimated and evaluated **before** loss computation (thus, supervision only acts on the upcoming frames as in [59]). This way, ConfNet runs at 0.09 seconds (11 FPS) against the 0.02 (50 FPS) without adaptation on Titan Xp, measuring network computations only. The learning rate is set to 0.0001 during adaptation. Table 4 collects the outcome of this evaluation. We point out that WILD can not be deployed for MADNet and GANet since a meaningful cost volume is not available for the former or cannot be used straightforwardly for the latter. On the other hand, SELF would require $(\mathcal{D}_L, \mathcal{D}_R)$ for supervision, while MADNet and GANet provide only the former. By assuming networks as gray-boxes, we get rid of this issue at training time obtaining $\mathcal{D}_R$ as shown in Eq. 1. Concerning SGM, OTB performs in between WILD and SELF. Nevertheless, keeping continuous adaptation active on the whole sequence makes it outperform both. Concerning MADNet, SELF results more effective than OTB. Again, performing online adaptation makes OTB the best solution in this case as well. Finally, concerning GANet, learned measures perform worse than $\Delta_{(\mathcal{I}_L, \tilde{\mathcal{I}_R})}$ and DA. Online adaptation results crucial for OTB to obtain the best results. To conclude, Fig. 4 shows qualitative examples for the SGM algorithm.

On-the-Fly Learning with Black-Box Sensors. Finally we report, as qualitative results, the outcome obtained by learning on-the-fly a confidence measure on disparity map sourced by an Apple iPhone XS, without any pre-training. Figure 5 shows examples of acquired disparity and estimated confidence maps by ConfNet adapted online, detecting gross errors like on turtle's shell.

5 Conclusion

In this paper, we have introduced a novel self-supervised paradigm aimed at learning from scratch a confidence measure for stereo. We leverage few, principled cues from the input stereo pair and the estimated disparity in order to source supervision signals in place of disparity ground truth labels. Being such cues available during deployment in-the-wild, our solution is suited for continuous online adaptation on any black-box framework. Experimental results proved that our strategy is equivalent or superior to existing self-supervised approaches and, conversely to them, allows for further improvements during deployment by leveraging the online self-adaptation process.

Acknowledgments. We gratefully acknowledge the support of NVIDIA Corporation with the donation of the Titan Xp GPU used for this research.

References

1. Aleotti, F., Tosi, F., Zhang, L., Poggi, M., Mattoccia, S.: Reversing the cycle: self-supervised deep stereo through enhanced monocular distillation. In: European Conference on Computer Vision (ECCV) (2020)
2. Banz, C., Hesselbarth, S., Flatt, H., Blume, H., Pirsch, P.: Real-time stereo vision system using semi-global matching disparity estimation: architecture and FPGA-implementation. In: ICSAMOS, pp. 93–101 (2010)
3. Casser, V., Pirk, S., Mahjourian, R., Angelova, A.: Depth prediction without the sensors: leveraging structure for unsupervised learning from monocular videos. In: Thirty-Third AAAI Conference on Artificial Intelligence (AAAI 2019). AAAI Press (2019)
4. Chang, J.R., Chen, Y.S.: Pyramid stereo matching network. In: The IEEE Conference on Computer Vision and Pattern Recognition (CVPR). IEEE, June 2018
5. Chen, Z., Sun, X., Wang, L., Yu, Y., Huang, C.: A deep visual correspondence embedding model for stereo matching costs. In: The IEEE International Conference on Computer Vision (ICCV). IEEE, December 2015
6. Di Stefano, L., Marchionni, M., Mattoccia, S.: A fast area-based stereo matching algorithm. Image Vis. Comput. **22**(12), 983–1005 (2004)
7. Duggal, S., Wang, S., Ma, W.C., Hu, R., Urtasun, R.: Deeppruner: learning efficient stereo matching via differentiable patchmatch. In: IEEE International Conference on Computer Vision (ICCV), pp. 4384–4393. IEEE (2019)
8. Fu, Z., Fard, M.A.: Learning confidence measures by multi-modal convolutional neural networks. In: 2018 IEEE Winter Conference on Applications of Computer Vision (WACV), pp. 1321–1330. IEEE (2018)
9. Gehrig, S.K., Eberli, F., Meyer, T.: A real-time low-power stereo vision engine using semi-global matching. In: ICVS, pp. 134–143 (2009)
10. Geiger, A., Lenz, P., Urtasun, R.: Are we ready for autonomous driving? The KITTI vision benchmark suite. In: CVPR. IEEE (2012)
11. Godard, C., Mac Aodha, O., Brostow, G.J.: Unsupervised monocular depth estimation with left-right consistency. In: CVPR. IEEE (2017)
12. Godard, C., Mac Aodha, O., Brostow, G.J.: Digging into self-supervised monocular depth estimation. In: IEEE International Conference on Computer Vision (ICCV). IEEE (2019)

13. Gul, M.S.K., Bätz, M., Keinert, J.: Pixel-wise confidences for stereo disparities using recurrent neural networks. In: British Machine Vision Conference (BMVC). BMVA (2019)
14. Haeusler, R., Nair, R., Kondermann, D.: Ensemble learning for confidence measures in stereo vision. In: IEEE Conference on Computer Vision and Pattern Recognition (CVPR), pp. 305–312. IEEE (2013)
15. Hirschmuller, H.: Accurate and efficient stereo processing by semi-global matching and mutual information. In: IEEE Computer Society Conference on Computer Vision and Pattern Recognition, CVPR 2005, vol. 2, pp. 807–814. IEEE (2005)
16. Hirschmüller, H., Innocent, P.R., Garibaldi, J.: Real-time correlation-based stereo vision with reduced border errors. Int. J. Comput. Vision **47**(1–3), 229–246 (2002)
17. Honegger, D., Oleynikova, H., Pollefeys, M.: Real-time and low latency embedded computer vision hardware based on a combination of FPGA and mobile CPU. In: IROS. IEEE (2014)
18. Hu, X., Mordohai, P.: A quantitative evaluation of confidence measures for stereo vision. PAMI **34**(11), 2121–2133 (2012)
19. Ilg, E., Saikia, T., Keuper, M., Brox, T.: Occlusions, motion and depth boundaries with a generic network for disparity, optical flow or scene flow estimation. In: Ferrari, V., Hebert, M., Sminchisescu, C., Weiss, Y. (eds.) ECCV 2018. LNCS, vol. 11216, pp. 626–643. Springer, Cham (2018). https://doi.org/10.1007/978-3-030-01258-8_38
20. Kendall, A., et al.: End-to-end learning of geometry and context for deep stereo regression. In: The IEEE International Conference on Computer Vision (ICCV). IEEE, October 2017
21. Kim, S., Kim, S., Min, D., Sohn, K.: LAF-net: locally adaptive fusion networks for stereo confidence estimation. In: The IEEE Conference on Computer Vision and Pattern Recognition (CVPR). IEEE, June 2019
22. Kim, S., Min, D., Kim, S., Sohn, K.: Feature augmentation for learning confidence measure in stereo matching. IEEE Trans. Image Process. **26**(12), 6019–6033 (2017)
23. Kim, S., Min, D., Kim, S., Sohn, K.: Unified confidence estimation networks for robust stereo matching. IEEE Trans. Image Process. **28**(3), 1299–1313 (2019)
24. Kim, S., Min, D., Kim, S., Sohn, K.: Adversarial confidence estimation networks for robust stereo matching. IEEE Trans. Intell. Transp. Syst., 1–15 (2020, early access)
25. Liang, Z., et al.: Learning for disparity estimation through feature constancy. In: The IEEE Conference on Computer Vision and Pattern Recognition (CVPR). IEEE, June 2018
26. Luo, W., Schwing, A.G., Urtasun, R.: Efficient deep learning for stereo matching. In: IEEE Conference on Computer Vision and Pattern Recognition (CVPR), pp. 5695–5703. IEEE (2016)
27. Marin, G., Zanuttigh, P., Mattoccia, S.: Reliable fusion of ToF and stereo depth driven by confidence measures. In: Leibe, B., Matas, J., Sebe, N., Welling, M. (eds.) ECCV 2016. LNCS, vol. 9911, pp. 386–401. Springer, Cham (2016). https://doi.org/10.1007/978-3-319-46478-7_24
28. Mattoccia, S., Poggi, M.: A passive RGBD sensor for accurate and real-time depth sensing self-contained into an FPGA. In: 9th ICDSC (2015)
29. Mayer, N., et al.: A large dataset to train convolutional networks for disparity, optical flow, and scene flow estimation. In: The IEEE Conference on Computer Vision and Pattern Recognition (CVPR). IEEE, June 2016
30. Menze, M., Geiger, A.: Object scene flow for autonomous vehicles. In: Conference on Computer Vision and Pattern Recognition (CVPR). IEEE (2015)

31. Mostegel, C., Rumpler, M., Fraundorfer, F., Bischof, H.: Using self-contradiction to learn confidence measures in stereo vision. In: The IEEE Conference on Computer Vision and Pattern Recognition (CVPR). IEEE, June 2016
32. Park, M.G., Yoon, K.J.: Leveraging stereo matching with learning-based confidence measures. In: IEEE Conference on Computer Vision and Pattern Recognition (CVPR), pp. 101–109. IEEE (2015)
33. Park, M.G., Yoon, K.J.: Learning and selecting confidence measures for robust stereo matching. IEEE Trans. Pattern Anal. Mach. Intell. **41**(6), 1397–1411 (2018)
34. Poggi, M., Agresti, G., Tosi, F., Zanuttigh, P., Mattoccia, S.: Confidence estimation for ToF and stereo sensors and its application to depth data fusion. IEEE Sens. J. **20**(3), 1411–1421 (2019)
35. Poggi, M., Aleotti, F., Tosi, F., Mattoccia, S.: Towards real-time unsupervised monocular depth estimation on CPU. In: IEEE/JRS Conference on Intelligent Robots and Systems (IROS). IEEE (2018)
36. Poggi, M., Mattoccia, S.: Deep stereo fusion: combining multiple disparity hypotheses with deep-learning. In: International Conference on 3D Vision (3DV), pp. 138–147. IEEE (2016)
37. Poggi, M., Mattoccia, S.: Learning a general-purpose confidence measure based on o (1) features and a smarter aggregation strategy for semi global matching. In: International Conference on 3D Vision (3DV), pp. 509–518. IEEE (2016)
38. Poggi, M., Mattoccia, S.: Learning from scratch a confidence measure. In: British Machine Vision Conference (BMVC). BMVA (2016)
39. Poggi, M., Mattoccia, S.: Learning to predict stereo reliability enforcing local consistency of confidence maps. In: IEEE Conference on Computer Vision and Pattern Recognition (CVPR), pp. 2452–2461. IEEE (2017)
40. Poggi, M., Pallotti, D., Tosi, F., Mattoccia, S.: Guided stereo matching. In: IEEE/CVF Conference on Computer Vision and Pattern Recognition (CVPR). IEEE (2019)
41. Poggi, M., Tosi, F., Mattoccia, S.: Even more confident predictions with deep machine-learning. In: IEEE Conference on Computer Vision and Pattern Recognition Workshops, pp. 76–84. IEEE (2017)
42. Poggi, M., Tosi, F., Mattoccia, S.: Quantitative evaluation of confidence measures in a machine learning world. In: IEEE International Conference on Computer Vision (ICCV), pp. 5228–5237. IEEE (2017)
43. Poggi, M., Tosi, F., Mattoccia, S.: Learning a confidence measure in the disparity domain from o (1) features. Comput. Vis. Image Underst. **193**, 102905 (2020)
44. Rahnama, O., et al.: Real-time highly accurate dense depth on a power budget using an FPGA-CPU hybrid SOC. IEEE Trans. Circuits Syst. II Express Briefs **66**(5), 773–777 (2019)
45. Rahnama, O., Cavalleri, T., Golodetz, S., Walker, S., Torr, P.: R3SGM: real-time raster-respecting semi-global matching for power-constrained systems. In: 2018 International Conference on Field-Programmable Technology (FPT), pp. 102–109. IEEE (2018)
46. Scharstein, D., et al.: High-resolution stereo datasets with subpixel-accurate ground truth. In: Jiang, X., Hornegger, J., Koch, R. (eds.) GCPR 2014. LNCS, vol. 8753, pp. 31–42. Springer, Cham (2014). https://doi.org/10.1007/978-3-319-11752-2_3
47. Scharstein, D., Szeliski, R.: A taxonomy and evaluation of dense two-frame stereo correspondence algorithms. Int. J. Comput. Vision **47**(1–3), 7–42 (2002)
48. Schmid, K., Hirschmuller, H.: Stereo vision and IMU based real-time ego-motion and depth image computation on a handheld device. In: IEEE International Conference on Robotics and Automation (ICRA). IEEE (2013)

49. Schönberger, J.L., Sinha, S.N., Pollefeys, M.: Learning to fuse proposals from multiple scanline optimizations in semi-global matching. In: Ferrari, V., Hebert, M., Sminchisescu, C., Weiss, Y. (eds.) ECCV 2018. LNCS, vol. 11217, pp. 758–775. Springer, Cham (2018). https://doi.org/10.1007/978-3-030-01261-8_45
50. Schops, T., et al.: A multi-view stereo benchmark with high-resolution images and multi-camera videos. In: IEEE Conference on Computer Vision and Pattern Recognition (CVPR), pp. 3260–3269. IEEE (2017)
51. Seki, A., Pollefeys, M.: Patch based confidence prediction for dense disparity map. In: British Machine Vision Conference (BMVC), vol. 2, p. 4. BMVA (2016)
52. Song, X., Zhao, X., Hu, H., Fang, L.: Edgestereo: a context integrated residual pyramid network for stereo matching. In: 14th Asian Conference on Computer Vision (ACCV) (2018)
53. Spyropoulos, A., Komodakis, N., Mordohai, P.: Learning to detect ground control points for improving the accuracy of stereo matching. In: IEEE Conference on Computer Vision and Pattern Recognition (CVPR), pp. 1621–1628. IEEE (2014)
54. Spyropoulos, A., Mordohai, P.: Ensemble classifier for combining stereo matching algorithms. In: 2015 International Conference on 3D Vision, pp. 73–81. IEEE (2015)
55. Spyropoulos, A., Mordohai, P.: Correctness prediction, accuracy improvement and generalization of stereo matching using supervised learning. Int. J. Comput. Vision **118**(3), 300–318 (2016)
56. Tonioni, A., Poggi, M., Mattoccia, S., Di Stefano, L.: Unsupervised adaptation for deep stereo. In: The IEEE International Conference on Computer Vision (ICCV). IEEE, October 2017
57. Tonioni, A., Poggi, M., Mattoccia, S., Di Stefano, L.: Unsupervised domain adaptation for depth prediction from images. IEEE Trans. Pattern Anal. Mach. Intell. **42**(10), 2396–2409 (2020)
58. Tonioni, A., Rahnama, O., Joy, T., Di Stefano, L., Thalaiyasingam, A., Torr, P.: Learning to adapt for stereo. In: The IEEE Conference on Computer Vision and Pattern Recognition (CVPR). IEEE, June 2019
59. Tonioni, A., Tosi, F., Poggi, M., Mattoccia, S., Di Stefano, L.: Real-time self-adaptive deep stereo. In: IEEE Conference on Computer Vision and Pattern Recognition (CVPR). IEEE, June 2019
60. Tosi, F., Poggi, M., Benincasa, A., Mattoccia, S.: Beyond local reasoning for stereo confidence estimation with deep learning. In: Ferrari, V., Hebert, M., Sminchisescu, C., Weiss, Y. (eds.) ECCV 2018. LNCS, vol. 11210, pp. 323–338. Springer, Cham (2018). https://doi.org/10.1007/978-3-030-01231-1_20
61. Tosi, F., Poggi, M., Tonioni, A., Di Stefano, L., Mattoccia, S.: Learning confidence measures in the wild. In: British Machine Vision Conference (BMVC). BMVA, September 2017
62. Wang, Z., Bovik, A.C., Sheikh, H.R., Simoncelli, E.P.: Image quality assessment: from error visibility to structural similarity. IEEE Trans. Image Process. **13**(4), 600–612 (2004)
63. Watson, J., Aodha, O.M., Turmukhambetov, D., Brostow, G.J., Firman, M.: Learning stereo from single images. In: Vedaldi, A., Bischof, H., Brox, T., Frahm, J.M. (eds.) ECCV 2020. LNCS, vol. 12346. Springer, Cham (2020). https://doi.org/10.1007/978-3-030-58452-8_42
64. Yang, G., Song, X., Huang, C., Deng, Z., Shi, J., Zhou, B.: Drivingstereo: a large-scale dataset for stereo matching in autonomous driving scenarios. In: IEEE Conference on Computer Vision and Pattern Recognition (CVPR). IEEE (2019)

65. Yang, G., Zhao, H., Shi, J., Deng, Z., Jia, J.: SegStereo: exploiting semantic information for disparity estimation. In: Ferrari, V., Hebert, M., Sminchisescu, C., Weiss, Y. (eds.) ECCV 2018. LNCS, vol. 11211, pp. 660–676. Springer, Cham (2018). https://doi.org/10.1007/978-3-030-01234-2_39
66. Yin, Z., Darrell, T., Yu, F.: Hierarchical discrete distribution decomposition for match density estimation. In: IEEE Conference on Computer Vision and Pattern Recognition (CVPR), pp. 6044–6053. IEEE (2019)
67. Zbontar, J., LeCun, Y.: Stereo matching by training a convolutional neural network to compare image patches. J. Mach. Learn. Res. **17**(1–32), 2 (2016)
68. Zhang, F., Prisacariu, V., Yang, R., Torr, P.H.: GA-net: guided aggregation net for end-to-end stereo matching. In: IEEE Conference on Computer Vision and Pattern Recognition (CVPR), pp. 185–194. IEEE (2019)
69. Zhang, K., Lu, J., Lafruit, G.: Cross-based local stereo matching using orthogonal integral images. IEEE Trans. Circuits Syst. Video Technol. **19**(7), 1073–1079 (2009)
70. Zhang, Z., Cui, Z., Xu, C., Jie, Z., Li, X., Yang, J.: Joint task-recursive learning for semantic segmentation and depth estimation. In: Ferrari, V., Hebert, M., Sminchisescu, C., Weiss, Y. (eds.) ECCV 2018. LNCS, vol. 11214, pp. 238–255. Springer, Cham (2018). https://doi.org/10.1007/978-3-030-01249-6_15
71. Zhong, Y., Li, H., Dai, Y.: Open-world stereo video matching with deep RNN. In: Ferrari, V., Hebert, M., Sminchisescu, C., Weiss, Y. (eds.) ECCV 2018. LNCS, vol. 11206, pp. 104–119. Springer, Cham (2018). https://doi.org/10.1007/978-3-030-01216-8_7

Deep Surface Normal Estimation on the 2-Sphere with Confidence Guided Semantic Attention

Quewei Li[1], Jie Guo[1](✉), Yang Fei[1], Qinyu Tang[1], Wenxiu Sun[2], Jin Zeng[2], and Yanwen Guo[1](✉)

[1] State Key Laboratory for Novel Software Technology, Nanjing University, Nanjing 210023, China
{queweili,yangf,tangqinyu}@smail.nju.edu.cn, {guojie,ywguo}@nju.edu.cn
[2] SenseTime Research, Shenzhen, China
sunwenxiu@sensetime.com, jzeng2010@gmail.com

Abstract. We propose a deep convolutional neural network (CNN) to estimate surface normal from a single color image accompanied with a low-quality depth channel. Unlike most previous works, we predict the normal on the 2-sphere rather than the 3D Euclidean space, which produces naturally normalized values and makes the training stable. Although the depth information is beneficial for normal estimation, the raw data contain missing values and noises. To alleviate this problem, we employ a confidence guided semantic attention (CGSA) module to progressively improve the quality of depth channel during training. The continuously refined depth features are fused with the normal features at multiple scales with the mutual feature fusion (MFF) modules to fully exploit the correlations between normals and depth, resulting in high quality normals and depth with fine details. Extensive experiments on multiple benchmark datasets prove the superiority of the proposed method.

Keywords: Normal estimation · 2-sphere · Confidence guided semantic attention · Mutual feature fusion

1 Introduction

This paper aims to recover high-quality surface normal and depth from a single RGB-D image using a deep neural network. Recently, the availability of depth information has promoted a great enhancement in the applications of object recognition [10,31], sematic segmentation [4,13,20,28], 3D scene reconstruction [17,22,25], pose estimation [5,37], *etc.* From the depth channel, we can

Electronic supplementary material The online version of this chapter (https://doi.org/10.1007/978-3-030-58586-0_43) contains supplementary material, which is available to authorized users.

A. Vedaldi et al. (Eds.): ECCV 2020, LNCS 12369, pp. 734–750, 2020.
https://doi.org/10.1007/978-3-030-58586-0_43

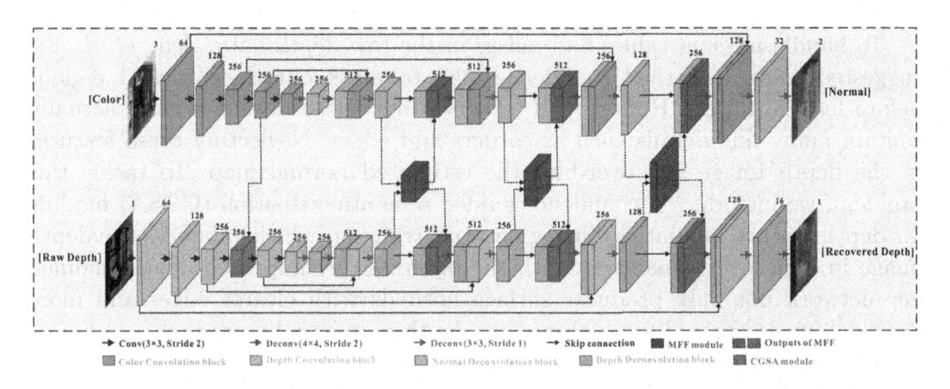

Fig. 1. The network architecture of our method (Color figure online)

easily obtain surface normals with least square optimization[24,29]. Unfortunately, depth images captured by low-cost depth sensors (*e.g.*, Microsoft Kinect and ASUS Xtion Pro) are notoriously corrupted by noises and contain missing/invalid values. These artifacts degrade the visual quality of both depth and normal images, influencing their usage in different tasks.

To improve the quality of normals and/or depth, a variety of methods have been proposed recently, among which the deep learning based solutions gain the most excellent results. These methods usually view the normal map as a conventional color image and predict the values directly in the 3D Euclidean space (*i.e.*, $\mathbb{R}^3$) [1,29,33,38]. In this way, the three components of any output normal are learned without any restriction. However, we know that the values of normal should lie on a unit 3D ball (*i.e.*, S^2). A straightforward strategy is to explicitly normalize the result before calculating the loss [38]. However, such a strategy still cannot guarantee that the final output is normalized and is sub-optimal since the gradients propagated backwards to the model are unconstrained [21]. In contrast to this strategy, we opt to predict surface normals on the 2-sphere by learning two independent parameters: azimuth angle θ and elevation angle ϕ. Both parameters are constrained: $\theta \in [-\pi, \pi]$ and $\phi \in [0, \pi]$. The benefit of this solution is two-fold. First, it ensures that the output normals are naturally normalized without any explicit normalization operation. Second, it makes the training stable since the regression gradients are constrained and well-behaved.

In this paper, we use RGB and depth images as input in the task of normal estimation. These two inputs are fed into two separate branches of our neural network and fused at multiple scales with a new mutual feature fusion (MFF) module, as shown in Fig. 1. As a main difference from previous methods [38,39], this module does not simply fuse features of different branches by concatenation but learns a mapping function from one feature map to the other. This fusion strategy can be viewed as using one branch as the guided filter to re-weight the feature maps in another branch via pixel-wise transformations, making a new form of hyper-network.

To handle missing values and noises in the raw depth data, Zeng *et al.* [38] suggested using a learned confidence map to mask out these invalid regions before feature fusion. However, we observe that these invalid regions actually contain many fine details such as corners and edges. Neglecting these features in the depth image will over-blur the estimated normal map. To tackle this problem, we introduce a confidence guided semantic attention (CGSA) module for depth feature inpainting during network training, which enhances the depth image in high-level semantics. With the progressively improved depth channel, our network not only produces surface normals with clearer edges and more details, but also provides a high-quality depth map as a by-product.

To summarize, the main contributions of our work are:

- a deep learning based solution that estimates surface normal on the 2-sphere with a well defined loss function,
- a CGSA module designed to enhance depth feature by exploiting high-level semantics, and
- the multi-scale MFF modules to effectively combine features from different branches.

2 Related Work

Per-pixel normal estimation has been extensively studied in the past years. Traditional methods like [24] estimated surface normals via least square optimization. Qi *et al.* [29] integrated traditional methods into a deep neural network architecture and jointly predicted depth and surface normals with the 3D geometric information.

With the emergence of deep learning, recent methods predict surface normals under the framework of deep neural network, most of which use a single RGB image as the input [2,8,19,40]. Wang *et al.* [33] proposed a network that integrates local, global, and vanishing point information to predict surface normals. Bansal *et al.* [1] proposed a skip connection network architecture to fuse features from different layers for normal estimation. Zhang *et al.* [41] designed a pattern-affinitive propagation network to predict surface normals and depth jointly by the affinity matrices. Due to the lack of geometric information in these RGB based methods, the details of the estimated results are not satisfactory, easily incurring over-blurriness or strange artifacts.

Compared with the RGB based methods, the study of RGB-D based surface normal estimation is far not enough. The 3D reconstruction based methods like [25] can be used for surface normal estimation but a sequence of RGB-D images are usually required for these methods. Zhang *et al.* briefly discussed normal estimation with RGB-D input in [39] and reported that their network produced better predictions with RGB input than the RGB-D input. More recently, Zeng *et al.* [38] proposed a hierarchical fusion network for surface normal estimation, which achieved the state-of-the-art performance. Notably, these works always treat invalid areas in the raw depth data as smooth plane that enables smooth transition in the corresponding areas of surface normals. However, as holes mainly exist along the boundaries of objects, the smooth transition will over-blur edges and erase details in the given scene.

To alleviate such a problem, we can first fill holes in the depth map in a preprocessing step with the guidance of the RGB image and further improve the performance of the estimated normals. Unfortunately, the accuracy of enhanced depth is still limited as single image depth enhancement itself is also a challenging problem [11,14,15,23,24,26,39]. Considering the strong correlations between depth and surface normals, we design a unified network to predict surface normals and enhance low-quality depth jointly. To fill large missing areas in the raw depth, we perform depth inpainting for high-level depth feature map by updating unreliable regions with patches in the reliable regions, which lessens the influence of big holes to a large extent and makes our method more stable.

3 Surface Normal Estimation on the 2-Sphere

The following sections detail our network to estimate per-pixel surface normal from a single RGB-D image. As illustrated in Fig. 1, the basic architecture of our network contains two autoencoders with skip connections (Sect. 3.1). One autoencoder maps a color image to a normal map with multi-scale features fused from the corresponding depth branch using the MFF modules (Sect. 3.2). The output normals are represented in the polar coordinate (Sect. 3.3) with a well defined loss function (Sect. 3.4), ensuring proper normalization and stable training. Another autoencoder progressively refines the raw depth image leveraging the color features and a CGSA module specifically designed for depth feature map inpainting (Sect. 4).

3.1 Network Architecture

Our network comprises two autoencoders, both of which are similar in their architectures. In the color branch of the first autoencoder, we adopt a modified VGG-16 network [32] by reducing channel numbers of the last two convolution blocks, *i.e.*, conv4 and conv5 in the original VGG-16, from 512 to 256. The raw depth branch of the second autoencoder is organized in a similar way to the color branch except that a CGSA module is inserted before the fourth block, *i.e.*, conv4, of the encoder as shown in Fig. 1 (dark orange box). Two decoders are symmetric to the encoders and are equipped with skip connections. Three MFF modules are used to fuse feature maps at different scales of the decoders. The last deconvolution layer with stride 1 and kernel size 3×3 outputs a 2-channel image (representing θ and ϕ of surface normal) and a single channel image (representing depth) in each decoder, respectively. We use ReLU activation function for all the (de)convolution layers except the last layer, which is equipped with the sigmoid activation function, to ensure that the output values fall in the range of [0,1].

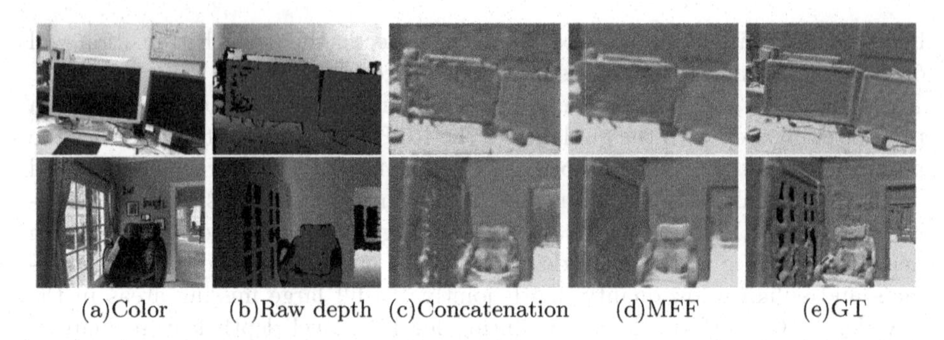

Fig. 2. Visual quality comparisons between different feature fusion strategies, *i.e.*, the simple concatenation operation and the mutual feature fusion (MFF) module

3.2 Multi-scale Mutual Feature Fusion

To fully utilize the close geometry relationship between depth and surface normals, some previous works resort to fuse information from different feature maps. The most widely used strategy, as adopted in [38,39], is by a simple concatenation operation. We observe that this is sub-optimal since the feature maps are from different domains such that they can not be properly handled with the same convolution operations. Zhang *et al.* [39] validated that such a strategy will make the network learn from the inaccurate and incomplete depth directly, lowering the influence of the color information. Consequently, as demonstrated in Fig. 2(c), the invalid or unreliable areas of the depth map will mislead the prediction of surface normals, leading to strange artifacts around the boundaries of depth holes.

Instead of simply concatenating features from different branches, we design a multi-scale fusion strategy in which the two autoencoders exchange features at multiple scales by several MFF modules. As illustrated in Fig. 3, each MFF module contains four conditional feature transform (CFT) blocks[1] (light blue boxes) and four convolution layers (dark blue arrays). The motivation of using the CFT blocks is to view the normal estimation as a conditional generative problem in which the depth image serves as an auxiliary feature. Similarly, the normal map is considered as an auxiliary feature in the task of depth inpainting. Such a strategy has been previously used in image dehazing [12], image translation [16] and Monte Carlo denoising [36]. All these methods including ours rely on some auxiliary features as a condition to address the ill-posed image generation problems.

Supposing $\mathcal{F}_a$ is an auxiliary image (or feature map) and $\mathcal{F}_s$ is the source image, the output of the CFT block at scale l is the target image $\mathcal{F}_t$ defined as

$$\mathcal{F}_t^l = \mathrm{CFT}(\mathcal{F}_s^l, \mathcal{F}_a^l | \gamma, \beta) = \gamma(\mathcal{F}_a^l) \otimes \mathcal{F}_s^l \oplus \beta(\mathcal{F}_a^l) \tag{1}$$

[1] We use four CFT blocks to improve the MFF module's ability of representing more complex feature transformations.

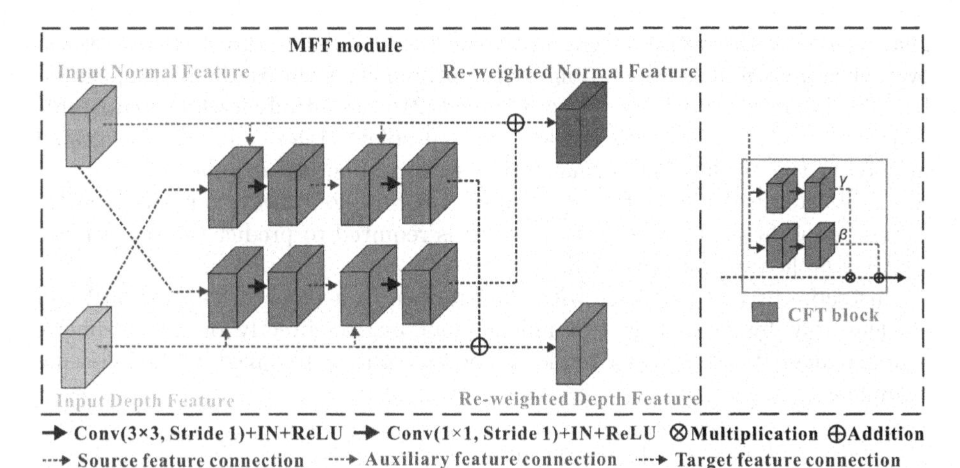

Fig. 3. The MFF module. IN represents the InstanceNorm layer and ReLU represents the ReLU activation function (Color figure online)

where $(\boldsymbol{\gamma}, \boldsymbol{\beta})$ is the modulation parameter pair with $\boldsymbol{\gamma}$ denoting the scaling operation matrix and $\boldsymbol{\beta}$ denoting the shifting operation matrix. $\otimes$ and $\oplus$ represent the element-wise multiplication and addition operation, respectively. Intuitively, the CFT block learns a mapping function that outputs the $(\boldsymbol{\gamma}, \boldsymbol{\beta})$ pair under some auxiliary feature conditions. Applying this mapping function to any source image yields a new target image.

In our MFF module, we utilize such conditioned mapping functions (*i.e.*, CFT blocks) mutually to generate parameters for pixel-wise transformation and modify the weights of each feature map in different branches at multiple scales. Specifically, the normal feature map acts as $\mathcal{F}_s$ and the depth feature map as $\mathcal{F}_a$ in the normal branch of the decoder while the roles exchange in the depth branch. Rather than simple concatenation, we re-weight the feature maps through additive and multiplicative interactions based on the conditioning representation. To better utilize feature information from high-level to low-level, we embed the MFF module at three scales in the decoder. As evidenced in Fig. 2(d), the MFF modules avoid the risk of incurring strange artifacts due to unreliable depth values. Note that masking out the invalid areas in the depth map, as suggested by Zeng *et al.* [38], will also remove these artifacts but lead to loss of details as compared in Fig. 6.

3.3 2-Sphere vs. 3D Euclidean Space

One of our important insights is that the 2-sphere space is more suitable for estimating surface normals than the 3D Euclidean space. Estimating surface normals in the 3D Euclidean space has several problems. First, the output 3-channel normals are learned without any restriction. Zeng *et al.* [38] performed a normalization before calculating the loss. However, this does not always guarantee that the outputs form unit vectors that indicate directionality. Moreover, such a normalized operation is sub-optimal since the gradients propagated backwards to the model are not constrained [21], potentially leading to unstable training or convergence. To tackle such a problem, Liao *et al.* [21] suggested modifying the traditional normalization with a spherical exponential function to enable stable training. However, since this function always outputs positive values, an additional classification branch is required to predict the sign values, which complicates the prediction.

Instead, our method deals with the problem in a more straightforward and efficient way by predicting two independent angles directly on the 2-sphere: azimuth angle θ and elevation angle ϕ. The two spaces are linked by the following formulas:

$$\begin{cases} \theta = \text{arctan2}\ (x, y) \\ \phi = \text{arctan2}\ (z, \sqrt{x^2 + y^2}) \end{cases} \tag{2}$$

where arctan2 is the two-dimension form of the arctan function. (x, y, z) is the Cartesian coordinate of the normal.

The benefit of predicting surface normals on the 2-sphere is at least two-fold. First of all, the 2-sphere is a naturally closed geometric manifold defined in the $\mathbb{R}^3$ that the output normal is expected to be normalized when getting back to the 3D Euclidean space. Second, without the need of any explicit normalization operation, the gradients escape from passing through the normalization layer for backward propagation, which makes the gradients constrained and enables stable training and easy convergence since the final layer is activated by the sigmoid function whose gradient is only determined by the constrained output. To train our model on the 2-sphere, we convert the ground-truth normals to the polar coordinate with Eq. 2. After prediction, we convert the learned θ and ϕ back to the 3D Euclidean space.

3.4 Loss Function

The loss function for our network is the L_1 norm which reflects the median angle difference between the predicted result and the ground truth. Denoting the input RGB and depth image by I_c and I_d, the output normal and depth image by T_n and T_d and the ground-truth/target normal and depth image by G_n and G_d, we aim at minimizing the distance between the ground truth and the output generated from the input RGB-D image. The loss function contains two components: one for surface normal and the other for depth. Though we represent the azimuth angle θ in the range of $[-\pi, \pi]$ (or in the range of $[0, 1]$ after normalization), $-\pi$ and π actually indicate the same direction on S^2. Considering this property, we define the "circle loss" operator as $\ominus$ for θ where

$$\|T_n^\theta \ominus G_n^\theta\|_1 = 2 \min\left(\|T_n^\theta - G_n^\theta\|_1, 1 - \|T_n^\theta - G_n^\theta\|_1\right). \tag{3}$$

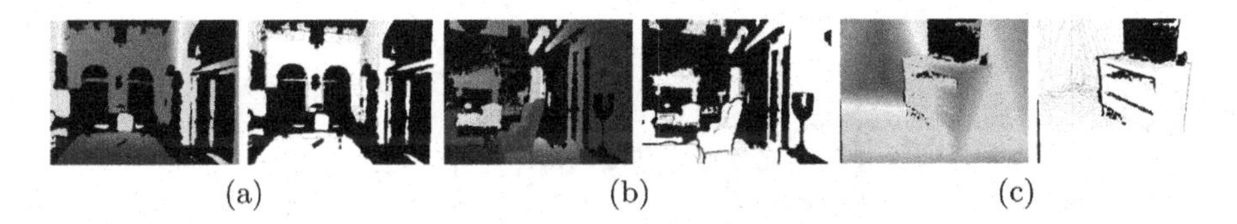

Fig. 4. Visualization of the raw depth images (the left image in each group) and the corresponding depth confidence maps (the right image in each group)

Then, our loss function is defined specifically as

$$\mathcal{L}(T_n, T_d, G_n, G_d | I_c, I_d) = \frac{1}{N}\sum_{i=1}^{N}(\|T^{\phi}_{n,i} - G^{\phi}_{n,i}\|_1 + \|T^{\theta}_{n,i} \ominus G^{\theta}_{n,i}\|_1 + \lambda\|T_{d,i} - G_{d,i}\|_1) \quad (4)$$

where N is the total pixel number, $\|\cdot\|_1$ denotes the L_1 norm and λ is a balanced factor between depth and surface normal.

4 Confidence Guided Semantic Attention

In this section, we introduce the CGSA module for depth feature inpainting. Although the depth information is becoming easier to obtain with the development of the RGB-D sensors, it is not fully exploited in the problem of surface normal estimation. The most important factor is that the depth data is not always reliable. Recently, Zeng *et al.* [38] proposed a confidence guided RGB-D fusion scheme to make use of limited geometric information from raw depth data for surface normal estimation. Their confidence map, which is also learned from a neural network, acts as a mask that masks out depth features with the low confidence before passing to the RGB branch. However, low confidence areas are mostly important scene details such as edges. Simply eliminating these areas will lead to the loss of details in the final results. Considering this, we propose a CGSA module to utilize spatial attention to recover missing values of the depth map conditioned on the valid patches from the reliable regions.

4.1 Confidence Map for Raw Depth

Before discussing the details of the CGSA module, we first introduce our definition of depth confidence map which will be used in the CGSA module. We assume that if there is no missing value in the neighboring region of a pixel i in the input depth image I_d and the variance of its local region is small, the depth confidence of i is high. Therefore, $C(I_d, i)$ is given by

$$C(I_d, i) = \begin{cases} 1, & \text{if } \widetilde{C}(I_d, i) > T_s \\ \widetilde{C}(I_d, i), & \text{otherwise} \end{cases} \quad (5)$$

$$\widetilde{C}(I_d, i) = \mathbb{I}(I_{d,i} > 0)\mathit{\Gamma}(I_{d,i}) \cdot \exp(-\frac{\sigma_i^2}{\gamma_d}) \quad (6)$$

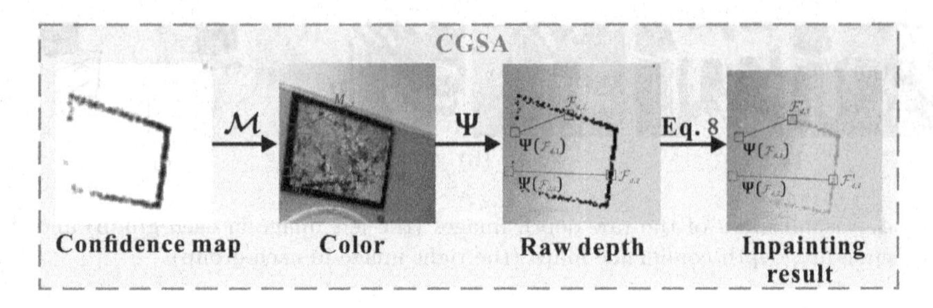

Fig. 5. The process of CGSA module. The confidence values of the confidence map in the white regions are 1 and the dark regions are lower than 1

with $\mathbb{I}(I_{d,i} > 0)$ being an indication function. γ_d is the controlling factor and T_s is a pre-defined threshold. $\Gamma(I_{d,i})$ represents the percentage of valid pixels in a neighboring region of i with the form

$$\Gamma(I_{d,i}) = \frac{\sum_{j \in \mathcal{N}(i)} \mathbb{I}(I_{d,j} > 0)}{|\mathcal{N}(i)|} \tag{7}$$

where $|\mathcal{N}(i)|$ returns the total number of the neighboring pixels. $\sigma_i^2 = \sum_{j \in \mathcal{N}(i)} ||I_{d,i} - I_{d,j}||^2$ represents the variance among the neighboring region $\mathcal{N}(i)$ of i. We use neighborhood of a 3×3 size of pixel i for all the scenes. Several examples of depth confidence maps are shown in Fig. 4. As seen, pixels with low confidence values always exist in large holes and along object boundaries. We down-sample the confidence map to match the spatial size of feature map in the CGSA module.

4.2 The CGSA Module

Considering the high correlation between RGB and depth, most existing methods [15,24,27,35] use RGB images to recover the missing depth values under the assumption that pixels with similar colors tend to have similar depth in a local region. Similarly, we perform depth feature inpainting to enhance the depth channel with the CGSA module. The workflow of CGSA is shown in Fig. 5, where the key point is to update each low confidence depth feature patch with the most similar high confidence depth feature patch.

More specifically, we first define the attention map $\mathcal{M}$ based on the calculated confidence map that $\mathcal{M}_i$ is set to 0 if $\hat{C}(i) = 1$ and set to 1 in other cases. Here, $\hat{C}$ denotes the down-sampled confidence map from C that $\hat{C}(i)$ returns the confidence value of patch i. We employ $\mathcal{M}$ on the color feature map $\mathcal{F}_c$ so that $\mathcal{F}_c$ is divided into the attention regions M_c (regions with low confidence) and the reference regions $\bar{M}_c$ (regions with confidence equal to 1). For each attention patch $M_{c,i}$, we find the closest-matching patch $\bar{M}_{c,i}$ in the reference regions. The relevant degree between these patches is measured with the squared L_2 distance defined as $||\bar{M}_{c,i} - M_{c,i}||_2^2$. By doing this, we actually find a mapping $\mathbf{\Psi}$ for every patch from the attention regions to the reference regions. We then apply $\boldsymbol{\Psi}$ to the depth feature map $\mathcal{F}_d$ so that each patch $\mathcal{F}_{d,i}$ in the low confidence regions couples with the most similar reference patch $\boldsymbol{\Psi}(\mathcal{F}_{d,i})$. Finally, we update $\mathcal{F}_{d,i}$ with the following scheme:

$$\mathcal{F}'_{d,i} = \hat{C}(i) \cdot \mathcal{F}_{d,i} + (1 - \hat{C}(i)) \cdot \boldsymbol{\Psi}(\mathcal{F}_{d,i}). \quad (8)$$

Considering that enhancing depth locally fails to fill large holes, we perform depth feature inpainting on high-level semantics, which avoids filling big holes directly at large scales and makes the module work more stable, efficiently improving the qualities of both the estimated surface normals and recovered depth. The detailed implementation of the CGSA module is provided in the supplemental material.

5 Experiments

5.1 Implementation Details

Datasets. We evaluate our method on three RGB-D datasets: NYUD-v2 [24], ScanNet [6] and Matterport3D [3]. NYUD-v2 dataset consists of RGB-D images collected from 464 different indoor scenes, among which 1449 images are provided with ground-truth normals and depth. We randomly choose 1200 of them for training and the remaining 249 images for testing. For the ScanNet and Matterport3D datasets, we use the ground-truth data provided by Zhang *et al.* [39] and follow their training and testing lists. Specifically, we use 105432 images for training and 12084 for testing of Matterport3D; 59743 for training and 7517 for testing of ScanNet. We convert all the normal maps in the training datasets from $\mathbb{R}^3$ to S^2 with Eq. 2 before training. After that, the training process carries on S^2. We train our network and test its performance on the three datasets, respectively. All the methods in comparison are tested with the same testing lists.

Training Details. Our network generally converges after 60 epochs. We implement it with PyTorch on four NVIDIA GTX 2080Ti GPUs. We use RMSprop optimizer and adjust the learning rate with the initial rate of $1e^{-3}$ and the power of 0.95 every 10 epochs. The hyper-parameters $\{\lambda, \gamma_d, T_s\}$ are set to $\{2.0, 0.2, 0.98\}$ according to validation on a 5% randomly split training data.

Evaluation Metrics. We adopt four metrics to evaluate the qualities of estimated normals: the mean of angle error (mean), the median of angle error (median), the root mean square error (rmse) and the pixel accuracy with angle difference with ground truth less than t_n where $t_n \in \{11.25°, 22.5°, 30°\}$ [29,38,40]. The qualities of the recovered depth are also evaluated with four metrics: the root mean square error (rmse), the mean relative error (rel), the mean log 10 error (log 10), and the pixel accuracy with $\max(\frac{T_d}{G_d}, \frac{G_d}{T_d})$ less than t_d where $t_d \in \{1.25, 1.25^2, 1.25^3\}$ [29,41].

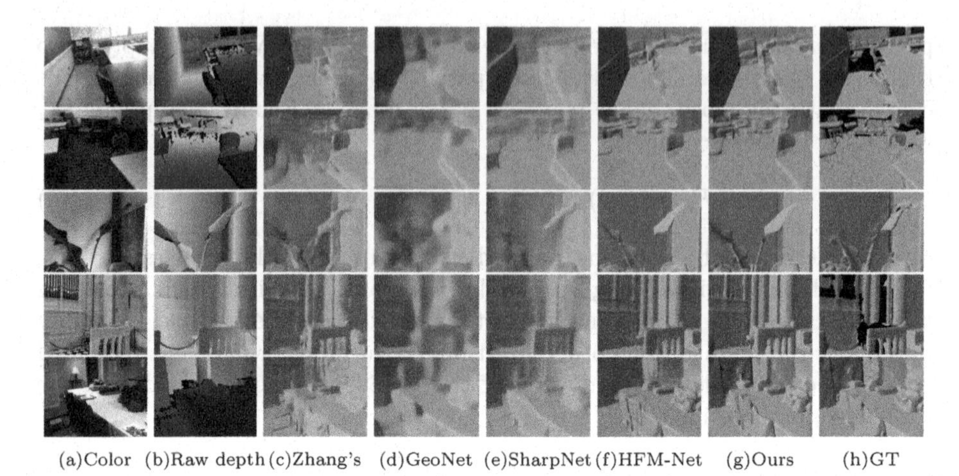

Fig. 6. Visual quality comparisons with the state-of-the-art surface normal estimation methods on ScanNet (the first and second rows), Matterport3D (the third and fourth rows) and NYUD-v2 (the last row) datasets

5.2 Comparisons with the State-of-the-Arts

In this section, we compare our network with the state-of-the-art normal and depth estimation methods.

Comparisons of Surface Normal Estimation. We compare the results of our method with some high-ranking surface normal estimation methods, including Zhang's network [40], GeoNet [29], SharpNet [30] and HFM-Net [38] with their public available pre-trained models fine-tuned on each dataset. Figure 6 shows the visual quality comparisons among all these methods. As seen, the results produced by Zhang's network and SharpNet have many unwanted details, *e.g.*, the highlights on the desk, due to the lack of depth information. GeoNet alleviates this problem by jointly learning depth and surface normals, incorporating geometric relation between them. However, it tends to generate over-blurred results. HFM-Net generally performs better than these previous methods, but still has the problem of detail losing, especially in the areas that have been masked out by the depth confidence map, *e.g.*, the edges of the desks in the first two scenes, the leaf stalk in the third scene and the lamp in the last scene. In comparison, our method achieves better visual effects that are very close to the ground truths and preserves most of scene details without introducing artifacts.

To further validate the accuracy of our method, we provide quantitative analysis for different datasets in Table 1. The best results are highlighted in bold. As seen, in all the cases, our method ranks first among these peer-reviewed methods according to the metrics mentioned above. It is worth noticing that our method achieves a significant improvement on the metrics of angle difference, especially in the case of $t_n = 11.25°$. This further proves the benefit of normal estimation on the 2-sphere since the 2-sphere is friendly to angle differences.

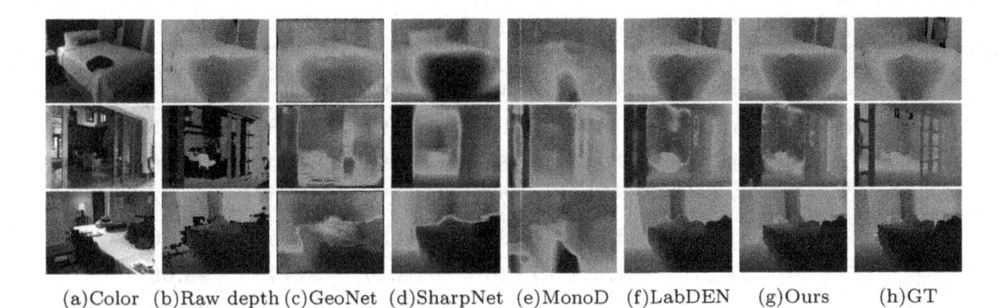

Fig. 7. Visual quality comparisons with the state-of-the-art depth estimation methods on ScanNet (the first row), Matterport3D (the second row) and NYUD-v2 (the third row) datasets

Table 1. Performance of surface normal estimation on the NYUD-v2, Matterport3D and ScanNet datasets. The last three columns are the different variants of the proposed method, *i.e.*, the model trained on $\mathbb{R}^3$, the model without CGSA (-CGSA) and the model without MFF (-MFF)

	Metrics	Zhang's	GeoNet	SharpNet	HFM-Net	Ours	$\mathbb{R}^3$	-CGSA	-MFF
NYUD-v2	mean	23.430	21.385	21.226	14.188	**12.172**	12.850	12.790	13.001
	median	14.446	12.810	14.084	6.827	**6.377**	7.222	7.595	7.193
	rmse	30.162	30.257	28.912	23.139	**19.152**	20.027	19.317	20.335
	11.25°	39.95	44.93	41.39	65.91	**69.41**	66.39	65.66	65.31
	22.5°	66.11	68.16	67.24	82.03	**85.90**	84.89	84.43	84.35
	30°	75.35	76.27	76.43	87.36	**90.40**	89.96	89.90	89.51
Matterport3D	mean	21.920	24.277	25.599	17.140	**14.687**	15.903	16.147	15.768
	median	11.039	15.975	18.319	6.483	**4.885**	5.759	6.336	6.010
	rmse	32.041	34.454	34.806	27.339	**25.308**	26.476	26.654	26.179
	11.25°	48.25	40.17	27.56	61.05	**69.74**	66.65	65.79	65.95
	22.5°	67.13	62.42	62.60	77.51	**82.55**	80.44	80.31	80.71
	30°	75.00	71.65	73.27	83.14	**86.73**	85.13	85.08	85.44
ScanNet	mean	23.306	23.289	23.977	14.590	**13.508**	14.205	14.390	14.425
	median	15.950	15.725	17.038	7.468	**6.739**	6.938	7.019	6.971
	rmse	31.371	29.902	31.974	23.638	**21.991**	23.024	22.933	23.093
	11.25°	40.43	46.41	27.95	65.65	**67.21**	65.70	65.18	65.24
	22.5°	63.08	64.04	63.91	81.21	**82.85**	81.72	81.27	81.21
	30°	71.88	76.78	75.74	86.21	**87.68**	86.75	86.38	86.29

Comparisons of Depth Estimation. As a by-product, our method also produces high-quality depth maps. We conduct comparisons with some popular depth estimation methods to verify the effectiveness and robustness of our proposed method. The methods in comparison are GeoNet [29], SharpNet [30], MonoD [9] and LabDEN [18]. Both visual comparisons in Fig. 7 and quantitative comparisons in Table 2 reveal that our method outperforms these previous methods, achieving the state-of-the-art performance in depth recovery.

Table 2. Performance of depth estimation on NYUD-v2 dataset

	Metrics	GeoNet	SharpNet	MonoD	LabDEN	Ours
NYUD-v2	rmse	0.106	0.104	0.197	0.028	**0.015**
	log10	0.121	0.150	0.198	0.030	**0.019**
	rel	0.283	0.278	0.610	0.069	**0.044**
	1.25	56.50	50.38	34.87	95.26	**97.80**
	1.25^2	84.98	75.16	58.20	98.74	**99.69**
	1.25^3	95.08	86.37	75.51	99.57	**99.96**

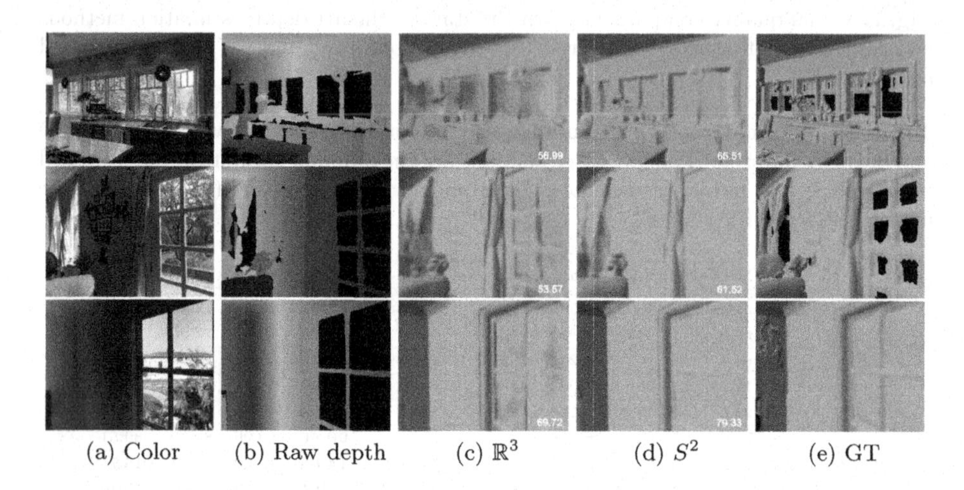

(a) Color (b) Raw depth (c) $\mathbb{R}^3$ (d) S^2 (e) GT

Fig. 8. Visual quality comparions between the $\mathbb{R}^3$ space and the S^2 space

5.3 Ablation Study

To validate the effectiveness of each module in our method, we conduct several ablation studies.

Comparisons between Different Spaces. We compare surface normal estimation on the 3D Euclidean space (*i.e.*, $\mathbb{R}^3$) and the 2-sphere space (*i.e.*, S^2) in Fig. 8. As expected, normal estimation on S^2 provides higher-quality results than directly regressing the Cartesian coordinate in the $\mathbb{R}^3$ space. It achieves an obvious improvement in accuracy in terms of angle differences. As reported in the bottom right corner of each image, our method on S^2 significantly surpasses that in $\mathbb{R}^3$ in the metric of the angle difference of $t_n = 11.25°$.

Effectiveness of the MFF Module. In Fig. 2 we show that the proposed MFF module is better than simple concatenation on fusing information from different branches. For simple concatenation, artifacts occur in the areas where the depth information is unreliable. It considers the inaccurate regions in the depth map as extra features and produces strange artifacts in these areas. The MFF module avoids this problem by not directly using the depth feature maps

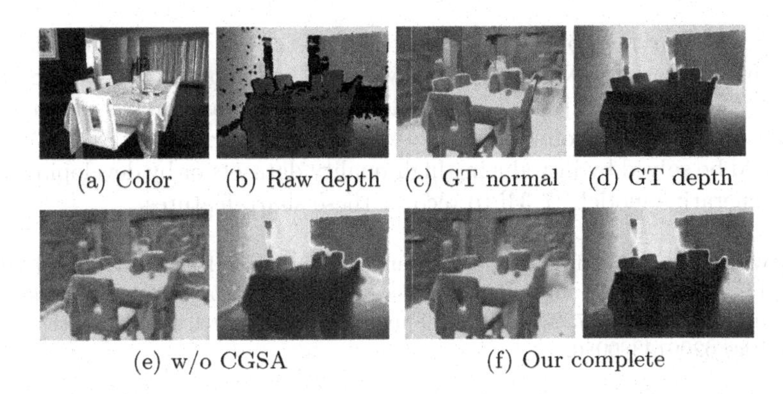

Fig. 9. Visual quality comparisons between with and w/o CGSA module

but re-weighting the normal feature maps via pixel-wise transformation based on conditioning representation, producing more stable and pleasing results.

Effectiveness of the CGSA Module. We verify the effectiveness of the CGSA module by removing it from our complete method. As shown in Fig. 9, without the depth inpainting procedure, wrong predictions appear in the large holes of the depth map. With the CGSA module enhancing depth feature map in high-level semantics, our complete method produces more plausible results in the regions where the raw depth is absent.

However, the enhanced depth values are not always reliable when our inpainting fails to find similar patches. As shown on the LCD screen of the first scene and the windows of the second scene in Fig. 2(c), these inaccurate depth values may affect the estimated normal values, leading to strange artifacts. The proposed MFF can avoid this as explained before. Nevertheless, CGSA is necessary to avoid wrong predictions in large holes of raw depth maps.

We also conduct quantitative analysis of different situations of our method on the above three datasets. The quantitative results in the last four columns of Table 1 show that our complete model consistently shows superior performance compared with other models.

6 Conclusion and Future Work

In this work, we prove the superiority of estimating surface normals on the naturally normalized 2-sphere than in the unconstrained $\mathbb{R}^3$ space. To improve the feature quality of the depth channel, we design a CGSA module to recover depth feature maps in high-level semantics. Our network fuses RGB-D features at multiple scales with the MFF modules, which organizes the two branches as a new form of hyper-network. Moreover, we design a loss function which is more suitable for the 2-sphere. Extensive experimental results verify that our method outperforms the state-of-the-art methods in providing high-quality surface normals with clearer edges and more details.

Although our method achieves the state-of-the-art performance, it suffers from some limitations. Notably, our network fails to capture some sharp details especially for distant objects, *e.g.*, the details on the wall. This is probably due to the low-quality of ground-truth normal maps in current datasets. We hope this would be solved by introducing high-quality datasets or by developing GAN-based generative models [7,34] to recover these sharp features.

Acknowledgement. The corresponding authors of this work are Jie Guo and Yanwen Guo. This research was supported by the National Natural Science Foundation of China under Grants 61772257 and the Fundamental Research Funds for the Central Universities 020914380080.

References

1. Bansal, A., Russell, B., Gupta, A.: Marr revisited: 2D–3D model alignment via surface normal prediction. In: CVPR (2016)
2. Li, B., Shen, C., Dai, Y., Van Den Hengel, A., He, M.: Depth and surface normal estimation from monocular images using regression on deep features and hierarchical CRFs. In: 2015 IEEE Conference on Computer Vision and Pattern Recognition (CVPR), pp. 1119–1127, June 2015. https://doi.org/10.1109/CVPR.2015.7298715
3. Chang, A., et al.: Matterport3D: learning from RGB-D data in indoor environments. International Conference on 3D Vision (3DV) (2017)
4. Cheng, Y., Cai, R., Li, Z., Zhao, X., Huang, K.: Locality-sensitive deconvolution networks with gated fusion for RGB-D indoor semantic segmentation. In: 2017 IEEE Conference on Computer Vision and Pattern Recognition (CVPR), pp. 1475–1483, July 2017. https://doi.org/10.1109/CVPR.2017.161
5. Zimmermann, C., Welschehold, T., Dornhege, C., Burgard, W., Brox, T.: 3D human pose estimation in RGBD images for robotic task learning. In: IEEE International Conference on Robotics and Automation (ICRA) (2018). https://lmb.informatik.uni-freiburg.de/projects/rgbd-pose3d/
6. Dai, A., Nießner, M., Zollöfer, M., Izadi, S., Theobalt, C.: Bundlefusion: real-time globally consistent 3D reconstruction using on-the-fly surface re-integration. ACM Trans. Graph. (TOG) **36**(4), 1 (2017)
7. Denton, E.L., Chintala, S., Szlam, A., Fergus, R.: Deep generative image models using a laplacian pyramid of adversarial networks. In: Cortes, C., Lawrence, N.D., Lee, D.D., Sugiyama, M., Garnett, R. (eds.) Advances in Neural Information Processing Systems 28, pp. 1486–1494. Curran Associates, Inc. (2015). http://papers.nips.cc/paper/5773-deep-generative-image-models-using-a-laplacian-pyramid-of-adversarial-networks.pdf
8. Eigen, D., Fergus, R.: Predicting depth, surface normals and semantic labels with a common multi-scale convolutional architecture. In: 2015 IEEE International Conference on Computer Vision (ICCV), pp. 2650–2658, December 2015. https://doi.org/10.1109/ICCV.2015.304
9. Godard, C., Mac Aodha, O., Brostow, G.J.: Unsupervised monocular depth estimation with left-right consistency. In: CVPR (2017)
10. Gupta, S., Girshick, R., Arbeláez, P., Malik, J.: Learning rich features from RGB-D images for object detection and segmentation. In: Fleet, D., Pajdla, T., Schiele, B., Tuytelaars, T. (eds.) ECCV 2014. LNCS, vol. 8695, pp. 345–360. Springer, Cham (2014). https://doi.org/10.1007/978-3-319-10584-0_23

11. Haefner, B., Quéau, Y., Möllenhoff, T., Cremers, D.: Fight ill-posedness with ill-posedness: single-shot variational depth super-resolution from shading. In: Proceedings of the IEEE Conference on Computer Vision and Pattern Recognition, pp. 164–174 (2018)
12. He, K., Sun, J., Tang, X.: Single image haze removal using dark channel prior. IEEE Trans. Pattern Anal. Mach. Intell. **33**(12), 2341–2353 (2011). https://doi.org/10.1109/TPAMI.2010.168
13. He, Y., Chiu, W., Keuper, M., Fritz, M.: STD2P: RGBD semantic segmentation using spatio-temporal data-driven pooling. In: 2017 IEEE Conference on Computer Vision and Pattern Recognition (CVPR), pp. 7158–7167, July 2017. https://doi.org/10.1109/CVPR.2017.757
14. Herrera, C.D., Kannala, J., Ladický, L., Heikkilä, J.: Depth map inpainting under a second-order smoothness prior. In: Kämäräinen, J.-K., Koskela, M. (eds.) SCIA 2013. LNCS, vol. 7944, pp. 555–566. Springer, Heidelberg (2013). https://doi.org/10.1007/978-3-642-38886-6_52
15. Hui, T.W., Loy, C.C., Tang, X.: Depth map super-resolution by deep multi-scale guidance. In: Proceedings of European Conference on Computer Vision (ECCV) (2016)
16. Isola, P., Zhu, J.Y., Zhou, T., Efros, A.A.: Image-to-image translation with conditional adversarial networks. In: The IEEE Conference on Computer Vision and Pattern Recognition (CVPR), July 2017
17. Izadi, S., et al.: Kinectfusion: real-time 3D reconstruction and interaction using a moving depth camera. In: UIST 2011 Proceedings of the 24th Annual ACM Symposium on User Interface Software and Technology, pp. 559–568. ACM, October 2011. https://www.microsoft.com/en-us/research/publication/kinectfusion-real-time-3d-reconstruction-and-interaction-using-a-moving-depth-camera/
18. Jeon, J., Lee, S.: Reconstruction-based pairwise depth dataset for depth image enhancement using CNN. In: Ferrari, V., Hebert, M., Sminchisescu, C., Weiss, Y. (eds.) ECCV 2018. LNCS, vol. 11220, pp. 438–454. Springer, Cham (2018). https://doi.org/10.1007/978-3-030-01270-0_26
19. Ladický, L., Zeisl, B., Pollefeys, M.: Discriminatively trained dense surface normal estimation. In: Fleet, D., Pajdla, T., Schiele, B., Tuytelaars, T. (eds.) ECCV 2014. LNCS, vol. 8693, pp. 468–484. Springer, Cham (2014). https://doi.org/10.1007/978-3-319-10602-1_31
20. Lee, S., Park, S.J., Hong, K.S.: RDFNet: RGB-D multi-level residual feature fusion for indoor semantic segmentation. In: 2017 IEEE International Conference on Computer Vision (ICCV), pp. 4990–4999 (2017)
21. Liao, S., Gavves, E., Snoek, C.G.M.: Spherical regression: learning viewpoints, surface normals and 3D rotations on n-spheres. In: The IEEE Conference on Computer Vision and Pattern Recognition (CVPR), June 2019
22. Litany, O., Bronstein, A.M., Bronstein, M.M., Makadia, A.: Deformable shape completion with graph convolutional autoencoders. In: 2018 IEEE/CVF Conference on Computer Vision and Pattern Recognition, pp. 1886–1895 (2017)
23. Liu, J., Gong, X., Liu, J.: Guided inpainting and filtering for kinect depth maps. In: Proceedings of the 21st International Conference on Pattern Recognition (ICPR 2012), pp. 2055–2058. IEEE (2012)
24. Silberman, N., Hoiem, D., Kohli, P., Fergus, R.: Indoor segmentation and support inference from RGBD images. In: Fitzgibbon, A., Lazebnik, S., Perona, P., Sato, Y., Schmid, C. (eds.) ECCV 2012. LNCS, vol. 7576, pp. 746–760. Springer, Heidelberg (2012). https://doi.org/10.1007/978-3-642-33715-4_54

25. Newcombe, R.A., Fox, D., Seitz, S.M.: Dynamicfusion: reconstruction and tracking of non-rigid scenes in real-time. In: 2015 IEEE Conference on Computer Vision and Pattern Recognition (CVPR), pp. 343–352, June 2015. https://doi.org/10.1109/CVPR.2015.7298631
26. Or-El, R., Rosman, G., Wetzler, A., Kimmel, R., Bruckstein, A.M.: RGBD-fusion: real-time high precision depth recovery. In: Proceedings of the IEEE Conference on Computer Vision and Pattern Recognition, pp. 5407–5416 (2015)
27. Park, J., Kim, H., Tai, Y.W., Brown, M.S., Kweon, I.: High quality depth map upsampling for 3D-TOF cameras. In: 2011 International Conference on Computer Vision, pp. 1623–1630, November 2011. https://doi.org/10.1109/ICCV.2011.6126423
28. Qi, X., Liao, R., Jia, J., Fidler, S., Urtasun, R.: 3D graph neural networks for RGBD semantic segmentation. In: ICCV (2017)
29. Qi, X., Liao, R., Liu, Z., Urtasun, R., Jia, J.: Geonet: geometric neural network for joint depth and surface normal estimation. In: The IEEE Conference on Computer Vision and Pattern Recognition (CVPR), June 2018
30. Ramamonjisoa, M., Lepetit, V.: Sharpnet: fast and accurate recovery of occluding contours in monocular depth estimation. In: The IEEE International Conference on Computer Vision (ICCV) Workshops (2019)
31. Ruizhongtai Qi, C., Liu, W., Wu, C., Su, H., Guibas, L.: Frustum pointnets for 3D object detection from RGB-D data, pp. 918–927, June 2018. https://doi.org/10.1109/CVPR.2018.00102
32. Simonyan, K., Zisserman, A.: Very deep convolutional networks for large-scale image recognition. arXiv preprint arXiv:1409.1556 (2014)
33. Wang, X., Fouhey, D.F., Gupta, A.: Designing deep networks for surface normal estimation. In: CVPR (2015)
34. Wu, J., Zhang, C., Xue, T., Freeman, B., Tenenbaum, J.: Learning a probabilistic latent space of object shapes via 3D generative-adversarial modeling. In: Advances in Neural Information Processing Systems, pp. 82–90 (2016)
35. Xie, J., Feris, R.S., Sun, M.: Edge-guided single depth image super resolution. IEEE Trans. Image Process. **25**(1), 428–438 (2016). https://doi.org/10.1109/TIP.2015.2501749
36. Xu, B., et al.: Adversarial monte carlo denoising with conditioned auxiliary feature. ACM Trans. Graph. (Proc. ACM SIGGRAPH Asia 2019) **38**(6), 224:1–224:12 (2019)
37. Yang, Z., Pan, J.Z., Luo, L., Zhou, X., Grauman, K., Huang, Q.: Extreme relative pose estimation for RGB-D scans via scene completion. In: The IEEE Conference on Computer Vision and Pattern Recognition (CVPR), June 2019
38. Zeng, J., et al.: Deep surface normal estimation with hierarchical RGB-D fusion. In: The IEEE Conference on Computer Vision and Pattern Recognition (CVPR), June 2019
39. Zhang, Y., Funkhouser, T.: Deep depth completion of a single RGB-D image. In: The IEEE Conference on Computer Vision and Pattern Recognition (CVPR) (2018)
40. Zhang, Y., et al.: Physically-based rendering for indoor scene understanding using convolutional neural networks. In: The IEEE Conference on Computer Vision and Pattern Recognition (CVPR) (2017)
41. Zhang, Z., Cui, Z., Xu, C., Yan, Y., Sebe, N., Yang, J.: Pattern-affinitive propagation across depth, surface normal and semantic segmentation. In: The IEEE Conference on Computer Vision and Pattern Recognition (CVPR), June 2019

AutoSTR: Efficient Backbone Search for Scene Text Recognition

Hui Zhang[1,2], Quanming Yao[2(✉)], Mingkun Yang[1], Yongchao Xu[1], and Xiang Bai[1(✉)]

[1] Department of Electronics and Information Engineering, Huazhong University of Science and Technology, Wuhan, China
[2] 4Paradigm Inc., Beijing, China
yaoquanming@4paradigm.com

Abstract. Scene text recognition (STR) is challenging due to the diversity of text instances and the complexity of scenes. However, no STR methods can adapt backbones to different diversities and complexities. In this work, inspired by the success of neural architecture search (NAS), we propose automated STR (AutoSTR), which can address the above issue by searching data-dependent backbones. Specifically, we show both choices on operations and the downsampling path are very important in the search space design of NAS. Besides, since no existing NAS algorithms can handle the spatial constraint on the path, we propose a two-step search algorithm, which decouples operations and downsampling path, for an efficient search in the given space. Experiments demonstrate that, by searching data-dependent backbones, AutoSTR can outperform the state-of-the-art approaches on standard benchmarks with much fewer FLOPS and model parameters. (Code is available at https://github.com/AutoML-4Paradigm/AutoSTR.git).

Keywords: Scene text recognition · Neural architecture search · Convolutional neural network · Automated machine learning

1 Introduction

Scene text recognition (STR) [28,49], which targets at recognizing text from natural images, has attracted great interest from both industry and academia due to its huge commercial values in a wide range of applications, such as identity authentication, digital financial system, and vehicle license plate recognition [1,9,41], etc. However, natural images are diverse, the large variations in size, fonts, background and layout all make STR still a very challenge problem [38]. A STR pipeline (Fig. 1) [38,48] usually consists of three modules: a rectification module, which aims at rectifying irregular text image to a canonical form before recognition; a feature sequence extractor, which employs a stack of convolutional layers to convert the input text image to a feature sequence; and a feature translator module, which is adopted to translate the feature sequence into a character sequence.

A. Vedaldi et al. (Eds.): ECCV 2020, LNCS 12369, pp. 751–767, 2020.
https://doi.org/10.1007/978-3-030-58586-0_44

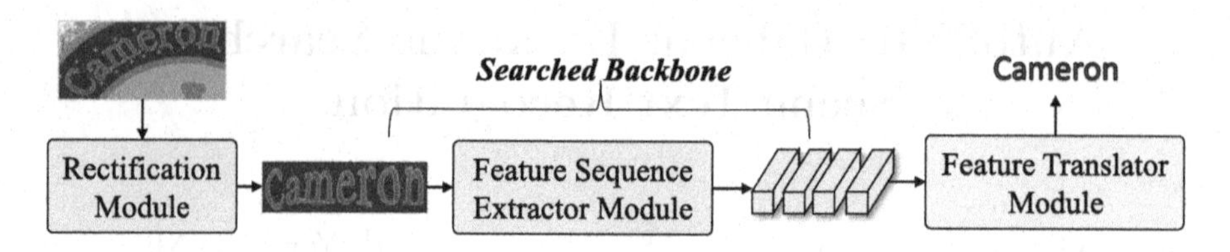

Fig. 1. Illustration of general structure of text recognition pipeline. Feature sequence extractor is searched in this paper.

In recent years, numerous methods [38,43,48] have successfully improved the text recognition accuracy via enhancing the performance of the rectification module. As for feature translator, inspired by some other sequence-to-sequence tasks, such as speech recognition [11] and machine translation [3], the translation module is also elaborately explored with both Connectionist Temporal Classification (CTC) based [36] and attention based methods [4,7,8,38,43]. In the contrast, the design of feature sequence extractor is relatively fewer explored. How to design a better feature sequence extractor has not been well discussed in the STR literature. However, text recognition performance can be greatly affected by feature sequence extractor. For example, the authors [38] can obtain significant performance gains by simply replacing feature extractor from vgg [39] to ResNet [14]. Furthermore, the feature sequence extractor bears heavy calculation and storage burden [21,43]. Thus, no matter for effectiveness or efficiency, the architecture of feature sequence extractor should be paid more attention to.

Besides, neural architecture search (NAS) [10,46] has also made a great success in designing data-dependent network architectures, of which the performances exceed the architectures crafted by human experts in many computer vision tasks, e.g., image classification [25,32], semantic segmentation [24] and object detection [6]. Thus, rather than adopt an off-the-shelf feature extractor (like ResNet [14]) from other tasks, a data-dependent architecture should be redesigned for a better text recognition performance.

In this paper, we present the first work, i.e., AutoSTR, on searching feature sequence extractor (backbone) for STR. First we design a domain-specific search space for STR, which contains both choices on operation for every convolution layer and constraints feature downsampling path. Since no existing NAS algorithms can handle the path constraint efficiently, we propose a novel two-step search pipeline, which decouples operation and downsampling path search. By optimizing the recognition loss with complexity regularization, we achieve a trade off between model complexity and recognition accuracy. Elaborate experiments demonstrate that, given a general text recognition pipeline, the searched sequence feature extractor can achieve state-of-the-art results with fewer FLOPS and parameters. The main contributions of this paper are summarized as follows:

- We discover that the architecture of the feature extractor, which is of great importance to STR, has not been well explored in the literature. This motivates

us to design a data-dependent backbone to boost text recognition performance, which is also the first attempt to introduce NAS into STR.
- We introduce a domain-specific search space for STR, which contains choices for downsampling path and operations. Then we propose a new search algorithm, which decouples operations and downsampling path for an efficient search in the space. We further incorporate an extra regularizer into the searching process, which helps effectively trade off the recognition accuracy with the model size.
- Finally, extensive experiments are carried on various benchmark datasets. Results demonstrate that, AutoSTR can discover data-dependent backbones, and outperform the state-of-the-art approaches with much fewer FLOPS and model parameters.

2 Related Works

2.1 Scene Text Recognition (STR)

As in Sect. 1, the general pipline (Fig. 1) of sequence-based STR methods [38,48], where a rectification module, a feature sequence extractor module and a feature translator module are involved. Currently, most works focus on improving rectification module or feature translator. Shi *et al.* [37,38] first introduce spatial transform network (STN) [17] for rectifying irregular text to a canonical form before recognition. Since then, [30,43,48] further push it forward and make the rectification module become a plug-in part. As for feature translator, CTC based and attention based decoder dominate this landscape for a long time [4,36,38]. Nevertheless, as another indispensable part, the feature sequence extractor has not been well discussed in the recent STR literature. As shown in Table 1, although different feature extractors are used in [36,43], they just follow the architecture proposed for other fundamental tasks, like image classification. But recently, some authors observe that the architectures of feature extractor have gaps between different tasks, like semantic segmentation [24] and object detection [6]. Therefore, these popular feature extractor might not be perfectly suitable for STR, and it is important to search a data-dependent architecture.

2.2 Neural Architecture Search (NAS)

Generally, there are two important aspects in neural architecture search (NAS) [10,46], i.e., the *search space* and *search algorithm*. The search space defines possibilities of all candidate architectures, and the search algorithm attempts to find suitable architectures from the designed space. Specifically, a proper space should explore domain information and cover good candidates, which are designed by humans. Besides, it cannot be too large, otherwise, no algorithms can efficiently find good architectures. Thus, usually designing a better search space by domain knowledge can make the searching process easier.

Table 1. Comparison with example text recognition and NAS methods that contain a downsampling path search. "feat.seq.extractor", "DS", "seq. rec.", "cls.","seg." and "grad. desc." means feature sequence extractor, downsampling, sequence recognition, classification, segmentation and gradient descent algorithm respectively.

	Model	feat. seq. extractor		Task	Search algorithm
		Operation	DS path		
Hand-designed	CRNN [36]	vgg	fixed	seq. rec.	—
	ASTER [38]	residual	fixed	seq. rec.	—
	SCRN [43]	residual	fixed	seq. rec.	—
NAS	DARTS [25]	searched	fixed	cls.	grad. desc.
	AutoDeepLab [24]	searched	one-dim	seg.	grad. desc.
	AutoSTR	searched	two-dim	seq. rec.	two-step

Classically, network architectures are treated as hyper-parameters, which are optimized by an algorithm like reinforcement learning [50] and evolution algorithms [42]. These methods are expensive since they need to train each sampled architecture fully. Currently one-shot neural architecture search (OAS) [5,25,32,45] have significantly reduced search time by sharing the network weights during the search progress. Specifically, these methods encode the whole search space into a supernet, which consists of all candidate architectures. Then, instead of training independent weights for each architecture, distinctive architecture inherits weights from the supernet. In this way, architectures can be searched by training the supernet once, which makes NAS much faster. However, as the supernet is a representation of the search space, a proper design of it is a non-trivial task. Without careful exploitation of domain information, OAS methods may not be even better than random search [22,35].

3 Methodology

3.1 Problem Formulation

As the input of STR is natural images, feature sequence extractor is constructed with convolutional layers. A convolutional layer $\mathcal{C}$ can be defined as $\mathcal{C}(X; o, s^h, s^w)$ (hereinafter referred to as $\mathcal{C}(X)$), where X is input tensor, o denotes convolution type, (e.g., 3×3 convolution, 5×5 depth-wise separable convolution), s^h and s^w represent its stride in height and width direction respectively. Therefore, a backbone $\mathcal{N}$ can be regarded as a stack of L convolution layers, i.e., $\mathcal{C}_L(\dots\mathcal{C}_2(\mathcal{C}_1(X)))$. After been processed by $\mathcal{N}$, X with input size (H, W) will be mapped into a feature map with a fixed size output to the feature translator module.

To determine a data-dependent backbone for STR, we need to search proper architectures, which is controlled by $\mathcal{S} \equiv \{(s_i^h, s_i^w)\}_{i=1}^L$ (for strides) and $\mathcal{O} \equiv \{o_i\}_{i=1}^L$ (for convolution type). Let $\mathcal{L}_{\text{tra}}$ measure the loss of the network on training dataset and $\mathcal{A}_{\text{val}}$ measure the quality of architecture on the validation set. We formulate the AutoSTR problem as:

$$\min_{\mathcal{S},\mathcal{O}} \mathcal{A}_{\text{val}}(\mathcal{N}(\boldsymbol{w}^*, \mathcal{S}, \mathcal{O})), \quad \text{s.t.} \quad \begin{cases} \boldsymbol{w}^* = \arg\min_{\boldsymbol{w}} \mathcal{L}_{\text{tra}}(\mathcal{N}(\boldsymbol{w}, \mathcal{S}, \mathcal{O})) \\ \mathcal{S} \in \mathcal{P} \end{cases}, \quad (1)$$

where $\mathcal{S}$ and $\mathcal{O}$ are upper-level variables representing architectures, and $\boldsymbol{w}$ as lower-level variable and $\mathcal{P}$ as constraint, i.e.,

$$\mathcal{P} \equiv \{\{(s_i^h, s_i^w)\}_{i=1}^L \mid {}^{H}/{\prod_{i=1}^L s_i^h} = c_1, {}^{W}/{\prod_{i=1}^L s_i^w} = c_2\}.$$

Specifically, c_1 and c_2 are two application dependent constants; and the constraint $\mathcal{P}$ is to ensure the output size of the searched backbone is aligned with the input size of the subsequent feature translator module.

3.2 Search Space

As explained in Sect. 2.2, the search space design is a key for NAS. Here, we design a two-level hierarchical search space for the AutoSTR problem in (1), i.e., the downsampling-path level and operation level, to represent the selection range of $\mathcal{S}$ and $\mathcal{O}$ respectively.

Downsampling-Path Level Search Space. Since the characters are horizontally placed in the rectified text image, following [36], we use CNN to exact a feature sequence to represent them. To reserve more discriminable features of the characters in compact text or in narrow shapes, a common way is to keep collapsing along the height axis until it reduces to 1, but compress less along the width axis to ensure that the length of final sequence is greater than the length of characters [36,38,48]. Specifically, the current mainstream methods are using the feature extractor proposed in ASTER [38]. The height of the input text image is unified to a fixed size, like 32. And to reserve more resolution along the horizontal axis in order to distinguish neighbor characters [4,36,38,43,48], the strides hyper-parameter s only selected from $\{(2,2),(2,1),(1,1)\}$, where $(2,2)$ appears twice and $(2,1)$ appears three times in the whole downsampling path to satisfy $\mathcal{P}$ with $\prod_{i=1}^L s_i^h = 32$ and $\prod_{i=1}^L s_i^w = 4$. Finally, a text image with size $(32, W)$ is mapped into a feature sequence with a length of ${}^{W}/{4}$.

The downsampling-path level search space is illustrated in Fig. 2. Our goal is to find an optimal path in this 3D-mesh, and the downsampling strategy in [38] is a spatial case in this search space. To the best of our knowledge, there are no suitable methods to search in such a constrained search space.

Operation Level Search Space. The convolutional layers of current text recognition networks usually share the same operation [7,38,48], such as 3×3 residual convolution. Instead of setting each o_i to be a fixed operation, we select a operator for each convolutional layer from a *choice block* with C parallel

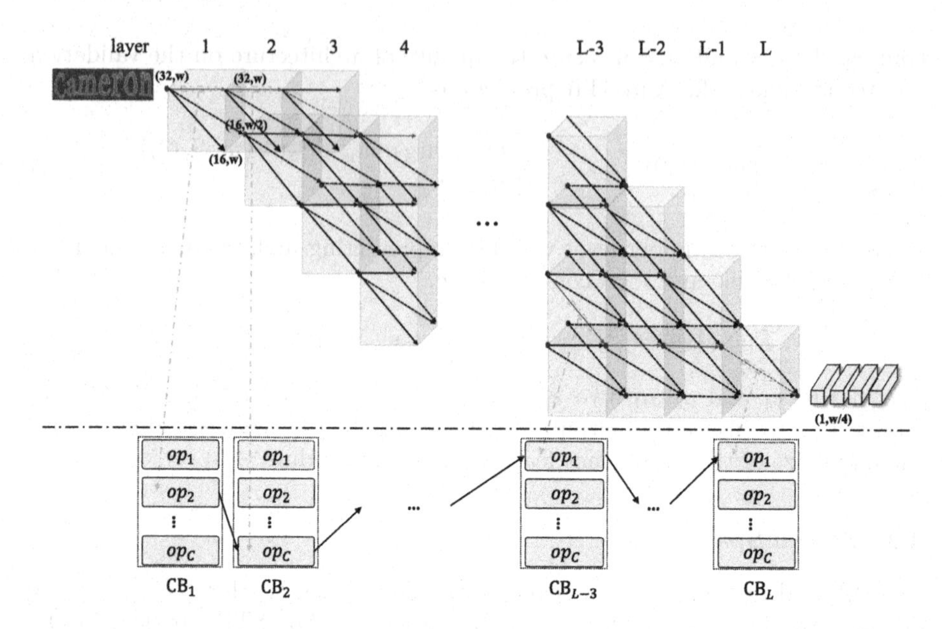

Fig. 2. Search space illustration. Top: a 3D-mesh representation of the downsampling-path level search space. Bottom: operation level search space, where "CB_i" donates the ith *choice block* and each block allows C choices for operations

operators, as illustrated in the bottom of Fig. 2. Then, we can obtain a deep convolutional network by stacking these *choice blocks*. Our choices on operations are inspired by MobileNetV2 [34], which uses lightweight depthwise convolutions to save FLOPS and the number of parameters. Thus, we build a set of mobile inverted bottleneck convolution layers (MBConv) with various kernel sizes $k \in \{3, 5\}$, expansion factors $e \in \{1, 3, 6\}$ and a skip-connect layer.

Table 2. Basic operation (i.e., op_i's) in the choice block (the bottom of Fig. 2), where "k" denotes kernel size and "e" denotes expansion factors.

MBConv(k:3 × 3,e:1)	MBConv(k:3 × 3,e:3)	MBConv(k:3 × 3,e:6)	MBConv(k:5 × 5,e:1)
MBConv(k:5 × 5,e:3)	MBConv(k:5 × 5,e:6)	Skip-Connect	

Complexity of the Search Space. When $L = 15$, there are 30030 possible downsampling paths in search space illustrated in Fig. 2. On operation level, if we allow it to be one of the seven operations in Table 2, then it leads to a total number $30030 \times 7^{15} \simeq 1.43 \times 10^{17}$ possible architectures for the backbone, which is prohibitively large. In the sequel, we show a two-step algorithm which can efficiently search through the space.

3.3 Search Algorithm

Since the combination of $\mathcal{S}$ and $\mathcal{O}$ generates a very space, directly optimizing the problem in (1) is a huge challenge. Motivated by the empirical observation that choices of the downsampling path and the operations are almost orthogonal with each other (see Sect. 4.6), in this section, we decouple the process of searching $\mathcal{S}$ and $\mathcal{O}$ into two steps for the backbone search. Specifically, in the first step, we fix the operation $\mathcal{O}$ as the 3×3 residual convolution (denote as $\hat{\mathcal{O}}$) and search for the downsampling path. In the second step, we search a convolution operation for every layer in the path.

Step 1: Search Downsampling Path. Let $\mathcal{L}_{\text{rec}}(\mathcal{N};\mathcal{D})$ measure the sequence cross-entropy loss [38] with predictions from a network $\mathcal{N}$ on a dataset $\mathcal{D}$, and $\mathcal{D}_{\text{tra}}$ (resp. $\mathcal{D}_{\text{val}}$) denotes the training (resp. validation) dataset. In this step, we search downsampling path when operations are fixed, and (1) becomes

$$\mathcal{S}^* = \arg\min_{\mathcal{S}\in\mathcal{P}} \mathcal{L}_{\text{rec}}(\mathcal{N}(\boldsymbol{w}^*, \mathcal{S}, \hat{\mathcal{O}}); \mathcal{D}_{\text{val}}), \tag{2}$$
$$\text{s.t. } \boldsymbol{w}^* = \arg\min_{\boldsymbol{w}} \mathcal{L}_{\text{rec}}(\mathcal{N}(\boldsymbol{w}, \mathcal{S}, \hat{\mathcal{O}}); \mathcal{D}_{\text{tra}}).$$

As in Sect. 3.2, downsampling path can only have two $(2, 2)$ and three $(2, 1)$ to satisfy $\mathcal{P}$, which are denoted as downsampling type A and B respectively. Note that some current NAS methods [25,45] use the same number of layers per convolution stage and have achieved good results. Using this reasonable prior knowledge, for a network with $L = 15$, we set downsampling at layers 1, 4, 7, 10, 13, and equally separate the network into five stages. Then the downsampling strategies can be divided into 10 types of typical paths: AABBB, ABABB, ABBAB, ABBBA, BAABB, BABAB, BABBA, BBAAB, BBABA, and BBBAA. We can do a small grid search in these typical paths to find a good path that is close to $\mathcal{S}^*$. Then by learning the skip-connect in searching step 2, we can reduce the number of layers for each convolutional stage.

Step 2: Search Operations. First, inspired by recent NAS methods [5,12] we associate the operation op_i^l at the lth layer with a hyperparameter α_i^l, which relaxes the categorical choice of a particular operation in the *choice block* (Fig. 2) to be continuous. Since op_i^l's influence both complexity and performance of the backbone [5,34,45], we introduce a regularizer, i.e.,

$$r(\boldsymbol{\alpha}) = [\log(\sum_{l=1}^{L}\sum_{j=1}^{C} \text{FLOPS}(\text{op}_j^l)) \cdot \alpha_j^l / \log \mathcal{G}]^{\beta}, \tag{3}$$

on the architecture $\boldsymbol{\alpha}$, where $\beta > 0$ and $\mathcal{G} > 1$ be application-specific constants. Then, (1) is transformed as

$$\boldsymbol{\alpha}^* = \arg\min_{\boldsymbol{\alpha}} r(\boldsymbol{\alpha}) \cdot \mathcal{L}_{\text{rec}}(\mathcal{N}(\boldsymbol{w}^*, \mathcal{S}^*, \boldsymbol{\alpha}), \mathcal{D}_{\text{val}}), \tag{4}$$
$$\text{s.t. } \boldsymbol{w}^* = \arg\min_{\boldsymbol{w}} \mathcal{L}_{\text{rec}}(\mathcal{N}(\boldsymbol{w}, \mathcal{S}^*, \boldsymbol{\alpha}); \mathcal{D}_{\text{tra}}).$$

We can see that when $\mathcal{L}_{\text{rec}}$ is the same, $r(\boldsymbol{\alpha})$ make architectures with less FLOPS favored. Thus, the regularizer can effectively trade off the accuracy with the model size. Finally, (4) can be solved by many existing NAS algorithms, such as DARTS [25], ProxylessNAS [5] and NASP [45]. Here, we adopt ProxylessNAS [5] as it consumes the GPU memory less.

3.4 Comparison with Other NAS Works

The search constraint $\mathcal{P}$ on the downsampling path for is new to NAS, which is specific to STR. Unfortunately, none of existing NAS algorithms can effectively deal with the AutoSTR problem in (1) here. It is the proposed search space (Sect. 3.2) and two-stage search algorithm (Sect. 3.3) that make NAS for STR possible (see Table 1). We notice that AutoDeepLab [24] also considers the downsampling path. However, it targets at in segmentation task and its unique decoding method is not suitable for our search space.

4 Experiments

4.1 Datasets

We evaluate our searched architecture on the following benchmarks that are designed for general STR. Note that the images in the first four datasets are regular while others are irregular.

- IIIT 5K-Words (**IIIT5K**) [31]: it contains 5,000 cropped word images for STR, 2,000 for validation and other 3,000 images for testing; all images are collected from the web.
- Street View Text (**SVT**) [40]: It is harvested from Google Street View. Its test set contains 647 word images, which exhibits high variability and often has low resolution. Its validation set contains 257 word images.
- ICDAR 2003 (**IC03**) [29]: it contains 251 full scene text images. Following [40], we discard the test images containing non-alphanumeric characters or have less than three characters. The resulting dataset contains 867 cropped images for testing and 1,327 images as validation dataset.
- ICDAR 2013 (**IC13**) [19]: it contains 1,015 cropped text images as test set and 844 images as validation set.
- ICDAR 2015 (**IC15**): it is the 4th Challenge in the ICDAR 2015 Robust Reading Competition [18], which is collected via Google Glasses without careful positioning and focusing. As a result, the dataset consists of a large proportion of blurred and multi-oriented images. It contains 1811 cropped images for testing and 4468 cropped text images as validation set.
- SVT-Perspective (**SVTP**): it is proposed in [33] which targets for evaluating the performance of recognizing perspective text and contains 645 test images. Samples in the dataset are selected from side-view images in Google Street View. Consequently, a large proportion of images in the datasets are heavily deformed by perspective distortion.

4.2 Implementation Details

The proposed method is implemented in PyTorch. We adopt ADADELTA [47] with default hyper-parameters (rho = 0.9, eps = 1e−6, weight decay=0) to minimize the objective function. When searching the downsampling path, we train 5 epochs for each typical path where the convolutional layers are equipped with default 3×3 convolution. In the operation searching step, we warm-up weights of each choice block by uniformly selecting operations for one epoch. And then use the method proposed in Sect. 3.3 to jointly train architecture parameters and weight parameters for two epochs. In the evaluation step, the searched architectures are trained on Synth90K [15] and SynthText [13] from scratch without finetuning on any real datasets. All models are trained on 8 NVIDIA 2080 graphics cards. Details of each module in Fig. 1 are as follows:

Rectification Module. Following [38], we use Spatial Transformer Network (STN) to rectify the input text image. The STN contains three parts: (1) Localization Network, which consists of six 3×3 convolutional layers and two fully connected layers. In this phase, 20 control points are predicted. Before fed into the localization network, the input image is resized to 32×64. (2) Grid Generator, which yields a sampling grid according to the control points. (3) Sampler, which generates a rectified text image with the sampling grid. The sampler produces a rectified image of size 32×100 from the input image of size 64×256 and sends it to the subsequent sequence feature extractor.

Feature Sequence Extractor Module. The feature sequence extractor dynamically changes during the search phase. For every test dataset, an architecture is searched. During the search phase, Synth90K [15] (90k) and SynthText [13] (ST) are used for training. The validation set of each target dataset is considered as the search validation set and is used to optimize the network structure. In order to prevent overfitting caused by too small validation set, we add extra COCO-Text training set to the search validation set. To control the complexity of the backbone, we only search those with less than 450M FLOPS which is close to that of the backbone used in ASTER. The maximum depth of our search network is 16 blocks, including a stem with 3×3 residual convolution and 15 choice blocks. It has a total of 5 convolutional stages, each stage has 3 choice blocks, the number filters of each stage are respectively 32, 64, 128, 256, 512 respectively.

Feature Translator Module. A common attention based decoder [38] is employed to translate the feature sequence to a character sequence. For simplicity, only left-to-right decoder is used. The module contains two layers of Bidirectional LSTM (BiLSTM) encoder (512 input units, 256 hidden units) and an attention based GRU Cell decoder (1 layer, 512 input units, 512 hidden units, 512 attention units) to model variable-length character sequence. The decoder yields 95 character categories at each step, including digits, upper-case and lower-case letters, 32 ASCII punctuation marks and an end-of-sequence symbol (EOS).

Table 3. Performance comparison on regular and irregular scene text datasets. "ST", "90k" are the training data of SynthText [27], Synth90k [15], and "extra" means extra real or synthetic data, respectively. The methods marked with "†" use the character box annotations. "ASTER (ours)" is the reproduced ASTER baseline method, whose difference is that only left-to-right translator is equipped.

Methods	Data	Regular				Irregular	
		IIIT5K	SVT	IC03	IC13	SVTP	IC15
Jaderberg et al. [16]	90k	–	80.7	93.1	90.8	–	–
CRNN [36]	90k	81.2	82.7	91.9	89.6	–	–
RARE [37]	90k	81.9	81.9	90.1	88.6	71.8	–
R^2AM [20]	90k	78.4	80.7	88.7	90.0	–	–
Yang et al. [44]	90k	–	–	–	–	75.8	–
Char-net [26]	90k	83.6	84.4	91.5	90.8	73.5	–
Liu et al. [27]	90k	89.4	87.1	94.7	94.0	73.9	–
AON [8]	ST + 90k	87.0	82.8	91.5	-	73.0	68.2
FAN† [7]	ST + 90k	87.4	85.9	94.2	93.3	–	70.6
EP [4]	ST + 90k	88.3	87.5	94.6	**94.4**	–	73.9
SAR [21]	ST + 90k	91.5	84.5	–	91.0	76.4	69.2
CA-FCN† [23]	ST + extra	92.0	86.4	–	91.5	–	–
ESIR [48]	ST + 90k	93.3	90.2	–	91.3	79.6	76.9
SCRN† [43]	ST + 90k	94.4	88.9	**95.0**	93.9	80.8	78.7
ASTER [38]	ST + 90k	93.4	89.5	94.5	91.8	78.5	76.1
ASTER (ours)	ST + 90k	93.3	89.0	92.4	91.5	79.7	78.5
AutoSTR	ST + 90k	**94.7**	**90.9**	93.3	94.2	**81.7**	**81.8**

4.3 Comparison with State of the Art

Recognition Accuracy. Following [2], all related works are compared in the unconstrained-lexicon setting. Equipped with the searched backbone, the whole framework is compared with other state-of-the-art methods, as shown in Table 3. AutoSTR achieves the best performance in IIIT5K, SVT, IC15, SVTP and get comparable results in IC03, IC13. It is worth noting that AutoSTR outperforms ASTER (ours) on IIIT5K, SVT, IC03, IC13, SVTP, IC15 by 1.4%, 1.9%, 0.9%, 2.7%, 2%, 3.3%, which domonstrate the effectiveness of AutoSTR. Although SCRN can achieve comparable performance with AutoSTR, its rectification module requires extra character-level annotations for more precise rectification. As a plug-in part, AutoSTR is expected to further improve the performance while been equipped with the rectification module of SCRN.

Memory and FLOPS. The comparison on FLOPS and memory size are in Fig. 3. We can see that, compared with the state-of-the-art methods, like SAR [21], CA-FCN [23], ESIR [48], SCRN [43], ASTER [38], the searched archi-

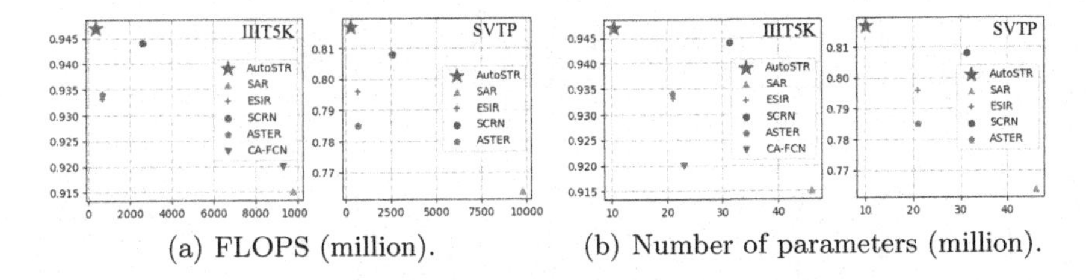

(a) FLOPS (million). (b) Number of parameters (million).

Fig. 3. Accuracy versus computational complexity and memory on IIIT5K and SVTP. Points closer to the top-left are better. Only methods with good performance, i.e., with accuracy greater than 90% on IIIT5K, are plotted.

tecture cost much less in FLOPS and memory size. Thus, AutoSTR is much more effective in mobile setting, where FLOPS and model size is limited.

4.4 Case Study: Searched Backbones and Discussion

Dataset-Dependency. In Fig. 4, we illustrate architectures of searched feature extractors on each test dataset to give some insights about the design of backbones. We observe some interesting patterns. The shallower convolutional stages (e.g., 1, 2) of the network, prefer larger MBConv operations (e.g., MBConv(k:5,e:6)) and do not have skip-connect layer. But in the deeper convolutional stages (e.g., 3, 4, 5), smaller MBConvs are employed and skip connections are learned to reduce the number of convolutional layers. Especially in the last convolutional stage, only one convolutional layer exists. The observed phenomenon is consistent with some manually designed network architecture, such as SCRN [43]. Specifically, in its first two stages, ResNet50 is used to extract features, and in the later stages, only a few convolutional layers are attached to quickly downsample feature map to generate feature sequences in the horizontal direction. This phenomenon may inspire us to design better text image feature extractor in the future.

Compactness. We compare our searched architectures with All MBConv(k:5,e:6) baseline model which choices blocks with the maximum number of parameters in each layer and uses ABABB downsampling strategy as shown in the right of Fig. 4. Comparing architectures in Fig. 4, we can see that our searched structure has less FLOPS and parameters, while maintain better accuracy, as shown in Table 4. Our searched architectures use less FLOPS and parameters, but exceeds the accuracy of the baseline model, which explains that the maximum number of parameters model (All MBConv(k:5,e:6) baseline) have lots of redundancy parameters, AutoSTR can remove some redundant layers and optimize the network structure.

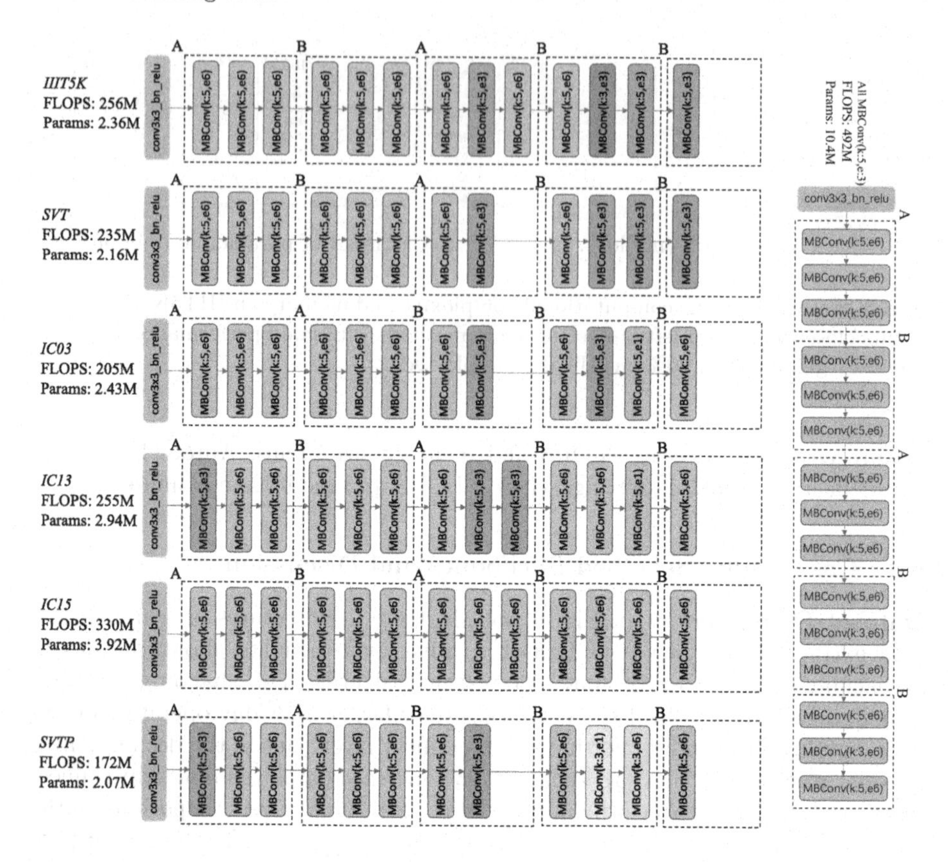

Fig. 4. Left: the searched architectures for AutoSTR in Table 3. Right: all MBConv(k:5,e:6) baseline architecture.

Table 4. Accuracies compared with the baseline of All MBConv(k:5,e:6).

Methods	IIIT5K	SVT	IC13	IC15
All MBConv(k:5,e:6)	94.5	90.4	92.3	81.1
AutoSTR	94.7	90.9	94.2	81.8

4.5 Comparison with Other NAS Approaches

Search Algorithm Comparison. From recent benchmarks and surveys [6, 22,35], the random search algorithm is a very strong baseline, we perform a random architecture search from the proposed search space. We choice 10 random architectures, train them from scratch, then test these random architectures on IIIT5K dataset. Random search takes about 15 × 4GPU days, while AutoSTR only costs 1.7 × 4GPU days in downsampling-path search step and 0.5 × 4GPU days in operation search step. The discovered architecture outperforms random

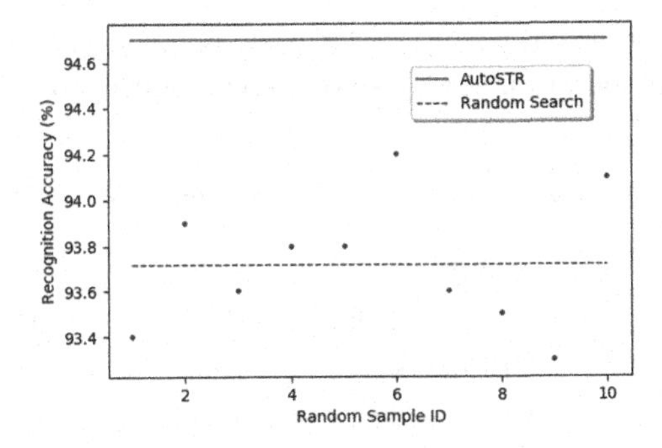

Fig. 5. Comparison to random search on IIIT5K dataset.

architectures in IIIT5K dataset by 0.5%–1.4% as in Fig. 5, which demonstrates AutoSTR is more effectiveness and efficiency.

Reusing Other Searched Architectures. NAS has been extensive researched on image classification task [5,25,45] and segmentation task [24], et al. We study whether those searched architectures are applicable here or not. Since the constraint in (1) cannot be directly satisfied, we manually tune the searched cell (from DARTS [25] and AutoDeepLab [24]) in ASTER. As can be seen from Table 5, the performance of the backbone from DARTS and AutoDeepLab is much worse. This further demonstrates direct reusing architecture searched from other tasks is not good.

Table 5. Comparison with DARTS and AutoDeepLab.

Backbones	IIIT5K	SVT	SVT	IC13	SVTP	IC15
ASTER [38]	93.3	89.0	92.4	91.5	79.7	78.5
DARTS [25]	90.6	83.9	91.3	88.3	76.1	73.5
AutoDeepLab [24]	93.0	87.2	91.8	91.0	77.5	76.6
AutoSTR	**94.7**	**90.9**	**93.3**	**94.2**	**81.7**	**81.8**

4.6 Ablation Study

Downsampling Path. In our proposed method, we decouple the searching problem in (1) into a two-step optimization problem as (2) and (4). This is based on an empirical assumption that a better feature downsampling path can provide a better startup for the operation searching problem, thus can get better architectures easier. We use two typical strategies in our downsampling path search space, i.e., AABBB and ABABB to search operations on IIIT5K datasets. As shown in Table 6, the optimal downsampling path will not be affected by the default operation (i.e. 3 $\times$ 3 residual convolution, MBConv(k:3,e:1)). Besides, a better downsampling strategy (i.e. ABABB) helps AutoSTR to find a better architecture in the operation search step, which confirms our assumption.

Table 6. Comparison of different downsampling path on IIIT5K dataset.

Downsample path	Default conv	Search step 1	Search step 2
AABBB	3×3 residual conv	92.5	93.9
	MBConv(k:3,e:1)	91.3	
ABABB	3×3 residual conv	**93.1**	**94.7**
	MBConv(k:3,e:1)	**92.0**	

Impact of the Regularizer. In (3), we introduce FLOPS into objective function as a regularization term. By adjusting β, we can achieve the trade off between the calculation complexity and accuracy, as shown in Table 7.

Table 7. Impact of the regularization on IIIT5K dataset.

β	0.0	0.3	0.6	0.9
Accuracy (%)	94.6	94.5	94.7	93.5
FLOPS (M)	319	298	256	149
Params (M)	3.82	3.40	2.36	1.32

5 Conclusion

In this paper, we propose to use neural architecture search technology finding data-dependent sequence feature extraction, i.e., the backbone, for the scene text recognition (STR) task. We first design a novel search space for the STR problem, which fully explore the prior from such a domain. Then, we propose a new two-step algorithm, which can efficiently search the feature downsampling path and operations separately. Experiments demonstrate that our searched backbone can greatly improve the capability of the text recognition pipeline and achieve the state-of-the-art results on STR benchmarks. As for the future work, we would like to extend the search algorithm to the feature translator.

Acknowledgments. The work is performed when H. Zhang was an intern in 4Paradigm Inc. mentored by Dr. Q. Yao. This work was partially supported by National Key R&D Program of China (No. 2018YFB1004600), to Dr. Xiang Bai by the National Program for Support of Top-notch Young Professionals and the Program for HUST Academic Frontier Youth Team 2017QYTD08.

References

1. Almazán, J., Gordo, A., Fornés, A., Valveny, E.: Word spotting and recognition with embedded attributes. IEEE Trans. Pattern Anal. Mach. Intell. **36**, 2552–2566 (2014)
2. Baek, J., et al.: What is wrong with scene text recognition model comparisons? Dataset and model analysis. In: International Conference on Computer Vision (2019)
3. Bahdanau, D., Cho, K., Bengio, Y.: Neural machine translation by jointly learning to align and translate. Technical report arXiv preprint arXiv:1409.0473 (2014)
4. Bai, F., Cheng, Z., Niu, Y., Pu, S., Zhou, S.: Edit probability for scene text recognition. In: IEEE Conference on Computer Vision and Pattern Recognition (2018)
5. Cai, H., Zhu, L., Han, S.: ProxylessNAS: direct neural architecture search on target task and hardware. In: International Conference on Learning Representations (2019)
6. Chen, Y., Yang, T., Zhang, X., Meng, G., Xiao, X., Sun, J.: DetNAS: backbone search for object detection. In: Advances in Neural Information Processing Systems (2019)
7. Cheng, Z., Bai, F., Xu, Y., Zheng, G., Pu, S., Zhou, S.: Focusing attention: towards accurate text recognition in natural images. In: IEEE Conference on Computer Vision and Pattern Recognition (2017)
8. Cheng, Z., Xu, Y., Bai, F., Niu, Y., Pu, S., Zhou, S.: Aon: towards arbitrarily-oriented text recognition. In: IEEE Conference on Computer Vision and Pattern Recognition (2018)
9. Dutta, K., Mathew, M., Krishnan, P., Jawahar, C.: Localizing and recognizing text in lecture videos. In: International Conference on Frontiers in Handwriting Recognition (2018)
10. Elsken, T., Metzen, J.H., Hutter, F.: Neural architecture search: a survey. J. Mach. Learn. Res. (2019)
11. Graves, A., Fernández, S., Gomez, F., Schmidhuber, J.: Connectionist temporal classification: labelling unsegmented sequence data with recurrent neural networks. In: International Conference on Machine Learning (2006)
12. Guo, Z., et al.: Single path one-shot neural architecture search with uniform sampling. Technical report arXiv preprint arXiv:1904.00420 (2019)
13. Gupta, A., Vedaldi, A., Zisserman, A.: Synthetic data for text localisation in natural images. In: IEEE Conference on Computer Vision and Pattern Recognition, pp. 2315–2324 (2016)
14. He, K., Zhang, X., Ren, S., Sun, J.: Deep residual learning for image recognition. In: IEEE Conference on Computer Vision and Pattern Recognition (2016)
15. Jaderberg, M., Simonyan, K., Vedaldi, A., Zisserman, A.: Synthetic data and artificial neural networks for natural scene text recognition. Technical report arXiv preprint arXiv:1406.2227 (2014)
16. Jaderberg, M., Simonyan, K., Vedaldi, A., Zisserman, A.: Reading text in the wild with convolutional neural networks. Int. J. Comput. Vis. **116**, 1–20 (2016). https://doi.org/10.1007/s11263-015-0823-z
17. Jaderberg, M., Simonyan, K., Zisserman, A., et al.: Spatial transformer networks. In: Advances in Neural Information Processing Systems (2015)
18. Karatzas, D., et al.: ICDAR 2015 competition on robust reading. In: International Conference on Document Analysis and Recognition (2015)

19. Karatzas, D., et al.: ICDAR 2013 robust reading competition. In: International Conference on Document Analysis and Recognition (2013)
20. Lee, C.Y., Osindero, S.: Recursive recurrent nets with attention modeling for OCR in the wild. In: IEEE Conference on Computer Vision and Pattern Recognition (2016)
21. Li, H., Wang, P., Shen, C., Zhang, G.: Show, attend and read: a simple and strong baseline for irregular text recognition. In: AAAI Conference on Artificial Intelligence (2019)
22. Li, L., Talwalkar, A.: Random search and reproducibility for neural architecture search. In: Uncertainty in Artificial Intelligence (2019)
23. Liao, M., et al.: Scene text recognition from two-dimensional perspective. In: AAAI Conference on Artificial Intelligence (2019)
24. Liu, C., et al.: Auto-deeplab: hierarchical neural architecture search for semantic image segmentation. In: IEEE Conference on Computer Vision and Pattern Recognition (2019)
25. Liu, H., Simonyan, K., Yang, Y.: DARTS: differentiable architecture search. In: International Conference on Learning Representations (2019)
26. Liu, W., Chen, C., Wong, K.Y.K.: Char-net: a character-aware neural network for distorted scene text recognition. In: AAAI Conference on Artificial Intelligence (2018)
27. Liu, Y., Wang, Z., Jin, H., Wassell, I.: Synthetically supervised feature learning for scene text recognition. In: European Conference on Computer Vision (2018)
28. Long, S., He, X., Yao, C.: Scene text detection and recognition: the deep learning era. Technical report arXiv preprint arXiv:1811.04256 (2018)
29. Lucas, S.M., et al.: ICDAR 2003 robust reading competitions: entries, results, and future directions. Int. J. Doc. Anal. Recogn. **7**, 105–122 (2005). https://doi.org/10.1007/s10032-004-0134-3
30. Luo, C., Jin, L., Sun, Z.: MORAN: a multi-object rectified attention network for scene text recognition. Pattern Recogn. **90**, 109–118 (2019)
31. Mishra, A., Alahari, K., Jawahar, C.: Top-down and bottom-up cues for scene text recognition. In: IEEE Conference on Computer Vision and Pattern Recognition (2012)
32. Pham, H., Guan, M.Y., Zoph, B., Le, Q.V., Dean, J.: Efficient neural architecture search via parameter sharing. In: International Conference on Machine Learning (2018)
33. Quy Phan, T., Shivakumara, P., Tian, S., Lim Tan, C.: Recognizing text with perspective distortion in natural scenes. In: IEEE Conference on Computer Vision and Pattern Recognition (2013)
34. Sandler, M., Howard, A., Zhu, M., Zhmoginov, A., Chen, L.C.: Mobilenetv 2: inverted residuals and linear bottlenecks. In: IEEE Conference on Computer Vision and Pattern Recognition (2018)
35. Sciuto, C., Yu, K., Jaggi, M., Musat, C., Salzmann, M.: Evaluating the search phase of neural architecture search. In: International Conference on Learning Representations (2020)
36. Shi, B., Bai, X., Yao, C.: An end-to-end trainable neural network for image-based sequence recognition and its application to scene text recognition. IEEE Trans. Pattern Anal. Mach. Intell. **39**, 2298–2304 (2017)
37. Shi, B., Wang, X., Lyu, P., Yao, C., Bai, X.: Robust scene text recognition with automatic rectification. In: IEEE Conference on Computer Vision and Pattern Recognition (2016)

38. Shi, B., Yang, M., Wang, X., Lyu, P., Yao, C., Bai, X.: Aster: an attentional scene text recognizer with flexible rectification. IEEE Trans. Pattern Anal. Mach. Intell. **41**, 2035–2048 (2019)
39. Simonyan, K., Zisserman, A.: Very deep convolutional networks for large-scale image recognition. Technical report arXiv preprint arXiv:1409.1556 (2014)
40. Wang, K., Babenko, B., Belongie, S.: End-to-end scene text recognition. In: International Conference on Computer Vision (2011)
41. Xie, F., Zhang, M., Zhao, J., Yang, J., Liu, Y., Yuan, X.: A robust license plate detection and character recognition algorithm based on a combined feature extraction model and BPNN. J. Adv. Transp. **2018**, 14 (2018)
42. Xie, L., Yuille, A.: Genetic CNN. In: IEEE International Conference on Computer Vision (2017)
43. Yang, M., et al.: Symmetry-constrained rectification network for scene text recognition. In: IEEE International Conference on Computer Vision (2019)
44. Yang, X., He, D., Zhou, Z., Kifer, D., Giles, C.L.: Learning to read irregular text with attention mechanisms. In: International Joint Conferences on Artificial Intelligence (2017)
45. Yao, Q., Xu, J., Tu, W.W., Zhu, Z.: Efficient neural architecture search via proximal iterations. In: AAAI Conference on Artificial Intelligence (2020)
46. Yao, Q., Wang, M.: Taking human out of learning applications: a survey on automated machine learning. Technical report arXiv preprint arXiv:1810.13306 (2018)
47. Zeiler, M.: Adadelta: an adaptive learning rate method. Technical report arXiv preprint arXiv:1212.5701 (2012)
48. Zhan, F., Lu, S.: ESIR: end-to-end scene text recognition via iterative image rectification. In: IEEE Conference on Computer Vision and Pattern Recognition (2019)
49. Zhu, Y., Yao, C., Bai, X.: Scene text detection and recognition: recent advances and future trends. Front. Comput. Sci. **10**, 19–36 (2016). https://doi.org/10.1007/s11704-015-4488-0
50. Zoph, B., Le, Q.V.: Neural architecture search with reinforcement learning. In: International Conference on Learning Representations (2017)

Mitigating Embedding and Class Assignment Mismatch in Unsupervised Image Classification

Sungwon Han[1], Sungwon Park[1], Sungkyu Park[1], Sundong Kim[2], and Meeyoung Cha[1,2](✉)

[1] Korea Advanced Institute of Science and Technology, Daejeon, South Korea
{lion4151,psw0416,shaun.park}@kaist.ac.kr

[2] Data Science Group, Institute for Basic Science, Daejeon, South Korea
{sundong,mcha}@ibs.re.kr

Abstract. Unsupervised image classification is a challenging computer vision task. Deep learning-based algorithms have achieved superb results, where the latest approach adopts unified losses from embedding and class assignment processes. Since these processes inherently have different goals, jointly optimizing them may lead to a suboptimal solution. To address this limitation, we propose a novel two-stage algorithm in which an embedding module for pretraining precedes a refining module that concurrently performs embedding and class assignment. Our model outperforms SOTA when tested with multiple datasets, by substantially high accuracy of 81.0% for the CIFAR-10 dataset (i.e., increased by 19.3 percent points), 35.3% accuracy for CIFAR-100-20 (9.6 pp) and 66.5% accuracy for STL-10 (6.9 pp) in unsupervised tasks.

1 Introduction

Deep learning-based algorithms have led to remarkable advances in various computer vision tasks thanks to their representative power [10,21,24]. However, these models often require active supervision from costly, high-quality labels. In contrast, unsupervised learning reduces labeling costs and, therefore, is more scalable [4,11,38].

Unsupervised image classification aims to determine the membership of each data point as one of the predefined class labels without utilizing any label information [18,39]. Since images are high dimensional objects, most existing methods focus on reducing dimensionality while discovering appropriate decision boundaries. The task of projecting high-dimensional data to lower dimensions is called embedding, and the task of identifying boundaries of dense groupings is called class assignment. Two methods are popularly used: 1) *a sequential method*, which separately trains to embed and assign samples to classes, and 2) *a joint method*, which simultaneously trains the samples in an end-to-end fashion.

Electronic supplementary material The online version of this chapter (https://doi.org/10.1007/978-3-030-58586-0_45) contains supplementary material, which is available to authorized users.

A. Vedaldi et al. (Eds.): ECCV 2020, LNCS 12369, pp. 768–784, 2020.
https://doi.org/10.1007/978-3-030-58586-0_45

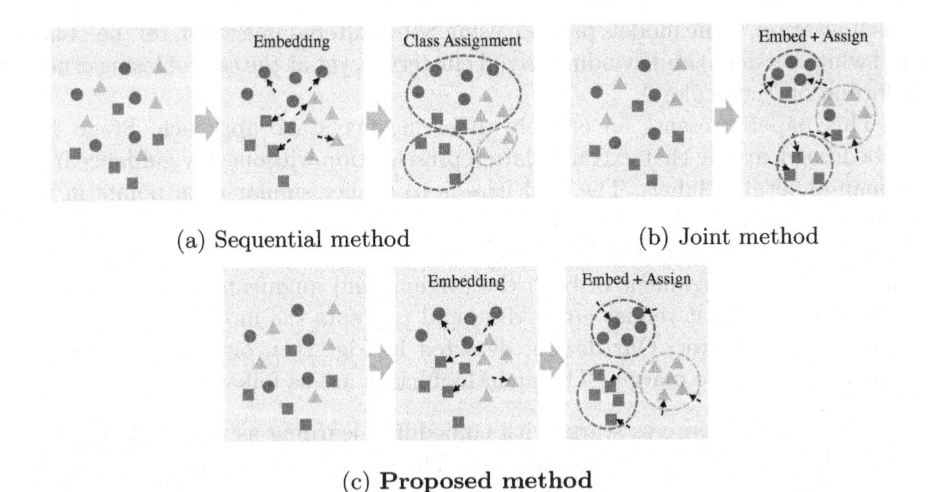

Fig. 1. Unsupervised image classification methods. (a) The sequential method embeds and assigns data points into classes separately, whereas (b) the joint method embeds and groups into classes together. (c) The proposed method first performs embedding learning as a pretraining step to find good initialization, then it jointly optimizes the embedding and class assignment processes. Our two-stage design introduces unique losses upon the pretraining step.

The sequential method, depicted in Fig. 1(a), utilizes embedding learning to represent visual similarity as a distance in the feature space [1,31]. Clustering algorithms such as k-means [25] and DBSCAN [8] are then applied. The embedding stage of this method solely reduces data dimensions, without knowledge of the immediately following class assignment, and hence may find representations that allow little separation between potential clusters.

The joint method, depicted in Fig. 1(b), simultaneously optimizes embedding and class assignment [4,39,41]. Some studies introduce the concept of clustering loss (e.g., k-means loss [40]) that is added to the embedding loss to yield enough separation between decision boundaries. Information Maximizing Self-Augmented Training (IMSAT) [15] and Invariant Information Clustering (IIC) [18] are new methods that show remarkable performance gain over conventional sequential methods. These models effectively extract invariant features against data augmentation by maximizing the mutual information.

Nevertheless, all of these models bear a common drawback that the goals for embedding and class assignment are inherently different. The former encodes high dimensional data points, whereas the latter identifies a proper class label for data points. Hence, jointly minimizing losses for these tasks may lead to what we identify as a "mismatched" result. A good example of a mismatch is when clusters are identified due to trivial traits such as colors rather than object shapes [4,18]. The gradient loss on the class assignment process in the early stage of training affects how images are grouped (i.e., by colors). To overcome this limitation, some models propose using Sobel-filtered images [4,18] (i.e., black and white versions) and avoiding trivial clustering, yet at the cost of losing crucial information (i.e., colors).

This paper presents an entirely different, two-stage approach. Stage 1 is embedding learning that extracts data representation without any guidance from human-annotated labels. The goal here is to gather similar data points in the embedding space and find a well-organized initialization. Stage 2 is class assignment and elaboratively minimizes two kinds of losses: (1) class assignment loss that considers assignment of both the original and augmented images and (2) embedding loss that refines embedding and prevents the model from losing its representation power. Our design, depicted in Fig. 1(c), outperforms existing baselines by a large margin. The main highlights are as follows:

- The two-stage process starts with embedding learning as a pretraining step, which produces a good initialization. The second aims to assign a class for each data point by refining its pretrained embedding. Our model successfully optimizes two objectives without falling into the mismatched state.
- Our method outperforms existing baselines substantially. With the CIFAR-10 dataset, we achieve an accuracy of 81.0%, whereas the best performing alternative reaches 61.7%.
- Extensive experiments and ablation studies confirm that both stages are critical to the performance gain. Comparison with the current state-of-the-art (SOTA) shows that our approach's most considerable benefit comes from the embedding learning initialization that gathers similar images nearby in the low-dimensional space.
- Our model can be adopted as a pretraining step for subsequent semi-supervised tasks. We discuss these implications in the experiments section.

Implementation details and codes are made available via GitHub[1] and the supplementary material.

2 Related Work

2.1 Unsupervised Embedding

Advances in unsupervised embedding can be discussed from three aspects: self-supervised learning, sample specificity learning, and generative models. Among them, self-supervised learning relies on auxiliary supervision. For instance, one may extract latent representations from images by expanding or rotating images [9] or solving a jigsaw puzzle made of input images [27]. Next, sample specificity learning considers every instance in the data as a single individual class [3,38]. The idea relies on the observation that deep learning can detect similarities between classes via supervised learning. By separating all data instances into the L2-normalized embedding space, this method gathers similar samples automatically due to its confined space. For example, Anchor Neighborhood

[1] Codes released at https://github.com/dscig/TwoStageUC.

Discovery (AND) [16] progressively discovers sample anchored neighborhoods to learn the underlying class decision boundaries iteratively. More recently, Super-AND [13] unifies some of the key techniques upon the AND model by maintaining invariant knowledge against small deformations [42], and by newly employing entropy-based loss. Finally, generative models learn to reconstruct the hidden data distribution itself without any labels. Thus, the model can generate new samples that likely belong to the input dataset [11,19]. Some research attempts have been made to use the generative model for deep embedding learning [29].

2.2 Unsupervised Classification

Class assignments can sequentially follow embedding or be optimized jointly. An example sequential setting is to apply principal component analysis before performing k-means clustering to relax the curse of dimensionality [7]. Another example is to use matrix factorization, which allows the derived matrix to learn a low-dimensional data representation before clustering [34]. For the face recognition task, one study embedded facial similarity features into Euclidean space and then clustered images based on the derived embedding [31]. Another approach is to utilize autoencoders. For example, one study stacked autoencoders in the deep networks to handle noisy input data [35]. Another study used a Boolean autoencoder as a form of non-linear representation [1]. Both studies showed a considerable improvement in classification tasks at that time.

The joint method is next, which considers embedding and class assignment simultaneously. Deep neural networks are used for the joint optimization of dimensionality reduction and clustering [40]. One study proposed the concept of deep embedded clustering (DEC), which learns latent representations and cluster assignments at the same time [39]. For the image clustering task, another study jointly updated clusters and their latent representations, derived by CNN via a recurrent process [41]. The deep adaptive clustering (DAC) model [5] computes the cosine similarities between pairs of hidden features on images via CNN. This model tackles the variant problem of the binary pairwise classification task, where the goal is to determine whether an image pair belongs to the same cluster. DeepCluster repeatedly learns and updates the features of CNN and the results of k-means clustering [4].

Most recently, a model called IIC (Invariant Information Clustering) has shown superior performance [18]. This model maximizes the mutual information of paired images. Because data augmentation does not deform the critical features of input images, this process can learn invariant features that persist in both the original and augmented data. The framework of the IIC is novel and straightforward. This algorithm is robust against any degeneracy in which one or more clusters have no allocated samples, or a single cluster dominates the others. Because of the entropy term in the mutual information, the loss cannot be minimized if a specific cluster dominates the others.

3 Model

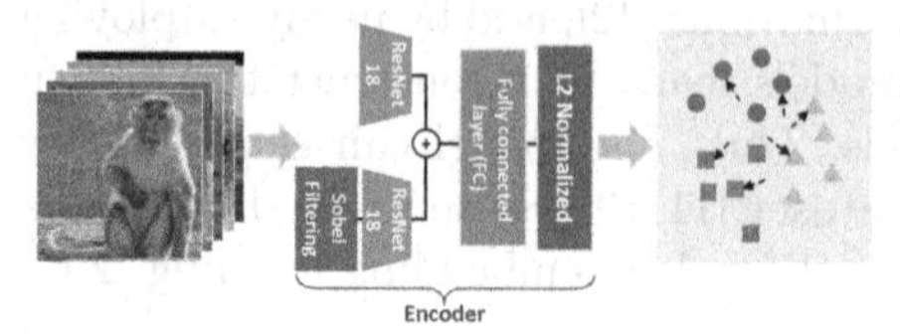

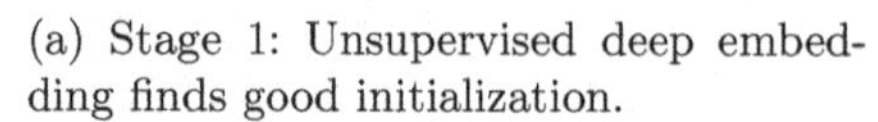

(a) Stage 1: Unsupervised deep embedding finds good initialization.

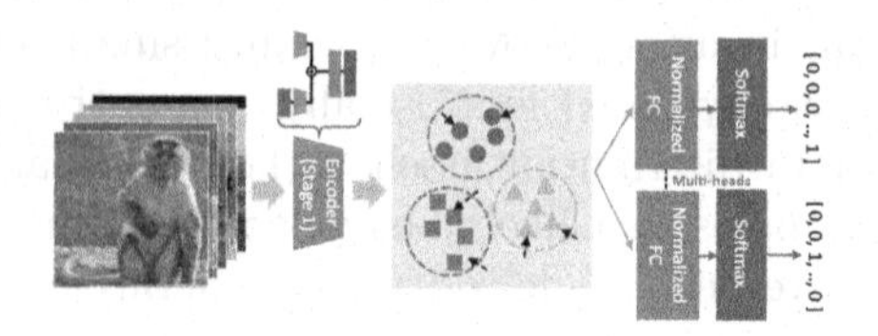

(b) Stage 2: Unsupervised class assignment with refining pretrained embeddings identifies consistent and dense groupings.

Fig. 2. Model illustration. The encoder projects input images to a lower dimension embedding sphere via deep embedding. The encoder is then trained to gather samples with similar visual contents nearby and separate them otherwise. Next, a multi-head normalized fully-connected layer classifies images by jointly optimizing the class assignment and embedding losses.

Problem Definition. Consider the number of underlying classes n_c and a set of n images $\mathcal{I} = \{\mathbf{x}_1, \mathbf{x}_2, ..., \mathbf{x}_n\}$. The objective of the unsupervised classification task is to learn a mapping f_θ that classifies images into the pre-defined n_c clusters without the use of any labels. Each stage of our method is described below.

3.1 Stage 1: Unsupervised Deep Embedding

The goal of this stage is to extract visually essential features without the use of any labels. An ideal embedding scenario should discriminate images of different classes and place them far apart in the embedding space while gathering similar images near each other. Since the model does not know beforehand which images are in the same class, the unsupervised embedding task is inherently challenging. Several advances have been made in this domain, including self-supervised learning, sample specificity learning, and generative models. Among them, we adopt Super-AND [13] to initialize the encoder. This model achieves high performance for unsupervised deep embedding tasks.

Super-AND extends AND [16], a powerful sample specificity learning method, by employing (1) data augmentation and (2) entropy-based loss. Total three losses are used for training: AND-loss (L_{and}), UE-loss (unification entropy loss, L_{ue}), and AUG-loss (augmentation loss, L_{aug}). Following the original AND algorithm, Super-AND considers every data occurrence as an individual class and separates the data in the L2-normalized embedding space. Then, the model groups the data points into small clusters by discovering the nearest neighborhood pairs, which is depicted in Fig. 2(a). The model runs iteratively to increase the number of identified subclasses by focusing on local clusters in each round. The subclass information is used for a self-supervised learning task to distinguish every local cluster. AND-loss considers each discovered neighborhood pair or remaining data instance as a single class to separate. This cross-entropy loss is written as:

$$L_{and} = -\sum_{i \in \mathcal{N}} \log(\sum_{j \in \tilde{\mathcal{N}}(\mathbf{x}_i) \cup \{i\}} \mathbf{p}_i^j) - \sum_{i \in \mathcal{N}^c} \log \mathbf{p}_i^i, \tag{1}$$

where $\mathcal{N}$ is the selected part of the neighborhood pair sets with its complement $\mathcal{N}^c$, $\tilde{\mathcal{N}}(\mathbf{x}_i)$ is the neighboring image i, and $\mathbf{p}_i^j$ represents the probability of i-th image being identified as j-th class.

UE-loss intensifies the concentration effect of AND-loss. UE-loss is defined as the entropy of the probability vector $\tilde{\mathbf{p}}_i$ except for instance itself. $\tilde{\mathbf{p}}_i$, which is computed from the softmax function, represents the similarity between instance i and the others in a probabilistic manner. The superscript j in $\tilde{\mathbf{p}}_i^j$ denotes the j-th component value of a given vector $\tilde{\mathbf{p}}_i$. By excluding the class of one's own, minimizing UE-loss makes nearby data occurrences attract each other—a concept that is contrary to the sample specificity learning. Jointly employing UE-loss with the AND-loss will enforce the overall neighborhoods to be separated while keeping similar neighborhoods to be placed closely. The UE-loss is calculated as follows:

$$L_{ue} = -\sum_i \sum_{j \neq i} \tilde{\mathbf{p}}_i^j \log \tilde{\mathbf{p}}_i^j. \tag{2}$$

Lastly, AUG-loss is defined to learn invariant image features against data augmentation. Since augmentation does not deform the underlying data characteristics, invariant features learned from the augmented data will still contain the class-related information. Model regards every augmentation instance as a positive sample and reduces the discrepancy between the original and augmented pair in embedding space. In Eq. 3, $\bar{\mathbf{p}}_i^j$ denotes the probability of wrong identification to class-j for the original i-th image, when $\bar{\mathbf{p}}_i^i$ describes that of correct identification to class-i for an augmented version of i-th image. Then, AUG-loss is defined as a cross entropy to minimize misclassification over batch instances.

$$L_{aug} = -\sum_i \sum_{j \neq i} \log(1 - \bar{\mathbf{p}}_i^j) - \sum_i \log \bar{\mathbf{p}}_i^i \tag{3}$$

The three losses are combined by weight manipulation on the UE-loss (Eq. 4). Weights for UE-loss $w(t)$ are initialized from 0 and increased gradually. The Super-AND model is trained by optimizing the total loss, and finally, the trained encoder becomes an initial point for the classification model in the next stage.

$$L_{stage1} = L_{and} + w(t) \times L_{ue} + L_{aug} \tag{4}$$

3.2 Stage 2: Unsupervised Class Assignment with Refining Pretrained Embeddings

The goal of this stage is to identify appropriate boundaries among classes. Unlike the first stage, an ideal class assignment requires not only ideal embedding, but also requires dense grouping to form decision boundaries with sufficient margins. This stage handles the given requirements by refining the initialized embeddings from the previous stage. Here, two kinds of losses are defined and used: class assignment loss and consistency preserving loss.

Mutual Information-Based Class Assignment. Mutual information quantifies mutual dependencies of two random variables [28], and measures how much two variables share the same kind of information. For example, if two variables are highly correlated or come from the same underlying distribution, their mutual information is significant, i.e., higher than zero. We can regard mutual information as the KL-divergence between the joint distribution and the product of its marginal distribution as follows:

$$I(x, y) = D_{KL}(p(x, y)||p(x)p(y)) \tag{5}$$

$$= \sum_{x \in \mathcal{X}} \sum_{y \in \mathcal{Y}} p(x, y) \log \frac{p(x, y)}{p(x)p(y)} \tag{6}$$

$$= H(x) - H(x|y). \tag{7}$$

The IIC (Invariant Information Clustering) model has been proposed for unsupervised classification to maximize the mutual information between the samples and the augmented set of samples [18]. This unique method trains the classifier with invariant features from data augmentation as follows: Suppose we have an image set $\mathbf{x}$ and a corresponding augmented image set $g(\mathbf{x})$ with a function g that geometrically transforms input images. The mapping f_θ classifies the images and generates the probability vector $y = f_\theta(\mathbf{x}), \hat{y} = f_\theta(g(\mathbf{x}))$ of all classes. Then, the model tries to find an optimal f_θ that maximizes the following terms:

$$\max_\theta I(y, \hat{y}) = \max_\theta (H(y) - H(y|\hat{y})). \tag{8}$$

By maximizing mutual information, we can prevent the clustering degeneracy, i.e., some clusters dominate the others, or there are no instances in a certain cluster. Since mutual information can be decomposed into two terms, maximizing $I(y, \hat{y})$ leads to maximizing $H(y)$ and minimizing $H(y|\hat{y})$. More specifically, $H(y)$ is maximized when every data sample is evenly assigned to every cluster; we can avoid degenerated solutions after optimization while consistent clusters are made with minimized $H(y|\hat{y})$.

We denote the joint probability distribution of y and $\hat{y}$ over the batch $\mathcal{B}$ to matrix $\mathbf{P}$ (i.e., $\mathbf{P} = \frac{1}{n} \sum_{i \in \mathcal{B}} f_\theta(x_i) \cdot f_\theta(g(x_i))^T$), where n is the size of batch $\mathcal{B}$). Using this matrix, we can easily define the objective function targeted to maximize the mutual information. By changing its sign, we define a class assignment loss L_{assign} (Eq. 9), where c and c' denote the class indices of the original and its augmented version, respectively. In the equation, $\mathbf{P}_{cc'}$ denotes the element at c-th row and c'-th column, where $\mathbf{P}_c$ and $\mathbf{P}_{c'}$ are the marginals over the rows and columns of the matrix.

$$L_{assign} = -\sum_{c} \sum_{c'} \mathbf{P}_{cc'} \cdot \log \frac{\mathbf{P}_{cc'}}{\mathbf{P}_{c'} \cdot \mathbf{P}_c} \tag{9}$$

Consistency Preserving on Embedding. We added an extra loss term, consistency preserving loss L_{cp}, to penalize any mismatch between original and augmented images. If the model only focuses on assigning class, during the process of dense grouping, embedding results can be easily modified just to match the number of final classes. By concurrently minimizing L_{cp}, our model can refine its embedding and avoid hasty optimization.

Assume that the $\mathbf{v}_i$ is the representation of an image $\mathbf{x}_i$ produced by the encoder in Stage 1. This $\mathbf{v}_i$ is projected into the normalized sphere, where the dot product can calculate the similarity of instances. Since augmented images have the same contents as the original ones, the similarity distance between them should be closer than other instances. We calculate $\hat{\mathbf{p}}_i^j$ $(i \neq j)$, the probability of given instance i classified as j-th instance, and $\hat{\mathbf{p}}_i^i$, the probability of being classified as its own i-th augmented instance (Eq. 10). The temperature value, τ, ensures that the label entropy distribution remains low [14]. Consistency preserving loss L_{cp} finally minimizes any misclassified cases over the batches (Eq. 11).

$$\hat{\mathbf{p}}_i^j = \frac{\exp(\mathbf{v}_j^\top \mathbf{v}_i/\tau)}{\sum_{k=1}^n \exp(\mathbf{v}_k^\top \mathbf{v}_i/\tau)}, \quad \hat{\mathbf{p}}_i^i = \frac{\exp(\mathbf{v}_i^\top \hat{\mathbf{v}}_i/\tau)}{\sum_{k=1}^n \exp(\mathbf{v}_k^\top \hat{\mathbf{v}}_i/\tau)} \tag{10}$$

$$L_{cp} = -\sum_i \sum_{j \neq i} \log(1 - \hat{\mathbf{p}}_i^j) - \sum_i \log \hat{\mathbf{p}}_i^i \tag{11}$$

The total unsupervised classification loss for the second stage is defined as follows (Eq. 12). λ is a hyper-parameter used for manipulating the weight of the consistency preserving loss term.

$$L_{stage2} = L_{assign} + \lambda \cdot L_{cp} \tag{12}$$

Normalized Fully-Connected Classifier. The commonly used fully-connected layer computes a weighted sum over the input. However, for feature vectors projected to a unit sphere space, as in the case of our encoder model, the conventional fully-connected layer with a bias term does not fit well. This is because the scale of the weights and bias obtained during training for each class can become pronounced to make drastic decisions after the softmax function. These variable sizes of weight vectors can lead to internal covariate shifts [17,30].

We introduce the normalized fully-connected layer (Norm-FC) without any bias term for classification. Each weight in Norm-FC becomes a prototype vector of each class, and images are classified by evaluating similarity in comparison with the class prototypes. Compared to the original classifier with the softmax function, Norm-FC is the layer whose weights $\mathbf{w}$ are L2-normalized (Eq. 13). To improve prediction confidence (i.e., reduce entropy in the distribution of model prediction), we additionally divide temperature $\tau_c < 1$ before the softmax function, as in Eq. 10. The temperature τ_c is critical for adjusting the concentration of feature vectors projected in unit sphere [36,38]. In the following equation, y_i^j is the j-th element of the classification probability vector for image $\mathbf{x}_i$, and $\mathbf{w}_j$

is the weight vector for class j in the fully-connected layer. Encoder with five Norm-FC classification heads is used for the second stage classifier.

$$y_i^j = \frac{\exp(\frac{\mathbf{w}_j}{||\mathbf{w}_j||} \cdot \mathbf{v}_i / \tau_c)}{\sum_k \exp(\frac{\mathbf{w}_k}{||\mathbf{w}_k||} \cdot \mathbf{v}_i / \tau_c)} \tag{13}$$

4 Experiments

We conducted extensive experiments and compared the model's performance against other baselines. We also examined how each of the two stages contributes to the performance gain. Last, we evaluated the performance after training with a dataset with scarce labels to analyze the relevance of the model to semi-supervised models. Implementation details such as model architecture and the reasoning behind hyper-parameter values such as τ_c, λ, and $w(t)$, can be found in supplementary material.

4.1 Image Classification Task

Datasets. Three datasets are used. (1) *CIFAR-10* [20] consists of a total of 60,000 images of 32×32 pixels and 10 classes, including airplanes, birds, and cats. (2) *CIFAR-100(20)* [20] is similar to CIFAR-10, but with 100 classes. Each class contains 600 images, which are then grouped into 20 superclasses. We use both CIFAR-100 and CIFAR-20, a version using 20 superclasses. (3) *STL-10* [6] contains 10 classes of 96×96 pixel images, based on 13,000 labeled images and 100,000 unlabeled images.

Evaluation. The Hungarian method [22] was used to achieve the best bijection permutation mapping between the unsupervised results and ground-truth labels. Then, the mapping was finally evaluated with top-1 classification accuracy. Given a network with five classification heads, we select the head with the lowest training loss for our final model.

Results. Table 1 compares the performances against the baselines. Across the three datasets, CIFAR-10, CIFAR-20, and STL-10, our model outperforms baselines substantially in classification accuracy, even when compared to the best performing sequential and joint classification algorithms. In the case of STL-10, we evaluated the model trained with both unlabeled and labeled datasets for a fair comparison with IIC. For CIFAR-100, whose results are omitted in the table, our model reports 28.1% for the lowest loss head accuracy as opposed to 20.3% for the IIC model (i.e., the current SOTA model).

4.2 Component Analyses

We conducted an ablation study by repeatedly testing the performance after removing or modifying each module. Such a step can confirm the efficacy of each component. We then interpret the model's predictions via qualitative analysis.

Table 1. Comparison of different unsupervised classification models. We report the accuracy (%) of our model from the head with having the lowest training loss. Baseline results are excerpted from the IIC paper [18].

Network	CIFAR-10	CIFAR-20	STL-10
Random network	13.1	5.9	13.5
k-means [37]	22.9	13.0	19.2
Autoencoder (AE) [2]	31.4	16.5	30.3
SWWAE [43]	28.4	14.7	27.0
GAN [29]	31.5	15.1	29.8
JULE [41]	27.2	13.7	27.7
DEC [39]	30.1	18.5	35.9
DAC [5]	52.2	23.8	47.0
DeepCluster [4]	37.4	18.9	33.4
ADC [12]	32.5	16.0	53.0
IIC [18]	61.7	25.7	59.6
Our Model	**81.0**	**35.3**	**66.5**

First Stage Ablations. We modify the first stage unsupervised embedding algorithm to various alternatives including state-of-the-art embedding methods [16,38], and measure the changes in the embedding quality and the final classification accuracy. Embedding quality is measured by the weighted k-NN classifier considering that the training labels are known [16]. For each test sample, we retrieve 200 nearest training samples based on the similarities in the embedding space and perform weighted voting for measuring its embedding quality. Table 2 describes the embedding quality and corresponding classification accuracy on CIFAR-10, which shows that the Super-AND algorithm outperforms other initialization alternatives in terms of the embedding quality and classification accuracy. This finding is evidence that the quality of pretrained embedding contributes to the final classification accuracy. We note that the Lemniscate [38] algorithm performs poorly in classification, although its embedding accuracy is reasonable. We speculate that this trend is due to the encoder's ambiguous knowledge learned only by separating every instance over the embedding space.

Second Stage Ablations. We modify (1) the network that trains the classifier and (2) the kind of classifier algorithm. For the former, we evaluated the classifier performance without the consistency preserving loss L_{cp}, without weight normalization, and without temperature in the softmax function. All experiments had an equal setting of a pretrained encoder from the first stage. Table 3 displays the comparison results, which demonstrate the significance of each component. We find temperature τ_c to be critical for correctly training the classifier, which can be explained by the topological characteristics of the normalized vector, i.e., it has a confined space, and the temperature is critical for amplifying certain signals.

Table 2. Comparison of deep embedding algorithms on CIFAR-10. The better encoder our model has, the more precise prediction the model can produce.

Initialization	Embedding quality (k-NN accuracy)	Class assignment quality (Top-1 classification accuracy)
Random	–	58.6
Lemniscate [38]	80.8	63.8
AND [16]	86.3	73.9
Super-AND [13]	**89.2**	**81.0**

Table 3. Test of alternative algorithms on CIFAR-10. Every component of the model contributes to a performance increase, and the two-stage design is superior than the sequential or joint versions.

Alternatives	Accuracy
Without L_{cp} in Eq. 12	58.2
Without weight normalization	56.0
Without temperature (i.e., $\tau_c = 1$)	19.5
Sequential: first stage + k-means	38.3
Sequential: first stage + Hierarchical clustering	49.9
Joint: first & second stage	68.4
Two-stage model with full features	**81.0**

We next test alternative algorithms, including sequential and joint versions of our model. For the sequential method, k-means and hierarchical clustering are adopted for the class assignment, while maintaining the pretrained encoder of the proposed model. The joint approach combines the losses from both stages and optimize concurrently. The comparison results in the table report a substantial drop in the accuracy of these variants, implying that the proposed two-stage paradigm contributes to a performance gain.

Qualitative Analysis. Figure 3 illustrates the visual interpretation of our model's prediction by Grad-CAM [32]. The blue framed images are the success cases and imply that our model can capture unique visual traits for each class, such as legs in the horse class, funnels in the ship class, wings in the airplane class, and horns in the deer class. The red framed images are failure cases, in contrast, and show that these cases are unable to detect key visual traits for the given class.

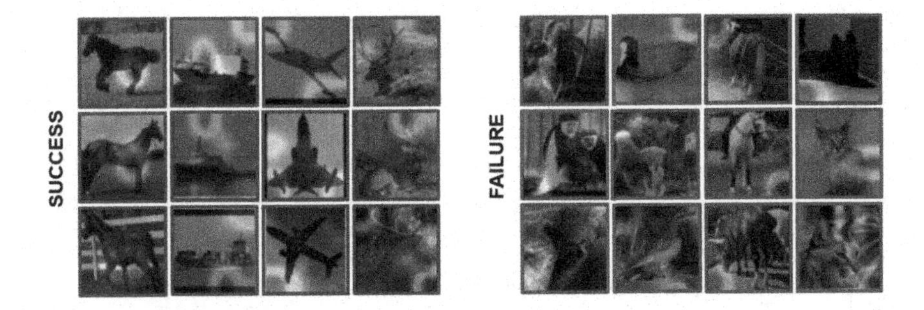

Fig. 3. Successes and failure cases from STL-10. Blue images are the success cases, where the highlighted part indicates how the model interpreted class traits based on the Grad-CAM visualization. Red images are the failure cases, where the model finds periphery areas to contain significant class traits. (Color figure online)

4.3 Improvement over SOTA

So far, we have demonstrated that the algorithm's superior performance. To understand what attributes to such novelty, we analyze the intermediate states of the proposed model against IIC, the current state-of-the-art. Figure 4(a) tracks the temporal changes in mutual information $I(y, \hat{y})$ in Eq. 7 by its two components: $H(y)$ and $H(y|\hat{y})$.

The figure shows that IIC gradually achieves higher $H(y)$, whereas our method starts with a high-value, thanks to the competitive advantage of the pretraining step. Note that a higher value indicates that data points are well divided across clusters. Next, IIC starts with a much larger $H(y|\hat{y})$ value compared to our model at epoch 0. Furthermore, while IIC gradually finds model instances with smaller $H(y|\hat{y})$ values, our method is more aggressive in identifying good clusters. Those drastic decreases are contributed by the consistency preserving loss, which enhances embedding refinement. Note that a lower value for this second term indicates that the cluster assignments of the original image and its augmentation version are closely related.

The visualization in Fig. 4(b) represents the corresponding clusters at different training epochs. Images appear well dispersed for our model at epoch 0. However, the lower $H(y)$ value for IIC is due to missing clusters, or those with zero matched images. This result implies that pretraining alleviates the mismatch between embedding and class assignments more effectively than IIC. The bottom row shows the per-class accuracy. The proposed model far exceeds IIC in the classification of images for every single class in all examined epochs.

We examined the confusion matrix between the ground truth labels and classification results in Fig. 5. The results are shown separately for our model trained after 300 epochs and IIC trained after 2,000 epochs. A perfect classification would only place items on the diagonal line. The figure shows that the proposed model eventually finds the right cluster for most images, although cluster assignments for some classes such as birds, cats, and dogs are error-prone.

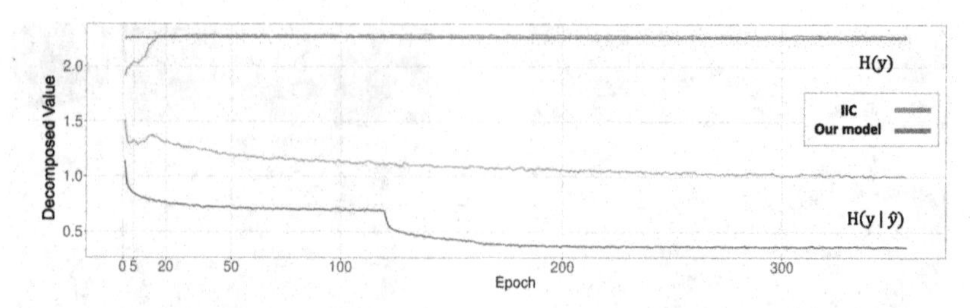

(a) Change of mutual information (i.e., $H(y) - H(y|\hat{y})$) across the epoch.

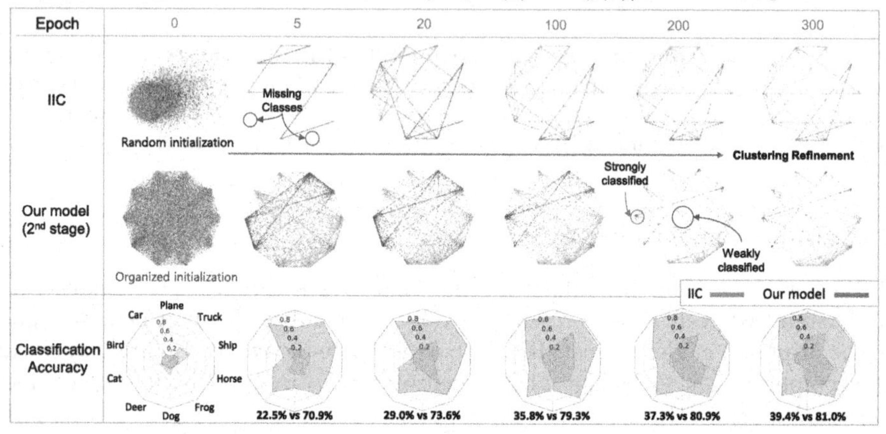

(b) Changes in classification results and the accuracy of each class. The color of each dot represents its original label.

Fig. 4. Performance difference between the two models, IIC and ours, as they are trained on CIFAR-10. (a) Mutual information $I(y, \hat{y})$ of IIC is smaller than that of our model across all epochs as it is associated with the quality of classification depicted in (b) the form of clustering and classification accuracy.

Nonetheless, the IIC model, even after six times longer training epochs, achieves less accurate performance compared to our model.

4.4 Implication on Semi-supervised Learning

Can unsupervised learning methods bring new insights into well-known tasks? As a practical implication, we next discuss the potential of the proposed approach to assist in classical image classification tasks with a small amount of labeled data. As an empirical test, we compare the classification accuracy of (semi-) supervised models against their variants that utilize our method in the pretraining step.

Datasets. Two datasets are chosen for evaluating semi-supervised learning tasks. (1) the *CIFAR-10* dataset and (2) the *SVHN* dataset [26], which is a real-world digit dataset with 10 classes of 32 × 32 pixel images. We used only 4,000 labels in CIFAR-10 and 1,000 labels in SVHN for training. The unlabeled

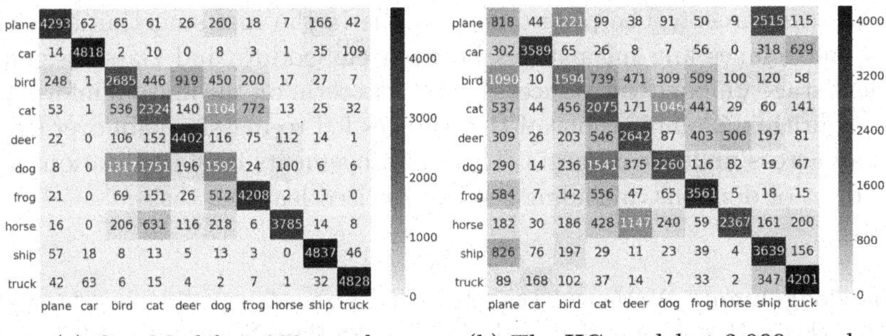

(a) Our Model at 300 epochs (b) The IIC model at 2,000 epochs

Fig. 5. Comparison of confusion matrices. The final confusion matrices on CIFAR-10 are given, where columns are the predicted class, and rows are the actual class. Classifications on the diagonal line show that our model produces superior results than IIC.

Table 4. Classification accuracy on partially labeled datasets. Applying our model as a pretraining to the existing semi-supervised methods such as Π-Model and Mean Teacher yields enhancing the classification accuracy.

Algorithms	CIFAR-10	SVHN
Supervised	79.7	87.7
Π-Model [23]	83.8	92.2
Mean Teacher [33]	84.2	93.5
Our Model + Supervised	**85.4**	**93.4**
Our Model + Π-Model	**85.8**	**93.6**
Our Model + Mean Teacher	**88.0**	**94.2**

dataset is also used by semi-supervised learning models, excluding the supervised model.

Results. For both CIFAR-10 and SVHN, our model surpasses the fully-supervised baselines. Table 4 shows that the proposed model can even match some of the semi-supervised learning algorithms. By applying a simple Π-model [23] or the mean teacher model [33] to our pretrained network, a semi-supervised model can obtain meaningful starting points that contribute to performance improvement.

5 Conclusion

Unsupervised image classification has endless potential. This study presented a new two-stage algorithm for unsupervised image classification, where an embedding module precedes the refining module that concurrently performs embedding and class assignment. The pretraining module in the first stage initializes data points and relaxes any mismatches between embedding and class assignment. The next stage uniquely introduces the L_{cp} loss term on the mutual information-based algorithm. Combinations of these stages led to massive gain over existing baselines across multiple datasets. These improvements have implications across a broad set of domains, including semi-supervised learning tasks.

Acknowledgement. We thank Cheng-Te Li and Yizhan Xu for their insights and discussions. This work was supported by the Institute for Basic Science (IBS-R029-C2) and the Basic Science Research Program through the National Research Foundation funded by the Ministry of Science and ICT in Korea (No. NRF-2017R1E1A1A01076400).

References

1. Baldi, P.: Autoencoders, unsupervised learning, and deep architectures. In: Proceedings of the ICML Workshop on Unsupervised and Transfer Learning, pp. 37–49 (2012)
2. Bengio, Y., Lamblin, P., Popovici, D., Larochelle, H.: Greedy layer-wise training of deep networks. In: Advances in Neural Information Processing Systems (NeurIPS), pp. 153–160 (2007)
3. Bojanowski, P., Joulin, A.: Unsupervised learning by predicting noise. In: Proceedings of the International Conference on Machine Learning (ICML), pp. 517–526 (2017)
4. Caron, M., Bojanowski, P., Joulin, A., Douze, M.: Deep clustering for unsupervised learning of visual features. In: Proceedings of the European Conference on Computer Vision (ECCV), pp. 132–149 (2018)
5. Chang, J., Wang, L., Meng, G., Xiang, S., Pan, C.: Deep adaptive image clustering. In: Proceedings of the IEEE International Conference on Computer Vision (ICCV), pp. 5879–5887 (2017)
6. Coates, A., Ng, A., Lee, H.: An analysis of single-layer networks in unsupervised feature learning. In: Proceedings of the International Conference on Artificial Intelligence and Statistics (AISTATS), pp. 215–223 (2011)
7. Ding, C., He, X.: K-means clustering via principal component analysis. In: Proceedings of the International Conference on Machine Learning (ICML), p. 29 (2004)
8. Ester, M., Kriegel, H.P., Sander, J., Xu, X., et al.: A density-based algorithm for discovering clusters in large spatial databases with noise. In: Proc. of the International Conference on Knowledge Discovery and Data Mining (SIGKDD), pp. 226–231 (1997)
9. Gidaris, S., Singh, P., Komodakis, N.: Unsupervised representation learning by predicting image rotations. arXiv preprint arXiv:1803.07728 (2018)
10. Goodfellow, I., Bengio, Y., Courville, A.: Deep Learning. MIT Press, Hoboken (2016)
11. Goodfellow, I., et al.: Generative adversarial nets. In: Advances in Neural Information Processing Systems (NeurIPS), pp. 2672–2680 (2014)

12. Haeusser, P., Plapp, J., Golkov, V., Aljalbout, E., Cremers, D.: Associative deep clustering: training a classification network with no labels. In: Brox, T., Bruhn, A., Fritz, M. (eds.) GCPR 2018. LNCS, vol. 11269, pp. 18–32. Springer, Cham (2019). https://doi.org/10.1007/978-3-030-12939-2_2
13. Han, S., Xu, Y., Park, S., Cha, M., Li, C.T.: A Comprehensive Approach to Unsupervised Embedding Learning based on AND Algorithm. arXiv preprint arXiv:2002.12158 (2020)
14. Hinton, G., Vinyals, O., Dean, J.: Distilling the knowledge in a neural network. arXiv preprint arXiv:1503.02531 (2015)
15. Hu, W., Miyato, T., Tokui, S., Matsumoto, E., Sugiyama, M.: Learning discrete representations via information maximizing self-augmented training. In: Proceedings of the International Conference on Machine Learning (ICML), pp. 1558–1567 (2017)
16. Huang, J., Dong, Q., Gong, S., Zhu, X.: Unsupervised deep learning by neighbourhood discovery. In: Proceedings of the International Conference on Machine Learning (ICML), pp. 2849–2858 (2019)
17. Ioffe, S., Szegedy, C.: Batch normalization: accelerating deep network training by reducing internal covariate shift. arXiv preprint arXiv:1502.03167 (2015)
18. Ji, X., Henriques, J.F., Vedaldi, A.: Invariant information clustering for unsupervised image classification and segmentation. In: Proceedings of the IEEE International Conference on Computer Vision (ICCV), pp. 9865–9874 (2019)
19. Kingma, D.P., Welling, M.: Auto-encoding variational bayes. arXiv preprint arXiv:1312.6114 (2013)
20. Krizhevsky, A.: Learning multiple layers of features from tiny images. Technical report. Citeseer (2009)
21. Krizhevsky, A., Sutskever, I., Hinton, G.E.: Imagenet classification with deep convolutional neural networks. In: Advances in Neural Information Processing Systems (NeurIPS), pp. 1097–1105 (2012)
22. Kuhn, H.W.: The Hungarian method for the assignment problem. Naval Res. Logist. Q. **2**(1–2), 83–97 (1955)
23. Laine, S., Aila, T.: Temporal ensembling for semi-supervised learning. arXiv preprint arXiv:1610.02242 (2016)
24. LeCun, Y., Bengio, Y., Hinton, G.: Deep learning. Nature **521**(7553), 436 (2015)
25. Lloyd, S.: Least squares quantization in PCM. IEEE Trans. Inf. Theory **28**(2), 129–137 (1982)
26. Netzer, Y., Wang, T., Coates, A., Bissacco, A., Wu, B., Ng, A.Y.: Reading digits in natural images with unsupervised feature learning (2011)
27. Noroozi, M., Favaro, P.: Unsupervised learning of visual representations by solving jigsaw puzzles. In: Leibe, B., Matas, J., Sebe, N., Welling, M. (eds.) ECCV 2016. LNCS, vol. 9910, pp. 69–84. Springer, Cham (2016). https://doi.org/10.1007/978-3-319-46466-4_5
28. Paninski, L.: Estimation of entropy and mutual information. Neural Comput. **15**(6), 1191–1253 (2003)
29. Radford, A., Metz, L., Chintala, S.: Unsupervised representation learning with deep convolutional generative adversarial networks. arXiv preprint arXiv:1511.06434 (2015)
30. Salimans, T., Kingma, D.P.: Weight normalization: a simple reparameterization to accelerate training of deep neural networks. In: Advances in Neural Information Processing Systems (NeurIPS), pp. 901–909 (2016)

31. Schroff, F., Kalenichenko, D., Philbin, J.: Facenet: a unified embedding for face recognition and clustering. In: Proceedings of the IEEE Conference on Computer Vision and Pattern Recognition (CVPR), pp. 815–823 (2015)
32. Selvaraju, R.R., Cogswell, M., Das, A., Vedantam, R., Parikh, D., Batra, D.: Grad-cam: visual explanations from deep networks via gradient-based localization. In: Proceedings of the IEEE International Conference on Computer Vision (ICCV), pp. 618–626 (2017)
33. Tarvainen, A., Valpola, H.: Mean teachers are better role models: weight-averaged consistency targets improve semi-supervised deep learning results. In: Advances in Neural Information Processing Systems (NeurIPS), pp. 1195–1204 (2017)
34. Trigeorgis, G., Bousmalis, K., Zafeiriou, S., Schuller, B.: A deep semi-NMF model for learning hidden representations. In: Proceedings of the International Conference on Machine Learning (ICML), pp. 1692–1700 (2014)
35. Vincent, P., Larochelle, H., Lajoie, I., Bengio, Y., Manzagol, P.A.: Stacked denoising autoencoders: learning useful representations in a deep network with a local denoising criterion. Journal of Machine Learning Research **11**, 3371–3408 (2010)
36. Wang, F., Xiang, X., Cheng, J., Yuille, A.L.: Normface: L2 hypersphere embedding for face verification. In: Proceedings of the ACM International Multimedia Conference, pp. 1041–1049 (2017)
37. Wang, J., Wang, J., Song, J., Xu, X.S., Shen, H.T., Li, S.: Optimized cartesian k-means. IEEE Trans. Knowl. Data Eng. **27**(1), 180–192 (2014)
38. Wu, Z., Xiong, Y., Yu, S.X., Lin, D.: Unsupervised feature learning via non-parametric instance discrimination. In: Proceedings of the IEEE Conference on Computer Vision and Pattern Recognition (CVPR), pp. 3733–3742 (2018)
39. Xie, J., Girshick, R., Farhadi, A.: Unsupervised deep embedding for clustering analysis. In: Proceedings of the International Conference on Machine Learning (ICML), pp. 478–487 (2016)
40. Yang, B., Fu, X., Sidiropoulos, N.D., Hong, M.: Towards k-means-friendly spaces: simultaneous deep learning and clustering. In: Proceedings of the International Conference on Machine Learning (ICML), pp. 3861–3870 (2017)
41. Yang, J., Parikh, D., Batra, D.: Joint unsupervised learning of deep representations and image clusters. In: Proceedings of the IEEE Conference on Computer Vision and Pattern Recognition (CVPR), pp. 5147–5156 (2016)
42. Ye, M., Zhang, X., Yuen, P.C., Chang, S.F.: Unsupervised embedding learning via invariant and spreading instance feature. In: Proceedings of the IEEE Conference on Computer Vision and Pattern Recognition (CVPR), pp. 6210–6219 (2019)
43. Zhao, J., Mathieu, M., Goroshin, R., Lecun, Y.: Stacked what-where auto-encoders. arXiv preprint arXiv:1506.02351 (2015)

Adversarial Training with Bi-directional Likelihood Regularization for Visual Classification

Weitao Wan[1], Jiansheng Chen[1](✉), and Ming-Hsuan Yang[2,3]

[1] Department of Electronic Engineering, Tsinghua University, Beijing, China
jschenthu@mail.tsinghua.edu.cn
[2] Department of EECS, UC Merced, Merced, USA
[3] Google Research, Mountain View, USA

Abstract. Neural networks are vulnerable to adversarial attacks. Practically, adversarial training is by far the most effective approach for enhancing the robustness of neural networks against adversarial examples. The current adversarial training approach aims to maximize the posterior probability for adversarially perturbed training data. However, such a training strategy ignores the fact that the clean data and adversarial examples should have intrinsically different feature distributions despite that they are assigned with the same class label under adversarial training. We propose that this problem can be solved by explicitly modeling the deep feature distribution, for example as a Gaussian Mixture, and then properly introducing the likelihood regularization into the loss function. Specifically, by maximizing the likelihood of features of clean data and minimizing that of adversarial examples simultaneously, the neural network learns a more reasonable feature distribution in which the intrinsic difference between clean data and adversarial examples can be explicitly preserved. We call such a new robust training strategy the adversarial training with bi-directional likelihood regularization (ATBLR) method. Extensive experiments on various datasets demonstrate that the ATBLR method facilitates robust classification of both clean data and adversarial examples, and performs favorably against previous state-of-the-art methods for robust visual classification.

Keywords: Adversarial training · Feature distribution · Optimization

1 Introduction

A key challenge for utilizing neural networks in visual classification is their vulnerability to adversarial examples, which has attracted increasing concerns in recent years [4,13,16,18]. Visual adversarial examples are crafted by adding small perturbations that are imperceptible to human eyes onto the clean data, causing the neural networks to produce wrong predictions. In addition, researches

Electronic supplementary material The online version of this chapter (https://doi.org/10.1007/978-3-030-58586-0_46) contains supplementary material, which is available to authorized users.

A. Vedaldi et al. (Eds.): ECCV 2020, LNCS 12369, pp. 785–800, 2020.
https://doi.org/10.1007/978-3-030-58586-0_46

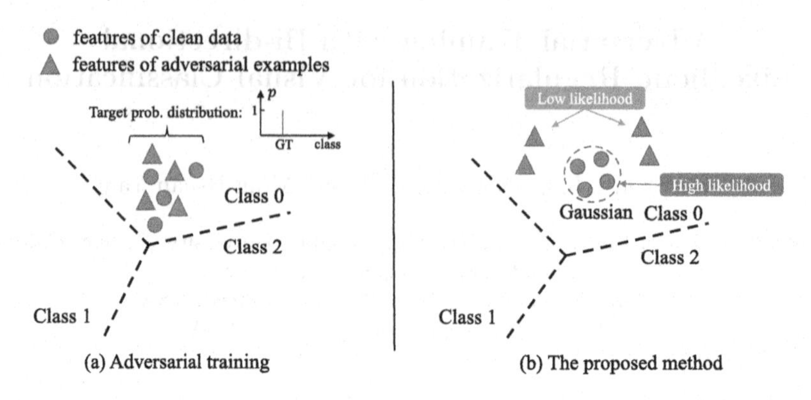

Fig. 1. Illustration of the expected feature space of (a) adversarial training and (b) the proposed ATBLR method. Adversarial examples are generated to resemble other classes. But existing adversarial training methods ignore their intrinsically different feature distribution and treat them equally with the clean data. The proposed method addresses this issue by optimizing not only the class probability distribution but also the likelihood of the feature distribution.

have demonstrated that adversarial examples can be transferable across different models [11,20], *i.e.* adversarial examples generated based on one model can successfully attack other models. As such, the existence of adversarial examples has become a serious threat to the safety of neural networks.

Improving the robustness of neural networks has become a critical issue in addition to increasing the classification accuracy. Numerous algorithms are proposed to address this issue, among which the most effective approaches are based on adversarial training [4,12]. The basic idea of adversarial training is to generate adversarial examples based on the latest model weights during training and feed them into the model for training. The adversarial examples are assigned the same class label as their source images. Madry *et al.* [12] propose a more generic form of adversarial training, which is formulated as a saddle-point optimization problem. However, the adversarial training only aims to optimize the posterior probability, without considering the feature distribution. The feature space of adversarial training is illustrated in Fig. 1(a). This paper focuses on the deepest features of neural networks, *e.g.* the output of the global average pooling layer after the last convolutional layer in ResNet [5]. Figure 1(a) shows the expected feature space of adversarial training but it is difficult to achieve in practice because existing adversarial training methods ignore the intrinsic difference between the feature distributions of the clean data and adversarial examples. For instance, a clean sample from class 0 is adversarially perturbed into class 1. Previous research [6] justifies that such an adversarial example contains highly predictive but non-robust features for class 1. As such, its features should follow a different distribution compared to the features of the clean data from class 0. However, the adversarial training scheme ignores its similarity to class 1 and forces the neural network to treat it the same way as the clean data from class 0 by assigning them with the same target class distribution, which is typically a one-hot distribution of the ground-truth (GT) class. This unreasonable underlying constraint in existing adversarial training methods leads to sub-optimal classification performance.

To address this issue, we propose to optimize the neural networks so that not only the clean data and its corresponding adversarial examples can be classified into the same class but also the their feature distributions are explicitly encouraged to be different. The proposed method is illustrated in Fig. 1(b). To achieve this, we explicitly learn the feature distribution of the clean data by incorporating the Gaussian Mixture Model into the network. More specifically, for the visual classification task, features belonging to each class correspond to one Gaussian component, of which the Gaussian mean is the trainable parameter updated by stochastic gradient descent and the Gaussian covariance matrix is reduced to identity matrix for simplicity. As such, the entire network can be trained end-to-end. Then we adopt the likelihood regularization term introduced in [21] to encourage the extracted features of clean data to follow the learned Gaussian Mixture distribution. We note that the likelihood regularization in this paper is intrinsically different from that in [21] because our method takes two different types of inputs, *i.e.* the clean data and adversarial examples, and optimizes the likelihood term towards different directions for these two inputs. For the clean data, the objective is to maximize the likelihood since we aim to learn its feature distribution through training. For the adversarial examples, since they should follow a distribution that is different from the one of clean data, the objective is to minimize the likelihood. The common objective for both the clean data and adversarial examples is the cross-entropy loss for the posterior probability and the target class. We refer to the proposed method as Adversarial Training with Bi-directional Likelihood Regularization (ATBLR). We present a comparison study in Fig. 3, Sect. 4.3 to demonstrate that the proposed bi-directional likelihood regularization leads to different feature distributions for the clean data and adversarial examples.

Our method can be implemented efficiently, without increasing the number of trainable parameters. The classification layer in a neural network is typically a fully-connected layer with $K \times C$ trainable parameters, in which K is the number of object classes and C is the dimension of the features. It outputs the class distribution based on the features. Our method replaces it with a Gaussian Mixture Model without adding extra trainable parameters. Since this paper is focused on the visual classification task, the deepest features belonging to each class can be assigned with one Gaussian component. As such, the GMM also requires $K \times C$ trainable parameters in total for the K Gaussian components when the covariance is reduced to identity matrix as aforementioned. The likelihood regularization, which is essentially the l_2 distance between features and the corresponding Gaussian mean, brings about very little computational overhead to the neural networks.

The main contributions of this paper are summarized as follows:

- We propose the bi-directional likelihood regularization on the conventional adversarial training method based on the learned feature distribution. Features of the clean data and adversarial examples are explicitly encouraged to follow different distributions.
- We improve both the robustness of neural networks and the classification performance on clean data without adding extra trainable parameters.
- We evaluate the proposed method on various datasets including MNIST [10], CIFAR-10 and CIFAR-100 [8] for different adversarial attacks. Experimental results show that the proposed algorithm performs favorably against the state-of-the-art methods.

2 Related Work

2.1 Adversarial Attacks

Adversarial examples are crated data with small perturbations that cause misclassification in neural networks [18]. Plenty of algorithms have been developed to generate adversarial examples.

Fast Gradient Sign Method (FGSM). Goodfellow *et al.* [4] propose the Fast Gradient Sign Method (FGSM) which uses a single-step perturbation along the gradient of the loss function $\mathcal{L}$ with respect to the input image x. The adversarial example x_{adv} is computed by $x_{adv} = x + \epsilon \cdot sign(\nabla_x \mathcal{L}(x, y))$. To perform a targeted attack, we replace the true label y with a wrong target label t and reverse the sign of the gradient by $x_{adv} = x - \epsilon \cdot sign(\nabla_x \mathcal{L}(x, t))$.

Basic Iterative Method (BIM). Kurakin *et al.* [9] extends the single-step approach to an iterative attack which updates the adversarial example at each iteration by the formulation of FGSM method and clips the resulting image to constrain it within the ϵ-ball from the original input x. The adversarial example is computed by $x^i_{adv} = clip_{x,\epsilon}(x^{i-1}_{adv} + \alpha \cdot sign(\nabla_x \mathcal{L}(x^{i-1}_{adv}, y))$, where α is the step size for each iteration.

Projected Gradient Descent (PGD). Madry *et al.* [12] discover that stronger attacks can be generated by starting the iterative search of the BIM method from a random initialization point within the allowed norm ball centered at the clean data. This method is called the Projected Gradient Descent (PGD) method.

Carlini & Wagner (C&W). Nicholas *et al.* [3] propose the C&W algorithm which is an optimization-based attack method. An auxiliary variable ω is introduced to reparameterize the variable for adversarial example by $x_{adv} = \frac{1}{2}(\tanh(\omega)+1)$ and solve $\min_\omega \|\frac{1}{2}(\tanh(\omega)+1) - x\|^2_2 + c \cdot f(\frac{1}{2}(\tanh(\omega)+1))$. The loss weight c is adjusted by binary search. And $f(x) = \max(\max\{Z(x)_i : i \neq t\} - Z(x)_t, -\kappa)$, in which $Z(x)_t$ is the logit for the target class t and the non-negative parameter κ controls the confidence for the adversarial example.

2.2 Defensive Methods

With the development of adversarial attack methods, the defense methods against them have attracted greater concerns in recent years. The defensive distillation approach [14] aims to train a substitute model with smoother gradients to increase the difficulty of generating adversarial examples. Nevertheless, it is not effective for the optimization-based attack methods such as [3]. Song *et al.* [17] propose to model the image distribution in the pixel space and restore an adversarial example to be clean by maximizing its likelihood. But it is difficult to effectively model the distribution in the pixel space where there is much noise and the dimension is much larger than that in the feature space. The adversarial training method [4,18] generates adversarial examples during training and use them as training data to improve the robustness against adversarial examples. However, this method is shown to be vulnerable to iterative attacks [9]. Tramer *et al.* [19] propose to improve the performance of adversarial training by generating adversarial examples using an ensemble of neural networks. Madry *et al.* [12] propose a more general framework for adversarial training and use random initialization before searching for adversarial examples to deal with iterative attack methods. Wong *et al.* [23] incorporate the linear programming into the training to minimize the loss for the worse case within the allowed perturbation around the clean data. However, the test accuracy on the clean data is severely compromised. Xie *et al.* [24] develop a network architecture which uses the non-local mean filter to remove the noises in the feature maps of adversarial examples. Song *et al.* [15] adopt the domain adaptation algorithms in adversarial training to learn domain-invariant representations across the clean domain and the adversarial domain. However, these methods are all based on the original adversarial training method and they do not address the issues concerning the feature distributions in the adversarial training which we discussed above.

3 Proposed Algorithm

3.1 Preliminaries

Adversarial Training. This paper focuses on the visual classification task. Suppose the number of object classes in the dataset is K. Denote the set of training samples as $\mathcal{D} = \{(x_i, y_i)\}_{i=1}^{N}$, in which $x_i \in \mathbb{R}^{H \times W \times 3}$ is the image, $y_i \in \{1, 2, ..., K\}$ is the class label and N is the number of training samples. Denote the one-hot label vector corresponding to label y_i as $\boldsymbol{y}_i$. Let $f_\theta(x) : \mathbb{R}^{H \times W \times 3} \to \mathbb{R}^K$ denote a neural network parameterized by θ. The network outputs the class probability distribution given an input image. Then the classification loss function for the training pair (x_i,y_i) is

$$\mathcal{L}_{cls}(x_i, y_i; \theta) = -\boldsymbol{y}_i \log f_\theta(x_i). \tag{1}$$

The adversarial training method [12] is formulated as a min-max optimization problem, which is expressed as

$$\min_{\theta} \max_{\|\delta_i\|_\infty \leq \epsilon} \frac{1}{N} \sum_{(x_i,y_i)\sim\mathcal{D}} \mathcal{L}_{cls}(x_i + \delta_i, y_i; \theta). \tag{2}$$

The maximizer of the inner problem can be approximately found by using k steps of the PGD attack or a single-step FGSM attack. The adversarial examples are crafted by adding the inner maximizer to the clean data. The min-max problem is solved by stochastic gradient descent by feeding the adversarial examples as inputs to the neural network.

3.2 Modeling the Feature Distribution

As discusses in Sect. 1, our motivation is to consider the difference between feature distributions of clean data and adversarial examples. We adopt an effective and tractable distribution to model the feature distribution, *i.e.* the Gaussian Mixture Model. For simplicity, the covariance matrix is reduced to the identity matrix. This is not only efficient but also beneficial for reducing the redundancy across different feature dimensions. Besides, we assume the prior distribution for each class is the constant $1/K$. For the visual classification task, features belonging to each class are assigned with one Gaussian component. Formally, denote the features at the deepest layer of the neural network by

$$\tilde{x}_i = h_\theta(x_i), \tag{3}$$

in which $h_\theta(\cdot)$ represents the feature extraction process in the neural network. As such, the posterior probability of the ground-truth class y_i is expressed by

$$p(y_i|\tilde{x}_i) = \frac{\mathcal{N}(\tilde{x}_i; \mu_{y_i})}{\sum_{k=1}^{K} \mathcal{N}(\tilde{x}_i; \mu_k)}, \tag{4}$$

in which μ_k is the Gaussian mean of class k and $\mathcal{N}(\cdot)$ is the density function of Gaussian distribution.

The computation in Eq. 4 can be implemented with a layer in the neural network, with the Gaussian means as its trainable parameters. This layer is deployed immediately after the deepest features of the neural network and outputs the class distribution. The entire network can be trained end-to-end and the Gaussian means are updated by gradient descent through back-propagation. Equipped with such a layer, the neural network can learn to not only predict class probabilities but also model the feature distribution.

3.3 Likelihood Regularization for Features

The adversarial training scheme in Eq. 2 adopts only the adversarial examples for training, without using the clean data. In this paper, we leverage both the clean data and the adversarial examples generated by the inner problem in Eq. 2 for training, with equal proportion. We train the neural networks equipped with the layer introduced in Sect. 3.2 to learn the feature distribution of clean data. In addition, we adopt the likelihood regularization [21] to maximize the likelihood of features of clean data. Formally, the likelihood regularization is defined as the negative log likelihood, which is given by

$$\mathcal{L}_{lkd} = -\frac{1}{N}\sum_{i=1}^{N} \log \mathcal{N}(h_\theta(x_i); \mu_{y_i}). \tag{5}$$

By ignoring the constant term and constant coefficient, we derive

$$\mathcal{L}_{lkd} = \frac{1}{N}\sum_{i} \|h_\theta(x_i) - \mu_{y_i}\|^2. \tag{6}$$

The likelihood regularization is weighted by a hyperparameter $\lambda > 0$ and added to the cross-entropy loss during training. Hence, the final objective function for the clean data is given by

$$\mathcal{L} = \frac{1}{N}\sum_{(x_i,y_i)\sim\mathcal{D}} (-\boldsymbol{y}_i \log f_\theta(x_i) + \lambda\|h_\theta(x_i) - \mu_{y_i}\|^2). \tag{7}$$

We note that this formulation is essentially different from the center loss [22] because the center loss does not consider modeling the feature distribution. However, the mapping function $f_\theta(\cdot)$ here contains the Gaussian Mixture Model and the posterior probability is generated based on the learned feature distribution.

By minimizing Eq. 7, the neural network not only learns to make classifications but also learns to model the feature distribution of clean data. For the adversarial training, the clean data and adversarial examples are assigned the same class label but their feature distributions should be different since adversarial examples are crafted to resemble the class other than the ground-truth one and researches [6] reveal that they contain highly predictive but non-robust features of other classes. As such, a more reasonable training approach should encourage the features of clean data and adversarial examples to be different. This can be achieved by introducing the regularization term. We propose to maximize the likelihood value for adversarial examples during training. Denote the adversarial examples generated by solving the inner maximization problem in Eq. 2 as $\{a_i\}_{i=1}^N$, in which $a_i = x_i + \arg\max_{\delta_i} \mathcal{L}_{cls}(x_i + \delta_i, y_i; \theta)$. We minimize the following loss for adversarial examples.

$$\mathcal{L}_{adv} = \frac{1}{N}\sum_{i}(-\boldsymbol{y}_i \log f_\theta(a_i) - \lambda\|h_\theta(a_i) - \mu_{y_i}\|^2). \tag{8}$$

The training scheme is illustrated in Fig. 2. It demonstrates two important modifications we make to the original adversarial training scheme. First, the original adversarial training is conducted on a discriminative model, which only considers the output probability distribution and maximizes the target probability. In contrast, our method explicitly models the feature distribution through end-to-end training. Second, we explicitly encourage different feature distributions by optimizing the likelihood regularization towards opposite directions for the clean data and adversarial examples. Our method facilitates a more reasonable feature distribution in the scope of robust classification and improves the classification accuracy of both clean data and various adversarial examples.

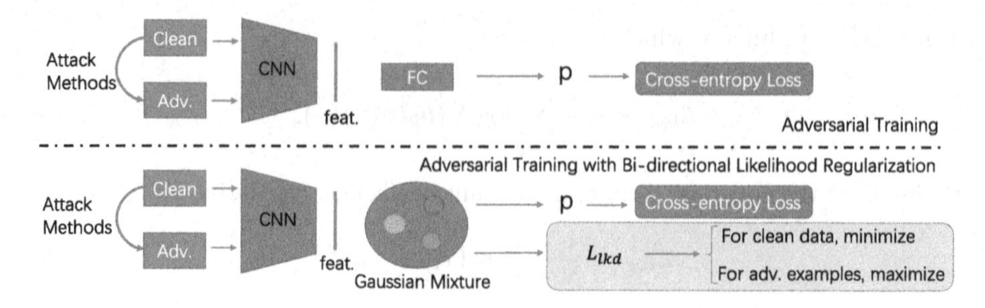

Fig. 2. Training schemes comparison. Top: the original adversarial training method [12]. Bottom: the proposed ATBLR method.

4 Experiments

To evaluate the robustness and generalization ability of the proposed method, we present experimental results on datasets including MNIST [10], CIFAR-10 and CIFAR-100 [8]. We report the natural accuracy, *i.e.* the accuracy of clean data, and that of adversarial examples. Following the widely adopted protocal [2,12, 24], we consider the adversarial attack methods including FGSM [4], PGD [12] and C&W [3]. We evaluate the robustness of our method under two different threat models.

- **White-Box Attack**: the attacker has access to all the information of the target classification model including the model architecture and model weights. The adversarial examples for testing are generated using gradient information of the target model.
- **Black-Box Attack**: the attacker has knowledge of the model architecture but has no access to its model weights. The adversarial examples for testing are generated using the gradient information of a substitute model, which is independently trained using the same architecture and training hyper-parameters as the target model.

Experiments are conducted with the Tensorflow [1] using the Nvidia TITAN X GPU. All the codes and trained models will be made available to the public.

4.1 MNIST

We apply the proposed ATBLR method to train robust models for image classification to compare with the baseline method, *i.e.* the adversarial training [12]. The MNIST dataset [10] is a handwritten digit dataset consisting of 10 classes including 60,000 images for training and 10,000 images for testing. We use the data augmentation method including mirroring and 28×28 random cropping after 2-pixel zero paddings on each side. The models are tested on different types of adversarial examples, including the FGSM [4], PGD [12] with varying steps and restarts, C&W [3] with $\kappa = 0$ and C&W with a high confidence parameter $\kappa = 50$ (denoted as C&W-hc method).

Table 1. Classification accuracy (%) on the MNIST dataset for clean data and adversarial attacks. The evaluation is conducted for both white-box and black-box attacks.

Testing Input	Steps	Restarts	Accuracy (%)	
			Adv. Training [12]	ATBLR (ours)
Clean	–	–	98.8	**99.3**
White-box Attack				
FGSM	–	–	95.6	**97.2**
PGD	40	1	93.2	**94.8**
PGD	100	1	91.8	**94.1**
PGD	40	20	90.4	**93.5**
PGD	100	20	89.3	**92.7**
C&W	40	1	94.0	**95.8**
C&W-hf	40	1	93.9	**96.3**
Black-box Attack				
FGSM	–	–	96.8	**98.4**
PGD	40	1	96.0	**97.7**
PGD	100	20	95.7	**97.6**
C&W	40	1	97.0	**98.8**
C&W-hf	40	1	96.4	**98.5**

Implementation Details. Following the practice in [12], we generate PGD attacks of 40 steps during training and use a network consisting of two convolutional layers with 32 and 64 filters respectively, followed by a fully connected layer of size 1024. The input images are divided by 255 so that the pixel range is $[0, 1]$. The l_∞ norm constraint of $\epsilon = 0.3$ is imposed on the adversarial perturbations and the step size for PGD attack is 0.01. The models are trained for 50 epochs using ADAM [7] optimizer with a learning rate of 0.001. The parameter λ in Eq. 7 and 8 which balances the trade-off between the classification loss and bi-directional likelihood regularization is set to 0.1. For the evaluation of the black-box attacks, we generate adversarial examples on an independently initialized and trained copy of the target network according to [12].

The experimental results on the MNIST dataset are presented in Table 1. The results show that the strongest attack is the PGD attack with multiple restarts. It can be observed that the proposed method not only improves the robustness against adversarial attacks but also improves the accuracy of the clean data. Moreover, the performance gain is achieved without introducing any extra trainable parameters, which validates the effectiveness of addressing the difference of feature distributions between clean data and adversarial examples.

Table 2. Classification accuracy (%) on the CIFAR-10 dataset for clean data and adversarial attacks. The PGD attacks for testing are generated with l_∞ norm constraint $\epsilon = 8$ and a step size of 2. †We re-run the code of PATDA with $\epsilon = 8$ since it reports the result for $\epsilon = 4$ which generates weaker adversarial attacks.

Method	Clean	White-box				Black-box
		FGSM	PGD-10	PGD-100	PGD-1000	PGD-1000
Network: ResNet-32						
Natural Training	92.73	27.54	0.32	0.11	0.00	3.03
Adv. Training [12]	79.37	51.72	44.92	43.44	43.36	60.22
IAAT [2]	83.26	52.05	44.26	42.13	42.51	60.26
PATDA† [15]	83.40	53.81	46.59	45.27	44.01	61.79
FD [24]	84.24	52.81	45.64	44.60	44.21	62.84
ATBLR (ours)	86.32	**58.60**	**50.18**	**48.56**	**47.88**	**64.38**
Network: WideResNet-32						
Natural Training	95.20	32.73	2.17	0.35	0.00	4.29
Adv. Training [12]	87.30	56.13	46.26	45.14	44.87	61.07
IAAT [2]	91.34	57.08	48.53	46.50	46.54	58.20
PATDA† [15]	84.63	57.79	49.85	48.73	48.04	58.53
FD [24]	86.28	57.54	49.26	46.97	46.75	59.31
ATBLR (ours)	92.12	**59.69**	**52.11**	**51.17**	**50.63**	**62.89**

4.2 CIFAR

We apply the proposed ATBLR method to train robust classification models on the CIFAR-10 and CIFAR-100 datasets and make comparisons with previous state-of-the-art methods. The CIFAR-10 dataset [8] consists of 32×32 pixel color images from 10 classes, with 50,000 training images and 10,000 testing images. The CIFAR-100 dataset [8] has 100 classes containing 50,000 training images and 10,000 testing images. We use the typical data augmentation method including mirroring and 32×32 random cropping after 4-pixel reflection paddings on each side. We use the network architectures of ResNet-32 [5] and WideRenset-32 [25] following Madry *et al.* [12] and Zhang *et al.* [26]. Our method is compared with the natural training and previous state-of-the-art training approaches designed to improve the robustness of classification models:

- Natural Training: Training with cross-entropy loss on the clean training data.
- Adversarial Training (Adv. Training) [12]: Training on the clean training data and the adversarial examples generated during training.
- Instance Adaptive Adversarial Training (IAAT) [2]: Training that enforces the sample-specific perturbation margin around every training sample.
- PGD-Adversairal Training with Domain Adaptation (PATDA) [15]: Adversarial training combined with domain adaptation algorithms.

- Feature Denoising (FD) [24]: Training that combines Adversarial Training and a network architecture with the non-local filters to remove the noise caused by the adversarial examples in feature space.

Implementation Details. During adversarial training, the adversarial examples are generated by PGD-10 attacks, *i.e.* 10 steps of PGD attack are conducted on the clean data for each training iteration. The step size for the PGD attack is set to 2 out of 255. A l_∞ norm constraint of $\epsilon = 8$ is imposed on the adversarial perturbations. The models are trained for 200 epochs using ADAM [7] optimizer with the learning rate of 0.001 and the batch size of 128. The parameter λ which balances the trade-off between the classification loss and bi-directional likelihood regularization is set to 0.02. We present more quantitative results in Sect. 4.4 to study the influence of the parameter λ. For evaluation in the white-box setting, the models are tested on (1) PGD-10 attacks with 5 random restarts, (2) PGD-100 attacks with 5 random restarts and (3) PGD-1000 attacks with 2 random restarts. For evaluation in the black-box setting, following the experimental setting of [2], the PGD-1000 attack with 2 random restarts is adopted.

The experimental results on the CIFAR-10 dataset are presented in Table 2. We observe that the proposed method improves the classification accuracy of both the clean data and adversarial examples. Compared with the original adversarial training, the results demonstrate that our method achieves large accuracy gain by considering the feature distribution differences and introducing the bi-directional likelihood regularization during training. Moreover, our method performs favorably against the Feature Denoising (FD) method [24], which is the previous state-of-the-art method. By switching from the network of ResNet-32 to its 10× wider variant, the classification performance is increased due to larger model capacity. Our method can increase the model's robustness for both the simple and complex models.

We present the experimental results on the CIFAR-100 dataset in Table 3. As shown by the results, the CIFAR-100 dataset is more challenging than the CIFAR-10 dataset. Nevertheless, we observe that the proposed ATBLR method consistently increases the robustness against adversarial examples and performs favorably against previous state-of-the-art methods.

4.3 Evolution of the Likelihood Regularization

During training, we propose to optimize different objective functions for clean data and adversarial examples, which are given by Eq. 7 and 8, respectively. Here we investigate the evolution of the values of $\mathcal{L}_{lkd}$ in the training progress to verify that the likelihood of clean data and adversarial examples is optimized to be different. We conduct the experiments on the CIFAR-10 dataset with the same network and training schemes as in Sect. 4.2. We evaluate and record the value of the likelihood regularization according to Eq. 6 for the clean data and adversarial examples, respectively, in each input batch. We compare two models. The first one is trained with the proposed ATBLR method and the second one is trained without optimizing the $\mathcal{L}_{lkd}$ during training.

Table 3. Classification accuracy (%) on the CIFAR-100 dataset for clean data and adversarial attacks. The PGD attacks for testing are generated with l_∞ norm constraint $\epsilon = 8$ and a step size of 2.

Method	Clean	White-box				Black-box
		FGSM	PGD-10	PGD-100	PGD-1000	PGD-1000
Network: ResNet-32						
Natural Training	74.88	4.61	0.02	0.01	0.00	1.81
Adv. Training [12]	55.11	26.25	20.69	19.68	19.91	35.57
IAAT [2]	63.90	27.13	18.50	17.17	17.12	35.74
PATDA [15]	59.40	27.33	20.25	19.45	19.08	35.81
FD [24]	65.13	26.96	21.14	20.39	20.06	36.28
ATBLR (ours)	67.34	**28.55**	**21.80**	**21.54**	**20.96**	**37.79**
Network: WideResNet-32						
Natural Training	79.91	5.29	0.01	0.00	0.00	3.22
Adv. Training [12]	59.58	28.98	26.24	25.47	25.49	38.10
IAAT [2]	68.80	29.30	26.17	24.22	24.36	35.18
PATDA [15]	64.24	28.35	24.51	23.45	23.08	35.81
FD [24]	67.13	29.54	27.15	25.69	25.14	37.95
ATBLR (ours)	70.39	**30.85**	**29.49**	**27.53**	**27.15**	**39.24**

The curves for the values of $\mathcal{L}_{lkd}$ are plotted in Fig. 3. We note that larger $\mathcal{L}_{lkd}$ indicates smaller likelihood value since $\mathcal{L}_{lkd}$ is essentially the negative logarithm of likelihood. In the left figure, as the training converges, the $\mathcal{L}_{lkd}$ of clean data (blue) is low, which means the network learns to model the feature distribution of the clean data. The $\mathcal{L}_{lkd}$ of adversarial examples (orange) is large, which is nearly twice that of the clean data. In contrast, the right figure shows that the $\mathcal{L}_{lkd}$ values of the clean data and adversarial examples are almost the same during training. The comparison verifies that the proposed method effectively encourages the features of clean data and adversarial examples to follow different distributions. The quantitative evaluation in Sect. 4.2 validates that introducing such a regularization during training is effective in improving the accuracy of both clean data and adversarial examples.

In addition, we observe in the left figure that the two curves do not separate until about 700 iterations. This phenomenon can be explained as below. The model parameters are randomly initialized and the likelihood of both types of inputs is low at the start since the features are far from their corresponding Gaussian mean. At the early training stage, the cross-entropy loss in Eq. 7 and 8 is dominating because the λ is small. The cross-entropy loss is driving the features to move closer to the corresponding Gaussian mean, thus decreasing the $\mathcal{L}_{lkd}$ for both types of inputs. As the cross-entropy loss becomes smaller, the likelihood regularization is having a larger impact. After a certain point of equilibrium, which is about 700 iterations in this experiment, the likelihood regularization

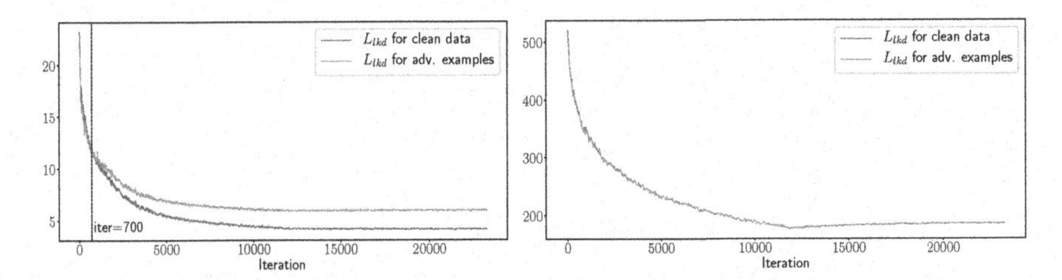

Fig. 3. Curves of the likelihood regularization for clean data and adversarial examples. Left: the model is trained using the proposed ATBLR method. **Right**: the model is trained without optimizing $\mathcal{L}_{lkd}$ but we record its value during training. The experiment is conducted on the CIFAR-10 dataset with ResNet-32. The first 60 epochs are shown here since the changes in the rest 140 epochs are not obvious.

term in Eq. 8 is slowing $\mathcal{L}_{lkd}$ from decreasing for adversarial examples. Finally, the training converges and the $\mathcal{L}_{lkd}$ value of adversarial examples is larger than that of clean data, which means that the clean data follows the learned GM distribution better and the adversarial examples follow a different distribution.

4.4 Hyper-parameter Analysis

We study the effect of choosing different values of λ in the proposed ATBLR method and compare the performance. We conduct the experiments on the CIFAR-10 and CIFAR-100 datasets using the ResNet-32 network.

The experimental results are presented in Fig. 4. The PGD attack is stronger than FGSM. Nevertheless, our method improves the classification performance for different types of attacks and the clean data. Comparing results of $\lambda = 0$ and the others, we conclude that the ATBLR method can improve the classification performance consistently for different values of the hyper-parameter λ. The results also demonstrate that it is disadvantageous to set a λ that is too large. This is reasonable considering that λ is a balancing coefficient for the classification loss and the likelihood regularization. Too large a λ will lead to the dominance of the likelihood regularization term. This damages the classification performance because the features of all the clean data tend to collapse into one point under this objective function. Nevertheless, our method makes steady improvements for different λ in a wide range. We choose $\lambda = 0.02$ for experiments in Sect. 4.2 based on this hyper-parameter study.

4.5 Adversaries for Training

In the previous experiments, following other works, we select PGD attacks with 10 steps as the adversarial examples for training. We investigate the effect of adopting other alternatives and present the results on the CIFAR-10 dataset in Table 4. The results show that the performance gain that our method achieves becomes larger when stronger attacks are used for training. For example, if

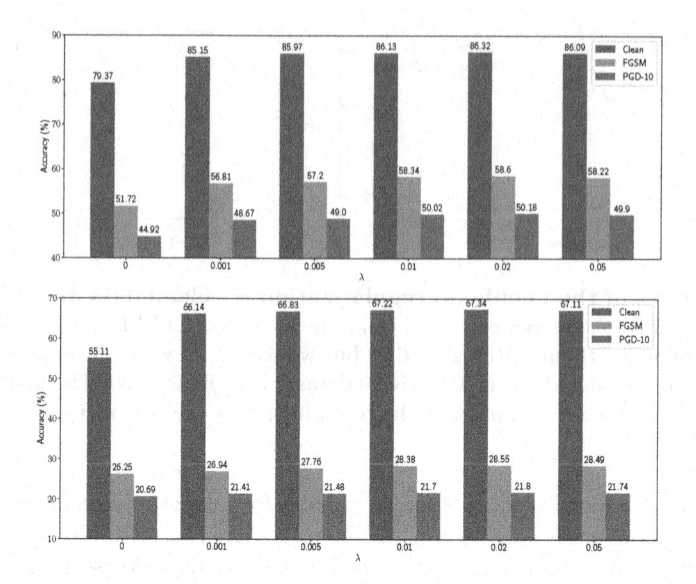

Fig. 4. Hyper-parameter study for λ on the CIFAR-10 dataset (top) and the CIFAR-100 dataset (bottom). $\lambda = 0$ denotes the original adversarial training.

Table 4. Classification accuracy (%) of the proposed ATBLR method/the original adversarial training when trained with different adversaries.

Model	Clean	FGSM	PGD-10	PGD-100
Training w/ FGSM	89.83/87.40	91.87/90.93	1.03/0.00	0.14/0.00
Training w/ PGD-10	86.32/79.37	58.60/51.72	50.18/44.92	48.56/43.44
Training w/ PGD-100	86.49/77.45	58.74/51.58	51.25/45.06	52.37/45.71

the training adversaries are switched from PGD-10 to the stronger PGD-100, the performance gain on clean data is increased from $86.32 - 79.37 = 6.95$ to $86.49 - 77.45 = 9.04$, likewise in other columns. This is expected because stronger adversarial examples have greater similarity to another class. As such, it is more favorable if they are encouraged to follow a feature distribution which is different from that of clean data of the original class.

5 Conclusion

In this paper, we propose a novel method for training robust classification models against adversarial attacks. In contrast to the previous adversarial training method which optimizes only the posterior class distribution, our method learns the feature distribution of clean data through end-to-end training. Furthermore, the intrinsic difference between feature distributions for clean data and adversarial examples is preserved by optimizing the likelihood regularization in opposite directions for these two types of inputs. Moreover, our method introduces no extra trainable parameters. Extensive experiments demonstrate that our method performs favorably against previous state-of-the-art methods in terms of the classification accuracy of both the clean data and adversarial examples.

Acknowledgements. This work was supported by the National Natural Science Foundation of China under Grant 61673234 and the program of China Scholarships Council (No. 201906210354). M.-H. Yang is supported in part by NSF CAREER Grant 1149783.

References

1. Abadi, M., et al.: TensorFlow: large-scale machine learning on heterogeneous systems (2015). https://www.tensorflow.org/. software available from tensorflow.org
2. Balaji, Y., Goldstein, T., Hoffman, J.: Instance adaptive adversarial training: improved accuracy tradeoffs in neural nets. arXiv preprint arXiv:1910.08051 (2019)
3. Carlini, N., Wagner, D.: Towards evaluating the robustness of neural networks. In: IEEE Symposium on Security and Privacy (SP) (2017)
4. Goodfellow, I.J., Shlens, J., Szegedy, C.: Explaining and harnessing adversarial examples. arXiv preprint arXiv:1412.6572 (2014)
5. He, K., Zhang, X., Ren, S., Sun, J.: Deep residual learning for image recognition. In: CVPR (2016)
6. Ilyas, A., Santurkar, S., Tsipras, D., Engstrom, L., Tran, B., Madry, A.: Adversarial examples are not bugs, they are features. In: NeurIPS (2019)
7. Kingma, D.P., Ba, J.: Adam: a method for stochastic optimization. arXiv preprint arXiv:1412.6980 (2014)
8. Krizhevsky, A., Hinton, G.: Learning multiple layers of features from tiny images. University of Toronto, Technical report (2009)
9. Kurakin, A., Goodfellow, I., Bengio, S.: Adversarial machine learning at scale. arXiv preprint arXiv:1611.01236 (2016)
10. LeCun, Y., Bottou, L., Bengio, Y., Haffner, P.: Gradient-based learning applied to document recognition. Proc. IEEE **86**, 2278–2324 (1998)
11. Liu, Y., Chen, X., Liu, C., Song, D.: Delving into transferable adversarial examples and black-box attacks. arXiv preprint arXiv:1611.02770 (2016)
12. Madry, A., Makelov, A., Schmidt, L., Tsipras, D., Vladu, A.: Towards deep learning models resistant to adversarial attacks. In: ICLR (2018)
13. Papernot, N., McDaniel, P., Jha, S., Fredrikson, M., Celik, Z.B., Swami, A.: The limitations of deep learning in adversarial settings. In: IEEE European Symposium on Security and Privacy (EuroS&P) (2016)
14. Papernot, N., McDaniel, P., Wu, X., Jha, S., Swami, A.: Distillation as a defense to adversarial perturbations against deep neural networks. In: IEEE Symposium on Security and Privacy (SP) (2016)
15. Song, C., He, K., Wang, L., Hopcroft, J.E.: Improving the generalization of adversarial training with domain adaptation. In: ICLR (2019)
16. Song, D., et al.: Physical adversarial examples for object detectors. In: 12th USENIX Workshop on Offensive Technologies (WOOT) (2018)
17. Song, Y., Kim, T., Nowozin, S., Ermon, S., Kushman, N.: Pixeldefend: leveraging generative models to understand and defend against adversarial examples. arXiv preprint arXiv:1710.10766 (2017)

18. Szegedy, C., et al.: Intriguing properties of neural networks. arXiv preprint arXiv:1312.6199 (2013)
19. Tramèr, F., Kurakin, A., Papernot, N., Goodfellow, I., Boneh, D., McDaniel, P.: Ensemble adversarial training: attacks and defenses. arXiv preprint arXiv:1705.07204 (2017)
20. Tramèr, F., Papernot, N., Goodfellow, I., Boneh, D., McDaniel, P.: The space of transferable adversarial examples. arXiv preprint arXiv:1704.03453 (2017)
21. Wan, W., Zhong, Y., Li, T., Chen, J.: Rethinking feature distribution for loss functions in image classification. In: CVPR (2018)
22. Wen, Y., Zhang, K., Li, Z., Qiao, Y.: A discriminative feature learning approach for deep face recognition. In: Leibe, B., Matas, J., Sebe, N., Welling, M. (eds.) ECCV 2016. LNCS, vol. 9911, pp. 499–515. Springer, Cham (2016). https://doi.org/10.1007/978-3-319-46478-7_31
23. Wong, E., Kolter, J.Z.: Provable defenses against adversarial examples via the convex outer adversarial polytope. In: ICML (2018)
24. Xie, C., Wu, Y., Maaten, L.V.D., Yuille, A.L., He, K.: Feature denoising for improving adversarial robustness. In: CVPR (2019)
25. Zagoruyko, S., Komodakis, N.: Wide residual networks. arXiv preprint arXiv:1605.07146 (2016)
26. Zhang, H., Yu, Y., Jiao, J., Xing, E.P., Ghaoui, L.E., Jordan, M.I.: Theoretically principled trade-off between robustness and accuracy. arXiv preprint arXiv:1901.08573 (2019)

Author Index

Printed in the United States
By Bookmasters